08763201
574.875 MAR
MARTONOSI
853279



BA

B.C.H.E. – LIBRARY
00126934

# The Enzymes of Biological Membranes

SECOND EDITION

Volume 3

Membrane Transport

*THE ENZYMES OF BIOLOGICAL MEMBRANES*
*Second Edition*

---

Volume 1: Membrane Structure and Dynamics
Volume 2: Biosynthesis and Metabolism
Volume 3: Membrane Transport
Volume 4: Bioenergetics of Electron and Proton Transport

# The Enzymes of Biological Membranes

## SECOND EDITION

## Volume 3

## Membrane Transport

Edited by

Anthony N. Martonosi

*State University of New York*
*Syracuse, New York*

Plenum Press • New York and London

Library of Congress Cataloging in Publication Data

Main entry under title:

The Enzymes of biological membranes.

Includes bibliographies and indexes.
Contents: v. 1. Membrane structure and dynamics— —v. 3. Membrane transport.
1. Membranes (Biology)—Collected works. 2. Enzymes—Collected works. I. Martonosi, Anthony, 1928–
QH601.E58 1984 574.87′5 84-8423
ISBN 0-306-41453-8 (v. 3)

© 1985 Plenum Press, New York
A Division of Plenum Publishing Corporation
233 Spring Street, New York, N.Y. 10013



Printed in the United States of America

# Contributors

*Giovanna Ferro-Luzzi Ames,* Department of Biochemistry, University of California at Berkeley, Berkeley, California 94720

*Robert Anholt,* The Salk Institute for Biological Studies, San Diego, California 92138

*Volkmar Braun,* Mikrobiologie II, Universität Tübingen, D-7400 Tübingen, West Germany

*Ernesto Carafoli,* Laboratory of Biochemistry, Swiss Federal Institute of Technology (ETH), CH-8092 Zurich, Switzerland.

*Sally E. Carty,* Department of Biochemistry and Biophysics, University of Pennsylvania School of Medicine, Philadelphia, Pennsylvania 19104

*Martin Crompton,* Department of Biochemistry, University College London, London WC1E 6BT, England

*Michael P. Czech,* Department of Biochemistry, University of Massachusetts Medical Center, Worcester, Massachusetts 01605

*L. D. Faller,* Center for Ulcer Research and Education, Wadsworth Veterans Administration Center, Los Angeles, California 90073; and School of Medicine, University of California at Los Angeles, Los Angeles, California 90024

*G. Gárdos,* National Institute of Haematology and Blood Transfusion, Budapest, Hungary

*I. M. Glynn,* Physiological Laboratory, University of Cambridge, Cambridge CB2 3EG, England

*Giuseppe Inesi,* Department of Biological Chemistry, University of Maryland School of Medicine, Baltimore, Maryland 21201

*Robert G. Johnson,* Department of Biochemistry and Biophysics, University of Pennsylvania School of Medicine, Philadelphia, Pennsylvania 19104

*Adam Kepes,* Laboratoire des Biomembranes, Institut Jacques Monod, 75251 Paris Cédex 05, France

*Maria A. Kukuruzinska,* Department of Biology and the McCollum-Pratt Institute, The Johns Hopkins University, Baltimore, Maryland 21218

*Robert Landick,* Department of Biological Sciences, Stanford University, Stanford, California 94305

*Jon Lindstrom,* The Salk Institute for Biological Studies, San Diego, California 92138

*Norman D. Meadow,* Department of Biology and the McCollum-Pratt Institute, The Johns Hopkins University, Baltimore, Maryland 21218

*Leopoldo de Meis,* Instituto de Ciencias Biomedicas, Departamento de Bioquimica, Universidade Federal do Rio de Janeiro, Rio de Janeiro 21910, Brasil

*Jean Meury,* Laboratoire des Biomembranes, Institut Jacques Monod, 75251 Paris Cédex 05, France

*Marek Michalak,* Bio Logicals, Ottawa, Ontario K1Y 4P1, Canada

*Mauricio Montal,* Departments of Biology and Physics, University of California at San Diego, La Jolla, California 92093

*Lorin J. Mullins,* Department of Biophysics, University of Maryland at Baltimore, Baltimore, Maryland 12101

*Dale L. Oxender,* Department of Biological Chemistry, The University of Michigan, Ann Arbor, Michigan 48109

*Jeffrey E. Pessin,* Department of Biochemistry, University of Massachusetts Medical Center, Worcester, Massachusetts 01605

*Meinrad Peterlik,* Department of General and Experimental Pathology, University of Vienna Medical School, A-1090 Vienna, Austria

*Aline Robin,* Laboratoire des Biomembranes, Institut Jacques Monod, 75251 Paris Cédex 05, France

*Saul Roseman,* Department of Biology and the McCollum-Pratt Institute, The Johns Hopkins University, Baltimore, Maryland 21218

*Terrone L. Rosenberry,* Department of Pharmacology, Case Western Reserve University, Cleveland, Ohio 44106

*G. Sachs,* Center for Ulcer Research and Education, Wadsworth Veterans Administration Center, Los Angeles, California 90073; and School of Medicine, University of California at Los Angeles, Los Angeles, California 90024

*B. Sarkadi,* National Institute of Haematology and Blood Transfusion, Budapest, Hungary

*Antonio Scarpa,* Department of Biochemistry and Biophysics, University of Pennsylvania School of Medicine, Philadelphia, Pennsylvania 19104

*Howard A. Shuman,* Department of Microbiology, College of Physicians and Surgeons, Columbia University, New York, New York 10032

*A. Smolka,* Center for Ulcer Research and Education, Wadsworth Veterans Administration Center, Los Angeles, California 90073; and School of Medicine, University of California at Los Angeles, Los Angeles, California 90024

*W. D. Stein,* Department of Biological Chemistry, Institute of Life Sciences, Hebrew University, Jerusalem, Israel

*Nancy A. Treptow,* Department of Microbiology, College of Physicians and Surgeons, Columbia University, New York, New York 10032

# Preface to the Second Edition

In the first edition of *The Enzymes of Biological Membranes,* published in four volumes in 1976, we collected the mass of widely scattered information on membrane-linked enzymes and metabolic processes up to about 1975. This was a period of transition from the romantic phase of membrane biochemistry, preoccupied with conceptual developments and the general properties of membranes, to an era of mounting interest in the specific properties of membrane-linked enzymes analyzed from the viewpoints of modern enzymology. The level of sophistication in various areas of membrane research varied widely; the structures of cytochrome *c* and cytochrome $b_5$ were known to atomic detail, while the majority of membrane-linked enzymes had not even been isolated.

In the intervening eight years our knowledge of membrane-linked enzymes expanded beyond the wildest expectations. The purpose of the second edition of *The Enzymes of Biological Membranes* is to record these developments. The first volume describes the physical and chemical techniques used in the analysis of the structure and dynamics of biological membranes. In the second volume the enzymes and metabolic systems that participate in the biosynthesis of cell and membrane components are discussed. The third and fourth volumes review recent developments in active transport, oxidative phosphorylation and photosynthesis.

The topics of each volume represent a coherent group in an effort to satisfy specialized interests, but this subdivision is to some extent arbitrary. Several subjects of the first edition were omitted either because they were extensively reviewed recently or because there was not sufficient new information to warrant review at this time. New chapters cover areas where major advances have taken place in recent years. As a result, the second edition is a fundamentally new treatise that faithfully and critically reflects the major transformation and progress of membrane biochemistry in the last eight years. For a deeper insight into membrane function, the coverage includes not only well-defined enzymes, but several membrane proteins with noncatalytic functions.

We hope that *The Enzymes of Biological Membranes* will catalyze the search for general principles that may lead to better understanding of the structure and function of membrane proteins. We ask for your comments and criticisms that may help us to achieve this aim.

My warmest thanks to all who contributed to this work.

Anthony N. Martonosi

*Syracuse, New York*

# Contents of Volume 3

27. *The Energetics of Active Transport*

*W. D. Stein*

I. Introduction ........ 1

II. Thermodynamics of Protein–Ligand Interactions ........ 3

A. The Thermodynamic Box 3 • B. Intrinsic and Apparent Affinities 5

III. Transport of the Simple Carrier—A Uni–Uni Isoenzyme ........ 7

A. Kinetics of the Simple Carrier 7 • B. Countertransport or Antiport—Coupling of Flows on the Simple Carrier 11

IV. The CoTransport Systems—Iso Bi–Bi Enzymes ........ 18

A. The Fundamental Equation of Cotransport 19 • B. Maximizing the Effectiveness of the Cotransporter 19 • C. Apparent Affinities for a Cotransport System 23

V. Primary Active Transport—Chemiosmotic Coupling ........ 27

A. Primary Active Transport vs. Secondary Active Transport 27 • B. The Kinetics of Primary Active Cotransport 28 • C. Maximizing the Effectiveness of the Primary Cotransporter 30 • D. Apparent Affinities of a Primary Cotransport System 30

VI. Conclusions ........ 31

References ........ 32

28. *The $Na^+,K^+$-Transporting Adenosine Triphosphatase*

*I. M. Glynn*

I. Introduction ........ 35

A. Definition 35 • B. Functions 35 • C. History 36

II. Purification of the $Na^+,K^+$-ATPase ........ 38

A. Membrane-Bound $Na^+,K^+$-ATPase 38 • B. Soluble $Na^+,K^+$-ATPase 39 • C. Criteria of Purity 39

III. Cation Fluxes ........ 40

A. An Outline Scheme 40 • B. The Six Flux Modes 40 • C. Compatibility of the Pump Model with Steady-State Flux Kinetics 47

IV. Catalytic Activities Not Associated with Ion Fluxes ........... 47
A. ATP–ADP Exchange 48 • B. Exchange of $^{18}O$ between Orthophosphate and Water 49 • C. Phosphatase Activity 50
V. Phosphorylation Studies ........................................ 52
A. Phosphorylation by ATP 52 • B. Phosphorylation by Orthophosphate 53 • C. Rephosphorylation of Newly Dephosphorylated Enzyme: Evidence for the Occlusion of $K^+$ Ions 54 • D. Mechanism of Hydrolysis of Phosphoenzyme 54 • E. Kinetic Studies 55
VI. Evidence for the Existence of Different Forms of the Dephosphoenzyme ........................................ 58
A. Differences in Reactivity to ATP and Orthophosphate 59 • B. Differences in Reactivity to Inhibitors 59 • C. Differences in Affinity for Nucleotides and Related Compounds 60 • D. Differences in the Pattern of Attack by Proteolytic Enzymes and in the Products of Proteolytic Digestion 61 • E. Differences in Intrinsic Fluorescence and in the Fluorescence of Probe Molecules 61 • F. Differences in Equilibrium Binding of $Na^+$ and $K^+$ Ions 65
VII. The Existence and Role of Occluded-Ion Forms of the $Na^+$,$K^+$-ATPase ........................................ 67
A. The Occluded-$K^+$ Form of Dephosphoenzyme 67 • B. The Occluded-$Na^+$ Form of Phosphoenzyme 70
VIII. Structure of the $Na^+$,$K^+$-ATPase ........................ 71
A. Number, Molecular Weight, and Ratio of Subunits 71 • B. Information from Electron Microscopy 72 • C. Structure of the α- and β-Subunits 73 • D. The Molecular Weight of $Na^+$,$K^+$-ATPase 75 • E. Is the Dimeric Nature of the $Na^+$,$K^+$-ATPase Relevant to Its Mechanism? 79 • F. The Role of Lipids 84
IX. Inhibitors ........................................ 85
A. Cardiac Glycosides 85 • B. Vanadate 88 • C. Oligomycin 90 • D. Thimerosal 91
X. Conclusion ........................................ 93
References ........................................ 94

29. *The Sarcoplasmic Reticulum Membrane*
*Marek Michalak*

I. Introduction ........................................ 115
II. Structure, Function, and Isolation of the Sarcotubular System .. 115
III. Protein Composition of the Sarcoplasmic Reticulum Membrane 117
IV. Ultrastructure and Asymmetry of the Sarcoplasmic Reticulum Membrane ........................................ 120
V. The ATPase of the Sarcoplasmic Reticulum Membrane ........ 122
A. Purification and Characterization of the ATPase 122 • B. Tryptic Fragments and Amino Acid Sequence of the ATPase 123 • C. ATPase–ATPase Interaction: An Oligomeric Form of the Enzyme 125 • D. Lipid–ATPase Interaction 128 • E. Reconstitution of the ATPase 132

VI. Phosphorylation of Sarcoplasmic Reticulum Proteins .......... 134
VII. Calcium Release by the Sarcoplasmic Reticulum ............. 134
VIII. Cardiac Sarcoplasmic Reticulum Membrane ................. 137
IX. Biosynthesis of the Sarcoplasmic Reticulum Membrane ....... 140
A. Biosynthesis of Sarcoplasmic Reticulum *in Vivo* 140 • B. Biosynthesis of Sarcoplasmic Reticulum *in Vitro* 141 • C. Synthesis of Sarcoplasmic Reticulum Proteins in Cell-Free Systems 142
X. Concluding Remarks .................................. 142
References ........................................... 143

*30. Kinetic Regulation of Catalytic and Transport Activities in Sarcoplasmic Reticulum ATPase*

*Giuseppe Inesi and Leopoldo de Meis*

I. Introduction ........................................ 157
A. $Ca^{2+}$ Uptake and ATP Hydrolysis 158 • B. $Ca^{2+}$ Activation of SR ATPase 160
II. Substrate Specificity ............................... 164
A. Phosphorylated Enzymes Intermediate 165 • B. Calcium Translocation and Phosphoenzyme Cleavage 168 • C. Interconversion of the $Ca^{2+}$-Binding Sites of High and Low Affinities 169 • D. The $Ca^{2+}$–$H^+$ Exchange 173 • E. Reversal of the $Ca^{2+}$ Pump and Coupled ATPase 174 • F. Enzyme Phosphorylation in the Absence of a Transmembrane $Ca^{2+}$ Gradient 177 • G. ATP Synthesis 179
III. Conformational Changes .............................. 183
IV. Conclusions ......................................... 184
References ........................................... 185

*31. Calcium-Induced Potassium Transport in Cell Membranes*

*B. Sarkadi and G. Gárdos*

I. Introduction ........................................ 193
II. Characteristics of the $Ca^{2+}$-Induced $K^+$ Transport in Human Red Cells ..................................... 194
A. Calcium Homeostasis in Red Cells—Induction of Rapid $K^+$ Transport 194 • B. Calcium Dependence and Activation Kinetics of $K^+$ Transport 195 • C. Effects of Metal Ions 199 • D. Membrane Potential and $Ca^{2+}$-Induced $K^+$ Transport 200 • E. Effects of Cellular Components and Drugs 205
III. $Ca^{2+}$-Induced $K^+$ Transport in Complex Cells ............... 208
A. Nonexcitable Cells 208 • B. Excitable Cells 212
IV. The Molecular Basis of $Ca^{2+}$-Induced $K^+$ Transport ........... 218
A. The Transport Molecule(s): Biochemical Characteristics 219 • B. The Transport Molecule(s): Physical Characteristics 221

V. Conclusions: The Prevalence and Significance of $Ca^{2+}$-Induced $K^+$ Transport ........ 222
References ........ 223

*32. Biochemistry of Plasma-Membrane Calcium-Transporting Systems*
*Ernesto Carafoli*

I. Introduction ........ 235
II. The $Ca^{2+}$-Pumping ATPase of the Plasma Membrane ........ 236
A. The Enzyme *in Situ* 236 • B. The Purified Enzyme 238
III. The Plasma Membrane $Na^+$–$Ca^{2+}$ Exchange ........ 243
References ........ 245

*33. The Calcium Carriers of Mitochondria*
*Martin Crompton*

I. Introduction ........ 249
II. The $Ca^{2+}$ Uniporter ........ 251
A. The Driving Force for $Ca^{2+}$ Accumulation 251 • B. The Molecular Nature of the Uniporter 253 • C. Factors Affecting Uniporter Activity 254
III. The $Na^+$-$Ca^{2+}$ Carrier ........ 255
A. The Binding of External Substrate Cations 255 • B. The Binding of Internal Cations 256 • C. The Transport Mechanism 257 • D. The Dependence of the $Na^+$-Induced Efflux of $Ca^{2+}$ on the Mitochondrial Energy State 259 • E. Physiological Effectors of the $Na^+$-$Ca^{2+}$ Carrier 260 • F. Tissue Distribution 262
IV. The $Na^+$-Independent Release of $Ca^{2+}$ ........ 262
A. The Reaction Mechanism 263 • B. The Transport System 264 • C. Factors Affecting Activity of the $Na^+$-Independent System 264 • D. Tissue Distribution 265
V. The Resolution of the Component Fluxes of the $Ca^{2+}$ Cycles ... 265
A. Lanthanides 265 • B. Ruthenium Red 266 • C. $Ca^{2+}$ Antagonists 266 • D. Other Distinctions 267
VI. The Capacity of Isolated Mitochondria to Retain Accumulated $Ca^{2+}$ ........ 267
A. The Dependence of $Ca^{2+}$-Induced Destabilization on Inorganic Phosphate and Redox State 267 • B. The Reversibility of $Ca^{2+}$-Induced Destabilization 268 • C. The Effects of Adenine Nucleotides and $Mg^{2+}$ 269
VII. $Ca^{2+}$ Recycling ........ 270
A. Properties of the $Ca^{2+}$ Cycles 272 • B. $Ca^{2+}$ Recycling in Heart Mitochondria 275 • C. $Ca^{2+}$ Recycling in Liver Mitochondria 277
VIII. Concluding Remarks ........ 278
References ........ 279

34. *Intestinal Phosphate Transport*
*Meinrad Peterlik*

I. Introduction ........................................ 287
II. Intestinal Sites and Modes of Phosphate Absorption ............ 288
A. $P_i$ Transport in Small and Large Intestine 288 • B. Influence of Vitamin D 289 • C. Transepithelial $P_i$ Transport: Transcellular Pathways 289 • D. Transepithelial $P_i$ Transport: Paracellular Pathway 295 • E. Absorption vs. Secretion 296
III. Active Transport of Inorganic Phosphate across the Brush-Border Membrane ........................................ 297
A. Transmucosal $P_i$ Transport in Intact Cells 298 • B. $Na^+$-Gradient-Driven $P_i$ Transport in Brush-Border Membrane Vesicles 299 • C. Attempts at Characterization of the $P_i$ Carrier 302
IV. Hormonal Regulation of Intestinal $P_i$ Transport ............... 303
A. Vitamin D 304 • B. Parathyroid Hormone and Cyclic AMP 308 • C. Calcitonin 309 • D. Glucocorticoids 309 • E. Insulin 310
V. Intestinal Phosphate Absorption in Health and Disease ......... 311
A. Efficiency of $P_i$ Absorption under Physiological Conditions 311 • B. Alteration of Phosphate Absorption in Human Disease 313
References ........................................ 315

35. *Ion Transport in Nerve Membrane*
*Lorin J. Mullins*

I. Introduction ........................................ 321
II. Methods for Transport Studies ............................ 322
A. Introduction 322 • B. Isotopes 322 • C. Analytical Measurements 322 • D. Electrode Measurements 323 • E. Optical Measurements 323
III. Active Transport ........................................ 323
A. Introduction to Na Fluxes 323 • B. Na-Transport Systems 324 • C. Potassium Fluxes 326 • D. Chloride Transport 326 • E. Introduction to Calcium Transport 326 • F. Magnesium Transport 329 • G. Hydrogen Ion Transport 331
IV. Physiological Integration of Ion Fluxes ..................... 331
References ........................................ 332

36. *The Molecular Basis of Neurotransmission: Structure and Function of the Nicotinic Acetylcholine Receptor*
*Robert Anholt, Jon Lindstrom, and Mauricio Montal*

I. Introduction ........................................ 335

II. Structural Aspects of the Acetylcholine Receptor .............. 337
A. Acetylcholine Receptor from Electric Organ 337 • B. Acetylcholine Receptor from Muscle 351 • C. Assembly and Degradation of Acetylcholine Receptors 353
III. Functional Aspects of the Acetylcholine Receptor ............. 360
A. Electrophysiology of the Postsynaptic Membrane 360 • B. Pharmacology of the Nicotinic Acetylcholine Receptor 363
IV. Structure–Function Correlations within the Acetylcholine Receptor Molecule: Reconstitution Studies as an Experimental Approach .. 371
A. Reconstitution of Acetylcholine Receptors in Model Membranes 371 • B. Functional Studies of Acetylcholine Receptors after Reconstitution in Model Membranes 376
V. Conclusions .......................................... 383
References .......................................... 384

37. *Structural Distinctions among Acetylcholinesterase Forms*
*Terrone L. Rosenberry*

I. Introduction .......................................... 403
II. Asymmetric Acetylcholinesterase Forms Contain Collagen-like Tail Structures .......................................... 404
A. Acetylcholinesterase from Fish Electric Organs Provides a Structural Model for the Asymmetric Forms 405 • B. Asymmetric Acetylcholinesterase Forms in Other Tissues Have Hydrodynamic and Aggregation Properties Similar to the Electric Organ Forms 412 • C. Asymmetric Acetylcholinesterases Appear To Be Localized in the Extracellular Basement Membrane Matrix in Skeletal Muscle 413
III. Globular Acetylcholinesterase Occurs as Soluble and Amphipathic Forms .......................................... 414
A. Human Erythrocyte Acetylcholinesterase Is an Amphipathic Form 415 • B. Comparison of Human Erythrocyte Acetylcholinesterase to Globular Acetylcholinesterases in Other Tissues That Bind Detergent 419
IV. Relationships among Acetylcholinesterase Forms .............. 419
A. Acetylcholinesterase Forms in Rat Diaphragm 420 • B. Biosynthesis of Acetylcholinesterase Forms 423
References .......................................... 424

38. *The Gastric H,K-ATPase*
*L. D. Faller, A. Smolka, and G. Sachs*

I. Introduction .......................................... 431
II. Localization of the H,K-ATPase within the Parietal Cell ....... 432
III. Discovery of a Pathway for the Transport of $K^+$ Salts in Membrane Vesicles Isolated from Secreting Tissues .......... 434
IV. Structure of the H,K-ATPase .............................. 437

V. Catalytic Properties of the H,K-ATPase .................... 439
VI. Model .................... 444
VII. Comparison with Other Transport ATPases .................... 445
VIII. Problems and Future Research .................... 446
References .................... 447

*39. $H^+$-Translocating ATPase and Other Membrane Enzymes Involved in the Accumulation and Storage of Biological Amines in Chromaffin Granules*

*Sally E. Carty, Robert G. Johnson, and Antonio Scarpa*

I. Introduction .................... 449
II. Isolation of Chromaffin Granules and Preparation of Chromaffin Ghosts .................... 450
III. The Composition of the Chromaffin Granule .................... 451
A. Membrane Proteins 451 • B. Soluble Proteins 454 • C. Lipids 455 • D. Storage Components 455 • E. Topography 456
IV. The Electrochemical $H^+$ Gradient .................... 458
A. Membrane Permeability to Ions 458 • B. Measurement of the $\Delta pH$ 459 • C. Measurement of the $\Delta\psi$ 461 • D. Measurement of the $\Delta\bar{\mu}_{H^+}$ 463
V. The $H^+$-Translocating ATPase .................... 463
A. Generation of a $\Delta\bar{\mu}_{H^+}$ 464 • B. Stoichiometry of $H^+$/ATP 465 • C. Reversal of the ATPase 467 • D. Physicochemical Properties of the ATPase 467 • E. Isolation and Reconstitution 469
VI. Amine Accumulation .................... 469
A. Physiochemical Properties 470 • B. Inhibitors 472 • C. Coupling to the $\Delta\mu_{H^+}$ 472 • D. Net Accumulation of Biogenic Amines 478
VII. The Electron-Transport Chain .................... 479
A. Organization 479 • B. Physicochemical Properties 481 • C. Physiologic Role 483
VIII. Other Transport Systems .................... 484
A. ATP Accumulation 484 • B. Ascorbate Accumulation 485 • C. Calcium Accumulation 486
IX. Conclusion: Biogenic Amine Transport into Other Organelles .. 488
References .................... 489

*40. Hexose Transport and Its Regulation in Mammalian Cells*

*Jeffrey E. Pessin and Michael P. Czech*

I. Introduction .................... 497
A. Facilitative D-Glucose Transport 498 • B. Assay Methodology 500
II. Human Erythrocyte D-Glucose Transport System .................... 501

A. Affinity Labeling 502 • B. Purification 503

III. Regulation of the D-Glucose Transporter in Cultured Cells ...... 504

IV. Insulin Regulation of D-Glucose Transport Activity ............ 506

A. Insulin Binding and D-Glucose Transport Activation 506 • B. Insulinomimetic Agents 507 • C. Insulin-Receptor Aggregation 509 • D. Structural Relationships between the D-Glucose Transporter and Insulin Receptor 510 • E. Identification of the Insulin-Sensitive D-Glucose Transporter 511

V. Mechanism of Insulin Activation of the D-Glucose Transporter ... 511

VI. Summary and Conclusions ................................ 513

References ................................................ 514

*41. The Bacterial Phosphoenolpyruvate:Sugar Phosphotransferase System*

*Norman D. Meadow, Maria A. Kukuruzinska, and Saul Roseman*

I. Introduction ............................................ 523

II. Enzyme I ............................................... 524

III. HPr ................................................... 528

IV. III$^{Glc}$ ............................................ 530

A. Isolation and Characterization 530 • B. Kinetic Studies with III$^{Glc}$ 533

V. Sugar Receptor (II-B) Proteins ........................... 537

A. Purification 537 • B. General Properties 538 • C. Organization of the Enzymes II-B in Membranes 540 • D. Exchange Transphosphorylation 540

VI. Regulation of the PTS .................................... 542

A. Introduction 542 • B. Regulation via Enzyme I 543 • C. Regulation via Acetate Kinase 543 • D. Regulation of the Activity of the Enzymes II 544 • E. Regulation of Methyl α-Glucoside Transport in Membrane Vesicles 546

VII. PTS Regulation of Non-PTS Systems: PTS-Mediated Repression 547

A. The *crr* and *iex* Genes 547 • B. *crr* Is the Structural Gene for III$^{Glc}$ 548 • C. Mechanism of Regulation of Non-PTS Transport Systems 550

VIII. Prospectus .............................................. 552

References ................................................ 553

*42. The Maltose–Maltodextrin-Transport System of Escherichia coli K–12*

*Howard A. Shuman and Nancy A. Treptow*

I. Introduction ............................................ 561

II. Maltose and Maltodextrin Catabolism in *Escherichia coli* K-12 .. 562

III. General Properties of Maltose and Maltodextrin Transport ...... 563

IV. Properties of the Individual Components ..................... 564

A. The LamB Protein 564 • B. The Maltose-Binding

Protein 566 • C. The MalF and MalK Proteins 566 • D. The MalG Protein 569
V. Interactions between the Maltose-Binding Protein and the Membrane Components ........ 569
VI. A Model for the Operation of the Maltose–Maltodextrin-Transport System ........ 571
A. Transport across the Outer Membrane 571 • B. Transport across the Cytoplasmic Membrane 572
References ........ 573

43. *Bacterial Amino-Acid-Transport Systems*
*Robert Landick, Dale L. Oxender and Giovanna Ferro-Luzzi Ames*

I. Introduction ........ 577
II. Classes of Transport Systems ........ 578
A. Multiplicity of Amino-Acid-Transport Systems 579 • B. Membrane-Bound Systems 585 • C. Periplasmic-BP-Dependent Systems 585
III. Nature of Protein Components of BP-Dependent Transport Systems ........ 587
A. Periplasmic Components 587 • B. Membrane Components 588
IV. Genetic and Physical Maps of the LIV-I and Histidine-Transport Genes ........ 588
V. Assembly of Transport Components ........ 590
VI. Energization and Reconstitution of Amino-Acid Transport ........ 593
A. Membrane-Bound, Osmotic-Shock-Resistant Systems: Energization and Reconstitution 594 • B. Energization of BP-Dependent Systems 595 • C. Reconstitution of BP-Dependent Amino-Acid Transport 596
VII. Possible Models for Amino-Acid Transport ........ 597
VIII. Regulation of Amino-Acid Transport ........ 598
A. Regulation of the Histidine-Transport System 599 • B. Regulation of the LIV-I-Transport System 600 • C. Control of Membrane Protein Synthesis 602
IX. Evolutionary Relationships among Periplasmic Systems ........ 602
References ........ 605

44. *The Iron-Transport Systems of Escherichia coli*
*Volkmar Braun*

I. Types of Transport Systems in *Escherichia coli* ........ 617
A. Uptake through the Outer Membrane 617 • B. Uptake through the Cytoplasmic Membrane 618
II. Peculiarities of the Iron-Transport Systems ........ 619

A. Requirement of Siderophores 619 • B. Receptor Proteins at the Cell Surface 622 • C. Functions of Genes Assigned to the Cytoplasmic Membrane; Release of Iron from the Siderophores; Modification of Siderophores 628 • D. The TonB and ExbB Functions 632 • E. Regulation of the Iron-Transport Systems 635

III. Iron Supply and Virulence of *Escherichia coli*. . . . . . . . . . . . . . . . 640
IV. Outlook . . . . . . . . . . . . . . . . . . . . . . . . . . . . . . . . . . . . 645
References . . . . . . . . . . . . . . . . . . . . . . . . . . . . . . . . . . . . 646

*45. Potassium Pathways in Escherichia coli*
*Adam Kepes, Jean Meury, and Aline Robin*

I. Introduction . . . . . . . . . . . . . . . . . . . . . . . . . . . . . . . . . . 653
II. Acquisition of Potassium-Free Cells: A Mechanical Disorganization of the Hydrophobic Continuum . . . . . . . . . . . . . . 654
III. Potassium Uptake and Its Mechanochemical Switch . . . . . . . . . . . 655
IV. Potassium–Potassium Exchange: A Metabolism-Dependent, Energy-Independent Process . . . . . . . . . . . . . . . . . . . . . . . . 658
V. The Effect of Thiol Reagents: A Reversible Opening of Potassium-Specific Channels . . . . . . . . . . . . . . . . . . . . . . . . 660
VI. The $K^+$ Channel is Specifically Controlled by Glutathione: It Is a GSH-Controlled Channel (GCC) . . . . . . . . . . . . . . . . . . . . 661
VII. Some Unsolved Problems . . . . . . . . . . . . . . . . . . . . . . . . . . 663
References . . . . . . . . . . . . . . . . . . . . . . . . . . . . . . . . . . . . 664

*Index* . . . . . . . . . . . . . . . . . . . . . . . . . . . . . . . . . . . . . 667

# Contents of Volume 1

1. *Electron Microscopy of Biological Membranes*
   *K. Mühlethaler and Frances Jay*

2. *Associations of Cytoskeletal Proteins with Plasma Membranes*
   *Carl M. Cohen and Deborah K. Smith*

3. *Cell Coupling*
   *Camillo Peracchia*

4. *Lipid Polymorphism and Membrane Function*
   *B. de Kruijff, P. R. Cullis, A. J. Verkleij, M. J. Hope,*
   *C. J. A. van Echteld, and T. F. Taraschi*

5. *Intrinsic Protein-Lipid Interactions in Biomembranes*
   *Jeff Leaver and Dennis Chapman*

6. *On the Molecular Structure of the Gramicidin Transmembrane Channel*
   *Dan W. Urry*

7. *Conventional ESR Spectroscopy of Membrane Proteins: Recent Applications*
   *Philippe F. Devaux*

8. *Saturation Transfer EPR Studies of Microsecond Rotational Motions in Biological Membranes*
   *David D. Thomas*

9. *Dye Probes of Cell, Organelle, and Vesicle Membrane Potentials*
*Alan S. Waggoner*

10. *Selective Covalent Modification of Membrane Components*
*Hans Sigrist and Peter Zahler*

11. *Calcium Ions, Enzymes, and Cell Fusion*
*Jack A. Lucy*

12. *Role of Membrane Fluidity in the Expression of Biological Functions*
*Juan Yguerabide and Evangelina E. Yguerabide*

13. *Rotational Diffusion of Membrane Proteins: Optical Methods*
*Peter B. Garland and Pauline Johnson*

*Index*

# Contents of Volume 2

14. *Ether-Linked Glycerolipids and Their Bioactive Species: Enzymes and Metabolic Regulation*
*Fred Snyder, Ten-ching Lee, and Robert L. Wykle*

15. *Fatty Acid Synthetases of Eukaryotic Cells*
*Salih J. Wakil and James K. Stoops*

16. *Properties and Function of Phosphatidylcholine Transfer Proteins*
*Karel W. A. Wirtz, Tom Teerlink, and Rob Akeroyd*

17. *Carnitine Palmitoyltransferase and Transport of Fatty Acids*
*Charles L. Hoppel and Linda Brady*

18. *Membrane-Bound Enzymes of Cholesterol Biosynthesis: Resolution and Identification of the Components Required for Cholesterol Synthesis from Squalene*
*James M. Trzaskos and James L. Gaylor*

19. *Membrane-Bound Enzymes in Plant Sterol Biosynthesis*
*Trevor W. Goodwin, C.B.E., F.R.S.*

20. *Glycosyltransferases Involved in the Biosynthesis of Protein-Bound Oligosaccharides of the Asparagine-N-Acetyl-D-Glucosamine and Serine (Threonine)-N-Acetyl-D-Galactosamine Types*
*Harry Schachter, Saroja Narasimhan, Paul Gleeson, George Vella, and Inka Brockhausen*

21. *Biosynthesis of the Bacterial Envelope Polymers Teichoic Acid and Teichuronic Acid*
*Ian C. Hancock and James Baddiley*

22. *The Major Outer Membrane Lipoprotein of Escherichia coli: Secretion, Modification, and Processing*
*George P. Vlasuk, John Ghrayeb, and Masayori Inouye*

23. *Anchoring and Biosynthesis of a Major Intrinsic Plasma Membrane Protein: The Sucrase–Isomaltase Complex of the Small-Intestinal Brush Border*
*Giorgio Semenza*

24. *Multifunctional Glucose-6-Phosphatase: A Critical Review*
*Robert C. Nordlie and Katherine A. Sukalski*

25. *The Beta Adrenergic Receptor: Elucidation of its Molecular Structure*
*Robert G. L. Shorr, Robert J. Lefkowitz, and Marc G. Caron*

26. *Ionic Channels and Their Metabolic Control*
*P. G. Kostyuk*

*Index*

# Contents of Volume 4


46. *The Enzymes and the Enzyme Complexes of the Mitochondrial Oxidative Phosphorylation System*
*Youssef Hatefi, C. Ian Ragan, and Yves M. Galante*

47. *Proton Diffusion and the Bioenergies of Enzymes in Membranes*
*Robert J. P. Williams*

48. *Relationships between Structure and Function in Cytochrome Oxidase*
*Mårten Wikström, Matti Saraste, and Timo Penttilä*

49. *$H^+$-ATPase as an Energy-Converting Enzyme*
*Toshiro Hamamoto and Yasuo Kagawa*

50. *The Proton-Translocating Membrane ATPase ($F_1F_0$) in Streptococcus faecalis (faecium)*
*Adolph Abrams*

51. *Cytochrome b of the Respiratory Chain*
*Henry R. Mahler and Philip S. Perlman*

52. *Cytochrome $b_5$ and Cytochrome $b_5$ Reductase from a Chemical and X-Ray Diffraction Viewpoint*
*F. Scott Mathews and Edmund W. Czerwinski*

53. *Iron–Sulfur Clusters in Mitochondrial Enzymes*
*Thomas P. Singer and Rona R. Ramsay*

54. *The Structure of Mitochondrial Ubiquinol: Cytochrome c Reductase*
*Hanns Weiss, Stephen T. Perkins, and Kevin Leonard*

55. *The Mechanism of the Ubiquinol:Cytochrome c Oxidoreductases of Mitochondrial and of Rhodopseudomonas sphaeroides*
*Antony R. Crofts*

56. *Functions of the Subunits and Regulation of Chloroplast Coupling Factor 1*
*Richard E. McCarty and James V. Moroney*

57. *Biosynthesis of the Yeast Mitochondrial $H^+$-ATPase Complex*
*Sangkot Marzuki and Anthony W. Linnane*

58. *Synthesis and Intracellular Transport of Mitochondrial Proteins*
*Matthew A. Harmey and Walter Neupert*

59. *Plasma-Membrane Redox Enzymes*
*F. L. Crane, H. Löw, and M. G. Clark*

60. *The ADP/ATP Carrier in Mitochondrial Membranes*
*Martin Klingenberg*

61. *Bacteriorhodopsin and Rhodopsin: Structure and Function*
*Yuri A. Ovchinnikov and Nazhmutdin G. Abdulaev*

*Index*

# 27

# The Energetics of Active Transport

W. D. Stein

## I. INTRODUCTION

This chapter aims to give an overview of the workings of active-transport systems. I shall not go much into the details of any particular transport system—these will be the subjects of later chapters of this treatise—but I shall want to present what seem to me to be the general principles involved in the coupling between two flows of transported substrates or the coupling between the flow of a transported substrate and the progress of the chemical reaction that drives this flow. The standpoint adopted in this chapter is that such coupling, and, hence, active transport itself, arises simply from the properties of the membrane carriers, and, in particular, that such carriers, by virtue of their role as carriers, exist in two conformations facing the two sides of the cell membrane. It is the redistribution of carrier between its two major forms that brings about transport and active transport. The study of active transport is, then, the exploration of the properties of membrane-carrier systems. I shall work step-by-step through the principles of ligand-carrier interactions, transport on the simple carrier, active transport by countertransport, and, finally, active transport by cotransport, and in this way succeed, I hope, in showing that a sound knowledge of carrier kinetics provides the basis for an understanding of active transport. Much has been written on the energetics of active transport, on the interaction between the driving and the driven substrates, on the conformational energetics of the transporters, and on the possible role of high-energy chemical substrates in active transport. But an emphasis on the simple carrier basis of active transport may provide a newer way of looking at these old problems.

The physical model that the reader should bear in mind, is of the carrier (or

---

*W. D. Stein* • Department of Biological Chemistry, Institute of Life Sciences, Hebrew University, Jerusalem, Israel.

transporter—these terms are used interchangeably) as being a protein molecule embedded in the cell membrane, and capable of existing in two conformations (Figure 1). In one conformation, the carrier has its binding sites for substrate(s) at one face of the membrane, e.g., the cytoplasmic surface. In the alternate conformation, the carrier has its binding sites for substrate(s) available at the other, e.g., extracellular, surface of the membrane. Flipping between these two conformations is the fundamental property of the transporter, from which all else follows. There will be rules that will govern the rates at which these flipping interconversions can occur. For instance, the flipping rate might be faster in one direction than in the other, the rate may be faster when substrate is bound to the transporter than when the transporter is free of substrate, or the flipping might occur only when two substrate molecules (the same or different) are bound to the carrier. But without flipping, there is no transport. This flipping is merely the well-studied phenomenon found for many protein molecules, that they are able to exist in two conformations and interconvert between these two conformations. The only novel feature of the carrier is that, in these two conformations, there is a shift in the availability of the substrate-binding sites from one side of the membrane to the other. What evidence is there that transporters exist in two conformations with binding sites alternately exposed to the two faces of the membrane? The strongest evidence comes from studies on the sodium pump, the sodium–potassium-activated ATPase found in animal-cell membranes (see Glynn, 1984). Karlish and Yates (1978) showed that the intrinsic tryptophan fluorescence of the purified enzyme (an indication of the local environment of these tryptophan residues and, hence, of the conformation of the transporter) was different according to whether the experimental conditions favor a conformation having high-affinity sites for sodium facing the cytoplasmic face of the membrane over a conformation with low-affinity sodium-binding sites facing the extracellular surface. Karlish and Pick (1981), using the ATPase reconstituted into phospholipid vesicles where the ionic composition of the media bathing the two surfaces of the membrane can be strictly controlled, showed that the transporter can be driven into one or other conformation by manipulation of the ionic gradients. The two conformations here differ both in the rate at which they are attacked by trypsin (present at the vesicle surface identified with the intracellular surface of the native-cell membrane) and in the products of this proteolytic digestion. Other reporters of conformational changes (reviewed by Glynn and Karlish, 1982) give comparable results. Another valuable study, in this case of the ATP–ADP exchange carrier of the mitochondrion (see Klingenberg, this volume), similarly showed that two conformations of this transporter, differing in their sensitivity to the inhibitors bongkrekate and atractylate, can be identified, and their interconversion can be manipulated by altering the gradients across the membrane of the transported substrates, ATP and ADP. For many other carrier systems (see Stein, 1981), kinetic evidence based on measurements of the transport process itself, points very strongly to the existence of two states of the carrier, their interconversion being the transport event. It seems clear that the essence of the carrier model is the existence of these two conformational states of the transporter, with the binding equilibria between substrate and carrier at one face of the membrane being physically separated from the binding equilibria at the other face of the membrane. The consequences of this "fundamental dogma" of our carrier world will be explored in the following sections.

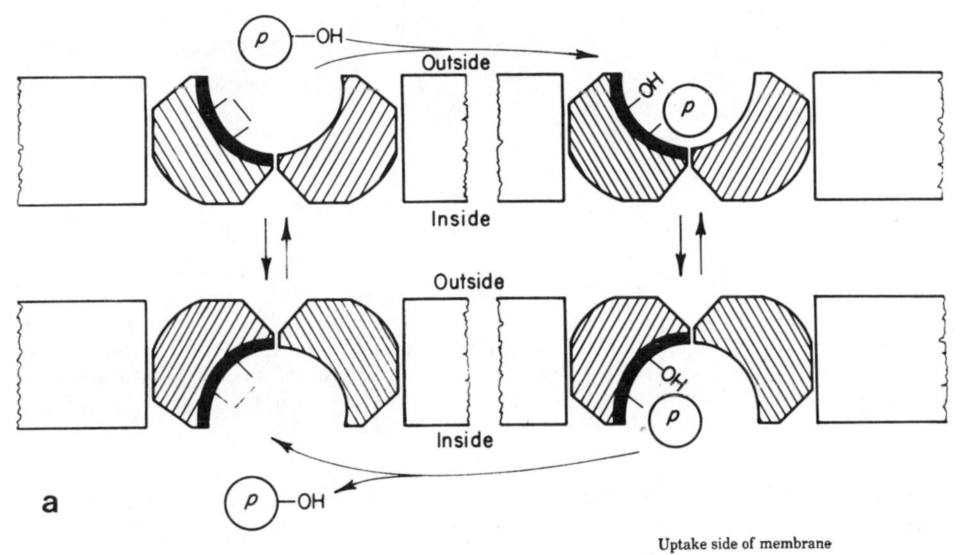


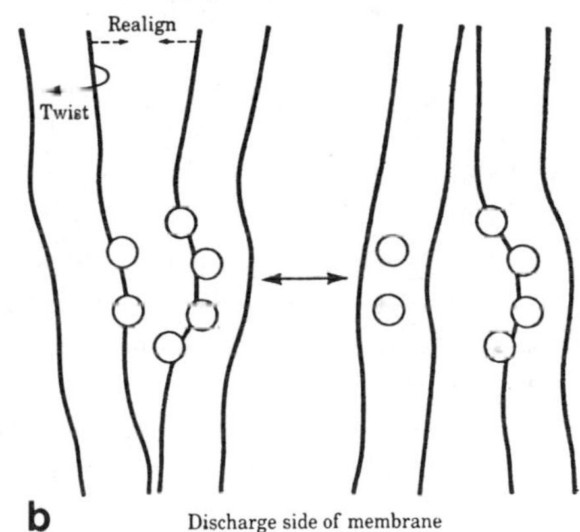


*Figure 1.* Twenty-five years of active transport models. (a) A hypothetical carrier molecule, embedded in a cell membrane (the hatched areas), capable of existing in two conformations, facing the inside and the outside of the cell, and free or in combination with the permeant. The ringed *P* is a phosphate anion and the substrate of the transporter. From Mitchell (1957). (b) A carrier whose binding sites for substrate (shown as amino acid side chains depicted as circles) are alternately available to the inside or outside of the cell, following a conformation change of the protein, and involving a twist of a polypeptide helix. From Tanford (1982).

## II. THERMODYNAMICS OF PROTEIN–LIGAND INTERACTIONS

### A. The Thermodynamic Box

Before I proceed to consider transport itself, I want to look at some simpler situations concerning the interaction between protein molecules (of any type) and the ligands that bind to them. Figure 2a shows the situation where a protein $E$ can bind two substrates $A$ and $B$, via intermediate complexes $EA$ or $EB$, to form the tertiary complex $EAB$. The overall reaction is $E + A + B = EAB$, and this is accompanied by a certain free energy change $\Delta G_{EAB}$. The free energy change must be the same whether $EAB$ is formed via the path containing $EA$ or via the path with $EB$. It could be that $A$ binds better to free $E$ than to $EB$, that is, with a lower investment of free energy, but then it must follow that $B$ binds better to free $E$ than it does to $EA$. Since

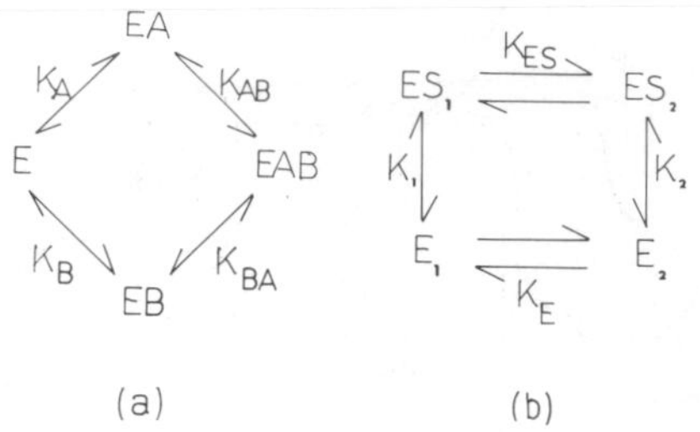

*Figure 2.* Two "thermodynamic boxes." (a) A protein $E$ interacting with two ligands $A$ and $B$ to form complexes $EA$ or $EB$, and then the tertiary complex $EAB$. (b) A protein $E$ capable of existing in two conformations $E_1$ and $E_2$, each capable of binding with the ligand $S$, to form $ES_1$ or $ES_2$. In this and subsequent figures, a pair of single-headed arrows represents interconvertible, conformational changes of the protein, while a double-headed arrow represents the equilibrium binding of ligand with the protein.

the dissociation constant for the formation of the various complexes $EA$, $EB$, and $EAB$ are simply related to the relevant free energy changes by the equation $\Delta G = RT \log_e K_{\text{diss}}$, it follows that the dissociation constants are connected by the equation

$$K_A K_{AB} = K_B K_{BA} \tag{1}$$

where the terms in $K$ are defined in Figure 2a.

To see the significance of this relation and to provide a basis for its use in subsequent discussions, consider the following "Gedanken experiment." Take a hypothetical, very unusual system where all four compounds at the corners of our "thermodynamic box" have the same standard state-free energy, that is, the energy of $(E + A + B)$ equals that of $EA + B$, $EB + A$, and finally $EAB$. Then all four complexes can be considered as being on the same horizontal plane of free energy (Figure 3a). Now stabilize one corner of the "box," that is, reduce the free energy of the form

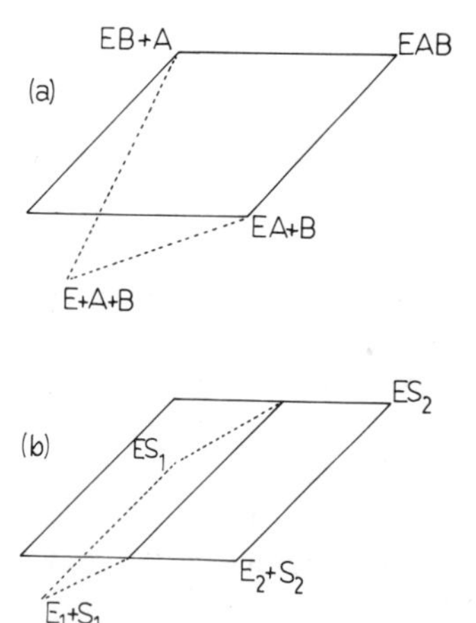


*Figure 3.* The thermodynamic boxes of Figure 2. (a) One corner of the box is stabilized. (b) One edge is stabilized.

represented at that corner, e.g., $E$ in Figure 2a. Formation of $EA$ is now more difficult than before, but so is the formation of $EB$. The rule of Eq. (1) holds. Alternatively, stabilize $EA$; the formation of $EA$ is easier, but the formation of $EAB$ from $EA$ is more difficult. The rule of Eq. (1) holds. By stabilizing or destabilizing any of the four states of this linked equilibrium, one will always alter the free energy relationships of two equilibria of the system. Stabilizing the binding of the first ligand is always at the expense of destabilizing the binding of the second ligand. It has frequently been argued that these linked-binding equilibria are at the basis of the effective coupling of cotransported substrates (see, for instance, Weber, 1975; Tanford, 1982). I shall now show that these effects are of some significance, but not as significant as those concerned with the transport equilibria themselves.

Figure 2b depicts the binding of a substrate $S$ to two different conformations of the protein $E$ ($E_1$ and $E_2$) with the formation of two different complexes $ES_1$ and $ES_2$. Once again, this is a thermodynamic "box." To go from $E_1$ to $ES_2$, via $ES_1$ or $E_2$, must involve the same overall free energy change with the result that the relevant dissociation constants are connected by the equation

$$K_E K_2 = K_{ES} K_1 \quad (2)$$

where the constants $K_E$ and $K_{ES}$ refer to the equilibrium constants of the conformation changes while the constants $K_1$ and $K_2$ refer to the dissociation constant of ligand binding, as depicted in the figure. Once again, we can perform a "Gedanken experiment" to see the effect of stabilizing one of the forms depicted by a corner of the "box." Thus, for instance, $E_1$ can be stabilized over $E_2$ (Figure 3a). $E_1$ is the prevalent form of the free protein, but it is more difficult to form $ES_1$ from $E_1$ than to form $ES_2$ from $E_2$; the affinity of $S$ for protein in state I is lower than that in state II. If a whole edge of the "box" is stabilized, that is, $E_1$ and $ES_1$ are stabilized equally over their counterparts in state II (Figure 3b), then the affinity of $S$ for the protein in either state I or state II is the same. An overall stabilizing of state I forms makes no difference to the intrinsic affinities of ligand binding.

## B. Intrinsic and Apparent Affinities

The above considerations apply to the intrinsic affinities of binding of ligand to protein. In many cases, however, one will be interested rather in the apparent affinities, which are the affinities empirically measured by studying the overall binding of ligand to protein. This distinction is of crucial importance when the protein exists in two conformations as do our transport proteins. In Figure 4, the protein exists in two conformations $E_1$ and $E_2$, and can bind to substrate $A$ in either conformation to form $EA_1$ or $EA_2$. The two conformations $E_1$ and $E_2$ can be of different standard state-free energies. Let $E_1$ be the more stable form, that is, of lower energy. Most of the free protein is in form $E_1$ and is available for binding to $A$ to form $EA_1$. The operative binding constant for formation of $EA_1$ is approximately given by the intrinsic affinity $K_1$. For binding to $E_2$, however, such is not at all the case. There will always be very little free $E_2$ available for interacting with $A$. To trap substantial amounts of protein

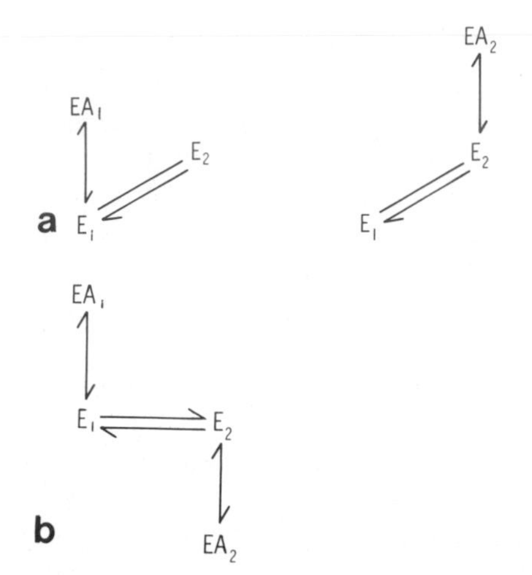


*Figure 4.* A ligand interacting with two forms of a protein. (a) The ligand $A$ interacts with the more stable form of the protein (left). It interacts with the less stable form (right). (b) The complex $EA_2$ is the stablest form, while the complex $EA_1$ is less stable.

in the form $EA_2$, enough $A$ must be present to pull the equilibrium between $E_1$ and $E_2$ over to the side of $E_2$. The apparent affinity for binding to form $EA_2$ is far lower than the intrinsic affinity of $A$ for $E_2$. Formally, we can readily derive the relationship between the intrinsic and apparent affinities. The free protein partitions between forms $E_1$ and $E_2$, such that $k_2/(k_1 + k_2)$ is in the form $E_1$ while $k_1/k_1 + k_2)$ is in the form of $E_2$ (where the terms in $k$ are the rate constants of the interconversion of $E$ forms, $k_1$ from $E_1$ to $E_2$ and $k_2$ from $E_2$ to $E_1$). The effective amount of $E$ present to complex with $A$ is diminished by these appropriate partition ratios, and so the effective dissociation constant of the $EA$ complex is *increased* by the reciprocal of these ratios. Thus,

$$K_1^{\text{apparent}} = K_1^{\text{intrinsic}} \times \frac{k_1 + k_2}{k_2} \tag{3a}$$

and

$$K_2^{\text{apparent}} = K_2^{\text{intrinsic}} \times \frac{k_1 + k_2}{k_1} \tag{3b}$$

Since $E_1$, we have postulated, is more stable than $E_2$ and, therefore, $k_2$ is larger than $k_1$, $K_1^{\text{app}}$ is smaller than $K_2^{\text{app}}$, and the former is far closer to the intrinsic affinity at that side. If we do an experiment in which we titrate the protein $E$ with $A$ in such a way that $A$ can combine only with form $E_1$ or with form $E_2$ (that is, if $E_1$ and $E_2$ face different sides of a membrane), then the amount of $A$ required to drive half of the protein into the form $EA$ (measured spectroscopically at equilibrium) will be far higher on the side at which the protein is less stable and, hence, less prevalent. The apparent half-saturation concentrations for this experiment will be given by Eqs. (3a and b). Such relations have been demonstrated directly for the sodium-pump system (Karlish, 1980).

Finally look at Figure 4b. This shows the protein $E$ able to combine with $A$ to form both $EA_1$ and $EA_2$. Take the case where the forms of $E$ are of equal energy, but where $EA_2$ is of lower free energy (that is, more stable) than $EA_1$. Here, the intrinsic affinity of $A$ for $E_2$ is higher than for $E_1$. If we add $A$ to $E$, the complex $EA_2$ is preferentially formed, with the protein as a whole driven into the state II, as $EA_2$. The protein will always be driven into the state in which it displays the highest affinity for ligand.

We can now drop the mask of innocence and state frankly that the two states of the protein that we have been considering are, in fact, the two conformations of the transporter protein, as its binding sites for ligand (substrate or permeant) face alternately one or the other side of the membrane. How do the relationships that have been established need to be modified when we consider the kinetic events of transmembrane transport, rather than merely the equilibria of binding, as we have discussed so far. Let us proceed to consider the kinetics of the simple carrier.

## III. TRANSPORT ON THE SIMPLE CARRIER—AN ISO UNI–UNI ENZYME

### A. Kinetics of the Simple Carrier

Figure 5 depicts the model of the simple carrier. Forms with subscript 1 face side 1 of the membrane, and those with subscript 2, face side 2. The rate constants for the various interconversions or bindings are depicted in the figure. There have been many

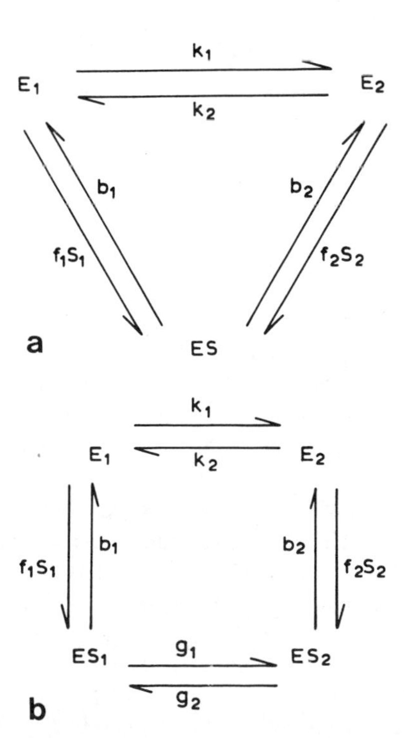


*Figure 5.* The simple carrier. (a) The "necessarily simplified" form of Stein and Lieb (1973), where only one form of the carrier–substrate complex is depicted. (b) The conventional form, with two carrier–substrate complexes. $E$ is the free carrier, $ES$, the carrier–substrate complex. Subscripts 1 refer to side 1 of the membrane, subscripts 2, to side 2. Rate constants in $f,b,g$, and $k$ refer to the indicated reaction steps.

Table 1. Kinetic Solutions for Various Forms of the Simple Carrier[a]

| Model | | Figure 5a | Figure 5b | Figure 5b, transport steps, rate limiting |
|---|---|---|---|---|
| $nR_{12}$ | = | $l/b_2 + l/k_2$ | $l/b_2 + l/k_2 + (l/g_1)((b_2 + g_2)/b_2)$ | $l/k_2 + l/g_1$ |
| $nR_{21}$ | = | $l/b_1 + l/k_1$ | $l/b_1 + l/k_1 + (l/g_2)((b_1 + g_1)/b_1)$ | $l/k_1 + l/g_2$ |
| $nR_{ee}$ | = | $l/b_1 + l/b_2$ | $l/b_1 + l/b_2 + (l/g_1)((b_2 + g_2)/b_2) + l/g_2)((b_1 + g_1)/b_1)$ | $l/g_1 + l/g_2$ |
| $nR_{oo}$ | = | $l/k_1 + l/k_2$ | $l/k_1 + l/k_2$ | $l/k_1 + l/k_2$ |
| $K$ | = | $k_1/f_1 + k_2/f_2$ | $k_1/f_1 + k_2/f_2 + b_1k_1/f_1g_1(= b_2k_2/f_2g_2)$ | $b_1k_1/f_1g_1 = b_2k_2/f_2g_2$ |
| Constraint | | $b_1f_2k_1 = b_2f_1k_2$ | $b_1f_2g_2k_1 = b_2f_1g_1k_2$ | $b_1f_2g_2k_1 = b_2f_1g_1k_2$ |

[a] $v_{1\rightarrow 2} = \frac{KS_1 + S_1S_2}{K^2R_{oo} + KR_{12}S_1 + KR_{21}S_2 + R_{ee}S_1S_2}$
where $R_{oo} + R_{ee} = R_{12} + R_{21}$. $n$ = Total number of carriers per unit area of membrane.

kinetic solutions of this model (which for the enzyme kineticist is known as the iso uni–uni model (Cleland, 1975), since one molecule of ligand binds each form of the enzyme and there is an isomerization between the different forms of the enzyme). Cleland has stressed that the two versions of the simple carrier depicted as (a) and (b) in Figure 5 are difficult to distinguish by kinetic tests. The four-state model has been convincingly demonstrated for the choline transport system of the human red blood cell by Devés and Krupka (1981), but it is often convenient to think in terms of the simpler three-stage model, and only the most subtle distinctions between them require one to make a choice. The kinetic solution that I find most useful (appropriate, of course, to both forms of Figure 5) was derived by Stein and Lieb (1973) and is as follows:

$$v_{12} = \frac{K\,S_1 + S_1S_2}{K^2R_{oo} + K\,R_{12}S_1 + K\,R_{21}S_2 + R_{ee}S_1S_2} \tag{4}$$

where $v_{12}$ is the unidirectional flux of substrate $S$ in the direction from face 1 to face 2 of the membrane, while $S_1$ and $S_2$ are the concentrations of $S$ at sides 1 and 2, respectively. The parameters in $K$ and $R$ are the defining parameters of the system, experimentally measurable from transport studies, and interpretable in terms of the fundamental rate constants of the models of Figure 5 as recorded in Table 1. $K$ is related to the intrinsic affinity of the substrate for the carrier. The terms in $R$ are combinations of rate constants and give the (maximal) rates of transport of substrate under various conditions that shall be specified below.

There are three fundamental types of transport experiment. The first of these is the "zero *trans*" experiment. Here, substrate is present at one face of the membrane only and the rate of transport from that face is measured as a function of the substrate concentration. In Eq. (4), put $S_2 = 0$. Simplifying, we get,

$$v_{12}^{zt} = \frac{(1/R_{12})S_1}{K\dfrac{R_{oo}}{R_{12}} + S_1} \tag{5}$$

where $v_{12}^{zt}$ is the zero *trans* flow of substrate. This is a simple Michaelis–Menten form with maximum velocity given by the reciprocal of the resistance parameter $R_{12}$ and a half-saturation concentration (which measures the *apparent affinity* of the ligand for the system at face 1) given by $K$ multiplied by a ratio of resistances. The appropriate parameters for the zero *trans* experiment in the opposite direction (from side 2 to side 1) are given by a similar equation with subscripts 1 and 2 interchanged. Thus, the asymmetry of the system given as the ratio of the maximum velocities in the two directions (*or* as the ratio of the half-saturation concentrations in the two directions) is given by the ratio $R_{12}/R_{21}$. This is a very important result, to which we shall return frequently in what follows, since the asymmetry of an active-transport system is of great relevance to its effectiveness of operation. The second type of experiment is the "infinite *trans*" experiment. The unidirectional flux of (isotopically-labeled) substrate is measured into a high concentration of nonlabeled substrate at the opposite face of the membrane. One measures, in effect, the exchange of labeled for unlabeled substrate. The unlabeled substrate is at such a high concentration that raising its concentration any further has no kinetically noticeable effect. To solve for this case, we return to Eq. (4) and put $S_2$ equal to infinity, so that only terms containing $S_2$ remain in the equation. Simplifying, we obtain,

$$v_{12}^{it} = \frac{(1/R_{ee})\, S_1}{K\dfrac{R_{21}}{R_{ee}} + S_1} \tag{6}$$

where $v_{12}^{it}$ is the infinite *trans* flux of labeled substrate. This, again, is a simple Michaelis–Menten form with maximum velocity given by the reciprocal of the resistance parameter $R_{ee}$, and a Michaelis coefficient given by $K$ times a ratio of resistances. The infinite *trans* experiment in the reverse direction has this same maximum velocity (since it is the same datum, with infinite concentration of substrate at both membrane faces), and a Michaelis coefficient given on interchanging subscripts 1 and 2. The asymmetry is, of course, given by $R_{12}/R_{21}$. The "apparent affinities," the Michaelis parameters, can be quite different for the two types of experiments, zero *trans* and infinite *trans*. If we do an experiment measuring flux from side 1 towards side 2 in zero *trans* conditions and in infinite *trans* conditions and compare the observable half-saturation concentrations or "apparent affinities," we can easily see that their ratio is given by $R_{oo}R_{ee}/R_{12}R_{21}$, or by rate constants pertaining only to the rate of interconversions of the various forms of the transporter. The differing stabilities of $E_1$ as compared with $E_2$, or of $ES_1$ as compared with $ES_2$, will completely determine the ratio of apparent affinities in the two types of experiment. Thus, the apparent affinity is determined by the distribution of carrier between its various forms, as these are shifted around by the gradient of substrate. A third very useful type of experiment is that in which one sets the substrate concentration equal on both sides of the membrane so that there is no net flow of matter, and measures the flux of substrate labeled only at one face of the membrane. To solve Eq. (4) for this case, one puts $S_1 = S_2$ and simplifies, obtaining, once again, a simple Michaelis–Menten form with maximum velocity of $1/R_{ee}$ and half-saturation concentration of $K$ multiplied by the ratio of $R_{oo}$ to $R_{ee}$.

As can be seen directly from Table 1, the four resistance parameters are connected by the relation that $R_{oo} + R_{ee} = R_{12} + R_{21}$, and, therefore, only three of them need to be specified to determine the fourth absolutely. Indeed, $R_{12}$, $R_{21}$, and $R_{ee}$ are the reciprocals of the maximum velocities of the various transport experiments, and with them determined, $R_{oo}$ can be calculated. With all four resistances available, the affinity parameter $K$ can be determined from each measured half-saturation parameter in turn. As five of these have been defined above, this gives five independent measures of $K$. Another three measures can be obtained by experimental procedures not discussed here (but see Stein, 1981). Thus, there are many ways of checking the validity of applying the simple carrier model to any given experimental situation. This approach, as well as Eqs. (4)–(6), have been verified for the uridine transport system of the human red cell (Cabantchik and Ginsburg, 1977; Plagemann *et al.*, 1982), the leucine transport system of that cell (Rosenberg, 1981; Lieb, 1982), and its cyclic AMP-transport system (Holman, 1978).

If $R_{12}/R_{21}$ is the measure of the functional asymmetry of the transport system, $R_{oo}/R_{ee}$ is a measure of the ability of the system to exchange substrate as compared with its ability to perform only net transport. This is because $R_{oo}$ is composed of the rate constant for interconversion of the unloaded forms of the transporter, while $R_{ee}$ is composed of the role constants for the interconversion of the loaded forms. There are many transport systems in which $R_{oo}$ is far bigger than $R_{ee}$, so that exchange transport is favored over net movements. Much of the effectiveness of the counter-transport systems, as we shall see, arises from this situation.

The affinity parameter $K$ is a rather complex measure of the affinity of ligand for transporter, averaged over the two faces of the membrane, and modified by the rates of interconversion of the different forms of the transporter. For the case where the transporter is in equilibrium with ligand, that is, where the interconversion steps are the rate-limiting ones, the value of $b_1$ is much greater than $g_1$, so that $K$ reduces to $b_1k_1/f_1g_1$ $(= b_2k_2/f_2g_2)$. The intrinsic dissociation constant for transporter with ligand is, at face 1, $b_1/f_1$, which is embedded in this value for $K$, but modified by the ratio of transporter interconversion rates $(k_1/g_1)$. The fact that $K$ can be written in terms of two equivalent forms arises from the "thermodynamic box" rules for the carrier model, as shown by Eq. (2).

It may come as a surprise to some readers that the simple carrier, which is in no way capable of causing the accumulation of substrate at one or other face of the membrane can yet be asymmetric. That the system cannot pump can be seen by calculating the net flux of substrate from side 1 to side 2, when the concentrations at side 1 and side 2 are equal. We use, of course, Eq. (4) which gives the unidirectional fluxes (from side 1 to side 2 as it is written; from side 2 to side 1 on interchanging the subscripts 1 and 2). The net flux is now the difference between these two unidirectional fluxes. The numerator of this difference term is $K(S_1 - S_2)$. When $S_1 = S_2$, this is zero and, therefore, no pumping occurs. That the system can be asymmetric follows merely from the possibly different stabilities of the various forms of the transporter. To see this, take again the case of the transporter for which transport steps are rate limiting (column 4 of Table 1). The asymmetry of the system, as we have seen, is given by the ratio $R_{12}/R_{21}$. If the different forms of the transporter are equally stable in form I as in form II, then $k_1 = k_2$, $g_1 = g_2$. Then (Table 1), $R_{12} = R_{21}$, and the system is symmetric. In any other case, the system is asymmetric. If, as is usually the case, the loaded transporter interconverts between its conformations more rapidly than the unloaded transporter, the resistance terms, dominated by the slowest step, are determined by $k_1$ and $k_2$. Thus, the asymmetry is determined by the differential stability of the two forms of the free transporter. If state I is more stable than state II, the rate constant $k_1$ being smaller than $k_2$, $R_{12}$ is smaller than $R_{21}$, and the system is asymmetric with a higher maximum velocity in the direction of transport from side 1 to side 2. By any experimental test, the system will look asymmetric, since it is, indeed, asymmetric. It will have a higher maximum velocity for zero *trans* transport in the 1 to 2 direction than in the other direction, the reason being that there are always more free transporters available at face 1 to carry substrate, this being the stable side. Associated with this higher maximum velocity in the zero *trans* experiment is the higher value of $K_m$, the Michaelis constant or the half-saturation concentration. There are more free transporters always present at face 1 and, therefore, a higher concentration of substrate will be required to complex with all of them. At the opposite face of the membrane, there will be fewer (unstable) transporters to complex with substrate. The maximum velocity of transport from that face will be low, but then so will the measured half-saturation concentration for this experiment. The Michaelis parameter measured from the side at which the transporter is less stable is closest to a true measure of the average affinity parameter $K$. (This is because, with interconversion of the unloaded transporter being slow, $K_m$ from that side is given by $K(k_1 + k_2)/k_2$, close to unity, while $K_m$ is far higher than unity for the opposite case). The simple carrier cannot pump, but it can operate as a valve (Krupka and Devés, 1979), preferring transport away from the side at which the transporter is more stable, but allowing net flow only in the direction of the concentration gradient.

Since we have seen that a system can be asymmetric in its transport parameters (apparent affinities and maximum velocities) and yet not be capable of pumping, it is clear that one can dissociate asymmetry from pumping. It is not sufficient for a system to be asymmetric in order for it to pump. We shall soon show that it is not even necessary for it to be asymmetric in order for it to pump. But it helps!

## B. Countertransport or Antiport—Coupling of Flows on the Simple Carrier

### 1. Kinetics of Countertransport

The first example of active transport will now be discussed. The transporter is a simple carrier, just as we have discussed these last few pages, but now it is one which

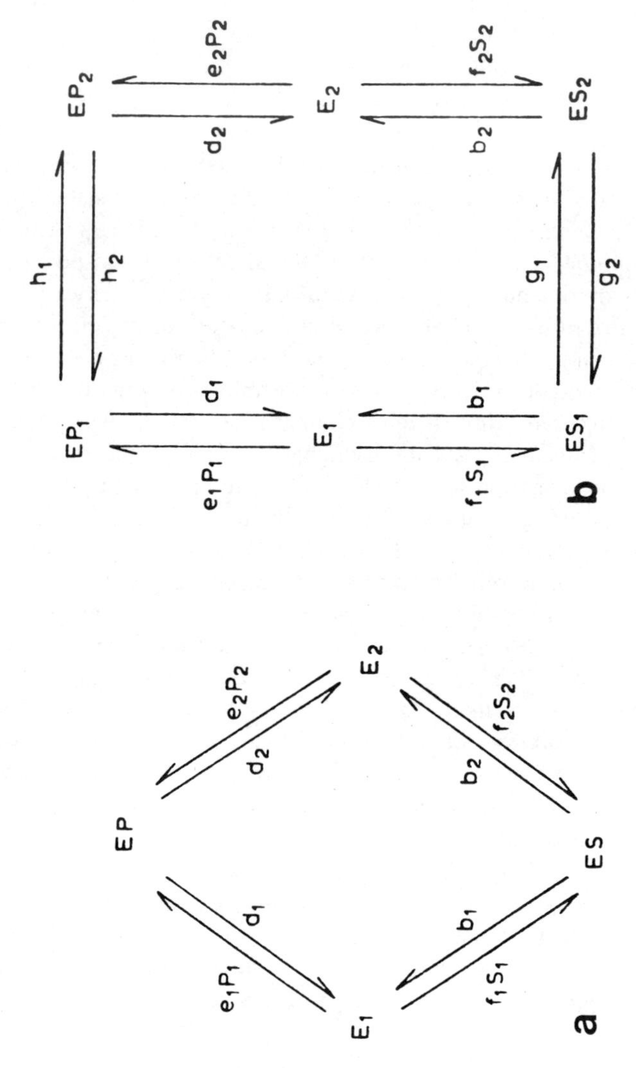


*Figure 6.* The simple countertransporter. A simple carrier capable of transporting either substrate $S$ or substrate $P$. (a) The one-complex form; (b) the two-complex form. Rate constants as indicated. The free carrier is incapable of undergoing a conformation change.

can bind to either of two substrates as in Figure 6. Instead of the free carrier being able to interconvert between the two conformations that face opposite sides of the membrane, interconversion is allowed only when the transporter has bound a transportable substrate. The pathway $E_1 = E_2$ is replaced by the pathway $EP_1 = EP_2$, where $P$ is the second substrate. Clearly, the system here is a simple carrier and all the considerations that we have obtained for the simple carrier also apply to this simple countertransport system. The system is described as being a countertransport, since the flow of $S$ from side 1 to side 2, for example, is accompanied by the interconversion of the carrier into the side-2 state, from which it will return by transporting substrate $S$ or $P$ in the counterdirection 2 to 1. The kinetics of this model are simply derived. Equation (7) describes how the unidirectional flux of $S$ from side 1 to side 2 ($v^s_{12}$) depends on the concentrations of $S$ and $P$ at sides 1 and 2:

$$v^s_{12} = \frac{\overline{S}_1\overline{S}_2 \dfrac{R_{pp}}{R_{ss}} + \overline{S}_1\overline{P}_2}{R_{ss}(\overline{P}_1 + \overline{P}_2 + \overline{P}_1\overline{P}_2) + R_{pp}(\overline{S}_1 + \overline{S}_2 + \overline{S}_1\overline{S}_2) + R_{ps}\overline{P}_1\overline{S}_2 + R_{sp}\overline{S}_1\overline{P}_2} \tag{7}$$

where the bars above the concentration term everywhere signifies the "reduced concentration." For example, $\overline{S}_1 = S_1/K^s_1$, where $K^s_1$ is the apparent dissociation constant for the formation of $ES$ at side 1, and correspondingly for $\overline{S}_2$, $\overline{P}_1$, and $\overline{P}_2$. The terms in $K$ and the terms in $R$, the resistances, are defined in Table 2. The unidirectional flux of $S$ from side 1 now depends on the concentrations of both $S$ and $P$. In particular,

*Table 2. Kinetic Solutions for Various Forms of the Simple Countertransporter*[a]

| Model | | Figure 6a | Figure 6b | Figure 6b, transport steps, rate limiting |
|---|---|---|---|---|
| $nR_{SP}$ | = | $l/b_2 + l/d_1$ | $l/b_2 + l/d_1 + (l/g_1)((b_2 + g_2)/b_2) + (l/h_2)((d_1 + h_1)/d_1)$ | $l/g_1 + l/h_2$ |
| $nR_{PS}$ | = | $l/b_1 + l/d_2$ | $l/b_1 + l/d_2 + (l/g_2)((b_1 + g_1)/b_1) + (l/h_1)((d_2 + h_2)/d_2)$ | $l/g_2 + l/h_1$ |
| $nR_{SS}$ | = | $l/b_1 + l/b_2$ | $l/b_1 + l/b_2 + (l/g_1)((b_2 + g_2)/b_2) + (l/g_2)((b_1 + g_1)/b_1)$ | $l/g_1 + l/g_2$ |
| $nR_{PP}$ | = | $l/d_1 + l/d_2$ | $l/d_1 + l/d_2 + (l/h_1)((d_2 + h_2)/d_2) + (l/h_2)((d_1 + h_1)/d_1)$ | $l/h_1 + l/h_2$ |
| $K^S_1$ | = | $b_1/f_1$ | $b_1g_2/f_1(g_1 + g_2)$ | $b_1g_2/f_1(g_1 + g_2)$ |
| $K^S_2$ | = | $b_2/f_2$ | $b_2g_1/f_2(g_1 + g_2)$ | $b_2g_1/f_2(g_1 + g_2)$ |
| $K^P_1$ | = | $d_1/e_1$ | $d_1h_1/e_1(h_1 + h_2)$ | $d_1h_1/e_1(h_1 + h_2)$ |
| $K^P_2$ | = | $d_2/e_2$ | $d_2h_2/e_2(h_1 + h_2)$ | $d_2h_2/e_2(h_1 + h_2)$ |
| Constraint | | $b_1d_2e_1f_2 = b_2d_1e_2f_1$ | $b_1d_2e_1f_2h_1g_2 = b_2d_1e_2f_1g_1h_2$ | $b_1d_2e_1f_2h_1g_2 = b_2d_1e_2f_1g_1h_2$ |

[a] $v^S_{1\to 2} = \dfrac{\overline{S}_1\overline{S}_2(R_{PP}/R_{SS}) + \overline{S}_1\overline{P}_2}{R_{SS}(\overline{P}_1 + \overline{P}_2 + \overline{P}_1\overline{P}_2) + R_{PP}(\overline{S}_1 + \overline{S}_2 + \overline{S}_1\overline{S}_2) + R_{PS}\overline{P}_1\overline{S}_2 + R_{SP}\overline{S}_1\overline{P}_2}$
where $R_{SS} + R_{PP} = R_{SP} + R_{PS}$. $n$ = Total number of countertransporters per unit area of membrane. Also, $\overline{X}_i = X_i/K^X_i$, $X_i$ being the concentration of $X$ at side $i$ and $K^S_1K^P_2 = K^S_2K^P_1$

with no transportable substrate at the *trans* face (side 2), there is no flux of $S$ from side 1. This is because, in Figure 6, there is no route for the return of transporter from its conformation in state II to state I. Hence, with no substrate at side 2, transporter will all be driven into state II *and remain in that state*. What is important to note is that there will be a net flow of substrate $S$ from side 1 to side 2 even when the concentrations of $S$ are identical on both sides of membrane, provided that $P$ is at a different concentration at the two sides of the membrane. Thus, in Eq. (7) put $S_1 = S_2$, and calculate the net flux of $S$ from side 1 to side 2 (by taking the difference between unidirectional fluxes $v^s_{12}$ and $v^s_{21}$). We obtain

$$\text{net } v^s_{12} = \frac{\overline{S}_1\overline{P}_2 - \overline{S}_2\overline{P}_1}{\text{denominator}} \tag{8}$$

The net flux will be zero when the numerator is zero or when $\overline{S}_1\overline{P}_2 = \overline{S}_2\overline{P}_1$, that is, when $S_1P_2/K^s_1\,K^p_2 = S_2P_1/K^s_2\,K^p_1$. But this expression reduces to $S_1P_2 = S_2P_1$, since the product $K^s_1K^p_2 = K^s_2K^p_1$ by the operation of our "thermodynamic box" rules on the equilibria of Figure 6. Thus, a net flow of $S$ continues until

$$S_1/S_2 = P_1/P_2 \tag{9}$$

The concentration gradient of $S$ is coupled to that of $P$ by the operation of the transporter. If, by some means, a concentration gradient of $P$, high at side 2, can be maintained by the cell, then the countertransport system of Figure 6 will bring about an active transport of $S$ from side 1 to side 2, until the concentration gradient of $S$ just balances that of $P$ (Eq. 9). Whatever the details of the kinetic steps involved in the conformation changes of the (counter-) transporter, the flows of $S$ and $P$ will be coupled and will tend to approach the equilibrium situation where $S_1/S_2 = P_1/P_2$. The concentrations of $S$ and $P$ will reach this limiting relation only if there is perfect coupling between the flows of $S$ and $P$, that is, there is no leakage of either $S$ or $P$ through any transport route other than the one given by Figure 6. In any real system, there will be departure from such strict coupling. It is convenient to make a formal distinction between two types of departure from this strict coupling. Let us distinguish between "leakage," which we define as movement of either $S$ or $P$ across the membrane by some path other than that involving the transporter $E$, and between "slippage," which we define as movement of either $S$ or $P$ on the transporter but uncoupled to the movement of the countersubstrate. Such slippage will occur if the transporter is able, to some extent, to interconvert between states I and II in the absence of substrate $S$ or $P$. Figure 7 shows such a countertransport system where slippage occurs. (Indeed this system is merely a simple carrier which can allow the movement of either $S$ or $P$ down their respective concentration gradients, as did the simple carrier of Figure 5, but which can transport both $S$ and $P$.) To the extent that the interconversion of free transporter is slow, that is, the step $E_1 \rightleftharpoons E_2$ is disallowed, so the coupling of flows of $S$ and $P$ becomes tighter. The rule for strict coupling of the flows of $S$ and $P$ and, hence, for the *effectiveness* of the countertransport system is that the rate constants $k_1$ and $k_2$ of the system be as small as possible in comparison with those describing the

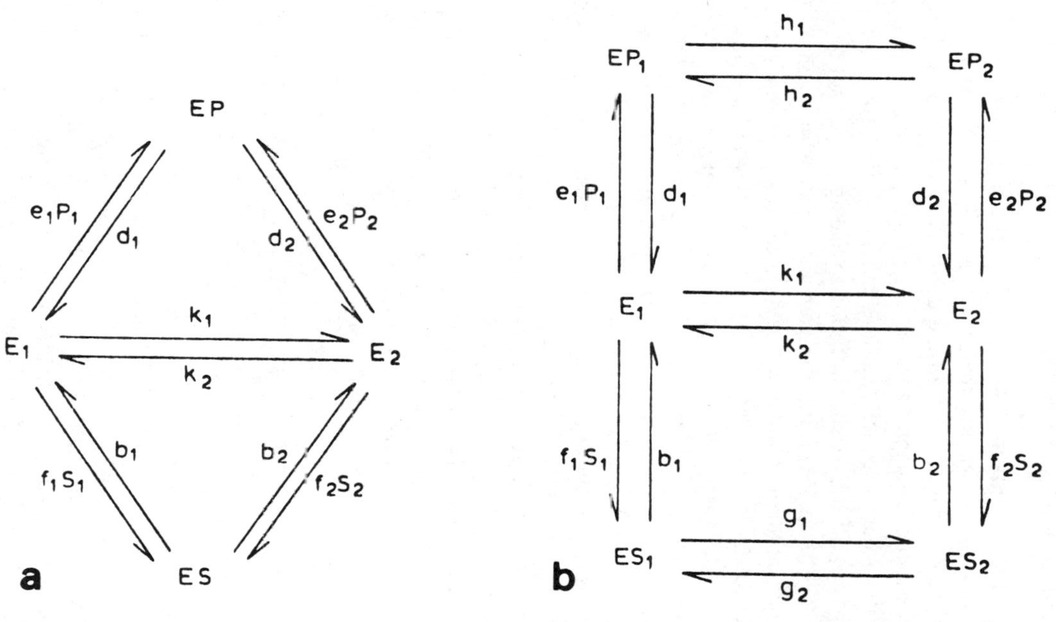


*Figure 7*. The countertransporter with slippage. As in Figure 6, but now the free carrier $E$ is capable of performing a conformation change allowing a looser coupling between the countertransport of $S$ and $P$. The system allows the net movement of $S$ or $P$ in the absence of a *trans* substrate.

interconversion rates of the loaded transporter, $g_1$, $g_2$ and $h_1$, $h_2$. Overcoming slippage does not at all involve the energetics of the active transport, but only the protein chemistry which allows a mechanism by which binding of substrate $S$ or $P$ triggers the rapid interconversion between the two states of the transporter.

## 2. *Apparent Affinities of a Countertransport System*

Energetics become important when we consider ways that a countertransport system might evolve so as to minimize the effects of leakage. As the argument I want to put forward now lies at the heart of all the considerations which will be taken up in subsequent sections of this chapter, I wish to go through it in some detail. Let us consider the situation where a concentration gradient of $P$ is somehow established while the concentration of $S$ approaches but, due to leakage, does not quite reach that given by Eq. (9). We term $P$, arbitrarily, the driving substrate and $S$, the driven substrate. With a concentration gradient of $S$ established ($P_2 > P_1$ leading to $S_2 > S_1$), due to the operation of the countertransport system, there will be a flow by leakage of $S$ down its concentration gradient through various sorts of pathways across the cell membrane, parallel with the countertransport system in question. $S$ will flow by simple diffusion, by the operation of any simple carriers present in the membrane, by the operation of other countertransport systems for which $S$ may be the driving substrate, and so on. The effect of these leakages will be minimal if the countertransport system in question evolves so as to maximize its effective rate of operation. It must move as fast as possible carrying $S$ up its concentration gradient (here in the direction 1 to 2); it must (1) move fast, and (2) be *occupied* in binding $S$. It will have to have a high effective affinity for $S$ at side 1. To see what factors satisfy these demands, we return to the transport Eq. (7), taking the case where $S$ has reached a high gradient ($S_2 > S_1$) arising from coupling to a high-concentration gradient of $P$ ($P_2 > P_1$). The net transport of $S$ in the 1 to 2 direction becomes, from Eq. (7),

$$\text{net } v_{12}^s = \frac{\overline{S}_1\overline{P}_2 - \overline{S}_2\overline{P}_1}{R_{ss}(\overline{P}_1 + \overline{P}_2 + \overline{P}_1\overline{P}_2) + (R_{pp} + R_{ps}\overline{P}_1)\overline{S}_2 + (R_{pp} + R_{sp}\overline{P}_2 + R_{pp}\overline{S}_2)\overline{S}_1} \tag{10}$$

The numerator of this expression is, of course, the very small distance that $\overline{S}_1\overline{P}_2$ is from $\overline{S}_2\overline{P}_1$ at the steady state, as a result of leakage, and gives the rate of transport of $S$ that is brought about the the action of the countertransport system in the face of the leakage. The expression on the right-hand side of Eq. (10) can be thought of as a Michaelis equation, with a maximal velocity being given by the reciprocal of the resistance term ($R_{pp} + R_{sp}\overline{P}_2 + R_{pp}\overline{S}_2$) and an effective half-saturation concentration ($K_m$) for $S$ at face 1, given by

$$K_m = \frac{K_1^s[R_{ss}(\overline{P}_1 + \overline{P}_2 + \overline{P}_1\overline{P}_2) + (R_{pp} + R_{ps}\overline{P}_1)\overline{S}_2]}{R_{pp} + R_{sp}\overline{P}_2 + R_{pp}\overline{S}_2} \tag{11}$$

This Michaelis constant $K_m$ determines the amount of transporter that is effectively occupied in transporting $S$ from side 1 to side 2. If this $K_m$ is small, so that the system has a high affinity for driven substrate $S$ at side 1, the transport rate will be maximized. Both at low $\overline{P}_2$ and $\overline{S}_2$ and at high $\overline{P}_2$ and $\overline{S}_2$, $K_m$ is given by $K_1^s$ multiplied by a ratio of resistance terms multiplied by $P_1/K^P$. Thus, if $K_m$ is to be as small as possible, $K_1^s$ must be as small as possible and $K_1^P$ must be as large as possible. This makes sense since it means that $S$, rather than $P$, moves preferentially from side 1. Our "thermodynamic box" rules give us the condition for a high intrinsic affinity of $S$ for the system relative to $P$ at side 1. Consider the model depicted in Figure 6. If state $E_1$ is destabilized over all other states, that is, the corner of the "box" at $E_1$ is at a higher energy than the other states, $S$ will bind to $E_1$ preferentially over $E_2$. On the other hand, it necessarily follows that $P$ also binds preferentially to $E_1$ rather than $E_2$. Thus, there is no effect on the competition between the driver $P$ and the driven substance $S$ at side 1. One can reduce the affinity of the transporter for $P$ in comparison with that for $S$, by destabilizing $EP$ with respect to $ES$, that is, by raising the energy of the $EP$ corner. But now, the affinity of $E_2$ for $P$ must equally be reduced, with the effect that the rate of return of transporter via $EP$ to the state I conformation will be reduced. The "thermodynamic box" makes serious constraints on the design of an effective transporter! We cannot increase the affinity of $S$ for $E$ at side 1 without increasing that of $P$ at this side, or reduce that of $P$ at one side without reducing that of $P$ at the other side. Both sets of changes will lower the overall pumping rate of $S$ from side 1. The effect of stabilizing the $EP$ conformation is less serious, however, since there is a substantial amount of $P$ present at side 2, driving the system from side 2 to side 1. An effective countertransport system will, therefore, use the fact that there is, indeed, a high concentration of the driving substrate and allow the system to have a poor affinity for $P$ (at both faces) by stabilizing the $EP$ form. $S$ will be preferred to $P$ at side 1, and the system will be maximally effective in transporting $S$ from side 1 to side 2, against the leak.

To summarize, a countertransport system in the absence of leakage or slippage will couple the transport of its substrates $S$ and $P$ until, at the steady state, $S_1/S_2 = P_1/P_2$. In the presence of leakage, however, this ideal distribution will not be achieved, the steady-state of $S_1/S_2$ being lower than $P_1/P_2$, owing to leakage of $S$ down its concentration gradient. A system will be most effective when it maximizes the rate of pumping of $S$ up its concentration gradient. It will do this best when it preferentially binds $S$ rather than $P$ at the side from which $S$ is being driven (we shall call this the "whence" side). One cannot achieve this by destabilizing the free transporter at this side, but only by stabilizing that form of the transporter which is bound to the driving substrate $P$. Although this stabilizing must also lower the affinity of $P$ at the side at which $S$ accumulates (which we shall call the "whither" side), there is already enough $P$ at this side to overcome this resulting low affinity. The energy in the gradient is used to drive the pumping of $S$, and a well-designed pump system uses the existence of the high concentration of the driving substrate to allow the driven substrate to have a relatively high affinity for the transporter.

Kinetically, one can characterize (and test) a countertransport system by performing the infinite *trans* experiments in infinite *trans* situations where either $S$ or $P$

only is present *trans*. There are eight such experiments, each of the two substrates flowing into either of the two *trans* substrates in either direction across the membrane. The four resistance parameters defined in Table 2 and the four affinity parameters can all be determined. They are connected by the interrelations listed at the foot of Table 2 and are, thus, heavily overdeterminable experimentally, providing an excellent set of tests for the applicability of the countertransport model to any real system.

## IV. THE COTRANSPORT SYSTEMS—ISO BI–BI ENZYMES

We now come to the cotransport systems. These are capable of simultaneously binding two transportable ligands, which often come from two quite different classes of chemical substances. The flows of the two ligands are coupled together by their joint flow on the transporter. We shall discuss first the kinetics of such transport systems, and then the question of how considerations of energetics could have controlled the evolution of such systems towards maximum effectiveness.

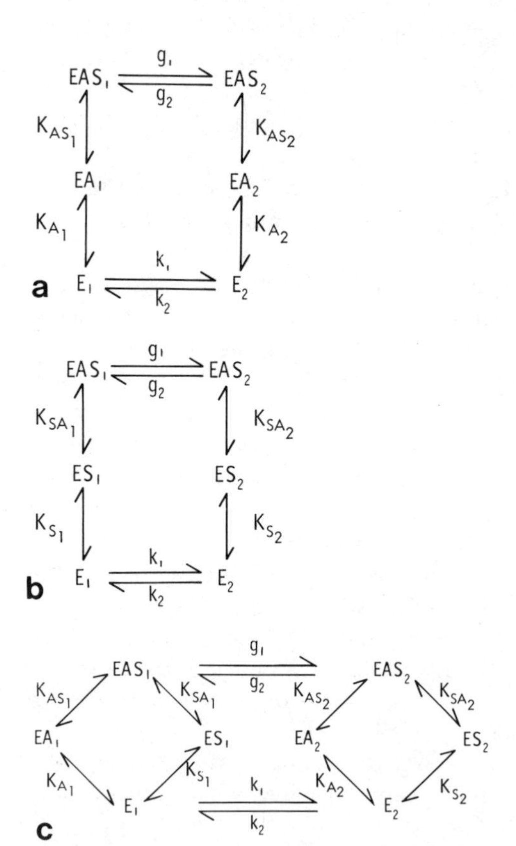


*Figure 8.* Three models for the simple cotransporter. (a) The driving substrate $A$ is the first to bind to the carrier. (b) The driven substrate $S$ is the first to bind. (c) The substrates bind in random order. The equilibrium constants for transporter–substrate interactions are the terms in $K_{X_i}$ depicted on each equilibrium.

## A. The Fundamental Equation of Cotransport

It was the work of Christensen, Crane, and Csaky in the mid-fifties (see the valuable review by Crane, 1977) that led to the realization that the active transport of sugars and amino acids was dependent on the presence of sodium ions and led Crane to formulate his "sodium-gradient" hypothesis which is the basis of our present understanding of these phenomena. The essence of this hypothesis is that the movement of sugar or amino acid occurs on a conventional carrier, but one which requires the binding (and transport) of sodium ions for transmembrane movement of the carrier–substrate complex. Only if both sodium and the metabolite are simultaneously bound to the carrier, can transport occur. The simplest such kinetic scheme for this cotransport model is depicted in Figure 8. Here we have a carrier capable of existing in two conformations $E_1$ and $E_2$ with substrate binding sites facing side 1 and side 2 of the membrane, respectively. The carrier is capable of binding, in turn, two substrates $A$ and $S$. In Figure 8a, $A$ is the first substrate bound at each face, Figure 8b shows the case where $S$ binds first, and Figure 8c shows where the binding order is random. $EA$ and $ES$ cannot cross the membrane, but after the formation of $EAS$, the carrier can undergo a conformation change, releasing $S$ and $A$ at side 2 of the membrane. The system is reversible. Kinetic analysis of any of the models of Figure 8 leads to the single solution

$$v_{12}^{s} = \frac{KS_1A_1 + S_1A_1S_2A_2}{K^2R_{oo} + KR_{12}S_1A_1 + KR_{21}S_2A_2 + R_{ee}S_1A_1S_2A_2} \tag{12}$$

where $v_{12}^{s}$ is the unidirectional flux of $S$ in the direction 1 to 2, when concentrations of $S$ and $A$ are $S_1$ and $A_1$ at side 1, and $S_2$ and $A_2$ at side 2. The terms in $K$ and $R$ are experimentally measurable parameters, defined in Table 3 in terms of the rate constants of the models of Figure 8. This equation is closely analogous to Eq. (4) which describes the simple carrier, except that in Eq. (12) we have the product of the two concentrations $S$ and $A$ in place of the single concentration term $S$ of Eq. (4). The reason for this substitution is quite obvious; it is the pair $S \times A$ that is the substrate for the cotransporter, just as $S$ was the substrate for the simple carrier. The parameters in $K$ and $R$ can be determined by the same set of "zero *trans*" and "infinite *trans*" experiments that are discussed for the simple carrier. In a way similar to that of the simple carrier, the cotransport models of Figure 8 can be tested and characterized. The only difference is that, as Table 3 makes clear, the resistance terms in $R$ in Eq. (12) (but not the parameter $K$) now contains terms for the concentration of the cosubstrate $A$, and it is this fact that must be considered.

## B. Maximizing the Effectiveness of the Cotransporter

Just as for the countertransporter system, we can consider how a cotransporter can evolve so as to be maximally effective. Equation (12) gives the unidirectional flux of $S$. To find the net flow in the 1 to 2 direction, subtract $v_{21}^{s}$ from $v_{12}^{s}$ and obtain, for the numerator, $K(S_1A_1 - S_2A_2)$. There is a net flow from side 1 to side 2 even if

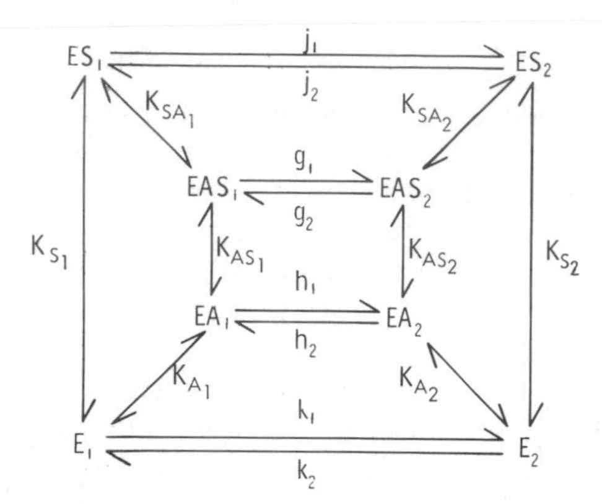


*Figure 9.* The simple cotransporter with slippage. As in Figure 8, but now the binary complexes in *EA* and *ES* are capable of performing a conformational change, allowing a looser coupling between the movement of *A* and *S*. *A* and *S* can be transported across the membrane alone, with no movement of the cosubstrate.

$S_1 = S_2$ and this net flow will continue, if all flows are strictly as on the model of Figure 8, until $S_1A_1 = S_2A_2$. At the steady state of an ideal cotransporter, then, the concentration gradients of the driven substrate, e.g., $S$, and the driving substrate $A$ are linked by the relation

$$S_1/S_2 = A_2/A_1 \tag{13}$$

This equation described the steady state of the ideal cotransporter, just as Eq. (9) described the steady state of the ideal countertransporter. We readily concede that no system is ideal. There is slippage in the cotransporter and this slippage is the movement of carrier when bound to only one of either of the two substrates. Figure 9 shows the kinetic scheme for a cotransporter with slippage. The most effective cotransporter is that in which the slippage paths are minimized, that is, where rate constants $h_1$, $h_2$, $j_1$, and $j_2$ are minimal. Again, as in the corresponding situation for the countertransporter, arranging that these rate constants are to be minimal involves solving the protein chemistry of a transporter. The system has to evolve until the only states of the transporter in which the fundamental transmembrane conformation changes occur are those of the free carrier and of the doubly-bound form, the so-called tertiary complex. This is not an energetics problem nor one of linked thermodynamic functions, but is akin to the absolute requirement that many enzymes have for a cofactor in order for them to function. Overcoming, or rather minimizing, the effect of leakage is, however, something that manipulation of the energetics of ligand binding will influence.

Consider the situation where design of the cotransporter is most critical, i.e., where the gradients of driven and driving substrates are almost those given by Eq. (13), except that $S_1/S_2$ is a little lower than the ideal value owing to leakage of $S$ down its concentration gradient (we assume $A_2 > A_1$ and hence $S_1 > S_2$). We want to maximize the rate of pumping of $S$ in the 2 to 1 direction, that is, the rate of net movement of $S$ as driven by the high ratio of $A_2$ to $A_1$. We solve Eq. (12) (and its analogous form with the subscripts 1 and 2 interchanged) for the net flow of $S$, putting $S_1A_1$ as approximately equal to $S_2A_2$ (on our assumption that we are near to the steady state defined by Eq. (13)). We write $SA$ for $S_1A_1(= S_2A_2)$ and obtain after simplifying, and realizing that $R_{oo} + R_{ee} = R_{12} + R_{21}$ (see Table 3)

$$\text{net } v_{21}^s = \frac{K\Delta(SA) \times 1/R_{ee}}{\left(K\dfrac{R_{oo}}{R_{ee}} + SA\right)(K + SA)} \tag{14}$$

where $\Delta(SA)$ is the small difference $(S_2A_2 - S_1A_1)$. This equation describes how the net transport of $S$ (and $A$) is affected by the parameters of the cotransport system and the prevailing value of the product $S \times A$, and tells us what has to be maximized for maximal effectiveness of the cotransport system in the face of leakage of $S$ down its concentration gradient. We consider two limiting cases. First, take the case where the prevailing level of $S \times A$ is high in comparison with the degree of saturation of the system with substrates. $SA$ is large in comparison with $K$ and with $K(R_{oo}/R_{ee})$ and we get the limiting value of the net flow as net $v_{12}^s = (K/SA) \cdot (\Delta SA/SA) \cdot (1/R_{ee})$. Pumping is a maximum when $K$ is large and when the resistance parameter $R_{ee}$ is small. $K$ is a parameter which (see Table 3) is embedded in the "thermodynamic box" describing the overall transport system. It is not affected by making the parameters of the system different on the two sides of the membrane or different for the driving and the driven substrates. If $K$ is high, both substrates bind poorly at both sides of the membrane and there is free carrier available for net transport without favoring one direction or substrate. No thermodynamic "trade-offs" or linked-function effects will alter $K$. The parameter $R_{ee}$ is, however, much affected by the details of the binding of the substrate in a very interesting fashion (Table 3). It is clear that the parameter $R_{ee}$ is smallest when the dissociation constants $K_{SA_2}$ and $K_{SA_1}$ are as small as possible. These dissociation constants describe the binding of $A$ to a transporter which has already bound $S$, that is, along the pathway where $S$ binds first, and $A$ binds second. Now, these dissociaton constants are smallest when they are indeed zero, that is, when the dissociation constants for binding of substrate $S$ first are infinite, namely, when it is $A$ that binds first. In particular, since we are discussing the situation where it is $A$ that is high, $K_{SA_2}$ is negligible in comparison with $A_2$, and it is $K_{SA_1}$ that should be minimized. We conclude that an effective cotransport system at high concentrations of $S \times A$ is one where the driving substrate ($A$) binds first in an ordered reaction at the whither side. The complex $ES_1$ should be heavily destabilized over the complex $EA_1$. In the "thermodynamic box" describing the interconversions of the various forms of the cotransporter at side 1 (see Figure 8c), the corner $ES_1$ should be of higher energy than other corners.

Consider the situation where the concentrations of $S$ and $A$ are low, so that the product $S \times A$ at each side of the membrane is low. The net transport of $S$ from side 2 to side 1 will then be: net $v_{21}^s = \Delta(SA)/K\,R_{oo}$. This net flow is a maximum when $K$ is a minimum, that is, both substrates at both sides of the membrane bind well to the cotransporter, a situation where manipulation of the differential energetics cannot affect matters and when $R_{oo}$ is a minimum. This latter situation, a minimum in $R_{oo}$, *is* one which the design of the cotransporter can effect. Referring to Table 3, we see

Table 3. *Kinetic Solutions for Various Forms of the Simple Cotransporter*[a]

| Model | | Figure 8a | Figure 8b | Figure 8c |
|---|---|---|---|---|
| $nR_{12}$ | = | $(l/k_2)(l + A_2/K_{A2}) + l/g_1$ | $l/k_2 + (l/g_1)(l + K_{SA_1}/A_1)$ | $(l/k_2)(l + A_2/K_{A_2}) + l/g_1)(l + K_{SA_1}/A_1)$ |
| $nR_{21}$ | = | $(l/k_1)(l + A_1/K_{A_1}) + l/g_2$ | $l/k_1 + (l/g_2)(l + K_{SA_2}/A_2)$ | $(l/k_1)(l + A_1/K_{A_1}) + (l/g_2)(l + K_{SA_2}/A_2)$ |
| $nR_{ee}$ | = | $l/g_1 + l/g_2$ | $(l/g_1)(l + K_{SA_1}/A_1) + (l/g_2)(l + K_{SA_2}/A_2)$ | $(l/g_1)(l + K_{SA_1}/A_1) + (l/g_2)(l + K_{SA_2}/A_2)$ |
| $nR_{oo}$ | = | $(l/k_1)(l + A_1/K_{A_1}) + (l/k_2)(l + k_2)/k_{A2}$ | $l/k_1 + l/k_2$ | $(l/k_1)(l + A_1/K_{A_1}) + (l/k_2)(l + k_2/k_{A2})$ |
| $K$ | = | $(k_1/g_1)K_{A_1} K_{AS_1} = (k_2/g_2)K_{A_2} K_{AS_2}$ | $(k_1/g_1)K_{S_1} K_{SA_1} = (k_2/g_2)K_{S_2}K_{SA_2}$ | $(k_1/g_1)K_{A_1} K_{AS_1} = (k_2/g_2)K_{A_2} K_{AS_2} = (k_1/g_1)K_{S_1} K_{SA_1} = (k_2/g_2)K_{S_2} K_{SA_2}$ |
| Constraint | | Above equation | Above equation | Above equation, which includes $K_{A_1} K_{AS_1} = K_{S_1} K_{SA_1}$ $K_{A_2} K_{AS_2} = K_{S_2} K_{SA_2}$ |

[a] $v^S_{1\to 2} = v^A_{1\to 2} = \dfrac{K\, S_1A_1 + S_1A_1S_2A_2}{K^2R_{oo} + KR_{12}S_1A_1 + KR_{21}S_2A_2 + R_{ee}S_1A_1S_2A_2}$
where $R_{oo} + R_{ee} = R_{12} + R_{21}$. $n$ = Total number of cotransporters per unit area of membrane. (The solutions given are for the situation where the transport steps are rate limiting).

that $R_{oo}$ will be minimal when the rate constants in $k$ are large and when the dissociation constants $K_{A_1}$ and $K_{A_2}$ are large. The latter can be affected by energetics. Since $A_2$ is (for the case we are considering) larger than $A_1$, it is, indeed, $K_{A_2}$ that is most critical and that must be as large as possible for an effective cotransporter. For $K_{A_2}$ to be large, the pathway of $S$ binding first at side 2 should be favored. The complex $EA_2$ should be destabilized and the corner $EA_2$ of the "thermodynamic box" should be raised.

It seems that a cotransport system will be most effective in minimizing the effects of leakage over a range of concentrations of the driving and driven substrates if it is designed so that the driven substrate is the first to bind at the whence side and the first to unbind at the whither side. This is achieved by destabilizing the nonfavored forms and, thus, ensuring ordered binding reactions at the two membrane faces. It is interesting that this "first-on, first-off" rule, derived here from first principles, seems to operate in the sodium–sugar cotransport system of the intestine (Hopfer and Groseclose, 1980), one of the few systems where the order of binding of the two substrates has been established. It should be pointed out that effective design of the cotransporter does not seem to involve linked functions of binding between the two substrates. It is not that binding of one substrate is traded-off against the binding of the other; the energetics manipulations are to ensure a particular order of binding, not to improve binding of one substrate over the other.

To appreciate how fixing the order of substrate binding can increase the effectiveness of the cotransporter, consider the following. Take first the situation of high substrate concentrations ($S \times A$) at both sides of the membrane. What limits the rate of pumping up the concentration gradient of the driven substrate is the rate of return of free transporter to the whence side. If the driven substrate binds first to the transporter at the whither side (where the concentration of the driven substrate is high), it will trap transporter in the conformation facing this side, reducing the amount available for return to the opposite side and, hence, reducing the rate of pumping up the gradient. If the driving substrate binds second at the side at which it is in high concentration, it cannot trap transporter at this face. Take the situation where the $S \times A$ product is low overall, although there *is* a concentration gradient of driving and, hence, of driven substrate. At the whence side, the concentration of the driving substrate is relatively high. If it binds second, it can pull the equilibrium between $E$, $S$, and itself, $A$ over into the $EAS$ form and reach the maximum velocity of cotransport. If it binds first, then no matter how much of it is present, the maximum velocity of transport from this side is never seen. (In the next section, we turn to a detailed description of how the maximum velocity of transport is affected by the order of binding of the cotransported substrates.) The cotransport system will be more effective if the substrate present at the higher concentration binds second.

### C. Apparent Affinities for a Cotransport System

A pump which is designed to operate effectively in a situation close to the steady state of a cotransport system is one in which there is a particular order of binding of the driven and driving substrates at the two faces of the membrane. Because, under these conditions, there is very little net movement of either substrate, this is a difficult situation to examine experimentally. Most experimental work has therefore been done in situations far from the steady state and this fact has tended to obscure the conclusions reached in the previous section. Take the case, usually considered, where the driven substrate is at the same concentration at both sides of the membrane, the driving substrate being present at a much higher consentration at one membrane face from which the driven substrate will be pumped. Clearly, under these conditions, the rate of net pumping of the driven substrate will be greatest if the transporter binds substrate most effectively at the whence side. The apparent affinity of transporter for substrate should be greater at the whence side, and should be low at the whither side. This is the traditional view of the design requirements of a cotransport system, and is indeed often considered to be the mechanism of action of a cotransport system. The change of affinity of transporter for substrate as the transporter interconverts between the two transmembrane conformations has been thought of as providing the driving force for the accumulation of the driven substrate. As we have seen, this view is quite erroneous. The force for accumulating substrate comes merely from the fact that the transporter is driven by the combined substrate pair $S \times A$, and it is effectively the $S \times A$ gradient that moves both $S$ and $A$. Whatever the design details, an ideal cotransporter will couple $S$ and $A$ together until the steady state of Eq. (13) is reached, while a real cotransporter existing in a world where leakage occurs, will have the requirement of a defined order of binding of its substrates, as is discussed in the previous section. The traditional view of design requirement (high apparent affinity at the whence side) is, of course, valid in a situation far from the steady state, and it is, thus, worthwhile to devote effort to understand what makes a cotransport system have a higher apparent affinity at the whence side. Again, it is traditionally thought that the change of apparent affinity for driven substrate, as the cotransporter changes conformation across the membrane, is brought about by linked-function interactions between the binding of driving and driven substrates. We shall see that such interactions do have a role to play, but that the apparent affinity changes will also be much affected by the local concentration of the driving substrate and by the energetics of the free transporter.

To understand what affects the apparent affinity of the cotransporter for the driven substrate, return to the fundamental cotransport equation, Eq. (12). We shall consider the simplest situation, where both the driving and the driven substrates are at zero concentration at one side of the membrane, the whither side. This is the zero *trans* experiment discussed in Section III-A. Since Eqs. (4) and (12), which describe transport on the simple carrier and on the cotransporter are so very similar, we can take directly the results derived in Section III-A as being valid for the present arguments. We can take the zero *trans* movement of substrates on the cotransport system as being described by Eq. (5) which expresses the zero *trans* movement of substrate on the simple carrier with the proviso that we replace substrate $S_1$ in Eq. (5) with the substrate product $S_1A_1$ for cotransport. The meanings of the parameters in $K$ and $R$ are, of course, different for the simple carrier (Table 1) and for the cotransporter (Table 3). The maximum velocity of cotransport is $l/R_{12}$ and the half-saturation concentration or Michaelis constant (giving the apparent affinity of the cotransporter) is $KR_{oo}/A_1R_{12}$, where $K$ and $R$ terms are as defined in Table 3. It is extremely important to note that the terms $R$ in Table 3 contain the substrate concentrations $A_1$ and $A_2$. Thus, the maximum rates and apparent affinities of the cotransport system will depend on the concentrations of the driving substrate. This we proceed to discuss in systematic fashion.

*Case 1. Ordered Reaction with* A *Binding First at Side 1*. $K_{S_1}$ is infinite (and $K_{SA_1}$ is zero). The maximum velocity of transport of $S$ is $l/R_{12}$, where $R_{12} = l/k_2 + l/g_1$. The maximum velocity does not depend at all on the concentration of driving substrate $A_2$. The apparent affinity of $S$ for the cotransporter at side 1 is given by the half-saturation concentration, $K_m$, as

$$K_m = \frac{K(l/k_1 + l/k_2 + A_1/k_1K_{A_1})}{A_1(l/k_2 + l/g_1)} \tag{15}$$

This apparent affinity depends strongly on the concentration of driving substrate $A_1$. At low values of $A_1$, $K_m$ tends to very high values (infinite in the limit of zero $A_1$). At very high levels of $A_1$, $K_m$ reduces, after simplification and substituting for $K$ from Table 3, to $K_{AS_1} \times k_2/(k_2 + g_1)$. What do these results mean? In the first place, the value of the apparent affinity for the driven substrate depends strongly on the concentration of the driving substrate at either face of the membrane. This is due merely to the fact that increasing concentrations of the driver bring about a redistribution of the transporter conformations, with more transporter being made available for cotransport of the driven substrate. Thus, by virtue of this redistribution effect, the transporter will have a low apparent affinity at the side where $A$ is low (that is, at the whither side) and a high apparent affinity for $S$ at the whence side. The limiting affinity of the cotransporter for the driven substrate at very high levels of the driver is fixed, however, by the intrinsic affinity of $S$ for the transporter, as described by the dissociation constant $K_{AS_1}$. It is only in this limited region of concentration of the cotransported substrates that the energetics of binding of $S$ and $A$ to the transporter come into play. It should be noted, too, that the higher the intrinsic affinity of $S$ for binding to the transporter, the lower is the intrinsic affinity of $A$ (by the "thermodynamic box" rules), and, hence, the higher the concentration of $A$ at which the effect of redistribution of transporter ceases to be the main determinant of the apparent affinity! Redistribution will, in many circumstances, be the main factor affecting the apparent affinity for the driven substrate. How the maximum velocity and apparent affinity of the transporter for the driven substrate depend on the concentration of the driver is shown in Figure 10a, where the conventional Lineweaver–Burk plot of l/velocity against l/substrate concentration is drawn for this case of ordered binding, with the driver binding first.

It is important to note that, even in that limited range of concentrations where the intrinsic affinity of the transporter for the driven substrate is a determinant of the apparent affinity, this increased affinity for the driven substrate is not necessarily at the expense of a decreased intrinsic affinity for the driving substrate. The relevant "thermodynamic box" describing the energetic interactions of the transporter is not merely that for the equilibrium between $E$, $EA$, $ES$, and $EAS$, but includes also the equilibria between the conformations of the free and loaded transporter on the two sides of the membrane. The rule that emerges from a consideration of the "thermodynamic box" as applied to the cotransporter is

$$(k_1/k_2)\, K_{A_1}K_{AS_1} = (g_1/g_2)\, K_{A_2}K_{AS_2} = (k_1/k_2)\, K_{S_1}K_{SA_1} = (g_1/g_2)\, K_{S_2}K_{SA_2} \quad (16)$$

Thus, $K_{AS_1}$ can be smaller than $K_{AS_2}$, that is, the cotransporter can have a higher intrinsic affinity for $S$ at side 1 than at side 2 if (1) $A$ binds more poorly to $E$ at side 1 than at side 2, (2) the free transporter is more stable at side 2 than at side 1 (when $k_1$ is bigger than $k_2$), or (3) the $EAS$ complex is more stable at side 1 than at side 2 (when $g_1$ is smaller than $g_2$). These latter effects depend on the energetic differences between the two states of the transporter and are not at all of the traditional type of "linked-function" effects, where the binding of one substrate deforms the ligand–protein complex so that the second substrate binds more or less poorly. We saw the importance of the energy difference between the two states of the carrier in determining the apparent affinity of the simple carrier for its substrate. Here, we see again the importance of such energy differences between two protein states in determining the apparent affinity of the cotransported substrates for their transporter. A high apparent affinity for the second substrate to be bound will arise if the free transporter is more stable on the whither side.

*Case 2. Ordered Reaction with* S *Binding First at Side 1.* Here, $K_{A_1}$ is infinite. The maximum velocity of transport is $(l/k_2 + l/g_1 + K_{SA_1}/g_1A_1)^{-1}$ and depends strongly on $A_1$. The apparent affinity of $S$ for the cotransporter is given by

$$K_m = \frac{K(l/k_1 + l/k_2)}{A_1(l/k_2 + l/g_1 + K_{SA_1}/g_1A_1)} \tag{17}$$

This, again, depends heavily on $A_1$. At very high values of $A_1$, the apparent affinity reaches very high values, $K_m$ tending to zero. At zero $A_1$, $K_m$ reaches a limiting value of $K_{S_1}(k_1 + k_2)/k_2$. The very fact of an unequal distribution of the driver $A$ leads to a difference in the apparent affinity of cotransporter for the driven substrate, the affinity, once again, being highest on the whence side. The limit of this effect is seen at very high concentrations of the driver and this limit is set by $K_{S_1}$, the intrinsic dissociation constant of $S$ with the cotransporter. It is only in this region of concentrations of the driving substrate that considerations of binding energetics play a role. Again, the tighter the binding of $S$ to the transporter at the whence side, the higher the concentration of driving substrate needed to show-up this effect. The reciprocal plot of velocity and substrate concentration for this case is given in Figure 10b.

Once again, an increased affinity of $S$ for $E$ at side 1, as compared with its affinity at side 2, can arise not only at the expense of a poor intrinsic affinity for $A$ at side 1, but also as a result of a differential stability of the states of the transporter. By the operation of our "box" rule, Eq. (16), $K_{S_1}$ will be lower than $K_{S_2}$ if $k_1$ is higher than $k_2$, that is, if the free transporter is more stable in that conformation in which its binding sites for substrate face side 2. This differential stability will arise from the energetic interactions between the amino-acid side chains of the transporter protein and not from some "linked-function" effect between binding of the driven and driving substrates, although these, too, can play a role in energetic considerations.

*Case 3. Random Order of Binding* S *and* A. This case is also dealt with simply. The maximum velocity term is just as in case 2. The half-saturation concentration or apparent affinity of $S$ for the cotransporter is given by

$$K_m = \frac{K(l/k_1 + l/k_2 + A_1/k_1K_{A_1})}{A\ (l/k_2 + l/g_1 + K_{SA_1}/g_1A_1)} \tag{18}$$

This will be strongly affected by the concentration of $A_1$, and, hence, determined by redistribution effects only, when $A_1$ is higher than $K_{SA_1}$ and lower than $K_{SA_1}$. Outside this range, the apparent $K_m$ reaches the two limiting values found for the ordered binding cases discussed just above. Again, if the energetics are set such that the intrinsic affinities of $S$ for the cotransporter (either for the free transporter $K_{S_1}$, or for the transporter already bound to $A$, $K_{AS_1}$) are high, then the range over which this effect of energetics rather than redistribution is dominant is accordingly small. The reciprocal plots for the random order case are given as Figures 10c and d, the former for the case where $K_{S_1}$ is larger than $K_{AS_1}$, the latter where the reverse is true.

We can conclude that in all cases of the cotransport model, the effects of redistribution of the transporter brought about by the mere presence of the driving substrate

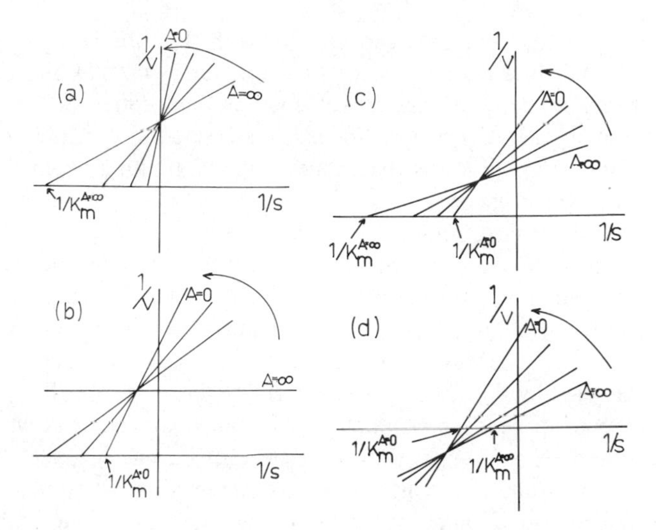

*Figure 10.* Kinetic analysis of the cotransporter. Schematic diagrams of plots of the reciprocal of the velocity of transport of $S$ ($1/v$) on the ordinate against the reciprocal of the concentration of $S$ ($1/S$) on the abscissa. The diagrams are for four cases. (a) An ordered reaction of ligand binding with $A$ binding first. (b) An ordered reaction with $S$ binding first. (c) A random order of binding where $S$ binds less avidly to the free transporter than to the form $EA$. (d) A random order of binding where $S$ binds more avidly to the free transporter than it does to $EA$. The arrows to the intersections of the lines with the abscissa give the reciprocals of the affinities of ligands with transporter as indicated. In each figure, the set of lines is for a set of different values of the concentration of the driving substrate $A$ decreasing from infinity to zero in counterclockwise progression.

are likely to be large. Affecting the intrinsic affinity of transporter for the driven substrate, at the expense of its affinity for the driving substrate or by virture of energetic differences between the two conformations of the transporter, does increase the effectiveness of transporter at the extreme range of concentrations of the driving substrate. The more the transporter affinity for the driven substrate is increased at the expense of that of the driver, the bigger is the range over which redistribution and net energetic effects are dominant. It should be stressed that even these second order effects of energetics are considerable only in situations far from the steady state; they are experimentally easier to handle but unlikely to be representative of the situation where the forces of evolution have selected among different designs of cotransport systems.

## V. PRIMARY ACTIVE TRANSPORT—CHEMI–OSMOTIC COUPLING

### A. Primary Active Transport vs. Secondary Active Transport

In many cases, the active transport of a substrate is linked not to the transport of a second substrate, but to the progress of a chemical reaction. These are the chemi–osmotic systems or the primary active transporters. It has become clear over the last two decades that these primary active transporters are, in principle, no way different from the secondary active (osmotic–osmotic) transporters that we have discussed. Kinetically, the same formal kinetic equations can equally well describe secondary as primary active transport. We shall see that, energetically, there is no real difference between them, and all that we have learned about counter- and cotransport is immediately applicable to primary active transport. Indeed, understanding the transport side of the primary active transports gives us the framework for thinking about the role of the chemical reaction. The properties of the primary active transport systems are discussed in detail in this and the next volume of this series.

In order to make explicit the connection between chemi–osmotic coupling and osmotic–osmotic coupling, we shall use precisely the same formal symbolism for the two processes in what follows. Thus, the chemical reaction considered will be written formally as $A_1 \rightleftharpoons A_2$, where $A_1$ is the reactant and $A_2$ the product of the chemical reaction ($A_1$ can be, for instance, ATP and $A_2$ is then ADP + Pi), this chemical reaction is, of course, reversible although with its equilibrium perhaps far in the direction of breakdown of $A_1$ to form $A_2$. The terms "reactant" and "product" are arbitrary; both are reactants or products depending on the distance from thermodynamic equilibrium. The conformation change which brings the binding site of the transporter for substrate $S$ to side 1 allows the chemical substrate $A_1$ to bind to the transporter. Thus, in conformation $E_1$, the transporter can bind and release $S$ at side 1, and can interact with reactant $A_1$. The conformation change to state $E_2$ brings the substrate $S$ to side 2 where the transporter can bind or release $S$, and is accompanied by the progress of the chemical reaction $A_1 \rightleftharpoons A_2$. In conformation $E_2$, the transporter can release (and interact with) the product $A_2$. The mechanism of coupling is (1) the restriction of the progress of the chemical reaction $A_1$ to $A_2$ to the simultaneous progress of the conformational change $E_1$ to $E_2$, and (2) the restriction of the access of $A_1$ to the enzyme-binding site in conformation $E_1$, and of $A_2$ to the enzyme-binding site in conformation $E_2$. Little is known of the details by which these linkages occur, but to understand this is to understand the heart of the problem of the mechanics of primary active transport. These rules that govern the rates of the conformation changes and the linked transport and chemical reaction events, and the rules that govern access of driven and driving substrates, or of reactants and products to their respective active centers are what determine coupling (Jencks, 1980). Much emphasis in research has been on describing and understanding the affinity changes for the driven substrate. These are subordinate, secondary problems.

## B. *The Kinetics of Primary Active Cotransport*

Since we have shown a perfect analogy between osmotic–osmotic coupling and chemi–osmotic coupling, the kinetic model for cotransport (Figure 8) will apply also to primary cotransport, without change. The movement of $S$ from side 1 to side 2 of the membrane is brought about by the conformation change from state 1 to state 2 of the transporter. This is obligatorily linked to the binding of $A_1$ to $E_1$ (or to $ES_1$, depending on the order of binding of $A$ and $S$ at side 1), and to the progress of the chemical reaction $A_1 \rightleftharpoons A_2$. We cannot take Eq. (12) as a description of the kinetics of primary cotransport, however, since this equation does not take into account the fact that the reactant undergoes a chemical transformation into the product, with a change in standard state-free energy. To take this fact into account, we shall have to consider the "thermodynamic box" relevant to the primary cotransporter. We consider the overall reaction of $E_1$ binding to $S$ and then to $A_1$ to yield $EAS_1$, followed by its transformation to $EAS_2$, the release of $S$ and $A_2$, and the return of free transporter to side 1. The free energy of binding of $S$ and $A_1$ to $E_1$, followed by the free energy of the transformation of $EAS_1$ to $EAS_2$, and the return of free transporter is not equal to the sum of free energies for the corresponding changes in the reverse direction, since there is also a free energy change after the $A_1$ to $A_2$ chemical transformation. This free energy change must be included within the "thermodynamic box." (For the transportable substrates, since there is no change in the nature of the substrate on its transport, from one side of the membrane to the other, there is no change in the standard state-free energy. This is not the case for the chemical reactants.) We replace Eq. (16) by the equations

$$(k_1/k_2)\, K_{A_1}\, K_{AS_1}\, K_{eq} = (g_1/g_2)\, K_{A_2}\, K_{AS_2} = (k_1/k_2)\, K_{S_1}\, K_{SA_1}\, K_{eq} = (g_1/g_2)\, K_{S_2}\, K_{SA_2} \quad (19a)$$

$$K_1 = (k_1/g_1)\, K_{A_1}\, K_{AS_1} = (k_1/g_1)\, K_{S_1}\, K_{SA_1} \quad (19b)$$

$$K_2 = (k_2/g_2)\, K_{A_2}\, K_{AS_2} = (k_2/g_2)\, K_{S_2}\, K_{SA_2} \quad (19c)$$

where, rather than a single parameter $K$, we have two parameters $K_1$ and $K_2$ connected by $K_2/K_1 = K_{eq}$, where $K_{eq}$ is the equilibrium constant for the chemical reaction $A_1$ going to $A_2$, written as the equilibrium concentration of products over that of reactant. We replace the kinetic Eq. (12) by the analogous equation

$$v_{12}^{s} = \frac{K_2 S_1 A_1 + S_1 A_1 S_2 A_2}{K_1 K_2 R_{oo} + K_2 R_{12} S_1 A_1 + K_1 R_{21} S_2 A_2 + R_{ee} S_1 A_1 S_2 A_2} \quad (20)$$

where $K_1$ and $K_2$ are defined as in Eq. (19), while the terms in $R$ are as defined in Table 3. The net transport of $S$ from side 1 to side 2 at the steady state is now given by $(K_2 S A_1 - K_1 S_2 A_2)$ divided by the denominator term. The net flow of $S$ is zero when

$$S_1/S_2 = (A_2/A_1)/K_{eq} \quad (21)$$

Thus, $A_1$ can be at the same absolute concentration as $A_2$, and yet there will be intense concentration of $S$ at side 2, provided that $K_{eq}$ is large, that is, the equilibrium of the chemical reaction is well over on the side of formation of products. The driving force for the transport of $S$ is no longer merely the ratio of concentrations of the driving substrate $A$, but rather the ratio of these concentrations multiplied by the equilibrium constant of the chemical reaction, that is, it depends on how far the given concentration ratio of the reactants to products is from the equilibrium value. In truth, precisely this ratio (that is, concentration ratio multiplied by equilibrium constant) is correct also for the cotransport of two substrates, but here the chemical equilibrium constant is unity since the transported substrates undergo no change in their chemical state in transport across the membrane. Equation (20) is the general form of the cotransport equation of which Eq. (12) is the special case.

## C. *Maximizing the Effectiveness of the Primary Cotransporter*

We can consider the primary cotransporter just as we did the two-substrate cotransporter. An equation equivalent to Eq. (14) is readily derived from Eq. (20). Precisely the same considerations of the situation near to the steady state show that minimizing the resistance parameters $R_{oo}$ and $R_{ee}$ will ensure a maximum rate of pumping of the cotransported substrate in conditions where the pump is working hardest, that is, at the highest obtainable gradient of the driven substrate. The minimizing of these resistances is simply a property of the rate constants of the interconversion rates of the transporter enzyme, and no considerations of "linked-energy" function are relevant. Slippage of the cotransporter can be defined as movement of substrate down its concentration gradient on the transporter, unlinked to the chemical reaction, and as progress of the chemical reaction via the transporter, but unlinked to simultaneous movement of the substrate. Slippage will be minimized by appropriate increase in the selectivity of the transport event, that is, in the tightening-up of the rules governing the movement of the transporter. Again, energy balances play no role.

## D. *Apparent Affinities of a Primary Cotransport System*

The apparent affinities of the cotransporter for the driven substrate are, in the case of the primary cotransporter, affected by precisely the same factors of transporter redistribution and energetics as in the case of the secondary cotransporter discussed in Section IV-C. However, the equilibrium constant of the reaction $A_1$ to $A_2$ must be taken into account. We can show that, even at equal concentrations of the reactant and products, that is when $A_1 = A_2$, there will be a substantial difference in the apparent affinities of the cotransporter for the driven substrate on the two sides of the membrane. We note that Eqs. (15), (16), and (18) contain the parameter $K$ in their numerators. For the primary cotransporter, this is replaced by $K_1$ defined in Eq. (19). The corresponding equation for the $K_m$ in the 2 to 1 direction will contain the corresponding parameter $K_2$. Since $K_1$ and $K_2$ are related through the equation $K_1/K_2 = K_{eq}$, the equilibrium constant for the chemical reaction enters directly into the relations determining the ratio of affinities of the transporter for the driven substrate at the two sides of the membrane. Thus, even if $A_1 = A_2$, since $K_1 = K_2$, the apparent affinities $K_{m_1}$ and $K_{m_2}$ will be very different. If, in any experimental situation, one wants to check whether the affinities for the driven substrate are intrinsically different at the two faces of the membrane, one first has to make sure that these effects of the equilibrium constant of the driven chemical reaction have been correctly taken into consideration. Since, for example, the reaction ATP to ADP + Pi has an equilibrium constant of some $10^8$, measurements purporting to be those of intrinsic affinities could be seriously distorted by failure to take into account the immense "effective concentration" of ATP.

Once again, there are ranges of concentrations of the driving reactants and products where it is an effective aid to the transport of the driven substrate to have its intrinsic affinity for the transporter set as high as possible on the whence side. This can be at the expense of the binding energy for the driving reactant, or it can be a result of the differential energy of the two forms of the transporter $E_1$ and $E_2$, or $EAS_1$ and $EAS_2$. In the case of the sodium pump (Glynn, this volume), $EAS_1$ is a high energy form (this is the E~PNa form described by investigators of the sodium pump) as compared with $EAS_2$ (the form E-PNa). In addition, $E_1$ is a low-energy form as compared with $E_2$. These two factors on their own would, by the operation of the "thermodynamic box" relations of Eq. (19), lead to $S$ having a lower affinity at side 1 than at side 2, side 1 being the cytoplasmic face of the transporter. It is clear, however, that, experimentally, the system is far more affine for sodium at the cytoplasmic face of the membrane than at the extracellular face. This higher affinity arises, by our "box" rules, from the fact that the equilibrium constant $K_{eq}$ for the reaction ATP to ADP + Pi is, of course, well over onto the side of the breakdown of ATP. A high value of $K_{eq}$ in the "box" Eq. (19) allows $K_{AS_1}$ to be smaller than $K_{AS_2}$, in spite of the conformation equlibria terms ($k_1/k_2$, $g_2/g_2$) being in a direction which would produce the opposite result. With $EAS_1$ destabilized compared with $EAS_2$, and $E_1$ stabilized with respect to $E_2$, the asymmetry of the sodium-pumping system is such that it has a much greater maximum velocity in the direction favoring sodium extrusion than in the opposite direction (the resistance term $R_{12}$ is far smaller than the term $R_{21}$). The situation of a high maximum velocity of sodium efflux in the face of a high apparent affinity for sodium at the cytoplasmic face is not one which could be achieved by a simple carrier (see Section III-A). It is achieved by the sodium pump by making use of the fact that the chemical reaction is well over to the side of ATP splitting, and the resulting high value of $K_{eq}$ in Eq. (19) allows an asymmetry in sodium-binding affinities, favoring an effective outward pumping system.

Potassium movements on the sodium pump are in the nature of countertransport, the transporter being brought to the extracellular face of the membrane by the cotransport of sodium and by the simultaneous progress of the reaction that splits ATP to ADP and Pi. The return of the transporter to the cytoplasmic face of the membrane with bound potassium goes through an occluded form. We have argued (Karlish *et al.*, 1982) that this occluded form is concerned with controlling the direction in which net pumping occurs and minimizing the rate of slippage. In this, as in any real system, detailed refinements of the control of the rates of transport in different directions and under the influence of the different ligands, substrates and reactants, will occur to maximize overall effectiveness of pumping in response to specific requirements of the biological system concerned.

## VI. CONCLUSIONS

We see that it is possible to put all carrier theory into a simple theoretical framework. The simple carrier, the cotransport of two substrates, and the cotransport of a substrate linked to the progress of a chemical reaction can all be handled by what is virtually the same formalism, the same fundamental kinetic equation. This similarity forces us to see the essential similarities in the three processes. All are driven by the progress of a thermodynamic event, the equilibration of substrate concentrations, or a chemical reaction. The effectiveness of transport is governed by rules which minimize the rate of certain reaction pathways, maximizing the rates of others. When working near to the steady state, that is, most slowly, the cotransport systems are most effective when the rates of loaded and unloaded transporter movements are maximized. Apparent affinities for the driven substrate play little role, but the order of binding of driving and driven substrate will be critical for maximal effectiveness. When working far from the steady state under conditions most often studied in the laboratory, the apparent affinity of the driven substrate for the transporter is a critical factor which must be maximized at the whence side. This increase in apparent affinity will often come about as a direct result of the mere presence of the driving substrate at a high concentration at the whence side, or by the presence of realistic concentrations of the reactant (chemi–osmotic coupling). It can also come about as a result of a differential stability of the two forms of the transporter that face opposite faces of the membrane. Finally, the differential intrinsic affinity of transporter for the driven substrate can arise from an energetic interaction on the transporter between driven and driving substrate, or between the reactant and the driven substrate. This last factor is often considered to be the sole or the most significant factor in increasing the effectiveness of a transport system. It could well turn out to have a relatively minor effect.

## REFERENCES


Cabantchik, Z. I., and Ginsburg, H., 1977, Transport of uridine in human red cells: Demonstration of a simple carrier-mediated process, *J. Gen. Physiol.* **69:**75–96.

Cleland, W. W., 1977, Determining the chemical mechanisms of enzyme-catalysed reactions by kinetic studies, *Adv. Enzymol.* **45:** 273–387.

Crane, R. K., 1977, The gradient hypothesis and other models of mediated active transport, *Rev. Physiol. Biochem. Pharmacol.* **78:**99–159.

Devés, R., and Krupka, R. M., 1981, Evidence for a two-state mobile carrier mechanism in erythrocyte choline transport: Effects of substrate analogs on inactivation of the carrier by *N*-ethylmaleimide, *J. Membr. Biol.* **61:**21–30.

Glynn, I. M., 1985, The $Na^+$, $K^+$-transporting adenosine triphosphate, in: *The Enzymes of Biological Membranes,* Vol. 3 (A. N. Martonosi, ed.), Plenum Press, New York, pp. 35–114.

Glynn, I. M., and Karlish, S. J. D., 1982, Conformational changes associated with $K^+$ transport by the $Na^+/K^+$-ATPase, in: *Membranes and Transport* (A. N. Martonosi, ed.), Plenum Press, New York, pp. 529–536.

Holman, G. D., 1978, Cyclic AMP transport in human erythrocyte ghosts, *Biochim. Biophys. Acta* **508:**174–183.

Hopfer, U., and Groseclose, R., 1980, The mechanism of $Na^+$-dependent D-glucose transport, *J. Biol. Chem.* **255:**4453–4462.

Jencks, W. P., 1980, The utilization of binding energy in coupled vectorial processes, *Adv. Enzymol.* **51:**75–106.

Karlish, S. J. D., 1980, Characterisation of conformational states in (Na,K)-ATPase labeled with fluorescein at the active site, *J. Bioenerg. Biomembr.* **12:**111–136.

Karlish, S. J. D., and Pick, U., 1981, Sidedness of the effects of sodium and potassium ions on the conformational state of the sodium–potassium pump, *J. Physiol. (London)* **312:**505–529.

Karlish, S. J. D., and Yates, D. W., 1978, Tryptophan fluorescence of ($Na^+$ + $K^+$)-ATPase as a tool for study of the enzyme mechanism, *Biochim. Biophys. Acta* **527:**111–130.

Karlish, S. J. D., Lieb, W. R., and Stein, W. D., 1982, Combined effects of ATP and phosphate on rubidium–rubidium exchange mediated by Na-K-ATPase reconstituted into phospholipid vesicles, *J. Physiol. (London)* **328:**333–350.

Klingenberg, M., 1985, The ADP/ATP carrier in mitochondrial membranes, in: *The Enzymes of Biological Membranes,* Vol. 4 (A. N. Martonosi, ed.), Plenum Press, New York, pp. 511–553.

Krupka, R. M., and Devés, R., 1979, The membrane valve: A consequence of asymmetrical inhibition of membrane carriers, *Biochim. Biophys. Acta* **550:**77–91.

Lieb, W. R., 1982, A kinetic approach to transport studies, in: *Red Cell Membranes—A Methodological Approach* (J. C. Ellory and J. D. Young, eds.), Academic Press, London, pp. 135–164.

Mitchell, P., 1957, A general theory of membrane transport from studies of bacteria, *Nature* **180:**134–136.

Plagemann, P. G. W., Wohlhueter, R. H., and Erbe, J., 1982, Nucleoside transport in human erythrocytes: A simple carrier with directional symmetry and differential mobility of loaded and empty carrier, *J. Biol. Chem.* **257:**12069–12074.

Rosenberg, R., 1981, L-leucine transport in human red blood cells: A detailed kinetic analysis, *J. Membr. Biol.* **62:**79–93.

Stein, W. D., 1981, Concepts of mediated transport, in: *Membrane Transport,* Chap. V (S. L. Bonting and J. J. H. H. M. de Pont, eds.), Elsevier, Amsterdam, pp. 123–157.

Stein, W. D., and Lieb, W. R., 1973, A necessary simplification of the kinetics of carrier transport, *Israel J. Chem.* **11:**325–339.

Tanford, C., 1982, Simple model for the chemical potential change of a transported ion in active transport, *Proc. Natl. Acad. Sci. USA* **79:**2882–2884.

Weber, G., 1975, Energetics of ligand binding to proteins, *Adv. Protein Chem.* **29:**1–83.

# 28

# The $Na^+,K^+$-Transporting Adenosine Triphosphatase

I. M. Glynn

## I. INTRODUCTION

### A. Definition

The $Na^+,K^+$-transporting adenosine triphosphatase ($Na^+,K^+$-ATPase), also often known as the sodium pump or sodium–potassium pump, is an enzyme, found in nearly all animal-cell membranes, that uses energy from the hydrolysis of intracellular ATP to transport $Na^+$ ions outwards and $K^+$ ions inwards. It may be thought of as having three substrates (ATP, intracellular $Na^+$ ions, and extracellular $K^+$ ions) and four products (ADP, orthophosphate, extracellular $Na^+$ ions, and intracellular $K^+$ ions.) Because more $Na^+$ ions are pumped out than $K^+$ ions are pumped in, the activity of the enzyme generates an outward movement of positive charge, and this outward current may also be considered a "product" of the reaction.

The $Na^+,K^+$-ATPase is now recognized to be one of a class of ion-transporting ATPases which also includes (1) the $Ca^{2+}$-transporting ATPases of cell membranes and of the sarcoplasmic reticulum, (2) the $H^+,K^+$-transporting ATPase responsible for acid secretion in the gastric mucosa, and (3) the $H^+$-transporting ATPases of fungal and bacterial membranes. The so-called $H^+$-transporting ATPases of mitochondria and chloroplasts are really ATP synthases running in reverse and do not resemble the enzymes listed above either in struture or mechanism.

### B. Functions

A detailed consideration of the functions of the $Na^+,K^+$-ATPase is beyond the scope of this chapter but a brief summary may be useful. The functions include:

---

I. M. Glynn • Physiological Laboratory, University of Cambridge, Cambridge CB2 3EG, England.

1. Maintenance of osmotic stability in animal cells. (The $Na^+$,$K^+$-ATPase ensures that the less penetrating $Na^+$ ion is largely extracellular and that the more penetrating $K^+$ ion is largely intracellular; because the pumped efflux of $Na^+$ is greater than the pumped influx of $K^+$, activity of the $Na^+$,$K^+$-ATPase also leads directly to a net loss of osmotically active particles, but the maintenance of osmotic stability does not depend on that inequality.)

2. Maintenance of the high-$K^+$ environment appropriate for many intracellular enzymes.

3. Maintenance of the transmembrane $Na^+$ and $K^+$ gradients that generate the resting membrane potential or provide energy for the action potentials of excitable cells.

4. Transport of sodium chloride across epithelia, both when the primary need is salt transport (thick ascending limb of the loop of Henle, salt glands of birds and fish, and possibly fish gills) and when the primary need is water transport (proximal kidney tubule, gall bladder, large intestine, amphibian urinary bladder, frog skin, and cuttle bone for controlling buoyancy).

5. Generation of a transmembrane electrochemical potential gradient for $Na^+$ ions which is then used to drive uphill movements of other substances, e.g., sugars, amino acids, $Ca^{2+}$, $Cl^-$, and $H^+$ ions, through appropriate coupled transport systems.

6. Generation of heat, accounting, probably, for more than 20% of the basal metabolic rate of adult mammals.

Why the sodium pump first evolved is a question that has often been asked but never satisfactorily answered. The need to maintain osmotic stability is perhaps the most likely explanation, though the fact that among the $K^+$-requiring enzymes there are several concerned with protein synthesis or with the transfer of phospho groups might suggest that $K^+$ accumulation was the primary function.

## C. *History*

The need for some process to maintain the $Na^+$ and $K^+$ gradients across the muscle membrane was pointed out by Overton in 1902, but it was not until 1941 that Dean developed the idea of an outwardly directed $Na^+$ pump as the explanation of the distribution of $Na^+$, $K^+$, and $Cl^-$ ions across that membrane. By 1941, it was known that muscles lost $K^+$ and gained $Na^+$ during activity or cold storage, and that the gradients could be restored during recovery at room temperature (Fenn and Cobb, 1936; Steinbach, 1940). Similar experiments on cold-stored red cells showed a similar restoration of the concentration gradients, provided that glycolysis could occur (Maizels and Patterson, 1940; Harris, 1941; Danowski, 1941). In red cells, it was clear that during recovery, the movements of $K^+$ ions as well as of $Na^+$ ions were uphill, since the (estimated) membrane potential (unlike that in muscle) was far from the $K^+$ equilibrium potential. At about the same time, experiments with tracers showed that the membranes of muscles and red cells were permeable to both $Na^+$ and $K^+$ ions, and this permeability suggested independently that the normal distribution of $Na^+$ and

$K^+$ across the cell membrane represented a balance between pumping and leaking (Cohn and Cohn, 1939; Heppel, 1940; Noonan *et al.*, 1940; Mullins *et al.*, 1941).

A link between $Na^+$ efflux and $K^+$ influx was indicated by the finding in muscle (Steinbach, 1951, 1952; Keynes, 1954), red cells (Harris and Maizels, 1951), and nerve (Hodgkin and Keynes, 1955) that $Na^+$ efflux was much reduced in $K^+$-free media. In muscle and nerve, removal of $K^+$ from the bathing medium hyperpolarizes the cell, but Hodgkin and Keynes (1955) showed that this hyperpolarization could not have caused the reduced $Na^+$ efflux in their experiments since a similar hyperpolarization produced by an applied current did not affect the outward movement of $Na^+$. In red cells, the membrane potential is "clamped" by the chloride ratio, yet glucose-dependent $Na^+$ efflux and $K^+$ influx varied in parallel when the external $K^+$ concentration was changed, and both disappeared when $K^+$ ions were absent from the bathing medium (Glynn, 1956). This parallelism and the promptness of the effects of changes in external $K^+$ concentration showed that, in physiological conditions, the active fluxes of $Na^+$ and $K^+$ are tightly coupled.

The dependence of active $Na^+$ and $K^+$ transport on glycolysis in mammalian red cells and on respiration in most other tissues pointed to ATP as the common energy source. Direct evidence for this role of ATP came from experiments in which ATP was incorporated into resealed red-cell ghosts (Gardos, 1954) or injected into cyanide-poisoned giant axons (Caldwell and Keynes, 1957). By 1957, therefore, it had been reasonably well established that the membranes of several types of cells possess a linked $Na^+$–$K^+$ pump, energized by ATP. It was known that several enzymes, particularly those involved in the transfer of phospho groups, were activated by $K^+$ ions, and it seemed possible that energy was fed into the active transport mechanism through the reactions of a $K^+$-activated enzyme situated on the cell surface (Glynn, 1956). It was also known that the red-cell membrane possessed ATPase activity (Clarkson and Maizels, 1952), but the dependence of that activity on $Na^+$ and $K^+$ ions had yet to be discovered. In the mid-1950's, cardiac glycosides, which had been shown to inhibit the red-cell sodium pump by Schatzmann in 1953, began to be used to define fluxes through the sodium pump mechanism and to count pump sites (Glynn, 1957).

In 1957, Skou published his classic paper demonstrating ($Na^+$ + $K^+$)-dependent ATP hydrolysis by particles of minced crab nerve; subsequently, he showed that this hydrolysis was inhibited by ouabain (Skou, 1960). The synergism between $Na^+$ and $K^+$ ions, the particulate nature of the preparation, and the sensitivity to ouabain suggested that the ($Na^+$ + $K^+$)-dependent ATPase might be identical with the $Na^+$,$K^+$ pump and this identity was subsequently proved by the following six lines of evidence:

1. There is a close parallelism between the properties of the $Na^+$,$K^+$ pump and the ($Na^+$ + $K^+$)-dependent ATPase in red cells (Post *et al.*, 1960; Dunham and Glynn, 1961).

2. There is a close parallelism between the distribution of pumping activity and ($Na^+$ + $K^+$)-dependent ATPase activity in different tissues and species (Bonting *et al.*, 1961; Bonting and Caravaggio, 1963; Tosteson, 1963; Baker and Simmonds, 1966).

3. Experiments with resealed red-cell ghosts show that ouabain-sensitive ATP

hydrolysis requires the presence of intracellular $Na^+$ and extracellular $K^+$ (Glynn, 1962; Laris and Letchworth, 1962; Whittam, 1962).

4. Increasing the concentration gradients of $Na^+$ and $K^+$ ions across the red-cell membrane to such an extent that the energy available from the hydrolysis of ATP is insufficient to drive the normal cycle leads to the backward running of the pump and the synthesis of ATP (Garrahan and Glynn, 1967e; Glynn and Lew, 1970).

5. Antibodies to a partially purified $Na^+,K^+$-ATPase preparation from pig kidney inhibited $Na^+$ and $K^+$ transport when trapped inside resealed red-cell ghosts (Jørgensen *et al.*, 1973).

6. Partially purified $Na^+,K^+$-ATPase prepared from brain or kidney and incorporated into the membranes of artificial lipid vesicles is capable of transporting $Na^+$ ions into the vesicles and $K^+$ ions out of them, with appropriate stoichiometry, when the vesicles are incubated in ATP-containing media (see Section III-B for references).

## II. PURIFICATION OF THE $Na^+,K^+$-ATPase

An excellent and detailed account of work on the purification of the $Na^+,K^+$-ATPase is included in the review recently published by Jørgensen (1982). Only a brief account is given here.

The starting material for purification is generally either (1) a tissue rich in epithelia specialized for sodium transport, e.g., the outer medulla of mammalian kidney (Kyte, 1971; Jørgensen and Skou, 1971; Jørgensen, 1974a,b), the salt glands of sea birds or salt-fed ducks (Hopkins *et al.*, 1976), or the rectal glands of dogfish (Hokin *et al.*, 1973; Perrone *et al.*, 1975; Skou and Esmann, 1979), or (2) a tissue rich in $Na^+,K^+$-ATPase associated with excitable cells, e.g., mammalian brain (Sweadner, 1979) or electric organ of electric eel (Dixon and Hokin, 1974; Cantley *et al.*, 1978b). For mammalian $Na^+,K^+$-ATPase, the best source is the dark red outer medulla of the kidney, where the basolateral membranes of the cells forming the thick ascending limbs of the loops of Henle are particularly rich in the enzyme (Fujita *et al.*, 1972; Beeuwkes and Rosen, 1975; Kyte, 1976a,b; Maunsbach *et al.*, 1978; Shaver and Stirling, 1978).

Homogenization of the tissue is followed by differential centrifugation to obtain a fraction rich in fragments of cell membrane, mostly in the form of vesicles, the so-called microsomal fraction.

### A. Membrane-Bound $Na^+,K^+$-ATPase

To obtain pure, or partially purified, membrane-bound $Na^+,K^+$-ATPase, the microsomal fraction is generally treated with detergents or with chaotropic agents such as sodium iodide. This treatment greatly increases the specific activity, partly by disrupting the vesicles so that ATP can gain access to the inner surface of the membrane and partly by selectively removing membrane proteins other than the $Na^+,K^+$-ATPase. The detergent-treated membranes are then subjected to density-gradient centrifugation of greater or lesser sophistication, depending on whether the aim is to obtain a small quantity of pure enzyme or a larger quantity of less pure enzyme.

With mammalian kidney, almost pure $Na^+$,$K^+$-ATPase can be prepared by the use of sodium dodecyl sulfate (SDS) in carefully controlled amounts, in the presence of ATP, which protects the enzyme against denaturation (Jørgensen, 1974a,b). SDS has also been used successfully in the purification of $Na^+$,$K^+$-ATPase from duck salt gland (Hopkins *et al.*, 1976), but it denatures the enzyme from dogfish rectal gland and eel electric organ, and it is unable to extract all of the unwanted membrane protein in preparations from mammalian brain. Deoxycholate has been used for purifying $Na^+$,$K^+$-ATPase from dogfish rectal gland (Skou and Esmann, 1979), and both deoxycholate and sodium iodide have been used in the purification of $Na^+$,$K^+$-ATPase from mammalian kidney and brain (Lane *et al.*, 1973; Nakao *et al.*, 1973; Hayashi *et al.*, 1977). Cantley *et al.* (1978b) have partially purified $Na^+$,$K^+$-ATPase from electric organ by density-gradient centrifugation of sonicated membranes using no chaotropic agents or detergents.

## B. *Soluble $Na^+$,$K^+$-ATPase*

The procedures described above yield a product which, even when it is believed to be substantially pure, consists of molecules aggregated together in small membranous fragments. To solubilize the enzyme, it is necessary to replace some of the lipid with detergent. Two approaches have been successful in yielding soluble $Na^+$,$K^+$-ATPase of high specific activity and high degrees of purity. In the first approach, detergents (usually nonionic) are used to solubilize crude preparations of $Na^+$,$K^+$-ATPase, and the enzyme is then separated from unwanted material by salt precipitation (Lane *et al.*, 1973; Dixon and Hokin, 1974; Perrone *et al.*, 1975) or by column chromatography (Dixon and Hokin, 1978; Hastings and Reynolds, 1979a,b; Esmann, 1980). In the second approach, $Na^+$,$K^+$-ATPase is purified in membrane-bound form and then solubilized with nonionic detergent (Brotherus *et al.*, 1981b).

## C. *Criteria of Purity*

An obvious criterion of purity is specific activity, but this alone is insufficient because the turnover number of a purified enzyme may be higher than the turnover number of the same enzyme under optimal conditions in its original environment. It is, therefore, customary to judge purity also by examination of the polypeptides that can be detected by gel electrophoresis after treatment of the enzyme with SDS. In 1971, Kyte showed that two polypeptide chains of molecular weight of about 100,000 and 40,000 copurified with the $Na^+$,$K^+$-ATPase, and much subsequent work has led to general agreement that these two polypeptides, usually referred to as $\alpha$ and $\beta$, together (probably) with a small polypeptide ($\gamma$) of molecular weight 10,000–12,000, constitute the only polypeptides in the enzyme. The nature and structure of these subunits will be discussed in Section VIII.

Examination of the polypeptides will not, in general, distinguish between active and damaged protein, and it is therefore useful to employ, as a third criterion, the number of phosphorylation sites or ouabain-binding sites per unit weight of enzyme.

## III. CATION FLUXES

### A. An Outline Scheme

The mechanism of the $Na^+,K^+$-ATPase is far from being fully understood, and such understanding as we have comes from experimental approaches of very different kinds. These approaches will be described in turn, though historically, of course, there has been a good deal of overlap. Before describing any of them, however, it will be convenient to outline the kind of model that has emerged from them, since it is often easier to follow an argument if one is aware of its destination. The model illustrated in Figure 1 is incomplete in several important ways, and parts of it are also controversial. Nevertheless, it can explain a large number of experimental observations, and is certainly useful as a set of working hypotheses.

The model is derived by grafting onto a chemical cycle [the so-called Albers–Post scheme (see Post *et al.*, 1972), based largely on studies of phosphorylation and dephosphorylation] hypotheses about accessibility to $Na^+$ and $K^+$ ions based partly on phosphorylation studies but mainly on studies of fluxes, of fluorescence changes, and of ion occlusion.

One molecule of ATP, one $Mg^{2+}$ ion, and three $Na^+$ ions are supposed to combine with high-affinity binding sites at the intracellular surface of the unphosphorylated enzyme in the $E_1$ conformation. Phosphorylation of a β-aspartyl carboxyl group in the enzyme is accompanied by the trapping of the $Na^+$ ions within the newly formed phosphoenzyme ($E_1P$). Release of ADP is followed by a spontaneous change of conformation (to $E_2P$), and this change makes the binding sites accessible to the cell exterior, and, at the same time, reduces their affinity for $Na^+$ so that the $Na^+$ ions are lost to the extracellular medium. Binding of two $K^+$ ions to high-affinity sites on the extracellular surface of $E_2P$ causes rapid hydrolysis, and this is accompanied by the trapping of the $K^+$ ions in the newly-formed dephosphoenzyme [$E_2(K)$]. Binding of ATP to a low-affinity site on $E_2(K)$ accelerates a spontaneous (but otherwise slow) change in conformation to $E_1K$, a form whose binding sites are accessible to the cell interior and have a low affinity for $K^+$. Release of $K^+$ ions to the intracellular medium completes the cycle. Further aspects of this model will be discussed later.

### B. The Six Flux Modes

With broken membranes, and with nearly all purified preparations of $Na^+,K^+$-ATPase (but cf. Forbush, 1982), it is impossible to distinguish between intracellular and extracellular ions, i.e., between some of the reactants and some of the products of the overall reaction. Important information about the working of the $Na^+,K^+$-ATPase is, therefore, available only from studies of the fluxes of $Na^+$ and $K^+$ ions across the membranes of intact or reconstituted cells, or across the membranes of

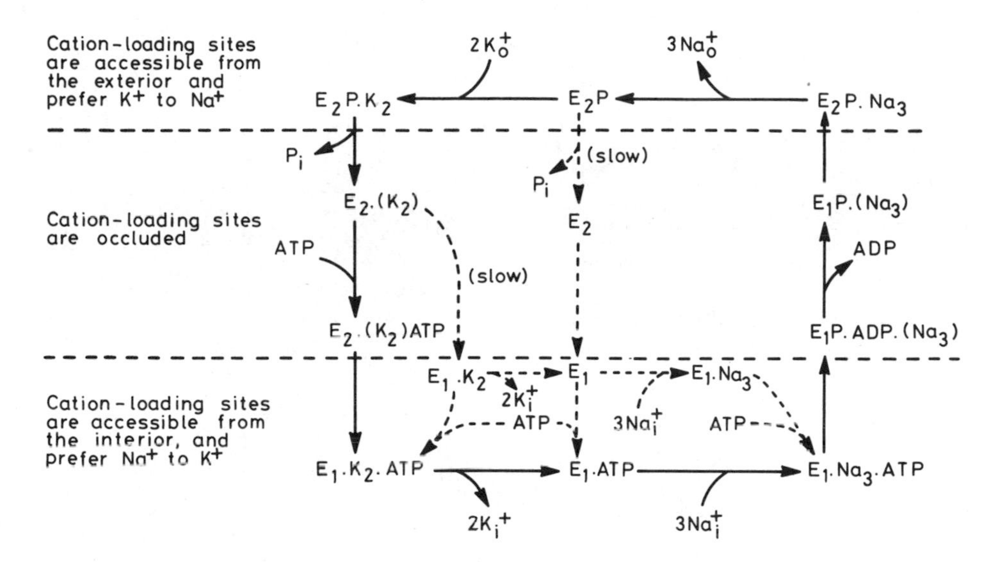


*Figure 1.* A scheme, modified from Karlish *et al.* (1978a,b), showing the proposed relations between ion movements, transfers of phospho groups, and conformational changes. The solid arrows show the cycle of events that are supposed to take place in physiological conditions. When the pump is forced to run backwards, each step is reversed and the order is reversed. $Na^+$–$Na^+$ exchange is explained by the alternate forward and backward running of the right-hand part of the scheme, and $K^+$–$K^+$ exchange by the alternate forward and backward running of the left-hand part of the scheme. Uncoupled $Na^+$ efflux is brought about by an outward movement through the right-hand part of the scheme, the cycle being completed via the central dotted pathway. All movements of ions across the membrane require the transfer of a phospho group as well as a change in conformation between the $E_1$ and $E_2$ forms of either phosphoenzyme or dephosphoenzyme. Combination with ATP, and release of ADP and orthophosphate take place at the intracellular surface of the membrane. For simplicity, $Mg^{2+}$ ions have been ignored. Although the sequence of cal events described by the scheme is the same as in recent versions of the Albers–Post scheme (see, for example, Taniguchi and Post, 1975), the relations between ion movements and conformational changes are different from those proposed by Post *et al.* (1973, 1975b). For further details see text. Note that the placing of the symbols "Na⁺" or "K⁺" within brackets is meant to imply that the ion is occluded within the enzyme. The same convention is used throughout the text.

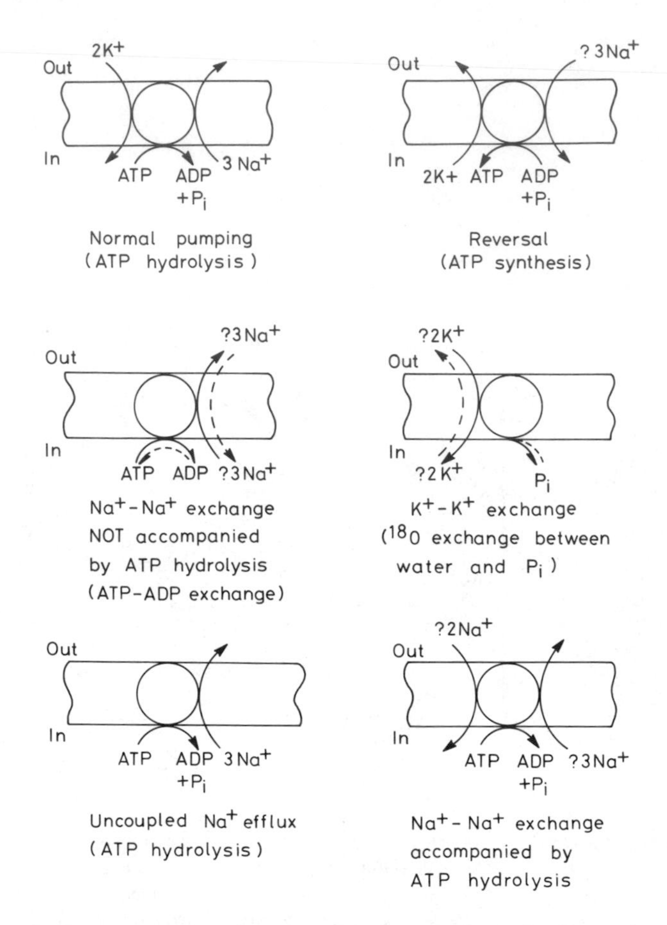


*Figure 2.* The six flux modes (see Section III-B). The diagrams for the first five modes are modified from Glynn and Karlish (1976).

vesicles prepared from them. Recently, it has also become possible to study fluxes catalyzed by $Na^+,K^+$-ATPase incorporated into the membranes of artificial lipid vesicles (Hilden and Hokin, 1975; Sweadner and Goldin, 1975; Goldin, 1977; Anner *et al.*, 1977; Karlish and Pick, 1981; Karlish and Stein, 1982a). This procedure is likely to be particularly useful for studying fluxes catalyzed by $Na^+,K^+$-ATPase modified in various ways (Jørgensen *et al.*, 1982b).

By suitable manipulation of the conditions, it is possible to make the $Na^+,K^+$-ATPase operate in six different modes (see Figure 2): (1) $Na^+$–$K^+$ exchange, the normal mode, (2) a reversed mode, (3) $Na^+$–$Na^+$ exchange *not* accompanied by ATP hydrolysis, (4) $K^+$–$K^+$ exchange, (5) uncoupled $Na^+$ efflux, i.e., $Na^+$ efflux not accompanied by the uptake of $Na^+$ or $K^+$, and (6) $Na^+$–$Na^+$ exchange accompanied by ATP hydrolysis. The belief that all six modes are brought about by the same system rests on the sensitivity of each of them to cardiac glycosides, on similarities in their responses to various physiological ligands, and on the fact that, with the exception of the reversed mode and of $Na^+$–$Na^+$ exchange accompanied by ATP hydrolysis, which have not been tested, all are inhibited by an antiserum to partially purified $Na^+$,$K^+$-ATPase from pig kidney (Glynn *et al.*, 1974).

### 1. $Na^+$–$K^+$ Exchange

Under physiological conditions, internal $Na^+$ is exchanged for external $K^+$, probably in a 3 : 2 ratio, at the expense of energy derived from the hydrolysis of ATP at the inner surface of the cell membrane. The process produces an outward movement of approximately one positive charge per cycle. Evidence concerning the coupling ratio, the electrogenic effect of the pump, and the effects, or lack of effects, of membrane potential on the pump, have been reviewed recently (Glynn, 1983) and will not be discussed here. In $Na^+$–$K^+$ exchange, ATP acts at both high-affinity and low-affinity intracellular sites (Glynn and Karlish, 1976) giving a biphasic activation curve similar to the activation curves found in studies of ATP hydrolysis by particulate $Na^+$,$K^+$-ATPase preparations (Neufeld and Levy, 1969; Kanazawa *et al.*, 1970; Robinson, 1976a).

### 2. Reversal

By arranging that the concentration gradients for $Na^+$ and $K^+$ are steeper than normal, and that the ratio [ATP]/([ADP] · [$P_i$]) is lower than normal, it is possible to make the red-cell sodium pump run backwards and synthesize ATP using energy derived from the downhill movements of the cations (Garrahan and Glynn, 1967e; Glynn and Lew, 1970; Lant *et al.*, 1970; Lew *et al.*, 1970). The rate of synthesis is roughly proportional to the external $Na^+$ concentration (Glynn and Lew, 1970), and synthesis is inhibited by external $K^+$ with a $K_{0.5}$ of about 1.3 mM ($[Na]_o$ = 150 mM). This is similar to the $K_{0.5}$ for stimulation of ouabain-sensitive $K^+$ influx by external $K^+$ under similar conditions (Glynn *et al.*, 1970), and it is also similar to the $K_{0.5}$ for activation of the forward running of the pump by external $K^+$ when normal cells are incubated in media containing 150 mM $Na^+$ (Glynn, 1956; Sachs and Welt, 1967). It is difficult to obtain accurate figures for the stoichiometry of the backward-running pump, but the synthesis of each molecule of ATP seems to be associated with the outward movement of 2–3 $K^+$ ions (Glynn and Lew, 1970).

Reversal of the pump has been demonstrated only in red cells and in resealed red-cell ghosts, but it would be difficult to demonstrate in other cells, and there is no reason to think that the $Na^+$,$K^+$-ATPase in other cells is any less reversible. Its relevance to an understanding of the mechanism of the pump is that reversibility of the overall system implies that no individual step in the normal cycle can be too far from equilibrium to be readily reversed by appropriate changes in the ligand concentrations.

### 3. $Na^+$–$Na^+$ Exchange Not Accompanied by ATP Hydrolysis

When red cells, squid axons, frog muscle, and probably other cells are incubated in high-$Na^+$, $K^+$-free media they show a ouabain-sensitive exchange of internal and external $Na^+$ ions (Caldwell *et al.*, 1960; Garrahan and Glynn, 1967a,c; Keynes and Steinhardt, 1968; Baker *et al.*, 1969b; Horowicz *et al.*, 1970; Sachs, 1970; De Weer, 1970; Sjodin, 1971; Beaugé and Ortiz, 1973; Kennedy and De Weer, 1976). The exchange is approximately one-for-one (Garrahan and Glynn, 1967a; De Weer *et al.*, 1979), and is electroneutral (Abercrombie and De Weer, 1978), but it shows a marked asymmetry in the affinity for $Na^+$ ions on the two sides of the membrane, the affinity being high inside and very low, $K_{0.5}$ of the order of 0.1 M, outside (Garrahan and Glynn, 1967c; Sachs, 1970; Garay and Garrahan, 1973). $K^+$ ions in the external solution inhibit $Na^+$–$Na^+$ exchange and make possible $Na^+$–$K^+$ exchange, with a similar (high) affinity for both effects, suggesting that the same external binding sites are involved. More surprisingly, intracellular $K^+$ has a facilitating effect on $Na^+$–$Na^+$ exchange (Garay and Garrahan, 1973; Sachs, 1981b). The explanation of this effect is not known, but Sachs suggests that $K^+$ ions may act by displacing MgATP from a low-affinity site at which it inhibits the enzyme (see also Banerjee and Wong, 1972; Robinson, 1977).

Although $Na^+$–$Na^+$ exchange of the kind being described is not accompanied by appreciable hydrolysis of ATP (Garrahan and Glynn, 1967d), it does not occur unless the cell contains both ADP (De Weer, 1970; Glynn and Hoffman, 1971) and ATP (Cavieres and Glynn, 1979); the ATP cannot be replaced by its nonphosphorylating $\beta,\gamma$-imido analogue. ADP acts with an affinity which is very low and independent of the ATP concentration, and ATP with an affinity which must be at least moderately high (Glynn and Hoffman, 1971; Kennedy and Hoffman, 1983).

Because ADP is not required for the forward running of the enzyme, it is attractive to suppose that its role in $Na^+$–$Na^+$ exchange is concerned with the inward movement of $Na^+$. It could act *either* by accepting a phospho group from $E_1P$ *or* (by analogy with the role of ATP in $Na^+$–$K^+$ exchange and $K^+$–$K^+$ exchange; see Section VII-A) by accelerating the conversion of $E_2P$ to $E_1P$ (see Figure 1). Evidence to be discussed later (see Section V-E; Klodos *et al.*, 1981) makes the second alternative less likely. In any event, the $Na^+$–$Na^+$ exchange must involve a step other than that in which the phospho group is transferred, because oligomycin (which is thought to inhibit the conversion of $E_1P$ to $E_2P$; see Section IX-C) inhibits $Na^+$–$Na^+$ exchange (Garrahan and Glynn, 1967d) but does not inhibit, and may stimulate, ATP–ADP exchange (Blostein, 1970). Recent work on the release of $Na^+$ ions occluded within the $E_1P$ form of the phosphoenzyme (Section VII-B) suggests strongly that ADP acts by accepting the phospho group, but experiments on ATP–ADP exchange are ambiguous. In resealed red-cell ghosts, ATP–ADP exchange has been shown to occur in conditions in which $Na^+$–$Na^+$ exchange occurs (Cavieres, 1980; Kaplan and Hollis, 1980), though it has not been possible to determine the stoichiometric relation between the two exchanges. In internally dialyzed squid giant axons, however, $Na^+$–$Na^+$ exchange appears not to be accompanied by the expected ATP–ADP exchange (De Weer *et al.*, 1983). This could be because, contrary to the hypothetical scheme shown in Figure 1, dephosphorylation of $E_1P$ is not necessary for $Na^+$ release, or, as suggested by De Weer *et al*. (1983), because in the conditions of their experiments the ATP formed from ADP is released sufficiently slowly from the enzyme for several $Na^+$ ions to be exchanged for each molecule of ATP synthesized. The fact that the axon contents outside the dialysis tube did not constitute a perfectly mixed compartment must also have reduced the sensitivity of their method of detecting ATP–ADP exchange.

The effects of $Mg^{2+}$ ions on $Na^+$–$K^+$ exchange are discussed by Flatman and Lew (1981).

### 4. $K^+$–$K^+$ Exchange

$K^+$–$K^+$ exchange has been studied in human red cells and resealed ghosts prepared from them (Glynn, 1956; Post and Sen, 1965; Glynn *et al.*, 1970, 1971; Sachs, 1972, 1980, 1981a; Simons, 1974, 1975; Eisner and Richards, 1983), and in artificial lipid vesicles with pig kidney $Na^+$,$K^+$-ATPase incorporated into their membranes (Karlish and Stein, 1982a,b; Karlish *et al.*, 1982). The exchange is roughly one-for-one, and the affinities are strikingly asymmetric with a high affinity outside and a low affinity inside (Simons, 1974; Sachs, 1981a). The exchange does not occur unless $Mg^{2+}$ ions, orthophosphate, and nucleotide are present (Glynn *et al.*, 1970, 1971), but ATP can be replaced by its nonphosphorylating $\beta$, $\gamma$-imido or methylene analogs (Simons, 1975) and also by ADP, deoxy-ATP, and CTP (Kaplan, 1982; D. E. Richards, J. L. Howland, and I. M. Glynn, unpublished results quoted by Glynn, 1982). The nucleotides act with a low affinity and without phosphorylating.

The need for orthophosphate suggests that the outward movement of $K^+$ involves a reversal of the normal hydrolytic step, and this and most of the other features of $K^+$–$K^+$ exchange can be accounted for by the left-hand part of the scheme in Figure 1. To explain the complex interactions of ATP and orthophosphate, however, it is necessary to assume that both can bind weakly to the same form of dephosphoenzyme (Sachs, 1981a, Karlish *et al.*, 1982; Eisner and Richards, 1983).

In the absence of both ATP and orthophosphate, Karlish and Stein (1982a) were able to detect very slow fluxes of $Rb^+$ ions in artificial lipid vesicles with $Na^+$,$K^+$-ATPase in their membranes. Because these fluxes could be inhibited by ouabain or vanadate, they were thought to occur through the $Na^+$,$K^+$-ATPase molecules. They can be reconciled with the scheme represented by the left-hand side of Figure 1 if it is assumed that, in the preparation used by Karlish and Stein, $Rb^+$ ions are able to escape to the exterior from $E_2$(Rb) at a very slow rate without phosphorylation by orthophosphate, and that this process is reversible.

### 5. Uncoupled $Na^+$ Efflux

When red cells or resealed ghosts are incubated in choline or $Mg^{2+}$ media lacking both $Na^+$ and $K^+$, there is a small ouabain-sensitive outward movement of $Na^+$ ions (Garrahan and Glynn, 1967a,b; Beaugé and Ortiz, 1973; Lew *et al.*, 1973). The efflux has a high affinity for $Na^+$ at the intracellular surface and is associated with the hydrolysis of ATP, one molecule of ATP being hydrolyzed for every 2–3 $Na^+$ ions ejected (Karlish and Glynn, 1974; Glynn and Karlish, 1976). The affinity for ATP is also very high when the concentration of orthophosphate is low (Glynn and Karlish, 1976; Karlish *et al.*, 1982), but is greatly reduced when the concentration of orthophosphate is high (Karlish and Glynn, 1974). The effect of orthophosphate on uncoupled $Na^+$ efflux is therefore different from the noncompetitive inhibition of $Na^+,K^+$-ATPase activity described by Hexum *et al.* (1970).

In the scheme of Figure 1, uncoupled $Na^+$ efflux is explained by a normal outward movement of $Na^+$ through the right-hand part of the cycle, followed by a slow hydrolysis of $E_2P$ (central dotted pathway). Two features of uncoupled $Na^+$ efflux remain unexplained, however: (1) the outward movement of $Na^+$ is not accompanied by a detectable outward movement of positive charge and seems to be coupled, in a way which is not understood, to a movement of anions (Dissing and Hoffman, 1983), and (2) the outward movement of $Na^+$ ions and the associated hydrolysis of ATP are both inhibited by extracellular $Na^+$ ions acting with a high affinity ($K_{0.5}$ = c. 1 mM; Garrahan and Glynn, 1967a,b; Glynn and Karlish, 1976). This effect is presumably related to the inhibitory effect that $Na^+$ ions in low concentrations have on the rate of hydrolysis of phosphoenzyme (Beaugé and Glynn, 1979b; Lee and Blostein, 1980) and on the rate of ATP–ADP exchange (Beaugé and Glynn, 1979b), but the nature of the high-affinity extracellular sites is not understood. They may be identical with the sites at which the binding of $Na^+$ is thought to alter, allosterically, the affinity for $K^+$ of the $K^+$-loading sites (see Sachs, 1967, 1974, 1977a; Garrahan and Glynn, 1967b; Priestland and Whittam, 1968; Cavieres and Ellory, 1975; Glynn and Karlish, 1976).

## 6. *$Na^+$–$Na^+$ Exchange Accompanied by ATP Hydrolysis*

In the experiments of Glynn and Karlish (1976), resealed red-cell ghosts containing ATP and no ADP were incubated in media lacking both $Na^+$ and $K^+$ and showed an efflux of $Na^+$ associated with a hydrolysis of ATP. As explained above, both the $Na^+$ efflux and the ATP hydrolysis were inhibited by low concentrations of extracellular $Na^+$. When $Na^+$ was added to the extracellular medium in high concentration, however, hydrolysis of ATP continued; and though Glynn and Karlish were unable, for technical reasons, to measure $Na^+$ fluxes in such media, Lee and Blostein (1980), using inside-out vesicles prepared from red-cell membranes, have shown that, under similar conditions, the hydrolysis of ATP is associated with both inward and outward movements of $Na^+$ ions. The $Na^+,K^+$-ATPase can therefore catalyze $Na^+$–$Na^+$ exchange of *two* kinds: (1) the familiar one-for-one exchange occurring in ADP-containing cells, unaccompanied by ATP hydrolysis and probably associated with ATP–ADP exchange, and (2) an exchange, smaller in magnitude, occurring only in cells lacking ADP, and associated with ATP hydrolysis. The explanation of the second kind of exchange is probably that extracellular $Na^+$ ions have a slight $K^+$-like effect in accelerating hydrolysis of $E_2P$ (Post *et al.*, 1972; Beaugé and Glynn, 1979b; Lee and Blostein, 1980), and that the catalyzing $Na^+$ ions are trapped transiently in the dephosphoenzyme and released to the interior as if they were $K^+$ ions.

This interpretation is supported by the experiments of Forgac and Chin (1981), who found that the exchange of $Na^+$ ions across the membranes of artificial lipid vesicles containing incorporated canine $Na^+,K^+$-ATPase was not one-for-one but led to a small net transport away from the surface at which ATP was being hydrolyzed. The observed stoichiometry of 0.5 $Na^+$ ions transported (net) per molecule of ATP hydrolyzed suggests that more than one kind of cycle must have been occurring, and it is possible that, as suggested by Blostein (1983), some of the hydrolysis of ATP under these conditions occurs via the hydrolysis of $E_1P$ rather than $E_2P$.

When $Na^+$,$K^+$-ATPase preparations are incubated with ATP in high-$Na^+$, $K^+$-free media, the hydrolysis of ATP is presumably associated with $Na^+$–$Na^+$ exchange through the pump molecules, though that exchange will not be detectable unless the enzyme is in the membrane of an intact cell or vesicle. With high ATP concentrations, $Na^+$–$K^+$ exchange can occur at a higher rate than $Na^+$–$Na^+$ exchange, and the addition of $K^+$ ions, therefore, accelerates ATP hydrolysis. With very low ATP concentrations, however, the slowness of the conversion of $E_2(K)$ to $E_1K$ makes $Na^+$–$K^+$ exchange slower than $Na^+$–$Na^+$ exchange, and the addition of $K^+$ ions, therefore, slows ATP hydrolysis. This, presumably, is the explanation of the inhibition of ATP hydrolysis by $K^+$ ions when the ATP concentration is very low, first described by Czerwinski *et al.* (1967).

### C. Compatibility of the Pump Model with Steady-State Flux Kinetics

In schemes such as that shown in Figure 1, outwardly transported $Na^+$ ions are released at the outer surface before extracellular $K^+$ ions are bound. The reaction mechanism is therefore a "ping-pong" mechanism in the Cleland (1970) sense. A feature of such mechanisms is that the apparent affinity for each reactant can be much affected by the concentrations of other reactants. Because flux studies failed to show effects of this kind (Baker and Stone, 1966; Hoffman and Tosteson, 1971; Garay and Garrahan, 1973, 1975; Garrahan and Garay, 1976; Chipperfield and Whittam, 1976; Sachs, 1977b), it seemed that the type of mechanism in which $Na^+$ and $K^+$ ions are transported consecutively could be ruled out. In 1979, however, Sachs pointed out that the existence of uncoupled $Na^+$ efflux complicated the analysis, and he showed that the experimental data are, in fact, compatible with a consecutive scheme. More recently, Beaugé and DiPolo (1979b, 1981, 1983), using squid axons, and Eisner and Richards (1981, 1982), using resealed red-cell ghosts, have shown that, under appropriate conditions, it is possible to demonstrate interactions between ATP and extracellular $K^+$ of the kind predicted by consecutive models (see also Garrahan *et al.*, 1982). Sachs (1980) has shown that the effects of oligomycin on the hydrolytic and transporting activities of the red-cell $Na^+$,$K^+$-ATPase strongly suggest that $Na^+$ ions are released to the outside before extracellular $K^+$ ions are bound. Garrahan *et al.* (1982), however, point out that not all of the observed interactions between $Na^+$ and ATP can be explained by the conventional scheme.

## IV. CATALYTIC ACTIVITIES NOT ASSOCIATED WITH ION FLUXES

We have seen that hydrolysis of ATP is associated with $Na^+$–$K^+$ exchange if $Na^+$ and $K^+$ ions are both present, and with uncoupled efflux of $Na^+$, or with the second type of $Na^+$–$Na^+$ exchange, if $Na^+$ ions are present without $K^+$ ions. In this section, we consider three other types of catalytic activity that are not associated (or not necessarily associated) with movements of ions across the membrane. They are (1) ATP–ADP exchange, (2) exchange of $^{18}O$ between orthophosphate and water, and (3) K-dependent phosphatase activity.

## A. ATP–ADP Exchange

$Na^+$,$K^+$-ATPase preparations can catalyze an exchange of isotope between ATP and $^{14}C$- or $^{3}H$-labeled ADP. The exchange occurs only in the presence of $Na^+$ ions (Fahn *et al.*, 1966b), and it is much faster if the enzyme is pretreated with N-ethylmaleimide, oligomycin, or equimolar concentrations of 2,3,dimercaptopropanol and arsenite (Fahn *et al.*, 1966a; Blostein, 1970; Siegel and Albers, 1967). Partly as a result of these observations, and partly because of observations on the extent to which the addition of ADP or $K^+$ leads to the disappearance of phosphoenzyme formed under different conditions (see Section V-A), all three procedures are thought to block the conversion of $E_1P$ (the phosphoenzyme supposed to be capable of reacting with ADP) into $E_2P$ (the $K^+$-sensitive form of phosphoenzyme); see Figure 1. At one time, it was thought that reducing the $Mg^{2+}$ concentration also slowed the conversion of $E_1P$ to $E_2P$, but Klodos and Skou (1975, 1977) have shown that the experiments that seemed to lead to that conclusion were complicated by unrecognized effects of the chelators used to reduce the $Mg^{2+}$ concentration.

The dependence of the rate of ATP–ADP exchange on $Na^+$ concentration is roughly hyperbolic if the enzyme has been pretreated to block the conversion of $E_1P$ to $E_2P$, but in the untreated enzyme it shows three phases, a steep but S-shaped rise between 0 and 2.5 mM, a slight fall between 2.5–10 mM, and a roughly linear rise between 10 and 150 mM (Wildes *et al.*, 1973; Beaugé and Glynn, 1979b). The similarity between this pattern and the pattern of activation of $Na^+$-ATPase activity by $Na^+$ ions, where the sidedness of the effects is known (Glynn and Karlish, 1976), led Beaugé and Glynn to conclude that (1) $Na^+$ ions in low concentrations stimulate at the inner surface by promoting phosphorylation, and inhibit at the outer surface by combining with high-affinity sites, and (2) $Na^+$ ions at higher concentrations reverse the inhibition by combining with extracellular low-affinity sites, converting $E_2P$ back to $E_1P$ (see Section V-A). Direct evidence that the effect of $Na^+$ ions in high concentrations depends on the filling of extracellular sites has now been provided by ingenious experiments in which ATP–ADP exchange was measured in resealed red-cell ghosts (Cavieres, 1980, 1983; Kaplan and Hollis, 1980; Kaplan, 1982, 1983).

The effects on ATP–ADP exchange of changing the concentrations of ATP, ADP, and $Mg^{2+}$ are complicated, but suggest that for the reaction leading to ATP formation *either* free ADP (rather than MgADP) must combine with $E_1P$, *or* $Mg^{2+}$ ions are inhibitory, or both (Robinson, 1976b; Beaugé and Glynn, 1979b). The hypothesis that free ADP rather than MgADP reacts with $E_1P$ fits well with the finding (Fukushima and Post, 1978) that $Mg^{2+}$ ions become firmly bound to the enzyme when it is phosphorylated.

## B. Exchange of $^{18}O$ between Orthophosphate and Water

In 1972, Skvortsevich *et al.* reported that $Na^+$,$K^+$-ATPase preparations can catalyze the exchange of $^{18}O$ between water and orthophosphate. Dahms and Boyer (1973) investigated the exchange using $Na^+$,$K^+$-ATPase from pig kidney outer medulla and from *Electrophorus* electric organ. With microsomal preparations under optimal conditions at 37°C, the exchange was very rapid, about 20 times the maximal rate of ouabain-sensitive ATP hydrolysis by the same preparations. Treatment of the preparations with deoxycholate to increase their hydrolytic activity caused a dramatic fall in the exchange activity. The exchange required the simultaneous presence of $Mg^{2+}$ and $K^+$ ions, and was inhibited by $Na^+$, ouabain, and *p*-chlormercuribenzoate, but not by oligomycin. Later work showed that $K^+$ can be replaced by $NH_4^+$, $Rb^+$, $Tl^+$, $Cs^+$, and, less effectively, by $Li^+$, and that $Mg^{2+}$ can be replaced by $Mn^{2+}$, $Co^{2+}$, $Sr^{2+}$, $Fe^{2+}$, and $Ba^{2+}$, though none of these divalent ions are nearly as effective as $Mg^{2+}$. $Ca^{2+}$, $Cd^{2+}$, $Zn^{2+}$, and $Cu^{2+}$ ions are ineffective (Shaffer *et al.*, 1978; Perez *et al.*, 1979). Nucleotide is not required, but ATP is able to reverse the inhibition by $Na^+$, and the exchange in the presence of ATP + $Na^+$ is peculiar in not being reduced by treatment with deoxycholate. Nonphosphorylating analogs of ATP are not capable of reversing the inhibition by $Na^+$.

Hydrolysis of ATP must, of course, lead to the incorporation of one atom of water oxygen into each molecule of orthophosphate formed, but the incorporation of $^{18}O$ was found to exceed that figure. The extra $^{18}O$ was not incorporated preferentially into the orthophosphate formed from cleavage of ATP, however, since the transfer of $^{18}O$ from [$^{18}O$]water into orthophosphate was shown to be equal to the transfer of $^{18}O$ from [$^{18}O$]orthophosphate into water.

Dahms and Boyer explained their observations by supposing that the exchange arises by the alternate formation and hydrolysis of phosphoenzyme by the reaction sequence (ignoring $Mg^{2+}$ ions):

$$E_2(K) + P_i \underset{k_{-1}}{\overset{k_1}{\rightleftharpoons}} E_2(K) \cdot P_i \underset{k_{-2}}{\overset{k_2}{\rightleftharpoons}} E_2PK + H_2O \qquad (1)$$

$Na^+$ ions prevent these reactions (see Section V-A) by converting $E_2(K)$ to $E_1Na$. In the presence of ATP, however, $E_1Na$ is phosphorylated and eventually generates $E_2(K)$ (see Figure 1). The effects of deoxycholate on the exchange of $^{18}O$ in the absence of $Na^+$ have not been explained.

If the reaction sequence summarized in Eq. (1) is responsible for $^{18}O$ exchange between water and orthophosphate, the pattern of incorporation of $^{18}O$ into orthophosphate will differ depending on the relative magnitudes of $k_2$ and $k_{-1}$. If $k_2$ is very much greater than $k_{-1}$, nearly every molecule of orthophosphate bound will exchange all of its oxygen atoms with water before it is released. If $k_2$ is very much less than $k_{-1}$, only one oxygen atom will be exchanged for each encounter of orthophosphate with the enzyme that leads to phosphorylation. Recently, Dahms and Miara (1983) have used the effect of $^{18}O$ on the $^{31}P$ NMR signal of orthophosphate to show that at physiological pH, at temperatures between 14 and 40°C, $k_2$ must be very much less than $k_{-1}$, that is to say, the bound orthophosphate is much more likely to dissociate from the enzyme than to phosphorylate it. Substitution of $Tl^+$, $Rb^+$, or $NH_4^+$ for $K^+$, or treatment with SDS, had no effect on the ratio of the rate constants, but when the pH was lowered to below 6.6 the ratio changed dramatically, as if dissociation became much slower than phosphorylation.

## C. Phosphatase Activity

$Na^+$,$K^+$-ATPase preparations are capable of catalyzing the release of orthophosphate from a number of non nucleotide substrates, e.g., *p*-nitrophenyl phosphate, acetyl phosphate, carbamyl phosphate, and umbelliferone phosphate, all of which contain phosphate groups with a moderate or high free energy of hydrolysis. The literature on this "phosphatase" activity of $Na^+$,$K^+$-ATPase is now very large and only a brief account will be given here (for further references see Glynn and Karlish, 1975; Robinson and Flashner, 1979; Schuurmans Stekhoven and Bonting, 1981; Cantley, 1981; Swann and Albers, 1975, 1978, 1979, 1980; Swann, 1983; Robinson *et al.*, 1978; Robinson, 1980, 1981, 1982).

The most striking feature of the phosphatase activity is that it requires the presence of $Mg^{2+}$ and $K^+$ ions but that, unlike the hydrolysis of ATP or other nucleoside triphosphates, it does not require, and may be inhibited by, $Na^+$ ions. Experiments on perfused squid axons by Brinley and Mullins (1968) and Mullins and Brinley (1969), and on red cells by Garrahan and Rega (1972), show that no significant fluxes of $Na^+$ or $K^+$ are associated with the hydrolysis of acetyl phosphate or of *p*-nitrophenyl phosphate.

An important finding by Nagai and Yoshida (1966) is that the simultaneous presence of $Na^+$ ions and ATP greatly increases phosphatase activity at low $K^+$ concentrations. This effect of $Na^+$ + ATP is associated with an increase in the apparent affinity for $K^+$ ions, and it almost certainly depends on phosphorylation (see Koyal *et al.*, 1971; Robinson, 1973; Swann and Albers, 1975; Glynn and Karlish, 1975). Recently, Blostein *et al.* (1979) and Drapeau and Blostein (1980) have shown that $Na^+$ and ATP acting together also produce a change in the location of the sites at which $K^+$ ions stimulate phosphatase activity. In the absence of $Na^+$ + ATP, $K^+$ ions act intracellularly and with a low affinity. In the presence of $Na^+$ + ATP, they act both extracellularly and intracellularly, and with a high affinity. The apparent contradiction with the earlier findings of Rega *et al.* (1970) is probably explained by the presence of small amounts of both $Na^+$ and ATP in the resealed red-cell ghosts used in the earlier experiments. These effects of $Na^+$ + ATP support the hypothesis of Post *et al.* (1972) that it is $E_2(K)$, the form of dephosphoenzyme that contains occluded $K^+$ ions (see Figure 1), that is responsible for the phosphatase activity. For, in the absence of $Na^+$ and ATP, $K^+$ ions promote $E_2(K)$ formation by combining at low-affinity intracellular sites (see Figure 1). In the presence of $Na^+$ + ATP, $K^+$ ions promote $E_2(K)$ formation by binding at high-affinity extracellular sites and accelerating the hydrolysis of $E_2P$. ATP, in the absence of $Na^+$ ions, would be expected to decrease the concentration of $E_2(K)$ (see Figure 1), and this fits well with the observation that, when ATP is present *without* $Na^+$, higher concentrations of $K^+$ are needed for activation of *p*-nitrophenyl phosphate hydrolysis (Nagai *et al.*, 1966; Garrahan *et al.*, 1970; Nagano *et al.*, 1973; Skou, 1974b; Gache *et al.*, 1976).

This interpretation of the effects of $Na^+$ + ATP on *p*-nitrophenylphosphatase activity is supported by the observation that oligomycin and N-ethylmaleimide, both of which are thought to block the conversion of $E_1P$ to $E_2P$, have no effect on *p*-nitrophenylphosphatase activity in the absence of $Na^+$, but prevent the stimulation of *p*-nitrophenylphosphatase activity by $Na^+$ + ATP or $Na^+$ + acetyl phosphate (Israel and Titus, 1967; Robinson, 1970; Garrahan *et al.*, 1970; Askari and Koyal, 1971).

There are, however, several unsolved problems. In the first place, the experiments of Drapeau and Blostein (1980) showing that in the presence of $Na^+$ + ATP the hydrolysis of *p*-nitrophenyl phosphate is stimulated by extracellular $K^+$ in low concentrations, also show that there is virtually no hydrolysis unless $K^+$ ions are also present at the intracellular surface. This is particularly puzzling since the intracellular $K^+$ ions seem to act with a high affinity (maximal effect with 0.2 mM $K^+$ in the presence of 10 mM $Na^+$).

Secondly, although it is true that full activation of the *p*-nitrophenylphosphatase activity by $K^+$ ions in low concentrations occurs only in the presence of $Na^+$ + ATP or $Na^+$ + acetyl phosphate, Skou (1974b) has shown that the addition of $Na^+$ alone is sufficient to cause a small high-affinity response to $K^+$. In the presence of $Na^+$, the $K^+$ activation curve becomes stepped, as if part of the enzyme were activated by $K^+$ ions with a high affinity and the rest with a low affinity. Activation of *p*-nitrophenylphosphatase activity at low $K^+$ concentrations by $Na^+$ in the absence of ATP has also been reported by Swann and Albers (1980).

Thirdly, even if $E_2(K)$ is the form of the $Na^+$,$K^+$-ATPase that is responsible for phosphatase activity, it is not clear how it acts, or even whether the mechanism is the same for all substrates. There is good evidence that acetyl phosphate can phosphorylate the enzyme in the presence of $Mg^{2+}$ + $K^+$ (without $Na^+$), though there is disagreement about the $K^+$ sensitivity of the phosphoenzyme formed, possibly resulting from differences in the conditions of phosphorylation (Sachs *et al.*, 1967; Bond *et al.*, 1971; Dudding and Winter, 1971; Swann and Albers, 1980). With *p*-nitrophenyl phosphate, however, no phosphorylation of the enzyme has been detected (Bond *et al.*, 1971), and it has been suggested that hydrolysis of *p*-nitrophenyl phosphate does not occur via a phosphorylated intermediate. This view is supported by the observations of Shaffer *et al.* (1978) who compared the exchange of $^{18}O$ between water and orthophosphate during the hydrolysis of *p*-nitrophenyl phosphate, acetyl phosphate, and ATP. With *p*-nitrophenyl phosphate, only one atom of oxygen was incorporated into orthophosphate for each molecule of *p*-nitrophenyl phosphate hydrolyzed (in other words there was no extra exchange), whereas with acetyl phosphate or with ATP, the extra exchanges were, respectively, 1.4–2.3 and 1.8–2.1 atoms of oxygen per molecule of substrate hydrolyzed. A difference between the mechanisms of hydrolysis of acetyl phosphate and *p*-nitrophenyl phosphate may account for the difference in the concentrations of $K^+$ ions necessary for activation, the $K_{0.5}$ for $K^+$ ions being 0.6–0.9 mM when acetyl phosphate is the substrate, and 2–5 mM when *p*-nitrophenyl phosphate is the substrate (Bader and Sen, 1966; Robinson, 1969, 1970, 1975, 1976a).

## V. PHOSPHORYLATION STUDIES

### A. Phosphorylation by ATP

It is now well established, mainly by the work of Post, Albers, and their colleagues, that $Na^+$,$K^+$-ATPase is phosphorylated on a β-aspartyl carboxyl group by ATP bound to a high-affinity site (Albers *et al.*, 1963; Post *et al.*, 1965; Siegel and Albers, 1967; Fahn *et al.*, 1968; Post and Kume, 1973; Degani *et al.*, 1974). Phosphorylation requires both $Mg^{2+}$ and $Na^+$ ions, and these must be present at the intracellular surface of the enzyme (Blostein and Chu, 1977; Blostein, 1979; Blostein *et al.*, 1979). The relation between the $Na^+$ concentration and the amount of phosphoenzyme formed in the steady state is complex (Siegel and Albers, 1967; Foster and Ahmed, 1976), though Mårdh and Post (1977) could find no evidence of cooperativity from measurements of initial phosphorylation rates. It is uncertain whether $Mg^{2+}$ binds with ATP, or at a separate site, or both (for discussion and references see Plesner and Plesner, 1981a; Cantley, 1981). Grisham (1979), however, has shown that $Mn^{2+}$ and a complex of $Cr^{2+}$ and ATP can bind simultaneously at nearby sites (see Section VIII-C.1). A number of other divalent cations, $Mn^{2+}$, $Ca^{2+}$, $Fe^{2+}$, $Ni^{2+}$, $Co^{2+}$, $Cu^{2+}$, and $Pb^{2+}$, can substitute for $Mg^{2+}$ as cofactors for phosphorylation, though they are less effective or ineffective in supporting ATP hydrolysis (Siegel *et al.*, 1973; Dahl and Hokin, 1974; Rossi *et al.*, 1978; Fukushima and Post, 1978; Siegel, 1979; Askari and Huang, 1981).

Although the free energy of hydrolysis of β-aspartyl phosphate is expected to be high, most of the phosphoenzyme formed by phosphorylating $Na^+$,$K^+$-ATPase by ATP in media containing $Na^+$ in low or moderate concentrations can transfer its phospho group to ADP only if the enzyme has been pretreated with either (1) sulfhydryl reagents such as N-ethyl maleimide (NEM) or a mixture of 2,3-dimercaptopropanol and arsenite (Fahn *et al.*, 1966a,b; Esmann and Klodos, 1983), or (2) the noncovalent inhibitors oligomycin or quercetin (Siegel and Albers, 1967; Blostein, 1970; Kuriki and Racker, 1976). Enzyme modified by treatment with α-chymotrypsin, under carefully controlled conditions, also yields a phosphoenzyme that can transfer its phospho group to ADP (Jørgensen *et al.*, 1982b). All of these treatments are thought to block the spontaneous conversion of the newly formed phosphoenzyme ($E_1P$) to a second form ($E_2P$) which is unable to react with ADP, but which can be hydrolyzed slowly spontaneously, and rapidly in the presence of $K^+$ ions (Post *et al.*, 1969, 1972). Various congeners of $K^+$, e.g., $Tl^+$, $Rb^+$, $Cs^+$, $NH_4^+$, and $Li^+$, also accelerate the hydrolysis of $E_2P$. Because $E_1P$ and $E_2P$ yield the same hydrolysis products after proteolytic digestion, they are thought to differ only in conformation (Post *et al.*, 1969).

As mentioned above, in connection with ATP–ADP exchange, the conversion of $E_1P$ to $E_2P$ was at one time thought to require the presence of $Mg^{2+}$ ions at high concentrations, but Klodos and Skou (1975, 1977) have shown that the experiments on which this belief is based were misinterpreted. In fact, the $Mg^{2+}$ ions which are required for the formation of $E_1P$ become very tightly bound, and are released only later in the cycle when $E_2P$ is hydrolyzed (Fukushima and Post, 1978).

If $Ca^{2+}$ is substituted for $Mg^{2+}$, phosphorylation by ATP leads to the formation of a phosphoenzyme which, in brain, is almost wholly ADP sensitive and $K^+$ insensitive (Tobin *et al.*, 1973, 1974), but which, in kidney, is partly ADP sensitive and partly $K^+$ sensitive (Fukushima and Post, 1978; Fukushima and Nakao, 1980). Unlike the phosphoenzyme formed in the presence of $Mg^{2+}$, that formed in the presence of $Ca^{2+}$ is able to lose its divalent ion readily to chelators and become unreactive to both ADP and $K^+$. Addition of $Ca^{2+}$ restores the original characteristics of the phosphoenzyme formed in the presence of $Ca^{2+}$; addition of $Mg^{2+}$ yields a phosphoenzyme that is $K^+$ sensitive and ADP insensitive.

An important finding is that the equilibrium between $E_1P$ and $E_2P$ can be displaced in favor of $E_1P$ by $Na^+$ ions in very high concentrations (Tobin *et al.*, 1973; Taniguchi and Post, 1975; Kuriki and Racker, 1976; Jørgensen and Karlish, 1980; Hara and Nakao, 1981). The very low affinity for $Na^+$ suggests that the sites at which the $Na^+$ ions act are the external sites from which $Na^+$ ions are released during the normal running of the pump, and to which extracellular ions bind during $Na^+$–$Na^+$ exchange and pump reversal. In $Na^+$,$K^+$-ATPase prepared from brain, the equilibrium between $E_1P$ and $E_2P$ is poised less far in favor of $E_2P$, and appreciable ADP-sensitive phosphoenzyme can be detected at low or moderate $Na^+$ concentrations without the use of measures to block the conversion of $E_1P$ to $E_2P$ (Tobin *et al.*, 1973; Mårdh, 1975a; Klodos and Nørby, 1979).

Besides the two forms of phosphoenzyme $E_1P$ and $E_2P$, sensitive, respectively, to ADP and to $K^+$ ions, two forms exist that are sensitive to neither. One is formed from $E_2P$ slowly at 0°C if the medium contains a high concentration of $Mg^{2+}$ ions, particularly if the $Na^+$ concentration is low (Post *et al.*, 1975a). Its physiological significance is unknown. The other is formed when a divalent cation chelator is added to enzyme phosphorylated by ATP + $Ca^{2+}$, as described above. It is unlikely to have any physiological significance since, normally, the divalent ion acting as a cofactor will be $Mg^{2+}$ rather than $Ca^{2+}$. Evidence for the existence of a fifth type of phosphoenzyme, containing trapped $Rb^+$, will be discussed later.

## B. Phosphorylation by Orthophosphate

Since the overall $Na^+$,$K^+$-ATPase reaction is reversible, phosphorylated intermediates should also be formed from orthophosphate under appropriate conditions.

Phosphorylation by orthophosphate was first detected in experiments in which membrane fragments were incubated with ouabain, orthophosphate, and $Mg^{2+}$, or with ouabain, orthophosphate, $Mg^{2+}$, and $K^+$ (Albers *et al.*, 1968; Lindenmayer *et al.*, 1968). $Na^+$ is inhibitory. Experiments with [$^{18}O$]orthophosphate show that no $^{18}O$ is incorporated into the C–O–P bridge oxygen atom in the phosphoenzyme; it follows that phosphorylation must occur by carboxylate attack on the phosphorus atom and displacement of a hydroxyl group (Dahms *et al.*, 1973).

Phosphorylation by orthophosphate has also been demonstrated in the absence of ouabain (Post *et al.*, 1973, 1975a; Taniguchi and Post, 1975; Kuriki *et al.*, 1976). Incubation of $Na^+$,$K^+$-ATPase with $Mg^{2+}$ and orthophosphate in the absence of $K^+$, leads to the formation of phosphoenzyme, part of which is insensitive both to $K^+$ and

ADP. When $K^+$ ions are also present during the phosphorylation, the steady-state yield of phosphoenzyme is smaller because the phosphoenzyme formed in the presence of $K^+$ is $K^+$ sensitive and rapidly hydrolyzed. The presence of a low concentration of $Na^+$ during the phosphorylation also leads to the formation of $K^+$-sensitive phosphoenzyme. The addition of $Na^+$ in high concentration, together with ADP, to $K^+$-sensitive phosphoenzyme leads to ATP synthesis, presumably by the route $E_2P \rightarrow E_1P \rightarrow$ ATP. Insensitive phosphoenzyme can also serve as the starting material for ATP synthesis.

Kuriki *et al.* (1976) have shown that the binding of $Mg^{2+}$ or of orthophosphate to dephosphoenzyme is associated with very large enthalpy changes.

The phosphoenzyme formed from orthophosphate is chemically identical with that formed from ATP (Post *et al.*, 1975a; Sen *et al.*, 1969; Siegel *et al.*, 1969; Bonting *et al.*, 1979). Furthermore, the maximum number of phospho groups that can be incorporated is the same whether they come from ATP or from orthophosphate, or partly from one and partly from the other (Schuurmans Stekhoven *et al.*, 1980).

## C. *Rephosphorylation of Newly Dephosphorylated Enzyme: Evidence for the Occlusion of $K^+$ Ions*

In 1972, Post *et al.* reported experiments in which they measured the rate of phosphorylation by ATP of enzyme that had just been dephosphorylated. They found that the rate of rephosphorylation differed depending on whether $Rb^+$ or $Li^+$ ions had been present during the hydrolysis, and that this was true even when the experiments were done in such a way that the conditions during rephosphorylation were identical. It follows that the hydrolysis products must have differed depending on whether $Rb^+$ or $Li^+$ had been present during the hydrolysis, and Post *et al.* suggest that this is because the catalyzing ions become occluded within the enzyme at the moment of hydrolysis and are released only later after a slow conformational change. Furthermore, since the enzyme became available for rephosphorylation sooner if higher ATP concentrations were used, they also suggest that ATP accelerates the change in conformation that precedes the release of the ions.

We shall see later (Section VII-A) that both of these hypotheses have been proved by later work. It is worth pointing out here, however, that newly formed dephosphoenzyme containing occluded $Rb^+$ can be phosphorylated by orthophosphate *even in the presence of a very high concentration of $Na^+$*, and that the phosphoenzyme produced is hydrolyzed much faster than is expected from the relative amounts of $Rb^+$ and $Na^+$ in the medium (see Figure 7 in Post *et al.*, 1975a). The rapid hydrolysis suggests that $Rb^+$ ions are still bound to the enzyme after its phosphorylation by orthophosphate. This is the evidence for the existence of the potassium-complexed phosphoenzyme postulated by Post *et al.* (1975a).

## D. *Mechanism of Hydrolysis of Phosphoenzyme*

When $Na^+$,$K^+$-ATPase from *Electrophorus* electric organ was allowed to hydrolyze ATP at 15°C in a medium containing [$^{18}O$]-$H_2O$, no $^{18}O$ was incorporated into the C–O–P bridge oxygen atom of the phosphoenzyme (Dahms *et al.*, 1973). This result, which was true of phosphoenzyme formed in the presence of $Na^+ + K^+$, or of $Na^+$ alone, shows that the phosphoenzyme is cleaved by water oxygen attack on the phosphorus atom. It rules out acyl transfer to some other group on the enzyme with subsequent hydrolysis by attack of water oxygen on the acyl carbon atom.

A surprising feature of these experiments is that some $^{18}O$ was incorporated into the phosphoryl oxygen atoms of the phosphoenzyme, and this cannot be accounted for by formation of phosphoenzyme from medium orthophosphate (which could have become labeled with $^{18}O$ either by water–orthophosphate exchange or by being formed by hydrolysis of ATP). Dahms *et al.* suggest that the release of orthophosphate from the enzyme may be slow, so that any phosphorylation of the enzyme by orthophosphate is likely to be by orthophosphate which has just been formed by hydrolysis of the phosphoenzyme. This newly formed orthophosphate will, necessarily, have acquired an oxygen atom from the $[^{18}O]$-$H_2O$.

## E. Kinetic Studies

The hypothesis that hydrolysis of ATP normally proceeds through $E_1P$ and $E_2P$ has received support from studies of both steady-state and transient kinetics (Kanazawa *et al.*, 1967, 1970; Neufeld and Levy, 1970; Post *et al.*, 1972; Fukushima and Tonomura, 1973, 1975; Tonomura and Fukushima, 1974; Mårdh and Zetterqvist, 1972, 1974; Mårdh, 1975a,b; Mårdh and Post, 1977; Mårdh and Lindahl, 1977; Blostein, 1968, 1975, 1979; Blostein and Whittington, 1973; Froehlich *et al.*, 1976; Lowe and Smart, 1977; Yamaguchi and Tonomura, 1977; Karlish *et al.*, 1978a; Hobbs *et al.*, 1980; Fukushima and Nakao, 1981; Hara and Nakao, 1981). There are also difficulties (Fukushima and Tonomura, 1973, but cf. Fukushima and Nakao, 1981; Kanazawa *et al.*, 1970; Tonomura and Fukushima, 1974; Yamaguchi and Tonomura, 1977; Klodos and Norby, 1979; Klodos *et al.*, 1981; Plesner and Plesner, 1981a,b; Plesner *et al.*, 1981; Hobbs *et al.*, 1983).

The transient kinetic studies most strongly supporting the orthodox hypothesis are those of Mårdh and his collaborators, who used sophisticated rapid-mixing devices that made it possible to obtain satisfactory results at 21°C. In experiments on $Na^+$,$K^+$-ATPase from bovine brain, Mårdh (1975a,b) showed that both phosphorylation by ATP and dephosphorylation in the presence of $K^+$ were more than sufficiently rapid to account for the overall rate of hydrolysis. When 10 mM KCl and excess unlabeled ATP were added simultaneously to enzyme previously phosphorylated by 100 μM$[\gamma^{32}P]$-ATP in the presence of $Na^+ + Mg^{2+}$, there were two phases of dephosphorylation: a fast phase with a rate constant of at least 230 $s^{-1}$, and a much slower phase. If the fast phase represented the breakdown of $E_2P$ and the slow phase the breakdown of $E_1P$, then there must have been about three times as much $E_2P$ as $E_1P$. When KCl was present before the addition of $[\gamma^{32}P]$-ATP, only the slow phase was seen, suggesting that under steady-state conditions, with 10 mM $K^+$ and 120 mM $Na^+$ present, nearly all of the phosphoenzyme was $E_1P$. The rate constant for the slow phase was 77 $sec^{-1}$, and the product of that rate constant and the amount of phosphoenzyme present agreed reasonably well with the steady-state rate of $(Na^+ + K^+)$-stimulated hydrolysis.

However, not all investigators have found good agreement between the steady-state rate of hydrolysis in the presence of $Na^+ + K^+$, and the product of (1) a measured rate constant for phosphoenzyme hydrolysis, and (2) the amount of phosphoenzyme present in the steady state. Kanazawa *et al.* (1970), working with ox brain enzyme at 0°C, found that with 0.6 mM $K^+$, the highest concentration at which the rate of dephosphorylation could be measured, the overall hydrolysis rate was about twice as high as that expected from the product of rate constant and concentration. A discrepancy of a similar kind was reported by Klodos and Nørby (1979), particularly if $Li^+$ were substituted for $K^+$.

A discrepancy of much greater magnitude was reported more recently by Klodos *et al.* (1981) and Plesner *et al.* (1981), who also worked with bovine brain preparations near 0°C. If overall hydrolysis occurs via the sequence

$$\underset{k_{-1}}{\rightleftharpoons} E_1P \underset{k_{-2}}{\overset{k_{+2}}{\rightleftharpoons}} E_2P \overset{k_3}{\rightarrow} \tag{2}$$

where $k_{-1}$, $k_{+2}$ etc. represent rate constants, then in the steady state:

$$k_{+2}[E_1P] = (k_{-2} + k_3)[E_2P].$$

When a large excess of labeled ATP, with or without ADP, was added to enzyme that had been phosphorylated with [$\gamma^{32}P$]-ATP in the presence of $Na^+$ and $Mg^{2+}$, the slow phase of dephosphorylation was unaffected by the presence of ADP; it follows that $k_{-2}$ must be much smaller than $k_3$ in the absence of $K^+$. From the observed loss of phosphoenzyme during the slow phase, and the relative amounts of $E_1P$ and $E_2P$ during the initial steady state, Klodos *et al.* and Plesner *et al.* were therefore able to calculate $k_{+2}$; it worked out at about 0.21 $s^{-1}$ (1°C), i.e., not much greater than the measured value of $k_3$, which was 0.13 $s^{-1}$ (1°C). Now, in the steady state in the absence of $K^+$, the greater part of the enzyme is in the $E_2P$ form, and the estimated rate of overall hydrolysis cannot therefore be less than 0.13 $E_T/2$, where $E_T$ is the total concentration of enzyme. When $K^+$ is present, the rate of overall hydrolysis cannot be greater than $k_{+2}[E_1P]$ and, hence, cannot be greater than 0.21 $E_T$. The ratio of hydrolysis rates at saturating levels of ATP in the presence and absence of $K^+$ ought not, therefore, to exceed $0.21\ E_T/(0.13\ E_T/2) = 3.2$. Yet, the observed ratio of the maximal hydrolysis rates in the presence and absence of $K^+$ was about 25 (Plesner and Plesner, 1981a,b; Plesner *et al.*, 1981).

Using the same procedure, Klodos *et al.* (1981) and Plesner *et al.* (1981) also calculated $k_{+2}$ from the published results of Mårdh (1975a) at 21°C. They obtained a figure of 7–8 $s^{-1}$, which is an order of magnitude less than the figure of 77 $s^{-1}$ obtained by Mårdh from direct measurements of the disappearance of $K^+$-insensitive phosphoenzyme (see above).

In view of these discrepancies, Klodos *et al.* abandon the hypothesis that the hydrolysis in the simultaneous presence of $Na^+$ and $K^+$ occurs via $E_1P$ and $E_2P$, and propose instead a cycle that does not involve acid-stable intermediates. A similar suggestion was made many years ago by Skou (1965, 1971), but that was to explain difficulties that have since been resolved by the discovery of the slow step that releases $K^+$ from the occluded $K^+$ form (see Section VII-A). A cycle that does not involve acid-stable intermediates does, of course, circumvent the difficulties discussed by Klodos *et al.*, but it raises other serious problems. It is generally agreed that the overall hydrolysis of ATP in the absence of $K^+$ does occur via phosphorylated intermediates, so if a different set of reactions accounts for the overall hydrolysis when $K^+$ ions are present, one would have two different pathways capable of moving $Na^+$ ions outwards. We shall see later that there is now rather strong evidence linking $K^+$ movements (in the absence of $Na^+$) with a sequence of reactions that involves the hydrolysis of $E_2P$; so if Klodos *et al.* are right, one would also appear to have two different pathways capable of moving $K^+$ ions inwards. It therefore seems more economical to retain the hypothesis that the acid-stable phosphorylated intermediates are involved in $Na^+$,$K^+$-ATPase activity and the associated $Na^+$–$K^+$ exchange, and to seek other ways of reconciling that hypothesis with the experimental data.

Estimates of rate constants based on observations of the rate of loss of labeled phosphoenzyme following the addition of excess unlabeled ATP will be too small if there is a pool of unphosphorylated enzyme with tightly-bound labeled ATP that can continue to phosphorylate. There is some evidence for the existence of tightly-bound ATP (Tonomura and Fukushima, 1974; Proverbio and Hoffman, 1977; Fukushima and Nakao, 1981; Mercer and Dunham, 1981; Froehlich *et al.*, 1983; Lowe and Reeve, 1983), but it is unlikely that the discrepancies described by the Aarhus workers can be entirely explained in this way. Another possible explanation of the discrepancies is that $K^+$ ions accelerate the rate of conversion of $E_1P$ to $E_2P$. This hypothesis might seem to be excluded by experiments of Tonomura and Fukushima (1974), Kuriki and Racker (1976) and Hara and Nakao (1981) showing that when $K^+$ ions were added to preformed phosphoenzyme the slow phase of the disappearance of the phosphoenzyme was not significantly faster than the disappearance in the absence of $K^+$. In the experiments of Mardh (1975), however, the rate constant of the slow phase was about an order of magnitude greater than the rate constant of the disappearance in the absence of $K^+$.

The work from Aarhus also has interesting implications for the part of the cycle involving $Na^+$ ions. When $Na^+$,$K^+$-ATPase was phosphorylated with [$\gamma$ $^{32}$P]-ATP in the presence of $Na^+$ and $Mg^{2+}$, and the formation of labeled phosphoenzyme was suddenly stopped by the addition of excess cold ATP, the disappearance of label from the enzyme followed a single exponential from the earliest time that could be measured. From this, and the fact that $k_{-2}$ is much smaller than $k_3$ (see above), Klodos *et al.* (1981) argue that (as well as being hydrolyzed via $E_2P$) $E_1P$ must be hydrolyzed directly, with a rate constant similar to $k_3$ (in the absence of $K^+$). The evidence is suggestive, though not quite compelling since, if the ratio of $k_3 : k_{+2}$ is, e.g., 1 : 3, the deviation from linearity on a semilogarithmic plot is detectable only with difficulty and at the earliest times. Direct hydrolysis of $E_1P$ has been suggested by Blostein (1979) to account for ATP hydrolysis unaccompanied by $Na^+$ efflux in experiments on inside-out vesicles prepared from red-cell membranes. That hydrolysis, however, occurred only when the concentration of $Na^+$ ions in contact with the cytoplasmic surface was very low. (See also Garrahan *et al.*, 1979).

Hara and Nakao (1981), using a pig kidney preparation at 0°C, investigated the transient kinetics of $Na^+,K^+$-ATPase at different $Na^+$ concentrations in the absence of $K^+$. They found that during steady-state turnover at $Na^+$ concentrations ranging from 0.12–1.2 M, the sum of $E_1P$ (defined as the fraction of phosphoenzyme that disappeared rapidly in the presence of ADP) and $E_2P$ (defined as the fraction that disappeared rapidly in the presence of $K^+$) was within a few percent of the total amount of phosphoenzyme. Increasing the $Na^+$ concentration displaced the equilibrium $E_1P \rightleftharpoons E_2P$ toward $E_1P$, the equilibrium constant varying with the third power of the $Na^+$ concentration. At concentrations up to 500 mM, the main effect of increasing $Na^+$ was to accelerate the conversion of $E_2P$ to $E_1P$. At higher concentrations, there was also significant slowing of the conversion of $E_1P$ to $E_2P$. It is uncertain whether $Na^+$ in concentrations less than 500 mM also slowed the conversion of $E_1P$ to $E_2P$, because of the technical difficulty of measuring the conversion rate when the concentration of $E_1P$ was low.

Hara and Nakao point out that an effect of $Na^+$ on the rate of conversion of $E_1P$ to $E_2P$ is to be expected if the conversion occurs not as a single step, but in such a way that one or two of the $Na^+$ ions are released at an intermediate stage. That would imply the existence of forms of phosphoenzyme intermediate between "$E_1P$" and "$E_2P$," though such forms need not exist in easily detectable amounts. If they do exist, and if one or more of them is sensitive both to ADP and to $K^+$ (or if forms separately sensitive to ADP and to $K^+$ are rapidly interconvertible), their presence in appreciable concentrations in some preparations might account for reports in the literature of experiments in which the sum of ADP-sensitive and $K^+$-sensitive phosphoenzyme exceeded the total phosphoenzyme. Yoda and Yoda (1982a), for example, in experiments with $Na^+,K^+$-ATPase from *Electrophorus* electric organ, found that with 10 mM $Na^+$, 90% of the phosphoenzyme behaved as $E_2P$, and with 1 M $Na^+$, virtually all of it behaved as $E_1P$, but with 50 mM $Na^+$, 70–80% was rapidly dephosphorylated in the presence of $K^+$, and over 90% was rapidly dephosphorylated in the presence of ADP. In similar experiments with shark enzyme, however, they found that the sum of $K^+$-sensitive phosphoenzyme and ADP-sensitive phosphoenzyme was always close to 100%. J. G. Nørby, I. Klodos, and N. O. Christiansen (personal communication) point out that the sum of $K^+$-sensitive phosphoenzyme and ADP-sensitive phosphoenzyme also exceeded 100% in the experiments of Kuriki and Racker (1976) on *Electrophorus* $Na^+,K^+$-ATPase.

## VI. EVIDENCE FOR THE EXISTENCE OF DIFFERENT FORMS OF DEPHOSPHOENZYME

The belief that the unphosphorylated enzyme can exist in two main stable conformations, $E_1$ in predominantly $Na^+$ media and $E_2$ in predominantly $K^+$ media, rests on seven independent lines of evidence, namely: (1) differences in reactivity to ATP and orthophosphate, (2) differences in affinity for nucleotides, (3) differences in reactivity to inhibitors, (4) differences in the pattern of attack by proteolytic enzymes and in the products of proteolytic digestion, (5) differences in intrinsic fluorescence and in the fluorescence of probe molecules, (6) differences in the equilibrium binding of $Na^+$ and $K^+$ ions, and (7) differences in the ability of the different forms to occlude $Rb^+$ (and presumably therefore $K^+$) ions. Although all seven lines of evidence support the notion that there are two main interconvertible conformations of dephosphoenzyme, stable respectively in $Na^+$ and $K^+$ media, not all of the experimental findings can be explained by just two conformations; it is, anyway, intrinsically unlikely that only two conformations can exist.

It will be convenient to consider the first six lines of evidence, in turn, in this section, and evidence from the occlusion of $Rb^+$ ions in Section VII. Several of the lines of evidence also give information about the rates of interconversion of the different forms.

## A. Differences in Reactivity to ATP and Orthophosphate

We have already seen that the reactivity of the enzyme to ATP and orthophosphate is greatly affected by the concentrations of $Na^+$ and $K^+$ ions, and Post *et al.* (1973) suggested that this is because the reactivity depends on whether the enzyme is stabilized in the $E_1$ form or the $E_2$ form. They were able to show that when phosphorylation was initiated by adding ATP + $Mg^{2+}$ + $Na^+$ to enzyme preincubated with $K^+$, the appearance of phosphoenzyme was delayed compared with its appearance when both $Na^+$ and $K^+$ were present initially. This delay they attributed to the time taken for a change in conformation from the $E_2$ form that existed in the $K^+$ medium to the $E_1$ form that could be phosphorylated by ATP. Recently, Hobbs *et al.* (1980), in experiments on *Electrophorus* enzyme using a rapid-flow technique, have been able to use the changes in reactivity of the enzyme to ATP to estimate the rate constants for the conversion in each direction. They found that, at 21°C and with 20 mM $K^+$, the pseudo-first-order rate constant for the conversion of $E_1$ to $E_2$ was 52–55 $sec^{-1}$, and this must have been nearly maximal since, with 30 mM $K^+$, it had increased to only 60 $sec^{-1}$. The rate constant for the conversion of $E_2$ to $E_1$ at 21°C and in the absence of nucleotides was about 6 $sec^{-1}$. Though slow, this is much faster than the rate (estimated by other methods) for mammalian $Na^+$,$K^+$-ATPase (see Section VI-B) perhaps because the *Electrophorus* enzyme has evolved to work at a lower temperature.

Although it is convenient to speak of the "$E_1$" form of the dephosphoenzyme, it is clear from the experiments of Mårdh and Post (1977) on the effects of changing the order of addition of ATP, $Na^+$, and $Mg^{2+}$ on the rate of phosphorylation of a kidney $Na^+$,$K^+$-ATPase that the enzyme must be able to change between subordinate conformations (within the class designated $E_1$) depending on whether $Na^+$ or ATP or both are bound to it.

## B. Differences in Reactivity to Inhibitors

If the conformations stabilized by $Na^+$, $K^+$, or other ligands have different affinities (or even different binding sites) for an inhibitor, then measurements of the extent or rate of loss of activity, or of the rate of binding of the inhibitor, under different conditions, can give information about changes in conformation. This approach will not be discussed further here, but examples are provided by work with N-ethylmaleimide (Banerjee *et al.*, 1972a,b; Hart and Titus, 1973a,b; Skou, 1974a; Schoot *et al.*, 1980; Winslow, 1981), and with a number of different ATP analogs which bind covalently to the enzyme (Patzelt-Wenczler and Mertens, 1981; Patzelt-Wenczler and Schoner, 1981).

## C. Differences in Affinity for Nucleotides and Related Compounds

In 1971, Hegyvary and Post and, independently, Nørby and Jensen, using flow dialysis, showed that $Na^+$,$K^+$-ATPase suspended in a $Na^+$ medium, lacking both $K^+$ and $Mg^{2+}$, bound ATP with a very high affinity ($K_d$ = 0.1–0.2 μM). The addition of $K^+$ led to a large drop in affinity (to a level at which binding could not easily be measured by flow dialysis), and this drop presumably reflected a change in conformation. $Tl^+$, $Rb^+$, $Cs^+$, and $NH_4^+$ acted like $K^+$. The effect of $K^+$ on nucleotide binding has been confirmed by more sensitive direct-binding experiments using a filtration technique, in a variety of tissues, with ADP as well as ATP (Kaniike *et al.*, 1973; Jensen and Ottolenghi, 1976; see also the excellent review by Nørby, 1983). The number and nature of nucleotide binding sites under different conditions will be discussed later (Section VIII-E.2).

Neither flow dialysis nor filtration allows binding to be measured rapidly enough for the measurements to be used to estimate the rates of the conformational changes. There exist, however, fluorescent formycin analogs of ATP and ADP, which are treated by the $Na^+$,$K^+$-ATPase very much like ATP and ADP, and which increase in fluorescence when they are bound to the enzyme (Karlish *et al.*, 1978a). Taking advantage of the change in fluorescence associated with the binding of formycin triphosphate (FTP) or formycin diphosphate (FDP), Karlish *et al.* (1978b; see also Glynn *et al.*, 1979) were able to use stopped-flow fluorimetry to measure the rates of net binding and net release of the nucleotides at room temperature and, hence, to infer the rates of the conformational changes.

When excess $Na^+$ was added to enzyme suspended in a medium containing low concentrations of $K^+$ and of FTP, the change in conformation ($E_2(K) \rightarrow E_1Na$) led to binding of FTP and an increase in fluorescence. The rate constant was estimated to be about 0.2 $s^{-1}$ at very low concentrations of FTP, and it increased linearly with FTP concentration at least up to 24 μM (beyond which the fluorescence of the free FTP made measurements impossible). This increase suggested that the binding of FTP at a low-affinity site on the $E_2$ form of the enzyme accelerated the otherwise remarkably slow conversion of $E_2$ to $E_1$. The slowness, the accelerating effect of nucleotide, and the fact that the nucleotide acted without phosphorylation, were all reminiscent of the behavior of the hypothetical occluded $K^+$ form of Post *et al.* (1972), and suggested strongly that the stable form of the dephosphoenzyme in $K^+$ media contains occluded $K^+$ ions and is identical with the form of the enzyme that exists transiently in high $Na^+$ media following the hydrolysis of $E_2P$ in the presence of $K^+$ ions.

When $K^+$ was added to enzyme suspended in media containing low concentrations of $Na^+$ and FTP, the change in conformation ($E_1Na \rightarrow E_2(K)$) led to release of bound FTP and a fall in fluorescence. The pseudo-first-order rate constant for the conversion varied roughly linearly with the $K^+$ concentration, and was about 100 $s^{-1}$ at 15 mM, the highest concentration at which measurements were practicable. Although the relation between the rate of the conformational change and the $K^+$ concentration implies that $K^+$ ions were acting at low-affinity sites ($K_{0.5}$ = c. 15 mM), concentrations of $K^+$ as low as 1 mM were sufficient to displace the equilibrium $E_1Na \rightleftharpoons E_2(K)$ well over to the right in similar conditions, i.e., low $Na^+$ and low FTP. To explain this paradox, Karlish *et al.* (1978a) suggested that $K^+$ ions bind rapidly to low-affinity sites on $E_1$, that the equilibrium $E_1K \rightleftharpoons E_2(K)$ is poised well to the right, and that the rate-limiting step in the formation of $E_2(K)$ is the conformational change rather than the binding of $K^+$.

## D. Differences in the Pattern of Attack by Proteolytic Enzymes and in the Products of Proteolytic Digestion

Unequivocal evidence that the $E_1$ and $E_2$ forms of dephosphoenzyme differ in conformation has been provided by experiments which showed two distinct patterns of attack by trypsin depending on whether the enzyme was in a $Na^+$ medium or a $K^+$ medium (Jørgensen, 1975a, 1977; Jørgensen and Klodos, 1978; Castro and Farley, 1979; Jørgensen and Anner, 1979; Jørgensen and Karlish, 1980, but cf. Koepsell, 1979). In $K^+$ media, the loss of activity with time followed a single exponential, and the α-polypeptide chain was cleaved at a single point to give two fragments of molecular weight 58,000 and 46,000. The loss of hydrolytic activity was paralleled by loss of the capacity to bind ATP and to be phosphorylated. In $Na^+$ media (and also in choline or Tris media), loss of activity followed a biphasic course. During the initial rapid phase, a very small peptide containing about 20 amino acids was split from the amino-terminal end of the α-polypeptide leaving an enzyme with a much reduced $V_{max}$ and other well-defined catalytic defects. Subsequently, there was a mono-exponential loss of activity, resulting from a second cleavage which left a 78,000 molecular weight fragment without enzymic activity. The relative magnitudes of the rate constants for the two exponential terms describing the biphasic loss of activity were found to be slightly different in Tris media and in $Na^+$ media (Jørgensen and Petersen, 1982), perhaps because $E_1$Tris and $E_1Na$ differ slightly in conformation.

Addition of ATP + $Mg^{2+}$ to enzyme in a $Na^+$ medium led to a conformation that was attacked by trypsin as if the enzyme (without ATP or $Mg^{2+}$) were in a $K^+$ medium. This implies that, so far as the accessibility of the peptide bonds to trypsin is concerned, the conformation of $E_2P$ must resemble that of $E_2(K)$. Experiments with chymotrypsin also showed a change in conformation of the unphosphorylated enzyme between $Na^+$ and $K^+$ media (Jørgensen *et al.*, 1982b).

## E. Differences in Intrinsic Fluorescence or in the Fluorescence of Probe Molecules

### 1. Intrinsic Fluorescence

The deductions drawn from the experiments with formycin nucleotides described above depend on the assumption that the observed changes in fluorescence reflect changes associated with the binding or release of nucleotide rather than changes in the fluorescence of bound nucleotide. The discovery that a similar pattern of behavior can be observed by measuring the intrinsic (tryptophan) fluorescence of the enzyme (Karlish and Yates, 1978) was, therefore, particularly welcome. By measuring changes in the intrinsic fluorescence of pig kidney $Na^+,K^+$-ATPase, Karlish and Yates were also able to show that ATP accelerates the rate of conversion of $E_2(K)$ to $E_1Na$, acting (like FTP) with a low affinity and without phosphorylation. At the highest concentration of ATP at which they were able to measure the rate (100 μM), the rate constant was about 17 $sec^{-1}$ (21°C), and they calculated that at saturating ATP concentrations, it would have been about 60 $sec^{-1}$. This large increase in rate constant must be mainly, if not wholly, responsible for the dramatic shift to the left of the equilibrium $E_1 \rightleftharpoons E_2$ that is seen when ATP, or its nonphosphorylating analogs, bind to the low-affinity ATP site (Beaugé and Glynn, 1980; Jørgensen and Karlish, 1980). The shift corresponds to a change in the equilibrium constant of between 2 and 3 orders of magnitude. Unfortunately, though changes in intrinsic fluorescence can be used to measure the relatively slow rates of conversion of $E_2(K)$ to $E_1K$ at low ATP concentrations, they are too small to be used to measure the faster change from $E_1Na$ to $E_2(K)$, even at low-$K^+$ concentrations.

Skou and Esmann (1980), working with $Na^+,K^+$-ATPase from the rectal gland of the spiny dogfish, found that choline and protonated Tris had a $Na^+$-like effect on the intrinsic fluorescence. Jørgensen and Petersen (1982), working with pig kidney $Na^+,K^+$-ATPase, found that the effect of Tris depended on its concentration; the higher the Tris concentration, the smaller the decrease in fluorescence caused by $Na^+$ and the greater the increase in fluorescence caused by $Rb^+$. With 150 mM Tris, $Na^+$ had no effect, and with 2 mM Tris, $Rb^+$ had no effect. The presumed explanation is that, in the absence of both $Na^+$ and $Rb^+$, the enzyme was in the $E_1$ form when the Tris concentration was 150 mM, and in the $E_2$ form when the Tris concentration was 2 mM. That explanation is supported by studies of the binding of ADP and AMPP(NH)P to a pig kidney $Na^+,K^+$-ATPase, which showed that both Tris and imidazole, at a concentration of 50 mM, had a $Na^+$-like effect (Jensen and Ottolenghi, 1983; Rempeters and Schoner, 1983).

### 2. *Differences in the Fluorescence of Probe Molecules*

*a. Formycin Nucleotides.* The use of formycin nucleotides has already been discussed in Section VI-C, dealing with nucleotide binding.

*b. Fluorescein Isothiocyanate.* Fluorescein isothiocyanate binds covalently to the α-polypeptide in the neighborhood of the ATP-binding site to give a labeled enzyme which shows a large drop in fluorescence when $K^+$ is added in excess (Karlish, 1980; Sen *et al.*, 1981). The labeled enzyme is unable to hydrolyze ATP because its high affinity ATP-binding site is blocked, but it can still be phosphorylated by orthophosphate, and it can still bind ouabain (Karlish, 1980). The effects of $Na^+$, $K^+$, $Mg^{2+}$, ouabain, and orthophosphate on the fluorescence have been investigated in some detail by Karlish (1980) and Hegyvary and Jørgensen (1981), using $Na^+,K^+$-ATPase from pig kidney. The fluorescence was highest in the presence of $Na^+$, and was only slightly less in the absence of ligands. $K^+$ (and the $K^+$ congeners, $Tl^+$, $Rb^+$, $Cs^+$, and $NH_4^+$) caused a large drop in fluorescence. $Mg^{2+}$ had no effect in the presence of

$Na^+$ (or in the absence of other ligands), but it reduced the quenching produced by $K^+$ and changed the relation between quenching and $K^+$ concentration from hyperbolic to sigmoid. The presence of $Na^+$ ions had a similar effect on the relation between quenching and $K^+$ concentration. It follows that, at least in the presence of $Mg^{2+}$ or $Na^+$, $K^+$ ions act at more than one site. The appearance of action at a single site in the absence of $Mg^{2+}$ and $Na^+$ may be the result of large differences between the affinities of sites all of which must be occupied for the change in conformation to occur.

The results just described suggest that it is possible to distinguish, by the intensity of their fluorescence, between four forms of the labeled dephosphoenzyme: $E_1$, $E_1Na$, $E_2(K)$, and $E_2Mg(K)$. The intensity of fluorescence of the phosphorylated enzyme $E_2P$ was only slightly greater than that of $E_2(K)$, whether or not $K^+$ ions were present. Binding of ouabain to the phosphoenzyme caused considerably more quenching, and since ouabain binds to the outer surface, whereas, the fluorescent label is near the ATP site at the inner surface, this implies that the conformations of $E_2PMg$ and $E_2PMg$-ouabain are different.

Because the quenching of the fluorescence of the dephosphoenzyme by $K^+$ is so large, it is possible to measure the rates of the interconversion between $E_1$ and $E_2$ forms in both directions. Karlish measured the rate of conversion of $E_1Na$ to $E_2(K)$ at $K^+$ concentrations up to 80 mM. At that concentration, the pseudo-first-order rate constant (in the absence of $Mg^{2+}$ and nucleotide) was nearly 200 $sec^{-1}$ (20°C), and Karlish estimated that the rate constant at saturating $K^+$ was about 290 $sec^{-1}$. The rate of the reverse transition ($E_2(K) \rightarrow E_1Na$) was measured both by Karlish (1980) and by Hegyvary and Jørgensen (1981), but there is a puzzling difference between their results. Karlish found a rate constant of 0.3 $sec^{-1}$ at 20°C and in the absence of $Mg^{2+}$ (1 mM EDTA). This is in reasonably good agreement with estimates from the experiments with formycin nucleotides or from the observations of intrinsic fluorescence (see above). Karlish did not investigate the effect of $Mg^{2+}$ on the rate. Hegyvary and Jørgensen obtained a much higher estimate for the rate constant (1.7 $sec^{-1}$ at 25°C and in the absence of $Mg^{2+}$), but this was reduced to about 0.4 $sec^{-1}$ by raising the $Mg^{2+}$ concentration to 4 mM. ATP had little effect on the rate of conversion of $E_2(K)$ to $E_1Na$, suggesting that the fluorescein also blocks the low-affinity ATP-binding site.

Vanadate, in the presence of $Mg^{2+}$, stabilizes the fluorescein–isothiocyanate-labeled enzyme in a low fluorescence form presumed to be $E_2(K)Mg$-vanadate (Karlish *et al.*, 1979).

*c. Eosin.* In 1980, Skou and Esmann described the use of eosin maleimide to study the effects of ligands on the conformation of $Na^+$,$K^+$-ATPase. They later discovered that the signals that they had been observing came from molecules that were not covalently bound, and that similar results could be obtained with eosin (Skou and Esmann, 1981).

In $K^+$ media, eosin binds to the enzyme with a low affinity and without much change in its fluorescence. In $Na^+$ media, it binds with a high affinity and its fluorescence changes in a way which suggests that its environment has become less polar. Both the change in affinity and the change in fluorescence point to a change in the conformation of the enzyme. The high-affinity site in $Na^+$ media seems to be identical with the high-affinity ATP site, since (1) ATP prevents the high-affinity binding of eosin, and (2) eosin is a competitive inhibitor of $Na^+$,$K^+$-ATPase activity. The stoichiometry of eosin binding varies with the $Na^+$ concentration, being one per phosphorylation site at 2-mM $Na^+$ and at 150-mM $Na^+$, but increasing to about 1.4 at 20-mM $Na^+$. The effects of $Na^+$ cannot, therefore, be wholly explained in terms of an equilibrium between $E_1$ and $E_2$ forms. An increase in the concentration of $H^+$ ions shifts the equilibrium in the same direction as the addition of $K^+$ ions (Skou and Esmann, 1980; Skou, 1982).

$Mg^{2+}$ causes a shift in the fluorescence curve similar to that produced by $Na^+$, but the maximal increase in fluorescence is greater and is associated with the binding of two molecules of eosin per phosphorylation site (Skou and Esmann, 1983). Binding at both sites is prevented by orthophosphate + ouabain, and Scatchard plots of the ouabain-sensitive binding show that the affinities for eosin of both sites are high, the precise values depending on the phosphate concentration (0.14 μM and 1.3 μM with 0.25 mM $P_i$). When $Mg^{2+}$ is present, the addition of $Na^+$ in high concentration has only a small effect on the binding affinity and does not alter the number of molecules of eosin bound. In contrast, the similar addition of $K^+$ greatly reduces the binding affinity, without changing the number of sites. In a 150-mM $K^+$, 5-mM $Mg^{2+}$ medium, binding is prevented by vanadate. Scatchard plots of the vanadate-sensitive binding show two sites, each with an affinity of about 17 μM.

Independent evidence for low-affinity eosin binding in the presence of $K^+$ has come from experiments in which ATP hydrolysis was measured at different $Na^+$ and $K^+$ concentrations (but with their sum held constant at 150 mM), and either (1) 1.5 μM, 10 μM, or 50 μM ATP, or (2) a fixed concentration of 1.5 μM ATP together with 0 μM, 10 μM, or 50 μM eosin (see Figure 10 in Skou and Esmann, 1981). When the results are plotted (not as absolute rates but relative to the maximal rate for each set of conditions), the curves are very similar, as if eosin can not only bind to the low-affinity ATP site but can also act like ATP in accelerating the release of occluded $K^+$ (see also Section VII-A). These results are important because they suggest that both of the eosin-binding sites are ATP-binding sites. Since two eosin molecules can be bound simultaneously, it would seem that two ATP-binding sites must exist per phosphorylation site. To explain why only one ATP-binding site seems to be detectable, Skou and Esmann suggest *either* that, for ATP-binding, the second site has a lower affinity (cf. Schuurmans Stekhoven *et al.*, 1981), *or* that, though two eosin molecules can bind simultaneously, steric hindrance prevents the binding of more than one ATP molecule at a time.

Skou and Esmann (1983) also used stopped-flow fluorimetry to measure some of the rates of interconversion between the different forms. Conversion of the $Mg^{2+}$ form to the $K^+$ form was very fast ($k$ not less than 330 $s^{-1}$ at 4°C), and in contrast to the conversion of the $Na^+$ form to the $K^+$ form (Skou, 1982), was slowed only slightly by oligomycin. Addition of 20 mM $Na^+$ to enzyme in the $Mg^{2+}$ form had only a slight effect on the rate of conversion to the $K^+$ form in the absence of oligomycin. When oligomycin, $Na^+$, and $Mg^{2+}$ were all present, however, the rate of conversion to the $K^+$ form was reduced by three orders of magnitude. Taken together, the estimates of numbers of binding sites, binding affinities, and of rates of conversion show that the enzyme must have different conformations depending on whether $Na^+$ alone, $Mg^{2+}$ alone, or $Na^+$ and $Mg^{2+}$ together are bound to it.

*d. 5-Iodoacetamidofluorescein.* All four of the fluorescence methods for studying conformational changes described so far (formycin nucleotides, intrinsic fluorescence, fluorescein isothiocyanate, and eosin) have serious disadvantages. Kapakos and Steinberg (1983) therefore investigated the behavior of 5-iodoacetamidofluorescein, arguing that because iodoacetate does not inhibit the enzyme, yet binds close to the phosphorylation site (Castro and Farley, 1979), its fluorescent derivative might signal conformational changes without interfering too much with the behavior of the enzyme. They found that about two molecules of the compound were incorporated per phosphorylation site, that ATPase and phosphatase activities were retained, that the kinetic parameters for $Na^+$, $K^+$, $Mg^{2+}$, ATP, and *p*-nitrophenyl phosphate were little changed, and that the fluorescence was increased by ATP, $Na^+$, and $Mg^{2+}$, and decreased by $K^+$. The compound therefore promises to be a useful tool.

## F. Differences in Equilibrium Binding of $Na^+$ and $K^+$ Ions

If $Na^+$ and $K^+$ stabilize, respectively, $E_1$ and $E_2$ forms of the dephosphoenzyme, it ought to be possible to detect conformation-dependent differences in the binding affinities of the dephosphoenzyme for the two ions. Because these binding affinities, particularly that for $Na^+$, are much smaller than the binding affinities for ATP or ouabain, the experimental difficulties in measuring them are much greater, and measurements only became possible when methods had been developed for preparing $Na^+$,$K^+$-ATPase of moderate degrees of purity.

Four techniques have been used for measuring equilibrium binding: (1) ion-selective electrodes (Hara and Nakao, 1979; Hastings and Skou, 1980), (2) forced dialysis (Cantley *et al.*, 1978a), (3) filtration (Yamaguchi and Tonomura, 1979, 1980a–c), and (4) centrifugation (Kaniike *et al.*, 1976; Matsui *et al.*, 1977; Matsui and Homareda, 1982; Homareda and Matsui, 1982; Jørgensen and Petersen, 1982; Jensen and Ottolenghi, 1983). Of these, centrifugation is the most sensitive because the proportion of enzyme to water is high in the pellet.

A further complication is the need to distinguish between specific and nonspecific binding. The distinction has generally been made on the basis that binding that is prevented by heat treatment or by the presence of excess competing ions is specific. Matsui and Homareda (1982), however, point out that though these methods will distinguish between ions bound at specific sites and ions trapped in the water in the pellet, both heat treatment and the presence of excess competing ions may also affect the nonspecific binding, including any trapping of ions that may be caused by the Donnan effect. They therefore prefer to use ouabain sensitivity as the criterion for specific binding, recognizing, however, that the enzyme may also bind ions in a way that is not prevented by ouabain.

For $K^+$ (or $Rb^+$) binding to normal enzyme, the picture that emerges from the various studies supports the hypothesis that $K^+$ ions are bound with a high apparent affinity to the $E_2$ form of the phosphoenzyme. The stoichiometry is less certain. Matsui and Homareda (1982), Cantley *et al.* (1978a), Yamaguchi and Tonomura (1979, 1980a–c), Jørgensen and Petersen (1982), and Matsui *et al.* (1983) all found that a maximum of close to two ions can be bound per α-subunit; Jensen and Ottolenghi (1983) found three, and Hastings and Skou (1980), four or five.

Equilibrium $Na^+$ binding has been investigated by Kaniike *et al.* (1976), Hara and Nakao (1979), Yamaguchi and Tonomura (1979, 1980a–c) and Matsui and his colleagues (see Matsui and Homareda, 1982). It is generally agreed that close to three $Na^+$ ions can be bound per α-subunit, but there is an important disagreement between the findings of Yamaguchi and Tonomura and of Matsui and Homareda. Although both groups of workers found that high concentrations of $Na^+$ could prevent the binding of $K^+$, and vice versa, the results of Yamaguchi and Tonomura seemed to show that three $Na^+$ and two $K^+$ ions could be bound (with a high affinity) simultaneously, whereas the results of Matsui and Homareda precluded simultaneous high-affinity binding. Matsui and Homareda suggest that the difference is a consequence of the lower precision of the filtration method used by Yamaguchi and Tonomura and of the inadequacy of the use of competing ions to define specific binding.

Magnesium decreases the affinity for $K^+$ binding, but does not affect the binding capacity. At a concentration of 5 mM, it completely prevents high-affinity binding of $Na^+$ (Homareda and Matsui, 1982). As would be expected from its effect on the $E_1 \rightleftharpoons E_2$ equilibrium, ATP decreases the affinity for $K^+$ ions and leaves the affinity for $Na^+$ ions unaffected (Yamaguchi and Tonomura, 1980b; Matsui and Homareda, 1982; Homareda and Matsui, 1982); ADP and AMPP(NH)P have a similar effect on $K^+$ binding (Homareda and Matsui, 1982). Oligomycin has little effect on $K^+$ binding in the absence of $Na^+$, but increases the inhibitory effect of $Na^+$. It causes a ten-fold increase in the affinity for $Na^+$ binding, with increased cooperativity (Matsui and Homareda, 1982; Homareda and Matsui, 1982). The insensitivity of $K^+$-binding to oligomycin in the absence of $Na^+$ is explained by the failure of oligomycin to slow the conversion of the enzyme to the $E_2(K)$ form in the absence of $Na^+$ ions, as demonstrated by the stopped-flow experiments of Skou and Esmann (1983) using eosin.

All the measurements of binding described so far were made with normal enzyme. Recently, Jørgensen and Petersen (1982) measured the binding of $Rb^+$ to enzyme treated with thimerosal, a sulfydryl reagent that blocks ATPase activity but does not block (and may stimulate) *p*-nitrophenylphosphatase activity (see Section IX-D). They found that the binding capacity for $Rb^+$ and the affinity for $Rb^+$ were both unchanged by the treatment; yet, in experiments under similar conditions, the addition of $Rb^+$ did not alter the intrinsic fluorescence of the thimerosal-treated enzyme. These observations make it awkward to explain the high apparent affinity for $Rb^+$ binding in terms of a combination of weak binding to $E_1$ and a conformational equilibrium between $E_1Rb$ and $E_2(Rb)$ poised far to the right. Jørgensen and Petersen suggest, instead, that the high-affinity binding is to the $E_1$ form of the enzyme. It is possible, however, that the thimerosal-treated enzyme does not show any fluorescence change on the addition of $Rb^+$ because it is already in the $E_2$ form. That hypothesis requires thimerosal-treated enzyme in Tris media of moderate concentration to behave like normal enzyme in Tris media of very low concentration. The fact that the addition of $Na^+$ leads to a decrease in fluorescence (Jørgensen and Petersen, 1982) makes the hypothesis plausible. Whatever the initial state of the thimerosal-treated enzyme, recent unpublished experiments by D. E. Richards and I. M. Glynn, showing that only a small fraction of the bound $Rb^+$ is released in 1 sec at 20°C, are incompatible with the notion that the thimerosal-treated enzyme with bound $Rb^+$ ions is in the $E_1$ conformation; it must be in a conformation in which the $Rb^+$ ions are occluded.

## VII. THE EXISTENCE AND ROLE OF OCCLUDED-ION FORMS OF THE $Na^+$,$K^+$-ATPase

Since, in the normal working of the $Na^+$,$K^+$-ATPase, $Na^+$ ions and $K^+$ ions are pumped through the enzyme molecule, it has always seemed likely that forms of the enzyme containing occluded $Na^+$ ions or occluded $K^+$ ions play a central role in cation transport in a literal as well as in a metaphorical sense. In the scheme of Figure 1, $K^+$ ions are occluded in the $E_2$ form of dephosphoenzyme and $Na^+$ ions are occluded in the $E_1P$ form of the phosphoenzyme. In this section, we shall consider the evidence for the existence of these occluded-ion forms and for the role that they play.

### A. The Occluded-$K^+$ Form of Dephosphoenzyme

We have seen that a transient occlusion of $K^+$ ions was postulated by Post *et al.* (1972) to account for differences in the rate of rephosphorylation of enzyme which had just been dephosphorylated in the presence of different congeners of $K^+$ (see Section V-C). We have also seen that the form of dephosphoenzyme that is stable in $K^+$ media resembles this hypothetical transient occluded $K^+$ form in that (1) both can be phosphorylated by orthophosphate but not by ATP, and (2) the conversion of both to the form that can be phosphorylated by ATP is slow unless accelerated by nucleotides, which act with a low affinity and without phosphorylating (see Sections VI-C and VI-E.1). The economical explanation of these resemblances is that the two forms are identical, which implies that the stable form of the dephosphoenzyme in $K^+$ media contains occluded $K^+$ ions.

Direct evidence for occlusion came from experiments in which enzyme suspended in $Na^+$-free media containing a low concentration of [$^{86}Rb$]-RbCl was forced through cation exchange columns at a rate that was slow enough for the resin to remove nearly all of the free $Rb^+$ ions yet was fast enough for the enzyme to emerge within less than 1 sec, i.e. within a period much smaller than the time constant for the conformational change $E_2(Rb) \rightarrow E_1Rb$ in the absence of nucleotides (Beaugé and Glynn, 1979a; Glynn and Richards, 1982). ($^{86}Rb$ was used rather than $^{42}K$ because $^{42}K$ is not available with a high enough specific activity.)

Experiments of this kind have shown that, if the conditions are such that the enzyme starts in the $E_2$ form (no nucleotide or $Na^+$), much more $^{86}Rb$ emerges from the column. By varying the flow rate, and therefore the time elapsing before the enzyme emerged, Glynn and Richards (1982) showed that the rate constant for release of occluded $Rb^+$ is 0.1–0.2 $sec^{-1}$ (21°C), in good agreement with estimates of the rate constant for the conformational change based on fluorescence measurements. The rate of release of occluded $Rb^+$ is not affected by the presence of $Na^+$ or $K^+$, but is greatly accelerated by high concentrations of ATP or ADP. (Recent unpublished experiments by D. E. Richards, J. L. Howland, and I. M. Glynn show that 30 μM eosin also accelerates the release of $Rb^+$.)

The relation between the amount of $Rb^+$ occluded and the $Rb^+$ concentration is sigmoid if $Mg^{2+}$ ions are present, but hyperbolic if they are rigorously exluded (2 mM CDTA). $Mg^{2+}$ does not affect the maximum amount of $Rb^+$ occluded, however, which was estimated to be close to 3 $Rb^+$ ions per phosphorylation site (or ouabain-binding site), a surprising figure in view of most of the equilibrium-binding data (see Section VI-F) and the stoichiometry of the pump.

Vanadate ($+Mg^{2+}$), at concentrations sufficient to prevent the change from $E_2(K)$ to $E_1K$ in enzyme labeled with fluorescein isothiocyanate (Karlish *et al.*, 1979), stabilizes the enzyme in the occluded-$Rb^+$ form. In the presence of ouabain + $Mg^{2+}$, however, no occlusion of $Rb^+$ ions can be detected. This effect of ouabain explains why Matsui and Homareda (1982) found that high-(apparent) affinity $K^+$ binding is ouabain sensitive. It also implies that the conformation of the ouabain-bound enzyme must differ from that of $E_2(Rb)$, despite the fact that it would be classified as an $E_2$ form on the basis of its reactivity with orthophosphate, its low affinity for ATP, and the low fluorescence of ouabain-bound, fluorescein-labeled enzyme.

Using their rapid ion-exchange technique, Glynn and Richards (1982) were also able to demonstrate the formation of the occluded-$Rb^+$ form of the enzyme when $Rb^+$ ions were allowed to catalyze the hydrolysis of phosphoenzyme generated by the addition of ATP to enzyme suspended in a high-$Na^+$ medium.

### 1. The Role of the Occluded-$K^+$ Form

The experiments just described show that the occluded-$Rb^+$ (and therefore presumably the occluded-$K^+$) form of the dephosphoenzyme can be reached by two different routes: addition of $Rb^+$ to dephosphoenzyme in a $Na^+$-free or low-$Na^+$ medium, or $Rb^+$-catalyzed hydrolysis of phosphoenzyme generated from ATP in a high-$Na^+$ medium. If we (1) accept the hypothesis of Karlish *et al.* (1978b) about the binding of $Rb^+$ to dephosphoenzyme, (2) ignore multiple binding sites and the role of $Mg^{2+}$ ions, and (3) assume that all reactions are reversible, the two routes coupled back-to-back may be written

$$Rb^+ + E_1 \rightleftharpoons E_1Rb \underset{\substack{\vdots \\ \text{accelerated} \\ \text{by ATP etc.}}}{\rightleftharpoons} E_2(Rb) \underset{P_i}{\rightleftharpoons} E_2PRb \rightleftharpoons E_2P + Rb^+ \quad (3)$$

There is good evidence (see especially Blostein and Chu, 1977) that the $K^+$ ions that catalyze the hydrolysis of $E_2P$ act extracellularly and at sites with a high affinity. The work with formycin nucleotides (Karlish *et al.*, 1978b) and the studies of intrinsic fluorescence (Beaugé and Glynn, 1980; Jørgensen and Karlish, 1980) strongly suggest that the $K^+$-binding sites on $E_1$ have a low affinity. There is also evidence that these sites are intracellular. In the first place, an intracellular location is suggested by observations on the ATPase activity and *p*-nitrophenylphosphatase activity of inside-out red-cell vesicles (Blostein and Chu, 1977; Blostein *et al.*, 1979) and on $Na^+$ fluxes in dialyzed squid axons (Beaugé and DiPolo, 1979b). Secondly, there is more direct evidence from experiments in which $Na^+$,$K^+$-ATPase was incorporated into tight artificial lipid vesicles and subjected to the actions of trypsin or vanadate (Karlish and Pick, 1981). The enzyme in artificial lipid vesicles may be oriented in either direction, but if ATP is present only in the suspending medium, the rate at which labeled $Na^+$ ions enter the vesicles reflects the activity solely of those enzyme molecules that are oriented with their ATP-binding sites facing outwards. By exposing the vesicles containing $Na^+$,$K^+$-ATPase to trypsin under different conditions, and then measuring ATP-dependent $Na^+$ uptake under standard conditions (which, where appropriate, included the presence of ionophores to allow $K^+$ ions to enter or leave the vesicles freely), Karlish and Pick showed that only when $K^+$ was present *outside* the vesicles did the action of trypsin follow the pattern characteristic of digestion of enzyme in the $E_2$ form. The inference is that only when $K^+$ ions were in contact with what was originally the intracellular surface did they lead to the formation of $E_2(K)$. The vanadate experiments took advantage of the facts that vanadate inhibits $Na^+$,$K^+$-ATPase activity by stabilizing the occluded-$K^+$ form of the enzyme (Karlish *et al.*, 1979; Glynn and Richards, 1982), and that the effect of vanadate is only slowly reversible (Cantley *et al.*, 1978a). By exposing vesicles containing $Na^+$,$K^+$-ATPase to vanadate under different conditions, removing the vanadate, and assaying the degree of inhibition under standard conditions, Karlish and Pick showed that only $K^+$ ions *outside* the vesicles were effective in promoting inhibition by vanadate. Again, the inference is that only $K^+$ ions at the original intracellular surface led to the formation of the occluded-$K^+$ form.

If the $K^+$-binding sites on $E_2P$ are extracellular and of high affinity, and those on $E_1$ are intracellular and of low affinity, the sequence of events represented by Eq. (3) is capable of catalyzing $K^+$ movements through the membrane with many of the features characteristic of the $K^+$ movements observed during normal activity of the pump, pump reversal, and $K^+$–$K^+$ exchange. Furthermore, detailed consideration of the rate constants of the individual reactions, so far as they are known, shows that the expected reaction rates are compatible with the observed fluxes (see Glynn and Karlish, 1982; Glynn and Richards, 1982).

If the sequence of reactions represented by Eq. (3) does account for $K^+$–$K^+$ exchange, the rate of the exchange at low-ATP concentrations should be limited by the rate of conversion of $E_2(K)$ to $E_1K$. Evidence that it is is provided by experiments, on $Na^+$,$K^+$-ATPase incorporated into artificial lipid vesicles, showing that when ATP is replaced by other nucleotides, there is a correlation between the effectiveness of the nucleotides in supporting $K^+$–$K^+$ exchange and their effectiveness in accelerating the release of occluded $Rb^+$ (D. E. Richards, J. L. Howland, and I. M. Glynn, unpublished work; Glynn, 1982).

Very recently, S. J. D. Karlish (personal communication) has shown that the very slow $Rb^+$–$Rb^+$ exchange catalyzed by $Na^+$,$K^+$-ATPase in artificial lipid vesicles in the absence of ATP and orthophosphate is accelerated by a rise in pH at the original intracellular surface. Under the conditions of his experiments, the rate-limiting step was thought to be the conversion of $E_2(Rb)$ to $E_1Rb$, so the result fits well with the observation of Skou and Esmann (1980) that a rise in pH displaces the equilibrium between the two forms of dephosphoenzyme toward the $E_1$ form.

There is, however, one set of recent experiments which suggests that the sequence of events represented by Eq. (3) may not give a complete picture. Attempts to demonstrate release of occluded $Rb^+$ by the addition of orthophosphate + $Mg^{2+}$ led to the surprising finding that only half of the occluded $Rb^+$ was released quickly; the rest was released at about the same rate as that observed in the absence of orthophosphate and $Mg^{2+}$ (Glynn *et al.*, 1983b). The most interesting explanation of the observations would be that the $Rb^+$ molecules occluded in the two halves of a dimeric enzyme behave differently (see Section VIII-E). Other possibilities are (1) that of two Rb ions occluded in a monomeric enzyme, or in one-half of a dimer, only one is released by phosphorylation, and (2) that the molecules containing occluded $Rb^+$ are not homogeneous.

## B. The Occluded-$Na^+$ Form of Phosphoenzyme

In 1971, Glynn and Hoffman put forward the hypothesis that $Na^+$ ions are trapped in the $E_1P$ form of phosphoenzyme and can be released to the interior only after the transfer of the phospho group to ADP, and to the exterior only after the conversion of $E_1P$ to $E_2P$ (see Figure 1). The basis of their hypothesis was that $Na^+$–$Na^+$ exchange requires ADP (which can act as an acceptor of the phospho group in $E_1P$) and is blocked by oligomycin (which is thought to prevent the conversion of $E_1P$ to $E_2P$).

To test for $Na^+$ occlusion directly, it is necessary first to generate $E_1P$ in circumstances in which the transfer of the phospho group back to ADP and the conformational change to $E_2P$ are both prevented, and then to see whether the $E_1P$ is capable of carrying $Na^+$ ions through a cation-exchange resin.

Experiments of this kind have recently been reported by Glynn and Richards (1983) and Glynn *et al.* (1983a; 1984). To prevent the dephosphorylation of $E_1P$ by ADP, they kept the ADP concentration very low by (1) working with low concentrations of ATP, (2) keeping the duration of exposure to ATP short, and (3) generally working at 0°C. To block the conversion of $E_1P$ to $E_2P$, they pretreated the enzyme with N-ethylmaleimide or with chymotrypsin. Jørgensen *et al.* (1982b) have recently shown that, under carefully controlled conditions, chymotrypsin modifies the enzyme so that the stable form of phosphoenzyme is $E_1P$. The most clear-cut results were obtained with the chymotrypsin-treated enzyme at 0°C. When phosphorylated by ATP, this enzyme was capable of carrying $Na^+$ ions through a cation-exchange column. Substitution of ADP for ATP, omission of $Mg^{2+}$, or omission of the chymotrypsin treatment, all prevented the effect. Addition of ADP to preformed $E_1P$ caused rapid release of the occluded $Na^+$. Provided the $Na^+$ concentration was high enough for the enzyme to be fully phosphorylated, about three $Na^+$ ions were occluded per molecule of phosphoenzyme. Experiments with N-ethylmaleimide-treated enzyme gave less complete results, but showed that occlusion of $Na^+$ can also occur at room temperature. These experiments provide strong evidence for the kind of mechanism represented in Figure 1, though, so far, it has not been possible to demonstrate the formation of the occluded $Na^+$ form by the alternative route, i.e. the combination of $Na^+$ ions with $E_2P$ followed by the conversion of $E_2P$ to $E_1P$. For this reason, and also because there is less information about the rate constants, the evidence for the $Na^+$ part of the cycle is weaker than the evidence for the $K^+$ part.

## VIII. STRUCTURE OF THE $Na^+$,$K^+$-ATPase

### A. Number, Molecular Weight, and Ratio of Subunits

The discovery of Kyte (1971) that gel electrophoresis of $Na^+$,$K^+$-ATPase solubilized in SDS reveals the existence of two kinds of subunit, a larger polypeptide (α) and a smaller glycopeptide (β), has been confirmed many times, though estimates of the molecular weights vary somewhat.

The most straightforward estimates come from sedimentation equilibrium measurements in the presence of detergent. For $Na^+$,$K^+$-ATPase from dogfish or shark rectal glands, Hastings and Reynolds (1979b) and Esmann *et al.* (1980) found figures of 106,000 and 37,000–40,000 for the α- and β-units, respectively. For $Na^+$,$K^+$-ATPase from pig kidney, Freytag and Reynolds (1981) found figures of 93,900 and 32,000. Using a different technique, which involved removing the SDS by dialysis and determining the molecular weights by centrifugation in the absence of detergent, Peters *et al.* (1981) found figures of 131,000 and 61,800 for the molecular weights of the α- and β-subunits of rabbit kidney $Na^+$,$K^+$-ATPase. The molecular weights have also been estimated in a variety of preparations by SDS gel electrophoresis (Kyte, 1972; Lane *et al.*, 1973; Dixon and Hokin, 1974; Jørgensen, 1974a, 1982; Giotta, 1976; Winter and Moss, 1979; Peterson and Hokin, 1981) and by gel filtration in SDS or guanidinium hydrochloride (Craig and Kyte, 1980). For the α-subunit, the estimates from gel electrophoresis range from 85,000 to 111,000; gel filtration (of a preparation from dog kidney) suggests a higher figure of 120,000. Because of the presence of carbohydrate, the analysis of the behavior of the peptide under electrophoresis is more complicated and uncertain (see Jørgensen, 1982). Kyte (1971) and Hokin (1974) obtained estimates of 40,000–60,000 for the molecular weight including the carbohydrate. Craig and Kyte (1980), using gel filtration, estimated the molecular weight of the β-subunit to be about 56,000; removal of the carbohydrate with glycosidase reduced this figure to 51,000. A figure of 38,000, much closer to the 37,000–40,000 suggested by the sedimentation equilibrium data, was obtained by Sabatini *et al.* (1982), who used SDS gel electrophoresis to estimate the molecular weight of the β-subunit of $Na^+$,$K^+$-ATPase from cultured cells in which glycosylation of the subunit had been prevented by tunicamycin.

Some of the disagreement between estimates no doubt reflects variation between

$Na^+,K^+$-ATPase from different tissues. Jørgensen (1982) has listed a number of known variations between subunits from different sources, including the particularly striking observation of Sweadner (1979) that the α-subunit from rat brain glial cells has about 20 amino acids less than the α-subunit from neurones in the same brain. There seems to be variation, too, in the carbohydrate content of different preparations (Munakata *et al.*, 1982). In $Na^+,K^+$-ATPase from the brine shrimp, *Electrophorus* electric organ, and lamb and rabbit kidney, some carbohydrate has also been found in the α-subunit (Churchill *et al.*, 1979; Peterson and Hokin, 1980; Peters *et al.*, 1981; Munakata *et al.*, 1982). Yamaguchi and Post (1982) recently published evidence suggesting that the α-subunit of $Na^+,K^+$-ATPase from the outer medulla of pig kidney is heterogeneous.

### 1. The α : β Ratio

Uncertainty about the molecular weights of the α- and β-subunits has led to uncertainty about their relative abundance, but recently evidence has hardened in favor of a 1 : 1 ratio (see Peters *et al.*, 1981; Peterson and Hokin, 1981). A 1 : 1 ratio is supported by experiments showing that α- and β-subunits disappear in parallel during crosslinking with Cu-*o*-phenanthrolene (Craig and Kyte, 1980), and by measurements of the ratio of amino-terminal residues for the α- and β-chains (R. Sauer, P. Slavin, and L. Cantley, unpublished work quoted by Cantley, 1981).

### 2. The γ-Subunit

Small peptides of molecular weight 10,000–12,000 have sometimes been reported in purified preparations of $Na^+,K^+$-ATPase (Hokin *et al.*, 1973; Reeves *et al.*, 1980) but, until recently, their status has been doubtful. The finding that ouabain, coupled to a suitable photoaffinity label, labels the γ-peptide as well the α-peptide (Forbush *et al.*, 1978; Rogers and Lazdunski, 1979a,b; Collins *et al.*, 1982) makes it unlikely that the γ-peptide is an impurity. So far, it has not proved possible to obtain a functioning enzyme by bringing together α-, β- and γ-subunits, so reconstitution cannot be used to decide whether the γ-subunit has a functional role.

## B. Information from Electron Microscopy

Purified $Na^+,K^+$-ATPase preparations have been studied by electron microscopy after thin sectioning, negative staining, and freeze-fracture (Van Winkle *et al.*, 1976; Deguchi *et al.*, 1977; Vogel *et al.*, 1977; Maunsbach *et al.*, 1979, 1983; Haase and Koepsell, 1979; Jackson *et al.*, 1980; Skriver *et al.*, 1980). Both negative staining and freeze-fracture show particles embedded asymmetrically in the membrane and projecting on both surfaces. The particles are more numerous and smaller after negative staining than after freeze-fracture (30–50 Å diameter compared with 100 Å), suggesting that the procedure used for negative staining may lead to the dissociation of the enzyme (Deguchi *et al.*, 1977; Haase and Koepsell, 1979). This interpretation is supported by rotary shadowing of the freeze-fracture particles which sometimes show a quadripartite structure (Haase and Koepsell, 1979). The size of the smaller units is compatible with a molecular weight of 170,000. Recently, some information has become available from electron microscopy, after negative staining, of the two-dimensional crystals of $Na^+$,$K^+$-ATPase that form very slowly in the presence of magnesium vanadate or magnesium phosphate (Hebert *et al.*, 1982; Jørgensen *et al.*, 1982a). The crystals differ depending on whether vanadate or phosphate is used, and the unit cell is thought to contain an α,β-unit in vanadate but an $(\alpha\beta)_2$-unit in phosphate. The significance of this is not clear.

## C. Structure of the α- and β-Subunits

Although the amino-acid composition of both subunits is known for several species (Kyte, 1972; Perrone *et al.*, 1975; Peters *et al.*, 1981), the high proportion of hydrophobic residues, about one-third of the total, prevents the sequencing of more than small parts of the polypeptide chains. In the α-subunit, the sequence at each end of the chain is known, together with the sequence in the immediate neighborhood of the aspartyl residue that provides the phosphorylation site (Post and Kume, 1973; Bastide *et al.*, 1973; Perrone *et al.*, 1975; Hopkins *et al.*, 1976; Castro and Farley, 1979; Cantley, 1981). It is significant that the sequence around the phosphorylation site, (Thr or Ser)-Asp-Lys, with a cysteine about four residues on the amino-terminal side of the aspartyl, is identical with that found in $Ca^{2+}$-ATPase from sarcoplasmic reticulum (Bastide *et al.*, 1973). In the β-subunit, only the amino-terminal sequence is known (Cantley, 1981).

Knowledge of the structure of the subunits, and of the arrangement of the polypeptides in the membrane, is therefore limited to what has been discovered by a combination of less direct methods. Because trypsin and chymotrypsin, under defined conditions, cut the α-chain at three particular sites (Jørgensen, 1975a, 1977; Giotta, 1975; Castro and Farley, 1979) all of which are accessible only at the intracellular surface (Giotta, 1975; Karlish and Pick, 1981), it is possible to divide the α-chain into four regions, and to know that the junctions between those regions are exposed at the cytoplasmic surface. By labeling individual residues in the chain with agents that attack exclusively from the cytoplasmic surface, from the lipid layer, or from the extracellular surface, and then identifying the label in one or other of the four regions, it is possible to build up a picture of the polypeptide chain as it winds backwards and forwards across the lipid layer.

Agents that attack from the cytoplasmic surface include [$\gamma^{32}$P]-ATP, [$^{32}$P]orthophosphate, fluorescein isothiocyanate (Karlish, 1980; Hegyvary and Jørgensen, 1981; Carilli *et al.*, 1982), and ATP attached to the photoaffinity label Cr(III)arylazido-β-alanyl (Munson, 1981). Agents that are thought to attack the portion of the chain that is embedded in the lipid include the photoactivable lipophilic substances, iodonaphthylazide (INA; Karlish *et al.*, 1977; Gitler and Bercovici, 1980; Jørgensen *et al.*, 1982a, 1983), adamantane diazirine (AD; Bayley and Knowles, 1980), and 3-trifluoromethyl-3-phenyldiazirine (TPD; Brunner *et al.*, 1980). Agents that attack the extracellular surface include a variety of derivatives of cardiac glycosides with photoactivable groups attached at different points to the glycoside molecule (Ruoho and Kyte, 1974; Forbush *et al.*, 1978; Rogers and Lazdunski, 1979a; Hall and Ruoho, 1980; Rossi *et al.*, 1980, 1982).

From investigations of this kind, it appears that the polypeptide chain forming the α-subunit probably crosses the membrane six times (see Figure 9 in Jørgensen, 1982). A possible mechanism for threading the polypeptide backwards and forwards through the membrane is discussed by Sabatini *et al.* (1982). The amino terminus must be free in the cytoplasm, since it is liberated at the cytoplasmic surface by the action of trypsin in $Na^+$ media, but the position of the carboxyl terminus is not known.

The exact disposition of the β-subunit is not known, but it must be accessible at the extracellular surface since it can be labeled with a ouabain-photoaffinity label (Hall and Ruoho, 1980).

Further information about structure has been obtained from studies of the inactivation of $Na^+$,$K^+$-ATPase by substances reacting with particular groups, and also from studies of the protective effects of ATP against these substances. Inhibition by 7-chloro-4-nitrobenzo-2-oxa-1,3,diazole (NBD-Cl), and protection against that inhibition by ATP suggests that there is a tyrosine or a cysteine residue near the ATP site (Cantley *et al.*, 1978b; Grosse *et al.*, 1978, 1979). From the absorption spectrum of the complex, Cantley *et al.* argue that the residue reacting with NBD-Cl is a tyrosine. The effects of N-acetylimidazole on the ATPase activity of erythrocyte membranes also point to the existence of a tyrosine residue near the ATP site (Masiak and D'Angelo, 1975). Inhibition by butanedione, and protection against that inhibition by ATP and ADP suggest that one, or more probably two, arginines are present at the ATP-binding sites (De Pont *et al.*, 1977; Grisham, 1979).

In interpreting the protective effect of ATP, it is important to remember that ATP may act by altering the conformation of the enzyme as well as by blocking access to its own binding site; see, for example, the work of Hara *et al.* (1981) on the effects of ATP on inhibition by 2-(β-D-ribofuranosyl)maleimide (showdomycin).

Studies of inhibition by various ATP analogs capable of reacting with sulfhydryl groups (thioinosine triphosphate (Sno*PPP*), its *S*-2,4-dinitrophenyl derivative, the disulfide ($(SnoPPP)_2$), and the β,γ-imido and β,γ-methylene analogs of ($(SnoPPP)_2$) led Patzelt-Wenczler and her colleagues to postulate the existence of two different types of sulfhydryl groups within the ATP-binding sites, a rapidly reacting group at the high-affinity ATP site and a slowly reacting group at the low-affinity ATP site (see Patzelt-Wenczler and Schoner, 1981; but cf. Koepsell *et al.*, 1982). A puzzling feature of inhibition by $(SnoPPP)_2$ (see Figure 2 in Patzelt-Wenczler and Schoner, 1981) is that, although the rates of both the fast and slow phases of inhibition increase in a similar way with inhibitor concentration, the fraction of activity inhibited by the end of the fast phase is not constant but increases markedly as the inhibitor concentration is increased. Koepsell *et al.* (1982), who found somewhat similar behavior with the ATP analog 6-[(3-carboxy-4-nitrophenyl)thiol]-9-β-D-ribofuranosylpurine 5′-triphosphate ($Nbs^6ITP$), suggested that the biphasic behavior with both $(SnoPPP)_2$ and $Nbs^6ITP$ was the result of very slow changes in the state of the enzyme, perhaps associated with changes in aggregation.

### 1. Spectroscopic Studies

Information of quite a different kind has been obtained by Grisham and his colleagues (see Grisham, 1982; Klevickis and Grisham, 1982) from measurements of $Mn^{2+}$ EPR, and of $^{205}Tl$, $^{7}Li$, and $^{31}P$ NMR. They also made use of $Cr(H_2O)_4ATP$ and $Co(NH_3)_4ATP$, the bidentate Cr(III), and Co(III) analogs of MgATP.

EPR measurements of $Mn^{2+}$ established that a single $Mn^{2+}$ ion was bound in the absence of ATP, and kinetic studies showed that the $Mn^{2+}$ detected by EPR could activate the ATPase reaction. A $Na^+$-binding site was identified by $^{205}Tl^+$ NMR (sic), 5.4 Å from the $Mn^{2+}$ site, and a $K^+$-binding site was identified by $^{7}Li$ NMR, 7.2 Å from the $Mn^{2+}$ site. Unfortunately, the nature of this $K^+$-binding site is obscure as it appeared to have similar affinities for $Li^+$ and $K^+$ (Grisham and Hutton, 1978). Using the dipolar interaction between $Cr^{3+}$ and $Mn^{2+}$ revealed in the $Mn^{2+}$ EPR spectra, O'Connor and Grisham (1980) calculated that a distance of 8.1 Å separated the $Mn^{2+}$ at the divalent cation-binding site of the enzyme from the $Cr^{3+}$ bound to the ATP. Measurements of $^{31}P$ NMR placed the β- and γ-phosphorus atoms of $Co(NH_3)_4ATP$ between the enzyme-bound $Mn^{2+}$ and the ATP-bound $Co^{3+}$, too far from the $Mn^{2+}$ to allow the formation of an inner-sphere complex but near enough for second-sphere coordination. Measurements of $^{31}P$ NMR in the presence of phosphate established the existence of a phosphate site 6.9 Å from the $Mn^{2+}$. Simultaneous binding of ATP and phosphate was shown to be possible.

Measurements of energy transfer between site-directed fluorescent probes have shown that the distance between the ATP site at the intracellular surface and the ouabain-binding site at the extracellular surface is probably in the range 60–80 Å (Fortes *et al.*, 1981; Carilli *et al.*, 1982). Both fluorescein isothiocyanate and 2′3,-0-(2,4,6 trinitrocyclohexadienylidine) adenosine 5′-triphosphate (TNP–ATP) have been used as probes for the ATP site, and anthroyl ouabain and several fluorescent derivatives of digitoxigenin have been used as probes for the ouabain-binding site. Fluorescence energy transfer between bound donor and acceptor cardiac glycoside molecules was studied by Fortes *et al.*, (1981) in $Na^+$,$K^+$-ATPase from *Electrophorus* electric organ and was found to be compatible with random distribution of the α,β-units in the plane of the membrane. Since the preparation showed ($Na^+$ + $K^+$)-dependent ATPase activity, Fortes *et al.*, argue that the $Na^+$- and $K^+$-pumping unit is the α,β-unit, not the $\alpha_2\beta_2$-dimer.

## D. The Molecular Weight of $Na^+$,$K^+$-ATPase

Five methods have been used to estimate the molecular weight of $Na^+$,$K^+$-ATPase: centrifugation, radiation inactivation, neutron scattering, the determination of phosphorylation sites or of ligand-binding sites, and crosslinking studies.

### 1. Analytical Centrifugation

Analytical centrifugation requires a pure solubilized preparation of $Na^+$,$K^+$-ATPase. Both sedimentation equilibrium analysis and sedimentation velocity analysis have been used, but results have been variable; Jørgensen (1982) gives a good account of the difficulties. Hastings and Reynolds (1979b) used sedimentation equilibrium analysis to determine the molecular weight of shark rectal gland enzyme solubilized with lubrol WX. They estimated the protein molecular weight to be 379,900, made up of two α-units and four β-units; attempts to dissociate the particles by increasing the detergent concentration led to loss of activity. Esmann *et al.* (1980) used sedimentation equilibrium analysis to determine the molecular weight of dogfish rectal gland $Na^+,K^+$-ATPase solubilized with octaethyleneglycol dodecyl monoether ($C_{12}E_8$). They obtained a figure for the protein molecular weight of 265,000; an earlier estimate using a similar procedure gave 276,000 (Esmann *et al.*, 1979). Higher concentrations of detergent caused dissociation into particles with a molecular weight of 139,000, but these particles had no enzymic activity. In contrast, Brotherus *et al.* (1981b), who used sedimentation velocity measurements to determine the molecular weight of pig kidney $Na^+,K^+$-ATPase immediately after solubilization in $C_{12}E_8$, obtained a maximum figure of 170,000 ± 9000, indicating the presence, predominantly, of α,β-units. Remarkably, the solubilized enzyme had 70–90% of the ATPase activity of the membrane-bound enzyme. It follows that, unless there was reassociation to form dimers during the ATPase assay, the α,β-particles must have been enzymically active, at least as judged by hydrolytic activity.

## 2. *Radiation Inactivation*

The great advantage of radiation inactivation for the determination of molecular weights of membrane proteins is that no purification is needed. The disadvantages are (1) that the crucial assumption of one-hit inactivation may not hold for membrane-bound proteins, (2) that the equation deriving the molecular weight from the radiation dose necessary to reduce activity to $1/e$ of its original value is derived empirically from studies of soluble proteins and may, therefore, give a systematic error when applied to membrane-bound proteins, and (3) that it is not known to what extent the estimated target size includes carbohydrate or even lipid. Radiation inactivation is, therefore, perhaps more useful for comparing the target sizes of the machinery responsible for the various partial reactions of which the $Na^+,K^+$-ATPase is capable than for determining the absolute molecular weight.

The earliest estimates of molecular weight, by Kepner and Macey (1968), gave a figure of about 300,000 for red-cell membrane $Na^+,K^+$-ATPase and about 190,000 for enzyme from kidney microsomes. Ellory *et al.* (1979) obtained a figure of 330,000 ± 30,000 for the $Na^+,K^+$-ATPase activity of red cells, but only 180,000 ± 28,000 for *p*-nitrophenylphosphatase or ouabain-binding activity. Ottolenghi *et al.* (1983), using kidney microsomes and brain microsomes, have compared the target sizes for $Na^+,K^+$-ATPase activity and for various partial reactions. The approximate equivalent molecular weights are: $Na^+,K^+$-ATPase activity, 264,000 (kidney), 261,000 ± 6000 (brain); $Na^+$-ATPase activity, 229,000 (kidney), 190,000 (brain); *p*-nitrophenylphosphatase activity, 229,000 (kidney), 196,000 ± 4000 (brain); ouabain binding in the presence of $Mg^{2+}$ + $Na^+$ + ATP, 145,000 (kidney), 154,000 (brain); vanadate binding, 145,000 (kidney); ATP binding, 145,000 (kidney). Cavieres and Ellory (1982) compared the effects of irradiation on $Na^+$-ATPase activity and ATP–ADP exchange activity of a partially purified preparation of pig kidney $Na^+$,$K^+$-ATPase, and obtained target sizes equivalent to molecular weights of 195,000 for $Na^+$-ATPase activity and 109,000 or 122,000 for ATP–ADP exchange activity. Richards *et al.* (1981), using a similar preparation, found that the target size of the mechanism responsible for occluding $Rb^+$ ions in a $Na^+$-free medium was surprisingly small, the equivalent molecular weight being only 40,000–60,000. Very recently, S. J. D. Karlish and E. Kempner (personal communication) have irradiated pig kidney $Na^+$,$K^+$-ATPase in a frozen (rather than a freeze-dried) state at −135°C, and then measured not only $Na^+$,$K^+$-ATPase activity and *p*-nitrophenylphosphatase activity, but also the magnitude of various fluxes after reconstitution of the irradiated enzyme in artificial lipid vesicles. From the target sizes for the various activities, they estimated the equivalent molecular weights to be: $Na^+$,$K^+$-ATPase, 188,000–195,000; *p*-nitrophenylphosphatase, 123,000; ATP-dependent $Na^+$–$K^+$ exchange, 199,000; (ATP + orthophosphate)-dependent $Rb^+$–$Rb^+$ exchange, 199,000; $Rb^+$–$Rb^+$ exchange in the absence of ATP and orthophosphate, 128,000. They also ran SDS gels of the irradiated samples and noted that the progressive loss of $Na^+$,$K^+$-ATPase activity was accompanied by the progressive loss of α-chains. The *relative* magnitudes of the molecular weights derived from the target sizes for $Na^+$,$K^+$-ATPase and *p*-nitrophenylphosphatase activities are not out of line with previous work, but the absolute magnitudes are much smaller (apart from the estimate by Kepner and Macey on kidney microsomes). The reason for the difference is not known, but the new results somewhat weaken the argument that irradiation inactivation data show that the enzyme must be a dimer.

### 3. Low-Angle Neutron Scattering Analysis

Very recently, Pachence *et al.* (1983) have reported the results of experiments in which the analysis of low-angle neutron scattering from a solution containing detergent-solubilized $Na^+$,$K^+$-ATPase from guinea-pig kidney was used to determine the molecular weight and the radius of gyration of the solubilized enzyme. The molecular weight was estimated to be 385,000–421,000 and the radius of gyration was 76.2Å. Pachence *et al.* concluded that the enzyme in the micelles was an $\alpha_2\beta_2$-dimer.

### 4. Molecular Weight from Phosphorylation or Ligand-Binding Data

It is now generally agreed that the $Na^+$,$K^+$-ATPase molecule contains equal numbers of phosphorylation sites, high-affinity ATP-binding sites, ouabain-binding sites, and vanadate-binding sites. The gram-molecular weight cannot, therefore, be less than the weight in grams of the amount of enzyme that contains one mole of each of the four kinds of site. Until a few years ago, the best preparations were thought to contain about $4 \times 10^{-6}$ moles of each kind of site per gram of protein, equivalent to a minimum molecular weight of 250,000. Given an α:β ratio of 1:1, this suggested that the $Na^+$,$K^+$-ATPase was an $\alpha_2\beta_2$-dimer, and it also implied that the enzyme showed "half-of-the-sites" behavior, since if one protomer was phosphorylated (or bound to, e.g., ouabain), the other could not be. The estimates of site concentration, however, were all based on measurements of protein by the Lowry method, and in 1980, Craig and Kyte showed that this method gave falsely high results for the $Na^+,K^+$-ATPase, when compared with the more reliable method of quantitative amino-acid analysis. This was confirmed both by Moczydlowski and Fortes (1981a) and by Peters *et al.* (1981), both groups finding that the Lowry method gave results about 34% too high. Reassessment of the site concentration by Moczydlowski and Fortes, who used TNP–ATP binding or ouabain binding to estimate the number of sites, and quantitative amino-acid analysis to estimate the protein, suggested a minimum molecular weight of 175,000 for enzyme from eel electric organ. A reassessment by Peters *et al.*, using phosphorylation to estimate the number of sites and quantitative amino-acid analysis to estimate the protein, suggested a minimum molecular weight of 179,000 for enzyme from rabbit kidney. Since even the best preparations probably contain small amounts of contaminant protein, or of denatured enzyme protein that may be incapable of being phosphorylated or of binding ligands, these figures are likely to be slightly too high. In any event, the notion that $Na^+,K^+$-ATPase behaves strictly as a "half-of-the-sites" enzyme must be given up. The results do not, of course, exclude the possibility that the enzyme is an $\alpha_2\beta_2$-dimer, or that interactions between the two halves of the dimer may be important, or even that the enzyme may display "half-of-the-sites" reactivity under other conditions.

### 5. *Crosslinking Studies*

We have seen that, although evidence from ultracentrifugation, radiation inactivation, neutron scattering, and electron microscopy points to a dimeric structure, none of this evidence is conclusive. A number of workers have attempted to get direct evidence for oligomeric structure from crosslinking studies. If the $Na^+,K^+$-ATPase is an $\alpha_2\beta_2$-dimer, it should be possible to demonstrate binding between the two $\alpha$-chains using an appropriate agent, such as cupric phenanthroline. Crosslinking to form $\alpha,\alpha$-dimers has in fact been achieved (Kyte, 1975; Giotta, 1976; Liang and Winter, 1977; Askari *et al.*, 1980; Huang and Askari, 1981) but, in view of the close packing of $Na^+,K^+$-ATPase molecules in the membrane, it is impossible, in most of the published experiments, to be certain that the crosslinking does not merely represent the binding of $\alpha$-chains from neighboring $\alpha,\beta$-molecules. The findings of Askari and Huang (1980) are, however, difficult to explain in this way. They chose conditions (0.25 mM $Cu^{2+}$, 1.25 mM *o*-phenanthroline) in which formation of crosslinked dimers was ATP dependent and stimulated by $Na^+$. Under these conditions, phosphorylation reached a steady level within 10 sec whereas the content of crosslinked $\alpha,\alpha$-dimer rose more slowly and reached a steady level within 1 min. SDS gel electrophoresis of the products showed that, when the reaction was complete, only 50% of the $\alpha$-chains had formed $\alpha,\alpha$-dimers. The implication would seem to be *either*, as Askari and Huang suggest, that the enzyme is a tetramer in which only two of the $\alpha$-chains can crosslink with each other, *or* that the enzyme is a dimer in which only one $\alpha$-chain at a time can crosslink with a near neighbor. In conditions in which $\alpha,\alpha$-crosslinkage was able to occur in the absence of phosphorylation, Askari and Huang found that all of the chains appeared to be able to crosslink to form $\alpha,\alpha$-dimers.

## E. Is the Dimeric Nature of the $Na^+$,$K^+$-ATPase Relevant to Its Mechanism?

In the mechanism shown in Figure 1, and in most of the discussion of mechanism so far, we have ignored the possibly dimeric nature of the enzyme. We must now repair that omission.

The earlier work on phosphorylation and ligand binding, together with the estimates of molecular weight obtained from ultracentrifugation and radiation inactivation, led to a number of hypotheses of which the central feature was the assumption that the $Na^+$,$K^+$-ATPase was a dimer with the two halves working out of phase in a "flip-flop" fashion (see Stein *et al.*, 1973; Repke and Schon, 1973; Repke and Dittrich, 1979). Hypotheses of this kind also appeared to provide a solution to the paradox that studies of the chemical changes involved in the $Na^+$,$K^+$-ATPase cycle pointed to a "ping-pong" mechanism, whereas studies of fluxes pointed to a "sequential" mechanism, both terms being used in the Cleland (1970) sense (see Hoffman and Tosteson, 1971; Garay and Garrahan, 1973). The realization that the phosphorylation and ligand-binding data do not reveal "half-of-the-sites" behavior, and that the existence of uncoupled efflux provides an alternative solution to the kinetic paradox (Sachs, 1979), removes two of the props of the "flip-flop" hypothesis, and makes it necessary to consider whether the dimeric (or oligomeric) nature of the enzyme has any role in its function. It is not possible to answer that question with any certainty, but there is evidence that the protomers can interact functionally. This evidence is of two kinds, which will be considered in turn.

### 1. Half-of-the-Sites Behavior

Although the experiments on phosphorylation and ligand binding that were originally thought to demonstrate "half-of-the-sites" behavior have had to be reassessed, more recent work in different conditions does appear to provide examples of such behavior.

1. Kudoh *et al.* (1979) showed that when ATP was added to pig kidney $Na^+$,$K^+$-ATPase in the presence of ouabain, there was rapid phosphorylation to the level observed in the absence of ouabain, followed by a slow fall to half that level. The slow fall was accompanied by the binding of ouabain, and in the final state the molar ratio of bound ouabain to phosphoenzyme was 2 : 1.

2. Askari and Huang (1981) and Askari *et al.* (1983) showed that when dog kidney $Na^+$,$K^+$-ATPase was phosphorylated by orthophosphate in the presence of $Mg^{2+}$ and ouabain, the molar ratio of bound ouabain to phosphoenzyme was 1 : 1. When $Ca^{2+}$ was substituted for $Mg^{2+}$, the binding capacity for ouabain was unaltered (though the affinity was much reduced) but the incorporation of $^{32}P$ was halved. Askari *et al.* suggest that $Ca^{2+}$ is able to replace $Mg^{2+}$ in only one of the protomers of a dimeric $Na^+$,$K^+$-ATPase. In further studies, Askari and Huang (1983) showed that the substitution of $Ca^{2+}$ for $Mg^{2+}$ halved the amount of $\alpha$,$\alpha$-cross linking that occurred when $Na^+$,$K^+$-ATPase was exposed to cupric phenanthroline in the presence of divalent cation, orthophosphate, and ouabain. With $Ca^{2+}$, only one-quarter of the $\alpha$-chains were involved in crosslinking (see Section VIII-D.4).

3. In a paper primarily concerned with monitoring changes between $E_1P$ and $E_2P$ using pig kidney $Na^+,K^+$-ATPase labeled with fluorescent sulfhydryl reagent, Taniguchi *et al.* (1982) describe an experiment in which the fluorescent enzyme was incubated with 20 μM [$\gamma^{32}P$]-ATP in a 16 mM $Na^+$ medium, containing also $Mg^{2+}$ and buffer. When the ATP was exhausted and the level of phosphoenzyme had fallen to zero, Taniguchi *et al.* added ouabain and observed that the level of phosphoenzyme rose again but reached only 50% of the level originally obtained with ATP. These results are obviously reminiscent of those described by Kudoh *et al.* (1979) and by Askari and Huang (1981), and suggest that phosphorylation by orthophosphate in the presence of $Mg^{2+}$, $Na^+$, ADP, and ouabain resembles phosphorylation by ATP in the presence of $Mg^{2+}$ and ouabain (or phosphorylation by orthophosphate in the presence of $Ca^{2+}$ and ouabain) in that only half of the phosphorylation sites are affected.

4. As already mentioned in Section VII-A.1, the addition of orthophosphate to pig kidney $Na^+,K^+$-ATPase containing occluded $Rb^+$ ions (in a Tris/$Mg^{2+}$ medium) led to the rapid loss of only half of the occluded $Rb^+$ (Glynn *et al.*, 1983b).

5. Observations on the phosphatase activity of thimerosal-treated enzyme, and of delipidated enzyme titrated with dioleoylphosphatidylcholine, provide further evidence for "half-of-the-sites" behavior, but are more conveniently considered later (see Section IX-D).

### 2. *The Number and Nature of the Nucleotide-Binding Sites*

The second kind of evidence for interaction between protomers in the dimeric (or oligomeric) enzyme comes from experiments to investigate the number and nature of the nucleotide-binding sites. From the evidence already discussed, it is clear that ATP has a dual role in $Na^+$–$K^+$ exchange, phosphorylating at a high-affinity site and accelerating the conformational change that releases $K^+$ ions to the interior at a low-affinity site. In this section, we consider the relationship between these sites, their number, whether they are interconvertible, whether they coexist, and, if they do coexist, whether they show cooperative behavior and whether they are on the same protomer. We also consider whether there are other nucleotide-binding sites.

The work referred to in Section VI-C, and also more recent work, using a more sensitive filtration method (Yamaguchi and Tonomura, 1980b; Schuurmans Stekhoven *et al.*, 1981) shows that, in media lacking $K^+$, the $Na^+,K^+$-ATPase has one high-affinity ATP-binding site per α-chain. The $K_d$ values vary slightly in different preparations, but are never far from the figure of 0.1–0.2 μM originally reported by Hegyvary and Post (1971) and Nørby and Jensen (1971) from measurements using flow dialysis. Similar estimates of the affinity of the high-affinity ATP-binding site have come from experiments in which ATP was made to compete with the fluorescent ATP analogs, formycin triphosphate (Karlish *et al.*, 1978a) or TNP–ATP (Moczydlowski and Fortes, 1981a).

Because of hydrolysis, the effect of $Mg^{2+}$ on ATP binding cannot be tested directly, but its effects on the binding of formycin diphosphate (Karlish *et al.*, 1978a),

AMPP(NH)P (Robinson and Flashner, 1979; Schuurmans Stekhoven *et al.*, 1981), and ADP (Nørby, 1983) show that it causes only a small reduction in the affinity of the high-affinity binding site. (In some of the experiments, low concentrations of $Mg^{2+}$ even caused a slight increase in the affinity.)

In $Na^+$-free media, $K^+$ ions at low concentrations reduce the affinity for ATP (Hegyvary and Post, 1971; Nørby and Jensen, 1971, 1974), for ADP (Kaniike *et al.*, 1973; Jensen and Ottolenghi, 1976) and for the formycin nucleotides (Karlish *et al.*, 1978a,b), presumably by converting the enzyme into the $E_2$ form. It is important to realize that the apparent binding affinity for ATP in the presence of a fixed concentration of $K^+$ is greater than the true affinity for ATP of the $E_2$ form of the enzyme; at low-$K^+$ concentrations, the discrepancy can be very large (see Beaugé and Glynn, 1980). The binding capacity for nucleotide is now thought not to be altered by $K^+$. $Mg^{2+}$ and $K^+$ together cause a much greater reduction in nucleotide-binding affinity than $K^+$ alone (Kaniike *et al.*, 1973; Moczydlowski and Fortes, 1981a).

A feature of ATP binding in the presence of $K^+$ ions is that the Scatchard plots are curved (concave upwards). This implies *either* that independent ATP-binding units are not homogeneous, *or* that there is negative cooperativity between two (or more) identical ATP-binding sites in each molecule.* Ottolenghi and Jensen (1983) excluded the first possibility by an ingenious experiment in which they compared the ATP-binding behavior of (1) normal enzyme, (2) enzyme in which all of the ouabain-binding sites had been filled with ouabain, and (3) enzyme in which only 81% of the ouabain-binding sites had been filled *after treatment with ouabain under conditions that gave a linear Scatchard plot for ouabain binding, indicating that all ouabain-binding sites were equally likely to bind ouabain*. If each α,β-protomer behaved independently, the ATP-binding behavior of the preparation in which 81% of the ouabain-binding sites had been filled would be predictable from the behavior of the normal enzyme and of the 100% ouabain-bound enzyme. In the event, the observed behavior deviated markedly from the predicted behavior. It follows that the ATP-binding sites are not independent but show negative cooperativity. Since the Scatchard plots of ATP binding in the absence of $K^+$ are straight lines, it also follows that the nucleotide-binding sites in the two protomers do not interact unless $K^+$ is present.

Interaction between the α,β-protomers can, nevertheless, occur in the absence of $K^+$, since Ottolenghi and Jensen (1983), confirming Hansen (1976), also showed that in $K^+$-free, $Na^+$ media, Scatchard plots for ouabain binding were curved, as if the ouabain-binding sites on the two protomers showed negative cooperativity. When the enzyme was treated with different concentrations of ouabain, however, the degree of inhibition of both $Na^+$,$K^+$-ATPase activity and *p*-nitrophenylphosphatase activity was always strictly proportional to the fraction of ouabain-binding sites that were filled,

* It might be thought that an enzyme which existed in two alternative forms, with high- and low-affinity substrate-binding sites respectively, could be regarded as an extreme example of negative cooperativity, and would therefore give a curved Scatchard plot. In fact, it is easy to show that such an enzyme gives a straight-line Scatchard plot, the apparent dissociation constant being a function of the true dissociation constants of the two forms and of the equilibrium constant for their interconversion.

irrespective of whether the treatment with ouabain took place in conditions in which filling was random or nonrandom. This implies that, despite the two kinds of interaction discussed above, each α,β-protomer seems to contribute the same amount of catalytic activity whether or not its partner is binding ouabain. If we accept Ottolenghi and Jensen's interpretation of their findings, we therefore reach the surprising but not impossible conclusion that, though there is evidence of cooperativity (under certain conditions) between two or more ATP-binding sites and (under other conditions) between two or more ouabain-binding sites, when the enzyme is catalyzing the hydrolysis of ATP in the presence of $Na^+$ and $K^+$ (or *p*-nitrophenyl phosphate in the presence of $K^+$) the two α,β-protomers appear to work independently.

If the analysis by Ottolenghi and Jensen (1983) is correct, the enzyme solubilized and dissociated into protomers by the procedure of Brotherus *et al.* (1981b) should not show a curved Scatchard plot for ATP binding in the presence of $K^+$, since interaction between binding sites could not occur. The experiment was tried by Jensen and Ottolenghi (1983), using $K^+$ concentrations up to 120 mM, and straight-line Scatchard plots were obtained. Interestingly, the effect of $K^+$ on the binding affinity for ATP was greatly reduced in this preparation, $K_d$ being 1.2 μM when $[K^+]$ was 120 mM. This reduction may reflect a shift in the poise of the $E_1 \rightleftharpoons E_2$ equilibrium. Unfortunately, for technical reasons, it is difficult to apply the usual criteria for investigating the conformational state of the solubilized enzyme.

### 3. *The Interconversion of High- and Low-Affinity ATP-Binding Sites*

A crucial question to which there is at present no unequivocal answer is: does the high-affinity site at which ATP phosphorylates become the low-affinity site at which ATP accelerates the conformational change, or can the two sites coexist in the same α,β-unit?

Moczydlowski and Fortes (1981a,b) investigated this problem using fluorescent triphenyl derivatives of adenine nucleotides. The compound TNP–ATP is not hydrolyzed by $Na^+,K^+$-ATPase but it binds to the enzyme with a high affinity, and the binding can be measured either by observing the fluorescence enhancement that accompanies it or by the use of $[^3H]$-TNP–ATP. In Tris media or $Na^+$ media, displacement of bound TNP–ATP by ATP led to estimates of about 1 μM (Tris medium) or 3 μM ($Na^+$ medium) for the enzyme–ATP dissociation constant, as if competition was occurring at the high-affinity site. In $K^+$ media, or in ($K^+$ + $Mg^{2+}$) media, displacement of TNP–ATP required higher concentrations of ATP ($K_d$ = 15–19 μM in a 20-mM $K^+$ medium; 70–120 μM in a 20-mM $K^+$, 4-mM $Mg^{2+}$ medium), as if competition was occurring at the low-affinity site. (All of these displacement experiments were done at 3°C.)

Independent evidence that TNP–ATP competes at both high- and low-affinity nucleotide-binding sites came from a study of its inhibitory effects. $Na^+$-ATPase activity was inhibited competitively; $Na^+,K^+$-ATPase activity was inhibited competitively if the ATP concentration was high, but noncompetitively (or uncompetitively) if the ATP concentration was low. The noncompetitive inhibition at low-ATP concentrations presumably reflects the inability of ATP in low concentrations to compete with the much more strongly-bound TNP–ATP at the low-affinity site. Competitive inhibition of *p*-nitrophenylphosphatase activity and inhibition of phosphorylation by orthophosphate in a high-$K^+$ medium, also suggest that TNP–ATP is able to compete at the low-affinity ATP site.

If TNP–ATP competes with a high affinity at both high- and low-affinity ATP sites, it ought to be easy to detect double binding if the two sites coexist. Since no more than one molecule of bound TNP–ATP per ouabain-binding site was ever detected, Moczydlowski and Fortes argue that the two sites cannot coexist in an $\alpha,\beta$-unit, and that the high-affinity site must become the low-affinity site in the second half of the cycle. A less likely possibility is that the two sites do coexist but that they cannot both bind TNP–ATP at the same time.

Although the findings of Moczydlowski and Fortes can all be accounted for in terms of the behavior of a single unit, they do not exclude the possibility that the enzyme is an $\alpha_2\beta_2$-dimer, or that, as suggested by the experiments of Ottolenghi and Jensen (1983), there may be interactions (between the two $\alpha,\beta$-protomers) which affect the ATP-binding sites. It seems to be necessary to postulate an interaction of this kind, too, to account for the results of some experiments of Koepsell *et al.* (1982) using the ATP analog $Nbs^6ITP$. At pH 7.4, this analog binds to the $Na^+$,$K^+$-ATPase rapidly and reversibly, and is hydrolyzed slowly. At pH 8.5, binding is followed by covalent-bond formation, probably in the neighborhood of the low-affinity site, and this produces irreversible inhibition. Under certain conditions (absence of $Na^+$ and $K^+$, and presence of 2 mM EDTA and pH 8.5), the rate of onset of irreversible inhibition was found to be *increased* by the presence of ATP at low or moderate concentrations. This implies that the enzyme must be able to combine with ATP and $Nbs^6ITP$ at the same time; it must, therefore, possess two (or more) ATP-binding sites. If, as seems likely, these sites are on different $\alpha$-chains it would follow that the binding of ATP to one chain alters the reactivity of $Nbs^6ITP$ bound to the other chain.

It is less easy to explain another observation of Koepsell *et al.*, (1982).They found that when $Na^+$,$K^+$-ATPase was treated with $Nbs^6ITP$, at pH 8.5, until 57% of the activity had been lost, and then washed and tested for $Na^+$,$K^+$-ATPase activity at different ATP concentrations, the Lineweaver–Burk plot was linear and the affinity for ATP was low. A possible explanation is that, when one half of the dimer has reacted with $Nbs^6ITP$, the other half turns over only very slowly unless ATP binds to the low-affinity site in the second half of the cycle.

Although the experiments of Moczydlowski and Fortes with TNP–ATP suggest that the high-affinity and the low-affinity nucleotide-binding sites cannot coexist in the same $\alpha,\beta$-unit, there are two recent reports suggesting that, under certain conditions, such coexistence can occur (see also the experiments on eosin binding discussed in Section VI-E.2.c).

The first is by Yamaguchi and Tonomura (1980b), who used a filtration method to measure ATP binding to a preparation of pig kidney $Na^+$,$K^+$-ATPase in a choline–imidazole–EDTA–glucose medium. They found that, in addition to one high-affinity site per $\alpha$-chain, they could detect at least two low-affinity sites. The low-affinity binding was not saturated at 1 mM ATP, but it cannot be dismissed as "unspecific" since it was prevented by 15 μM KC1, and the bound ATP could be displaced completely by AMPP(NH)P.

The second report was by Schuurmans Stekhoven *et al.* (1981), who used a filtration method and AMPP(NH)P to study the effects of $Mg^{2+}$ ions on nucleotide binding. Working with a rabbit kidney $Na^+,K^+$-ATPase in a sucrose–imidazole medium, they found that if $Mg^{2+}$ ions were absent, they could detect one high-affinity binding site ($K_d$ = 3.4 μM) per α-chain, and no low-affinity binding (but see Yamaguchi and Tonomura, 1980b). The addition of $Mg^{2+}$ ions led to curvature of the Scatchard plots, which was attributed to the appearance of a second binding site of low affinity ($K_d$ = c. 150 μM) on each unit. A curious feature of these experiments, however, is that the capacity of the low-affinity sites, estimated by extrapolating the binding curves to infinite nucleotide concentration, increased (to a maximum of about one per unit) as the $Mg^{2+}$ ion concentration was increased from 0.5 mM to 4mM. As Nørby (1983) has pointed out, this cannot readily be explained in terms of equilibrium binding.

At present, there seems to be no obvious way of reconciling these different findings, nor is it possible to say what the relations are between the $Mg^{2+}$-dependent low-affinity sites described by Schuurmans Stekhoven *et al.* (1981), the low-affinity sites seen in the absence of $Mg^{2+}$ by Yamaguchi and Tonomura (1980b), the sites at which ATP competes with a low affinity to displace TNP–ATP (Moczydlowski and Fortes, 1981a,b) and the low-affinity sites at which ATP, in the presence or absence of $Mg^{2+}$, accelerates the conformational change that releases occluded $K^+$.

## F. The Role of Lipids

An aspect of the $Na^+,K^+$-ATPase that we have not yet discussed is the role of the lipids in which the protein portion of the enzyme is embedded. There is a large literature on the lipid requirements of the $Na^+,K^+$-ATPase, and only a brief summary will be given here (for reviews of the subject and for references, see Jørgensen 1975b, 1982; Schuurmans Stekhoven and Bonting, 1981).

Pure membrane-bound preparations of $Na^+,K^+$-ATPase contain many molecules of cholesterol, and of both neutral and acidic phospholipids, for each molecule of $Na^+,K^+$-ATPase; but it is clear that much of this lipid is unnecessary for the enzyme's function. To investigate the lipid requirements of the enzyme, and the role of the lipids in enzymic activity, four main approaches have been used. The oldest and most straightforward approach is to delipidate the enzyme with detergents or organic solvents and then to see which lipids are able to restore activity. A drawback of this method is that delipidation may damage the protein; one may, therefore, be investigating the lipid requirements for recovery rather than for activity of the undamaged enzyme. The second approach is to treat $Na^+,K^+$-ATPase preparations with enzymes that attack, specifically, different classes of lipids, and then to examine the remaining activity. A drawback of this method is that one cannot be sure that all of the susceptible lipid is, in fact, accessible to the attacking enzyme. The third approach is to use spin-labeled lipids, or other physical methods (see Brotherus *et al.*, 1981a) to distinguish between lipids in the bilayer and lipids at the protein interface. The fourth approach is to vary the fluidity or thickness of the membrane by changes in temperature or lipid composition, and to see how these changes affect the $Na^+$,$K^+$-ATPase.

Application of these methods has led to the following general conclusions:

1. The lipid requirements are most stringent for ($Na^+$ + $K^+$)-dependent ATPase activity and less stringent for the various partial reactions; indeed, ATP can be bound with a high affinity even by completely delipidated enzyme.
2. Activation of delipidated enzyme by lipid shows positive cooperativity.
3. There seems to be no absolute requirement for any particular phospholipid. For maximal activity, however, negatively-charged phospholipids seem to be necessary, and the use of spin-labeled phospholipids shows that the protein possesses sites which bind negatively-charged species preferentially. There is some recent evidence suggesting a specific role for sulfatides (Hansson *et al.*, 1978; Zambrano *et al.*, 1981; Gonzalez and Zambrano, 1983).
4. The fluidity and thickness of the lipid bilayer are important in determining enzymic activity (see Johannsson *et al.*, 1981).

## IX. INHIBITORS

A general review of inhibitors of $Na^+$,$K^+$-ATPase has been published by Schwartz *et al.* (1975). In this section, we consider only cardiac glycosides, vanadate, oligomycin, and thimerosal. Information about a number of other inhibitors (arsenite, N-ethylmaleimide, butanedione, NBD-C1, N-acetylimidazole, showdomycin, quercetin, fluorescein isothiocyanate, eosin, and various ATP analogs) is given in earlier sections.

### A. Cardiac Glycosides

The cardiac glycosides, of which ouabain is the most widely used because it is the most water soluble, seem to be specific inhibitors of $Na^+$,$K^+$-ATPase. They are, therefore, invaluable for investigating the mechanism and distribution of the enzyme, and work with them has given rise to a vast literature. This section will contain only a brief review; further information, including references to earlier reviews, may be found in the article by Akera (1981).

Under appropriate conditions, the cardiac glycosides inhibit: (1) all six flux modes (Schatzmann, 1953; Matchett and Johnson, 1954; Glynn, 1957; Caldwell and Keynes, 1959; Garrahan and Glynn, 1967a,c,e; Baker *et al.*, 1969b; Glynn *et al.*, 1970; Beaugé and Ortiz, 1973; Lew *et al.*, 1973; Lee and Blostein, 1980), (2) ($Na^+$ + $K^+$)-dependent and $Na^+$-dependent ATP hydrolysis (Post *et al.*, 1960; Skou, 1960; Dunham and Glynn, 1961, (3) ATP–ADP exchange (Fahn *et al.*, 1966b), (4) $^{18}O$ exchange between orthophosphate and water (Dahms and Boyer, 1973), (5) *p*-nitrophenylphosphatase activity (Judah *et al.*, 1962), (6) the binding of ATP (Hansen *et al.*, 1971; Hegyvary and Post, 1971), (7) the equilibrium binding of $Na^+$ and $K^+$ ions (Matsui and Homareda, 1982), and (8) the occlusion of $Rb^+$ ions (Glynn and Richards, 1983).

The sensitivity of the different processes to cardiac glycosides appears to differ, but this is almost certainly because, at low glycoside concentrations, the rate of binding is slow and very sensitive to the incubation conditions (see below).

Because ouabain sensitivity tends to be used as a criterion that the $Na^+,K^+$-ATPase is involved, there is some danger, particularly in work on intact cells, that an activity of the $Na^+,K^+$-ATPase not inhibited by ouabain would not be recognized. The possibility that some of the ouabain-insensitive fluxes observed in intact cells may involve the $Na^+,K^+$-ATPase has been discussed by Glynn and Karlish (1975) and by Lew and Beaugé (1979).

### 1. Requirements for Inhibition

Cardiac glycosides act on the $Na^+,K^+$-ATPase only from the extracellular surface (Caldwell and Keynes, 1959; Hoffman, 1966; Hilden and Hokin, 1975), and inhibition occurs when one molecule is bound per $\alpha,\beta$-protomer (see Ottolenghi and Jensen, 1983). Little is known of the receptor site, but experiments using glycosides modified so that they can form covalent bonds show that the bound glycoside molecule must be close to the outer faces of the $\alpha$-, $\beta$-, and $\gamma$-subunits (Ruoho and Kyte, 1974; Hegyvary, 1975; Forbush *et al.*, 1978; Rogers and Lazdunski, 1979a; Hall and Ruoho, 1980; Rossi *et al.*, 1980, 1982).

Enzyme–glycoside interaction is described by a single reversible equilibrium (Baker and Willis, 1970; Hansen, 1971; but cf. Taniguchi and Iida, 1972). The association rate is first order with respect to both enzyme and glycoside concentrations, dissociation follows a monoexponential course, and the measured equilibrium constant is not significantly different from the quotient of dissociation and association rate constants (Baker and Willis, 1970; Barnett, 1970; Tobin and Sen, 1970; Hansen, 1971; Akera and Brody, 1971; Tobin and Brody, 1972; Tobin *et al.*, 1972; Schönfeld *et al.*, 1972; Erdmann and Schoner, 1973; Lindenmayer and Schwartz, 1973; Wallick and Schwartz, 1974).

In all circumstances, the rate of onset of inhibition is limited by the rate of binding, and this varies greatly depending on the ligands present. Combinations of ligands most effective at promoting binding are (1) ATP + $Mg^{2+}$ + $Na^+$, and (2) orthophosphate + $Mg^{2+}$ (Matsui and Schwartz, 1968; Schwartz *et al.*, 1968; Albers *et al.*, 1968; Lindenmayer *et al.*, 1968). Both combinations are thought to act by generating $E_2P$, which combines with the glycoside to form a noncovalent complex that is resistant to dephosphorylation by $K^+$ ions, but which, in certain circumstances, can be hydrolyzed slowly leaving the glycoside bound to the dephosphoenzyme (see Sen *et al.*, 1969). $E_1P$ combines poorly, if at all, with cardiac glycosides (Sen *et al.*, 1969; Yoda and Yoda, 1982b).

Although there is no doubt that phosphorylation promotes ouabain binding (Forbush and Hoffman, 1979), binding of cardiac glycosides also occurs, at a slow rate, in the absence of phosphorylation, provided that $Mg^{2+}$ ions are present and that $Na^+$ and $K^+$ ions are either absent, or present only in low concentrations (Schwartz *et al.*, 1968; Sen *et al.*, 1969).

The effects of $Na^+$ and $K^+$ on ouabain binding are complicated (see reviews by Akera (1981); Schuurmans Stekhoven and Bonting, 1981) even when the effects on the two sides of the membrane are distinguished (Bodemann and Hoffman, 1976a,b). $Na^+$ ions probably prevent ouabain binding in the presence of $Mg^{2+}$, or of $Mg^{2+}$ + orthophosphate, by holding the enzyme in the $E_1$ form. The protective effect of extracellular $K^+$ ions on inhibition by low concentrations of glycoside (Glynn, 1957) is probably at least partly the result of hydrolysis of the phosphoenzyme, catalyzed by the extracellular $K^+$ ions.

Cardiac glycosides differ widely in their binding affinity and hence in their potency (Repke and Portius, 1966). Studies of both binding affinity and inhibitory activity, in relation to structure, show that certain features of the molecule are critical, namely: (1) the unsaturated lactone ring (which may be five- or six-membered) attached in the correct configuration at $C_{17}$ (Glynn, 1957; Dunham and Glynn, 1961), (2) the *cis* configuration of the AB and CD ring junctions in the steroid nucleus (Erdmann and Schoner, 1974), (3) the presence of a hydroxyl group at $C_{14}$ (Repke, 1965; Wilson *et al.*, 1970; Naidoo *et al.*, 1974), and (4) the presence of an appropriate sugar at $C_3$ (Erdmann and Schoner, 1974). Repke and Portius (1966) argue that an important feature of the lactone ring is the carbonyl group conjugated with a double bond, since the erythrophleum alkaloid cassaine, which includes that structure but lacks a lactone ring, has a digitalis-like effect. Although aglycones are generally less convenient to use than glycosides because they are less water-soluble, *K*-strophanthidin is sometimes used because it dissociates from its complex with the enzyme fast enough for the inhibition to be readily reversible.

With a particular cardiac glycoside and fixed conditions, the degree of inhibition of $Na^+$,$K^+$-ATPase depends on the source of the enzyme. The variation derives mainly from differences in the dissociation rate of the enzyme-inhibitor complex (Tobin and Brody, 1972; Tobin *et al.*, 1972; Erdmann and Schoner, 1973). $Na^+$,$K^+$-ATPase from rat heart has a particularly low affinity, but this cannot be a general feature of rat $Na^+$,$K^+$-ATPase because the brain enzyme is highly sensitive to cardiac glycosides (Tobin and Brody, 1972; Lin and Akera, 1978).

## 2. *Mechanism of Inhibition*

In circumstances in which phosphorylation can occur, the cardiac glycosides inhibit by rendering the phosphoenzyme insensitive both to $K^+$ ions and to ADP. Karlish (1980), using fluorescein-labeled enzyme in the presence of $Mg^{2+}$ ions, found that either ouabain or orthophosphate in optimal concentrations (or a mixture of both in suboptimal concentrations) could quench the fluorescence to about the same extent as $K^+$ ions, as if, so far as the fluorescein was concerned, $MgE_2(K)$, $MgE_2P$, $MgE_2$ ouabain, and $MgE_2P$ ouabain all had the same conformation. The experiments of Castro and Farley (1979), who used tryptic digestion to investigate the conformation of the enzyme in the presence of ouabain, are also compatible with the hypothesis that the phosphoenzyme–ouabain complex is similar in conformation to the $E_2(K)$ form. Hegyvary and Jørgensen (1981), however, using fluorescein-labeled enzyme, found that $MgE_2P$ ouabain had a significantly lower fluorescence than $E_2(K)$, $MgE_2(K)$, or $MgE_2P$, as if it had a different conformation producing a different environment for the fluorescein.

In the absence of phosphorylation, cardiac glycosides stabilize the enzyme in a form which, judging from Karlish's (1980) experiments with fluorescein-labeled enzyme, resembles $E_2(K)$ but which, judging from the $Rb^+$ occlusion experiments of Glynn and Richards (1983) is not able to occlude $K^+$ ions. It is not known whether this is because the $K^+$-binding sites are freely accessible, or because they are occluded but empty.

### 3. The Cardiotonic Effect

A detailed consideration of the mechanism by which cardiac glycosides increase the force of contraction of cardiac muscle is beyond the scope of this chapter. There seems to be little doubt that inhibition of $Na^+,K^+$-ATPase in the cardiac muscle membrane can strengthen contraction, but there is no general agreement that that is the only or even the main mechanism (see Noble, 1980). It is tempting to suppose that inhibition of the $Na^+,K^+$-ATPase increases the force of contraction by raising the intracellular $Na^+$ concentration, thus increasing $Ca^{2+}$ entry through the $Na^+$–$Ca^{2+}$ exchange mechanism (Glynn, 1969; Baker *et al.*, 1969a), but even that is uncertain.

### 4. Endogenous Cardiac-Glycoside-Like Activity

The specificity and potency of the effects of cardiac glycosides on $Na^+,K^+$-ATPase, the existence in toad skin of substances closely related to the cardiac glycosides, the accumulated, if inconclusive, evidence for a "natriuretic" hormone, and a supposed analogy with morphine and the endorphins have led a number of investigators to look for endogenous cardiac-glycoside-like principles in extracts of animal tissues. Although a number of claims have been made (Fishman, 1979; Haupert and Sancho, 1979; Godfraind and Hernandez, 1981; Gruber, *et al.*, 1983; Haupert, 1983; Whitmer *et al.*, 1983; Chipperfield, 1983), no endogenous principle has yet been identified. One difficulty is that the extraction procedures may themselves produce or concentrate substances which inhibit $Na^+,K^+$-ATPase but which have no physiological role (see Schwartz *et al.*, 1982; Kracke, 1983). The recent report by Hamlyn *et al.* (1982; see also MacGregor *et al.*, 1981; Poston *et al.*, 1981) that an inhibitor of $Na^+,K^+$-ATPase may be detected in the plasma of subjects with essential hypertension, and to a slight extent in normal subjects, is obviously of great interest, though whether the inhibitor is either the cause or the result of the high blood pressure remains to be determined (see Glynn and Rink, 1982).

## B. Vanadate

In 1977, Cantley *et al.* identified as orthovanadate the $Na^+,K^+$-ATPase inhibitor found in certain commercial preparations of ATP. Vanadate is thought to inhibit by combining at the orthophosphate-discharge site and stabilizing the enzyme in the $E_2$ form (see below). It acts intracellularly (Cantley *et al.*, 1978c; Beaugé and DiPolo, 1979a), provided that $Mg^{2+}$ (or $Mn^{2+}$) ions are also present at the intracellular face of the enzyme (Cantley *et al.*, 1978a; Beaugé *et al.*, 1980).

Under optimal conditions (high $Mg^{2+}$ and $K^+$, no ATP or $Na^+$), one molecule of vanadate is bound with a high affinity ($K_d$ = c.4 μM) per ouabain-binding site. A second molecule can bind with a lower affinity (Cantley *et al.*, 1978a), but the binding of a single molecule is sufficient to inhibit $Na^+$,$K^+$-ATPase activity. The high-affinity binding is accompanied by the trapping of one $Mg^{2+}$ ion per molecule of bound vanadate, and the vanadate-bound enzyme is unable to bind ATP (Smith *et al.*, 1980). Both binding and release of vanadate are slow, the time constant for release being several minutes at 37°C and many minutes at 0°C. The rate of release is unaffected by ATP. It is slightly increased by $K^+$ ions, which act with a high affinity ($K_{0.5}$ = 0.5 mM in a choline-magnesium medium), and is greatly increased by $Na^+$ ions, which act with a low affinity ($K_{0.5}$ = 250 mM in a choline–magnesium medium) (Cantley *et al.*, 1978a). The binding affinity is increased by $Mg^{2+}$ ions, $K^+$ ions, ouabain, and dimethylsulfoxide, and is decreased by $Na^+$ ions, ATP, AMPP(NH)P, *p*-nitrophenylphosphate, orthophosphate, and oligomycin (Cantley *et al.*, 1978a; Bond and Hudgins, 1981; Robinson and Mercer, 1981). Some of these effects can be explained by the known effects of the ligands on the poise of the $E_1 \rightleftharpoons E_2$ equilibrium, but $Mg^{2+}$ ions must have a specific role, and orthophosphate and *p*-nitrophenylphosphate probably both compete directly with vanadate. The action of $K^+$ ions will be discussed later.

Following the interpretation of the effects of vanadate on acid and alkaline phosphatases (see Lopez *et al.*, 1976), Cantley *et al.* (1978a) suggested that vanadate binds to form a trigonal bipyramidal structure analogous to the transition state that is thought to exist transiently during the hydrolysis of the phosphoenzyme. This implies that vanadate binds to the site that normally releases orthophosphate, a conclusion that is compatible with the intracellular location of the site and with the competitive effects of orthophosphate, *p*-nitrophenylphosphate, and ATP.

Since orthophosphate is normally released from $E_2P$, vanadate should stabilize $E_2$. Evidence that it does is provided by (1) studies of the effects of trypsin (Cantley *et al.*, 1978a), (2) measurements of the fluorescence of enzyme labeled with fluorescein isothiocyanate (Karlish *et al.*, 1979), (3) measurements of $Rb^+$ occlusion (Glynn and Richards, 1982), (4) studies of $Na^+$ and $K^+$ or $Rb^+$ fluxes in squid axons or red cells (Beaugé and DiPolo, 1979a; Beaugé, 1979; Beaugé *et al.*, 1980), and (5) phosphorylation studies showing that vanadate inhibits the enzyme by forming a complex that cannot be phosphorylated by ATP (Hobbs *et al.*, 1983). The stabilization is partly the result of ATP being kept off the low-affinity site at which it accelerates the conformational change $E_2(K) \rightarrow E_1K$ and partly an effect of vanadate on the equilibrium in the absence of nucleotide.

This effect of vanadate on the equilibrium in the absence of nucleotide is not fully understood. From their measurements of the rates of release of occluded $Rb^+$ in the absence of nucleotides, Glynn and Richards (1982) argued that vanadate could not much reduce the rate of conversion of $E_2(Rb)$ to $E_1Rb$, but they ignored the possibility that the rate of conversion might be greatly reduced but that the vanadate, acting like a feeble phosphate, might allow a slow release of $Rb^+$ *directly* from $E_2(Rb)$. There is, therefore, no evidence to decide which rate constant is changed when vanadate changes the poise of the $E_1 \rightleftharpoons E_2$ equilibrium in the absence of nucleotides.

$Na^+$,$K^+$-ATPase treated with trypsin in a $Na^+$ medium for a period long enough for the rapid phase of inhibition to be complete was found to be very much less sensitive to vanadate (Beaugé and Glynn, 1978; Beaugé, 1979). This effect is presumably related to the shift to the left in the poise of the $E_1 \rightleftharpoons E_2$ equilibrium that results from the splitting off of the small polypeptide from the amino-terminal end of the α-chain (Jørgensen and Karlish, 1980).

The effects of $K^+$ ions on inhibition by vanadate are complex. Intracellular $K^+$ ions promote vanadate binding (and the resulting inhibition) by displacing the $E_1 \rightleftharpoons E_2$ equilibrium to the right (Karlish *et al.*, 1979; Robinson and Mercer, 1981; Beaugé and Berberian, 1983). Whether extracellular $K^+$ ions are needed depends on the conditions. Vanadate, at low concentrations, inhibits $Na^+$–$K^+$ exchange, $K^+$–$K^+$ exchange, and uncoupled $Na^+$ efflux, but not $Na^+$–$Na^+$ exchange or pump reversal (Beaugé *et al.*, 1980). Hydrolysis of *p*-nitrophenylphosphate is also inhibited even when $K^+$ ions are present only at the intracellular surface (Robinson and Mercer, 1981; Beaugé and Berberian, 1983). It follows that inhibition by vanadate does not require extracellular $K^+$ ions when $Na^+$ ions are absent from the extracellular face of the enzyme. When the $Na^+$ concentration at the extracellular face is high, however, vanadate in low concentrations is able to inhibit only if the extracellular medium also contains $K^+$ ions, or ions of one of the congeners of $K^+$ (Grantham and Glynn, 1979; Beaugé and DiPolo, 1979a; Beaugé and Berberian, 1983). The order of effectiveness, $Rb^+ > K^+ > Cs^+ > NH_4^+ > Li^+$, is the same as the order of effectiveness in promoting hydrolysis. It is tempting to suppose that $K^+$ ions re required to displace $Na^+$ ions from external sites at which they accelerate vanadate release (perhaps by converting the $E_2$ form to a form analogous to $E_1P$). However, it is difficult to understand why the concentration of $K^+$ ions required should be higher than that necessary to saturate $K^+$ influx (see Beaugé *et al.*, 1980), unless the release of vanadate from the $Na^+$-loaded enzyme is so fast that $Na^+$ loading for only a very small fraction of the time is sufficient to lower the apparent affinity for vanadate.

The inhibition of $Na^+$–$Na^+$ exchange by vanadate at high concentrations probably involves the binding of vanadate at the second (low-affinity) site, but nothing is known of the mechanism of inhibition.

Because vanadium at low concentrations is widely distributed in animal tissues (see Post *et al.*, 1979), it has been suggested that intracellular vanadate might be involved in the physiological control of the sodium–potassium pump. The finding that intracellular vanadium exists predominantly not as vanadate but in a relatively ineffective tetravalent form makes that hypothesis less attractive (Cantley and Aisen, 1979; Grantham and Glynn, 1979; Macara *et al.*, 1980; Heinz *et al.*, 1982).

A useful review of the chemistry of vanadium with particular reference to its biochemical effects has been published by Rubinson (1981).

## C. Oligomycin

Oligomycin is a potent, though generally incomplete, inhibitor of $Na^+$–$K^+$ exchange, pump reversal, and $Na^+$–$Na^+$ exchange unaccompanied by ATP hydrolysis (Glynn, 1963; Garrahan and Glynn, 1967d,e). $K^+$–$K^+$ exchange is also inhibited, but only if $Na^+$ is present inside the cells (Sachs, 1980). In particulate preparations, $Na^+,K^+$-ATPase activity and $Na^+$-ATPase activity are inhibited (Järnefelt, 1962;

Glynn, 1963; Jöbsis and Vreman, 1963; Van Groningen and Slater, 1963; Whittam *et al.*, 1964; Inturrisi and Titus, 1968; Robinson, 1974), but phosphorylation by ATP and the exchange of $^{18}O$ between water and orthophosphate are unaffected (Whittam *et al.*, 1964; Blostein and Whittington, 1973; Dahms and Boyer, 1973). ATP–ADP exchange is either unaffected or stimulated (Fahn *et al.*, 1966a,b; Blostein, 1970). Phosphatase activity, in the absence of $Na^+$ and ATP, is not inhibited (Israel and Titus, 1967), but oligomycin prevents the stimulating effect of $Na^+$ + ATP (Garrahan *et al.*, 1970; Robinson, 1970, 1980; Askari and Koyal, 1971).

All of these observations can be explained by supposing that oligomycin blocks the conversion of $E_1P$ to $E_2P$.

Using Cleland's (1970) rules to interpret the results of a detailed study of the inhibition by oligomycin of the $Na^+$,$K^+$ pump in red cells, Sachs (1980) concluded that oligomycin must bind preferentially to the enzyme in the $E_1P$ form. Phosphorylation cannot be essential for oligomycin binding, however, nor can the effect of oligomycin be restricted to the blocking of the conversion of $E_1P$ to $E_2P$. Robinson (1982), using tryptic digestion, showed that oligomycin held the dephosphoenzyme in the $E_1$ form even in the presence of $K^+$; and Matsui and Homareda (1982) and Homareda and Matsui (1982) found that oligomycin caused a large increase in equilibrium $Na^+$ binding, and increased the inhibitory effect of $Na^+$ on equilibrium $K^+$ binding (see Section VI-F). Skou and Esmann (1983), using stopped-flow fluorimetry and eosin to monitor conformational changes, found that when $K^+$ was added to enzyme suspended in a medium containing $Mg^{2+}$ and $Na^+$, the presence of oligomycin reduced the rate of conversion of the enzyme to the $K^+$ form by three orders of magnitude (see Section VI-E.2.c; Skou, 1982). It may be this last effect, rather than trapping of the enzyme in the $E_1P$ form, that is responsible for the inhibition by oligomycin of $K^+$–$K^+$ exchange when the cells contain $Na^+$.

Robinson (1980) found that pretreatment of $Na^+$,$K^+$-ATPase with nonionic detergents made the phosphatase activity sensitive to oligomycin even in the absence of $Na^+$ and ATP. He suggested that the detergents had this effect because they displace the $E_1K \rightleftharpoons E_2K$ equilibrium to the left.

There are two reports of activating effects of oligomycin. Askari and Koyal (1971), using a rat brain $Na^+$,$K^+$-ATPase, found that low (but not high) concentrations of oligomycin stimulated *p*-nitrophenylphosphatase activity in the presence of $Na^+$ and low concentrations of $K^+$. Blostein and Whittington (1973), using red-cell membranes from sheep with high- or low-$K^+$ red cells, found that oligomycin could stimulate $Na^+$-ATPase activity if either the ATP concentration or the $Na^+$ concentration were very low. In neither the phosphatase experiments nor the $Na^+$-ATPase experiments is the stimulation understood, though various possibilities are discussed by the respective authors.

## D. Thimerosal

Ethylmercurithiosalicylate (thimerosal) and certain short-chain alkylmercury compounds were first used to investigate $Na^+$,$K^+$-ATPase by Askari and his colleagues (see Askari *et al.*, 1979). They are interesting because it is possible that they produce their effects by abolishing subunit interactions.

Under appropriate conditions, thimerosal inhibits $Na^+$,$K^+$-ATPase activity, $Na^+$–$K^+$ exchange, $Na^+$–$Na^+$ exchange unaccompanied by ATP hydrolysis, and uncoupled $Na^+$ efflux; it has little effect on ATP–ADP exchange, $Na^+$-dependent phosphorylation, $K^+$-dependent dephosphorylation, or $K^+$-dependent phosphatase activity measured in high-$K^+$, $Na^+$-free media (Henderson and Askari, 1976, 1977; Henderson *et al.*, 1979; Askari *et al.*, 1979). All the effects of thimerosal can be reversed with dithiothreitol. Studies with [$^{203}$Hg]ethylmercury showed that the kind of selective inhibition described above occurs when about 14 of the 24 sulfhydryl groups capable of reacting with the mercurial have done so (Henderson and Askari, 1977). All of the 14 groups are on the α-subunit, and crosslinking studies suggest that the $\alpha_2\beta_2$-structure of the enzyme is not altered by the thimerosal treatment (Askari *et al.*, 1979).

Hansen *et al.* (1979) found that by working with higher concentrations of thimerosal, but at 0°C instead of 37°C, they could modify the enzyme so that $Na^+$,$K^+$-ATPase activity was more than 90% inhibited, while phosphatase activity was stimulated by up to 50% above control levels. The number of ouabain-binding sites was reduced by the mercurial, but the properties of the sites that did bind were normal.

Since both $Na^+$,$K^+$-ATPase activity and phosphatase activity could be completely inhibited by ouabain, it was possible to calculate the molar activity of the enzyme for each process. Remarkably, whereas the $Na^+$,$K^+$-ATPase activity per ouabain-binding site was found to be much lower than normal, the phosphatase activity (measured in a medium containing 150 mM $K^+$ and no $Na^+$) was twice normal; this doubling of molar activity occurred whether the conditions of thimerosal treatment led to an absolute increase or an absolute decrease in the phosphatase activity. Furthermore, the doubling of phosphatase activity was accompanied by a change in its properties. Though normal phosphatase activity is inhibited by $Na^+$ ions and, at low $K^+$ concentrations is stimulated by ATP + $Na^+$, the phosphatase activity of the thimerosal-treated enzyme was unaffected by $Na^+$ ions in low or moderate concentrations, whether or not ATP was present. Hansen *et al.* point out that, in its insensitivity to $Na^+$ and ATP, the phosphatase activity of thimerosal-treated enzyme resembles the phosphatase activity of delipidated enzyme treated with dioleoylphosphatidylcholine in amounts sufficient to saturate only 1% of available sites, so that the enzyme consists only of molecules lacking lipid, and molecules in which only one unit is relipidated (Ottolenghi, 1979). The phosphatase activity of normal enzyme, on the other hand, resembles that of fully relipidated enzyme.

Both the thimerosal experiments and the relipidation experiments suggest that (1) though both α,β-protomers possess $K^+$-dependent phosphatase activity, in the normal enzyme only one protomer can act at a time, and (2) the effects of ATP + $Na^+$ on phosphatase activity involve subunit interaction (Hansen *et al.*, 1979). The hypothesis that thimerosal abolishes subunit interaction had been proposed earlier to explain the loss of cooperativity observed in kinetic studies of the thimerosal-treated enzyme (Askari *et al.*, 1979; but cf. Mone and Kaplan, 1983). If thimerosal does act in this way, it is of particular interest that the thimerosal-treated enzyme is still able to occlude

$Rb^+$ ions (D. E. Richards and I. M. Glynn, unpublished results, discussed in Section VI-F).

Mone and Kaplan (1983) have shown that thimerosal-treated $Na^+$,$K^+$-ATPase from dog kidney outer medulla shows significant ATPase activity in the absence of alkali-metal ions. This activity has a low affinity for ATP ($K_{0.5} > 70$ μM) and is abolished by treatment with dithiothreitol. It is not known whether a similar activity exists in partially relipidated enzyme.

Robinson (1983) has shown that treatment of dog kidney $Na^+$,$K^+$-ATPase with thimerosal increases $Na^+$-ATPase activity in high-$Na^+$, low-$Mg^{2+}$ media, perhaps by increasing the efficacy of $Na^+$ ions in activating the hydrolysis of $E_2P$.

Yet another feature of thimerosal inhibition is that the properties of the thimerosal-treated enzyme differ depending on the $Na^+$ and $K^+$ concentrations of the medium in which the enzyme is treated (Jensen *et al.*, 1979). Enzyme that had been treated in a 230-mM $Na^+$, $K^+$-free medium possessed equal numbers of ouabain-binding sites and ATP-binding sites. Enzyme that had been treated in an 80-mM $Na^+$, 1-M $K^+$ medium possessed twice as many ouabain-binding sites as ATP-binding sites. Jensen *et al.* also showed that the residual $Na^+$,$K^+$-ATPase activity of thimerosal-treated enzyme is attributable to enzyme molecules with a low turnover and a moderately low affinity for ATP ($K_{0.5} = 10.4$ μM).

## X. CONCLUSION

The evidence that the $Na^+$,$K^+$-ATPase is identical with the $Na^+$,$K^+$ pump in cell membranes is overwhelming. We have seen that there is also strong evidence of various kinds that phosphorylation of the $Na^+$,$K^+$-ATPase by ATP, followed by a change in conformation, can lead to the outward movement of $Na^+$ ions, and that the hydrolysis of the phosphoenzyme, followed by a change in conformation, can lead to the inward movement of $K^+$ ions. It is economical to assume that in the normal running of the pump the two steps occur in succession, as in the scheme of Figure 1; but, though that scheme has been remarkably successful in explaining observations and making predictions, it also has serious weaknesses. In the first place, it is difficult to reconcile with some of the kinetic data. Secondly, it ignores the evidence that the pump is probably a dimer, and that, at least in some conditions, the two halves can influence each other's behavior. Thirdly, it lacks molecular detail. The first two weaknesses may be related, since it is possible that it is interaction between the two protomers that complicates the kinetics. Whether information about molecular detail will, ultimately, come from X-ray diffraction analysis of crystalline $Na^+$,$K^+$-ATPase, or from analysis of the nucleotide sequence of a messenger RNA, or in some other way, it would be rash to predict. What is clear now is the width of the gap between the functional studies and the structural studies of the enzyme, despite the progress in both that has taken place in the last few years. A textbook of zoology popular in England earlier this century included a chapter with the odd title: "The Functions and Their Organs" (Woodger, 1924). For the $Na^+$,$K^+$-ATPase molecule, that chapter has hardly begun to be written.

ACKNOWLEDGMENTS

I am grateful to Dr. Y. Hara and Dr. D. E. Richards for reading large parts of this chapter in manuscript and for many helpful comments.

## REFERENCES

Abercrombie, R. F., and De Weer, P., 1978, Electric current generated by squid giant axon sodium pump: External K and internal ADP effects. *Am. J. Physiol.* **235**(1):C63–C68.

Akera, T., 1981, Effects of cardiac glycosides on $Na^+,K^+$-ATPase, in: *Handbook of Experimental Pharmacology,* Vol. 56/I (K. Greeff, ed.), Springer-Verlag, Berlin, pp. 287–336.

Akera, T., and Brody, T. M., 1971, Membrane adenosine triphosphatase: The effect of potassium on the formation and dissociation of the ouabain–enzyme complex, *J. Pharmacol. Exp. Ther.* **176**:545–557.

Albers, R. W., Fahn, S., and Koval, G. J., 1963, The role of sodium ions in the activation of *Electrophorus* electric organ adenosine triphosphatase, *Proc. Natl. Acad. Sci. USA* **50**:474–481.

Albers, R. W., Koval, G. J., and Siegel, G. J., 1968, Studies on the interaction of ouabain and other cardio-active steroids with sodium–potassium-activated adenosine triphosphatase, *Mol. Pharmacol.* **4**:324–336.

Anner, B. M., Lane, L. K., Schwartz, A., and Pitts, B. J. R., 1977, A reconstituted $Na^+ + K^+$ pump in liposomes containing purified $(Na^+ + K^+)$-ATPase from kidney medulla, *Biochim. Biophys. Acta* **467**:340–345.

Askari, A., and Huang, W-H., 1980, $Na^+,K^+$-ATPase: Half of the subunits cross-linking reactivity suggests an oligomeric structure containing a minimum of four catalytic subunits, *Biochem. Biophys. Res. Commun.* **93**:448–453.

Askari, A., and Huang, W-H., 1981, $Na^+,K^+$-ATPase: $(Ca^{2+}$ + ouabain)-dependent phosphorylation by $P_i$, *FEBS Lett.* **126**:215–218.

Askari, A., and Huang, W-H., 1983, $Na^+,K^+$-ATPase: relation of quaternary conformational transitions to function, *Current Topics in Membranes and Transport* 19 (in press).

Askari, A., and Koyal, D., 1971, Studies on the partial reactions catalyzed by the $(Na^+ + K^+)$-activated ATPase. II. Effects of oligomycin and other inhibitors of the ATPase on the *p*-nitrophenylphosphatase, *Biochim. Biophys. Acta* **225**:20–25.

Askari, A., Huang, W., and Henderson, G. R., 1979, Na,K-ATPase: Functional and structural modifications induced by mercurials, in: *Na,K-ATPase: Structure and Kinetics* (J. C. Skou and J. G. Nørby, eds.), Academic Press, London, pp. 205–215.

Askari, A., Huang, W-H., and Antieau, J. M., 1980, $Na^+,K^+$-ATPase: Ligand-induced conformational transitions and alterations in subunit interactions evidenced by cross-linking studies, *Biochemistry* **19**:1132–1140.

Askari, A., Huang, W-H., and McCormick, P. W., 1983, $(Na^+ + K^+)$-dependent adenosine triphosphatase: Regulation of inorganic phosphate, magnesium ion and calcium ion interactions with the enzyme by ouabain, *J. Biol. Chem.,* **258**:3453–3460.

Bader, H., and Sen, A. K., 1966, $(K^+)$-dependent acyl phosphatase as part of the $(Na^+ + K^+)$-dependent ATPase of cell membranes, *Biochim. Biophys. Acta* **118**:116–123.

Baker, E., and Simmonds, W. J., 1966, Membrane ATPase and electrolyte levels in marsupial erythrocytes, *Biochim. Biophys. Acta* **126**:492–499.

Baker, P. F., and Stone, A. J., 1966, A kinetic method for investigating hypothetical models of the sodium pump, *Biochim. Biophys. Acta* **126**:321–329.

Baker, P. F., and Willis, J. S., 1970, Potassium ions and the binding of cardiac glycosides to mammalian cells, *Nature* **226**:521–523.

Baker, P. F., Blaustein, M. P., Hodgkin, A. L., and Steinhardt, R. A., 1969a, The influence of calcium on sodium efflux in squid axons, *J. Physiol.* **200**:431–468.

Baker, P. F., Blaustein, M. P., Keynes, R. D., Manil, J., Shaw, T. I., and Steinhardt, R. A., 1969b, The ouabain-sensitive fluxes of sodium and potassium in squid giant axons, *J. Physiol.* **200**:459–496.

Banerjee, S. P., and Wong, S. M. E., 1972, Effect of potassium on sodium-dependent adenosine diphosphate–adenosine triphosphate exchange activity in kidney microsomes, *J. Biol. Chem.* **247:**5409–5413.

Banerjee, S. P., Wong, S. M. E., Khanna, V. K., and Sen, A. K., 1972a, Inhibition of sodium- and potassium-dependent adenosine triphosphatase by *N*-ethylmaleimide. I. Effects on sodium-sensitive phosphorylation and potassium-sensitive dephosphorylation, *Mol. Pharmacol.* **8:**8–17.

Banerjee, S. P., Wong, S. M. E., and Sen, A. K., 1972b, Inhibition of sodium- and potassium-dependent adenosine triphosphatase by *N*-ethylmaleimide. II. Effects on sodium-activated transphosphorylation, *Mol. Pharmacol.* **8:**18–29.

Barnett, R. E., 1970, Effect of monovalent cations on the ouabain inhibition of the sodium and potassium ion activated adenosine triphosphatase, *Biochemistry* **9:**4644–4648.

Bastide, F., Meissner, G., Fleischer, S., and Post, R. L., 1973, Similarity of the active site of phosphorylation of the adenosine triphosphatase for transport of sodium and potassium ions in kidney to that for transport of calcium ions in the sarcoplasmic reticulum of muscle, *J. Biol. Chem.* **248:**8385–8391.

Bayley, H., and Knowles, J. R., 1980, Photogenerated reagents for membranes: Selective labeling of intrinsic membrane proteins in the human erythrocyte membrane, *Biochemistry* **19:**3883–3892.

Beaugé, L. A., 1979, Vanadate–potassium interactions in the inhibition of Na,K-ATPase, in: *Na,K-ATPase: Structure and Kinetics* (J. C. Skou and J. G. Nørby, eds.), Academic Press, London, pp. 373–387.

Beaugé, L. A., and Berberian, G., 1983, The effects of several ligands on the potassium–vanadate interaction in the inhibition of the Na,K-ATPase and the Na-K pump, *Biochim. Biophys. Acta,* **727:**336–350.

Beaugé, L. A., and DiPolo, R., 1979a, Vanadate selectively inhibits the $K^+_o$-activated $Na^+$ efflux in squid axons, *Biochim. Biophys. Acta* **551:**220–223.

Beaugé, L. A., and DiPolo, R., 1979b, Sidedness of the ATP-$Na^+$-$K^+$ interactions with the $Na^+$ pump in squid axons, *Biochim. Biophys. Acta* **553:**495–500.

Beaugé, L. A., and DiPolo, R., 1981, The effects of ATP on the interactions between monovalent cations and the sodium pump in dialysed squid axons, *J. Physiol.* **314:**457–480.

Beaugé, L. A., and DiPolo, R., 1983, Sidedness of cations and ATP interactions with the sodium pump, *Curr. Top. Membr. Trans.* **19:**643–647.

Beaugé, L. A., and Glynn, I. M., 1978, Commercial ATP containing traces of vanadate alters the response of ($Na^+$ + $K^+$)-ATPase to external potassium, *Nature* **272:**551–552.

Beaugé, L. A., and Glynn, I. M., 1979a, Occlusion of K ions in the unphosphorylated sodium pump, *Nature* **280:**510–512.

Beaugé, L. A., and Glynn, I. M., 1979b, Sodium ions, acting at high-affinity extracellular sites, inhibit sodium ATPase activity of the sodium pump by slowing dephosphorylation, *J. Physiol.* **289:**17–31.

Beaugé, L. A., and Glynn, I. M., 1980, The equilibrium between different conformations of the unphosphorylated sodium pump: Effects of ATP and of potassium ions, and their relevance to potassium transport, *J. Physiol.* **299:**367–383.

Beaugé, L. A., and Ortiz, O., 1973, Na fluxes in rat red blood cells in K-free solutions, *J. Membr. Biol.* **13:**165–184.

Beaugé, L. A., Cavieres, J. D., Glynn, I. M., and Grantham, J. J., 1980, The effects of vanadate on the fluxes of sodium and potassium ions through the sodium pump, *J. Physiol.* **301:**7–23.

Beeuwkes, R., and Rosen, S., 1975, Renal Na,K-ATPase. Optical localization and X-ray microanalysis, *J. Histochem. Cytochem.* **23:**828–839.

Blostein, R., 1968, Relationships between erythrocyte membrane phosphorylation and adenosine triphosphate hydrolysis, *J. Biol. Chem.* **243:**1957–1965.

Blostein, R., 1970, Sodium activated adenosine triphosphatase activity of the erythrocyte membrane, *J. Biol. Chem.* **245:**270–275.

Blostein, R., 1975, $Na^+$ ATPase of the mammalian erythrocyte membrane. Reversibility of phosphorylation at O°, *J. Biol. Chem.* **250:**6118–6124.

Blostein, R., 1979, Side-specific effects of sodium on (Na,K)-ATPase, *J. Biol. Chem.* **254:**6673–6677.

Blostein, R., 1983, Sidedness of sodium interactions with the sodium pump in the absence of $K^+$, *Curr. Top. Membr. Trans.* **19:**649–652.

Blostein, R., and Chu, L., 1977, Sidedness of (sodium, potassium)-adenosine triphosphatase of inside-out red cell membrane vesicles. Interactions with potassium, *J. Biol. Chem.* **252:**3035–3043.

Blostein, R., and Whittington, E. S., 1973, Studies of high potassium and low potassium sheep erythrocyte membrane sodium-adenosine triphosphatase: Interactions with oligomycin, adenosine triphosphate, sodium, and potassium, *J. Biol. Chem.* **248:**1772–1777.

Blostein, R., Pershadsingh, H. A., Drapeau, P., and Chu, L., 1979, Side-specificity of alkali cation interactions with Na,K-ATPase: Studies with inside-out red cell membrane vesicles, in: *Na,K-ATPase: Structure and Kinetics* (J. C. Skou and J. G. Nørby, eds.), London, Academic Press, pp. 233–245.

Bodemann, H. H., and Hoffman, J. F., 1976a, Side-dependent effects of internal versus external Na and K on ouabain binding to reconstituted human red blood cell ghosts, *J. Gen. Physiol.* **67:**497–525.

Bodemann, H. H., and Hoffman, J. F., 1976b, Comparison of the side-dependent effects of Na and K on orthophosphate-, UTP- and ATP-promoted ouabain binding to reconstituted human red blood cell ghosts, *J. Gen. Physiol.* **67:**527–545.

Bond, G. H., and Hudgins, P. M., 1981, Dog kidney ($Na^+$,$K^+$)-ATPase is more sensitive to inhibition by vanadate than human red cell ($Na^+$,$K^+$)-ATPase, *Biochim. Biophys. Acta* **646:**479–482.

Bond, G. H., Bader, H., and Post, R. L., 1971, Acetyl phosphate as a substitute for ATP in ($Na^+$ + $K^+$)-dependent ATPase, *Biochim. Biophys. Acta* **241:**57–67.

Bonting, S. L., and Caravaggio, L. L., 1963, Studies on Na : K activated ATPase. V. Correlation of enzyme activity with cation flux in six tissues, *Arch. Biochem. Biophys.* **101:**37–46.

Bonting, S. L., Simon, K. A., and Hawkins, N. M., 1961, Studies on sodium-potassium-activated adenosine triphosphatase. I. Quantitative distribution in several tissues of the cat, *Arch. Biochem. Biophys.* **95:**416–423.

Bonting, S. L., Schuurmans Stekhoven, F. M. A. H., Swarts, H. G. P., and de Pont, J. J. H. H. M., 1979, The low-energy phosphorylated intermediate of Na,K-ATPase, in: *$Na^+$,$K^+$-ATPase: Structure and Kinetics* (J. C. Skou and J. G. Nørby, eds.), Academic Press, London, pp. 317–330.

Brinley, F. J., and Mullins, L. J., 1968, Sodium fluxes in internally dialyzed squid axons, *J. Gen. Physiol.* **52:**181–211.

Brotherus, J. R., Griffith, O. H., Brotherus, M. O., Jost, P. C., and Silvius, J. R., 1981a, Lipid–protein multiple binding equilibria in membranes, *Biochemistry* **20:**5261–5267.

Brotherus, J. R., Moller, J. V., and Jorgensen, P. L., 1981b, Soluble and active renal Na,K-ATPase with maximum protein molecular mass 170,000 ± 9000 Daltons; formation of larger units by secondary aggregation, *Biochem. Biophys. Res. Commun.* **100:**146–154.

Brunner, J., Senn, H., and Richards, F. M., 1980, 3-Trifluoromethyl-3-phenyldiazirine. A new carbene generating group for photolabeling reagents, *J. Biol. Chem.* **255:**3313–3318.

Caldwell, P. C., and Keynes, R. D., 1957, The utilization of phosphate bond energy for sodium extrusion from giant axons, *J. Physiol.* **137:**12P.

Caldwell, P. C., and Keynes, R. D., 1959, The effect of ouabain on the efflux of sodium from a squid axon, *J. Physiol.* **148:**8–9P.

Caldwell, P. C., Hodgkin, A. L., Keynes, R. D., and Shaw, T. I., 1960, The effects of injecting "energy-rich" phosphate compounds on the active transport of ions in the giant axons of *Loligo*, *J. Physiol.* **152:**561–590.

Cantley, L. C., 1981, Structure and mechanism of the ($Na^+$,$K^+$)-ATPase, *Curr. Top. Bioenerg.* **11:**201–237.

Cantley, L. C., and Aisen, P., 1979, The fate of cytoplasmic vanadium. Implications on (Na,K)-ATPase inhibition, *J. Biol. Chem.* **254:**1781–1784.

Cantley, L. C., Josephson, L., Warner, R., Yanagisawa, M., Lechene, C., and Guidotti, G., 1977, Vanadate is a potent (Na,K)-ATPase inhibitor found in ATP derived from muscle, *J. Biol. Chem.* **252:**7421–7423.

Cantley, L. C., Cantley, L. G., and Josephson, L., 1978a, A characterization of vanadate interactions with the (Na,K)-ATPase. Mechanistic and regulatory implications, *J. Biol. Chem.* **253:**7361–7368.

Cantley, L. C., Gelles, J., and Josephson, L., 1978b, Reaction of (Na-K)-ATPase with 7-chloro-4-nitrobenzo-2-oxa-1,3-diazole: Evidence for an essential tyrosine at the active site, *Biochemistry* **17:**418–425.

Cantley, L. C., Resh, M., and Guidotti, G., 1978c, Vanadate inhibits the red cell ($Na^+$,$K^+$)ATPase from the cytoplasmic side, *Nature* **272:**552–554.

Carilli, C. T., Farley, R. A., Perlman, D. M., and Cantley, L. C., 1982, The active site structure of $Na^+$- and $K^+$-stimulated ATPase. Location of a specific fluorescein isothiocyanate reactive site, *J. Biol. Chem.* **257:**5601–5606.

Castro, J., and Farley, R. A., 1979, Proteolytic fragmentation of the catalytic subunit of the sodium and potassium adenosine triphosphatase. Alignment of tryptic and chymotryptic fragments and location of sites labelled with ATP and iodoacetate, *J. Biol. Chem.* **254:**2221–2228.

Cavieres, J. D., 1980, Extracellular sodium stimulates ATP–ADP exchange by the sodium pump, *J. Physiol.* **308:**57P.

Cavieres, J. D., 1983, Ouabain-sensitive ATP–ADP exchange and Na-ATPase of resealed red cell ghosts, *Curr. Top. Membr. Trans.*, **19:**677–681.

Cavieres, J. D., and Ellory, J. C., 1975, Allosteric inhibition of the sodium pump by external sodium, *Nature* **255:**338–340.

Cavieres, J. D., and Ellory, J. C., 1982, The target size for ATP–ADP exchange and Na-ATPase activities of a purified Na,K-ATPase preparation, *J. Physiol.* **332:**120*P*.

Cavieres, J. D., and Glynn, I. M., 1979, Sodium–sodium exchange through the sodium pump: The roles of ATP and ADP, *J. Physiol.* **297:**637–645.

Chipperfield, A. R., 1983, Stimulation and inhibition by plasma of ouabain-sensitive sodium efflux in human red blood cells *Curr. Top. Membr. Trans.* **19:**1013–1016.

Chipperfield, A. R., and Whittam, R., 1976, The connexion between the ion-binding sites of the sodium pump, *J. Physiol.* **260:**371–385.

Churchill, L., Peterson, G. L., and Hokin, L. E., 1979, The large subunit of (sodium + potassium)-activated adenosine triphosphatase from the electroplax of *Electrophorus electricus* is a glycoprotein, *Biochem. Biophys. Res. Commun.* **90:**488–490.

Clarkson, E. M., and Maizels, M., 1952, Distribution of phosphatases in human erythrocytes, *J. Physiol.* **116:**112–128.

Cleland, W. W., 1970, Steady-state kinetics, in: *The Enzymes,* Vol. 2 (P. D. Boyer, ed.), Academic Press, New York, pp. 1–65.

Cohn, W. E., and Cohn, E. T., 1939, Permeability of red corpuscles of the dog to sodium ion, *Proc. Soc. Exp. Biol. Med.* **41:**445–449.

Collins, J. H., Forbush, B., Lane, L. K., Ling, E., Schwarz, A., and Zot, A., 1982, Purification and characterization of an ($Na^+$ + $K^+$)-ATPase proteolipid labeled with a photoaffinity derivative of ouabain, *Biochim. Biophys. Acta* **686:**7–12.

Craig, W. S., and Kyte, J., 1980, Stoichiometry and molecular weight of the minimum asymmetric unit of canine renal sodium and potassium ion-activated adenosine triphosphatase, *J. Biol. Chem.* **255:**6262–6269.

Czerwinski, A., Gitelman, H. J., and Welt, L. G., 1967, A new member of the ATPase family, *Am. J. Physiol.* **213:**786–792.

Dahl, J. L., and Hokin, L. E., 1974, The sodium-potassium adenosinetriphosphatase, *Annu. Rev. Biochem.* **43:**327–356.

Dahms, A. S., and Boyer, P. D., 1973, Occurrence and characteristics of $^{18}O$ exchange reactions catalyzed by sodium- and potassium-dependent adenosine triphosphatases, *J. Biol. Chem.* **248:**3155–3162.

Dahms, A. S., and Miara, J. E., 1983, $^{31}P(^{18}O)$-NMR kinetic analysis of the oxygen-18 exchange reaction between inorganic phosphate and water catalyzed by the ($Na^+$,$K^+$)-ATPase, *Curr. Top. Membr. Trans.* **19:**371–375.

Dahms, A. S., Kanazawa, T., and Boyer, P. D., 1973, Source of the oxygen in the C-O-P linkage of the acyl phosphate in transport adenosine triphosphatases, *J. Biol. Chem.* **248:**6592–6595.

Danowski, T. S., 1941, The transfer of potassium across the human blood cell membrane, *J. Biol. Chem.* **139:**693–705.

Dean, R. B., 1941, Theories of electrolyte equilibrium in muscle, *Biol. Symp.* **3:**331–348.

Degani, C., Dahms, A. S., and Boyer, P. D., 1974, Characterization of acyl phosphate in transport ATPases by a borohydride reduction method, *Ann. N. Y. Acad. Sci.* **242:**77–79.

Deguchi, N., Jørgensen, P. L., and Maunsbach, A. B., 1977, Ultrastructure of the sodium pump. Comparison of thin sectioning, negative staining and freeze-fracture of purified, membrane-bound ($Na^+$,$K^+$)-ATPase, *J. Cell Biol.* **75:**619–634.

De Pont, J. J. H. H. M., Schoot, B. M., Van Prooijen-Van-Eeden, A., and Bonting, S. L., 1977, An essential arginine residue in the ATP-binding centre of ($Na^+$ + $K^+$)-ATPase, *Biochim. Biophys. Acta* **482:**213–227.

De Weer, P., 1970, Effects of intracellular 5′ADP and orthophosphate on the sensitivity of sodium efflux from squid axon to external sodium and potassium, *J. Gen. Physiol.* **56:**583–620.

De Weer, P., Kennedy, B. G., and Abercrombie, R. F., 1979, Relationship between the Na : K exchanging and Na : Na exchanging modes of operation of the sodium pump, in: *Na,K-ATPase: Structure and Kinetics* (J. C. Skou and J. G. Nørby, eds.), Academic Press, London, pp. 503–515.

De Weer, P., Breitwieser, G. E., Kennedy, B. G., and Smith, H. G., 1983, ADP–ATP exchange in internally dialysed squid giant axons, *Curr. Top. Membr. Trans.* **19:**665–669.

Dissing, S., and Hoffman, J. F., 1983, Anion-coupled Na efflux mediated by the Na : K pump in human red blood cells, *Curr. Top. Membr. Trans.* **19:**693–695.

Dixon, J. F., and Hokin, L. E., 1974, Studies in the characterization of the sodium–potassium adenosine triphosphatase. Purification and properties of the enzyme from the electric organ of *Electrophorus electricus, Arch. Biochem.* **163:**749–758.

Dixon, J. F., and Hokin, L. E., 1978, A simple procedure for the preparation of highly purified (sodium + potassium) adenosine triphosphatase from the rectal salt gland of *Squalus acanthias* and the electric organ of *Electrophorus electricus, Anal. Biochem.* **86:**378–385.

Drapeau, P., and Blostein, R., 1980, Interactions of $K^+$ with (Na,K)-ATPase: Orientation of $K^+$-phosphatase sites studied with inside-out red cell membrane vesicles, *J. Biol. Chem.* **255:**7827–7834.

Dudding, W. F., and Winter, C. G., 1971, On the reaction sequence of the $K^+$-dependent acetyl phosphatase activity of the $Na^+$ pump, *Biochim. Biophys. Acta* **241:**650–660.

Dunham, E. T., and Glynn, I. M., 1961, Adenosine triphosphatase activity and the active movements of alkali metal ions, *J. Physiol.* **156:**274–293.

Eisner, D. A., and Richards, D. E., 1981, The interaction of potassium ions and ATP on the sodium pump of resealed red cell ghosts, *J. Physiol.* **319:**403–418.

Eisner, D. A., and Richards, D. E., 1982, Inhibition of the sodium pump by inorganic phosphate in resealed red cell ghosts, *J. Physiol.* **326:**1–10.

Eisner, D. A., and Richards, D. E., 1983, Stimulation and inhibition by ATP and orthophosphate of the potassium–potassium exchange in resealed red cell ghosts, *J. Physiol.* **335:**495–506.

Ellory, J. C., Green, J. R., Jarvis, S. M., and Young, J. D., 1979, Measurement of the apparent molecular volume of membrane-bound transport systems by radiation inactivation, *J. Physiol.* **295:**10–11*P*.

Erdmann, E., and Schoner, W., 1973, Ouabain-receptor interactions in ($Na^+$ + $K^+$)-ATPase preparations from different tissues and species. Determination of kinetic constants and dissociation constants, *Biochim. Biophys. Acta* **307:**386–398.

Erdmann, E., and Schoner, W., 1974, Ouabain-receptor interactions in ($Na^+$ + $K^+$)-ATPase preparations. IV. The molecular structure of different cardioactive steroids and other substances and their affinity to the glycoside receptor, *Naunyn-Schmiedebergs Arch. Pharmacol.* **283:**335–356.

Esmann, M., 1980, Concanavalin A-Sepharose purification of soluble Na,K-ATPase from rectal glands of the spiny dogfish, *Anal. Biochem.* **108:**83–85.

Esmann, M., and Klodos, I., 1983, Sulphydryl groups of Na,K-ATPase: Effects of *N*-ethyl-maleimide on phosphorylation from ATP in the presence of $Na^+$ + $Mg^{2+}$, *Curr. Top. Membr. Trans.* **19:**349–352.

Esmann, M., Skou, J. C., and Christiansen, C., 1979, Solubilization and molecular weight determination of Na,K-ATPase from rectal glands of *Squalus Acanthias, Biochim. Biophys. Acta* **567:**410–420.

Esmann, M., Christiansen, C., Karlsson, K-A., Hansson, G. C., and Skou, J. C., 1980, Hydrodynamic properties of solubilized ($Na^+$ + $K^+$)-ATPase from rectal glands of *Squalus acanthias, Biochim. Biophys. Acta* **603:**1–12.

Fahn, S., Hurley, M. R., Koval, G. J., and Albers, R. W., 1966a, Sodium–potassium-activated adenosine triphosphatase of *Electrophorus* electric organ. II. Effects of *N*-ethylmaleimide and other sulfhydryl reagents, *J. Biol. Chem.* **241:**1890–1895.

Fahn, S., Koval, G. J., and Albers, R. W., 1966b, Sodium–potassium-activated adenosine triphosphatase of *Electrophorus* electric organ. I. An associated sodium-activated transphosphorylation, *J. Biol. Chem.* **241:**1882–1889.

Fahn, S., Koval, G. J., and Albers, R. W., 1968, Sodium–potassium-activated adenosine triphosphatase of *Electrophorus* electric organ. V. Phosphorylation by adenosine triphosphate-$^{32}P$, *J. Biol. Chem.* **243:**1993–2002.

Fenn, W. O., and Cobb, D. M., 1936, Electrolyte changes in muscle during activity, *Am. J. Physiol.* **115:**345–356.

Fishman, M. C., 1979, Endogenous digitalis-like activity in mammalian brain, *Proc. Natl. Acad. Sci. USA* **76**:4661–4663.

Flatman, P. W., and Lew, V. L., 1981, The magnesium dependence of sodium-pump-mediated sodium–potassium and sodium–sodium exchange in intact human red cells, *J. Physiol.* **315**:421–446.

Forbush, B., 1982, Characterization of right-side-out membrane vesicles rich in (Na,K)-ATPase and isolated from dog kidney outer medulla, *J. Biol. Chem.* **257**:12678–12684.

Forbush, B., and Hoffman, J. F., 1979, Evidence that ouabain binds to the same large polypeptide chain of dimeric Na,K-ATPase that is phosphorylated by $P_i$, *Biochemistry* **18**:2308–2315.

Forbush, B., Kaplan, J. H., and Hoffman, J. F., 1978, Characterization of a new photoaffinity derivative of ouabain: labeling of the large polypeptide and of a proteolipid component of the Na,K-ATPase, *Biochemistry* **17**:3667–3676.

Forgac, M., and Chin, G., 1981, $K^+$-independent active transport of $Na^+$ by the ($Na^+$ + $K^+$)-stimulated adenosine triphosphatase, *J. Biol. Chem.* **256**:3645–3646.

Fortes, P. A. G., Moczydlowski, E. G., Yagi, A., and Lee, J. A., 1981, Na, K-ATPase structure and mechanism studied with site-directed fluorescent probes, *Proc. VIIth International Biophysics Congress*, p. 66. IUPAB.

Foster, D., and Ahmed, K., 1976, $Na^+$-dependent phosphorylation of rat brain ($Na^+$ + $K^+$)-ATPase. Possible non-equivalent activation sites for $Na^+$, *Biochim. Biophys. Acta* **429**:258–273.

Freytag, J. W., and Reynolds, J. A., 1981, Polypeptide molecular weights of the ($Na^+$,$K^+$)-ATPase from porcine kidney medulla, *Biochemistry* **20**:7211–7214.

Froehlich, J. P., Albers, R. W., Koval, G. J., Goebel, R., and Berman, M., 1976, Evidence for a new intermediate state in the mechanism of ($Na^+$ + $K^+$)-adenosine triphosphatase, *J. Biol. Chem.* **251**:2186–2188.

Froehlich, J. P., Hobbs, A. S., and Albers, R. W., 1983, Evidence for parallel pathways of phosphoenzyme formation in the mechanism of ATP hydrolysis by *Electrophorus* (Na,K)-ATPase, *Curr. Top. Membr. Trans.* **19**:513–535.

Fujita, M., Ohta, H., Kawai, K., Matsui, H., and Nakao, M., 1972, Differential isolation of microvillous and basolateral plasma membranes from intestinal mucosa: Mutually exclusive distribution of digestive enzymes and ouabain-sensitive ATPase, *Biochim. Biophys. Acta* **274**:336–347.

Fukushima, Y., and Nakao, M., 1980, Changes in affinity of $Na^+$- and $K^+$-transport ATPase for divalent cations during its reaction sequence, *J. Biol. Chem.* **255**:7813–7819.

Fukushima, Y., and Nakao, M., 1981, Transient state in the phosphorylation of sodium- and potassium-transport adenosine triphosphatase by adenosine triphosphate, *J. Biol. Chem.* **256**:9136–9143.

Fukushima, Y., and Post, R. L., 1978, Binding of divalent cation to phosphoenzyme of sodium- and potassium-transport adenosine triphosphatase, *J. Biol. Chem.* **253**:6853–6862.

Fukushima, Y., and Tonomura, Y., 1973, Two kinds of high energy phosphorylated intermediate, with and without bound ADP, in the reaction of $Na^+$-$K^+$-dependent ATPase, *J. Biochem. (Tokyo)* **74**:135– 142.

Fukushima, Y., and Tonomura, Y., 1975, The pre-steady state of $Na^+$-$K^+$-dependent ATPase after addition of $Na^+$ ions. Transition of the phosphorylated intermediate from an ADP-sensitive to an ADP-insensitive form, *J. Biochem. (Tokyo* **78**:749–755.

Gache, C., Rossi, B., and Lazdunski, M., 1976, ($Na^+$,$K^+$)-activated adenosine triphosphatase of axonal membranes, cooperativity and control. Steady-state analysis, *Eur. J. Biochem.* **65**:293–306.

Garay, R. P., and Garrahan, P. J., 1973, The interaction of sodium and potassium with the sodium pump in red cells, *J. Physiol.* **231**:297–325.

Garay, R. P., and Garrahan, P. J., 1975, The interaction of adenosine triphosphate and inorganic phosphate with the sodium pump in red cells, *J. Physiol.* **249**:51–67.

Gardos, G., 1954, Akkumulation der Kaliumionen durch menschliche Blutkörperchen, *Acta Physiol. Hung.* **6**:191–199.

Garrahan, P. J., and Garay, R. P., 1976, The distinction between simultaneous and sequential models for sodium and potassium transport, *Curr. Top. Membr. Trans.* **8**:29–97.

Garrahan, P. J., and Glynn, I. M., 1967a, The behavior of the sodium pump in red cells in the absence of external potassium, *J. Physiol.* **192**:159–174.

Garrahan, P. J., and Glynn, I. M., 1967b, The sensitivity of the sodium pump to external sodium, *J. Physiol.* **192**:175–188.

Garrahan, P.J., and Glynn, I. M., 1967c, Factors affecting the relative magnitudes of the sodium : potassium and sodium : sodium exchanges catalysed by the sodium pump, *J. Physiol.* **192:**189–216.

Garrahan, P. J., and Glynn, I. M., 1967d, The stoichiometry of the sodium pump, *J. Physiol.* **192:**217–235.

Garrahan, P. J., and Glynn, I. M., 1967e, The incorporation of inorganic phosphate into adenosine triphosphate by reversal of the sodium pump, *J. Physiol.* **192:**237–257.

Garrahan, P. J., and Rega, A. F., 1972, Potassium activated phosphatase from human red blood cells. The effects of *p*-nitrophenylphosphate on cation fluxes, *J. Physiol.* **233:**595–617.

Garrahan, P. J., Pouchan, M. I., and Rega, A. F., 1970, Potassium activated phosphatase from human red blood cells. The effects of adenosine triphosphate, *J. Membr. Biol.* **3:**26–42.

Garrahan, P. J., Horenstein, A. H., and Rega, A. F., 1979, The interaction of ligands with the Na,K-ATPase during Na-ATPase activity, in: *Na,K-ATPase: Structure and Kinetics* (J. C. Skou and J. G. Nørby, eds.), Academic Press, London, pp. 261–274.

Garrahan, P. J., Rossi, R. C., and Rega, A. F., 1982, The interaction of $K^+$,$Na^+$, $Mg^{2+}$, and ATP with the (Na,K)ATPase, *Ann. N. Y. Acad. Sci.,* **402:**239–251.

Giotta, G. J., 1975, Native ($Na^+$ + $K^+$)-dependent adenosine triphosphatase has two trypsin-sensitive sites, *J. Biol. Chem.* **250:**5159–5164.

Giotta, G. J., 1976, Quaternary structure of ($Na^+$ + $K^+$)-dependent adenosine triphosphatase, *J. Biol. Chem.* **251:**1247–1252.

Gitler, C., and Bercovici, T., 1980, Use of lipophilic photoactivatable reagents to identify the lipid-embedded domain of membrane proteins, *Ann. N. Y. Acad. Sci.* **346:**199–211.

Glynn, I. M., 1956, Sodium and potassium movements in human red cells, *J. Physiol.* **134:**278–310.

Glynn, I. M., 1957, The action of cardiac glycosides on sodium and potassium movements in human red cells, *J. Physiol.* **136:**148–173.

Glynn, I. M., 1962, Activation of adenosinetriphosphatase activity in a cell membrane by external potassium and internal sodium, *J. Physiol.* **160:**18–19*P*.

Glynn, I. M., 1963, "Transport adenosinetriphosphatase" in electric organ. The relation between ion transport and oxidative phosphorylation, *J. Physiol.* **169:**452–465.

Glynn, I. M., 1969, The effects of cardiac glycosides on metabolism and ion fluxes, in: *Digitalis* (C. Fisch and B. Surawicz, eds.), Grune & Stratton, New York, pp. 30–42.

Glynn, I. M., 1982, Occluded-ion forms of the Na,K-ATPase, *Ann. N. Y. Acad. Sci.* **402:**287–288.

Glynn, I. M., 1984, The electrogenic sodium pump, in: *Electrogenic Transport: Fundamental Principles and Physiological Implications* (M. P. Blaustein and M. Lieberman, eds.), Raven Press, New York, pp. 33–48.

Glynn, I. M., and Hoffman, J. F., 1971, Nucleotide requirements for sodium–sodium exchange catalysed by the sodium pump in human red cells, *J. Physiol.* **218:**239–256.

Glynn, I. M., and Karlish, S. J. D., 1975, The sodium pump, *Annu. Rev. Physiol.* **37:**13–55.

Glynn, I. M., and Karlish, S. J. D., 1976, ATP hydrolysis associated with an uncoupled sodium flux through the sodium pump: Evidence for allosteric effects of intracellular ATP and extracellular sodium, *J. Physiol.* **256:**465–496.

Glynn, I. M., and Karlish, S. J. D., 1982, Conformational changes associated with $K^+$ transport by the $Na^+$/$K^+$-ATPase, in: *Membranes and Transport,* Vol. I (A. N. Martinosi, ed.), Plenum Press, New York, pp. 529–536.

Glynn, I. M., and Lew, V. L., 1970, Synthesis of adenosine triphosphate at the expense of downhill cation movements in intact human red cells, *J. Physiol.* **207:**393–402.

Glynn, I. M., and Richards, D. E., 1982, Occlusion of rubidium ions by the sodium-potassium pump: Its implications for the mechanism of potassium transport, *J. Physiol.* **330:**17–43.

Glynn, I. M., and Richards, D. E., 1983, The existence and role of occluded-ion forms of Na,K-ATPase, *Curr. Top. Membr. Trans.* **19:**625–638.

Glynn, I. M., and Rink, T. J., 1982, Hypertension and inhibition of the sodium pump: A strong link but in which chain? *Nature* **300:**576–577.

Glynn, I. M., Lew, V. L., and Lüthi, V., 1970, Reversal of the potassium entry mechanism in red cells, with and without reversal of the entire pump cycle, *J. Physiol.* **207:**371–391.

Glynn, I. M., Hoffman, J. F., and Lew, V. L., 1971, Some "partial reactions" of the sodium pump, *Phil. Trans. R. Soc. Lond. B.* **262:**91–102.

Glynn, I. M., Karlish, S. J. D., Cavieres, J. D., Ellory, J. C., Lew, V. L., and Jørgensen, P. L., 1974, The effects of an antiserum to $Na^+$,$K^+$-ATPase on the ion transporting and hydrolytic activities of the enzyme, *Ann. N. Y. Acad. Sci.* **242:**357–371.

Glynn, I. M., Karlish, S. J. D., and Yates, D. W., 1979, The use of formycin nucleotides to investigate the mechanism of Na,K-ATPase, in: *Na,K-ATPase: Structure and Kinetics* (J. C. Skou and J. G. Nørby, eds.), Academic Press, London, pp. 101–113.

Glynn, I. M., Hara, Y., and Richards, D. E., 1983a, Trapping of sodium ions by a phosphorylated form of the sodium-potassium pump (Na,K-ATPase), *J. Physiol.*, **339:**56–57*P*.

Glynn, I. M., Howland, J. L., and Richards, D. E., 1983b, Orthophosphate plus magnesium causes the rapid release of only 50% of rubidium ions occluded in the unphosphorylated Na, K-ATPase, *J. Physiol.* **343:**94*P*.

Glynn, I. M., Hara, Y., and Richards, D. E., 1984, The occlusion of sodium ions within the mammalian sodium-potassium pump: its role in sodium transport. *J. Physiol.* **351:**531–547

Godfraind, T., and Hernandez, G. C., 1981, Properties of a digitalis-like factor extracted from guinea-pig brain, *Arch. Int. Pharmacodyn. Ther.* **250:**316–317.

Goldin, S. M., 1977, Active transport of sodium and potassium ions by the sodium and potassium ion-activated adenosine triphosphatase from renal medulla. Reconstitution of the purified enzyme into a well-defined *in vitro* transport system, *J. Biol. Chem.* **252:**5630–5642.

Gonzalez, E., and Zambrano, F., 1983, Possible role of sulphatide in $K^+$-activated phosphatase activity, *Biochim. Biophys. Acta* **728:**66–72.

Grantham, J. J., and Glynn, I. M., 1979, Renal Na,K-ATPase: Determinants of inhibition by vanadium, *Am. J. Physiol.* **236**(6)**:**F530–F535.

Grisham, C. M., 1979, Characterization of essential arginine residues in sheep kidney ($Na^+$ + $K^+$)-ATPase, *Biochem. Biophys. Res. Commun.* **88:**229–236.

Grisham, C. M., 1982, Ion-transporting ATPases. Characterizing structure and function with paramagnetic probes, in: *Membranes and Transport,* Vol. 1 (A. N. Martonosi, ed.), Plenum, New York, pp. 585–592.

Grisham, C. M., and Hutton, W., 1978, Lithium-7 NMR as a probe of monovalent cation sites at the active ($Na^+$ + $K^+$)-ATPase from kidney, *Biochem. Biophys. Res. Commun.* **81:**1406–1411.

Grosse, R., Eckert, K., Malur, J., and Repke, K. R. H., 1978, Analysis of function-related interactions of ATP, sodium and potassium ions with $Na^+$ and $K^+$ transporting ATPase studied with a thiol reagent as tool, *Acta Biol. Med. Ger.* **37:**83–96.

Grosse, R., Rapoport, T., Malur, J., Fischer, J., and Repke, K. R. H., 1979, Mathematical modelling of ATP, $K^+$ and $Na^+$ interactions with ($Na^+$ + $K^+$)-ATPase occurring under equilibrium conditions, *Biochim. Biophys. Acta* **550:**500–514.

Gruber, K. A., Whitaker, J.M., and Buckalew, V. M., 1983, Immunochemical approaches to the isolation of an endogenous digoxin-like factor. *Curr. Top. Membr. Trans.* **19:**917–921.

Haase, W., and Koepsell, H., 1979, Substructure of membrane-bound $Na^+$-$K^+$-ATPase protein, *Pflügers Arch.* **381:**127–135.

Hall, C., and Ruoho, A., 1980, Ouabain-binding-site photoaffinity probes that label both subunits of $Na^+$,$K^+$-ATPase, *Proc. Natl. Acad. Sci. USA* **77:**4529–4533.

Hamlyn, J. M., Ringel, R., Schaeffer, J., Levinson, P. D., Hamilton, B. P., Kowarski, A. A., and Blaustein, M. P., 1982, A circulating inhibitor of ($Na^+$ + $K^+$)ATPase associated with essential hypertension, *Nature* **300:**650–652.

Hansen, O., 1971, The relationship between *g*-strophanthin-binding capacity and ATPase activity in plasma membrane fragments from ox brain, *Biochim. Biophys. Acta* **233:**122–132.

Hansen, O., 1976, Nonuniform population of *g*-strophanthin-binding sites of ($Na^+$ + $K^+$)-activated AT-Pase. Apparent conversion to uniformity by $K^+$, *Biochim. Biophys. Acta* **433:**383–392.

Hansen, O., Jensen, J., and Norby, J. G., 1971, Mutual exclusion of ATP, ADP, and *g*-strophanthin binding to NaK-ATPase, *Nature* **234:**122–124.

Hansen, O., Jensen, J., and Ottolenghi, P., 1979, Na,K-ATPase: The uncoupling of its ATPase and *p*-nitrophenyl phosphatase activities by thimerosal, in: *Na,K-ATPase: Structure and Kinetics* (J. C. Skou and J. G. Nørby, eds.), Academic Press, London, pp. 217–226.

Hansson, C. G., Karlsson, K-A., and Samuelsson, B. E., 1978, The identification of sulfatides in human erythrocyte membrane and their relation to sodium–potassium dependent adenosine triphosphatase, *J. Biochem. (Tokyo)* **83:**813–819.

Hara, Y., and Nakao, M., 1979, Detection of sodium binding to $Na^+,K^+$-ATPase with a sodium sensitive electrode, in: *Cation Flux across Biomembranes* (Y. Mukohata and L. Packer, eds.), Academic Press, New York, pp. 21–28.

Hara, Y., and Nakao, M., 1981, Sodium ion discharge from pig kidney $Na^+,K^+$-ATPase. $Na^+$-dependency of the $E_1P \rightleftharpoons E_2P$ equilibrium in the absence of KC1, *J. Biochem. (Tokyo)* **90:**923–931.

Hara, S., Hara, Y., Nakao, T., and Nakao, M., 1981, Ligand-dependent reactivity of $(Na^+ + K^+)$-ATPase with showdomycin, *Biochim. Biophys. Acta* **644:**53–61.

Harris, J. E., 1941, The influence of the metabolism of human erythrocytes on their potassium content, *J. Biol. Chem.* **141:**579–595.

Harris, E. J., and Maizels, M., 1951, The permeability of human red cells to sodium, *J. Physiol.* **113:**506–524.

Hart, W. M., and Titus, E. O., 1973a, Isolation of a protein component of sodium-potassium transport adenosine triphosphatase containing ligand-protected sulfhydryl groups, *J. Biol. Chem.* **248:**1365–1371.

Hart, W. M., and Titus, E. O., 1973b, Sulfhydryl groups of sodium–potassium transport adenosine triphosphatase. Protection by physiological ligands and exposure by phosphorylation, *J. Biol. Chem.* **248:**4674–4681.

Hastings, D. F., and Reynolds, J. A., 1979a, Non-ionic detergent solubilized Na,K-ATPase from shark rectal glands—molecular weight and peptide stoichiometry of the active complex, in: *Na,K-ATPase: Structure and Kinetics* (J. C. Skou and J. G. Nørby, eds.), Academic Press, London, pp. 15–20.

Hastings, D. F., and Reynolds, J. A., 1979b, Molecular weight of $(Na^+,K^+)$ATPase from shark rectal gland, *Biochemistry* **18:**817–821.

Hastings, D., and Skou, J. C., 1980, Potassium binding to the $(Na^+ + K^+)$-ATPase, *Biochim. Biophys. Acta* **601:**380–385.

Haupert, G. T., 1983, Endogenous glycoside-like substances, *Curr. Top. Membr. Transport* **19:**843–855.

Haupert, G. T., and Sancho, J. M., 1979, Sodium transport inhibitor from bovine hypothalamus, *Proc. Natl. Acad. Sci. USA* **76:**4658–4660.

Hayashi, Y., Kimimura, M., Homareda, H., and Matsui, H., 1977, Purification and characteristics of $(Na^+,K^+)$-ATPase from canine kidney by zonal centrifugation in sucrose density gradient, *Biochim. Biophys. Acta* **482:**185–196.

Hebert, H., Jørgensen, P. L., Skriver, E., and Maunsbach, A. B., 1982, Crystallization patterns of membrane-bound $(Na^+ + K^+)$-ATPase, *Biochim. Biophys. Acta* **689:**571–574.

Hegyvary, C., 1975, Covalent labeling of the digitalis-binding component of plasma membranes, *Mol. Pharmacol.* **11:**588–594.

Hegyvary, C., and Jørgensen, P. L., 1981, Conformational changes of renal sodium plus potassium ion-transport adenosine triphosphatase labelled with fluorescein, *J. Biol. Chem.* **256:**6296–6303.

Hegyvary, C., and Post, R. L., 1971, Binding of adenosine triphosphate to sodium and potassium ion-stimulated adenosine triphosphatase, *J. Biol. Chem.* **246:**5234–5240.

Heinz, A., Rubinson, K. A., and Grantham, J. J., 1982, The transport and accumulation of oxyvanadium compounds in human erythrocytes *in vitro, J. Lab. Clin. Med.* **100:**593–612.

Henderson, G. R., and Askari, A., 1976, Transport ATPase: Thimerosal inhibits the $Na^+,K^+$-dependent ATPase activity without diminishing the $Na^+$-dependent ATPase activity, *Biochem. Biophys. Res. Commun.* **69:**499–505.

Henderson, G. R., and Askari, A., 1977, Transport ATPase: Further studies on the properties of the thimerosal-treated enzyme, *Arch. Biochem. Biophys.* **182:**221–226.

Henderson, G. R., Huang, W., and Askari, A., 1979, Transport ATPase—the different modes of inhibition of the enzyme by various mercury compounds, *Biochem. Pharmacol.* **28:**429–433.

Heppel, L. A., 1940, The diffusion of radioactive sodium into the muscles of potassium-deprived rats, *Am. J. Physiol.* **128:**449–454.

Hexum, T., Samson, F. E., and Himes, R. H., 1970, Kinetic studies of $(Na^+ + K^+ + Mg^{2+})$-ATPase, *Biochim. Biophys. Acta* **212:**322–331.

Hilden, S., and Hokin, L. E., 1975, Active potassium transport coupled to active sodium transport in vesicles reconstituted from purified sodium and potassium ion-activated adenosine triphosphatase from the rectal gland of *Squalus acanthias, J. Biol. Chem.* **250:**6296–6303.

Hobbs, A. S., Albers, R. W., and Froehlich, J. P., 1980, Potassium-induced changes in phosphorylation and dephosphorylation of $(Na^+ + K^+)$-ATPase observed in the transient state, *J. Biol. Chem.* **255:**3395–3402.

Hobbs, A. S., Froehlich, J. P., and Albers, R. W., 1983, Inhibition by vanadate of the reactions catalyzed by the $Na^+$ plus $K^+$-stimulated ATPase: A transient state kinetic characterization, *J. Biol. Chem.* **255:**3724–3727.

Hodgkin, A. L., and Keynes, R. D., 1955, Active transport of cations in giant axons from *Sepia* and *Loligo*, *J. Physiol.* **128:**28–60.

Hoffman, J. F., 1966, The red cell membrane and the transport of sodium and potassium, *Am. J. Med.* **41:**666–680.

Hoffman, P. G., and Tosteson, D. C., 1971, Active sodium and potassium transport in high potassium and low potassium sheep red cells, *J. Gen. Physiol.* **58:**438–466.

Hokin, L. E., 1974, Purification and properties of the (sodium + potassium)-activated adenosine triphosphatase and reconstitution of sodium transport, *Ann. N. Y. Acad. Sci.* **242:**12–23.

Hokin, L. E., Dahl, J. L., Deupree, J. D., Dixon, J. F., Hackney, J. F., and Perdue, J. F., 1973, Studies on the characterization of the sodium–potassium transport adenosine triphosphatase. X. Purification of the enzyme from rectal gland of *Squalus acanthias*, *J. Biol. Chem.* **248:**2593–2605.

Homareda, H., and Matsui, H., 1982, Interaction of sodium and potassium ions with $Na^+$,$K^+$-ATPase. II. General properties of ouabain-sensitive $K^+$ binding, *J. Biochem (Tokyo)* **92:**219–231.

Hopkins, B. E., Wagner, H., and Smith, T. W., 1976, Sodium- and potassium-activated adenosine triphosphatase of the nasal salt gland of the duck *(Anas platyrhynchos)*. Purification characterization, and $NH_2$-terminal amino acid sequence of the phosphorylating polypeptide, *J. Biol. Chem.* **251:**4365–4371.

Horowicz, P., Taylor, J. W., and Waggoner, D. M., 1970, Fractionation of sodium efflux in frog sartorius muscles by strophanthidin and removal of external sodium, *J. Gen. Physiol.* **55:**401–425.

Huang, W-H., and Askari, A., 1981, Phosphorylation-dependent cross-linking of the α-subunits in the presence of $Ca^{2+}$ and *o*-phenanthroline, *Biochim. Biophys. Acta* **645:**54–58.

Inturrisi, C. E., and Titus, E., 1968, Kinetics of oligomycin inhibition of sodium and potassium-activated adenosine triphosphatase from beef brain, *Mol. Pharmacol.* **4:**591–599.

Israel, Y., and Titus, E. O., 1967, A comparison of microsomal ($Na^+$ + $K^+$)-ATPase with $K^+$-acetylphosphatase, *Biochim. Biophys. Acta* **139:**450–459.

Jackson, R. L., Verkleij, A. J., Van Zoelen, E. J. J., Lane, L. K., Schwartz, A., and Van Deenen, L. L. M., 1980, Asymmetric incorporation of $Na^+$,$K^+$-ATPase into phospholipid vesicles, *Arch. Biochem. Biophys.* **200:**269–278.

Järnefelt, J., 1962, Properties and possible mechanism of the $Na^+$ and $K^+$-stimulated microsomal adenosine triphosphatase, *Biochim. Biophys. Acta* **59:**643–654.

Jensen, J., and Ottolenghi, P., 1976, Adenosine diphosphate binding to sodium-plus-potassium ion-dependent adenosine triphosphatase. The role of lipid in nucleotide-potassium ion interplay, *Biochem. J.* **159:**815–817.

Jensen, J., and Ottolenghi, P., 1983, Binding of $Rb^+$ and ADP to a potassium-like form of Na,K-ATPase, *Curr. Top. Membr. Trans.* **19:**223–227.

Jensen, J., Nørby, J. G., and Ottolenghi, P., 1979, Is there a relationship between ATP-binding capacity and enzyme activity in thimerosal-treated Na,K-ATPase? in: *Na,K-ATPase: Structure and Kinetics* (J.C. Skou and J. G. Nørby, eds.), Academic Press, London, pp. 227–230.

Jöbsis, F. F., and Vreman, H. J., 1963, Inhibition of a $Na^+$ and $K^+$ stimulated adenosinetriphosphatase by oligomycin, *Biochim. Biophys. Acta* **73:**346–348.

Johannsson, A., Smith, G. A., and Metcalfe, J. C., 1981, The effect of bilayer thickness on the activity of ($Na^+$ pl $K^+$)-ATPase, *Biochim. Biophys. Acta* **641:**416–421.

Jørgensen, P. L., 1974a, Purification and characterization of ($Na^+$ + $K^+$)-ATPase. III. Purification from the outer medulla of mammalian kidney after selective removal of membrane components by SDS, *Biochim. Biophys. Acta* **356:**36–52.

Jørgensen, P. L., 1974b, Purification and characterization of ($Na^+$ + $K^+$)-ATPase. IV. Estimation of the purity and of the molecular weight and polypeptide content per enzyme unit in preparations from the outer medulla of the rabbit kidney, *Biochim. Biophys. Acta* **356:**53–67.

Jørgensen, P. L., 1975a, Purification and characterization of ($Na^+$ + $K^+$)-ATPase. V. Conformational changes in the enzyme. Transitions between the Na-form and the K-form studied with tryptic digestion as a tool, *Biochim. Biophys. Acta* **401:**399–415.

Jørgensen, P. L., 1975b, Isolation and characterization of the components of the sodium pump, *Q. Rev. Biophys.* **7:**239–274.

Jørgensen, P. L., 1977, Purification and characterization of ($Na^+$ + $K^+$)-ATPase. VI. Differential tryptic modification of catalytic functions of the purified enzyme in presence of NaCl and KCl, *Biochim. Biophys. Acta* **466:**97–108.

Jørgensen, J. C., 1982, Mechanism of the $Na^+$,$K^+$ pump. Protein structure and conformations of the pure ($Na^+$ + $K^+$)-ATPase, *Biochim. Biophys. Acta* **694:**27–68.

Jørgensen, P. L., and Anner, B. M., 1979, Purification and characterization of ($Na^+$ + $K^+$)-ATPase. VIII. Altered $Na^+$ : $K^+$ transport ratio in vesicles reconstituted with purified ($Na^+$ + $K^+$)-ATPase that has been selectively modified with trypsin in presence of NaCl, *Biochim. Biophys. Acta* **555:**485–492.

Jørgensen, P. L., and Karlish, S. J. D., 1980, Defective conformational response in a selectively trypsinized ($Na^+$ + $K^+$)-ATPase studied with tryptophan fluorescence, *Biochim. Biophys. Acta* **597:**305–317.

Jørgensen, P. L., and Klodos, I., 1978, Purification and characterization of ($Na^+$ + $K^+$)-ATPase. VII. Tryptic degradation of the Na-form of the enzyme protein resulting in selective modification of dephosphorylation reactions of the ($Na^+$ + $K^+$)-ATPase, *Biochim. Biophys. Acta* **507:**8–16.

Jørgensen, P. L., and Petersen, T., 1982, High-affinity 86Rb-binding and structural changes in the α-subunit of $Na^+$,$K^+$-ATPase as detected by tryptic digestion and fluorescence analysis, *Biochim. Biophys. Acta* **705:**38–47.

Jørgensen, P. L., and Skou, J. C. 1971, Purification and characterization of ($Na^+$ + $K^+$)-ATPase. I. Influence of detergents on the activity of ($Na^+$ + $K^+$)-ATPase in preparations from the outer medulla of rabbit kidney, *Biochim. Biophys. Acta* **233:**366–380.

Jørgensen, P. L., Hansen, O., Glynn, I. M., and Cavieres, J. D., 1973, Antibodies to pig kidney ($Na^+$ + $K^+$)-ATPase inhibit the $Na^+$ pump in human red cells provided they have access to the inner surface of the cell membrane, *Biochim. Biophys. Acta* **291:**795–800.

Jørgensen, P. L., Karlish, S. J. D., and Gitler, C., 1982a, Evidence for the organization of the transmembrane segments of (Na,K)-ATPase based on labeling lipid-embedded and surface domains of the α-subunit, *J. Biol. Chem.* **257:**7435–7442.

Jørgensen, P. L., Skriver, E., Hebert, H., and Maunsbach, A. B., 1982b, Structure of the Na,K-pump: Crystallization of pure membrane-bound Na,K-ATPase and identification of functional domains of the α-subunit, *Ann. N. Y. Acad. Sci.* **402:**207–224.

Jørgensen, P. L., Karlish, S. J. D., and Gitler, C., 1983, Organization of the transmembrane segments of Na,K-ATPase. Labeling of lipid embedded and surface domains of the α-subunit and its tryptic fragments with [$^{125}$I]-iodonaphthylazide, [$^{32}$P]-ATP, and photolabeled ouabain, *Curr. Top Membr. Trans.* **19:**127–130.

Judah, J. D., Ahmed, K., and McLean, A. E. M., 1962, Ion transport and phosphoproteins of human red cells, *Biochim. Biophys. Acta* **65:**472–480.

Kanazawa, T., Saito, M., and Tonomura, Y., 1967, Properties of a phosphorylated protein as a reaction intermediate of the Na + K sensitive ATPase, *J. Biochem. (Tokyo)* **61:**555–566.

Kanazawa, T., Saito, M., and Tonomura, Y., 1970, Formation and decomposition of a phosphorylated intermediate in the reaction of $Na^+$-$K^+$ dependent ATPase, *J. Biochem. (Tokyo)* **67:**693–711.

Kaniike, K., Erdmann, E., and Schoner, W., 1973, ATP binding to ($Na^+$ + $K^+$)-activated ATPase, *Biochim. Biophys. Acta* **298:**901–905.

Kaniike, K., Lindenmayer, G. M., Wallick, E. T., Lane, L. K., and Schwartz, A., 1976, Specific sodium-22 binding to a purified sodium + potassium adenosine triphosphatase. Inhibition by ouabain, *J. Biol. Chem.* **251:**4794–4796.

Kapakos, J. G., and Steinberg, M., 1982, Fluorescent labeling of (Na + K)-ATPase by 5-iodoacetamidofluorescein, *Biochim. Biophys. Acta* **693:**493–496.

Kaplan, J. H., 1982, Sodium pump mediated ATP–ADP exchange. The sided effects of sodium and potassium ions, *J. Gen. Physiol.* **80:**915–937.

Kaplan, J. H., 1983, Na pump catalysed ATP : ADP exchange in red blood cells: The effects of intracellular and extracellular Na and K ions, *Curr. Top. Membr. Trans.* **19:**671–675.

Kaplan, J. H., and Hollis, R. J., 1980, External Na dependence of ouabain-sensitive ATP–ADP exchange initiated by photolysis of intracellular caged-ATP in human red cell ghosts, *Nature* **288:**587–589.

Karlish, S. J. D., 1980, Characterization of conformational changes in (Na,K)ATPase labeled with fluorescein at the active site, *J. Bioenerg. Biomembr.* **12:**111–135.

Karlish, S. J. D., and Glynn, I. M., 1974, An uncoupled efflux of sodium ions from human red cells, probably associated with Na-dependent ATPase activity, *Ann. N. Y. Acad. Sci.* **242:**461–470.

Karlish, S. J. D., and Pick, U., 1981, Sidedness of the effects of sodium and potassium ions on the conformational state of the sodium-potassium pump, *J. Physiol.* **312:**505–529.

Karlish, S. J. D., and Stein, W. D., 1982a, Passive rubidium fluxes mediated by Na-K-ATPase reconstituted into phospholipid vesicles when ATP- and phosphate-free, *J. Physiol.* **328:**295–316.

Karlish, S. J. D., and Stein, W. D., 1982b, Effects of ATP or phosphate on passive rubidium fluxes mediated by Na-K-ATPase reconstituted into phospholipid vesicles, *J. Physiol.* **328:**317–331.

Karlish, S. J. D., and Yates, D. W., 1978, Tryptophan fluorescence of ($Na^+$ + $K^+$)-ATPase as a tool for study of the enzyme mechanism, *Biochim. Biophys. Acta* **527:**115–130.

Karlish, S. J. D., Jørgensen, P. L., and Gitler, C., 1977, Identification of a membrane-embedded segment of the large polypeptide chain of ($Na^+$,$K^+$)ATPase, *Nature* **269:**715–717.

Karlish, S. J. D., Yates, D. W., and Glynn, I. M., 1978a, Elementary steps of the ($Na^+$ + $K^+$)-ATPase mechanism, studied with formycin nucleotides, *Biochim. Biophys. Acta* **525:**230–251.

Karlish, S. J. D., Yates, D. W., and Glynn, I. M., 1978b, Conformational transitions between $Na^+$-bound and $K^+$-bound forms of ($Na^+$ + $K^+$)-ATPase, studied with formycin nucleotides, *Biochim. Biophys. Acta* **525:**252–264.

Karlish, S. J. D., Beaugé, L. A., and Glynn, I. M., 1979, Vanadate inhibits ($Na^+$ + $K^+$)ATPase by blocking a conformational change of the unphosphorylated form, *Nature* **282:**333–335.

Karlish, S. J. D., Lieb, W. R., and Stein, W. D., 1982, Combined effects of ATP and phosphate on rubidium exchange mediated by Na-K-ATPase reconstituted into phospholipid vesicles, *J. Physiol.* **328:**333–350.

Kennedy, B. G., and De Weer, P., 1976, Relationship between Na : K and Na : Na exchange by the sodium pump of skeletal muscle, *Nature* **268:**165–167.

Kennedy, B. G., Lunn, G., and Hoffman, J. F., 1983, Effect of intracellular adenine nucleotides on sodium pump catalyzed Na : Na and Na : K exchanges, *Curr. Top. Memb. Trans.* **19:**683–686.

Kepner, G. R., and Macey, R. I., 1968, Membrane enzyme systems. Molecular size determinations by radiation inactivation, *Biochim. Biophys. Acta* **163:**188–203.

Keynes, R. D., 1954, The ionic fluxes in frog muscle, *Proc. R. Soc. B.* **142:**359–382.

Keynes, R. D., and Steinhardt, R. A., 1968, The components of the Na efflux in frog muscle, *J. Physiol.* **198:**581–600.

Klevickis, C., and Grisham, C. M., 1982, Phosphorus-31 nuclear magnetic resonance studies of the conformation of an adenosine 5′-triphosphate analogue at the active site of ($Na^+$ + $K^+$)-ATPase from kidney medulla, *Biochemistry* **21:**69–79.

Klodos, I., and Nørby, J. G., 1979, Effect of $K^+$ and $Li^+$ on intermediary steps in the Na,K-ATPase reaction, in: *Na,K-ATPase: Structure and Kinetics* (J. C. Skou and J. G. Nørby, eds.), Academic Press, London, pp. 331–342.

Klodos, I., and Skou, J. C., 1975, The effect of $Mg^{2+}$ and chelating agents on intermediary steps of the reaction of $Na^+$,$K^+$-activated ATPase, *Biochim. Biophys. Acta* **391:**474–485.

Klodos, I., and Skou, J. C., 1977, The effect of chelators on $Mg^{2+}$, $Na^+$-dependent phosphorylation of ($Na^+$ + $K^+$)-activated ATPase, *Biochim. Biophys. Acta* **481:**667–679.

Klodos, I., Nørby, J. G., and Plesner, I. W., 1981, The steady-state kinetic mechanism of ATP hydrolysis catalysed by membrane-bound ($Na^+$ + $K^+$)-ATPase from ox brain. II. Kinetic characterization of phosphointermediates, *Biochim. Biophys. Acta* **643:**463–482.

Koepsell, H., 1979, Conformational changes of membrane-bound ($Na^+$ + $K^+$)-ATPase as revealed by trypsin digestion, *J. Membr. Biol.* **48:**69–94.

Koepsell, H., Hulla, F. W., and Fritzsch, G., 1982, Different classes of nucleotide binding sites in the ($Na^+$ + $K^+$)-ATPase studied by affinity labeling and nucleotide-dependent SH-group modifications, *J. Biol. Chem.* **257:**10733–10741.

Koyal, D., Rao, S. N., and Askari, A., 1971, Studies on the partial reactions catalyzed by the ($Na^+$ + $K^+$)-activated ATPase. I. Effects of simple anions and nucleotide triphosphates on the alkali-cation specificity of the *p*-nitrophenylphosphatase, *Biochim. Biophys. Acta* **225:**11–19.

Kracke, G. R., 1983, Absence of ouabain-like activity of the Na,K-ATPase inhibitor in guinea pig brain extract. *Curr. Top. Membr. Trans.* **19:**927–930.

Kudoh, F., Nakamura, S., Yamaguchi, M., and Tonomura, Y., 1979, Binding of ouabain to $Na^+K^+$-dependent ATPase during the ATPase reaction. Evidence for a dimer structure of the ATPase, *J. Biochem. (Tokyo)* **86:**1023–1028.

Kuriki, Y., and Racker, E., 1976, Inhibition of ($Na^+$ + $K^+$)-adenosine triphosphatase and its partial reactions by quercetin, *Biochemistry* **15:**4951–4956.

Kuriki, Y., Halsey, J., Biltonen, R., and Racker, E., 1976, Calorimetric studies of the interactions of magnesium and phosphate with ($Na^+$ + $K^+$)ATPase: Evidence for a ligand induced conformational change in the enzyme, *Biochemistry* **15:**4956–4961.

Kyte, J., 1971, Purification of the sodium- and potassium-dependent adenosine triphosphatase from canine renal medulla, *J. Biol. Chem.* **246:**4157–4165.

Kyte, J., 1972, Properties of the two polypeptides of sodium- and potassium-dependent adenosine triphosphatase, *J. Biol. Chem.* **247:**7642–7649.

Kyte, J., 1975, Structural studies of sodium and potassium ion-activated adenosine triphosphatase. The relationship between molecular structure and the mechanism of active transport, *J. Biol. Chem.* **250:**7443–7449.

Kyte, J., 1976a, Immunoferritin determination of the distribution of ($Na^+$ + $K^+$)ATPase over the plasma membranes of renal convoluted tubules. I. Distal segment, *J. Cell Biol.* **68:**287–303.

Kyte, J., 1976b, Immunoferritin determination of the distribution of ($Na^+$ + $K^+$)ATPase over the plasma membranes of renal convoluted tubules. II. Proximal segment, *J. Cell Biol.* **68:**304–318.

Lane, L. K., Copenhaver, J. H., Lindenmayer, G. E., and Schwartz, A., 1973, Purification and characterization of, and [$^3$H]ouabain binding to the transport adenosine triphosphatase from outer medulla of canine kidney, *J. Biol. Chem.* **248:**7197–7200.

Lant, A. F., Priestland, R. N., and Whittam, R., 1970, The coupling of downhill ion movements associated with reversal of the Na pump in human red cells, *J. Physiol.* **207:**291–301.

Laris, P. C., and Letchworth, P. E., 1962, Cation influence on inorganic phosphate production in human erythrocytes, *J. Cell. Comp. Physiol.* **60:**229–234.

Lee, K. H., and Blostein, R., 1980, Red cell sodium fluxes catalysed by the sodium pump in the absence of $K^+$ and ADP, *Nature* **285:**338–339.

Lew, V. L., and Beaugé, L. A., 1979, Passive cation fluxes in red cell membranes, in: *Membrane Transport in Biology,* (G. Giebisch, D. C. Tosteson, and H. H. Ussing, eds.), Springer-Verlag, Berlin, pp. 81–115.

Lew, V. L., Glynn, I. M., and Ellory, J. C., 1970, Net synthesis of ATP by reversal of the sodium pump, *Nature* **225:**865–866.

Lew, V. L., Hardy, M. A., and Ellory, J. C., 1973, The uncoupled extrusion of $Na^+$ through the $Na^+$ pump, *Biochim. Biophys. Acta* **323:**251–266.

Liang, S-M., and Winter, C. G., 1977, Digitonin-induced changes in subunit arrangement in relation to some *in vitro* activities of the ($Na^+$,$K^+$)-ATPase, *J. Biol. Chem.* **252:**8278–8284.

Lin, M. H., and Akera, T., 1978, Increased ($Na^+$,$K^+$)-ATPase concentrations in various tissues of rats caused by thyroid hormone treatment, *J. Biol. Chem.* **253:**723–726.

Lindenmayer, G. E., and Schwartz, A., 1973, Nature of the transport adenosine triphosphatase digitalis complex. IV. Evidence that sodium–potassium competition modulates the rate of ouabain interaction with ($Na^+$ + $K^+$) adenosine triphosphatase during enzyme catalysis, *J. Biol. Chem.* **248:**1291– 1300.

Lindenmayer, G. E., Laughter, A. H., and Schwartz, A., 1968, Incorporation of inorganic phosphate-32 into a $Na^+$,$K^+$-ATPase preparation: Stimulation by ouabain, *Arch. Biochem. Biophys.* **127:**187–192.

Lopez, V., Stevens, T., and Lindquist, R. N., 1976, Vanadium ion inhibition of alkaline phosphatase-catalyzed phosphate ester hydrolysis, *Arch. Biochem. Biophys.* **175:**31–38.

Lowe, A. G., and Reeve, L. A., 1983, Pre-steady state hydrolysis of ATP and enzyme phosphorylation in the Na,K-ATPase, *Curr. Top. Membr. Trans.* **19:**577–580.

Lowe, A. G., and Smart, J. W., 1977, The pre-steady-state hydrolysis of ATP by porcine brain ($Na^+$ + $K^+$)-dependent ATPase, *Biochim. Biophys. Acta* **481:**695–705.

Macara, I. G., Kustin, K., and Cantley, L. C., 1980, Glutathione reduces cytoplasmic vanadate. Mechanism and physiological implications, *Biochim. Biophys. Acta* **629:**95–106.

MacGregor, G. A., Fenton, S., Alaghband-Zadeh, J., Markandu, N., Roulston, J. E., and De Wardener, H., 1981, Evidence for a raised concentration of a circulating sodium transport inhibitor in essential hypertension, *Br. Med. J.* **283:**1355–1357.

Maizels, M., and Patterson, J. H., 1940, Survival of stored blood after transfusion, *Lancet* **2:**417–420.

Mårdh, S., 1975a, Bovine brain $Na^+$,$K^+$-stimulated ATP phosphohydrolase studied by a rapid-mixing technique. $K^+$-stimulated liberation of [$^{32}$P]orthophosphate from [$^{32}$P]phosphoenzyme and resolution of the dephosphorylation into two phases, *Biochim. Biophys. Acta* **391:**448–463.

Mårdh, S., 1975b, Bovine brain $Na^+$,$K^+$-stimulated ATP phosphohydrolase studied by a rapid-mixing technique. Detection of a transient [$^{32}$P]phosphoenzyme formed in the presence of potassium ions, *Biochim. Biophys. Acta* **391:**464–473.

Mårdh, S., and Lindahl, S., 1977, On the mechanism of sodium- and potassium-activated adenosine triphosphatase. Time course of intermediary steps examined by computer simulation of transient kinetics, *J. Biol. Chem.* **252:**8058–8061.

Mårdh, S., and Post, R. L., 1977, Phosphorylation from adenosine triphosphate of sodium- and potassium-activated adenosine triphosphatase. Comparison of enzyme-ligand complexes as precursors to the phosphoenzyme, *J. Biol. Chem.* **252:**633–638.

Mårdh, S., and Zetterqvist, O., 1972, Phosphorylation of bovine brain $Na^+$,$K^+$-stimulated ATP phosphohydrolase by adenosine [$^{32}$P]triphosphate studied by a rapid-mixing technique, *Biochim. Biophys. Acta* **255:**231–238.

Mårdh, S., and Zetterqvist, O., 1974, Phosphorylation and dephosphorylation reactions of bovine brain ($Na^+$-$K^+$)-stimulated ATP phosphohydrolase studied by a rapid-mixing technique, *Biochim. Biophys. Acta* **350:**473–483.

Masiak, S. J., and D'Angelo, G., 1975, Effects of *N*-acetylimidazole on human erythrocyte ATPase activity. Evidence for a tyrosyl residue at the ATP-binding site of the ($Na^+$ + $K^+$)-dependent ATPase, *Biochim. Biophys. Acta* **382:**83–91.

Matchett, P. A., and Johnson, J. A., 1954, Inhibition of Na and K transport in frog sartorii in the presence of ouabain, *Fed. Proc.* **13:**384.

Matsui, H., and Homareda, H., 1982, Interaction of sodium and potassium ions with $Na^+$,$K^+$-ATPase. I. Ouabain-sensitive alternative binding of three $Na^+$ or two $K^+$ to the enzyme, *J. Biochem. (Tokyo)* **92:**193–217.

Matsui, H., and Schwartz, A., 1968, A mechanism of cardiac glycoside inhibition of the ($Na^+$ + $K^+$)-dependent ATPase from cardiac tissue, *Biochim. Biophys. Acta* **151:**655–663.

Matsui, H., Hayashi, Y., Homareda, H., and Kimimura, M., 1977, Ouabain sensitive $^{42}$K binding to $Na^+$,$K^+$-ATPase purified from canine kidney outer medulla, *Biochem. Biophys. Res. Commun.* **75:**373–380.

Matsui, H., Hayashi, Y., Homareda, H., and Taguchi, M., 1983, Stoichiometrical binding of ligands to less than 160 K Daltons of $Na^+$,$K^+$-ATPase, *Curr. Top. Membr. Trans.* **19:**145–148.

Maunsbach, A. B., Deguchi, N., and Jørgensen, P. L., 1978, Ultrastructure of purified Na,K-ATPase, in: *FEBS Symp. A4, Membrane Proteins,* Vol. 45 (P. Nicholls, J. V. Moller, P. L. Jørgensen, and A. J. Moody, eds.), Pergamon, New York, pp. 173–181.

Maunsbach, A. B., Skriver, E., and Jørgensen, P. L., 1979, Ultrastructure of purified Na,K-ATPase membranes, in: *Na,K-ATPase: Structure and Kinetics* (J. C. Skou and J. G. Nørby, eds.), Academic Press, London, pp. 3–13.

Maunsbach, A. B., Skriver, E., and Jørgensen, P. L., 1983, Electron microscope analysis of protein distribution in purified, membrane-bound Na,K-ATPase, *Curr. Top. Membr. Trans.* **19,** in press.

Mercer, R. W., and Dunham. P. B., 1981, Membrane-bound ATP fuels the Na/K pump. Studies on membrane-bound glycolytic enzymes on inside-out vesicles from human red cell membranes, *J. Gen. Physiol.* **78:**547–568.

Moczydlowski, E. G., and Fortes, P. A. G., 1981a, Characterization of 2′3′-*O*-(2,4,6-trinitrocyclohexadienylidene) adenosine 5′-triphosphate as a fluorescent probe of the ATP site of sodium and potassium transport adenosine triphosphatase. Determination of nucleotide binding stoichiometry and ion-induced changes in affinity for ATP, *J. Biol. Chem.* **256:**2346–2356.

Moczydlowski, E. G., and Fortes, P. A. G., 1981b, Inhibition of sodium and potassium adenosine triphosphatase by 2′3′-*O*-(2,4,6-trinitrocyclohexadienylidene) adenine nucleotides. Implications for the structure and mechanism of the Na : K pump, *J. Biol. Chem.* **256:**2357–2366.

Mone, M.D. and Kaplan, J.H., 1983, Cation activation of Na,K-ATPase after treatment with thimerosal, *Curr. Top. Membr. Trans.* **19:**465–469.

Mullins, L. J., and Brinley, F. J., 1969, Potassium fluxes in dialyzed squid axons, *J. Gen. Physiol.* **53:**704–740.

Mullins, L. J., Fenn, W. O., Noonan, T. R., and Haege, L., 1941, Permeability of erythrocytes to radioactive potassium, *Am. J. Physiol.* **135:**93–101.

Munakata, H., Schmid, K., Collins, J. H., Zot, A., Lane, L. K., and Schwartz, A., 1982, The α and β subunits of lamb kidney Na,K-ATPase are both glycoproteins, *Biochem. Biophys. Res. Commun.* **107:**229–231.

Munson, K. B.; 1981, Light dependent inactivation of ($Na^+$ + $K^+$)-ATPase with a new photoaffinity reagent, chromium arylazido-β-alanyl ATP, *J. Biol. Chem.* **256:**3223–3230.

Nagai, K., and Yoshida, H., 1966, Biphasic effects of nucleotides on potassium dependent phosphatase, *Biochim. Biophys. Acta* **128:**410–412.

Nagai, K., Izumi, F., and Yoshida, H., 1966, Studies on potassium dependent phosphatase: Its distribution and properties, *J. Biochem (Tokyo)* **59:**295–303.

Nagano, K., Fujihara, Y., Hara, Y., and Nakao, M., 1973, ATP as a modulator of $Na^+$,$K^+$,-ATPase, in: *Organization of Energy-Transducing Membranes* (M. Nakao and L. Packer, eds.), University of Tokyo Press, Tokyo, pp. 47–61.

Naidoo, B. K., Witty, T. R., Remers, W. A., and Besch, H. R., 1974, Cardiotonic steroids: I. Importance of 14 β-hydroxy in digitoxigenin, *J. Pharmcol. Sci.* **63:**1391–1394.

Nakao, T., Nakao, M., Nagai, F., Kawai, K., Fujihara, Y., Hara, Y., and Fujita, M., 1973, Purification and some properties of Na,K-transport ATPase. II. Preparations with high specific activity obtained using aminoethyl cellulose chromatography, *J. Biochem. (Tokyo)* **73:**781–791.

Neufeld, A. H., and Levy, H. M., 1969, A second ouabain-sensitive Na dependent ATPase in brain microsomes, *J. Biol. Chem.* **244:**6493–6497.

Neufeld, A. H., and Levy, H. M., 1970, The steady state level of phosphorylated intermediate in relation to the two sodium-dependent adenosine triphosphatases of calf brain microsomes, *J. Biol. Chem.* **245:**4962–4967.

Noble, D., 1980, Mechanisms of action of therapeutic levels of glycosides, *Cardiovasc. Res.* **14:**495–514.

Noonan, T. R., Fenn, W. O., and Haege, L., 1940, The distribution of injected radioactive potassium in rats, *Am. J. Physiol.* **132:**474–488.

Nørby, J. G., 1983, Ligand interactions with the substrate site of Na,K-ATPase: Nucleotides, vanadate and phosphorylation, *Curr. Top. Membr. Trans.* **19:**281–314.

Nørby, J. G., and Jensen, J., 1971, Binding of ATP to brain microsomal ATPase. Determination of the ATP-binding capacity and the dissociation constant of the enzyme-ATP complex as a function of $K^+$ concentration, *Biochim. Biophys. Acta* **233:**104–116.

O'Connor, S. E., and Grisham, C. M., 1980, Distance determination at the active site of kidney ($Na^+$ + $K^+$)-ATPase by Mn(II) ion electron paramagnetic resonance, *FEBS Lett.* **118:**303–307.

Ottolenghi, P., 1979, The relipidation of delipidated Na,K-ATPase. An analysis of complex formation with dioleoylphosphatidylcholine and with dioleoylphophatidylethanolamine, *Eur. J. Biochem.* **99:**113–131.

Ottolenghi, P., and Jensen, J., 1983, The $K^+$-induced apparent heterogeneity of high-affinity nucleotide-binding sites in ($Na^+$ + $K^+$)-ATPase can only be due to the oligomeric structure of the enzyme, *Biochim. Biophys. Acta* **727,** in press.

Ottolenghi, P., Ellory, J. C., and Klein, R., 1983, Radiation inactivation analysis of the partial reactions of NaK activated ATPase, *Curr. Top. Membr. Trans.* **19,** in press.

Overton, E., 1902, Beitrage zur allgemeinen Muskel- und Nervenphysiologie. II. Mittheilung. Ueber die Unentbehrlichkeit von Natrium- (oder Lithium-) Ionen fur den Contractionsact des Muskels, *Pflugers Arch. Ges. Physiol.* **92:**346–386.

Pachence, J. M., Schoenborn, B. P., and Edelman, I. S., 1983, Low angle neutron scattering analysis of Na/K-ATPase in detergent solution, *Biophys. J.* **41:**370a.

Patzelt-Wenczler, R., and Mertens, W., 1981, Effects of cations on high-affinity and low-affinity ATP-binding sites of (Na,K)-ATPase as studied by disulfides of thioinosine triphosphate and its analogue, *Eur. J. Biochem.* **121:**197–202.

Patzelt-Wenczler, R., and Schoner, W., 1981, Evidence for two different reactive sulfhydryl groups in the ATP-binding sites of ($Na^+$ + $K^+$)-ATPase, *Eur. J. Biochem.* **114:**79–87.

Perez, B., Miara, J., and Dahms, A. S., 1979, Probes at the medium and intermediate water oxygen exchange reactions of the Na,K-ATPase, in: *Na,K-ATPase: Structure and Kinetics* (J. C. Skou and J. G. Nørby, eds.), Academic Press, London, pp. 343–358.

Perrone, J. R., Hackney, J. F., Dixon, J. F., and Hokin, L. E., 1975, Molecular properties of purified (sodium and potassium)-activated adenosine triphosphatases and their subunits from the rectal gland of *Squalus acanthias* and the electric organ of *Electrophorus electricus*, *J. Biol. Chem.* **250:**4178–4184.

Peters, W. H. M., Du Pont, J. J. H. H. M., Koppers, A., and Bonting, S. L., 1981, Studies on ($Na^+$ + $K^+$)-activated ATPase. XLVII. Chemical composition, molecular weight and molar ratio of the subunits of the enzyme from rabbit kidney outer medulla, *Biochim. Biophys. Acta* **641:**55–70.

Peterson, G. L., and Hokin, L. E., 1980, Improved purification of brine-shrimp *(Artemia saline)* ($Na^+$ + $K^+$)-activated adenosine triphosphatase and amino-acid and carbohydrate analogues of the isolated subunits, *Biochem. J.* **192:**107–118.

Peterson, G. L., and Hokin, L. E., 1981, Molecular weight and stoichiometry of the sodium- and potassium-activated adenosine triphosphatase subunits, *J. Biol. Chem.* **256:**3751–3761.

Plesner, L., and Plesner, I. W., 1981a, The steady-state kinetic mechanism of ATP hydrolysis catalyzed by membrane-bound ($Na^+$ + $K^+$)-ATPase from ox brain. I. Substrate identity, *Biochim. Biophys. Acta* **643:**449–462.

Plesner, I. W., and Plesner, L., 1981b, The steady state kinetic mechanism of ATP hydrolysis catalyzed by membrane-bound ($Na^+$ + $K^+$)-ATPase from ox brain. IV. Rate-constant determination, *Biochim. Biophys. Acta* **648:**231–246.

Plesner, I. W., Plesner, L., Nørby, J. G., and Klodos, I., 1981, The steady-state kinetic mechanism of ATP hydrolysis catalyzed by membrane-bound ($Na^+$ + $K^+$)-ATPase from ox brain. III. A minimal model, *Biochim. Biophys. Acta* **643:**483–494.

Post, R. L., and Kume, S., 1973, Evidence for an aspartyl phosphate residue at the active site of sodium and potassium ion transport adenosine triphosphatase, *J. Biol. Chem.* **248:**6993–7000.

Post, R. L., and Sen, A. K., 1965, An enzymatic mechanism of active sodium and potassium transport, *J. Histochem.* **13:**105–112.

Post, R. L., Merritt, C. R., Kinsolving, C. R., and Albright, C. D., 1960, Membrane adenosine triphosphate-dependent sodium and potassium transport across kidney membrane, *J. Biol. Chem.* **240:**1437–1445.

Post, R. L., Sen, A. K., and Rosenthal, A. S., 1965, A phosphorylated intermediate in adenosine triphosphate-dependent sodium and potassium transport across kidney membranes, *J. Biol. Chem.* **240:**1437–1445.

Post, R. L., Kume, S., Tobin, T., Orcutt, B., and Sen, A. K., 1969, Flexibility of an active centre in sodium-plus-potassium adenosine triphosphatase, *J. Gen. Physiol.* **54:**306*s*–326*s*.

Post, R. L., Hegyvary, C., and Kume, S., 1972, Activation by adenosine triphosphate in the phosphorylation kinetics of sodium and potassium ion transport adenosine triphosphatase, *J. Biol. Chem.* **247:**6530–6540.

Post, R. L., Kume, S., and Rogers, F. N., 1973, Alternating paths of phosphorylation of the sodium and potassium ion pump of plasma membranes, in: *Mechanisms in Bioenergetics* (G. F. Azzone, S. Ernster, E. Papa, N. Quagliariello, and N. Siliprandi, eds.), Academic Press, New York, pp. 203–218.

Post, R. L., Toda, G., and Rogers, F. N., 1975a, Phosphorylation by inorganic phosphate of sodium plus potassium ion transport adenosine triphosphatase. Four reactive states, *J. Biol. Chem.* **250:**691–701.

Post, R. L., Toda, G., Kume, S., and Taniguchi, K., 1975b, Synthesis of adenosine triphosphate by way of potassium-sensitive phosphoenzyme of sodium, potassium adenosine triphosphatase, *J. Supramolec. Struct.* **3:**479–497.

Post, R. L., Hunt, D., Walderhaug, M. O., Perkins, R. C., Park, J. H., and Beth, A. H., 1979, Vanadium compounds in relation to inhibition of sodium and potassium adenosine triphosphatase, in: *Na,K-ATPase: Structure and Kinetics* (J. C. Skou and J. G. Nørby, eds.), Academic Press, London, pp. 389–401.

Poston, L., Sewell, R. B., Wilkinson, S. P., Richardson, P. J., Williams, R., Clarkson, E. M., MacGregor, G. A., and De Wardener, H. E., 1981, Evidence for a circulating sodium transport inhibitor in essential hypertension, *Br. Med. J.* **282:**847–849.

Priestland, R. N., and Whittam, R., 1968, The influence of external sodium ions on the sodium pump in erythrocytes, *J. Physiol.* **109:**369–374.

Proverbio, F., and Hoffman, J. F., 1977, Membrane compartmentalized ATP and its preferential use by the Na,K-ATPase of human red cell ghosts, *J. Gen. Physiol.* **69:**605–632.

Reeves, A. S., Collins, J. H., and Schwartz, A., 1980, Isolation and characterization of (Na,K)-ATPase proteolipid, *Biochem. Biophys. Res. Commun.* **95:**1591–1598.

Rega, A. F., Garrahan, P. J., and Pouchan, M. I., 1970, Potassium activated phosphatase from human red blood cells. The asymmetrical effects of $K^+$, $Na^+$, $Mg^{++}$ and adenosine triphosphate, *J. Membr. Biol.* **3:**14–25.

Rempeters, G., and Schoner, W., 1983, Imidazole chloride and Tris chloride substitute for sodium chloride in inducing high affinity Ado*PP*[NH]*P* binding to ($Na^+$ + $K^+$)-ATPase, *Biochim. Biophys. Acta* **727:**13–21.

Repke, K. R. H., 1965, Effect of digitalis on membrane adenosine triphosphatase of heart muscle, in: *Drugs and Enzymes,* Vol. 4 (B. B. Brodie and J. Gillette, eds.), Proc. 2nd Int. Pharmacol. Meet. Prague, 1963, Pergamon Press, Oxford, and Czechoslovak Medical Press, pp. 65–87.

Repke, K. R. H., and Dittrich, F., 1979, Subunit–subunit interaction: Determinant of reactivity and cooperativity of Na,K-ATPase, in: *Na,K-ATPase: Structure and Kinetics* (J. C. Skou and J. G. Nørby, eds.), Academic Press, London, pp. 487–500.

Repke, K. R. H., and Portius, H. J., 1966, Analysis of structure activity relationships in cardioactive compounds on the molecular level, in: *Scientiae Pharmaceuticae—I,* Proc. 25th Congr. Pharmaceut. Sci., Prague, 1965, pp. 39–57.

Repke, K. R. H., and Schon, R., 1973, Flip-Flop model of (Na,K)-ATPase function, *Acta Biol. Med. Ger.* **31:**K19–K30.

Richards, D. E., Ellory, J. C., and Glynn, I. M., 1981, Radiation inactivation of ($Na^+$ + $K^+$)-ATPase. A small target size for the $K^+$-occluding mechanism, *Biochim. Biophys. Acta* **648:**284–286.

Robinson, J. D., 1969, Kinetic studies on a brain microsomal adenosine triphosphatase. III. Potassium-dependent phosphatase activity, *Biochemistry* **8:**3348–3355.

Robinson, J. D., 1970, Phosphatase activity stimulated by $Na^+$ plus $K^+$: Implications for the ($Na^+$ plus $K^+$)-dependent adenosine triphosphatase, *Arch. Biochem. Biophys.* **139:**164–171.

Robinson, J. D., 1973, Cation sites of the ($Na^+$ + $K^+$)-dependent ATPase. Mechanisms for $Na^+$-induced changes in $K^+$ affinity of the phosphatase activity, *Biochim. Biophys. Acta* **321:**662–670.

Robinson, J. D., 1974, Nucleotide and divalent cation interactions with the ($Na^+$ + $K^+$)-dependent ATPase, *Biochim. Biophys. Acta* **341:**232–247.

Robinson, J. D., 1975, Functionally distinct classes of $K^+$ sites on the ($Na^+$ + $K^+$)-dependent ATPase, *Biochim. Biophys. Acta* **384:**250–264.

Robinson, J. D., 1976a, Substrate sites of the ($Na^+$ + $K^+$)-dependent ATPase, *Biochim. Biophys. Acta* **429:**1006–1019.

Robinson, J. D., 1976b, The ($Na^+$ + $K^+$)-dependent ATPase; mode of inhibition of ADP/ATP exchange activity by $MgCl_2$, *Biochim. Biophys. Acta* **440:**711–722.

Robinson, J. D., 1980, Sensitivity of the ($Na^+$ + $K^+$)-ATPase to state-dependent inhibitors: Effects of digitonin and Triton X-100, *Biochim. Biophys. Acta* **598:**543–553.

Robinson, J. D., 1981, Substituting manganese for magnesium alters certain reaction properties of the ($Na^+$ + $K^+$)-ATPase, *Biochim. Biophys. Acta* **642:**405–417.

Robinson, J. D., 1982, Tryptic digestion of the (Na + K)-ATPase is both sensitive to and modifies $K^+$ interactions with the enzyme, *J. Bioenerg. Biomembr.* **14:**319–333.

Robinson, J. D., 1983, Kinetic studies on the ($Na^+$ + $K^+$)-dependent ATPase. Evidence for coexisting sites for $Na^+$, $K^+$ and $Mg^{2+}$, *Biochim. Biophys. Acta* **727:**63–69.

Robinson, J. D., and Flashner, M. S., 1979, The ($Na^+$ + $K^+$)-activated ATPase; enzymatic and transport properties, *Biochim. Biophys. Acta* **549:**145–176.

Robinson, J. D., and Mercer, R. W., 1981, Vanadate binding to the (Na + K)-ATPase, *J. Bioenerg. Biomembr.* **13:**205–218.

Robinson, J. D., Flashner, M. S., and Marin, G. K., 1978, Inhibition of the ($Na^+$ + $K^+$)-dependent ATPase by inorganic phosphate, *Biochim. Biophys. Acta* **509:**419–428.

Rogers, T. B., and Lazdunski, M., 1979a, Photoaffinity labeling of the digitalis receptor in the (sodium + potassium)-activated adenosinetriphosphatase, *Biochemistry* **18:**135–140.

Rogers, T. B., and Lazdunski, M., 1979b, Photoaffinity labelling of a small protein component of a purified ($Na^+ + K^+$)ATPase, *FEBS Lett.* **98:**373–376.

Rossi, B., Gache, C., and Lazdunski, M., 1978, Specificity and interactions at the cationic sites of the axonal ($Na^+,K^+$)-activated adenosinetriphosphatase, *Eur. J. Biochem.* **85:**561–570.

Rossi, B., Vuilleumier, P., Gache, C., Balerna, M., and Lazdunski, M., 1980, Affinity labeling of the digitalis receptor with *p*-nitrophenyltriazene-ouabain, a highly specific alkylating agent, *J. Biol. Chem.* **255:**9936–9941.

Rossi, B., Ponzio, G., and Lazdunski, M., 1982, Identification of the segment of the catalytic subunit of ($Na^+$, $K^+$)ATPase containing the digitalis binding site, *EMBO J.* **1:**859–861.

Rubinson, K. A., 1981, Concerning the form of biochemically active vanadium, *Proc. R. Soc. B.* **212:**65–84.

Ruoho, A., and Kyte, J., 1974, Photoaffinity labeling of the ouabain-binding site on ($Na^+ + K^+$)adenosinetriphosphatase, *Proc. Natl. Acad. Sci. USA* **71:**2352–2356.

Sabatini, D., Golman, D., Sabban, E., Sherman, J., Morimoto, T., Kreibich, G., and Adesnik, M., 1982, Mechanisms for the incorporation of protein into the plasma membranes, *Cold Spring Harbor Symp. Quant. Biol.* **XLVI:**807–818.

Sachs, J. R., 1967, Competition effects of some cations on active potassium transport in the human red blood cell, *J. Clin. Invest.* **46:**1433–1441.

Sachs, J. R., 1970, Sodium movements in the human red blood cell, *J. Gen. Physiol.* **56:**322–341.

Sachs, J. R., 1972, Recoupling the Na-K pump, *J. Clin. Invest.* **51:**3244–3247.

Sachs, J. R., 1974, Interaction of external K, Na, and cardioactive steroids with Na-K pump of the human red blood cell, *J. Gen. Physiol.* **63:**123–143.

Sachs, J. R., 1977a, Kinetics of the inhibition of the Na-K pump by external sodium, *J. Physiol.* **264:**449–470.

Sachs, J. R., 1977b, Kinetic evaluation of the Na-K pump reaction mechanisms, *J. Physiol.* **273:**489–514.

Sachs, J. R., 1979, A modified consecutive model for the Na,K-pump, in: *Na,K ATPase: Structure and Kinetics* (J. C. Skou and J. G. Nørby, eds.), Academic Press, London, pp. 463–473.

Sachs, J. R., 1980, The order of release of sodium and addition of potassium in the sodium–potassium pump reaction mechanism, *J. Physiol.* **302:**219–240.

Sachs, J. R., 1981a, Mechanistic implications of the potassium–potassium exchange carried out by the sodium–potassium pump, *J. Physiol.* **316:**263–277.

Sachs, J. R., 1981b, Internal potassium stimulates the sodium–potassium pump by increasing cell ATP concentration, *J. Physiol.* **319:**515–528.

Sachs, J. R., and Welt, L. G., 1967, The concentration dependence of active K transport in the human red blood cell, *J. Clin. Invest.* **46:**65–76.

Sachs, S., Rose, J. D., and Hirschowitz, B. I., 1967, Acetyl phosphatase in brain microsomes: A partial reaction of $Na^+ + K^+$-ATPase, *Arch. Biochem. Biophys.* **119:**277–281.

Schatzmann, H. J., 1953, Herzglykoside als Hemmstoffe fur den aktiven Kalium und Natrium Transport durch die Erythrocytenmembran, *Helv. Physiol. Acta* **11:**346–354.

Schönfeld, W., Schön, R., Menke, K. H., and Repke, K. R. H., 1972, Identification of conformational states of transport ATPase by kinetic analysis of ouabain binding, *Acta Biol. Med. Germ.* **28:**935–956.

Schoot, B. M., Van Emst-de Vries, S. E., Van Haard, P. M. M., De Pont, J. J. H. H. M., and Bonting, S. L., 1980, Studies on ($Na^+ + K^+$)-activated ATPase. XLVI. Effect on cation-induced conformational changes on sulfhydryl group modification, *Biochim. Biophys. Acta* **602:**144–154.

Schuurmans Stekhoven, F. M. A. H., and Bonting, S. L., 1981, Transport adenosine triphosphatases: Properties and functions, *Physiol. Rev.* **61:**1–76.

Schuurmans Stekhoven, F. M. A. H., Swarts, H. G. P., De Pont, J. J. H. H. M., and Bonting, S. L., 1980, Studies in ($Na^+ + K^+$)-activated ATPase. XLIV. Single phosphate incorporation during dual phosphorylation by inorganic phosphate and adenosine triphosphate, *Biochim. Biophys. Acta* **597:**100–111.

Schuurmans Stekhoven, F. M. A. H., Swarts, H. G. P., De Pont, J. J. H. H. M., and Bonting, S. L., 1981, Studies on ($Na^+ + K^+$)-activated ATPase. XLV. Magnesium induces two low-affinity non-phosphorylating nucleotide binding sites per molecule, *Biochim. Biophys. Acta* **649:**533–540.

Schwartz, A., Matsui, H., and Laughter, A. H., 1968, Tritiated digoxin binding to ($Na^+ + K^+$)-activated adenosine triphosphatase: Possible allosteric site, *Science* **160:**323–325.

Schwartz, A., Lindenmayer, G. E., and Allen, J. C., 1975, The sodium-potassium adenosine triphosphatase: pharmacological, physiological, and biochemical aspects, *Pharmacol. Rev.* **27**:3–134.

Schwartz, A., Whitmer, K., Grupp, G., Grupp, I., Adams, R. J., and Lee, S-W., 1982, Mechanism of action of digitalis: Is the Na,K-ATPase the pharmacological receptor? *Ann. N. Y. Acad. Sci.* **402**:253–270.

Sen, A. K., Tobin, T., and Post, R. L., 1969, A cycle for ouabain inhibition of sodium- and potassium-dependent adenosine triphosphatase, *J. Biol. Chem.* **244**:6596–6604.

Sen, P. C., Kapakos, J. G., and Steinberg, M., 1981, Modification of ($Na^+$ + $K^+$)-dependent ATPase by fluorescein isothiocyanate: Evidence for the involvement of different amino groups at different pH values, *Arch. Biochem. Biophys.* **211**:652–661.

Shaffer, E., Azari, J., and Dahms, A. S., 1978, Properties of the Pi–oxygen exchange reaction catalyzed by ($Na^+$,$K^+$)-dependent adenosine triphosphatase, *J. Biol. Chem.* **253**:5696–5706.

Shaver, J. L., and Stirling, C., 1978, Ouabain binding to renal tubules of the rabbit, *J. Cell Biol.* **76**:278–292.

Siegel, G. J., 1979, Revised enzyme reaction model for Na,K-ATPase incorporating consecutive and simultaneous reactions with $Na^+$ and $K^+$, in: *Na,K-ATPase: Structure and Kinetics* (J. C. Skou and J. G. Nørby, eds.), Academic Press, London, pp. 287–299.

Siegel, G. J., and Albers, R. W., 1967, Sodium–potassium activated adenosine triphosphatase of *Electrophorus* electric organ. IV. Modification of responses to sodium and potassium by arsenite plus 2,3-dimercaptopropanol, *J. Biol. Chem.* **242**:4972–4975.

Siegel, G. J., Koval, G. J., and Albers, R. W., 1969, Sodium–potassium-activated adenosine triphosphatase. VI. Characterization of the phosphoprotein formed from orthophosphate in the presence of ouabain, *J. Biol. Chem.* **244**:3264–3269.

Siegel, G. J., Fogt, S. K., and Iyengar, S., 1973, Characteristics of lead ion-stimulated phosphorylation of *Electrophorus electricus* electroplax ($Na^+$ + $K^+$)-adenosine triphosphatase and inhibition of ATP–ADP exchange, *J. Biol. Chem.* **253**:7207–7211.

Simons, T. J. B., 1974, Potassium : potassium exchange catalysed by the sodium pump in human red cells, *J. Physiol.* **237**:123–155.

Simons, T. J. B., 1975, The interaction of ATP-analogues possessing a blocked γ-phosphate group with the sodium pump in human red cells, *J. Physiol.* **244**:731–739.

Sjodin, R. A., 1971, The kinetics of Na extrusion in striated muscle as functions of the external sodium and potassium ion concentrations, *J. Gen. Physiol.* **57**:164–187.

Skou, J. C., 1957, The influence of some cations on an adenosine triphosphatase from peripheral nerves, *Biochim. Biophys. Acta* **23**:394–401.

Skou, J. C., 1960, Further investigations on a $Mg^{++}$ + $Na^+$-activated adenosine triphosphatase, possibly related to the active linked transport of $Na^+$ and $K^+$ across the nerve membrane, *Biochim. Biophys. Acta* **42**:6–23.

Skou, J. C., 1965, Enzymatic basis for active transport of $Na^+$ and $K^+$ across cell membrane, *Physiol. Rev.* **45**:596–617.

Skou, J. C., 1971, Sequence of steps in the (Na + K)-activated enzyme system in relation to sodium and potassium transport, *Curr. Top. Bioenerg.* **4**:357–398.

Skou, J. C., 1974a, Effect of ATP on the intermediary steps of the reaction of the ($Na^+$ + $K^+$)-dependent enzyme system. I. Studied by the use of *N*-ethylmaleimide inhibition as a tool, *Biochim. Biophys. Acta* **339**:234–245.

Skou, J. C., 1974b, Effect of ATP on the intermediary steps of the reaction of the ($Na^+$ + $K^+$)-dependent enzyme system. III. Effect on the *p*-nitrophenylphosphatase activity of the system, *Biochim. Biophys. Acta* **339**:258–273.

Skou, J. C., 1982, The effect of pH, of ATP and of modification with pyridoxal 5-phosphate on the conformational transition between the $Na^+$-form and the $K^+$-form of the ($Na^+$ + $K^+$)-ATPase, *Biochim. Biophys. Acta* **688**:369–380.

Skou, J. C., and Esmann, M., 1979, Preparation of membrane-bound and of solubilized ($Na^+$ + $K^+$)-ATPase from rectal glands of *Squalus acanthias*. The effect of preparative procedures on purity, specific and molar activity, *Biochim. Biophys. Acta* **567**:436–444.

Skou, J. C., and Esmann, M., 1980, Effects of ATP and protons on the Na : K selectivity of the ($Na^+$ + $K^+$)-ATPase studied by ligand effects on intrinsic and extrinsic fluorescence, *Biochim. Biophys. Acta* **601**:386–402.

Skou, J. C., and Esmann, M., 1981, Eosin, a fluorescent probe of ATP binding to the (Na$^+$ + K$^+$)-ATPase, *Biochim. Biophys. Acta* **647**:232–240.

Skou, J. C., and Esmann, M., 1983, Effect of magnesium ions on the high-affinity binding of eosin to the (Na$^+$ + K$^+$)-ATPase, *Biochim. Biophys. Acta* **727**:101–107.

Skriver, E., Maunsbach, A. B., and Jørgensen, P. L., 1980, Ultrastructure of Na,K-transport vesicles reconstituted with purified renal Na,K-ATPase, *J. Cell Biol.* **86**:746–754.

Skvortsevich, E. G., Panteleeva, N. S., and Pisareva, L. N., 1972, The reaction of oxygen isotope exchange in the system of Na$^+$-K$^+$-dependent ATPase, *Proc. Acad. Sci. USSR* **206**:240.

Smith, R. L., Zinn, K., and Cantley, L. C., 1980, A study of the vanadate-trapped state of the (Na,K)-ATPase. Evidence against interacting nucleotide site models, *J. Biol. Chem.* **255**:9852–9859.

Stein, W. D., Lieb, W. R., Karlish, S. J. D., and Eilam, Y., 1973, A model for the active transport of sodium and potassium ions as mediated by a tetrameric enzyme, *Proc. Natl. Acad. Sci. USA* **70**:275–278.

Steinbach, H. B., 1940, Sodium and potassium in frog muscle, *J. Biol. Chem.* **133**:695–701.

Steinbach, H. B., 1951, Sodium extrusion from isolated frog muscle, *Am. J. Physiol.* **167**:284–287.

Steinbach, H. B., 1952, On the sodium and potassium balance of isolated frog muscles, *Proc. Natl. Acad. Sci. USA* **38**:451–455.

Swann, A. C., 1983, (Na$^+$ + K$^+$)-ATPase of mammalian brain: Effects of temperature on cation and ATP interactions regulating phosphatase activity, *Arch. Biochem. Biophys.* **221**:148–157.

Swann, A. C., and Albers, R. W., 1975, Sodium + potassium-activated ATPase of mammalian brain; regulation of phosphatase activity, *Biochim. Biophys. Acta* **382**:437–456.

Swann, A. C., and Albers, R. W., 1978, Sodium and potassium ion dependent adenosine triphosphatase of mammalian brain; interactions of magnesium ions with the phosphatase site, *Biochim. Biophys. Acta* **523**:215–227.

Swann, A. C., and Albers, R. W., 1980, (Na$^+$ + K$^+$)-ATPase of mammalian brain: Differential effects on cation affinities of phosphorylation by ATP and acetylphosphate, *Arch. Biochem. Biophys.* **203**:422–427.

Sweadner, K. J., 1979, Two molecular forms of (Na$^+$ + K$^+$)-stimulated ATPase in brain. Separation and difference in affinity for strophanthidin, *J. Biol. Chem.* **254**:6060–6067.

Sweadner, K. J., and Goldin, S. M., 1975, Reconstitution of active ion transport by the sodium and potassium ion-stimulated adenosine triphosphatase from canine brain, *J. Biol. Chem.* **250**:4022–4024.

Taniguchi, K., and Iida, S., 1972, Two apparently different ouabain binding sites of (Na$^+$ + K$^+$)-ATPase, *Biochim. Biophys. Acta* **288**:98–102.

Taniguchi, K., and Iida, S., 1972, Two apparently different ouabain binding sites of (Na$^+$ + K$^+$)-ATPase, *Biochim. Biophys. Acta* **288**:98–102.

Taniguchi, K., and Post, R. L., 1975, Synthesis of adenosine triphosphate and exchange between inorganic phosphate and adenosine triphosphate in sodium and potassium ion transport adenosine triphosphatase, *J. Biol. Chem.* **250**:3010–3018.

Taniguchi, K., Suzuki, K., and Iida, S., 1982, Conformational change accompanying transition of ADP-sensitive phosphoenzyme to potassium-sensitive phosphoenzyme of (Na$^+$,K$^+$)-ATPase modified with *N*-[*p*-(2-benzimidazolyl)phenyl]maleimide, *J. Biol. Chem.* **257**:10659–10667.

Tobin, T., and Brody, T. M., 1972, Rates of dissociation of enzyme–ouabain complexes and $K_{0.5}$ values in (Na$^+$ + K$^+$) adenosine triphosphatase from different species, *Biochem. Pharmacol.* **21**:1553–1560.

Tobin, T., and Sen, A. K., 1970, Stability and ligand sensitivity of ($^3$H)ouabain binding to (Na$^+$ + K$^+$)-ATPase, *Biochim. Biophys. Acta* **198**:120–131.

Tobin, T., Henderson, R., and Sen. A. K., 1972, Species and tissue differences in the rate of dissociation of ouabain from (Na$^+$ + K$^+$)-ATPase, *Biochim. Biophys. Acta* **274**:551–555.

Tobin, T., Akera, T., Baskin, S. I., and Brody, T. M., 1973, Calcium ion and sodium- and potassium-dependent adenosine triphosphatase: Its mechanism of inhibition and identification of the *E*1-P intermediate, *Mol. Pharmacol.* **9**:336–349.

Tobin, T., Akera, T., and Brody, T. M., 1974, Studies on the two phosphoenzyme conformations of Na$^+$ + K$^+$ ATPase, *Ann. N. Y. Acad. Sci.* **242**:120–132.

Tonomura, Y., and Fukushima, Y., 1974, Kinetic properties of phosphorylated intermediates in the reaction of Na$^+$,K$^+$-ATPase, *Ann. N. Y. Acad. Sci.* **242**:92–105.

Tosteson, D. C., 1963, Active transport, genetics, and cellular evolution, *Fed. Proc.* **22**:19–26.

Van Groningen, H. E. M., and Slater, E. C., 1963, The effect of oligomycin on the ($Na^+ + K^+$)-activated Mg-ATPase of brain microsomes and erythrocyte membrane, *Biochim. Biophys. Acta* **73:**527–530.

Van Winkle, W. B., Lane, L. K., and Schwartz, A., 1976, The subunit fine structure of isolated, purified $Na^+$,$K^+$-adenosine triphosphatase, *Exp. Cell Rev.* **100:**291–296.

Vogel, F., Meyer, H. W., Grosse, R., and Repke, K. R. H., 1977, Electron microscopic visualization of the arrangement of the two protein compounds of ($Na^+ + K^+$)-ATPase, *Biochim. Biophys. Acta* **470:**497–502.

Wallick, E. T., and Schwartz, A., 1974, Thermodynamics of the rate of binding of ouabain to the sodium, potassium adenosine triphosphatase, *J. Biol. Chem.* **249:**5141–5147.

Whitmer, K. R., Epps, D., and Schwartz, A., 1983, An endogenous inhibitor of $Na^+$,$K^+$-ATPase: "Endodigin," *Curr. Top. Membr. Trans.* **19,** in press.

Whittam, R., 1962, The asymmetrical stimulation of a membrane adenosine triphosphatase in relation to active cation transport, *Biochem. J.* **84:**110–118.

Whittam, R., Wheeler, K. P., and Blake, A., 1964, Oligomycin and active transport reactions in cell membranes, *Nature* **203:**720–724.

Wildes, R. A., Evans, H. J., and Chiu, J., 1973, Effects of cations on the adenosine diphosphate–adenosine triphosphate exchange reaction catalyzed by rat brain microsomes, *Biochim. Biophys. Acta* **307:**162–168.

Wilson, W. E., Sivitz, W.I., and Hanna, L. T., 1970, Inhibition of calf brain membranal sodium-and potassium-dependent adenosine triphosphatase by cardioactive sterols. A binding site model, *Mol. Pharmacol.* **6:**449–459.

Winslow, J. W., 1981, The reaction of sulfhydryl groups of sodium and potassium ion-activated adenosine triphosphatase with *N*-ethylmaleimide. The relationship between ligand-dependent alterations of nucleophilicity and enzymatic conformational states, *J. Biol. Chem.* **256:**9522–9531.

Winter, C. G., and Moss, A. J., 1979, Ultracentrifugal analysis of the enzymatically active fragments produced by digitonin action on Na,K-ATPase, in: *Na,K-ATPase: Structure and Kinetics* (J. C. Skou and J. G. Nørby, eds.), Academic Press, London, pp. 25–32.

Woodger, J. H., 1924, *A Textbook of Morphology and Physiology for Medical Students,* Oxford University Press, Oxford.

Yamaguchi, M., and Post, R. L., 1982, Inhomogeneity of alpha subunits of (Na,K)ATPase from renal outer medulla, *Fed. Proc.* **41:**673.

Yamaguchi, M., and Tonomura, Y., 1977, Kinetic studies on the ADP–ATP exchange reaction catalyzed by $Na^+$,$K^+$-dependent ATPase. Evidence for the K. S. T. mechanism with two enzyme–ATP complexes and two phosphorylated intermediates of high-energy type, *J. Biochem. (Tokyo)* **81:**249–260.

Yamaguchi, M., and Tonomura, Y., 1979, Simultaneous binding of three $Na^+$ and two $K^+$ ions to $Na^+$,$K^+$-dependent ATPase and changes in its affinities for the ions induced by the formation of a phosphorylated intermediate, *J. Biochem. (Tokyo)* **86:**509–523.

Yamaguchi, M., and Tonomura, Y., 1980a, Binding of monovalent cations to $Na^+$,$K^+$-dependent ATPase purified from porcine kidney. I. Simultaneous binding of three sodium and two potassium or rubidium ions to the enzyme, *J. Biochem. (Tokyo)* **88:**1365–1375.

Yamaguchi, M., and Tonomura, Y., 1980b, Binding of monovalent cations to $Na^+$,$K^+$-dependent ATPase purified from porcine kidney. II. Acceleration of transition from a $K^+$-bound form to a $Na^+$-bound form by binding of ATP to a regulatory site of the enzyme, *J. Biochem. (Tokyo)* **88:**1377–1385.

Yamaguchi, M., and Tonomura, Y., 1980c, Binding of monovalent cations to $Na^+$,$K^+$-dependent ATPase purified from porcine kidney. III. Marked changes in affinities for monovalent cations induced by formation of an ADP-insensitive but not an ADP-sensitive phosphoenzyme, *J. Biochem. (Tokyo)* **88:**1387–1397.

Yoda, A., and Yoda, S., 1982a, Formation of ADP-sensitive phosphorylated intermediate in the electric eel Na,K-ATPase preparation, *Mol. Pharmacol.* **22:**693–699.

Yoda, A., and Yoda, S., 1982b, Interaction between ouabain and the phosphorylated intermediate of Na,K-ATPase, *Mol. Pharmacol.* **22:**700–705.

Zambrano, F., Morales, M., Fuentes, N., and Rojas, M., 1981, Sulfatide role in the sodium pump, *J. Membr. Biol.* **63:**71–75.

29

# The Sarcoplasmic Reticulum Membrane

Marek Michalak

## I. INTRODUCTION

The sarcoplasmic reticulum is a closed membranous system surrounding myofibrils in muscle cells. It regulates the level of free $Ca^{2+}$ in the cytoplasm, thereby regulating the contraction and relaxation of muscle cells. Aspects of the sarcoplasmic reticulum and of excitation–contraction coupling have been reviewed recently (MacLennan and Holland, 1975, 1976; Ebashi, 1976; Tada *et al.*, 1978a; Fabiato and Fabiato, 1979; Hasselbach, 1979; de Mais and Vianna, 1979; Inesi, 1981; Ikemoto, 1982; Moller *et al.*, 1982; Berman, 1982). This review is complementary to Chapter 30 and describes work on the sarcoplasmic reticulum that is not directly related to the mechanism of ATP-dependent $Ca^{2+}$ transport.

## II. STRUCTURE, FUNCTION, AND ISOLATION OF THE SARCOTUBULAR SYSTEM

The sarcotubular system is composed of the transverse tubular system (T-system) and the sarcoplasmic reticulum (Porter and Palade, 1957; Franzini-Armstrong, 1972, 1980). The T-system consists of tubular invaginations of the surface membrane, which enter muscle cells perpendicular to the muscle fibers. They are organized so that in mammalian skeletal muscle two T-tubules pass each sarcomere at each A-I junction. The sarcoplasmic reticulum is composed of two continuous elements, the terminal cisternae, the portion of the sarcoplasmic reticulum membrane directly facing the T-tubules, and the longitudinal reticulum overlying the remainder of the myofibril (Fran-

*Marek Michalak* • Bio Logicals, Ottawa, Ontario KIY 4P1, Canada.

zini-Armstrong, 1980). The cisternal elements contain a filamentous interior matrix, whereas, the longitudinal elements are relatively free of internal content. The two terminal cisternae elements of the sarcoplasmic reticulum and one tubule of the T-system come into close opposition to form the triad (Franzini-Armstrong, 1972, 1980). The junction between sarcoplasmic reticulum and T-tubules forms the junctional gap with a width of 10–20 nm. Periodic densities (feet) cross the junctional gap joining sarcoplasmic reticulum and T-tubule membranes (Franzini-Armstrong, 1980). The feet are presumably strongly attached to the sarcoplasmic reticulum since they tend to remain with that membrane during fractionation procedures (Campbell *et al.*, 1980).

It is now well established that the release and subsequent reaccumulation of $Ca^{2+}$ by sarcoplasmic reticulum triggers contraction and relaxation of myofibrillar bundles (Ebashi and Endo, 1968; Fabiato and Fabiato, 1979). The relaxing effect of sarcoplasmic reticulum results from its very rapid and high-affinity $Ca^{2+}$ transport system which can reduce sarcoplasmic $Ca^{2+}$ to less than micromolar concentration within milliseconds. This $Ca^{2+}$ uptake occurs by way of $Ca^{2+}$-dependent ATPase which is present in all regions of the sarcoplasmic reticulum structure except for the junctional face (Meissner, 1975; Jorgensen *et al.*, 1982b). The $Ca^{2+}$-dependent ATP hydrolysis results in stoichiometric $Ca^{2+}$ transport; 2 moles of $Ca^{2+}$ are transported for every mole of ATP hydrolyzed. Clear evidence that $Ca^{2+}$ transport is mediated by the ATPase came from reconstitution experiments in which vesicles formed from the purified ATPase and phospholipids were shown to carry out ATP-dependent accumulation of $Ca^{2+}$ (Racker, 1972).

Different procedures for the purification of saroplasmic reticulum vesicles have been used in different laboratories (Martonosi, 1968; MacLennan, 1970; Ikemoto *et al.*, 1971a; McFarland and Inesi, 1971; de Meis and Hasselbach, 1971). Since the intact sarcoplasmic reticulum is composed of different regions, it would be expected that microsomal fractions would not be homogeneous. Sarcoplasmic reticulum has been fractionated into light and heavy components by a variety of procedures (Meissner, 1975; Caswell *et al.*, 1976; Lau *et al.*, 1977; Louis *et al.*, 1980; Campbell *et al.*, 1980). From biochemical and morphological analyses of the fractions obtained, it was concluded that the light fraction originated in elements of the longitudinal reticulum, while the heavy fraction originated in the terminal cisternae. The major difference among the fractions was in their content of calsequestrin, one of the $Ca^{2+}$-binding proteins of the sarcoplasmic reticulum membrane (MacLennan and Wong, 1971). It was proposed that the matrix material present inside vesicles of the heavy fraction was composed of calsequestrin. Campbell *et al.* (1980) found that vesicles of the heavy sarcoplasmic reticulum fraction had evenly spaced densities over some part of their surface. These were identified as the feet, originally occupying the junctional gap in the triads (Ikemoto *et al.*, 1976). That structure was most obvious in cases where intact triads were isolated (Lau *et al.*, 1977). These findings strongly support the idea that certain functions of the sarcoplasmic reticulum are restricted to specific regions of this membrane system. Lau *et al.* (1977), while separating the membrane fraction, found that [$^3$H]ouabain used as a surface membrane marker was concentrated in a heavy microsomal fraction containing complexes of T-tubules and sarcoplasmic reticulum (triad). Subsequently, heavy fractions were disrupted in a French pressure cell and separated into a transverse tubule-enriched fraction and a fraction enriched in heavy sarcoplasmic reticulum. According to recent reports (Scales and Sabbadini, 1979; Rosenblatt *et al.*, 1981), a membrane fraction, presumably of T-tubule origin, can also be isolated without a second round of mechanical disruption.

## III. PROTEIN COMPOSITION OF THE SARCOPLASMIC RETICULUM MEMBRANE

About 60–70% of the protein of the sarcoplasmic reticulum is accounted for by an intrinsic protein, the ($Ca^{2+}$ + $Mg^{2+}$)-dependent ATPase of molecular weight approximately 110,000 (Martonosi and Halpin, 1971; MacLennan *et al.*, 1971; Inesi, 1972; Meissner *et al.*, 1973). This protein has been localized in all regions of the sarcoplasmic reticulum except the junctional face of the terminal cisternae (Meissner, 1975; Jorgensen *et al.*, 1982b).

In addition to the ATPase, two extrinsic protein components were isolated from sarcoplasmic reticulum by MacLennan and Wong (1971) and Ostwald and MacLennan (1974). These acidic, $Ca^{2+}$-binding proteins, have been referred to as calsequestrin (MacLennan and Wong, 1971) and the high-affinity $Ca^{2+}$-binding protein (Ostwald and MacLennan, 1974).

The mobility of calsequestrin in SDS-gel electrophoresis is dependent on electrophoretic conditions of ionic composition and pH (MacLennan and Wong, 1971; Ikemoto *et al.*, 1971a; Meissner *et al.*, 1973; Michalak *et al.*, 1980). In SDS–polyacrylamide gels run according to Weber and Osborn (1969), calsequestrin moves with a mobility corresponding to a molecular weight of 44,000 (MacLennan and Wong, 1971), whereas, in gels run according to Laemmli (1970), it moves with a mobility corresponding to 63,000 (Meissner *et al.*, 1973). Michalak *et al.* (1980) developed a two-dimensional SDS-gel electrophoretic system using two different pH values to identify calsequestrin in skeletal muscle. Campbell *et al.* (1983a) used the same system to identify calsequestrin in cardiac muscle. Proteins with the same mobility in the two gel systems fell on a diagonal line, whereas, calsequestrin fell far from the diagonal and was purified in the process.

Calsequestrin is very acidic being composed of 37% glutamic and aspartic acids and only 9% lysine and arginine. It is a glycoprotein, containing three glucosamine and five mannose residues per 63,000 daltons (Jorgensen *et al.*, 1977). It binds [$^{125}$I] concanavalin A only weakly (Michalak *et al.*, 1980; Campbell and MacLennan, 1981), and it is not sensitive to digestion with endo-β-acetylglucosaminidase (Endo H) due to its low mannose content (Campbell and MacLennan, 1981). Calsequestrin binds nearly 1 μmole of $Ca^{2+}$/mg of protein with a dissociation constant in absence of salt of about 50 μM, and in the presence of physiological salt of about 800 μM (MacLennan and Wong, 1971; Ostwald and MacLennan, 1974; Ikemoto *et al.*, 1971a). Upon binding $Ca^{2+}$, calsequestrin undergoes conformation changes which include an increase in helical content. At a high-$Ca^{2+}$ concentration, calsequestrin becomes insoluble (MacLennan and Wong, 1971; Ostwald *et al.*, 1974). Precipitation of calsequestrin with calcium phosphate or calcium chloride has been used to purify the protein from skeletal (Ikemoto *et al.*, 1971a) and cardiac sarcoplasmic reticulum (Campbell *et al.*, 1983a).

Calsequestrin is exclusively localized in the interior of the sarcoplasmic reticulum (MacLennan and Wong, 1971; Stewart and MacLennan, 1974; MacLennan, 1975; Martonosi and Fortier, 1974; Meissner, 1975; Hidalgo and Ikemoto, 1977; Michalak *et al.*, 1980). Vesicles of light sarcoplasmic reticulum, originating in longitudinal tubules, are free of calsequestrin, whereas, heavy vesicles originating in the terminal cisternae are rich in this acidic protein (Meissner, 1975; Campbell *et al.*, 1980; Louis, *et al.*, 1980). Jorgensen *et al.* (1979), using immunofluorescent staining, localized calsequestrin only near the interface between the I- and A-band regions of rat skeletal muscle sarcomeres, where the terminal cisternae of the sarcoplasmic reticulum are localized.

The high-affinity $Ca^{2+}$-binding protein, of molecular weight 55,000, is less acidic than calsequestrin (MacLennan *et al.*, 1972). The protein binds about 1 mole of $Ca^{2+}$ per mole with a dissociation constant of about 4 μM (Ostwald and MacLennan, 1974). Recent studies (Michalak *et al.*, 1980) on the localization of the high-affinity $Ca^{2+}$-binding protein showed that, like calsequestrin, it is localized in the interior of sarcoplasmic reticulum vesicles. It is not labeled by a cycloheptaamylase–fluorescamine complex, and it is not sensitive to trypsin digestion in intact sarcoplasmic reticulum vesicles (Michalak *et al.*, 1980). The high-affinity $Ca^{2+}$-binding protein was found to be present in a rather constant ratio with the ATPase in light (longitudinal tubules) and heavy (terminal cisternae) sarcoplasmic reticulum fractions (Michalak *et al.*, 1980). The high-affinity $Ca^{2+}$-binding protein seems to be concentrated, however, in a fraction of the sarcotubular system corresponding to the T-tubules (Lau *et al.*, 1977; Michalak *et al.*, 1980).

The function of the high-affinity $Ca^{2+}$-binding protein in the sarcoplasmic reticulum membrane is not obvious. Its low amount and low $Ca^{2+}$-binding capacity suggest that it could bind only a fraction of the amount of $Ca^{2+}$ bound by calsequestrin. Its high affinity for $Ca^{2+}$ is comparable to that of troponin (Ostwald and MacLennan, 1974) and is characteristic of a regulatory protein. However, the $Ca^{2+}$ concentration inside of the sarcoplasmic reticulum, where the high-affinity $Ca^{2+}$-binding protein is located, is high and might obviate a regulatory role. The high-affinity $Ca^{2+}$-binding protein might act as a $Ca^{2+}$ receptor or concentrator in the transport process or play some role in communication between the sarcoplasmic reticulum and the T-system membrane, where it is apparently present in relatively high amounts.

Michalak *et al.* (1980) and Campbell and MacLennan (1981) have used [$^{125}$I]concanavalin A binding as a test for glycoproteins in vesicles of the sarcoplasmic reticulum. Four glycoproteins of molecular weight 160,000, 63,0000, 60,000, and 53,000 have been identified. The 63,000-dalton glycoprotein is calsequestrin and the 53,000-dalton glycoprotein is a major intrinsic protein of skeletal muscle sarcoplasmic reticulum (Michalak *et al.*, 1980; Campbell and MacLennan, 1981). The glycoprotein is not tightly associated with the ATPase (MacLennan 1970; Campbell and MacLennan, 1981). The 53,000-dalton glycoprotein can be solubilized with low concentrations of deoxycholate in the presence of 1 M KCl. It remains soluble but oligomeric upon removal of the detergent. This behavior of the sarcoplasmic reticulum glycoprotein resembles that observed for cytochrome *c* and cytochrome $b_5$ (Capaldi and Vanderkooi, 1972).

The glycoprotein contains 52 mole percent polar amino acids, about 22 mole percent aspartic and glutamic acid residues, and about 12 mole percent arginine and lysine residues (Campbell and MacLennan, 1981). It also contains two high-mannose sugar chains of the composition $(Man)_9$:$(GlcNAc)_2$ (Campbell and MacLennan, 1981). This is similar to the carbohydrate composition of the dolichol oligosaccharide (Liu *et al.*, 1979) that donates $(Glc)_3$:$(Man)_9$:$(GlcNAc)_2$ to proteins. Campbell and MacLennan (1981) digested the glycoprotein with Endo H, an enzyme which splits the chitobiosyl core of *N*-linked glycoproteins with high specificity for high-mannose sugar chains (Arakawa and Muramatsu, 1974). The molecular weight was reduced by about 4000, which agrees with the molecular weight of two oligosaccharides composed of $(Man)_9$:$(GlcNAc)_2$. The 160,000- and 60,000-dalton glycoproteins were also found to be sensitive to Endo H digestion. Calsequestrin, which was shown earlier to be a low-mannose glycoprotein (Jorgensen *et al.*, 1977) was not affected by Endo H digestion (Campbell and MacLennan, 1981).

A substantial part of the glycoprotein is exposed on the cytoplasmic side of the sarcoplasmic reticulum, since it is highly labeled with a cycloheptaamylase–fluorescamine complex (Michalak *et al.*, 1980). Labeling of the sarcoplasmic reticulum with the nonpenetrating fluorescent sulfhydryl probe *N*-(Iodoacetylaminoethyl)-5-naphthylamine-1-sulfonic acid (Hudson and Weber, 1973) showed that there are also SH groups of the glycoprotein exposed to the cytoplasmic side of the membrane (Campbell and MacLennan, 1981). In other membrane systems such as the endoplasmic reticulum (Boulan *et al.*, 1978) and rod outer segment disks (Clark and Molday, 1979), oligosaccharide chains are localized exclusively on the luminal side of the membrane. Similarly, the sugar chains in the 53,000-dalton glycoprotein of the sarcoplasmic reticulum were shown by Endo H digestion of intact and detergent-treated vesicles to be localized on the luminal side of the sarcoplasmic reticulum membrane (Campbell and MacLennan, 1981). Thus, the glycoprotein is clearly a transmembrane protein.

The 53,000-dalton glycoprotein has been found in both light and heavy sarcoplasmic reticulum fractions in a rather constant ratio with the ATPase, but was diminished in the membrane fraction corresponding to the T-system (Michalak *et al.*, 1980). The glycoprotein has also been found in vesicles reconstituted following deoxycholate or Triton X-100 solubilization of the sarcoplasmic reticulum (Meissner and Fleischer, 1974; Repke *et al.*, 1976; Michalak *et al.*, 1980).

It is of interest that the 53,000-dalton, intrinsic glycoprotein of the sarcoplasmic reticulum membrane might be comparable to the intrinsic glycoprotein associated with the $Na^+$, $K^+$-ATPase (Kyte, 1972). The $Na^+$, $K^+$-ATPase is composed of two different polypeptides, a 100,000-dalton protein and a 53,000-dalton glycoprotein in a molar ratio of 1 : 1 (Craig and Kyte, 1980). The function of this glycoprotein is unknown, but it was suggested that it might play a role in the transport of cations rather than in ATPase activity (Pennington and Hokin, 1979). The 53,000-dalton glycoprotein of the $Na^+$, $K^+$-ATPase, in contrast to the 53,000-dalton glycoprotein of the sarcoplasmic reticulum, could not be separated from the ATPase by treatment with a low level of detergent.

Sarzala *et al.* (1974), Ikemoto *et al.* (1976), and Hidalgo and Ikemoto (1977) have reported that a 30,000-dalton protein in the sarcoplasmic reticulum is an intrinsic glycoprotein. However, this protein did not bind any [$^{125}$I]concanavalin A (Michalak *et al.*, 1980, Campbell and MacLennan, 1981). Hidalgo and Ikemoto (1977) and Ikemoto *et al.* (1976) suggested that the 30,000-dalton intrinsic protein of the sarcoplasmic reticulum was a component of the feet, originally occupying the junctional gap in the triads (Franzini-Armstrong, 1980). It was later shown by Campbell *et al.* (1980) that the heavy fraction of the sarcoplasmic reticulum contained the 30,000- and 34,000-dalton proteins, whereas, the light fraction did not. The 34,000-dalton protein was identified by Campbell *et al.* (1980) as a portion of the "feet" since extraction of the vesicles with 0.6 M KCl resulted in a loss of the 34,000-dalton protein and in the disappearance of the visible portion of the feet. The separation of the 34,000-dalton and the 30,000-dalton proteins was not achieved and it was not possible to determine the relationship between the feet and those two proteins.

Another minor component of the sarcoplasmic reticulum is proteolipid which was isolated by MacLennan *et al.* (1972). This protein has not been well characterized and may, indeed, consist of more than one component. Proteolipid has been partially sequenced by Ohnoki and Martonosi (1980). Recently, Knowles *et al.* (1980) and Collins *et al.* (1982) showed that rabbit skeletal muscle sarcoplasmic reticulum contains at least two types of proteolipid. Although the proteolipid is an intrinsic protein and fractionates with the ATPase, it is not essential to ATPase or $Ca^{2+}$ uptake functions (MacLennan *et al.*, 1980). Proteolipids have been found in membrane systems other than the sarcoplasmic reticulum (Folch-Pi and Stoffyn, 1972; Schlesinger, 1981).

## IV. ULTRASTRUCTURE AND ASYMMETRY OF THE SARCOPLASMIC RETICULUM MEMBRANE

Electron microscopic analysis of the sarcoplasmic reticulum membrane using either freeze-fracture, negative staining, or thin sectioning shows it to be a highly asymmetric membrane. The isolated sarcoplasmic reticulum consists of closed spherical vesicles with a diameter of 80–150 nm. Freeze-fracture of the sarcoplasmic reticulum membrane (Deamer and Baskin, 1969; Baskin, 1971) has shown the presence of 9-nm intramembranous particles highly concentrated on the cytoplasmic leaflet. The occurrence of similar particles in reconstituted ATPase preparations (MacLennan *et al.*, 1971; Packer *et al.*, 1974; Jilka *et al.*, 1975; Malan *et al.*, 1975; Wang *et al.*, 1979) suggests that they represent the ATPase. In reconstituted sarcoplasmic reticulum vesicles, the intramembranous particles were found to be symmetrically distributed in the two membrane leaflets (MacLennan *et al.*, 1971; Packer *et al.*, 1974), probably as a result of random insertion of the ATPase during the reformation of membranes. The concentration of particles on the cytoplasmic leaflet of intact sarcoplasmic reticulum membrane is consistent with X-ray data that indicate an asymmetrical protein distribution across the membrane (Dupont *et al.*, 1973), Herbette *et al.*, 1977; Herbette and Blasie, 1980; Fleischer *et al.*, 1979). These 9-nm intramembranous particles increase with an increase in ATPase activity during muscle differentiation (Boland *et al.*, 1974; Tillack *et al.*, 1974).

Electron microscopic observation of negatively-stained membranes of sarcoplasmic reticulum permits a view of the external surface of the membrane, where a part of the ATPase molecule would be assumed to be exposed. Particles 4 nm in size are present on the surface of negatively-stained sarcoplasmic reticulum vesicles (Inesi and Asai, 1968; Ikemoto *et al.*, 1968, 1971b). These particles are also present in vesicles formed from the purified ATPase and represent cytoplasmic extensions of the ATPase molecule (Migala *et al.*, 1973; Thorley-Lawson and Green, 1973; Stewart and MacLennan, 1974). The appearance of 4-nm surface particles in the sarcoplasmic reticulum membrane correlates well with the appearance of ATPase activity during development of skeletal muscle (Michalak and Sarzala, 1975; Sarzala *et al.*, 1975a, b; Martonosi, 1975).

Stewart *et al.* (1976) have shown that part of the ATPase polypeptide is exposed to the outer surface of sarcoplasmic reticulum membrane by demonstrating the binding of antibodies against the ATPase and its proteolytic fragments to the membrane. Hasselbach and Elfvin (1967) earlier demonstrated the binding of Hg-phenylazoferritin to the membrane of sarcoplasmic reticulum, suggesting that some of the SH groups of the ATPase polypeptide are exposed to the external surface of the membrane.

The number of 4-nm surface particles appears to be greater than that of the 9-nm intramembranous particles in reconstituted ATPase vesicles and in sarcoplasmic reticulum membrane (Jilka *et al.*, 1975; Scales and Inesi, 1976). It was suggested that the discrepancy in the number of both types of particles may be due to the formation of oligomers containing three or four hydrophobic portions of the ATPase that appear as single 9-nm particles.

Saito *et al.* (1978) visualized the asymmetry of proteins in the sarcoplasmic reticulum by staining of thin sections with tannic acid. Electron-dense stained material was localized only on the outer, cytoplasmic surface of the sarcoplasmic reticulum membrane. This was also shown to be true for thin sections of tannic acid-stained intact muscle cells. The symmetric appearance of this dense material on both sides of the membrane was observed in vesicles reconstituted from solubilized sarcoplasmic reticulum (Saito *et al.*, 1978; Wang *et al.*, 1979). However, Fleischer *et al.* (1979), Wang *et al.* (1979), and Herbette *et al.* (1981b) have shown asymmetric staining of reconstituted sarcoplasmic reticulum vesicles when the phospholipid:protein ratio was higher than 88 moles/mole of ATPase.

There is also evidence for an asymmetric distribution of phospholipids between the two halves of the sarcoplasmic reticulum membrane bilayer. Hasselbach and Migala (1975) used fluorescamine to show that aminophospholipids are present mainly on the outer monolayer. However, fluorescamine was added to the sample as a solution in an organic solvent, which might perturb the lipid bilayer. Hidalgo and Ikemoto (1977) used a stable, water-soluble complex of fluorescamine with cycloheptaamylase which was shown not to penetrate the membrane. They also concluded that most of the phosphatidylethanolamine was situated in the cytoplasmic half of the lipid bilayer. This was confirmed by Vale (1977) who used 2,4,6-trinitrobenzenesulfonate as a label. Sarzala and Michalak (1978) used both 2,4,6-trinitrobenzenesulfonate and phospholipases to show that phosphatidylethanolamine as well as phosphatidylserine were in the cytoplasmic monolayer.

## V. THE ATPase OF THE SARCOPLASMIC RETICULUM MEMBRANE

The ATPase is an asymmetric, intrinsic, transmembrane protein of the sarcoplasmic reticulum. At present, most experimental evidence suggests that the ATPase protein is solely responsible for $Ca^{2+}$ transport into sarcoplasmic reticulum (MacLennan and Holland, 1975; Hasselbach, 1979; Ikemoto, 1982). The ATPase which couples the hydrolysis of ATP to the active transport of $Ca^{2+}$ is phosphorylated by ATP (Kanazawa *et al.*, 1971; Meissner, 1973). There are two high-affinity $Ca^{2+}$-binding sites and one phosphorylation site per ATPase molecule and for each ATP hydrolyzed; two $Ca^{2+}$ are transported per phosphorylation site (Kanazawa *et al.*, 1971; Meissner, 1973).

This review will not deal with a detailed description of current knowledge on the ATPase reaction mechanism since this is the subject of another review article in this volume. However, understanding of the transport mechanisms requires a detailed knowledge of the ATPase structure and assembly within the membrane. In this section of the article, we focus on the structure of the ATPase polypeptide (sequence and disposition within the membrane), protein–protein and protein–lipid interactions of the ATPase, and the current status of the reconstitution of the ATPase.

### A. Purification and Characterization of the ATPase

Various methods employing different detergents have been used for the solubilization of the sarcoplasmic reticulum membrane and the isolation of the ATPase (MacLennan and Holland, 1975, 1976). The first procedure for the isolation of the ATPase (MacLennan, 1970) consists in solubilization of the membrane with deoxycholate and fractionation with ammonium acetate. Ikemoto *et al.* (1971a) used Triton X-100 to solubilize the sarcoplasmic reticulum membrane and gel-exclusion chromatography to isolate the ATPase. Warren and Metcalf (1976) isolated the ATPase by solubilization of the membrane with deoxycholate followed by centrifugation of the solution into a sucrose gradient to remove deoxycholate. Le Maire *et al.* (1976a,b) used two nonionic detergents, Tween 80 and dodecyloctaxyethylene glycol monoether ($C_{12}E_8$), to solubilize the ATPase. Banerjee *et al.* (1979) used a nonionic detergent octylglucoside, in the presence of high-salt concentrations, to isolate the ATPase from sarcoplasmic reticulum.

The molecular weight of the sarcoplasmic reticulum ATPase has been estimated to range from 102,000–119,000 based on either SDS-gel electrophoresis (MacLennan, 1970; McFarland and Inesi, 1971; Louis and Shooter, 1972; Meissner *et al.*, 1973;

Thorley-Lawson and Green, 1973; Warren *et al.*, 1974a) or analytical ultracentrifugation (Rizzolo *et al.*, 1976; Le Maire *et al.*, 1976a,b). The amino acid composition of the ATPase (MacLennan *et al.*, 1971; Martonosi and Halpin, 1971; Louis and Shooter, 1972; Meissner *et al.*, 1973; Thorley-Lawson and Green, 1973) shows that the protein contains approximately 43% polar residues compared to 47% for an average water-soluble protein (Capaldi and Vanderkooi, 1972). The relatively high content of polar residues is probably related to the fact that a large portion of the protein is located outside the lipid phase of the membrane. Allen and Green (Allen, 1977, 1980a,b; Allen and Green, 1978; Allen *et al.*, 1980a,b; Green *et al.*, 1980) have sequenced large portions of the ATPase. They have calculated that the percentage of polar residues in the region of the ATPase where they obtained long stretches of sequences was 48%, whereas, it was only 18% in the nonsequenced residues which are very likely associated with the lipid phase of the membrane. A relatively high proportion of acidic amino acids (Glx + Asx = 19%) were distributed almost equally between sequenced and nonsequenced residues. Nearly all the tryptophan residues (18 out of 19) were located in the intramembranous part of the ATPase, while most of the cysteine residues were associated with the part of the protein external to the lipid bilayer. This was supported by the fact that, in the intact membrane, most of the cysteine residues of the ATPase could be modified by 5,5′-dithio-*bis*-2-nitrobenzoic acid (Murphy, 1976; Thorley-Lawson and Green, 1977). Out of a total of 27 cysteine residues, six were present as disulfides (Thorley-Lawson and Green, 1977). Out of 51 lysine residues per 110,000 daltons of the ATPase, 41 residues were located in the sequenced polar segments (Allen *et al.*, 1980a). Murphy (1977) showed that one of the lysine residues located in the large tryptic fragment of the ATPase was present next to the phosphorylation site of the enzyme and they identified this residue as the "essential" lysine. The studies by Hidalgo (1980) show that chemical modification of lysine residues of the ATPase with fluorescamine resulted in inhibition of ATPase activity either by blocking phosphorylated intermediate formation or its decomposition.

## B. Tryptic Fragments and Amino Acid Sequence of the ATPase

Tryptic digestion of the intact sarcoplasmic reticulum membranes results in cleavage of the ATPase at specific sites (Thorley-Lawson and Green, 1973; Migala *et al.*, 1973; Stewart and MacLennan, 1974). Active sites in these isolated fragments have been postulated because of their ability to be phosphorylated by [$^{32}$P]-ATP (catalytic site), by their ability to confer $Ca^{2+}$ selective conduction channels in black lipid films or bilayers (Shamoo, 1978), and by their ability to bind dicyclohexylcarbodiimide (DCCD), an inhibitor of ATP hydrolysis and $Ca^{2+}$ uptake (Pick and Racker, 1979) or fluorescein 5′-isothiocyanate, an inhibitor of ATP binding (Pick and Bassilian, 1980; Pick, 1981; Mitchinson *et al.*, 1982). All of the tryptic fragments contain distinct hydrophilic and hydrophobic regions, and they remain associated with the membrane. However, solubilization and separation of the peptide fragments can be achieved in the presence of SDS (Thorley-Lawson and Green, 1975; Rizzolo *et al.*, 1976, Stewart *et al.*, 1976; Shamoo *et al.*, 1976). When tryptic hydrolysis was limited to one cleavage site, yielding two major species of molecular weight 45,000 and 55,000, there was no loss of ATPase or of $Ca^{2+}$ transport activity. Loss of $Ca^{2+}$ transport followed the cleavage of the 55,000-dalton fragment to 25,000- and 30,000-dalton fragments. The phosphorylation site resides in the 55,000- and in the 30,000-dalton fragment. $Ca^{2+}$ conduction activity was associated with the 25,000-dalton tryptic fragment (Shamoo and MacLennan, 1974; Shamoo, 1978). DCCD, an inhibitor of ATP hydrolysis and $Ca^{2+}$ uptake, was found to bind competitively with $Ca^{2+}$ to a site that could be located in the 25,000-dalton fragment (Pick and Racker, 1979). This suggests that the 25,000-dalton fragment might be a site of $Ca^{2+}$ binding and translocation. The fluorescein 5′-isothiocyanate-binding site was located in the 45,000-dalton fragment (Pick and Bassilian, 1980; Pick, 1981; Mitchinson *et al.*, 1982)

It was of interest that, even after extensive treatment with trypsin, which removes about 30% of the protein from the sarcoplasmic reticulum, ATPase activity was still maintained (Ikemoto *et al.*, 1971b) or even increased (Saito *et al.*, 1978). Very extensive tryptic digestion leads to loss of ATPase activity and loss of surface particles but not to loss of intramembranal particles (Stewart and MacLennan, 1974). The intramembranous segment of the ATPase was relatively resistant to digestion with trypsin and gave rise to large aggregated, relatively hydrophobic peptides (Stewart and MacLennan, 1974, 1975; Allen, 1977). However, considerable progress has been made toward establishing the primary sequences of the ATPase from proteolytic fragments that do become water soluble (Allen and Green, 1978; Allen, 1980a,b; Allen *et al.*, 1980a,b; Green *et al.*, 1980). More than 50% of the ATPase protein has been sequenced. Sequences of three fragments containing 116, 122, and 298 residues have been aligned in the ATPase, and a 32-residues $NH_2$-terminal end and an 8-residues COOH-terminal end have been sequenced. The $NH_2$-terminal Met residue is acetylated (Tong, 1977). Klip *et al.* (1980) have shown that trypsin cleavages occur specifically between Arg and Ala residues and no peptides were lost during digestion and purification procedures. The localization of known sequences of the ATPase has been shown by Klip *et al.* (1980) and Allen *et al.* (1980b). A 116-residue sequence has been localized around the 25,000- to 30,000-dalton trypsin cleavage point, while a 298-residue sequence has been localized around the 30,000- to 45,000-dalton trypsin cleavage point. The position of a 112-residue sequence within the tryptic fragments is unknown, but it could be located only within the 45,000-dalton fragment because of size constraints.

Green *et al.* (1980) digested the ATPase protein extensively while still in membranous form where hydrophobic, membrane-associated regions would be inaccessible to digestion, and showed that peptides released under these conditions were all located in the five known sequences. Thus, the five known sequences appear to be composed of part of the ATPase molecule residing on the aqueous phase on the membrane surface, whereas, remaining peptides would correspond to the membrane-associated material. It was suggested from the labeling of each tryptic fragment with a hydrophobic photoactivated azide (Green *et al.*, 1980), from their similar content of tryptophan, and from their firm anchorage in the membrane (Stewart and MacLennan, 1974; Rizzolo *et al.*, 1976; Thorley-Lawson and Green, 1973) that each has an extensive membrane-associated segment(s).

Reithmeier *et al.* (1980) and Reithmeier and MacLennan (1981) have suggested that the $NH_2$ terminus of the ATPase has a cytoplasmic location. Reithmeier and MacLennan (1981) labeled the cysteine residues of the ATPase with $^3H$-labeled *N*-ethylmaleimide and isolated the $NH_2$-terminal peptide (residues 1–32) containing a cysteine residue at position 12 (Allen and Green, 1978). They showed that Cys-12 was labeled. Labeling of this SH group was blocked when the membrane was pretreated with the nonpenetrating sulfhydryl reagent glutathione maleimide. They demonstrated that all cysteine residues labeled with [$^3H$]-NEM, including Cys-12, were localized on the cytoplasmic surface of the membrane. The location of the $NH_2$ terminus of the ATPase on the cytoplasmic surface of the membrane was also suggested by studies on the cell-free synthesis of the sarcoplasmic reticulum ATPase. The enzyme is found without an $NH_2$-terminal signal sequence implying that this region would not cross the membrane (Reithmeier *et al.*, 1980; Mostov *et al.*, 1981). A cytoplasmic location for the COOH-terminal peptide is tentative and is based on the view that, during biosynthesis, it would be left on the cytoplasmic surface at the time of chain termination.

The aspartate residue, which was phosphorylated during ATP hydrolysis was located at position 26 at the 298-residue sequence fragment. De Meis *et al.* (1980) suggested that the aspartate residue may be situated in a border region between the hydrophilic and hydrophobic portions of the ATPase.

Available data suggest that the known cytoplasmic sequences are separated by a nonsequenced gap of about 100 amino acid residues. Klip *et al.* (1980), Allen *et al.* (1980b), and MacLennan and Reithmeier, (1982) have suggested that each large gap might correspond to a loop of the ATPase chain through the core of a phospholipid bilayer, i.e., to a total of eight crosses of the membrane. A probable folding pattern of the ATPase peptide chain through the sarcoplasmic reticulum membrane has been suggested by MacLennan *et al.* (1980). The $NH_2$ terminal 32 amino acids lie in the cytoplasm. The chain enters the hydrophobic region of the membrane and passes through. Since it is unlikely that a peptide would turn within a hydrophobic environment (Kennedy, 1978), it would be expected to turn within the aqueous phase in the lumen of the membrane. Then, the chain would reenter the hydrophobic phase of the membrane, passing through in a regular structure to create the cytoplasmic portion of the peptide. This process is repeated probably four times, so that the chain stitches back and forth eight times through the membrane. There is no clear evidence that the ATPase chain extends to the luminal surface. However, Hidalgo and Ikemoto (1977) compared the extent of labeling of the ATPase with a cycloheptaamylase–fluorescamine complex in intact sarcoplasmic reticulum vesicles and in the purified ATPase, which was isolated as a lipoprotein complex forming "leaky" vesicles. The purified ATPase was labeled about two times higher, indicating that a part of the ATPase molecule might be accessible from the inside of the vesicles.

## C. ATPase–ATPase Interaction: An Oligomeric Form of the Enzyme

Evidence from several independent experimental approaches has suggested an oligomeric structure for the ATPase of the sarcoplasmic reticulum. Physicochemical evidence for an interaction between ATPase molecules in the native state has come mainly from freeze-fracture studies on sarcoplasmic reticulum (Jilka *et al.*, 1975; Scales and Inesi, 1976; Le Maire *et al.*, 1981), crosslinking studies on proteins of sarcoplasmic reticulum (Louis and Shooter, 1972; Murphy, 1976; Baskin and Hanna, 1979; Hebdon *et al.*, 1979; Bailin, 1980), studies on detergent solubilized ATPase (Le Maire *et al.*, 1976a,b, 1978; Jorgensen *et al.*, 1978; Moller *et al.*, 1980; Murphy *et al.*, 1982), and spectroscopic studies on the ATPase (Vanderkooi *et al.*, 1977; Thomas and Hidalgo, 1978; Kiring *et al.*, 1978; Hidalgo *et al.*, 1978; Hoffman *et al.*, 1980; Burkli and Cherry, 1981).

The number of intramembranous particles observed after freeze-fracture of sarcoplasmic reticulum membrane is about 5000–6000 $\mu m^2$ of membrane surface (Jilka *et al.*, 1975; Scales and Inesi, 1976). However, Jilka *et al.* (1975) showed that the number of surface particles counted after negative staining of the vesicles was about 16,000/$\mu m^2$ in the sarcoplasmic reticulum membrane and about 22,000/$\mu m^2$ in purified ATPase vesicles. Scales and Inesi (1976) estimated that the number of surface particles in freeze-etched specimens of sarcoplasmic reticulum vesicles was about 21,000 $\mp$ 3900 surface particles/$\mu m^2$. These experiments suggested that the intramembranous particles represent 3–4 ATPase peptides which are associated in their intramembranous portion but not in the part projecting from the surface. It should be emphasized that it is very difficult to obtain reliable estimates of the number of surface particles per unit membrane of sarcoplasmic reticulum or ATPase vesicles. However, it was clear that the number of intramembranous particles observed by freeze-fracture were fewer than the number of ATPase polypeptides present. The large size of the particles (9 nm) was in agreement with the view that the particles contained a self-associated ATPase. Hydrodynamic measurements (Le Maire *et al.*, 1976a,b) and freeze-fracture studies (Le Maire *et al.*, 1981) showed that deoxycholate-solubilized ATPase was an asymmetric, elongated particle with a length of approximately 11 nm and with the major part of the mass located at one end.

Louis and Shooter (1972) observed formation of dimers, trimers, tetramers, pentamers, and hexamers of the ATPase as a function of time, using the crosslinking reagent suberimidate. Murphy (1976) observed the formation of a tetrameric ATPase by air oxidation of SH groups in the presence of $Cu^{2+}$-1,10 phenanthroline. In contrast, Baskin and Hanna (1979) showed that most of the protein was in a dimer form after a short exposure to $Cu^{2+}$-1,10 phenanthroline. Hebdon *et al.*, (1979), using the same reagent, found that a large fraction of the ATPase remained in a monomeric state. Bailin (1980) used 1,5-difluoro-2,4-dinitrobenzene to study the formation of oligomers of the ATPase. He found that monomer in ATPase was the predominant species present. The results of these studies are controversial and do not necessarily prove an oligomeric form of the ATPase in the membrane. Random molecular collision might account for the various crosslinking results. A significant correlation between formation of oligomers of the ATPase and changes in enzymatic activity has not been observed.

In order to define the smallest functional unit of the ATPase, characterization of the aggregation state of the transport protein in detergent-solubilized ATPase has been undertaken. Le Maire *et al.* (1976a,b) demonstrated ATP hydrolysis by the ATPase in the presence of nonionic detergents ($C_{12}E_8$ and Tween 80) and showed high enzymatic activities when the solubilized enzyme was in a form corresponding to the molecular weight of a trimer or a tetramer. However, an oligomeric form was not a prerequisite for maintenance of full enzymatic activity since ATP hydrolysis could also be demonstrated (over a shorter time period) by solubilization of the ATPase in a monomeric form in deoxycholate (Le Maire *et al.*, 1976a,b; Jorgensen *et al.*, 1978) or $C_{12}E_8$ (Moller *et al.*, 1980). ATPase activity could be maintained for hours in the presence of deoxycholate provided that solubilization was carried out in high sucrose, high protein concentration, and a high concentration of KCl. The activity could be preserved for days in $C_{12}E_8$ if the solubilization medium contained 20% glycerol (Dean and Tanford, 1977, 1978). Examination of the kinetics of the enzyme has indicated that basic aspects of ATP hydrolysis such as the presence of high-affinity sites for ATP and $Ca^{2+}$ were retained by both the $C_{12}E_8$ and deoxycholate-solubilized monomers of the ATPase (Jorgensen *et al.*, 1978; Moller *et al.*, 1980). However, this did not preclude the possibility that the ATPase is present in associated form in the membrane. The ATPase complex could be held together by hydrophobic forces and be dissociated by detergent without any major conformational changes. A number of observations suggest different aggregational states of the membranous ATPase (Yates and Duance, 1976; Moller *et al.*, 1980). Moller *et al.* (1980) showed that the enzyme activity of membranous ATPase in the presence of an intermediate concentration of ATP (Tada *et al.*, 1978b; Tong, 1980) could be analyzed in terms of negative cooperativity. This could be caused by protein–protein interaction and was not observed after solubilization of the ATPase into monomeric form. In contrast to the oligomeric or membranous ATPase, the monomeric ATPase was rapidly inactivated by removal of $Ca^{2+}$ from the high-affinity sites (Moller *et al.*, 1982) and the monomer had a different SH reactivity than was observed for oligomeric or membranous ATPase (Andersen *et al.*, 1980, Moller *et al.*, 1982). In addition, oligomerization of detergent-solubilized ATPase could be detected by the presence of an immobilized component in the EPR spectra of maleimide nitroxide probes, covalently attached to the ATPase (Andersen *et al.*, 1980). This was observed for the membranous ATPase as well (Andersen *et al.*, 1980). All the above observations are consistent with the presence of the ATPase in a self-associated, oligomeric form in the membrane.

Vanderkooi *et al.* (1977) analyzed the interaction between ATPase molecules in reconstituted ATPase vesicles by measuring the efficiency of energy transfer between two populations of ATPase molecules, one labeled with *N*-iodoacetyl-*N*′-(5-sulfo-1-naphthyl)-ethylene-diamine serving as energy donor and the other with iodoacetamidofluorescein serving as energy acceptor. They found that the efficiency of energy transfer was not affected by dilution of the lipid phase of the reconstituted vesicles suggesting that ATPase molecules exist as an oligomer. Studies on the incorporation of dicyclohexylcarbodiimide (Pick and Racker, 1979) or fluorescein 5′-isothiocyanate (Pick and Karlish, 1980; Pick, 1981) also favored an oligomeric structure for the ATPase. It has been shown that the incorporation of approximately 1 mole of dicyclohexylcarbodiimide/4 moles of ATPase and 1 mole of fluorescein 5′-isothiocyanate/2 moles of ATPase (Pick and Karlish, 1980; Pick, 1981) leads to almost complete inhibition of the ATPase activity. In contrast to Pick's (1981) results, Mitchinson *et al.* (1982) and Andersen *et al.* (1982) showed that a higher level of incorporation of fluorescein 5′-isothiocyanate was required (up to 1 mol/mol) to lead to an inhibition of enzymatic activity. Their results do not support an oligomeric form of the ATPase but do not exclude the possibility of a dimeric functional unit for $Ca^{2+}$ transport.

The results presented above indicate that the ATPase of the sarcoplasmic reticulum might be present as an oligomeric structure (dimers to tetramers) in the membrane. Evidence also indicates that protein–protein interaction is probably a characteristic feature of transport proteins. A self-associated state of the $Na^+$, $K^+$ ATPase in the membrane was considered probable on the basis of crosslinking experiments (Giotta, 1976a,b; Liang and Winter, 1977) and detergent solubilization of the protein (Rizzolo *et al.,* 1976; Hastings and Reynolds, 1979; Brotherus *et al.,* 1981). The minimal molecular weight complex that could be obtained for Band 3 after solubilization in Triton X-100 corresponded to a dimer (Clarke, 1975). This was supported by flash photolysis (Nigg and Cherry, 1979) and crosslinking studies (Peters and Richards, 1977). The ATP–ADP translocase, after solubilization, was also found to be a dimer (Hackenberg and Klingenberg, 1980).

It is still difficult to say which of the observed ATPase–ATPase interactions represents the physiological state of the molecule. Formation of nonspecific ATPase oligomers resulting in complexes of varying shapes and sizes cannot be ruled out. Further studies are required using a variety of different physicochemical techniques to clarify the aspect of self-association of the ATPase peptide and its functional implications.

## D. Lipid–ATPase Interaction

The purified ATPase contains about 530 mg of phospholipids per g of protein and its composition is virtually unchanged from that of the intact sarcoplasmic reticulum (MacLennan *et al.,* 1971; Owens *et al.,* 1972). The major classes of phospholipids present are phosphatidylcholine and phosphatidylethanolamine but lower concentration of phosphatidylserine and sphingomyelin are also detectable. Removal of phospholipds from the ATPase of the sarcoplasmic reticulum by phospholipase treatment (Martonosi *et al.,* 1968; Fiehn and Hasselbach, 1970; Meissner and Fleischer, 1972), by use of detergent followed by sucrose density centrifugation (Warren *et al.,* 1974a), by polyethylene glycol fractionation (Dean and Tanford, 1977), by gel filtration (Hardwicke and Green, 1974; Knowles *et al.,* 1976; Le Maire *et al.,* 1976a,b, 1978), or by DEAE-cellulose chromatography (Andersen *et al.,* 1980) all lead to inactivation of the ATPase. Addition of phospholipids to the depleted sarcoplasmic reticulum vesicles restores the ATPase activity (Martonosi *et al.,* 1968). In contrast, enzymatic delipidation of sarcoplasmic reticulum vesicles with phospholipase A2, followed by albumin washes leads to a very stable ATPase preparation (Svaboda *et al.,* 1979). High-ATPase activity as well as $Ca^{2+}$ transport could be obtained in systems containing either one of the two major phospholipid classes in sarcoplasmic reticulum, phosphatidylcholine, or phospholidylethanolamine (Warren *et al.,* 1974a,b; Knowles *et al.,* 1976; Zimniak and Racker, 1978), but other phospholipid classes may serve as well (Bennett *et al.,*

1980). Under some conditions, detergent alone could substitute for phospholipids to support ATPase activity (Dean and Tanford, 1978; Melgunov and Alimara, 1980; Nestrucke-Goyk and Hasselbach, 1981). Melgunov and Alimara (1980) showed that the efficiency of nonionic detergent in reactivation of lipid-depleted ATPase was closely related to their hydrophile–lipophile balance; the more lipophilic detergents being the best activators. Although the critical role of one or two phospholipid molecules in the ATPase activity cannot be excluded, it appears that the hydrophobic environment (hydrocarbon chains of phospholipids or nonionic detergent) might be of primary importance for ATPase activity. This hydrophobic environment or specific phospholipid molecules might also play a role in mediating self-association of the ATPase (Marcelja, 1976; Le Maire *et al.*, 1978).

Katz *et al.* (1982) investigated the effect of different fatty acids on $Ca^{2+}$ influx and efflux in the sarcoplasmic reticulum membrane. At low concentrations, fatty acids caused an inhibition of the initial $Ca^{2+}$ uptake by sarcoplasmic reticulum vesicles. The extent of inhibition varied with chain length and unsaturation in a series of $C_{14}$-$C_{20}$ fatty acids. They also observed a disparity among the inhibition potencies of fatty acids with different degrees of saturation but similar chain length and between stereoisomers. Thus, the inhibitory effect of the fatty acids might have resulted from specific interaction with a region of the phospholipid membrane intimately related to the ATPase.

Studies on enzymatically delipidated ATPase preparations show that mono-unsaturated fatty acids were more effective as ATPase reactivating agents, compared to the corresponding saturated fatty acids (Svaboda *et al.*, 1979; The *et al.*, 1981). This is in agreement with observations of Nakamura *et al.* (1976) and Nakamura and Martonosi (1980) revealing that replacement of membrane phospholipids with saturated diacylglycerophosphocholine markedly reduced phosphoprotein levels and the rate of phosphoprotein formation. These results suggest an important role of unsaturated fatty acids in the activity of the sarcoplasmic reticulum ATPase. A general finding that the ATPase is more active when membrane phospholipids contain polyunsaturated than saturated fatty acyl chains (Hidalgo *et al.*, 1978; Quinn *et al.*, 1980; Bennett *et al.*, 1980) favors the view that the phospholipid structure most favorable for ATPase function is a fluid rather than a gel structure.

The sarcoplasmic reticulum membrane also contains a very low level of cholesterol and other neutral lipids which can be removed without any loss of ATPase activity (Drabikowski *et al.*, 1972). However, the ATPase activity is inhibited by incorporation of cholesterol into the membrane (Warren *et al.*, 1975; Madden *et al.*, 1979, 1981). Madden and Quinn (1979) suggested that cholesterol might be increasing the viscosity of bulk lipid causing enzymatic inhibition. However, Warren *et al.* (1975) observed inhibition of ATPase by cholesterol in a lipid-depleted preparation suggesting a direct interaction of cholesterol with the ATPase polypeptide. Borchman *et al.* (1982) showed that the cholesterol content of slow twitch muscle sarcoplasmic reticulum was three times greater than that for fast twitch muscle sarcoplasmic reticulum. They suggested that the greater cholestrol content as well as greater sphingomyelin and phospatidylcholine ratio observed in slow twitch muscle contributed to decreased bilayer fluidity and decreased ATPase activity.

The insertion of protein into the lipid bilayer reduces the mobility of the hydrocarbon chains of the lipids. It has been observed, using spin-label probes, that relatively immobile and mobile components exist in phospholipids or fatty acids incorporated into the membrane (Hidalgo *et al.*, 1976; Hesketh *et al.*, 1976). According to Warren and Metcalf (1976), the immobilized lipid forms a single layer of about 30 phospholipid in direct contact with the ATPase. Warren *et al.* (1974b) showed, by plotting the ATPase activity as a function of lipid–protein ratio, that there was a critical lipid–protein ratio that represents the minimum lipid content that will support maximal activity of the ATPase. The critical lipid content required for the enzyme activity was approximately 30 moles lipid per mole protein. This was observed when cholate (Warren *et al.*, 1974a,b), octyl glucoside (Bennett *et al.*, 1980), Triton X-100, Tween 80, and $C_{12}E_8$ (Le Maire *et al.*, 1976a,b) were used. Warren *et al.* (1974b) have called these 30 molecules of lipid per molecule of ATPase the "lipid annulus." The presence of a lipid annulus has been supported by studies with spin-labeled lipid probes which showed that the same number of lipid molecules (30) were being immobilized regardless of the composition of the bilayer. Hesketh *et al.* (1976) showed, in addition, that the physical properties of the annulus lipids (examined using spin-label techniques) were different from the physical properties of the rest of the bilayer, and that it was the physical properties of the lipid annulus that appeared to modulate enzyme activity. They also pointed out that resolution of the annulus by electron-spin resonance techniques provides an upper limit for the rate of exchange of annulus lipids into the free bilayer of about $10^5$ $sec^{-1}$. Thus, annular lipids can exchange rapidly into the bilayer and appear immobilized only by comparison with the extremely fast diffusion rates of lipids in the free bilayer. Bennett *et al.* (1980) showed that all the annulus lipids contributed equally to the modulation of enzyme activity. However, when dioleoylphosphatidic acid (negatively charged) and dioleoylphosphatidylcholine (zwitterionic) were used to study lipid-ATPase interactions in the membrane, the lateral segregation of these lipids was observed. About 80% of the total lipid pool consisted of dioleoylphosphatidic acid, whereas, the lipid annulus consisted principally of dioleoylphosphatidylcholine (Bennett *et al.*, 1980). Bennett *et al.* (1978) investigated the disposition of the annulus phospholipids between the two leaflets of the bilayer. The 30 lipids per ATPase protein were resistant to digestion with phospholipase D. In the outer monolayer, 15 moles of lipid per mole ATPase remained undigested. This meant that the annulus phospholipids were distributed approximately equally between the inner and outer leaflets of the membrane bilayer.

Experimental evidence against the lipid annulus model has been obtained by Moore *et al.* (1978), who used fluorescence depolarization of diphenylhexatriene to indicate changes in the microviscosity of sarcoplasmic reticulum preparations of varying lipid content. They could not interpret their data by two noninteracting lipid domains of different viscosity at physiological temperatures. Only below 25°C could this data be fitted by a model representing an independent domain of restricted lipid and a domain corresponding to bulk, bilayer lipid. ESR and NMR studies (Nakamura and Ohnishi, 1975; Davis *et al.*, 1976; Rice *et al.*, 1979) showed an extensive restriction of lipid mobility by the ATPase. The ATPase was able to restrict the motion of as many lipid molecules as were present in the native sarcoplasmic reticulum.

Rice *et al.* (1979) and Chapman *et al.* (1979) showed that exchange of lipid molecules between "bound" and "free" environments is fast ($\geqslant 10^3$ sec$^{-1}$), and suggested that lipid molecules may exchange during the time scale of the turnover time of ATPase. Their results did not support the lipid annulus model. Fleischer and coworkers (Fleischer *et al.*, 1979; Seelig *et al.*, 1981; Herbette *et al.*, 1981a,b) studied the lipid interaction with the ATPase in the reconstituted sarcoplasmic reticulum vesicles. Seelig *et al.* (1981) have reconstituted sarcoplasmic reticulum ATPase with a single lipid environment. 1,2-Dioleoyl-*sn*-glycero-3-phosphocholine and 1,2-dielaidoyl-*sn*-glycero-3-phosphocholine were employed and selectively deuterated in either the 9,10 or the 2,2 positions of the fatty acyl chains. The deuterium and phosphorus spectra of deuterium-labeled lipids showed that an interaction of the protein led to only a minor reduction in the rate of the internal modes of phospholipid motions. Their data, in addition to results of McLaughlin *et al.* (1981), provided no evidence for a strong, long-lived interaction between the membrane protein and the lipid. Nevertheless, it cannot be ruled out completely that a small number of lipid molecules might be tightly bound to the membrane proteins.

Alterations in temperature have been used to study the role of lipids in ATPase activity. Changes in the physical state of lipid and protein as a function of temperature are best characterized in reconstituted preparations of the ATPase. Endogenous lipids of sarcoplasmic reticulum have been exchanged in such preparations with saturated synthetic lipids such as dipalmitoyllecithin (DPL) or dimyristoyllecithin (DML) (Hidalgo *et al.*, 1976; Hesketh *et al.*, 1976; Gomez-Fernandez *et al.*, 1979, 1980; Hoffman *et al.*, 1980). In pure DPL preparations, phase transitions occur at 41°C and are accompanied by an abrupt change in lipid fluidity. For DPL–ATPase complexes, a phase transition was observed between approximately 29–40°C (Gomez-Fernandez *et al.*, 1980). These data were interpreted as follows: the protein-rich areas melt at 29°C, while the crystalline pure lipid melts gradually between 29–40°C (Moller *et al.*, 1982). The ATPase activity was stimulated because of a decrease in viscosity of the protein microenvironment. About 40°C, the lipid was fluid and the protein might have been homogenously distributed in the plane of the membrane. Similar results were obtained in the DML–ATPase complex (Nakamura *et al.*, 1976; Gomez-Fernandez *et al.*, 1979; Hoffman *et al.*, 1980) and a similar interpretation might be applied. Changes in the phase transition for native sarcoplasmic reticulum were less conspicuous than those detected in the synthetic lipids–ATPase complex. It was not expected that a phase transition like that observed with synthetic lipids would be observed for sarcoplasmic reticulum due to its heterogeneous nature and the presence of a large fraction of unsaturated fatty acid chains. Spin-label studies indicated that the character of the thermal motion of sarcoplasmic reticulum lipid changes in the same temperature interval at which there is a transition in the activation energy for ATPase activity (Inesi *et al.*, 1973; Lee *et al.*, 1974; Hidalgo *et al.*, 1976). From NMR studies, it has been concluded that a transition of lipid motion occurs between 10–30°C (Davis *et al.*, 1976) involving changes in the mobility of *N*-methyl protons of the phosphatidylcholine head groups.

The influence of temperature on the activity of the ATPase might also occur independently of changes in lipid fluidity. Dean and Tanford (1978) observed a break in the Arrhenius plot of ATPase activity not only for the lipid–ATPase complex but also for monomers of the ATPase, which were almost completely delipidated. Anzai *et al.* (1978) have reported a break in the Arrhenius plots of ATPase activity at 18°C for native sarcoplasmic reticulum as well as for preparations of dioleoyllecithin or egg lecithin-exchanged ATPase. The rotational motion of protein-bound probes, detected by saturation transfer ESR (Kiring *et al.*, 1978; Thomas and Hidalgo, 1978; Kaizu *et al.*, 1980) or by flash photolysis techniques (Hoffman *et al.*, 1980); Burkli and Cherry, 1981) shows a change in activation energy at 15–20°C. There is also some evidence that both the spin-label and photodichroism probes showed an intramolecular motion of the ATPase as well as rotational motion of the whole protein (Hidalgo *et al.*, 1978; Burkli and Cherry, 1981).

It has to be concluded that much more work is needed to clarify the precise nature of temperature-dependent changes in lipid and/or ATPase motion in the sarcoplasmic reticulum membrane and its functional implications.

## E. Reconstitution of the ATPase

Studies of phospholipid and fatty acid interaction with the ATPase (Section V-D) used assays only for $Ca^{2+}$-dependent ATPase activity. However, the lipid–ATPase interaction should also support accumulation of $Ca^{2+}$. The reconstitution procedure first developed by Racker (1972) permitted studies of incorporation of the ATPase into phospholipid vesicles under conditions where the vesicles would accumulate $Ca^{2+}$.

Racker (1972) solubilized the ATPase with cholate and then removed the detergent by dialysis. As a result, the ATPase was incorporated into bimolecular lipid and was able to catalyze ATP-dependent accumulation of $Ca^{2+}$. The complete $Ca^{2+}$ transport unit consisted of phospholipids, ATPase, and proteolipid. Knowles and Racker (1975) found that the presence of phosphatidylcholine and phosphatidylethanolamine was essential for $Ca^{2+}$-transport activity. In addition, Knowles *et al.* (1975) showed that acetylated phosphatidylethanolamine, in which the amine group was blocked, did not support $Ca^{2+}$ uptake in reconstitution experiments unless alkylamines were added. This led Racker and his co-workers to suggest that the presence of the free amine group in the membrane was essential for reconstitution of an active $Ca^{2+}$-transporting ATPase.

Warren *et al.* (1974b) showed that reconstitution of the ATPase with synthetic dioleoylphosphatidylcholine was sufficient to create a $Ca^{2+}$-accumulating vesicle. Later, Bennett *et al.* (1980) showed that $Ca^{2+}$ uptake was sensitive to small changes in the ratios of lipid detergent and protein.

Meissner and Fleischer (1973, 1974) were able to reconstitute $Ca^{2+}$ transport with deoxycholate solubilized intact sarcoplasmic reticulum or purified ATPase without added phospholipids. Reconstitution of functional vesicles of sarcoplasmic reticulum was dependent on the temperature of dialysis, pH, and the composition of the dialysis buffer, as well as on the deoxycholate concentration used initially for solubilization of the membrane. Reconstituted vesicles contained the ATPase, phospholipids, some intrinsic glycoprotein, and probably some proteolipid. Meissner and Fleischer (1974) showed that reconstituted vesicles have an elevated ATPase activity and retained approximately 25% and 50% of $Ca^{2+}$ uptake and loading capacity, respectively.

Repke *et al.* (1976) reconstituted the sarcoplasmic reticulum membrane under conditions similar to those used by Meissner and Fleischer. They recovered approximately 80% of the $Ca^{2+}$ uptake velocity and up to 90% of the $Ca^{2+}$ storage capacity. The reconstituted sarcoplasmic reticulum vesicles obtained by Repke *et al.* (1976) consisted of the 100,000-dalton ATPase and more than 80% of the 59,000-dalton protein present in intact sarcoplasmic reticulum as well as 30,000- and 37,000-dalton proteins. Michalak *et al.* (1980) showed that the 53,000-dalton glycoprotein was present in vesicles reconstituted from solubilized sarcoplasmic reticulum. Most likely, Repke's 59,000-dalton protein was the same as the 53,000-dalton glycoprotein.

The presence of proteolipid in reconstituted sarcoplasmic reticulum was reported by Racker (1972) and Racker and Eyton (1975). They suggested that it might constitute a $Ca^{2+}$ uptake channel (Racker, 1972). However, the ATPase can be stripped of proteolipid and, after reconstitution, still catalyze $Ca^{2+}$ transport (MacLennan *et al.*, 1980; Knowles *et al.*, 1980). These experiments indicate a lack of any essential role for the proteolipid in providing a channel for the $Ca^{2+}$ uptake process.

Reconstituted membranes have been prepared with a lipid content similar to that of the original sarcoplasmic reticulum membrane, i.e., 60% protein and 40% phospholipid (Meissner and Fleischer, 1974). Wang *et al.* (1979) have developed the methodology to reconstitute membranes of defined lipid content containing either lower or higher lipid content than the original sarcoplasmic reticulum membrane. Herbette *et al.* (1981a) measured ATP-induced $Ca^{2+}$ accumulation and rates of $Ca^{2+}$ uptake of the reconstituted sarcoplasmic reticulum membrane system over an extended range of lipid to protein molar ratios. Estimates of the initial $Ca^{2+}$ uptake rates and efficiencies of reconstituted sarcoplasmic reticulum were found to be comparable to those of isolated sarcoplasmic reticulum controls (light sarcoplasmic reticulum fraction) at comparable lipid:protein molar ratios. By contrast, reconstituted sarcoplasmic reticulum vesicles at low lipid:protein molar ratios showed a poorer functional behavior which might have been related to an altered structural organization of these membranes. It was also shown that reconstituted sarcoplasmic reticulum vesicles with a lipid:protein molar ratio greater than 88 were highly asymmetric, and that the ATPase activation was similar to native sarcoplasmic reticulum vesicles (Wang *et al.*, 1979; Fleischer *et al.*, 1979; Herbette *et al.*, 1981a,b). Herbette *et al.* (1981b) have proposed that the ATPase penetrated into the lipid bilayer to different degrees in native and reconstituted sarcoplasmic reticulum membrane (lipid:protein molar ratio greater than 88). The majority of the ATPase, whose vectorial distribution in the reconstituted sarcoplasmic reticulum membrane profile was identical to that of native sarcoplasmic reticulum, may have been more deeply embedded in the lipid bilayer than in the native sarcoplasmic reticulum.

The results of reconstitution experiments indicate that the ATPase complexed with phospholipids exhibits the requirements of an active $Ca^{2+}$ transport unit. However, $Ca^{2+}$ transport activity in reconstituted sarcoplasmic reticulum vesicles is not restored to the level exhibited by the original vesicles. This might be due to the presence of "leaky" reconstituted vesicles unable to permit an effective accumulation of $Ca^{2+}$ or the fact that additional minor component(s) essential for $Ca^{2+}$ transport are lost during the process of reconstitution.

## VI. PHOSPHORYLATION OF SARCOPLASMIC RETICULUM PROTEINS

Studies on the phosphorylation of sarcoplasmic reticulum membrane have been focused mainly on cardiac muscle sarcoplasmic reticulum (Tada and Katz, 1982). Less extensive studies have been devoted to phosphorylation of proteins of skeletal muscle sarcoplasmic reticulum.

As part of the reaction mechanism of ATP-dependent $Ca^{2+}$ transport the ATPase is autophosphorylated forming a phosphoprotein intermediate. Phosphorylation occurs at carboxyl group of an aspartic acid residue (Martonosi, 1967; Yamamoto and Tonomura, 1968; Yamamoto *et al.*, 1971; Degani and Boyer, 1973). Phosphoprotein levels in the steady state depend on the concentration of $Ca^{2+}$ and ATP and parallel those of the $Ca^{2+}$-dependent ATPase activity.

Campbell and Shamoo (1980) found that several proteins were phosphorylated when phosphorylation of the skeletal muscle sarcoplasmic reticulum membrane was carried out in the presence of EGTA, but in the absence of cAMP and exogenous protein kinase. These were proteins of molecular weight 64,000, 42,000, and 20,000. The 64,000-dalton phosphoprotein was more concentrated in the heavy sarcoplasmic reticulum vesicles and its phosphorylation was inhibited by dantrolene, an inhibitor of $Ca^{2+}$ release.

Campbell and MacLennan (1982) examined the effect of calmodulin on phosphorylation of skeletal muscle sarcoplasmic reticulum proteins. They found that an endogenous calmodulin-dependent kinase in skeletal muscle sarcoplasmic reticulum specifically catalyzed the phosphorylation of 60,000-and 20,000-dalton proteins. Chiesi and Carafoli (1982) also demonstrated calmodulin-dependent phosphorylation of 57,000-, 35,000-, and 20,000-dalton proteins of the sarcoplasmic reticulum membrane from fast skeletal muscle.

## VII. CALCIUM RELEASE BY THE SARCOPLASMIC RETICULUM

The mechanism of $Ca^{2+}$ release from sarcoplasmic reticulum is not well understood. The release of $Ca^{2+}$ occurs at rates 1000-fold faster than the rate of $Ca^{2+}$ uptake (Martonosi, 1972). These rates suggest involvement of an open channel rather than an enzymatic process in $Ca^{2+}$ release.

The physiological stimulus for $Ca^{2+}$ release is the depolarization of the surface membrane by an action potential that is conducted into the muscle cell through the T-tubules (Huxley, 1971; Bastin and Nakajima, 1974). $Ca^{2+}$ release from sarcoplasmic reticulum is studied either in single-skinned fibers (Constantin and Podolsky, 1967; Endo *et al.*, 1970) or in isolated sarcoplasmic reticulum vesicles (Inesi and Malan, 1976; Katz *et al.*, 1976, 1977; Miyamoto and Racker, 1981, 1982). Skinned fibers are more advantageous for such studies because the overall structure of the sarcoplasmic reticulum and T-tubules is not altered while the cytoplasmic medium becomes accessible to chemical modification. Major questions in $Ca^{2+}$ release in these partially resolved systems are the nature of the $Ca^{2+}$-release channel, the role of the ATPase and the ways in which the $Ca^{2+}$-release channel controls its opening and closing.

The autoradiographic experiments of Winegrad (1968, 1970) strongly suggest that $Ca^{2+}$ is released from the terminal cisternae of the sarcoplasmic reticulum. Recently, Somlyo *et al.* (1981) have determined, by electron-probe analysis of ultrathin cryosections of frog muscle, that $Ca^{2+}$ is specifically released from the terminal cisternae of sarcoplasmic reticulum. They suggested that the charge compensation during $Ca^{2+}$ release was achieved by movement of protons into the sarcoplasmic reticulum and/or by the movement of organic co- or counterions.

Campbell *et al.* (1980) and Campbell and Shamoo (1980) have shown that chloride-induced $Ca^{2+}$ release from isolated heavy sarcoplasmic reticulum vesicles (corresponding to terminal cisternae) can be partially blocked by dantrium, in contrast to vesicles corresponding to longitudinal sarcoplasmic reticulum which lack the dantrium-sensitive component of chloride-induced $Ca^{2+}$ release. Dantrium has been shown to inhibit $Ca^{2+}$ release induced by T-tubule depolarization, a finding which adds to the evidence that $Ca^{2+}$ release occurs from the terminal cisternae. Miyamoto and Racker (1981, 1982) supported these observations showing that heavy sarcoplasmic reticulum vesicles, presumably derived from the terminal cisternae of the sarcotubular system, will release $Ca^{2+}$ in response to $Ca^{2+}$ concentrations in the physiological range, whereas, $Ca^{2+}$-induced $Ca^{2+}$ release from the longitudinal sarcoplasmic reticulum is less dramatic. These observations suggest again that $Ca^{2+}$-release channels might reside specifically in the terminal cisternae of sarcoplasmic reticulum and that the T-system elements (found in the heavy sarcoplasmic reticulum fraction) might also play an important role in $Ca^{2+}$-induced $Ca^{2+}$ release. Miyamoto and Racker (1981, 1982) provided evidence that $Ca^{2+}$ release could be induced by physiological concentrations of $Ca^{2+}$ since Endo (1977) pointed out that results on $Ca^{2+}$-induced $Ca^{2+}$ release were obtained in the presence of $Ca^{2+}$ concentrations too high to be considered physiologically relevant.

The influence of an interaction between the T-system and the sarcoplasmic reticulum during $Ca^{2+}$ release has been suggested by Eisenberg and Eisenberg (1982). They reported the formation of "pillars" between the terminal cisternae of the sarcoplasmic reticulum and the T-tubules upon depolarization of the muscle. The fact that "pillar" density is higher in depolarized muscle than in resting fibers strongly suggests that their formation may be a morphological correlate of the process linking electrical excitation with $Ca^{2+}$ release. The above results do not, of course, rule out a completely different distribution of the $Ca^{2+}$-release channels, for example, one involving the whole sarcoplasmic reticulum membrane.

Other stimuli than $Ca^{2+}$ have been shown to induce $Ca^{2+}$ release. Caffeine (Endo, 1977) and chloride (Stephenson and Podolsky, 1977) will induce $Ca^{2+}$ release from the sarcoplasmic reticulum. Chloride-induced $Ca^{2+}$ release is believed to result from changes in an electrical potential across the membrane. However, such a potential across the sarcoplasmic reticulum membrane was neither measured, nor reliably calculated. $Ca^{2+}$-induced $Ca^{2+}$ release from sarcoplasmic reticulum has been found to be accelerated by $\gamma,\beta$-methylene adenosine triphosphate, an unhydrolyzable ATP analogue (Ogawa and Ebashi, 1976). Beirao and de Meis (1976) also reported an ADP activation of $Ca^{2+}$ efflux. Nucleotide binding to the channel for $Ca^{2+}$ release might play a significant role in the inactivation and activation of the channel.

Several investigators have studied passive permeability of the sarcoplasmic reticulum for different ions. Passive $Ca^{2+}$ efflux from sarcoplasmic reticulum vesicles has been measured by dilution of vesicles passively loaded with $Ca^{2+}$ (Meissner and McKinley, 1976; Nagasaki and Kasai, 1980; Campbell *et al.*, 1980; Millman, 1980; Bennett and Dupont, 1981) or by adding EGTA to vesicles loaded through the operation of the sarcoplasmic reticulum ATPase (Katz *et al.*, 1977; Hidalgo, 1980).

Jilka *et al.* (1975) have shown that the passive $Ca^{2+}$ permeability of phospholipid vesicles is much less than the $Ca^{2+}$ permeability of sarcoplasmic reticulum membranes or of phospholipid vesicles reconstituted with the ATPase. The permeability to sucrose, $Na^+$, choline$^+$, or $SO_4^{2-}$ was similarly enhanced by the presence of the protein. The rate of $Ca^{2+}$ efflux in this system was 1000-fold lower than the rate of release necessary to elevate the sarcoplasmic concentration to levels that generate contraction of the fibrillar systems. It could only account for the rate of release observed in a muscle fiber in a relaxed state. The role of the sarcoplasmic reticulum ATPase in passive $Ca^{2+}$ efflux was ruled out by Jilka *et al.* (1975). Feher and Briggs (1982) studied $Ca^{2+}$ efflux from sarcoplasmic reticulum as a function of the $Ca^{2+}$ load of membrane vesicles. They showed that passive $Ca^{2+}$ efflux was very slow compared to $Ca^{2+}$ uptake or to physiological $Ca^{2+}$ release and it was not regulated by the external $Ca^{2+}$ concentration. The ATPase was not involved as a carrier of these passive effluxes of $Ca^{2+}$.

$Ca^{2+}$ release might require or be influenced by the movement of other ions across the sarcoplasmic reticulum membrane. McKinley and Meissner (1978) presented evidence for two populations of sarcoplasmic reticulum, type I and II, the latter being far less permeable to $K^+$ and $Na^+$. On the basis of the influence of an inside negative membrane potential on the permeability of type I and II vesicles, they suggested that the structure of the $K^+$,$Na^+$ permeable channel of the sarcoplasmic reticulum might be changed by a membrane potential. This $K^+$,$Na^+$ channel might be responsible for a compensation of the charge during $Ca^{2+}$ movement across the membrane. They followed those studies by investigating the $H^+$ permeability of sarcoplasmic reticulum membranes (Meissner and Young, 1980; Meissner, 1981). They used a fluorescent dye as an indicator of membrane potential, and concluded that both type I and type II sarcoplasmic reticulum vesicles were permeable to $H^+$, both being significantly more permeable to $H^+$ than sarcoplasmic reticulum phospholipid vesicles.

The relation between a proton gradient generated across the sarcoplasmic reticulum membrane during $Ca^{2+}$ uptake and $Ca^{2+}$ release has been studied by Shoshan *et al.* (1981) and MacLennan *et al.* (1982). They found that increasing the pH of the bathing solution of chemically skinned muscle fibers beyond a threshold of 0.2 pH units leads to $Ca^{2+}$ release, and that an increase of 0.5 units above loading pH leads to maximal $Ca^{2+}$ release. A decrease in pH never led to $Ca^{2+}$ release. Since, in the pH range 6.5–7.5, the threshold for $Ca^{2+}$ release was always 0.2 pH units above the loading pH, they suggested that a pH gradient rather than an absolute, external pH was critical for control of $Ca^{2+}$ release. They showed, in addition, that dicyclohexylcarbodiimide blocks $Ca^{2+}$ release by pH elevation but not caffeine-induced $Ca^{2+}$ release. This might suggest that these two stimuli operate at different points in a chain of events leading to $Ca^{2+}$ release or that they have different triggering mechanisms.

Shoshan and co-workers (Shoshan *et al.*, 1980; Shoshan and MacLennan, 1981; MacLennan *et al.*, 1982) studied the effects of quercetin on $Ca^{2+}$ uptake and release by the sarcoplasmic reticulum membrane. They have shown that quercetin is a potent inhibitor of ATP-dependent $Ca^{2+}$ uptake, by blocking decomposition of a phosphorylated form of the ATPase. In this stabilized, phosphorylated form of the ATPase, reversal of the $Ca^{2+}$ pump is inhibited. In skinned muscle fibers, release of $Ca^{2+}$ was observed in the presence of quercetin. Moreover, the measured rate of tension development was enhanced 4- to 7-fold and the relaxation was inhibited (Shoshan *et al.*, 1980). On these bases, they suggested that the physiologically relevant $Ca^{2+}$-release channels are separate from the $Ca^{2+}$-uptake channels in the ATPase. Isolated sarcoplasmic reticulum vesicles will release $Ca^{2+}$ when the pH is elevated in the presence of EGTA (Shoshan *et al.*, 1981). Eighty-five percent of this release is quercetin sensitive, however, suggesting an involvement of the ATPase in this process but not in the relevant $Ca^{2+}$-release channel (Shoshan and MacLennan, 1981). This suggests that proton gradient-sensitive $Ca^{2+}$ release is a component of the more complex skinned fiber system but not of the more resolved sarcoplasmic reticulum vesicles. It might suggest that the pH sensitivity is localized in T-tubules (MacLennan *et al.*, 1982).

The precise mechanism of $Ca^{2+}$ release from the sarcoplasmic reticulum membrane and the role of various membrane proteins in this process needs further studies.

## VIII. CARDIAC SARCOPLASMIC RETICULUM MEMBRANE

Several reviews published in the last few years have dealt with cardiac sarcoplasmic reticulum (Tada *et al.*, 1978a, 1982; Van Winkle and Entman, 1979; Katz, 1979; Inesi, 1981; Winegrad, 1982; Tada and Katz, 1982). Here, attention is focused on general properties of cardiac sarcoplasmic reticulum with an emphasis on the role of phosphorylation in the regulation of ATPase and $Ca^{2+}$-transport activities.

The sarcotubular system of the myocardium is less developed than that of the skeletal muscle. It consists of the sarcoplasmic reticulum, which forms networks around the myofibrillar bundles and T-system tubules (Sommer and Johnson, 1979; Inesi, 1981). The portion of the sarcoplasmic reticulum membrane that is located in close apposition to the T-system is called interior junctional sarcoplasmic reticulum membrane, while that portion in close apposition to the sarcolemma is called peripheral junctional sarcoplasmic reticulum membrane. In mammalian heart, the T-system is less well developed in atrial fibers than in ventricular fibers, and no T-system is present in Purkinje fibers (Suzuki and Sugi, 1982). The cardiac sarcoplasmic reticulum vesicles, isolated by differential centrifugation and/or sucrose density centrifugation (Jones *et al.*, 1979; Jones and Cala, 1981), retain major characteristics of the sarcoplasmic reticulum in that $Ca^{2+}$ is actively transported in an ATP-dependent manner (Tada *et al.*, 1978b).

Cardiac sarcoplasmic reticulum vesicles are not so highly purified and so appear to contain a more complex protein composition than skeletal vesicles (Jones *et al.*, 1979). The major protein of the cardiac sarcoplasmic reticulum is the ATPase enzyme of about 100,000 daltons, consisting of up to 40% of total proteins (Suko and Hasselbach, 1976; Affolter *et al.*, 1976; Jones and Cala, 1981). The ATPase of cardiac sarcoplasmic reticulum has been isolated (Levitsky *et al.*, 1976; Van Winkle *et al.*, 1978) and shown to be 100% $Ca^{2+}$ dependent. The maximum specific activity of the cardiac ATPase is approximately 30–40% of that of the fast skeletal enzyme (Van Winkle *et al.*, 1978).

Campbell *et al.* (1983a) have purified calsequestrin and identified the 53,000-dalton glycoprotein in cardiac sarcoplasmic reticulum. A combination of both dark blue staining of calsequestrin with the cationic carbocyanine dye stains (Campbell *et al.*, 1983b) and application of two-dimensional gel electrophoresis, let them identify calsequestrin in the cardiac sarcoplasmic reticulum. Unlike skeletal muscle calsequestrin, cardiac calsequestrin was sensitive to Endo H digestion, indicating that it contained a "high-mannose" oligosaccharide (Campbell *et al.*, 1983a). Cardiac calsequestrin bound 300 μmoles of $Ca^{2+}$ per mg protein. Thirty-two percent of the amino-acid residues in the protein were glutamic and aspartic acid (Campbell *et al.*, 1983a). Calsequestrin has also been identified in cardiac muscle by indirect immunofluorescence labeling and shown to be localized in those regions of the cardiac cells where sarcoplasmic reticulum is localized (Campbell *et al.*, 1983a). Using the immunofluorescein-labeling technique, Jorgensen *et al.* (1982a) showed that the cardiac ATPase was uniformly distributed in the free cardiac sarcoplasmic reticulum membrane.

Cardiac sarcoplasmic reticulum has been found to form two classes of phosphoproteins. The ATPase forms a phosphoprotein intermediate by incorporating the terminal phosphate of ATP into hydroxylamine labile acylphosphate (Shigekawa *et al.*, 1976; Tada *et al.*, 1979; Tada and Katz, 1982). In skeletal muscle sarcoplasmic reticulum, phosphorylation occurs at the carboxyl group of an aspartic acid residue. The second type of phosphoprotein in cardiac sarcoplasmic reticulum is formed upon reaction with a cAMP-dependent protein kinase system. In this reaction, the terminal phosphate of ATP is incorporated mainly into a serine residue in a 22,000-dalton protein that exhibits the stability characteristics of a phosphoester (Kirchberger *et al.*, 1974; Tada *et al.*, 1978b; Le Peuch *et al.*, 1980; Bidlack and Shamoo, 1980). The 22,000-dalton phosphoprotein of cardiac sarcoplasmic reticulum has been named phospholamban.

Other lower molecular weight proteins of cardiac sarcoplasmic reticulum are also phosphorylated by the cAMP-dependent protein kinase system, including a phosphoprotein of 6000 (Le Peuch *et al.*, 1979; Bidlack and Shamoo, 1980) or 7000 daltons (Jones *et al.*, 1979; Le Peuch *et al.*, 1980). An 11,000-dalton phosphoprotein that appears after addition of Triton X-100 and boiling of the phosphorylated sarcoplasmic reticulum solubilized in SDS may be a monomer of phospholamban (Le Peuch *et al.*, 1980). Kirchberger and Antonetz (1982) showed that boiling of SDS-denatured, phosphorylated sarcoplasmic reticulum vesicles results in a shift to the 11,000- and 5500-dalton forms of phospholamban. They suggested that the 5500-dalton phosphoprotein represents the monomeric form of phospholamban. Phosphorylation of phospholamban has also been shown in a calmodulin-dependent protein kinase system (Le Peuch *et al.*, 1979; Kranias *et al.*, 1980; Louis and Maffitt, 1982). Louis *et al.* (1982) and Louis and Jarvis (1982) proposed a model for the structure of phospholamban suggesting that the protein is composed of three subunits of 11,000, 8000, and 4000 daltons. In this model, only the 11,000-dalton subunit would be phosphorylated in the presence of either calmodulin or cAMP-dependent protein kinase. They suggested that smaller-sized subunits that were produced when phosphorylated sarcoplasmic reticulum was boiled in the presence of Triton X-100 or SDS would be derived from partial or total dissociation of the 22,000-dalton phosphorylated phospholamban.

Phosphorylation of phospholamban has been shown to have a marked stimulatory effect on active $Ca^{2+}$ transport and ATPase activity (Tada *et al.*, 1978a,b, 1982; Van Winkle and Entman, 1979; Jones *et al.*, 1979; Tada and Katz, 1982. The rates of oxalate- or phosphate-supported $Ca^{2+}$ uptake by cardiac sarcoplasmic reticulum are more than doubled after phosphorylation of phospholamban. Protein kinase-catalyzed phosphorylation of phospholamban is capable of stimulating $Ca^{2+}$ transport even in the absence of oxalate (Will *et al.*, 1976). Tada *et al.* (1979, 1980) found that cAMP-dependent protein kinase stimulated ATPase activity as well as $Ca^{2+}$ uptake with the expected coupling stoichiometry of two. Katz *et al.* (1975) reported an increase in EGTA-induced $Ca^{2+}$ efflux from cardiac sarcoplasmic reticulum after phosphorylation of the membrane with cAMP-dependent protein kinase. Tada *et al.* (1980) showed that phospholamban phosphorylation can enhance the rate of decomposition of the phosphorylated intermediate of the ATPase relative to that of phosphorylated intermediate formation when the latter is inhibited at low-ATP concentrations. In addition, Hicks *et al.* (1979) showed that phospholamban phosphorylation alters the $Ca^{2+}$ sensitivity of the ATPase and influences an interaction between the two $Ca^{2+}$-binding sites of the enzyme in a manner that reduces the positive cooperativity for $Ca^{2+}$ which is found when phospholamban is in its phosphorylated form. Since phosphorylation of phospholamban in cardiac sarcoplasmic reticulum is accompanied by significant changes in both the $Ca^{2+}$-dependent and kinetic properties of the ATPase (Katz, 1979, Tada *et al.*, 1982; Tada and Katz, 1982), it can be assumed that phospholamban is closely associated with the ATPase protein.

Molecular models have been presented in which phosphorylated phospholamban serves as an activator (Tada *et al.*, 1978b) or derepressor (Hicks *et al.*, 1979; Katz, 1980) of the ATPase. A protein–protein interaction has been suggested to occur between phospholamban and the ATPase (Hicks *et al.*, 1979; Tada *et al.*, 1979, 1980), although an involvement of membrane lipids is also possible. Although a part of phospholamban is located on the cytoplasmic surface of sarcoplasmic reticulum membranes (Le Peuch *et al.*, 1980), a large portion of the molecule appears to be embedded within the membrane (Bidlack and Shamoo, 1980; Le Peuch *et al.*, 1980). Hicks *et al.* (1979) compared the $Ca^{2+}$ dependence of cardiac sarcoplasmic reticulum vesicles before and after cAMP-dependent protein kinase phosphorylation with that of a comparable preparation from rabbit fast skeletal muscle to show whether dephospho-phospholamban acts as an inhibitor of the ATPase, or whether phosphorylated phospholamban is an activator. Their finding, that the $Ca^{2+}$ sensitivity of phosphorylated cardiac sarcoplasmic reticulum vesicles is similar to that of skeletal sarcoplasmic reticulum vesicles which appear to lack phospholamban, suggests that the dephospho form of phospholamban is an inhibitor of cardiac ATPase.

A regulatory mechanism of $Ca^{2+}$ transport involving phosphorylation of a protein similar to phospholamban may also operate in platelets (Haslem *et al.*, 1978) and smooth muscle (Kimura *et al.*, 1977; Bhalla *et al.*, 1978; Nishikori and Maeno, 1979).

## IX. BIOSYNTHESIS OF THE SARCOPLASMIC RETICULUM MEMBRANE

The biosynthesis of the sarcoplasmic reticulum has been studied extensively *in vivo*, in tissue culture, and in cell-free translation systems (MacLennan *et al.*, 1978; Martonosi, 1980; Martonosi, 1982).

### A. Biosynthesis of Sarcoplasmic Reticulum in Vivo

The ATPase and $Ca^{2+}$ transport activities are very low at the earlier stages of development of skeletal muscle. The ATPase content of the whole muscle and of isolated sarcoplasmic reticulum increases more then 20-fold with parallel changes in $Ca^{2+}$ transport, $Ca^{2+}$-sensitive ATPase activity and steady-state concentrations of the phosphoenzyme intermediate (Martonosi *et al.*, 1972, 1977; Boland *et al.*, 1974; Sarzala *et al.*, 1975a,b). There is also an increase in total surface area of sarcoplasmic reticulum during development (Boland *et al.*, 1974). The density of 9-nm freeze-fracture and 4-nm negative-staining particles increases during development (Boland *et al.*, 1974; Martonosi, 1975; Sarzala *et al.*, 1975a,b). The increase in ATPase concentration was reflected as an increase in the protein:lipid ratio of isolated sarcoplasmic reticulum membranes (Boland *et al.*, 1974; Sarzala *et al.*, 1975a,b; Boland and Martonosi, 1976). The change in ATPase concentration determined by gel electrophoresis, electron microscopy, and analysis of the active site concentration of ATPase after specific labeling with [$^{32}$P]-ATP parallels the change in $Ca^{2+}$ transport activity throughout development (Martonosi *et al.*, 1977, 1980). Calsequestrin and the high-affinity $Ca^{2+}$-binding protein seems to be present in sarcoplasmic reticulum at early stages of muscle development (Sarzala *et al.*, 1975a,b; Martonosi, 1975; Zubrzycka *et al.*, 1979).

The phospholipid composition of isolated chicken sarcoplasmic reticulum remained relatively constant between 10–50 days of development (Boland *et al.*, 1974; Boland and Martonosi, 1976). There were, however, marked changes in the fatty acid composition which maintained fluidity of the developing membrane rather constant (Boland *et al.*, 1977). In developing rabbit muscle, the phosphatidylcholine content of sarcoplasmic reticulum increased during the first week after birth with a decrease in phosphatidylethanolamine and other phospholipids (Sarzala *et al.*, 1975a,b; Zubrzycka *et al.*, 1979).

## B. The Biosynthesis of Sarcoplasmic Reticulum in Vitro

In tissue culture systems, differentiation of muscle cells takes place in a manner similar to that *in vivo*. Myoblasts proliferate and fuse to form mononucleated myotubes, followed by accumulation of contractile protein and sarcoplasmic reticulum.

Biosynthesis of the sarcoplasmic reticulum protein has been extensively studied by MacLennan and co-workers (Holland and MacLennan, 1976; Zubrzycka and MacLennan, 1976; Jorgensen *et al.*, 1977; Zubrzycka *et al.*, 1983) and Martonosi and co-workers (Martonosi *et al.*, 1977, Martonosi, 1980; Ha *et al.*, 1979; Wu *et al.*, 1981; Roufa *et al.*, 1981). MacLennan and co-workers have compared the process of biosynthesis of the intrinsic (ATPase and the 53,000-dalton glycoprotein) and extrinsic (calsequestrin and the high-affinity $Ca^{2+}$-binding protein) proteins of sarcoplasmic reticulum membrane in differentiating primary culture of rat skeletal muscle. The biosynthetic pattern of the ATPase and the 53,000-dalton glycoprotein is indistinguishable (Holland and MacLennan, 1976; Zubrzycka *et al.*, 1983). The initiation of the synthesis of those proteins takes place at about the same time as the beginning of cell fusion. Similar observations for ATPase synthesis measured by active-site labeling with [$^{32}$P]-ATP, were reported for primary cultures of chicken myoblasts (Ha *et al.*, 1979). Holland (1979) showed that the rate of synthesis of the ATPase in embryonic chicken heart cells increases steadily during the entire culture period without the initial lag phase seen in skeletal muscle. This may suggest that cell proliferation may not be a required prelude to differentiation and ATPase synthesis in cardiac muscle.

The synthesis of intrinsic proteins of sarcoplasmic reticulum is not coordinated with the biosynthesis of calsequestrin (Zubrzycka and MacLennan, 1976; Michalak and MacLennan, 1980) or the high-affinity $Ca^{2+}$-binding protein (Michalak and MacLennan, 1980), which are turned on about 20 hr earlier. The pattern of biosynthesis of the two extrinsic proteins of sarcoplasmic reticulum is almost identical even though calsequestrin is glycosylated and the high-affinity $Ca^{2+}$-binding protein is not. Almost identical rates of synthesis of those proteins does not correlate with a relatively low concentration of the high-affinity $Ca^{2+}$-binding protein in mature membrane. The turnover of the high-affinity $Ca^{2+}$-binding protein (half-life about 10 hr; Michalak and MacLennan, 1980) is greater than that of calsequestrin (half-life about 33 hr; Zubrzycka and MacLennan, 1976), which may account for the differences in the amounts of these protein in the mature membrane. In addition, the two extrinsic sarcoplasmic reticulum proteins may be distributed in different regions of the sarcotubular system, since the high-affinity $Ca^{2+}$-binding protein has been shown to be enriched in preparations of T-tubules (Michalak *et al.*, 1980) and calsequestrin is found mainly in the heavy sarcoplasmic reticulum fraction.

The above biochemical findings are in agreement with morphological studies using direct immunofluorescein as a test for the appearance of the sarcoplasmic reticulum protein in cultured cells (Jorgensen *et al.*, 1977). Jorgensen *et al.* (1979) found that the ATPase appears in numerous foci throughout the cytoplasm in the earliest stages of its synthesis. By contrast, calsequestrin appears, in its earliest stages of synthesis, in a sharply defined, brightly staining, perinuclear area. Calsequestrin is glycosylated and the sugar composition suggests that it is a highly processed carbohydrate chain. It was postulated by MacLennan and co-workers (Zubrzycka and MacLennan, 1976; Jorgensen *et al.*, 1977; Michalak and MacLennan, 1980; Reithmeier *et al.*, 1980) that calsequestrin enters the luman of the rough endoplasmic reticulum where it is glycosylated and passes through the Golgi region where sugars are processed.

Ha *et al.* (1979), Wu *et al.* (1981), and Roufa *et al.* (1981) have evaluated a role for $Ca^{2+}$ in the synthesis of sarcoplasmic reticulum proteins. They observed an inhibition of synthesis of several muscle specific proteins, including the ATPase of the sarcoplasmic reticulum in a culture medium of relatively low $Ca^{2+}$ concentration where cell fusion was prevented. Martonosi (Martonosi *et al.*, 1978, Martonosi, 1980; Martonosi, 1982) has presented a hypothesis in which changes in intracellular free $Ca^{2+}$ concentration influence the rate of synthesis of the ATPase and other sarcoplasmic reticulum proteins by $Ca^{2+}$-dependent regulation of the transcription or processing of the relevant classes of mRNA. Direct evidence to support this hypothesis is still required.

### C. The Synthesis of Sarcoplasmic Reticulum Proteins in Cell-Free Systems

It was first shown by Greenway and MacLennan (1978) and confirmed by Chyn *et al.* (1979) that the ATPase and calsequestrin are synthesized on membrane-bound polysomes. Free polysomes are essentially inactive in synthesis of these proteins. The electrophoretic mobility, isoelectric point, and tryptic map of the *in vitro* translated product are similar to authentic ATPase isolated from mature muscle or from tissue culture (Chyn *et al.*, 1979; Reithmeier *et al.*, 1980). The $NH_2$-terminal methionine group of the ATPase derived from initiator methionyl tRNA Met, was acetylated during translation (Reithmeier *et al.*, 1980). These observations suggest that the ATPase is synthesized without an $NH_2$-terminal signal sequence. Recent observations suggest cotranslational insertion of the ATPase into microsomal membranes (Mostov *et al.*, 1981). The results of structural studies of the ATPase suggest that there are a number of transmembrane segments in the molecule. Since the 30 $NH_2$-terminal amino acids of the ATPase are relatively hydrophilic, the $NH_2$-terminal segment of the molecule remains on the cytoplasmic side of the membrane (Reithmeier and MacLennan, 1981).

Calsequestrin is synthesized on membrane-bound polysomes in a precursor form (66,000 daltons), which is processed cotranslationally to mature calsequestrin (63,000 daltons) in the presence of pancreatic microsomes (Reithmeier *et al.*, 1980). This suggests that calsequestrin follows a biosynthetic pathway similar to secreted proteins.

## X. CONCLUDING REMARKS

This review summarizes studies on the structural organization of the sarcoplasmic reticulum membrane in relation to its function. The $Ca^{2+}$-transporting ATPase of sarcoplasmic reticulum, similar to other transport proteins, is critically dependent upon its lipid environment for enzyme activity. The lipids interacting with protein should contain "fluid" fatty acid chains. There is an increasing evidence that the ATPase is self-associated in the membrane in oligomeric form. The polypeptide chain of the ATPase passes several times through the hydrophobic region of the membrane and the $NH_2$- and COOH-termini of the enzyme are localized on the cytoplasmic side of the membrane. The sarcoplasmic reticulum, like the $Na^+,K^+$-ATPase, contains a 53,000-dalton glycoprotein which is not essential to the ATPase activity.

The membrane of sarcoplasmic reticulum provides one of the most suitable systems for examining the organization, active transport, and biogenesis of biological membranes.

ACKNOWLEDGMENTS

I would like to thank very much Professor D. H. MacLennan for his valuable suggestions and criticism, useful discussion, and constant encouragement. I sincerely thank Lorraine G. Beehler for her skillful secretarial assistance.

## REFERENCES

Affolter, H., Chiesi, M., Dabrowska, R., and Carafoli, E., 1976, Calcium regulation in heart cells. The interaction of mitochondria and sarcoplasmic reticulum with troponin-bound calcium, *Eur. J. Biochem.* **67:**389.

Allen, G., 1977, On the primary structure of the $Ca^{2+}$ ATPase of sarcoplasmic reticulum, in: *FEBS 11th Meeting Copenhagen, Membrane Proteins,* Vol. 45 (P. Nicholls, Y. V. Moller, P. L. Jorgensen, and A. Y. Moody, eds.), Pergamon Press, New York, p. 159.

Allen, G., 1980a, The primary structure of the calcium ion-transporting adenosine triphosphatase of rabbit skeletal sarcoplasmic reticulum, *Biochem. J.* **187:**545.

Allen, G., 1980b, Primary structure of the calcium ion-transporting adenosine triphosphatase of rabbit skeletal sarcoplasmic reticulum, *Biochem. J.* **187:**565.

Allen, G., and Green, N. M., 1978, Primary structures of cysteine-containing peptides from the calcium ion-transporting adenosine triphosphatase of rabbit sarcoplasmic reticulum, *Biochem. J.* **173:**393.

Allen, G., Bottomley, R. C., and Trinnaman, B. J., 1980a, Primary structure of the calcium ion transporting adenosine triphosphatase from rabbit skeletal sarcoplasmic reticulum, *Biochem. J.* **187:**577.

Allen, G., Trinnaman, B. J., and Green, N. M., 1980b, The primary structure of the calcium ion-transporting adenosine triphosphatase protein of rabbit skeletal sarcoplasmic reticulum, *Biochem. J.* **187:**591.

Andersen, J. P., Le Maire, M., and Moller, J. V., 1980, Properties of detergent-solubilized and membranous ($Ca^{2+}$ + $Mg^{2+}$) activated ATPase from sarcoplasmic reticulum as studied by sulfhydryl reactivity and ESR spectroscopy. Effect of protein–protein interactions, *Biochim. Biophys. Acta* **603:**84.

Andersen, P., Moller, J. V., and Jorgensen, P. L., 1982, The functional unit of sarcoplasmic reticulum $Ca^{2+}$-ATPase. Active site titration and fluorescence measurements, *J. Biol. Chem.* **257:**8300.

Anzai, K., Kiring, Y., and Shimizu, H., 1978, Temperature induced change in the $Ca^{2+}$-dependent ATPase activity and in the state of the ATPase protein of sarcoplasmic reticulum membrane, *J. Biochem.* **84:**815.

Arakawa, M., and Muramatsu, T., 1974, Endo-β-*N*-acetylglucosaminidases acting on the carbohydrate moieties of glycoproteins. The differential specificities of the enzyme from *Streptomyces griseus* and *Diplococcus pneumoniae, J. Biochem.* **76:**307.

Bailin, G., 1980, Crosslinking of sarcoplasmic reticulum ATPase protein with 1,5-difluoro 2,4-dinitrobenzene, *Biochim. Biophys. Acta* **624:**511.

Banerjee, R., Epstein, M., Kandrach, M., Zimniak, P., and Racker, E., 1979, A new method of preparing $Ca^{2+}$-ATPase from sarcoplasmic reticulum: Extraction with octyl-glucoside, *Membr. Biochem.* **2:**283.

Baskin, R. J., 1971, Ultrastructure and calcium transport in crustacean muscle microsomes, *J. Cell Biol.* **48:**49.

Baskin, R. J., and Hanna, S., 1979, Cross-linking of the ($Ca^{2+}$ + $Mg^{2+}$) ATPase protein, *Biochim. Biophys. Acta* **576:**61.

Bastin, J., and Nakajima, S., 1974, Action potential in the transverse tubules and its role in the activation of skeletal muscle, *J. Gen. Physiol.* **63:**257.

Beirao, P. S., and de Meis, L. 1976, ADP-activated calcium ion exchange in sarcoplasmic reticulum vesicles, *Biochim. Biophys. Acta* **433:**520.

Bennett, N., and Dupont, Y., 1981, Evidence for a calcium gated cation channel in sarcoplasmic reticulum vesicles, *FEBS Lett.* **128:**269.

Bennett, J. P., McGill, K. A., and Warren, G. B., 1978, Transbilayer disposition of phospholipid annulus surrounding a calcium transport protein, *Nature* **274:**823.

Bennett, J. P., McGill, K. A., and Warren, G. B., 1980, The role of lipids in the functioning of a membrane protein: The sarcoplasmic reticulum pump, *Curr. Top. Membr. Trans.* **14:**127.

Berman, M. C., 1982, Energy coupling and uncoupling of active calcium transport by sarcoplasmic reticulum membranes, *Biochim. Biophys. Acta* **694:**95.

Bhalla, R. C., Webb, R. C., Singh, D., and Brock, T., 1978, Role of cyclic AMP in rat aortic microsomal phosphorylation and calcium system, *Am. J. Physiol.* **234:**H508.

Bidlack, J. M., and Shamoo, A. F., 1980, Adenosine 3′, 5′-monophosphate-dependent phosphorylation of a 6,000 and 22,000 dalton protein from cardiac sarcoplasmic reticulum, *Biochim. Biophys. Acta* **632:**310.

Boland, R., and Martonosi, A., 1976, The lipid composition and $Ca^{2+}$-transport function of sarcoplasmic reticulum (SR) membranes during development *in vivo* and *in vitro,* in: *Function and Biosynthesis of Lipids* (N. G. Bazan, R. R. Brenner, and N. M. Guisto, eds.), Plenum, New York and London, p. 233.

Boland, R., Martonosi, A., and Tillack, T. W., 1974, Developmental changes in the composition and function of sarcoplasmic reticulum, *J. Biol. Chem.* **249:**612.

Boland, R., Chyn, T., Roufa, D., Reyes, E., and Martonosi, A., 1977, The lipid composition of muscle cells during development, *Biochim. Biophys. Acta* **489:**349.

Borchman, D., Simon, R., and Bicknell-Brown, E., 1982, Variation in the lipid composition of rabbit muscle sarcoplasmic reticulum membrane with muscle type, *J. Biol. Chem.* **257:**14136.

Boulan, E. R., Kreibich, G., and Sabatini, D. D., 1978, Spatial orientation of glycoproteins in membranes of rat liver rough microsomes. I. Localization of lectin binding sites in microsomal membranes, *J. Cell Biol.* **78:**894.

Brotherus, J. R., Moller, J. V., and Jorgensen, P. L., 1981, Soluble and active renal Na, K-ATPase with maximum protein molecular mass 170,000 ± 9,000 daltons; formation of larger units by secondary aggregation, *Biochem. Biophys. Res. Commun.* **100:**146.

Burkli, A., and Cherry, R. J., 1981, Rotational motion and flexibility of $Ca^{2+}$ + $Mg^{2+}$-dependent adenosine 5′-triphosphatase in sarcoplasmic reticulum membranes, *Biochemistry* **20:**138.

Campbell, K. P., and MacLennan, D. H., 1981, Purification and characterization of the 53,000 dalton glycoprotein from the sarcoplasmic reticulum, *J. Biol. Chem.* **256:**4626.

Campbell, K. P., and MacLennan, D. H., 1982, A calmodulin-dependent protein kinase system from skeletal muscle sarcoplasmic reticulum, *J. Biol. Chem.* **257:**1238.

Campbell, K. P., and Shamoo, A. E., 1980, Phosphorylation of heavy sarcoplasmic reticulum vesicles: Identification and characterization of three phosphorylated proteins, *J. Membr. Biol.* **56:**241.

Campbell, K. P., Franzini-Armstrong, C., and Shamoo, A. E., 1980, Further characterization of light and heavy sarcoplasmic reticulum vesicles. Identification of the "sarcoplasmic reticulum feet" associated with heavy sarcoplasmic reticulum vesicles, *Biochim. Biophys. Acta* **602:**97.

Campbell, K. P., MacLennan, D. H., Jorgensen, A. O., and Mintzer, M. C., 1983a, Purification and characterization of calsequestrin from canine cardiac sarcoplasmic reticulum and identification of the 53,000-dalton glycoprotein, *J. Biol. Chem.* **258:**1197.

Campbell, K. P., MacLennan, D. H., and Jorgensen, A. O., 1983b, Staining of the $Ca^{2+}$-binding proteins, calsequestrin, calmodulin, troponin C, and S-100, with the cationic carbocyanine dye "stains all," *J. Biol. Chem.* **258:**11267.

Capaldi, R. A., and Vanderkooi, G., 1972, The low polarity of many membrane proteins, *Proc. Natl. Acad. Sci. USA* **69:**930.

Caswell, H. A., Lou, Y. H., and Brunschwig, J-P., 1976, Ouabain-binding vesicles from skeletal muscle, *Arch. Biochem. Biophys.* **176:**417.

Chapman, D., Gomez-Fernandez, J. C., and Goni, F. M., 1979, Intrinsic protein-lipid interactions: Physical and biochemical evidence, *FEBS Lett.* **98:**211.

Chiesi, M., and Carafoli, E., 1982, The regulation of $Ca^{2+}$-transport by fast skeletal muscle sarcoplasmic reticulum, *J. Biol. Chem.* **257:**984.

Chyn, T. L., Martonosi, A. G., Morimoto, T., and Sabatini, D. D., 1979, *In vitro* synthesis of the $Ca^{2+}$ transport ATPase by ribosomes bound to sarcoplasmic reticulum membranes, *Proc. Natl. Acad. Sci. USA* **76:**1241.

Clark, S. P., and Molday, R. S., 1979, Orientation of membrane glycoproteins in sealed rod outer segment disks, *Biochemistry* **18:**5868.

Clarke, S., 1975, The size and detergent binding of membrane proteins, *J. Biol. Chem.* **250:**5459.

Collins, J. H., Zot, A. S., and Kranias, E. G., 1982, Isolation of two proteolipids from rabbit muscle sarcoplasmic reticulum, *Prep. Biochem.* **12:**255.

Constantin, L. L., and Podolsky, R. J., 1967, Depolarization of the internal membrane system in the activation of frog skeletal muscle, *J. Gen. Physiol.* **50:**1101.

Craig, W. S., and Kyte, J., 1980, Stoichiometry and molecular weight of the minimum asymmetric units of canine renal sodium and potassium ion-activated adenosine triphosphatase, *J. Biol. Chem.* **255:**6262.

Davis, D. G., Inesi, G., and Gulik-Krzywicki, T., 1976, Lipid molecular motion and enzyme activity in sarcoplasmic reticulum membranes, *Biochemistry* **15:**1271.

Deamer, D. W., and Baskin, R. Y., 1969, Ultrastructure of sarcoplasmic reticulum preparations, *J. Cell Biol.* **42:**296.

Dean, W. L., and Tanford, C., 1977, Reactivation of lipid-depleted $Ca^{2+}$-ATPase by a nonionic detergent, *J. Biol. Chem.* **252:**3551.

Dean, W. L., and Tanford, C. 1978, Properties of a delipidated detergent activated $Ca^{2+}$ ATPase, *Biochemistry* **17:**1683.

Degani, C., and Boyer, P. D., 1973, A borohydride reduction method for characterization of the acyl phosphate linkage in proteins and its application to sarcoplasmic reticulum adenosine triphosphatase, *J. Biol. Chem.* **244:**3733.

de Meis, L., and Hasselbach, W., 1971, Acetyl phosphate as substrate for $Ca^{2+}$ uptake in skeletal muscle microsomes, inhibition by alkali ions, *J. Biol. Chem.* **246:**4759.

de Meis, L., and Vianna, A. L., 1979, Energy interconversion by the $Ca^{2+}$-dependent ATPase of the sarcoplasmic reticulum, *Annu. Rev. Biochem.* **48:**275.

de Meis, L., Martins, O. B., and Alvez, E. W., 1980, Role of water, hydrogen ion and temperature on the synthesis of adenosine triphosphate by the sarcoplasmic reticulum adenosine triphosphatase in the absence of a calcium ion gradient, *Biochemistry* **19:**4252.

Drabikowski, W., Sarzala, M. G., Wroniszewska, A., Lagwinska, E., and Drzewiecka, B., 1972, Role of cholesterol in the $Ca^{2+}$ uptake and ATPase activity of fragmented sarcoplasmic reticulum, *Biochim. Biophys. Acta* **274:**158.

Dupont, Y., Harrison, S. G., and Hasselbach, W., 1973, Molecular organization in the sarcoplasmic reticulum membrane studied by X-ray diffraction, *Nature* **244:**555.

Ebashi, S., 1976, Excitation-contraction coupling, *Annu. Rev. Physiol.* **38:**293.

Ebashi, S., and Endo, M., 1968, Calcium ion and muscle contraction, *Prog. Biophys. Mol. Biol.* **18:**123.

Eisenberg, B. R., and Eisenberg, R. S., 1982, The T-SR junction in contracting single skeletal muscle fibers, *J. Gen. Physiol.* **79:**1.

Endo, M., 1977, Calcium release from the sarcoplasmic reticulum, *Physiol. Rev.* **57:**71.

Endo, M., Tanake, M., and Ogawa, T., 1970, Calcium induced release of calcium from the sarcoplasmic reticulum of skinned skeletal muscle fibers, *Nature* **228:**34.

Fabiato, A., and Fabiato, F., 1979, Calcium and cardiac excitation–contraction coupling, *Annu. Rev. Physiol.* **41:**473.

Feher, J. J., and Briggs, F. N., 1982, The effect of calcium load on the calcium permeability of sarcoplasmic reticulum, *J. Biol. Chem.* **257:**10191.

Fiehn, W., and Hasselbach, W., 1970, The effect of phospholipase A on the calcium transport and the role of unsaturated fatty acids in ATPase activity of sarcoplasmic vesicles, *Eur. J. Biochem.* **13:**510.

Fleischer, S., Wang, C-T., Hymel, L., Seeling, J., Brown, M. F., Herbette, L., Scarpa, A., McLaughlin, A. C., and Blasie, J. K., 1979, Structural studies of the sarcoplasmic reticulum membrane using the reconstitution approach, in: *Function and Molecular Aspects of Biomembrane Transport* (E. Quagliariello, F. Palmieri, S. Papa, and M. Klingenberg, eds.), Biomedical Press, Elsevier, North-Holland, p. 465.

Folch-Pi, J., and Stoffyn, P. J., 1972, Proteolipids from membrane systems, *Ann. N.Y. Acad. Sci.* **195:**86.

Franzini-Armstrong, C., 1972, Membrane systems in muscle fibers, in: *The Structure and Function of Muscle,* Vol. 1 (G. H. Bourne, ed.), Academic Press, New York, p. 532.

Franzini-Armstrong, C., 1980, Structure of sarcoplasmic reticulum, *Fed. Proc.* **39:**2403.

Giotta, G. J., 1976a, Distribution of the quaternary structure of ($Na^+$ + $K^+$)-dependent adenosine triphosphatase by Triton X-100, *Biochem. Biophys. Res. Commun.* **71:**776.

Giotta, G. J., 1976b, Quaternary structure of ($Na^+$ + $K^+$)-dependent adenosine triphosphatase, *J. Biol. Chem.* **251:**1247.

Gomez-Fernandez, J. C., Goni, F. M., Bach, D., Restall, C. J., and Chapman, D., 1979, Protein–lipid interactions. A study of ($Ca^{2+}$ + $Mg^{2+}$) ATPase reconstituted with synthetic phospholipids, *FEBS Lett.* **98:**224.

Gomez-Fernandez, J. C., Goni, F. M., Bach, D., Restall, C. J., and Chapman, D., 1980, Biophysical studies of ($Ca^{2+}$ + $Mg^{2+}$)-ATPase reconstituted systems, *Biochim. Biophys. Acta* **598:**502.

Green, N. M., Allen, G., and Hebdon, G. M., 1980, Structural relationship between the calcium and magnesium-transporting ATPase of sarcoplasmic reticulum and the membrane, *Ann. N.Y. Acad. Sci.* **358:**149.

Greenway, D. C., and MacLennan, D. H., 1978, Assembly of the sarcoplasmic reticulum. Synthesis of calsequestrin and the $Ca^{2+}$ + $Mg^{2+}$ adenosine triphosphatase on membrane-bound polyribosomes, *Can. J. Biochem.* **56:**452.

Ha, D-B., Boland, R., and Martonosi, A., 1979, Synthesis of the calcium transport ATPase of sarcoplasmic reticulum and other muscle proteins during development of muscle cells *in vivo* and *in vitro, Biochim. Biophys. Acta* **585:**165.

Hackenberg, H., and Klingenberg, M., 1980, Molecular weight and hydrodynamic parameters of the adenosine 5′-diphosphate-adenosine 5′-triphosphate carrier in Triton X-100, *Biochemistry* **19:**548.

Hardwicke, P. M. D., and Green, N. M., 1974, The effect of delipidation on the adenosine triphosphatase of sarcoplasmic reticulum. Electron microscopy and physical properties, *Eur. J. Biochem.* **42:**183.

Haslem, R. J., Davidson, M. M. L., Davis, T., Lynham, J. A., and McCleuagham, M. O., 1978, Regulation of blood platelet function by cyclic nucleotides, *Adv. Cycl. Nuc. Res.* **9:**533.

Hasselbach, W., 1979, The sarcoplasmic calcium pump. A model of energy transduction in biological membranes, in: *Topics in Current Chemistry,* Vol. 78 (F. L. Boschke, ed.), Springer-Verlag, Berlin, Heidelberg, New York, p. 1.

Hasselbach, W., and Elfvin, L-G., 1967, Structural and chemical asymmetry of the calcium transporting membranes of the sarcotubular system as revealed by electron microscopy, *J. Ultrastruct. Res.* **17:**598.

Hasselbach, W., and Migala, A., 1975, Arrangement of proteins and lipids in the sarcoplasmic reticulum membrane, *Z. Naturforsch.* C30:681.

Hastings, D. F., and Reynolds, J. A., 1979, Molecular weight of ($Na^+$,$K^+$) ATPase from shark rectal gland, *Biochemistry* **18:**817.

Hebdon, G. M., Cunningham, L. W., and Green, N. M., 1979, Cross-linking experiments with the adenosine triphosphatase of sarcoplasmic reticulum, *Biochem. J.* **179:**135.

Herbette, L., and Blasie, J. K., 1980, Static and time resolved structural studies on isolated sarcoplasmic reticulum membrane, in: *Calcium-Binding Proteins: Structure and Function* (F. L. Siegel, E. Carafoli, R. H. Kretsinger, D. H. MacLennan, and R. H. Wasserman, eds.), Elsevier, North-Holland, p. 115.

Herbette, L., Marquardt, J., Scarpa, A., and Blasie, J., 1977, A direct analysis of lamellar X-ray diffraction from hydrated oriented multilayers of fully functional sarcoplasmic reticulum, *Biophys. J.* **20:**245.

Herbette, L., Scarpa, A., Blasie, J. K., Bauer, D. R., Wang, C-T., and Fleischer, S., 1981a, Functional characteristics of reconstituted sarcoplasmic reticulum membranes as a function of the lipid-to-protein ratio, *Biophys. J.* **36:**27.

Herbette, L., Scarpa, A., Blasie, J. K., Wang, C-T., Saito, A., and Fleischer, S., 1981b, Comparison of the profile structures of isolated and reconstituted sarcoplasmic reticulum membranes, *Biophys. J.* **36:**47.

Hesketh, T. R., Smith, G. A., Houslay, M. D., McGill, K. A., Birdsall, N. J. M., Metcalfe, J. C., and Warren, G. B., 1976, Annular lipids determine the ATPase activity of a calcium transport protein complexed with dipalmitoyllecithin, *Biochemistry* **15:**4145.

Hicks, M. J., Shigekawa, M., and Katz, A. M., 1979, Mechanism by which cyclic adenosine 3', 5'-monophosphate dependent protein kinase stimulates calcium transport in cardiac sarcoplasmic reticulum, *Circ. Res.* **44:**384.

Hidalgo, C., 1980, Inhibition of calcium transport in sarcoplasmic reticulum after modification of highly reactive amino groups, *Biochem. Biophys. Res. Commun.* **92:**757.

Hidalgo, C., and Ikemoto, N., 1977, Disposition of proteins and aminophospholipids in the sarcoplasmic reticulum membrane, *J. Biol. Chem.* **252:**8446.

Hidalgo, C., Ikemoto, N., and Gergely, J., 1976, Role of phospholipids in Ca-dependent ATPase of sarcoplasmic reticulum. Enzymetic and ESR studies with phospholipid-replaced membranes, *J. Biol. Chem.* **251:**4224.

Hidalgo, C., Thomas, D. D., and Ikemoto, N., 1978, Effect of the lipid environment on protein motion and enzymatic activity of the sarcoplasmic reticulum calcium ATPase, *J. Biol. Chem.* **253:**6879.

Hoffman, W., Sarzala, M. G., Gomez-Fernandez, J. C., Goni, F. M., Restall, C. J., Chapman, D., Heppeler, G., and Kreutz, U., 1980, Protein rotational diffusion and lipid structure of reconstituted systems of $Ca^{2+}$-activated adenosine triphosphatase, *J. Mol. Biol.* **141:**119.

Holland, P. C., 1979, Biosynthesis of the $Ca^{2+}$ and $Mg^{2+}$-dependent adenosine triphosphatase of sarcoplasmic reticulum in cell cultures of embryonic chick heart, *J. Biol. Chem.* **254:**7604.

Holland, P. C., and MacLennan, D. H., 1976, Assembly of sarcoplasmic reticulum. Biosynthesis of the adenosine triphosphatase in rat skeletal muscle cell culture, *J. Biol. Chem.* **251:**2030.

Hudson, E. N., and Weber, G., 1973, Synthesis and characterization of two fluorescent sulfhydryl reagents, *Biochemistry* **12:**4154.

Huxley, A. F., 1971, The activation of striated muscle and its mechanical response, *Proc. R. Soc. Lond. Ser. B* **178:**1.

Ikemoto, N., 1982, Structure and function of the calcium pump protein of sarcoplasmic reticulum, *Annu. Rev. Physiol.* **44:**297.

Ikemoto, N., Bhatnagar, G. M., and Gergely, J., 1971a, Fractionation of solubilized sarcoplasmic reticulum, *Biochem. Biophys. Res. Commun.* **44:**1540.

Ikemoto, N., Sreter, F. A., and Gergely, J., 1971b, Structural features of the vesicles of FSR-lack functional role in $Ca^{2+}$ uptake and ATPase activity, *Arch. Biochem. Biophys.* **147:**571.

Ikemoto, N., Sreter, F. A., Nakamura, A., and Gergely, J., 1968, Tryptic digestion and localization of calcium uptake and ATPase activity in fragments of sarcoplasmic reticulum, *J. Ultrastruct. Res.* **23:**216.

Ikemoto, N., Cucchiaro, J., and Garcia, A. M., 1976, A new glycoprotein factor of the sarcoplasmic reticulum, *J. Cell Biol.* **70:**290a.

Inesi, G., 1972, Active transport of calcium ion in sarcoplasmic membranes, *Annu. Rev. Biophys. Bioeng.* **1:**19.

Inesi, G., 1981, The sarcoplasmic reticulum of skeletal and cardiac muscle, in: *Cell and Muscle Motility,* Vol. 1 (R. M. Dowben and J. W. Shay, eds.), Plenum Publishing Corporation, New York, p. 63.

Inesi, G., and Asai, H., 1968, Trypsin digestion of fragmented sarcoplasmic reticulum, *Arch. Biochem. Biophys.* **126:**469.

Inesi, G., and Malan, N., 1976, Mechanisms of calcium release in sarcoplasmic reticulum. Mini review, *Life Sci.* **18:**773.

Inesi, G., Millman, M., and Eletr, S., 1973, Temperature-induced transitions of function and structure in sarcoplasmic reticulum membranes, *J. Mol. Biol.* **81:**483.

Jilka, R. L., Martonosi, A. N., and Tillack, T. W., 1975, Effect of the purified ($Mg^{2+}$ + $Ca^{2+}$)-activated ATPase of sarcoplasmic reticulum upon the passive $Ca^{2+}$ permeability and ultrastructure of phospholipid vesicles, *J. Biol. Chem.* **250:**7511.

Jones, L. R., and Cala, S. E., 1981, Biochemical evidence for functional heterogeneity of cardiac sarcoplasmic reticulum vesicles, *J. Biol. Chem.* **256:**11809.

Jones, L. R., Besch, H. R., Fleming, J. W., McConnaughey, M. M., and Watanabe, A. M., 1979, Separation of vesicles of cardiac sarcolemma from vesicles of cardiac sarcoplasmic reticulum, *J. Biol. Chem.* **254:**530.

Jorgensen, A. O., Kalnins, V. I., Zubrzycka, E., and MacLennan, D. H., 1977, Assembly of the sarcoplasmic reticulum. Localization by immunofluorescence of sarcoplasmic reticulum proteins in differentiating rat skeletal muscle cell cultures, *J. Cell Biol.* **74:**287.

Jorgensen, K. E., Lind, K. E., Roigaard-Peterson, H., and Moller, J. V., 1978, The functional unit of calcium-plus-magnesium-ion-dependent adenosine triphosphatase from sarcoplasmic reticulum, *Biochem. J.* **169:**489.

Jorgensen, A. O., Kalnins, V., and MacLennan, D. H., 1979, Localization of sarcoplasmic reticulum proteins in rat skeletal muscle by immunofluorescence, *J. Cell Biol.* **80:**372.

Jorgensen, A. O., Shen, A. C-Y., Daly, P., and MacLennan, D. H., 1982a, Localization of $Ca^{2+}$ + $Mg^{2+}$-ATPase of the sarcoplasmic reticulum in adult rat papillary muscle, *J. Cell Biol.* **93:**883.

Jorgensen, A. O., Shen, A. C.-Y., MacLennan, D. H., and Tokuyasu, K. T., 1982b, Ultrastructural localization of the $Ca^{2+}$ + $Mg^{2+}$-dependent ATPase of sarcoplasmic reticulum in rat skeletal muscle by ferritin labeling of ultrathin frozen sections, *J. Cell Biol.* **92:**409–416.

Kaizu, T., Kirino, Y., and Shimizu, H., 1980, A saturation transfer electron spin resonance study on the break in the arrhenius plot for the rotational motion of $Ca^{2+}$-dependent adenosine triphosphatase molecules in purified and lipid replaced preparations of rabbit skeletal muscle sarcoplasmic reticulum, *J. Biochem.* **88:**1837.

Kanazawa, T., Yamada, S., Yamamoto, T., and Tonomura, Y., 1971, Reaction mechanism of the $Ca^{2+}$-dependent ATPase of sarcoplasmic reticulum. V. Vectorial requirements for calcium reactions of ATPase: Formation and decomposition of a phosphorylated intermediate and ATP-formation from ADP and the intermediate, *J. Biochem.* **70:**95.

Katz, A. M., 1979, Role of the contractile proteins and sarcoplasmic reticulum in the response of the heart to catecholamines: A historical review, *Adv. Cycl. Nuc. Res.* **11:**303.

Katz, A. M., 1980, Relaxing effects of catecholamines in the heart, *Trends Pharmacol. Sci.* **1:**434.

Katz, A. M., Tada, M., and Kirchberger, M. A., 1975, Control of calcium transport in the myocardium by the cyclic AMP protein kinase system, *Adv. Cycl. Nuc. Res.* **5:**453.

Katz, A. M., Dunnett, J., Repke, D. I., and Hasselbach, W., 1976, Control of calcium permeability in the sarcoplasmic reticulum, *FEBS Lett.* **67:**208.

Katz, A. M., Repke, D. I., Fudyma, G., and Shigekawa, M., 1977, Control of calcium efflux from sarcoplasmic reticulum vesicles by external calcium, *J. Biol. Chem.* **252:**4210–4214.

Katz, A. M., Nesh-Adler, P., Watras, J., Messineo, F. C., Takenaka, H., and Louis, C. F., 1982, Fatty acid effects on calcium influx and efflux in sarcoplasmic reticulum vesicles from rabbit skeletal muscle, *Biochim. Biophys. Acta* **687:**17.

Kennedy, S. J., 1978, Structures of membrane proteins, *J. Membr. Biol.* **42:**265.

Kimura, M., Kimura, I., and Kobayashi, S., 1977, The activation of cyclic 3′,5′-adenosine monophosphate dependent protein kinase on sarcoplasmic reticulum fractions of various smooth muscles and its related novel relaxants, *Biochem. Pharmacol.* **26:**994.

Kirchberger, M. A., and Antonetz, T., 1982, Phospholamban: Dissociation of the 22,000 molecular weight protein of cardiac sarcoplasmic reticulum into 11,000 and 5,500 molecular weights forms, *Biochem. Biophys. Res. Commun.* **105:**152.

Kirchberger, M. A., Tada, M., and Katz, A. M., 1974, Adenosine 3′:5′-monophosphate-dependent protein kinase catalyzed phosphorylation reaction and its relationship to calcium transport in cardiac sarcoplasmic reticulum, *J. Biol. Chem.* **249:**6166.

Kiring, Y., Okkume, T., and Shimizu, H., 1978, Saturation transfer electron study on the rotational diffusion of calcium and magnesium dependent adenosine triphosphatase in sarcoplasmic reticulum membranes, *J. Biochem.* **84:**111.

Klip, A., Reithmeier, R. A. F., and MacLennan, D. H., 1980, Alignment of the major tryptic fragments of the adenosine triphosphatase from sarcoplasmic reticulum, *J. Biol. Chem.* **255:**6562.

Knowles, A. F., and Racker, E., 1975, Properties of a reconstituted calcium pump, *J. Biol. Chem.* **250:**3538.

Knowles, A. F., Kandrach, A., Racker, E., and Khorana, H. G., 1975, Acetyl phosphatidylethanolamine in the reconstitution of ion pumps, *J. Biol. Chem.* **250:**1809.

Knowles, A. F., Eyton, E., and Racker, E., 1976, Phospholipid protein interactions in the $Ca^{2+}$ adenosine triphosphatase of sarcoplasmic reticulum, *J. Biol. Chem.* **251:**5161.

Knowles, A., Zimniak, P., Alfonso, M., Zimnik, A., and Racker, E., 1980, Isolation and characterization of proteolipids from sarcoplasmic reticulum, *J. Membr. Biol.* **55:**233.

Kyte, T., 1972, Properties of the two polypeptides of sodium and potassium dependent adenosine triphosphatase, *J. Biol. Chem.* **247**:7642.

Laemmli, U. K., 1970, Cleavage of structural proteins during the assembly of the head of bacteriophage T4, *Nature* **227**:680.

Lau, Y. H., Caswell, A. H., and Brunschwig, J. P., 1977, Isolation of transverse tubules by fractionation of triad junctions of skeletal muscle, *J. Biol. Chem.* **252**:5565.

Lee, A. G., Birdsall, N. J. M., Metcalf, J. C., Toon, P. A., and Warren, G. B., 1974, Clusters in lipid bilayers and the interpretation of thermal effects in biological membranes, *Biochemistry* **13**:3699.

Le Maire, M., Jorgensen, K. E., Roigaard-Peterson, H., and Moller, J. V., 1976a, Properties of deoxycholate solubilized sarcoplasmic reticulum $Ca^{2+}$-ATPase, *Biochemistry* **15**:5805.

Le Maire, M., Moller, J. V., and Tanford, C., 1976b, Retention of enzyme activity by detergent solubilized sarcoplasmic reticulum $Ca^{2+}$ ATPase, *Biochemistry* **15**:2336.

Le Maire, M., Lind, K. F., Jorgensen, K. E., Roigaard-Peterson, H., and Moller, J. V., 1978, Enzymatically active $Ca^{2+}$ ATPase from sarcoplasmic reticulum membrane, solubilized by nonionic detergents. Role of lipid for aggregation of the protein, *J. Biol. Chem.* **253**:7051.

Le Maire, M., Moller, J. V., and Gulik-Krzywicki, T., 1981, Freeze-fracture study of water-soluble, standard proteins and of detergent solubilized forms of sarcoplasmic reticulum $Ca^{2+}$ ATPase, *Biochim. Biophys. Acta* **643**:115.

Le Peuch, C. J., Haieck, J., and Demaille, J. G., 1979, Concerted regulation of cardiac sarcoplasmic reticulum calcium transport by cyclic adenosine monophosphate dependent and calcium-colmodulin dependent phosphorylation, *Biochemistry* **19**:3368.

Le Peuch, C. J., Le Peuch, D. A. M., and Demaille, J. G., 1980, Phospholamban, activator of the cardiac sarcoplasmic reticulum calcium pump. Physiochemical properties and diagonal purification, *Biochemistry* **19**:3368.

Levitsky, D. O., Aliev, M. K., Kuzmin, A. V., Levchenko, T. S., Smirna, V. N., and Chazov, E. I., 1976, Isolation of calcium pump system and purification of calcium ion-dependent ATPase from heart muscle, *Biochim. Biophys. Acta* **443**:468.

Liang, S-M., and Winter, C. G., 1977, Digitonin-induced changes in subunit arrangement in relation to some *in vitro* activities of the $(Na^+,K^+)$ ATPase, *J. Biol Chem.* **252**:8278.

Liu, T., Stetson, B., Turco, S. J., Hubbard, S. C., and Robbins, P. W., 1979, Arrangement of glucose residues in the lipid linked oligosaccharide precursor of asparaginyl oligosaccharide, *J. Biol. Chem.* **254**:4554.

Louis, C. F., and Jarvis, B., 1982, Affinity labeling of calmodulin-binding components in canine cardiac sarcoplasmic reticulum, *J. Biol. Chem.* **257**:15187.

Louis, C. F., and Maffitt, M., 1982, Characterization of calmodulin-mediated phosphorylation of cardiac muscle sarcoplasmic reticulum, *Arch. Biochem. Biophys.* **218**:109.

Louis, C. F., and Shooter, E. M., 1972, The proteins of rabbit skeletal muscle sarcoplasmic reticulum, *Arch. Biochem. Biophys.* **153**:641.

Louis, C. F., Nash-Adler, P. A., Fudyma, G., Shigekawa, M., Akowitz, A., and Katz, A. M., 1980, A comparison of vesicles derived from terminal cisternae and longitudinal tubules of sarcoplasmic reticulum isolated from rabbit skeletal muscle, *Eur. J. Biochem.* **111**:1.

Louis, C. F., Maffitt, M., and Jarvis, B., 1982, Factors that modify the molecular size of phospholamban, the 23,000-dalton cardiac sarcoplasmic reticulum phosphoprotein, *J. Biol. Chem.* **257**:15182.

MacLennan, D. H., 1970, Purification and properties of an adenosine triphosphatase from sarcoplasmic reticulum, *J. Biol. Chem.* **245**:4508.

MacLennan, D. H., 1975, Resolution of the calcium transport system of sarcoplasmic reticulum, *Can. J. Biochem.* **53**:251.

MacLennan, D. H., and Holland, P. C., 1975, Calcium transport in sarcoplasmic reticulum, *Annu. Rev. Biophys. Bioenerg.* **4**:377.

MacLennan, D. H., and Holland, P. C., 1976, The calcium transport ATPase of sarcoplasmic reticulum, in: *The Enzymes of Biological Membranes,* Vol. 3 (A. Martonosi, ed.), Plenum Press, New York, p. 221.

MacLennan, D. H., and Reithmeier, R. A. F., 1982, The structure of the $Ca^{2+}/Mg^{2+}$-ATPase of sarcoplasmic reticulum, in: *Membranes and Transport,* Vol. 1 (A. Martonosi, ed.), Plenum Publishing Corporation, New York, p. 567.

MacLennan, D. H., and Wong, P. T. S., 1971, Isolation of a calcium sequestering protein from sarcoplasmic reticulum, *Proc. Natl. Acad. Sci. USA* **68:**1231.

MacLennan, D. H., Seeman, P., Iles, G. H., and Yip, C. C., 1971, Membrane formation by the adenosine triphosphatase of sarcoplasmic reticulum, *J. Biol. Chem.* **246:**2702.

MacLennan, D. H., Yip, C. C., Iles, G. H., and Seeman, P., 1972, Isolation of sarcoplasmic reticulum proteins, *Cold Spring Harbor Symp. Quant. Biol.* **37:**460.

MacLennan, D. H., Zubrzycka, E., Jorgensen, A. O., and Kalnins, I., 1978, Assembly of the sarcoplasmic reticulum, in: *The Molecular Biology of Membranes* (S. Fleischer, Y. Hatefi, D. H. MacLennan, and A. Tzagoloff, eds.), Plenum Press, New York, p. 304.

MacLennan, D. H., Reithmeier, R. A. F., Shoshan, V., Campbell, K. P., LeBel, D., Herrmann, T. R., and Shamoo, A. F., 1980, Ion pathways in proteins of the sarcoplasmic reticulum, *Ann. N.Y. Acad. Sci.* **358:**138.

MacLennan, D. H., Shoshan, V., and Wood, D. S., 1982, Studies of $Ca^{2+}$ release from sarcoplasmic reticulum, *Ann. N.Y. Acad. Sci.* **402:**400.

Madden, T. D., and Quinn, P. J., 1979, Arrhenius discontinuities of $Ca^{2+}$-ATPase activity are related to changes in membranes lipid fluidity of sarcoplasmic reticulum, *FEBS Lett.* **107:**110.

Madden, T. D., Chapman, D., and Quinn, P. J., 1979, Cholesterol modulates activity of calcium dependent ATPase of the sarcoplasmic reticulum, *Nature* **279:**538.

Madden, T. D., King, M. D., and Quinn, P. J., 1981, The modulation of $Ca^{2+}$-ATPase activity of sarcoplasmic reticulum membrane cholesterol. The effect of enzyme coupling, *Biochim. Biophys. Acta* **641:**265.

Malan, N., Sabbadini, R., Scales, D., and Inesi, G., 1975, Functional and structural roles of sarcoplasmic reticulum protein components, *FEBS Lett.* **60:**122.

Marcelja, S., 1976, Lipid-mediated protein interaction in membranes, *Biochim. Biophys. Acta* **455:**1.

Martonosi, A., 1967, The role of phospholipids in the ATPase activity of skeletal muscle microsomes, *Biochem. Biophys. Res. Commun.* **29:**753.

Martonosi, A., 1968, Sarcoplasmic reticulum. IV. Localization of microsomal adenosine triphosphatase, *J. Biol. Chem.* **243:**71.

Martonosi, A., 1972, Biochemical and clinical aspects of sarcoplasmic reticulum function, *Curr. Top. Membr. Trans.* **3:**83.

Martonosi, A., 1975, Membrane transport during development of animals, *Biochim. Biophys. Acta* **415:**311.

Martonosi, A., 1980, Calcium pumps, *Fed. Proc.* **39:**2401.

Martonosi, A., 1982, The development of sarcoplasmic reticulum membranes, *Annu. Rev. Physiol.* **44:**337.

Martonosi, A., and Fortier, F., 1974, The effect of anti-ATPase antibodies upon the $Ca^{2+}$ transport of sarcoplasmic reticulum, *Biochem. Biophys. Res. Commun.* **60:**382.

Martonosi, A., and Halpin, R., 1971, Sarcoplasmic reticulum. X. The protein composition of sarcoplasmic reticulum membranes, *Arch. Biochem. Biophys.* **144:**66.

Martonosi, A., Donley, J., and Halpin, R. A., 1968, Sarcoplasmic reticulum. III. The role of phospholipids in the adenosine triphosphatase activity and $Ca^{2+}$ transport, *J. Biol. Chem.* **253:**61.

Martonosi, A., Boland, R., and Halpin, R. A., 1972, The biosynthesis of sarcoplasmic reticulum membranes and the mechanism of calcium transport, *Cold Spring Harbor Symp. Quant. Biol.* **37:**455.

Martonosi, A., Roufa, D., Boland, R., Reyes, E., and Tillack, T. W., 1977, Development of sarcoplasmic reticulum in cultured chicken muscle, *J. Biol. Chem.* **252:**318.

Martonosi, A., Chyn, T. L., and Schibeci, A., 1978, The calcium transport of sarcoplasmic reticulum, *Ann. N.Y. Acad. Sci.* **307:**148.

McFarland, B. H., and Inesi, G., 1971, Solubilization of sarcoplasmic reticulum with Triton X-100, *Arch. Biochem. Biophys.* **145:**456.

McKinley, D., and Meissner, G., 1978, Evidence for a $K^+$, $Na^+$ permeable channel in sarcoplasmic reticulum, *J. Membr. Biol.* **44:**159–186.

McLaughlin, A. C., Herbette, L., Blasie, J. K., Wang, C. T., Hymel, L., and Fleischer, S., 1981, $^{31}P$-NMR studies of oriented multilayers formed from isolated sarcoplasmic reticulum and reconstituted sarcoplasmic reticulum. Evidence that "boundary-layer" phospholipid is not immobilized, *Biochim. Biophys. Acta* **643:**1.

Meissner, G., 1973, ATP and $Ca^{2+}$ binding by the $Ca^{2+}$ pump protein of sarcoplasmic reticulum, *Biochim. Biophys. Acta* **298:**906.

Meissner, G., 1975, Isolation and characterization of two types of sarcoplasmic reticulum vesicles, *Biochim. Biophys. Acta* **389:**51.

Meissner, G., 1981, Calcium transport and monovalent cation and proton fluxes in sarcoplasmic reticulum, *J. Biol. Chem.* **256:**636.

Meissner, G., and Fleischer, S., 1972, The role of phospholipid in $Ca^{2+}$ stimulated ATPase activity of sarcoplasmic reticulum, *Biochim. Biophys. Acta* **255:**19.

Meissner, G., and Fleischer, S., 1973, $Ca^{2+}$ uptake in reconstituted sarcoplasmic reticulum vesicles, *Biochem. Biophys. Res. Commun.* **59:**913.

Meissner, G., and Fleischer, S., 1974, Dissociation and reconstitution of functional sarcoplasmic reticulum vesicles, *J. Biol. Chem.* **249:**302.

Meissner, G., and McKinley, D., 1976, Permeability of sarcoplasmic reticulum membrane: The effect of changed ionic environments on $Ca^{2+}$ release, *J. Biol. Chem.* **255:**6814.

Meissner, G., and Young, G., 1980, Proton permeability of sarcoplasmic reticulum vesicles, *J. Biol. Chem.* **255:**6814.

Meissner, G., Conner, G., and Fleischer, S., 1973, Isolation of sarcoplasmic reticulum by zonal centrifugation and purification of $Ca^{2+}$ pump and $Ca^{2+}$ binding proteins, *Biochim. Biophys. Acta* **298:**246.

Melgunov, V. I., and Alimara, E. I., 1980, The dependence for reactivation of lipid-depleted $Ca^{2+}$-ATPase of sarcoplasmic reticulum by non-ionic detergents on their hydrophile/lipophile balance, *FEBS Lett.* **121:**235.

Michalak, M., and MacLennan, D. H., 1980, Assembly of the sarcoplasmic reticulum. Biosynthesis of the high affinity calcium binding protein in rat skeletal muscle cell cultures, *J. Biol. Chem.* **255:**1327.

Michalak, M., and Sarzala, M. G., 1975, Ultrastructure of sarcoplasmic reticulum membrane during development of rabbit skeletal muscle, *Ann. Med. Sect. Pol. Acad. Sci.* **20:**93.

Michalak, M., Campbell, P. K., and MacLennan, D. H., 1980, Localization of the high affinity calcium binding protein and an intrinsic glycoprotein in sarcoplasmic reticulum membranes, *J. Biol. Chem.* **255:**1317.

Migala, A., Agostini, B., and Hasselbach, W., 1973, Tryptic fragmentation of the calcium transport system in the sarcoplasmic reticulum, *Z. Naturforsch.* **28:**178.

Millman, M. S., 1980, A thermal transition of passive calcium efflux in fragmented sarcoplasmic reticulum, *Membr. Biochem.* **3:**271.

Mitchinson, C., Wilderspin, A. F., Trinnaman, B. J., and Green, N. M., 1982, Identification of a labelled peptide after stoichiometric reaction of fluorescein isothiocyanate with the $Ca^{2+}$-dependent adenosine triphosphatase of sarcoplasmic reticulum, *FEBS Lett.* **146:**87.

Miyamoto, H., and Racker, E., 1981, Calcium induced calcium release at terminal cisternae of skeletal sarcoplasmic reticulum, *FEBS Lett.* **133:**235.

Miyamoto, H., and Racker, E., 1982, Mechanism of calcium release from skeletal sarcoplasmic reticulum, *J. Membr. Biol.* **66:**193.

Moller, J. V., Lind, K. E., and Andersen, J. P., 1980, Enzyme kinetics and substrate stabilization of detergent solubilized and membranous ($Ca^{2+}$ + $Mg^{2+}$)-activated ATPase from sarcoplasmic reticulum. Effect of protein–protein interactions, *J. Biol. Chem.* **255:**1912.

Moller, J. V., Andersen, J. P., and le Maire, M., 1982, The sarcoplasmic reticulum $Ca^{2+}$-ATPase, *Mol. Cell. Biochem.* **42:**83.

Moore, B. M., Lentz, B. R., and Meissner, G., 1978, Effects of sarcoplasmic reticulum $Ca^{2+}$ ATPase on phospholipid bilayer fluidity: Boundry lipid, *Biochemistry* **17:**5248.

Mostov, K. E., De Foor, P., Fleischer, S., and Blobel, G., 1981, Co-translational membrane integration of calcium pump protein without signal sequence cleavage, *Nature* **292:**87.

Murphy, A. J., 1976, Sulfhydryl group modification of sarcoplasmic reticulum membranes, *Biochemistry* **15:**4492.

Murphy, A. J., 1977, Sarcoplasmic reticulum adenosine triphosphatase: Labeling of an essential lysyl residue with pyridoxyl-5′-phosphate, *Arch. Biochem. Biophys.* **180:**114.

Murphy, A. J., Pepitone, M., and Highsmith, S., 1982, Detergent-solubilized sarcoplasmic reticulum ATPase. Hydrodynamic and catalytic properties, *J. Biol. Chem.* **257:**3551.

Nagasaki, K., and Kasai, M., 1980, Magnesium permeability of sarcoplasmic reticulum vesicles monitored in terms of chlortetracycline fluorescence, *J. Biochem.* **87:**709.

Nakamura, H., and Martonosi, A. N., 1980, Effect of phospholipid substitution on the mobility of protein bound spin labels in sarcoplasmic reticulum, *J. Biochem.* **87**:525.

Nakamura, H., Jilka, R. L., Boland, R., and Martonosi, A. N., 1976, Mechanism of ATP hydrolysis by sarcoplasmic reticulum and the role of phospholipids, *J. Biol. Chem.* **25**:5414.

Nakamura, M., and Ohnishi, S., 1975, Organization of lipids in sarcoplasmic reticulum membranes and $Ca^{2+}$-dependent ATPase activity, *J. Biochem.* **78**:1039.

Nestruck-Goyke, A. C., and Hasselbach, W., 1981, Preparative isolation of Apo ($Ca^{2+}$-ATPase) from sarcoplasmic reticulum and the reactivation by lysophosphatidylcholine of $Ca^{2+}$-dependent ATP hydrolysis and partial-reaction steps of the enzyme, *Eur. J. Biochem.* **114**:339.

Nigg, E. A., and Cherry, R. J., 1979, Dimeric association of band 3 in the erythrocyte membrane demonstrated by protein diffusion measurements, *Nature* **277**:493.

Nishikori, K., and Maeno, H., 1979, Close relationship between adenosine 3′ : 5′-monophosphate-dependent endogenous phosphorylation of a specific protein and stimulation of calcium uptake in rat uterine microsomes, *J. Biol. Chem.* **254**:6099.

Ogawa, Y., and Ebashi, S., 1976, Ca-releasing action of γ,β-methylene adenosine triphosphate on fragmented sarcoplasmic reticulum, *J. Biochem.* **80**:1179.

Ohnoki, S., and Martonosi, A., 1980, Purification and characterization of the proteolipid of rabbit sarcoplasmic reticulum, *Biochim. Biophys. Acta* **626**:170.

Ostwald, T. J., and MacLennan, D. H., 1974, Isolation of a high affinity calcium-binding protein from sarcoplasmic reticulum, *J. Biol. Chem.* **249**:974.

Ostwald, T. J., MacLennan, D. H., and Dorrington, K. J., 1974, Effects of cation binding on the conformation of calsequestrin and the high affinity calcium binding protein of sarcoplasmic reticulum, *J. Biol. Chem.* **249**:5867.

Owens, K., Ruth, R. C., and Waglicki, W. B., 1972, Lipid composition of purified fragmented sarcoplasmic reticulum of the rabbit, *Biochim. Biophys. Acta* **288**:479.

Packer, L., Mehard, C. W., Meissner, G., Zahler, W. L., and Fleischer, S., 1974, The structural role of lipids in mitochondrial and sarcoplasmic reticulum membranes. Freeze-fracture electron microscopy studies, *Biochim. Biophys. Acta* **254**:9754.

Pennington, J., and Hokin, L. E., 1979, Effects of wheat germ agglutinin on the coupled transports of sodium and potassium in reconstituted (Na,K)-ATPase liposomes, *J. Biol. Chem.* **254**:9754.

Peters, K., and Richards, F. M., 1977, Chemical cross-linking reagents and proteins in studies of membrane structure, *Annu. Rev. Biochem.* **46**:523.

Pick, U., 1981, Dynamic interconversions of phosphorylated and non-phosphorylated intermediates of the $Ca^{2+}$ ATPase from sarcoplasmic reticulum followed in a fluorescein-labeled enzyme, *FEBS Lett.* **123**:131.

Pick, U., and Bassilian, S., 1980, Modification of the ATP binding site of the $Ca^{2+}$-ATPase from sarcoplasmic reticulum by fluorescein isothiocyanate, *FEBS Lett.* **123**:127.

Pick, U., and Karlish, S. J. D., 1980, Indications for an oligomeric structure and for conformation changes in sarcoplasmic reticulum $Ca^{2+}$-ATPase labeled selectively with fluorescence, *Biochim. Biophys. Acta* **626**:255.

Pick, U., and Racker, E., 1979, Inhibition of the ($Ca^{2+}$) ATPase from sarcoplasmic reticulum by dicyclohexyl carbodiimide. Evidence for location of the $Ca^{2+}$ binding site in a hydrophobic region, *Biochemistry* **18**:108.

Porter, K. P., and Palade, G. E., 1957, Studies on the endoplasmic reticulum. III. Its form and distribution in striated muscle cells, *J. Biophys. Biochem. Cytol.* **3**:269–319.

Quinn, P. J., Gomez, R., and Madden, T. D., 1980, Modification of membrane lipids of sarcoplasmic reticulum to probe the influence of bilayer fluidity on $Ca^{2+}$ activated ATPase activity, *Biochem. Soc. Trans.* **8**:38.

Racker, E., 1972, Reconstitution of a calcium pump with phospholipids and a purified $Ca^{++}$-adenosine triphosphatase from sarcoplasmic reticulum, *J. Biol. Chem.* **247**:8798.

Racker, E., and Eyton, E., 1975, A coupling factor from sarcoplasmic reticulum required for the translocation of $Ca^{2+}$-ions in a reconstituted $Ca^{2+}$-ATPase pump, *J. Biol. Chem.* **250**:7533.

Reithmeier, R. A. F., and MacLennan, D. H., 1981, The $NH_2$ terminus of the ($Ca^{2+}$ + $Mg^{2+}$) adenosine triphosphatase is located on the cytoplasmic surface of the sarcoplasmic reticulum membrane, *J. Biol. Chem.* **256**:5957.

Reithmeier, R. A. F., de Leon, S., and MacLennan, D. H., 1980, Assembly of the sarcoplasmic reticulum. Cell-free synthesis of the $Ca^{2+}$ + $Mg^{2+}$ adenosine triphosphatase and calsequestrin, *J. Biol. Chem.* **255:**11839.
Repke, D. I., Spivak, J. C., and Katz, A. M., 1976, Reconstitution of an active calcium pump in sarcoplasmic reticulum, *J. Biol. Chem.* **251:**3169.
Rice, D. M., Meadows, M. D., Scheinman, A. O., Goni, F. M., Gomez-Fernandez, J. C., Moscarello, M. A., Chapman, D., and Oldfield, E., 1979, Protein–lipid interactions. A nuclear magnetic resonance study of sarcoplasmic reticulum $Ca^{2+}$, $Mg^{2+}$-ATPase, lipophilin, and proteolipid apoprotein-lecithin systems and a comparison with the effects of cholesterol, *Biochemistry* **18:**5893.
Rizzolo, L., Le Maire, M., Reynolds, J. A., and Tanford, C., 1976, Molecular weights and hydrophobicity of the polypeptide chain of sarcoplasmic reticulum calcium (II) adenosine triphosphatase and its primary tryptic fragments, *Biochemistry* **15:**3433.
Rosemblatt, M., Hidalgo, C., Vergara, C., and Ikemoto, N., 1981, Immunological and biochemical properties of transverse tubule membranes isolated from rabbit skeletal muscle, *J. Biol. Chem.* **256:**8140.
Roufa, D., Wu, F. S., and Martonosi, A., 1981, The effect of $Ca^{2+}$ ionophores upon the synthesis of proteins in cultured skeletal muscle, *Biochim. Biophys. Acta* **674:**225.
Saito, A., Wang, C-T., and Fleischer, S., 1978, Membrane asymmetry and enhanced ultrastructural detail of sarcoplasmic reticulum revealed with the use of tannic acid, *J. Cell Biol.* **79:**601.
Sarzala, M. G., and Michalak, M., 1978, Studies on the heterogeneity of sarcoplasmic reticulum vesicles, *Biochim. Biophys. Acta* **513:**221.
Sarzala, M. G., Zubrzycka, E., and Drabikowski, W., 1974, Characterization of the constituents of sarcoplasmic reticulum membrane, in: *Calcium Binding Proteins* (W. Drabikowski, M. Strzelecka-Golaszewska, and E. Carafoli, eds.), Elsevier, Amsterdam, p. 317.
Sarzala, M. G., Zubrzycka, E., and Michalak, M., 1975a, Comparison of some features of undeveloped and mature sarcoplasmic reticulum, in: *Calcium Transport in Contraction and Secretion* (E. Carafoli, F. Clementi, W. Drabikowski, and A. Margreth, eds.), Amsterdam, North-Holland, p. 329.
Sarzala, M. G., Pilarska, M., Zubrzycka, E., and Michalak, M., 1975b, Changes in the structure, composition and function of sarcoplasmic reticulum membrane during development, *Eur. J. Biochem.* **57:**25.
Scales, D. J. and Inesi, G., 1976, Assembly of ATPase protein in sarcoplasmic reticulum membranes, *Biophys. J.* **16:**735.
Scales, D. J., and Sabbadini, R. A., 1979, Microsomal T system. A sterological analysis of purified microsomes derived from normal and dystrophic skeletal muscle, *J. Cell Biol.* **83:**33.
Schlesinger, M. J., 1981, Proteolipids, *Annu. Rev. Biochem.* **50:**193.
Seelig, J., Tamm, L., Hymel, L., and Fleischer, S., 1981, Deuterium and phosphorus nuclear magnetic resonance and fluorescence depolarization studies of functional reconstituted sarcoplasmic reticulum membrane vesicles, *Biochemistry* **20:**3922.
Shamoo, A. E., 1978, Inophorous properties of the 20,000 dalton fragment of ($Ca^{2+}$ + $Mg^{2+}$)-ATPase in phosphatidylcholine: Cholesterol membranes, *J. Membr. Biol.* **43:**227.
Shamoo, A. E., and MacLennan, D. H., 1974, A $Ca^{++}$ dependent and selective ionophore as part of the $Ca^{++}$ + $Mg^{++}$ dependent adenosine triphosphatase of sarcoplasmic reticulum, *Proc. Natl. Acad. Sci. USA* **71:**3522.
Shamoo, A. E., Ryan, T. E., Stewart, P. S., and MacLennan, D. H., 1976, Localization of ionophore activity in a 50,000 dalton fragment of the adenosine triphosphatase of sarcoplasmic reticulum, *J. Biol. Chem.* **251:**4147.
Shigekawa, M., Finegan, J. A. M., and Katz, A. M., 1976, Calcium transport ATPase of canine cardiac sarcoplasmic reticulum. A comparison with that of rabbit fast skeletal muscle sarcoplasmic reticulum, *J. Biol. Chem.* **251:**6894.
Shoshan, V., and MacLennan, D. H., 1981, Quercetin interaction with the $Ca^{2+}$ + $Mg^{2+}$ ATPase of sarcoplasmic reticulum, *J. Biol. Chem.* **256:**887.
Shoshan, V., Campbell, K. P., MacLennan, D. H., Frodis, W., and Britt, B. A., 1980, Quercetin inhibits $Ca^{2+}$ uptake but not $Ca^{2+}$ release by sarcoplasmic reticulum skinned muscle fibers, *Proc. Natl. Acad. Sci. USA* **77:**4435.
Shoshan, V., MacLennan, D. H., and Wood, D. S., 1981, A proton gradient controls a calcium release channel in sarcoplasmic reticulum, *Proc. Natl. Acad. Sci. USA* **78:**4828.

Somlyo, A. V., Gonzalez-Serratos, H., Shuman, H., McClellan, G., and Somlyo, A. P., 1981, Calcium release and ionic changes in the sarcoplasmic reticulum of tetamized muscle: An electron-probe study, *J. Cell Biol.* **90:**577.

Sommer, J., and Johnson, E., 1979, Ultrastructure of cardiac muscle, in: *Handbook of Physiology,* Vol. 1 (R. Berne, J. Sperelakis, and S. Geiger, eds.), Am. Physiol. Soc., Bethesda, Maryland, p. 113.

Stephenson, E. W., and Podolsky, R. J., 1977, Influence of magnesium on chloride induced calcium release in skinned muscle fibers, *J. Gen. Physiol.* **69:**17.

Stewart, P. S., and MacLennan, D. H., 1974, Surface particles of sarcoplasmic reticulum membrane, Structural features of the adenosine triphosphatase, *J. Biol. Chem.* **249:**985.

Stewart, P. S., and MacLennan, D. H., 1975, Isolation and characterization of tryptic fragments of the sarcoplasmic reticulum adenosine triphosphatase, *Ann. N.Y. Acad. Sci.* **264:**326.

Stewart, P. S., MacLennan, D. H., and Shamoo, A. E., 1976, Isolation and characterization of tryptic fragments of the adenosine triphosphatase of sarcoplasmic reticulum, *J. Biol. Chem.* **251:**712.

Suko, J., and Hasselbach, W., 1976, Characterization of cardiac sarcoplasmic reticulum ATP–ADP phosphate exchange and phosphorylation of the calcium transport adenosine triphosphatase, *Eur. J. Biochem.* **64:**123.

Suzuki, S., and Sugi, H., 1982, Mechanisms of intracellular calcium translocation in muscle, in: *The Role of Calcium in Biological Systems,* Vol. 1 (L. J. Aughileri and A. M. Tuffet-Aughileri, eds.), CRC Press, Boca Raton, Florida, p. 201.

Svaboda, G., Fritzsche, J., and Hasselbach, W., 1979, Effects of phospholipase $A_2$ and albumine on the calcium dependent ATPase and the lipid composition of sarcoplasmic reticulum, *Eur. J. Biochem.* **95:**77.

Tada, M., and Katz, A. M., 1982, Phosphorylation of sarcoplasmic reticulum and sarcolemma, *Annu. Rev. Physiol.* **44:**401.

Tada, M., Yamamoto, T., and Tonomura, Y., 1978a, Molecular Mechanism of Active calcium transport by sarcoplasmic reticulum, *Physiol. Rev.* **58:**1.

Tada, M., Ohmori, F., Kinoshite, N., and Abe, H., 1978b, Cyclic AMP regulation of active calcium transport across membranes of sarcoplasmic reticulum: Role of the 22,000-dalton protein phospholamban, *Adv. Cycl. Nuc. Res.* **9:**355.

Tada, M., Ohmori, F., Yamada, M., and Abe, H., 1979, Mechanism of the stimulation of $Ca^{2+}$ dependent ATPase of cardiac sarcoplasmic reticulum by adenosine 3′-5′-monophosphate dependent protein kinase. Role of the 22,000 dalton protein, *J. Biol. Chem.* **254:**319.

Tada, M., Yamada, M., Ohmori, F., Kuzuya, T., Inui, M., and Abe, H., 1980, Transient state kinetic studies of $Ca^{2+}$ dependent ATPase and calcium transport by cardiac sarcoplasmic reticulum. Effect of cyclic AMP-dependent protein kinase catalyzed phosphorylation of phospholamban, *J. Biol. Chem.* **255:**198.

Tada, M., Yamamoto, M., Kadoma, M., Inui, M., and Ohmori, F., 1982, Calcium transport by cardiac sarcoplasmic reticulum and phosphorylation of phosphalamban, *Mol. Cell. Biochem.* **46:**73.

The, R., Husseini, S. H., and Hasselbach, W., 1981, Synthetic monoacylphospholipids as reactivators of the calcium dependent ATPase of enzymatically delipidated sarcoplasmic reticulum, *Eur. J. Biochem.* **118:**223.

Thomas, D. D., and Hidalgo, C., 1978, Rotational motion of the sarcoplasmic reticulum $Ca^{2+}$-ATPase *Proc. Natl. Acad. Sci. USA* **75:**5488.

Thorley-Lawson, D. A., and Green, N. M., 1973, Studies on the location and orientation of proteins in the sarcoplasmic reticulum, *Eur. J. Biochem.* **40:**403.

Thorley-Lawson, D. A., and Green, N. M., 1975, Separation and characterization of tryptic fragments from the adenosine triphosphatase of sarcoplasmic reticulum, *Eur. J. Biochem.* **59:**193.

Thorley-Lawson, D. A., and Green, N. M., 1977, The reactivity of the thiol groups of the adenosine triphosphatase of sarcoplasmic reticulum and their location on tryptic fragments of the molecule, *Biochem. J.* **167:**739.

Tillack, T. W., Boland, R., and Martonosi, A., 1974, The ultrastructure of developing sarcoplasmic reticulum, *J. Biol. Chem.* **249:**624.

Tong, S. W., 1977, The aceylated $NH_2$-terminus of Ca ATPase from rabbit skeletal muscle sarcoplasmic reticulum: A common $NH_2$ termined acetylated methionyl sequence, *Biochem. Biophys. Res. Commun.* **74:**1242.

Tong, S. W., 1980, Studies on the structure of the calcium dependent adenosine triphosphatase from rabbit skeletal muscle sarcoplasmic reticulum, *Arch. Biochem. Biophys.* **203:**780.

Vale, M. G. P., 1977, Localization of the amino phospholipids in sarcoplasmic reticulum membranes revealed by trinitrobenzenesulfonate and fluorodinitrobenzene, *Biochim. Biophys. Acta* **417:**39.

Vanderkooi, J. M., Ierokomas, A., Nakamura, H., and Martonosi, A., 1977, Fluorescence energy transfer between $Ca^{2+}$ transport ATPase molecules in artificial membranes, *Biochemistry* **16:**1262.

Van Winkle, W. B., and Entman, M. L., 1979, Minireview. Comparative aspects of cardiac and skeletal muscle sarcoplasmic reticulum, *Life Sci.* **25:**1189.

Van Winkle, W. B., Phitts, B. J. R., and Entman, M. L., 1978, Rapid purification of canine cardiac sarcoplasmic reticulum $Ca^{2+}$ ATPase, *J. Biol. Chem.* **253:**8671.

Wang, C-T., Saito, A., and Fleischer, S., 1979, Correlation of ultrastructure of reconstituted sarcoplasmic reticulum membrane vesicles with variation in phospholipid to protein ratio, *J. Biol. Chem.* **254:**9209.

Warren, G. B., and Metcalf, J. G., 1976, The molecular architecture of a reconstituted calcium pump, in: *Structural and Kinetic Approach to Plasma Membrane Functions* (C. Nicolau and A. Parat, eds.), Springer-Verlag, Berlin, p. 188.

Warren G. B., Toon, P. A., Birdsall, N. J. M., Lee, A. G., and Metcalf, J. C., 1974a, Reversible lipid titrations of the activity of pure adenosine triphosphatase lipid complexes, *Biochemistry* **13:**5501.

Warren, G. B., Toon, P. A., Birdsall, N. J. M., Lee, A. G., and Metcalfe, J. C., 1974b, Reconstitution of a calcium pump using defined membrane components, *Proc. Natl. Acad. Sci. USA* **71:**622.

Warren, G. B., Houslay, M. D., Metcalf, J. C., and Birdsall, N. J. M., 1975, Cholesterol is excluded from the phospholipid annulus surrounding an active calcium transport protein, *Nature* **255:**684.

Weber, K., and Osborn, M., 1969, The reliability of molecular weight determinations by dodecyl sulfate-polyacrylamide gel electrophoresis, *J. Biol. Chem.* **244:**4406.

Will, H., Blanck, J., Smettan, G., and Wollenberger, A., 1976, A quench flow kinetic investigation of calcium ion accumulation by isolated cardiac sarcoplasmic reticulum. Dependence of initial velocity on free calcium ion concentration and influence of preincubation with a protein kinase, MgATP and cyclic AMP, *Biochim. Biophys. Acta* **449:**295.

Winegrad, S., 1968, Intracellular calcium movements of frog skeletal muscle during recovery from tetanus, *J. Gen. Physiol.* **51:**65.

Winegrad, S., 1970, The intracellular site of calcium activation of contraction in frog skeletal muscle, *J. Gen. Physiol.* **55:**77.

Winegrad, S., 1982, Calcium release from cardiac sarcoplasmic reticulum, *Annu. Rev. Physiol.* **44:**451.

Wu, F. S., Park, Y. C., Roufa, D., and Martonosi, A., 1981, Selective stimulation of the synthesis of an 80,000-dalton protein by calcium ionophores, *J. Biol. Chem.* **256:**5309.

Yamamoto, T., and Tonomura, Y., 1968, Reaction mechanism of the $Ca^{++}$-dependent ATPase of sarcoplasmic reticulum from skeletal muscle. II. Intermediate formation of phosphoryl protein, *J. Biochem.* **64:**789.

Yamamoto, T., Yoda, A., and Tonomura, Y., 1971, Reaction mechanism of the $Ca^{2+}$-dependent ATPase of sarcoplasmic reticulum from skeletal muscle. IV. Hydroxomate formation from a phosphorylated intermediate and 2-hydroxy-5-nitrobenzyl hydroxylamine, *J. Biochem.* **69:**807.

Yates, D. W., and Duance, V. C., 1976, The binding of nucleotides and bivalent cations to the calcium and magnesium ion-dependent adenosine triphosphatase from rabbit muscle sarcoplasmic reticulum, *Biochem. J.* **159:**719.

Zimniak, P., and Racker, E., 1978, Electrogenicity of $Ca^{2+}$-transport catalyzed by the $Ca^{2+}$-ATPase from sarcoplasmic reticulum, *J. Biol. Chem.* **253:**4631.

Zubrzycka, E., and MacLennan, D. H., 1976, Assembly of the sarcoplasmic reticulum. Biosynthesis of calsequestrin in rat skeletal muscle cell cultures, *J. Biol. Chem.* **251:**7733.

Zubrzycka, E., Michalak, M., Kosk-Kosicka, O., and Sarzala, M. G., 1979, Properties of microsomal subfractions isolated from developing rabbit skeletal muscle, *Eur. J. Biochem.* **93:**113.

Zubrzycka, E., Campbell, K. P., MacLennan, D. H., and Jorgensen, A. O., 1983, Biosynthesis of intrinsic sarcoplasmic reticulum proteins during differentiation of the myogenic cell line, L6, *J. Biol. Chem.*, **258:**4576.

# 30

# Kinetic Regulation of Catalytic and Transport Activities in Sarcoplasmic Reticulum ATPase

Giuseppe Inesi and Leopoldo de Meis

## I. INTRODUCTION

The cytoplasmic $Ca^{2+}$ concentration is generally several orders of magnitude lower than that of extracellular fluids. This gradient must be maintained by energy-dependent transport mechanisms located in the cell membrane. In addition, intracellular membranous structures may form compartments which serve as $Ca^{2+}$ sinks and reservoirs inside the cell. In some tissues, in which specific functional controls require rapid $Ca^{2+}$ delivery to, and sequestration from, the cytoplasm, these intracellular systems are developed to a highly differentiated and prominent degree. A prime example is the sarcoplasmic reticulum (SR) of fast striated muscles. The membrane of the sarcoplasmic reticulum can be isolated in vesicular form from muscle homogenates by differential centrifugation. When suspended in a medium containing ATP and $Mg^{2+}$, these vesicles reduce the $Ca^{2+}$ concentration of the medium from $10^{-4}$ M to less than $10^{-7}$ M, which is the level found in the cytosol of living muscle cells in a resting state. This was discovered, independently, by Hasselbach and Makinose (1962) and Ebashi and Lipman (1962).

The mechanism of ATP-dependent $Ca^{2+}$ sequestration by SR vesicles consists of transmembrane transport coupled to hydrolytic cleavage of the ATP terminal phosphate, and results in formation of a $Ca^{2+}$ gradient across the SR membranes.

*Giuseppe Inesi* • Department of Biological Chemistry, University of Maryland School of Medicine, Baltimore, Maryland 21201. *Leopoldo de Meis* • Instituto de Ciencias Biomedicas, Departamento de Bioquimica, Universidade Federal do Rio de Janeiro, Rio de Janeiro 21910, Brasil.

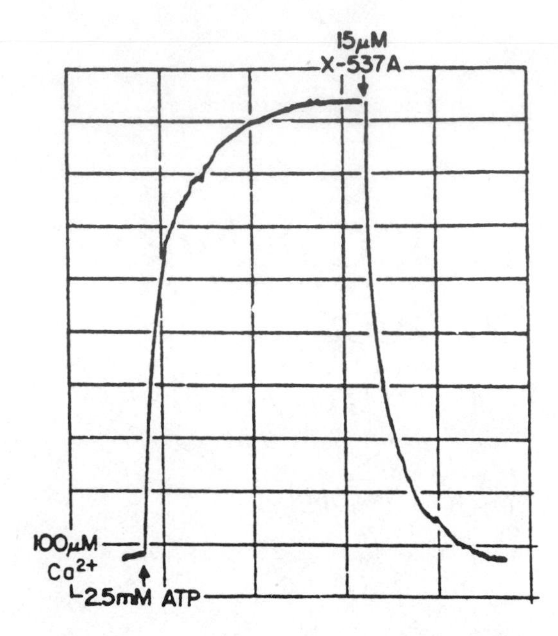


*Figure 1.* $Ca^{2+}$ uptake and subsequent release of $Ca^{2+}$ obtained by adding ATP and the ionophore X-537 A to SR vesicles (Scarpa *et al.*, 1972).

If the passive permeability of the SR membrane to $Ca^{2+}$ is increased by the addition of divalent cation ionophores (Scarpa *et al.*, 1972), the accumulated $Ca^{2+}$ is rapidly released by facilitated diffusion and the gradient collapses (Figure 1). On the other hand, under suitable conditions, the actual process of $Ca^{2+}$ transport can be reversed and the very same ATPase involved in $Ca^{2+}$ transport is able to catalyze the synthesis of ATP from ADP and $P_i$ in a process coupled with the release of $Ca^{2+}$ through reversal of the pump (Barlogie *et al.*, 1971; Makinose and Hasselbach, 1971; Makinose, 1972).

It is clear that the SR ATPase is able to interconvert different forms of energy. During the process of $Ca^{2+}$ accumulation, the chemical energy derived from the hydrolysis of ATP is used to build up a $Ca^{2+}$ concentration gradient across the vesicles membrane, and in the reverse process, the energy derived from the $Ca^{2+}$ gradient is used for synthesis of ATP.

This chapter will describe a number of sequential reactions which were recently found to be integral parts of the catalytic and transport cycle of SR ATPase in the forward and reverse directions. The experimental systems used for such studies are (1) SR vesicles permitting net $Ca^{2+}$ accumulation simultaneously with ATP utilization, (2) vesicles rendered "leaky" by addition of ionophores or by partial denaturation, and (3) purified ATPase. Systems (2) and (3) sustain $Ca^{2+}$-dependent ATP hydrolysis, but are unable to maintain a transmembrane $Ca^{2+}$ gradient.

## A. $Ca^{2+}$ Uptake and ATP Hydrolysis

Two different ATPase activities are often obtained with sarcoplasmic reticulum vesicles: the low-rate $Mg^{2+}$-dependent ("basal") ATPase activity which is observed in the presence of EGTA to remove contaminant $Ca^{2+}$ from the assay medium, and the high-ATPase activity which is correlated with $Ca^{2+}$ transport. The latter requires both $Ca^{2+}$ and $Mg^{2+}$ for full activity and is calculated by subtracting the "basal" ATPase from the total activity measured in the presence of $Mg^{2+}$ and $Ca^{2+}$ (Hasselbach and Makinose, 1962; Hasselbach, 1964, 1978; Tada *et al.*, 1978).

The rates of $Ca^{2+}$ uptake and ATP hydrolysis by the $Ca^{2+}$ + $Mg^{2+}$-dependent ATPase vary with the $Ca^{2+}$ concentration in the medium, half-maximal activity being attained in both cases in the presence of 0.1–2.0 μM $Ca^{2+}$ (Martonosi, 1971; Hasselbach, 1978; Tada *et al.*, 1978; Ikemoto, 1982). In most experimental conditions, two calcium ions are accumulated by the vesicles for each molecule of ATP hydrolyzed.

Maximal levels of calcium uptake by SR vesicles are reached in a few seconds following addition of ATP (Figure 1), after which the activity of the pump is reduced due to inhibition by the high concentration of $Ca^{2+}$ accumulated inside the vesicles. Such an inhibition can be removed by the use of oxalate or phosphate in order to precipitate the accumulated calcium and reduce the intravesicular $Ca^{2+}$ activity. Thereby, calcium uptake and ATPase activity can be prolonged to allow useful measurements in steady-state conditions (Hasselbach and Makinose, 1962; Martonosi and Feretos, 1964; Weber *et al.*, 1966).

Measurements in the initial phase of the reaction require rapid kinetic methods. One approach to this problem is based on the use of stopped-flow mixing and chromogenic $Ca^{2+}$ indicators. Thus, the initial velocity of calcium uptake can be monitored by double-wavelength differential measurements of light absorption changes undergone by these indicators (Figure 2). The turnover number (10–20 $sec^{-1}$ at 25°C) for the

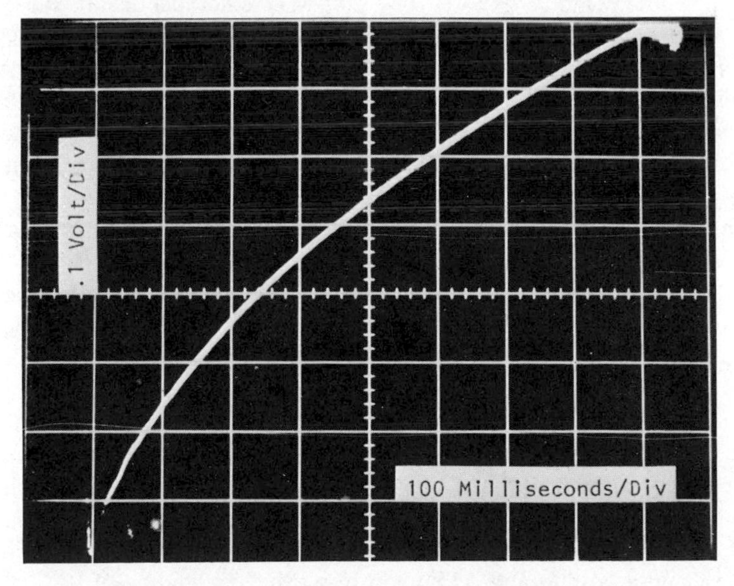


*Figure 2.* ATP-dependent calcium uptake by SR vesicles, measured by double-wavelength (660 vs. 690 mm) photometery in a stopped-flow apparatus. The reaction was followed by monitoring light-absorption changes undergone by the calcium-sensitive dye Arsenazo III (Inesi, 1981).

$Ca^{2+}$ pump was first obtained by this method (Inesi and Scarpa, 1972). Furthermore, the initial velocity of substrate, e.g., ATP, hydrolysis which is coupled to $Ca^{2+}$ uptake can be followed by measuring $H^+$ production in the presence of appropriate $H^+$ indicators (Verjovski-Almeida *et al.*, 1978). Finally, utilization of substrates yielding chromogenic moieties upon hydrolysis can be monitored directly (Figure 3).

Another approach is based on the use of rapid-quench methods (Martonosi *et al.*, 1974; Froehlich and Taylor, 1975; Kurzmack and Inesi, 1977). Rapid quenching is a very powerful approach, the advantages of which can be maximized by the choice of suitable quench reagents (Chiesi and Inesi, 1979). Acid quenchers permit kinetic measurements of phosphoenzyme formation and $P_i$ production. On the other hand, quenching with agents that prevent calcium occupancy of specific activating sites, e.g., EGTA and $La^{3+}$, stops the reaction without causing leak of accumulated calcium; therefore, movements of calcium can be measured directly by the use of radioactive tracers and filtration methods.

## B. $Ca^{2+}$ Activation of SR ATPase

Owing to the high affinity of SR ATPase for $Ca^{2+}$, the dependence of enzyme activity on $Ca^{2+}$ cannot be demonstrated unless calcium introduced into the assay medium as a contaminant of the reagents is complexed by an appropriate chelating agent.

EGTA (ethylene glycol *bis* (β-aminoethyl ether) *N, N', N', N'*-tetraacetic acid) is a suitable chelating agent, owing to its much greater affinity for $Ca^{2+}$ than for $Mg^{2+}$. Although the values given in the literature for the Ca-EGTA dissociation constant at neutral pH ($5 \times 10^{-7} - 4 \times 10^{-6}$ M) are not identical (Schmid and Reilley, 1957; Schwartzenbach *et al.*, 1957; Holloway and Reilley, 1960; Ebashi, 1961; Ogawa, 1968; de Meis and Hasselbach, 1971; Allen *et al.*, 1977), they are sufficiently low to allow the use of EGTA for reduction of the $Ca^{2+}$ concentration in the medium below levels producing activation of the pump. Addition of increasing concentrations of $CaCl_2$ then yields a Ca-EGTA buffer for control of the free $Ca^{2+}$ concentration and stepwise activation of the SR vesicles in the presence of total calcium concentrations sufficiently high to support the activity of the pump for experimentally useful times. It is noteworthy that EGTA does not penetrate the SR membrane (Weber *et al.*, 1966), and its $Ca^{2+}$ buffering effect is limited to the medium outside the vesicles. A very informative discussion on calculations of $Ca^{2+}$ buffers and free $Ca^{2+}$ concentrations in various conditions and in the presence of other ligands is given by Fabiato and Fabiato (1979).

Generally, it is reported in the literature that activation of the SR pump displays a sigmoidal dependence on the $Ca^{2+}$ concentration in the medium, with half-maximal saturation at approximately 0.5 μM $Ca^{2+}$ (de Meis, 1971; The and Hasselbach, 1972a,b; Gattass and de Meis, 1975; Vianna, 1975; Suko and Hasselbach, 1976; Neet and Green, 1977). Such a sigmoidal dependence, as well as the 2 : 1 molar ratio observed between calcium transport and ATP hydrolysis, suggest binding of two calcium ions to the same enzyme unit. It is then clear that the enzyme must bind calcium before being able to utilize ATP. In fact, calcium binding to SR vesicles in the absence of

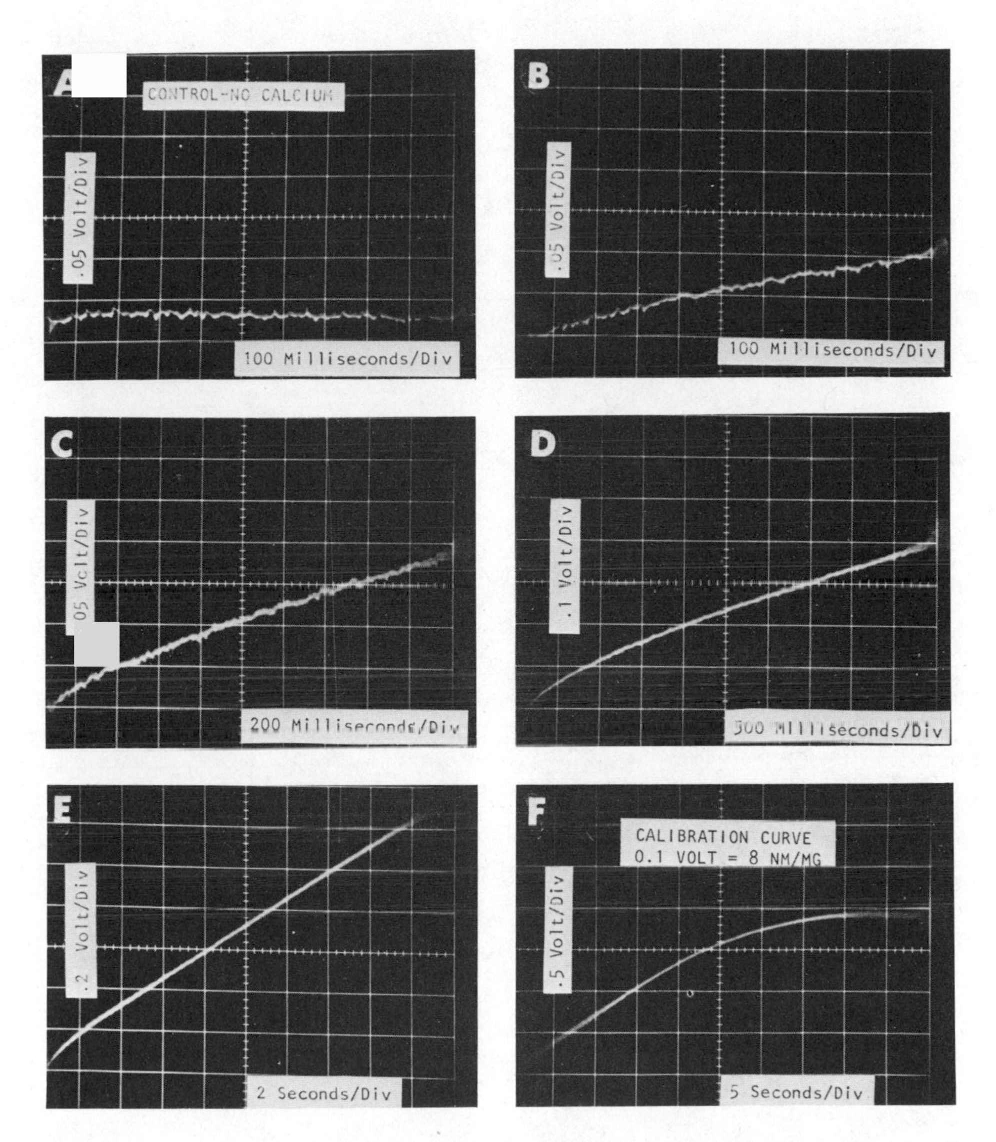


*Figure 3.* Hydrolysis of the chromogenic substrate furylacryloylphosphate used by SR vesicles to sustain calcium transport. The activity was measured by double-wavelength (350 vs. 370 mm) photometry in a stopped-flow apparatus, taking advantage of the differential absorption of furylacryloate anion released on furylacryloylphosphate utilization (Kurzmack *et al.*, 1981).

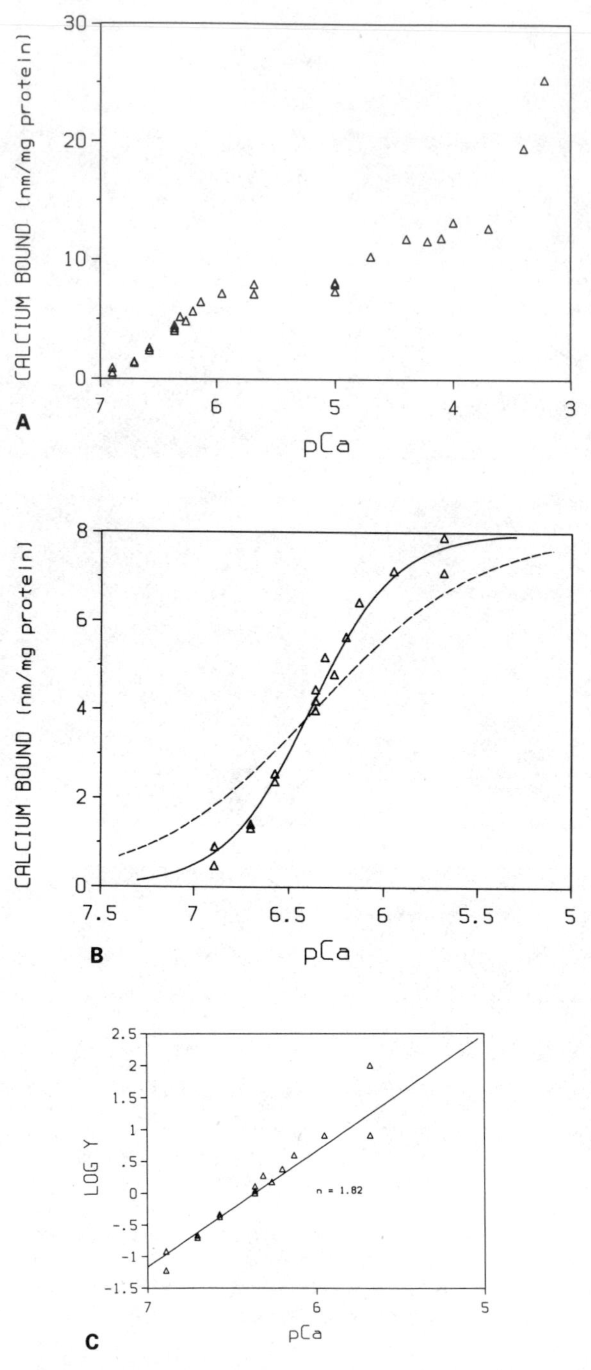


*Figure 4.* Calcium binding to SR vesicles in the absence of ATP. (A) Graph showing various classes of binding sites. (B,C) Graphs showing the cooperative character of the high-affinity binding (Inesi *et al.*, 1980a).

ATP can be measured in equilibrium conditions. These measurements reveal various classes of binding sites (Carvalho, 1966; Chevallier and Butow, 1971; Fiehn and Migala, 1971; Meissner *et al.*, 1973; Ikemoto, 1974, 1975; Chiu and Haynes, 1977).

Of the various classes, the high-affinity class ($K_d < 1$ μM) can be attributed to binding sites on the ATPase protein (Meissner *et al.*, 1973), which are exposed on the outer surface of the vesicles. This class of binding sites is involved in activation of the enzyme. Measurement of binding at such low concentrations presents serious difficulties, and most of the early studies report few experimental points in the $Ca^{2+}$ concentration range in which this class is saturated. More recent studies (Ikemoto, 1975; Kalbitzer *et al.*, 1978; Inesi *et al.*, 1980a) indicate that calcium binding to high-affinity sites occur through a cooperative mechanism which is fully consistent with the $Ca^{2+}$ concentration dependence for activation of ATP hydrolysis and $Ca^{2+}$ transport (Figure 4). The high-affinity calcium-binding sites correspond to 8–10 nmol/mg vesicles protein, in preparations of SR vesicles yielding 4–5 nmole of phosphorylated enzyme intermediate (catalytic site)/mg protein.

Ample experimental evidence based on EPR spectroscopy of spin-labeled SR (Coan and Inesi, 1977; Champeil *et al.*, 1978), kinetics of —SH reactivity (Ikemoto *et al.*, 1978; Murphy, 1978), and measurements of intrinsic fluorescence (Dupont and Leigh, 1978) indicates that calcium binding to high-affinity sites is accompanied by a protein conformational change. This change which is fully reversible was attributed by Inesi *et al.* (1980a) to cooperative interactions and sequential binding as in:

$$E + Ca^{2+} \rightleftharpoons E.Ca \rightleftharpoons E'.Ca + Ca^{2+} \rightleftharpoons E''.Ca_2. \qquad (1)$$

where $K_3 \gg K_1$, and 2 is a conformational transition induced by the initial calcium binding.

Dupont (1982) studied, at low temperature (0–12°C), the kinetics of both $Ca^{2+}$ binding and changes in protein intrinsic fluorescence. His observations are consistent with a sequential binding mechanism, whereby, $Ca^{2+}$ has very fast access to a site of low apparent affinity ($K_d \sim 25 \times 10^{-6}$ M). Occupation of this site induces a slow conformational change which reveals a second site of high affinity. Based on the data reported, Dupont proposed a more general scheme (Figure 5) involving sequential

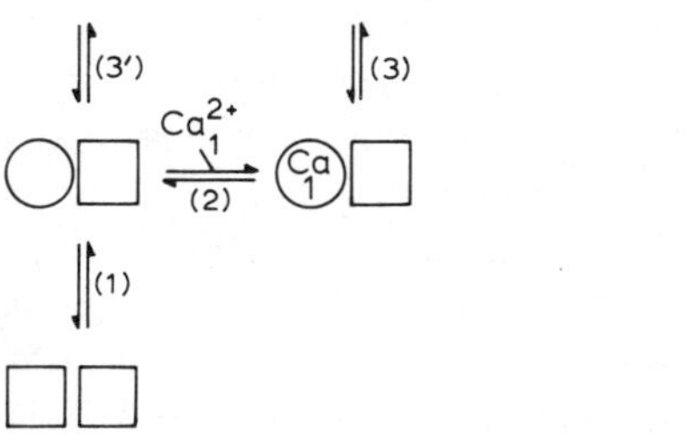

*Figure 5.* Diagram for calcium binding to two interacting binding domains (DuPont, 1982).

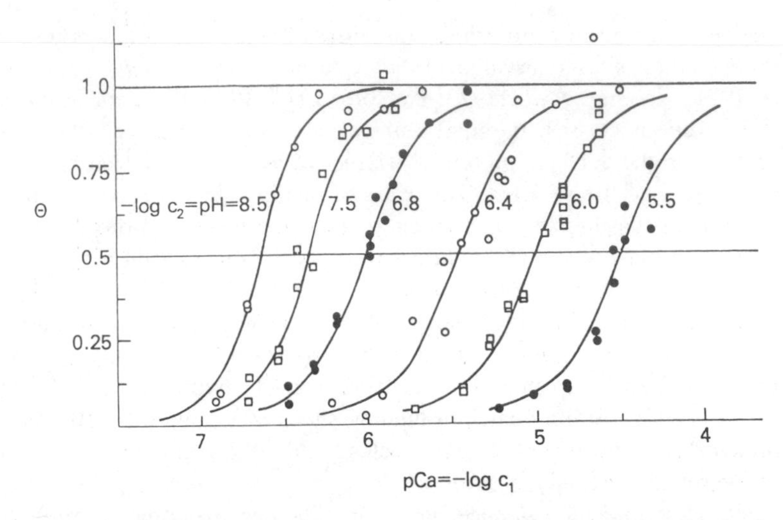


*Figure 6.* $Ca^{2+}$ concentration dependence of high-affinity calcium binding to SR ATPase in the absence of ATP, at varying pH. The experimental points are fitted according to a model based on competitive binding of $Ca^{2+}$ or $H^+$ to interacting binding domains (Hill and Inesi, 1982).

conformational changes. Presently, it is not clear to what extent the cooperative binding mechanism should be attributed to interaction of different polypeptides, as opposed to interaction of binding domains within one polypeptide chain.

An interesting feature of the calcium-binding mechanism is its pH dependence, as it was observed that the affinity of the enzyme for $Ca^{2+}$ is increased by high pH, and reduced by low pH (Meissner, 1973; Verjovski-Almeida and de Meis, 1977; de Meis and Tume, 1977; Beil *et al.*, 1977). Watanabe *et al.* (1981) measured directly calcium binding to SR ATPase as a function of pH in equilibrium conditions, and found that both the affinity and the extent of cooperativity increase as the $H^+$ concentrations are reduced. A family of calcium-binding curves (Figure 6), as a function of $Ca^{2+}$ concentration at various pH, was fitted satisfactorily (Hill and Inesi, 1982) assuming $Ca^{2+}$ and $H^+$ competition for interacting sites. Displacement of 1 $H^+$ per 1 $Ca^{2+}$ bound to each site was in fact measured directly (Chiesi and Inesi, 1980).

## II. SUBSTRATE SPECIFICITY

In early studies, it was found that the nucleotides ATP, ITP, GTP, CTP, and UTP were suitable substrates for SR ATPase, and were able to drive the $Ca^{2+}$ pump (Hasselbach, 1964; Makinose and The, 1965; Makinose, 1966). Later, it was found that a large variety of pseudosubstrates could also support active transport of $Ca^{2+}$; among them are acetylphosphate, *p*-nitrophenylphosphate, carbamylphosphate, methylumbelliferylphosphate, and furylacryloylphosphate (de Meis, 1969; Friedman and Makinose, 1970; Pucell and Martonosi, 1971; de Meis and Hasselbach, 1971; Inesi, 1971; Kurzmack *et al.*, 1981; Nakamura and Tonomura, 1978; Rossi *et al.*, 1979). The affinity of the enzyme for ATP is one or more orders of magnitude higher than for other substrates (de Meis, 1969b; de Meis and de Mello, 1973; Verjovski-Almeida *et al.*, 1978; Rossi *et al.*, 1979). Hasselbach (1978) reported that the molar ratio of $Ca^{2+}$ transport to substrate hydrolysis is two when nucleotides or *p*-nitrophenylphosphate are used as substrates. On the other hand, Rossi *et al.* (1979) found a coupling ratio of one when *p*-nitrophenylphosphate, methylumbelliferylphosphate, or furylacryloylphosphate were used. Of all these substrates, ATP is the most specific inasmuch as it is utilized at very low concentrations and high rates. The true substrate appears to be the ATP–Mg complex (Weber *et al.*, 1966; Yamamoto and Tonomura, 1967; Vianna, 1975).

The ATP concentration dependence of the ATPase reaction has been studied first by measuring the hydrolytic reaction, e.g., release of $P_i$, in steady-state conditions. A complex dependence was found in this case, with a first activation obtained with ATP concentrations ranging between 0.5–50 μM, and a further rise of activity at higher concentrations (Inesi *et al.*, 1967; Yamamoto and Tonomura, 1967; de Meis, 1971; The and Hasselbach, 1972a,b; de Meis and de Mello, 1973; Froehlich and Taylor, 1975; Vianna, 1975; Yates and Duance, 1976; Dupont, 1977; Ribeiro and Vianna, 1978; Scofano *et al.*, 1979; Taylor and Hattan, 1979). The observed dependence cannot be fitted according to simple Michaelis–Menten kinetics. It is possible that the effect of high ATP is related to activation of partial reactions which follow the initial enzyme phosphorylation of ATP and contribute to rate limitation with respect to the enzyme turnover. Consistent with this suggestion, it is found that the ATP concentration dependence of the initial enzyme phosphorylation is limited to the lower concentration range (Kanazawa *et al.*, 1971; Froehlich and Taylor, 1975, 1976; Verjovski-Almeida *et al.*, 1978; Scofano *et al.*, 1979), but the hydrolytic cleavage of the phosphoenzyme formed in the initial part of the reaction is activated by high-ATP concentrations (de Meis and de Mello, 1973; Froehlich and Taylor, 1975; de Souza and de Meis, 1976; Verjovski-Almeida and Inesi, 1979).

It is clear that the ATP concentration dependence ($K_m$) for the ATPase, which is a multistep reaction, is related but does not reflect directly the concentration dependence of ATP binding to the enzyme. Direct measurements of ATP binding are quite difficult from the experimental point of view. Measurements by Pang and Briggs (1977) and Dupont (1977) have detected nucleotide-binding sites in the $10^{-5}$ range, and suggest the presence of lower-affinity sites which are titrated with disturbing experimental scatter at high-ATP concentrations.

## A. Phosphorylated Enzyme Intermediate

Formation of a phosphorylated enzyme intermediate following addition of ATP to SR ATPase was first suggested by the occurrence of ATP–ADP exchange (Ebashi and Lipman, 1962; Hasselbach and Makinose, 1962). The exchange is revealed by formation of [$^{14}$C]-ATP following incubation of ATP and [$^{14}$C]-ADP with the enzyme, and is attributed to reversal of enzyme phosphorylation by ATP:

$$\begin{array}{ccccc} \text{ATP} & + \text{ E} \rightleftharpoons \text{E-P} + & \text{ADP} \\ \updownarrow & & \updownarrow \\ [^{14}\text{C}]\text{-ATP} & & [^{14}\text{C}]\text{-ADP} \end{array}$$

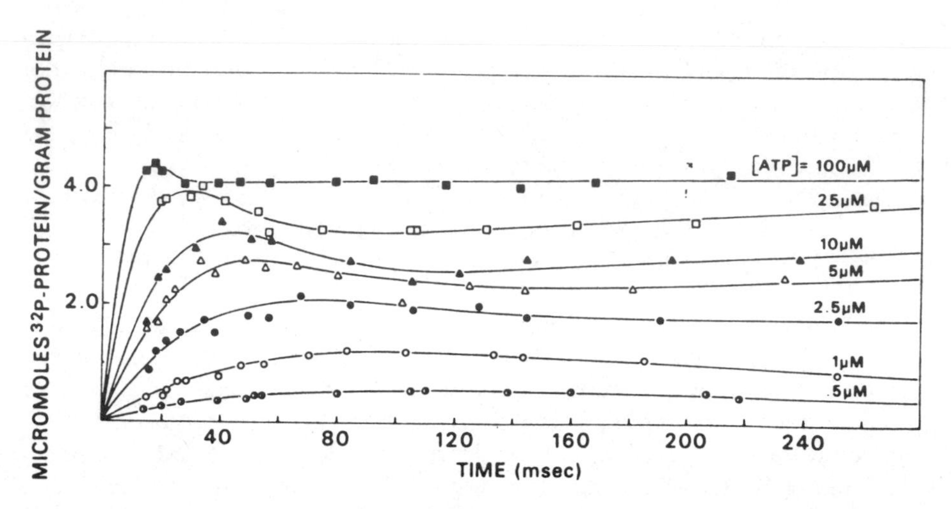


*Figure 7.* ATP-concentration dependence of ATPase phosphorylation with ATP in the presence of saturating $Ca^{2+}$ (Froehlich and Taylor, 1975).

It should be emphasized that this exchange is specifically $Ca^{2+}$-dependent (Ebashi and Lipman, 1962; Inesi and Almendares, 1968), and it is, therefore, related to the ATPase activity rather than to nonspecific enzymes contaminating the SR preparation.

Enzyme phosphorylation was demonstrated directly by incorporation of [$\gamma$-$^{32}$P]-ATP terminal phosphate and acid quenching (Makinose, 1967, 1969; Yamamoto and Tonomura, 1967, 1968; Martonosi, 1969; Inesi *et al.*, 1970). Phosphorylation involves an aspartyl residue at the active site (Bastide *et al.*, 1973; Degani and Boyer, 1973).

Time resolution of the phosphorylation reaction requires fast-mixing techniques owing to the rapidity of the reaction. A phosphorylation rate constant of approximately $10^2$ $sec^{-1}$ is obtained in rapid kinetic experiments (Figure 7; Froehlich and Taylor, 1975; Verjovski-Almeida *et al.*, 1978) in which ATP is added to enzyme *preincubated with* $Ca^{2+}$. On the other hand, if ATP and $Ca^{2+}$ are added to enzyme *deprived of* $Ca^{2+}$, i.e., exposed to EGTA, a slower phosphorylation curve is obtained (Sumida *et al.*, 1978; Scofano *et al.*, 1979; Inesi *et al.*, 1980a). This is due to a time limit imposed by the $Ca^{2+}$-induced activation of the enzyme.

The reverse rate constant of the phorphorylation reaction, i.e., $Ca^2$.E-P + ADP $\rightharpoondown$ $Ca_2$.E.ATP, appears to be 2–3 times faster than the forward constant (Froehlich and Taylor, 1975; Verjovski-Almeida *et al.*, 1978; Pickart and Jencks, 1982). The level of phosphoenzyme measured following addition of ATP represents the steady-state levels of different intermediate species in the enzyme cycle. The phosphoenzyme reaches significant levels owing to a favorable ratio between the rate constants of the reaction causing formation, and those causing disappearance of the intermediate.

In optimal conditions with respect to ATP and $Ca^{2+}$ concentrations, temperature, and pH, maximal steady-state levels of phosphoenzyme range between 3–5 nmol/mg protein when SR vesicles are used (Makinose, 1967; Froehlich and Taylor, 1975;

*Table 1. Levels and $P_i$ Production Obtained with Phosphorylation by ATP or ITP*[a]

| | | | ATP | | ITP | |
|---|---|---|---|---|---|---|
| Vesicles | $CaCl_2$ (mM) | Temperature (°C) | Phosphoenzyme (µmoles E-P/g prot.) | ATPase (µmoles $P_i$/mg min) | Phosphoenzyme (µmoles/g prot.) | ITPase (µmoles $P_i$/mg per min) |
| Intact (back inhibition) | 0.1 | 30 | 3.8 ± 0.12 | 0.8 ± 0.07 | 3.4 ± 0.2 | 0.28 ± 0.05 |
| Leaky vesicles | 0.1 | 30 | 3.5 ± 0.15 | 4.2 ± 0.10 | 1.0 ± 0.1 | 0.60 ± 0.07 |
| Leaky vesicles | 10.0 | 30 | 3.6 ± 0.11 | 0.7 ± 0.05 | 3.7 ± 0.1 | 0.15 ± 0.01 |
| Leaky vesicles | 0.1 | 0 | 3.9 ± 0.13 | 0.02 ± 0.05 | 3.6 ± 0.10 | 0.02 ± 0.04 |

[a] The assay medium was 1 mM [$\gamma$-$^{32}$P]-ATP or [$\gamma$-$^{32}$P]-ITP, 10 mM $MgCl_2$, 30 mM Tris-maleate buffer (pH 7.0), and the $CaCl_2$ concentration shown in the table. For details see de Souza and de Meis (1976), Verjovski-Almeida and de Meis (1977), and Masuda and de Meis (1977).

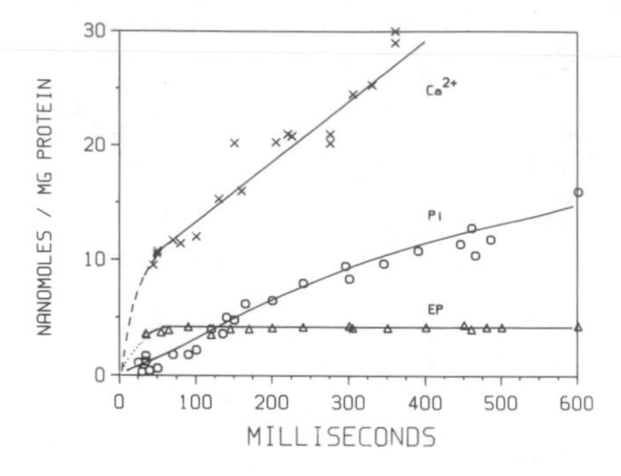


*Figure 8.* Time resolution of phosphoenzyme formation, calcium translocation, and $P_i$ production following addition of ATP to ATPase saturated with $Ca^{2+}$ (Inesi, 1981).

Verjovski-Almeida *et al.,* 1978). However, at variance with the constant phosphoenzyme levels obtained with saturating ATP, variable steady-state levels of phosphoenzyme are obtained with saturating ITP or GTP (de Souza and de Meis, 1976; Verjovski-Almeida and de Meis, 1977; de Meis and Boyer, 1978; Ronzani *et al.,* 1979). When calcium binding is limited to the sites in the high-affinity state, the level of phosphoenzyme obtained with ITP or GTP is lower than that obtained with ATP. This indicates that in the steady state obtained with ITP or GTP, a significant portion of the enzyme is in a nonphosphorylated form (Table 1).

## B. Calcium Translocation and Phosphoenzyme Cleavage

The first detectable event following enzyme phosphorylation is internalization of calcium bound to the high-affinity sites. This phenomenon can be detected by rapid-quench techniques, and is manifested as an early burst of calcium uptake; i.e., bound calcium becomes unavailable for rapid equilibration with EGTA or for displacement by $La^{3+}$ added to the medium (Kurzmack *et al.,* 1977; Chiesi and Inesi, 1979). The early burst of calcium internalization is related to the phosphoenzyme with a stoichiometric ratio of 2 : 1 (Figure 8). It is apparent that the calcium involved in the initial burst is not immediately released into the lumen of the vesicles, but is rather maintained in an "occluded" state during rate-limiting steps related to evolution of the enzymes through the catalytic cycle (Takakuwa and Kanazawa, 1979; Dupont, 1980). Finally, $Ca^{2+}$ is released inside the vesicles and the phosphoenzyme undergoes hydrolytic clevage. Steady-state $P_i$ production and calcium uptake then take place, still maintaining a stoichiometric ratio of two calcium ions for 1 mole of $P_i$ (Figure 8). This ideal ratio of two is reduced by unfavorable experimental conditions such as high pH, the use of certain pseudosubstrates, or partial SR denaturation.

When the $Ca^{2+}$ concentration inside the vesicles is raised to levels higher than $10^{-3}$ M as a consequence of several pump cycles, the activity is inhibited. This "back inhibition" (Makinose and Hasselbach, 1965; Weber *et al.,* 1966; Weber, 1971a,b; Yamada and Tonomura, 1972) is abolished if the permeability of the membrane is artificially increased by the use of ionophores, detergents, or diethyl ether. In this case, the vesicles become leaky and are unable to retain the $Ca^{2+}$ pumped in by the ATPase. The "back inhibition" can also be prevented with the use of phosphate or oxalate. These anions form calcium salts of low solubility within the vesicles. Thus, they act as calcium-sequestering agents which maintain the free calcium concentration low and constant in the vesicle lumen. "Back inhibition" can be reproduced with leaky vesicles or soluble preparations of the ATPase, simply by raising the $Ca^{2+}$ concentration in the assay medium to levels expected to be found in the lumen of intact vesicles following incubation with ATP (de Meis and Carvalho, 1974; Ikemoto, 1974, 1975; de Meis and Sorenson, 1975; Carvalho *et al.*, 1976; de Souza and de Meis, 1976).

## C. *Interconversion of the $Ca^{2+}$-Binding Sites of High and Low Affinities*

The active transport of $Ca^{2+}$ has been attributed to an ATP-dependent conversion of the $Ca^{2+}$-binding sites from a state of high to one of low affinity (Hasselbach, 1964). According to this model, the binding site would initially face the outer surface of the vesicles and, due to its low $K_s$, would bind $Ca^{2+}$ even when the $Ca^{2+}$ concentration in the medium is less than 1 μM. In the process of ATP utilization, the ATPase would undergo a conformational change and the same $Ca^{2+}$-binding site would now face the lumen of the vesicles. This would be associated with an increase of the $K_d$ from micromolar to millimolar, permitting release of bound $Ca^{2+}$ into the lumen of the vesicles. Following hydrolysis of the phosphorylated intermediate, the enzyme would return to its original high-affinity state in order to initiate a new cycle of $Ca^{2+}$ transport (Figure 9; de Meis and Vianna, 1979). This model, based on transformation of the binding sites from high to low affinity, is consistent with the experiments of Ikemoto (1975, 1976) and Watanabe *et al.* (1981) who, in fact, observed $Ca^{2+}$ dissociation from the enzyme upon addition of ATP. The released $Ca^{2+}$ was again bound to the enzyme following hydrolysis of ATP. In these experiments, enzyme preparations assembled in "leaky" vesicles were used in order to avoid net accumulation of the dissociated $Ca^{2+}$ inside the vesicles. The enzyme was incubated in a medium containing a low $Ca^{2+}$ concentration, sufficient to allow binding only to the high-affinity sites. Upon addition of a small amount of ATP, part of the bound $Ca^{2+}$ was released into the medium. Following exhaustion of the added ATP and decay of the phosphorylated enzyme intermediate, $Ca^{2+}$ was bound again by the enzyme (Figure 10).

It is apparent that, *in the presence of $Ca^{2+}$*, the *dephosphorylated* enzyme is drawn into the high-affinity state ($E.Ca_2$ in Figure 9) from which it can be displaced only with expenditure of energy, i.e., by utilization of ATP. Then, *following phosphorylation* by ATP, the phosphoenzyme shifts to a low-affinity state, whereby, $Ca^{2+}$

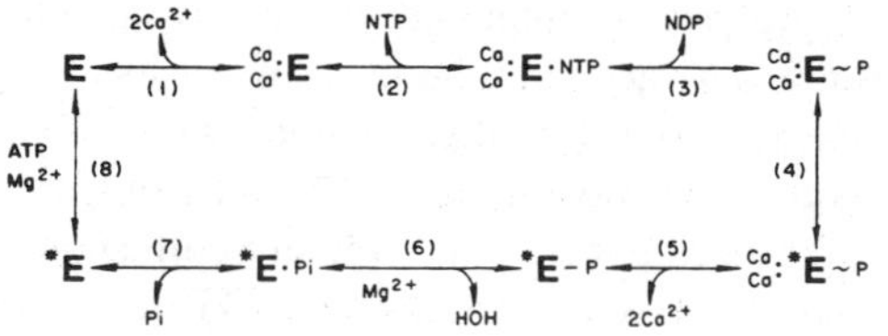

*Figure 9.* The reaction scheme proposed by de Meis and Vianna (1979).

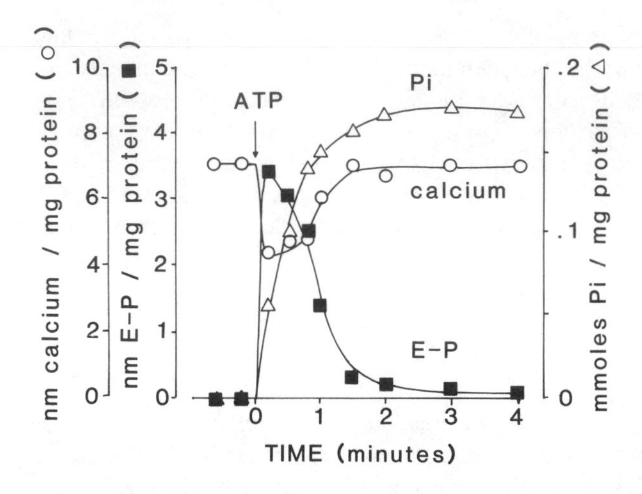


*Figure 10.* Phosphoenzyme formation, calcium release, and $P_i$ production following addition of ATP to leaky vesicles (Watanabe *et al.*, 1981).

is released. Saturation of the low-affinity sites in the phosphoenzyme occurs when the intravesicular $Ca^{2+}$ is increased to levels higher than $10^{-3}$ M. In this case, the net forward flux of ATP is inhibited ("back inhibition").

The regulatory function of the high- and low-affinity states of the calcium sites can also be observed with detergent-solubilized enzyme or "leaky" vesicles which do not maintain a transmembrane $Ca^{2+}$ gradient. In studies of ATP fluxes as a function of $Ca^{2+}$ in the medium, it was shown with these systems that the ATPase activity is stimulated by micromolar $Ca^{2+}$ and inhibited by millimolar $Ca^{2+}$ (Inesi *et al.*, 1967; Ikemoto, 1974).

It is of interest that, even though the hydrolytic activity is inhibited by high $Ca^{2+}$, the phosphorylated enzyme intermediate formed with ATP remains at approximately the same level in the presence of either micromolar or millimolar $Ca^{2+}$ concentrations, because in either case, the rate of phosphorylation is much higher than that of phosphoenzyme cleavage. These levels nearly match the maximal number of catalytic sites. On the other hand, when GTP or ITP (Table 1, Figure 11) are used as substrates, the phosphoenzyme level is rather low in the presence of micromolar $Ca^{2+}$ owing to a rather slow rate of phosphorylation, but increases to levels equal to those obtained with ATP in the presence of millimolar $Ca^{2+}$ (de Souza and de Meis, 1976; Verjovski-Almeida and de Meis, 1977; de Meis and Boyer, 1978; Scofano *et al.*, 1979). A similar effect is observed when the temperature is decreased from 30°C to 0°C (Masuda and de Meis, 1977). When leaky vesicles are incubated in a medium containing a low $Ca^{2+}$ concentration sufficient to saturate only the high-affinity binding site, it is found that, as the temperature is decreased from 25°C to 2°C, the rates of ATP and ITP hydrolysis are inhibited, the steady-state level of phosphoenzyme *obtained with ATP* does not vary, and the level of phosphoenzyme *obtained with ITP* increases reaching the same level as that obtained with ATP (Table 1). It can be safely concluded that both low temperatures and binding of $Ca^{2+}$ to the low-affinity sites inhibit hydrolysis of the phosphoenzyme without a corresponding inhibition of phosphoenzyme formation with ATP. A detailed study of GTP utilization by SR ATPase was published by Ronzani *et al.* (1979).

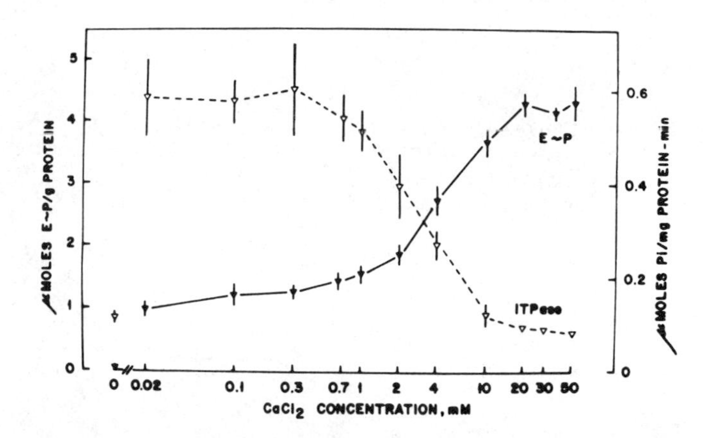


*Figure 11.* Effect of high $Ca^{2+}$ concentrations on the steady level of phosphoenzyme formed with ITP, and on the ITP flux (de Souza and de Meis, 1976).

Experimental determination of the apparent association constant for binding of calcium to the ATPase before and after phosphorylation, yields the evidence required to demonstrate that the mechanism of free energy transduction is operated through direct coupling between calcium-binding sites and phosphorylation sites. Calcium binding to the enzyme *before phosphorylation* can be measured in equilibrium conditions as explained above, and found to occur with an apparent $K_a = 10^6 \ M^{-1}$. This constant remains unchanged in the presence of ATP analogues, i.e., AMP–PNP, which do not yield significant levels of phosphorylation (Inesi *et al.*, 1980a). On the other hand, *following phosphorylation* of the catalytic site by ATP, calcium is internalized by the ATPase protein (a step leading to its transfer across the membrane) and then released from the binding sites (Ikemoto, 1975, 1976; Watanabe *et al.*, 1981). Therefore, it is agreed that the binding sites must undergo a reduction in their affinity for calcium as a consequence of phosphorylation. However, direct measurement of calcium binding to the phosphoenzyme in equilibrium conditions has not been possible, owing to the tendency of the phosphoenzyme to undergo hydrolysis.

A special and very interesting case is presented by vanadate. Following studies on the $Na^+$-$K^+$ ATPase (Cantley *et al.*, 1977, 1978; Smith *et al.*, 1980), it was found that vanadate inhibits the $Ca^{2+}$ ATPase by acting as an analogue of phosphate (Inesi *et al.*, 1980b; Dupont and Bennett, 1982; Pick, 1982). It is then found that vanadate inhibits high-affinity calcium binding. This effect is analogous to that of ATP; however, while the effect of ATP can be studied only in steady state owing to the tendency of ATP to undergo hydrolysis, vanadate is stable and its effect can be studied by equilibrium experimentation (Figure 12). It is likely that vanadate functions as a stable stereo-analogue of a phosphate transition state of the phosphorylation reaction owing to its bipyramidal structure. Stabilization of vanadate at the site may also be gained by acceptance of electrons from neighboring oxygens into its *d* orbitals. Thereby, vanadate, in analogy to ATP, is able to overcome the energy barrier presented by the calcium-bound enzyme ($E.Ca_2$), and to form a complex with the catalytic site. The resulting inhibition of high-affinity calcium binding demonstrates that the effective

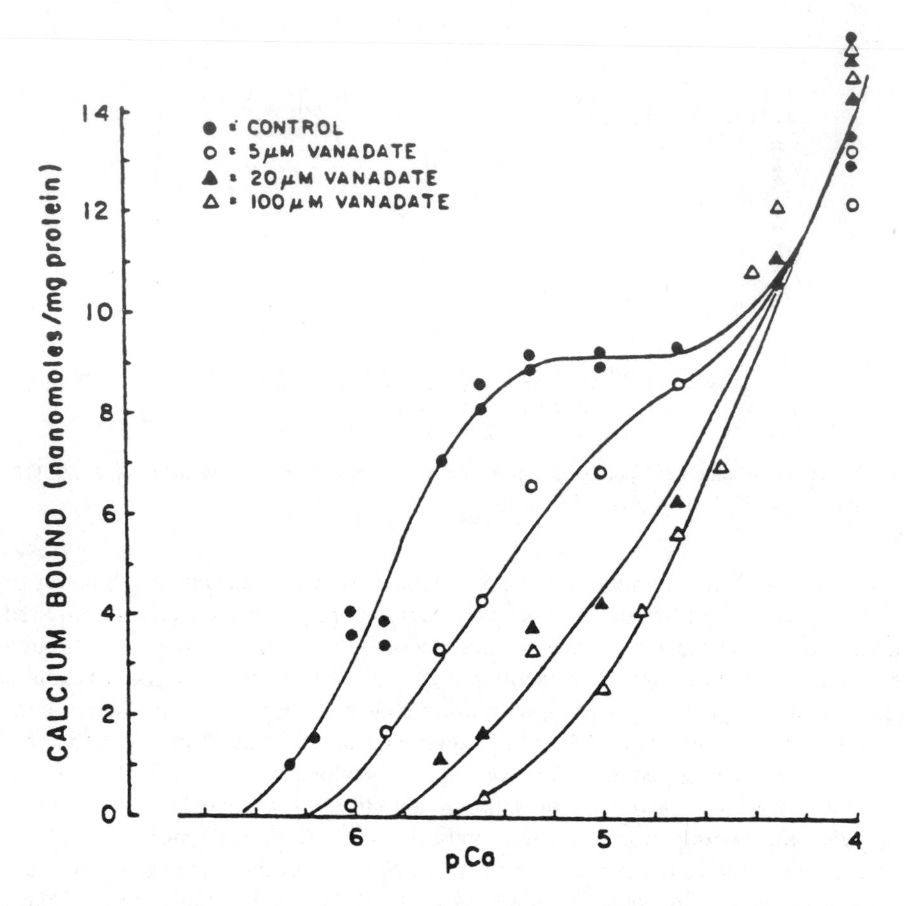


*Figure 12.* Vanadate inhibition of calcium binding to SR ATPase. The inhibition is specific for the high-affinity sites, and does not involve the lower affinity, nonspecific sites. Binding was measured at equilibrium in the absence of ATP (Inesi *et al.*, 1984).

mechanism for conversion of the enzyme from the high-calcium affinity state to the low-calcium affinity state, is interaction of a phosphate moiety (or its analogue) with the catalytic site. In the case of ATP, the phosphorylation reaction is a means for reacting the phosphate moiety with the catalytic site in the presence of calcium, whereby, free energy transduction is manifested by a decrease in affinity of the binding sites, and dissociation of $Ca^{2+}$ against a concentration gradient. The subsequent phosphohydrolase reaction is simply the means to allow the enzyme to recycle.

It will be explained below that, as opposed to vanadate, $P_i$ does not lower the calcium affinity of the enzyme *in the presence of* $Ca^{2+}$, but it does *in the absence of* $Ca^{2+}$. This is due to an unfavorable free energy barrier posed by the calcium–enzyme complex, as opposed to the free enzyme. This barrier is overcome by stabilization factors related to the structure of vanadate (see above) or by the phosphorylation potential of ATP.

## D. The $Ca^{2+}$–$H^+$ Exchange

An important feature of the calcium-binding mechanism is that, in the presence of $H^+$ concentrations permitting saturation of appropriate residues, association of one mole of calcium to specific high-affinity sites is accompanied by release of one mole of $H^+$ (Chiesi and Inesi, 1980). Furthermore, the calcium-binding dependence of $Ca^{2+}$ concentration is displaced by changes in $H^+$ concentration with a pattern consistent with $Ca^{2+}$ and $H^+$ competition, as well as cooperative interactions of binding domains (Hill and Inesi, 1982). The question is then whether the $Ca^{2+}$–$H^+$ competition is relevant to enzyme activation. Such a relevance can be demonstrated unambiguously by measuring the formation of phosphorylated enzyme intermediate following addition of ATP in the presence of fixed $Ca^{2+}$ and various $H^+$ concentrations (Inesi and Hill, 1983). In perfect agreement with the binding curves of Figure 6, it is found that the enzyme is phosphorylated only when calcium occupancy, rather than $H^+$ occupancy of activating sites is predicted (Figure 13).

Progressing further with the evolution of the catalytic and transport cycle, it is found that, in the presence of $Ca^{2+}$ concentrations sufficiently high to overcome $H^+$ competition, maximal phosphorylation of the enzyme and the related burst of calcium translocation occur following addition of ATP, even in the presence of high-$H^+$ concentrations (Figure 14). On the other hand, a reduced $P_i$ production is noted at pH 8.0 as compared with pH 6.0 (Figure 14). This is not due to an inhibitory effect of high pH on hydrolytic $P_i$ cleavage. In fact, in kinetic studies of phosphoenzyme formed by incubation of the enzyme with $P_i$ (Inesi and Hill, 1983), the hydrolytic rate constant is found to be approximately the same at pH 6.8 and 8.0 (Figure 14). Therefore, it is apparent that the inhibition of $P_i$ production noted at pH 8.0 is due to slow progression of the nonprotonated (as compared to protonated) enzyme species through a step following release of $Ca^{2+}$ inside the vesicles and preceding the hydrolytic reaction.

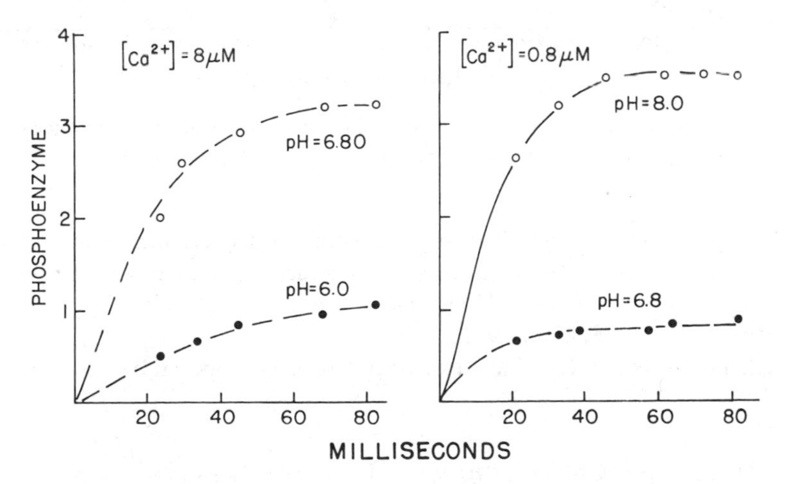


*Figure 13.* Enzyme phosphorylation following addition of ATP to SR vesicles, in the presence of various $Ca^{2+}$ and $H^+$ concentrations (Inesi and Hill, 1983).

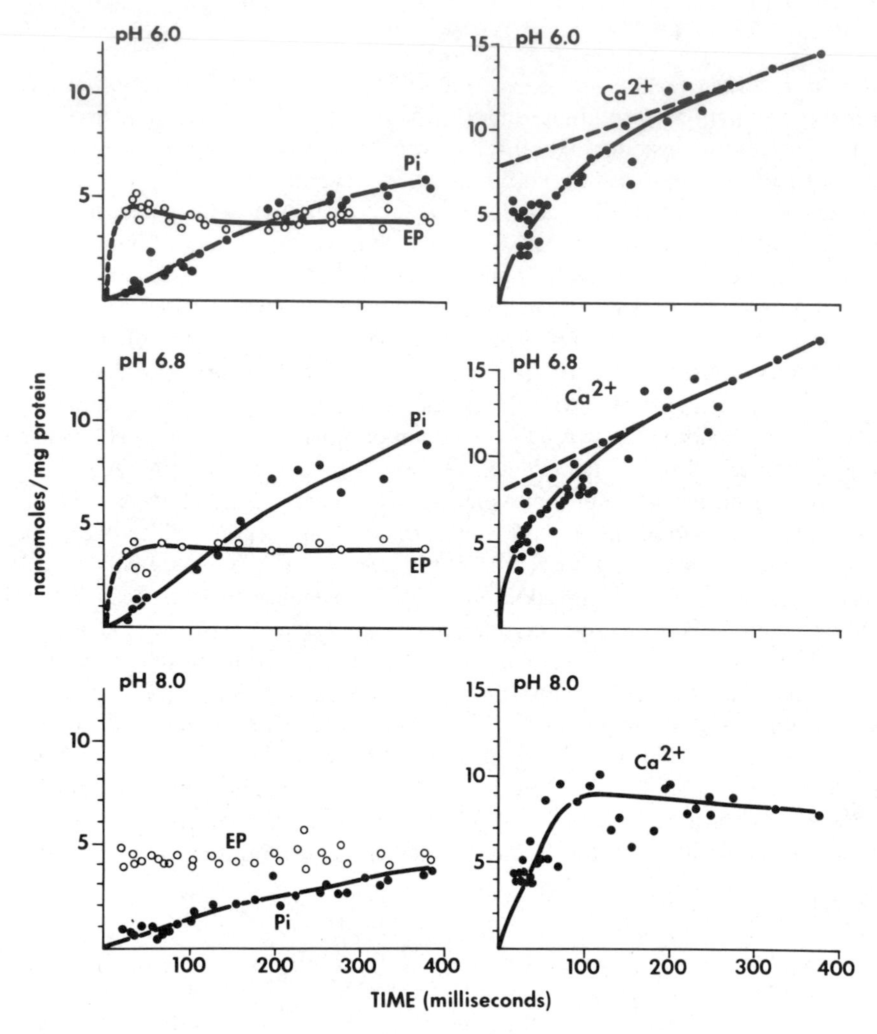


*Figure 14.* Enzyme phosphorylation, calcium translocation, and $P_i$ production following addition of ATP to SR vesicles, in the presence of *saturating* $Ca^{2+}$ and various $H^+$ concentrations (Inesi and Hill, 1983).

All these experimental observations suggest a transport cycle including exchange of two $Ca^{2+}$ with two $H^+$ and operated by two interacting binding domains within one ATPase unit (Carvalho, 1972; de Meis and Tume, 1977; Madeira, 1978; Chiesi and Inesi, 1980; Veno and Sekine, 1981).

In addition to $H^+$, a $K^+$–$Ca^{2+}$ exchange has been proposed by Chiu and Haynes (1980).

## E. Reversal of the $Ca^{2+}$ Pump and Coupled ATPase

When loaded (with $Ca^{2+}$), SR vesicles are placed in a medium containing EGTA (to lower the $Ca^{2+}$ concentration in the outside medium), $P_i$, and ADP; the $Ca^{2+}$

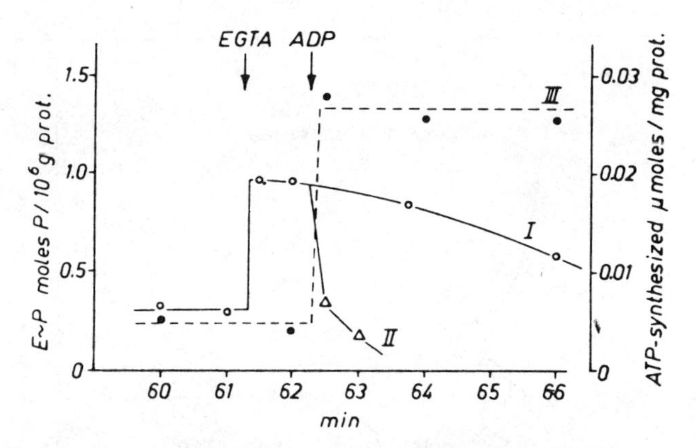


Figure 15. Formation of (I) phosphoenzyme from $P_i$ following addition of EGTA to calcium-loaded vesicles. Subsequent addition of ADP produces breakdown of the (II) phosphoenzyme and formation of (III) ATP (Makinose, 1972).

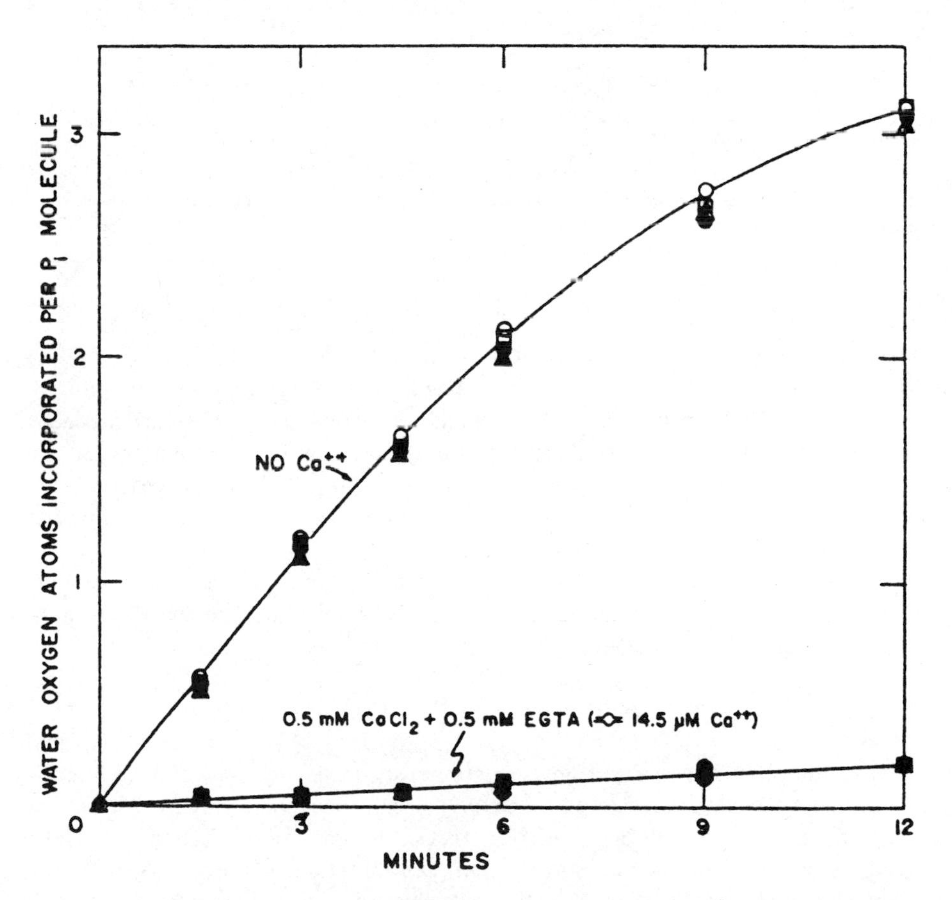


Figure 16. The $P_i$–HOH exchange sustained by SR vesicles in the absence of $Ca^{2+}$ (Kanazawa and Boyer, 1973).

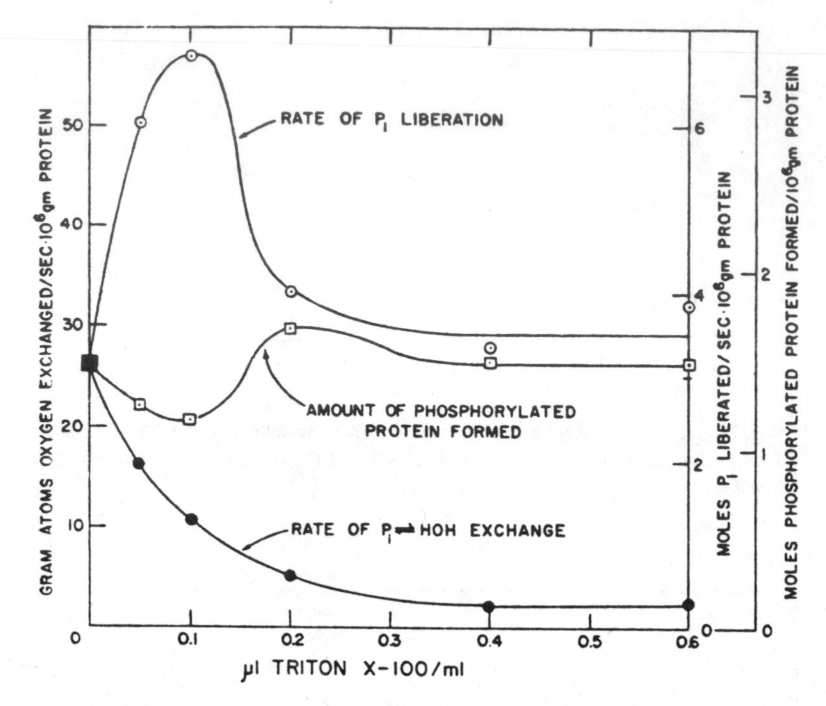


*Figure 17.* Effect of Triton X-100 on phosphoenzyme formation, $P_i$–HOH exchange, and ATP hydrolysis by SR vesicles. The measurements were obtained in reaction mixtures optimized for each reaction (Kanazawa and Boyer, 1973).

pump and coupled ATPase undergo repeated reverse cycles as long as the $Ca^{2+}$ concentration inside the vesicles remains sufficiently high (Barlogie *et al.*, 1971; Makinose, 1971, 1972). Each reverse cycle begins with enzyme phosphorylation by $P_i$ (Makinose, 1972; Yamada *et al.*, 1972; Yamada and Tonomura, 1973), and proceeds to $Ca^{2+}$ efflux from the lumen of the vesicles to the outside medium, and phosphoryl transfer from the phosphoenzyme to ADP (Barlogie *et al.*, 1971; Hasselbach, 1978; Makinose, 1972; de Meis, 1976). One of the original experiments demonstrating the reversal of the pump is shown in Figure 15.

If ADP is omitted from the assay medium, the fast efflux of $Ca^{2+}$ coupled with synthesis of ATP is impaired and an equilibrium between phosphoenzyme, dephosphoenzyme, and $P_i$ is established:

$$E + P_i \rightleftharpoons {}^*E - P + H_2O$$

At equilibrium, the reaction flows continuously in the forward and reverse directions leading to incorporation of water oxygen into phosphate (Figure 16) as first reported by Kanazawa and Boyer (1973). These authors found that this $P_i$–HOH exchange is inhibited by the addition of small concentrations of the detergent Triton X-100 which render the vesicles leaky without impairing the ATPase activity or phosphorylation of the enzyme by [$\gamma$-$^{32}$P]-ATP (Figure 18).

Another index of the reversal of the $Ca^{2+}$ pump is the ATP–$P_i$ exchange reaction. Makinose (1971) observed that when vesicles were incubated in the medium described in Figure 17, $Ca^{2+}$ was accumulated by the vesicles up to a steady level at which time the ATP driven $Ca^{2+}$ influx was balanced by $Ca^{2+}$ efflux through reversal of the pump. As soon as net $Ca^{2+}$ uptake ceased, a steady rate of exchange between $^{32}P_i$ and the ATP terminal phosphate began. As with the $P_i$–HOH exchange, the ATP–$P_i$ exchange was abolished when the vesicles were rendered leaky.

## F. *Enzyme Phosphorylation in the Absence of a Transmembrane $Ca^{2+}$ Gradient*

Phosphorylation of SR ATPase involves an aspartyl residue at the catalytic site (Bastide *et al.*, 1973; Degani and Boyer, 1973), and it is expected that formation of such an acylphosphate requires a sizable free energy input. This is provided by ATP in the forward direction of the enzyme cycle. Conversely, in enzyme phosphorylation with $P_i$, the free energy source was initially identified with the potential of the transmembrane $Ca^{2+}$ gradient (Makinose, 1971, 1972; Yamada *et al.*, 1972; Yamada and Tonomura, 1973). However, it became soon apparent that the $P_i$ reaction occurs even in the absence of a transmembrane gradient. This was demonstrated directly by the use of "leaky" vesicles (Masuda and de Meis, 1973; de Meis and Masuda, 1974). In fact, when the experimental conditions were optimized (pH 6, high $Mg^+$, no $K^+$ or $Na^+$), the observed phosphoenzyme levels were found to be stoichiometrically equivalent to those obtained with ATP. Experimentation was then extended to systems including not only "leaky vesicles," but also purified ATPase (Kanazawa, 1975; Knowles and Racker, 1975; de Meis, 1976; Beil *et al.*, 1977; Boyer *et al.*, 1977; Rauch *et al.*,

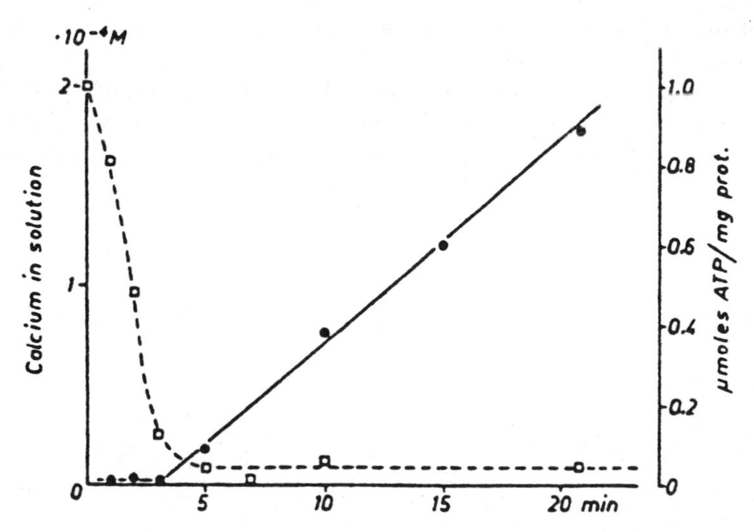


*Figure 18.* Calcium uptake (dashed line) and ATP formation (solid line) following addition of SR vesicles to a medium containing 20 mM $P_i$, 5 mM ATP, 2 mM ADP, 7 mM $MgCl_2$, and 0.2 mM $CaCl_2$ (Makinose, 1971).

1977; Hasselbach, 1978; Punzengruber *et al.*, 1978; Kolassa *et al.*, 1979; Prager *et al.*, 1979; Martin and Tanford, 1981; Lacapere *et al.*, 1981). A commonly accepted reaction scheme (Punzengruber *et al.*, 1978) involves formation of a ternary complex by independent binding of $Mg^{2+}$ and $P_i$ to the enzyme (Figure 19). In this scheme $K_1$, $K_2$, $K_3$, and $K_4$ are binding constants, and $K_5$ is the equilibrium constant for the phosphorylation reaction.

A requirement for enzyme phosphorylation with $P_i$, as well as for $P_i$–HOH exchange (Figure 15), is that $Ca^{2+}$ can be dissociated from the enzyme; therefore, the $Ca^{2+}$ concentration in the medium must be kept below $10^{-8}$ M to avoid calcium binding to the high-affinity sites. In fact, by varying the $Ca^{2+}$ concentration within the $10^{-7}$–$10^{-8}$ M range, it can be shown that part of the enzyme reacts with $P_i$ and part with ATP in proportion to the saturation of the high-affinity sites with $Ca^{2+}$ (Masuda and de Meis, 1973; de Meis, 1976; Beil *et al.*, 1977; Prager *et al.*, 1979). If $P_i$ and EGTA are added simultaneously to enzyme pre-incubated with $Ca^{2+}$, the phosphorylation reaction occurs following a lag period of approximately 100 msec, as compared to the immediate reaction observed when $P_i$ is added to enzyme pre-incubated with EGTA (Figure 20). This lag phase has been attributed to slow transformation of the enzyme following $Ca^{2+}$ dissociation from the high-affinity sites (Rauch *et al.*, 1977; Chaloub *et al.*, 1979; Guimaraes-Motta and de Meis, 1980), suggesting that the reactive species is *E, rather than E (see scheme in Figure 7). Alternatively, it could be related to reversal of the $Ca^{2+}$-induced enzyme transition (Inesi *et al.*, 1980a).

The reaction of SR ATPase with $P_i$ is clearly affected by pH (Masuda and de Meis, 1973; Beil *et al.*, 1977). This effect was attributed to the pH dependence of $P_i$ ionization (Beil *et al.*, 1977). However, pH effects are not observed in the presence of organic solvents such as dimethylsulfoxide, glycerol, and dimethylformamide. In this case, a marked reduction of the $P_i$ concentration required for the phosphorylation reaction is noted, independent of pH and of $P_i$ ionization. Therefore, the influence of pH, when observed, must also be attributed to ionization of protein residues which may be overcome by solvation effects (de Meis *et al.*, 1980).

In experimental conditions most commonly used in studies of SR ATPase (pH 6.8, 80 mM KCl, and 10 mM $MgCl_2$), the equilibrium constant for this reaction is approximately one, and the forward and reverse rate constants approximately 60 $sec^{-1}$ (Inesi *et al.*, 1983).

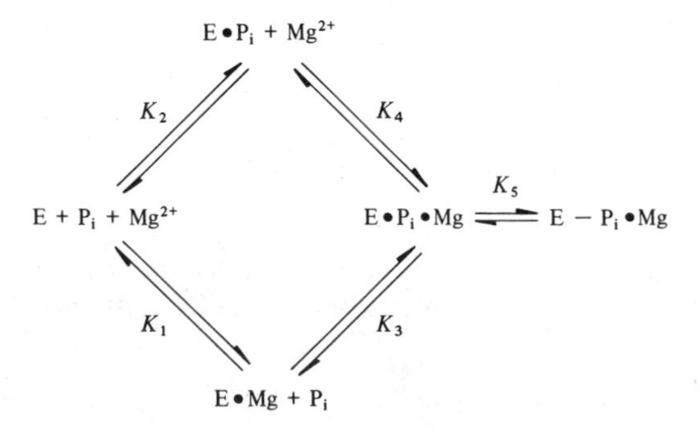


*Figure 19.* The scheme for the reaction of $P_i$ with SR ATPase in the presence of $Mg^{2+}$ and in the absence of $Ca^{2+}$ as originally proposed by Punzengruber *et al.* (1978).

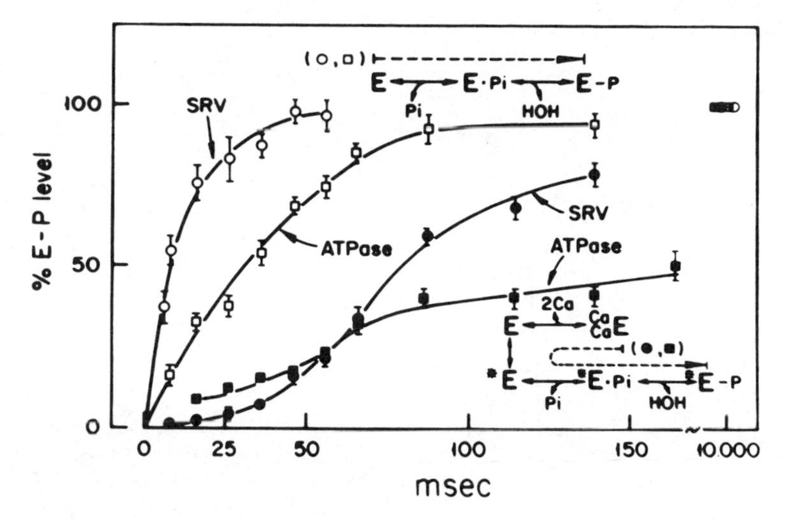


*Figure 20.* Phosphoenzyme formation following addition of $P_i$ to enzyme preincubated with EGTA (empty symbols), or $P_i$ and EGTA to enzyme preincubated with $Ca^{2+}$ (filled symbols). A lag period is observed in the latter case (Guimaraes-Motta and de Meis, 1980).

## G. ATP Synthesis

A basic difference between the phosphoenzyme formed by reacting $P_i$ with SR vesicles in the presence of a $Ca^{2+}$ transmembrane gradient (Makinose, 1972; Yamada *et al.*, 1972; Yamada and Tonomura, 1973), as compared to the absence of a $Ca^{2+}$ gradient (Masuda and de Meis, 1973), is that the former reacts with ADP to form ATP, while the latter does not (de Meis and Sorenson, 1975; Knowles and Racker, 1975; de Meis, 1976). This difference can be overcome simply by the addition of high (millimolar) concentrations of $Ca^{2+}$ to saturate the binding sites of the phosphoenzyme on the low-affinity state (Knowles and Racker, 1975; de Meis and Tume, 1977), an effect that could be predicted based on the $Ca^{2+}$ (millimolar) dependence of ATP–$P_i$ exchange in the absence of transmembrane gradient (Figure 22; de Meis and Carvalho, 1974).

Even though the phosphoenzyme obtained with $P_i$ in the absence of $Ca^{2+}$ is not reactive to ADP, such a phosphoenzyme is, in fact, an intermediate species in the ATPase catalytic cycle (*E-P in Figure 9). For instance, it can be demonstrated that while the enzyme does not react with $P_i$ in most reaction mixtures containing micromolar $Ca^{2+}$, this $Ca^{2+}$ inhibition can be bypassed if ITP or low ATP concentrations are added to sustain steady-state ATPase activity of leaky vesicles, e.g., unable to form a $Ca^{2+}$ gradient and to reduce the $Ca^{2+}$ concentration in the medium. In these conditions, it is found that a sizable portion of the phosphoenzyme steady-state level is formed by incorporation of $^{32}P_i$ (de Meis and Masuda, 1974; Carvalho *et al.*, 1976). This indicates that during progression of the ATPase cycle, an enzyme species which is reactive to $P_i$ and is able to sustain HOH–$P_i$ exchange (de Meis and Boyer, 1978) is present in significant concentrations. Its presence can be better detected in conditions in which the $Ca^{2+}$-induced transition is somewhat slow, as in the presence of acetylphosphate, ITP, or low concentrations of ATP (Figure 21).

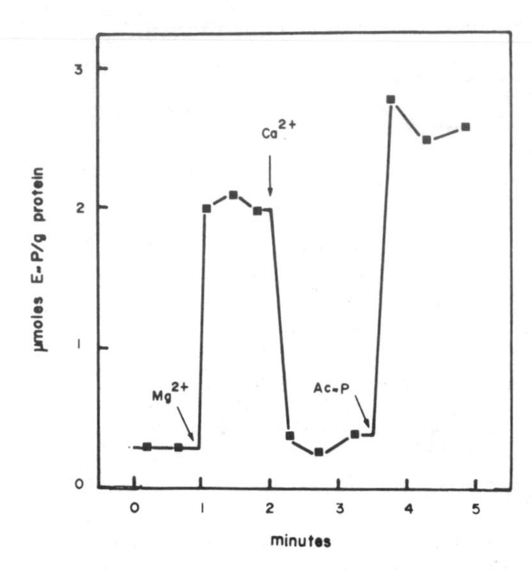


*Figure 21.* Phosphorylation of ATPase with $P_i$. The reaction is dependent on $Mg^{2+}$, but is inhibited by $Ca^{2+}$. In the presence of $Ca^{2+}$, the enzyme can be phosphorylated with acetylphosphate (de Meis and Masuda, 1974).

In similar experiments with leaky vesicles or purified ATPase (de Meis and Carvalho, 1974; Sorenson and de Meis, 1977; Carvalho *et al.*, 1976; Plank *et al.*, 1979), it is possible to obtain complete $P_i$–ATP exchange if the $Ca^{2+}$ concentration in the medium is raised to levels (millimolar) permitting occupancy of the calcium-binding sites in the low-affinity state and rise of the $^*E\text{-}P.Ca_2$ steady-state levels. The occurrence of $P_i$–ATP exchange in this conditions (Figure 22) demonstrates that reversal of the catalytic cycle and formation of ATP can be obtained even in the absence of a $Ca^{2+}$ gradient.

It is interesting that, in analogy to the calcium-binding sites in the high-affinity state (Watanabe *et al.*, 1981), the calcium binding in the low-affinity state, as well as the $Ca^{2+}$-concentration dependence of ATP formation are affected by pH shifts (de Meis and Tume, 1977; Verjovski-Almeida and de Meis, 1977). In fact, the $Ca^{2+}$-concentration in the reaction mixture can be adjusted to suitable levels to permit phosphorylation with $P_i$ at pH 5.0, and then obtain ATP formation simply by the addition of ADP and a pH jump to 8.0, without changing the $Ca^{2+}$ concentration (de Meis and Tume, 1977; Ratkje and Shamoo, 1980). In this case, the pH jump increases the affinity of the calcium sites to obtain binding at the $Ca^{2+}$ concentrations already present in the medium, thereby permitting ATP formation (Figure 23).

Changes in temperature can also affect the $Ca^{2+}$ dependence of the enzyme reaction with $P_i$, and the reaction of phosphoenzyme with ADP to form ATP. The former reaction is inhibited at low temperature, and no phosphoenzyme is formed with $P_i$ at 0°C. On the other hand, the $Ca^{2+}$ concentration required to render the phosphoenzyme reactive to ADP at pH 8.0 is three orders of magnitude lower at 0°C than at 30°C (de Meis *et al.*, 1980). It is likely that this effect is due to a shift of the equilibrium between different forms of the enzyme, permitting significant levels of phosphoenzyme in a high-affinity state, which then yields $ADP.E \sim P.Ca_2$ and $ATP + E.Ca_2$ in the presence of micromolar $Ca^{2+}$.

Phosphorylation of the enzyme with $P_i$ and formation of ATP can be affected by the presence of organic solvents in the reaction medium. In a comparative study of several organic solvents, de Meis *et al.* (1980) found that the $P_i$ concentration dependence of enzyme phosphorylation with $P_i$ is very much reduced in the presence of solvents such as dimethylsulfoxide, while subsequent phosphoryl transfer from the phosphoenzyme to ADP is inhibited. However, the ADP sensitivity is acquired again by the phosphoenzyme if the concentration of dimethylsulfoxide is reduced by dilution of the reaction mixture with aqueous media. Based on these experiments, de Meis *et al.* (1980) have suggested that solvation of substrates and appropriate residues at the active site may be involved in determining free energy changes in phosphorylation reactions within the ATPase cycle.

A detailed study of the effect of perturbations related to $Ca^{2+}$ and $H^+$ concentrations, temperature, and water activity was recently conducted by de Meis and Inesi

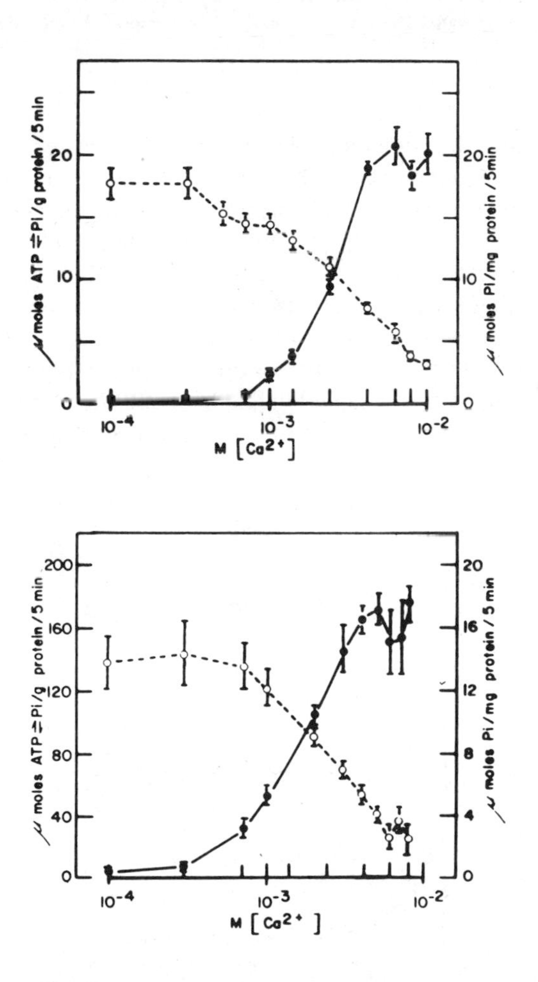


*Figure 22.* Inhibition of ATP flux ($P_i$ production) and increase of ATP–$P_i$ exchange by high- (mM) $Ca^{2+}$ concentrations. Solubilized SR vesicles (top) Leaky SR vesicles (bottom) (de Meis and Carvalho, 1974).

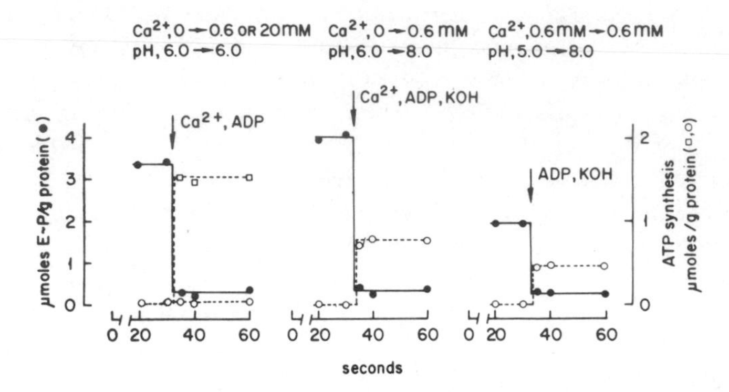


*Figure 23.* Changes in the phosphoenzyme level and formation of ATP, following changes in $Ca^{2+}$ and/or $H^+$ concentrations upon addition of ADP. The phosphoenzyme was obtained by incubating SR ATPase previous to the addition of ADP. The $Ca^{2+}$ and/or $H^+$ concentration changes are indicated on top of the figure (de Meis and Tume, 1977).

(1972). In this study, the commonly observed requirement for calcium occupancy of low-affinity sites was interpreted as a step permitting equilibrium between a phosphoenzyme with low phosphorylation potential and binding sites in a low-affinity state and a phosphoenzyme with high phosphorylation potential and binding sites in a high-affinity state. It was postulated that low- and high-affinity binding sites are an expression of two interconverting states of the phosphoenzyme. At alkaline pH and low temperature, proton dissociation from enzyme residues transforms the sites from the low- to the high-affinity state even in the absence of calcium. It is then apparent that both $Ca^{2+}$ and $H^+$ interact with the same binding site, and the effect of such interactions has a profound influence on both the forward and reverse directions of the enzyme cycle.

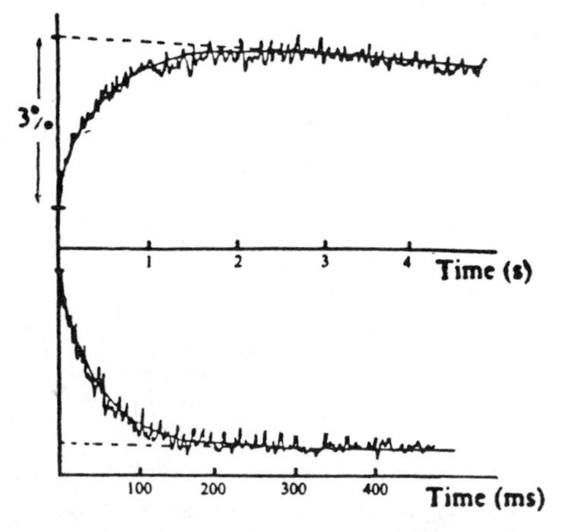


*Figure 24.* Stopped-flow records of intrinsic fluorescence changes associated with binding or release of $Ca^{2+}$, as originally published by DuPont and Leigh (1978).

## III. CONFORMATIONAL CHANGES

Several lines of evidence indicate that protein conformational changes are operative in at least two steps of the transport mechanism: (1) cooperative calcium binding and activation of the enzyme, and (2) internalization and reduction in affinity of the calcium-binding sites following enzyme phosphorylation with ATP.

The calcium-induced conformational change is demonstrated by measurements of intrinsic fluorescence (Dupont and Leigh, 1978; Ikemoto *et al.*, 1978; Pick and Karlish, 1980), ESR spectroscopy of spin-labeled enzyme (Coan and Inesi, 1977), and kinetics of –SH reactivity (Murphy, 1978). An example of the reversible change of intrinsic fluorescence induced by $Ca^{2+}$ on SR ATPase is shown in Figure 24.

Structural changes following enzyme phosphorylation with ATP are suggested by the kinetics of –SH titration (Murphy, 1978) and experiments with fluorescent labels (Ikemoto, 1982). In addition, it was shown that phosphorylation of membranous ATPase in the presence of low concentrations of detergent produces reversible dis-

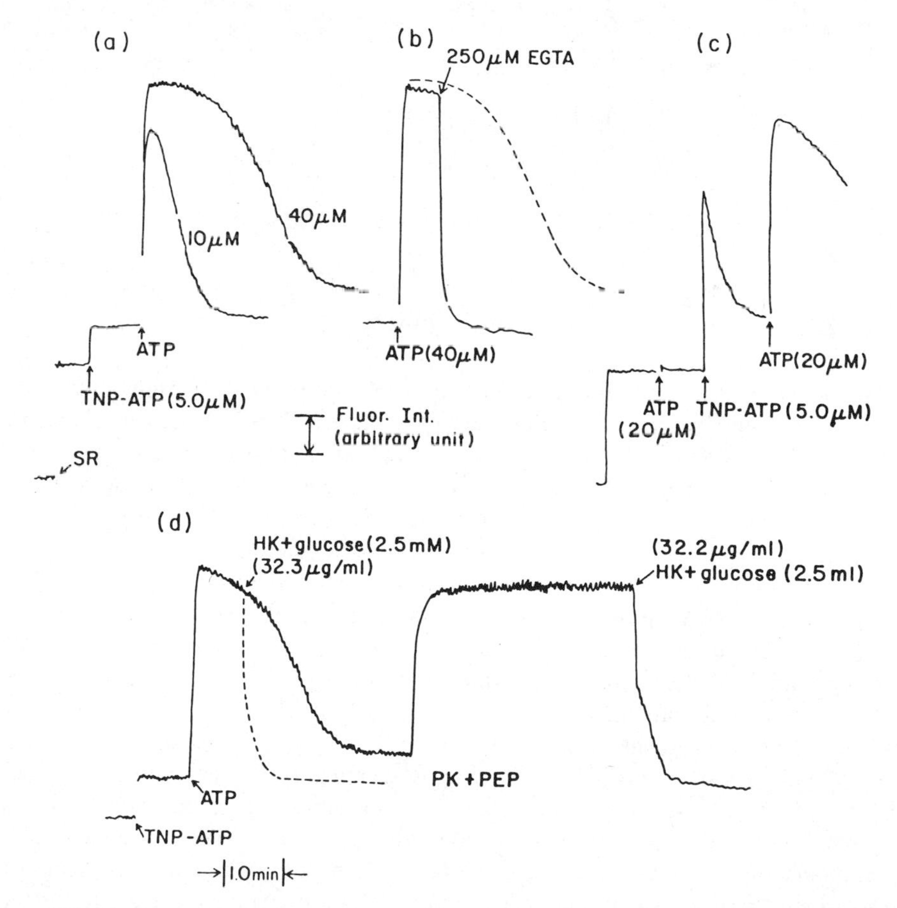


*Figure 25.* Fluorescence enhancement of bound TNP–ATP, following enzyme phosphorylation with ATP in the presence of $Ca^{2+}$ (Watanabe and Inesi, 1982b).

sociation of the ATPase units, as demonstrated by measurements of light scattering and fluorescence energy transfer (Watanabe and Inesi, 1982a). Finally, the fluorescence intensity of the substrate analogue 2′,3′-*O*-(2,4,6-trimitrophenyl) adenosine 5′-triphosphate was found to be very much enhanced (Figure 25) by enzyme phosphorylation (Watanabe and Inesi, 1982b).

## IV. CONCLUSIONS

Equilibrium and kinetic characterization of the partial reactions of the SR ATPase cycle has already yielded a number of equilibrium and rate constants. Approximate values obtained (pH 6.8, 80 mM KCl, 25°C) from references cited throughout this chapter may be listed as follows:

$$E + Ca_{out} \rightleftharpoons E.Ca : K_{app} \cong 10^6\ M^{-1}$$

$$E + ATP \rightleftharpoons E.ATP : K_{app} \cong 10^5\ M^{-1}$$

$$E.Ca.ATP \rightleftharpoons ADP.E \sim P.Ca : k_{for} \cong 100\ sec^{-1};\ k_{rev} \cong 300\ sec^{-1}$$

$$ADP.E \sim P.Ca \rightleftharpoons E \sim P.Ca + ADP : K_{app} = 10^{-3} - 10^{-4}\ M$$

$$E\text{-}P.Ca \rightleftharpoons E\text{-}P + Ca_{in}^{2+} : K_{app} = 10^{-3} - 10^{-2}\ M$$

$$E\text{-}P \rightleftharpoons E.P_i : K_{for} \cong 60\ sec^{-1};\ k_{rev} \cong 60\ sec^{-1}$$

$$E.P_i \rightleftharpoons E + P_i : K_{app} \cong 10^{-2}\ M$$

Overall turnover: 10–20 $sec^{-1}$

While the values listed above are derived from experimental measurements, several comments must be made regarding their role in the ATPase reaction mechanism. First, it is generally accepted that each reaction cycle involves two calcium ions which are bound with a cooperative mechanism and slow kinetics to interacting domains of the enzyme (Inesi *et al.*, 1980a; Dupont, 1982). The kinetics of ATP binding and dissociation also appear to be slow, as recently discussed by Pickart and Jencks (1982). A transition between an ADP sensitive (R $\sim$ $P.Ca_2$) and ADP insensitive ($E\text{-}P.Ca_2$) phosphoenzyme has been postulated by Shigekawa and Dougherty (1978a,b); however, the related constants are not known. Finally, a transition between a low-affinity (*E) and a high-affinity (E) state for the enzyme has been proposed (de Meis and Vianna, 1979), but its equilibrium constant has not been measured directly. In spite of these limitations, transient (Froehlich and Taylor, 1975; Inesi *et al.*, 1980a) and steady-state (Haynes, 1983; Inesi and Hill, 1983) kinetic measurements have been fitted satisfactorily by computer modeling, using both experimentally determined and assumed constants, and based on various modifications of the reaction scheme given above.

Explanation of bioenergetic transductions is generally formulated in terms of transmembrane potentials, covalent interactions, or conformational changes. It is clear that all these factors are pertinent to the calcium pump of SR vesicles. In the first place, the $Ca^{2+}$ concentration transmembrane potential can be expressed by equations defining its thermodynamic relationship with the chemical potential of ATP. In addition, experimental characterization of intermediate steps is necessary to clarify the mechanism of enzyme catalysis coupled to active transport. For instance, direct measurements of calcium binding reveal a change in orientation and a reduction in the affinity of the binding sites for calcium following enzyme phosphorylation. This demonstrates a primary and self-contained mechanism of transport within the ATPase protein, as well as the relationship between the covalent interaction of the enzyme with ATP and its ability to bind or dissociate $Ca^{2+}$. The observed reduction of the calcium association constant requires a free energy input consistent with that derived from utilization of ATP, and with that required to overcome the maximal $Ca^{2+}$ gradient generated by the pump. Furthermore, spectroscopic experimentation indicates that intermediate steps, such as the enzyme activation by calcium binding and the phosphorylation reaction, are accompanied by changes in the conformation of the ATPase protein. It is likely that these conformational changes are operative in the catalytic and transport mechanism.

The SR $Ca^{2+}$ pump has inspired several discussions regarding mechanisms of energy translocation in biological systems (Jencks, 1980; Hammes, 1982; Hill and Eisenberg, 1981; Tanford, 1982a,b). Preliminary diagrams of the standard free energy changes in the sequential reactions of the SR ATPase cycle have been published (Inesi *et al.*, 1980a), providing a basis to obtain free energy diagrams for diverse experimental conditions by simply correcting the second order equilibrium constants for the appropriate concentrations of reagents or products. Undoubtedly, the favorable yield of detailed experimental information on functional and structural parameters of this system will contribute to the development of important generalizations.

## REFERENCES


Allen, D., Blinks, J., and Prendergast, F., 1977, Aequorin luminescence: Relation of light emission to calcium concentration—A calcium-independent component, *Science* **195:**996–998.

Barlogie, B., Hasselbach, W., and Makinose, M., 1971, Activation of calcium efflux by ADP and inorganic phosphate, *FEBS Lett.* **12:**267–268.

Bastide, F., Meissner, G., Fleischer, S., and Post, R. L., 1973, Similarity of the active site of phosphorylation of the ATPase for transport of sodium and potassium ions in kidney to that for transport of calcium ions in sarcoplasmic reticulum of muscle, *J. Biol. Chem.* **248:**8385–8391.

Beil, F., Chak, D., and Hasselbach, W., 1977, Phosphorylation from inorganic phosphate and ATP synthesis of sarcoplasmic membranes, *Eur. J. Biochem.* **81:**151–164.

Boyer, P., de Meis, L., and Carvalho, M., 1977, Dynamic reversal of enzyme carboxyl group phosphorylation as the basis of the oxygen exchange catalyzed by sarcoplasmic reticulum adenosine triphosphatase, *Biochemistry* **16:**136–140.

Cantley, L., Josephson, L., Warner, R., Yanagisawa, M., Lechene, C., and Guidotti, G., 1977, Vanadate is a potent (Na,K)-ATPase inhibitor found in ATP derived from muscle, *J. Biol. Chem.* **252:**7421–7423.

Cantley, L. C., Cantley, L. G., and Josephson, L., 1978, A characterization of vanadate interactions with the (Na,K)-ATPase. Mechanistic and regulatory implications, *J. Biol. Chem.* **253:**7361–7368.

Carvalho, A., 1966, Binding of cations by microsomes from rabbit skeletal muscle, *J. Cell. Physiol.* **67:**73–84.

Carvalho, A., 1972, Binding and release of cations by sarcoplasmic reticulum before and after removal of lipid, *Eur. J. Biochem.* **27:**491–502.

Carvalho, M., de Souza, D., and de Meis, L., 1976, On a possible mechanism of energy conservation in sarcoplasmic reticulum membrane, *J. Biol. Chem.* **251:**3629–3636.

Chaloub, R., Guimaraes-Motta, H., Verjovski-Almeida, S., de Meis, L., and Inesi, G., 1979, Sequential reactions in $P_i$ utilization for ATP synthesis by sarcoplasmic reticulum, *J. Biol. Chem.* **254:**9464–9468.

Champeil, P., Bastide, F., Taupin, C., and Gary-Bobo, C. M., 1976, Spin labelled sarcoplasmic reticulum vesicles: $Ca^{2+}$-induced spectral change, *FEBS Letters* **63:**270–272.

Chevallier, J., and Buton, R., 1971, Calcium binding to the sarcoplasmic reticulum of rabbit skeletal muscle, *Biochemistry* **10:**2733–2737.

Chiesi, M., and Inesi, G., 1979, The use of quench reagents for resolution of single transport cycles in sarcoplasmic reticulum, *J. Biol. Chem.* **254:**10370–10377.

Chiesi, M., and Inesi, G., 1980, Adenosine triphosphate dependent fluxes of manganese and hydrogen ions in sarcoplasmic reticulum vesicles, *Biochemistry* **19:**2912–2918.

Chiu, V., and Haynes, D., 1977, High and low affinity $Ca^{++}$ binding to sarcoplasmic reticulum—use of a high-affinity fluorescent calcium indicator, *Biophys. J.* **18:**3–22.

Chiu, V., and Haynes, D., 1980, Rapid kinetic studies of active $Ca^{++}$ transport in sarcoplasmic reticulum, *J. Membr. Biol.* **56:**219–239.

Coan, C., and Inesi, G., 1977, Calcium dependent effect of ATP on spin-labeled sarcoplasmic reticulum, *J. Biol. Chem.* **252:**3044–3049.

Degani, C., and Boyer, P., 1973, A borohydride reduction method for characterization of the acyl phosphate linkage in proteins and its application to sarcoplasmic reticulum adenosine triphosphatase, *J. Biol. Chem.* **248:**8222–8226.

de Meis, L., 1969, $Ca^{2+}$ uptake and acetyl phosphatase of skeletal muscle microsomes, *J. Biol. Chem.* **244:**3733–3739.

de Meis, L., 1971, Allosteric inhibition by alkali ions of the $Ca^{2+}$ uptake and adenosine triphosphatase activity of skeletal muscle microsomes, *J. Biol. Chem.* **246:**4764–4773.

de Meis, L., 1976, Regulation of steady state level of phosphoenzyme ATP synthesis in sarcoplasmic reticulum vesicles during reversal of the $Ca^{2+}$ pump, *J. Biol. Chem.* **251:**2055–2062.

de Meis, L., and Hasselbach, W., 1971, Acetylphosphate as substrate for $Ca^{2+}$ uptake in skeletal muscle microsomes, *J. Biol. Chem.* **246:**4759–4763.

de Meis, L., and Boyer, P., 1978, Induction by nucleotide triphosphate hydrolysis of a form of sarcoplasmic reticulum ATPase capable of medium phosphate–oxygen exchange in presence of calcium, *J. Biol. Chem.* **253:**1556–1559.

de Meis, L., and Carvalho, M., 1974, Role of the $Ca^{++}$ concentration gradient in the adenosine 5′-triphosphate-inorganic phosphate exchange catalyzed by sarcoplasmic reticulum, *Biochemistry* **13:**5032–5038.

de Meis, L., and de Mello, M. C. F., 1973, Substrate regulation of membrane phosphorylation and calcium transport in the sarcoplasmic reticulum, *J. Biol. Chem.* **248:**3691–3701.

de Meis, L., and Inesi, G., 1982, ATP synthesis by sarcoplasmic reticulum ATPase following $Ca^{++}$, pH, temperature, and water activity jumps, *J. Biol. Chem.* **257:**1289–1294.

de Meis, L., and Masuda, H., 1974, Phosphorylation of the sarcoplasmic reticulum membrane by orthophosphate through two different reactions, *Biochemistry* **13:**2057–2062.

de Meis, L., and Sorenson, M., 1975, ATP–$P_i$ exchange and membrane phosphorylation in sarcoplasmic reticulum vesicles: Activation by silver in the absence of a $Ca^{++}$ concentration gradient, *Biochemistry* **14:**2739–2744.

de Meis, L., and Tume, R., 1977, A new mechanism by which an $H^+$ concentration gradient drives the synthesis of adenosine triphoshate, pH jump, and adenosine triphosphate synthesis by the $Ca^{++}$-dependent adenosine triphosphatase of sarcoplasmic reticulum, *Biochemistry* **16:**4455–4463.

de Meis, L., and Vianna, A., 1979, Energy interconversion by the $Ca^{++}$-dependent ATPase of the sarcoplasmic reticulum, *Annu. Rev. Biochem.* **48:**275–292.

de Meis, L., Martins, O., and Alves, E., 1980, Role of water, $H^+$ and temperature on the synthesis of ATP by the sarcoplasmic reticulum ATPase in the absence of a $Ca^{++}$ gradient, *Biochemistry* **19:**4252–4261.

de Souza, D., and de Meis, L., 1976, Calcium and magnesium regulation of phosphorylation by ATP and ITP in sarcoplasmic reticulum vesicles, *J. Biol. Chem.* **251:**6355–6359.

Dupont, Y., 1977, Kinetics and regulation of sarcoplasmic reticulum ATPase, *Eur. J. Biochem.* **72:**185–190.

Dupont, Y., 1980, Occlusion of divalent cations in the phosphorylated calcium pump of sarcoplasmic reticulum, *Eur. J. Biochem.* **109:**231–238.

Dupont, Y., 1982, Low-temperature studies of the sarcoplasmic reticulum calcium pump mechanism of calcium binding, *Biochim. Biophys. Acta* **688:**75–87.

Dupont, Y., and Bennett, N., 1982, Vanadate inhibition of the $Ca^{++}$-dependent conformational change of the sarcoplasmic reticulum $Ca^{++}$-ATPase, *FEBS Lett.* **139:**237–240.

Dupont, Y., and Leigh, J., 1978, Transient kinetics of sarcoplasmic reticulum Ca + Mg ATPase studied by fluorescence, *Nature* **273:**396–398.

Ebashi, S., 1961, Calcium binding activity of vesicular relaxing factor, *J. Biochem. (Tokyo)* **50:**236–244.

Ebashi, S., and Lipmann, F., 1962, Adenosine triphosphate-linked concentration of calcium ions in a particulate fraction of rabbit muscle, *J. Cell Biol.* **14:**389–400.

Fabiato, A., and Fabiato, F., 1979, Calculator programs for computing the composition of the solutions containing multiple metals and ligands used for experiments in skinned muscle cells, *J. Physiol. (Paris)* **75:**463–505.

Fiehn, W., and Migala, A., 1971, Calcium binding to sarcoplasmic membranes, *Eur. J. Biochem.* **20:**245–248.

Friedman, Z., and Makinose, M., 1970, Phosphorylation of skeletal muscle microsomes by acetylphosphate, *FEBS Letters* **11:**69–72.

Froehlich, J., and Taylor, E., 1976, Transient state kinetic effects of calcium ion on sarcoplasmic reticulum adenosine triphosphatase, *J. Biol. Chem.* **251:**2307–2315.

Gattass, C., and de Meis, L., 1975, $Ca^{++}$-dependent inhibitory effects of $Na^{+}$ and $K^{+}$ on $Ca^{++}$-transport in sarcoplasmic reticulum vesicles, *Biochim. Biophys. Acta* **389:**506–515.

Guimaraes-Motta, H., and de Meis, L., 1980, Pathway for ATP synthesis by sarcoplasmic reticulum ATPase, *Arch. Biochem. Biophys.* **203:**395–403.

Hammes, G., 1982, Unifying concept for the coupling between ion pumping and ATP hydrolysis or synthesis, *Proc. Natl. Acad. Sci. USA* **79:**6881–6884.

Hasselbach, W., 1964, Relating factor and the relaxation of muscle, *Progress in Biophysics and Mol. Biol.* **14:**167–222.

Hasselbach, W., 1978, The reversibility of the sarcoplasmic calcium pump, *Biochim. Biophys. Acta* **515:**23–53.

Hasselbach, W., and Makinose, M., 1962, ATP and active transport, *Biochem. Biophys. Res. Commun.* **7:**132–136.

Haynes, D., 1983, Computer modeling of the $Ca^{++}$-$Mg^{++}$-ATPase pump of skeletal sarcoplasmic retic ulum, *Biophys. J.* **41:**233.

Hill, T., and Eisenberg, E., 1981, Can free energy transduction be localized at some crucial part of the enzymatic cycle?, *Quart. Rev. Biophys.* **14:**1–49.

Hill, T., and Inesi, G., 1982, Equilibrium cooperative binding of calcium and protons by sarcoplasmic reticulum ATPase, *Proc. Natl. Acad. Sci. USA* **79:**3978–3982.

Holloway, J., and Reilley, C., 1960, Metal chelate stability constants of aminopolycarboxylate ligands, *Anal. Chem.* **32:**249–256.

Ikemoto, N., 1974, The calcium binding sites involved in the regulation of the purified adenosine triphosphatase of the sarcoplasmic reticulum, *J. Biol. Chem.* **249:**649–651.

Ikemoto, N., 1975, Transport and inhibitory calcium binding sites on the ATPase enzyme isolated from the sarcoplasmic reticulum, *J. Biol. Chem.* **250:**7219–7224.

Ikemoto, N., 1976, Behavior of the $Ca^{++}$ transport sites linked with the phosphorylation reaction of ATPase purified from the sarcoplasmic reticulum, *J. Biol. Chem.* **251:**7275–7277.

Ikemoto, N., 1982, Structure and function of the calcium pump protein of sarcoplasmic reticulum, *Annu. Rev. Physiol.* **44:**297–317.

Ikemoto, N., Morgan, T., and Yamada, S., 1978, $Ca^{++}$ controlled conformational states in the $Ca^{++}$ transport enzyme of sarcoplasmic reticulum, *J. Biol. Chem.* **253:**8027–8033.

Inesi, G., 1981, The sarcoplasmic reticulum of skeletal and cardiac muscle, in: *Cell and Muscle Motility*, Vol. 1 (R. Dowben and J. Shay, eds.), Plenum Publishing Corporation, New York, pp. 63–97.

Inesi, G., and Almendares, J., 1968, Interaction of fragmented sarcoplasmic reticulum with $^{14}C$-ADP, $^{14}C$-ATP, and $^{32}P$-ATP. Effect of Ca and Mg, *Arch. Biochem. Biophys.* **126:**733–735.

Inesi, G., and Scarpa, A., 1972, Fast kinetics of adenosine triphosphate $Ca^{2+}$ uptake by frequented sarcoplasmic reticulum, *Biochemistry* **11:**356–359.

Inesi, G., and Hill, T., 1983, The calcium and proton dependence of sarcoplasmic reticulum ATPase, *Biophys. J.*, in press.

Inesi, G., Goodman, J. J., and Watanabe, S., 1967, Effect of diethyl ether on the ATPase activity and calcium uptake of fragmented sarcoplasmic reticulum of rabbit skeletal muscle, *J. Biol. Chem.* **242:**4637–4643.

Inesi, G., Lewis, D., and Murphy, A., 1984, Interdependence of $H^+$, $Ca^{2+}$ and $P_i$ (or vanodate) sites in sarcoplasmic reticulum ATPase, *J. Biol. Chem.* (in press).

Inesi, G., Maring, E., Murphy, A., and McFarland, B., 1970, A study of the phosphorylated intermediate in sarcoplasmic reticulum ATPase, *Arch. Biochem. Biophys.* **138:**285–294.

Inesi, G., Kurzmack, M., Coan, C., and Lewis, D., 1980a, Cooperative calcium binding and ATPase activation in sarcoplasmic reticulum vesicles, *J. Biol. Chem.* **255:**3025–3031.

Inesi, G., Kurzmack, M., Nakamoto, R., de Meis, L., and Bernhard, S., 1980b, Uncoupling of calcium control and phosphohydrolase activity in sarcoplasmic reticulum vesicles, *J. Biol. Chem.* **255:**6040–6043.

Inesi, G., Watanabe, T., Coan, C., and Murphy, A., 1983, The mechanism of sarcoplasmic reticulum ATPase, *Ann. N.Y. Acad. Sci.* **402:**515–534.

Inesi, G., Lewis, D., and Murphy, A. J., 1984, Interdependence of $H^+$, $Ca^{2+}$, and $P_i$ (or vanadate) sites in sarcoplasmic reticulum ATPase, *J. Biol. Chem.* **259:**996–1003.

Jencks, W., 1980, The utilization of binding energy in coupled vectorial processes, *Adv. Enzymol.* **51:**75–106.

Kalbitzer, H., Stehlik, D., and Hasselbach, W., 1978, Binding of calcium and magnesium to sarcoplasmic-reticulum vesicles as studied by manganese electron-paramagnetic resonance, *Eur. J. Biochem.* **82:**245–255.

Kanazawa, T., 1975, Phosphorylation of solubilized sarcoplasmic reticulum by orthophosphate and its thermodynamic characteristics—the dominant role of entropy in the phosphorylation, *J. Biol. Chem.* **250:**113–119.

Kanazawa, T., and Boyer, P., 1973, Occurrence and characteristics of a rapid exchange of phosphate–oxygen catalyzed by sarcoplasmic reticulum vesicles, *J. Biol. Chem.* **248:**3163–3172.

Kanazawa, T., Yamada, S., Yamamoto, T., and Tonomura, Y., 1971, Reaction mechanism of the $Ca^{++}$-dependent ATPase of sarcoplasmic reticulum from skeletal muscle: V. Vectorial requirements for calcium and magnesium ions of three partial reations of ATPase: Formation and decomposition of a phosphorylated intermediate ATP-formation from ADP and the intermediate, *J. Biochem. (Tokyo)* **70:**95–123.

Knowles, A., and Racker, E., 1975, Formation of adenosine triphosphate from $P_i$ and adenosine diphosphate by purified $Ca^{2+}$-adenosine triphosphatase, *J. Biol. Chem.* **250:**1949–1951.

Kolassa, N., Punzengruber, C., Suko, J., and Makinose, M., 1979, Mechanism of calcium-independent phosphorylation of sarcoplasmic reticulum ATPase by orthophosphate, *FEBS Lett.* **108:**495–500.

Kurzmack, M., and Inesi, G., 1977, The initial phase of calcium uptake and ATPase activity of sarcoplasmic reticulum vesicles, *FEBS Lett.* **74:**35–37.

Kurzmack, M., Verjovski-Almeida, S., and Inesi, G., 1977, Detection of an initial burst of calcium translocation in sarcoplasmic reticulum, *Biochem. Biophys. Res. Commun.* **78:**772–776.

Kurzmack, M., Inesi, G., Tal, N., and Bernhard, S., 1981, Transient-state kinetic studies on the mechanism of furylacryloylphosphatase-coupled calcium ion transport with sarcoplasmic reticulum adenosine triphosphatase, *Biochemistry* **20:**486–491.

Lacapere, J., Gingold, M., Champeil, P., and Guillain, F., 1981, Sarcoplasmic reticulum ATPase phosphorylation from inorganic phosphate in the absence of a calcium gradient: Steady state and fluorescence studies, *J. Biol. Chem.* **256:**2302–2306.

Madeira, V., 1978, Proton gradient formation during transport of $Ca^{++}$ by sarcoplasmic reticulum, *Arch. Biochem. Biophys.* **185:**316–325.

Makinose, M., 1966, Die Nucleosid-triphosphat, nucleosid-diphosphat-transphosphorylase-aktivitat der Vesikel des sarkoplasmatischen reticulums, *Biochem. Zeitschrift* **345:**80–86.

Makinose, M., 1967, Gibt es zwei phosphorylierte intermediate des aktiven calcium-transportes in den membranen des sarcoplasmatischen reticulums?, *Pfluger's Arch. Ges. Physiol.* **294:**82–83.

Makinose, M., 1969, The phosphorylation of the membrane protein of the sarcoplasmic vesicles during active calcium transport, *Eur. J. Biochem.* **10:**74–82.

Makinose, M., 1971, Calcium efflux dependent formation of ATP from ADP and orthophosphate by the membranes of the sarcoplasmic vesicles, *FEBS Lett.* **12:**269–270.

Makinose, M., 1972, Phosphoprotein formation during osmo-chemical energy conversion in the membrane of the sarcoplasmic reticulum, *FEBS Lett.* **25:**113–115.

Makinose, M., and Hasselbach, W., 1965, Der einfluss von oxalat auf den calcium-transport isolierter vesikel des sarkoplasmatischen reticulum, *Biochem. Zeitschrift* **343:**360–382.

Makinose, M., and Hasselbach, W., 1971, ATP synthesis by the reverse of the sarcoplasmic calcium pump, *FEBS Lett.* **12:**271–272.

Makinose, M., and The, R., 1965, Calcium-akkumulation und nucleosidtriphosphat-spaltung durch die vesikel des sarkoplasmatischen reticulum, *Biochem. Zeitschrift* **343:**383–393.

Martin, D., and Tanford, C., 1981, Phosphorylation of ($Ca^{++}$)-ATPase by inorganic phosphate: van't Hoff analysis of enthalpy changes, *Biochemistry* **20:**4597–4603.

Martonosi, A., 1969, Sarcoplasmic reticulum. VII. Properties of a phosphoprotein intermediate implicated in calcium transport, *J. Biol. Chem.* **244:**613–620.

Martonosi, A., 1971, The structure and function of sarcoplasmic reticulum membranes, in: *Biomembranes,* Vol. 1 (L. Manson, ed.), Plenum Press, New York, pp. 191–256.

Martonosi, A., and Feretos, R., 1964, Sarcoplasmic reticulum: I. The uptake of $Ca^{++}$ by sarcoplasmic reticulum fragments, *J. Biol. Chem.* **239:**648–658.

Martonosi, A., Lagwinska, E., and Oliver, M., 1974, Elementary processes in the hydrolysis of ATP by sarcoplasmic reticulum membranes, *Ann. N.Y. Acad. Sci.* **227:**549–567.

Masuda, H., and de Meis, L., 1973, Phosphorylation of the sarcoplasmic reticulum membrane by orthophosphate. Inhibition by calcium ions, *Biochemistry* **12:**4581–4585.

Masuda, H., and de Meis, L., 1977, Effect of temperature on the $Ca^{++}$ transport ATPase of sarcoplasmic reticulum, *J. Biol. Chem.* **252:**8567–8571.

Meissner, G., 1973, ATP and $Ca^{++}$ binding by the $Ca^{++}$ pump protein of sarcoplasmic reticulum, *Biochim. Biophys. Acta* **298:**906–926.

Meissner, G., Conner, G., and Fleischer, S., 1973, Isolation of sarcoplasmic reticulum by zonal centrifugation and purification of $Ca^{2+}$-pump and $Ca^{2+}$-binding proteins, *Biochim. Biophys. Acta* **298:**246–269.

Murphy, A., 1978, Effects of divalent cations and nucleotides on the reactivity of the sulfhydryl groups of sarcoplasmic reticulum membranes, *J. Biol. Chem.* **253:**385–389.

Nakamura, Y., and Tonomura, Y., 1978, Reaction mechanism of *p*-nitrophenylphosphatase of sarcoplasmic reticulum, *J. Biochem. (Tokyo)* **83:**571–583.

Neet, K., and Green, N., 1977, Kinetics of the cooperativity of the $Ca^{++}$-transporting adenosine triphosphatase of sarcoplasmic reticulum and the mechanism of the ATP interaction, *Arch. Biochem. Biophys.* **178:**588–597.

Ogawa, Y., 1968, The apparent binding constant of glycoletherdiaminetetraacetic acid for calcium at neutral pH, *J. Biochem. (Tokyo)* **64:**255–257.

Pang, D., and Briggs, F., 1977, Effect of calcium and magnesium on binding of beta, gamma-methylene ATP to sarcoplasmic reticulum, *J. Biol. Chem.* **252:**3262–3266.

Pick, U., 1982, The interaction of vanadate ions with the Ca-ATPase from sarcoplasmic reticulum, *J. Biol. Chem.* **257:**6111–6119.

Pick, U., and Karlish, S., 1980, Indications for an oligomeric structure and for conformational changes in sarcoplasmic reticulum $Ca^{++}$-ATPase labeled selectively with fluorescein, *Biochim. Biophys. Acta* **626:**255–261.

Pickart, C., and Jencks, W., 1982, Slow dissociation of ATP from the calcium ATPase, *J. Biol. Chem.* **257:**5319–5322.

Plank, B., Hellman, G., Punzengruber, C., and Suko, J., 1979, ATP–$P_i$ and ITP–$P_i$ exchange by cardiac sarcoplasmic reticulum, *Biochim. Biophys. Acta* **550:**259–268.

Prager, R., Punzengruber, C., Kolassa, N., Winkler, F., and Suko, J., 1979, Ionized and bound calcium inside isolated sarcoplasmic reticulum of skeletal muscle and its significance in phosphorylation of adenosine triphosphatase by orthophosphate, *Eur. J. Biochem.* **97:**239–250.

Pucell, A., and Martonosi, A., 1971, Sarcoplasmic reticulum: XIV. Acetylphosphate and carbamylphosphate as energy sources for $Ca^{++}$ transport, *J. Biol. Chem.* **246:**3389–3397.

Punzengruber, C., Prager, R., Kolassa, N., Winkler, F., and Suko, J., 1978, Calcium gradient-dependent and calcium gradient-independent phosphorylation of sarcoplasmic reticulum by orthophosphate, *Eur. J. Biochem.* **92:**349–359.

Ratkje, S., and Shamoo, A., 1980, ATP synthesis by $Ca^{++}$ + $Mg^{++}$-ATPase in detergent solution at constant $Ca^{++}$ levels, *Biophys. J.* **30:**523–530.

Rauch, B., Chak, D., and Hasselbach, W., 1977, Phosphorylation by inorganic phosphate of sarcoplasmic membranes, *Z. Naturforsch. Part C* **32:**828–834.

Ribeiro, J., and Vianna, L., 1978, Allosteric modification by $K^+$ of the ($Ca^{++}$ + $Mg^{++}$)-dependent ATPase of sarcoplasmic reticulum, *J. Biol. Chem.* **253:**3153–3157.

Ronzani, N., Migala, A., Hasselbach, W., 1979, Comparison between ATP-supported and GTP-supported phosphate turnover of the calcium-transporting sarcoplasmic reticulum membranes, *Eur. J. Biochem.* **101:**593–606.

Rossi, B., Leone, F., Gache, C., and Lazdunski, M., 1979, Psuedosubstrates of the sarcoplasmic $Ca^{++}$-ATPase as tools to study the coupling between substrate hydrolysis and $Ca^{++}$ transport, *J. Biol. Chem.* **254:**2302–2307.

Scarpa, A., Baldassare, J., and Inesi, G., 1972, The effect of calcium ionophores on fragmented sarcoplasmic reticulum, *J. Gen. Physiol.* **60:**735–749.

Schmid, R., and Reilley, C., 1957, New complexon for titration of calcium in the presence of magnesium, *Anal. Chem.* **29:**264–268.

Schwartzenbach, G., Senn, H., and Anderegg, G., 1957, Komplexone. XXIX. Ein grosse Chelateffekt besonderer Azt., *Helv. Chim. Acta* **40:**1186–1900.

Scofano, H., Vieyra, A., and de Meis, L., 1979, Substrate regulation of the sarcoplasmic reticulum ATPase: Transient kinetic studies, *J. Biol. Chem.* **254:**10227–10231.

Shigekawa, M., and Dougherty, J., 1978a, Reaction mechanism of $Ca^{++}$-dependent ATP hydrolysis by skeletal muscle sarcoplasmic reticulum in the absence of added alkali metal salts. II. Kinetic properties of the phosphoenzyme formed at the steady state in high $Mg^{++}$ and low $Ca^{++}$ concentrations, *J. Biol. Chem.* **253:**1451–1457.

Shigekawa, M., and Dougherty, J., 1978b, Reaction mechanism of $Ca^{++}$-dependent ATP hydrolysis by skeletal muscle sarcoplasmic reticulum in the absence of added alkali metal salts. III. Sequential occurrence of ADP-sensitive and ADP-insensitive phosphoenzymes, *J. Biol. Chem.* **253:**1458–1464.

Smith, R., Zinn, K., and Cantley, L., 1980, A study of the vanadate-trapped state of the (Na,K)-ATPase. Evidence against interacting nucleotide site models, *J. Biol. Chem.* **255:**9852–9859.

Sorenson, M., and de Meis, L., 1977, Effects of anions, pH and magnesium on calcium accumulation and release by sarcoplasmic reticulum vesicles, *Biochim. Biophys. Acta* **465:**210–223.

Suko, J., and Hasselbach, W., 1976, Characterization of cardiac sarcoplasmic reticulum ATP–ADP exchange and phosphorylation of the calcium transport ATPase, *Eur. J. Biochem.* **64:**123–130.

Sumida, M., Wang, T., Mandel, F., Froehlich, J., and Schwartz, A., 1978, Transient kinetics of $Ca^{++}$ transport of sarcoplasmic reticulum, *J. Biol. Chem.* **253:**8772–8777.

Tada, M., Yamamoto, T., and Tonomura, Y., 1978, Molecular mechanism of active calcium transport by sarcoplasmic reticulum, *Physiol. Rev.* **58:**1–79.

Takakuwa, Y., and Kanazawa, T., 1979, Slow transition of phoshoenzyme from ADP-sensitive to ADP-insensitive forms in solubilized $Ca^{++}$, $Mg^{++}$-ATPase of sarcoplasmic reticulum: Evidence for retarded dissociation of $Ca^{++}$ from the phosphoenzyme, *Biochem. Biophys. Res. Commun.* **88:**1209–1216.

Tanford, C., 1982a, Steady state of an ATP-driven calcium pump: Limitations on kinetic and thermodynamic parameters, *Proc. Natl. Acad. Sci. USA* **79:**6161–6165.

Tanford, C., 1982b, Mechanism of active transport: Free energy dissipation and free energy transduction, *Proc. Natl. Acad. Sci. USA* **79:**6527–6531.

Taylor, J., and Hattan, D., 1979, Biphasic kinetics of ATP hydrolysis by calcium-dependent ATPase of the sarcoplasmic reticulum of skeletal muscle, *J. Biol. Chem.* **254:**4402–4407.

The, R., and Hasselbach, W., 1972a, The modification of the reconstituted sarcoplasmic ATPase by monovalent cations, *Eur. J. Biochem.* **30:**318–324.

The, R., and Hasselbach, W., 1972b, Properties of the sarcoplasmic ATPase reconstituted by oleate and lysolecithin after lipid depletion, *Eur. J. Biochem.* **28:**357–363.

Veno, T., and Sekine, T., 1981, A role of $H^+$ flux in active $Ca^{++}$ transport in sarcoplasmic reticulum vesicles. I. Effect of an artificially imposed $H^+$ gradient on $Ca^{++}$ uptake, *J. Biochem. (Tokyo)* **89:**1239–1246.

Verjovski-Almeida, S., and de Meis, L., 1977, pH-induced changes in the reactions controlled by the low- and high-affinity $Ca^{++}$-binding sites in sarcoplasmic reticulum, *Biochemistry* **16**:329–334.

Verjovski-Almeida, S., and Inesi, G., 1979, Fast-kinetic evidence for an activating effect of ATP on the $Ca^{++}$ transport of sarcoplasmic reticulum ATpase, *J. Biol. Chem.* **254**:18–21.

Verjovski-Almeida, S., Kurzmack, M., and Inesi, G., 1978, Partial reactions in the catalytic and transport cycle of sarcoplasmic reticulum ATPase, *Biochemistry* **17**:5006–5013.

Vianna, A., 1975, Interaction of calcium and magnesium in activating and inhibiting the nucleoside triphosphatase of sarcoplasmic reticulum vesicles, *Biochim. Biophys. Acta* **410**:389–406.

Watanabe, T., and Inesi, G., 1982a, Structural effects of substrate utilization on the ATPase chains of sarcoplasmic reticulum, *Biochemistry* **21**:3254–3259.

Watanabe, T., and Inesi, G., 1982b, The use of 2′,3′-*O*-(2,4,6-trinitrophenyl) adenosine 5′-triphoshate for studies of nucleotide interaction with sarcoplasmic reticulum vesicles, *J. Biol. Chem.* **257**:11510–11516.

Watanabe, T., Lewis, D., Nakamoto, R., Kurzmack, M., Fronticelli, C., and Inesi, G., 1981, Modulation of calcium binding in sarcoplasmic reticulum adenosinetriphosphatase, *Biochemistry* **20**:6617–6625.

Weber, A., 1971a, Regulatory mechanisms of the calcium transport system of fragmented rabbit sarcoplasmic reticulum. I. The effect of accumulated calcium on transport and adenosine triphosphate hydrolysis, *J. Gen. Physiol.* **57**:50–63.

Weber, A., 1971b, Regulatory mechanisms of the calcium transport system of fragmented rabbit sarcoplasmic reticulum. II. Inhibition of outflux in calcium-free media, *J. Gen. Physiol.* **57**:64–70.

Weber, A., Herz, R., and Reiss, I., 1966, Study of the kinetics of calcium transport by isolated fragmented sarcoplasmic reticulum, *Biochem. Zeitschrift* **345**:329–369.

Yamada, S., and Tonomura, Y., 1972, Phosphorylation of the $Ca^{++}$-$Mg^{++}$-dependent ATPase of the sarcoplasmic reticulum coupled with cation translocation, *J. Biochem. (Tokyo)* **71**:1101–1104.

Yamada, S., and Tonomura, Y., 1973, Reaction mechanism of the $Ca^{++}$-dependent ATPase of sarcoplasmic reticulum from skeletal muscle. IX. Kinetic studies on the conversion of osmotic energy to chemical energy in the sarcoplasmic reticulum, *J. Biochem. (Tokyo)* **74**:1091–1096.

Yamada, S., Sumida, M., and Tonomura, Y., 1972, Reaction mechanism of the $Ca^{++}$-dependent ATPase of sarcoplasmic reticulum from skeletal muscle. VIII. Molecular mechanism of the conversion of osmotic energy to chemical energy in the sarcoplasmic reticulum, *J. Biochem. (Tokyo)* **72**:1537–1548.

Yamamoto, T., and Tonomura, Y., 1967, Reaction mechanism of the $Ca^{++}$-dependent ATPase of sarcoplasmic reticulum from skeletal muscle. I. Kinetic studies, *J. Biochem. (Tokyo)* **62**:558–575.

Yamamoto, T., and Tonomura, Y., 1968, Reaction mechanism of the $Ca^{++}$-dependent ATPase of sarcoplasmic reticulum from skeletal muscle. II. Intermediate formation of phosphoryl protein, *J. Biochem. (Tokyo)* **64**:137–145.

Yates, D., and Duance, V., 1976, The binding of nucleotides and bivalent cations to the calcium-and-magnesium ion-dependent adenosine triphosphatase from rabbit muscle sarcoplasmic reticulum, *Biochem. J.* **159**:719–728.

# 31

# Calcium-Induced Potassium Transport in Cell Membranes

B. Sarkadi and G. Gárdos

## I. INTRODUCTION

Studies in the 1930-1940s, directed toward the elucidation of the connection between cell metabolism and ion–water content of human red cells, noted extreme alterations in $K^+$ leakage under various experimental conditions. In lead-poisoned red cells (Ørskov, 1935), and in NaF-treated cells, Wilbrandt (1937, 1940) observed an increase in net $K^+$ efflux and a concomitant cell shrinkage. It was first demonstrated by Gárdos (1956, 1958a,b, 1959) that, in metabolically depleted red cells, the enhanced $K^+$ efflux took place only when $Ca^{2+}$ ions were present in the suspending media. Since that time, it has been firmly established that rapid $K^+$ transport in red cells is triggered by a specific interaction of $Ca^{2+}$ ions with the intracellular membrane surface (see Section II-B), and the process, often noted in the literature as the "Gárdos phenomenon," has become a model system to entertain numerous membrane physiologists, biochemists, and biophysicists. Training courses in membrane biology use this easily reproducible phenomenon to illustrate specificity, side-dependent activation, and selectivity of natural transport processes. Established research workers, deeply involved in the investigation of complex phenomena in complex cellular systems, from time to time return to the red-cell $Ca^{2+}$-induced $K^+$ transport* to reveal new and important aspects of this process. The $Ca^{2+}$-induced $K^+$ transport in red cells gave new insights into the coupling of ion movements to changes in membrane potential, and into the

* The notation $Ca^{2+}$-induced (-activated, -stimulated, etc.) $K^+$-transport system (pathway, channel, etc.) is used without actually referring to a given molecular mechanism (but see Section IV).

B. Sarkadi and G. Gárdos • National Institute of Haematology and Blood Transfusion, Budapest, Hungary.

problem of side-dependent triggering and gating of ionic channels. In the meantime, as it generally occurs with red cell membrane phenomena, the phenomenon has turned out to be present in many animal cell membranes. There is a renewed interest in finding the physiological relevance and significance of the $Ca^{2+}$-induced $K^+$ transport in muscle and nerve, as well as in non-excitable cells, and a reopening of the search for clarifying the molecular mechanism of this transport system. The present review intends to summarize the information currently available about the basic mechanism and the general physiological aspects of the $Ca^{2+}$-induced $K^+$ transport in plasma membranes. Here, we concentrate on the recent advances in this field; for the historical background we refer to the reviews by Lew and Ferreira (1977, 1978), Putney (1979), Passow (1981), Parker (1981), and Schwarz and Passow (1983).

## II. CHARACTERISTICS OF THE $Ca^{2+}$-INDUCED $K^+$ TRANSPORT IN HUMAN RED CELLS

### A. Calcium Homeostasis in Red Cells—Induction of Rapid $K^+$ Transport

The calcium concentration in normal human red cells is about 15–20 μmoles/liter of cells and, as most of this calcium is membrane bound, cytoplasmic calcium concentration is only a fraction of a micromole (Harrison and Long, 1968; Simons, 1982). The permeability of the red-cell membrane for calcium is extremely low (the rate of passive calcium influx is in the order of $10^{-6}$ mole/liter of cells per hour (see Lew, 1974; Szász *et al.*, 1978b,c; Lew and Beaugé, 1979), and red cells possess a powerful calcium extrusion pump which has a maximum transport capacity of $5–10 \times 10^{-3}$ mol/liter of cells per hour. The calcium pump becomes activated at micromolar cellular calcium concentrations and the calcium-dependent binding of calmodulin, a cytoplasmic regulatory protein, further increases both the calcium affinity and the transport capacity of the pump (the characteristics and regulation of the red-cell calcium pump are discussed in detail in the reviews by Schatzmann, 1975, 1982; Lew and Ferreira, 1978; Larsen and Vincenzi, 1979; Roufogalis, 1979; Sarkadi and Tosteson, 1979; Sarkadi, 1980; Sarkadi *et al.*, 1982).

The careful balance of the alkali cation pumps and leaks, as outlined by Tosteson and Hoffman (1960), allows red cells to conserve their nonequilibrium cation distribution during their 4-month-long lifetime in spite of their lack of an oxidative metabolism. Glycolysis provides enough ATP to support the energy needs of the $Na^+,K^+$ and $Ca^{2+}$ pumps, working at a low rate. In the case of the $Na^+,K^+$ pump, the density of the pump protein in the membrane is so low that, although under physiological conditions the pump is working with about 50% of its maximum capacity, the need for ATP is less than 1 mmole/liter of cells per hour. In contrast, the $Ca^{2+}$ pump is hardly activated at physiological cellular calcium concentrations and is working only with less than 1% of its maximum transport rate. The huge reserve capacity of the $Ca^{2+}$ extrusion can be regarded as a safety device to avoid the harmful effects of even a small increase in cellular calcium. These effects include the unfavorable changes in red cell shape and plasticity, inhibition of glycolytic enzymes, and the $Na^+$,$K^+$ pump (Weed *et al.*, 1969; Dunn, 1974; Szász *et al.*, 1974; Lorand *et al.*, 1976; Sarkadi *et al.*, 1976, 1977; Parker *et al.*, 1978; Parker, 1981), as well as the opening of a selective $K^+$ transport pathway.

In the course of red cell transport studies, several experimental conditions were reported to incresase the $K^+$ permeability of the membrane. Some of these conditions, such as the incubation of the cells in low ionic strength media and the effect of some SH blocking agents, produce a nonselective increase in the monovalent ion permeability (LaCelle and Rothstein, 1966; Cotterrell and Whittam, 1970; Knauf and Rothstein, 1971), while other treatments result in a selective rise of $K^+$ movement and a concomitant shrinkage of red cells in isotonic NaCl media. As listed below in the approximate order of discovery, a selective, 50- to 500-fold increase in the $K^+$ permeability of the red cell membrane was found to be evoked by micromolar concentrations of lead salts (Ørskov, 1935; Grigarzik and Passow, 1958; Lindemann and Passow, 1960; Riordan and Passow, 1971; Passow, 1981), by NaF with a maximum effect at approximately 40 mM (Wilbrandt, 1937, 1940; Davson, 1941; Lepke and Passow, 1960, 1968), by ATP depletion with iodoacetate and inosine (Gárdos, 1956, 1958a,b 1959), by prolonged incubation of the cells in substrate-free media or with iodoacetate (Blum and Hoffman, 1970, 1971; Kregenow and Hoffman, 1972), by the addition of propranolol and several related compounds (Ekman *et al.*, 1969; Manninen, 1970), or by the divalent cation ionophore A23187 (Reed, 1973, 1976; Gárdos *et al.*, 1975b; Kirkpatrick *et al.*, 1975). Gárdos (1956, 1958a,b, 1959) first showed that in ATP-depleted cells, the effect entirely depends on the presence of calcium ions; calcium-chelating agents such as EDTA or EGTA completely inhibit the increase in $K^+$ transport. It is by now generally accepted that all the promoters act by allowing a specific membrane–metal interaction to open a $K^+$ conductance pathway. Lead can replace calcium in this activation, thus provoking the effect in itself, while in the other cases, the effectors increase cellular calcium concentration and/or increase the sensitivity of the $K^+$ transport system for calcium ions.

## B. Calcium Dependence and Activation Kinetics of $K^+$ Transport

The exclusive role of *intracellular calcium* in producing a selective increase in red cell $K^+$ permeability was indicated by Hoffman (1962), Whittam (1968), Lew (1970), and Romero and Whittam (1971), but first clearly demonstrated by Blum and Hoffman (1972). By using red cell ghosts, they showed that internal calcium is much more effective in inducing $K^+$ transport than external calcium, and while external EGTA cannot revert the action of incorporated calcium, internal EGTA completely abolishes any calcium-induced response. In red cells, the role of ATP depletion in evoking rapid $K^+$ transport is probably three-fold. ATP depletion (1) increases the permeability of the membrane for $Ca^{2+}$, (2) decreases the ability of the calcium pump to extrude cellular calcium, and (3) increases the susceptibility of the $K^+$ transport system for calcium (Lew, 1971a, 1974; Lew and Ferreira, 1976; Szász *et al.*, 1978a). By using the ionophore A23187, it could be shown in intact red cells that the basis of rapid $K^+$ efflux was an increase in cellular calcium (Reed, 1976; Sarkadi *et al.*,

1976; Lew and Ferreira, 1976), although a direct effect of the ionophore on the calcium affinity of $K^+$ transport has also been indicated (Lew and Ferreira, 1976, Yingst and Hoffman, 1984). In propranolol-treated cells, a rapid, transient increase in cellular calcium was reported (Porzig, 1975, Szász *et al.*, 1977). Based on these findings, it is widely accepted that a calcium-sensitive receptor is present at the internal surface of the red cell membrane and the activation of a selective $K^+$ transport is based on the binding of calcium to this receptor.

An important question is the *sensitivity* of the receptor *to calcium* under various conditions. In ghost preparations, in which unbuffered calcium was entrapped (Hoffman, 1962), in intact red cells loaded with calcium by ATP depletion (Romero and Whittam, 1971), by salicylate (Bürgin and Schatzmann, 1979) or by a short exposure to A23187 (Sarkadi *et al.*, 1976; Gárdos *et al.*, 1977), as well as in red cells incubated with low concentrations of A23187 (Lew and Ferreira, 1976), the intracellular calcium concentration producing half-maximum activation of $K^+$ transport ($K_{Ca}$) was in the range of $10^{-4}$–$10^{-3}$ M. As shown by various techniques (Schatzmann, 1973; Ferreira and Lew, 1977), human red cells contain a low-affinity, high-capacity calcium buffer, which binds 50–70% of intracellular calcium in a range between $10^{-6}$ and $10^{-2}$ M. Thus, the concentration of free calcium for half-maximum $K^+$-transport activation ($K_{Ca\ free}$) can be estimated to be around 50–300 μM. In contrast to this, when red cell ghosts were loaded with a Ca-EGTA buffer (Simons, 1976a), $K_{Ca\ free}$ was between 0.5 and 1.0 μM, and calcium concentrations higher than 70 μM inhibited $K^+$ transport (Simons, 1976b). In ghosts containing arsenazo III, a calcium indicator and calcium buffer dye, the value of $K_{Ca\ free}$ was 5–10 μM (Yingst and Hoffman, 1978, 1981, 1984). As shown in red cell ghosts (Schatzmann, 1973) and in inside-out membrane vesicles (Sarkadi *et al.*, 1979; Al-Jobore and Roufogalis, 1981), the use of Ca-chelator buffers increases the apparent calcium affinity of the red cell calcium pump by two orders of magnitude, and a similar effect may occur in the case of $K^+$ transport as well. The apparent $K_{Ca}$ of rapid $K^+$ transport is also in the micromolar range if red cells are incubated in the presence of high concentrations (≈10 μM) of A23187, or when the cells are depleted from ATP and/or magnesium (Lew and Ferreira, 1976; Szász *et al.*, 1980). The same high $Ca^{2+}$ sensitivity is found in calcium-loaded red cells when propranolol is present (Gárdos *et al.*, 1977). Lew and Bookchin (1980) reported that sickle cells, in spite of their elevated cellular calcium level, do not show a quinine-sensitive increased $K^+$ release, thus, the transport system is in a "$Ca^{2+}$-refractory state." Addition of A23187 (even in the presence of external EGTA, when the cells are continuously depleted from calcium) increases quinine-sensitive $K^+$ efflux and, thus, sensitizes the system to intracellular calcium. Whether this $Ca^{2+}$-insensitive state of sickle cells is a specific feature in this disease, or is similar to the low-affinity state observed in normal red cells (Lew and Ferreira, 1976; Gárdos *et al.*, 1977), is yet to be answered. In calcium-loaded ghosts, the presence of electron-donor systems significantly increases the apparent calcium affinity of the $K^+$-transport system (Garcia-Sancho *et al.*, 1979). NaF and triose-reducton probably also sensitizes the cellular receptors for calcium ions (Lew and Ferreira, 1976). All these findings suggest that the sensitivity of rapid $K^+$ transport for calcium in human red cells is affected by the actual experimental conditions and that this variable calcium sensitivity may be involved in the physiological regulation as well as in the pathological alterations of membrane $K^+$ transport.

An additional problem in evaluating the calcium sensitivity of $K^+$ transport is an *all-or-none* type response observed both in ATP-depleted or calcium-loaded cells (Gárdos, 1958a,b; Riordan and Passow, 1973; Lew, 1974; Szász *et al.*, 1974; Yingst and Hoffman, 1984). The experiments showed that, under suboptimal conditions, only a certain fraction of cellular potassium is lost, or only a certain fraction of the cells loses substantial amounts of $K^+$. The first case may occur if rapid $K^+$ transport is spontaneously inactivated, while the other indicates a heterogeneity either in calcium concentration, or in the calcium-sensitivity of the $K^+$-transport system in various cells (see below).

1. A time-dependent spontaneous inactivation of the $Ca^{2+}$-induced $K^+$ transport in human red cells can be convincingly excluded in most cases. ATP-depleted red cells, which lose $K^+$ and shrink in a NaCl medium, rapidly reswell when placed in a medium containing predominantly KCl (Szász *et al.*, 1974), implying that the $K^+$ leak is continuously open in such cells. Rapid $K^+$ transport, as measured by tracer $K^+$ techniques, is present in ghosts or in intact cells long after the induction of $K^+$ conductance by calcium (Brown and Lew, 1981). In intact cells, with working metabolism, active calcium extrusion may terminate rapid $K^+$ transport, as it is probably in the case when *Amphiuma* cells are exposed to calcium by electrode micropuncture (Lassen *et al.*, 1976). Under experimental conditions, when intracellular calcium concentrations show transient changes, e.g., in the presence of low concentrations of A23187 ionophore, an oscillation of $K^+$ permeability was observed (Vestergaard-Bogind and Bennekou, 1982).

2. *Heterogeneity* of the *calcium concentration* in red cell or ghost preparations may certainly be involved in the all-or-none response. When ATP-depleted cells or arsenazo III-containing ghosts (preincubated in NaCl media in the presence of $Ca^{2+}$) were separated according to their densities, the calcium content in the heavier cells (which lost more $K^+$ and shrunk more than the others) was significantly higher than in the lighter cell fractions (Yingst and Hoffman, 1984). Since net $K^+$ efflux hyperpolarizes the cell membrane (see Section II-D) and increases $Ca^{2+}$ influx (Szász *et al.*, 1978b), this change in calcium concentration can be partly a consequence instead of the cause of a heterogeneous $K^+$ response. This explanation is supported by the findings of Lew and Ferreira (1976), who failed to see any heterogeneity in different density red cell populations in the absence of net $K^+$ loss. Moreover, Lew and Simonsen (1981) and Simonsen *et al.* (1982) convincingly documented the uniform distribution of calcium in cells treated with various concentrations of ionophore A23187.

3. Even if a heterogeneous $Ca^{2+}$ distribution is present, a threshold-type all-or-none activation of the $K^+$-transport systems is suggested by the responses observed. Experiments with A23187- or propranolol-treated cells show that all the red cells contain enough $K^+$ channels to produce a rapid response under favorable conditions. (As shown by Heinz and Passow (1980), the different sensitivity of the cells to extracellular $K^+$ may also cause heterogeneous response.) A heterogeneity in the calcium sensitivity of the $K^+$-transport system could be convincingly demonstrated in inside-out red cell membrane vesicle preparations (see below).

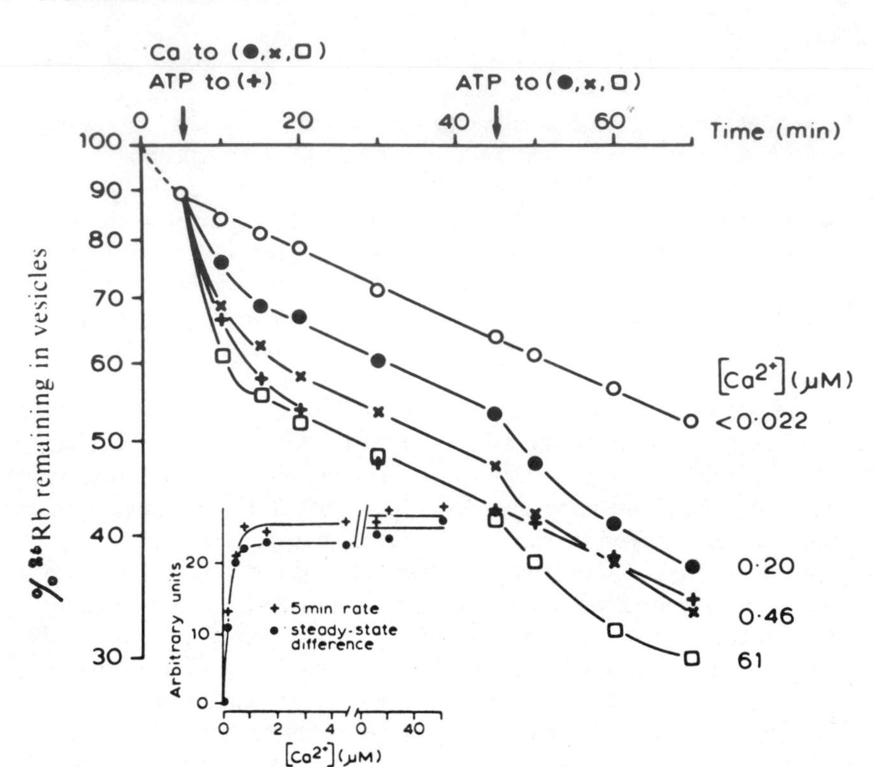


*Figure 1.* Ca-sensitive $^{86}$Rb loss from one-step inside-out vesicles prepared from human red cells, as a function of the free calcium concentration. The insert shows the threshold distribution curve plotted against $Ca^{2+}$. At increasing free-calcium concentrations, progressively more vesicles release their $^{86}$Rb at the same maximum relative rate. This indicates an all-or-nothing activation for each channel with different threshold $Ca^{2+}$ sensitivity. For the experimental details see the original paper. From Lew *et al.* (1982).

A promising experimental system for studying transport phenomena in red cells is the preparation of sealed, *inside-out* membrane *vesicles* (IOVs). In IOVs, the cytoplasmic side of the membrane is exposed to the incubation media and the transport effects of various agents can be directly estimated. The Steck-type IOVs (Steck, 1974), prepared by a multistep procedure, failed to show a $Ca^{2+}$-induced $K^+$ transport (Grinstein and Rothstein, 1978), or the calcium sensitivity and the $K^+$,$Na^+$ selectivity of the process was very much reduced (Sze and Solomon, 1979). In our laboratory, we had Tris-HCl IOV preparations in which a $Ca^{2+}$-induced response for $K^+$ transport could be evoked by the addition of a hemoglobin-free protein concentrate of the red cell cytoplasm (Sarkadi *et al.*, 1980). We could not further purify a protein fraction responsible for this effect, and studies in various laboratories with various IOV preparations failed to find a restoration of $Ca^{2+}$-induced $K^+$ transport by cytoplasmic proteins in "silent" IOVs [a possible irreversible inhibition of the transport system by low $K^+$ media, (see Section II-C), could explain some of the negative results]. Recently, the group of Dr. Lew in Cambridge developed a one-step IOV preparation, in which $Ca^{2+}$-induced $K^+$ transport was preserved (Lew *et al.*, 1980, 1982). Studies with such vesicles indicate that each red cell contains about 100–200 $Ca^{2+}$-activated $K^+$-transport "channels," and all these individual channels respond to $Ca^{2+}$ in an all-or-none fashion, with considerable differences in their calcium sensitivity (Figure 1; Garcia-Sancho *et al.*, 1982; Lew *et al.*, 1982). These studies were carried out in media containing EGTA-Ca buffers and no data are available as yet for a possible variability in the calcium sensitivity of $K^+$ transport caused by cellular factors, drugs, chelators, or by the ionophore A23187 in these IOVs.

## C. Effects of Metal Ions

### 1. Monovalent Cations

Blum and Hoffman (1971) first showed that $Ca^{2+}$-induced $K^+$ transport (measured as tracer $K^+$ efflux) in ATP-depleted red cells was sensitive to changes in extracellular $K^+$ concentrations ($K_o$; Figure 2). In these experiments, $K^+$ transport was maximum at about 2 mM $K_o$, while both smaller and greater external potassium concentrations substantially decreased the transport rate. The stimulating effect of increased $K_o$ between 0 and 3 mM was subsequently shown in various laboratories by measuring equilibrium exchange of tracer $K^+$ (Gárdos *et al.*, 1975a), or net $K^+$ transport (Knauf *et al.*, 1974, 1975). A time-dependent irreversible inactivation of the $Ca^{2+}$-induced $K^+$ transport by $K^+$-free media has also been reported (Heinz and Passow, 1980).

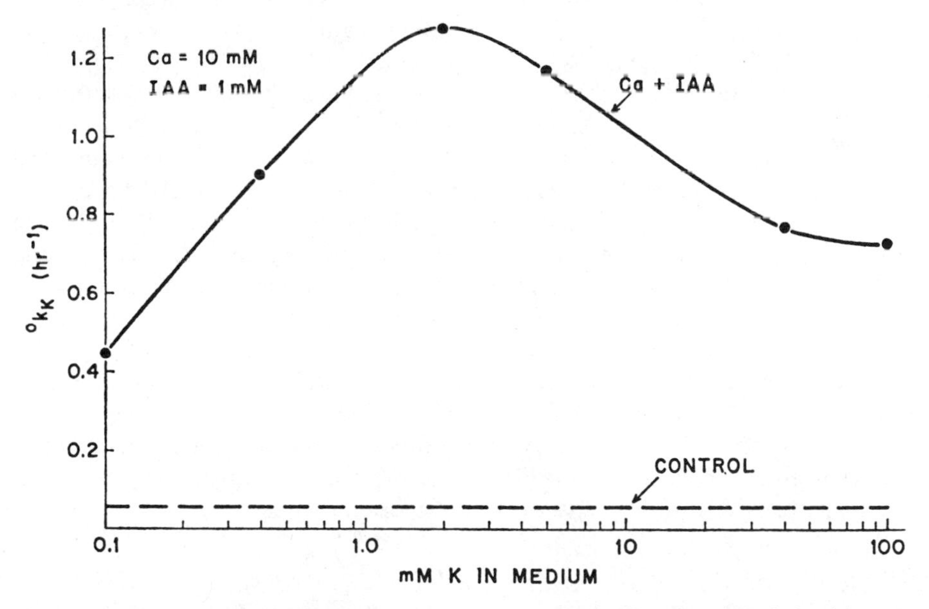


*Figure 2.* Effect of external $K^+$ concentration on the rate of $K^+$ efflux from ATP-depleted human red cells. The $K^+$ concentration in the medium was varied by replacing $Na^+$ with $K^+$. The fluxes were measured at 37°C, pH 7.15, $Ca^{2+}$. concentration was 10 mM, iodoacetamide (IAA) was added in 1 mM final concentration. From Blum and Hoffman (1971).

The inhibitory effect of high $K_o$, however, was not observed in propranolol-treated cells (Gárdos *et al*, 1975a), or in ghosts loaded with Ca-EGTA buffers (Simons, 1976b). This discrepancy may be explained by the inhibitory action of high $K_o$ on calcium influx (Lew, 1974; Porzig, 1977), which may cause an apparent inhibition of $K^+$ transport in ATP-depleted cells (see Lew and Ferreira, 1978). Recent data by Yingst and Hoffman (1984) obtained with arsenazo III + $Ca^{2+}$-loaded ghosts indicate a direct inhibition of $Ca^{2+}$-induced $K^+$ transport by high $K_o$.

Increased intracellular $K^+$ concentrations ($K_i$) were reported to activate $Ca^{2+}$-induced $K^+$ transport in a nonsaturable manner (Simons, 1976b). $K^+$ in the activation of transport can be replaced by $Rb^+$, but not by any other monovalent or divalentcation. According to the experiments of Simons (1976b), the apparent transport rate for monovalent ions, relative to $K^+$, is the following: $Rb^+(1.5) > K^+(1.0) > Cs^+(0.05) \gg Na^+, Li^+$. Intracellular $Na^+$ inhibits the $K^+$ transport system and $Na_i$ inhibition is more pronounced at low $K_o$ (Riordan and Passow, 1973; Knauf *et al.*, 1975; Simons, 1976b; Hoffman and Blum, 1977; Yingst and Hoffman, 1984). The curve for the rate constant of $K^+$ movement as a function of $K_i$ becomes sigmoid in $Na^+$-containing ghosts (Simons, 1976b). All these data suggest that $Ca^{2+}$-induced $K^+$ transport is modified by specific interactions of monovalent cations with both sides of the cell membrane.

### 2. Divalent and Trivalent Cations

$Ca^{2+}$ ions are specifically required to activate rapid $K^+$ transport in red cells and only $Pb^{2+}$ and $Sr^{2+}$ ions were convincingly shown to replace $Ca^{2+}$ in this role (see Lew and Ferreira, 1978; Passow, 1981). The reported effect of $Mg^{2+}$ in evoking rapid $K^+$ transport in NaF-poisoned cells (Lepke and Passow, 1960; Lindemann and Passow, 1960) is probably due to $Ca^{2+}$ contamination and an increase in the $Ca^{2+}$ sensitivity of the $K^+$-transport system under these conditions. In fact, in inside-out membrane vesicles, Mg was shown to inhibit $Ca^{2+}$-induced $K^+$ transport, probably by competing with $Ca^{2+}$ at the specific receptors (Figure 3; Garcia-Sancho *et al.*, 1982). The reported effect of vanadium salts in producing selective rapid $K^+$ transport in red cells (Siemon *et al.*, 1982) is probably also based on $Ca^{2+}$ contamination in the media, on an increased sensitivity of the membrane receptors to Ca, and on the displacement of calcium ions by vanadium from chelator substances (Fuhrmann *et al.*, 1984).

Externally added trivalent lanthanum ions do not affect $Ca^{2+}$-induced $K^+$ transport in $Ca^{2+}$-loaded red cells when intracellular calcium is present in unlimiting concentrations, but considerably inhibit $K^+$ transport in ATP-depleted or propranolol-treated cells (Szász *et al.*, 1978d). This effect is probably based on lanthanum inhibition of calcium influx or calcium redistribution in these cells. Intracellular lanthanum directly interferes with $Ca^{2+}$-activation of $K^+$ transport (Szász *et al.*, 1978c,d).

## D. Membrane Potential and $Ca^{2+}$-induced $K^+$ Transport

In propranolol-treated red cells, Ekman *et al.* (1969) and Manninen (1970), and in ATP-depleted cells, Blum and Hoffman (1971) observed a rapid and temporary accumulation of externally added tracer $K^+$ during $Ca^{2+}$-induced net $K^+$ efflux. This

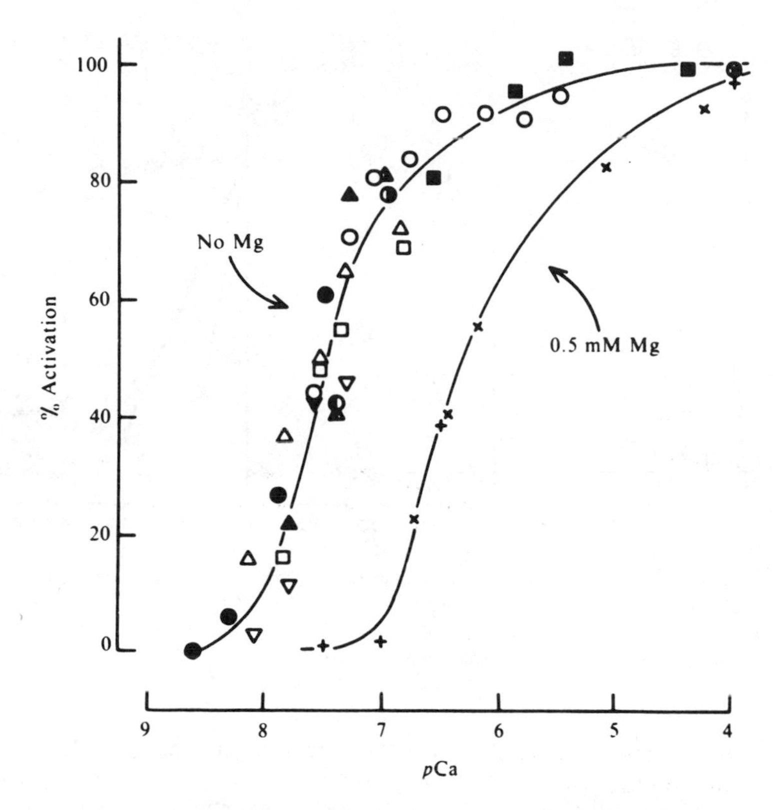


*Figure 3.* Activation of $^{86}$Rb transport in one-step inside-out red-cell membrane vesicles. Squares represent data obtained in isotope release experiments, while all other symbols correspond to uptake experiments. The curve on the left was obtained in the absence of $Mg^{2+}$, and the one on the right with 0.5 mM $Mg^{2+}$. For the details of measurements and calculations see the original paper. From Garcia-Sancho *et al.* (1982).

tracer $K^+$ (or $Rb^+$) influx was inhibited by increasing extracellular $K^+$ concentration. Both groups concluded that the "countertransport"-like phenomenon was a strong indication for the carrier-mediation of $Ca^{2+}$-induced $K^+$ transport. Models assuming that the "loaded" carrier crosses the membrane much faster than the "unloaded" one, can well explain countertransport phenomena and, indeed, the validity of these assumptions was shown in the functioning of some non-electrolyte transport systems. However, in ion-transport processes, the effect of membrane potential may considerably modify the flux components. In the 1960s, membrane potential in human red cells was believed to be exclusively governed by $Cl^-$ ion distribution, as isotope $Cl^-$ fluxes indicated a permeability of the red cell membrane for this anion of about $10^{-4}$ cm/sec (Tosteson, 1959; Gunn *et al.,* 1973; 1979; Knauf, 1979). Even a thousand-fold increase in the $K^+$ permeability of the membrane, resulting in a change of $P_K$ from about $10^{-10}$–$10^{-7}$ cm/sec, should not affect membrane potential (calculated to be about $-10$mV, inside negative) in this case. However, by direct microelectrode measurements in the giant red cells of *Amphiuma,* Lassen (1972) found a membrane

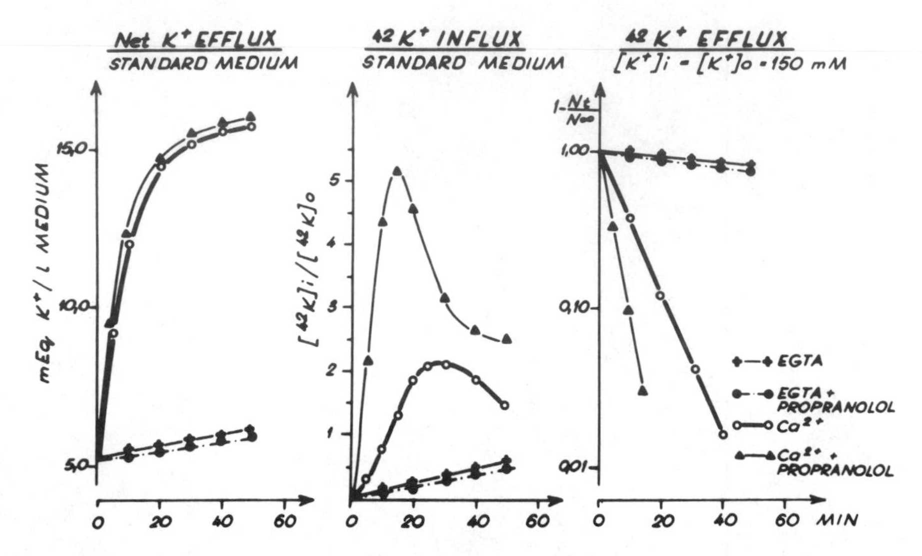


*Figure 4.* Effect of propranolol on the $Ca^{2+}$-dependent $K^+$ transport in phosphate ester-depleted human red cells. Standard medium, 145 mM NaCl, 5mM KCl, 10 mM Tris-HCl, pH 7.4, EGTA, 1 mM; $Ca^{2+}$, 2 mM; propranolol, 0.5 mM. Propranolol does not affect net $K^+$ efflux but increases the rate of $^{42}K$ equilibration and tracer $K^+$ influx. From Gárdos *et al.* (1975a).

resistance four orders of magnitude higher than predicted by tracer $Cl^-$ flux measurements. In human red cells treated with the $K^+$-selective ionophore valinomycin, Hunter (1971) showed that net $K^+$ movement was limited by net $Cl^-$ movement, and the $Cl^-$ conductance of the membrane was in the range of $10^{-8}$ cm/sec, that is four orders of magnitude smaller than previously thought. Based on these data, in a theoretical approach, Glynn and Warner (1972) proposed a model to explain the countertransport phenomena during $Ca^{2+}$-induced $K^+$ transport. By supposing that $Ca^{2+}$-induced $K^+$ permeability is higher than net $Cl^-$ permeability, tracer $K^+$ accumulation could be ascribed to a hyperpolarization of the red cell membrane in low $K_o$ media. The membrane potential, which develops under such conditions (maximum 60–70 mV, inside negative), would slow down net $K^+$ efflux to the rate allowed by net $Cl^-$ movement, while the potential difference would induce a rapid influx of tracer $K^+$. Calculations by Glynn and Warner (1972), by applying the Goldman theory for changes in membrane potential and using variable ratios for $Cl^-$ and $K^+$ permeabilities, gave computer-simulated curves which could be fitted to the experimental data of Manninen (1970). This approach stimulated further research in the field of net cation and anion transport of red cells, eventually fully confirming the original idea by Glynn and Warner. A series of experiments reported by our group (Gárdos *et al.*, 1975a) may illustrate the experimental set-up used for distinguishing the effects of membrane potential and other factors during $Ca^{2+}$-induced $K^+$ transport. As shown in Figure 4, parallel measurements of net $K^+$ efflux into low-$K^+$ media, tracer $K^+$ influx from the same media, and tracer $K^+$ efflux under $K^+$ equilibrium exchange conditions, allow independent estimation of the $K^+$ permeability (from $K^+$ exchange at zero membrane potential) and the anion conductance (reflected in net $K^+$ efflux and tracer $K^+$ influx). Based on such experiments, the direct and indirect effects of various agents could be differentiated (see Section II-E.3), and models describing these phenomena could be tested. Schubert and Sarkadi (1977) used a model in which the $K^+$-transport system was treated as a simple, nonsaturable diffusion channel. Tracer and net $K^+$ movements were measured and the value of net $Cl^-$ permeability was calculated. The obtained value for $P_{Cl,\ net}$ ($1.6 \times 10^{-8}$ cm/sec) is in good agreement with the values computed from valinomycin experiments (Hunter, 1971, 1977; Knauf *et al.*, 1977). Based on the obtained $P_K$ and $P_{Cl}$ values, a computer simulation of tracer $K^+$ influx could be closely fitted to the experimental data (Schubert and Sarkadi, 1977). Experiments with various inhibitors of red-cell anion transport, or by exchanging chloride with more slowly penetrating anions, showed a decrease in net $K^+$ efflux and an increase in tracer $K^+$ uptake, as predicted by the models implicating the role of membrane potential (Hoffman and Knauf, 1973; Gárdos *et al.*, 1975a).

By direct microelectrode measurements in *Amphiuma* red cells, a hyperpolarization of the membrane during $K^+$ transport induced by insertion of the electrode in high $Ca^{2+}$ media (Lassen *et al.*, 1973, 1976), or by the A23187 ionophore plus $Ca^{2+}$ (Gárdos *et al.*, 1976) could be demonstrated. In the relatively small human erythrocytes, microelectrode measurements are unreliable, thus, several indirect methods were applied to assess the changes in membrane potential. The use of carbocyanine dyes for measuring membrane potential in red cells (Hoffman and Laris, 1974; Hladky and Rink, 1976), is questionable in the case of $Ca^{2+}$-induced $K^+$ transport, as most of such dyes inhibit rapid $K^+$ transport (Simons, 1976c, 1979). Still, by using this method (dye inhibition of $K^+$ movement was reduced by external $K^+$, and dye distribution and microelectrode measurements were directly compared), Pape (1982) could evaluate $Ca^{2+}$ induced membrane potential changes in *Amphiuma* and in human red cells. He showed a considerable membrane hyperpolarization when $Ca^{2+}$ influx and $K^+$ channel openings were evoked by an elevation of extracellular pH.

An indirect, but dependable method for estimating membrane potential in red cells was worked out by Macey *et al.* (1978). In this technique, the cells are incubated in unbuffered media and the membrane is rendered permeable for $H^+$ ions by the addition of the protonophore CCCP. Changes in the extracellular pH under these conditions directly reflect the changes in membrane potential. In this system, a hyperpolarization caused by $Ca^{2+}$-induced $K^+$ transport can be easily demonstrated (Figure 5). By using Macey's technique, Vestergaard-Bogind and Bennekou (1982) found an A23187 + $Ca^{2+}$-induced oscillation in membrane potential and in $K^+$ efflux rates at low ionophore concentrations, while a permanent hyperpolarization occurred at higher A23187 concentrations or in ATP-depleted cells. The oscillation is probably the result of a reversible, temporary interaction of calcium ions with the specific intracellular membrane receptors of the $K^+$ transport in cells where active calcium extrusion can partially compensate $Ca^{2+}$ influx. All the above-mentioned studies support the view that, during $Ca^{2+}$-induced $K^+$ transport, a hyperpolarization of the red cell membrane occurs and the normal, approximately $-10$ mV, practically $Cl^-$-distribution potential in the cells is shifted to a largely $K^+$-distribution potential of $-60$ to $-70$ mV. They also indicate that this is a conductive transport pathway and carries

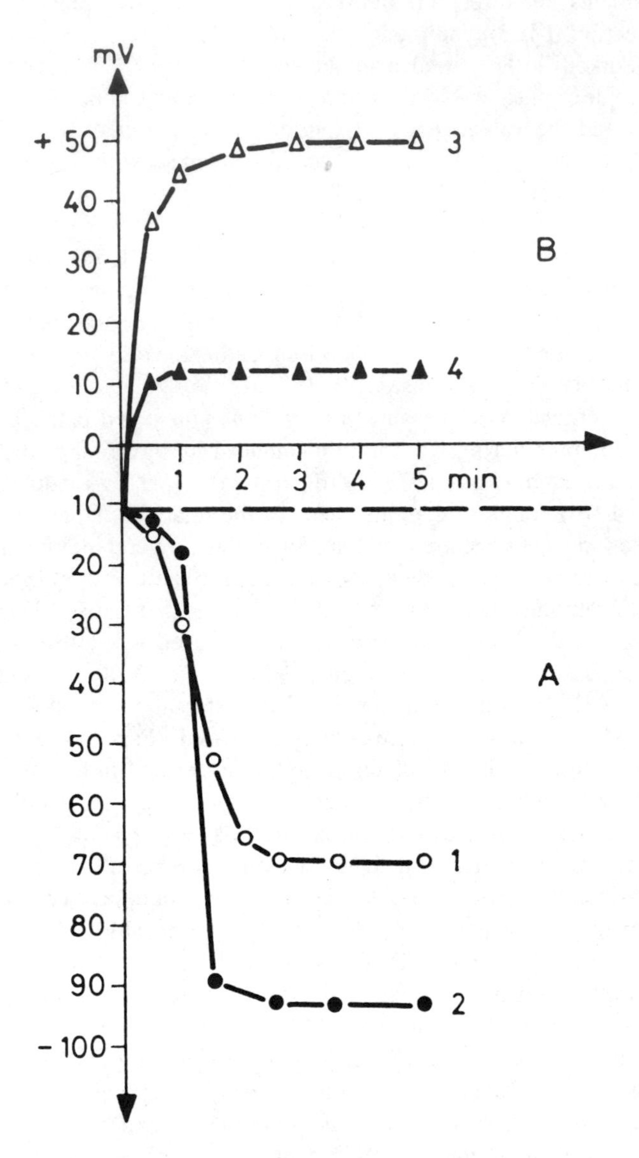


*Figure 5.* Membrane potential changes in human red cells assessed by the method of Macey *et al.* (1978). Unbuffered media contained 20 μM CCCP to allow rapid $K^+$ equilibration and external pH was measured. The membrane potential was calculated as described by Macey *et al.* (1978). (A) 0.155 M NaCl; (B) 0.3 M sucrose. Additions: (1) 0.5 mM propranolol, 0.5 mM $CaCl_2$, (2) as in (1), + 0.5 mM SITS, (3) none, (4) 0.5 mM SITS. Temperature, 20°C; Hematocrit, 15%. From Szász *et al.* (1978b).

a net $K^+$ movement. In the experiments of Szönyi (1960) and Gárdos and Szász (1968), in $HCO_3^-$-containing media, the absorption of $CO_2$ and the increase in the pH of the media showed good correlation with $K^+$ efflux, and the authors suggested the presence of an electrically silent $K^+$–$H^+$ exchange. However, all the further studies indicated that such an exchange is the consequence of membrane hyperpolarization and does not significantly contribute to the $K^+$ movement. There is no apparent membrane potential dependence of the $Ca^{2+}$-induced $K^+$ permeability of red cells in the experiments mentioned above.

## E. Effects of Cellular Components and Drugs

### 1. Metabolic Intermediates

As we already mentioned, the ATP content of red cells affects $Ca^{2+}$-induced $K^+$ transport in several ways: A reduction in cellular ATP stops calcium pumping, increases calcium influx, and probably increases the affinity of the $K^+$ transport system for $Ca^{2+}$. The role of 2,3-DPG, a glycolytic intermediate present in high concentrations in red cells, in modifying $Ca^{2+}$-induced $K^+$ transport was indicated by Gárdos (1966a,b, 1967) and Gárdos and Szász (1973), while the group of Passow searched for the effects of carbohydrate derivatives, such as triose-reducton, in this process (Passow, 1963; Passow and Vielhauer, 1966). As most of these experiments were carried out in cells where changes in ATP and/or calcium concentrations also occurred, the direct effect of these intermediates on $K^+$ transport is questionable. The most probable basis of such effects is calcium chelation and the modification of the calcium affinity of the $K^+$ transport system. The role of NADH in maintaining optimum conditions for $Ca^{2+}$-induced $K^+$ transport was indicated by Gárdos (1956) and Gárdos and Szász (1968). As shown by Garcia-Sancho *et al.* (1979), the addition of artificial electron-donor systems (such as ascorbate + phenazine methosulfate) increases the sensitivity of the red-cell $K^+$ transport to calcium, and a similar effect is produced when the natural compound SH-glutathione, NADH, or NADPH is incorporated into resealed ghosts. These findings suggest a role of the actual redox state of some membrane components in $Ca^{2+}$-induced $K^+$ transport. An involvement of membrane phospholipases in inducing $Ca^{2+}$-dependent $K^+$ transport was also indicated (Grey and Gitelman, 1979).

### 2. Calmodulin

The regulatory role of calmodulin, a heat-stable, low molecular weight, acidic calcium-binding protein has been demonstrated in several $Ca^{2+}$-activated cellular functions (see Cheung, 1982). Experiments with calmodulin-antagonist drugs, which also inhibit red-cell $Ca^{2+}$-induced $K^+$ transport (Lackington and Orrego, 1981), point to a possible involvement of calmodulin in this process. However, the effect of such drugs cannot be regarded specific for calmodulin inhibition as they seem to block various calcium-induced, and especially, lipid-dependent enzyme actions (Vincenzi, 1981; Roufogalis, 1981; Sarkadi *et al.*, 1982). We have to note that Plishker *et al.* (1980) found an activation of $Ca^{2+}$-induced $K^+$ transport by calmodulin inhibitors when such drugs inhibited calcium pump and, thus, increased calcium influx into red cells. Calmodulin binding and release, respectively, may evoke an alternating activation of the $Ca^{2+}$ pump and the $Ca^{2+}$-induced $K^+$ transport in a model proposed by Lassen *et al.* (1980). However, until this time, any effort to show a direct involvement of calmodulin in $Ca^{2+}$-induced $K^+$ transport was unsuccessful. In "silent" inside-out red-cell membrane vesicles, a cytoplasmic protein concentrate partially restored $Ca^{2+}$-induced $K^+$ transport, but a boiled extract (containing the heat-stable calmodulin and stimulating calcium pumping) was ineffective in this respect (Sarkadi *et al.*, 1980). In several other laboratories, no cytoplasmic protein could restore $Ca^{2+}$-induced $K^+$ transport in IOVs (Grinstein and Rothstein, 1978; Sze and Solomon, 1979). In one-step IOVs, which retain $Ca^{2+}$-induced $K^+$ transport, calmodulin had no effect on this transport system (Garcia-Sancho *et al.*, 1982), although active calcium pumping in these IOVs was significantly stimulated by calmodulin (V. L. Lew, personal communication).

### 3. Drug Effects

Several compounds are known to be pharmacological *promoters* of the red-cell $Ca^{2+}$-induced $K^+$ transport. These agents include various metabolic inhibitors, propranolol, and related compounds, as well as the ionophore A23187 (see Section II-A), and probably all of them have multiple effects. NaF, e.g., blocks glycolysis and, thus, decreases ATP level, which in turn allows calcium entry into red cells. However, the concentration curve for NaF induction of $K^+$ transport has a bell shape with a maximum at about 40 mM NaF, distinctly higher concentrations than required for a complete inhibition of glycolysis (Wilbrandt, 1937, 1940; Gárdos and Straub, 1957; Gárdos, 1972). Propranolol and pronethalol, in concentrations of $10^{-6}$–$10^{-7}$ M, are known inhibitors of β-adrenergic receptors, but the induction of $Ca^{2+}$-dependent $K^+$ transport requires them in concentrations of $2–5 \times 10^{-4}$ M. The β-receptor blocking effect is clearly distinguishable from the induction of $K^+$ movement, as, e.g., Trasicor, a compound closely related to propranolol and having similar potency in β-blocking, has practically no effect on red-cell $K^+$ transport (Sarkadi and Gárdos, unpublished results). The membrane action of propranolol and pronethalol is probably related to their local anesthetic properties and involves the liberation of membrane-bound calcium as well as a perturbation of the lipid structure (see Porzig, 1975; Szász *et al.*, 1977). It is worth noting that propranolol and related drugs are potent inhibitors of calmodulin action in the concentrations required to provoke $Ca^{2+}$-induced $K^+$-transport (Volpi *et al.*, 1981). Both A23187 (which evokes rapid $K^+$ transport by increasing calcium influx), and propranolol were shown to increase the sensitivity of the $K^+$-transport system for calcium (Lew and Ferreira, 1976; Gárdos *et al.*, 1977).

Table 1 lists some of the drugs which were shown to *inhibit* $Ca^{2+}$-induced $K^+$ transport in red cells, and summarizes their suggested mechanisms of action. These inhibitors are nonspecific, at least in the sense that they also inhibit several other transport or enzymatic functions. However, under well-defined conditions, the concentrations blocking $Ca^{2+}$-induced $K^+$ transport can be distinguished from those required to inhibit other systems. The most promising compounds in this respect are oligomycin and quinine. Both drugs were shown to inhibit all types of $Ca^{2+}$- or $Pb^{2+}$-

*Table 1. Inhibitors of $Ca^{2+}$-Induced $K^+$ Transport in Human Red Cells*

| Inhibitor | Possible mechanism of action | Order of $K_i$ | References |
|---|---|---|---|
| Oligomycin | Unknown, relation to inhibition of electron transport systems? | 2–3 μg/ml (?) (mixture of A,B,C) | Blum and Hoffman (1971), Riordan and Passow (1971), Gárdos *et al.* (1975a) |
| Quinine, quinidine, carbocyanine dyes | Reaction with the external $K^+$ (activator, modifier?) site | 5–100 μM (external K reduces inhibition) | Armando-Hardy *et al.* (1975), Reichstein and Rothstein (1981), Simons (1976c), Simons (1979) |
| Ethacrinate, mersalyl, PCMB | Oxidation of SH groups in the transport protein | 0.1–2.0 mM | Gárdos *et al.* (1975a) |
| Chlorobutanol, chlorpromazine, promethazine, trifluoperazine | Affecting lipid–protein interactions, also calmodulin inhibitory effects may be involved? | 10–200 μM | Gárdos *et al.* (1975a), Szász and Gárdos (1974), Szász *et al.* (1978a), Gárdos *et al.* (1976) |
| TEA, TMA | Blocking $K^+$ channels? | 5–20 mM | Gárdos *et al.* (1975a), Simons (1976c) |
| Cetiedil | Blocking $K^+$ transport irreversibly | 0.1 mM | Berkowitz and Orringer (1981) |
| Ouabain | Indirect, probably sparing ATP by blocking the $Na^+$, $K^+$ pump | As for the $Na^+$, $K^+$ pump | Blum and Hoffman (1971), Blum and Hoffman (1970), Lew (1971b), Lew (1974) |
| SITS, DIDS, dipyridamole | Indirect, by blocking inorganic anion transport | As for anion transport | Gárdos *et al.* (1975a), Hoffman and Knauf (1973), Macey *et al.* (1978), Eaton *et al.* (1980) |

induced $K^+$ transports in red cells, and in the case of quinine, the probable site of action has also been revealed. The basis of transport inhibition is an interaction with the externally located $K^+$-modifying site (see Table 1). It is important to note that both oligomycin and quinine are hydrophobic compounds and they probably reach sites which are in a hydrophobic environment. The relative specificity of quinine and quinidine inhibition of $Ca^{2+}$-induced $K^+$ transport could be used to identify this process in various complex cells (see Section III). Cetiedil, an *in vitro* antisickling agent, was shown to block $Ca^{2+}$-induced $K^+$ transport irreversibly. This compound does not inhibit passive $Ca^{2+}$ movement or anion transport in red cells but slightly enhances $Na^+$ transport (Berkowitz and Orringer, 1981). The action of ouabain on $Ca^{2+}$-induced $K^+$ transport is seen only under certain specific conditions, when cellular ATP levels and calcium concentrations are modified by the inhibition of the pump (Lew and Ferreira, 1978). In fact, ouabain can be used in various tissues to differentiate movements through the pump and the $Ca^{2+}$-induced pathway, respectively. The inhibitors of anion transport slow down net $K^+$ movement and increase membrane hyperpolarization, but they do not affect $Ca^{2+}$-induced $K^+$ transport under equilibrium-exchange conditions.

## *III. $Ca^{2+}$-INDUCED $K^+$ TRANSPORT IN COMPLEX CELLS*

In the early 1970's several investigators reported the activation of a selective $K^+$ transport mechanism by increased intracellular calcium concentrations in various cell types. Reviews by Meech (1976, 1978), Lew and Ferreira (1978), and Putney (1979) summarized these findings and gave new initiative for a further investigation of such phenomena in single cells as well as in complex tissues. As mentioned by Putney (1979), the Gárdos phenomenon, which was earlier suspected of being a "red herring" characteristic only of "dying" erythrocytes, by now is thought to be a widespread control system in animal cell membranes. Studies in complex cells are certainly complicated by the fact that both calcium influx from the extracellular space and calcium liberation from intracellular stores can produce a significant increase in cytoplasmic calcium levels, and this level is regulated by various intracellular binding and transport processes (Carafoli and Crompton, 1978). Still, $Ca^{2+}$-induced $K^+$ transport seems to be an important membrane process in numerous cell types. In the following, we summarize some of the recent discoveries in this field and consider especially those data which emphasize the physiological importance, or may help in the understanding of the molecular nature of this transport system.

### *A. Nonexcitable Cells*

#### *1. Animal Red Cells*

Results from studies directed to investigate the presence of a $Ca^{2+}$-induced $K^+$ transport in various animal red cells showed that among mammals, rat, coypu, and guinea pig red cells have a large $Ca^{2+}$-dependent increase in their $K^+$ transport, similar to that found in human red cells. In contrast, in sheep, cow, or goat red cells, this phenomenon is either inactive or absent (Jenkins and Lew, 1973). As shown by Brown *et al.* (1978), $Ca^{2+}$-induced $K^+$ transport is present in fetal sheep red cells but absent from reticulocytes or red cells of the adult sheep. Dog red cells, which lack a $Na^+$,$K^+$ pump and regulate their volume mostly by a $Na^+/Ca^{2+}$ countertransport system (Parker *et al.*, 1975; Parker, 1978) were shown to exhibit a $Ca^{2+}$-stimulated increase in $K^+$ permeability. In metabolically depleted dog red cells, $Ca^{2+}$ evokes a rapid net $K^+$ uptake which is inhibited by quinine or oligomycin (Richhardt *et al.*, 1979). In a recent communication, Parker (1983) described the reinvestigation of the hemolytic effects of potassium salts in dog red cells, first reported by Davson (1942), and showed that the basis of this effect is a $Ca^{2+}$-induced, quinine-sensitive $K^+$ transport. In $Na^+$-free media, dog red cells rapidly accumulate $Ca^{2+}$ which, in turn, initiates a specific increase in $K^+$ permeability. The rate of the colloid–osmotic type hemolysis produced by this $K^+$ influx depends on the nature of the anion present in the incubation medium. The more permeable the anion, the faster is the salt influx and the hemolytic response.

In the nucleated red cells of *Amphiuma,* Lassen *et al.* (1973, 1976) first demonstrated a $Ca^{2+}$-induced hyperpolarization of the membrane which could be related to an increase in $K^+$ permeability. Membrane potential and $K^+$ transport measurements in *Amphiuma* red cells were highly stimulating for the characterization of net cation and anion permeabilities in red cells (see Section II-D). In fish and chicken nucleated red cells Marino *et al.* (1981) showed the presence of an A23187 ionophore + $Ca^{2+}$-induced, quinine-sensitive rapid $K^+$ transport. It is important to note that $K^+$ efflux in these cells is compensated by a ouabain-insensitive $Na^+$ influx, thus, the selectivity of the $Ca^{2+}$-dependent increase in cation permeability is strongly reduced (or $Na^+$ transport is activated by a parallel or consecutive mechanism during $Ca^{2+}$-induced $K^+$ transport).

### 2. Leukocytes and Platelets

In *lymphocytes,* the role of an increase in cytoplasmic calcium concentration during blast transformation is well documented. An early calcium influx or rapid calcium liberation from cellular stores seems to have a basic function in initiating cell multiplication (Whitney and Sutherland, 1972; Freedman *et al.*, 1975; Freedman, 1979; Tsien *et al.*, 1982; Deutsch and Price, 1982). Rapid changes in $K^+$ transport under these conditions have also been reported, and these effects may be consistent with a $Ca^{2+}$-dependent activation of a leak $K^+$ transport and a parallel or consequent activation of the Na,K pump (Quastel and Kaplan, 1970; Segel *et al.*, 1979; Holian *et al.*, 1979). Lymphoblasts or lymphocytes, when placed into hypotonic media, regulate their volume by a rapid $K^+$ loss (Roti-Roti and Rothstein, 1973, Doljanski *et al.*, 1974) which was indicated to be a $Ca^{2+}$-dependent phenomenon (Grinstein *et al.*, 1982a). The effect does not require the presence of extracellular calcium but is strongly inhibited by $Ca^{2+}$-depletion of the cells and can be instantaneously restored by the addition of A23187 + $Ca^{2+}$. In the presence of cholinergic agents or A23187 + $Ca^{2+}$, the $K^+$ permeability of isotonic lymphocytes increases significantly (Szász *et al.*, 1980; Grinstein *et al.*, 1982a) and both these effects and the volume-induced increase in $K^+$ transport are blocked by quinine, oligomycin, chlorobutanol, or trifluoperazine (Figure 6; Szász *et al.*, 1982; Grinstein *et al.*, 1982a; Sarkadi *et al.*, 1983).

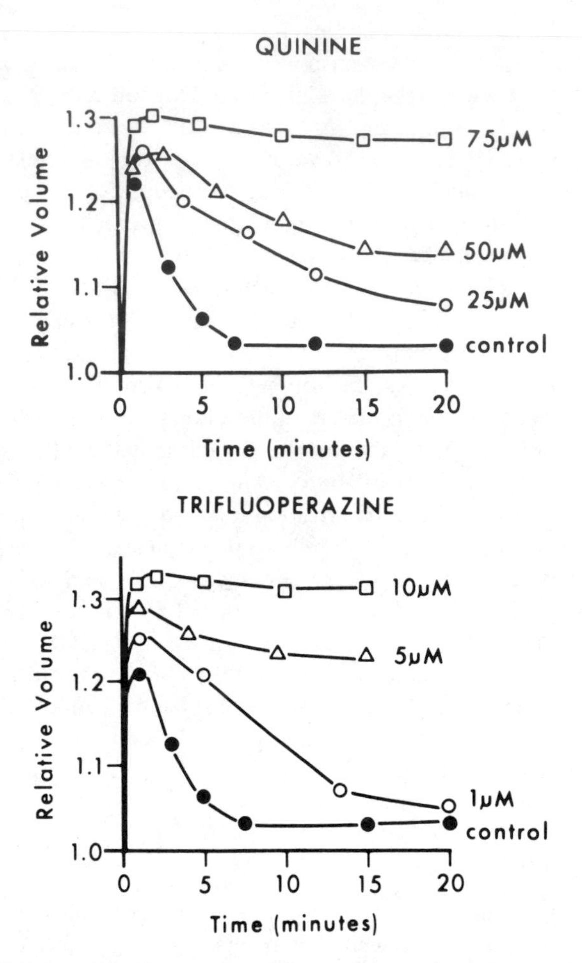


*Figure 6.* Effect of quinine and trifluoperazine on the regulatory volume decrease in human peripheral blood lymphocytes. Cells were hypotonically challenged (0.67 × isotonic) in media containing the indicated concentrations of the drugs, and their volume was recorded with a Coulter counter. For further details see the original paper. From Grinstein *et al.* (1982a).

In *neutrophils,* the secretion of lysosomal enzymes is most probably a $Ca^{2+}$-induced phenomenon, and the secretory effects produced by chemotactic peptides or A23187 + $Ca^{2+}$ are accompanied by an increase in $K^+$ movement in these cells (Naccache *et al.,* 1979). The ionic events in neutrophil chemotaxis and secretion are discussed in detail by Sha'afi and Naccache (1981).

As shown by direct microelectrode measurements, the slow hyperpolarization in activated *macrophages* is based on a $Ca^{2+}$-induced increase in $K^+$ permeability of the cell membrane (Oliveira-Castro and Dos Reis, 1981).

An A23187 + $Ca^{2+}$-induced, quinine-, chlorobutanol-, and chlorpromazine-sensitive (but ouabain-insensitive) increase in $K^+$ ($Rb^+$) uptake has been demonstrated in human *platelets* by Szász, Sarkadi, and Gárdos (unpublished experiments).

### 3. Cultured Cells

Recent studies by Valdeolmillos *et al.* (1982) demonstrated the presence of a $Ca^{2+}$-induced selective $K^+$ transport in Ehrlich ascites tumor cells. A23187 + $Ca^{2+}$, propranolol, or artificial electron donor systems induce a net $K^+$ loss which is balanced by net $Na^+$ influx. The initial effect of $Ca^{2+}$ is probably an increase in $K^+$ conductance, which may stimulate $Na^+$ uptake through membrane hyperpolarization. The effect of $Ca^{2+}$ on $K^+$ movement is inhibited by quinine (but not by oligomycin), and has a transient character similar to that observed in some other nucleated cells (see below).

Cultured mouse *fibroblasts* (L-cells) exhibit repeated hyperpolarizing responses, and chemical, mechanical, or electrical stimuli also produce a hyperpolarization which is probably connected to the initiation of endocytic activity or cell motility. The hyperpolarization response is based on a $Ca^{2+}$-induced increase in the $K^+$ permeability of the membrane (Nelson and Henkart, 1979; Okada *et al.*, 1979), which is blocked by quinine, quinidine, or TEA analogues (Okada *et al.*, 1982). The latter authors showed that blockers of $Ca^{2+}$ channels inhibit, while $Ca^{2+}$-pump inhibitors facilitate, hyperpolarization in intact cells. These drugs have no effect on hyperpolarization in calcium-injected L-cells. Okada *et al.* (1982) concluded that the transient increase in $K^+$ permeability and the consequent hyperpolarization are induced by transient elevations in cytoplasmic calcium concentrations through a balance between $Ca^{2+}$ channels and active calcium pumping.

### 4. Secretory Tissues

In several cell types, the second messenger of the hormonal regulation of secretion seems to be the intracellular calcium (Rasmussen and Goodman, 1977). A response generally present in stimulation is an increase in the $K^+$ permeability of the membrane, apparently controlled by the cytoplasmic ionized calcium concentration. The source of this calcium may be either the extracellular environment or some intracellular $Ca^{2+}$ stores (for review see Putney, 1979).

In *salivary glands* and in *lacrimal glands*, it has been shown that calcium is an obligatory intermediate during muscarinic or α-adrenergic stimulation of secretion and $K^+$ movement (Batzri *et al.*, 1973; Schramm and Selinger, 1975; Putney, 1978, 1979). The two phases (a rapid initial and a sustained phase), generally observed in the secretagogue-induced $K^+$ loss, probably reflect a rapid $Ca^{2+}$ release from cellular stores and a following $Ca^{2+}$ influx from the extracellular space, respectively (Putney, 1976). In rat parotid gland, a $Ca^{2+}$-dependent desensitization or inactivation of the $K^+$ transport was also observed, which may be due to an enzymatic control of this transport system (Putney *et al.*, 1977, 1978; Parod and Putney, 1978). It is important to mention that changes in cellular calcium in secretory cells probably also affect other ion-transport systems (especially $Na^+$ movements), and that the transient changes in membrane potential are mainly due to the consecutive activation of these transport pathways. By using more or less specific inhibitors of the various ionic fluxes, it has been indicated that $Ca^{2+}$-activated $K^+$ transport and $Na^+$ fluxes occur through separate pathways (Parod and Putney, 1978). A $Ca^{2+}$-induced, quinine- and trifluoperazine-sensitive $K^+$ efflux has recently been described in rat submandibular glands (Kurtzer and Roberts, 1982). It is worth mentioning that apparently there is no adrenergic control in the exocrine pancreas and the Gárdos effect is probably also absent, that is, an increase in cellular calcium in this tissue does not provoke $K^+$ efflux (Schramm and Selinger, 1975). In secretory tissues, certain hormonal responses liberate cyclic AMP and this compound may affect $Ca^{2+}$-induced $K^+$ transport either by increasing $Ca^{2+}$ influx and liberation, or by modifying the $K^+$ transport system itself. In the present review, we can not discuss these effects in detail but mention some of the molecular characteristics in Section IV (for review see Rasmussen and Goodman, 1977).

In the *endocrine β-cells* from mouse pancreas, a calcium-activated, quinine-inhibited $K^+$ channel has been described and shown to be the predominant way to control membrane potential. Addition of glucose in these cells induces a membrane depolarization, while $Ca^{2+}$ release from cellular stores, such as mitochondria, hyperpolarizes the membrane by increasing $K^+$ permeability (Matthews, 1975; Atwater and Biegelman, 1976; Atwater *et al.*, 1979).

### 5. Liver Cells

In hepatocytes, which react to α-adrenergic agents, adenine nucleotides, and peptides like angiotensin II or vasopressin, by a membrane hyperpolarization and a net $K^+$ loss (Craig, 1958; Haylett and Jenkinson, 1972), such a response was shown to be connected to $Ca^{2+}$-induced increase in $K^+$ permeability of the membrane (Van Rossum, 1970; Haylett, 1976; Burgess *et al.*, 1979). Ca is released mostly from intracellular stores, and the nature of the $K^+$ permeability increase is transient; an inactivation of the transport system is observed (Burgess *et al.*, 1981). In isolated hepatocytes, the α-adrenoceptor-activated increase in $K^+$ permeability could be mimicked by the addition of A23187 ionophore + $Ca^{2+}$, and all these effects were inhibited by quinine (Burgess *et al.*, 1979). According to these authors, quinine did not inhibit the changes in cellular calcium level, thus, probably acting directly on the K-transport system as also indicated in red cells (see Section II-E). Further studies by the above group (Burgess *et al.*, 1981) showed that adrenergic- or A23187 + $Ca^{2+}$-triggered K loss in hepatocytes was specifically inhibited by a bee venom toxin, apamin. Apamin blocked $Ca^{2+}$-induced $K^+$ transport in hepatocytes in nanomolar concentrations, while it was ineffective in red cells even if applied in high concentrations and to both sides of the membrane. Rat hepatocytes, in contrast to several other mammalian liver cells, probably lack $Ca^{2+}$-induced $K^+$ transport (Burgess *et al.*, 1981).

## B. Excitable Cells

In cells which show complex voltage-dependent changes in their membrane permeability, the investigation of a $Ca^{2+}$-induced $K^+$ transport is inherently more difficult than in nonexcitable cells. The first demonstration of this transport system in an invertebrate nerve (Meech and Strumwasser, 1970) and in cat spinal neurons (Krnjevic and Lisiewicz, 1972; Feltz *et al.*, 1972) initiated an extensive search for similar phenomena and for their function and molecular mechanism. By now, it is generally accepted that, in addition to the classical voltage-sensitive $K^+$ channel producing a rapid hyperpolarization after $Na^+$-dependent depolarization in most excitable tissues, a $Ca^{2+}$-dependent $K^+$ current is also present in variable extent and importance in various cells. In the past few years, a real flood of papers documented the increasing interest in this topic, and exciting new techniques provided valuable information concerning the $Ca^{2+}$-induced $K^+$ transport pathway. Without the aim of even a nearly complete coverage, we summarize some of the data available in this respect below.

### 1. Invertebrate Excitable Tissues

By pressure injection of calcium salts into *Aplysia* and *Helix neurons* (Meech and Strumwasser, 1970; Meech, 1974), an external $K^+$-inhibited hyperpolarization of the cell membrane was observed which could be closely related to the so-called posttetanic hyperpolarization in these cells. In the firing neurons, as shown in cells injected with aequorin (Stinnakre and Tauc, 1973), or arsenazo III (Thomas and Gorman, 1977; Gorman and Thomas, 1978), the action potential increases $Ca^{2+}$ influx, and intracellular $Ca^{2+}$ ($Ca_i$) produces a hyperpolarization of the membrane due to an increased $K^+$ conductance. EGTA injection abolishes $Ca^{2+}$-dependent after-hyperpolarization (Meech, 1974, 1976).

Under voltage-clamp conditions, it has been demonstrated that the currents evoked by a depolarizing pulse consist of transitory inward $Na^+$ and $Ca^{2+}$ movements, and the prolonged outward current is carried by $K^+$ (Meech and Standen, 1975; Gorman and Hermann, 1979). It is important to note that the injection of Ca EGTA buffers provokes an increased $K^+$ conductance at free calcium levels of $10^{-7}$–$10^{-6}$ M, while unbuffered calcium has to be introduced in concentrations higher than $10^{-3}$ M to produce the same response (Meech, 1976). The 1000-fold higher effectivity of buffered calcium is partially explained by calcium "sinks" in these cells, but a higher apparent sensitivity of the $K^+$-transport system to buffered than to unbuffered $Ca^{2+}$, as indicated in red cells, can also play a role.

As shown in *Aplysia* pacemaker neurons under voltage-clamp conditions (Gorman and Hermann, 1979), the efflux of $K^+$ is a linear function of the $Ca^{2+}$-activated outward current, and while $Ca^{2+}$ can be substituted (in a decreasing order of effectivity) by divalent $Cd^{2+}$, $Hg^{2+}$, $Sr^{2+}$, $Mn^{2+}$, or $Fe^{2+}$ ions, $Ba^{2+}$ injection inhibits $K^+$ current. Tetraethylammonium (TEA) in high concentration (above 50 mM) inhibits $Ca^{2+}$-induced $K^+$ transport in molluscan neurons (Meech and Strumwasser, 1970), although the sensitivity of this system to TEA is much less pronounced than that of the depolarization-induced K movements (Hermann and Gorman, 1981).

In the pacemaker function of *Aplysia* neurons, the membrane oscillations are mostly produced by a voltage-dependent $Ca^{2+}$ influx and a $Ca_i$ + voltage-dependent increase in $K^+$ permeability. When active calcium extrusion and internal sequestration mechanisms reduce the cytoplasmic calcium concentration, $K^+$ current decreases, and the closure of $K^+$ channels results in a slow depolarization, initiating a $Ca^{2+}$ influx and, thus, starting the cycle again (Gorman *et al.*, 1982).

A recent development in these studies has been the demonstration of cyclic AMP-dependent modulation of the $Ca^{2+}$-induced $K^+$ transport in molluscan neurons. Greengard (1978) first advocated that cAMP-dependent protein phosphorylation may regulate neuronal activity, and indeed, such a regulation was shown by experiments in which the catalytic subunit of the cAMP-dependent protein kinase (Kaczmarek *et al.*, 1980) or its specific protein inhibitor (Adams and Levitan, 1982) were injected into *Aplysia* neurons. DePeyer *et al.* (1982) recently reported that in perfused *Helix* neurons the intracellular addition of the catalytic subunit of the cAMP-dependent protein kinase selectively increased $Ca^{2+}$-induced $K^+$ current, possibly by increasing the affinity of the $K^+$ transport system for $Ca^{2+}$ ions. The interdependent actions of $Ca^{2+}$ ions and the catalytic subunit of the cAMP-dependent kinase suggest that these effectors may interact in the regulation of a single type of ionic channels.

In crustacean *muscle fibers,* the presence of a $Ca^{2+}$-induced $K^+$ current has also been indicated (Mounier and Vassort, 1975; Meech, 1976). In *Limulus* and *Balanus* (barnacle) photoreceptors, the adaptation to light by a decreased electrical activity is probably the result of a $Ca_i$-mediated $K^+$ current producing prolonged membrane hyperpolarization (Lisman and Brown 1972; Hanani and Shaw, 1977). In some invertebrate eyes, e.g., scallop, the photoreceptors respond to light with a hyperpolarization, which was indicated to be the result of $Ca_i$-induced $K^+$ current (Gorman and McReynolds, 1974).

## 2. Vertebrate Neurons

*In cat spinal motoneurons,* Krnjevic and Lisiewicz (1972) first demonstrated the direct involvement of intracellular calcium in the activation of a $K^+$ current, which is probably responsible for the long-lasting depolarization in these cells (Krnjevic *et al.*, 1975). In *frog motoneurons,* Barrett and Barrett (1976) showed the presence of $Ca^{2+}$-induced $Mn^{2+}$- or $Co^{2+}$-inhibited slow after-hyperpolarization, produced by a $K^+$ selective current. This transport is less sensitive to TEA than the fast, $Ca^{2+}$-independent $K^+$ current in the same tissue. It is interesting to note that motoneurons which innervate slow muscles have longer after-hyperpolarization than those innervating fast muscles, thus, the $Ca^{2+}$-induced $K^+$ current may play a key role in the regulation of the motoneuron discharge frequency (Meech, 1978).

In the *toad sympathetic neurons,* after-hyperpolarization is probably a result of the functioning of various $K^+$ channels, one of which is $Ca_i$ and voltage dependent. Both an increase in cellular calcium and a positive shift in the membrane potential activate this $K^+$ current (Adams *et al.*, 1982a,b). Caffeine and theophylline potentiate $Ca^{2+}$-induced $K^+$ current in these cells, probably by releasing $Ca^{2+}$ from cellular stores (Kuba and Nishi, 1976). According to Adams *et al.* (1982a), the voltage sensitivity of the $K^+$ channels is independent of the voltage sensitivity of the $Ca^{2+}$ channels in the membrane. The authors used the "patch-clamp" method to study the characteristics of the $K^+$ channel. In this technique, intact cells or isolated cell membranes can be attached to the electrodes and the currents registered under voltage-clamp conditions. Specific sides of the patched membrane can be exposed to changes in the incubation media and, thus, the sidedness of the ionic and drug effects studied. Adams *et al.*

(1982a) showed that cytoplasmic calcium evokes a single-channel, short-lifetime $K^+$ conductance which is inhibited by external TEA. The maximum single-channel conductance was about 100 p*S* in these experiments.

A similar, $Ca^{2+}$-induced $K^+$ conductance was reported to be present in guinea pig *myenteric neurons* (North, 1973; Grafe *et al.*, 1980), and a basic role of $Ca^{2+}$ entry during action potential was indicated (Morita *et al.*, 1982). Certain neurotransmitters reduce $Ca^{2+}$ influx and, as a consequence, they decrease the conductance for $K^+$ through the $Ca^{2+}$-activated channels (Grafe *et al.*, 1980). The $K^+$-dependent after-hyperpolarization is eliminated in $Ca^{2+}$-free solutions, or by the extracellular addition of $Co^{2+}$ ions, inhibiting $Ca^{2+}$ entry. The $K^+$ response is prolonged by high external $Ca^{2+}$ concentrations, and by the addition of TEA or caffeine (Morita *et al.*, 1981). TEA in low concentrations causes only a partial inhibition of $K^+$ transport, but increases the driving force for $Ca^{2+}$ entry, while caffeine probably decreases intracellular calcium sequestration (Kuba, 1980).

In *hippocampal CA1 pyramidal cells,* an epileptiform burst after-hyperpolarization was indicated to be a calcium-induced $K^+$ potential. This action potential-induced hyperpolarization was inhibited if the media contained low $Ca^{2+}$ and higher $Co^{2+}$ concentrations. The phenomenon was suggested to play an important role in the regulation of the electrical activity of these cells (Alger and Nicoll, 1980; Nicoll and Alger, 1981).

In cultured *neuroblastoma cells,* which provide a useful model system for studying ionic currents in mammalian excitable cells, a prolonged after-hyperpolarization was identified as $Ca^{2+}$-induced $K^+$ current (Moolenaar and Spector, 1979a, 1979b; Boonstra *et al.*, 1981). High $Ca^{2+}$ concentrations enhance this TEA-resistant, long-lasting hyperpolarization, which considerably reduces the firing frequency. Thus, $Ca^{2+}$-induced $K^+$ transport regulates nerve activity also in these cells. In voltage-clamped neuroblastoma cells, the above authors showed that $K^+$ conductance is activated both by cellular calcium and membrane depolarization. A recent report by Yellen (1982), by using the patch-clamp technique in neuroblastoma cells, described the presence of a nonselective, calcium-activated cation channel (almost equally permeable to $Na^+$, $K^+$, $Li^+$, or $Cs^+$ ions) in this tissue. The single-channel conductance in this case was about 22 p*S*. The author also mentioned unpublished experiments in which he showed the presence of $Ca^{2+}$-activated $K^+$-selective channels in the same membrane preparations. The single-channel conductance in this case was about ten times higher (that is about 200 p*S*).

Also by using neuroblastoma cell lines, Hugues *et al.* (1982a) showed that $Ca^{2+}$-induced $K^+$ current is specifically blocked by nanomolar concentrations of apamin (see also guinea pig hepatocytes). The receptor for apamin binding could be destroyed by external proteolytic digestion, thus, the binding site is probably a protein, located on the cell surface. The number of the apamin-binding sites, in correlation with the $Ca^{2+}$-induced $K^+$ current, was significantly increased by the differentiation of the neuroblastoma cells, and apamin binding was reduced by external divalent and monovalent cations. The maximum number of the $Ca^{2+}$-activated $K^+$ channels was calculated to be about one-fifth of the fast $Na^+$ channels. We mention here that apamin binding was also examined by this group in synaptosomes where the biochemical and ion-interacting studies gave similar results and, moreover, by a covalent-labeling method, the apamin receptor protein could be isolated (Hugues *et al.*, 1982b; see Section IV).

The *chromaffin cells,* isolated from *adrenal medulla,* provide another useful and widely applied model system for studying mammalian excitable tissues. These cells, which originate from the neural crest of the embryo, respond to specific stimuli and have $Na^+$ and $K^+$ permeability systems similar to those in nerve cells. In chromaffin cells, by using the patch-clamp method, Marty (1981) showed the presence of an intracellular $Ca^{2+}$-induced selective $K^+$ current, which could be distinguished from other currents in this membrane. The $Ca^{2+}$-activated appearance of $K^+$ current with large single-channel conductance ($\approx$180 p*S*) did not require the presence of any soluble cytoplasmic substances in the isolated membrane patches, and the channels at low calcium concentrations were sensitive to changes in membrane potential.

### 3. Cardiac Purkinje Fibers

In this tissue, the action potential has a plateau phase which is probably caused by a slowly developing inward $Ca^{2+}$ current, generated by the fast initial $Na^+$ inward current. Rapid repolarization involves a transient outward current, while the termination of the plateau is most probably the result of a slowly developing, $Ca^{2+}$-induced outward $K^+$ current (Bassingthwaite *et al.*, 1976). Isenberg (1975) showed that $Ca^{2+}$ injection into Purkinje fibers, due to the development of a $K^+$ current, shortens the plateau, whereas, intracellular EGTA prolongs the depolarization phase. Calcium antagonists, such as divalent $Mn^{2+}$ or trivalent $La^{3+}$ ions inhibit both inward $Ca^{2+}$ and outward $K^+$ currents (Kass and Tsien, 1976). According to Siegelbaum *et al.* (1977), transient outward current was also inhibited when D600 (methoxy-verapamil, a calcium channel blocker) or $Mn^{2+}$ was added to the medium. By using a microelectrode voltage-clamp technique, Siegelbaum and Tsien (1980) showed that transient outward current is in fact a $Ca^{2+}$-stimulated $K^+$ current. TEA or 4-aminopyridine (4-AP) inhibit this current (Isenberg, 1978), but the investigation of $Ca^{2+}$ and $K^+$ currents is very difficult in their overlapping presence. A recent study by Marban and Tsien (1982) showed that Purkinje fibers could be loaded with $Cs^+$ ions by using the ionophore nystatin. Since $Cs^+$ ions do not ride the $K^+$ channels, they abolish transient outward current. These investigations provided further indications for the independence of $Ca^{2+}$ current and $Ca^{2+}$-induced $K^+$ transport in cardiac fibers.

### 4. Smooth Muscle

In this contractile tissue, the investigation of the $Ca^{2+}$-induced $K^+$ transport in the cell membrane is complicated by the presence of other $Ca^{2+}$-dependent phenomena, especially those involved in muscle contraction. Variation in the surface receptors and in other membrane features of different smooth muscles is also abundant. A relatively well-characterized system is the guinea pig *tenia coli,* in which catecholamines inhibit contraction. This effect has been indicated to depend on the hyperpolarization of the cell membrane, produced by a $Ca^{2+}$-dependent $K^+$ current (Brading *et al.*, 1969; Bülbring and Tomita, 1977; Walsh and Singer, 1980). Apamin was shown to prevent adrenergic and other agents (such as ATP) to inhibit contraction in guinea pig visceral muscle (Banks *et al.*, 1979).

The inhibitory effect of catecholamines on the *uterus muscle* contraction in certain species is probably also mediated by $Ca^{2+}$-induced $K^+$ permeability pathway (Marshall, 1977). By using a double sucrose-gap technique, Moronneau and Savineau (1980) studied the $Ca^{2+}$-induced outward $K^+$ current in rat uterine muscle, and found a fast and a slow component of this $K^+$ current, respectively. Depolarization activates these pathways, and they are both blocked by $Mn^{2+}$ (which inhibits $Ca^{2+}$ entry), or by TEA. The fast component was shown to be specifically inhibited by 4-AP. In the majority of smooth muscles, where catecholamines induce contraction, the complex

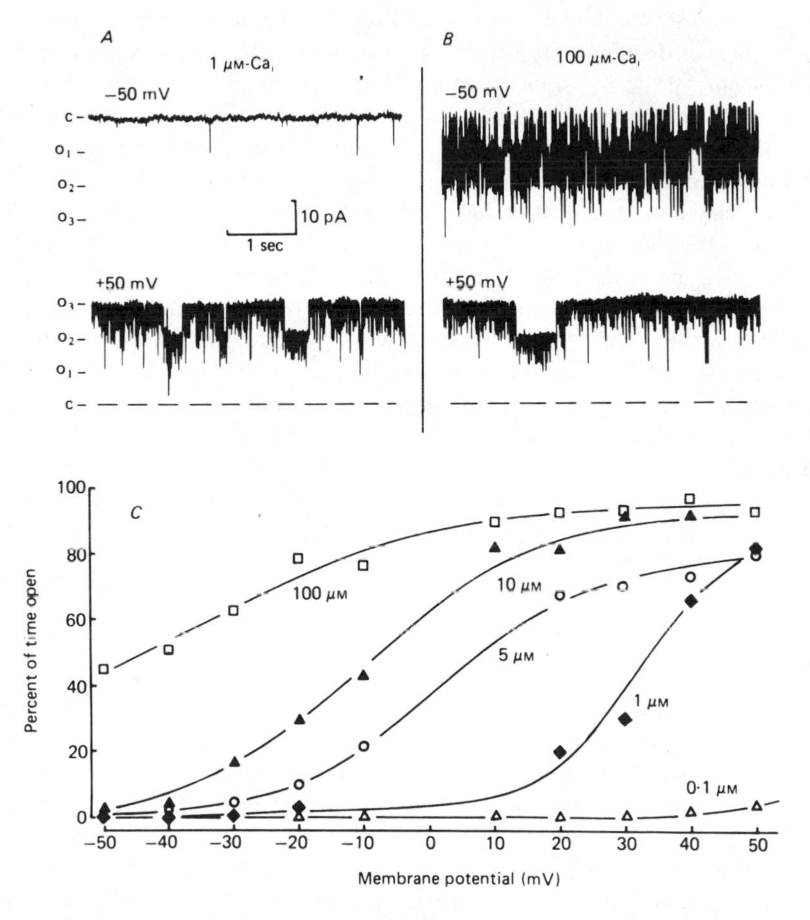


*Figure 7.* $Ca^{2+}$-induced $K^+$ channels in excised patches of cultured muscle cell surface membranes. Effects of $Ca^+$ concentration and membrane potential on channel activity. (A and B) Records of membrane currents at two membrane potentials, -50 mV (upper traces) and +50 mV (lower traces). (A) 1 μM $Ca_i$, (B) 100 μM $Ca_i$. Channel currents are downward-going at -50 mV and upward-going at +50 mV. (C) Plot of percentage of time channels were open (percent of time open) against membrane potential for the indicated $Ca_i$. From Barrett *et al.*, (1982).

action of $Ca_i$ on $Na^+$ and $K^+$ fluxes has been indicated, and these effects have not been properly distinguished as yet (Meech, 1978).

### 5. Skeletal Muscle

In this complex excitable tissue, a $Ca^{2+}$-induced $K^+$ transport was indicated to be present by the studies of Fink and Lüttgau (1976) and Barrett *et al.* (1981), but the phenomenon could be reliably studied only under special experimental conditions with special methods. By using the patch-clamp technique, Pallotta *et al.* (1981) demonstrated $Ca^{2+}$-activated selective $K^+$ channels in cultured rat myotubule membranes. As analyzed in detail in a recent paper by the same group (Barrett *et al.*, 1982), the opening frequency and the effective open times for $K^+$ channels depend on both cytoplasmic $Ca^{2+}$ concentration and the membrane potential. Depolarization facilitates channel opening, but at high calcium levels, the channels are almost continuously open even at hyperpolarized membrane potentials (Figure 7). The single-channel conductance is independent from cytoplasmic calcium concentration or voltage, and has a value between 100–300 p*S*, depending on the temperature. It has to be noted that cytoplasmic soluble components were not required to obtain these $Ca^{2+}$-induced $K^+$ channels, but such factors may modify their *in vivo* characteristics.

Another effective approach for studying ion conductance in muscle membranes was applied by Latorre *et al.* (1982). They isolated T-tubule membrane vesicles from rabbit skeletal muscle, and then fused these vesicles with a planar bilayer membrane. In these experiments, a $Ca^{2+}$-induced selective $K^+$ conductance was observed, again with a single-channel conductance of about 200 p*S*. Voltage and $Ca^{2+}$ ions affected this conductance similarly to that found in patch-clamped muscle membranes (Barrett *et al.*, 1982). Divalent ions such as $Mg^{2+}$, $Ba^{2+}$, or $Co^{2+}$ did not evoke $K^+$ conductance, and TEA inhibited the $Ca^{2+}$-induced response when applied to the "external" side. Such studies, when combined with thorough biochemical investigations are most promising for clarifying the molecular nature of the $Ca^{2+}$-induced $K^+$ transport.*

## IV. THE MOLECULAR BASIS OF $Ca^{2+}$-INDUCED $K^+$ TRANSPORT

The first question which arises in this context is whether the $Ca^{2+}$-induced $K^+$ transport pathways described in various tissues represent the same molecular machinery or whether just the phenomenology of their behavior is similar. It is impossible to give a definite answer to this question at present but here we favor a "unitarian" approach, that is, suppose that the basic molecular mechanism is common in these pathways. We may suppose that in human red cells, only pathological states or the artifactual conditions devised by experimental biochemists provoke the activation of

* $Ca^{2+}$-induced $K^+$ transport, producing membrane hyperpolarization, was also shown to be present in single-cell organisms such as *Paramecium* (Eckert, 1977; Brehm *et al.*, 1978; Satow and Kung, 1980).

this system. In contrast, in various complex cells, this transport pathway is used for important functions during cell life (see Section V).

If we try to pack all the available information into one molecular model, we find some characteristics which easily suit such a treatment, while other features do not seem to fit into the picture. Without fabricating a "Procrustes-bed" model, in the following, we summarize these "common" and "specific" characteristics of the $Ca^{2+}$-induced $K^+$ transport in various cell membranes.

## A. The Transport Molecule(s)—Biochemical Characteristics

The molecular basis of the $Ca^{2+}$-induced $K^+$ transport is most probably a membrane-bound *polypeptide (protein)*, which has *hydrophobic* characteristics. The high *selectivity* of this transport system is one of the features which advocates the involvement of a specific peptide structure. In most tissues, as discussed in Sections II and III, the $Ca^{2+}$-induced pathway carries only $K^+$ and $Rb^+$ ions, while $Cs^+$, $Na^{2+}$, $Ca^{2+}$, etc. ions are not transported. In some tissues, such as the nucleated red cells of fish and chicken (Marino *et al.*, 1981), Ehrlich ascites cells (Valdeolmillos *et al.*, 1982), or secretory tissues (see Putney, 1979), a parallel increase of $Na^+$ movement with the $Ca^{2+}$-induced $K^+$ transport has also been observed. However, chances are high that this is not due to a decreased selectivity of the $Ca^{2+}$-induced $K^+$ transport, but separate pathways are activated for $Na^+$ movement. An observation which suggests a heterogeneity in the transport characteristics of the $Ca^{2+}$-induced $K^+$ transports in different tissues is, that in muscle T-tubule membranes, when incorporated into planar bilayers, the pathway is blocked by $Rb^+$ ions (R. Latorre, personal communication).

The role of proteins in the $Ca^{2+}$-induced $K^+$ transport is shown by the *inhibition* of this pathway *by SH-blocking* reagents (Table 1) and by its *specific modulation* by internal $Ca^{2+}$, $K^+$, $Na^+$, and external $K^+$ *ions*, as described in Section II-C. This side-dependent regulation by ion binding is generally characteristic for membrane spanning transport proteins, such as the $Na^+$,$K^+$ pump. These features, the $K^+$-"countertransport" and the ouabain inhibition of the $Ca^{2+}$-induced $K^+$ transport observed in their experiments, prompted Blum and Hoffman (1971) to suggest that the molecule responsible for this transport process is an altered, $Ca^{2+}$-inhibited conformation of the $Na^+$,$K^+$ pump. The validity of this assumption is challenged by the following arguments: (1) ouabain inhibition is not a general feature of this transport system but is indirect, and appears only under specific conditions (Lew, 1971b, 1974; Lew and Ferreira, 1978), (2) $Ca^{2+}$-activation of $K^+$ transport and $Ca^{2+}$ inhibition of the $Na^+$,$K^+$ pump show different kinetics (Yingst and Hoffman, 1984), (3) in several tissues, a parallel functioning of the $Ca^{2+}$-induced $K^+$ transport and the $Na^+$,$K^+$ pump has been observed (Putney, 1979), (4) $Ca^{2+}$-induced $K^+$ transport is present in mammalian red cells in which $Na^+$,$K^+$ pump is absent (Richhardt *et al.*, 1979; Parker, 1983), and (5) the isolated and reconstituted $Na^+$,$K^+$ pump does not exhibit a $Ca^{2+}$-induced $K^+$ transport (Karlish *et al.*, 1981). The hypothesis that the $Ca^{2+}$-induced $K^+$ transport would represent an altered functioning of the calcium pump (Lassen *et al.*, 1980) can be also strongly criticized by a number of arguments, such as the absence of lanthanum inhibition of the $Ca^{2+}$-induced $K^+$ transport in the presence of saturating cellular $Ca^{2+}$-concentrations (Szász *et al.*, 1978d), the lack of effect of quinine or oligomycin on the calcium pump (Sarkadi *et al.*, 1977, 1980), etc. The role of calmodulin, or other cytoplasmic proteins in the functioning of $Ca^{2+}$-induced $K^+$ transport in red cells has been suggested, but not satisfactorily demonstrated as yet (see Section II-E.2).

It has to be confessed that we hardly know anything about the nature of the membrane protein responsible for $Ca^{2+}$-induced $K^+$ transport, and still lack a decent technique for an affinity-binding type isolation of this molecule. Careful biochemistry, such as antibody characterization, protease digestion, covalent modification, and separation, has hardly been done on this system until now. A good "tagging" with a specific ligand would be of great importance to facilitate this work. Quinine or oligomycin, although strong inhibitors of this system in most tissues, are required in much higher concentrations than would suggest a stoichiometric interaction with the transport system. Apamin inhibits $Ca^{2+}$-induced $K^+$ transport in several complex cells in nanomolar concentrations (Banks *et al.*, 1979; Burgess *et al.*, 1981; Hugues *et al.*, 1982a), and indeed, an irreversible labeling with this toxin allowed the isolation of an apamin-receptor protein with a molecular weight of 28,000 from synaptosome membranes (Hugues *et al.*, 1982b). However, since apamin does not block $Ca^{2+}$-induced $K^+$ transport in red cells (Burgess *et al.*, 1982) or in lymphocytes (S. Grinstein, personal communication), the general application of this method is not possible. It is also not known whether the apamin-binding protein is the transport molecule itself or at least a part of it, and this question can be answered only by isolation and incorporation studies. The variable effect of apamin raises the question again; do the $Ca^{2+}$-induced $K^+$ transport systems with different sensitivity to apamin represent similar molecular entities? If so, they are certainly different with respect to an extracellularly located apamin-binding unit or subunit, which when liganded with apamin, blocks $K^+$ transport.

An important characteristic of this transport system is its specific activation by calcium ions. In most tissues, only $Pb^{2+}$ and $Sr^{2+}$ ions can substitute for calcium in this role, although some activation of $K^+$ transport was indicated by $Cd^{2+}$ and $Hg^{2+}$ ions in *Aplysia* neurons (Gorman and Hermann, 1979). The question, whether the system is activated by the binding of one or more $Ca^{2+}$ ions, is still open. A slight sigmoidicity of the curve of the $Ca^{2+}$-dependence of $K^+$ transport activation was observed by Garcia-Sancho *et al.* (1982) in red cell IOVs, but the reproducibility of this finding has not been reinforced as yet. The findings of Barrett *et al.* (1982) on the $Ca^{2+}$-dependence of the $K^+$ current in patch-clamped muscle membranes are consistent with a model supposing the binding of two or more $Ca^{2+}$ to open the $K^+$ channel. $Mg^{2+}$, $Ba^{2+}$, $Mn^{2+}$, and $Co^{2+}$ ions inhibit calcium activation in most tissues (Figure 3).

In red cells, the complex actions of the monovalent cations $Na^+$ and $K^+$ in the regulation of $Ca^{2+}$-induced $K^+$ transport are well established (Section II-C), but in other tissues, there are no detailed studies available in this respect. The absence of external $K^+$ in contrast to red cells in some tissues does not seem to block $Ca^{2+}$-induced $K^+$ transport (Grinstein *et al.*, 1982). The possible involvement of a cAMP-dependent protein phosphorylation in the modulation of the calcium sensitivity of the $K^+$ transport (DePeyer *et al.*, 1982) indicates a complex regulation of this transport in the intact cells.

The *hydrophobic nature* of the $Ca^{2+}$-induced $K^+$ transport system is indicated by (1) the strongly membrane-bound character of the transport pathway which allows it to survive membrane preparation, (2) modification of the $Ca^{2+}$-affinity by hydrophobic molecules such as propranolol or the ionophore A23187 (see Section II-B), and (3) inhibition of the transport system by various alcohols such as heptanol and chlorobutanol, and by other hydrophobic compounds (Table 1). Based on experiments in which a $Ca_i$-induced breakdown of phospholipids and the formation of 1,2-diacylglycerol was shown in human red cells (Allan and Michell, 1975), the role of this process in the initiation of the $Ca^{2+}$-induced $K^+$ transport was suggested (Lew and Ferreira, 1978). Allan and his colleagues, in their follow-up work (Allan and Michell, 1977; Allan and Thomas, 1981a,b) could clearly separate the effects caused by diacylglycerol accumulation from the activation of the $K^+$ pathway. Grey and Gitelman (1979), by observing a correlation between $Ca^{2+}$-activated phospholipases and $Ca^{2+}$-stimulated $K^+$ transport under various conditions, also advocated the role of membrane lipid changes in the induction of rapid $K^+$ transport. Although a modifying role of membrane lipids on $Ca^{2+}$-induced $K^+$ transport is likely to be present, these slowly reversible changes can hardly play a significant role in the actual gating and transport process.

## B. Physical Characteristics

Most of the data available suggest that the $Ca^{2+}$-induced $K^+$ transport has the physical characteristics of a *conducting channel* in the cell membranes. The indications that the system would represent a $K^+$–$H^+$ exchange (Szönyi, 1960; Gárdos and Szász, 1968) or a carrier-mediated $K^+$–$K^+$ exchange (Manninen, 1970; Blum and Hoffman, 1971) in the red cell membrane were based on findings which can be well explained by changes in membrane potential, due to a low net anion permeability of the membrane (see Section II-D). Experiments in both non-excitable and excitable cells show that the $Ca^{2+}$-induced $K^+$ transport does not represent an electrically silent, obligatory exchange system.

The channel-type behavior of the $Ca^{2+}$-induced $K^+$ transport is suggested by the following observations: (1) kinetic data in red cells favor a model of $K^+$ diffusion channel (Schubert and Sarkadi, 1977), (2) the apparent low density and high $K^+$-transport capacity of the system in red cells (see Lew and Ferreira, 1978; Lew *et al.*, 1982) gives a transference number 2–3 orders of magnitude higher than that of the $Na^+$,$K^+$ pump, and this is characteristic for channels rather than for carriers, (3) patch-clamp experiments in red cells (Hamill, 1981) and in various complex cell membranes (Pallotta *et al.*, 1981; Adams *et al.*, 1982a; Yellen, 1982; Marty, 1981; Barrett *et al.*, 1982) showed $Ca^{2+}$-activated single-channel conductances with high conductivity for $K^+$, and (4) a bilayer incorporation study by Latorre *et al.* (1982) showed the presence of the same type of highly conductive, $Ca^{2+}$-activated $K^+$ channels in muscle T-tubule membranes. (For a recent review on conductivity and selectivity of potassium channels see Latorre and Miller, 1983).

In several tissues, a voltage sensitivity of the $Ca^{2+}$-induced $K^+$ transport was reported. As shown by direct measurements, (inside) positive membrane potentials facilitate the opening of the $K^+$ channels and/or increase the effective opening times (Adams *et al.*, 1982a; Barrett *et al.*, 1982, Latorre *et al.*, 1982). In all the cases, this effect was overcome by high calcium concentrations at the cytoplasmic side of the membrane. This finding indicates that it is the $Ca^{2+}$-receptor interaction which is modified by the actual membrane potential, and depolarization in fact may increase the local $Ca^{2+}$ concentration at the binding site and increase the apparent $Ca^{2+}$ affinity of the system (Gorman and Thomas, 1980; Barrett *et al.*, 1982).

Based on the above considerations, the most likely mechanism responsible for $Ca^{2+}$-induced $K^+$ transport is a *$Ca^{2+}$-gated, $K^+$-specific conductive channel,* which carries $K^+$ ions in the direction of the electrochemical potential gradient.

## V. CONCLUSIONS—THE PREVALENCE AND SIGNIFICANCE OF CALCIUM-INDUCED $K^+$ TRANSPORT

Intracellular $Ca^{2+}$-induced rapid $K^+$ transport is a phenomenon present in a wide range of animal cell membranes, probably from the unicellular organisms to the mammalian brain tissues. The transport process was first recognized in human red cells, and its biochemical and biophysical characteristics were extensively studied in the past 30–40 years. There is no known physiological function of $Ca^{2+}$-induced $K^+$ transport in human red cells; but increased intracellular calcium concentrations, especially when combined with a metabolic exhaustion of the cells, may provoke an activation of this pathway. Sheer stress in the capillaries significantly increases calcium influx into erythrocytes (Larsen *et al.*, 1981) and, in sickle cells (Eaton *et al.*, 1973) or in red cells with other genetic abnormalities (Feig and Bassilian, 1974; Wiley and Gill, 1976), intracellular calcium is constantly elevated. Oxidants or other hemolytic agents inhibit cell metabolism and calcium extrusion and, thus, also increase cellular calcium levels (Shalev *et al.*, 1981). Under these pathologic conditions, an increase in the rate of $K^+$ efflux, dehydration, and shrinkage of red cells were reported (Glader *et al.*, 1974; Wiley, 1981; Orringer and Parker, 1973; Parker *et al.*, 1978). In certain forms of sickle cell anemia, $K^+$ loss and dehydration are probably the direct causes of cell sequestration and destruction (Hellerstein and Bunthrarungoj, 1974; Masys *et al.*, 1974) and, in fact, some antisickling drugs act through the inhibition of the $Ca^{2+}$-induced $K^+$ transport (see Berkowitz and Orringer, 1981, 1983). Quinine, an inhibitor of this pathway, was also shown to prevent cell dehydration in anemia with pyruvate kinase deficient red cells (Koller *et al.*, 1979). $Ca^{2+}$-induced $K^+$ transport has been suggested to be an important factor in the sequestration and destruction of senescent erythrocytes, although no convincing evidence is available in this respect as yet (see Parker *et al.*, 1978).

As discussed in Section III of this review, in complex cells, the role of $Ca^{2+}$-induced $K^+$ transport was implicated in cell volume regulation, initiation of cell multiplication, exocrine and endocrine secretion, cellular activation by hormones and neurotransmitters, regulation of nerve firing and muscle contractile activity by the modification of the action potential, cell motility, etc. If only a fraction of this list proves to be valid, this phenomenon has a basic physiological importance in the regulation of cellular functions. Further research concerning the physiological, biochemical, and biophysical features of the $Ca^{2+}$-induced $K^+$ transport may soon change the situation characterized by Lew and Beaugé (1979). "We know nothing about the actual nature of the $K^+$ pathway but disguise our ignorance with lively controversies."

ACKNOWLEDGMENTS

The authors wish to thank Drs. Ilma Szász and Sergio Grinstein for their helpful comments on the manuscript.

## Note Added in Proof

A recent publication (*Cell Calcium,* Volume 4, Number 5–6, 1983) was devoted to the problem of $Ca^{2+}$-activated $K^+$ channels. Petersen and Maruyama have published a review article on the role of this process in secretion (*Nature* **307**:693–696, 1984).

## REFERENCES

Adams, P. R., Constanti, A., Brown, D. A., and Clark, R. B., 1982a, Intracellular $Ca^{2+}$ activates a fast voltage-sensitive $K^+$ current in vertebrate sympathetic neurons, *Nature* **296**:746–749.

Adams, P. R., Brown, D. A., and Constanti, A., 1982b, M-currents and other potassium currents in bullfrog sympathetic neurones, *J. Physiol. (London)* **330**:537–572.

Adams, W. B., and Levitan, I. B., 1982, Intracellular injection of protein kinase inhibitor blocks the serotonin-induced increase in $K^+$ conductance in Aplysia neuron R 15, *Proc. Natl. Acad. Sci. USA* **79**:3877–3880.

Alger, B. E., and Nicoll, R. A., 1980, Epileptiform burst after-hyperpolarization: Calcium-induced potassium potential in hippocampal pyramidal cells, *Science* **210**:1122–1124.

Al-Jobore, A., and Roufogalis, B. D., 1981, Influence of EGTA on the apparent $Ca^{2+}$-affinity of $Mg^{2+}$-dependent, $Ca^{2+}$-stimulated ATPase in the human erythrocyte membrane, *Biochim. Biophys. Acta* **645**:1–9.

Allan, D., and Michell, R. H., 1975, Accumulation of 1,2-diacylglycerol in the plasma membrane may lead to echinocyte transformation of erythrocytes, *Nature* **258**:348–349.

Allan, D., and Michell, R. H., 1977, Calcium ion-dependent diacylglycerol accumulation in erythrocytes is associated with microvesiculation but not with efflux of potassium ions, *Biochem. J.* **166**:495– 499.

Allan, D., and Thomas, P., 1981a, $Ca^{2+}$-induced biochemical changes in human erythrocytes and their relation to microvesiculation, *Biochem. J.* **198**:433–440.

Allan, D., and Thomas, P., 198lb, The effects of $Ca^{2+}$ and $Sr^{2+}$ on $Ca^{2+}$-sensitive biochemical changes in human erythrocytes and their membranes, *Biochem. J.* **198**:441–445.

Armando-Hardy, M., Ellory, J. C., Ferreira, H. G., Fleminger, S., and Lew, V. L., 1975, Inhibition of the calcium-induced increase in the potassium permeability of human red blood cells by quinine, *J. Physiol. (London)* **250**:32–33P.

Atwater, I., and Biegelman, P. M., 1976, Dynamic characteristics of electrical activity in pancreatic β cells. Effects of calcium and magnesium, *J. Physiol. (Paris)* **72**:769–786.

Atwater, I., Dawson, C. M., Ribalet, B., and Rojas, E., 1979, Potassium permeability activated by intracellular calcium ion concentration in the pancreatic B-cell, *J. Physiol. (London)* **288**:575–588.

Baker, P. F., 1972, Transport and metabolism of calcium ions in nerve, *Prog. Biophys. Mol. Biol.* **24**:177–223.

Banks, B. E. C., Brown, C., Burgess, G. M., Burnstock, G., Claret, M., Cocks, T. M., and Jenkinson, D. H., 1979, Apamin blocks certain neurotransmitter induced increases in potassium permeability, *Nature* **282**:417–419.

Barrett, E. F., and Barrett, J. N., 1976, Separation of two voltage sensitive potassium currents and demonstration of a tetrodotoxin-resistant calcium current in frog motoneurons, *J. Physiol. (London)* **255**:737–774.

Barrett, J. N., Barrett, E. F., and Dribin, L. B., 1981, Calcium-dependent slow potassium conductance in rat skeletal myotubules. *Dev. Biol.* **82**:258–266.

Barrett, J. N., Magleby, K. L., and Pallotta, B. S., 1982, Properties of single calcium-activated potassium channels in cultured rat muscle, *J. Physiol (London)* **331**:211–230.

Bassingthwaite, J. B., Fry, C. H., and McGuigan, J. A. S., 1976, Relationship between intracellular calcium and outward current in mammalian ventricular muscle; a mechanism for the control of action potential duration? *J. Physiol. (London)* **262**:15–37.

Batzri, S., Selinger, Z., Schramm, M., and Robinovitch, M. R., 1973, Potassium release mediated by the epinephrine L-receptor in rat parotid slices. Properties and relation to enzyme secretion, *J. Biol. Chem.* **248**:361–368.

Berkowitz, L. R., and Orringer, E. P., 1981, Effect of cetiedil, an *in vitro* antisickling agent on erythrocyte membrance cation permeability, *J. Clin. Invest.* **68**:1215–1220.

Blum, R. M., and Hoffman, J. F., 1970, Carrier mediation of Ca-induced K transport and its inhibition in red blood cells, *Fed. Proc.* **29**:663a.

Blum, R. M., and Hoffman, J. F., 1971, The membrane locus of Ca-stimulated K transport in energy-depleted human red blood cells, *J. Membr. Biol.* **6**:315–328.

Blum, R. M., and Hoffman, J. F., 1972, Ca-induced K transport in human red cells: Localization of the Ca-sensitive site to the inside of the membrane, *Biochem. Biophys. Res. Commun.* **46**:1146–1151.

Bodemann, H., and Passow, H., 1972, Factors controlling the resealing of the membrane of human erythrocyte ghosts after hypotonic hemolysis, *J. Membr. Biol.* **8**:1–26.

Bookchin, R. M., Lew, V. L., Nagel, R. L., and Raventos, C., 1981, Increase in potassium and calcium transport in human red cells infected with *Plasmodium* falciparum *in vitro, J. Physiol. (London)* **312**:65.

Boonstra, J., Mummery, C. L., Tertoole, L. G., Vandersa, P. T., and Delaat, S. W., 1981, Characterization of $^{42}$K and $^{86}$Rb transport and electrical membrane properties of exponentially growing neuroblastoma cells, *Biochim. Biophys. Acta* **643**:89–100.

Brading, A., Bülbring, E., and Tomita, T., 1969, The effect of sodium and calcium on the action potential of the smooth muscle of the guinea-pig taenia coli, *J. Physiol. (London)* **200**:637–654.

Brehm, P., Dunlap, K., and Eckert, R., 1978, Calcium-dependent repolarization in *Paramecium, J. Physiol. (London)* **274**:639–654.

Brown, A. M., and Lew, V. L., 1981, Lack of time-dependent inactivation of the Ca-sensitive K channel of red cells, *J. Physiol (London)* **320**:122P.

Brown, A. M., Ellory, J. C., Young, J. D., and Lew, V. L., 1978, A calcium activated potassium channel present in foetal red cells of sheep but absent from reticulocytes and mature red cells, *Biochim. Biophys. Acta* **511**:163–175.

Bülbring, E., and Tomita, T., 1977, Calcium-requirement of the L action of catecholamines on guinea-pig tenia coli, *Proc. Roy. Soc. Biol. Sci.* **197**:271–284.

Burgess, G. M., Claret, M., and Jenkinson, D. H., 1979, Effects of catecholamines, ATP and ionophore A23187 on potassium and calcium movements in isolated hepatocytes, *Nature* **279**:544–546.

Burgess, G. M., Claret, M., and Jenkinson, D. H., 1981, Effects of quinine and apamin on the calcium-dependent potassium permeability of mammalian hepatocytes and red cells, *J. Physiol. (London)* **317**:67–90.

Bürgin, H., and Schatzmann, H. J., 1979, The relation between net calcium, alkali cation and chloride movements in red cells exposed to salicylate, *J. Physiol. (London)* **287**:15–32.

Carafoli, E., and Crompton, M., 1978, The regulation of intracellular calcium, in: *Current Topics in Membranes and Transport,* Vol. 10 (F. Bronner and A. Kleinzeller, eds.), Academic Press, New York, pp. 151–216.

Cheung, W. Y., 1982, Calmodulin—an overview. *Fed. Proc.* **41**:2253–2257.

Cotterrell, D., and Whittam, R., 1970, An increase in potassium efflux in human red cells associated with reversing the sign of the membrane potential, *J. Physiol. (London)* **210:**136–137P.

Craig, A. B., 1958, Observations on epinephrine and glucagon-induced glycogenolysis and potassium loss in the isolated perfused frog liver, *Am. J. Physiol.* **193:**425–430.

Davson, H., 1941, The effect of some metabolic poisons on the permeability of the rabbit erythrocyte to potassium, *J. Cell. Comp. Physiol.* **18:**173–185.

Davson, H., 1942, The haemolytic action of potassium salts, *J. Physiol. (London)* **101:**265–283.

dePeyer, J. E., Cachelin, A. B., Levitan, I. B., and Reuter, H., 1982, $Ca^{2+}$-activated $K^+$ conductance in internally perfused snail neurons is enhanced by protein phosphorylation, *Proc. Natl. Acad. Sci. USA* **79:**4207–4211.

Deutsch, C., and Price, M. A., 1982, Cell calcium in human peripheral blood lymphocytes and the effect of mitogen, *Biochim. Biophys. Acta* **687:**211–218.

Doljanski, F., Ben-Sasson, S., Reich, M., and Groves, N. B., 1974, Dynamic osmotic behavior of chick blood lymphocytes. *J. Cell Physiol.* **84:**215–224.

Dunn, M. J., 1974, Red blood cell calcium and magnesium: Effects upon sodium and potassium transport and cellular morphology, *Biochim. Biophys. Acta* **352:**97–116.

Eaton, J. W., Skelton, T. D., Swofford, H. S., Kolpin, C. E., and Jacob, H. S., 1973, Elevated erythrocyte calcium in sickle cell disease, *Nature* **246:**105–106.

Eaton, J. W., Branda, R. F., Hadland, C., and Dreher, K., 1980, Anion channel blockade—effects upon erythrocyte membrane calcium response, *Am. J. Hematol.* **9:**391–399.

Eckert, E., 1977, Genes, channels and membrane currents in *Paramecium, Nature* **368:**104–105.

Ekman, A., Manninen, V., and Salminen, S., 1969, Ion movements in red cells treated with propranolol, *Acta Physiol. Scand.* **75:**333–344.

Feig, S. A., and Bassilian, S., 1974, Abnormal RBC Ca metabolism in hereditary spherocytosis, *Blood* **44:**937.

Feltz, A., Ronjevic, K., and Lisiewicz, A., 1972, Intracellular free $Ca^{2+}$ and membrane properties of motoneurones, *Nature New Biol.* **237:**179–181.

Ferreira, H. G., and Lew, V. L., 1976, Use of ionophore A23187 to measure cytoplasmic Ca buffering and activation of the Ca pump by internal Ca, *Nature* **259:**47–49.

Ferreira, H. G., and Lew, V. L., 1977, Passive Ca transport and cytoplasmic Ca buffering in intact red cells, in: *Membrane Transport in Red Cells* (J. C. Ellory and V. L. Lew, eds.), Academic Press, New York, pp. 53–92.

Fink, R., and Lüttgau, H. C., 1976, An evaluation of the membrane constants and the potassium conductance in metabolically exhausted muscle fibres, *J. Physiol. (London)* **263:**215–238.

Freedman, M. H., 1979, Early biochemical events in lymphocyte activation I. Investigation of the nature and significance of early calcium fluxes observed in mitogen-induced T and B lymphocytes, *Cell. Immunol.* **44:**290–313.

Freedman, M. H., Raff, M. C., and Gomperts, B., 1975, Induction of increased calcium uptake in mouse T lymphocytes by concanavalin A and its modulation by cyclic nucleotides, *Nature* **255:**378–380.

Fuhrmann, G. F., Hüttermann, F., and Knauf, P. A., 1984, The mechanism of vanadium action on selective $K^+$-permeability in human erythrocytes, *Biochim. Biophys. Acta* **769:**130–140.

Garcia-Sancho, J., Sanchez, A., and Herreros, B., 1979, Stimulation of monovalent cation fluxes by electron donors in the human red cell membrane, *Biochim. Biophys. Acta* **556:**118–130.

Garcia-Sancho, J., Sanchez, A., and Herreros, B., 1982, All-or-none response of the $Ca^{2+}$-dependent $K^+$ channel in inside-out vesicles, *Nature* **296:**744–746.

Gárdos, G., 1956, The permeability of human erythrocytes to potassium, *Acta Physiol. Acad. Sci. Hung.* **10:**185–189.

Gárdos, G., 1958a, Effect of ethylenediamine-tetraacetate on the permeability of human erythrocytes, *Acta Physiol. Acad. Sci. Hung.* **14:**1–5.

Gárdos, G., 1958b, The function of calcium in the potassium permeability of human erythrocytes, *Biochim. Biophys. Acta* **30:**653–654.

Gárdos, G., 1959, The role of calcium in the potassium permeability of human erythrocytes, *Acta Physiol. Acad. Sci. Hung.* **15:**121–125.

Gárdos, G., 1966a, The role of 2,3-diphosphoglyceric acid in the potassium transport of human erythrocytes, *Experientia* **22:**308.

Gárdos, G., 1966b, The mechanism of ion transport in human erythrocytes I. The role of 2,3-diphosphoglyceric acid in the regulation of potassium transport, *Acta Biochim. Biophys. Acad. Sci. Hung.* **1:**139–148.

Gárdos, G., 1967, Studies on potassium permeability changes in human erythrocytes, *Experientia* **23:**19.

Gárdos, G., 1972, Ion transport across the erythrocyte membrane, *Haematologia* **6:**237–247.

Gárdos, G., and Straub, F. B., 1957, Über die Rolle Der Adenosintriphosphorsaüre (ATP) in der K-Permeabilitat der menschlichen roten Blutkorperchen, *Acta Physiol. Acad. Sci. Hung.* **12:**1–8.

Gárdos, G., and Szász, I., 1968, The mechanism of ion transport in human erythrocytes II. The role of histamine in regulation of cation transport, *Acta Biochim. Biophys. Acad. Sci. Hung.* **3:**13–27.

Gárdos, G., and Szasz, I., 1973, Studies on the leak cation transport of human erythrocytes, in: *Erythrocytes, Thrombocytes, Leukocytes* (E. Gerlach, K. Moser, E. Deutsch, and W. Wilmanns, eds.), Georg Thieme, Stuttgart, pp. 31–33.

Gárdos, G., Szász, I., and Sarkadi, B., 1975a, Mechanism of Ca-dependent K transport in human red cells, *FEBS Proc.* **35:**167–180.

Gárdos, G., Sarkadi, B., and Szász, I., 1975b, Effect of the Ca-ionophore A23187 on the K transport of human red cells, *Abst. Vol. 5th Int. Cong. Biophys.* p. 100.

Gárdos, G., Lassen, U. V., and Pape, L., 1976, Effect of antihistamines and chlorpromazine on the calcium-induced hyperpolarization of the *Amphiuma* red cell membrane, *Biochim. Biophys. Acta* **448:**599–606.

Gárdos, G., Szász, I., and Sarkadi, B., 1977, Effect of intracellular calcium on the cation transport processes in human red cells, *Acta Biol. Med. Germ.* **36:**823–829.

Glader, B. E., Fortier, N., Albala, M. M., and Nathan, D. G., 1974, Congenital anemia associated with dehydrated erythrocytes and increased potassium loss, *New Engl. J. Med.* **291:**491–496.

Glynn, I. M., and Warner, A. E., 1972, Nature of the calcium dependent potassium leak induced by (+)-propranolol, and its possible relevance to the drug's antiarrhythmic effect, *Br. J. Pharmacol.* **44:**271–278.

Gorman, A. L. F., and Hermann, A., 1979, Internal effects of divalent cations on potassium permeability in molluscan neurones, *J. Physiol. (London)* **296:**393–410.

Gorman, A. L. F., and McReynolds, J. S., 1974, Control of membrane $K^+$ permeability in a hyperpolarizing photoreceptor: Similar effects of light and metabolic inhibition, *Science* **185:**620–621.

Gorman, A. L. F., and Thomas, M. V., 1978, Changes in intracellular concentration of free calcium ions in a pace-maker neurone, measured with the metallochromic dye Arsenazo III, *J. Physiol. (London)* **275:**357–376.

Gorman, A. L. F., and Thomas, M. V., 1980, Potassium conductance and internal calcium accumulation in a molluscan neurone, *J. Physiol. (London)* **308:**287–313.

Gorman, A. L. F., Hermann, A., and Thomas, M. V., 1982, Ionic requirements for membrane oscillations and their dependence on the calcium concentration in a molluscan pace-maker neurone, *J. Physiol. (London)* **327:**185–217.

Grafe, P., Mayer, C. J., and Wood, J. D., 1980, Synaptic modulation of calcium-dependent potassium conductance in myenteric neurons in the guinea-pig, *J. Physiol. (London)* **305:**235–248.

Greengard, P., 1978, Phosphorylated proteins as physiological effectors, *Science* **199:**146–152.

Grey, J. E., and Gitelman, H. J., 1979, Phospholipase participates in the calcium-induced potassium efflux of human erythrocytes, *Fed. Proc.* **38:**1127.

Grigarzik, H., and Passow, H., 1958, Versuche zum mechanismus der bleiwirkung auf die kalium-permeabilität roter blutköperchen, *Pflugers Arch.* **267:**73–92.

Grinstein, S., and Rothstein, A., 1978, Chemically induced cation permeability in red cell membrane vesicles. The sidedness of the response and the proteins involved, *Biochim. Biophys. Acta* **508:**236–245.

Grinstein, S., DuPre, A., and Rothstein, A., 1982a, Volume regulation by human lymphocytes—role of calcium, *J. Gen. Physiol.* **79:**849–868.

Grinstein, S., Clarke, C. A., DuPre, A., and Rothstein, A., 1982b, Volume-induced increase of anion permeability in human lymphocytes, *J. Gen. Physiol.* **80:**801–823.

Gunn, R. B., 1979, Transport of anions across red cell membranes, in: *Membrane Transport in Biology,* Vol. 2 (G. Giebisch, D. C. Tosteson, and H. H. Ussing, eds.), Springer Verlag, Berlin, pp. 59–80.

Gunn, R. B., Dalmark, M., Tosteson, D. C., and Wieth, I. O., 1973, Characteristics of chloride transport in human red blood cells, *J. Gen. Physiol.* **61:**185–206.

Hamill, O. P., 1981, Potassium channel currents in human red blood cells, *J. Physiol. (London)* **314:**125P.

Hanani, M., and Shaw, C., 1977, A potassium contribution of the response of the barnacle photoreceptor, *J. Physiol. (London)* **270:**151–163.

Harrison, D. G., and Long, C., 1968, The calcium content of human erythrocytes, *J. Physiol. (London)* **199:**367–381.

Haylett, D. G., 1976, The effects of sympathomimetic amines on $^{45}Ca$ efflux from liver slices, *J. Pharmacol.* **57:**158–160.

Haylett, D. G., and Jenkinson, D. H., 1972, Effects of noradrenaline on potassium efflux, membrane potential and electrolyte levels in tissue slices prepared from guinea pig liver, *J. Physiol. (London)* **225:**721–750.

Heinz, A., and Passow, H., 1980, Role of external potassium in the calcium-induced potassium efflux from human red blood cell ghosts, *J. Membr. Biol.* **57:**119–131.

Hellerstein, S., and Bunthrarungoj, T., 1974, Erythrocyte composition in sickle cell anemia, *J. Lab. Clin. Med.* **83:**611–624.

Hermann, A., and Gorman, A. L., 1981, Effects of tetraethylammonium on potassium currents in a molluscan neuron, *J.. Gen. Physiol.* **78:**87–110.

Hladky, S. B., and Rink, T. J., 1976, Potential difference and the distribution of ions across the human red blood cell membrane: A study of the mechanism by which the fluorescent cation, diS-$C_3$(5) reports membrane potential, *J. Physiol. (London)* **263:**287–319.

Hoffman, J. F., 1962, Cation transport and structure of the red cell plasma membrane, *Circulation* **26:**1201–1213.

Hoffman, J. F., and Blum, R. M., 1977, On the nature of the transport pathway used for $Ca^{2+}$-dependent $K^+$ movement in human red blood cells, in: *Membrane Toxicity* (M. W. Miller and A. E. Shamoo, eds.), Plenum Press, New York, pp. 381–404.

Hoffman, J. F., and Knauf, P. A., 1973, The mechanism of the increased K transport induced by Ca in human red blood cells, in: *Erythrocytes, Thrombocytes, Leukocytes* (E. Gerlach, K. Moser, E. Deutch, and W. Wilmanns, eds.), Georg Thieme, Stuttgart, pp. 66–70.

Hoffman, J. F., and Laris, P. C., 1974, Determination of membrane potentials in human and *Amphiuma* red blood cells by means of a fluorescent probe, *J. Physiol. (London)* **239:**519–552.

Holian, A., Deutsch, C. J., Holian, S. K., Daniele, R. P., and Wilson, D. F., 1979, Lymphocyte response to phytohemagglutinin: Intracellular volume and intracellular $K^+$, *J. Cell. Physiol.* **98:**137–144.

Howland, J. L., 1974, Abnormal potassium conductance associated with muscular dystrophy, *Nature* **251:**724–725.

Hugues, M., Romey, G., Duval, D., Vincent, J. P., and Lazdunski, M., 1982a, Apamin as a selective blocker of the calcium-dependent potassium channel in neuroblastoma cells: Voltage-clamp and biochemical characterization of the toxin receptor, *Proc. Natl. Acad. Sci. USA* **79:**1308–1312.

Hugues, M., Schmid, H., and Lazdunski, M., 1982b, Identification of a protein component of the $Ca^{2+}$-dependent $K^+$ channel by affinity labeling with apamin, *Biochem. Biophys. Res. Commun.* **107:**1557–1582.

Hunter, M. J., 1971, A quantitative estimate of the non-exchange-restricted chloride permeability of the human red cell, *J. Physiol (London)* **218:**49P.

Hunter, M. J., 1977, Human erythrocyte anion permeabilities measured under conditions of net charge transfer, *J. Physiol. (London)* **268:**35–49.

Isenberg, G., 1975, Is potassium conductance of cardiac Purkinje fibres controlled by $[Ca^{2+}]_i$? *Nature* **253:**273–274.

Isenberg, G., 1978, The positive dynamic current of the cardiac Purkinje fibre is not a chloride but a potassium current, *Pflugers Arch.* **377:**R5.

Jenkins, D. M. G., and Lew, V. L., 1973, Ca uptake by ATP depleted red cells from different species with and without associated increase in K permeability, *J. Physiol. (London)* **234:**41–42P.

Kaczmarek, L. K., Jennings, K. R., Strumwasser, F., Nairn, A. C., Walter, V., Wilson, F. D., and Greengard, P., 1980, Microinjection of catalytic subunit of cyclic AMP-dependent protein kinase enhances calcium action potential of bag cell neurons in cell culture, *Proc. Natl. Acad. Sci. USA* **77:**7487–7491.

Karlish, S. J. D., Ellory, J. C., and Lew, V. L., 1981, Evidence against $Na^+$-pump mediation of $Ca^{2+}$-activated $K^+$ transport and diuretic-sensitive $Na^+/K^+$ cotransport, *Biochim. Biophys. Acta* **646:**353–355.

Kass, R. S., and Tsien, R. W., 1976, Control of action potential duration by calcium ions in cardiac Purkinje fibers, *J. Gen. Physiol.* **67:**599–617.

Kirkpatrick, F. H., Hillman, D. G., and LaCelle, P. L., 1975, A23187 and red cells: Changes in deformability, $K^+$, $Mg^{2+}$, $Ca^{2+}$, and ATP, *Experientia* **31:**653–654.

Knauf, P. A., 1979, Erythrocyte anion exchange and the Band III protein: Transport kinetics and molecular structure, in: *Current Topics in Membranes and Transport,* Vol. 12 (F. Bronner and A. Kleinzeller, eds.), Academic Press, New York, pp. 251–365.

Knauf, P. A., and Rothstein, A., 1971, Chemical modifications of membranes: I. Effects of sulfhydryl and amino reactive reagents on anion and cation permeability of human red blood cell, *J. Gen. Physiol.* **58:**190–210.

Knauf, P. A., Riordan, J. R., Schuhmann, B., and Passow, H., 1974, Effects of external potassium on calcium-induced potassium leakage from human red blood cell ghosts, in: *Membranes: Comparative Biochemistry and Physiology of Transport* (L. Bolis, K. Bloch, S. E. Luria, and F. Lynen, eds.), North Holland, Amsterdam, pp. 305–309.

Knauf, P. A., Riordan, J. R., Schuhmann, B., Wood-Guth, I., and Passow, H., 1975, Calcium-potassium stimulated net potassium efflux from human erythrocyte ghosts, *J. Membr. Biol.* **25:**1–22.

Knauf, P. A., Fuhrmann, G. F., Rothstein, S., and Rothstein, A., 1977, The relationship between anion exchange and net anion flow across the human red blood cell membrane, *J. Gen. Physiol.* **69:**363–386.

Koller, C. A., Orringer, E. P., and Parker, J. C., 1979, Quinine protects pyruvate kinase deficient red cells from dehydration, *Am. J. Hematol.* **7:**193–199.

Kregenow, F. M., and Hoffman, J. F., 1972, Some kinetic and metabolic characteristics of calcium-induced potassium transport in human red cells, *J. Gen. Physiol.* **60:**406–429.

Krnjevic, K., and Lisiewicz, A., 1972, Injection of calcium ions into spinal motoneurones, *J. Physiol. (London)* **225:**363–390.

Kuba, K., 1980, Release of calcium ions linked to the activation of potassium conductance in a caffeine-treated sympathetic neurone, *J. Physiol. (London)* **298:**251–269.

Kuba, K., and Nishi, S., 1976, Rhythmic hyperpolarizations and depolarization of sympathetic ganglion cells induced by caffeine, *J. Neurophysiol.* **39:**547–563.

Kurtzer, R., and Roberts, M. L., 1982, Calcium-dependent $K^+$ efflux from rat submandibular gland. The effects of trifluoperazine and quinidine, *Biochim. Biophys. Acta* **693:**479–484.

LaCelle, P. L., and Rothstein, A., 1966, The passive permeability of the red blood cell to cations, *J. Gen. Physiol.* **50:**171–188.

Lackington, I., and Orrego, F., 1981, Inhibition of calcium-activated potassium conductance of human erythrocytes by calmodulin inhibitory drugs, *FEBS Lett.* **133:** 103–106.

Larsen, F. L., and Vincenzi, F. F., 1979, Calcium transport across the plasma membrane: Stimulation by calmodulin, *Science* **204:**306–309.

Larsen, F. L., Katz, S., Roufogalis, B. D., and Brooks, D. E., 1981, Physiological sheer stresses enhance the $Ca^{2+}$ permeability of human erythrocytes, *Nature* **294:** 667–668.

Lassen, U. V., 1972, Membrane potential and membrane resistance of red cells, in: *Oxygen Affinity and Red Cell Acid Base Status* (M. Rorth and P. Astrup, eds.), Munksgaard, Copenhagen, pp. 291–304.

Lassen, U. V., Pape, L., and Vestergaard-Bogind, B., 1973, Membrane potential of *Amphiuma* red cells: Effect of calcium, in: *Erythrocytes, Thrombocytes, Leukocytes* (E. Gerlach, K. Moser, E. Deutsch, and W. Wilmanns, eds.), Georg Thieme, Stuttgart, pp. 33–36.

Lassen, U. V., Pape, L., and Vestergaard-Bogind, B., 1976, Effect of calcium on the membrane potential of Amphiuma red cells, *J. Membr. Biol.* **26:**51–70.

Lassen, U. V., Pape, L., and Vestergaard-Bogind, B., 1980, Calcium related transient changes in membrane potential of red cells, in: *Membrane Transport in Erythrocytes* (U. V. Lassen, H. H. Ussing, and J. O. Wieth, eds.), Munksgaard, Copenhagen, pp. 255–273.

Latorre, R., and Miller, C., 1983, Conduction and selectivity in potassium channels, *J. Membr. Biol.* **71:**11–30.

Latorre, R., Vergara, C., and Hidalgo, C., 1982, Reconstitution in planar lipid bilayers of $Ca^{2+}$-dependent $K^+$ channel from transverse tubule membranes isolated from rabbit skeletal muscle, *Proc. Natl. Acad. Sci. USA* **79:**805–809.

Lepke, S., and Passow, H., 1960, Die Wirkung von Erdalkalimetallionen auf die Kationpermeabilitat fluoridvergifteter Erythrocyten, *Pflugers Arch.* **271:**473–487.

Lepke, S., and Passow, H., 1968, Effects of fluoride on potassium and sodium permeability of the erythrocyte membrane, *J. Gen. Physiol.* **51:**365–372.

Lew, V. L., 1970, Effect of intracellular calcium on the potassium permeability of human red cells, *J. Physiol. (London)* **206:**35–36P.

Lew, V. L., 1971a, On the ATP-dependence of the $Ca^{2+}$-induced increase in $K^+$ permeability observed in human red cells, *Biochim. Biophys. Acta* **233:**827–830.

Lew, V. L., 1971b, Effect of ouabain on the $Ca^{++}$-dependent increase in $K^+$ permeability in ATP depleted guinea-pig red cells, *Biochim. Biophys. Acta* **249:**236–239.

Lew, V. L., 1974, On the mechanism of the Ca-induced increase in K permeability observed in human red cell membranes, in: *Comparative Biochemistry and Physiology of Transport* (L. Bolis, K. Bloch, S. E. Luria, and F. Lynen, eds.), North Holland, Amsterdam, pp. 310–316.

Lew, V. L., and Beaugé, L., 1979, Passive cation fluxes in red cell membranes, in: *Membrane Transport in Biology,* Vol. 2 (G. Giebisch, D. C. Tosteson, and H. H. Ussing, eds.), Springer Verlag, Berlin, pp. 81–116.

Lew, V. L., and Bookchin, R. M., 1980, A $Ca^{2+}$-refractory state of the $Ca^{2+}$-sensitive $K^+$ permeability mechanism in sickle cell anaemia red cells, *Biochim. Biophys. Acta* **602:**196–200.

Lew, V. L., and Ferreira, H. G., 1976, Variable Ca sensitivity of a K selective channel in intact red cell membranes, *Nature* **263:**336–338.

Lew, V. L., and Ferreira, H. G., 1977, The effect of Ca on the K permeability of red cells, in: *Membrane Transport in Red Cells* (J. C. Ellory and V. L. Lew, eds.), Academic Press, New York, pp. 93–100.

Lew, V. L., and Ferreira, H. G., 1978, Calcium transport and the properties of a Ca-activated potassium channel in red cell membranes, in: *Current Topics in Membranes and Transport,* Vol. 10 (F. Bronner and A. Kleinzeller, eds.), Academic Press, New York, pp. 217–277.

Lew, V. L., and Simonsen, L. O., 1981, A23187-induced $^{45}Ca$ flux kinetics reveal uniform ionophore distribution and cytoplasmic calcium buffering in ATP-depleted human red cells, *J. Physiol. (London)* **316:**6–7P.

Lew, V. L., Muallem, S., and Seymour, C. A., 1980, One-step vesicles from mammalian red cells, *J. Physiol. (London)* **307:**36–37P.

Lew, V. L., Muallem, S., and Seymour, C. A., 1982, Properties of the $Ca^{2+}$ activated $K^+$ channel in one-step inside-out vesicles from human red cell membranes, *Nature* **296:**742–744.

Lindemann, B., and Passow, H., 1960, Kaliumverlust und ATP-Zerfall in bleivergifteten Menschenerythrocyten, *Pflugers Arch.* **271:**369–373.

Lisman, J. E., and Brown, J. E., 1972, The effects of intracellular iontophoretic injection of calcium and sodium ions on the light response of *Limulus* ventral photoreceptors, *J. Gen. Physiol.* **59:**701–719.

Lorand, L., Weissmann, L. B., Epel, D. L., and Lorand, J. B., 1976, Role of the intrinsic transglutaminase in the $Ca^{2+}$ mediated crosslinking of erythrocyte proteins, *Proc. Natl. Acad. Sci. USA* **73:**4479–4481.

Macey, R. I., Adorante, J. S., and Orme, F. W., 1978, Erythrocyte membrane potentials determined by hydrogen ion distribution, *Biochim. Biophys. Acta* **512:**284–295.

Manninen, V., 1970, Movements of sodium and potassium ions and their tracers in propranolol-treated red cells and diaphragm muscle, *Acta Physiol. Scand. Suppl.* **355:**1–37.

Marban, E., and Tsien, R. W., 1982, Effects of nystatin-mediated intracellular ion substitution on membrane currents in calf Purkinje fibres, *J. Physiol. (London)* **329:**569–587.

Marino, D., Sarkadi, B., Gardos, G., and Bolis, L., 1981, Calcium-induced alkali cation transport in nucleated red cells, *Mol. Physiol.* **1:**295–300.

Marshall, J. M., 1977, Modulation of smooth muscle activity by catecholamines, *Fed. Proc.* **36:**2450–2455.

Marty, A., 1981, Ca-dependent K-channels with large unitary conductance in chromaffin cell membranes, *Nature* **291:**497–500.

Masys, D. R., Bromberg, P. A., and Balcerzak, S. P., 1974, Red cells shrink during sickling, *Blood* **44:**885–890.

Matthews, E. K., 1975, Calcium and stimulus-secretion coupling in pancreatic islet cells, in: *Calcium Transport in Contraction and Secretion* (E. Carafoli, ed.), North Holland, Amsterdam, pp. 203–210.

Meech, R. W., 1974, The sensitivity of *Helix aspersa* neurones to injected calcium ions, *J. Physiol. (London)* **237:**259–277.

Meech, R. W., 1976, Intracellular calcium and the control of membrane permeability, in: *Calcium in Biological Systems, Symp. Soc. Exp. Biol. Med.* **30:**161–191.

Meech, R. W., 1978, Calcium-dependent potassium activation in nervous tissues, *Annu. Rev. Biophys. Bioeng.* **7**:1–18.

Meech, R. W., and Standen, N. B., 1975, Potassium activation in *Helix aspersa* under voltage clamp: A component mediated by calcium influx, *J. Physiol. (London)* **249**:211–239.

Meech, R. W., and Strumwasser, F., 1970, Intracellular calcium injection activates potassium conductance in *Aplysia* nerve cells, *Fed. Proc.* **29**:834.

Mironneau, J., and Savineau, J. P., 1980, Effects of calcium ions on outward membrane currents in rat uterine smooth muscle, *J. Physiol. (London)* **302**:411–425.

Moolenaar, W. H., and Spector, I., 1979a, The calcium action potential and a prolonged calcium-dependent after-hyperpolarization in mouse neuroblastoma cells, *J. Physiol. (London)* **292**:297–306.

Moolenaar, W. H., and Spector, I., 1979b, The calcium current and the activation of a slow potassium conductance in voltage-clamped mouse neuroblastoma cells, *J. Physiol. (London)* **292**:307–323.

Morita, K., North, R. A., and Tokimasa, T., 1982, The calcium-activated potassium conductance in guinea-pig myenteric neurones, *J. Physiol. (London)* **329**:341–354.

Mounier, Y., and Vassort, G., 1975, Evidence for a transient potassium membrane current dependent on calcium influx in crab muscle fibre, *J. Physiol. (London)* **251**:609–625.

Naccache, P. H., Volpi, M., Shawell, H. J., Becker, E. L., and Sha'afi, R. I., 1979, Chemotactic factor-induced release of membrane calcium in rabbit neutrophils, *Science* **203**:461–463.

Nelson, P. G., and Henkart, M. I., 1979, Oscillatory membrane potential changes in cells of mesenchymal origin: The role of an intracellular regulation system, *J. Exp. Biol.* **81**:49–61.

Nicoll, R. A., and Alger, B. E., 1981, Synaptic excitation may activate a calcium-dependent potassium conductance in hippocampal pyramidal cells, *Science* **212**:957–958.

North, R. A., The calcium-dependent slow after-hyperpolarization in myenteric plexus neurones with tetrodotoxin-resistant action potentials, *Br. J. Pharmacol.* **49**:709–711.

Okada, Y., Tsuchiya, W., and Inouye, A., 1979, Oscillations of membrane potential in L cells. IV. Role of intracellular $Ca^{2+}$ in hyperpolarizing excitability, *J. Membr. Biol.* **47**:357–376.

Okada, Y., Tsuchiya, W., and Yada, T., 1982, Calcium channel and calcium pump involved in oscillatory hyperpolarizing responses of L-strain mouse fibroblasts, *J. Physiol. (London)* **327**:449–461.

Oliveira-Castro, G. M., and Dos Reis, G. A., 1981, Electrophysiology of phagocytic membranes III. Evidence for a calcium-dependent potassium permeability change during slow hyperpolarizations in activated macrophages, *Biochim. Biophys. Acta* **640**:500–511.

Orringer, E. P., and Parker, J. C., 1973, Ion and water movements in red blood cells, in: *Progress in Hematology,* Vol. 8 (E. B. Brown, ed.), Grune and Stratton, New York, pp. 1–23.

Ørskov, S. L., 1935, Untersuchungen über den einfluss von kohlensaure und blei auf die permeabilität der blutkörperchen für kalium und rubidium, *Biochem. Z.* **279**:250–261.

Pallotta, B. S., Magleby, K. L., and Barrett, J. N., 1981, Single channel recordings of $Ca^{2+}$-activated $K^+$ currents in rat muscle cell culture, *Nature* **293**:471–474.

Pape, L., 1982, Effect of extracellular $Ca^{2+}$, $K^+$, and $OH^-$ on erythrocyte membrane potential as monitored by the fluorescent probe 3,3-dipropylthiodicarbocyanine, *Biochim. Biophys. Acta* **686**: 225–232.

Parker, J. C., 1978, Sodium and calcium movements in dog red blood cells, *J. Gen. Physiol.* **71**:1–17.

Parker, J. C., 1981, Effects of drugs on calcium related phenomena in red blood cells, *Fed. Proc.* **40**:2872–2876.

Parker, J. C., 1983, Hemolytic action of potassium salts on dog red blood cells, *Am. J. Physiol.,* **244**:C313–317.

Parker, J. C., Gitelman, H. J., Glosson, P. S., and Leonard, D. L., 1975, The role of calcium in volume regulation by dog red blood cells, *J. Gen. Physiol.* **65**:84–96.

Parker, J. C., Orringer, E. P., and McManus, T. J., 1978, Disorders of ion transport in red blood cells: *Physiology of Membrane Disorders* (T. E. Andreoli, J. F. Hoffman, and D. D. Fanestil, eds.), Plenum, New York, pp. 773–800.

Parod, R. J., and Putney, J. W., Jr., 1978, Role of calcium in the receptor-mediated control of potassium permeability in the rat lacrimal gland, *J. Physiol. (London)* **281**:371–381.

Passow, H., 1963, Metabolic control of passive cation permeability in human red cells, in: *Cell Interface Reactions* (H. D. Brown, ed.), Scholar's Library, New York, pp. 57–107.

Passow, H., 1981, Selective enhancement of potassium efflux from red blood cells by lead, in: *The Functions of Red Blood Cells: Erythrocyte Pathobiology* (D. F. Wallach, ed.), Alan R. Liss, New York, pp. 80–104.

Passow, H., and Vielhauer, E., 1966, Die wirkung von trioseredukton auf die kalium und natriumpermeabilität roter blutkörperchen, *Pflugers Arch.* **288:**1–14.

Plishker, G. A., Appel, S. H., Dedman, J. R., and Means, A. R., 1980, Phenothiazine inhibition of calmodulin stimulates Ca-dependent K-efflux in human red blood cells, *Fed. Proc.* **39:**1713.

Porzig, H., 1975, Comparative study of the effects of propranolol and tetracaine on cation movements in resealed human red cell ghosts, *J. Physiol. (London)* **249:**27–50.

Porzig, H., 1977, Studies on the cation permeability of human red cell ghosts, *J. Membr. Biol.* **31:**317–349.

Putney, J. W., Jr., 1976, Stimulation of $^{45}Ca$ influx in rat parotid gland by carbachol, *J. Pharmacol. Exp. Ther.* **199:**526–537.

Putney, J. W., Jr., 1978, Ionic millieu and control of K permeability in rat parotid gland, *Am. J. Physiol.* **235:**C180–C187.

Putney, J. W., Jr., 1979, Stimulus-permeability coupling: Role of calcium in the receptor regulation of membrane permeability, *Pharmacol. Rev.* **30:**209–245.

Putney, J. W., Jr., Parod, R. J., and Marier, S. H., 1977, Control by calcium of protein discharge and membrane permeability to potassium in the rat lacrimal gland, *Life Sci.* **2:**1905–1912.

Putney, J. W., Jr., van de Walle, C. M., and Leslie, B. A., 1978, Stimulus-secretion coupling in the rat lacrimal gland, *Am. J. Physiol.* **235:**C188–C198.

Quastel, M. R., and Kaplan, J. G., 1970, Early stimulation of potassium uptake in lymphocytes treated with PHA, *Exp. Cell. Res.* **63:**230–233.

Rasmussen, H., and Goodman, D. P. H., 1977, Relationship between calcium and cyclic nucleotides in cell activation, *Physiol. Rev.* **57:**421–509.

Reed, P. W., 1973, Calcium-dependent potassium efflux from rat erythrocytes incubated with antibiotic A23187, *Fed. Proc.* **32:**635.

Reed, P. W., 1976, Effects of the divalent cation ionophore A23187 on potassium permeability of rat erythrocytes, *J. Biol. Chem.* **251:**3489–3494.

Reichstein, E., and Rothstein, A., 1981, Effects of quinine on $Ca^{2+}$-induced $K^+$ efflux from human red blood cells, *J. Membr. Biol.* **59:**57–63.

Richhardt, H. W., Fuhrmann, G. F., and Knauf, P. A., 1979, Dog red blood cells exhibit a Ca-stimulated increase in K permeability in the absence of (Na,K) ATPase activity, *Nature* **279:**248–250.

Riordan, J. R., and Passow, H., 1971, Effects of calcium and lead on potassium permeability of human erythrocyte ghosts, *Biochim. Biophys. Acta* **249:**601–605.

Riordan, J. R., and Passow, H., 1973, The effects of calcium and lead on the potassium permeability of human erythrocytes and erythrocyte ghosts, in: *Comparative Physiology* (L. Bolis, K. Schmidt-Nielsen, and S. H. P. Maddrell, eds.), North Holland, Amsterdam, pp. 543–581.

Romero, P. J., and Whittam, R., 1971, The control by internal calcium of membrane permeability to sodium and potassium, *J. Physiol. (London)* **214:**481–507.

Roti-Roti, L. W., and Rothstein, A., 1973, Adaptation of mouse leukemic cells (L5178Y) to anisotonic media, *Exp. Cell. Res.* **79:**295–310.

Roufogalis, B. D., 1979, Regulation of calcium translocation across the red blood cell membrane, *Can. J. Physiol. Pharmacol.* **57:**1331–1349.

Roufogalis, B. D., 1981, Phenothiazine antagonism of calmodulin: A structurally nonspecific interaction, *Biochem. Biophys. Res. Commun.* **98:**607–613.

Sarkadi, B., 1980, Active calcium transport in human red cells, *Biochem. Biophys. Acta* **604:**159–190.

Sarkadi, B., and Tosteson, D. C., 1979, Active cation transport in human red cells, in: *Membrane Transport in Biology,* Vol. 2 (G. Giebisch, D. C. Tosteson, and H. H. Ussing, eds.), Springer Verlag, Berlin, pp. 117–160.

Sarkadi, B., Szász, I., and Gárdos, G., 1976, The use of ionophores for rapid loading of human red cells with radioactive cations for cation pump studies, *J. Membr. Biol.* **26:**357–370.

Sarkadi, B., Szász, I., Gerlóczi, A., and Gárdos, G., 1977, Transport parameters and stoichiometry of active calcium ion extrusion in intact human red cells, *Biochim. Biophys. Acta* **464:**93–107.

Sarkadi, B., Schubert, A., and Gárdos, G., 1979, Effects of calcium-EGTA buffers on active calcium transport in inside-out red cell membrane vesicles, *Experientia* **35:**1045–1047.

Sarkadi, B., Szebeni, J., and Gárdos, G., 1980, Effects of calcium on cation transport processes in inside-out red cell membrane vesicles, in: *Membrane Transport in Erythrocytes* (U. V. Lassen, H. H. Ussing, and J. O. Wieth, eds.), Munksgaard, Copenhagen, pp. 220–235.

Sarkadi, B., Enyedi, A., Nyers, A., and Gárdos, G., 1982, The function and regulation of the calcium pump in the erythrocyte membrane, *Ann. N.Y. Acad. Sci.* **402:**329–348.

Sarkadi, B., Grinstein, S., Mack, E., and Rothstein, A., 1983, An anion conductance pathway is involved in regulatory volume decrease in human lymphocytes, *Biophys. F.* **41:**188a.

Satow, Y., and Kung, C., 1980, Ca-induced K outward current in *Paramecium tetraurelia, J. Exp. Biol.* **88:**293–303.

Schatzmann, H. J., 1973, Dependence on calcium concentration and stoichiometry of the calcium pump in human red cells, *J. Physiol. (London)* **235:**551–569.

Schatzmann, H. J., 1975, Active calcium transport and $Ca^{2+}$-activated ATPase in human red cells in: *Current Topics in Membranes and Transport,* Vol. 6 (F. Bronner and A. Kleinzeller, eds.), Academic Press, New York, pp. 125–168.

Schatzmann, H. J., 1982, The plasma membrane calcium pump of erythrocytes and other animal cells, in: *Membrane Transport of Calcium* (E. Carafoli, ed.), Academic Press, New York, pp. 41–108.

Schramm, M., and Selinger, Z., 1975, The functions of cyclic AMP and calcium as alternative second messengers in parotid gland and pancreas, *J. Cycl. Nuc. Res.* **1:**181–192.

Schubert, A., and Sarkadi, B., 1977, Kinetic studies on the calcium-dependent potassium transport in human red blood cells, *Acta Biochim. Biophys. Acad. Sci. Hung.* **12:**207–216.

Schwarz, W., and Passow, H., 1983, $Ca^{2+}$-activated $K^+$ channels in erythrocytes and excitable cells, *Ann. Rev. Physiol.* **45:**359–374.

Segel, G. B., Simon, W., and Lichtman, M. A., 1979, Regulation of sodium and potassium transport in phytohemagglutinin-stimulated human blood lymphocytes, *J. Clin. Invest.* **64:**834–841.

Sha'afi, R. I., and Naccache, P. H., 1981, Ionic events in neutrophil chemotaxis, in: *Advances in Inflammation Research,* Vol. 2 (G. Weissmann, ed.), Raven Press, pp. 115–148.

Shalev, O., Leida, M. N., Hebbel, R. P., Jacob, H. S., and Eaton, J. W., 1981, Abnormal erythrocyte calcium homeostasis in oxidant-induced hemolytic disease, *Blood* **58:**1232–1238.

Siegelbaum, S. A., and Tsien, R. W., 1980, Calcium-activated transient outward current in calf cardiac Purkinje fibres, *J. Physiol. (London)* **299:**485–506.

Siegelbaum, S. A., Tsien, R. W., and Kass, R. S., 1977, Role of intracellular calcium in the transient outward current of calf Purkinje fibres, *Nature* **269:**611–613.

Siemon, H., Schneider, H., and Fuhrmann, G. F., 1982, Vanadium increases selective $K^+$ permeability in human erythrocytes, *Toxicology* **22:**271–278.

Simons, T. J. B., 1976a, The preparation of human red cell ghosts containing calcium buffers, *J. Physiol. (London)* **256:**209–225.

Simons, T. J. B., 1976b, Calcium-dependent potassium exchange in human red cell ghosts, *J. Physiol. (London)* **256:**227–244.

Simons, T. J. B., 1976c, Carbocyanine dyes inhibit Ca-dependent K efflux from human red cell ghosts, *Nature (London)* **264:**467–469.

Simons, T. J. B., 1979, Actions of a carbocyanine dye on calcium-dependent potassium transport in human red cell ghosts, *J. Physiol. (London)* **288:**481–507.

Simons, T. J. B., 1982, A method for estimating free Ca within human red blood cells, with an application to the study of their Ca-dependent K permeability, *J. Membr. Biol.* **66:**235–247.

Simonsen, L. O., Gomme, J., and Lew, V. L., 1982, Uniform ionophore A23187 distribution and cytoplasmic calcium buffering in intact human red cells, *Biochim. Biophys. Acta* **692:**431–440.

Steck, T. L., 1974, Preparation of impermeable inside-out and right-side-out vesicles from erythrocyte membrane, in: *Methods in Membrane Biology,* Vol. 2 (E. D. Korn, ed.), Plenum Press, New York, pp. 245–282.

Stinnakre, J., and Tauc, L., 1973, Calcium influx in active *Aplysia* neurones detected by injected aequorin, *Nature, New Biol.* **242:**113–115.

Szász, I., and Gárdos, G., 1974, Mechanism of various drug effects on the $Ca^{2+}$-dependent $K^+$-efflux from human red blood cells, *FEBS Lett.* **44:**213–216.

Szász, I., Sarkadi, B., and Gárdos, G., 1974, Erythrocyte parameters during induced $Ca^{2+}$-dependent rapid $K^+$-efflux: Optimum conditions for kinetic analysis, *Haematologia* **8:**143–151.

Szász, I., Sarkadi, B., and Gárdos, G., 1977, Mechanism of $Ca^{2+}$-dependent selective rapid $K^+$-transport induced by propranolol in red cells, *J. Membr. Biol.* **35:**75–93.

Szász, I., Sarkadi, B., and Gárdos, G., 1978a, Effects of drugs on calcium-dependent rapid potassium transport in calcium-loaded intact red cells, *Acta Biochim. Biophys. Acad. Sci. Hung.* **13:**133–141.

Szász, I., Sarkadi, B., and Gárdos, G., 1978b, Mechanism for passive calcium transport in human red cells, *Acta Biochim. Biophys. Acad. Sci. Hung.* **13:**239–249.

Szász, I., Sarkadi, B., and Gárdos, G., 1978c, Changes in the $Ca^{2+}$-transport processes of red cells during storage in ACD, *Brit. J. Haematol.* **39:**559–568.

Szász, I., Sarkadi, B., Schubert, A., and Gárdos, G., 1978d, Effects of lanthanum on calcium-dependent phenomena in human red cells, *Biochim. Biophys. Acta* **512:**331–340.

Szász, I., Sarkadi, B., and Gárdos, G., 1980, Calcium sensitivity of calcium-dependent functions in human red blood cells, in: *Advances in Physiological Sciences* Vol. 6 (S. R. Hollán, G. Gárdos, and B. Sarkadi, eds.), Pergamon Press, Akadémiai Kiadó, Budapest, pp. 211–221.

Szász, I., Sarkadi, B., and Gárdos, G., 1982, Operation of a Ca-dependent K(Rb)-transport in human lymphocytes, *Haematologia* **15:**83–89.

Sze, H., and Solomon, A. K., 1979, Calcium-induced potassium pathways in sided erythrocyte membrane vesicles, *Biochim. Biophys. Acta* **554:**180–194.

Szönyi, S., 1960, Wirkung von Fluorid auf die Verteilung von Kalium und Natrium sowie auf die $Co_2$-Bindung in menschlichen Blut, *Acta Physiol. Acad. Sci. Hung.* **17:**9–13.

Thomas, M. V., and Gorman, A. L. F., 1977, Internal calcium changes in a bursting pace-maker neuron measured with arsenazo III., *Science* **196:**531–533.

Tosteson, D. C., 1959, Halide transport in red blood cells, *Acta Physiol. Scand.* **46:**19–41.

Tosteson, D. C., and Hoffman, J. F., 1960, Regulation of cell volume by active cation transport in high and low potassium sheep red cells, *J. Gen. Physiol.* **44:**169–194.

Tsien, R. Y., Pozzan, T., and Rink, T. J., 1982, T-cell mitogens cause early changes in cytoplasmic free $Ca^{2+}$ and membrane potential in lymphocytes, *Nature* **295:**68–71.

Valdeolmillos, M., Garcia-Sancho, J., and Herreros, B., 1982, $Ca^{2+}$-dependent $K^+$ transport in the Ehrlich ascites tumor cells, *Biochim. Biophys. Acta* **685:**273–278.

van Rossum, G. D. V., 1970, Relation of intracellular $Ca^{2+}$ to retention of $K^+$ by liver slices, *Nature (London)* **225:**638–639.

Vestergaard-Bogind, B., and Bennekou, P., 1982, Calcium-induced oscillations in $K^+$ conductance and membrane potential of human erythrocytes mediated by the ionophore A23187, *Biochim. Biophys. Acta* **688:**37–44.

Vincenzi, F. F., 1981, Calmodulin pharmacology, *Cell Calcium* **2:**387–409.

Volpi, M., Sha'afi, R. I., and Feinstein, M. B., 1981, Antagonism of calmodulin by local anesthetics—inhibition of calmodulin-stimulated calcium transport of erythrocyte inside-out membrane vesicles, *Mol. Pharmacol.* **20:**363–370.

Walsh, J. V., and Singer, J. J., 1980, Penetration-induced hyperpolarization as evidence for $Ca^{2+}$-activation of $K^+$ conductance in isolated smooth muscle cells, *Am. J. Physiol.* **239:**182–189.

Weed, R. I., LaCelle, P. L., and Merrill, E. M., 1969, Metabolic dependence of red cell deformability, *J. Clin. Invest.* **48:**795–809.

Whitney, R. B., and Sutherland, R. M., 1972, Enhanced uptake of calcium by transforming lymphocytes, *Cell. Immunol.* **5:**137–147.

Whittan, R., 1968, Control of membrane permeability to potassium in red blood cells, *Nature (London)* **219:**610.

Wilbrandt, W., 1937, A relation between the permeability of red cell and its metabolism, *Trans. Faraday Soc.* **33:**956–959.

Wilbrandt, W., 1940, Die Abhängigkeit der Ionenpermeabilität der Erythrozyten vom glykolytischen Stoffwechsel, *Pflugers Arch.* **243:**519–536.

Wiley, J. S., 1981, Increased erythrocyte cation permeability in thalassemia and conditions of marrow stress, *J. Clin. Invest.* **67:**917–922.

Wiley, J. S., and Gill, F. M., 1976, Red cell calcium leak in congenital hemolytic anemia with extreme microcytosis, *Blood* **47:**197–210.

Yellen, G., 1982, Single $Ca^{2+}$-activated nonselective cation channels in neuroblastoma, *Nature* **296:**357–359.

Yingst, D. R., and Hoffman, J. F., 1978, Changes of intracellular $Ca^{++}$ as measured by arsenazo III in relation to the K permeability of human erythrocyte ghosts, *Biophys. J.* **23:**463–471.

Yingst, D. R., and Hoffman, J. F., 1981, Effect of intracellular Ca on inhibiting the Na-K pump and stimulating Ca-induced K transport in resealed human red cell ghosts, *Fed. Proc.* **40:**543.

Yingst, D. R., and Hoffman, J. F., 1984, Ca-induced K transport in human red blood cell ghosts containing arsenazo III: Transmembrane interactions of Na, K, and Ca and the relationship to the functioning Na-K pump, *J. Gen. Physiol.,* **83:**19–45.

32

# *Biochemistry of Plasma Membrane Calcium Transporting Systems*

*Ernesto Carafoli*

## I. INTRODUCTION

Plasma membranes contain three $Ca^{2+}$-transporting systems, a specific channel, a pumping ATPase, and a $Na^{+}$–$Ca^{2+}$ exchanger. The first system, which mediates the downhill influx of $Ca^{2+}$ into cells, has been known since 1958, when $Ca^{2+}$ action potentials, implying a $Ca^{2+}$ component in plasma membrane, were first recorded by Fatt and Ginsborg (1958) in crayfish muscle fiber membranes. In the years that followed, the observation was extended to several other excitable plasma membranes, and is now commonly attributed to the existence of a specific $Ca^{2+}$-conducting channel, different from the well-known $Na^{+}$ channel. The traditional tool for studying the $Ca^{2+}$ channel has been the recording of $Ca^{2+}$-dependent electrical currents in intact, or nearly intact, tissue preparations. It has emerged from a large number of electrophysiological studies of this type that the density of $Ca^{2+}$ channels in most plasma membranes is vanishingly low, much lower, for example, than that of the $Na^{+}$ or $K^{+}$ channels. As a result, biochemical studies of the $Ca^{2+}$ channel are faced with almost insurmountable difficulties, and are very scarce. Recent developments, particularly on isolated vesicles of heart sarcolemma (see for example Rinaldi *et al.*, 1981), hold some promises, but it will certainly be a while before a reasonably detailed report on the biochemistry of the $Ca^{2+}$ channel can be written.

Of the other two plasma-membrane $Ca^{2+}$ transporting systems, one, the $Ca^{2+}$-pumping ATPase, has now been studied biochemically for a number of years, and decisive progress has recently been made toward understanding it in molecular terms. The other system, the $Na^{+}$–$Ca^{2+}$ exchanger, originally studied only with electro-

*Ernesto Carafoli* • Laboratory of Biochemistry, Swiss Federal Institute of Technology (ETH), CH-8092 Zurich, Switzerland.

physiological tools, has recently become amenable to penetrating biochemical investigations and has also witnessed considerable advancement. This chapter offers a concise survey of the biochemical knowledge of these two systems.

## II. THE $Ca^{2+}$-PUMPING ATPase OF THE PLASMA MEMBRANE

### A. The Enzyme in Situ

The first indication for the involvement of an ATP-dependent system in the pumping of $Ca^{2+}$ out of cells came from the work of Dunham and Glynn (1961) on erythrocytes. The existence of a $Ca^{2+}$-specific ATPase in the membrane of erythrocytes was conclusively documented in 1966 by Schatzmann, who showed that it could indeed mediate the uphill transport of $Ca^{2+}$ out of the cells. Originally, the enzyme was considered typical of nonexcitable cells, and for a long time the plasma membrane of choice for its study has been almost exclusively that of erythrocytes. In the last 3 or 4 years, however, it has become clear that $Ca^{2+}$-transporting ATPases occur in nonexcitable as well as in excitable plasma membranes (see Penniston, 1983, for a review), and the starting material for their study now ecompasses a large number of plasma-membrane types. Comprehensive reviews on the enzyme have appeared (Roufogalis, 1979; Sarkadi, 1980; Schatzmann, 1982), and the reader is referred to them, particularly to the excellent review by Schatzmann, for detailed information on the enzyme *in situ*. In a recent development, the $Ca^{2+}$-ATPase has been isolated and purified, first from erythrocytes (Niggli *et al.*, 1979), and then from heart plasma membranes (Caroni and Carafoli, 1981) and from the synaptosomal membrane (Hakim *et al.*, 1982). A $Ca^{2+}$-ATPase preparation has been obtained also from liver plasma membranes, but its degree of purification and its relationship to $Ca^{2+}$ pumping have not been determined (Lotersztain *et al.*, 1981). A recent review summarizes the information on the purified enzyme (Carafoli and Zurini, 1982).

The $Ca^{2+}$-ATPase can be considered as a high-affinity, low-capacity enzyme. It interacts with $Ca^{2+}$ with an affinity expressed by a $K_m$ of less than 1 $\mu$M (however, see below), and pumps $Ca^{2+}$ out of cells with a $V_{max}$ of the order of 0.5 nmol per mg of membrane protein per second. This rather low maximal rate of $Ca^{2+}$ pumping is adequate for the demands of nonexcitable cells, but probably not for those of excitable cells, which undergo periodic increases in the influx of $Ca^{2+}$ from the extracellular spaces and, thus, require a high-capacity pumping system to dispose of it rapidly and efficiently. This is probably the evolutionary rationale for the development, in the latter cell types, of a predominant parallel route for $Ca^{2+}$ ejection, the $Na^{+}$–$Ca^{2+}$ exchanger (see next section).

Work on erythrocyte membranes has established that the $Ca^{2+}$-ATPase becomes phosphorylated by ATP (Knauf *et al.*, 1974) and thus belongs to the aspartyl-phosphate-forming, $E_1$–$E_2$ class of ion-motive ATPases. However, no conclusive demonstration has been provided that the phosphorylated intermediate is actually formed on an aspartyl residue of the enzyme protein. The ATP hydrolysis, and the associated transport of $Ca^{2+}$, are inhibited by vanadate. Barrabin *et al.* (1980) have shown the inhibition to be enhanced by $K^{+}$ and $Mg^{2+}$, which increase the affinity of vanadate for the enzyme, to a $K_i$ of 1.5 μM. The inhibition of the ATPase by vanadate is unaffected by $Ca^{2+}$ up to a concentration of 50 μM (Barrabin *et al.*, 1980), but Rossi *et al.* (1981) have shown that 1 mM external $Ca^{2+}$ relieves partially the inhibition of $Ca^{2+}$ transport in resealed erythrocyte ghosts by vanadate. Both the ATPase (Schatzmann and Rossi, 1971; Bond and Green, 1971) and the associated $Ca^{2+}$ transport in red-cell ghosts (Romero, 1981) are stimulated by $Na^+$ and $K^+$.

The affinity of the ATPase enzyme for its two substrates, ATP and $Ca^{2+}$, is a complex matter. Agreement now seems to exist that the enzyme contains two ATP-interacting sites, one with low and one with high affinity (Richards *et al.*, 1978; Mualem and Karlish, 1979; Stieger and Luterbacher, 1981). The high-affinity site is the catalytic site ($K_m$, 2.5 μM); the low-affinity site accelerates the hydrolysis at the catalytic site ($K_m$, about 200 μM). As for $Ca^{2+}$, the very high affinity mentioned above ($K_m < 1$ μM) has recently come into question, since experiments have been reported (Sarkadi *et al.*, 1979) where it has been shown that the high $Ca^{2+}$ affinity requires the presence of EGTA (the latter is normally present as EGTA-$Ca^{2+}$ buffer). The interpretation of the EGTA effect is not clear, and it has been proposed (Sarkadi *et al.*, 1979) that the enzyme has two $Ca^{2+}$ sites, one for $Ca^{2+}$ alone, one for $Ca^{2+}$ and $Ca^{2+}$-EGTA. Both sites must be occupied for $Ca^{2+}$ transport to occur.

The measurement of the $Ca^{2+}$–ATP stoichiometry of the ATPase requires tightly sealed membranes to minimize $Ca^{2+}$ leaks, subtraction of the $Mg^{2+}$-ATPase, i.e., the activity seen prior to the addition of $Ca^{2+}$, and complete inhibition of the $Na^+$, $K^+$-ATPase. In most laboratories, a stoichiometry approaching one has been measured (see Schatzmann, 1982, for a review of the problem), but some (Quist and Roufogalis, 1975, 1977; Sarkadi *et al.*, 1977) have concluded for a stoichiometry of two $Ca^+$ per ATP. The question is not yet decided, although recent experiments on the purified reconstituted ATPase (Niggli *et al.*, 1981a; Clark and Carafoli, 1984; see below) have provided strong support for a 1 : 1 stoichiometry.

As mentioned above, the reaction cycle of the $Ca^{2+}$-ATPase involves the formation of a hydroxylamine-sensitive phosphorylated intermediate, most likely an aspartyl-phosphate. The $Ca^{2+}$-dependent phosphorylation of the membrane protein corresponding to the ATPase ($M_r$, about 140,000) is very rapid ($t_{1/2}$ at 0°C of the order of 10 sec; Rega and Garrahan, 1975; Garrahan and Rega, 1978; Schatzmann and Bürgin, 1978) and, thus, permits the differentiation of the $Ca^{2+}$-ATPase from other phosphorylatable proteins of the (erythrocyte) membrane. Work in several laboratories (Rega and Garrahan, 1975; Garrahan and Rega, 1978; Schatzmann and Bürgin, 1978; Bürgin and Schatzmann, 1979; Richards *et al.*, 1978; Mualem and Karlish, 1979, 1980; Szasz *et al.*, 1978; Rega and Garrahan, 1980; see Schatzmann, 1982, for a comprehensive review) has established the principal characteristics of the transport cycle which can be summarized in the following scheme:

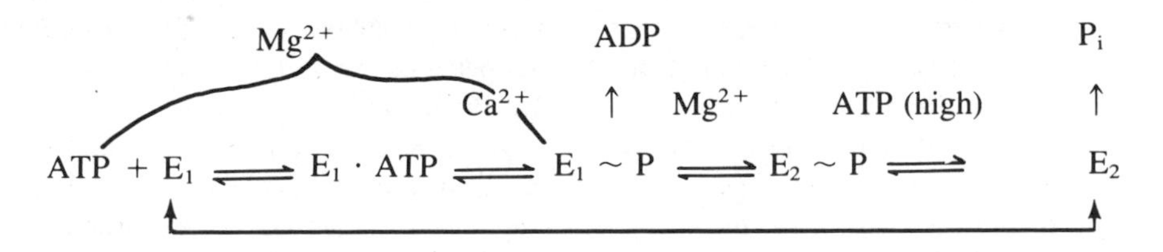

$E_1$ and $E_2$ represent two different conformational states of the ATPase. The phosphorylation of the enzyme requires $Ca^{2+}$, but not $Mg^{2+}$, indicating that the substrate of the ATPase is not $Mg^{2+}$-ATP. $Mg^{2+}$, however, accelerates the $Ca^{2+}$-dependent phosphorylation and increases the steady-state decay of the phosphoenzyme, possibly resulting from the increase in the affinity of the enzyme for free ATP, or from a shift of the reaction $E_1 \cdot ATP \rightarrow E_1 \sim P$ to the right. $Mg^{2+}$ promotes the decay of the phosphoprotein only if the latter is formed in its absence. If the phosphoenzyme is formed in the presence of both $Mg^{2+}$ and $Ca^{2+}$, then its decay is equally rapid in the presence and in the absence of $Mg^{2+}$. This suggests that $Mg^{2+}$ is not required for the hydrolysis of the phosphorylated intermediate, but for a prior step, which favors the transition of the enzyme to a form where the accessibility of the phosphorylated intermediate with water is favored. The hydrolytic step of $E_2 \sim P \rightarrow E_2 + P_i$ is favored also by the low-affinity interaction of ATP with the enzyme.

Simultaneous work by Gopinath and Vincenzi (1977) and Jarrett and Penniston (1977) has established that the $Ca^{2+}$-ATPase of erythrocyte membranes is stimulated by calmodulin, and subsequent work by Lynch and Cheung (1979) and Niggli *et al.* (1979a) has shown that the stimulation is due to direct interaction. Only nanomolar amounts of calmodulin are required for the stimulation of the ATPase and the accompanying $Ca^{2+}$ transport (Vincenzi and Larsen, 1980). In a series of studies, Scharff (1972, 1976, 1978) and Scharff and Foder (1978) have defined two states of the ATPase, determined by the absence of calmodulin (A state, low $Ca^{2+}$ affinity) or by its presence (B state, high $Ca^{2+}$ affinity).

## B. The Purified Enzyme

With the multiplication of the reports on $Ca^{2+}$-dependent ATPases from sources different from erythrocytes, it has become clear that the sensitivity to calmodulin is a marker for the enzyme (see Penniston, 1983, for a review). At least two exceptions, however, apparently exist, liver (Lotersztain *et al.*, 1981; Kraus-Friedmann *et al.*, 1982) and corpus luteum (Verma and Penniston, 1981). It is certainly fortunate that the $Ca^{2+}$-ATPase of erythrocytes, the membrane on which the attempts to purify the enzyme have traditionally been made, is so sensitive to calmodulin, and associated with it in a way that easily permits its detachment. This has permitted Niggli *et al.* (1979) to devise a simple and highly effective method, based on a calmodulin affinity column, for the isolation and purification of the enzyme. In the procedure of Niggli *et al.* (1979), erythrocyte ghosts were washed free of calmodulin with EDTA, dissolved in Triton X-100, and applied to a Sepharose 4-B calmodulin affinity chromatography column in the presence of phosphatidyl serine. The latter was added following the discovery by Ronner *et al.* (1977) that the erythrocyte $Ca^{2+}$-ATPase is specifically stabilized by acidic phospholipids. Elution of the column with EDTA yielded a protein that had $Ca^{2+}$-stimulated ATPase activity, formed a phosphoenzyme when incubated with $Ca^{2+}$ and ATP, and mediated the ATP-dependent pumping of $Ca^{2+}$ when reconstituted into liposomes (Niggli *et al.*, 1981a; Figure 1). The EDTA-eluted protein had an $M_r$ of 138,000, interacted with $Ca^{2+}$ with a $K_m$ of about 0.4 $\mu$M, and was vanadate-sensitive. It, thus, repeated the properties of the $Ca^{2+}$-ATPase of the erythrocyte membrane, except for the fact that its activity was insensitive to calmodulin. The sensitivity to calmodulin was, on the other hand, preserved in a preparation of

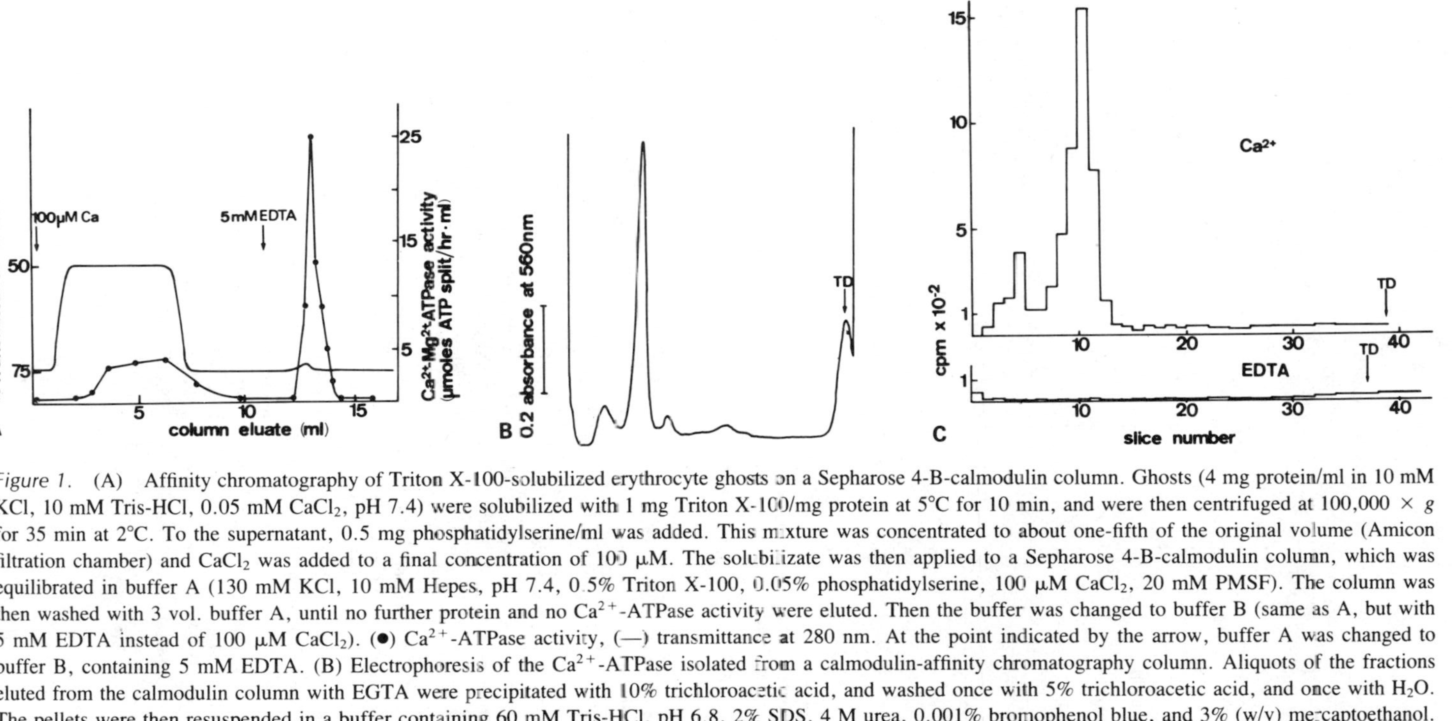


*Figure 1.* (A) Affinity chromatography of Triton X-100-solubilized erythrocyte ghosts on a Sepharose 4-B-calmodulin column. Ghosts (4 mg protein/ml in 10 mM KCl, 10 mM Tris-HCl, 0.05 mM $CaCl_2$, pH 7.4) were solubilized with 1 mg Triton X-100/mg protein at 5°C for 10 min, and were then centrifuged at 100,000 × *g* for 35 min at 2°C. To the supernatant, 0.5 mg phosphatidylserine/ml was added. This mixture was concentrated to about one-fifth of the original volume (Amicon filtration chamber) and $CaCl_2$ was added to a final concentration of 100 µM. The solubilizate was then applied to a Sepharose 4-B-calmodulin column, which was equilibrated in buffer A (130 mM KCl, 10 mM Hepes, pH 7.4, 0.5% Triton X-100, 0.05% phosphatidylserine, 100 µM $CaCl_2$, 20 mM PMSF). The column was then washed with 3 vol. buffer A, until no further protein and no $Ca^{2+}$-ATPase activity were eluted. Then the buffer was changed to buffer B (same as A, but with 5 mM EDTA instead of 100 µM $CaCl_2$). (●) $Ca^{2+}$-ATPase activity, (—) transmittance at 280 nm. At the point indicated by the arrow, buffer A was changed to buffer B, containing 5 mM EDTA. (B) Electrophoresis of the $Ca^{2+}$-ATPase isolated from a calmodulin-affinity chromatography column. Aliquots of the fractions eluted from the calmodulin column with EGTA were precipitated with 10% trichloroacetic acid, and washed once with 5% trichloroacetic acid, and once with $H_2O$. The pellets were then resuspended in a buffer containing 60 mM Tris-HCl, pH 6.8, 2% SDS, 4 M urea, 0.001% bromophenol blue, and 3% (w/v) mercaptoethanol. The mixture was heated for 4 min in a boiling-water bath and then applied to 8% polyacrylamide gels (10 µg protein stained with Coomassie blue). TD, tracking dye. (C) Formation of a $Ca^{2+}$- and ATP-dependent phosphorylated intermediate of the purified $Ca^{2+}$-ATPase. The isolated ATPase was incubated at a concentration of 20 µg/ml in a medium containing 130 mM KCl, 20 mM Tris-HCl, pH 7.4, 12 µM $mgCl_2$, 2 µM [γ-$^{32}$P]-ATP (specific radioactivity 5 µCi/nmol) and 50 µM $CaCl_2$ (top) or 500 µM EDTA (bottom). The reaction was stopped after 15 sec at 0°C by adding 1 vol. ice-cold solution containing 10% trichloroacetic acid, 0.1 mM $K_2$ATP, and 0.8 mM phosphoric acid. After centrifugation for 5 min at 1300*g*, at 4°C, the pellet was first washed with 2 vol. 5% trichloroacetic acid, 0.1 mM $K_2$ATP, 1 mM phosphoric acid, then with 2 vol. water. The enzyme was resuspended in 0.3 ml buffer containing 100 mM sodium phosphate, pH 6, 1% SDS, bromophenol blue (as the tracking dye), and 10% glycerol. Gel electrophoresis was performed with 20 µg enzyme on 5% SDS-polyacrylamide gels, at pH 6, the gels were frozen in dry ice, and pairs of 1-mm slices were incubated overnight with shaking in 1 ml 0.5% SDS, 10 mM Tris-HCl, pH 9, at 40°C. After addition of 10 ml Instagel (Packard), the vials were counted in a scintillation counter.

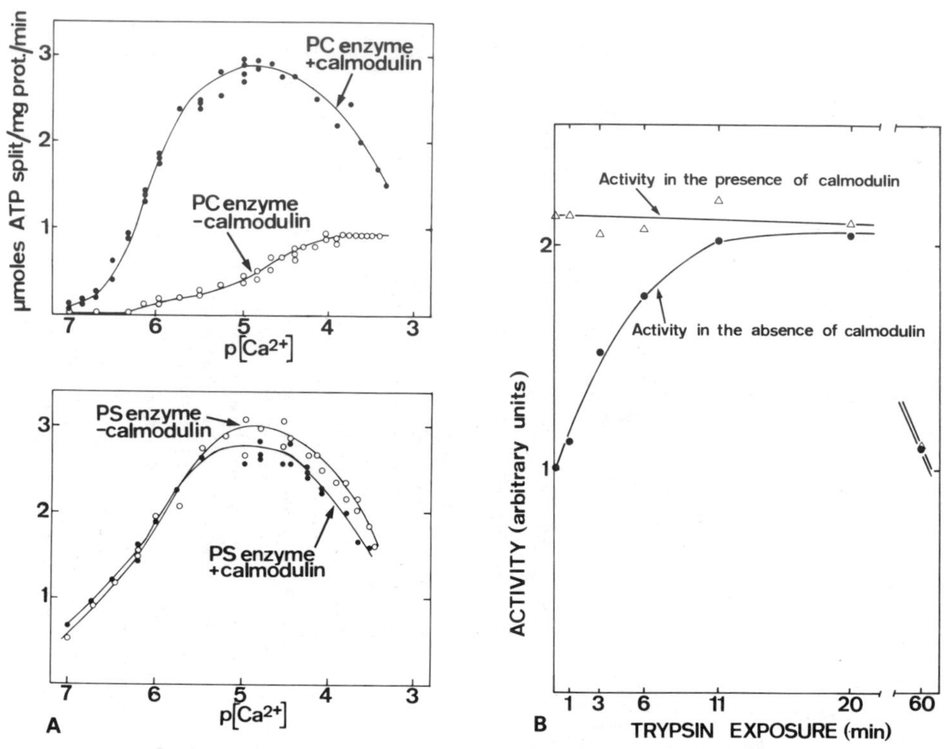


*Figure 2.* (A) High- and low-affinity forms of the purified $Ca^{2+}$-ATPase in the presence of phosphatidylcholine (PC) and phosphatidylserine (PS). The ATPase was solubilized from the membrane in 0.4% Triton X-100 as indicated in the legend to Figure 1. The procedure for isolation is described in Niggli *et al.* (1979). After several washes with $Ca^{2+}$-containing buffer, the medium was changed to a PC-containing buffer (130 mM KCl, 20 mM Hepes, pH 7.4, 1 mM $MgCl_2$, 100 μM $CaCl_2$, 2 mM dithiothreitol, 0.05% Triton X-100, 0.05% PC). After several washings of the column with this buffer, the medium was changed to a buffer of the same composition, but with 2 mM EDTA (potassium salt) instead of $CaCl_2$. Fractions of the ATPase-containing eluate were reconstituted in liposomes consisting of either PC (top) or PS (bottom). The reconstitution was performed by the Biobeads SM-2 Triton-removal method. The assay of the ATPase with a coupled enzyme assay system is described in Niggli *et al.* (1979). The free $Ca^{2+}$ concentrations indicated on the abscissae were obtained by adding different amounts of $CaCl_2$ to the medium containing 500 μM HEDTA, and were calibrated with a $Ca^{2+}$ electrode. The amount of reconstituted protein added was 2–3 μg, that of calmodulin 5 μg. Final volume, 1 ml; temperature, 37°C. Redrawn from Niggli *et al.* (1981a). (B) Activation of the purified $Ca^{2+}$-ATPase by trypsin. Conditions for the solubilization and the isolation of the ATPase as in (A). 7μg of trypsin for the time indicated. The reaction was stopped by adding a solution consisting of 30 mM Na Na-phosphate, pH 7.0, 30% (vol./vol.) glycerol, 7.5% (wt./vol.) sodium dodecyl-sulfate, 10 mM dithiothreitol followed by 5 min boiling. Activity was determined as in (A), in the absence or presence of saturating amounts of calmodulin.

the ATPase obtained by Gietzen *et al.* (1980) essentially by the method of Niggli *et al.* (1979), but in the absence of phosphatidyl serine. This indicated that the sensitivity of the ATPase to calmodulin was determined by the phospholipid environment. Niggli *et al.* (1981a,b) then carried out a detailed study of the effect of phospholipids on the ATPase, and established that the purified enzyme was shifted to the high-affinity state seen in the presence of calmodulin (the B state of the terminology of Scharff) by acidic phospholipids or long-chain polyunsaturated fatty acids (Figure 2A). Furthermore, following earlier findings by Taverna and Hanahan (1980) on erythrocyte ghosts, Niggli *et al.* (1981b) established that the purified ATPase could be shifted to the high-affinity form in the absence of calmodulin and acidic phospholipids by a controlled proteolytic treatment with trypsin (Figure 2B). The logical suggestion at this point (Carafoli and Zurini, 1982) was that calmodulin, acidic phospholipids or fatty acids, and limited proteolysis somehow increased the accessibility of the active site of the enzyme, as indicated in the scheme of Figure 3. Very interestingly (Adunyah *et al.*, 1982), a number of so-called "anticalmodulin drugs" were found to interfere not only with the activation of the purified ATPase by calmodulin, but also by acidic phospholipids and limited proteolysis. This indicates compellingly that these anticalmodulin drugs, in addition to interacting with calmodulin, also interact directly with the ATPase molecule.

The purified $Ca^{2+}$-ATPase was reconstituted into liposomes by Niggli *et al.* (1981a). The reconstituted system pumped $Ca^{2+}$ into $Ca^{2+}$-tight liposomes with a stoichiometry to ATP of one (Niggli *et al.*, 1981a; Clark and Carafoli, 1983). Using the reconstituted system it was also shown (Niggli *et al.*, 1982) that the ATPase functions as an electroneutral carrier, which exchanges two $H^+$ per one $Ca^{2+}$.

Controlled proteolysis by trypsin has been used by Graf *et al.* (1982) and Zurini *et al.* (1983) to map zones of functional interest in the purified enzyme molecule. It has, thus, been established that the active site resides in a limit polypeptide of $M_r$ about 76,000, and that calmodulin is bound to a polypeptide of $M_r$ 90,000 and to one

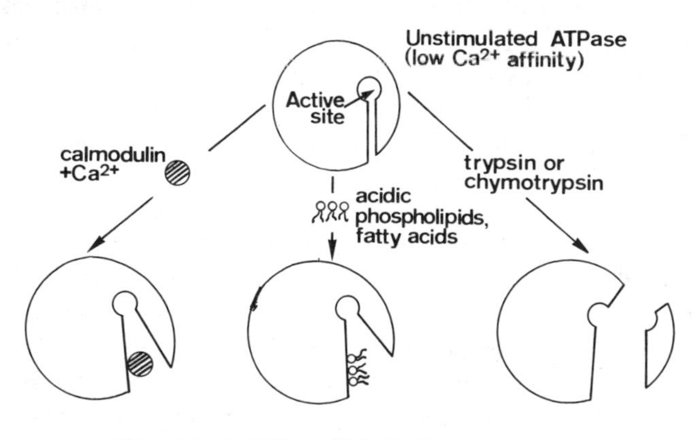


*Figure 3.* A scheme of the possible mechanism of activation of the $Ca^{2+}$-ATPase by calmodulin, acidic phospholipids, and controlled proteolysis.

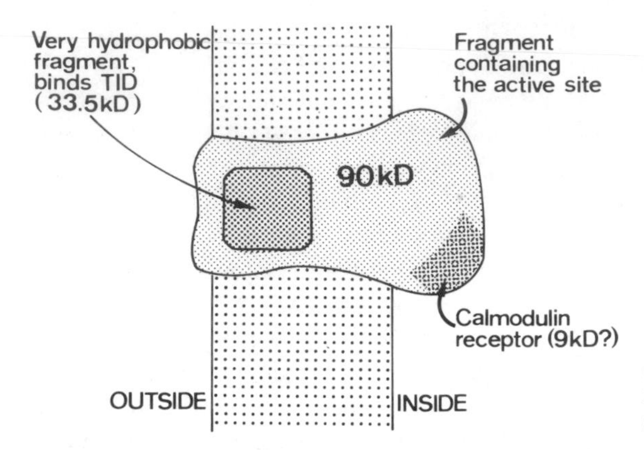


*Figure 4.* A hypothetical model of the functional architecture of the $Ca^{2+}$-ATPase in the erythrocyte membrane.

or more fragments of $M_r$ between 25,000–28,000. Labeling of the proteolyzed ATPase with the radioactive hydrophobic probe ((3-trifluoromethyl-3-(m-[$^{125}$I]iodophenyl) diazirine (TID)) has shown unusually high levels of radioactivity in a limited polypeptide of $M_r$ about 33,500 which may, thus, contain most of the intramembrane regions of the molecule. A summary of the results of the experiments on controlled proteolysis is presented in the scheme of Figure 4.

As mentioned above, $Ca^{2+}$-ATPases have now been documented in a number of plasma membranes different from that of erythrocyte (see the recent review by Penniston, 1983). Of special interest among them are those of the corpus luteum and of liver, since they apparently lack calmodulin sensitivity. The liver enzyme has now been isolated (albeit not purified to homogeneity) by Lotersztain *et al.* (1981) from sodium cholate-solubilized liver plasma membranes using concanavalin A–Ultragel chromatography. Famulski and Carafoli (1982) and Kraus-Friedmann *et al.* (1982) have documented its role in the transport of $Ca^{2+}$. Other interesting properties of the liver enzyme are its nucleotide specificity, which is less strict than that of the related erythrocyte system, its extremely high affinity for $Ca^{2+}$ ($K_m$ in the nanomolar range), and its mediocre sensitivity to vanadate. It appears that the liver $Ca^{2+}$-transporting ATPase may be of a type at least partially different from that of the other plasma-membrane $Ca^{2+}$-transporting ATPases, of which the erythrocyte enzyme is the best-known representative.

Also of special interest are the $Ca^{2+}$-ATPases of synaptosomal and heart plasma membrane (Hakim *et al.*, 1982; Caroni and Carafoli, 1981), since their demonstration has disposed of the once prevailing concept that the only system for ejecting $Ca^{2+}$ from excitable cells is the $Na^+$–$Ca^{2+}$ exchanger. The $Ca^{2+}$-ATPase of heart sarcolemma has been purified to homogeneity (Caroni and Carafoli, 1981b), and shown to repeat the properties of the purified erythrocyte enzyme. It has been reconstituted into liposomes (Caroni *et al.*, 1982, 1984) and shown to pump $Ca^{2+}$ with a stoichiometry to ATP of 1 : 1. It differs from the erythrocyte enzyme in one respect; it is modulated by a kinase–phosphatase system (Caroni and Carafoli, 1981b) which, however, does not act on the ATPase molecule proper, but probably on some accessory protein associated with it.

Regulatory systems have been described for $Ca^{2+}$-ATPases also in other plasma membranes. Ghiisen and Van Os (1982) have demonstrated stimulation of ATP-dependent $Ca^{2+}$ transport in basolateral plasma membranes of intestinal cells by 1,25-dihydroxy vitamin $D_3$, Pershadsingh and McDonald (1979) have shown that direct addition of insulin inhibits the $Ca^{2+}$-ATPase of adipocyte plasma membranes, and Davis and Blas (1981), Galo *et al.* (1981), and Davis *et al.* (1982) have described stimulatory effects by thyroid hormone. Soloft and Sweet (1982) have reported inhibition of the ATPase in myometrial plasma membranes by oxytocin.

## III. THE PLASMA MEMBRANE $Na^+$–$Ca^{2+}$ EXCHANGE

In many cells, the ejection of $Ca^{2+}$ has been shown to depend on the presence of $Na^+$ in the external medium, indicating that the exit of $Ca^{2+}$ is coupled to the entrance of $Na^+$. Experiments in the late 1960's by Reuter and Seitz on heart, and by Blaustein, Baker, Hodgkin, and their co-workers on the giant axon of the squid, have offered direct support to the indication, and have led to the discovery of an exchange diffusion carrier that mediates the movements of $Na^+$ and $Ca^{2+}$ across plasma membranes. It has now become clear that the system predominates quantitatively over the $Ca^{2+}$-ATPase in the ejection of $Ca^{2+}$ in the two tissues originally studied, and probably also in other excitable plasma membranes. But it has recently emerged that it operates also in most, if not all, nonexcitable cells (erythrocytes may be the only exception), where the role of the $Ca^{2+}$-ATPase is most likely predominant (see Blaustein and Nelson, 1982, for a recent review).

If the movement of $Na^+$ into the cell down its steep electrochemical gradient provides the energy for the uphill extrusion of $Ca^{2+}$, it follows that the latter would cease, and even be reversed, when the trans-plasma membrane electrochemical gradient of $Na^+$ is reduced or eliminated. This can indeed be experimentally verified; internal $Ca^{2+}$ increases in isolated heart muscle preparations upon lowering of the extracellular $Na^+$ (Wilbrandt and Koller, 1948; Lüttgau and Niedergerke, 1958) and decreases promptly again when the preparation is placed in a $Na^+$-rich medium. Since the electrochemical gradient of $Na^+$ is not the only pulling force in the movement of $Ca^{2+}$ (the exchange is electrogenic, see below), the potential across the plasma membrane will contribute a factor that adds algebraically to the chemical gradient of $Na^+$ in determining the direction of the $Ca^{2+}$ movements. In cells like heart, where the magnitude and the sign of the trans-plasma membrane electrical potential vary during the physiological cycle, the $Na^+$–$Ca^{2+}$ exchanger most likely mediates both the efflux and the influx of $Ca^{2+}$.

The basic observation made by the groups mentioned above was that the rate of $Ca^{2+}$ efflux from cells depended largely on the presence and amount of $Na^+$ in the external medium (Reuter and Seitz, 1968; Blaustein and Hodgkin, 1968, 1969). Raising the intracellular $Na^+$ in dialyzed axons, or lowering its external concentration, reversed the reaction, i.e., promoted the influx of $Ca^{2+}$ into the axon (Baker *et al.*, 1967a,b). The efflux of $Na^+$ from axons was partially dependent on external $Ca^{2+}$ (Baker *et al.*, 1967a,b).

One important problem related to the suggestion that the $Na^+$–$Ca^{2+}$ exchange is the main system for the ejection of $Ca^{2+}$ from (excitable) cells and, thus, for the maintenance of the very low free $Ca^{2+}$ concentration normally found in the cytoplasm is whether the energy content of the electrochemical $Na^+$ gradient is adequate. Blaustein and Hodgkin (1969) observed that an electroneutral exchange could hardly account for the submicromolar concentration of $Ca^{2+}$ in axons, and argued in favor of an electrogenic, three $Na^+$ per one $Ca^{2+}$ exchange. Indirect support for the electrogenic operation of the exchanger came from the observation that the efflux of $Ca^{2+}$ from axons or barnacle muscle fibers is a sigmoidal function of the external $Na^+$ concentration (Blaustein, 1974; Russel and Blaustein, 1974; Blaustein, 1977), fitting a Hill equation with a Hill coefficient of between 2.6 and 3.0. Direct measurements of $Ca^{2+}$ and $Na^+$ fluxes in axons (Blaustein and Russel, 1975) also supported a 3 : 1 stoichiometry. In heart, data on the relationship between external $Ca^{2+}$ and $Na^+$ concentrations, $Ca^{2+}$ efflux, and tension were originally interpreted in terms of a two $Na^+$ for one $Ca^{2+}$ stoichiometry (Reuter and Seitz, 1968; Glitsch *et al.*, 1970; Jundt *et al.*, 1975; Chapman and Ellis, 1977). A problem with the electroneutral operation of the carrier in heart is that the minimal level of intracellular $Ca^{2+}$ that could be attained in this case is 1 μM or more, i.e., a level that would cause permanent activation of contraction.

A recent technological development that has clarified the matter of the stoichiometry of the exchanger in heart, and has permitted other significant advancements, has been the introduction of heart plasma-membrane vesicles (Jones *et al.*, 1979). Essential kinetic parameters of the $Na^+$–$Ca^{2+}$ exchanger have, thus, been determined (Reeves and Sutko, 1979; Pitts, 1979; Caroni *et al.*, 1980), among them the affinity of the exchanger for $Na^+$ ($K_m$, about 20 mM) and for $Ca^{2+}$ ($K_m$, 1.5–5.0 μM), and the maximal velocity of $Ca^{2+}$ transport (about 20 nmoles per mg of sarcolemmal protein per second). It has also been established conclusively that the heart exchanger operates electrogenically, exchanging at least three $Na^+$ per one $Ca^{2+}$. Pitts (1979) has measured directly $Na^+$ and $Ca^{2+}$ fluxes, and has extrapolated his results to a three $Na^+$ per one $Ca^{2+}$ stoichiometry. Reeves and Sutko (1980) and Caroni *et al.* (1980) have used as a tool the movements of lipophilic cations during the operation of the exchanger, and have found that the interior of the vesicles becomes negative as $Na^+$ leaves them in exchange for $Ca^{2+}$; i.e., more than two $Na^+$ are exchanged for one $Ca^{2+}$. Philipson and Nishimoto (1980) and Bers *et al.* (1980) have shown that the $Ca^{2+}$ uptake driven by the efflux of $Na^+$ becomes faster if a positive potential is established inside the vesicles, and slower if a negative inside potential is established.

One interesting aspect of the $Na^+$–$Ca^{2+}$ exchanger is its response to ATP, which has been observed in squid axons, barnacle muscle fibers, and heart. The first observation is that by Baker and Glitsch (1973), who reported that the efflux of $Ca^{2+}$ from axons, including the $Na^+$-dependent portion, was reduced when the cytoplasmic ATP concentration was lowered. Their data, and later results by Di Polo (1974) and Blaustein (1977), indicated that lowering cytoplasmic ATP decreased the affinity of the exchanger for $Ca^{2+}$ and $Na^+$, without affecting the maximal rate of exchange. Of great interest is the observation by Di Polo (1977) that only hydrolyzable analogues of ATP (2-β-methylene ATP and 2-deoxy-ATP) were able to replace the latter in promoting $Na^+$-dependent $Ca^{2+}$ efflux from axons. This suggests the involvement of a phosphorylation step, a possibility that has been supported by very recent experiments by Caroni and Carafoli (1984). They have found that the exchange activity of heart sarcolemmal vesicles is depressed by an endogenous protein phosphatase which requires $Ca^{2+}$ and calmodulin, and enhanced by an endogenous protein kinase which is also dependent on $Ca^{2+}$ and calmodulin. The kinase, however, has higher affinity for calmodulin and $Ca^{2+}$ than the phosphatase. The $Na^{+}$–$Ca^{2+}$ exchange, then, appears to be modulated by a phosphorylation–dephosphorylation process, in analogy to what is shown for the $Ca^{2+}$-ATPase (see above). Since the $Ca^{2+}$-ATPase is stimulated by a controlled proteolytic treatment (see above), it is of interest that controlled proteolysis also stimulates the $Na^{+}$–$Ca^{2+}$ exchanger (Philipson and Nishimoto, 1982).

## REFERENCES

Adunyah, E. S., Niggli, V., and Carafoli, E., 1982, The anticalmodulin drugs trifluoperazine and R24571 remove the activation of purified erythrocyte $Ca^{2+}$ ATPase by acidic phospholipids and by controlled proteolysis, *FEBS Lett.*, **143:**65–68.

Baker, P. F., and Glitsch, H. G., 1973, Does metabolic energy participate directly in the sodium-dependent extrusion of calcium from squid giant axons? *J. Physiol. (London)* **233:**44P.

Baker, P. F., Blaustein, M. P., Manil, J., and Steinhardt, R. A., 1967a, A ouabain-insensitive, calcium-sensitive sodium efflux from giant axons of Loligo, *J. Physiol., (London)* **191:**100P.

Baker, P. F., Blaustein, M. P., Hodgkin, A. L., and Steinhardt, R. A., 1967b, The effect of sodium concentration on calcium movements in giant axons of Loligo forbesi, *J. Physiol. (London)* **192:**43P.

Barrabin, H., Garrahan, P. J., and Rega, A. F., 1980, Vanadate inhibition of the $Ca^{2+}$-ATPase of human red cell membranes, *Biochim. Biophys. Acta* **600:**796–804.

Bers, D. M., Philipson, K. D., and Nishimoto, A. Y., 1980, Sodium-calcium exchange and sidedness of isolated cardiac sarcolemmal vesicles, *Biochim. Biophys. Acta* **601:**358–371.

Blaustein, M. P., 1974, The interrelationship between sodium and calcium fluxes across cell membranes, *Rev. Physiol. Biochem. Pharmacol.* **70:**33–82.

Blaustein, M. P., 1977, Effect of internal and external cations and of ATP on sodium–calcium and calcium–calcium exchange in squid axons, *Biophys. J.* **20:**79–111.

Blaustein, M. P., and Hodgkin, A. L., 1968, The effect of cyanide on calcium efflux in squid axons, *J. Physiol. (London)* **198:**46P.

Blaustein, M. P., and Hodgkin, A. L., 1969, The effect of cyanide on the efflux of calcium from squid axons, *J. Physiol. (London)* **200:**497–527.

Blaustein, M. P., and Nelson, M., 1982, $Na^{+}$–$Ca^{2+}$ exchange: Its role in the regulation of cell calcium, in: *Membrane Transport of Calcium* (E. Carafoli, ed.), Academic Press, New York, pp. 217–236.

Blaustein, M. P., and Russel, J. M., 1975, Sodium–calcium exchange and calcium–calcium exchange in internally dialyzed squid giant axons, *J. Membr. Biol.* **22:**285–312.

Bond, G. H., and Green, J. W., 1971, Effects of monovalent cations on the ($Mg^{2+}$ + $Ca^{2+}$)-dependent ATPase of red cell membranes, *Biochim. Biophys. Acta* **241:**393–398.

Bürgin, H., and Schatzmann, H. J., 1979, The relation between net calcium, alkali cation, and chloride movements in red cells exposed to salicylate, *J. Physiol. (London)* **287:**15–32.

Carafoli, E., and Zurini, M., 1982, The calcium pumping ATPase of plasma membranes. Purification, reconstitution, and properties, *BBA Rev. Bioenerg.* **683:**279–301.

Caroni, P., and Carafoli, E., 1981a, Regulation of the $Ca^{2+}$-pumping ATPase of heart sarcolemma by a phosphorylation–dephosphorylation process, *J. Biol. Chem.* **253:**9371–9373.

Caroni, P., and Carafoli, E., 1981b, The $Ca^{2+}$-pumping ATPase of heart sarcolemma: Characterization, calmodulin dependence, and partial purification, *J. Biol. Chem.* **256:**3263–3270.

Caroni, P., and Carafoli, E., 1984, The regulation of the $Na^{+}/Ca^{2+}$ exchanger of heart sarcolemma, *Eur. J. Biochem.*, in press.

Caroni, P., Reinlib, L., and Carafoli, E., 1980, Charge movements during $Na^+$–$Ca^{2+}$ exchange in heart sarcolemmal vesicles, *Proc. Natl. Acad. Sci. USA* **77:**6354–6358.

Caroni, P., Zurini, M., and Clark, A., 1982, The calcium pumping ATPase of heart sarcolemma, *Ann. N.Y. Acad. Sci.* **402:**402–421.

Caroni, P., Zurini, M., Clark, A., and Carafoli, E., 1984, Further characterization and reconstitution of the purified $Ca^{2+}$-pumping ATPase of heart sarcolemma, *J. Biol. Chem.*, in press.

Chapman, R. A., and Ellis, D., 1977, The effects of manganese ions on the concentration of the frog's heart, *J. Physiol. (London)* **272:**331–354.

Clark, A., and Carafoli, E., 1984, The stoichiometry of the $Ca^{2+}$-pumping ATPase of erythrocytes, *Cell Calcium,* in press.

Davis, P. J., and Blas, S. D., 1981, *In vitro* stimulation of human red blood cell $Ca^{2+}$ ATPase by thyroid hormone, *Biochem. Biophys. Res. Commun.* **99:**1073–1080.

Davis, P. J., Davis, F. B., and Blas, S. D., 1982, Studies on the mechanism of thyroid hormone stimulation *in vitro* of human red cell $Ca^{2+}$ ATPase activity, *Life Sci.* **30:**675–682.

Di Polo, R., 1974, Effect of ATP on the calcium efflux in dialyzed squid giant axons, *J. Gen. Physiol.* **64:**503–517.

Di Polo, R., 1977, Characterization of the ATP-dependent calcium efflux in dialyzed squid giant axons, *J. Gen. Physiol.* **69:**795–813.

Dunham, E. T., and Glynn, L. M., 1961, Adenosine triphosphatase activity and the active movements of alkali metal ions, *J. Physiol. (London)* **156:**274–293.

Famulski, K., and Carafoli, E., 1982, Calcium transporting activities of membrane functions isolated from the post-mitochondrial supernatant of rat liver, *Cell Calcium* **3:**263–281.

Fatt, P., and Ginsborg, B. L., 1958, The ionic requirements for the production of action potentials in crustacean muscle fibers, *J. Physiol. (London)* **142:**516–543.

Galo, M. G., Uñates, L. E., and Farias, R. N., 1981, Effect of membrane fatty acid composition on the action of thyroid hormones on ($Ca^{2+}$ + $Mg^{2+}$)-adenosine triphosphatase from rat erythrocyte, *J. Biol. Chem.* **256:**7113–7114.

Garrahan, P. J., and Rega, A. F., 1978, Activation of partial reactions of the $Ca^{2+}$-ATPase from human red cells by $Mg^{2+}$ and ATP, *Biochim. Biophys. Acta* **513:**59–65.

Ghiisen, W. E. J. M., and Van Os, C. H., 1982, 1α, 25-Dihydroxy-vitamin $D_3$ regulates ATP-dependent calcium transport in basolateral plasma membranes of rat enterocytes, *Biochim. Biophys. Acta* **689:**170–172.

Gietzen, K., Tejcka, M., and Wolf, H. U., 1980, Calmodulin affinity chromatography yields a functional purified erythrocyte ($Ca^{2+}$ + $Mg^{2+}$)-dependent adenosine triphosphatase, *Biochem. J.* **189:**81–88.

Glitsch, H. G., Reuter, H., and Scholz, H., 1970, The effect of the internal sodium concentration on calcium fluxes in isolated guinea-pig auricles, *J. Physiol. (London)* **209:**25–43.

Gopinath, R. M., and Vincenzi, F. F., 1977, Phosphodiesterase protein activator mimics red blood cell cytoplasmic activator of ($Ca^{2+}$-$Mg^{2+}$) ATPase, *Biochem. Biophys. Res. Commun.* **77:**1203–1209.

Graf, E., Verma, A. K., Gorski, J. P., Lopaschuk, G., Niggli, V., Zurini, M., Carafoli, E., and Penniston, J. T., 1982, Molecular properties of calcium pumping ATPase from human erythrocytes, *Biochemistry* **21:**4511–4516.

Hakim, G., Itano, T., Verma, A. K., and Penniston, J. T., 1982, Purification of the $Ca^{2+}$ and $Mg^{2+}$-requiring ATPase from rat brain synaptic plasma membrane, *Biochem. J.* **207:**225–231.

Jarrett, H. W., and Penniston, J. T., 1977, Partial purification of the $Ca^{2+}$-$Mg^{2+}$ ATPase activator from human erythrocytes: Its similarity to the activator of $3^1$ : $5^1$-cyclic nucleotide phosphodiesterase, *Biochem. Biophys. Res. Commun.* **77:**1210–1216.

Jones, L. R., Besch, H. R., Jr., Fleming, J. W., McConnaughey, M. M., and Watanabe, A. M., 1979, Separation of vesicles of cardiac sarcolemma from vesicles of cardiac sarcoplasmic reticulum, *J. Biol. Chem.* **254:**530–539.

Jundt, H., Portzig, H., Reuter, H., and Stucki, J. W., 1975, The effect of substances releasing intracellular calcium ions on sodium-dependent calcium efflux from guinea-pig auricles, *J. Physiol. (London)* **246:**229–253.

Knauf, P. A., Proverbio, F., and Hoffmann, J. F., 1974, Electrophoretic separation of different phosphoproteins associated with Ca-ATPase and Na, K-ATPase in human red cell ghosts, *J. Gen. Physiol.* **63:**324–336.

Kraus-Friedmann, N., Biber, J., Murer, H., and Carafoli, E., 1982, Calcium uptake in isolated hepatic plasma membrane vesicles, *Eur. J. Biochem.* **129:**7–12.

Lotersztain, S., Hanoune, J., and Pecker, F., 1981, A high affinity calcium-stimulated magnesium-dependent ATPase in rat liver plasma membranes: Dependence on an endogenous protein activator distinct from calmodulin, *J. Biol. Chem.* **256:**11209–11215.

Lüttgau, H. C., and Niedergerke, R., 1958, The antagonism between Ca and Na ions in frog's heart, *J. Physiol. (London)* **143:**486–505.

Lynch, Th. J., and Cheung, W. J., 1979, Human erythrocyte $Ca^{2+}$-$Mg^{2+}$-ATPase: Mechanism of stimulation by $Ca^{2+}$, *Arch. Biochem. Biophys.* **194:**165–170.

Mualem, S., and Karlish, S. J. D., 1979, Is the red cell calcium pump regulated by ATP? *Nature* **277:**238–240.

Mualem, S., and Karlish, S. J. D., 1980, Regulatory interaction between calmodulin and ATP on the red cell $Ca^{2+}$ pump, *Biochim. Biophys. Acta* **597:**631–636.

Niggli, V., Ronner, P., Carafoli, E., and Penniston, J. T., 1979a, Effects of calmodulin on the ($Ca^{2+}$-$Mg^{2+}$) ATPase partially purified from erythrocyte membranes, *Arch. Biochem. Biophys.* **198:**124–130.

Niggli, V., Penniston, J. T., and Carafoli, E., 1979b, Purification of the ($Ca^{2+}$ + $Mg^{2+}$)-ATPase from human erythrocyte membranes using a calmodulin affinity column, *J. Biol. Chem.* **254:**9955–9958.

Niggli, V., Adunyah, E. S., Penniston, J. T., and Carafoli, E., 1981a, Purified ($Ca^{2+}$ + $Mg^{2+}$) ATPase of the erythrocyte membrane. Reconstitution and effect of calmodulin and phospholipids, *J. Biol. Chem.* **256:**395–401.

Niggli, V., Adunyah, E. S., and Carafoli, E., 1981b, Acidic phospholipids, unsaturated fatty acids, and limited proteolysis mimic the effect of calmodulin on the purified erythrocyte $Ca^{2+}$-ATPase, *J. Biol. Chem.* **256:**8588–8592.

Niggli, V., Sigel, E., and Carafoli, E., 1982, The purified $Ca^{2+}$ pump of human erythrocyte membranes catalyzes an electroneutral $Ca^{2+}$ : $H^+$ exchange in reconstituted liposomal systems, *J. Biol. Chem.* **257:**2350–2356.

Penniston, J. T., 1983, Plasma membrane $Ca^{2+}$ ATPases as active $Ca^{2+}$ pumps, in: *Calcium and Cell Function,* Vol. 4 (W. Y. Cheung, ed.), Academic Press, New York, pp. 100–149.

Pershadsingh, H. A., and McDonald, J. M., 1979, Direct addition of insulin inhibits a high affinity $Ca^{2+}$-ATPase in isolated adipocyte plasma membranes, *Nature* **281:**495–497.

Philipson, K. D., and Nishimoto, A. Y., 1980, $Na^+$–$Ca^{2+}$-exchange is affected by membrane potential in cardiac sarcolemmal vesicles, *J. Biol. Chem.* **255:**6880–6882.

Philipson, K. D., and Nishimoto, A. Y., 1982, Stimulation of $Na^{2+}$–$Ca^{2+}$ exchange in cardiac sarcolemmal vesicles by proteinase pretreatment, *Am. J. Physiol.* **243:**c191–c195.

Pitts, B. J. R., 1979, Stoichiometry of sodium–calcium exchange in cardiac sarcolemmal vesicles, *J. Biol. Chem.* **254:**6232–6235.

Quist, E. E., and Roufogalis, B. D., 1975, Determination of the stoichiometry of the calcium pump in human erythrocytes using lanthanum as a selective inhibitor, *FEBS Lett.* **50:**135–139.

Quist, E. E., and Roufogalis, B. D., 1977, Association of (Ca + Mg)-ATPase activity with ATP-dependent Ca uptake in vesicles prepared from human erythrocytes, *J. Supramol. Struct.* **6:**375–381.

Reeves, J. P., and Sutko, J. L., 1979, Sodium–calcium exchange in cardiac membrane vesicles, *Proc. Natl. Acad. Sci. USA* **76:**590–594.

Reeves, J. P., and Sutko, J. L., 1980, Sodium–calcium exchange activity generates a current in cardiac membrane vesicles, *Science* **208:**1461–1464.

Rega, A. F., and Garrahan, P. J., 1975, Calcium ion-dependent phosphorylation of human erythrocyte membranes, *J. Membr. Biol.* **22:**313–327.

Rega, A. F., and Garrahan, P. J., 1980, Effects of calmodulin on the phosphoenzyme of the $Ca^{2+}$-ATPase of human red cell membranes, *Biochim. Biophys. Acta* **596:**487–489.

Reuter, H., and Seitz, N., 1968, The dependence of $Ca^{2+}$ efflux from cardiac muscle on temperature and external ion composition, *J. Physiol. (London)* **195:**451–470.

Richards, D. E., Rega, A. F., and Garrahan, P. J., 1978, Two classes of site for ATP in the $Ca^{2+}$-ATPase from human red cell membranes, *Biochim. Biophys. Acta* **511:**194–201.

Rinaldi, M. L., Le Peuch, C. J., and Demaille, J. G., 1981, The epinephrine-induced activation of the cardiac slow $Ca^{2+}$ channel is mediated by the cAMP-dependent phosphorylation of calciductin, a 23,000 $M_r$ sarcolemmal protein, *FEBS Lett.* **129:**277–281.

Romero, P. J., 1981, Active calcium transport in red cell ghosts resealed in dextran solutions, *Biochim. Biophys. Acta* **649**:404–418.

Ronner, P., Gazzotti, P., and Carafoli, E., 1977, A lipid requirement for the ($Ca^{2+}$ + $Mg^{2+}$)-activated ATPase of erythrocyte membranes, *Arch. Biochem. Biophys.* **179**:578–583.

Rossi, J. P. F. C., Garrahan, P. J., and Rega, A. F., 1981, Vanadate inhibition of active calcium transport across red cell membranes, *Biochim. Biophys. Acta* **648**:145–150.

Roufogalis, B. D., 1979, Regulation of calcium translocation across the red blood cell membrane, *Can. J. Physiol. Pharmacol.* **57**:1331–1339.

Russel, J. M., and Blaustein, M. P., 1974, Calcium efflux from barnacle muscle fibers: Dependence on external cations, *J. Gen. Physiol.* **63**:144–167.

Sarkadi, B., 1980, Active transport of calcium in human red cells, *BBA Rev. Biomembr.* **604**:159–190.

Sarkadi, B., Szasz, I., Gerloczy, A., and Gardos, G., 1977, Transport parameters and stoichiometry of active calcium ion extrusion in intact human red cells, *Biochim. Biophys. Acta* **464**:93–107.

Sarkadi, B., Schubert, A., and Gardos, G., 1979, Effects of calcium-EGTA buffers on active calcium transport in inside-out red cell membrane vesicles, *Experientia* **35**:1045–1047.

Scharff, O., 1972, The influence of calcium ions on the preparation of the ($Ca^{2+}$ + $Mg^{2+}$)-activated membrane ATPase in human red cells, *Scand. J. Clin. Lab. Invest.* **30**:313–320.

Scharff, O., 1976, $Ca^{2+}$ activation of membrane-bound ($Ca^{2+}$ + $Mg^{2+}$)-dependent ATPase from human erythrocytes prepared in the presence or absence of $Ca^{2+}$, *Biochim. Biophys. Acta* **443**:206–218.

Scharff, O., 1978, Stimulating effects of monovalent cations on activator-dissociated and activator-associated states of $Ca^{2+}$-ATPase in human erythrocytes, *Biochim. Biophys. Acta* **512**:309–317.

Scharff, O., and Foder, B., 1978, Reversible shift between two states of $Ca^{2+}$ ATPase in human erythrocytes mediated by $Ca^{2+}$ and a membrane-bound activator, *Biochim. Biophys. Acta* **50**:67–77.

Schatzmann, H. J., 1966, ATP-dependent $Ca^{++}$ extrusion from human red cells, *Experientia* **22**:364–368.

Schatzmann, H. J., 1982, The plasma membrane calcium pump of erythrocytes and other animal cells, in: *Membrane Transport of Calcium* (E. Carafoli, ed.), Academic Press, New York, pp. 41–108.

Schatzmann, H. J., and Bürgin, H., 1978, Calcium in human blood red cells, *Ann. N.Y. Acad. Sci.* **307**:125–147.

Schatzmann, H. J., and Rossi, G. L., 1971, ($Ca^{2+}$ + $Mg^{2+}$)-activated membrane ATPases in human red cells and their possible relations to cation transport, *Biochim. Biophys. Acta* **241**:379–392.

Soloft, M. S., and Sweet, P., 1982, Oxytocin inhibition of ($Ca^{2+}$ + $Mg^{2+}$)-ATPase activity in rat myometrial plasma membranes, *J. Biol. Chem.* **257**:10687–10693.

Stieger, J., and Luterbacher, S., 1981, Some properties of the purified ($Ca^{2+}$ + $Mg^{2+}$) ATPase from human red cell membranes, *Biochim. Biophys. Acta* **641**:270–275.

Szasz, I., Hasitz, M., Sarkadi, B., and Gardos, G., 1978, Phosphorylation of the calcium pump intermediate in intact red cells, isolated membranes, and inside-out vesicles, *Mol. Cell. Biochem.* **22**:147–152.

Taverna, R. D., and Hanahan, D. J., 1980, Modulation of human erythrocyte $Ca^{2+}/Mg^{2+}$ ATPase activity by phospholipase $A_2$ and proteases. A comparison with calmodulin, *Biochem. Biophys. Res. Commun.* **94**:652–659.

Verma, A. K., and Penniston, J. T., 1981, A high affinity $Ca^{2+}$-stimulated and $Mg^{2+}$-dependent ATPase in rat corpus luteum plasma membrane fractions, *J. Biol. Chem.* **256**:1269–1275.

Vincenzi, F. F., and Larsen, F. L., 1980, The plasma membrane calcium pump: Regulation by a soluble $Ca^{2+}$-binding protein, *Fed. Proc.* **39**:2427–2431.

Wilbrandt, W., and Koller, H., 1948, Die Calciumwirkung an Froschherzen als Funktion des Ionengleichgewichts zwischen Zellmembrane und Umgebung, *Helv. Physiol. Pharmacol. Acta* **6**:208–221.

Zurini, M., Krebs, J., Penniston, J. T., and Carafoli, E., 1983, Controlled proteolysis of the purified $Ca^{2+}$-ATPase of the erythroenzyme, submitted.

# 33

# The Calcium Carriers of Mitochondria

Martin Crompton

## I. INTRODUCTION

Until quite recently, the phenomenon of mitochondrial $Ca^{2+}$ transport was enigmatic. The generally held belief that mitochondria may provide an intracellular sink for $Ca^{2+}$ evolved naturally from extensive work in the early 1960's (Lehninger *et al.*, 1967; Carafoli and Lehninger, 1971), which established that mitochondria from very nearly all sources examined accumulate massive amounts of $Ca^{2+}$ under appropriate *in vitro* conditions. Indeed, more recent studies have demonstrated that mitochondria *in situ* do sequester $Ca^{2+}$ introduced into the cytoplasm by microinjection (salivary gland cells; Rose and Lowenstein, 1975) or by electrical stimulation (nerve cells; Brinley *et al.*, 1977). Subsequent studies concentrated on the kinetic properties of the transport system responsible for $Ca^{2+}$ uptake and its relation to the general features of mitochondrial energy transduction (Bygrave, 1977). These studies interpreted the profound capacity for $Ca^{2+}$ uptake in chemiosomotic terms, whereby the mitochondrial inner membrane potential ($\Delta\psi$)* provides the driving force for $Ca^{2+}$ accumulation, and revealed the relatively low affinity of the transport system for external $Ca^{2+}$. These features in themselves, i.e., massive capacity and low affinity, favored the sink concept in implying that the capacity for $Ca^{2+}$ uptake would be largely unused except in the event of some cellular emergency involving a rise in cytoplasmic $Ca^{2+}$. Moreover, intramitochondrial $Ca^{2+}$ was considered inert biochemically. Consequently, mito-

* $\psi_e$-$\psi_i$ as conventionally used in mitochondrial studies. The subscripts *i* and *e* refer to intramitochondrial and extramitochondrial, respectively.

*Martin Crompton* • Department of Biochemistry, University College London, London WC1E 6BT, England.

chondrial $Ca^{2+}$ transport was frequently viewed as a mere appendix of mainstream $Ca^{2+}$ metabolism, apparent only when something went wrong.

The last few years, however, have witnessed a drastic revision of these concepts, as a result principally of two complementary lines of investigation. First, it has become apparent that mitochondria contain distinct carriers for $Ca^{2+}$ uptake and release, and that the distribution of $Ca^{2+}$ across the inner membrane is established by steady-state recycling via these independent systems (Crompton *et al.*, 1976a; Puskin *et al.*, 1976). The extreme magnitude of the driving force for $Ca^{2+}$ uptake is thereby rationalized, not in terms of the potential it provides for $Ca^{2+}$ accumulation, but in ensuring that the influx reaction proceeds irreversibly so that the $Ca^{2+}$ distribution is governed by steady-state recycling. The basic properties of the $Ca^{2+}$ cycles allow mitochondria to act, in principle, as regulators of intramitochondrial free $Ca^{2+}$ (Denton and McCormack, 1980) and as precise buffers of extramitochondrial free $Ca^{2+}$ (Nicholls, 1978a). Moreover, $Ca^{2+}$ cycling permits the $Ca^{2+}$ distribution across the mitochondrial inner membrane to be altered via modification of the kinetic properties of the $Ca^{2+}$-transport systems (Kessar and Crompton, 1981). In other words, the level of intramitochondrial $Ca^{2+}$ is no longer seen, necessarily, as a passive response to changes in cytoplasmic $Ca^{2+}$, but may be controlled independently.

Second, the reasons for such control are now apparent. It has become clear that mitochondrial $Ca^{2+}$ is not biochemically inert, but that several key enzymes of oxidative metabolism are extremely sensitive to $Ca^{2+}$ (Denton and McCormack, 1980). Thus, intramitochondrial $Ca^{2+}$ may need to be controlled as precisely as cytoplasmic free $Ca^{2+}$. In addition, the possible function of mitochondria as $Ca^{2+}$ donors to the cytoplasm has received much attention (Exton, 1980; Williamson *et al.*, 1981).

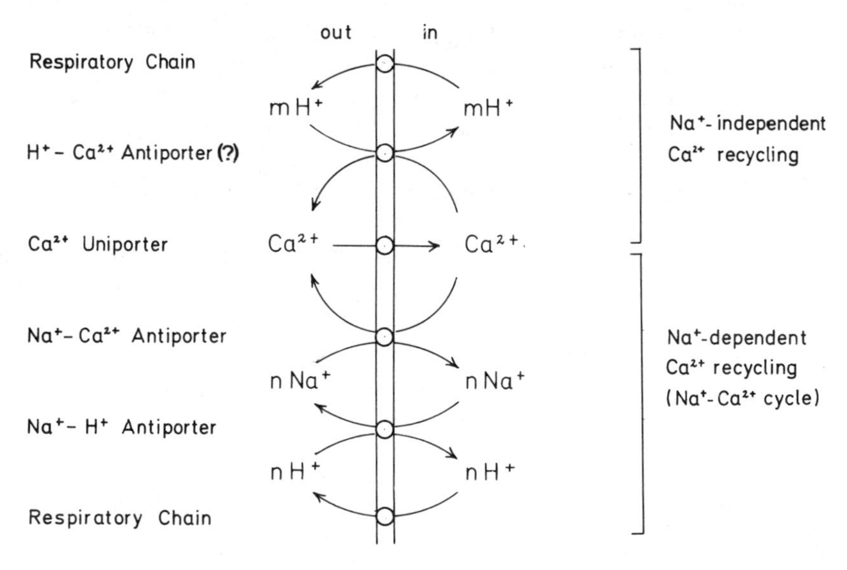


*Figure 1.* $Ca^{2+}$ recycling by $Na^+$-dependent and $Na^+$-independent mechanisms across the mitochondrial inner membrane. The quantities $n$ and $m$ refer to the stoichiometries of $Na^+$–$Ca^{2+}$ and $H^+$–$Ca^{2+}$ exchange, respectively.

The mitochondrial $Ca^{2+}$ cycle is depicted in Figure 1. $Ca^{2+}$ influx is catalyzed by a single system, the $Ca^{2+}$ uniporter, whereas $Ca^{2+}$ efflux occurs via two distinct systems. In mitochondria from many tissues, however, the activity of the $Na^+$-independent efflux system is negligible in comparison to the activity of the $Na^+$ dependent system and, in these cases, the term $Na^+$–$Ca^{2+}$ cycle for recycling via the uniporter and $Na^+$-dependent system will be used to indicate this fact. The basic role of the cycle is to establish the steady-state distribution of $Ca^{2+}$ across the inner membrane, and it is timely to draw together those aspects of mitochondrial $Ca^{2+}$ transport that bear directly on this function. These comprise the kinetic properties of the component $Ca^{2+}$ carriers and the factors that influence their activities, either directly by modification of these properties, or indirectly by changing the driving forces for ion movement. Accordingly, this article considers these particular features, and how they may influence the properties of the cycle as a whole, together with the basic evidence used to construct the cycle.

## *II. THE $Ca^{2+}$ UNIPORTER*

### *A. The Driving Force for $Ca^{2+}$ Accumulation*

The nature of the driving force for $Ca^{2+}$ influx contains implications not only for the mechanism of the transport process but also the principles on which control of mitochondrial $Ca^{2+}$ fluxes may be achieved.

Fundamental observations, made over 20 years ago, clearly showed that $Ca^{2+}$ uptake is an alternative use of respiratory energy (Lehninger *et al.*, 1967). Whereas $Ca^{2+}$ uptake driven by ATP hydrolysis is inhibited by oligomycin, uptake driven by respiration is not. Both processes are inhibited by uncoupling agents. The demonstration that $Ca^{2+}$ accumulation may be driven in the absence of respiratory chain or ATPase activity by diffusion potentials generated by gradients of $H^+$ (in the presence of protonophores), $SCN^-$, or $K^+$ (in the presence of valinomycin) indicated an electrogenic mode of transport (Selwyn *et al.*, 1970; Scarpa and Azzone, 1970). In support of this, Åkerman (1978a) observed $Ca^{2+}$ diffusion potentials in respiration-inhibited mitochondria that were abolished by ruthenium red. These observations rationalized the process in chemiosmotic terms, as proposed earlier by Mitchell (1966).

The question of the precise charge transfer during the net reaction catalyzed by the $Ca^{2+}$ carrier has received considerable experimental attention, since it bears directly on the physiological reversibility of the process with $\Delta\psi$ equal to 150–200 mV (Mitchell and Moyle, 1969; Nicholls, 1974; Rottenberg, 1975). Two basic approaches have been adopted, both of which indicate a positive charge stoichiometry of two per $Ca^{2+}$ ion translocated.

First, the uptake of $Ca^{2+}$ driven by $K^+$-diffusion potentials in respiration-inhibited mitochondria is accompanied by the release of two $K^+$ per $Ca^{2+}$ (Azzone *et al.*, 1976; Akerman, 1978a; Fiskum *et al.*, 1979). An analogous approach, in which the $H^+$–$Ca^{2+}$ stoichiometry is determined during respiration-supported $Ca^{2+}$ uptake, is ambiguous since it does not distinguish between $H^+$ ejected by the respiratory chain and possible $H^+$ translocation by the $Ca^{2+}$ carrier, e.g., one $Ca^{2+}$–one $H^+$ antiport. Nevertheless, with one exception (Moyle and Mitchell, 1977), such measurements have revealed that two $H^+$ are released per $Ca^{2+}$ taken up (Crompton and Heid, 1978; Pfeiffer *et al.*, 1978; Vercesi *et al.*, 1978; Williams and Fry, 1978; Fiskum *et al.*, 1979). In the exceptional case, cotransport of $Ca^{2+}$ with inorganic phosphate was proposed, but the procedures on which these conclusions were based have been questioned (Crompton *et al.*, 1978b; Fiskum *et al.*, 1979; Saris and Åkerman, 1980).

Secondly, at relatively low values of $\Delta\psi$, the $Ca^{2+}$ accumulation ratio is consistent with charge uncompensated $Ca^{2+}$ flux via the $Ca^{2+}$ carrier even when $\Delta\psi$ is opposed by an equivalent pH gradient, i.e., when $\Delta\psi \simeq -2.3$ RT/F $\Delta$pH;* this effectively excludes the possibility of one $Ca^{2+}$-one $H^+$ antiport, in which case zero accumulation would be predicted (Heaton and Nicholls, 1976). Rigorous interpretation of the $Ca^{2+}$ accumulation ratio is precluded by the fact that most internal $Ca^{2+}$ is bound to a variable extent depending on the experimental conditions (Chappell *et al.*, 1963; Puskin *et al.*, 1976 and references therein), and a quantitative correlation can only be made between $\Delta\psi$ and the free $Ca^{2+}$ gradient. One approach to this problem involves the use of $Mn^{2+}$ as a paramagnetic $Ca^{2+}$ analogue and the estimation of internal binding by EPR (Chappell *et al.*, 1963). In a revealing application of this technique, Puskin *et al.* (1976) showed that the free $Mn^{2+}$ closely approached a Nernstian distribution when $\Delta\psi$ was low (<90mV). That is, at relatively low $\Delta\psi$

$$\Delta\psi \simeq 2.3 \frac{RT}{2F} \log \frac{[Mn^{2+}]_i}{[Mn^{2+}]_e}$$

An alternative approach was used by Crompton and Heid (1978) to examine the degree to which the distribution of $Ca^{2+}$ obeyed the Nernst relation. In this case, $[Ca^{2+}]_i$ was eliminated as a variable by determining $[Ca^{2+}]_e$ that was required to maintain total internal $Ca^{2+}$ constant as $\Delta\psi$ was varied. In the presence of acetate and oligomycin, a Nernstian distribution was obeyed when $\Delta\psi$ was below 80 mV. A somewhat similar technique was used by Nicholls (1978a) with analogous conclusions if reasonable assumptions were made about the internal activity coefficient for $Ca^{2+}$.

It is abundantly clear, however, that at high values of $\Delta\psi$, $Ca^{2+}$ and $Mn^{2+}$ deviate greatly from a Nernstian distribution (Puskin *et al.*, 1976; Nicholls, 1978a; Azzone *et al.*, 1976). The reason for this behavior is almost surely the existence of independent efflux systems for $Ca^{2+}$. The attainment of near-equilibrium by the uniporter requires that unidirectional efflux via this system is fast relative to $Ca^{2+}$ efflux via the independent $Ca^{2+}$-efflux systems, a condition that may not apply at high $\Delta\psi$ (Nicholls, 1978a; Nicholls and Crompton, 1980). At relatively high $\Delta\psi$ and, therefore, high internal $Ca^{2+}$ (or $Mn^{2+}$), the uniporter may be essentially saturated with internal cation. In this condition, the rate of unidirectional efflux via the uniporter will be essentially zero-order with respect to internal $Ca^{2+}$ and will be determined by the

* The pH difference across the mitochondrial inner membrane ($pH_i - pH_e$).

dependence of unidirectional efflux on $\Delta\psi$. Further increase in $\Delta\psi$ will decrease the rate of unidirectional efflux so that the fraction of total $Ca^{2+}$ efflux catalyzed by the independent systems will become significant, and the uniport reaction will be displaced from equilibrium.

This interpretation (Nicholls and Crompton, 1980; Nicholls and Åkerman, 1982) predicts that non-Nernstian distributions of $Ca^{2+}$ or $Mn^{2+}$ will be affected significantly by the activity of the independent efflux processes which, in liver, may be driven by $\Delta pH$ (Section IV). Consistent with this, Puskin *et al.* (1976) observed that the internal free $Mn^{2+}$ of liver mitochondria in the steady state decreased (from 17.3 mM to 2.4 mM) when $\Delta pH$ was increased (from 0.48 to 1.68) even though both $\Delta\psi$ and external $Mn^{2+}$, and presumably unidirectional influx via the uniporter, increased.

In summary, data obtained with a variety of techniques are consistent with a uniport mechanism of $Ca^{2+}$ entry, in which $Ca^{2+}$ flux is driven solely by its own conjugate driving force, $\Delta\bar{\mu}_{Ca^{2+}}$.* The data taken together are very difficult to interpret in an alternative manner. A definitive answer may be provided by investigations with the isolated reconstituted carrier; indeed, such studies support the uniport model (Section II-B). The immediate implication of the uniport model is that this system must operate far from equilibrium *in vivo* and catalyze unidirectional flux only. This topic is considered further in Section VII.

## B. The Molecular Nature of the Uniporter

There is evidence for the involvement of a glycoprotein in the uniport reaction. Mitochondrial glycoproteins with $Ca^{2+}$-binding properties have been isolated by several groups (Gomez-Puyou *et al.*, 1972; Sottocasa *et al.*, 1972; Miranova *et al.*, 1982). The glycoprotein isolated by Sottocasa, Carafoli, and co-workers (Carafoli and Sottocasa, 1974) is $H_2O$ soluble (glutamate and aspartate accounting for 35% of its amino acid residues). It binds $Ca^{2+}$ to two classes of sites with relatively high affinity ($K_d$ = 0.1–1.0 μM, 2–3 moles of $Ca^{2+}$/mole) and low affinity ($K_d \approx$ 10 μM, 21 moles of $Ca^{2+}$/mole). About 22% of the glycoprotein content of rat liver mitochondria is released by osmotic rupture of the outer membrane; the remainder is solubilized by sonication (about 50%) and treatment with chaotropic agents (about 28%). There are two lines of evidence that this glycoprotein is involved in $Ca^{2+}$ transport. First, the impaired ability of osmotically shocked mitochondria (mitoplasts) to accumulate $Ca^{2+}$ is improved by addition of purified glycoprotein (Sandri *et al.*, 1979). Second, antibodies raised against the glycoprotein inhibit $Ca^{2+}$ uptake by mitoplasts and also by mitochondria (Panfili *et al.*, 1976); in the latter case, prolonged preincubation is required presumably to allow antibody access to the intermembrane space. In later studies, Panfili *et al.* (1980) reported that the inhibition of $Ca^{2+}$ uptake by antiglycoprotein antibodies is not associated with impairment of respiration. Moreover, the antibodies inhibit $Ca^{2+}$ efflux by uniport reversal when $\Delta\psi$ is collapsed by addition of acetoacetate or oxaloacetate (Panfili *et al.*, 1980; Section V-C).

* Electrochemical gradient of $Ca^{2+}$ across the inner membrane.

More recently, Miranova *et al.* (1982) described the purification of two fractions (40K glycoprotein and 2K peptide) from ethanol extracts of heart mitochondria which exhibit selective $Ca^{2+}$-transport properties when reconstituted into black lipid membranes. These membranes developed near-Nernstian potentials with applied $[Ca^{2+}]$ gradients. The $Ca^{2+}$-ionophoric activity was inhibited by ruthenium red (1–10 μM).

Jeng and Shamoo (1980) reported the isolation of a $Ca^{2+}$-binding protein from heart mitochondria ($K_d^{Ca} \approx 5$ μM, mol. wt. 3000). This protein transports $Ca^{2+}$ electrogenically through bulk organic phases in a manner blocked by ruthenium red.

## C. Factors Affecting Uniporter Activity

### 1. Extramitochondrial $[Ca^{2+}]$ and $\Delta\psi$

$Ca^{2+}$ entry driven solely by $\Delta\bar{\mu}_{Ca^{2+}}$ implies a strict dependence of the rate of unidirectional influx on $\Delta\psi$. Although true unidirectional fluxes have not been resolved, the rate of net $Ca^{2+}$ influx exhibits a linear dependence on $\Delta\psi$ when $\Delta\psi$ is sufficiently high to displace the reaction far from equilibrium (Åkerman, 1978b; Crompton and Kessar, 1981). This ohmic relation has an important physiological consequence in that it is notably insensitive, and fluctuations in $\Delta\psi$ would need to be unphysiologically large, with severe repercussions on oxidative phosphorylation, to cause meaningful changes in the rate of unidirectional $Ca^{2+}$ influx.

The $\Delta\psi$-dependence of $Ca^{2+}$ influx has precluded precise determination of the $[Ca^{2+}]$ required for half maximal velocity ($K_{0.5}^{Ca}$). The $\Delta\psi$ is depressed increasingly as external $[Ca^{2+}]$ and the rate of influx increase (Åkerman, 1978b; Lötscher *et al.*, 1980a), and this leads to an underestimate of $V_{max}$ and, consequently, of $K_{0.5}^{Ca}$. The maximal rates of $Ca^{2+}$ uptake *in vitro* generally exceed 350 nmol of $Ca^{2+}$/mg protein · min at 25°C (Crompton *et al.*, 1976b; Scarpa and Graziotti, 1973; Vercesi *et al.*, 1978; Bragadin *et al.*, 1979; McMillin-Wood *et al.*, 1980). In a KCl-based medium containing $Mg^{2+}$ (0.5–5.0 mM), the $[Ca^{2+}]$-dependence of uniporter activity is strongly sigmoidal with $K_{0.5}^{Ca} > 30$ μM (Zoccarato and Nicholls, 1982; Crompton *et al.*, 1976b; Vinogradov and Scarpa, 1973; Scarpa and Graziotti, 1973). The essential point is that with external $Ca^{2+}$ in the physiological range, the uniporter operates considerably below $K_{0.5}^{Ca}$ so that an increase in cytosolic free $Ca^{2+}$ will cause a proportionally greater increase in the rate of unidirectional influx. The implications of this are considered in Section VII-A.

### 2. Hormones

Numerous hormones have been reported to influence the capacity of mitochondria to take up and retain $Ca^{2+}$. In assessing this phenomenon, however, it is important to distinguish between flux changes that reflect modification in the kinetic properties of the uniporter and those caused by changes in the driving force for $Ca^{2+}$ uptake, most obviously in $\Delta\psi$. In the latter case, changes in $Ca^{2+}$ flux observed *in vitro* need not necessarily be expressed *in vivo*. Conventional *in vitro* conditions have usually included high external $Ca^{2+}$ (to facilitate measurement) and, consequently, unphysiologically high net rates of $Ca^{2+}$ influx, so that $\Delta\psi$ is partly collapsed to an extent dependent on respiratory chain activity. As outlined in Sections VI and VII, it is unlikely that changes in $\Delta\psi$ *in vivo* significantly affect $Ca^{2+}$ fluxes, or vice versa.

These considerations may apply to the observations that liver pretreatment with glucagon causes enhanced rates of $Ca^{2+}$ uptake by isolated mitochondria coexistent with increased respiratory chain activity (Friedman, 1980; Hughes and Barritt, 1978; Yamazaki *et al.*, 1980; Titheradge and Coore, 1976; Halestrap, 1978), whereas during steady-state recycling (at low external $[Ca^{2+}]$) no changes are observed (Brand and DeSelincourt, 1980).

The rates of $Ca^{2+}$ uptake by liver and heart mitochondria are also increased after tissue pretreatment with adrenaline (Taylor *et al.*, 1980; Kessar and Crompton, 1981). The activation is mimicked by $\alpha_1$-adrenergic agonists and it is blocked by $\alpha$-adrenergic (but not $\beta$-adrenergic) antagonists and by trifluoperazine (Reinhardt *et al.*, 1980), an inhibitor of calmodulin action. Regarding the latter observation, Blackmore *et al.* (1981) drew attention to the action of trifluoperazine at the plasma membrane as an $\alpha$-adrenergic antagonist. These observations indicate that the activation is mediated by an $\alpha_1$-adrenergic mechanism.

The phenylephrine-induced activation in liver is accompanied by increased respiratory chain activity which precludes an immediate discrimination between a direct effect on uniporter activity and an indirect effect caused by increased $\Delta\psi$. However, the action of the $\alpha_1$-adrenergic agonist, methoxamine, on heart mitochondrial $Ca^{2+}$ fluxes is less ambiguous. In this case, the activation of respiration-dependent $Ca^{2+}$ uptake is not associated with significant changes in respiration or in $\Delta\psi$ when succinate is the respiratory substrate, and it was concluded that the kinetic properties of the uniporter are modified by methoxamine pretreatment (Kessar and Crompton, 1981). In agreement with this, measurements of the rate of $Ca^{2+}$ uptake vs. $\Delta\psi$ (manipulated by partial uncoupling) revealed that uniporter conductance (nmole $Ca^{2+}$/mg protein · mV) is increased by about 40% by preadministration of adrenaline to perfused heart (Crompton and Kessar, 1981).

## III. THE $Na^+$-$Ca^{2+}$ CARRIER

### A. The Binding of External Substrate Cations

The $Na^+$-$Ca^{2+}$ carrier catalyzes $Ca^{2+}$ efflux on addition of $Na^+$, $Li^+$, $Ca^{2+}$, or $Sr^{2+}$. Each of these four reactions has been instrumental to a greater or lesser degree in formulating concepts of the carrier mechanism and properties in relation to physiological function. The reactions are most commonly investigated after accumulation of varying amounts of $Ca^{2+}$ and inhibition of uniporter activity by ruthenium red to prevent reuptake of the $Ca^{2+}$ released.

$Na^+$ induces $Ca^{2+}$ efflux more effectively than $Li^+$. In heart mitochondria, about 8 mM $Na^+$ and 15 mM $Li^+$ elicit half-maximal velocity of $Ca^{2+}$ efflux, and the $V_{max}$ with $Na^+$ is about threefold greater than that obtained with $Li^+$ (Crompton *et al.*, 1976a). Both relations are sigmoidal, indicating the involvement of more than one $Na^+$ or $Li^+$ per reaction cycle. No significant release of $Ca^{2+}$ is caused by $K^+$, $Rb^+$, or $Cs^+$ (Carafoli *et al.*, 1974; Crompton *et al.*, 1980), although $K^+$ and $Rb^+$ activate $Na^+$-induced $Ca^{2+}$ efflux (Section III-E.2). The release of intramitochondrial $Ca^{2+}$ ($^{45}Ca^{2+}$) in the presence of ruthenium red is also induced by extramitochondrial $Ca^{2+}$ and $Sr^{2+}$, but not by $Mg^{2+}$ (20 mM), $Ba^{2+}$, or $Mn^{2+}$ (both at 100 μM). The $Ca^{2+}$-induced efflux of $^{45}Ca^{2+}$ is due to 1 : 1 exchange between internal and external $Ca^{2+}$ (Crompton *et al.*, 1977). Both the $Ca^{2+}$–$Ca^{2+}$ exchange and the $Sr^{2+}$-induced efflux of $Ca^{2+}$ (presumably $Sr^{2+}$–$Ca^{2+}$ exchange) exhibit hyperbolic kinetics with respect to external cation, the $K_s^{Ca}$* and $K_s^{Sr}$ values being about 2 μM and 5 μM, respectively (Hayat and Crompton, 1982).

Two lines of evidence indicate that the effluxes of $Ca^{2+}$ induced by monovalent and by divalent cations are catalyzed by the same system. First, the $Na^+$-dependent and $Ca^{2+}$-dependent effluxes of $Ca^{2+}$ show a similar sensitivity to $La^{3+}$ inhibition; in both cases, about 0.3 nmole $La^{3+}$/mg protein yields 50% inhibition (Crompton *et al.*, 1977). Secondly, $Na^+$ and $Ca^{2+}$ compete as inducers of efflux of internal $Ca^{2+}$. This may be investigated by discriminating between the $Na^+$-induced and $Ca^{2+}$-induced effluxes of $Ca^{2+}$ when these occur simultaneously. Since the $Ca^{2+}$-induced efflux of $Ca^{2+}$ involves $Ca^{2+}$–$Ca^{2+}$ exchange, this component may be resolved by the initial rate of $^{45}Ca^{2+}$ entry in exchange for $^{40}Ca^{2+}$ (Crompton *et al.*, 1977). Alternatively, the initial rate of $^{40}Ca^{2+}$ entry in exchange for internal $Sr^{2+}$ may be determined (Hayat and Crompton, 1982). The latter technique is advantageous since it permits continuous spectrophotometric monitoring of the exchange. Both techniques have shown that the initial velocity of $Ca^{2+}$–$Ca^{2+}$ and $Ca^{2+}$–$Sr^{2+}$ exchange is strongly inhibited by external $Na^+$. The spectrophotometric assay revealed that this inhibition is competitive with respect to external $Ca^{2+}$ and that the $K_i^{Na}$ value, about 2.5 mM, is similar to the $K_s^{Na}$ value determined from the $[Na^+]$ dependence of the $Na^+$-induced efflux of $Ca^{2+}$; i.e., the affinities of the transport system for $Na^+$ as an inducer (of $Ca^{2+}$ efflux) and as a competitive inhibitor (of $Ca^{2+}$–$Sr^{2+}$ exchange) are approximately the same.

These observations support the concept that the $Na^+$-binding sites and the binding sites involved in $Ca^{2+}$ translocation are located on the same carrier, and that occupation of these sites by $Na^+$ and external $Ca^{2+}$ is mutually exclusive. As stated in Section III-E.2, there is now evidence that, in addition, $Ca^{2+}$ may bind to separate regulatory sites on the carrier. It is important to note that the above data relate to conditions under which the $Ca^{2+}$-regulatory sites were fully occupied, i.e., $[Ca^{2+}]_e > 3$ μM. These conditions presumably ensured the absence of complications due to $La^{3+}$ or (varied) $Ca^{2+}$ binding to the regulatory sites.

## B. *The Binding of Internal Cations*

Numerous observations indicate that the $Ca^{2+}$ released on addition of $Na^+$ in the presence of ruthenium red originates from the matrix space. Thus, $Na^+$ induces the release of essentially all the $Ca^{2+}$ preaccumulated via the uniporter (Crompton *et al.*,

* $K_s$ refers to the dissociation constants for the binding of ions as substrates.

1976a), whereas EGTA and lanthanides, which would be predicted to chelate or displace superficially bound $Ca^{2+}$, cause no significant release (Crompton *et al.*, 1979; Hayat and Crompton, 1982). Steady-state recycling of $Ca^{2+}$ via the uniport and efflux reactions leads to partial uncoupling of respiration, consistent with energy-dissipating transmembrane ion fluxes (Crompton *et al.*, 1976a, 1978a). More recent studies have elegantly demonstrated $Na^+$-induced release of matrix $Ca^{2+}$ by using the activity of $Ca^{2+}$-sensitive matrix dehydrogenases as indices of matrix free $Ca^{2+}$ (McCormack and Denton, 1980; Denton *et al.*, 1980; Hansford, 1981; Hansford and Castro, 1981).

Investigations of the binding of internal $Ca^{2+}$ to the $Na^+$-$Ca^{2+}$ carrier is impeded by the experimental inaccessibility of the matrix space and the fact that, to date, $Na^+$-$Ca^{2+}$ carrier activity in inverted submitochondrial particles has not been demonstrated. Recently, however, Coll *et al.* (1982) have developed a null-titration technique with the $Ca^{2+}$ ionophore A23187 to estimate matrix free $Ca^{2+}$. With this technique, these investigators have obtained an apparent $K_m$ value for internal $Ca^{2+}$ of about 6 μM. Application of the same techniques by Hansford and Castro (1982) has provided a similar value (3–4 μM). It is important to bear in mind, however, that $Na^+$ and $Ca^{2+}$ may compete for internally facing binding sites on the carrier, i.e., as for external sites, above. In this case, the apparent affinity for internal $Ca^{2+}$ would depend on internal $[Na^+]$ and hence on ΔpH which establishes the transmembrane distribution of $Na^+$. Studies in this laboratory (unpublished) have revealed inhibition of $Na^+$-$Ca^{2+}$ carrier activity in heart mitochondria by high external $[Na^+]$, typically with >80 mM $Na^+$ in the absence of permanent anions (when ΔpH is high) and with >25 mM Na in the presence of 1 mM $P_i$ (when ΔpH is low).

## C. The Transport Mechanism

The reaction catalyzed by the mitochondrial $Na^+$-$Ca^{2+}$ carrier is not established. Nevertheless, its kinetic behavior indicates simple models that are consistent with its physiological function.

According to the concepts developed over the last few years, the fundamental function of the carrier is to extrude $Ca^{2+}$ against $\Delta\bar{\mu}_{Ca^{2+}}$. Accordingly, evidence that strongly indicates the passive nature of $Ca^{2+}$ entry via the uniporter (Section II-A) is equally indicative of active $Ca^{2+}$ extrusion by the $Na^+$-$Ca^{2+}$ carrier in the presence of ruthenium red and $Na^+$, provided that these agents do not decrease $\Delta\psi$. This proviso has been confirmed in mitochondria from heart and brain (Nicholls, 1978b; Affolter and Carafoli, 1980) and also liver (Puskin *et al.*, 1976; although in this case it is uncertain whether the $Na^+$-$Ca^{2+}$ carrier was operative).

The capacity of the $Na^+$-$Ca^{2+}$ carrier to extrude $Ca^{2+}$ against $\Delta\bar{\mu}_{Ca^{2+}}$ may be accounted for by a direct exchange between $Na^+$ and $Ca^{2+}$, so that $\Delta\bar{\mu}_{Na^+}$* provides the driving force for $Ca^{2+}$ efflux (Crompton *et al.*, 1976a, 1977). However, direct examination of whether $Ca^{2+}$ efflux is indeed associated with a stoichiometric uptake of $Na^+$ is prevented by the presence of the $Na^+$–$H^+$ antiporter (Mitchell and Moyle,

* Electrochemical gradient of $Na^+$ across the inner membrane.

1967; Brierley, 1976). Any $Na^+$ that enters would be able to exchange with $H^+$. The relative rapidity of the $Na^+$–$H^+$ exchange, about seven-fold greater than the $Na^+$-induced efflux of $Ca^{2+}$, would allow $Na^+$–$H^+$ exchange to approach equilibrium irrespective of whether $Na^+$ is a substrate for exchange with $Ca^{2+}$ or not; this prediction is borne out experimentally (Crompton and Heid, 1978).

The $Na^+$-induced release of $Ca^{2+}$ from respiring heart mitochondria in the presence of ruthenium red is accompanied in the early stages by the uptake of two $H^+$ per $Ca^{2+}$ released (Crompton and Heid, 1978). These data are consistent with a sequential operation of $n$ $Na^+$–$Ca^{2+}$ and $Na^+$–$H^+$ exchanges as shown in Figure 2a (Crompton *et al.*, 1976a, 1977).

It is important to note that the stoichiometry two $H^+$–one $Ca^{2+}$ does not necessarily mean that the value of $n$ (Figure 2a) is 2, since any charge imbalance in the $Na^+$–$Ca^{2+}$ exchange, i.e., with $n > 2$, would be compensated by an equivalent respiratory expulsion of $H^+$. The data may be equally well explained by the fluxes shown in Figure 2b, in which $Na^+$ simply activates $m$ $H^+$–$Ca^{2+}$ exchange and $Na^+$–$H^+$ antiport is not involved. In this case, $\Delta\bar{\mu}_{H^+}$* would provide the immediate driving force for $Ca^{2+}$ extrusion, but since the $Na^+$-$H^+$ antiporter operates close to equilibrium, i.e., $\Delta\bar{\mu}_{Na^+} \simeq \Delta\bar{\mu}_{H^+}$ the driving forces for $Ca^{2+}$ efflux in the two models are about the same. The lack of a known selective inhibitor of the $Na^+$–$H^+$ antiporter prevents direct discrimination between these two models. However, kinetic data are accommodated with fewer assumptions about possible carrier conformations by the model in Figure 2a. The competitive interrelation between external $Na^+$ and external $Ca^{2+}$ may be accounted for as follows, where $x$ is the carrier (with $n = 3$; Hayat and Crompton, 1982):

$$\text{Ca}\cdot x \rightleftarrows x \rightleftarrows x\cdot\text{Na} \rightleftarrows x\cdot\text{Na}_n$$

Figure 2b would require additional steps, e.g.,

$$\text{Ca}\cdot x \rightleftarrows x \rightleftarrows x\cdot\text{Na} \rightleftarrows x\cdot\text{Na}_n \rightleftarrows \text{H}\cdot x\cdot\text{Na}_n \rightleftarrows \text{H}_m\cdot x\cdot\text{Na}_n$$

in which $H^+$ may only bind to the $Na^+$ form of the carrier at physiological pH values. In addition, this model would need to account for the fact that alkali metal ions also affect the translocation step. For example, although $Li^+$ may substitute for $Na^+$ in inducing $Ca^{2+}$ efflux, the $V_{max}$ with $Li^+$ is only 30% of that obtained with $Na^+$ (Crompton *et al.*, 1976a).

Figure 2a is also favored by analogy with numerous secondary active transport systems of plasma membranes that utilize $\Delta\bar{\mu}_{Na^+}$, including $Na^+$–$Ca^{2+}$ exchange (Blaustein, 1974). The analogy is perhaps strengthened by the fact that the mitochondrial and plasma membrane $Na^+$-$Ca^{2+}$ carriers have two common features, namely energy dependence and $K^+$ activation, as described below.

* Electrochemical gradient of $H^+$ across the mitochondrial inner membrane ($= 2.3\ RT\ \Delta pH + zF\Delta\psi$).

### D. The Dependence of the $Na^+$-Induced Efflux of $Ca^{2+}$ on the Mitochondrial Energy State

Both the $Na^+$-induced and $Li^+$-induced effluxes of $Ca^{2+}$ are activated by coupled respiration (Crompton *et al.*, 1976a, 1977; Affolter and Carafoli, 1980; Heffron and Harris, 1981). Kinetic studies indicate that the $V_{max}$ of $Na^+$-induced $Ca^{2+}$ efflux is increased about 2.5-fold by energization (Crompton *et al.*, 1977). A plausible explanation is that the exchange is electrogenic (i.e., $n > 2$; Figure 2a,) by analogy with the $Na^+$–$Ca^{2+}$ exchange of heart sarcolemma, which is electrogenic (Pitts, 1974; Caroni *et al.*, 1980; Reeves and Sutko, 1980; Philipson and Nishimoto, 1980) with the stoichiometry three $Na^+$–one $Ca^{2+}$ (Pitts, 1979). On the other hand, the 1 : 1 $Ca^{2+}$–$Ca^{2+}$ and $Sr^{2+}$–$Ca^{2+}$ exchanges catalyzed by the mitochondrial carrier are not affected by the mitochondrial energy state in agreement with their implied electroneutrality (Crompton *et al.*, 1977).

Affolter and Carafoli (1980) detected no significant depression of $\Delta\psi$ during $Na^+$-induced efflux of $Ca^{2+}$ from heart mitochondria and concluded that the exchange was electroneutral. However, any charge imbalance in the reaction may have been effectively compensated by the activity of the respiratory chain, oxidizing succinate. The capacity for compensating charge ($q^+$) displacement during succinate oxidation (100–150 nmol succinate/mg protein$^{-1}$/min$^{-1}$ at 25° (Denton *et al.*, 1980; Kessar and Crompton, 1981) would be about 600–900 $q^+$/mg · min (assuming 6 $q^+$/2 $e$; Wikström and Krab, 1980) which is 40 to 60-fold greater than the rate of inward positive charge flux predicted for three $Na^+$–one $Ca^{2+}$ exchange ($V_{max} \simeq 15$ nmole $Ca^{2+}$/mg$^{-1}$/min$^{-1}$ at $> 2$ μM $[Ca^{2+}]_e$; Crompton *et al.*, 1976a).

A further factor that may be related to the energy dependence of $Na^+$-induced efflux of $Ca^{2+}$ is suggested by the reported inhibition by oligomycin (Harris and Heffron, 1982). This was interpreted to reflect a requirement for inorganic phosphate generated by an intramitochondrial ATPase; however, this is inconsistent with earlier

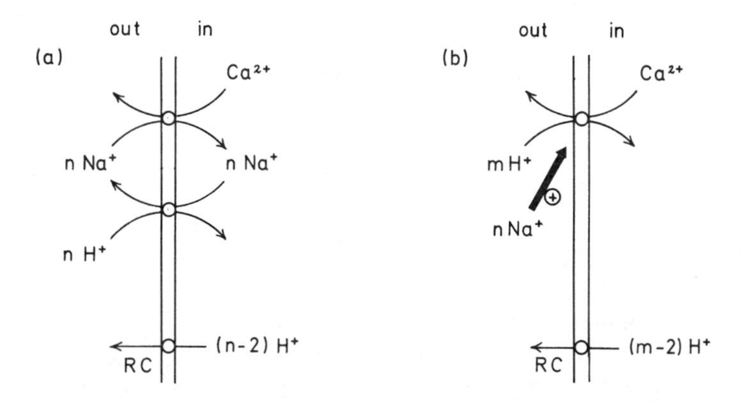


*Figure 2.* Mechanisms proposed for the $Na^+$-induced efflux of $Ca^{2+}$ from mitochondria. RC refers to the respiratory chain and ⊕→ to activation by $Na^+$.

reports that oligomycin inhibits when phosphate is supplied (Heffron and Harris, 1981). Superficially at least, the effect would appear to indicate a possible requirement for intramitochondrial ATP. In this connection, the $Na^+$–$Ca^{2+}$ exchange of nerve axolemma is stimulated by intracellular ATP (Baker, 1972; Di Polo, 1974; Reinlib *et al.*, 1981). Unfortunately, the oligomycin sensitivity of the mitochondrial system has not been confirmed (Saris and Åkerman, 1980).

## E. Physiological Effectors of the $Na^+$-$Ca^{2+}$ Carrier

### 1. Monovalent Cations

The relation between [$Na^+$] and carrier activity in muscle and other tissues is sigmoidal with an inflexion at 5–8 mM $Na^+$ (Crompton *et al.*, 1978a; Hayat and Crompton, 1982). These values are close to the activity of $Na^+$ in the cytosol of muscle of several species. Studies with $Na^+$-sensitive microelectrodes indicate that the $Na^+$ activity in skeletal muscle (6.3–7.7 mM; Lee and Armstrong, 1974) and cardiac muscle (5.7 mM; Lee and Fozzard, 1975) is much lower than would be predicted if all the tissue $Na^+$ were present in free solution. In principle, therefore, the [Na] sensitivity of the $Na^+$-$Ca^{2+}$ carrier is well poised for a sensitive response to changes in cytosolic $Na^+$. However, changes in cytosolic [$Na^+$] during depolarization are far too small to affect the $Na^+$-$Ca^{2+}$ significantly. The importance of this factor for subsarcolemmal mitochondria of heart is not known (McMillin-Wood *et al.*, 1980).

The $Na^+$-$Ca^{2+}$ carrier of cardiac mitochondria is strongly activated by extramitochondrial $K^+$ at constant ionic strength. $K^+$ is not a substrate for exchange, since it is ineffective in the absence of an external exchangeable cation ($Na^+$, $Ca^{2+}$, or $Sr^{2+}$). About 18 mM $K^+$ is sufficient to half-maximally activate $Na^+$- and $Ca^{2+}$-induced effluxes of $Ca^{2+}$. The $Na^+$-$Ca^{2+}$ carrier of nerve axolemma is also activated by $K^+$ under conditions in which the membrane potential is not changed (Sjödin and Abercrombie, 1978). Kinetic studies of the mitochondrial carrier show that external $K^+$ markedly promotes the binding of external $Na^+$ and $Sr^{2+}$ (as a $Ca^{2+}$ analogue), i.e., the apparent dissociation constants for $Na^+$ and $Sr^{2+}$ are decreased about threefold and nine-fold, respectively, with little change in $V_{max}$ (Hayat and Crompton, 1982). It appears, therefore, that $K^+$ may stabilize the conformation of external substrate-binding sites.

### 2. Divalent Cations

Hansford and Castro (1981) observed a small inhibition (23%) of $Na^+$–$Ca^{2+}$ exchange by 1 mM $Mg^{2+}$. A similar degree of inhibition has been obtained in this laboratory (27–31% with 1 mM $Mg^{2+}$ and 10 mM $Na^+$; M. Crompton, unpublished data). The comparatively weak effect of $Mg^{2+}$ on the $Na^+$-$Ca^{2+}$ carrier, with respect to its effect on the uniporter, is evident also from the $Mg^{2+}$-induced displacement of the steady-state distribution of $Ca^{2+}$ across the inner membrane in the direction of decreased $[Ca^{2+}]_i$ (Hansford, 1981; Denton *et al.*, 1980).

The competitive interrelation between the binding of external $Na^+$ and external $Ca^{2+}$ to sites involved in cation translocation (substrate-binding sites) has been mentioned in Section III-A. Kinetic data are consistent with a model in which the apparent dissociation constant for external $Ca^{2+}$ is increased by the factor $(1 + [Na^+]/K_i^{Na^+})^3$, where the value of $K_i^{Na+}$ is about 2.5 mM (Hayat and Crompton, 1982)*. Accordingly, as little as 3 mM $Na^+$, for example, suffices to raise the apparent dissociation constant for external $Ca^{2+}$ by about tenfold, i.e., to 20–40 μM. Thus, at physiological $[Na^+]$ and $[Ca^{2+}]$, the occupation of external substrate (product) binding sites by $Ca^{2+}$ will be negligible. Nevertheless, in heart mitochondria, the initial rate of $Na^+$-induced efflux of $Ca^{2+}$ is strongly inhibited by external $Ca^{2+}$ over the range 0–2 μM. Kinetic studies indicate that this inhibition is noncompetitive with respect to $Na^+$, and that it is partial, yielding a maximal decrease in $V_{max}$ of about 70% irrespective of $[Na^+]$. These features led to the proposal that the $Na^+$-$Ca^{2+}$ carrier of cardiac mitochondria may contain regulatory (inhibitory) binding sites for $Ca^{2+}$, distinct from the substrate-binding sites (Hayat and Crompton, 1982). These sites yield half-maximal inhibition at about 0.7 μM external $Ca^{2+}$ and show cooperative interactions ($n = 2$). These properties may confer significant sensitivity to external $Ca^{2+}$ under physiological conditions.

### 3. *Hormones*

Recent studies indicate that the $Na^+$-$Ca^{2+}$ carrier of liver mitochondria responds to tissue pretreatment with certain hormones (Goldstone and Crompton, 1982; Goldstone *et al.*, 1983). Under optimal assay conditions, the $Na^+$-$Ca^{2+}$ carrier activity is increased two- to threefold by administration (2 min) of 1 μM adrenaline, 10 nM glucagon, or 100 μM cAMP to perfused rat liver. The adrenaline-induced activation was mimicked by the β-agonist, isoprenaline, but the α-agonist, phenylephrine, was ineffective. The degree of activation is markedly dependent on internal $Ca^{2+}$ load, being evident only with <10 nmole $Ca^{2+}$/mg protein, which indicates that isoprenaline may induce an increase in the apparent affinity for internal $Ca^{2+}$. An alternative explanation is that isoprenaline causes a decrease in intramitochondrial buffering power for $Ca^{2+}$ so that, at constant $Ca^{2+}$ load, more internal $Ca^{2+}$ is free. This possibility appears unlikely since neither the $Na^+$-independent efflux of $Ca^{2+}$ nor efflux induced by the ionophore A23187 was affected by isoprenaline treatment (Goldstone *et al.*, 1983). In addition, Coll *et al.* (1982), using a null-point titration technique to estimate internal free $Ca^{2+}$, observed that glucagon did not change the fraction of total $Ca^{2+}$ that was free.

These findings reopen the controversy concerning the role of cAMP in the release of mitochondrial $Ca^{2+}$. Borle (1974) and Matlib and O'Brien (1974) first reported that the addition of cAMP to liver, kidney, and adrenal medulla mitochondria *in vitro* induced release of mitochondrial $Ca^{2+}$. More recently, Arshad and Holdsworth (1980) reported a decreased capacity of liver mitochondria to retain accumulated $Ca^{2+}$ *in vitro* during oxidation of palmitoyl CoA in the presence of high cAMP concentration (>38 μM). Other workers, however, have failed to observe any effect of cAMP added

* $K_i$ refers to the dissociation constants for the binding of ions as inhibitors.

to isolated mitochondria *in vitro* (Scarpa *et al.*, 1976; Schotland and Mela, 1977; Goldstone *et al.*, 1983).

Goldstone *et al.* (1983) investigated further the role of cAMP by use of the fact that the capacity of glucagon and β-adrenergic agonists to raise tissue cAMP requires a relatively reduced tissue redox state. The capacity of glucagon to activate the $Na^+$-$Ca^{2+}$ carrier correlated with its ability to increase cAMP (manipulated by the lactate–pyruvate ratio of the perfusion medium). On the other hand, the isoprenaline-induced activation was independent of tissue redox state, and was unaffected by depression of cAMP changes to below detectable limits.

### F. Tissue Distribution

The $Na^+$-induced efflux of $Ca^{2+}$ was first reported with mitochondria from heart (Carafoli *et al.*, 1974) and neurohypophysis (Thorn *et al.*, 1975). Subsequent work from several laboratories indicated the presence of this system in mitochondria from many mammalian tissues. The most active mitochondrial preparations have been obtained from rat heart and bovine adrenal cortex ($V_{max}$ = 12–27 nmole $Ca^{2+}$/mg protein · min at 25°C; Crompton *et al.*, 1976a 1978a; Hayat and Crompton, 1982; Nicholls, 1978b). Rather lower activities have been observed in mitochondria from rabbit red and white skeletal muscle, bovine parotid gland, and brown fat ($V_{max}$ = 4–6 nmole $Ca^{2+}$/mg protein · min; Crompton *et al.*, 1978a; Al-Shaikhaly *et al.*, 1979). Early experiments failed to detect significant $Na^+$-$Ca^{2+}$ carrier activity in liver, kidney cortex, ileum muscle, lung (Crompton *et al.*, 1978a), and blowfly flight muscle (Crompton, unpublished). However, later studies revealed low activities (1–2 nmoles $Ca^{2+}$/mg protein · min) in liver, kidney cortex and medulla, and lung (Nedergaard and Cannon, 1980; Haworth *et al.*, 1980; Heffron and Harris, 1981). Goldstone and Crompton (1982) attributed early difficulties in detecting significant activity in liver to two factors. First, $Na^+$-induced release from liver mitochondria in the presence of ruthenium red diminishes markedly as $Ca^{2+}$ efflux progresses. Second, the hormonal state of the animal on sacrifice may influence the activity considerably, as discussed in the previous section.

## IV. THE $Na^+$-INDEPENDENT RELEASE OF $Ca^{2+}$

The search for $Na^+$-independent release mechanisms came with the growing appreciation that whereas the need to reconcile a uniport mechanism of $Ca^{2+}$ entry with a high $\Delta\psi$ required an independent process for $Ca^{2+}$ release, no $Na^+$-$Ca^{2+}$ carrier activity could be detected in mitochondria from certain tissues, notably liver. The problem was explicitly exposed in the studies of Puskin *et al.* (1976) and Gunter *et al.* (1978) who observed that the gradients of free $Mn^{2+}$ across the inner membrane of liver mitochondria approached electrochemical equilibrium at low $\Delta\psi$, but at high $\Delta\psi$, the $Mn^{2+}$ distribution was less than that predicted. In particular, partial inhibition of the uniporter with ruthenium red shifted the final $Mn^{2+}$ distribution further from electrochemical equilibrium (rather than merely slowing the approach to the same distribution). These authors postulated the existence of an independent, ruthenium red-insensitive route for active $Ca^{2+}$ extrusion. A similar conclusion was reached by Nicholls (1978a) who observed increased disparity between predicted (on the basis $\Delta\bar{\mu}_{Ca} \simeq 0$) and measured accumulation ratios when the uniporter activity was limited by low external $Ca^{2+}$ (<0.8 μM) or by $Mg^{2+}$. Increased divergence of $Ca^{2+}$ from the predicted distribution on partial inhibition of the uniporter by $Mg^{2+}$ and ruthenium red was reported also by Azzone *et al.* (1976) and Pozzan *et al.* (1977).

The classification of liver mitochondria as $Na^+$-insensitive is now known to be incorrect (Section III-F). Moreover, the studies of Puskin and Nicholls were conducted in the presence of $Na^+$, although it is possible that $Na^+$-$Ca^{2+}$ carrier activity was negligible at the $[Na^+]$ used (Section III-B). Nevertheless, mitochondria, particularly from liver and kidney, do exhibit significant rates of $Ca^{2+}$ release in the absence of $Na^+$, when the uniporter is blocked by ruthenium red (Crompton *et al.*, 1978a). Ruthenium red does not decrease the magnitude of $\Delta\psi$, rather a slight increase occurs in parallel with $Ca^{2+}$ release (Puskin *et al.*, 1976; Nicholls, 1978b), and it is evident that these mitochondria contain a $Na^+$-independent mechanism for active $Ca^{2+}$ extrusion, in addition to the $Na^+$-dependent mechanism.

## A. The Reaction Mechanism

There is evidence that the $Na^+$-independent system of liver mitochondria may exchange $Ca^{2+}$ and $H^+$. Åkerman (1978c) reported $Ca^{2+}$ release from liver mitochondria on acidification of the external medium in the presence of ruthenium red. The rate of $Ca^{2+}$ release was roughly proportional to the $Ca^{2+}$ content over the range 10–100 nmole $Ca^{2+}$/mg protein. Increase in $\Delta pH$ caused a decrease in $\Delta\psi$, as predicted, but efflux via uniport reversal was presumably prevented by inclusion of ruthenium red. The $Ca^{2+}$ release was substantially inhibited by inclusion of about 2 nmole $La^{3+}$/mg protein; this sensitivity may be compared to that of the $Na^+$-independent system of liver mitochondria (about 10 nmole $La^{3+}$/mg protein for 50% inhibition; Crompton *et al.*, 1979).

Fiskum and Cockrell (1978) observed $Ca^{2+}$ uptake in respiration- and uniporter-inhibited mitochondria when a pH gradient, acid inside, was imposed via the ionophore dianemycin. These authors provided a persuasive correlation between this process and that responsible for ruthenium red-induced release of $Ca^{2+}$ from respiring mitochondria. Both processes were very largely depressed in mitochondria from ascites tumor cells, whereas $\Delta pH$-induced $Ca^{2+}$ uptake via the ionophore A23187 (which reflects the magnitude of $\Delta pH$) was equivalent in the two mitochondrial types. Later work (Fiskum and Lehninger, 1979) showed that $Ca^{2+}$ release from liver mitochondria is associated with the uptake of two $H^+$ per $Ca^{2+}$. Bernardi and Azzone (1979), however, detected no significant ruthenium red-insensitive uptake by liver mitochondria on imposing a pH gradient with nigericin.

Whether or not $H^+$ exchanges directly with $Ca^{2+}$, or indirectly via an ionic intermediary, has not been established. Wehrle and Pedersen (1979) and Lötscher *et al.* (1979) reported that purified inverted submitochondrial particles accumulate $Ca^{2+}$ upon energization in the presence of inorganic phosphate. While this is consistent with a mechanism involving $Ca^{2+}$–phosphate symport (coupled with phosphate-$H^+$ symport to give $Ca^{2+}$–$H^+$ antiport overall), studies with intact mitochondria have provided no support for this conclusion, since they have involved either collapse of $\Delta\psi$ (Rugolo *et al.*, 1981), a more rapid release of phosphate than $Ca^{2+}$ (Roos *et al.*, 1980), or an inhibition of $Ca^{2+}$ release by phosphate (Zoccarato and Nicholls, 1982).

## B. The Transport System

The molecular identity of the $Na^+$-independent system has not been resolved. Roman *et al.* (1979) suggested that long-chain fatty acids may act as $Ca^{2+}$–$H^+$ ionophores independently of their uncoupling action, since their ability to release $Ca^{2+}$, but not uncoupling activity, was blocked by 120 mM $Na^+$.

Puskin *et al.* (1976) first suggested that $Na^+$-independent $Ca^{2+}$ efflux may reflect a degree of nonselectivity in a cation carrier. It is unlikely that this is the $Na^+$-$Ca^{2+}$ carrier, as outlined in Section V-A. A potential candidate is the $Na^+$–$H^+$ antiporter. Kinetic studies indicate that $Ca^{2+}$ is a competitive inhibitor of this system with respect to $Na^+$ ($K_i^{Ca^{2+}} \simeq 0.24$ mM; Crompton and Heid, 1978). Although such affinity for $Ca^{2+}$ is low, the affinity of the $Na^+$-independent system for internal $Ca^{2+}$ may also be low, since it requires >80 nmoles of internal $Ca^{2+}$/mg protein for saturation (Zoccarato and Nicholls, 1982; Fiskum and Cockrell, 1978; Åkerman, 1978c). Both the $Na^+$-independent system and the $Na^+$-$H^+$ antiporter are relatively insensitive to lanthanide inhibition (Crompton *et al.*, 1979; Crompton and Heid, 1978). In addition, simultaneous reconstitution of $Na^+$–$H^+$ exchange and $Ca^{2+}$–$H^+$ exchange activities into phospholipid vesicles has been reported ($K_m^{Ca^{2+}} \simeq 0.4$ mM; Dubinsky *et al.*, 1979).

## C. Factors Affecting Activity of the $Na^+$-Independent System

The $Na^+$-independent release of $Ca^{2+}$ from mitochondria of several tissues is inhibited by ADP and ATP (Harris, 1979; Heffron and Harris, 1981). There is some evidence that the liver system is affected by factors in the circulation, since its activity in mitochondria from perfused liver is about half that in mitochondria from freshly excised liver (Goldstone and Crompton, 1982). Coll *et al.* (1982) have provided evidence that the $Na^+$-independent system may be activated by an α-adrenergic mechanism. Pretreatment of liver with phenylephrine decreased the apparent $K_m$ for internal $Ca^{2+}$ from about 10 μM to 5 μM. However, phenylephrine also caused a two-fold decrease in the proportion of total intramitochondrial $Ca^{2+}$ that was free, so that the response of the $Na^+$-independent system to total intramitochondrial $Ca^{2+}$ was unchanged (in agreement with Goldstone *et al.*, 1983).

Gunter *et al.* (1978) reported that addition of small amounts of uncouplers, insufficient to cause measurable uncoupling of respiration, inhibited the release of $Ca^{2+}$ from liver mitochondria in the presence of ruthenium red. This may have reflected the energy dependence of the $Na^+$-dependent system (Section III-D) since the media contained $Na^+$. In contrast, other work suggests that the $Na^+$-independent release of $Ca^{2+}$ by heart and brain mitochondria is increased by antimycin *a* or partial uncoupling (Crompton *et al.*, 1976a, 1977; Zoccarato and Nicholls, 1982). A recurring question concerning the interpretation of such experiments is whether ruthenium red is an effective inhibitor of the uniporter in the deenergized state and whether uniport reversal contributes to $Ca^{2+}$ efflux under these conditions (Pozzan *et al.*, 1977). This query arises from the observation that $Ca^{2+}$ efflux on complete uncoupling, when uniport reversal would be predicted, is not markedly inhibited by ruthenium red (Gunter *et al.*, 1978; Vasington *et al.*, 1972). Gunter *et al.* (1978) raised the possibility that the uniporter may "close" on deenergization. Whatever the explanation, it is clear that the uniporter is "open" and is inhibited by ruthenium red, when the membrane polarity is reversed (positive inside) by $K^+$-diffusion potentials (Rigoni *et al.*, 1980).

### D. Tissue Distribution

$Na^+$-independent release of $Ca^{2+}$ is evident in all mitochondria that have been examined. In some mitochondria, e.g., heart, it is relatively inactive at moderate $Ca^{2+}$ loads, i.e., <0.5 nmole $Ca^{2+}$/mg protein · min with 10–20 nmole internal $Ca^{2+}$/mg protein, which amounts to <2% of the maximal rate of $Na^+$-induced efflux (Hayat and Crompton, 1982). In liver, kidney, and brain, however, roughly equivalent activities of the $Na^+$-dependent and independent activities have been reported (2–5 nmole $Ca^{2+}$/mg protein · min; Haworth *et al.*, 1980; Heffron and Harris, 1981; Goldstone and Crompton, 1982; Zoccarato and Nicholls, 1982).

## V. THE RESOLUTION OF THE COMPONENT FLUXES OF THE $Ca^{2+}$ CYCLES

The concept of steady-state $Ca^{2+}$ recycling via independent systems has been derived largely from studies that demonstrate that the component reactions are indeed catalyzed by separate systems. It is instructive, therefore, to outline criteria by which individual reactions may be resolved.

### A. Lanthanides

The ability of lanthanides to substitute for $Ca^{2+}$ in biological systems is well documented. Mela (1969; Mela and Chance, 1969) first reported that $La^{3+}$ and $Ho^{3+}$ are potent inhibitors of the uniporter; 50–100 pmoles of lanthanide/mg of protein sufficed for 50% inhibition. In a careful study, Reed and Bygrave (1974) established that complete $La^{3+}$ inhibition reflected occupation of <1 pmole uniporter/mg of protein; put into perspective, this comprises <1% of the amount of individual cytochromes and the adenine nucleotide carrier. These measurements indicated a molecular activity of the uniporter of >1400 $S^{-1}$ at 0°C. These authors noted that $La^{3+}$ inhibits competitively with respect to $Ca^{2+}$ ($K_i^{La^{3+}} \simeq 2 \times 10^{-8}$ M) in agreement with earlier work (Scarpa and Azzone, 1970).

The relative potency of different lanthanides was investigated by Tew (1977), who reported that $Sm^{3+}$ and $Nd^{3+}$ were the best inhibitors of a range of ionic sizes from $Lu^{3+}$ to $Pr^{3+}$. A rather different pattern of inhibition was seen by Crompton *et al.* (1979), who determined the amount of lanthanide required for 50% inhibition of uniporter activity ($I_{0.5}$). $La^{3+}$ was the weakest inhibitor in both liver and heart mitochondria, and the potency increased markedly (18- to 20-fold) with decreasing ionic radius to $Dy^{3+}$. The $I_{0.5}^{Dy^{3+}}$ value was 1–2 pmole/mg protein in agreement with the earlier conclusion of Reed and Bygrave (1974) of the scarcity of the uniporter.

In contrast, the $Na^+$-$Ca^{2+}$ carrier is much less sensitive to lanthanide inhibition (Crompton *et al.*, 1979). Moreover, the pattern of inhibition is opposite to that displayed by the uniporter, larger lanthanides being the best inhibitors, e.g., $I_{0.5}^{La^{3+}} \simeq 400$ pmole/mg protein).

The $Na^+$-independent efflux of $Ca^{2+}$ from both heart and liver mitochondria is even less sensitive to lanthanides (Crompton *et al.*, 1979). For example, the $I_{0.5}^{La^{3+}}$ value is about 10–15 nmole/mg protein (when nonspecific effects may well apply), which is about 30 times less sensitive than the $Na^+$-$Ca^{2+}$ carrier. The $Na^+$-$H^+$ antiporter is similarly resistant to lanthanide inhibition ($I_{0.5}^{La^{3+}} \simeq 25$ nmoles/mg protein; Crompton and Heid, 1978).

The use of the lanthanide series, therefore, has provided clear evidence that the uniporter, the $Na^+$-$Ca^{2+}$ carrier, and the system responsible for $Na^+$-independent $Ca^{2+}$ efflux are distinct. Lanthanides do not discriminate, however, between the $Na^+$-independent system and the $Na^+$–$H^+$ antiporter.

## B. Ruthenium Red

Inhibition of the uniporter by ruthenium red was first observed by Moore (1971). Using purified ruthenium red, Reed and Bygrave (1974) showed that 50% inhibition is attained with about 70 pmoles ruthenium red/mg protein (unpurified samples may require 5- to 10-fold this amount; Moore, 1971; Vasington *et al.*, 1972). These investigations also showed that ruthenium red inhibits noncompetitively with respect to $Ca^{2+}$ ($K_i \simeq 3 \times 10^{-8}$ M). No deleterious effects on mitochondrial energy transduction have been observed with [ruthenium red] just sufficient to block the uniporter, although higher amounts may inhibit respiration (Vasington *et al.*, 1972; Puskin, *et al.*, 1976; Nicholls, 1978b). No effects of ruthenium red on the $Na^+$-$Ca^{2+}$ carrier have been noted (Crompton *et al.*, 1976a). The apparent selectivity of ruthenium red for the uniporter may reflect its glycoprotein content (Section II-B).

## C. $Ca^{2+}$ Antagonists

Vaghy *et al.* (1982) have demonstrated unambiguously that a range of $Ca^{2+}$-antagonist drugs act as selective inhibitors of the $Na^+$-$Ca^{2+}$ carrier of heart mitochondria. The most potent inhibitors of those tested were diltiazem, prenylamine, and fendipine which yielded 50% inhibition at 7 μM, 12 μM, and 13 μM, respectively. These findings are potentially of great utility since means are now available for selective suppression of either the uniporter (by ruthenium red and, at appropriate concentrations, lanthanides) or the $Na^+$-$Ca^{2+}$ carrier.

### D. Other Distinctions

A number of other criteria provide means of discriminating between the component carriers. Briefly, these are as follows:

1. $K^+$ greatly activates all known exchanges of the $Na^+$-$Ca^{2+}$ carrier (Section III-E.1), but has little or no effect on the $Na^+$-independent process of heart mitochondria (substitution of 120 mM choline chloride by 120 mM KCl inhibits by 60–80%; M. Crompton, unpublished observations). There is no evidence that $K^+$ activates $Na^+$–$H^+$ exchange (Brierley, 1976).

2. Antibodies raised against the $Ca^{2+}$-binding glycoprotein (Section II-B) inhibit the uniporter, but have no effect on the $Na^+$-$Ca^{2+}$ carrier (Panfili *et al.*, 1981).

3. Extramitochondrial $Ca^{2+}$ is a partial noncompetitive inhibitor of the $Na^+$-$Ca^{2+}$ carrier in heart (half-maximal inhibition at 0.7 μM $Ca^{2+}$; Section III-E.2), whereas extramitochondrial $Ca^{2+}$ is a very much weaker, competitive inhibitor of $Na^+$–$H^+$ exchange ($K_i \simeq 240$ μM; Section V-A).

## VI. THE CAPACITY OF ISOLATED MITOCHONDRIA TO RETAIN ACCUMULATED $Ca^{2+}$

In principle, maintenance of the $Ca^{2+}$ uniport reaction far from equilibrium by steady-state recycling allows the $Ca^{2+}$ distribution across the inner membrane to be changed solely by modification of the kinetic properties of component carriers of the cycle, i.e., without obligatory change in $\Delta\psi$, and, therefore, without perturbing other basic mitochondrial functions dependent on $\Delta\psi$, in particular, oxidative phosphorylation. In practice, however, this theoretical capacity for selective control of $Ca^{2+}$ distribution will be limited by the extent to which the generation, utilization, and dissipation of $\Delta\psi$ are not adversely affected by changes in intramitochondrial $Ca^{2+}$. It becomes important, therefore, to evaluate the consequences of increased $Ca^{2+}$ load on mitochondrial function. The positive effects of increased intramitochondrial $Ca^{2+}$ on the generation (via activation of intramitochondrial dehydrogenases) and, possibly, utilization of $\Delta\psi$ ($\Delta\tilde{\mu}_{H^+}$) are outlined in Section VII. However, it is now clear that when the intramitochondrial $Ca^{2+}$ load exceeds certain limits, the efficiency of energy transduction becomes severely impaired. Moreover, these limits are influenced by a somewhat bewildering set of factors.

### A. The Dependence of $Ca^{2+}$-Induced Destabilization on Inorganic Phosphate and Redox State

The accumulation of >50 nmoles of $Ca^{2+}$/mg of protein by liver mitochondria in the presence of inorganic phosphate (0.5–2.0 mM) is followed by time-dependent swelling, uncoupling of succinate oxidation, dissipation of $\Delta\psi$, oxidation of pyridine nucleotides, and losses of intramitochondrial $K^+$, $Mg^{2+}$, adenine nucleotides, and preaccumulated $Ca^{2+}$ (Harris and Heffron, 1982; Beatrice *et al.*, 1980; Nicholls and Brand, 1980; Chappell *et al.*, 1963; Siliprandi *et al.*, 1975; Nicholls and Scott, 1980).

It appears that swelling, loss of $\Delta\psi$, and the inability to retain $Ca^{2+}$ may be promoted by addition of oxaloacetate or acetoacetate; this may reflect oxidation of pyridine nucleotides, since the effects are reversed by β-hydroxybutyrate (Nicholls and Brand, 1980; Lehninger *et al.*, 1978; Beatrice *et al.*, 1980). In a revealing study, Nicholls and Brand (1980) related the maintenance of $\Delta\psi$ by liver mitochondria to the quantity of $Ca^{2+}$ accumulated in the presence of 2 mM phosphate. When the intramitochondrial pyridine nucleotides were maintained in an oxidized state by inclusion of acetoacetate, dissipation of $\Delta\psi$ commenced with $Ca^{2+}$ loads >25 nmoles/mg protein. When the pyridine nucleotides were maintained reduced by β-hydroxybutyrate, as much as 160 nmoles $Ca^{2+}$/mg protein was accumulated without adverse effects on $\Delta\psi$.

The deleterious effects are promoted by inorganic phosphate. Thus, replacement of phosphate with lactate prevents the swelling and loss of preaccumulated $Ca^{2+}$ from liver mitochondria on pyridine nucleotide oxidation (Wolkowicz and McMillin-Wood, 1980). It appears that phosphate acts intramitochondrially since it is ineffective when entry is prevented with *N*-ethylmaleimide (Beatrice *et al.*, 1980). Heart mitochondria are more resistant to $Ca^{2+}$ load than liver mitochondria, and may accumulate up to 100 nmoles $Ca^{2+}$/mg protein in the presence of 2 mM phosphate with impunity, although increased phosphate (20 mM) is deleterious (Palmer and Pfeiffer, 1981).

## B. The Reversibility of $Ca^{2+}$-Induced Destabilization

As stated above, the capacity of mitochondria to retain relatively large amounts of $Ca^{2+}$ *in vitro* is markedly decreased by an oxidized state of intramitochondrial pyridine nucleotides and by phosphate. These two factors may operate synergistically and predispose mitochondria to $Ca^{2+}$-induced deleterious effects. However, it has been shown in several cases that these effects are reversible. Lehninger *et al.* (1978) first demonstrated that with relatively low phosphate (0.2 mM) and in the presence of 5 mM $Mg^{2+}$ (which may exert a protective effect; Beatrice *et al.*, 1980), the release of $Ca^{2+}$ induced by pyridine nucleotide oxidation (oxaloacetate addition) is reversed on subsequent addition of β-hydroxybutyrate. Under these conditions, irreversible changes do not occur. Under similar conditions, with $Mg^{2+}$ replaced by oligomycin (which may also "protect"; Hofstetter *et al.*, 1981), the release of $Ca^{2+}$ on pyridine nucleotide oxidation is associated with a stoichiometric uptake of $H^+$, i.e., two $H^+$–one $Ca^{2+}$, from a saline suspension medium (Fiskum and Lehninger, 1979). The selective use of $H^+$ as a countercation is not consistent with a general loss of inner membrane permeability characteristics. Richter and co-workers (Hofstetter *et al.*, 1981; Lötscher *et al.*, 1980b) used organic hydroperoxides as an alternative means of oxidizing intramitochondrial pyridine nucleotides via the combined action of glutathione peroxidase and glutathione reductase. After the accumulation of large amounts of $Ca^{2+}$ (100 nmoles/mg protein) in the absence of added phosphate (when, presumably, the matrix pH would be very alkaline), hydroperoxide-induced oxidation is followed by hydrolysis, and nicotinamide is released from the mitochondria together with $Ca^{2+}$. However, general mitochondrial integrity is preserved as judged by the retention of malate dehydrogenase activity, impermeability to added ferricyanide, and the restoration of $\Delta\psi$ on addition of EDTA.

There is evidence that the impairment of energy transduction after accumulation of relatively large amounts of $Ca^{2+}$ (>60 nmoles $Ca^{2+}$/mg of protein) by liver mitochondria reflects activity of a membrane-bound phospholipase $A_{II}$ (Pfeiffer *et al.*, 1979; Beatrice *et al.*, 1980; Palmer and Pfeiffer, 1981). The presence of a $Ca^{2+}$-dependent phospholipase $A_{II}$ in mitochondria has been demonstrated (Nachbaur *et al.*, 1972; Waite and Sisson, 1971). Simultaneous determinations revealed that the progressive collapse of $\Delta\psi$, swelling, and uncoupling were accompanied by the release of fatty acids (Pfeiffer *et al.*, 1979); indeed, many earlier studies had reported the release of free fatty acids during mitochondrial swelling (Lehninger *et al.*, 1967). In the presence of phosphate (or other potentiators of $Ca^{2+}$-induced destabilization, e.g., diamide, *N*-ethylmaleimide), the fatty acids released were predominantly polyunsaturated reflecting preferential hydrolysis at carbon-2 of the glycerol skeleton. Further products, therefore, include *1*-acyllysophospholipids. Either lysophospholipids or free fatty acids (or derived metabolites, e.g., acyl CoA) may cause the deleterious effects (Pfeiffer *et al.*, 1979; Beatrice *et al.*, 1980).

In support of the "phospholipase theory," a variety of local anaesthetics known to inhibit phospholipase $A_{II}$ (Waite and Sisson, 1971) also inhibit the $Ca^{2+}$-induced destabilization in the presence of oxaloacetate and phosphate, as described above, and other "promoters" including palmitoyl CoA, *N*-ethylmaleimide, and diamide (Beatrice *et al.*, 1980). In this context, it is relevant to note the studies of Mela (1969), which showed that the uptake of $Ca^{2+}$ and attendant intramitochondrial alkalinization are promoted by local anaesthetics. Mela observed that local anaesthetics exerted a considerably greater effect on $\Delta pH$ than on $Ca^{2+}$ uptake, and concluded that the intramitochondrial buffering capacity may have been decreased. Alternatively, the preferential action of local anaesthetics on $\Delta pH$ may indicate that they retard $\Delta pH$ dissipation by maintaining the low $H^+$ permeability of the inner membrane.

It is perhaps unreasonable to assume that "promoters" of Ca-induced destabilization of such diversity exert their effects at a common site. Pfeiffer *et al.* (1979) stress that the level of lysophospholipids may be determined by cyclic deacylation and reacylation reactions, and that the effects are consistent with inhibition of reacylation, in addition to activation of phospholipase by $Ca^{2+}$. Such a cycle could, in principle, account for the reversibility of the phenomena noted above.

## C. *The Effects of Adenine Nucleotides and* $Mg^{2+}$

The impairment of mitochondrial function *in vitro* by high internal $Ca^{2+}$ is largely abolished by inclusion of ADP or ATP and $Mg^{2+}$ in the suspension medium (Nicholls and Scott, 1980; Coelho and Vercesi, 1980; Nicholls and Brand, 1980; Beatrice *et al.*, 1980; Leblanc and Clauser, 1974; Sordahl, 1974). Conversely, phosphoenol pyruvate, which exchanges with internal adenine nucleotides, potentiates the adverse effects of $Ca^{2+}$ (Roos *et al.*, 1978). Although the stabilizing effects of adenine nucleotides have been appreciated for many years (Chappell *et al.*, 1963; Lehninger *et al.*, 1967; Rossi and Lehninger, 1964), the biochemical basis remains unresolved. Hofstetter *et al.* (1981) reported that the hydrolysis of intramitochondrial NAD (in the presence of hydroperoxide) is accompanied by the covalent binding of a hydrolysis product, perhaps ADP-ribose, to a protein of the inner membrane; this protein, they postulated, was involved in $Ca^{2+}$ release. These authors ascribed the action of ATP to inhibition of pyridine nucleotide hydrolysis. Hunter and Haworth (1979) concluded from studies of the osmotic response of heart mitochondria preexposed to $Ca^{2+}$ that relatively high internal $Ca^{2+}$ induces permeability to low molecular weight solutes (<1000), and that this channel is inhibited by ADP, NADH, and $Mg^{2+}$.

Regarding the role of $Mg^{2+}$, work from several laboratories indicates a modification in ion permeability properties of the inner membrane on $Mg^{2+}$ depletion (Packer *et al.*, 1966; Brierley, 1976; Kun *et al.*, 1969). In particular, it appears that passive permeability to monovalent cations is increased by $Mg^{2+}$ loss (Settlemire *et al.*, 1968). Perhaps significantly, $Mg^{2+}$ is lost from mitochondria *in vitro* by a respiration-dependent process that is potentiated by inorganic phosphate (Zoccarato *et al.*, 1981; Crompton *et al.*, 1976c) and diamide (Siliprandi *et al.*, 1975), two potentiators of $Ca^{2+}$-induced destabilization. However, *in vitro* net $Mg^{2+}$ efflux only occurs with $[Mg^{2+}]_e < 2$ mM; above this value, net influx occurs (Crompton *et al.*, 1976c; Brierley, 1976). That is, zero net flux *in vitro* is attained at close to physiological extramitochondrial $[Mg^{2+}]$.

In summary, irrespective of the underlying mechanism(s), it is clear that mitochondria are able to retain large amounts of $Ca^{2+}$ (>100 nmoles/mg protein) *in vitro* for prolonged periods with no observable deleterious effects when supplemented with physiological concentrations of ATP, ADP, and $Mg^{2+}$, and that their presence overcomes the adverse effects of inorganic phosphate and other "potentiators" (Nicholls and Scott, 1980; Nicholls and Brand, 1980). The proposal that $Na^+$ itself may be deleterious and induce phospholipase activation (Harris and Heffron, 1982) was based on experiments that did not discriminate between the $Ca^{2+}$ efflux induced by $Na^+$ and by mitochondrial destabilization. Indeed, evidence that mitochondrial integrity is not perturbed by $Na^+$ is considerable. Thus, mitochondria maintain a high $\Delta\psi$ for prolonged periods in the presence of $Na^+$ (20–75 mM) and $Ca^{2+}$ (>60 nmoles/mg protein; Nicholls and Brand, 1980; Beatrice *et al.*, 1980). $Ca^{2+}$ efflux in the presence of ruthenium red and $Na^+$ is not associated with a decrease in $\Delta\psi$ (Nicholls, 1978b; Affolter and Carafoli, 1980) or uncoupling of respiration (Crompton *et al.*, 1976a, 1978a). The presence of high $[Na^+]$ does not cause swelling (Hunter and Haworth, 1979) or loss of internal $K^+$ (Crompton *et al.*, 1976c). $Ca^{2+}$ efflux on the $Na^+$-$Ca^{2+}$ carrier may also be induced by $Sr^{2+}$ (Crompton *et al.*, 1977; Hayat and Crompton, 1982), which does not activate phospholipase (Pfeiffer *et al.*, 1979; Beatrice *et al.*, 1980). $Na^+$-$Ca^{2+}$ carrier activity is most active in mitochondria that show the greatest resistance to $Ca^{2+}$ load, e.g., heart, and, in these, it is half-saturated by low internal $Ca^{2+}$ (5–8 nmoles $Ca^{2+}$/mg protein; Hansford and Castro, 1982; Coll *et al.*, 1982) relative to that required for destabilization (>100 nmoles $Ca^{2+}$/mg protein; Palmer and Pfeiffer, 1981).

## VII. $Ca^{2+}$ RECYCLING

In previous sections, evidence has been outlined that mitochondria contain the catalytic machinery for $Ca^{2+}$ recycling, i.e., distinct $Ca^{2+}$ carriers that catalyze either net $Ca^{2+}$ influx or net $Ca^{2+}$ efflux under physiological conditions of high $\Delta\psi$. It is important to note that whereas influx is associated with the extrusion of two $H^+$–one $Ca^{2+}$, the efflux of $Ca^{2+}$ (in the presence of ruthenium red) is accompanied by the reuptake of two $H^+$-one $Ca^{2+}$ (Section III-C). Thus, the net ionic fluxes of the influx and efflux segments of the cycle are the precise converse of each other, enabling the two segments to be apposed in a mutually compatible manner (Crompton and Heid, 1978). $Ca^{2+}$ recycling is thereby integrated into the $H^+$ circuit, and is classically chemiosmotic in being driven by $\Delta\bar{\mu}_{H^+}$. Whereas influx dissipates $\Delta\psi$, efflux dissipates $\Delta pH$ (plus $\Delta\psi$ in the event that the $Na^+$-$Ca^{2+}$ carrier catalyzes an electrogenic exchange); i.e., the complete cycle dissipates $\Delta\bar{\mu}_{H^+}$, as illustrated schematically in Figure 3. The number of $H^+$ that enter for each complete $Ca^{2+}$ cycle ($n$) will be equal to the stoichiometry of the $Ca^{2+}$ antiport systems (e.g., $n$ $Na^+$–$Ca^{2+}$).

The energy dissipation associated with $Ca^{2+}$- (recycling) dependent $H^+$ recycling is evident as a component of state-4 respiration. Stucki and Ineichen (1974) observed that about 20% of state-4 respiration of liver mitochondria was inhibited by ruthenium red. State-4 respiration of heart, adrenal cortex, and brain mitochondria is activated by $Na^+$ in the presence of $Ca^{2+}$, and the $Na^+$-induced increment is abolished by ruthenium red (Crompton *et al.*, 1976a, 1978a). The energy drain, however, is small. In the presence of $Mg^{2+}$ and $<1$ μM external $Ca^{2+}$, the rate of $Ca^{2+}$ recycling ($<1$ nmole $Ca^{2+}$/mg protein · min) by the $Na^+$-$Ca^{2+}$ cycle (with $n = 3$) would give rise to a respiratory increment of $<0.3$ ng atoms $O_2$/mg protein · min (with nine $H^+$ extruded per $O_2$ atom reduced by electron transport from NADH; Wikström and Krab, 1980). If the mitochondrial content of heart, for example, is taken to correspond to about 110 mg of mitochondrial protein/gm wet weight (calculated from the cytochrome *c* contents of heart, i.e., 50 nmoles/gm wet weight; Nishiki *et al.*, 1978), and of heart mitochondria, i.e., 0.44 nmoles/mg protein (Williams, 1968), then $Ca^{2+}$ recycling would consume about 35 ng atoms $O_2$/gm wet weight · min. This amounts to $<0.2\%$ of the total $O_2$ consumption by the tissue, e.g., 20 μg atoms $O_2$/gm · min in rat hearts perfused under 80 cm $H_2O$ (Erecinska and Wilson, 1982).

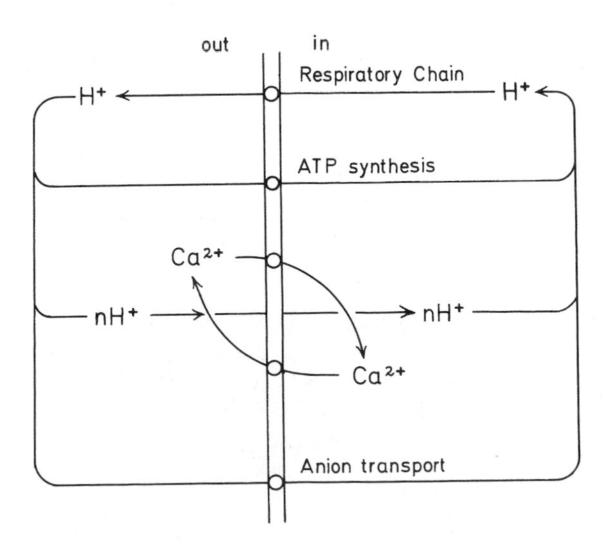


*Figure 3.* Integration of the $Ca^{2+}$ cycle within the $H^+$ circuit of the mitochondrial inner membrane.

## A. Properties of the $Ca^{2+}$ Cycles

The establishment of the $Ca^{2+}$ distribution across the inner mitochondrial membrane by steady-state recycling means that this distribution may be controlled kinetically by modifying the kinetic properties of the carriers, and without change in $\Delta\bar{\mu}_{H^+}$. Indeed, if a large span of the respiratory chain is close to thermodynamic equilibrium with the cytosolic ATP pool (Erecinska and Wilson, 1982), changes in $\Delta\bar{\mu}_{H^+}$ would have a greater influence on ATP synthesis than on $Ca^{2+}$ distribution, which is far removed from equilibrium. Modest changes in $\Delta\psi$ have a relatively small effect on $Ca^{2+}$ distribution during steady-state recycling (Puskin *et al.*, 1976; Nicholls, 1978a). In the case of the $Na^+$–$Ca^{2+}$ cycle, a minimal perturbation by $\Delta\psi$ is predicted from the rather insensitive ohmic response of unidirectional influx via the uniporter and the fact that $Ca^{2+}$ extrusion by the $Na^+$-$Ca^{2+}$ carrier is also energy-dependent, so that a change in $\Delta\psi$ affects influx and efflux in the same manner.

Fine control of the $Ca^{2+}$ distribution by changed carrier activities has been demonstrated. Partial inhibition of the uniporter by $Mg^{2+}$ leads to a decrease in the steady-state ratio $[Ca^{2+}]_i/[Ca^{2+}]_e$ (Nicholls, 1978a; Denton *et al.*, 1980). A similar shift in the steady-state $Ca^{2+}$ distribution is caused by activation of efflux with increased $[Na^+]$ (Crompton *et al.*, 1976a; Kessar and Crompton, 1981). Moreover, the effects of $Mg^{2+}$ and $Na^+$ are roughly additive in accordance with their independent actions (Denton *et al.*, 1980). In order to analyze the effects of possible physiological regulators of the cycle, however, it is necessary to consider the basic kinetic behavior of the cycle, and how this behavior is influenced by the level of extramitochondrial free $Ca^{2+}$.

At relatively low extramitochondrial free $Ca^{2+}$, when uniporter activity is low, steady-state recycling will be attained at submaximal activities of $Ca^{2+}$ efflux, i.e., when internal $Ca^{2+}$ is insufficient to saturate the antiporter(s). If external $Ca^{2+}$ is increased, then uniporter activity will increase, and internal $Ca^{2+}$ will rise until the consequent increase in antiport activity matches that of the uniporter; i.e., a new steady state will be attained with increased external $Ca^{2+}$, increased internal $Ca^{2+}$, and increased recycling rate. These conditions undoubtedly appertained in the studies of Denton *et al.* (1980). In these, heart mitochondria were subjected to increases in free external $Ca^{2+}$, stabilized by $Ca^{2+}$ buffers, and the resultant changes in internal free $Ca^{2+}$ were monitored by the response of $Ca^{2+}$-sensitive dehydrogenases in the matrix space, i.e., α-oxoglutarate dehydrogenase and pyruvate dehydrogenase. In the presence of $Na^+$ and $Mg^{2+}$, an increase in external free $Ca^{2+}$ from 0.1 to 1 μM caused a 5- to 7-fold activation. This behavior was confirmed by Hansford (1981) and Kessar and Crompton (1981). Since these enzymes when either isolated or assayed in intact mitochondria in the presence of uncoupling agents (to abolish $Ca^{2+}$ gradients) respond to free $Ca^{2+}$ over the range 0.1–10.0 μM (McCormack and Denton, 1980), it was reasonable to conclude that the increase in intramitochondrial free $Ca^{2+}$ lay within the same range. This conclusion was eventually borne out by Coll *et al.* (1982) who, by using a null-point titration technique to estimate internal free $Ca^{2+}$, reported a three-fold increase in α-oxoglutarate dehydrogenase activity as matrix free $Ca^{2+}$ was increased from 0.5 to 3.0 μM.

In this condition, the action of the cycle is to relay an increase in external free $Ca^{2+}$ into an increase in free $Ca^{2+}$ in the mitochondrial matrix. The precise characteristics of the relay will depend on the kinetic responses of the uniporter and antiporter to changes in external and internal free $Ca^{2+}$, respectively. In the case of heart mitochondria, considered here, the Hill coefficients for the interaction of $Ca^{2+}$ with the uniporter and the $Na^+$-$Ca^{2+}$ carrier are, respectively, 1.5–2.0 (Crompton *et al.*, 1976b; Saris and Åkerman, 1980) and 1 (Hayat and Crompton, 1982; Coll *et al.*, 1982). It follows that an increase in external $Ca^{2+}$ will require a proportionally larger increase in internal $Ca^{2+}$ for reattainment of steady-state recycling. In other words, the transition from one steady state (*a*) to another (*b*), with higher external free $Ca^{2+}$, will involve amplification in the steady-state distribution of $Ca^{2+}$, i.e.

$$\frac{([Ca^{2+}]_i/[Ca^{2+}]_e)_b}{([Ca^{2+}]_i/[Ca^{2+}]_e)_a} > 1$$

Moreover, the degree of amplification will increase as internal $Ca^{2+}$ exceeds the $K_m$ value for $Ca^{2+}$ of the $Na^+$-$Ca^{2+}$ carrier, and it becomes less sensitive to changes in internal free $Ca^{2+}$. The $K_m$ value of the $Na^+$-$Ca^{2+}$ carrier for internal $Ca^{2+}$ is, therefore, an important parameter in relation to the physiological performance of the cycle as a whole. According to Coll *et al.* (1982) and Hansford and Castro (1982) this value is about 3–6 μM. It is clear that amplification requires that the uniporter, on the other hand, remain highly sensitive to $Ca^{2+}$ irrespective of the level of extramitochondrial free $Ca^{2+}$; this requirement is satisfied by the relatively high $K_m$ value of the uniporter for $Ca^{2+}$ (Section II-C.1). It appears, therefore, that some degree of amplification is inherent in the relay properties of the $Na^+$-$Ca^{2+}$ cycle.

An essential feature of the cycle in this condition is that steady-state recycling is attained at internal $Ca^{2+}$ contents insufficient to cause intramitochondrial precipitation of calcium phosphate since, if this happened, internal free $Ca^{2+}$ would not vary with changes in external free $Ca^{2+}$. Under these conditions, $Na^+$-$Ca^{2+}$ carrier activity would be constant irrespective of $Ca^{2+}$ load. Invariant $Na^+$-$Ca^{2+}$ carrier activity with respect to $Ca^{2+}$ load will also occur if internal free $Ca^{2+}$ is sufficient to saturate the carrier. There is evidently, therefore, a limiting internal free $Ca^{2+}$ above which the amplified relay behavior does not apply. The limiting condition may be stated $v_{uniporter}$* < maximal efflux rate. Expressed in this way, it is apparent that the condition under which the restriction is satisfied will depend on the relative activities of the uniporter and the $Ca^{2+}$ efflux systems. Since the $Na^+$-$Ca^{2+}$ carrier is very active in heart mitochondria (in comparison with mitochondria from other sources), these mitochondria will accommodate this restriction at relatively high external free $Ca^{2+}$. Indeed, heart mitochondria suspended in media containing 0.5–1.0 mM $Mg^{2+}$, 10–15 mM $Na^+$, and 5–20 mM phosphate satisfy this restriction even when the external free $Ca^{2+}$ is increased to 1–3 μM, as judged by the responses of internal dehydrogenases (Denton *et al.*, 1980; Hansford, 1981).

* $v_{uniporter}$ refers to the velocity of the uniport reaction.

*In vitro* studies with liver and brain mitochondria have revealed a quite different kinetic behavior of the $Ca^{2+}$ cycle under conditions where the limiting condition ($v_{uniporter}$ < maximal efflux rate) does not apply. Nicholls (1978a; Nicholls and Scott, 1980) drew attention to the capacity of mitochondria to buffer extramitochondrial $Ca^{2+}$ very precisely independently of $Ca^{2+}$ load when this exceeds certain limits. Thus, brain mitochondria *in vitro* increased their matrix $Ca^{2+}$ content from 20 to 200 nmoles/mg protein, while maintaining extramitochondrial $Ca^{2+}$ at about 0.35 μM (± 10%). This perfect buffering capacity follows if sufficient $Ca^{2+}$ is accumulated to exceed the solubility product of calcium phosphate, so that internal free $Ca^{2+}$ and, hence, the efflux rate is clamped constant, independently of $Ca^{2+}$ load (Zoccarato and Nicholls, 1982). In the steady state, when influx equals efflux, an equally precise value is imposed on uniporter activity and hence on external free $Ca^{2+}$ on which uniporter activity depends. The mitochondria provides a means of sequestering $Ca^{2+}$ to a level that exceeds the solubility product of calcium phosphate. The transition from one state to another is accompanied only by a change in the amount of precipitate and the free internal $Ca^{2+}$, free external $Ca^{2+}$ and recycling rate are unchanged.

The concept of perfect $Ca^{2+}$ buffering imposed via constant internal free $Ca^{2+}$ is based on the observation that the rate of $Ca^{2+}$ efflux from liver mitochondria in the presence of 3 mM phosphate does not vary over the range 12–80 nmoles internal $Ca^{2+}$/mg protein (Zoccarato and Nicholls, 1982). Substitution of phosphate by acetate leads to much greater rates of efflux, which increase as the $Ca^{2+}$ load is increased over the same range. In principle, perfect $Ca^{2+}$ buffering would result also if the internal free $Ca^{2+}$ greatly exceeds the $K_m$ value of the efflux systems (Nicholls and Crompton, 1980). It seems unlikely, however, that internal free $Ca^{2+}$ increases sufficiently in the presence of phosphate for this condition to be satisfied, at least by liver mitochondria. In liver mitochondria, a significant proportion of $Ca^{2+}$ efflux occurs via the $Na^+$-independent system, which has a relatively low affinity for internal $Ca^{2+}$ (Zoccarato and Nicholls, 1982; Nicholls and Åkerman, 1982).

The above outline states possible modes of behavior of the $Ca^{2+}$ cycle. The domain in which regulation of internal free $Ca^{2+}$ is achieved and the domain of perfect buffering adjoin at the boundary condition, $v_{uniporter}$ = maximal efflux rate, and the "position" of the boundary, i.e., free external [$Ca^{2+}$], will be determined by the relative activities of the influx and efflux systems. In the case of heart mitochondria, the boundary appears to be outside the normal physiological range of cytosolic $Ca^{2+}$ (strictly, the time-average, cytosolic $Ca^{2+}$ as discussed in Section VII-B). In the cases of liver and brain mitochondria *in vitro,* which display a lower capacity for $Ca^{2+}$ extrusion, the boundary occurs at lower extramitochondrial free [$Ca^{2+}$], i.e., 0.3–0.8 μM $Ca^{2+}$ in the presence of 1–3 mM $Mg^{2+}$ (Nicholls, 1978a; Nicholls and Scott, 1980; Becker *et al.*, 1980). The two domains do not overlap and the two operational modes are mutually exclusive simultaneously. Since each mode would appear to provide the basis for distinct function of the cycle, namely regulation of intramitochondrial free $Ca^{2+}$ on the one hand and regulation of extramitochondrial free $Ca^{2+}$ on the other, their applicability *in vivo* may be assessed in terms of their possible contribution to cell metabolism. The following sections utilize this approach with heart and liver mitochondria.

### B. $Ca^{2+}$ Recycling in Heart Mitochondria

Several key matrix dehydrogenases in mitochondria from heart (and other tissues, including liver, kidney, brain, white fat, and brown fat) are acutely sensitive to $Ca^{2+}$ (reviewed by Denton and McCormack, 1980). $Ca^{2+}$ promotes the conversion of pyruvate dehydrogenase to the active form by activating pyruvate dehydrogenase phosphate phosphatase and inhibiting pyruvate dehydrogenase kinase. The affinities of NAD-linked isocitrate dehydrogenase and α-oxoglutarate dehydrogenases for their respective substrates are markedly increased by $Ca^{2+}$. When isolated, these enzymes are all affected by an increase in free $Ca^{2+}$ over the range 0.1–10.0 μM. In addition, Yamada *et al.* (1980, 1981) reported that about 1 μM free $Ca^{2+}$ promotes the dissociation of the inhibitor protein from $F_1$-ATPase in skeletal muscle, thereby activating the enzyme. These observations have fueled the concept that the major role of the $Na^+$–$Ca^{2+}$ cycle may be the control of matrix free $Ca^{2+}$ in accordance with the regulatory requirements of these enzymes (Denton and McCormack, 1980). The capacity of the $Na^+$–$Ca^{2+}$ cycle to relay changes in sarcoplasmic $Ca^{2+}$ to the mitochondrial matrix may be limited by its slow response relative to the rapidity of the beat-to-beat changes in sarcoplasmic $Ca^{2+}$, so that these large-scale changes may be transmitted as a largely damped-out ripple (Crompton, 1980; McCormack and Denton, 1981). Robertson *et al.* (1982) have calculated that $<$10 pmole of $Ca^{2+}$ may enter porcine heart mitochondria during each beat. It seems likely, therefore, that the cycle may respond to the time-average cytoplasmic free $Ca^{2+}$.

Denton and McCormack (1981) proposed that the passive relay of changes in time-average cytoplasmic free $Ca^{2+}$ to the mitochondrial matrix may be one factor that allows oxidative metabolism in heart to adjust to the demands imposed by increased mechanical activity, when cytoplasmic $Ca^{2+}$ increases. This function will be facilitated by the amplification inherent in the relay response of the $Na^+$–$Ca^{2+}$ cycle (Section VII-A). Additional amplification may be introduced by two factors. First, the presence of external regulatory (inhibitory) sites for $Ca^{2+}$ on the heart mitochondrial $Na^+$-$Ca^{2+}$ carrier, which show cooperation interactions and confer half-maximal inhibition of efflux at about 0.7 μM free $Ca^{2+}$, will render the rate of $Ca^{2+}$ efflux sensitive to changes in sarcoplasmic free $Ca^{2+}$ (Hayat and Crompton, 1982). Second, the uniporter may be activated by an $\alpha_1$-adrenergic mechanism. Kessar and Crompton (1981) assessed the degree to which $\alpha_1$-adrenergic activation of the uniporter *per se* may amplify the relay response of the cycle by using α-oxoglutarate dehydrogenase activity as an index of intramitochondrial free $Ca^{2+}$. Under steady-state recycling conditions, uniporter activation yielded a maximal increase of about 50% in α-oxoglutarate dehydrogenase activity. During the action of adrenaline on heart, the sarcoplasmic free $Ca^{2+}$ attained at each beat increases (Tsien 1977; Allen and Kurihara, 1980; Fabiato, 1981), and some increase in time-average free $Ca^{2+}$ will occur. The opposing actions of an increase in time-average free $Ca^{2+}$ on the uniporter (as a substrate) and on the antiporter (as an inhibitor) and the $\alpha_1$-adrenergic activation of uniporter conductance may operate in concert to yield increased intramitochondrial free $Ca^{2+}$, with consequent activation of oxidative metabolism, as outlined in Figure 4 (Hayat and Crompton, 1982).

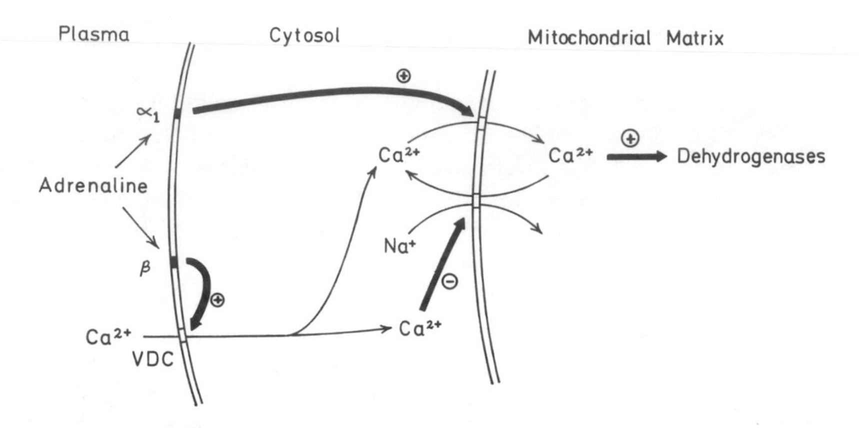


*Figure 4.* Mechanisms proposed for the adrenergic control of the $Na^+$–$Ca^{2+}$ cycle of cardiac mitochondria. VDC refers to the voltage-dependent channel of the sarcolemma. Activation is represented by $\overset{\oplus}{\rightarrow}$ and inhibition by $\overset{\ominus}{\rightarrow}$.

The mediation of uniporter activation via an $\alpha_1$-adrenergic mechanism is interesting since the mechanical response of heart to adrenaline is brought about by β-adrenergic mechanisms, involving modification of $Ca^{2+}$ transport by the sarcolemma and sarcoplasmic reticulum and of the contractile proteins (Tsien, 1977). Nevertheless, $\alpha_1$-adrenergic receptors have been identified in heart (Scholz, 1980; Williams and Lefkowitz, 1978), and these may be involved in the stimulation of glucose transport (Keely *et al.*, 1977) and the activation of phosphofructokinase (Patten *et al.*, 1982). These features led to the proposed coordinated α-adrenergic control of glucose uptake and glycolysis in heart (Patten *et al.*, 1982). Conceivably, the α-adrenergic action on the $Ca^{2+}$ uniporter extends this control to include the regulation of pyruvate oxidation via changes in mitochondrial free $Ca^{2+}$.

In summary, current data favor the concept that the prime role of the $Na^+$–$Ca^{2+}$ cycle in heart is the control of intramitochondrial free $Ca^{2+}$. Two features of the cycle effectively prevent the mitochondria from acting as cytoplasmic $Ca^{2+}$ buffers and opposing the work of the sarcolemma and sarcoplasmic reticulum in changing cytoplasmic free $Ca^{2+}$, on which heart performance depends, i.e., the slow kinetic response of the cycle, and the relatively high limiting extramitochondrial $Ca^{2+}$ that must be exceeded for effective buffering of extramitochondrial $Ca^{2+}$. In addition, the α-adrenergic activation of the uniporter is incompatible with cytosolic $Ca^{2+}$ buffering by mitochondria, since it would imply that cytosolic $Ca^{2+}$ is decreased during adrenergic action, in contrast to the increase that actually occurs. Similarly, inhibition of the $Na^+$-$Ca^{2+}$ carrier by cytosolic $Ca^{2+}$ would have no obvious value in terms of cytosolic $Ca^{2+}$ buffering, merely enabling the steady state to be reestablished more quickly following a change in cytosolic $Ca^{2+}$. Both phenomena, however, are readily explicable in terms of amplification of changes in intramitochondrial free $Ca^{2+}$.

The critical question of whether the cycle *in vivo* sets the intramitochondrial free $Ca^{2+}$ within the range required for dehydrogenase regulation (as it does *in vitro;*

Section VII-A) has not been answered definitively. Hansford and Castro (1981) and Coll *et al.* (1982) reported that essentially full activation of α-oxoglutarate dehydrogenase by $Ca^{2+}$ is attained with 2–3 nmoles $Ca^{2+}$/mg protein *in vitro*. Heart mitochondria isolated in the presence of EGTA, when some $Ca^{2+}$ losses may be predicted, typically contain 6–8 nmoles $Ca^{2+}$/mg protein (M. Crompton, unpublished observation); similar values have been reported for the $Ca^{2+}$ content of liver mitochondria after rapid density-gradient separation in the presence of EGTA (Reinhardt *et al.*, 1982). While these comparisons reveal a discrepancy, this is not large and underlines the need for a further investigation of this aspect.

## C. $Ca^{2+}$ Recycling in Liver Mitochondria

In the case of liver mitochondria, the relatively low activities of the $Ca^{2+}$ efflux systems demonstrable *in vitro* lead to effective $Ca^{2+}$ buffering at relatively low external free $Ca^{2+}$ (Becker *et al.*, 1980; Becker, 1980). In addition, the mitochondria may be kinetically competent to bring about changes in cytoplasmic $Ca^{2+}$ within the time period predicted from the physiological response, e.g., glycogenolysis. Particular attention has been directed to the role of mitochondria during the stimulation of hepatic glycogenolysis by α-adrenergic agonists, vasopressin and angiotensin. Extensive evidence indicates that this response reflects activation of phosphorylase *b* kinase by a rise in cytosolic free $Ca^{2+}$ (Exton, 1980; Williamson *et al.*, 1981; Murphy *et al.*, 1980). Moreover, observed losses of mitochondrial $Ca^{2+}$ after tissue treatment with these agonists have led to the concept that mitochondria may provide part, at least, of the cytosolic $Ca^{2+}$ (Taylor *et al.*, 1980; Murphy *et al.*, 1980; Blackmore *et al.*, 1979; Babcock *et al.*, 1979; Barritt *et al.*, 1981). A treatment of this controversial topic is outside the scope of this article (Williamson *et al.*, 1981). Nevertheless, the view that mitochondria may control cytoplasmic $Ca^{2+}$ fits naturally with the concept of their proposed role as buffers of cytoplasmic $Ca^{2+}$, and a brief comment is appropriate.

Perfect buffering *in vitro* is attained when internal free $Ca^{2+}$ equilibrates with internal deposits of calcium phosphate (Nicholls and Åkerman, 1982; Zoccarato and Nicholls, 1982) which are in the amorphous, rather than crystalline, state (Greenawalt *et al.*, 1964; Tew *et al.*, 1980). Under these conditions, $Ca^{2+}$ losses would be achieved at the expense of the deposits and internal free $Ca^{2+}$ would be unchanged, at least initially. In this connection, Babcock *et al.* (1979) used chlortetracycline fluorescence as an index of intramitochondrial free $Ca^{2+}$ in hepatocytes and reported a lag phase of about 1 min following addition of norepinephrine, before significant decrease in fluorescence occurred.

A change in the $Ca^{2+}$-buffering properties of liver mitochondria in line with increased cytoplasmic free $Ca^{2+}$ would be brought about by either activation of the $Ca^{2+}$ efflux process or inhibition of the uniporter. In this respect, the α-adrenergic (phenylephrine)-induced two-fold increase in the apparent affinity of the $Na^+$-independent system for internal $Ca^{2+}$ may be important (Coll *et al.*, 1982). These observations do not conflict with those of Goldstone *et al.* (1983) showing that phenylephrine pretreatment does not change the response of this system to total intramitochondrial $Ca^{2+}$ since, under the experimental conditions used (limiting inorganic phosphate), phenylephrine also caused a two-fold decrease in the proportion of total intramitochondrial $Ca^{2+}$ that was free (Coll *et al.*, 1982). It would seem important to investigate this activation further under conditions where internal free $Ca^{2+}$ may be independent of the hormonal state of the mitochondria, e.g., at nonlimiting concentrations of intramitochondrial phosphate.

Liver mitochondrial $Ca^{2+}$ is also mobilized in response to glucagon (Foden and Randle, 1978; Blackmore *et al.*, 1979; Taylor *et al.*, 1980; Baddams *et al.*, 1983). In this case, it appears that activation of the $Na^{+}$-$Ca^{2+}$ carrier may be responsible (Goldstone *et al.*, 1983). Again, an increase in the affinity of the system for internal $Ca^{2+}$ was indicated.

In summary, separate mechanisms may exist for the control of the $Na^{+}$-$Ca^{2+}$ carrier (β-adrenergic, glucagon) and the $Na^{+}$-independent systems (α-adrenergic) of liver mitochondria, and these may provide means for the hormonally induced losses of mitochondrial $Ca^{2+}$. However, further studies are necessary to evaluate the capacity of these modifications to produce meaningful increases in cytosolic $Ca^{2+}$. Our current understanding of the $Ca^{2+}$ cycles indicates that they amplify relatively small changes in external $Ca^{2+}$ into relatively large changes in intramitochondrial $Ca^{2+}$ (Section VII-A); in other words, the system appears to be wrongly "geared" for the regulation of cytosolic free $Ca^{2+}$. As stressed by Williamson *et al.* (1981; Coll *et al.*, 1982), this behavior reflects the sigmoïdal kinetics of the uniporter, i.e., high sensitivity to changes in cytosolic free $Ca^{2+}$. One might predict, therefore, that modification of the kinetic behavior of the uniporter, rather than the efflux systems, would be more appropriate for establishing a change in cytosolic free $Ca^{2+}$.

## VIII. CONCLUDING REMARKS

The general features of mitochondrial $Ca^{2+}$ transport bear a broad resemblance to those of plasma-membrane $Ca^{2+}$ transport, in that influx is passive down the $Ca^{2+}$ electrochemical gradient, and active Ca extrusion in mitochondria from many tissues occurs predominantly by $Na^{+}$–$Ca^{2+}$ exchange (there is no recognized mitochondrial equivalent of the $Ca^{2+}$-ATPase, however). In spite of this analogy, the mitochondrial carriers may operate somewhat differently than their plasma membrane counterparts. Many, at least, of the plasma membrane $Ca^{2+}$ channels are acutely regulated by voltage-dependent mechanisms or by linkage to the occupation of a hormone receptor, so that influx may be minimal in resting cells and greatly increased on stimulation to bring about the appropriate $Ca^{2+}$-sensitive response. In contrast, there is no evidence that the mitochondrial $Ca^{2+}$ uniporter exists in anything but an "open" conformation, so that $Ca^{2+}$ cycling presumably occurs continuously at significant rates across the inner membrane.

This feature has certain implications for the mode of operation and function of the cycle *in vivo*. It is easily accommodated into the concept that the cycle provides one means of controlling oxidative metabolism, via matrix free $Ca^{2+}$, since this is not an all-or-none response, but a continuous process requiring adjustment according to the cellular demand for ATP. It is equally well accommodated by the concept that mitochondria may function as $Ca^{2+}$ donors to the cytosol under certain conditions, since the "resting" state would require an active uniporter to enable mitochondrial $Ca^{2+}$ retention. Both functions imply that the cycle operates by readjustment from one steady state to another.

It appears, therefore, that the functional role of mitochondrial $Ca^{2+}$ transport is realized via the steady-state properties of the cycle as a whole. Although it is true that the behavior of the cycle in establishing a steady-state gradient of $Ca^{2+}$ will be determined by the kinetic properties of the individual carriers, these carriers function as a discrete operational entity, and it is the relation between their properties that will be paramount in establishing its steady-state properties and its function. For example, the relative activities of the influx and efflux systems and their relative affinities for $Ca^{2+}$ are critical factors in the overall response of the cycle to changes in external free $Ca^{2+}$, as exemplified by *in vitro* studies with heart and liver mitochondria. It follows that some degree of caution is necessary in extrapolating from recycling experiments *in vitro,* when native carrier properties may be incompletely expressed, to the presumed function of the cycle *in vivo*. A clearer understanding will undoubtedly emerge when diverse approaches coalesce into a consistent pattern. Above all, there is a clear need to establish techniques for the measurement of internal free $Ca^{2+}$ to allow unambiguous expression of cycle performance in establishing the gradient of free $Ca^{2+}$. The identification of factors that control the reactions of the cycle should prove informative. The ubiquitous nature of mitochondrial $Ca^{2+}$ transport in mammalian tissues may well render this approach particularly incisive, since it is barely conceivable that the $Ca^{2+}$ cycle responds to the same regulatory influences irrespective of tissue type. The elucidation of specific control factors and their relation to other cellular metabolic events should enable the specific roles of the $Ca^{2+}$ cycle in cell physiology to be clarified further.

## REFERENCES


Affolter, H., and Carafoli, E., 1980, The $Ca^{2+}/Na^{+}$ antiporter of heart mitochondria operates electroneutrally, *Biochem. Biophys. Res. Commun.* **95:**193–196.

Åkerman, K. E. O., 1978a, Charge transfer during valinomycin-induced $Ca^{2+}$ uptake in rat liver mitochondria, *FEBS Lett.* **93:**293–296.

Åkerman, K. E. O., 1978b, Changes in membrane potential during calcium ion influx and efflux across the mitochondrial membrane, *Biochim. Biophys. Acta* **502:**359.

Åkerman, K. E. O., 1978c, Effect of pH and $Ca^{2+}$ on the retention of $Ca^{2+}$ by rat liver mitochondria, *Arch. Biochem. Biophys.* **189:**256–262.

Allen, D. G., and Kurihara, S., 1980, Calcium transients in mammalian ventricular muscle, *Eur. Heart J.* **1**(Suppl. A)**:**5–15.

Al-Shaikhaly, M. H. M., Nedergaard, J., and Cannon, B., 1975, Sodium induced calcium release from mitochondria in brown adipose tissue, *Proc. Natl. Acad. Sci. USA* **76:**2350–2352.

Arshad, J. H., and Holdsworth, E. S., 1980, Stimulation of calcium efflux from rat liver mitochondria by adenosine 3′, 5′-cyclic monophosphate, *J. Membr. Biol.* **57:**207–212.

Azzone, G. F., Pozzan, T., Massari, S., Bragadin, M., and Dell'Antone, P., 1976, Proton electrochemical potential in steady state rat liver mitochondria, *Biochim. Biophys. Acta* **459:**96–109.

Babcock, D. F., Chen, J. J., Yip, B. P., and Lardy, H. A., 1979, Evidence for mitochondrial localisation of the hormone-responsive pool of $Ca^{2+}$ in isolated hepatocytes, *J. Biol. Chem.* **254:**8117–8120.

Baddams, H. M., Chang, L. B. F., and Barritt, G. J., 1983, Evidence that glucagon acts on the liver to decrease mitochondrial calcium stores, *Biochem. J.* **210**:73–77.

Baker, P. F., 1972, Transport and metabolism of calcium ions in nerve, *Prog. Biophys. Mol. Biol.* **24**:177–233.

Barritt, G. J., Parker, J. C., and Wadsworth, J. C., 1981, A kinetic analysis of the effects of adrenalin on calcium distribution in isolated rat liver parenchymal cells, *J. Physiol. (London)* **312**:29–55.

Beatrice, M. C., Palmer, J. W., and Pfeiffer, D. R., 1980, The relationship between mitochondrial membrane permeability, membrane potential and the retention of $Ca^{2+}$ by mitochondria, *J. Biol. Chem.* **255**:8663–8671.

Becker, G. L., 1980, Steady state regulation of extramitochondrial $Ca^{2+}$ by rat liver mitochondria, *Biochim. Biophys. Acta* **591**:234–239.

Becker, G. L., Fiskum, G., and Lehninger, A. L., 1980, Regulation of free $Ca^{2+}$ by liver mitochondria and endoplasmic reticulum, *J. Biol. Chem.* **255**:9009–9012.

Bernardi, P., and Azzone, G. F., 1979, ΔpH Induced calcium fluxes in rat liver mitochondria, *Eur. J. Biochem.* **102**:555–562.

Blackmore, P. F., Dehaye, J., and Exton, J. H., 1979, Studies on the α-adrenergic activation of hepatic glucose output. The role of mitochondrial $Ca^{2+}$ release in α-adrenergic activation of phosphorylase in perfused rat liver, *J. Biol. Chem.* **254**:6945–6950.

Blackmore, P. F., El-Refai, M. F., Dehaye, J. P., Strickland, W. G., Hughes, B. P., and Exton, J. H., 1981, Blockade of hepatic α-adrenergic receptors and responses by chlorpromazine and trifluoperazine, *FEBS* Lett. **123**:245–248.

Blaustein, M. P., 1974, The interrelationship between sodium and calcium fluxes across cell membranes, *Rev. Physiol. Biochem. Pharmacol.* **70**:33–82.

Borle, A. B., 1974, Cyclic AMP stimulation of calcium efflux from kidney, liver and heart mitochondria, *J. Membr. Biol.* **16**:221–236.

Bragadin, M., Pozzan, T., and Azzone, G. F., 1979, Kinetics of $Ca^{2+}$ carrier in rat liver mitochondria, *Biochemistry* **18**:5972–5978.

Brand, M. D., and DeSelincourt, C., 1980, Effects of glucagon and $Na^+$ on the control of extramitochondrial free $Ca^{2+}$ concentration by mitochondria from liver and heart, *Biochem. Biophys. Res. Commun.* **92**:1377–1382.

Brierley, G. P., 1976, The uptake and extrusion of monovalent cations by isolated heart mitochondria, *Mol. Cell. Biochem.* **10**:41–63.

Brinley, F. J., Tiffert, T., Scarpa, A., and Mullins, L. J., 1977, Intracellular calcium buffering capacity in isolated squid axons, *J. Gen. Physiol.* **70**:355–384.

Bygrave, F. L., 1977, Mitochondrial calcium transport, in: *Current Topics in Bioenergetics* (D. R. Sanadi, ed.), Academic Press, New York, pp. 260–318.

Carafoli, E., and Lehninger, A. L., 1971, A survey of the interaction of calcium ions with mitochondria from different tissues and species, *Biochem. J.* **122**:681–690.

Carafoli, E., and Sottocasa, G., 1974, The $Ca^{2+}$ transport system of the mitochondrial membrane and the problem of the $Ca^{2+}$ carrier, in: *Dynamics of Energy-Transducing Membranes* (L. Ernster, R. W. Estabrook, and Slater, eds.), Elsevier, Amsterdam, pp. 455–469.

Carafoli, E., Tiozzo, R., Lugli, G., Crovetti, F., and Kratzing, C., 1974, The release of calcium from heart mitochondria by sodium, *J. Mol. Cell. Cardiol.* **6**:361–371.

Caroni, P., Reinlib, L., and Carafoli, E., 1980, Charge movements during the $Na^+$–$Ca^{2+}$ exchange in heart sarcolemmal vesicles, *Proc. Natl. Acad. Sci. USA* **77**:6354–6358.

Chappell, J. B., Cohn, M., and Greville, G. D., 1963, The accumulation of divalent cations by isolated mitochondria, in: *Energy-Linked Functions of Mitochondria* (B. Chance, ed.), Academic Press, New York, pp. 219–231.

Coelho, R. L. C., and Vercesi, A. E., 1980, Retention of $Ca^{2+}$ by rat liver and rat heart mitochondria. Effect of phosphate, $Mg^+$ and NAD(P) redox state, *Arch. Biochem. Biophys.* **204**:141–147.

Coll, K. E., Joseph, S. K., Corkey, B. E., and Williamson, J. R., 1982, Determination of the matrix free $Ca^{2+}$ concentration and kinetics of $Ca^{2+}$ efflux in liver and heart mitochondria, *J. Biol. Chem.* **257**:8696–8704.

Crompton, M., 1980, The sodium ion/calcium ion cycle of cardiac mitochondria, *Biochem. Soc. Trans.* **8**:261–262.

Crompton, M., and Heid, I., 1978, The cycling of calcium, sodium and protons across the inner membrane of cardiac mitochondria, *Eur. J. Biochem.* **91:**599–608.

Crompton, M., and Kessar, P., 1981, The activation of the calcium uniporter of cardiac mitochondria by α-adrenergic agonists, in: *Vectorial Reactions in Electron and Ion Transport in Mitochondria and Bacteria* (F. Palmieri, ed.), Elsevier, Amsterdam, pp. 261–264.

Crompton, M., Capano, M. and Carafoli, E., 1976a, The sodium-induced efflux of calcium from heart mitochondria. A possible mechanism for the regulation of mitochondrial calcium, *Eur. J. Biochem.* **69:**453–462.

Crompton, M., Sigel, E., Salzman, M., and Carafoli, E., 1976b, A kinetic study of the energy-linked influx of $Ca^{2+}$ into heart mitochondria, *Eur. J. Biochem.* **69:**429–434.

Crompton, M., Capano, M., and Carafoli, E., 1976c, Respiration dependent efflux of $Mg^{2+}$ from heart mitochondria, *Biochem. J.* **154:**735–742.

Crompton, M., Kunzi, M., and Carafoli, E., 1977, The calcium induced and sodium induced effluxes of calcium from heart mitochondria. Evidence for a sodium-calcium carrier, *Eur. J. Biochem.* **79:**549–558.

Crompton, M., Moser, R., Ludi, H., and Carafoli, E., 1978a, The interrelations between the transport of sodium and calcium in mitochondria of various mammalian tissues, *Eur. J. Biochem.* **82:**25–31.

Crompton, M., Hediger, M., and Carafoli, E., 1978b, The effect of inorganic phosphate on calcium influx into rat heart mitochondria, *Biochem. Biophys. Res. Commun.* **80:**540–546.

Crompton, M., Heid, I., Baschera, C., and Carafoli, E., 1979, The resolution of calcium fluxes in heart and liver mitochondria using the lanthanide series, *FEBS Lett.* **104:**352–354.

Crompton, M., Heid, I., and Carafoli, E., 1980, The activation by potassium of the sodium-calcium carrier of cardiac mitochondria, *FEBS Lett.* **115:**257–259.

Denton, R. M., and McCormack, J. G., 1980, On the role of the calcium transport cycle in heart and other mammalian mitochondria, *FEBS Lett.* **119:**1–8.

Denton, R. M., and McCormack, J. G., 1981, Calcium ions, hormones and mitochondrial metabolism, *Clin. Sci.* **61:**135–140.

Denton, R. M., McCormack, J. G., and Edgell, N. J., 1980, Role of calcium ions in the regulation of intramitochondrial metabolism. Effects of $Na^{+}$, $Mg^{2+}$ and ruthenium red on the $Ca^{2+}$ stimulated oxidation of oxoglutarate and on pyruvate dehydrogenase activity in intact rat heart mitochondria, *Biochem. J.* **190:**107–117.

Di Polo, R., 1974, Effect of ATP on the calcium efflux in dialyzed squid axons, *J. Gen. Physiol.* **64:**503–517.

Dubinsky, W., Kandrack, A., and Racker, E., 1979, Resolution and reconstitution of a calcium transporter from bovine heart mitochondria, in: *Membrane Bioenergetics* (C. P. Lee, G. Schatz, and L. Ernster, eds.), Addison-Wesley, Reading, Massachusetts, pp. 267–280.

Erecinska, M., and Wilson, M., 1982, Regulation of cellular energy metabolism, *J. Membr. Biol.* **70:** 1–16.

Exton, J. H., 1980, Mechanisms involved in α-adrenergic phenomena. Role of calcium ions in the actions of catecholamines in liver and other tissues, *Am. J. Physiol.* **238:**E3–E12.

Fabiato, A., 1981, Myoplasmic free calcium concentration reached during the twitch of an intact isolated cardiac cell and during calcium-induced release of calcium from the sarcoplasmic reticulum of a skinned cardiac cell from the adult rat or rabbit ventricle, *J. Gen. Physiol.* **78:**447–497.

Fiskum, G., and Cockrell, R. S., 1978, Ruthenium red sensitive and insensitive calcium transport in rat liver and Ehrlich ascites tumour cell mitochondria, *FEBS Lett.* **92:**125–128.

Fiskum, G., and Lehninger, A. L., 1979, Regulated release of $Ca^{2+}$ from respiring mitochondria by $Ca^{2+}/2H^{+}$ antiport, *J. Biol. Chem.* **254:**6236–6239.

Fiskum, G., Reynafarje, B., and Lehninger, A. L., 1979, The electric charge stoichiometry of respiration-dependent $Ca^{2+}$ uptake by mitochondria, *J. Biol. Chem.* **254:**6288–6295.

Foden, S., and Randle, P. J., 1978, Calcium metabolism in rat hepatocytes, *Biochem. J.* **170:**615–625.

Friedman, N., 1980, Studies on the mechanism of the glucagon elicited changes in mitochondrial $Ca^{2+}$ uptake, *J. Supramol. Struct.* **13**(Suppl. 4)**:**107.

Goldstone, T. P., and Crompton, M., 1982, Evidence for β-adrenergic activation of $Na^{+}$-dependent efflux of $Ca^{2+}$ from isolated liver mitochondria, *Biochem. J.* **204:**369–371.

Goldstone, T. P., Duddridge, R. J., and Crompton, M., 1983, The activation of $Na^{+}$-dependent efflux of $Ca^{2+}$ from liver mitochondria by glucagon and β-adrenergic agonists, *Biochem. J.* **210:**463–472.

Gomez-Puyou, A., Tuena de Gomez-Puyou, M., Becker, G., and Lehninger, A. L., 1972, An insoluble $Ca^{2+}$ binding factor from rat liver mitochondria, *Biochem. Biophys. Res. Commun.* **47:**814–819.

Greenawalt, J. W., Rossi, C. S., and Lehninger, A. L., 1964, Effect of active accumulation of calcium and phosphate ions on the structure of rat liver mitochondria, *J. Cell. Biol.* **23:**21–38.

Gunter, T. E., Gunter, K. K., Puskin, J. S., and Russell, P. R., 1978, Efflux of $Ca^{2+}$ and $Mn^{2+}$ from rat liver mitochondria, *Biochemistry* **17:**339–345.

Halestrap, A. P., 1978, Stimulation of the respiratory chain of rat liver mitochondria between cytochrome $c_1$ and cytochrome *c* by glucagon treatment of rats, *Biochem. J.* **172:**399–405.

Hansford, R. G., 1981, Effect of micromolar concentrations of free $Ca^{2+}$ ions on pyruvate dehydrogenase interconversion in intact rat heart mitochondria, *Biochem. J.* **194:**721–732.

Hansford, R. G., and Castro, F., 1981, Effect of micromolar concentrations of free calcium ions on the reduction of heart mitochondrial NAD(P) by 2-oxoglutarate, *Biochem. J.* **198:**525–533.

Hansford, R. G., and Castro, F., 1982, Intramitochondrial and extramitochondrial free calcium ion concentration of suspensions of heart mitochondria with low, plausibly physiological contents of total calcium, *J. Bioenerg. Biomembr.* **14:**361–376.

Harris, E. J., 1979, Modulation of $Ca^{2+}$ efflux from heart mitochondria, *Biochem. J.* **178:**673–680.

Harris, E. J., and Heffron, J. J. A., 1982, The stimulation of the release of $Ca^{2+}$ from mitochondria by sodium ions and its inhibition, *Arch. Biochem. Biophys.* **218:**513–520.

Haworth, R. A., Hunter, D. R., and Berkoff, H. A., 1980, $Na^+$ releases $Ca^{2+}$ from liver, kidney and lung mitochondria, *FEBS Lett.* **110:**216–218.

Hayat, L. H., and Crompton, M., 1982, Evidence for the existence of regulatory sites for $Ca^{2+}$ on the $Na^+/Ca^{2+}$ carrier of cardiac mitochondria, *Biochem. J.* **202:**509–518.

Heaton, G. M., and Nicholls, D. G., 1976, The calcium conductance of the inner membrane of rat liver mitochondria and the determination of the calcium electrochemical gradient, *Biochem. J.* **156:**635–646.

Heffron, J. J. A., and Harris, E. J., 1981, Stimulation of calcium ion efflux from liver mitochondria by sodium ions and its response to ADP and energy state, *Biochem. J.* **194:**925–929.

Hofstetter, W., Mühleback, T., Lötscher, H. R., Winterhalter, K., and Richter, C., 1981, ATP prevents both hydroperoxide induced hydrolysis of pyridine nucleotides and release of calcium in rat liver mitochondria, *Eur. J. Biochem.* **117:**361–367.

Hughes, B. P., and Barritt, G. J., 1978, Effects of glucagon and $N^6$, $O^2$-dibutyryladenosine 3′ : 5′-cyclic monophosphate on calcium transport in isolated rat liver mitochondria, *Biochem. J.* **176:**295–304.

Hunter, P. R., and Haworth, R. A., 1979, The $Ca^{2+}$-induced membrane transitions in mitochondria. III. Transitional $Ca^{2+}$ release, *Arch. Biochem. Biophys.* **195:**468–477.

Jeng, A. Y., and Shamoo, A. E., 1980, The electrophoretic properties of a $Ca^{2+}$ carrier isolated from calf heart inner mitochondrial membranes, *J. Biol. Chem.* **255:**6904–6912.

Keely, S. L., Corbin, J. D., and Lincoln, T., 1977, Alpha adrenergic involvement in heart metabolism: Effects on adenosine cyclic 3′,5-monophosphate, adenosine cyclic 3′,5′-monophosphate dependent protein kinase, guanosine cyclic 3′,5′-monophosphate and glucose transport, *Mol. Pharmacol.* **13:**964–975.

Kessar, P., and Crompton, M., 1981, The α-adrenergic-mediated activation of $Ca^{2+}$ influx into cardiac mitochondria, *Biochem. J.* **200:**379–388.

Kun, E., Kearney, E. B., Wiedemann, I., and Lee, N. M., 1969, Regulation of mitochondrial metabolism by specific cellular substances. II. The nature of stimulation of mitochondrial glutamate metabolism by a cytoplasmic component, *Biochemistry* **8:**4443–4449.

Leblanc, P., and Clauser, H., 1974, ADP and $Mg^{2+}$ requirement for $Ca^{2+}$ accumulation by hog heart mitochondria, *Biochim. Biophys. Acta* **347:**87–101.

Lee, C. O., and Armstrong, W. McD. 1974, State and distribution of potassium and sodium ions in frog skeletal muscle, *J. Membr. Biol.* **15:**331–362.

Lee, C. O., and Fozzard, H. A., 1975, Activities of potassium and sodium ions in rabbit heart muscle, *J. Gen. Physiol.* **65:**695–708.

Lehninger, A. L., Carafoli, E., and Rossi, C. S., 1967, Energy linked ion movements in mitochondrial systems, *Adv. Enzymol.* **29:**259–320.

Lehninger, A. L., Vercesi, A., and Bababunmi, E. A., 1978, Regulation of $Ca^{2+}$ release from mitochondria by the oxidation-reduction state of pyridine nucleotides, *Proc. Natl. Acad. Sci. USA* **75:**1690–1694.

Lötscher, H.-R., Schwerzmann, K., and Carafoli, E., 1979, The transport of $Ca^{2+}$ in a purified population of inside-out vesicles from rat liver mitochondria, *FEBS Lett.* **99:**194–198.

Lötscher, H.-R., Winterhalter, K. H., Carafoli, E., and Richter, C., 1980a, The energy state of mitochondria during the transport of $Ca^{2+}$, *Eur. J. Biochem.* **110:**211–216.

Lötscher, H.-R., Winterhalter, K. H., Carafoli, E., and Richter, C., 1980b, Hydroperoxide induced loss of pyridine nucleotides and release of $Ca^{2+}$ from rat liver mitochondria, *J. Biol. Chem.* **255:**9325–9330.

Matlib, A., and O'Brien, J. P., 1974, Adenosine 3′ : 5′-cyclic monophosphate stimulation of calcium efflux, *Biochem. Soc. Trans.* **2:**997–1000.

McCormack, J. G., and Denton, R. M., 1980, The role of calcium ions in the regulation of intramitochondrial metabolism. Properties of the $Ca^{2+}$-sensitive dehydrogenases within intact uncoupled mitochondria from the white and brown adipose tissue of the rat, *Biochem. J.* **190:**95–105.

McCormack, J. G., and Denton, R. M., 1981, The activation of pyruvate dehydrogenase in the perfused rat heart by adrenaline and other inotropic agents, *Biochem. J.* **194:**639–643.

McMillin-Wood, J., Wolkowicz, P. E., Chu, A., Tate, C. A., Goldstone, M. A., and Entman, M. L., 1980, Calcium uptake by two preparations of mitochondria from heart, *Biochim. Biophys. Acta* **591:**251–265.

Mela, L., 1969, Inhibition and activation of calcium transport in mitochondria. Effect of lanthanides and local anaesthetic drugs, *Biochemistry* **8:**2481–2486.

Mela, L., and Chance, B., 1969, Calcium carrier and the "high affinity calcium binding site" in mitochondria, *Biochem. Biophys. Res. Commun.* **35:**556–669.

Miranova, G. D., Tutjana, V. S., Pronevitch, L. A., Trofimenko, N. T., Miranov, G. P., Grigorjev, P. A., and Kondrashora, M., 1982, Isolation and properties of $Ca^{2+}$ transporting glycoprotein and peptide from beef heart mitochondria, *J. Bioenerg. Biomembr.* **14:**213–219.

Mitchell, P., 1966, *Chemiosmotic Coupling in Oxidative and Photosynthetic Phosphorylation,* Glynn Research, Bodmin, U.K.

Mitchell, P., and Moyle, J., 1967, Respiration-driven proton translocation in rat liver mitochondria, *Biochem. J.* **105:**1147–1162.

Mitchell, P., and Moyle, J., 1969, Estimation of membrane potential and pH difference across the cristae membrane of rat liver mitochondria, *Eur. J. Biochem.* **7:**471–484.

Moore, C. L., 1971, Specific inhibition of mitochondrial $Ca^{2+}$ transport by ruthenium red, *Biochem. Biophys. Res. Commun.* **42:**298–305.

Moyle, J., and Mitchell, P., 1977, The lanthanum sensitive calcium phosphate porter of rat liver mitochondria, *FEBS Lett.* **77:**136–145.

Murphy, E., Coll, K. E., Rich, T. L., and Williamson, J. R., 1980, Hormonal effects on calcium homeostasis in isolated hepatocytes, *J. Biol. Chem.* **255:**6600–6608.

Nachbaur, J., Colbeau, A., and Vignais, P. M., 1972, Distribution of membrane confined phospholipases in the rat hepatocyte. *Biochem. Biophys. Res. Commun.* **274:**426–446.

Nedergaard, J., and Cannon, B., 1980, Effects of monovalent cations on $Ca^{2+}$ transport in mitochondria; a comparison between brown fat and liver mitochondria, *Acta Chem. Scand.* **B34:**149–151.

Nicholls, D. G., 1974, The influence of respiration and ATP hydrolysis on the proton-electrochemical gradient across the inner membrane of rat liver mitochondria as determined by ion distribution, *Eur. J. Biochem.* **50:**305–315.

Nicholls, D. G., 1978a, The regulation of extramitochondrial free $Ca^{2+}$ ion concentration by rat liver mitochondria, *Biochem. J.* **176:**463–474.

Nicholls, D. G., 1978b, Calcium transport and proton electrochemical gradient in mitochondria from guinea pig cerebral cortex and rat heart, *Biochem. J.* **170:**511–522.

Nicholls, D. G., and Åkerman, K. E. O., 1982, Mitochondrial calcium transport, *Biochim. Biophys. Acta* **683:**57–88.

Nicholls, D. G., and Brand, M. D., 1980, The nature of the calcium ion efflux induced in rat liver mitochondria by the oxidation of endogenous nicotinamide nucleotides, *Biochem. J.* **188:**113–118.

Nicholls, D. G., and Crompton, M., 1980, Mitochondrial calcium transport, *FEBS Lett.* **111:**261–268.

Nicholls, D. G., and Scott, L. D., 1980, The regulation of brain mitochondrial calcium-ion transport, *Biochem. J.* **186:**833–839.

Nishiki, K., Erecinska, M., and Wilson, D. F., 1978, Energy relationships between cytosolic metabolism and mitochondrial respiration in rat heart, *Am. J. Physiol.* **234:**c73–c81.

Packer, L., Utsumi, K., and Mustafa, M. G., 1966, Oscillatory states of mitochondria 1. Electron and energy transfer pathways, *Arch. Biochem. Biophys.* **117:**381–393.

Palmer, J. W., and Pfeiffer, D. R., 1981, The control of $Ca^{2+}$ release from heart mitochondria, *J. Biol. Chem.* **256:**6742–6750.

Panfili, E., Sandri, G., Sottocasa, G. L., Lunazzi, G., and Liut, G., 1976, Specific inhibition of mitochondrial $Ca^{2+}$ transport by antibodies directed to the $Ca^{2+}$-binding glycoprotein, *Nature* **264:**185–186.

Panfili, E., Sottocasa, G. L., Sandri, G., and Liut, G., 1980, The $Ca^{2+}$ binding glycoprotein as the site of metabolic regulation of mitochondrial $Ca^{2+}$ movements, *Eur. J. Biochem.* **105:**205–210.

Panfili, E., Crompton, M., and Sottocasa, G. L., 1981, Immunochemical evidence of the independence of the $Ca^{2+}/Na^{+}$ antiporter and the electrophoretic $Ca^{2+}$ uniporter in heart mitochondria, *FEBS Lett.* **123:**30–32.

Patten, G. S., Filsell, O. H., and Clark, M. G., 1982, Epinephrine regulation of phosphofructokinase in perfused rat heart. A calcium-ion dependent mechanism mediated via α-receptors, *J. Biol. Chem.* **257:**9480–9486.

Pfeiffer, D. R., Kaufman, R. F., and Lardy, H. A., 1978, Effects of *N*-ethylmaleimide on the limited uptake of $Ca^{2+}$, $Mn^{2+}$ and $Sr^{2+}$ by rat liver mitochondria, *J. Biol. Chem.* **253:**4165–4171.

Pfeiffer, D. R., Schmid, P. C., Beatrice, M. C., and Schmid, H. O., 1979, Intramitochondrial phospholipase activity and the effects of $Ca^{2+}$ plus *N*-ethylamaleimide on mitochondrial function, *J. Biol. Chem.* **254:**11485–11494.

Philipson, K. O., and Nishimoto, A. Y., 1980, $Na^{+}$–$Ca^{2+}$ exchange is affected by membrane potential in cardiac sarcolemmal vesicles, *J. Biol. Chem.* **255:**6880–6882.

Pitts, I. R., 1979, Stoichiometry of sodium–calcium exchange in cardiac sarcolemmal vesicles, *J. Biol. Chem.* **254:**6232–6235.

Pozzan, T., Bragadin, M., and Azzone, G. F., 1977, Disequilibrium between steady-state $Ca^{2+}$ accumulation ratio and membrane potential in mitochondria: Pathway and role of $Ca^{2+}$ efflux, *Biochemistry* **16:**5618–5625.

Puskin, J. S., Gunter, T. E., Gunter, K. K., and Russell, P. R., 1976, Evidence for more than one $Ca^{2+}$ transport mechanism in mitochondria, *Biochemistry* **15:**3834–3842.

Reed, K. C., and Bygrave, F. L., 1974, Inhibition of mitochondrial calcium transport by lanthanides and ruthenium red, *Biochem. J.* **140:**143–153.

Reeves, J. P., and Sutko, J. L., 1980, Sodium–calcium exchange activity generates a current in cardiac membrane vesicles, *Science* **208:**1461–1463.

Reinhardt, P. H., Taylor, W. M., and Bygrave, F. L., 1982, A procedure for the rapid preparation of mitochondria from rat liver, *Biochem. J.* **204:**731–735.

Reinlib, L., Caroni, P., and Carafoli, E., 1981, Studies on heart sarcolemmal vesicles of opposite orientation and the effect of ATP on the $Na^{+}/Ca^{2+}$ exchange, *FEBS Lett.* **126:**74–76.

Rigoni, F., Mathien-Shire, Y., and Deana, R., 1980, Effect of ruthenium red on calcium efflux from rat liver mitochondria, *FEBS Lett.* **120:**255–258.

Robertson, S. P., Potter, J. P., and Rouslin, W., 1982, The $Ca^{2+}$ and $Mg^{2+}$ dependence of $Ca^{2+}$ uptake and respiratory function of porcine heart mitochondria. Probable physiological significance during the cardiac contraction–relaxation cycle, *J. Biol. Chem.* **257:**1743–1751.

Roman, I., Gmaj, P., Nowicka, C., and Angielski, S., 1979, Regulation of $Ca^{2+}$ efflux from kidney and liver mitochondria by unsaturated fatty acids and $Na^{+}$ ions, *Eur. J. Biochem.* **102:**615–623.

Roos, I., Crompton, M., and Carafoli, E., 1978, The effect of phosphoenol pyruvate on the retention of $Ca^{2+}$ by liver mitochondria, *FEBS Lett.* **94:**418–421.

Roos, I., Crompton, M., and Carafoli, E., 1980, The role of inorganic phosphate in the release of $Ca^{2+}$ from rat liver mitochondria, *Eur. J. Biochem.* **110:**319–325.

Rose, B., and Lowenstein, W. R., 1975, Calcium ion distribution in cytoplasm visualized by aequorin: Diffusion in cytosol restricted by energised sequestering, *Science* **190:**1204–1206.

Rossi, C. S., and Lehninger, A. L., 1964, Stoichiometry of respiratory stimulation, accumulation of $Ca^{2+}$ and phosphate and oxidative phosphorylation in rat liver mitochondria, *J. Biol. Chem.* **239:**3971–3980.

Rottenberg, H., 1975, Measurement of transmembrane electrochemical proton gradients, *Bioenergetics* **7:**61–74.

Rugolo, M., Siliprandi, D., Siliprandi, N., and Toninello, A., 1981, Parallel efflux of $Ca^{2+}$ and Pi in energised rat liver mitochondria, *Biochem. J.* **200:**481–486.

Sandri, G., Sottocasa, G., Panfili, E., and Liut, G., 1979, The ability of the mitochondrial $Ca^{2+}$-binding glycoprotein to restore $Ca^{2+}$ transport in glycoprotein depleted rat liver mitochondria, *Biochim. Biophys. Acta* **558:**214–220.

Saris, N., and Åkerman, K. E. O., 1980, Uptake and release of bivalent cations in mitochondria, in: *Current Topics in Bioenergetics,* Vol. 10 (D. R. Sanadi, ed.), Academic Press, New York, pp. 103–179.

Scarpa, A., and Azzone, G. F., 1970, The mechanism of ion translocation in mitochondria, *Eur. J. Biochem.* **12:**328–335.

Scarpa, A., and Graziotti, P., 1973, Mechanisms for intracellular calcium regulation in heart, *J. Gen. Physiol.* **62:**756–772.

Scarpa, A., Malmström, K., Chiesi, M., and Carafoli, E., 1976, On the problem of the release of mitochondrial calcium by cyclic AMP, *J. Membr. Biol.* **29:**205–206.

Scholz, H., 1980, Effects of beta and alpha-adrenergic activators and adrenergic transmitter releasing agents on the mechanical activity of the heart, in: *Adrenergic Activators and Inhibitors,* Part 1 (L. Szekeres, ed.), Springer-Verlag, Heidelberg, pp. 651–733.

Schotland, J., and Mela, L., 1977, Role of cyclic nucleotides in the regulation of mitochondrial calcium uptake and efflux kinetics, *Biochem. Biophys. Res. Commun.* **75:**920–924.

Selwyn, H. J., Dawson, A. P., and Dunnett, S. J., 1970, Calcium transport in mitochondria, *FEBS Lett.* **10:**1–5.

Settlemire, C. T., Hunter, G. R., and Brierley, G. P., 1968, Ion transport in heart mitochondria XIII. The effect of ethylenediaminetetraacetate on monovalent ion uptake, *Biochim. Biophys. Acta* **162:**487–499.

Siliprandi, D., Toninello, A., Zoccarato, F., Rugolo, M., and Siliprandi, N., 1975, Synergic action of calcium ions and diamide on mitochondrial swelling, *Biochem. Biophys. Res. Commun.* **66:**956– 961.

Sjödin, R. A., and Abercrombie, R. F., 1978, The influence of external cations and membrane potential on Ca-activated Na efflux in Myxicola giant axons, *J. Gen. Physiol.* **71:**453–466.

Sordahl, L. A., 1974, Effects of magnesium, ruthenium red and the antibiotic ionophore A-23187 on initial rates of calcium uptake and release by heart mitochondria, *Arch. Biochem. Biophys.* **167:**104–115.

Sottocasa, G. L., Sandri, G., Panfili, E., de Bernard, B., Gazzotti, P., Vasington, F., and Carafoli, E., 1972, Isolation of a soluble $Ca^{2+}$ binding glycoprotein from ox-liver mitochondria, *Biochem. Biophys. Res. Commun.* **47:**808–813.

Stucki, J. W., and Ineichen, E. A., 1974, Energy dissipation by calcium recycling and the efficiency of calcium transport in rat liver mitochondria, *Eur. J. Biochem.* **48:**365–375.

Taylor, W. M., Prpic, V., Exton, J. H., and Bygrave, F. L., 1980, Stable changes to calcium fluxes in mitochondria isolated from rat livers perfused with α-adrenergic agonists and with glucagon, *Biochem. J.* **188:**443–450.

Tew, W. P., 1977, Use of the coulombic interactions of the lanthanide series to identify two classes of $Ca^{2+}$ binding sites in mitochondria, *Biochem. Biophys. Res. Commun.* **78:**624–630.

Tew, W. P., Mahle, C., Benavides, J., Howard, J. E., and Lehninger, A. L., 1980, Synthesis and characterisation of phosphocitric acid, a potent inhibitor of hydroxyapatite crystal growth, *Biochemistry* **19:**1983–1988.

Thorn, N. A., Russel, J. T., and Robinson, I. C. A. F., 1975, Factors affecting intracellular concentrations of free calcium ions in neurosecretory nerve endings, in: *Calcium Transport in Contraction and Secretion* (E. Carafoli, F. Clementi, W. Drabikowski, and A. Margreth, eds.), North-Holland/Elsevier, Amsterdam, pp. 261–270.

Titheradge, M. A., and Coore, H. G., 1976, Hormonal regulation of liver mitochondrial pyruvate carrier in relation to glucoenogenesis and lipogenesis, *FEBS Lett.* **72:**73–78.

Tsien, R. W., 1977, Cyclic AMP and contractile activity in heart, in: *Advances in Cyclic Nucleotide Research* (P. Greengard and G. A. Robison, eds.), Raven Press, New York, pp. 363–421.

Vaghy, P. L., Johnson, D. J., Matlib, M., Wang, T., and Schwartz, A., 1982, Selective inhibition of $Na^{+}$-induced $Ca^{2+}$ release from heart mitochondria by diltiazem and $Ca^{2+}$ antagonist drugs, *J. Biol. Chem.* **257:**6000–6002.

Vasington, F. D., Gazzotti, P., Tiozzo, R., and Carafoli, E., 1972, The effect of ruthenium red on $Ca^{2+}$ transport and respiration in rat liver mitochondria, *Biochim. Biophys. Acta* **256:**43–54.

Vercesi, A., Reynafarje, B., and Lehninger, A. L., 1978, Stoichiometry of $H^+$ ejection and $Ca^{2+}$ uptake coupled to electron transfer in rat heart mitochondria, *J. Biol. Chem.* **253:**6379–6385.

Vinogradov, A., and Sarpa, A., 1973, The initial velocities of calcium uptake by rat liver mitochondria, *J. Biol. Chem.* **248:**5527–5531.

Waite, M., and Sisson, P., 1971, Partial purification and characterisation of the phospholipase $A_2$ from rat liver mitochondria, *Biochemistry* **10:**2377–2383.

Wehrle, J. P., and Pedersen, P. L., 1979, Phosphate transport in rat liver mitochondria, *J. Biol. Chem.* **254:**7265–7295.

Wikström, M., and Krab, K., 1980, Respiration-linked $H^+$ translocation in mitochondria: Stoichiometry and mechanism, in, *Current Topics in Bioenergetics,* Vol. 10 (D. R. Sanadi, ed.), Academic Press, New York, pp. 51–101.

Williams, J. N., 1968, A comparative study of cytochrome ratios in mitochondria from organs of the rat, chicken and guinea pig, *Biochim. Biophys. Acta* **162:**175–181.

Williams, A. J., and Fry, C. H., 1978, Calcium–proton exchange in cardiac and liver mitochondria, *FEBS Lett.* **97:**288–292.

Williams, R. S., and Lefkowitz, R. J., 1978, Alpha adrenergic receptors in rat myocardium. Identification by binding [$^3$H]-dehydro ergocryptine, *Circ. Res.* **43:**721–727.

Williamson, J. R., Cooper, R. H., and Hoek, J. B., 1981, Role of calcium in the hormonal regulation of liver metabolism, *Biochim. Biophys. Acta* **639:**243–295.

Wolkowicz, P. E., and McMillin-Wood, J. M., 1980, Dissociation between mitochondrial calcium ion release and pyridine nucleotide oxidation, *J. Biol. Chem.* **255:**10348–10353.

Yamada, E. W., Shiffman, F. H., and Huzel, N. J., 1980, $Ca^{2+}$-regulated release of an ATPase inhibitor protein from submitochondrial particles derived from skeletal muscle of the rat, *J. Biol. Chem.* **255:**267–273.

Yamada, E. W., Huzel, N. J., and Dickison, J. C., 1981, Reversal by uncouplers of oxidative phosphorylation and by $Ca^{2+}$ of the inhibition of mitochondrial ATPase activity by the ATPase inhibition protein of rat skeletal muscle, *J. Biol. Chem.* **256:**10203–10207.

Yamazaki, R. K., Mickey, D. L., and Storey, M., 1980, Rapid action of glucagon on hepatic mitochondrial calcium metabolism and respiratory rates, *Biochim. Biophys. Acta* **592:**1–12.

Zoccarato, F., and Nicholls, D. G., 1982, The role of phosphate in the regulation of the independent $Ca^{2+}$ efflux pathway of liver mitochondria, *Eur. J. Biochem.* **127:**333–338.

Zoccarato, F., Rugolo, M., Siliprandi, D., and Siliprandi, N., 1981, Correlated effluxes of adenine nucleotides, $Mg^{2+}$ and $Ca^{2+}$ induced in rat liver mitochondria by external $Ca^{2+}$ and phosphate, *Eur. J. Biochem.* **114:**195–199.

# 34

# Intestinal Phosphate Transport

Meinrad Peterlik

## I. INTRODUCTION

Functions of inorganic phosphate ($P_i$) differ widely in the vertebrate organism. This can be illustrated, e.g., by the role of phosphate in soft tissue on the one hand, where it serves mainly as an inorganic precursor of a large variety of essential organic phosphocompounds (phospholipids, nucleotides, metabolic intermediates, phosphoproteins, etc.), and by its occurrence in the skeletal system on the other hand, where it exists in crystalline form as hydroxyapatite together with calcium. $P_i$, in mineralized bone, makes up as much as 85% of total body phosphate. The remaining 15% is accounted for by soft-tissue $P_i$ and is about equally distributed between muscle and other nonosseous tissue (Bringhurst and Potts, 1979).

Since such divergent forms as intracellular $P_i$ in soft tissue as well as $P_i$ deposited in bone both require the maintenance of a certain plasma concentration for proper function, a homeostatic system has to be envisaged that would include the coordinate regulation of absorption and excretion as well as the exchange of soluble with crystalline skeletal $P_i$, and which would thereby eventually maintain plasma $P_i$ within certain limits. In fact, plasma concentrations of $P_i$, although they exhibit variations to some degree, ranging, e.g., in healthy adults between 0.81 and 1.34 mM (Wertheim *et al.*, 1954), and thus might appear not to be regulated as strictly as those of its counterpart calcium, are nevertheless by no means fortuitous but rather the result of an intricate interplay of hormonally controlled transport mechanisms in intestine, kidney, and bone. Homeostasis of phosphate, just as of calcium, is brought about in general by regulatory mechanisms involving mainly the actions of vitamin D and parathyroid hormone on their respective target organs (for overview see Bronner and Peterlik, 1981).

Because of the importance of inorganic phosphate for a variety of cellular and organ functions, the organism must have the ability to conserve $P_i$ in efficient ways.

---

*Meinrad Peterlik* • Department of General and Experimental Pathology, University of Vienna Medical School, A-1090 Vienna, Austria.

This is achieved at the level of excretion by adaptation of tubular reabsorption in the kidney (Steele, 1977), as well as in the intestine, where $P_i$ conservation is aided by specific epithelial transport processes that are under the control of vitamin D. Thereby, an adequate supply of $P_i$ can be secured according to the needs of the organism.

Impairment of intestinal phosphate absorption by whatever reason inevitably leads to a state of phosphate depletion which is characterized by major pathological alterations such as skeletal abnormalities, cardiac and central nervous dysfunction, as well as disturbances of renal and electrolyte metabolism among others (Massry, 1978). On the other hand, uncontrolled uptake of phosphate from the intestine, when not matched by renal excretion, is the cause of the hyperphosphatemia encountered as a major complication of chronic kidney failure.

The far-reaching consequences that any perturbation of intestinal $P_i$ absorption can have on the whole organism certainly warrant intensive research on the basic modes of $P_i$ transfer across the gut wall. This review is an attempt to summarize our present knowledge on the specificity, localization, development, and hormonal control of pathways of epithelial $P_i$ transport in the intestine.

## II. INTESTINAL SITES AND MODES OF PHOSPHATE ABSORPTION

A number of investigations have been carried out in the past on the mode of $P_i$ absorption, e.g., active vs. passive transport, at different sites of the intestine. The results of these studies are not readily comparable because they differ widely not only in the experimental methods employed which include various *in vivo* and *in vitro* techniques, but also with respect to experimental animals, their vitamin D status, and conditions under which $P_i$ transport was measured, such as $P_i$ concentration and electrolyte composition of perfusion or incubation buffers. Of these studies, only a few (which in addition were limited to the small intestine of a single species, either chicks or rats) have evaluated $P_i$ absorption with respect to different anatomical locations in a truly comparative manner. However, by collation of all available data on $P_i$ absorption, including also those on the large intestine and those obtained from other species, a reasonable inference can be made on some general features of $P_i$ absorption along the intestine.

### A. $P_i$ Transport in Small and Large Intestine

In the chick and the rat, inorganic phosphate can be absorbed along the entire length of the intestine, though it is quite obvious that the small intestine (MacHardy and Parsons, 1956; Harrison and Harrison, 1961; Hurwitz and Bar, 1972; Wasserman and Taylor, 1973; Walling, 1978; Peterlik and Wasserman, 1978a) surpasses the large bowel by far in its absorptive capacity (Skadhauge and Thomas, 1979; Lee *et al.*, 1980). This can be explained by the fact that the colon in contrast to the small intestine lacks the possibility of active $P_i$ transfer in the lumen-to-blood direction and, thus, any translocation is possible only within the limitations imposed on passive movements of $P_i$ by the "tightness" of the colonic epithelium (for discussion see Section II-D).

Considerable differences in $P_i$ absorptive capacity have been described in the various portions of the small intestine. There is, however, compelling evidence that $P_i$ is least effectively cleared from the ileum of chicks (Peterlik and Wasserman, 1978a; Blahos and Care, 1981), rats (Harrison and Harrison, 1961; Walling, 1977, 1978), and sheep (Ben-Ghadalia *et al.*, 1975). In these species, phosphate is most readily absorbed in the proximal small intestine, where higher rates of $P_i$ transfer are consistently observed in the jejunum as compared to the duodenum. In humans, the jejunum was also defined as the intestinal site most active in $P_i$ absorption (Juan *et al.*, 1976; Walton and Gray, 1979).

## B. Influence of Vitamin D

Another aspect concerns the influence of vitamin D on the rate of $P_i$ absorption in relation to intestinal location. In absorption measurements *in vivo* (Hurwitz and Bar, 1972) as well as in studies utilizing everted gut sacs from the small intestine of 4-week-old chicks (Peterlik and Wasserman, 1975, 1978a), the highest response to vitamin D repletion was observed in the jejunum, while the duodenum, showing an already high level of $P_i$ transfer in the vitamin D-deficient state, could be less stimulated; again, the lowest rate of $P_i$ transfer was found in the ileum and it rose only marginally in response to vitamin D. Also in the rat, vitamin D-related increments of $P_i$ transport exhibit the same variations with respect to specific intestinal sites. This could be inferred already from data reported by Harrison and Harrison (1961) on the influence of vitamin D on $P_i$ concentrative transfer in everted intestinal segments, and was later confirmed through direct evidence from $P_i$ flux measurements in short-circuited gut segments (Walling, 1977, 1978).

In the colon, $P_i$ transport was not found to be inducible by the active vitamin D metabolite 1,25-dihydroxyvitamin $D_3$ (Lee *et al.*, 1980).

## C. Transepithelial $P_i$ Transport: Transcellular Pathways

It is obvious that the lack of effect by vitamin D on colonic $P_i$ transport cannot be due at all to some unresponsiveness of the epithelium towards vitamin D, since at the same time the sterol is able to induce calcium transport in the distal intestine (Lee *et al.*, 1980). This suggests that only the active component of transepithelial $P_i$ transport, which apparently does not exist in the colon, may be subject to the action of vitamin D. In fact, early investigations on intestinal $P_i$ absorption were very much focused on the question of whether $P_i$ is transferred across the gut wall mainly by active or passive means, and, consequently, which mode of transfer was affected by vitamin D (Morgan, 1969; Avioli and Birge, 1977). The data of MacHardy and Parsons (1956) obtained from *in situ* perfusion of rat intestinal segments appeared to suggest that $P_i$ leaves the gut lumen only by nonsaturable transfer, thus favoring the assumption that $P_i$ absorption was solely the result of passive diffusion out of the intestinal lumen. In perfusion studies of human jejunum and ileum, Juan *et al.* (1976) and Walton and Gray (1979) observed a linear concentration dependence of $P_i$ absorption when the test solution contained 5–16 mM $P_i$. There is no doubt that, particularly at high intraluminal concentrations, $P_i$ is absorbed mainly by simple diffusion. However, other laboratories (Jacobi *et al.*, 1958; Pfleger *et al.*, 1958; Harrison and Harrison, 1961, 1963; Taylor, 1974; Chen *et al.*, 1974) convincingly demonstrated that small intestinal segments from various species (rat, chick, guinea pig) are capable also of concentrative transepithelial $P_i$ transfer which depends on cellular energy, shows considerable temperature sensitivity, and operates only in the presence of $Na^+$. Thus, in addition to transfer by diffusion, active transmural $P_i$ transport had definitely to be considered to exist in the small intestine.

### 1. *Active and Passive Transport Across Enterocyte Membranes: Theoretical Considerations*

As for epithelia in general, two principal pathways of solute transfer can be defined in the intestine. The paracellular path involves transfer across specialized intercellular structures, so-called "tight junctions", thereby enabling the direct exchange of a solute between the lumen and the serosal extracellular space (for details see Section II-D). The transcellular route in the lumen-to-blood direction encompasses the successive transfer of a solute (1) across the brush-border membrane of enterocytes (absorptive, columnar cells), (2) migration through the cellular interior to the contralateral aspect of the cell, and (3) exit across the basolateral membrane of the enterocyte into the interstitial serosal space. (The arrangement of transport compartments and separating membranes is illustrated in Figure 1.) Owing to the specific localization of the paracellular shunt between two adjacent cells (see also Figure 1), this pathway can display only characteristics of diffusional transfer, whereas the transcellular route is a composite of active and passive transmembrane movements alike, depending on the concentration and potential differences prevailing at the particular aspects of the enterocytic cell membrane.

Since an electrical potential difference (PD) in a two-compartmental system causes an asymmetrical distribution of any ion on both sides of the separating structure, a certain contribution therefrom to the concentrative transfer of an ion across either a single cell membrane or even an entire epithelium can be expected. Available data for the small intestine from *in vivo* and *in vitro* measurements in experimental animals and in humans (see Wasserman *et al.*, 1961, 1966; Schultz, 1979 for review) indicate a relatively small PD, only 3 mV (mucosa side negative) on the average, with variations between 1 and 8 mV (Figure 1). Using a form of the Nernst equation describing the mutual dependence of PDs and concentration gradients (PD [volts] $= -0.059/n \times \log c_1/c_2$, where $n$ is the number of electrical charges of the ion, and $c_1$ and $c_2$ are the concentrations in the respective, e.g., mucosal and serosal, compartments), one can calculate that the highest reported PD of 8 mV would be responsible for a serosa/mucosa concentration gradient of 1.87 in case of $HPO_4^{2-}$, or of 1.37, if only $H_2PO_4^-$ would be transferred across the small-intestinal epithelium. Any values obtained for serosa/mucosa $P_i$ concentration gradients exceeding those above would be indicative of the involvement of an active transfer step in the course of mucosa-to-serosa $P_i$ transport, though no conclusion can be drawn therefrom on its exact location. It had to be determined, therefore, whether the mucosal, i.e., brush-border, or the serosal,

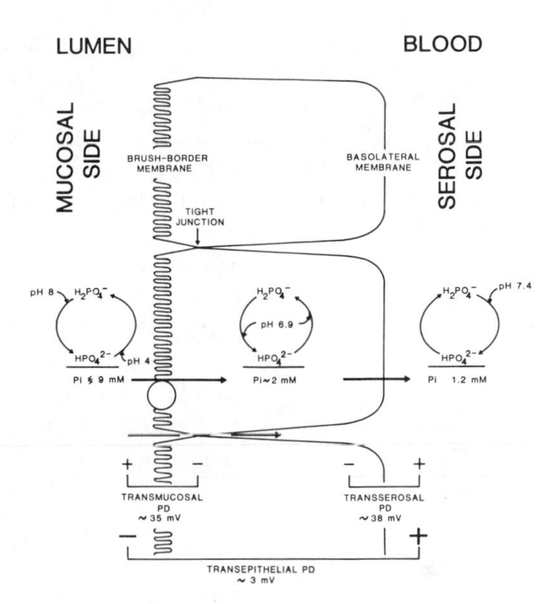


*Figure 1.* Compartmental model of absorptive $P_i$ transport under *in vivo* conditions.

i.e., basolateral, membrane, which $P_i$ has to traverse sequentially in the course of its transcellular migration, carries an active transport mechanism responsible for generation of serosa-to-mucosa $P_i$ gradients.

At any rate, the particular mode by which $P_i$ is actually transferred across the mucosal and serosal membrane is determined by both the individual PDs and the concentration gradients prevailing at these membranes.

a. *Relevant $P_i$ Concentrations in Transport Compartments.* A key role in these processes is played by the intracellular steady-state $P_i$ concentration since this determines largely whether $P_i$ can leave or enter one of the extracellular compartments along or against a concentration gradient under actual *in vivo* conditions. The concentration of intracellular $P_i$ available for exchange with the extracellular compartments has been estimated to about 2–3 mM (Morgan, 1969). Since the p$K$ of the dissociation reaction $H_2PO_4^- = H^+ + HPO_4^{2-}$ is 6.9 (Walser, 1961), and an identical value has been calculated for the intracellular pH of small-intestinal absorptive cells (Jackson and Kutcher, 1977), both phosphate anion species certainly coexist within the enterocyte in a 1 : 1 equilibrium (Figure 1).

Considering the importance of extra/intracellular concentration gradients for the mode by which transmembrane $P_i$ transfer is eventually achieved, actual *in vivo* concentrations on both sides of the gut wall are of particular interest.

The luminal $P_i$ concentration is of course expected to depend widely on the amount of $P_i$ consumed with the diet and, in addition, will show diurnal variations depending on eating habits. In the fasting state, the minimal $P_i$ concentration in the lumen is determined by the $P_i$ content of the endogenous secretions which equally amounts to 0.9 mM in gastric and pancreatic juice as well as in bile (Demand *et al.*, 1968; Lewis, 1973). Assuming that, in healthy adults, the average $P_i$ intake of 30 mmoles/day

(Bringhurst and Potts, 1979) is contained in an average fluid volume of 3400 ml entering the duodenum in 24 hr (Soergel and Hofmann, 1972), the mean $P_i$ concentration in the gut lumen would be approximately 9 mM. Although peak concentrations certainly exceed this value, one has to take into account that this figure pertains to total $P_i$, and that free phosphate available for absorption may represent only a small fraction thereof due to protein binding and complexing with calcium and magnesium ions.

The pH profile of the small intestine (Meldrum *et al.*, 1972), which changes from slightly acidic to more alkaline values in the proximal-to-distal direction, induces a change in the relative proportion of mono- and divalent $P_i$ anions. Accordingly, with the exception of the upper duodenum, where neutralization of acidic gastric contents is not complete and $H_2PO_4^-$ is therefore expected to be present in a larger proportion, in the greater part of the small intestine, the equilibrium is shifted to the $HPO_4^{2-}$ form, which at the average intraluminal pH of 7.5 may represent 80% of "ionized" inorganic phosphate (Figure 1).

The ionic composition of the interstitial fluid which bathes the serosal side of the epithelium is generally believed to be equal to that of plasma. Hence, the mean plasma value of 1.2 mM total $P_i$ (Wertheim *et al.*, 1954; Walser, 1961) is valid also for the serosal compartment. A normal pH of 7.4 could shift the dissociation equilibrium of phosphate anions to near 80% $HPO_4^{2-}$.

*b. Transmembrane Potential Differences.* From measurements of membrane potentials in enterocytes of various species, it seems that the transmucosal PD in small intestinal cells in general is about −35 mV, inside negative (Rose and Schultz, 1971; Schultz, 1979; Cremaschi *et al.*, 1982). Considering the fact that enterocytes even exhibit a variation in their membrane potential between −18 and −70 mV associated with their differentiation along the villus axis (Tsuchiya and Okada, 1982), an average transmucosal PD of −35 mV seems a reasonable assumption. Given an average transepithelial PD of 3 mV, mucosa side negative, this would imply that the transserosal PD differs only slightly from the transmucosal value (Figure 1).

*c. A Model of Actual Transmembrane $P_i$ Fluxes.* The above estimates of the intracellular exchangeable $P_i$ concentration and of the transmembrane PDs have been used to construct a model which would allow prediction of certain essential features of $P_i$ transfer across the intestinal epithelium (Figure 1).

The direction of the transmucosal PD, which is negative with respect to the interior of the cell, opposes in principle the entry of both anions, $HPO_4^{2-}$ and $H_2PO_4^-$, from the lumen into the epithelial cells. Actually, according to the Nernst equation (see before) a PD of −35 mV would require the concentration of $HPO_4^{2-}$ in the lumen to be 15 times higher than within the enterocyte to allow for passive out-of-lumen movement. If $H_2PO_4^-$ is the prevailing phosphate anion, an outside/inside concentration gradient of at least 4 would be necessary to overcome the opposing force of the PD. From this point of view, the supposed active transport mechanism ought to reside at the brush-border membrane of absorptive intestinal cells to overcome the restraints which the existence of a PD of this magnitude puts on $P_i$ entry, and to mediate $P_i$ transport against the transmucosal electrochemical gradient.

Similar considerations pertain to the mode of $P_i$ transfer across the basolateral membrane, except that the direction and the degree of the transserosal PD would favor $P_i$ diffusion out of the cell into the interstitial space, even if outside $HPO_4^{2-}$ or $H_2PO_4^-$ rose to 15, or fourfold higher than intracellular levels.

### 2. *Active and Passive Transport across Enterocyte Membranes: Experimental Validation*

The "everted gut sac" technique, originally introduced by Wilson and Wiseman (1954), was used to study unidirectional $P_i$ movements across both the mucosal and serosal boundaries of chick small-intestinal epithelium (Peterlik and Wasserman, 1975, 1977, 1978a). The results of this study verified the predictions based on the model of transcellular $P_i$ transport (Figure 1), that $P_i$ enters the cell from the lumen mainly by active transport and can leave it on the basolateral side on a diffusional pathway.

*a. $P_i$ Entry into the Epithelium.* Selected experimental conditions allowed total suppression of $P_i$ net movement across the serosal surface of everted gut segments. Thus, $P_i$ transfer across the mucosal surface of the epithelium could be studied under steady-state conditions at its contralateral boundary. $P_i$ influx was determined by analysis of $^{32}P_i$ tracer fluxes, while $P_i$ movement in the opposite direction was calculated from the appearance of endogenous stable $P_i$ in the outside solution.

Not unexpectedly, endogenous $P_i$ leaked into the outside (mucosal) solution, and this process was increased by incubation under anaerobic conditions. Conversely, $P_i$ influx was suppressed at the same time. Apparently, cellular energy is necessary for $P_i$ uptake from the lumen. In agreement with the requirements for active carrier-mediated transport, this flux exhibits saturation kinetics with a $K_m$ of 0.2 mM $P_i$. In combination, these experimental data provide firm proof for the existence of a high-affinity, active $P_i$ transport system at the epithelial brush border.

It is important to note that active transmucosal $P_i$ transport proceeds in the absence of extracellular calcium and is greatly stimulated when vitamin D-deficient chicks are repleted with the hormone. In addition, this pathway was thereby identified as the only vitamin D-regulated transfer step at the brush-border membrane (Figure 2).

*b. $P_i$ Exit across the Serosal Epithelial Boundary.* Unlike the mucosal side, the basolateral surface of the epithelium is shielded from direct exposure to the serosal buffer compartment by the serosal muscular and connective tissue layers; therefore, $P_i$ fluxes out of enterocytes into the serosal interstitium cannot be measured directly. However, some indirect clues on possible modes of $P_i$ exit could be obtained.

In fact, it is obvious that $P_i$ transferred across the serosal surface originates in principle from two distinct intracellular pools, one of which comprises the bulk $P_i$ initially taken up from the gut lumen (transport pool), while the other one may contain endogenous metabolic $P_i$ in an unspecified intracellular location (Figure 2).

On incubation of everted sacs from chick jejunum, endogenous $P_i$ is immediately released into the serosal compartment. The magnitude of this flux, which certainly receives a contribution also from nonepithelial cells, is not changed when $P_i$ net absorption across the contralateral mucosal border is induced by vitamin D. In addition, this type of $P_i$ flux from endogenous sources is in itself not influenced by vitamin D, does not require cellular energy and, hence, confirms the predicted $P_i$ leak across the serosal epithelial border.

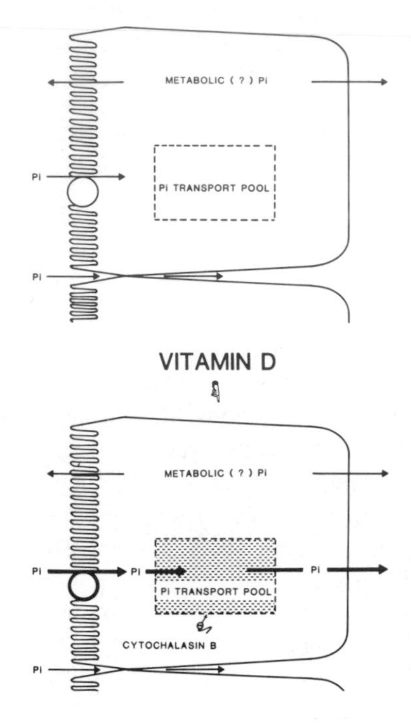


*Figure 2.* Induction of intestinal $P_i$ absorption by vitamin D. Upper part, $P_i$ fluxes in the absence of vitamin D; lower part, $P_i$ fluxes stimulated by vitamin D.

When the serosal concentration is adjusted to 4.0 mM $P_i$, net exchange at this side of the epithelium is zero. From measurements of $P_i$ equilibrium fluxes under this condition, no evidence could be obtained that reentry of $P_i$ into the cellular compartment is affected by vitamin D.

Considering the magnitude of the transserosal PD (Figure 1), it can be calculated that the intracellular pool in equilibrium with 4.0 mM serosal $P_i$ contains approximately 0.6 mM $P_i$. This is less than half of the estimate for intracellular exchangeable $P_i$ (Morgan, 1969; Figure 1) and might be explained by the fact that endogenous $P_i$ in this particular pool, which is most probably of metabolic origin, is not identical, and apparently does also not mix with $P_i$ contained in the "transport pool" specified below.

### 3. *Intracellular Transport*

The term "channelized transport" (Kowarski and Schachter, 1969; Morgan, 1969) was coined to indicate a specific mode of intracellular migration. $P_i$ taken up from the lumen does not readily mix with intracellular metabolic $P_i$ and is straightforwardly transported to the basolateral membrane of the enterocyte (Figure 2). This assumption is based on the following observations. First, in isolated rat or chick small intestine, only a small percentage of initially intraluminal $P_i$ is incorporated into organic phosphocompounds, while almost the entire amount of $P_i$ transferred into the cellular interior can be recovered as nonmetabolized $P_i$ from the cytoplasm (Kowarski and Schachter, 1969; Wasserman and Taylor, 1973). Second, there are obvious discrepancies in intracellular specific activity, depending on whether radiolabeled $P_i$ was introduced from the mucosal or serosal side. In chick as well as in rat small intestine, higher specific labeling of intracellular $P_i$ was achieved when $^{32}P_i$ was added to the mucosal rather than to the serosal solution (Kowarski and Schachter, 1969; Morgan, 1969). Further evidence for "channelized transport" of $P_i$ in a discrete "transport pool" came from investigations on a possible influence of vitamin D on particular pathways of transepithelial $P_i$ transport (Peterlik and Waserman, 1977, 1978a). In everted jejunum of vitamin D-deficient chicks, the amount of $P_i$ which is actively transferred into the cell matches the flux in the opposite direction so that net change is about zero. Repletion with vitamin D enhances the brush-border active transport system, thereby initiating net absorption from the lumen. This leads, after a characteristic lag time of approximately 15 min, to the appearance of bulk mucosal $P_i$ in the serosal compartment. Under the same experimental conditions, calcium is transferred in about 5 min (Fuchs and Peterlik, 1979b). This can be interpreted that not all solutes taken up simultaneously from the lumen move freely to the serosal aspect of the plasma membrane within the same time interval. The migration of some, like of $P_i$, might proceed in a different manner due to their designation to a specific transport pool from where they are transferred by "channelized transport" to the basolateral membrane for their final discharge into the serosal interstitial space (Figure 2).

It should be noted that this specific intracellular pathway cannot be utilized in the vitamin D-deficient state, even though some $P_i$, as mentioned above, reaches the intracellular space from the lumen. This is just enough to compensate for the leak in the opposite direction so that $P_i$ is not available for the transport pool (Figure 2).

At present, intracellular transport of $P_i$ is considered to proceed through a "black box" since nothing is known of its specific location and whether particular cellular organelles, membrane systems, or transport proteins are involved. The existence of a specific $P_i$-binding protein comparable to the vitamin D-dependent calcium-binding protein (Wasserman and Taylor, 1966), which could be implicated in vitamin D-induced intracellular migration of $P_i$, is unlikely, since Wasserman and Taylor (1973) found no evidence for any particular $P_i$-binding activity in the cytoplasm of intestinal cells from vitamin D-replete chicks.

A hint on the participation of cytoskeletal structures in intracellular $P_i$ transport can be taken from the effect of inhibitors of microfilamentous function on transcellular $P_i$ transport (Fuchs and Peterlik, 1979a). Cytochalasin B does not inhibit active $P_i$ entry into the cell but particularly blocks its transcellular migration induced by vitamin D (Figure 2) at a concentration which is known to cause disruption of microfilamentous structures, although another mechanism of action cannot be excluded.

## D. Transepithelial $P_i$ Transport: Paracellular Pathway

Until now, it has been tacitly assumed that all transepithelial $P_i$ transport is exclusively absorptive, i.e., in the lumen-to-blood direction, and proceeds thereby only on a transcellular route. This is probably due to the fact that the first step in the $P_i$ absorptive process had been theoretically and experimentally identified as an active transport mechanism at the brush-border membrane, and this could lead to the belief that intraluminal $P_i$ is entirely ferried on the transcellular route. However, from a theoretical point of view, the paracellular route should be considered as an alternative transepithelial pathway.

It is inherent in the location of this pathway through the "tight junctions" of adjacent epithelial cells that the exchange of small ions and molecules between the lumen and the serosal extracellular space can only be by passive means. Although the transepithelial PD is positive with respect to the serosal side and would thus favor the diffusion of anions across the epithelium (Figure 1), it remains doubtful whether this shunt path could be efficiently used for $P_i$ absorption. Although a definite change in the permeability of the so-called "tight junctions" along the intestine provides the basis for classification of the small intestine as "leaky" and the large bowel as "tight" epithelium, this pertains only to the permselectivity of cations. It is believed that "fixed" negative charges, which facilitate the permeation of cations, would inhibit anion transfer even in leaky epithelia regardless of the ionic radius which, as a matter of fact, in case of $P_i$ anions could be an additional obstacle for their migration on the shunt pathway. Actually, Peterlik and Wasserman (1978a) were unable to demonstrate any significant leak of $P_i$ from the serosal into the mucosal compartment of everted chick small intestine even if the imposed serosa/mucosa concentration gradient was sufficiently high to overcome the opposing PD. This provides evidence that a substantial secretory backflux of $P_i$ occurs under no circumstances, neither on the transcellular nor on the paracellular route. This would have been expected if transfer of $P_i$ on the shunt pathway would be feasible at least to some extent. In another study, Fuchs and Peterlik (1979b, 1980a) confirmed that only a minute amount of $P_i$ could enter the serosal compartment from the lumen on a diffusion pathway. This type of transfer, which by exclusion of other possibilities was tentatively assigned to the shunt pathway, is not influenced by vitamin D. Actually, it may constitute the only route of $P_i$ absorption in the vitamin D-deficient state (Figure 2).

## E. Absorption vs. Secretion

In the preceding sections, transepithelial $P_i$ transport in the intestine has been treated only with respect to absorption, i.e., in the lumen-to-blood (mucosa-to-serosa) direction. Secretory pathways in the opposite direction were disregarded because it remains doubtful whether a substantial backflux of $P_i$ into the gut lumen on any other than a paracellular route can occur at all. When intestinal loops are perfused *in vivo* with a $P_i$-free solution, secretion of $P_i$ into the perfusion fluid is minimal (MacHardy and Parsons, 1956). This might be explained by several facts. First, the transepithelial PD, as already pointed out, does not favor passive movements of anions in the blood-to-lumen direction. Second, even if the electromotive force of the PD is counteracted by an appropriate serosa/mucosa concentration gradient, which, however, is unlikely to occur *in vivo* (concentrations shown in Figure 1), evidence for the existence of efficient secretory pathways is inconclusive. Paracellular transfer is certainly inefficient as reasoned before, and whether the transcellular route is feasible for secretory $P_i$ transfer remains to be established. A prerequisite for this would be the existence of an active transport mechanism at the basolateral membrane of the enterocyte to mediate

$P_i$ reentry against an electrochemical gradient. A low-affinity transport system has been detected in isolated intestinal cells which differs from the high-affinity system of the brush border with respect to its insensitivity to $Na^+$ and its dependence on $Ca^{2+}$ (Avioli *et al.*, 1981; Birge and Avioli, 1981). It is possible that this transport system resides not only at the brush border but also at the basolateral membrane. Since it can be activated by vitamin D, it is conceivable that its genuine function is primarily in facilitating $P_i$ diffusion out of the cell. It remains to be established whether this transport mechanism, if activated by an appropriate energy source, could mediate the transfer of $P_i$ also into the opposite direction.

Passive efflux of $P_i$ from the cell across the brush border has been shown to occur to some degree (Peterlik and Wasserman, 1978a; Section II-C-2). In the same series of experiments, it was also shown that serosal $P_i$, though it was taken up by epithelial cells, cannot be exchanged with the lumen. This apparent compartmentalization of $P_i$ having entered the cell from the basolateral side probably also underlies the observations by Kowarski and Schachter (1969) and Morgan (1969) of different labeling of intracellular $P_i$ by serosal and mucosal $^{32}P_i$, and certainly constitutes a severe limitation for secretory $P_i$ transport in the intestine on the transcellular route.

## III. ACTIVE TRANSPORT OF INORGANIC PHOSPHATE ACROSS THE BRUSH-BORDER MEMBRANE

Past studies on $P_i$ transport in the intestine were done primarily with the intention to understand the events underlying the stimulatory action of vitamin D on $P_i$ transfer across the gut wall. Although this effect of the sterol has been known for a long time (Nicolaysen, 1937b), it was believed that increased transfer of $P_i$ was somehow a consequence of vitamin D-induced calcium absorption (Nicolaysen, 1937c; Harrison and Harrison, 1961). In a number of studies, however, evidence to the contrary was presented as it was convincingly demonstrated that vitamin D-stimulated $P_i$ transfer could proceed without calcium moving in parallel (Kowarski and Schachter, 1969; Wasserman and Taylor, 1973; Taylor, 1974; Chen *et al.*, 1974; Birge and Miller, 1977). There is now unanimous agreement that the transepithelial transfer of calcium and inorganic phosphate involves separate pathways which, in consequence, ought to be stimulated independently by vitamin D through different mechanisms of action.

Once the independent action of vitamin D on $P_i$ absorption had been established, it became clear very soon that the sterol increased unidirectional $P_i$ flux in the lumen-to-blood direction in rat and chick small intestine (Kowarski and Schachter, 1969; Wasserman and Taylor, 1973) and had no effect on the opposite flux. Peterlik and Wasserman (1975, 1977, 1978a) then studied the influence of vitamin D on transmucosal and transserosal $P_i$ unifluxes and were able to pinpoint its action on a single transmembrane flux, namely, the uptake step across the enterocyte brush-border membrane. The intrinsic properties of this particular transfer system turned out to be the essential determinants of the features of intestinal transport of inorganic phosphate, viz., requirement for cellular energy, $Na^+$ dependence, stimulation by vitamin D, and independence from extracellular calcium.

## A. Transmucosal $P_i$ Transport in Intact Cells

In everted gut sacs from vitamin D-deficient chicks, $P_i$ transfer across the brush-border membrane is reduced by inhibitors and uncouplers of respiration, and, in agreement with the requirements for an active carrier-mediated transport system, shows saturation kinetics of the Michaelis–Menten type with an apparent $K_m$ of 0.2 mM $P_i$ (Peterlik and Wasserman, 1975, 1977, 1978a; Fuchs and Peterlik, 1979b). Repletion of deficient animals with vitamin $D_3$ or with its biologically active metabolite, 1,25-dihydroxyvitamin $D_3$, increases the maximal velocity of active transmucosal $P_i$ transport more than twofold. As $K_m$ was not changed by vitamin D, it has to be concluded that the sterol aids the absorption of $P_i$ solely by increasing the number of functional carrier complexes without affecting their affinity towards $P_i$.

Vitamin D stimulation of a high-affinity $Na^+$-dependent $P_i$ transport system at the luminal aspect of the small-intestinal epithelium to initiate transepithelial $P_i$ transfer is a mechanism common to all species so far investigated. As another general feature, all the brush-border $P_i$ transport mechanisms, irrespective of animal species or location in the small intestine, show a remarkable uniformity in their affinity towards $P_i$ ($K_m$ 0.1–0.2 mM $P_i$) which, notably, is not changed by vitamin D. No effect of the sterol on transport kinetics other than on $V_{max}$ has been reported so far. In chicks, this mode of vitamin D action has been additionally observed in embryonic small intestine maintained in organ culture (Peterlik, 1978a,b; Cross and Peterlik, 1982b). Walling (1978) presented evidence that administration of 1,25-dihydroxyvitamin $D_3$ [1,25-$(OH)_2D_3$] to vitamin D-deficient rats leads to stimulation of active, $Na^+$-dependent $P_i$ influx across the brush border in the jejunum as measured under short-circuit conditions in an Ussing chamber. The $K_m$ of this transport system was found to be invariably at 0.1 mM $P_i$, whereas its $V_{max}$ increased about twofold upon vitamin D action (Kabakoff *et al.*, 1982). In ileal explants of the same species, Birge and Miller (1977) described a saturable transport system which is regulated by vitamin D solely through an effect on its $V_{max}$. $K_m$ (0.125 mM) is not changed after 25 hydroxyvitamin $D_3$, as a biologically active metabolite of vitamin $D_3$, is added to the culture system. Danisi and Straub (1980) reported that, in the rabbit, $Na^+$-dependent phosphate transport can be demonstrated only in the duodenum. The $V_{max}$ but not the $K_m$ of this system was reduced when the animals were treated with disodium ethane-hydroxy-1,1-diphosphonate to make them vitamin D-deficient by blocking the endogenous synthesis of 1,25-$(OH)_2D_3$. Administration of this metabolite restored the $V_{max}$ to its normal value (Danisi *et al.*, 1980).

Since $Na^+$-dependence of intestinal $P_i$ transport has been reported previously by several authors (Harrison and Harrison, 1963; Taylor, 1974; Birge and Miller, 1977) a more detailed study was carried out on the influence of extracellular $Na^+$ on the kinetics of active $P_i$ uptake (Peterlik 1978a,b; Fuchs and Peterlik, 1979b). In the absence of $Na^+$, $P_i$ movement into the epithelium is reduced to the level of nonsaturable flux, which is equally low in vitamin D-deficient and replete chicks. Although saturable $P_i$ uptake basically depends on the presence of $Na^+$ in the extracellular fluid, the interaction of $Na^+$ with the carrier seems to be different in vitamin D-independent and vitamin D-induced $P_i$ active transport. While in the former, a change in $K_m$ suggests an increased carrier affinity for $P_i$ as a result of $Na^+$ influence, the vitamin D-induced transport system, in contrast, displays a $Na^+$-dependent increase of its maximal velocity which may indicate a higher mobility of the carrier–$P_i$ complex within the cell membrane.

In addition to the $Ca^{2+}$-independent, $Na^+$-dependent high-affinity $P_i$ transport system extensively described before, Birge and co-workers presented evidence for the existence of a $Ca^{2+}$-dependent, $Na^+$-insensitive low-affinity $P_i$ transport system in isolated chick small-intestinal cells (Birge and Avioli, 1981; Avioli *et al.*, 1981). $P_i$ uptake via this mechanism can be augmented by biologically active vitamin D metabolites such as 25-hydroxyvitamin $D_3$. A close association of this additional $P_i$ absorptive mechanism with a $Ca^{2+}$-stimulated alkaline phosphatase can be inferred from the identity of enzyme and transport activities with respect to ionic dependence, pH optimum, as well as insensitivity to L-phenylalanine (see also Section III-C.3). From the distribution of the $Ca^{2+}$-activated phosphatase at different membrane sites of the enterocyte, a brush-border localization of the associated $P_i$ transport system is strongly suggested, although both activities may reside to a certain degree on the basolateral membrane as well. Knowledge of the exact localization of this second $P_i$ absorptive mechanism will certainly help to clarify its role in transepithelial $P_i$ transport.

## B. $Na^+$-Gradient-Driven $P_i$ Transport in Brush-Border Membrane Vesicles

Studies with isolated membrane vesicles clearly established $P_i$ active transport at the brush border as a $Na^+$-coupled transfer step and brought further confirmation for the specific mode of vitamin D action on its kinetic properties. Berner *et al.* (1976) first demonstrated accumulation of $P_i$ by rat duodenal brush border vesicles in the presence of a $Na^+$ gradient directed to their interior. Uphill flux of $P_i$ clearly ceased with the $Na^+$ gradient dissipating so that initial accumulation exceeded the equilibrium value. This "overshoot" phenomenon is commonly believed to indicate a "secondary active" mode of transport of a solute inasmuch as its translocation is coupled to a concomitant flux of $Na^+$ and, consequently, derives its energy from an appropriately directed electrochemical $Na^+$ gradient. Since vesicle preparations in this study were derived only from normal, vitamin D-replete rats, no information was provided on particular effects of vitamin D on $Na^+$-coupled $P_i$ transport in the small intestine.

### 1. Transport Kinetics and Vitamin D

As an isolation procedure for brush-border vesicles from chick small intestine (Max *et al.*, 1978) became available, Fuchs and Peterlik (1980b) and, independently, Matsumoto *et al.* (1980) studied the effect of vitamin D on $Na^+$-coupled $P_i$ translocation across the brush-border membrane. In addition, Hildmann *et al.* (1982) provided information on vitamin D-induced $Na^+$-dependent $P_i$ uptake by rabbit duodenal brush-border membranes. In general, vesicles derived from the duodenum or jejunum of animals repleted with vitamin D showed higher rates of $Na^+$-gradient-driven $P_i$ accumulation than those from the vitamin D-deficient control group.

Although secondary active $P_i$ transport is abrogated at equal sodium concentrations on both sides of the vesicular membrane (since, by the absence of an appropriate $Na^+$ gradient, the system is deprived of its driving force), $P_i$ uptake into vesicles still occurs with a higher velocity than in the total absence of $Na^+$.

It is noteworthy that a stimulatory effect of vitamin D can still be observed even when the carrier operates in the nonenergized state. Two important conclusions can be drawn from this observation with respect to the nature of vitamin D action on $Na^+$-coupled $P_i$ transport. First, the action of the sterol is not exclusively directed at the extent of the $Na^+$ gradient, though, as will be discussed later, an effect like this can at least partially account for the increase in $P_i$ transfer observed after vitamin D repletion (see Section IV-A.2). Second, the effect of vitamin D, which then is necessarily on the $P_i$ transport system itself, probably involves a change in the number of functional carrier sites and/or in their mobility within the plasma membrane. That this is the quintessential type of vitamin D action on intestinal $Na^+$-coupled $P_i$ transport is equally evident from studies on the kinetics of $Na^+$-gradient-driven $P_i$ accumulation by isolated brush-border vesicles. The more than twofold increase in its $V_{max}$ as a result of vitamin D treatment, with the carrier affinity for $P_i$ unchanged at $K_m$ 0.1 mM, can hardly be interpreted in a different way.

If $Na^+$ in the outside medium is totally replaced by, e.g., choline, $P_i$ uptake into vesicles is further incapacitated, now proceeding in a nonsaturable mode which is not influenced by vitamin D. This substantiates the previous notion that the passive permeability of $P_i$ across the brush-border membrane is not subject to regulation by vitamin D (see Section II-C.2).

### 2. *Segmental Pattern*

When $P_i$ uptake was compared in vesicles from different portions of the small intestine, the vitamin D-independent mode of $P_i$ transfer declined in the direction from duodenum to ileum. Stimulation by vitamin D was highest in the jejunum and least in the ileum (Peterlik *et al.*, 1981b, 1982a). Thus, it turned out that the variations in the level of basal $P_i$ uptake and in the extent of its stimulation by vitamin D, which have been previously observed in intact epithelium in different sections of the small intestine (see Section II-A and B), reflect the actual endowment of the corresponding brush-border membrane with $Na^+$-activated $P_i$ transport sites.

### 3. *Influence of pH, $Na^+$:$P_i$ Coupling Ratio, Electroneutral vs. Electrogenic Transport*

Another important aspect of intestinal $P_i$ absorption concerns its dependence on the hydrogen ion concentration. Berner *et al.* (1976) reported that rat duodenal vesicles accumulate more $P_i$ at pH 6 than at 7.4. Additional observations indicate that transport of $Na^+$ and $P_i$ is coupled in a 1 : 1 ratio, and that the cotransport of both ions is electroneutral at pH 7.4. Taken together, these data suggest that the rat intestinal $P_i$ transport system preferentially accepts the monovalent $H_2PO_4^-$ anion which is translocated on the carrier together with one $Na^+$ ion. Danisi *et al.* (1982), utilizing brush-border membrane vesicles from rabbit duodenum, evaluated the influence of potential changes on $Na^+$-dependent $P_i$ transport at different pH and came to somewhat different conclusions. $Na^+$–$P_i$ cotransport, which is clearly potential-sensitive at pH 6, still responds to changes in vesicular membrane potential at pH 7.4. These data conform to the assumption that the transport system equally accepts $P_i$ in its mono- and divalent forms together with a single $Na^+$ ion, and that at acidic pH the $Na^+$–$P_i$ carrier bears a positive charge.

### 4. *Specificity of $P_i$ Translocation*

The $Na^+$-dependent transport system has not been directly tested for its anion specificity except that it was studied for the acceptance of $P_i$ in the mono- and divalent forms (Berner *et al.*, 1976; Danisi *et al.*, 1982). Studies on anion transfer across the intestinal brush-border membrane in general (Liedtke and Hopfer, 1977) have revealed that various anions, such as acetate, chloride, nitrate, and thiocyanate, are preferentially translocated by the completely different mode of proton/anion symport. The possibility that these anions would thus compete with $P_i$ for the $Na^+$ cotransport system seems therefore very unlikely. Lücke *et al.* (1981), utilizing brush-border vesicles from rat small intestine, showed that $P_i$ lacks the ability to interact with the $Na^+$-linked sulfate transport system. Conversely, it may be assumed that sulfate will not compete with the $P_i$ carrier since it is specifically translocated by its own transport complex.

Arsenate is probably the only anion able to compete with $P_i$ for transfer. Arsenate inhibits $P_i$ transport in everted chick ileum (Taylor, 1974) and also blocks $Na^+$-gradient-driven $P_i$ uptake by isolated brush-border vesicles from rat duodenum and chick jejunum (Berner *et al.*, 1976; Fuchs and Peterlik, 1980b). This proves that arsenate intercedes with $P_i$ transport actually at the membrane level and not through interference with oxidative metabolism. The latter possibility was never completely excluded in experiments with intact cells. Similar to $P_i$ uptake by renal vesicles (Hoffmann *et al.*, 1976), $Na^+$-dependent uptake into rat duodenal vesicles was competitively inhibited by arsenate (Berner *et al.*, 1976). Diffusional $P_i$ uptake was not affected by arsenate in rat and chick brush-border membrane vesicles (Berner *et al.*, 1976; Fuchs and Peterlik, 1980b).

Another aspect of specificity concerns the obvious relationship between $Na^+$-gradient-driven $P_i$ and D-glucose transport. With regard to the corresponding systems at the brush-border membrane of renal tubules, the issue was raised whether mutual inhibition by $P_i$ or D-glucose (Corman *et al.*, 1978; DeFronzo *et al.*, 1976; Harter *et al.*, 1974; Knight *et al.*, 1980) could be explained by the interaction of one solute with the transfer mechanism of the other or whether there is competition of the two transport systems for the $Na^+$ gradient, the only energy source available to both of them. Another facet of the relation between $Na^+$–$P_i$ and $Na^+$–D-glucose transport, particularly in the small intestine, is that both translocation mechanisms have been shown to be stimulated by vitamin D treatment of vitamin D-deficient chicks (Peterlik *et al.*, 1981b). Although, in isolated brush-border membrane vesicles from chick jejunum inhibition of $Na^+$-dependent D-glucose uptake by $P_i$ and vice versa could be demonstrated (Peterlik *et al.*, 1981a), it is unlikely that $P_i$ and D-glucose are transferred on the same carrier. Specific inhibitors like arsenate for $P_i$ transport and phloridizin for D-glucose transfer blocked only one system and had no effect on the other (Fuchs and Peterlik, unpublished observations). It is assumed therefore that both transport mechanisms can compete for the $Na^+$ gradient. This would imply also a certain tie between the vitamin D-dependent fractions of both $P_i$ and D-glucose transport since they could be jointly stimulated by an action of the sterol on the $Na^+$ gradient (for discussion see Section IV-A.2).

## C. Attempts at Characterization of the $P_i$ Carrier

### 1. Sensitivity to Inhibitors

Information on molecular characteristics of $P_i$ carriers in general is only slowly emerging. In a preliminary approach, inhibitors with known effects on $P_i$ translocation, particularly in mitochondria, were tested for their ability to interact with intestinal transmucosal $P_i$ transfer in everted chick jejunum (Peterlik, 1978a; Fuchs and Peterlik, unpublished observations). In these studies, $P_i$ transport was inhibited by the SH-group blocker *N*-ethylmaleimide as well as by similar agents like *p*-chloromercuribenzoate, mersalyl, and ethacrynic acid. It is noteworthy that vitamin D-related $P_i$ transport exhibits less sensitivity to these inhibitors since it cannot be reduced to the same level as vitamin D-independent $P_i$ transport. Whether this reflects some alteration of properties of the carrier or a change in its membrane topography induced by vitamin D (for discussion see Section IV-A.2) remains unknown.

### 2. Phosphate Binding; Protein Phosphorylation

In a very promising approach, Fanestil and co-workers (Kessler and Fanestil, 1981; Kessler *et al.*, 1982) have isolated a phosphate-binding proteolipid from renal brush border. High-affinity binding of $P_i$ is prevented by arsenate and 2,4-dinitrofluorobenzene, which also inhibit $Na^+$-driven $P_i$ uptake in renal brush-border vesicles. The $P_i$-binding site is not affected by sulfhydryl reagents. It remains to be seen whether a similar protein, which is likely a part of the $P_i$ translocation system, can also be isolated from the intestinal brush-border membrane.

The importance of protein phosphorylation and dephosphorylation for the operation of $Na^+$-gradient-driven $P_i$ translocation can be inferred from the findings of Hammerman and Hruska (1982) that the inhibitory action of cyclic AMP on $Na^+$-coupled $P_i$ transport in renal brush borders is associated with phosphorylation of certain proteins. With respect to intestinal brush borders, Wasserman and his colleagues (Wasserman and Brindak, 1979; Wasserman *et al.*, 1981) could demonstrate a subtle vitamin D-related change in the phosphorylation pattern of intestinal brush-border membranes, involving possibly only one protein. This so-called Q protein, which is phosphorylated with [$\gamma$-$^{32}$P]-ATP or with $^{32}P_i$ in a cyclic-AMP-independent manner, shows different electrophoretic mobility depending on the vitamin D status of the chick. It is entirely possible that vitamin $D_3$ treatment changes only the electrophoretic mobility of a single phosphorylated protein or, alternatively, represses the synthesis of one protein while simultaneously inducing a second one. Although it remains to be seen whether these proteins are involved in $P_i$ transfer at all, it is conceivable that a change in the pattern of protein phosphorylation could be responsible for the shift in the rate of $Na^+$-coupled $P_i$ transport associated with vitamin D action. The ineffectiveness of cyclic AMP on phosphorylation of these particular proteins would correspond to its inability to modulate intestinal $P_i$ transport (see Section IV-B).

### 3. *Possible Role of Alkaline Phosphatase in $P_i$ Translocation*

Moog and Glazier (1972) originally observed a close association between alkaline phosphatase and $P_i$ accumulation in mouse and chick small-intestinal rings, particularly when $P_i$ could have been derived through the activity of the enzyme from the β-glycerophosphate added to the incubation medium. The relationship between enzyme activity and $P_i$ transport showed considerable variability in two strains of chicks when phosphate was present as inorganic salt only. Nevertheless, a role of alkaline phosphatase in $P_i$ transfer has been considered possible, mainly because the enzyme is known to possess a $P_i$-binding site and could thus function as $P_i$ acceptor in the translocation process, and, in addition, the alkaline phosphatase activity of the intestinal brush border like $P_i$ transport is enhanced by vitamin D treatment (Norman *et al.*, 1970). However, there are reports denying such a relationship. Taylor (1974) could not find any influence of L-phenylalanine, an inhibitor of alkaline phosphatase activity, on $P_i$ transport in everted chick ileum, no matter whether it was derived from vitamin D-deficient or replete animals. Peterlik and Wasserman (1980) followed the time course of vitamin D induction of both alkaline phosphatase and active transmucosal $P_i$ transfer. In this study, increased enzyme activity lagged significantly behind vitamin D induction of $P_i$ transport. This would exclude any identity between the enzyme and the transport system.

From studies on diverse mechanisms of $P_i$ uptake in isolated intestinal cells, Birge and co-workers concluded that the relationship between alkaline phosphatase and $P_i$ transport is not relevant for the L-phenylalanine-sensitive $Mg^{2+}$-activated enzyme and the $Ca^{2+}$-independent high-affinity $Na^+$–$P_i$ cotransport system, but rather concerns a L-phenylalanine-insensitive, $Ca^{2+}$-stimulated alkaline phosphatase residing at the plasma membrane and a low-affinity $Ca^{2+}$-dependent $P_i$ transport system with optimal function at alkaline pH (Avioli *et al.*, 1981; Birge and Avioli, 1981).

More recent studies reassumed the original suggestion by Moog and Glazier (1972) that alkaline phosphatase through its hydrolytic action would facilitate the transfer of $P_i$ derived from organic phosphocompounds. Infusion of 1,25-$(OH)_2D_3$ into the vasculature of intestinal loops enhanced lumen-to-serosa transport of $^{32}P_i$ when the lumen was perfused with [γ-$^{32}$P]-ATP. This type of $P_i$ transport was inhibited by pretreatment with the alkaline phosphatase inhibitor levamisole (Bachelet *et al.*, 1982). The conclusions of these authors were essentially supported by the finding of Letellier *et al.* (1982) that alkaline phosphatase activity was required for $P_i$ uptake by isolated brush-border vesicles when organic phosphate was present as sole phosphate source.

## IV. HORMONAL REGULATION OF INTESTINAL $P_i$ TRANSPORT

There can be no doubt that the rate of $Na^+$-gradient-driven transport of $P_i$ across the brush border determines the extent of $P_i$ absorption from the lumen. Consequently, this transfer step would be a logical target for the action of those hormones that play a role in the regulation of intestinal $P_i$ absorption.

## A. Vitamin D

As for vitamin D, ample evidence for this assumption has already been presented in the preceding sections. The term vitamin D is used in this review (as elsewhere) for vitamin $D_3$ itself [cholecalciferol, 9,10-seco-5,7,10(19)cholestatrien-3-ol] as it is for any structurally related and biologically active compound. In this respect, it is well known that vitamin $D_3$ itself has little biological activity and, therefore, has to undergo a series of hydroxylations to be converted into more active metabolites. This aspect of vitamin D action has been extensively reviewed (Fraser, 1980).

### 1. In Vivo and in Vitro Actions of Vitamin $D_3$ Derivatives

*a. Vitamin $D_3$.* The capability of this compound to raise intestinal $P_i$ absorption, when administered to intact, vitamin D-deficient animals, has been extensively demonstrated. Like other typical responses to vitamin $D_3$, the effect on intestinal $P_i$ transport is abolished upon nephrectomy (Chen *et al.*, 1974) or when renal synthesis of the most active metabolite, 1,25-$(OH)_2D_3$, is blocked otherwise, e.g., by feeding a strontium-supplemented diet (Peterlik and Wasserman, 1978b, 1980). This suggests that elevation of $P_i$ absorption from the intestine, like any other biological activity of vitamin $D_3$, is largely due to its conversion into more active 1,25-$(OH)_2D_3$. Intrinsic bioactivity of vitamin $D_3$ is thereby not excluded, since the sterol has been shown to be fully active *in vitro,* viz., in organ culture of embryonic chick intestine (Corradino, 1973a) where further metabolism is highly unlikely to occur (Corradino, 1973b). Nevertheless, when added to the culture medium, vitamin $D_3$ exhibits its typical bioactivity which includes the ability to induce intestinal $P_i$ absorption (Corradino, 1973a; Peterlik, 1978a,b; Cross and Peterlik, 1982b).

*b. 25-Hydroxyvitamin $D_3$.* This metabolite, which is the hepatic hydroxylation product of vitamin $D_3$, has been used as a model compound for vitamin D action on intestinal $P_i$ transport in the *in vitro* studies of Birge and co-workers. 25-Hydroxyvitamin $D_3$ raises $P_i$ uptake effectively in rat ileal explants (Birge and Miller, 1977) as well as in isolated intestinal cells (Birge and Avioli, 1981; Avioli *et al.*, 1981; Karsenty *et al.*, 1982). Since 25-hydroxyvitamin $D_3$ is inactive in nephrectomized rats (Chen *et al.* 1974), its *in vivo* activity apparently depends also on further conversion to 1,25-$(OH)_2D_3$.

*c. 1,25-Dihydroxyvitamin $D_3$.* The discovery of the renal synthesis of a particularly active compound which was identified as 1,25-dihydroxyvitamin $D_3$ (Fraser and Kodicek, 1970) marked the end of the successful search for the ultimate metabolite of vitamin $D_3$ that was responsible for its bioactivity in the intact organism. 1,25-$(OH)_2D_3$ has been shown to restore intestinal $P_i$ transport to normal levels in nephrectomized rats (Chen *et al.*, 1974; Walling and Kimberg, 1975; Walling *et al.*, 1976, 1977) or in strontium-inhibited chicks (Peterlik and Wasserman, 1977, 1978b, 1980).

In cultured embryonic small intestine, 1,25-$(OH)_2D_3$ is about 300 times more effective than vitamin $D_3$ in raising intestinal $P_i$ transport (Cross and Peterlik, 1982b).

*d. 24R,25-Dihydroxyvitamin $D_3$ and 1,24R,25-Trihydroxyvitamin $D_3$.* 25-Hydroxyvitamin $D_3$ is hydroxylated in the kidney either on C-1 or, alternatively, on C-24. Although, through the latter reaction, considerable amounts of 24R,25-dihydroxyvitamin $D_3$ are produced under physiological conditions, the function of this metabolite within the organism is still a matter of intensive research. In vitamin D-deficient rats, 24R,25-dihydroxyvitamin $D_3$ is about ten-fold less effective than 1,25-$(OH)_2D_3$ in raising intestinal $P_i$ transport. Its activity is lost in nephrectomized animals, suggesting that this vitamin D metabolite as well acts *in vivo* through its 1$\alpha$-hydroxylated metabolite. In fact, 1,24R,25-hydroxyvitamin $D_3$ is about equally effective as 1,25-$(OH)_2D_3$ in both sham-operated and nephrectomized rats (Walling *et al.*, 1977).

### 2. Mechanisms of Vitamin D Action on Intestinal $P_i$ Transport

The vitamin D compounds tested so far have all shown a certain effectiveness in stimulating intestinal $P_i$ transport. As a result of numerous studies, 1,25-$(OH)_2D_3$ must be viewed as being more active than any other related compound in evoking any of the great number of vitamin D responses. Quantitative differences in bioactivity among vitamin D compounds are believed to arise from different degrees of receptor binding. 1,25-$(OH)_2D_3$ excels all other vitamin D steroids in affinity binding to a specific receptor protein that has been found in the cytoplasm of all target cells. This apparently determines the relative biopotency of these structurally related substances, since binding to the receptor is an essential step in the mechanism of action; vitamin D like any other steroid hormone alters the genetic expression of its target cells (for review see Lawson, 1978). Recently, however, some attention has been given to an alternative mode of action by which vitamin D could change the lipid composition of the plasma membrane without prior interaction with the genome leading to stimulation of calcium flux across the brush-border membrane of enterocytes (Fontaine *et al.*, 1981).

*a. Genomic Action of Vitamin D.* Evidence has been accumulated to indicate that the stimulatory effect of vitamin D on $P_i$ transport in the intestine is due to its genomic action with subsequent modulation of protein synthesis. In this respect, it has been demonstrated that actinomycin D blunts the response of the $P_i$ absorptive system when vitamin $D_3$ is injected into vitamin D-deficient rats (Ferraro *et al.*, 1976). In addition, actinomycin D as well as $\alpha$-amanitin, another inhibitor of mRNA synthesis, both depress the action of vitamin $D_3$ or of 1,25-$(OH)_2D_3$ on $P_i$ transport in cultured embryonic small intestine (Peterlik, 1978a,b; Cross and Peterlik, 1982b). Furthermore, the action of vitamin D compounds on the $P_i$ transport system *in vivo* and *in vitro* can be effectively blocked by cycloheximide. When this inhibitor of protein synthesis was administered to vitamin D-deficient chicks, no effect of 1,25-$(OH)_2D_3$ on $P_i$ transfer from the lumen of everted gut sacs could be observed (Peterlik and Wasserman, 1977, 1980).

Since multiple dosing of an adequate amount of cycloheximide is necessary to achieve an optimal effect in intact animals (Norman *et al.*, 1970), this might explain the failure to demonstrate any inhibition of vitamin D-induced $P_i$ uptake by brush-border vesicles after a single dose of cycloheximide (Matsumoto *et al.*, 1980). Addition of the inhibitor to cultures of rat ileal explants or of embryonic chick duodenum prevents the stimulatory action of 25-hydroxyvitamin $D_3$ or of vitamin $D_3$, respectively, on $P_i$ transport (Birge and Miller, 1977; Peterlik, 1978a,b).

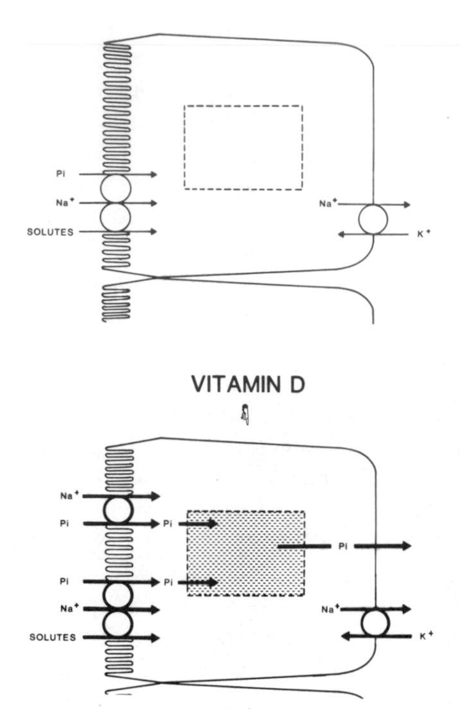


*Figure 3.* Mechanisms of vitamin D action. Upper part, $Na^+$-gradient-driven transport of $P_i$ (and other solutes) and $Na^+$, $K^+$-ATPase activity in absence of vitamin D; lower part, vitamin D-related increase in number of $Na^+$–$P_i$ carriers, electrochemical $Na^+$ gradient, and $Na^+$, $K^+$-ATPase.

*b. Possible Vitamin Effect on Synthesis of Carrier Proteins.* In terms of kinetics, vitamin D acts on the intestinal $Na^+$–$P_i$ cotransport system solely by raising its maximal velocity, $V_{max}$. This observation has been made before in whole-tissue studies and was later confirmed in isolated brush-border membrane vesicles (see Section III-B.1). Since there is no evidence that vitamin D has any influence on the average vesicular volume (Fuchs and Peterlik, 1980b; Matsumoto *et al.*, 1980) and, consequently, also not on the absorptive surface of isolated vesicles, an increase in $V_{max}$ would thus imply that the number of carrier sites has been increased (Figure 3). This, of course, could occur through activation of latent, preexisting carrier sites or via synthesis of new carrier sites under the direction of vitamin D. With respect to this second possibility, the question arises, then, whether vitamin D augments only the synthesis of carrier proteins that are already being genetically expressed in the vitamin D-deficient state, or whether the genomic action of the sterol results in the expression of new carrier proteins. There is, in fact, some evidence for the latter possibility from measurements of $P_i$ transport kinetics at different $Na^+$ concentrations in everted gut sacs from chick jejunum (Fuchs and Peterlik, 1979b). It has already been mentioned (see Section III-A) that, in the vitamin D-deficient state, increasing $Na^+$ concentrations raise the affinity of the carrier for $P_i$ as reflected by an approximately ten-fold change in $K_m$; maximal $P_i$ binding ($K_m$ 0.2 mM) is not observed until $Na^+$ reaches its extracellular concentration of 143 mM. In contrast, the $K_m$ of the vitamin D-induced transport system, which equally is 0.2 mM, is much less affected by $Na^+$ so that a high extent of $P_i$ binding is observed even at low $Na^+$ concentrations. In other words, a relatively higher $Na^+$ concentration is necessary to achieve the same extent of $P_i$ binding to the carrier in the vitamin D-deficient state as after vitamin D repletion; consequently, more $Na^+$ ions per $P_i$ are transferred in vitamin D-independent as compared to vitamin D-dependent transport. Preliminary evidence from studies with isolated brush-border vesicles tends to confirm the assumption that the two carriers differ in their properties with respect to $Na^+$-$P_i$ coupling (unpublished observations).

*c. Does Vitamin D Activate Nonfunctional Carrier Sites?* As long as there is no definite proof for vitamin D-directed synthesis of carrier proteins, alternative possibilities have to be considered, one of which could be a change in the composition or in the arrangement of lipids and proteins in certain domains of the brush-border membrane. Conceivably, that could occur through transcriptional control by vitamin D of the metabolism and/or of the assembly of membrane components. By a change in the microenvironment, the arrangement of a nonfunctional carrier complex within the membrane bilayer could be altered so that participation in $Na^+$–$P_i$ translocation across the membrane becomes possible. The assumption of such a "membranotropic" action of vitamin D can be derived from several observations. Goodman *et al.* (1972) have shown that vitamin $D_3$, in fact, alters the lipid composition of the microvillar membrane. O'Doherty (1979) has observed an effect of 1,25-$(OH)_2D_3$ on intestinal phospholipid metabolism. Finally, Norman *et al.* (1981) provided evidence that 1,25-$(OH)_2D_3$ changes the architecture of the brush-border membrane proteins. In light of the preceding discussion, it remains to be seen how changes in a domain that the carrier is embedded in could be conveyed to the carrier itself so that its binding affinity for $P_i$ and $Na^+$ is altered.

*d. Vitamin D Action on the Electrochemical $Na^+$ Gradient* Apart from an effect on the carrier itself, it is conceivable that vitamin D could accelerate $Na^+$–$P_i$ cotransport through appropriate action on the energy source of this system, the transmucosal electrochemical $Na^+$ gradient. In fact, firm evidence has been obtained that 1,25-$(OH)_2D_3$ could modify passive $Na^+$ fluxes across the brush-border membrane as well as active $Na^+$ extrusion via the basolateral $Na^+$,$K^+$-ATPase in a way that would lead to hyperpolarization of the enterocyte and, accordingly, to an increase of the transmucosal $Na^+$ gradient.

Experiments with isolated brush-border vesicles from chick small intestine revealed that luminal $Na^+$ entry is principally possible via conductive $Na^+$ flux and $Na^+$–$H^+$ antiport (Peterlik *et al.*, 1982b; Fuchs and Peterlik, 1983). 1,25-$(OH)_2D_3$ restricts $Na^+$ movement on both pathways (Fuchs and Peterlik, manuscript in preparation). The resultant reduction in $Na^+$ permeability of the brush-border membrane, which is observed in the jejunum and ileum though not in the duodenum, is equivalent to an increment of the transmucosal $Na^+$ electrochemical gradient. This could be utilized as additional driving force for $Na^+$-dependent transport processes (Figure 3). This assumption was validated by the observation that in those sections of the gut which show reduced luminal $Na^+$ permeability after vitamin D dosage, $Na^+$-dependent transport, not only of $P_i$, but also of D-glucose and of several amino acids, increases in parallel (Peterlik *et al.*, 1981b, 1982a; Fuchs *et al.*, 1982).

In a search for other pathways of epithelial $Na^+$ transport which possibly could be influenced by vitamin D, Cross and Peterlik (1983) observed a vitamin D-related increment in $Na^+$, $K^+$-ATPase activity in embryonic chick small intestine in organ culture. Elevated $^{86}Rb^+$ uptake as indication for enhanced $Na^+$–$K^+$ exchange was visible to a varying degree also along the gut of 4-week-old chicks. Since $Na^+$ pumping across the basolateral membrane mediated by the $Na^+$, $K^+$-ATPase is ultimately responsible for the maintenance of the $Na^+$ gradient between the extra- and intracellular space, the mechanism by which vitamin D could influence any $Na^+$-gradient-driven transport process by an action on the sodium pump is highly apparent (Figure 3).

Although it became evident through these studies that vitamin D can definitively elevate $Na^+$–$P_i$ transport via an effect on the electrochemical $Na^+$ gradient, this has to be viewed only as an accessory mechanism of vitamin D action on intestinal $P_i$ transport, because there are certain situations where a link between vitamin D-induced $Na^+$–$P_i$ transport and a change in $Na^+$ permeability (or vice versa) cannot be demonstrated. Vitamin D-related $Na^+$–$P_i$ cotransport can be observed in the duodenum without any change in luminal $Na^+$ permeability, and stimulation of $Na^+$, $K^+$-ATPase activity can occur, e.g., in undifferentiated embryonic small intestine, without a concomitant rise in vitamin D-dependent $Na^+$-linked $P_i$ transport. That necessarily leaves the purported action on the number and possibly also properties of carrier complexes as the major instrument of vitamin D to raise $Na^+$-dependent $P_i$ transport at the brush-border membrane of enterocytes (Figure 3).

## B. Parathyroid Hormone and Cyclic AMP

Parathyroid hormone (PTH) is the main regulator of renal tubular $P_i$ reabsorption. Underlying the well-documented phosphaturic response to elevated plasma levels of the hormone is its cyclic-AMP-mediated inhibitory action on the $Na^+$–$P_i$ transport system of the tubular brush-border membrane. By analogy, a similar effect of PTH on the intestinal $Na^+$–$P_i$ system is conceivable. However, despite an intensive search, no conclusive evidence has so far been obtained for a direct effect of PTH on intestinal $P_i$ transport. Walling (1978) reported that $P_i$ movements across rat small-intestinal epithelium mounted in an Ussing chamber were not changed when PTH was added to the incubation solution. The hormone was also not able to increase intracellular cyclic AMP levels in the intestinal epithelium. When dibutyryl cyclic AMP was added to the bathing solution, again no effect on $P_i$ fluxes was observed. This inability of the nucleotide to change active $P_i$ transport was confirmed by measurements of unstimulated and vitamin D-stimulated $Na^+$-dependent $P_i$ transport in everted chick jejunum (Fuchs and Peterlik, 1980a). Although vitamin $D_3$ clearly raises the intracellular concentration of cyclic AMP in organ-cultured embryonic chick duodenum (Corradino, 1974), $P_i$ accumulation appears to be independently affected by the sterol (Corradino, 1979b). In another study, only inconsistent effects on $P_i$ uptake were observed when dibutyryl cAMP was added to the organ culture of embryonic chick jejunum at different stages of development (Cross and Peterlik, 1980). Depression of basal $P_i$ uptake by the cyclic AMP analogue was observed at a developmental stage when the embryonic intestine was refractory to vitamin D action (Cross and Peterlik,

1982b); in contrast, basal $P_i$ uptake by cultured jejunum seemed unaffected just prior to hatching, while at this time, vitamin D-induced $P_i$ uptake was reduced by cyclic AMP. The results of these studies certainly warrant the conclusion that the similarities between the renal and the intestinal $P_i$ transport system, which undoubtedly exist with respect to their localization at the respective brush border, their $Na^+$ dependence and their common sensitivity to vitamin D (Kinne, 1982), cannot be extended to a regulatory role of PTH and/or cAMP in both systems.

The earlier report by Lifshitz *et al.* (1969), that parathyroidectomy reduces active $P_i$ absorption from isolated intestinal loops in the rat, while administration of a parathyroid extract restores $P_i$ transport to normal, must be viewed in light of the now well-known fact that PTH increases the renal synthesis of 1,25-$(O)_2D_3$ which, in turn, acts on the intestinal $P_i$ transport system.

## C. Calcitonin

Under experimental conditions, calcitonin lowers serum levels of both calcium and phosphate in mammals (Hirsch and Munson, 1969). The hypophosphatemic response in the rat might be partially mediated by an effect of the peptide hormone on intestinal $P_i$ absorption (Tanzer and Navia, 1973). In man, intravenous infusion of salmon calcitonin in healthy volunteers produced net secretion of water and electrolytes in the jejunum and concomitantly reduced net absorption of calcium and phosphate (Juan *et al.*, 1976). Serum $P_i$ levels, however, were not affected. The mechanism by which calcitonin influences $P_i$ fluxes has not yet been elucidated. Although it is obvious from these studies that calcitonin can inhibit intestinal $P_i$ absorption under certain conditions, a physiological importance of these findings has not yet been established.

## D. Glucocorticoids

There is clinical and experimental evidence that prolonged administration of high doses of glucocorticoids reduces the absorption of calcium from the intestinal lumen. In experimental animals, glucocorticoid treatment produces a significant reduction of $P_i$ absorption in the small intestine as well (Ferraro *et al.*, 1976; Fox *et al.*, 1981). It is still a matter of controversy whether this is due to an alteration of vitamin D metabolism resulting in lower circulating levels of the active metabolite, 1,25-$(OH)_2D_3$, or whether the steroids have an inhibitory effect directly on the intestinal absorptive mechanisms.

In the experiments of Fox *et al.* (1981), inhibition of $P_i$ absorption due to betamethasone treatment of chicks was not noted when the animals were placed on experimental diets low in phosphorus or calcium. It is believed that these conditions provide a stimulus for the production of 1,25-$(OH)_2D_3$ (Hughes *et al.*, 1975) resulting in an adaptive increase of $P_i$ absorption (Peterlik and Wasserman, 1977, 1980). This may counteract the inhibitory action of betamethasone, indicating that metabolism of vitamin D and, consequently, its control over intestinal $P_i$ absorption is still intact in these glucocorticoid-treated animals. It appears, therefore, that betamethasone acts upon basal, i.e., vitamin D-independent $P_i$ transfer. In contrast, a direct inhibitory action of glucocorticoids on $P_i$ uptake could not be substantiated by the *in vitro* experiments of Corradino (1979a,b). In organ culture of embryonic chick duodenum, hydrocortisone was shown to stimulate basal $P_i$ uptake and even to potentiate the vitamin D effect thereon. The exact mechanism by which hydrocortisone affects $P_i$ absorption from the intestine has, therefore, still to be clarified. The obvious differences between *in vivo* and *in vitro* activites of glucocorticoids could be due to the fact that hydrocortisone was present in the organ culture for only a short period relative to the duration of the betamethasone treatment of intact animals. Reduction of $P_i$ absorption, therefore, might be secondary to a general debilitating effect of corticosteroids on the function of the intestinal mucosa.

## E. Insulin

Insulin so far has not been recognized as a phosphate-regulating hormone. Nevertheless, experiments with embryonic chick small intestine maintained in organ culture have revealed its potential to modulate intestinal $P_i$ absorption in two ways (Cross and Peterlik, 1980, 1981, 1982a): First, insulin was shown to facilitate the induction of $Na^+$-dependent $P_i$ uptake by vitamin D without any effect of its own on the transport system. Second, under certain experimental conditions, insulin may depress basal $P_i$ uptake without interfering with vitamin D action.

### 1. Precocious Induction of Vitamin D-Related $P_i$ Uptake

Even undifferentiated enterocytes, which exclusively constitute the small-intestinal epithelium, e.g., on day 15 of embryonic development, have the potential for $Na^+$-dependent $P_i$ uptake (Cross and Peterlik, 1982b), although their responses to vitamin D in organ culture do not include the elevation of this particular transport system. Only immediately before hatching and, hence, coincidental with the appearance of the highest possible number of differentiated absorptive cells in the epithelium can $Na^+$-dependent $P_i$ transport be induced by vitamin D.

Insulin, when present in the culture medium in physiological concentrations, permits precocious induction of $P_i$ uptake when the organ culture of immature gut segments, otherwise refractory to the sterol, has been primed with vitamin D. The insulin effect develops within 1 hr and is, therefore, probably too short in its duration to allow any effect of insulin directly on the transport system. By this permissive effect, insulin allows the mechanism of action of vitamin D on $P_i$ uptake to proceed even in immature epithelium.

### 2. Long-Term Action on Vitamin D-Independent $P_i$ Transport

*a. Effect in Low Concentration Range.* Prolonged exposure of the small intestine during culture to insulin concentrations equivalent to circulating plasma levels depresses basal $Na^+$-dependent $P_i$ uptake. The inhibitory response is observed in immature as well as in differentiated epithelium. The latter remains fully responsive to vitamin D also in the presence of insulin. It can, therefore, be concluded that the peptide hormone does not interfere with the steroid hormone's action (Cross and Peterlik, 1981, 1982a).

*b. Somatomedin-Like Effect of Insulin in the High Concentration Range.* The long-term action of insulin on basal $P_i$ uptake is complicated by the fact that, in differentiated small intestine, high culture concentrations of the hormone tend to reverse the inhibitory effect described before. This can take the form of an outright stimulation above control levels (Cross and Peterlik, 1981, 1982a).

When gut segments are cultured in the presence of vitamin D and high insulin, both hormones stimulate $P_i$ uptake in an additive manner.

The effects observed with supraphysiological concentrations of insulin are commonly interpreted as being mediated through the somatomedin receptor which has a low affinity also for insulin.

## V. INTESTINAL PHOSPHATE ABSORPTION IN HEALTH AND DISEASE

At this point, it is certainly evident that our understanding of the mechanisms regulating intestinal $P_i$ absorption is still far from being complete albeit a considerable amount of data has been accumulated in recent years. Nevertheless, if available experimental evidence is pieced together, a general though sketchy picture of the total $P_i$ absorptive process is clearly emerging.

### A. Efficiency of $P_i$ Absorption under Physiological Conditions

Fractional absorption of dietary phosphate is the resultant of two principal modes of $P_i$ transfer across the gut wall. First, uptake by simple diffusion (MacHardy and Parsons, 1956; Juan *et al.*, 1976; Walton and Gray, 1979) as nonsaturable transport is unlimited in its capacity and can hardly be adjusted to homeostatic control. The amount of $P_i$ transferred on this pathway is proportionally related to the actual intraluminal $P_i$ concentration and, therefore, reflects the variations in the amount of ingested $P_i$. On the other hand, the second mode of uptake via the $Na^+$-gradient-driven transport system as a saturable process imposes limitations on the amount of $P_i$ that can be absorbed, though these can be aligned through endocrine control to the increased need of the organism (see below). In addition, transfer by this high-affinity system could result in almost complete extraction of $P_i$ from luminal contents and thereby guarantee maximal efficiency of $P_i$ uptake even at very low intraluminal concentrations.

It is obvious that net absorption of $P_i$ depends on the actual proportion of uncontrolled diffusional and controlled active absorption in the intestine, which may exhibit individual variations and certainly may differ to a great extent among various species.

In man, a mean fractional absorption of 60% (Coburn *et al.*, 1977) from the average daily dietary intake of 800–900 mg P ensures that phosphate is absorbed in moderate excess over the minimal daily requirement of 400 mg P (Bringhurst and

Potts, 1979). This, however, does not prevent hypoabsorption of $P_i$ from occuring under certain conditions (see below). Unlike in man, $P_i$ consumption just matches the requirement in the rat (Nicolaysen, 1937a). This again indicates the importance of the controlled mode of intestinal $P_i$ absorption in maintaining $P_i$ homeostasis at normal. In cattle, even adaptive control of $P_i$ absorption may not be sufficient to procure an adequate amount of $P_i$, particularly when availability of $P_i$ from feed is low. Phosphate depletion resulting from hypoabsorption poses a serious problem in cattle farming in certain areas in the world (Lobao *et al.*, 1982).

### 1. Uncontrolled Passive Absorption

As long as there is no evidence to the contrary, it must be assumed that within the limits set by the modest junctional permeability of $P_i$ and by the transepithelial potential difference, passive diffusion of $P_i$ out of the lumen is not subject to any regulation. As far as vitamin D is concerned, the passive permeability of the luminal plasma membrane of enterocytes is apparently not altered, as can be inferred from evaluation of diffusional $P_i$ uptake by isolated brush-border vesicles (Fuchs and Peterlik, 1980b; Matsumoto *et al.*, 1980), nor is passive transepithelial transfer on a presumably junctional pathway eased by the sterol (Fuchs and Peterlik, 1979b, 1980a).

The chance that a fraction of $P_i$ is passively absorbed is certainly highest in the upper duodenum, where, at the onset of the absorptive process, total $P_i$ concentration is definitely highest and the phosphate anion equilibrium is shifted by the relatively acidic pH to the monovalent form, so that $P_i$ can passively enter the absorptive cells as long as the intraluminal ionic concentrations of $H_2PO_4^-$ or $HPO_4^{2-}$ are not lowered to less than 4 or 15 mM, respectively (see Section II-C).

Passive $P_i$ absorption is, in all likelihood, responsible for the fact that plasma $P_i$ levels can change in a certain relation to dietary $P_i$ intake.

$P_i$ absorption from the large intestine, which, in the absence of active transport (Lee *et al.*, 1980), is through diffusion across the gut wall, only seems to be of minor importance for net absorption of phosphate (Innes and Nicolaysen, 1937).

### 2. Controlled Active Absorption

As intestinal contents are moved from the upper duodenum farther distally, uncontrolled absorption becomes the less important of the two modes of $P_i$ transport across the gut wall. Passive transfer of $P_i$ may have already reduced the total amount of $P_i$ available for absorption and, conversely, the threshold concentration for passive entry is raised from 4 mM for $H_2PO_4^-$ to 15 mM for the divalent $HPO_4^{2-}$ which is now preferentially formed at relatively alkaline pH (see Section II-C).

In general, conditions prevailing in almost the entire small intestine favor the absorption of $P_i$ by an active transport that is adjustable to the needs of the organism through hormonal control and thereby is an integral part in the endocrine regulation of $P_i$ homeostasis. In this respect, $1,25\text{-}(OH)_2D_3$ is by far the most efficient regulator. Whether other steroids, e.g., glucocorticoids, or peptide hormones such as calcitonin, insulin, or somatomedins exert relevant control under physiological conditions has yet to be established.

*a. Adaptive Control by Nutritional Requirements.* Rizzoli *et al.* (1977) demonstrated that intestinal $P_i$ absorption in normal rats maintained on a diet adequate in vitamin $D_3$, responds to a change in circulating levels of 1,25-$(OH)_2D_3$. These authors thereby stressed the importance of the active vitamin $D_3$ metabolite for the modulation of the absorptive process under normal conditions, which would include the adjustment to dietary needs for phosphorus.

It is well known that plasma concentrations of 1,25-$(OH)_2D_3$ are inversely related to the dietary intake of calcium and phosphorus (Hughes *et al.*, 1975) and consequently mediate the respective adaptive change of the calcium absorptive process. Not unexpectedly, a similar response of intestinal $P_i$ absorption could be demonstrated. When experimental animals are placed on a diet adequately supplemented with vitamin $D_3$ but low in either phosphorus or calcium, the rate of $P_i$ active transfer across the brush border rises within 3–5 days to maximal levels (Peterlik and Wasserman, 1977, 1980; Fox *et al.*, 1981).

*b. Control of Absorption by Degree of Enterocyte Differentiation.* Since the so-called columnar cells represent the great majority of epithelial cells in the small intestine, it is consequently assumed that they function as absorptive cells. However, it cannot be excluded that other cells, e.g., goblet cells, take part in absorptive processes, particularly since all intestinal cells originate from the same undifferentiated type of enterocyte (Cheng and Leblond, 1974), which is endowed already with $Na^+$-dependent transport systems for, importantly, $P_i$ and also for D-glucose (Cross and Peterlik, 1982b). In the same study, these authors made the observation that as long as the epithelium consists only of undifferentiated cells, $Na^+$-dependent $P_i$ transport cannot be augmented by vitamin D sterols. Sensitivity of $P_i$ uptake to the steroid hormone is apparently conveyed very late in the process of differentiation. These findings may have some bearing on the mode of $P_i$ absorption along the villi of mature small intestine. The ontogenetic process of epithelial differentiation is replicated in essence, as undifferentiated crypt cells, though within a shorter time span, develop into fully differentiated absorptive cells along their migration to the villus tip. One may conclude, therefore, that vitamin D-related $P_i$ absorption occurs preferentially in the upper part of the villi and is less probable in the crypt region. By the same token, one may also speculate that insulin, by facilitating the expression of vitamin D-dependent $P_i$ transport in less differentiated cells (see Section IV-E.1), could thereby modify the control imposed on $P_i$ absorption by the degree of enterocyte differentiation along the villus axis.

## B. *Alteration of Phosphate Absorption in Human Disease*

Elevated plasma levels of $P_i$ are frequently encountered in patients with end-stage chronic renal failure. This seems paradoxical at first, since these patients have circulating levels of 1,25-$(OH)_2D_3$ near zero (Haussler *et al.*, 1976) and, accordingly, are expected to absorb less $P_i$ than normal due to the lack of vitamin D-related $P_i$ absorption. This must certainly be the case (Coburn *et al.*, 1977), particularly in light of the fact that vitamin D deficiency is usually associated with malabsorption of $P_i$ leading to hypophosphatemia of some degree (see below). However, $P_i$ absorption in these patients can still proceed in the uncontrolled, diffusional mode and might even be elevated in the uremic state (Marcinowska-Suchowierska *et al.*, 1981). Since absorption in excess to nutritional requirements cannot be corrected through urinary excretion of $P_i$ due to the loss of renal function, hyperphosphatemia is bound to develop in these patients, particularly when their dietary intake of $P_i$ is high.

From this example, it may be deduced that the $P_i$ requirements of the organism could be met even in case of malfunctioning vitamin D-dependent transmural $P_i$ transfer. However, the importance of this mode of absorption for $P_i$ homeostasis under physiological conditions is highlighted by the fact that vitamin D deficiency in general leads to hypophosphatemia in experimental animals as well as in man (Bronner, 1976; Massry, 1978). Impairment of intestinal $P_i$ absorption is likely to occur regardless of whether the vitamin D deficiency encountered in a particular case is of the nutritional type or, more frequently, is the consequence of a severe disturbance of vitamin D metabolism associated with a great number of gastrointestinal, hepatobiliary, and renal diseases as well as endocrine dysfunctions like hypoparathyroidism and diabetes (Haussler and McCain, 1977).

Reduced absorption of $P_i$ from the intestine certainly contributes to the hypophosphatemia frequently encountered in chronic alcoholism (Massry, 1978). Those patients are generally malnourished due to their drinking habits and, accordingly, their dietary intake of $P_i$ is already low. In addition, efficient extraction of $P_i$ from the lumen to meet the minimal requirements, which normally is achieved by vitamin D-dependent active absorption (see before), is hindered (Farrington *et al.*, 1979; Long *et al.*, 1978) because of impaired production of active vitamin D metabolites associated with alcoholic fatty liver or liver cirrhosis (Sonnenberg *et al.*, 1977) in many alcoholics.

Vitamin D-resistant rickets represents a unique case of a hypophosphatemic disease. It is characterized by massive loss of $P_i$ in the urine, severe hypophosphatemia, and rickets which are resistant to treatment with active vitamin D sterols. The etiological basis is a genetic defect concerning the PTH-sensitive part of the $Na^+$-$P_i$ cotransport system of the renal brush border (Tenenhouse and Scriver, 1978, 1979). It has long been debated whether the intestinal $Na^+$-linked $P_i$ transport system is also affected by the same genetic defect. X-linked hypophosphatemic rickets have been found to occur in the mouse. The hemizygous male *Hyp*/Y mouse is a suitable animal model for this human disease. Studies with everted gut sacs from *Hyp* mice suggested a defective intestinal $P_i$ transport system (O'Doherty *et al.*, 1977). Supportive evidence came from a previous study showing that biopsy specimens from the small intestine of patients with familial vitamin D-resistant hypophosphatemia exhibit an apparently reduced capability of $P_i$ accumulation (Short *et al.*, 1973). However, Tenenhouse *et al.* (1981), by comparing $P_i$ uptake by isolated brush-border vesicles derived from the small intestine of *Hyp*/Y and normal mice, found no evidence for defective intestinal $Na^+$-dependent $P_i$ absorption in this disease.

ACKNOWLEDGMENTS

Investigations carried out in the author's laboratory at the University of Vienna Medical School, which are cited in this review, were supported by Grants Nos. 3031 and 4422 of the Fonds zur Förderung der Wissenschaften and by the Anton Dreher

Memorial Fund donated to the University of Vienna Medical School. The author thanks Dr. Heide S. Cross for critical reading and Mrs. E. Gindel for excellent secretarial help in preparation of the manuscript.

## REFERENCES

Avioli, L. V., and Birge, S. J., 1977, Controversies regarding intestinal phosphate transport and absorption, in: *Phosphate Metabolism* (S. G. Massry and E. Ritz, eds.), Plenum Press, New York, pp. 507–513.

Avioli, R. C., Miller, R. A., and Birge, S. J., 1981, Characterization of phosphate uptake in isolated chick intestinal cells, *Min. Electrolyte Metabol.* **5:**287–295.

Bachelet, M., Lacour, B., and Ulmann, A., 1982, Early effects of $1\alpha$,25-dihydroxy-vitamin $D_3$ on phosphate absorption. A role for alkaline phosphatase? *Min. Electrolyte Metabol.* **8:**261–266.

Ben-Ghadalia, D., Tagari, H., Zamwel, S., and Bondi, A., 1975, Solubility and net exchange of calcium, magnesium and phosphorus in digesta flowing along the gut of the sheep, *Brit. J. Nutr.* **33:**87–94.

Berner, W., Kinne, R., and Murer, H., 1976, Phosphate transport into brush-border membrane vesicles isolated from rat small intestine, *Biochem. J.* **160:**467–474.

Birge, S. J., and Avioli, R. C., 1981, Intestinal phosphate transport and alkaline phosphatase activity in the chick, *Am. J. Physiol.* **240:**E384–E390.

Birge, S. J., and Miller, R., 1977, Role of phosphate in the action of vitamin D on the intestine, *J. Clin. Invest.* **60:**980–988.

Blahos, J., and Care, A. D., 1981, The jejunum is the site of maximal rate of intestinal absorption of phosphate in chicks. *Physiol. Bohemoslov.* **30:**157–159.

Bringhurst, F. R., and Potts, J. T., 1979, Calcium and phosphate distribution, turnover, and metabolic actions, in: *Endocrinology*, Vol. 2 (L. J. DeGroot, G. F. Cahill, Jr., L. Martini, D. H. Nelson, W. D. Odell, J. T. Potts, Jr., E. Steinberger, and A. I. Winegrad, eds.), Grune & Stratton, New York, pp. 551–592.

Bronner, F., 1976, Vitamin D deficiency and rickets, *Am. J. Clin. Nutr.* **29:**1307–1314.

Bronner, F., and Peterlik, M. (eds), 1981, *Calcium and Phosphate Transport Across Biomembranes*, Academic Press, New York.

Chen, T. C., Castillo, L., Korycka-Dahl, M., and DeLuca, H. F., 1974, Role of vitamin D metabolites in phosphate transport of rat intestine, *J. Nutr.* **104:**1056–1060.

Cheng, H., and Leblond, C. P., 1974, Origin, differentiation and renewal of the four main epithelial cell types in the mouse small intestine, *Am. J. Anat.* **141:**461–562.

Coburn, J. W., Brickman, A. S., Hartenbower, D. L., and Norman, A. W., 1977, Intestinal phosphate absorption in normal and uremic man: Effects of 1,25$(OH)_2$-vitamin $D_3$ and $1\alpha$(OH)-vitamin $D_3$, in: *Phosphate Metabolism* (S. G. Massry and E. Ritz, eds.), Plenum Press, New York, pp. 549–557.

Corman, B., Touvay, C., Poujeol, P., and deRouffignac, C., 1978, Glucose-mediated inhibition of phosphate reabsorption in rat kidney, *Am. J. Physiol.* **235:**F430–F439.

Corradino, R. A., 1973a, Embryonic chick intestine in organ culture: A unique system for the study of the intestinal calcium absorptive mechanism, *J. Cell. Biol.* **58:**64–78.

Corradino, R. A., 1973b, Embryonic chick intestine in organ culture: Response to vitamin $D_3$ and its metabolites, *Science* **179:**402–405.

Corradino, R. A., 1974, Embryonic chick intestine in organ culture: Interaction of adenylate cyclase and vitamin $D_3$-mediated calcium absorptive mechanism, *Endocrinology* **94:**1607–1614.

Corradino, R. A., 1979a, Embryonic chick intestine in organ culture: Hydrocortisone and vitamin D-mediated processes, *Arch. Biochem. Biophys.* **192:**302–310.

Corradino, R. A., 1979b, Hydrocortisone and vitamin $D_3$ stimulation of $^{32}P_i$-phosphate accumulation by organ-cultured chick embryo duodenum, *Horm. Metabol. Res.* **11:**519–523.

Cremaschi, D., James, P. S., Meyer, G., Peacock, M. A., and Smith, M. W., 1982, Membrane potentials of differentiating enterocytes, *Biochim. Biophys. Acta* **688:**271–274.

Cross, H. S., and Peterlik, M., 1980, Role of differentiation and hormonal effectors (vitamin $D_3$, insulin, cyclic nucleotides) in phosphate transport by embryonic intestine, *Hoppe Seyler's Z. Physiol. Chem.* **361:**1275.

Cross, H. S., and Peterlik, M., 1981, Effects of vitamin D and insulin on phosphate transport in the differentiating chick small intestine, in: *Calcium and Phosphate Transport Across Biomembranes* (F. Bronner and M. Peterlik, eds.), Academic Press, New York, pp. 293–296.

Cross, H. S., and Peterlik, M., 1982a, Hormonal regulation of phosphate transport in the differentiating chick small intestine, in: *Regulation of Phosphate and Mineral Metabolism* (S. G. Massry, J. M. Letteri, and E. Ritz, eds.), Plenum Press, New York, pp. 127–135.

Cross, H. S., and Peterlik, M., 1982b, Differential response of enterocytes to vitamin D during embryonic development: Induction of intestinal inorganic phosphate, D-glucose and calcium uptake, *Horm. Metabol. Res.* **14:**649–652.

Cross, H. S., and Peterlik, M., 1983, Vitamin D stimulates $(Na^+-K^+)$ATPase activity in chick small intestine, *FEBS Lett.* **153:**141–145.

Danisi, G., and Straub, R. W., 1980, Unidirectional influx of phosphate across the mucosal membrane of rabbit small intestine, *Pfluegers Arch.* **385:**117–122.

Danisi, G., Bonjour, J.-Ph., and Straub, R. W., 1980, Regulation of Na-dependent phosphate influx across the mucosal border of duodenum by 1,25-dihydroxycholecalciferol, *Pfluegers Arch.* **388:**227–232.

Danisi, G., Murer, H., and Straub, R. W., 1982, Effect of pH on rabbit intestinal phosphate transport, *Experientia* **38:**712.

DeFronzo, R. A., Goldberg, M., and Agus, Y., 1976, The effects of glucose and insulin on renal electrolyte transport, *J. Clin. Invest.* **58:**83–90.

Demand, H. A., Weichmann, K., and Berg, G., 1968, Some effects of epinephrine on histamine-induced human gastric secretion, *Gastroenterology* **55:**272–276.

Farrington, K., Epstein, O., Varghese, Z., Newman, S. P., Moorhead, J. F., and Sherlock, S., 1979, Effect of oral 1,25-dihydroxycholecalciferol on calcium and phosphate malabsorption in primary biliary cirrhosis, *Gut* **20:**616–619.

Ferraro, C., Ladizesky, M., Cabrejas, M., Montoreano, R., and Mautalen, C., 1976, Intestinal absorption of phosphate: Action of protein synthesis inhibitors and glucocorticoids in the rat, *J. Nutr.* **106:**1752–1756.

Fontaine, O., Matsumoto, T., Goodman, D. B. P., and Rasmussen, H., 1981, Liponomic control of $Ca^{2+}$ transport: Relationship to mechanism of action of 1,25-dihydroxyvitamin $D_3$, *Proc. Natl. Acad. Sci. USA* **78:**1751–1754.

Fox, J., Bunnett, N. W., Farrar, A. R., and Care, A. D., 1981, Stimulation by low phosphorus and low calcium diets of duodenal absorption of phosphate in betamethasone-treated chicks, *J. Endocrinol.* **88:**147–153.

Fraser, D. R., 1980, Regulation of the metabolism of vitamin D, *Physiol. Rev.* **60:**551–613.

Fraser, D. R., and Kodicek, E., 1970, Unique biosynthesis by kidney of a biologically active vitamin D metabolite, *Nature* **228:**764–766.

Fuchs, R., and Peterlik, M., 1979a, Vitamin D-induced transepithelial phosphate and calcium transport by chick jejunum: Effect of microfilamentous and microtubular inhibitors, *FEBS Lett.* **100:**357–359.

Fuchs, R., and Peterlik, M., 1979b, Pathways of phosphate transport in chick jejunum: Influence of vitamin D and extracellular sodium, *Pfluegers Arch.* **381:**217–222.

Fuchs, R., and Peterlik, M., 1980a, Intestinal phosphate transport, in: *Phosphate and Minerals in Health and Disease* (S. G. Massry, E. Ritz, and H. Jahn, eds.), Plenum Press, New York, pp. 381–390.

Fuchs, R., and Peterlik, M., 1980b, Vitamin D-induced phosphate transport in intestinal brush border membrane vesicles, *Biochem. Biophys. Res. Commun.* **93:**87–92.

Fuchs, R., and Peterlik, M., 1983, Effect of vitamin D on transmucosal sodium fluxes, *Calc. Tissue Int.* **35**(Suppl.)**:**A50.

Fuchs, R., Cross, H. S., and Peterlik, M., 1982, Effect of vitamin D on intestinal sodium and sodium-dependent transport, in: *Vitamin D: Chemical, Biochemical and Clinical Endocrinology of Calcium Metabolism* (A. W. Norman, K. Schaefer, D. v. Herrath, and H.-G. Grigoleit, eds.), Walter de Gruyter Verlag, Berlin and New York, pp. 305–307.

Goodman, D. B. P., Haussler, M. R., and Rasmussen, H., 1972, Vitamin $D_3$ induced alteration of microvillar membrane lipid composition, *Biochem. Biophys. Res. Commun.* **46:**80–86.

Hammerman, M. R., and Hruska, K. A., 1982, Cyclic AMP-dependent protein phosphorylation in canine renal brush-border membrane vesicles is associated with decreased phosphate transport, *J. Biol. Chem.* **257:**992–999.

Harrison, H. E., and Harrison, H. C., 1961, Intestinal transport of phosphate: Action of vitamin D, calcium and potassium. *Am. J. Physiol.* **201:**1007–1012.

Harrison, H. E., and Harrison, H. C., 1963, Sodium, potassium, and intestinal transport of glucose, 1-tyrosine, phosphate and calcium, *Am. J. Physiol.* **205:**107–111.

Harter, H. R., Mercado, A., Rutherford, W. E., Rodriguez, H., Slatopolsky, E., and Klahr, S., 1974, Effects of phosphate depletion and parathyroid hormone on renal glucose absorption. *Am. J. Phsyiol.* **227:**1422–1427.

Haussler, M. R., and McCain, T. A., 1977, Basic and clinical concepts related to vitamin D metabolism and action, *New Engl. J. Med.* **297:**1041–1050.

Haussler, M. R., Baylink, D. J., Hughes, M. R., Brumbaugh, P. F., Wergedal, J. E., Shen, F. H., Nielsen, R. L., Counts, S. J., Bursac, K. M., and McCain, T. A., 1976, The assay of 1α,25-dihydroxyvitamin $D_3$: Physiology and pathologic modulation of circulating hormone levels, *Clin. Endocrinol.* **5:**151s–165s.

Hildmann, B., Storelli, C., Danisi, G., and Murer, H., 1982, Regulation of $Na^+$-$P_i$ cotransport by 1,25-dihydroxyvitamin $D_3$ in rabbit duodenal brush-border membrane, *Am. J. Physiol.* **242:** G533–G539.

Hirsch, P. F., and Munson, P. L., 1969, Thyrocalcitonin, *Physiol. Rev.* **49:**548–622.

Hoffmann, N., Thees, M., and Kinne, R., 1976, Phosphate transport by isolated renal brush border vesicles, *Pflügers Arch.* **362:**147–156.

Hughes, M. R., Brumbaugh, P. F., Haussler, M. R., Wergedal, J. E., and Baylink, D. J., 1975, Regulation of serum 1α,25-dihydroxyvitamin $D_3$ by calcium and phosphate in the rat, *Science* **190:**578–580.

Hurwitz, S., and Bar, A., 1972, Site of vitamin D action in chick intestine, *Am. J. Physiol.* **222:**761–767.

Innes, J. R. M., and Nicolaysen, R., 1937, The assimilation of the Steenblock–Black diet in normal and vitamin D-deficient rats with and without caecum, *Biochem. J.* **31:**101–104.

Jackson, M. J., and Kutcher, L. M., 1977, The three-compartment system for transport of weak electrolytes in the small intestine, in: *Intestine Permeation* (M. Kramer, and F. Lankbach, eds.), Excerpta Medica, Amsterdam and Oxford, pp. 65–73.

Jacobi, H., Rummel, W., and Pfleger, K., 1958, Dic Bcziehungen zwischen Phosphatdurchtritt und Glucoseresorption, *Naunyn-Schmiedeberg's Arch. Exp. Pathol. Pharmakol.* **234:**404–413.

Juan, D., Liptak, P., and Gray, T. K., 1976, Absorption of inorganic phosphate in human jejunum and its inhibition by salmon calcitonin, *J. Clin. Endocrinol. Metabol.* **43:**517–522.

Kabakoff, B., Kendrick, N. C., and DeLuca, H. F., 1982, 1,25-Dihydroxyvitamin $D_3$-stimulated active uptake of phosphate by rat jejunum, *Am. J. Physiol.* **243:**E470–E475.

Karsenty, G., Ulmann, A., Lacour, B., Picrandrei, E., and Drueke, T., 1982, Early stimulation by $1,25\text{-}(OH)_2D_3$ of $^{33}P_i$ uptake by isolated enterocytes, in: *Vitamin D: Chemical, Biochemical and Clinical Endocrinology of Calcium Metabolism* (A. W. Norman, K. Schaefer, D. v. Herrath, and H.-G. Grigoleit, eds.), Walter de Gruyter Verlag, Berlin and New York, pp. 345–347.

Kessler, R. J., and Fanestil, D. D., 1981, Identification of a phosphate-binding proteolipid in kidney brush border, in: *Calcium and Phosphate Transport Across Biomembranes* (F. Bronner and M. Peterlik, eds.), Academic Press, New York, pp. 123–126.

Kessler, R. J., Vaughn, D. A., and Fanestil, D. D., 1982, Phosphate binding proteolipid from brush-border, *J. Biol. Chem.* **257:**14311–14317.

Kinne, R., 1982, Calcium and phosphate transport across renal plasma membranes: Concepts, problems and future developments, in: *Calcium and Phosphate Transport Across Biomembranes* (F. Bronner and M. Peterlik, eds.), Academic Press, New York, pp. 105–110.

Knight, T. F., Senekjian, H. O., Sansom, S., and Weinman, E. J., 1980, Influence of D-glucose on phosphate absorption in the rat proximal tubule, *Min. Electrolyte Metabol.* **4:**37–42.

Kowarski, S., and Schachter, D., 1969, Effects of vitamin D on phosphate transport and incorporation into mucosal constituents of rat intestinal mucosa, *J. Biol. Chem.* **244:**211–217.

Lawson, D. E. M., 1978, Biochemical responses of the intestine to vitamin D, in: *Vitamin D* (D. E. M. Lawson, ed.), Academic Press, London, pp. 167–200.

Lee, D. B. N., Walling, M. W., Gafter, U., Silis, V., and Coburn, J. W., 1980, Calcium and inorganic phosphate transport in rat colon. Dissociated response to 1,25-dihydroxyvitamin $D_3$, *J. Clin. Invest.* **65:**1326–1331.

Letellier, M., Plante, G. E., Briere, N., and PetitClerc, C., 1982, Participation of alkaline phosphatase in the active transport of phosphates in brush border membrane vesicles, *Biochem. Biophys. Res. Commun.* **108:**1394–1400.

Lewis, K. O., 1973, The nature of the copper complexes in bile and their relationship to the absorption and excretion of copper in normal subjects and in Wilson's disease, *Gut* **14:**221–232.

Liedtke, C. M., and Hopfer, U., 1977, Anion transport in brush border membranes isolated from rat small intestine, *Biochem. Biophys. Res. Commun.* **76:**579–585.

Lifshitz, F., Harrison, H. C., and Harrison, H. E., 1969, Influence of parathyroid function upon the *in vitro* transport of calcium and phosphate by the rat intestine, *Endocrinology* **84:**912–917.

Lobao, A. O., Vitti Marcondes, D. M. S. S., Lemos, J. W., Peixoto Escubedo, M. I. B., de Oliveira, A. A. D., and Binnerts, W. T., 1982, The use of phosphorus-32 in the diagnosis of phosphorus deficiency in sheep, in: *The Use of Isotopes to Detect Moderate Mineral Imbalances in Farm Animals,* IAEA-TECDOC-267 (A Technical Document issued by the International Atomic Energy Agency), Vienna, 1982, pp. 33–48.

Long, R. G., Varghese, Z., Skinner, R. K., Wills, M. R., and Sherlock, S., 1978, Phosphate metabolism in chronic liver disease, *Clin. Chim. Acta* **87:**353–358.

Lücke, H., Stange, G., and Murer, H., 1981, Sulfate-sodium cotransport by brush-border membrane vesicles isolated from rat ileum, *Gastroenterology* **80:**22–30.

MacHardy, G. J. R., and Parsons, D. S., 1956, The absorption of inorganic phosphate from the small intestine of the rat, *Quart. J. Exp. Physiol.* **41:**398–409.

Marcinowska-Suchowierska, E., Lorenc, R. S., and Gray, T. K., 1981, Ileal calcium and phosphate absorption in chronic renal failure, in: *Calcium and Phosphate Transport across Biomembranes* (F. Bronner and M. Peterlik, eds.), Academic Press, New York, pp. 167–170.

Massry, S. G., 1978, The clinical syndrome of phosphate depletion, in: *Homeostasis of Phosphate and Other Minerals* (S. G. Massry, E. Ritz, and A. Rapado, eds.), Plenum Press, New York, pp. 301–312.

Matsumoto, T., Fontaine, O., and Rasmussen, H., 1980, Effect of 1,25-dihydroxyvitamin D-3 on phosphate uptake into chick intestinal brush border membrane vesicles, *Biochim. Biophys. Acta* **599:**13–23.

Max, E. E., Goodman, D. B. P., and Rasmussen, H., 1978, Purification and characterization of chick intestine brush border membrane. Effects of 1$\alpha$(OH)vitamin $D_3$ treatment, *Biochim. Biophys. Acta* **511:**224–239.

Meldrum, S. J., Watson, B. W., and Riddle, H. C., 1972, pH Profile of gut as measured by radiotelemetry capsule, *Brit. Med. J.* **2:**104–106.

Moog, F., and Glazier, H. S., 1972, Phosphate absorption and alkaline phosphatase activity in the small intestine of the adult mouse and of the chick embryo and hatched chick, *Comp. Biochem. Physiol.* **42A:**321–336.

Morgan, D. B., 1969, Calcium and phosphorus transport across the intestine, in: *Malabsorption* (R. H. Girdwood and A. N. Smith, eds.), Williams and Wilkins, Baltimore, pp. 74–91.

Nicolaysen, R., 1937a, A note on the calcium and phosphate requirement of rachitic rats, *Biochem. J.* **31:**105–106.

Nicolaysen, R., 1937b, Studies upon the mode of action of vitamin D. II. The influence on the faecal output of endogenous calcium and phosphorus in the rat, *Biochem. J.* **31:**107–121.

Nicolaysen, R., 1937c, Studies on the mode of action of vitamin D. V. The absorption of phosphates from isolated loops of the small intestine in the rat, *Biochem. J.* **31:**1086–1088.

Norman, A. W., Mircheff, A. K., Adams, T. H., and Spielvogel, A., 1970, Studies on the mechanism of action of calciferol. III. Vitamin D-mediated increase of intestinal brush border alkaline phosphatase activity, *Biochim. Biophys. Acta* **215:**348–359.

Norman, A. W., Putkey, J. A., and Nemere, I., 1981, Vitamin D-mediated intestinal calcium transport: Analysis of the complexity of the process, in: *Calcium and Phosphate Transport across Biomembranes* (F. Bronner and M. Peterlik, eds.), Academic Press, New York, p. 263–268.

O'Doherty, P. J. A., 1979, 1,25-Dihydroxyvitamin $D_3$ increases the activity of the intestinal phosphatidylcholine deacylation–reacylation cycle, *Lipids* **14:**75–77.

O'Doherty, P. J. A., DeLuca, H. F., and Eicher, E. M., 1977, Lack of effect of vitamin D and its metabolites on intestinal phosphate transport in familial hypophosphatemia of mice, *Endocrinology* **101:**1325–1330.

Peterlik, M., 1978a, Vitamin D-dependent phosphate transport by chick intestine: Inhibition by low sodium and *N*-ethylmaleimide, in: *Homeostasis of Phosphate and Other Minerals* (S. G. Massry, E. Ritz, and A. Rapado, eds.), Plenum Press, New York, pp. 149–159.

Peterlik, M., 1978b, Phosphate transport by embryonic chick duodenum: Stimulation by vitamin $D_3$, *Biochim. Biophys. Acta* **514:**164–171.

Peterlik, M., and Wasserman, R. H., 1975, Basic features of the vitamin D-dependent phosphate transport by chick jejunum *in vitro, Fed. Proc.* **34:**887.

Peterlik, M., and Wasserman, R. H., 1977, Effect of vitamin $D_3$ and 1,25-dihydroxyvitamin $D_3$ on intestinal transport of phosphate, in: *Phosphate Metabolism* (S. G. Massry and E. Ritz, eds.), Plenum Press, New York, pp. 323–332.

Peterlik, M., and Wasserman, R. H., 1978a, Effect of vitamin D on transepithelial phosphate transport in chick intestine, *Am. J. Physiol.* **234:**E379–388.

Peterlik, M., and Wasserman, R. H., 1978b, Stimulatory effect of 1,25-dihydroxycholecalciferol-like substances from *Solanum malacoxylon* and *Cestrum diurnum* on phosphate transport in chick jejunum, *J. Nutr.* **108:**1673–1679.

Peterlik, M., and Wasserman, R. H., 1980, Regulation by vitamin D of intestinal phosphate absorption, *Horm. Metabol. Res.* **12:**216–219.

Peterlik, M., Fuchs, R., and Cross, H. S., 1981a, Phosphate transport in intestine: Cellular pathways and hormonal regulation, in: *Calcium and Phosphate Transport across Biomembranes* (F. Bronner and M. Peterlik, eds.), Academic Press, New York, pp. 173–179.

Peterlik, M., Fuchs, R., and Cross, H. S., 1981b, Stimulation of D-glucose transport: A novel effect of vitamin D on intestinal membrane transport, *Biochim. Biophys. Acta* **649:**138–142.

Peterlik, M., Cross, H. S., and Fuchs, R., 1982a, Vitamin D and intestinal transport of phosphate, D-glucose and sodium, in: *Electrolyte and Water Transport Across Gastrointestinal Epithelia* (M. R. Case, A. Garner, L. A. Turnberg and J. A. Young, eds.), Raven Press, New York, pp. 305–308.

Peterlik, M., Cross, H. S., and Fuchs, R., 1982b, Effects of vitamin D on pathways of epithelial sodium transport in chick small intestine, *Naunyn-Schmiedeberg's Arch. Pharmacol.* **321**(Suppl.)**:**R57.

Pfleger, K., Rummel, W., and Jacobi, H., 1958, Phosphatdurchtritt am isolierten Darm unter Dinitrophenol und Thyroxin, *Biochem. Z.* **330:**303–309.

Rizzoli, R., Fleisch, H., and Bonjour, J.-P., 1977, Role of 1,25-dihydroxyvitamin $D_3$ on intestinal phosphate absorption in rats with a normal vitamin D supply, *J. Clin. Invest.* **60:**639–647.

Rose, R. C., and Schultz, S. G., 1971, Studies on the electrical potential profile across rabbit ileum, *J. Gen. Physiol.* **57:**639–663.

Schultz, S. G., 1979, Transport across small intestine, in: *Membrane Transport in Biology* (G. Giebisch, D. C. Tosteson, and H. H. Ussing, eds.), Springer-Verlag, Berlin, Heidelberg, New York, pp. 749–780.

Short, E. M., Binder, H. J., and Rosenberg, L. E., 1973, Familial hypophosphatemic rickets: Defective transport of inorganic phosphate by intestinal mucosa, *Science* **179:**700–702.

Skadhauge, E., and Thomas, D. H., 1979, Transepithelial transport of $K^+$, $NH_4^+$, inorganic phosphate and water by hen (Gallus domesticus) lower intestine (colon and coprodeum) perfused luminally *in vivo, Pfluegers Arch.* **379:**237–243.

Soergel, K. H., and Hofmann, A. F., 1972, Absorption, in: *Pathophysiology* (E. D. Frohlich, ed.), J. B. Lippincott, Philadelphia and Toronto, pp. 423–453.

Sonnenberg, A.,v. Lilienfeld-Toal, H., Sonnenberg, G. E., Rohner, H. G., and Strohmeyer, G., 1977, Serum 25-hydroxyvitamin $D_3$ levels in patients with liver disease, *Acta Hepato-Gastroenterol.* **24:**256–258.

Steele, T. H., 1977, Independence of phosphate homeostasis from parathyroid function in the phosphate-depleted rat, in: *Phosphate Metabolism* (S. G. Massry and E. Ritz, eds.), Plenum Press, New York, pp. 183–192.

Tanzer, F. S., and Navia, J. M., 1973, Calcitonin inhibition of intestinal phosphate absorption, *Nature* **242:**221–222.

Taylor, A. N., 1974, *In vitro* phosphate transport in chick ileum: Effect of cholecalciferol, calcium, sodium and metabolic inhibitors, *J. Nutr.* **104:**489–494.

Tenenhouse, H. S., and Scriver, C. R., 1978, The defect in transcellular transport of phosphate in the nephron is located in brush-border membranes in X-linked hypophosphatemia (*Hyp* mouse model), *Can. J. Biochem.* **56:**640–646.

Tenenhouse, H. S., and Scriver, C. R., 1979, Renal brush border membrane adaptation to phosphorus deprivation in the *Hyp*/Y mouse, *Nature* **281**:225–227.

Tenenhouse, H. S., Fast, D. K., and Scriver, C. R., 1981, Effect of 1,25$(OH)_2D_3$ on renal and intestinal transport of phosphate anion in *Hyp* mouse, in: *Calcium and Phosphate Transport across Biomembranes* (F. Bronner and M. Peterlik, eds.), Academic Press, New York, pp. 221–224.

Tsuchiya, W., and Okada, Y., 1982, Membrane potential changes associated with differentiation of enterocytes in the rat intestinal villi in culture, *Dev. Biol.* **94**:284–290.

Walling, M. W., 1977, Intestinal Ca and phosphate transport: Differential response to vitamin $D_3$ metabolites, *Am. J. Physiol.* **233**:E488–494.

Walling, M. W., 1978, Intestinal inorganic phosphate transport, in: *Homeostasis of Phosphate and Other Minerals* (S. G. Massry, E. Ritz, and A. Rapado, eds.), Plenum Press, New York, pp. 131–147.

Walling, M. W., and Kimberg, D. V., 1975, Effects of 1α,25-dihydroxyvitamin $D_3$ and *Solanum glaucophyllum* on intestinal calcium and phosphate transport and on plasma, Ca, Mg, and P levels in the rat, *Endocrinology* **97**:1567–1576.

Walling, M. W., Kimberg, D. V., Wasserman, R. H., and Feinberg, R. R., 1976, Duodenal active transport of calcium and phosphate in vitamin D-deficient rats: Effect of nephrectomy, *Cestrum diurnum*, and 1α,25-dihydroxyvitamin $D_3$, *Endocrinology* **98**:1130–1134.

Walling, M. W., Hartenbower, D. L., Coburn, J. W., and Norman, A. W., 1977, Effects of 1α,25-, 24R,25-, and 1α,24R,25-hydroxylated metabolites of vitamin $D_3$ on calcium and phosphate absorption by duodenum from intact and nephrectomized rats, *Arch. Biochem. Biophys.* **182**:251–257.

Walser, M., 1961, Ion association. VI. Interactions between calcium, magnesium, inorganic phosphate, citrate and protein in normal human plasma, *J. Clin. Invest.* **40**:723–730.

Walton, J., and Gray, T. K., 1979, Absorption of inorganic phosphate in the human small intestine, *Clin. Sci. Mol. Med.* **56**:407–412.

Wasserman, R. H., and Brindak, M. E., 1979, The effect of cholecalciferol on the phosphorylation of intestinal membrane proteins, in: *Vitamin D. Basic Research and Its Clinical Application* (A. W. Norman, K. Schaefer, D.v. Herrath, H.-G. Grigoleit, J. W. Coburn, H. F. DeLuca, E. B. Mawer, and T. Suda, eds.), Walter de Gruyter Verlag, Berlin and New York, pp. 703–710.

Wasserman, R. H., and Taylor, A. N., 1966, Vitamin $D_3$-induced calcium-binding protein in chick intestine mucosa, *Science* **152**:791–793.

Wasserman, R. H., and Taylor, A. N., 1973, Intestinal absorption of phosphate in the chick: Effect of vitamin $D_3$ and other parameters, *J. Nutr.* **103**:586–599.

Wasserman, R. H., Kallfelz, F. A., and Comar, C. L., 1961, Active transport of calcium by rat duodenum *in vivo*, *Science* **133**:883–884.

Wasserman, R. H., Taylor, A. N., and Kallfelz, F. A., 1966, Vitamin D and transfer of plasma calcium to intestinal lumen in chicks and rats, *Am. J. Physiol.* **211**:419–423.

Wasserman, R. H., Brindak, M. E., and Fullmer, C. S., 1981, Calcium-binding protein (CaBP) and other vitamin D-responsive proteins, in: *Calcium and Phosphate Transport across Biomembranes* (F. Bronner and M. Peterlik, eds.), Academic Press, New York, pp. 279–287.

Wertheim, A. R., Eurman, G. H., and Kalinsky, H. J., 1954, Changes in serum inorganic phosphorus during intravenous glucose tolerance tests: In patients with primary (essential) hypertension, other disease states, and in normal man, *J. Clin. Invest.* **33**:565–571.

Wilson, T. H., and Wiseman, G., 1954, The use of sacs of everted small intestine for the study of the transference of substances from the mucosal to the serosal surface, *J. Physiol.* **123**:116–125.

# 35

# Ion Transport in Nerve Membrane

Lorin J. Mullins

## I. INTRODUCTION

It has been recognized that since the Na electrochemical gradient in nerve is the energy source for flow of bioelectric currents, some restorative process is necessary to maintain the Na gradient. This entity, the Na pump, is primarily responsible for the conversion of the chemical free energy of the ATP molecule into an alternate store of free energy, the electrochemical gradient of $Na^+$ ions.

This review will be concerned with those sorts of ion movements that are either energized directly by ATP or those secondarily so energized when they draw on the Na electrochemical gradient. The function of the Na pumping has undergone a revision in our thinking mainly because it is now recognized that $Na^+$ can play a regulatory role in affecting the entry of Ca during depolarization, and presumably can control other processes that at present are poorly understood.

Ion transport in nerve has also been used to study ion transport generally since it can yield transport information that may be difficult or impossible to obtain in, for example, epithelia. Thus, giant nerve fibers of the squid and *Myxicola* have contributed information far beyond the sort of information that may be particularly important for axonal function (for reviews see Mullins, 1979; DeWeer, 1975; Baker, 1970, 1972).

---

*Lorin J. Mullins* • Department of Biophysics, University of Maryland at Baltimore, Baltimore, Maryland 21201.

## II. METHODS FOR TRANSPORT STUDIES

### A. Introduction

The development of methods for the measurement of ion and molecule fluxes across the plasma membrane has been of critical importance in furthering our understanding of membrane transport. At first sight, it might seem that isotope measurements would fulfill all the requirements for transport measurement. Unfortunately, this is not so since (1) some substances do not have isotopes, (2) some cells are sufficiently small so that one cannot resolve ion fluxes from extracellular space mixing ($Ca^{2+}$ movement in cardiac cells), and (3) internal complexation sometimes confuses isotopic measurements.

For reasons cited above, it has proved important to develop methods that measure net fluxes (the glass electrode for $H^+$), electric current measurement under a voltage clamp (net $Na^+$ movement), or optical measurements (aequorin light emission in the presence of $Ca^{2+}$). It is useful to note that a $Ca^{2+}$ influx measurement with isotopes may not reflect true influx if extracellular $Ca^{2+}$ binding absorbs much of the isotope. Similarly, a voltage-clamp measurement of current may have this contaminated with outward currents, while an electrode or optical signal sensitive to $Ca^{2+}$ must cope with intracellular sequestration or buffering.

### B. Isotopes

Most of our information about ion and molecule transport across the cell membrane comes from the use of isotopes of the substance being followed; these measurements allow the unequivocal assignment of influx and efflux of a substance. Such estimates, while doing much to make clear the movements of $Na^+$ and $K^+$ both in their diffusional and active transport aspects, have led to some difficulties in interpreting ion movements of $H^+$ (no isotopic movements are possible) and of $Ca^{2+}$ (binding and intracellular complexation are the predominant mode of dealing with $Ca^{2+}$ influx).

### C. Analytical Measurements

Considerations such as those detailed above have led to the idea that a measurement of the net gain of a substance may be more easily interpreted than an isotopic measurement. Such measurements, in the case of $Ca^{2+}$, have proved indispensable to understanding net $Ca^{2+}$ flux balance (Requena *et al.*, 1979) and the movements of ions that follow excitation of nerve cells. As an example, the change of membrane potential of a squid axon during an action potential is from a resting level of about $-60$ mV to a value of $+40$ mV or a difference of 100 mV. With a membrane capacitance of 1 $\mu F/cm^2$, the change would require a charge transfer of 0.1 volt $\times$ $10^{-6}$ coulomb/volt, $10^{-7}$ coulomb, or dividing by the Faraday, $10^{-12}$ $mole/cm^2$. Analytical values for Na gain per impulse show 3.5 $pmoles/cm^2$ imp or substantially more than the theoretical requirement for $Na^+$ gain.

### D. Electrode Measurements

In the case of $H^+$, its concentration in cells is so low ($10^{-7}$ M) that special methods for its measurement have always been required. The near-universal use of a glass electrode that is selectively permeable to $H^+$ is a testimonial to the need for measurement and to the difficulty with alternate measuring methods. It has proved possible to reduce $H^+$-sensing electrodes to sizes that allow the insertion of the electrode into cells of reasonable size and to make measurements that are reasonable (Thomas, 1971). More recently, use has been made of Na-sensitive glass electrodes for the intracellular measurement of such ions and for the detection of net fluxes of Na (Thomas, 1972).

A somewhat different approach to ion-specific electrodes has been the development of a series of ion-specific ion-exchange compounds that can be introduced into the tip of a microelectrode and, hence, used to detect concentration changes in $Mg^{2+}$, $Ca^{2+}$, $Cl^-$, and a variety of other ions. These measurements complement measurements of analytical changes in the case of the substances involved.

### E. Optical Measurements

Complementary to measurements with ion-specific electrodes is the use of dyes that change their absorbance when they complex with $Ca^{2+}$, $H^+$, or other ions. In the case of squid giant axons, measurements with dyes are not exactly analogous to measurements with electrodes since the former measure an "average" optical absorbance while the latter measure a local ion concentration. In the case of ions such as $Mg^{2+}$ the concentration of which in axoplasm is millimolar, there is undoubtedly no difference between the "average" (or optical) concentration and the local (surrounding the electrode) concentration, but in the case of both $H^+$ and $Ca^{2+}$, where concentrations are approximately 30–50 nM, there may be very large differences indeed between these two concentration-measuring locales.

## III. ACTIVE TRANSPORT

### A. Introduction to Na Fluxes

Because of the widespread distribution of the Na/K pump in cells, it has been the subject of intensive study by investigators interested in red cells, muscle, nerve, and epithelial secretory cells. A parallel effort by biochemists to understand $Na^+$,$K^+$-ATPase has resulted in our having a detailed knowledge of the pump and the steps involved in its activation. During the course of such investigations, it has also become clear that the Na/K pump is not the only mechanism capable of producing an Na efflux from cells as a variety of co- or countertransport systems are able to do just this. In particular, in squid axons, Baker *et al.*, (1969a,b) have shown that if $Na_i$ is elevated and $Na_o$ made zero then there is a large $Ca_o$-sensitive but ouabain-insensitive Na efflux that is presumably an exchange of $Na_i$ for $Ca_o$.

Because the literature on Na transport from axons is so extensive and has been reviewed extensively, only certain newer studies are referred to here and these concern (1) the electrogenic nature of the Na/K pump, and (2) the sensitivity of the pump to conditions that cause it to undergo abnormal modes of operation.

There is substantial agreement that the Na,K pump in red cells operates with a stoichiometry of 3 Na : 2 K : 1 ATP, and as a result of this unequal charge transfer (one third of the Na efflux should appear as electric current), the pump is electrogenic. The electrogenicity of the Na,K pump has been amply confirmed in all cells where it has been examined.

## B. Na-Transport Systems

The concentration of Na inside cells is about 1/15th to 1/20th that of the external solution bathing the cell, and there is furthermore usually an electric potential difference between the cell interior and the external solution such that cations would be attracted to the cell interior. Thus, it is apparent that some energy-consuming process must be involved in the maintenance of the steady-state $[Na]_i$.

Early studies with muscle (Steinbach, 1941) showed that a muscle could be kept in K-free solutions overnight in the refrigerator and that such muscles would gain substantial quantities of $Na^+$ and lose almost equivalent quantities of $K^+$. Furthermore, if K-containing solutions were reimposed on the fibers and they were warmed, the muscles would recover the $K^+$ they lost and extrude the $Na^+$ so that the suggestion was strong that there was a mechanism in the fiber membrane that could exchange $Na^+$ and $K^+$.

Early studies with nerve (Hodgkin and Keynes, 1955), showed that Na efflux as measured with isotopes was enhanced by the presence of $K_o$, and was inhibited by drugs such as ouabain or substances that interfered with the generation of ATP such as dinitrophenol. Thus, the idea emerging from these and other studies was that there was a mechanism in the nerve membrane that exchanged $Na^+$ for $K^+$. Since work was being done, it was also assumed that ATP was the substrate that provided the energy for this process although the actual studies with squid axons showed that phosphoarginine was a more effective substance than ATP in restoring Na efflux.

The first assumptions about Na/K pumping were that the exchange was electroneutral since measurements of membrane potential then undergoing intensive study appeared to favor a view that they could be explained solely on the basis of diffusion potentials; the demonstration by Kernan (1962) that the membrane potential of frog muscle fibers undergoing $Na^+$ unloading could exceed plausible values for the $K^+$ equilibrium potential provided some convincing evidence that Na/K exchange was not electroneutral and that, in fact, the electrogenicity of the $Na^+$ pump was a general phenomenon in nerve and indeed all other cells where $Na^+$ transport has been adequately studied.

While the electrogenicity of the Na pump in nerve fibers is at times difficult to demonstrate, especially if the membrane resistance is low and the $Na^+$ efflux small, it has been so demonstrated by DeWeer and Geduldig (1978); the current and ion flux produced by pumping is compared by Abercrombie and DeWeer (1978). Thus, they are both the direct demonstration of an increase in membrane potential and a demonstration of an extra current both of which are eliminated by cardioactive drugs.

The quantitative aspects of Na/K pumping, namely a specification of the coupling ratio Na/K, remains less satisfactory mainly because there are so many sources of $Na^+$ efflux and because $K^+$ influx is quite sensitive to membrane potential. The sorts of measurements that have been made are the ATP-dependent $K^+$ influx and $Na^+$ efflux (Brinley and Mullins, 1968), and these show that the coupling ratio increases with increases in $[Na]_i$ to values well in excess of three, while conventional coupling ratios obtained from studies with red cells show a rather constant 3:2 ratio. The use of ouabain in squid axons is less clear cut than it is in some other cells since it does not reduce $Na^+$ efflux to levels expected for passive diffusion; one is then forced to aim comparisons of the change in flux upon application of the inhibitor a procedure that can lead to large error.

It would appear that precise measurements of the future will utilize voltage-clamp methods to hold the membrane potential constant and that the change in current through the clamp will be used to measure Na-pump current when glycosides are applied. Such a measurement will have to be coupled to a measurement of $K^+$ influx $\pm$ glycoside to yield active $K^+$ influx. One then will have pump current ($Na^+$ current—$K^+$ current) and an independent measure of $K^+$ current so that stoichiometry can be measured.

Measurements utilizing purely the flux changes produced by compounds such as ouabain are not satisfactory because of the finding (Brinley and Mullins, 1968; Beaugé and Mullins, 1976) that ouabain, while totally inhibiting $Na^+$ pump efflux, induces apparently a new $Na^+$ efflux that requires ATP and a monovalent cation outside the fiber.

### 1. Na Influx

One of the predictions of the Hodgkin–Huxley analysis is that at the resting membrane potential there should be an influx of Na in the steady state that is given by the infinite time values of the product $m^3h$. This means that there should be an influx of Na that is dependent on membrane potential and capable of being inhibited by (tetrodotoxin) TTX. Such an inhibition has actually been identified (Baker *et al.*, 1969b) and, quantitatively, this influx is about 50% of the total Na influx. There is also a component of Na influx that depends on ATP (Brinley and Mullins, 1967), and this is presumably an inward movement that is coupled to the extrusion of ions such as $Mg^{2+}$and $Ca^{2+}$that have been shown to depend on the Na electrochemical gradient for their extrusion. The question arises as to why Na-gradient-dependent ion extrusion should have a requirement for ATP; this question remains unanswered.

### 2. Na Efflux

In addition to its producing an increase in $Na^+$ influx in squid axons of about 15 pmoles/cm$^2$ per sec, ATP also energizes a $Na^+$ efflux of about 35–50 pmoles/cm$^2$ per sec depending on values for $Na_i$ and $K_o$. The details of the $Na^+$ efflux from axons under a variety of conditions have been previously reviewed (Mullins, 1979) and are not covered here.

## C. Potassium Fluxes

In addition to the need for a $K^+$ reaccumulating mechanism that will compensate for the $K^+$ loss involved in the generation of each action potential in nerve, there is also a regulatory function that external $K^+$ concentrations exert on the operation of the Na/K pump. A recent study (Abercrombie and DeWeer, 1978) showed quite clearly both the regulatory role of $K_o$ and of $ADP_i$ on the generation of Na pump current. It had been known from earlier studies (DeWeer, 1970) that a rise in ADP would suppress Na/K exchange in favor of Na/Na exchange and, indeed, that while the Na pump is normally quite insensitive to changes in membrane potential, a low ratio of ATP/ADP makes it actually sensitive to hyperpolarization (Brinley and Mullins, 1971). These studies indicate that one can go from an electrogenic mechanism (the Na/K pump) to a nonelectrogenic one (Na/Na exchange) by changes in ATP/ADP ratio and that $K_o$ can change the extent to which the Na pump operates in a particular mode.

## D. Chloride Transport

Although chloride ions appear to be passively distributed in red blood cells and frog skeletal muscle, there is increasing recognition that these ions are not passively distributed in a variety of tissues including nerve. Other experiments show that a Cl/ $HCO_3$ exchange is a factor involved in the regulation of intracellular pH.

The original studies of $Cl_i$ of squid axoplasm were those of Keynes (1963) and Deffner (1961) and these gave 108 and 137 mM, while Brinley and Mullins (1965) found 131 mM. These values are almost three times higher than the values expected for a Donnan distribution and a membrane potential of the axon of $-70$ mV.

Flux studies by Russell (1976) showed that there was a requirement for ATP if maximal values for Cl influx were to be realized, and most recently (Russell, 1983) it has been possible to show that Cl influx in squid axons is a Na-gradient-dependent process with 3 $Cl^-$, $2Na^+$, and 1 $K^+$ moving per cycle of the carrier.

## E. Introduction to Calcium Transport

The recognition of the important regulatory role that $Ca^{2+}$ plays in the release of transmitter from nerve terminals as well as the role it plays in allowing its own entry via $Ca^{2+}$ channels has stimulated intensive study of Ca movement in nerve over the past 10 years. We are indebted to DiPolo (1973a,b) for introducing the internal dialysis technique to $Ca^{2+}$ studies with results that have had wide applicability. These studies, supplemented by aequorin measurements and arsenazo III measurements, have led to a large body of experimental information that as yet is not completely sorted out (for recent reviews see Requena and Mullins, 1979; DiPolo and Beaugé, 1980; DiPolo, 1983; Requena, 1983).

### 1. Ca Influx

It is of some interest to compare the experimentally measured values of Ca influx under a variety of conditions. From DiPolo (1979a,b) we can note that with physiological values of $Ca_i$, $Ca_o$, and $Na_i$, $Ca^{2+}$ influx is 140 $\pm$ 12 fmoles/cm$^2$ per sec in the presence of ATP and 85 $\pm$ 7 in its absence. This yields an ATP-sensitive flux of

55 fmoles; by comparison, if $Na_i$ is made zero, but ATP is present, $Ca^{2+}$ influx is 87 ± 8 fmoles so that the ATP-sensitive and the $Na_i$-sensitive fluxes are identical (55 or 53 fmoles).

In a more recent study (DiPolo *et al.*, 1982), a $Ca^{2+}$ influx in the presence of ATP is given as 117.3 ± 5.5 fmoles/$cm^2$ per sec and this fractionated between a TTX-sensitive $Ca^{2+}$ influx (82 fmoles), a $Na_i$-dependent Ca influx (24 fmoles), and an influx that is insensitive to either TTX or $Na_i$ and is 12 fmoles. One concludes that most of the $Ca^{2+}$ influx is via $Na^+$ channels, and the ATP- and $Na_i$-sensitive fluxes are identical or that Na/Ca exchange is via an ATP-sensitive mechanism. This sensitivity is not absolute since a very large $Ca^{2+}$ influx can be produced by depolarization if $Na_i$ is finite (Mullins and Requena, 1981) even in the absence of ATP.

### 2. Ca Efflux

The first measurements of $Ca^{2+}$ efflux were by Hodgkin and Keynes (1957), who used the microinjection of $^{45}Ca$ into squid axons in order to initiate measurements. These studies gave a variety of useful results but they emphasized the lack of control that an experimenter has over efflux measurement. The introduction of dialysis methods (DiPolo, 1973a,b) overcame these limitations but imposed new ones since now the total control of $Ca_i$ necessarily was in the hands of the experimenter, and since reagent contamination is substantial, EGTA buffers are a uniform necessity.

Efflux measurements did establish that external Na was a stimulant of $Ca^{2+}$ efflux with a $K_{1/2}$ of about 40 mM, that ATP enhanced $Ca^{2+}$ efflux that was dependent on $Na_o$, and that if $Ca_i$ were low enough, then $Ca^{2+}$ efflux was unaffected by removing $Na_o$. The measurements also established that $Ca^{2+}$ efflux at normal values for $Ca_i$ was in the range of 50–100 fmoles/$cm^2$ per sec, a value close to that measured for $Ca^{2+}$ influx (see above). Since $Ca^{2+}$ efflux was thought to be an Na/Ca exchange, it was puzzling that ATP had a clear enhancing action on $Ca^{2+}$ efflux. Note that it also enhanced Ca influx so that if one supposed that the $Ca^{2+}$ efflux to be an ATP-driven Ca pump (DiPolo, 1979b), then some other explanation was necessary for the ATP-dependent Ca influx. Since Ca efflux is clearly a function of $Ca_i$, $Na_i$, $Na_o$, $Ca_o$, $E_m$, and ATP, it is difficult to sort out the variable in the equation and to be certain that one can divide $Ca^{2+}$ efflux into separate ATP-driven pumps and ATP-dependent Na/Ca exchange.

### 3. Internal Ca Concentration

One of the difficulties with $Ca^{2+}$ efflux measurements was an uncertainty about the value for ionized Ca in axoplasm. A measurement of this, using aequorin and arsenazo III as indicators (DiPolo *et al.*, 1976), established a value between 20 and 50 nM. Other values obtained using Ca-sensitive electrodes have been in excess of 100 nM but there are reasons (below) for supposing that these electrodes are insufficiently sensitive for the measurements that have been undertaken.

### 4. Ca Entry with Depolarization

The first application of aequorin injection in squid axons was by Baker *et al.* (1971) who showed that $Ca^{2+}$ entry was greatly enhanced by depolarizing the axon.

Now, Mullins and Brinley (1975), have shown that Ca efflux was enhanced by increasing membrane potential and decreased by depolarization, so that rather than supposing that depolarization opened Ca channels in the squid axon membrane, one could also imagine that the Na/Ca exchange was either electrogenic or that the carrier itself had a charge such that depolarization led to more carrier availability.

A test of the hypothesis that Ca entry with depolarization *is* Na/Ca exchange was made by Mullins and Requena (1981) who showed that there was no Ca entry with depolarization if $[Na]_i$ were reduced to low levels and that the response was greatly enhanced by raising internal Na concentration. These are the changes expected if Ca entry with depolarization is rate-limited by the availability of internal Na. A further study (Mullins *et al.*, 1983) of this effect showed that Ca entry is half maximal if $Na_i$ is 25 mM, while it is virtually undetectable if $Na_i$ is 18 mM. The Hill coefficient of the curve relating Ca entry to $Na_i$ is 7.

### 5. *Effect of ATP on Ca Fluxes*

It was noticed quite early in the study of Ca efflux from dialyzed axons by DiPolo (1974) that Ca efflux was larger from the axon if ATP was present. For technical reasons, $Ca_i$ was set rather high (300 nM) and, under these conditions, ATP doubled Ca efflux (the absolute flux increment was 140 fmoles/cm$^2$ per sec). A subsequent study (DiPolo, 1977) showed that if $[Ca]_i$ were 60 nM, then 95% of the Ca efflux would disappear if ATP were removed from the dialysis solution. The absolute magnitude of the Ca efflux dependent on ATP was of the order of 45 fmoles/cm$^2$ per sec.

When the effects of ATP on Ca influx were examined (DiPolo, 1979a), it was found that ATP produced a 55 *fmole/cm*$^2$*s* increment in Ca influx when $[Ca]_i$ was 60 nM (as above), and removing *either* ATP or $Na_i$ reduced influx to the same lower level. Thus, a conclusion from these studies would be that ATP has an equal effect in promoting an increment in the influx and in the efflux of Ca.

A complicating factor in our analysis is that Brinley *et al.* (1975) showed that the Ca efflux from dialyzed axons was reduced to low levels when $Na_o$ was removed if $Ca_i$ were high (300 nM), but that there was no effect on Ca efflux upon making $Na_o$ zero if $Ca_i$ were 50 nM or less. Two approaches are possible: (1) defining Ca efflux in the absence of $Na_o$ as via another mechanism (the uncoupled Ca pump), or (2) supposing that when $Ca_i$ is low and, hence, the demands on the Na/Ca carrier are minimal, ions other than Na can have some ability to promote Ca extrusion especially if the affinity of the carrier for such ions is promoted by the addition of ATP to the system.

Discriminating between the two hypotheses outlined above has proved difficult for a variety of reasons. One approach (DiPolo and Beaugé, 1981) has been to show that vanadate has a large inhibitory effect on Ca efflux if $Na_o$ is absent and little effect if it is present; unfortunately, since ATP is supposed by hypothesis to make the Na/Ca carrier more able to accept non-$Na^+$ ions, this experiment is of little use. Another approach (DiPolo and Beaugé, 1982) has been to show that the Ca efflux in the absence of $Na_o$ has a smaller sensitivity to internal pH change than does the Ca efflux in the presence of $Na_o$. Again, because of the kinetic complexity of a carrier that must bind multiple $H^+$ and $Na^+$ along with Ca, it can be argued that we are, in fact, dealing with the same carrier in two different states.

In another study (DiPolo *et al.*, 1982) it has been possible to show that Ca influx is reduced to 33% or less of control values by TTX, suggesting that most Ca enters via the Na channels. Since, in the steady state, Ca influx must equal Ca efflux, this finding suggests that with total fluxes in both directions of the order of 100 fmoles/$cm^2$ per sec, only 30 fmoles is carrier-mediated Na/Ca and the rest leak and enter via Na channels. One of the interesting measurements of this paper (DiPolo *et al.*, 1982, Figures 6 and 7) shows that there is an increase of about 80 fmoles/$cm^2$ per sec upon depolarizing an axon with 100 mM K seawater when $Ca_i$ is 100 nM, $Na_i$ is 60, and ATP is 2 mM. A repeat of this experiment with $Na_i = 0$ shows by contrast a 40 fmoles decrease in Ca influx (presumably because Ca entry via Na channels is decreased because depolarization decreases the steady-state value of the Hodgkin–Huxley parameter *h* and also decreases the driving force moving Na inward). A repeat of this sort of experiment with all conditions the same but with $Na_i = 90$ mM shows that a change from $Ca_i = 100$ nM to 600 nM leads to a seven-fold increase in the Ca entry with depolarization. This surprising finding is totally at variance with all studies of Ca entry with depolarization as these show either that there is no effect of prior Ca entry (a high $Ca_i$) on subsequent Ca entry or that Ca entry is depressed by previous depolarizations and a high $Ca_i$. The finding suggests that what is at fault is the method for measuring Ca influx by dialysis. In fact, the method has not been subjected to a rigorous test for adequacy. It is necessary to appreciate that for the method to work, Ca entering the axon has to diffuse around 200 μm in axoplasm, a medium with strong Ca buffering properties and past a variety of organelles with significant Ca-concentrating properties. It is known that Ca counts cannot be measured if EGTA is omitted from the dialysis medium and that beyond 300 μM EGTA there is no further sequestering of Ca by this agent; this is not the same as showing that all $^{45}Ca$ counts are being collected. It may well be that if Ca influx is very small and the concentration change in axoplasm is small that Ca is collected, but Ca fluxes have been measured at the 10,000 fmole/$cm^2$s level and these may well lead to a loss of substantial fractions of the entering Ca. According to this analysis, the effect of raising the $[Ca]_i$ would be that the cold Ca supplied by the dialysis system would then improve the collection of entering $^{45}Ca$ by exchanging with it. This scheme implies that there is a reservoir of stored $^{45}Ca$ counts somewhere in the axoplasm that can be drained of its counts by increasing the cold Ca in axoplasm.

It has been demonstrated in intact axons (Baker and McNaughton, 1976) that the injection of EGTA leads to a loss of $Ca_o$-dependent Na efflux. They suggested that the complexation of some heavy metal might be responsible for the effect. DiPolo (1979a) showed that in dialyzed axons Ca influx measured with ATP was constant at 40 fmoles/$cm^2$ per sec when $Ca_i$ was varied from 0–800 nM but that it rose with increases in $Ca_i$ when $Na_i = 90$ mM. This is the basis for a claim that Ca influx is enhanced by $Ca_i$. Against this finding is that of Abercrombie and Sjodin (1980) that in *Myxicola* axons Ca influx was independent of $Ca_i$ in the range of 16–560 nM and was also independent of EGTA in the range of 0.01–10 mM.

## F. Magnesium Transport

DeWeer (1976) has shown that the efflux of magnesium from squid giant axons is virtually unaffected by a strong depolarization of the membrane from its normal value of −60 to −30 mV. Repetitive stimulation of the axon on the other hand showed a small but definite increase in magnesium efflux which amounts to 10 fmoles/$cm^2$ per sec, suggesting that magnesium ions have some small ability to pass through either sodium or potassium channels.

A study of magnesium inward movement in perfused squid axons (Rojas and Taylor, 1975) found that for voltage-clamp pulses up to a duration of 5 msec, the entry was about 20 fmoles/$cm^2$ per impulse. The external magnesium concentration was 55 mM, whereas internal Mg is closer to 5 mM; the results with stimulation suggest that influx and efflux with stimulation are of a similar order of magnitude.

Original measurements of magnesium movement in squid axons were made by Baker and Crawford (1972) which showed that magnesium efflux was sensitive to metabolic inhibition with cyanide and that it was sensitive to external sodium concentration. Since intracellular magnesium is a mandatory requirement for the formation of MgATP, the substrate for the Na/K transport system, its regulation in the nerve fiber, and indeed in other cells, is an important consideration. It has been shown (DeWeer, 1970) that high intracellular magnesium concentrations inhibit Na efflux and in some unpublished measurements, Mullins and Brinley have found that complexing intracellular magnesium with 1,2-cyclohexane diaminetetraacetic acid (CDTA), a compound similar to EDTA, virtually stopped the Na/K pump. DeWeer (1976) has exploited the sensitivity of the Na/K pump to intracellular magnesium to determine that at a level of free ion between 3–4 mM there is no change in the normal Na efflux of the squid axon; hence, he infers that these values represent free magnesium. Analysis of the total Mg of the fiber was 6.7 mM, and since ATP binds Mg very strongly and ATP is of the order of 3 mM, this would leave 3.7 mM for free magnesium in agreement with the above estimates. Although the Mg concentration in the hemolymph was found to be 44 mM, the fraction of this that is ionized is unknown. The study of DeWeer also confirmed the notion that ATP is essential for the efflux of Mg since the injection of apyrase reduced Mg efflux about ten-fold. External Na was also essential for the efflux process. Thus, these findings are similar to those for Na/Ca exchange.

Rojas and Taylor (1975) studied Mg influx in perfused squid axons and found it to be about 0.016 pmole/$cm^2$ per sec as compared with estimates in intact axons (Baker and Crawford, 1972) of 0.16 *p*mole. Since the internal perfusion solution contains no Na, and since it is likely that Mg movement is by an Na/Mg exchange, the difference between these values is understandable. In dialyzed axons (Mullins *et al.*, 1977), it was possible to show that Mg efflux is a complex function of $Na_o$, $Na_i$, $Mg_i$, and $ATP_i$. Mg efflux can equal that measured in the absence of ATP if $Mg_i$ is high enough. $Na_i$ is an inhibitor of Mg efflux and $Na_o$ a promoter. In a companion study (Mullins and Brinley, 1978), magnesium influx in dialyzed squid axons was also measured. The results showed that influx is increased about 2.5-fold (from 1–2.5 *p*mole/$cm^2$/sec) by the addition of ADP to the dialysis fluid and that influx can be further increased by increasing $Na_i$.

A summary of the results with magnesium fluxes is that they appear to be driven by the Na gradient and substantially accelerated by the presence of ATP in the fiber. Since both magnesium influx and efflux are stimulated by ATP, it is not entirely clear that ATP is used as a conventional substrate rather than as a substance that catalyzes Mg flux.

### G. Hydrogen Ion Transport

Because of the dissociation of water into hydrogen and hydroxyl ions, it is not feasible to use isotopic techniques to measure hydrogen ion movement, and recourse must be made to net flux measurements either by a glass electrode or ion-exchange electrode or via dyes and spectrophotometry.

The first such measurements in nerve fibers were those of Caldwell (1958). Caldwell established a normal value for $pH_i$ of 7.3 and showed that exposure of the nerve fiber to $CO_2$ produced a prompt and sustained acidification. These early studies were followed 20 years later by Thomas (1974) who designed recessed tip microelectrodes of pH-sensitive glass, and observed in snail neurons that the $CO_2$ or $NH_4$ produced, respectively, acidification and alkalinization of the cell and that these changes were transient.

This work was followed by that of Boron and DeWeer (1976) who showed that the internal pH of 7.32 was susceptible to manipulation with $CO_2$ or $NH_4$, and that the transient nature of the pH change could be explained by assuming that membrane hydrolysis led to a rapid movement of either $CO_2$ itself or $NH_3$ followed by a slower movement of the ions $NH_4^+$ and $HCO_3^-$. In spite of this, a theoretical study showed that the simultaneous passive movements of $CO_2$ and $HCO_3$ cannot explain the results of $CO_2$ treatment of the nerve fiber and an active proton extrusion in exchange for Na is required. A detailed study of intracellular pH and regulation is given by Roos and Boron (1981). The prevailing current view is that intracellular regulation of hydrogen ion concentration is carried out by both $Na_o$- and $Cl_i$-dependent processes with the principal energy for $H^+$ extrusion coming from the Na gradient. Fluxes are of the order of two hydrogen ions per sodium and per chloride thus making the process effectively electroneutral.

In addition to physiological variables such as $CO_2$ controlling intracellular pH, there is now abundant evidence in nerve cells and in squid axons that the entry of calcium in connection with various physiological activities results in an intracellular acidification as a result of the buffering of the calcium that enters. Inatracellular buffers apparently exchange protons for calcium ions. This means that the rapid entry of Ca is followed by an acidification and that the acidification is likely to produce physiological effects on the calcium transport mechanisms. A review of pH transients in squid axons is given by DeWeer (1978).

## IV. PHYSIOLOGICAL INTEGRATION OF ION FLUXES

One of the conclusions that seem apparent from an examination of ion fluxes in nerve and their dependence either on substrate or on the Na electrochemical gradient is that gradient control makes it possible for the nerve fiber to respond to previous bioelectric activity (stored as a form of memory in extra Na concentration in the fiber) by, for example, increasing the entry of another important ion ($Ca^{2+}$) by virtue of there being an extraordinarily large increase in $Ca^{2+}$ entry with depolarization for a trivial increase in $Na_i$ (of the order of 1 mM).

At the same time, this memory in the form of an enhanced $Na_i$ is itself a transient since it will stimulate the Na pump to increase its efflux rate, and $Na_i$ will return to control values where Ca entry with depolarization is much more modest.

While this situation is satisfactory, it by no means exhausts the complexity of the response of a nerve fiber to previous stimulation. The Ca entry that is generated by extra $Na_i$ leads to most of the entering $Ca^{2+}$ being buffered by Ca-absorbing systems that apparently exchange Ca for $H^+$. In turn, this leads to a decreased Ca entry and to the need for the extrusion or reabsorption of protons if the Na/Ca exchange is to recover its original properties. While $Ca_i$ declines promptly upon the cessation of depolarization (as a result of the action of buffers + Ca extrusion), the change in pH is more long lasting (Ahmed and Connor, 1980).

## REFERENCES


Abercrombie, R. F., and DeWeer, P., 1978, The electric current generated by the sodium pump of squid giant axons: Effects of external potassium and internal ADP, *Am. J. Physiol.* **4:**C63–68.

Abercrombie, R. F., and Sjodin, R. A., 1980, Calcium efflux from Myxicola giant axons: Effects of extracellular calcium and intracellular EGTA, *J. Physiol. (London)* **306:**175–191.

Ahmed, Z., and Connor, J. A., 1980, Intracellular pH changes induced by calcium influx during electrical activity in molluscan neurons, *J. Gen. Physiol.* **75:**403–426.

Baker, P. F., 1970, Sodium–calcium exchange across nerve cell membrane, in: *Calcium and Cellular Function* (A. W. Cuthbert, ed.), Macmillan, London, pp. 96–107.

Baker, P. F., 1972, Transport and metabolism of calcium ions in nerve, *Prog. Biophys. Mol. Biol.* **24:**179–223.

Baker, P. F., and Crawford, A. C., 1972, Mobility and transport of magnesium in squid giant axons, *J. Physiol. (London)* **227:**855–874.

Baker, P. F., and McNaughton, P. A., 1976, Kinetics and energetics of calcium efflux from intact squid giant axons, *J. Physiol. (London)* **259:**103–144.

Baker, P. F., Blaustein, M. P., Hodgkin, A. L., and Steinhardt, R. A., 1969a, The influence of calcium on sodium efflux in squid axons, *J. Physiol. (London)* **200:**431–458.

Baker, P. F., Blaustein, M. P., Keynes, R. D., Manil, J., Shaw, T. I., and Steinhardt, R. A., 1969b, The ouabain-sensitive fluxes of sodium and potassium in squid giant axons, *J. Physiol. (London)* **200:**459–496.

Baker, P. F., Hodgkin, A. L., and Ridgway, E. B., 1971, Depolarization and calcium entry in squid axons, *J. Physiol. (London)* **218:**709–755.

Beaugé, L. A., and Mullins, L. J., 1976, Strophanthidin-induced sodium efflux, *Proc. R. Soc. Lond. B.* **194:**279–284.

Boron, W. F., and DeWeer, P., 1976, Intracellular pH transients in squid giant axons caused by $CO_2$, $NH_3$, and metabolic inhibitors, *J. Gen. Physiol.* **67:**91–112.

Brinley, F. J., Jr., and Mullins, L. J., 1965, Ion fluxes and transference numbers in squid axons, *J. Neurophysiol.* **28:**526–544.

Brinley, F. J., Jr., and Mullins, L. J., 1967, Sodium extrusion by internally dialyzed squid axons, *J. Gen. Physiol.* **50:**2303–2331.

Brinley, F. J., Jr., and Mullins, L. J., 1968, Sodium fluxes in internally dialyzed squid axons, *J. Gen. Physiol.* **52:**181–211.

Brinley, F. J., Jr., and Mullins, L. J., 1971, The fluxes of sodium and potassium across the squid axon membrane under conditions of altered membrane potential, *Fed. Proc.* **30:**255.

Brinley, F. J., Jr., Spangler, S. G., and Mullins, L. J., 1975, Calcium and EDTA fluxes in dialyzed squid axons, *J. Gen. Physiol.* **66:**223–250.

Caldwell, P. C., 1958, Studies on the internal pH of large muscle and nerve fibres, *J. Physiol. (London)* **142:**22–62.

Deffner, G. G. J., 1961, The dialyzable free organic constituents of squid blood; a comparison with nerve axoplasm, *Biochim. Biophys. Acta* **47:**378–388.

DeWeer, P., 1970, Effects of intracellular adenosine-5′-diphosphate and orthophosphate on the sensitivity of sodium efflux from squid axons to external sodium and potassium, *J. Gen. Physiol.* **56:**583–620.

DeWeer, P., 1975, Aspects of the recovery processes in nerve, in: *Physiology, Vol. 3: Neurophysiology* (C. C. Hunt, ed.), University Park Press, Baltimore, pp. 232–278.

DeWeer, P., 1976, Axoplasmic free Mg levels and Mg extrusion from squid axons, *J. Gen. Physiol.* **68:**159–178.

DeWeer, P., 1978, Intracellular pH transients induced by $CO_2$ or $NH_3$[1,2], *Resp. Physiol.* **33:**41–50.

DeWeer, P., and Geduldig, D., 1978, Contribution of sodium pump to resting potential of squid giant axon, *Am. J. Physiol. Soc.* **235:**C55–C62.

DiPolo, R., 1973a, Sodium dependent calcium influx in dialyzed barnacle muscle fibers, *Biochim. Biophys. Acta* **298:**279–283.

DiPolo, R., 1973b, Calcium efflux from internally dialyzed squid giant axons, *J. Gen. Physiol.* **62:**575–589.

DiPolo, R., 1974, Effect of ATP on the calcium efflux in dialyzed squid axons, *J. Gen. Physiol.* **64:**503–517.

DiPolo, R., 1977, Characterization of the ATP-dependent calcium efflux in dialyzed squid giant axons, *J. Gen. Physiol.* **69:**795–814.

DiPolo, R., 1979a, Calcium influx in internally dialyzed squid giant axons, *J. Gen. Physiol.* **73:**91–113.

DiPolo, R., 1979b, Physiological role of ATP-driven calcium pump in squid axon, *Nature* **278:**271–273.

DiPolo, R., and Beaugé, L., 1983, *Annu. Rev. Physiol.*, vol. 45.

DiPolo, R., and Beaugé, L., 1980, Mechanisms of calcium transport in the giant axon of the squid and their physiological role, *Cell Calcium* **1:**147–169.

DiPolo, R., and Beaugé, L., 1981, The effects of vanadate on calcium transport in dialyzed squid axons. Sidedness of vanadate-cation interactions, *Biochim. Biophys. Acta* **645:**229–236.

DiPolo, R., and Beaugé, L., 1982, The effect of pH on $Ca^{2+}$ extrusion mechanisms in dialyzed squid axons, *Biochim. Biophys. Acta* **688:**237–245.

DiPolo, R., Requena, J., Brinley, F. J., Jr., Mullins, L. J., Scarpa, A., and Tiffert, T., 1976, Ionized calcium concentrations in squid axons, *J. Gen. Physiol.* **67:**433–467.

DiPolo, R., Rojas, H., and Beauge, L., 1982, Ca entry at rest and during prolonged depolarization in dialyzed squid axons, *Cell Calcium* **3:**19–41.

Hodgkin, A. L., and Keynes, R. D., 1955, Active transport of cations in giant axons from *Sepia* and *Loligo*, *J. Physiol. (London)* **128:**28–60.

Hodgkin, A. L., and Keynes, R. D., 1957, Movements of labelled calcium in squid giant axons, *J. Physiol. (London)* **138:**253–281.

Kernan, R. P., 1962, Membrane potential changes during sodium transport in frog sartorius muscle, *Nature* **193:**986–987.

Keynes, R. D., 1963, Chloride in the giant squid axon, *J. Physiol. (London)* **169:**690–705.

Mullins, L. J., 1979, Transport across axon membranes, in: *Membrane Transport in Biology,* Vol. II (D. C. Tosteson, ed.), Springer-Verlag, New York, pp. 161–210.

Mullins, L. J., and Brinley, F. J., Jr., 1975, Sensitivity of calcium efflux from squid axons to changes in membrane potential, *J. Gen. Physiol.* **65:**135–152.

Mullins, L. J., and Brinley, F. J., Jr., 1978, Magnesium influx in dialyzed squid axons, *J. Membr. Biol.* **43:**243–250.

Mullins, L. J., and Requena, J., 1981, The "late" Ca channel, *J. Gen. Physiol.* **78:**683–700.

Mullins, L. J., Brinley, F. J., Jr., Spangler, S. G., and Abercrombie, R. F., 1977, Magnesium efflux in dialyzed squid axons, *J. Gen. Physiol.* **69:**389–400.

Mullins, L. J., Tiffert, T., Vassort, G., and Whittembury, J., 1983, Effects of internal sodium and hydrogen ions and of external calcium ions and membrane potential on calcium entry in squid axons, *J. Physiol. (London)* **338:**319.

Requena, J., 1983, Ca transport and regulation in nerve fibers. *Annu. Rev. Biophys.* **12:**237–258.

Requena, J., and Mullins, L. J., 1979, Calcium movement in nerve fibres, *Quart. Rev. Biophys.* **12:**371–460.

Requena, J., Mullins, L. J., and Brinley, F. J., Jr., 1979, Calcium content and net fluxes in squid giant axons, *J. Gen. Physiol.* **73:**327–342.

Rojas, E., and Taylor, R. E., 1975, Simultaneous measurements of magnesium, calcium and sodium influxes in perfused squid giant axons under membrane potential control, *J. Physiol. (London)* **252:**1–27.

Roos, A., and Boron, W. F., 1981, Intracellular pH, *Physiol. Rev.* **61:**296–434.

Russell, J. M., 1976, ATP-dependent chloride influx into internally dialyzed squid giant axons, *J. Membr. Biol.* **28:**335.

Russell, J. M., 1983, Cation coupled chloride influx in squid axon, *J. Gen. Physiol.* **81:**909–925.

Steinbach, H. B., 1941, Chloride in the giant axons of the squid, *J. Cell. Comp. Physiol.* **17:**57.

Thomas, R. C., 1971, Na microelectrodes with sensitive glass inside the tip, in: *Ion Selective Microelectrodes* (N. C. Herbert and R. Khuri, eds.), Marcel Dekker, New York.

Thomas, R. C., 1972, Intracellular sodium activity and the sodium pump in snail neurons, *J. Physiol. (London)* **220:**55–71.

Thomas, R. C., 1974, Intracellular pH of snail neurones measured with a new pH-sensitive glass microelectrode, *J. Physiol. (London)* **238:**159–180.

# 36

# The Molecular Basis of Neurotransmission: Structure and Function of the Nicotinic Acetylcholine Receptor

*Robert Anholt, Jon Lindstrom, and Mauricio Montal*

> One thing I have learned in a long life: that all our science, when measured against reality, is primitive and childlike; and yet it is the most precious thing we have.
>
> Albert Einstein

## I. INTRODUCTION

Acetylcholine was identified as the first neurotransmitter as a result of elegant experiments by Otto Loewi in 1921 who demonstrated that the vagus nerve liberates a substance that has an inhibitory effect on the rate of the heartbeat of an isolated frog heart. Loewi showed that this "vagus substance" could be transferred from the fluid filling the heart onto another heart and there reproduce the same inhibitory effect. He coined the term "humoral transmission" to describe this activity (Loewi, 1921). Subsequent experiments identified the "vagus substance" as acetylcholine. It became clear

*Robert Anholt and Jon Lindstrom* • The Salk Institute for Biological Studies, San Diego, California 92138. *Mauricio Montal* • Departments of Biology and Physics, University of California, San Diego, La Jolla, California 92093. *Present address of R.A.:* Department of Neuroscience, The Johns Hopkins University School of Medicine, Baltimore, Maryland 21205.

that acetylcholine, in addition to its influence on the heart, exerted a variety of pharmacologically distinct effects, which were classified by Sir Henry Dale (1934) as "muscarinic" and "nicotinic" actions, because some were mimicked best by muscarine and others by nicotine. Acetylcholine receptors with muscarinic ligand-binding properties are characterized by prolonged responses of slow onset which are mediated through nucleotide cyclases (for short reviews see Sokolovsky and Bartfai, 1981; Hartzell, 1982), whereas acetylcholine receptors with nicotinic-binding properties are characterized by rapid responses in which ligand binding regulates the opening and closing of a cation-specific channel through a conformational alteration in the molecule. Acetylcholine receptors at the neuromuscular junctions of striated muscle and at the synapses of fish electric organs (which are phylogenetically related to muscle tissue) are the best-studied nicotinic acetylcholine receptors and the subjects of this review.

A detailed description of the role of acetylcholine as chemical transmitter at the synaptic junction evolved in the 1950's and subsequent years as a result of the pioneering work by, among others, Sir Bernhard Katz, John Eccles, and Stephen Kuffler. These early electrophysiological experiments showed that the binding of acetylcholine to a receptor on the postsynaptic membrane results in an increase in the membrane conductance for cations which leads to depolarization of the muscle cell membrane, an event that represents the initial trigger for muscle contraction (see Katz, 1966). David Nachmansohn, in 1953, suggested that the postsynaptic receptor for acetylcholine might be a protein, which after binding the neurotransmitter would undergo a change in conformation leading to the formation of a transmembrane conductance pathway across the membrane. The collective electrophysiological, pharmacological, and structural data that have accumulated during the last three decades have validated this hypothesis.

Detailed electrophysiological and pharmacological studies of neuromuscular transmission, primarily using preparations of frog muscle, preceded the isolation and biochemical characterization of the acetylcholine receptor by nearly two decades. These studies generated a wealth of information about the selectivity, conductance amplitude, lifetime, and voltage dependence of the acetylcholine receptor channel and led to the identification and pharmacological characterization of numerous agonists, e.g., carbamylcholine and suberyldicholine, antagonists, e.g., *d*-tubocurarine, and noncompetitive blockers, e.g., local anesthetics.

Although major advances had already been made towards a detailed electrophysiological and pharmacological description of the cholinergic response by the beginning of the 1960's, the biochemical isolation and subsequent structural studies on the nicotinic acetylcholine receptor became feasible only in the late 1960's after the identification by Lee and collaborators in Taiwan of certain elapid snake neurotoxins as very specific, high-affinity probes for the acetylcholine-binding site (Lee, 1970). These toxins, such as α-bungarotoxin (Chang and Lee, 1963; Mebs *et al.*, 1971; Clark *et al.*, 1972; Ravdin and Berg, 1979) or cobratoxin (*Naja naja siamensis* toxin III; Karlsson *et al.*, 1971; Walkinshaw *et al.*, 1980) are small basic proteins ($M_r \sim 8000$) which can be conveniently iodinated to high specific activities and which have proven invaluable for the localization (Fertuck and Salpeter, 1974, 1976), purification (reviewed by Karlin, 1980; Changeux, 1981), and quantitation (Schmidt and Raftery,

1973; Lindstrom *et al.*, 1981a) of the acetylcholine receptor. Biochemical studies of the acetylcholine receptor benefited further from the recognition that the electric organs of electric eels (*Electrophorus electricus*) and rays (*Torpedo californica, marmorata* and *nobiliana,* and *Narcine brasiliensis* and *japonica*) are extremely rich sources of acetylcholine receptor (for reviews see Lindstrom, 1978; Karlin, 1980; Changeux, 1981). The ontogeny of the electric organ resembles the development of muscle, but contractile proteins are present only transiently during the development of the electric organ (Mellinger *et al.*, 1978; Fox and Richardson, 1979; Krenz *et al.*, 1980). Thousands of synapses develop per electrocyte in contrast to the formation of a single synapse per muscle fiber. Thus, the electric organ becomes specialized for the generation of large electrical discharges through acetylcholine-activated channels (Bennett *et al.*, 1961; Mellinger *et al.*, 1978). Acetylcholine receptors can be obtained in 1000-fold larger quantities from electric organ tissue (~100 mg/kg) than from muscle (<0.5 mg/kg). As a result, most structural information on the acetylcholine receptor has been obtained from studies on electric organ receptor. Recent studies on receptors purified from mammalian muscle indicate extensive overall similarities between the acetylcholine receptor from *Torpedo* electric organ and the acetylcholine receptor from mammalian muscle which confirms the validity of the *Torpedo* receptor as a model system for receptors from other sources (Conti-Tronconi *et al.*, 1982e).

In spite of the extensive progress in the structural and functional characterization of the acetylcholine receptor, a number of crucial questions remain unanswered. These questions include the three-dimensional structure of the receptor molecule, the nature of the interactions between its subunits, the regulation of gene expression and subunit assembly during synthesis of the acetylcholine receptor, and a complete structural description at the molecular level of the functional events that underlie neurotransmitter-regulated gating of the transmembrane cation channel. Recently, progress towards solutions of some of these problems has been made by several major achievements such as cloning of genes which encode the receptor subunits, production of a large number of monoclonal antibodies against different determinants on the surface of the receptor molecule, and development of reconstituted systems which allow the incorporation of functional acetylcholine receptors in model membranes. In this chapter, we will describe the principal structural and functional aspects of the acetylcholine receptor with particular emphasis on their interrelationships.

## II. STRUCTURAL ASPECTS OF THE ACETYLCHOLINE RECEPTOR

### A. Acetylcholine Receptor from Electric Organ

#### 1. Subunit Composition and Molecular Size of the Purified Acetylcholine Receptor

Acetylcholine receptors were initially isolated from fish electric organ by affinity chromatography using neurotoxins or agonist analogues conjugated to agarose. Purification of receptors by affinity chromatography yielded preparations with specific activities ranging between 5–10 μmoles of [$^{125}$I]α-bungarotoxin-binding sites/g protein. This corresponds to one toxin binding site per molecular weight of 100,000–125,000 (reviewed by Karlin, 1980; Changeux, 1981).

Proteolysis of acetylcholine receptors by tissue proteases during homogenization of the electric organ generated initially considerable controversy with respect to the subunit composition of the purified acetylcholine receptor. Proteolytic nicking of the acetylcholine receptor during the initial homogenization and subsequent purification causes the breakdown products of previously intact subunits to run with the tracking dye preventing their visualization on the gel (Lindstrom *et al.*, 1980a; Huganir and Racker, 1980). Proteolytic degradation of the receptor can be prevented by the inclusion of EDTA to inhibit $Ca^{2-}$dependent proteases, iodoacetamide to block thiolproteases, and phenylmethanesulfonyl fluoride to inactivate serine proteases during the initial steps of the isolation procedure. Acetylcholine receptor preparations from *Torpedo* prepared under conditions that prevent proteolysis appear as four polypeptide bands on SDS–polyacrylamide gels with *apparent* molecular weights of 40,000, 50,000, 57,000, and 64,000 (Figure 1; Karlin *et al.*, 1975; Hucho *et al.*, 1978; Vandlen *et al.*, 1979; Lindstrom *et al.*, 1979a). Since the apparent molecular weight values vary slightly among different laboratories depending on the species of ray and the gel system used, the subunits are indicated as α, β, γ, and δ, respectively. As we will discuss in detail below, elucidation of the complete primary structures of all four subunits has made it clear that the real molecular weights of the subunits are more similar than would appear from their migration on polyacrylamide gels (Noda *et al.*, 1982, 1983a,b; Claudio *et al.*, 1983). A similar subunit pattern has been found in the case of receptor from the electric eel, although here proteolysis tends to be a more severe problem than in the case of *Torpedo,* because the receptor in eel electric organ is present at ten-fold lower concentration (Lindstrom *et al.*, 1980b).

Figure 2 presents a schematic representation of the four receptor subunits. All four receptor subunits are glycopeptides (4–7% carbohydrate by weight; Vandlen *et al.*, 1979; Lindstrom *et al.*, 1979a). The four chains are similar in amino acid composition being particularly rich in aspartic acid and glutamic acid, and having a low content of arginine, lysine, and histidine (Vandlen *et al.*, 1979; Lindstrom *et al.*, 1979a; see also Noda *et al.*, 1982, 1983a,b). Hence, the receptor is an acidic protein with an isoelectric point of 4.5–4.8 (Brockes *et al.*, 1975). Amino acid analysis has also indicated the presence of phosphothreonine and phosphoserine residues (Vandlen *et al.*, 1979). Protein kinase (Gordon *et al.*, 1977a, 1980; Teichberg *et al.*, 1977; Saitoh and Changeux, 1980; Huganir and Greengard, 1983) and phosphatase (Gordon *et al.*, 1979) activities have been identified in fractions of receptor-rich electric organ membranes, and phosphorylation of some (Gordon *et al.*, 1977b; Saitoh and Changeux, 1980; Smilowitz *et al.*, 1981; Huganir and Greengard, 1983) or all (Davis *et al.*, 1982) of the four subunits has been demonstrated. Evidence suggests that the principal phosphorylation sites are located on the cytoplasmic side of the membrane (Figure 2; Wennogle *et al.*, 1981; Davis *et al.*, 1982). Methylation of the acetylcholine receptor has also been reported (Kloog *et al.*, 1980; Flynn *et al.*, 1982). Thus far, the functional significance of phosphorylation or methylation of the acetylcholine receptor remains unknown.

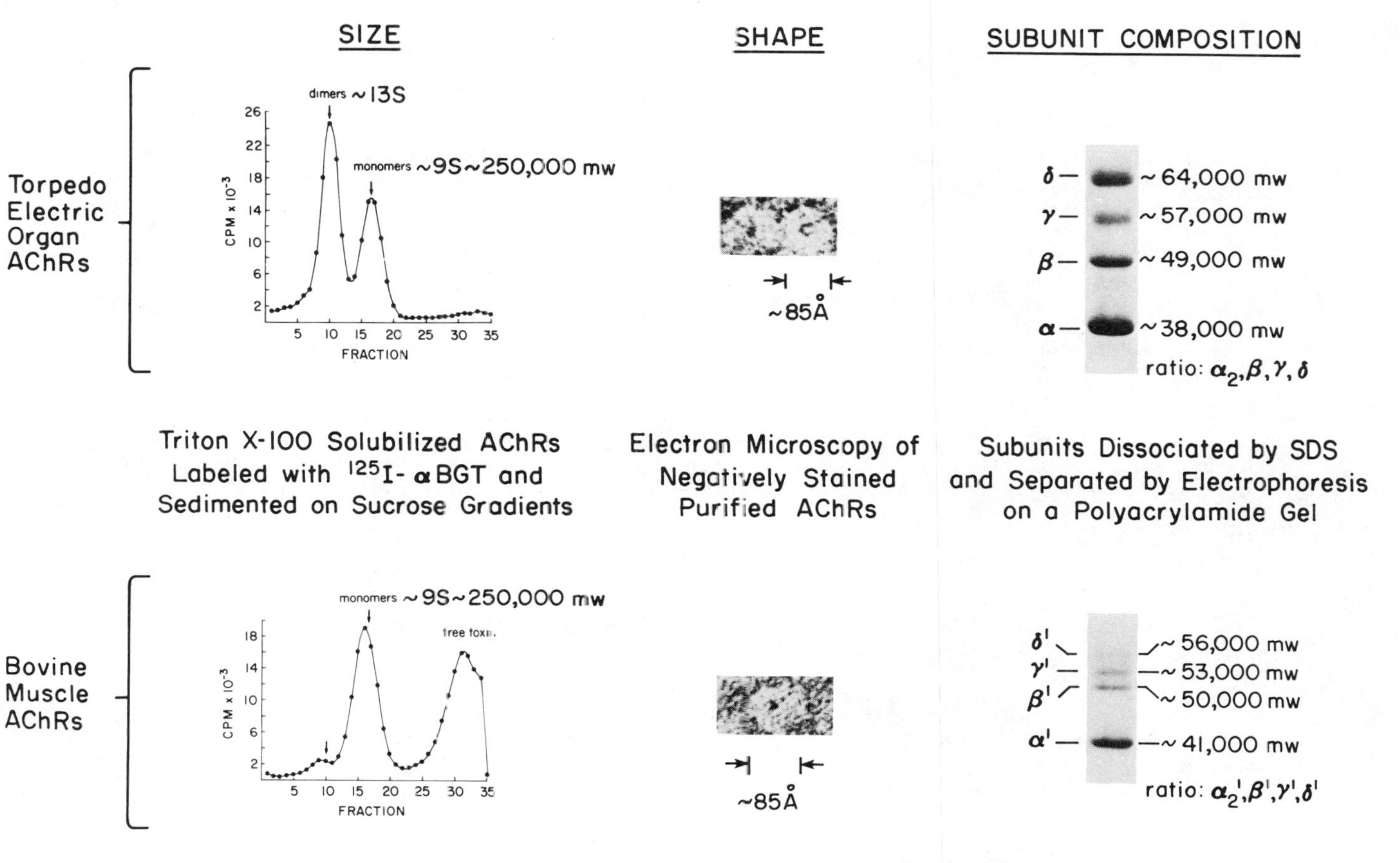


*Figure 1.* Structural similarities between acetylcholine receptors from fish electric organs and mammalian muscle. AChR indicates the acetylcholine receptor, [$^{125}$I]-αBGT (iodinated α-bungarotoxin) and SDS (the detergent, sodium dodecyl sulfate). Notice that most receptors from *Torpedo* electric organ are present as dimers, whereas, receptors from bovine muscle appear only as monomers, although a small peak is discernible on the gradients (indicated by the arrow) at a position expected for a 13 S dimeric form. Whether this peak is a small aggregation artifact or a remnant of a proteolytically damaged dimer remains to be determined. This figure is reproduced from Lindstrom (1984).

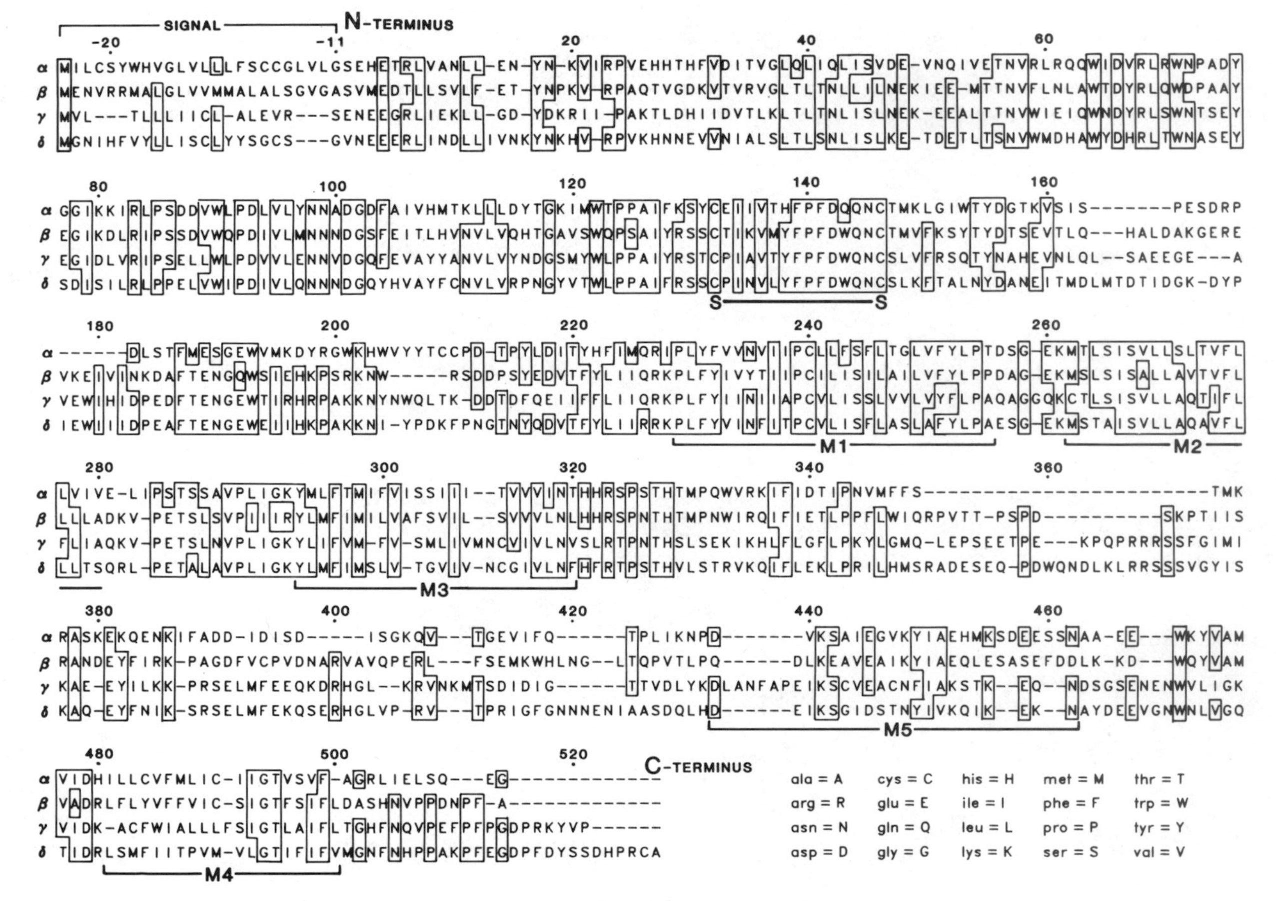
SIGNAL
N-TERMINUS
C-TERMINUS
M1
M2
M3
M4
M5
S—S
ala = A
arg = R
asn = N
asp = D
cys = C
glu = E
gln = Q
gly = G
his = H
ile = I
leu = L
lys = K
met = M
phe = F
pro = P
ser = S
thr = T
trp = W
tyr = Y
val = V

Figure 2.

Careful determination of the molecular weight of the acetylcholine receptor complex by sedimentation of solubilized receptor under conditions in which the contribution of the detergent was compensated with $D_2O$ has yielded a molecular-weight value of 250,000 daltons (Reynolds and Karlin, 1978). This value is close to molecular-weight estimates obtained by different techniques, such as osmometry (Martinez-Carrion *et al.*, 1975) and SDS–gel electrophoresis of crosslinked receptors (Hucho *et al.*, 1978). The molecular weight of the receptor is consistent with the presence of two α-subunits for each of β, γ, and δ, and implies the presence of two α-bungarotoxin binding sites per complex. This subunit stoichiometry has been confirmed directly by quantitation of the individual chains extracted from polyacrylamide gels after electrophoresis in SDS (Lindstrom *et al.*, 1979a) and by sequence analysis of both membrane-bound and Triton-solubilized receptor, in which N-terminal amino acids are released from the four chains in molar ratios of 2 : 1 : 1 : 1 (Figure 2; Raftery *et al.*, 1980; Conti-Tronconi *et al.*, 1982a). Sucrose-gradient centrifugation of purified acetylcholine receptors in the presence of monoclonal antibodies directed against determinants on the α-subunit has also demonstrated the presence of two α-subunits per receptor complex for *Torpedo*, eel, and bovine acetylcholine receptors (Conti-Tronconi *et al.*, 1981b; Tzartos *et al.*, 1983).

In the native membrane of *Torpedo* electric organ, the acetylcholine receptor occurs predominantly as a dimer, formed by a disulfide bridge between the δ-subunits of adjacent monomers (Chang and Bock, 1977; Hucho *et al.*, 1978; Hamilton *et al.*, 1979). On sucrose gradients, receptor monomers sediment as a 9 S species and dimers have a sedimentation coefficient of 13 S (Figure 1; Chang and Bock, 1977). Treatment with reducing agents (Chang and Bock, 1977; Hucho *et al.*, 1978; Hamilton *et al.*, 1979; Anholt *et al.*, 1980) or trypsin (Chang and Bock, 1977; Wennogle *et al.*, 1981) converts dimers into monomers. Tryptic mapping of the δ-subunit suggests that the intermolecular δ–δ disulfide bond is formed on the cytoplasmic side of the membrane (Wennogle *et al.*, 1981). Dimer formation has not yet been demonstrated for receptors from eel electroplaque (Lindstrom and Patrick, 1974) or mammalian muscle (Berg *et*

←

*Figure 2.* Amino acid sequences of the subunits of the acetylcholine receptor from *Torpedo californica*. The amino acid sequences presented were deduced by Noda *et al.* (1983b) from the corresponding nucleotide sequences of cloned cDNAs. The alignment for maximum homology is according to Noda *et al.* (1983b). The boxes indicate the occurrence of identical amino acids at homologous positions. Notice that each subunit possesses a signal peptide. In this alignment the proposed active site disulfide would be formed between cysteines -132 and -146 of the α subunit, and the proposed δ–δ disulfide might be formed via the cysteine at position 526. The four transmembranous hydrophobic α helical domains proposed by Noda *et al.* (1983b), Claudio *et al.* (1983), and Devillers-Thiery *et al.* (1983) are designated M1, M2, M3 and M4. In addition, Fairclough *et al.* (1984) and Guy (1984) have proposed a fifth transmembranous domain in each subunit, indicated in the figure as M5. This domain consists of an amphipathic α helix which contains charged amino acids on one side and apolar amino acids on the other side, such that alignment of these corresponding amphiphatic domains of the five subunits would form a transmembrane channel with a hydrophilic core (Fairclough *et al.*, 1984; Guy, 1984). In this scheme the N-termini would be located on the synaptic side of the acetylcholine receptor, whereas the C-termini would be intracellular.

*al.*, 1972; Shorr *et al.*, 1981), but efficient proteolysis around the disulfide bond may eliminate dimers during the initial steps of the isolation procedure.

Immunological crossreactivity between subunits detected by monoclonal antibodies revealed the existence of homologous domains between the subunits (Tzartos and Lindstrom, 1980; Tzartos *et al.*, 1981). Sequencing of the first 54 N-terminal amino acids of all four subunits of *Torpedo* receptor showed 35–50% homology between the polypeptides with closest correspondence between the γ- and δ-subunits (Figure 2; Raftery *et al.*, 1980). A similar degree of homology was found for the four receptor subunits of eel receptor (Conti-Tronconi *et al.*, 1982a) and bovine receptor (Conti-Tronconi *et al.*, 1982e). In addition, comparison of the sequences of eel subunits with the corresponding subunits from *Torpedo* revealed a 46–71% homology between these two species (Conti-Tronconi *et al.*, 1982a,b). These sequence homologies suggest that acetylcholine receptor subunits were derived by a process of duplication and reduplication of a primordial subunit at some time during evolution before the divergence of primitive vertebrates into chondrichthyes and teleosts approximately 400 million years ago. The complete amino acid sequences of all four subunits have been deduced from the nucleotide sequences of cDNA clones (Figure 2; Noda *et al.*, 1982, 1983a,b; Claudio *et al.*, 1983; Devillers-Thiery *et al.*, 1983). Although there are similarities in amino acid sequence throughout the lengths of the four subunits, Noda *et al.* (1983a,b) identified three highly conserved regions one of which they postulated to contain the site to which affinity labels of acetylcholine attach to the α subunits. In addition, they tentatively assigned a role to cysteine 500 on the δ chain in the formation of receptor dimers (Noda *et al.*, 1983a). The molecular weights of the α (50,116 daltons), β (53,681 daltons), γ (56,601 daltons) and δ (57,565 daltons) subunits are closer than would have been expected from their apparent molecular weights on acrylamide gels in sodium dodecyl sulfate (40,000, 50,000, 57,000 and 64,000 daltons, respectively). Claudio *et al.* (1983) suggested that the γ subunit may cross the membrane four times with both the N-terminal and C-terminal protruding from the synaptic side of the membrane. Four putative transmembrane segments have also been identified in each of the other subunits (Noda *et al.*, 1983b). Several invariant proline residues are evident at positions where the peptides are proposed to bend into the membrane. An inverted repeat of the amino acid sequence around proline 302 of the α chain is especially intriguing (Noda *et al.*, 1982). In addition, Fairclough *et al.* (1982) and Guy (1984) have proposed a fifth transmembrane segment, which consists of an amphipatic α helix containing hydrophilic amino acids on one side and apolar ones on the other side. Alignment of these corresponding amphipathic domains of the five subunits might then form a transmembrane channel with a hydrophilic interior (Fairclough *et al.*, 1984; Guy, 1984). According to this model the N-termini of the subunits would protrude from the synaptic side of the membrane whereas the C-termini would be intracellular (Fig. 2). Noda *et al.* (1982) suggested that the acetylcholine binding site is formed in the vicinity of cysteines 128 and 142 of the α subunit, which could form the disulfide bond which has to be reduced prior to affinity alkylation of the acetylcholine receptor (described in detail below). In addition, descriptions of cDNA clones containing partial sequences of the α subunit have been reported (Giraudat *et al.*, 1982; Sumikawa *et al.*, 1982b) and Ballivet *et al.*, (1982) have also reported the cloning of the gene encoding the γ subunit.

## 2. Shape and Subunit Topography of the Membrane-Bound Acetylcholine Receptor

Initial demonstrations that all subunits are transmembrane polypeptides came from investigations of the accessibility of receptor subunits to proteolytic enzymes (Wennogle and Changeux, 1980; Strader and Raftery, 1980) or iodinating agents (St. John *et al.*, 1982) acting from the external and internal side of the membrane as well as binding of antibodies to saponin and/or alkali treated membranes (Tarrab-Hadzai *et al.*, 1978; Strader *et al.*, 1979; Froehner, 1981). The subunits are intimately associated and can only be dissociated under denaturing conditions, such as treatment with SDS. They do not dissociate in 8 M urea, guanidine hydrochloride, Triton X-100, or sodium cholate (reviewed by Karlin, 1980; Changeux, 1981). Although extensive proteolysis leads to the disappearance from polyacrylamide gels of distinct peptide bands greater than 10,000 daltons, the nicked chains hang tightly together in such a way that the complexes retain both α-bungarotoxin binding and cation channel activity (Lindstrom *et al.*, 1980a; Huganir and Racker, 1980). Despite the loss of some antigenic determinants as indicated from binding studies with monoclonal antibodies (Einarson *et al.*, 1982; Gullick and Lindstrom, 1983a), such proteolytically nicked receptors still retain their characteristic appearance as observed by electron microscopy (described below; Lindstrom *et al.*, 1980a) as well as a sedimentation coefficient close to the value of native acetylcholine receptor (Huganir and Racker, 1980; Conti-Tronconi *et al.*, 1982c).

Negative staining of purified acetylcholine receptors in solution (Cartaud *et al.*, 1978; Holtzman *et al.*, 1982) and negative staining or deep-etching of acetylcholine receptor-enriched electric organ membranes (Cartaud *et al.*, 1978; Schiebler and Hucho, 1978; Klymkowsky and Stroud, 1979, Heuser and Salpeter, 1979; Zingsheim *et al.*, 1980, 1982a,b; Anholt *et al.*, 1982; St. John *et al.*, 1982) followed by electron microscopic observation shows the acetylcholine receptors as characteristic 80- to 90-Å doughnut-shaped particles containing an approximately 20-Å wide central pit (Figure 1). This central portion can be stained with tannic acid and may be continuous with the ion channel (Potter and Smith, 1977; Sealock, 1982a,b). In native electric organ membranes, these particles can often be seen as extensive double rows interdigitated with particle-free areas (Heuser and Salpeter, 1979). Immunoelectron microscopy using subunit-specific antibodies labeled with gold has demonstrated that all four polypeptides of the acetylcholine receptor are contained within these doughnut-shaped particles (Klymkowsky and Stroud, 1979; Kistler and Stroud, 1981). Digital image reconstruction of negatively stained electron micrographs revealed a highly asymmetrical density contour map of horseshoe-shaped unit structures containing three principal stain excluding areas (Zingsheim *et al.*, 1980, 1982a,b; Kistler and Stroud, 1981). X-ray diffraction data suggest that acetylcholine receptor molecules traverse the membrane extending 15 ± 5 Å from the bilayer on the cytoplasmic side and protruding 55 ± 5 Å from the membrane on the synaptic side with an overall length normal to the plane of the membrane of 110 Å (Ross *et al.*, 1977; Klymkowsky and Stroud, 1979). These structural dimensions are consistent with low-angle neutron scattering data (Wise *et al.*, 1979).

The center-to-center distance between individual rosettes in acetylcholine receptor

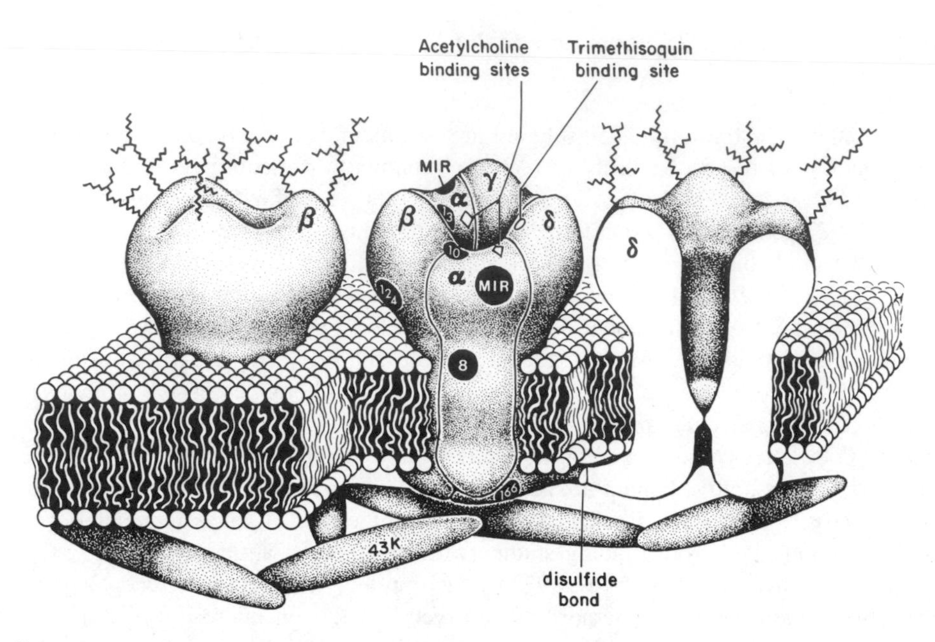


*Figure 3.* A structural model of the acetylcholine receptor. The figure presented is a modification of a model by Kistler *et al.* (1982). Notice that the bulk of the macromolecule protrudes from the synaptic side of the membrane. All subunits are glycopeptides and traverse the membrane at least once. Receptor monomers are linked as dimeric units via a disulfide bond between δ-subunits. The α-subunits are shown to contain the acetylcholine-binding sites (Weill *et al.*, 1974), the MIR (Tzartos and Lindstrom, 1980), and binding sites for monoclonal antibodies 10 and 13 (Tzartos and Lindstrom, 1980; see Table 1). The position of monoclonal antibody 124 at a sterically masked location is tentative. The binding site of monoclonal antibody 8 near the membrane interphase is based on Gullick *et al.* (1981) and the intracellular position of the antigenic determinant to which monoclonal antibody 166 binds is according to Anderson *et al.*, (1983) and Gullick and Lindstrom (1983a). A local anesthetic-binding site has been indicated on the extracellular surface of the δ-subunit according to Wennogle *et al.* (1981). Subunit topography is indicated as proposed by Kistler *et al.* (1982). Relative dimensions are based on Ross *et al.* (1977) and Wise *et al.* (1979). It should be noted that the shapes of the subunits are merely schematic and hypothetical. The positions of antigenic regions and ligand-binding sites are, similarly, only schematic illustrations and are not intended to indicate precise topographical locations.

dimers was determined to be 96 Å, both by electron microscopy and by neutron scattering (Wise *et al.*, 1981a). The contact point between the individual doughnuts in the dimer defines the position of the δ-subunits and can therefore serve as a reference point for the topographical localization of the individual subunits within the rosette. The δ-subunit has been localized opposite the gap in the horseshoe-shaped density contour map of the receptor monomer (Zingsheim *et al.*, 1982b). Several plausible, although not yet conclusive, models have been suggested for the topography of the subunits relative to the δ–δ disulfide bond (Figure 3; Wise *et al.*, 1981b; Kistler *et al.*, 1982; Zingsheim *et al.*, 1982a). Treatment of receptors in the membrane with the crosslinking reagent diamide leads to the formation of higher oligomers in which the monomeric units are linked by disulfide bonds alternately between pairs of δ-chains and pairs of β-chains (Hamilton *et al.*, 1979). High-resolution electron microscopy has indicated that in trimers formed in this way, the disulfide bond between the δ-subunits is separated from the disulfide bond between the β-subunits by an angle in the range of 50–80° (Wise *et al.*, 1981b). Labeling of the α-subunits with biotin-conjugated neurotoxin complexed with avidin (Holtzman *et al.*, 1982) and studies using monoclonal antibodies against the α-subunits (Stroud and Lindstrom, unpublished observations) indicate the angle between the two α-subunits to be 150 ± 30°. In addition, the toxin-binding sites have been localized by an analysis of the difference in mass between receptors labeled with α-bungarotoxin and unlabeled receptors after image averaging of negatively stained electron micrographs of each preparation (Zingsheim *et al.*, 1982a). Two regions were identified in the horseshoe-shaped density map which showed a significant increase in mass upon binding of α-bungarotoxin. One of these two toxin-binding regions was localized adjacent to the site of the δ–δ disulfide bond and the other region was ~50 Å away from the δ-subunit. Both toxin-binding sites appeared to be diametrically opposed within the rosette separated by a distance of ~50 Å. From these studies, it seems feasible that the subunits are topographically distributed around the central pit in such a way that both α-subunits are adjacent to the β-subunit on one side and separated by the γ- and δ-subunits on the other side (Figure 3; Wise *et al.*, 1981b; Kistler *et al.*, 1982; Zingsheim *et al.*, 1982a,b). This model is consistent with results from crosslinking experiments in which extensive crosslinking between the subunits has been reported, especially between the α- and δ-subunits (Witzemann *et al.*, 1979; Schiebler *et al.*, 1980; Oswald and Changeux, 1982), but in which crosslinking between the two α-subunits has never been observed (reviewed by Karlin, 1980). Studies in which [$^{125}$I]α-bungarotoxin was crosslinked to the receptor by ultraviolet light irradiation indicate that, in addition to the α-subunit, the δ-subunit and, to a lesser extent, the γ-subunit contribute to the binding sites for α-bungarotoxin (Oswald and Changeux, 1982). Although these binding sites overlap with the ligand-binding sites, additional groups on the neurotoxin can interact with the receptor enabling the formation of a particularly stable, high-affinity toxin–receptor complex. Kistler *et al.*, (1982) and Fairclough *et al.* (1984) have suggested that the binding sites for α-bungarotoxin are located on the top of the synaptic crest of the acetylcholine receptor about 55 Å above the plane of the membrane, implying that agonist-induced channel gating is accompanied by major conformational alterations which are propagated along the molecule.

### 3. Antigenic Structure of the Acetylcholine Receptor

The critical role which snake neurotoxins have played in the initial biochemical characterization of the acetylcholine receptor and the advent of hybridoma technology have stimulated an interest in the development of sets of monoclonal antibodies against different antigenic determinants on the acetylcholine receptor molecule. Such antibodies could form a library of toxin-like, specific high-affinity reagents for sites other than the agonist-binding sites and would be valuable both as structural and functional probes. Several laboratories have described the production of monoclonal antibodies against the acetylcholine receptor (Gomez *et al.*, 1979; Tzartos and Lindstrom, 1980; Tzartos *et al.*, 1981, 1983; Lennon *et al.*, 1980; James *et al.*, 1980; Mochly-Rosen and Fuchs, 1981) including monoclonal antibodies against the acetylcholine-binding sites of the receptor (Gomez *et al.*, 1979; Jameas *et al.*, 1980; Mochly-Rosen and

Fuchs, 1981). These latter antibodies have, however, limited potential since they duplicate the applications of neurotoxins.

Lindstrom and collaborators have developed a library of approximately 150 monoclonal antibodies against acetylcholine receptors. Initially, Tzartos and Lindstrom (1980) described the production of 17 monoclonal antibodies against *Torpedo* acetylcholine receptor and found that the majority of these antibodies competed for binding to the same region on the acetylcholine receptor (Table 1). They termed this region, which dominates the immunogenicity of the acetylcholine receptor, the "main immunogenic region" (MIR). More than half of the antibodies in sera from animals immunized with purified *Torpedo* acetylcholine receptor are directed against this region (Tzartos and Lindstrom, 1980). The MIR is located on the extracellular surface of the α-subunits and is distinct from the acetylcholine-binding site so that antibodies bound to the MIR still allow the binding of neurotoxins (Figure 3; Tzartos and Lindstrom, 1980). Subsequently, a main immunogenic region was also identified on receptors from eel (Tzartos *et al.*, 1981) and several mammalian species, including bovine (Tzartos *et al.*, 1983) and human receptor (Tzartos *et al.*, 1982, 1983), as well as the amphibian *Rana pipiens* (Sargent *et al.*, 1984). The MIR appears to be highly conserved in evolution (Tzartos and Lindstrom, 1980; Tzartos *et al.*, 1981) and may, therefore, fulfill an essential function, which has as yet not been identified. Antibodies bound to the MIR do not affect the neurotransmitter-regulated permeability response (Lindstrom *et al.*, 1981b) and the origin of its strong immunogenicity is not clear. Denaturation of the acetylcholine receptor in SDS obliterates the MIR so that crossreaction of antibodies to the MIR with denatured α-subunits is limited (Tzartos and Lindstrom, 1980; Tzartos *et al.*, 1981; Gullick *et al.*, 1981). It has been speculated that the MIR may represent a recognition site for interactions with components of the basal lamina (Tzartos *et al.*, 1981).

In addition to the MIR, approximately 30 other antigenic regions have been identified on the acetylcholine receptor. These have been distinguished by competitive binding between monoclonal antibodies (Tzartos and Lindstrom, 1980; Tzartos *et al.*, 1981, 1982) and by their ability to bind to specific sets of overlapping proteolytic fragments obtained from the isolated subunits (Gullick *et al.*, 1981; Gullick and Lindstrom, 1982a, 1983a). Some of these monoclonal antibodies react specifically with receptors from a unique species, whereas, others crossreact extensively between receptors from different species. A significant number of monoclonal antibodies crossreact between different receptor subunits illustrating the homologies and evolutionary relationships between these polypeptides (Tzartos and Lindstrom, 1980; Tzartos *et al.*, 1981, 1983; Gullick *et al.*, 1981; Gullick and Lindstrom, 1982a, 1983a).

Some monoclonal antibodies have been found to bind poorly to membrane-bound acetylcholine receptor although they bind well to detergent-solubilized receptor. Some of these antibodies can be categorized as antibodies which bind to regions which are sterically hindered by the close proximity of receptors at the synaptic junction or regions close to the membrane interface (Gullick *et al.*, 1981). A large number of monoclonal antibodies, which were raised against denatured acetylcholine receptor subunits, were found to be directed against that portion of the receptor which protrudes into the cytoplasmic compartment (Figure 3; Anderson *et al.*, 1983). This observation

*Table 1. Anti-Acetylcholine Receptor Antibody Producing Hybridoma Clones*[a]

| Clone number | Concentration: Electric organs: Torpedo | Electric organs: Eel | Muscles: Rat | Muscles: Fetal calf | Muscles: Mouse | Muscles: $BC_3H1$ cells | Muscles: Human | Muscles: Chicken | Subunit specificity |
|---|---|---|---|---|---|---|---|---|---|
| 1 | 36,800 | 0 | 0 | 0 | 0 | 0 | 0 | 0 | α, MIR |
| 2 | 31,300 | 0 | 0 | 0 | 0 | 0 | 0 | 0 | α, MIR |
| 3 | 13,000 | 0 | 0 | 0 | 0 | 0 | 0 | 0 | α[b] |
| 4 | 13,200 | 0 | 0 | 0 | 0 | 0 | 0 | 0 | α, MIR |
| 5 | 41,700 | 0 | 0 | 0 | 0 | 0 | 0 | 0 | α[b] |
| 6 | 47,700 | 16,200 | 51.1 | 176 | 88 | 292 | 59.7 | 87.7 | α, MIR |
| 7 | 5,280 | 12 | 0 | 3.8 | 0.3 | 9.6 | 3.5 | 0 | γ/δ |
| 8 | 10,500 | 1660 | 6 | 2.6 | 0 | 85 | 0 | 0 | α[c] |
| 9 | 312 | 0 | 0 | 0 | 0 | 0 | 0 | 0 | δ |
| 10 | 221 | 0 | 0 | 0 | 0 | 0 | 0 | 0.6 | α/β[d] |
| 11 | 16,200 | 8.3 | 0.9 | 4.2 | 0.8 | 8.4 | 0.6 | 0 | β |
| 12 | 18,300 | 0 | 1.6 | 2.8 | 6.3 | 0 | 0 | 0 | α, MIR |
| 13 | 422 | 0 | 0 | 0 | 0 | 0 | 0 | 0 | α[d] |
| 14 | 41,800 | 0 | 2 | 4.7 | 16.3 | 0 | 0 | 0 | α, MIR |
| 16 | 2,750 | 2.5 | 1.8 | 5.4 | 30.6 | 0.6 | 0.4 | 0 | α, MIR |
| 17 | 2,820 | 2270 | 0.2 | 6.2 | 2.7 | 7.3 | 1.1 | 0 | α, MIR |
| 19 | 923 | 0 | 0 | 0 | 0 | 0 | 0 | 3 | α[b] |

[a] Reaction with acetylcholine receptor from various species (nmoles of $[^{125}I]$α-bungarotoxin-labeled receptor bound/liter). Adapted from Tzartos and Lindstrom (1980).
[b] Monoclonal antibodies 3, 5, and 19 compete for the same region on the α-subunit, which is distinct from the MIR (Tzartos and Lindstrom, 1980; Gullick *et al.*, 1981).
[c] Postulated to bind near the membrane interphase (Gullick *et al.*, 1981; see Figure 3).
[d] Binds to the extracellular surface of the acetylcholine receptor and can noncompetitively inhibit function (Lindstrom *et al.*, 1981b).

is consistent with the data of Froehner (1981) who showed that many of the antibodies in antireceptor sera raised against denatured receptor were directed at the cytoplasmic surface.

In addition, sedimentation of antibody–receptor complexes on sucrose gradients has been informative for a description of their binding to the receptor surface (Conti-Tronconi *et al.*, 1981a,b; Lindstrom *et al.*, 1981b; Tzartos *et al.*, 1983). The MIRs on the α-subunits are oriented in such a way that an antibody bound to one α-subunit cannot bind to the other α-subunit on the same receptor monomer, but is still able to bind to an α-subunit on a different receptor monomer. Antibodies against the MIR can aggregate acetylcholine receptors because each receptor monomer contains two α-subunits and each immunoglobulin molecule contains two antigen-binding sites. In contrast, some antibodies directed against different determinants on the α-subunits are not able to crosslink receptors and have been proposed to form an intramolecular crosslink between the two α-subunits within the receptor monomer forming a distinct complex on sucrose gradients (Conti-Tronconi *et al.*, 1981b). At an excess of antibody, noncrossreactive antibodies against the β-, γ-, and δ-subunits can crosslink only two receptor monomers since there is only one antigenic site on each monomer.

Thus far, two monoclonal antibodies have been reported which act as noncompetitive inhibitors of the agonist-induced opening of the cation channel (Lindstrom *et al.*, 1981b). One of these monoclonal antibodies crossreacts with α- and β-subunits, whereas the other was found to react with the α-subunit of the acetylcholine receptor (Figure 3). Unfortunately, the low affinity of these antibodies precluded more detailed mapping. These preliminary observations, however, raise optimistic expectations for the use of monoclonal antibodies as functional probes for the identification of regions on the acetylcholine receptor molecule that are directly involved in conformational transitions triggered by the binding of the neurotransmitter.

### 4. *Affinity Labeling of Acetylcholine Receptor Subunits*

The first successful attempts to correlate specific components of the receptor molecule with functional roles were undertaken by A. Karlin and co-workers. They found that tritiated affinity labels for the ligand-binding site, such as the agonist bromoacetylcholine (Damle *et al.*, 1978), or the antagonist 4-(*N*-maleimido)benzyl trimethylammonium iodide (MBTA; Weill *et al.*, 1974; Damle and Karlin, 1978; Figure 4), attach to the α-subunit of the acetylcholine receptor which, therefore, appears to form at least part of the acetylcholine-binding site. In order to label the acetylcholine receptor with bromoacetylcholine or MBTA, a disulfide close to the acetylcholine binding site has to be reduced in order to supply a free sulfhydryl with which the affinity reagent can react. As mentioned above, this disulfide bond has been postulated to be formed between cysteines 128 and 142 on the α subunits (Noda *et al.*, 1982). Reduction of this disulfide occurs readily in the case of one of the α-subunits, but with difficulty for the other α-subunit. Hence, in early studies which employed relatively mild reducing conditions before alkylation with the affinity label, only one of the α-subunits became labeled (reviewed by Karlin, 1980). Later it was shown that the other α subunit could also be affinity alkylated using stronger reducing conditions (Wolosin *et al.*, 1980). Several laboratories have investigated the role of active-site

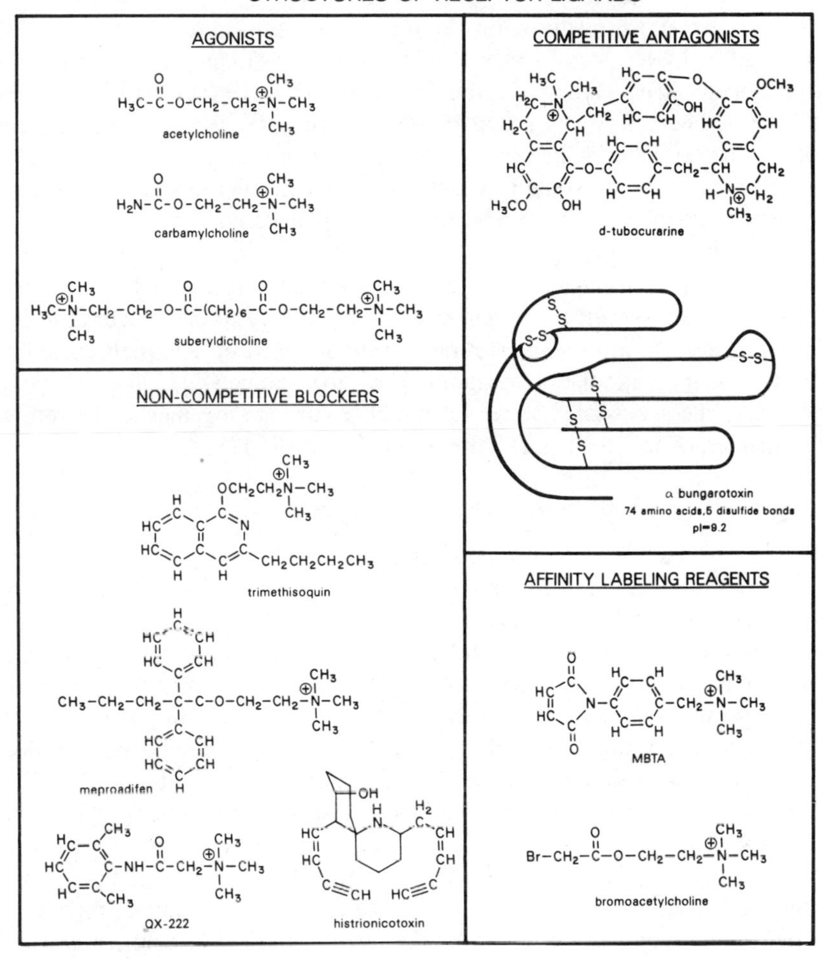


*Figure 4.* Chemical structures of some frequently used agonists, affinity-labeling reagents, competitive antagonists, and noncompetitive blockers of the acetylcholine receptor. Notice the structural similarities between the aminated local anesthetics and acetylcholine, and the hydrophobic nature of the noncompetitive blockers and, to a lesser extent, *d*-tubocurarine.

thiol groups in receptor function using reagents other than affinity labels. Alkylation of reactive thiols, in general, results in a lowered affinity of the acetylcholine receptor for agonists (Rang and Ritter, 1971; Ben-Haim *et al.*, 1973; Suárez-Isla and Hucho, 1977; Moore and Raftery, 1979; Damle and Karlin, 1980; Walker *et al.*, 1981a). Whether the difference in reducibility of the active site disulfide bonds of the two α-subunits reflects a conformational difference between the two α-subunits, a difference in primary sequence, or selective posttranslational modification is not known. Sequencing data indicate that the two α-subunits are identical at least with respect to the first 54 amino acids from the N-terminal (Devillers-Thiery *et al.*, 1979; Raftery *et al.*, 1980; Conti-Tronconi *et al.*, 1982a,b).

The δ-subunit of the receptor has been labeled by a tritiated photoaffinity derivative of the local anesthetic trimethisoquin (Figure 4; Oswald *et al.*, 1980; Saitoh *et al.*, 1980; Oswald and Changeux, 1981a) and by ultraviolet light-induced photolysis of the hallucinogen phencyclidine (Oswald and Changeux, 1981b). These compounds are members of a heterogeneous group of noncompetitive blockers of the acetylcholine receptor, the net effect of which is to stabilize the receptor in a "desensitized" conformation, i.e., a refractory state in which the cation channel remains closed even in the presence of bound agonist. Acetylcholine receptor desensitization and the effect of noncompetitive blockers will be discussed in detail below.

No specific functional roles have yet been assigned to the β- and γ-subunits, nor is it known to what extent the different subunits contribute to the structure of the ion channel. The fact that all subunits are transmembrane proteins and their considerable sequence homology led to the speculation that all five subunits may line up across the membrane like the staves of a barrel so that all contribute together to the formation of a transmembrane ion channel (Figure 3; Kistler *et al.*, 1982).

### 5. *Interactions between the Acetylcholine Receptor and the 43,000 Protein*

Evidence which has accumulated over the last five years indicates that acetylcholine receptor dimers may be immobilized at the synaptic junction through association with a cytoskeletal component commonly referred to as the "43,000 protein" and sometimes indicated as the ν-peptide. This 43,000-dalton protein is a prominent component of acetylcholine receptor-enriched electric organ membranes (Sobel *et al.*, 1977, 1978; Hamilton *et al.*, 1979; Neubig *et al.*, 1979; Moore *et al.*, 1979; Lindstrom *et al.*, 1980c; St. John *et al.*, 1982; Barrantes, 1982) and can be removed by treatment of the membranes with alkaline pH (Neubig *et al.*, 1979; Moore *et al.*, 1979; Elliott, 1979; Barrantes *et al.*, 1980) or lithium diiodosalicylate (Neugebauer and Zingsheim, 1982). Receptors in alkali-treated membranes devoid of the 43,000 protein were found to be more sensitive to proteolysis (Klymkowsky *et al.*, 1980; Wennogle and Changeux, 1980) and heat inactivation (Saitoh *et al.*, 1979), and their rotational mobility was greatly increased (Barrantes *et al.*, 1980; Lo *et al.*, 1980; Bartholdi *et al.*, 1981). Furthermore, removal of the 43,000 protein from the membranes by alkaline extraction is correlated with the disappearance of electron-dense cytoplasmic condensations observed on the inner face of the electric organ membrane (Cartaud *et al.*, 1981; Sealock, 1982a). In addition, several investigators have reported the localization of the 43,000 protein on the cytoplasmic side of the membrane using monoclonal antibodies against this protein (Froehner *et al.*, 1983; Nghiem *et al.*, 1983). A protein which crossreacts immunologically with the 43,000-dalton protein has been localized under the postsynaptic membrane of rat diaphragm (Froehner *et al.*, 1981), and subsynaptic densities attached to acetylcholine receptors similar to those observed under postsynaptic membranes from *Torpedo* electric organ have been identified on the inside of postsynaptic membranes of mouse sternomastoid muscle (Sealock, 1982b). In addition to electron microscopic studies, lactoperoxidase-catalyzed iodination of electric organ membranes before and after treatment with saponin indicates that the 43,000 protein is located on the cytoplasmic surface of the postsynaptic membranes (St. John *et al.*, 1982; Figure 3). Only a study by Conti-Tronconi *et al.* (1982c) is at variance with these observations.

They conclude that the 43,000 protein may be present both on the cytoplasmic and the synaptic surface of the electric organ membranes. However, this investigation relied on proteolytic degradation of membrane components by externally added or membrane-entrapped trypsin and since, in this study, the δ–δ disulfide bond was found to be susceptible to tryptic cleavage from either side of the membrane, it is likely that incompletely sealed membranes can account for the aberrant results.

The 43,000 protein has an amino acid composition which is distinct from actin (Strader *et al.*, 1980) which has also been localized at the neuromuscular junction (Hall *et al.*, 1981). Furthermore, there are indications that the 43,000 protein is rich in sulfhydryl groups which may be involved in its aggregation (Sobel *et al.*, 1978; Hamilton *et al.*, 1979; Criado and Barrantes, 1982). Two-dimensional electrophoresis and tryptic fingerprinting has resolved the 43,000 band into a group of three distinct components designated $\nu_1$, $\nu_2$, and $\nu_3$ (Gysin *et al.*, 1981). The $\nu_1$ component was found to be exclusively membrane-bound, whereas $\nu_2$ and $\nu_3$ were found to be also prominent cytoplasmic proteins. Crosslinking of electric organ membranes with *p*-azidophenacyl bromide, which resulted in the selective crosslinking of the $\nu_1$ component, revealed after extraction with Triton X-100 a filamentous network reminiscent of the cytoskeleton underlying the erythrocyte cell membrane (Cartaud *et al.*, 1982). It is not yet clear whether the $\nu_2$ and $\nu_3$ proteins are contaminants or whether they participate in anchoring acetylcholine receptors at the synaptic junction. Creatine kinase activity associated with the $\nu$ proteins of *Torpedo* electric organ membranes has also been reported (Barrantes *et al.*, 1983).

In addition to the 43,000 protein, a 51,000-dalton protein has been identified under the postsynaptic membrane of *Torpedo* electroplax and frog *cutaneous pectoris* muscle, which is related to, but distinct from, proteins present in intermediate filaments, and which appears as amorphous electron-dense material coincident with acetylcholine receptors at synaptic as well as extrasynaptic areas (Burden, 1982). Thus, it is evident that the acetylcholine receptors at the synaptic junction are immobilized by a complex cytoskeletal architecture. In addition, there is evidence which suggests that components of the extracellular matrix may also interact with acetylcholine receptors (Sanes *et al.*, 1978; Fambrough *et al.*, 1982).

## B. Acetylcholine Receptor from Muscle

Mammalian skeletal muscle contains acetylcholine receptors at a concentration of approximately 0.3 pmole $[^{125}I]\alpha$-bungarotoxin-binding sites per gram of tissue (Lindstrom and Lambert, 1978). Denervation of the muscle may increase the receptor concentration to about 6 pmoles of $[^{125}]\alpha$-bungarotoxin binding sites per gram tissue (Berg and Hall, 1975a,b). This concentration is still several hundred-fold lower than the acetylcholine receptor concentration in *Torpedo* electric organ (Karlin, 1980; Changeux, 1981). The unfavorably low amount of acetylcholine receptor relative to the protease content of muscle has made the isolation of intact acetylcholine receptor from muscle extracts a particularly challenging problem (Froehner *et al.*, 1977a,b; Shorr *et al.*, 1978, 1981; Merlie *et al.*, 1978; Nathanson and Hall, 1979; Stephenson *et al.*, 1981; Gotti *et al.*, 1982; Einarson *et al.*, 1982; Momoi and Lennon, 1982).

Early indications that acetylcholine receptors from muscle may be structurally similar to electric organ receptors came from studies on experimental autoimmune myasthenia gravis. This experimental disease serves as a model for the human autoimmune neuromuscular disorder "myasthenia gravis," which is characterized by severe muscular weakness as a result of a reduction in the amount of active acetylcholine receptors due to an enhanced rate of receptor degradation ("antigenic modulation"; Merlie *et al.*, 1976, 1979a,b; Kao and Drachman, 1977; Heinemann *et al.*, 1977, 1978; Stanley and Drachman, 1978; Reiness *et al.*, 1978) and complement-mediated focal lysis of the endplate (Engel *et al.*, 1976, 1977a,b, 1979; Sahashi *et al.*, 1978). Myasthenic symptoms can be induced in experimental animals by immunization with purified acetylcholine receptor from *Torpedo* electric organ as a consequence of structural similarities and concomitant immunological crossreactivity between this receptor and the animal's own muscle receptors (reviewed by Lindstrom, 1979; Vincent, 1980; Drachman, 1981; Lindstrom and Engel, 1981). Lindstrom *et al.* (1978) observed that immunization of rats with either of the four purified component polypeptides of *Torpedo* acetylcholine receptor caused experimental autoimmune myasthenia gravis, indicating the presence on rat muscle receptor of crossreactive determinants against all four subunits of *Torpedo* receptor. It was, however, not until 1982 that intact acetylcholine receptors showing four polypeptide bands on SDS–polyacrylamide gels which corresponded immunochemically to the four subunits of receptor from *Torpedo* were successfully purified from muscle tissue (Einarson *et al.*, 1982). Previously, the purification of muscle receptors from rat (Froehner *et al.*, 1977a,b; Nathanson and Hall, 1979), cat (Shorr *et al.*, 1978, 1981), fetal calf (Merlie *et al.*, 1978), and human muscle (Stephenson *et al.*, 1981) had been reported. Some laboratories observed only a single subunit similar in molecular weight to the α-subunit of *Torpedo* receptor (Merlie *et al.*, 1978; Shorr *et al.*, 1978, 1981), whereas other laboratories reported the presence of 2–5 subunits (Boulter and Patrick, 1977; Nathanson and Hall, 1979, 1980; Merlie and Sebbane, 1981; Kemp *et al.*, 1980). In one study, rat muscle receptor was labeled *in situ* with a photoaffinity derivative of α-bungarotoxin (Nathanson and Hall, 1980). The label attached to five polypeptides, which corresponded in size to similar polypeptides observed by the same investigators for the purified rat receptor (Nathanson and Hall, 1979). Another study, which reported the purification of acetylcholine receptors from the nonfusing murine $BC_3H1$ smooth muscle cell line, observed four subunits, three of which resembled the α-, β-, and δ-subunits of *Torpedo* receptor in molecular weight (Boulter and Patrick, 1977). Antisera and monoclonal antibodies have detected antigenic determinants characteristic of α-, β-, γ-, and δ-subunits from *Torpedo* on acetylcholine receptors from eel electric organ (Lindstrom 1979), eel muscle (Gullick and Lindstrom, 1983b), cattle (Conti-Tronconi *et al.*, 1981b; Einarson *et al.*, 1982; Tzartos *et al.*, 1983), rats (Lindstrom *et al.*, 1979b; Einarson *et al.*, 1982), human (Tzartos *et al.*, 1982, 1983), *Rana pipiens,* and *Xenopus laevis* (Sargent *et al.*, in preparation). An MIR has been found in all cases except *Xenopus laevis* (Sargent *et al.*, 1984).

The purification of intact acetylcholine receptor from mammalian muscle was reported by Einarson *et al.* (1982) and Gotti *et al.* (1982). Fetal calf muscle was in each case used as a relatively rich source for acetylcholine receptor. Purified bovine receptor appeared by electron microscopy as doughnut-shaped particles which were identical in shape and size to acetylcholine receptor rosettes from electric organ, and was found to be composed of four glycoprotein subunits, which corresponded immunochemically to the four subunits from *Torpedo* receptor (Figure 1; Einarson *et al.*, 1982). In addition, bovine receptor and *Torpedo* receptor were shown to have a similar amino acid composition (Gotti *et al.*, 1982). Similarly, purification of acetylcholine receptors from eel muscle (Gullick and Lindstrom, 1983b; Lindstrom *et al.*, 1983) and from rat (Einarson *et al.*, 1982), under conditions that minimize proteolytic damage to the receptor, revealed four subunits which correspond immunochemically to the α-, β-, γ-, and δ-subunits of *Torpedo* receptor. Although proteolytic digestion of electric organ and muscle receptors with V8 protease revealed no extensive homology in the peptide maps, probing of the resulting fragment patterns with monoclonal antibodies indicated structural homology between the corresponding subunits (Gullick and Lindstrom, 1982b).

Thus far, muscle receptors like the acetylcholine receptors from eel electric organ have been observed on sucrose gradients only as a light 9 S form corresponding in size to the monomeric form of *Torpedo* acetylcholine receptor (Figure 1; Berg *et al.*, 1972; Shorr *et al.*, 1981; Lo *et al.*, 1981; Tzartos *et al.*, 1983; Gotti *et al.*, 1982; Momoi and Lennon, 1982). Dimerization of the acetylcholine receptor from *Torpedo* may represent a species-specific modification rather than a universal structural feature. Alternatively, it is possible that efficient proteolysis around the δ–δ disulfide bond prevents the isolation of intact dimers from eel electric organ and mammalian muscle.

The lowest molecular weight peptide of mammalian acetylcholine receptor can be affinity labeled with tritiated MBTA, indicating that this peptide, like the α-subunit from electric organ receptor, forms part or all of the acetylcholine-binding site (Shorr *et al.*, 1981; Gullick and Lindstrom, 1982b; Momoi and Lennon, 1982). Studies with monoclonal antibodies directed against α-subunits suggest that bovine muscle receptors contain two α-subunits per receptor monomer, each of which carries a main immunogenic region (Conti-Tronconi *et al.*, 1981b; Tzartos *et al.*, 1983). Furthermore, cDNA encoding the α-subunit of bovine muscle as well as a genomic DNA segment containing the corresponding human gene have been cloned by Noda *et al.* (1983c). Nucleotide sequence analysis of these cDNAs revealed 97, 80, and 81% sequence homology between the α subunits for the human/cow, human/torpedo, and cow/torpedo counterparts, respectively. Thus, biochemical, immunochemical, and structural similarities between acetylcholine receptors from electric organ and muscle indicate that these receptors consist of a fundamentally similar monomeric $\alpha_2\beta\gamma\delta$ structure and justify the assumption that acetylcholine receptors from electric organ represent a valid model for acetylcholine receptors at the mammalian neuromuscular junction.

## C. Assembly and Degradation of Acetylcholine Receptors

### 1. Junctional and Extrajunctional Acetylcholine Receptors

In the early stages of embryonic development, prior to innervation of the muscle, acetylcholine receptors can be found at low density scattered over the entire muscle fiber (Hartzell and Fambrough, 1973). During synapse formation, clusters of acetylcholine receptors develop at the nascent endplate (Diamond and Miledi, 1962; Bevan and Steinbach, 1977; Anderson and Cohen, 1977; Frank and Fischbach, 1979; Steinbach, 1981; Pumplin and Fambrough, 1982). The stimuli which direct this clustering of receptors still remain obscure. Evidence suggests that as yet undefined factors synthesized by the nerve or the basal lamina overlying the postsynaptic membrane may play a role in the regulation of the distribution of receptors along the muscle fiber (Sanes *et al.*, 1978; Burden *et al.*, 1979; Pumplin and Fambrough, 1982). Junctional acetylcholine receptors located directly under the nerve terminal turn over slowly (half-lifetime ~5–9 days). These acetylcholine receptors differ in this respect from their extrajunctional counterparts, which have a half-life time of approximately 20 hr (Berg and Hall, 1975a,b; Chang and Huang, 1975; Sakmann and Brenner, 1978; Steinbach *et al.*, 1979; Michler and Sakmann, 1980; Reiness and Weinberg, 1981; Pumplin and Fambrough, 1982). Verdenhalven *et al.* (1982) have made the attractive suggestion that receptor degradation might normally be initiated by proteolytic nicking of the cytoplasmic surface of the receptor and proposed that association of the 43,000 protein might protect junctional receptors from proteolysis and thereby slow their degradation rate. Acetylcholine receptors are immobilized at the synaptic junction (Axelrod *et al.*, 1976, 1978) and the high density of junctional receptors at the tips of junctional folds of the postsynaptic membrane persists after denervation of the muscle (Birks *et al.*, 1960; Hartzell and Fambrough, 1972; Porter and Barnard, 1975). Denervation of the muscle leads, however, to an increase in the number of extrajunctional receptors (reviewed by Fambrough, 1979).

In addition to their differences in metabolic stability, junctional and extrajunctional receptors can be distinguished on the basis of the lifetime of the open state of the neurotransmitter-controlled cation channel. The mean channel open time of the junctional receptor is in the order of ~1 ms and extrajunctional receptors appear to have a 2- to 3-fold longer mean channel open time (Sakmann and Brenner, 1978; Michler and Sakmann, 1980). Junctional and extrajunctional acetylcholine receptors differ further with respect to their sensitivity for *d*-tubocurarine (Beranek and Vyskocil, 1967) and their isoelectric point, which is slightly more acidic in the case of the junctional receptor (Brockes *et al.*, 1975). Some antigenic differences between junctional and extrajunctional receptors have also been reported (Weinberg and Hall, 1979). However, no differences have been detected in subunit molecular weights or peptide maps between junctional and extrajunctional receptors (Nathanson and Hall, 1979; Sumikawa *et al.*, 1982a). These observations are consistent with the hypothesis that these two forms of receptor arise from posttranslational modification of proteins encoded by the same gene. Saitoh and Changeux (1981) have correlated changes in the state of phosphorylation of the acetylcholine receptor during maturation of the electric organ of *Torpedo marmorata* with a change in the isoelectric point of the receptors. These studies generate the attractive suggestion that junctional and extrajunctional receptors might be interconvertible through phosphorylation (Saitoh and Changeux, 1981). However, no direct link has yet been demonstrated between the state of phosphorylation of the acetylcholine receptor and its mean channel open time, sensitivity to curare, or metabolic stability. Furthermore, alterations in the metabolic turnover rate and in channel-gating properties occur at different times during the development of the neuromuscular junction in rat embryos (Dennis *et al.*, 1981). Studies on cloned acetylcholine receptor genes may eventually resolve the question whether junctional and extrajunctional receptors arise by covalent modification of a single gene product or whether they are coded for by separate genes.

## 2. Synthesis and Assembly of Acetylcholine Receptor Subunits

Figure 5 presents a schematic illustration of the major steps in the synthesis of the acetylcholine receptor. Synthesis, transport, and degradation of both acetylcholine receptor and acetylcholinesterase in muscle cells in culture have been reviewed by Rotundo and Fambrough (1982) who show that the regulation of synthesis and destruction of these two molecules are quite distinct, but that they probably share a common transport mechanism from the rough endoplasmic reticulum to the cell surface. *In vitro* translation studies using mRNA from *Torpedo* electric organ have demonstrated that each of the receptor subunits is synthesized from a separate mRNA (Mendez *et al.*, 1980; Anderson and Blobel, 1981). In these studies, the *in vitro* synthesized receptor peptides were identified by immunoprecipitation with subunit-specific antisera and it was demonstrated that each of the component polypeptides of the receptor could be labeled with the initiator amino acid [$^{35}$S]*N*-formylmethionine (Anderson and Blobel, 1981). The primary translation products identified in these *in vitro* studies differed in electrophoretic mobility from the subunits of acetylcholine receptor purified from electric organ. This has been attributed to the absence of glycosylation, since incubation with pancreatic rough microsomes during translation gave rise to glycosylated forms of each of the four subunits with apparent molecular weights which were close to those of the subunits of purified receptor (Anderson and Blobel, 1981).

Integration of *in vitro* synthesized subunits in the membranes of dog pancreatic microsomes proceeds independently. The cytoplasmic portions of the *in vitro* synthesized subunits are, like mature receptor, sensitive to proteolysis by trypsin and the trypsin-sensitive cytoplasmic domains have been reported to increase in proportion to the apparent molecular weights of the subunits ($\alpha \sim 5000$ mol. wt., $\beta \sim 13{,}000$ mol. wt., $\gamma \sim 1500$ mol. wt., and $\delta \sim 21{,}000$ mol. wt.; Anderson and Blobel, 1981).

Anderson *et al.* (1982) have demonstrated that synthesis of the δ-subunit of the acetylcholine receptor involves a 21 amino acid N-terminal signal sequence of predominantly hydrophobic amino acids. Similar signal sequences of 24 amino acids for the α-subunit (Noda *et al.*, 1982; Barnard *et al.*, 1982; Sebbane *et al.*, 1983) as well as for the β-subunit (Noda *et al.*, 1983a) and 17 amino acids for the γ-subunit (Claudio *et al.*, 1983) have also been reported. In addition, Noda *et al.* (1983c) have reported that the protein coding sequence of the human gene which encodes the α-subunit precursor is divided by eight introns into nine exons, which seem to correspond to different structural and functional domains of the subunit precursor molecule such that the proposed transmembrane domains, the intracellular domain, the extracellular domain containing the acetylcholine-binding site, and the leader peptide are encoded in separate exons. The exact mechanisms through which the message is processed so

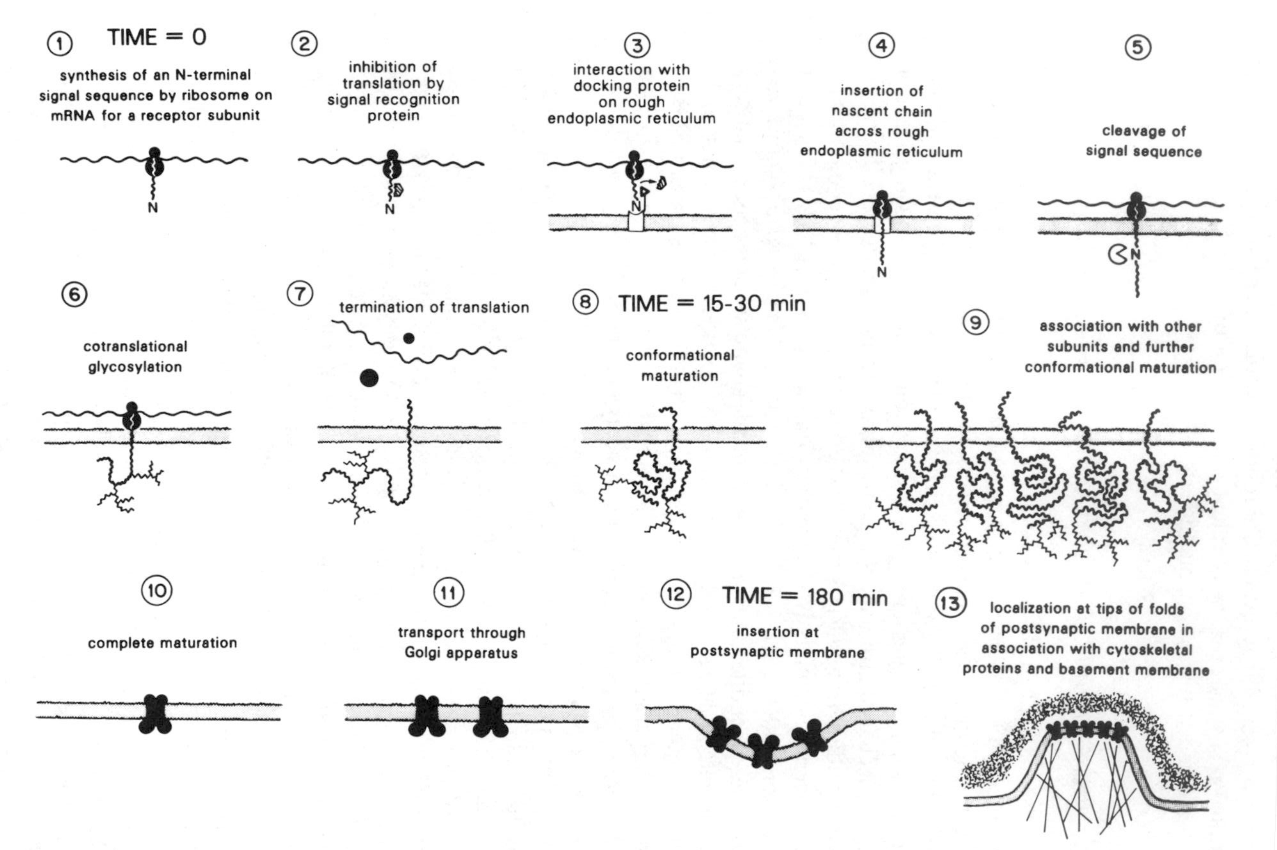


*Figure 5.* A schematic summary of the major steps involved in the biosynthesis and assembly of the acetylcholine receptor. Modified from Lindstrom (1984). Although the involvement of signal recognition protein (SRP) has been demonstrated for the integration of the δ-subunit into the membrane of the endoplasmic reticulum (Anderson *et al.*, 1982), the postulated participation of the docking protein in this process is based on analogy with other systems (Meyer *et al.*, 1982). It is not known how many times each of the receptor subunits traverses the membrane. Neither is the mechanism of their assembly understood.

that the exons are correctly linked together in the translated product remain to be elucidated.

A protein of molecular weight 250,000 consisting of 6 subunits and obtained from an 0.5 M salt wash of microsomal membranes was demonstrated to be essential for the transmembrane insertion of membrane proteins and for the translocation of secretory proteins across the endoplasmic reticulum (Walter and Blobel, 1981a,b; Walter *et al.*, 1981). This protein has been termed "signal recognition protein" (SRP) and was found to be essential for the translocation of the nascent δ-chain across the microsomal membrane (Anderson *et al.*, 1982). SRP appears to cause an arrest in chain elongation by interacting with the signal sequence. This elongation arrest is released after attachment of the polysomes which synthesize the membrane protein to microsomal membranes (Walter and Blobel, 1981b). The displacement of SRP from the signal sequence of the nascent polypeptide chain after attachment of the polysomes to the microsomal membrane appears to be mediated by an integral protein of the endoplasmic reticulum referred to as the "docking protein" (Meyer *et al.*, 1982). Removal of SRP by the docking protein allows continuation of the synthesis of the membrane protein and its transmembrane insertion via an as yet incompletely understood mechanism (Figure 5).

Although *in vitro* synthesized components of the acetylcholine receptor are inserted in microsomal membranes, they do not assemble into functional complexes able to bind [$^{125}$I]α-bungarotoxin (Anderson and Blobel, 1981). This may, at least in part, be due to the low concentration of each of the *in vitro* synthesized subunits and the low probability with which multiple subunits would be present in the same microsomal membrane vesicle. Sumikawa *et al.* (1981) microinjected mRNA from *Torpedo marmorata* into *Xenopus* oocytes and demonstrated that oocytes injected with this mRNA synthesized intact acetylcholine receptor capable of binding α-bungarotoxin and opening cation-specific channels in response to acetylcholine (Barnard *et al.*, 1982). This suggested that posttranslational modifications, which can take place in the oocytes but not in a cell-free translation system, are important for the assembly of mature acetylcholine receptors. The receptors synthesized by the oocytes were present only as a 9 S form, suggesting that dimer formation may be a process that uniquely occurs in the electric organ (Sumikawa *et al.*, 1981). Mishina *et al.* (1984) coupled cloned cDNAs of all four acetylcholine receptor subunits to a simian virus 40 vector and, after amplification of the messages in COS monkey cells, injected the resulting mRNAs into *Xenopus* oocytes. They observed that injection of a mixture of mRNAs encoding all four subunits of the acetylcholine receptor resulted in the synthesis of functional nicotinic cholinergic receptors in the oocytes. From studies in which mRNAs coding for some of the subunits were selectively deleted during injection, they deduced that all four subunits are required for the assembly of a functional acetylcholine receptor, although only the α-subunit appeared indispensable for α-bungarotoxin binding activity.

Glycosylation of the newly synthesized polypeptides appears to be important for the acquisition of neurotoxin-binding activity. Presence of tunicamycin, an inhibitor of protein glycosylation on asparaginyl residues, during the biosynthesis of the acetylcholine receptor prevents the maturation of the neurotoxin-binding site (Merlie *et al.*, 1982) and, in addition, enhances the rate of receptor degradation (Prives and

Olden, 1980). Such under-glycosylated receptors functioned, but had altered ligand-binding affinity (Prives and Bar-Sagi, 1982). Studies on the biosynthesis of acetylcholine receptors in the murine $BC_3H1$ muscle cell line using anti-α-bungarotoxin antiserum in conjunction with anti-acetylcholine receptor serum have shown that synthesis takes place on the rough endoplasmic reticulum (Merlie *et al.*, 1981), and that the acquisition of neurotoxin-binding activity takes place over a period of 15–30 min after synthesis of the α-subunit concomitant with subunit assembly (Merlie and Sebbane, 1981). Monoclonal antibodies against the different subunits may provide powerful tools for the investigation of subunit assembly after their cotranslational insertion in the membrane. It is not clear how synthesis of the subunits is directed in such a way that they are synthesized in the correct proportions for the ultimate formation of an $\alpha_2\beta\gamma\delta$ complex. In $BC_3H1$ cells, three times more α-subunits are synthesized than are assembled into mature receptor (Merlie *et al.*, 1982). However, the turnover of acetylcholine receptors in $BC_3H1$ cells is more rapid than receptor turnover in innervated muscle, and synthesis of receptor subunits in the correct proportions may proceed more efficiently *in vivo* than in this murine muscle cell line.

Mature acetylcholine receptors are transported via the Golgi apparatus to the plasma membrane where their insertion occurs probably through fusion of microsomal membrane vesicles with the plasma membrane (Fambrough and Devreotes, 1978; Rotundo and Fambrough, 1982; Figure 5). As a result, that portion of the receptors which faced the lumen of the microsomal membrane becomes exposed on the synaptic side of the plasma membrane. The question whether a distinction exists between junctional and extrajunctional receptors at the level of the microsomal membrane vesicles and how the site of insertion of receptors at the plasma membrane is determined remains still unresolved. It is conceivable that glycosylation in the rough endoplasmic reticulum (probably on amino residues) and in the Golgi (probably on serine residues; Hanover and Lennarz, 1981) may be involved in targeting the acetylcholine receptors for insertion at junctional and extrajunctional sites. Differences in glycosylation might also account partially for differences in antigenicity and isoelectric point between junctional and extrajunctional receptors. In addition, muscle activity suppresses the synthesis of extrajunctional receptors and, therefore, appears to be an important factor in the regulation of acetylcholine receptor metabolism (Reiness and Hall, 1977; Linden and Fambrough, 1979).

### 3. *Degradation of Acetylcholine Receptors*

Labeling of acetylcholine receptors of cultured muscle cells with radioactive amino acids (Merlie *et al.*, 1976; Gardner and Fambrough, 1979) or with $[^{125}I]$α-bungarotoxin permitted studies of the rate of receptor degradation through measurements of the release of the radioisotope in the culture medium (reviewed by Fambrough, 1979; Rotundo and Fambrough, 1982). In addition, the degradation rate of junctional receptors labeled with $[^{125}I]$α-bungarotoxin in live rats has been found to have a turnover time of 8–9 days (Fumagalli *et al.*, 1982b). Although the precise mechanism which initiates receptor degradation is unknown, endocytosis appears to be the rate-limiting step (schematically represented in Figure 6; Devreotes and Fambrough, 1975, 1976; Fambrough, 1979). Crosslinking of the acetylcholine receptors by antibodies enhances

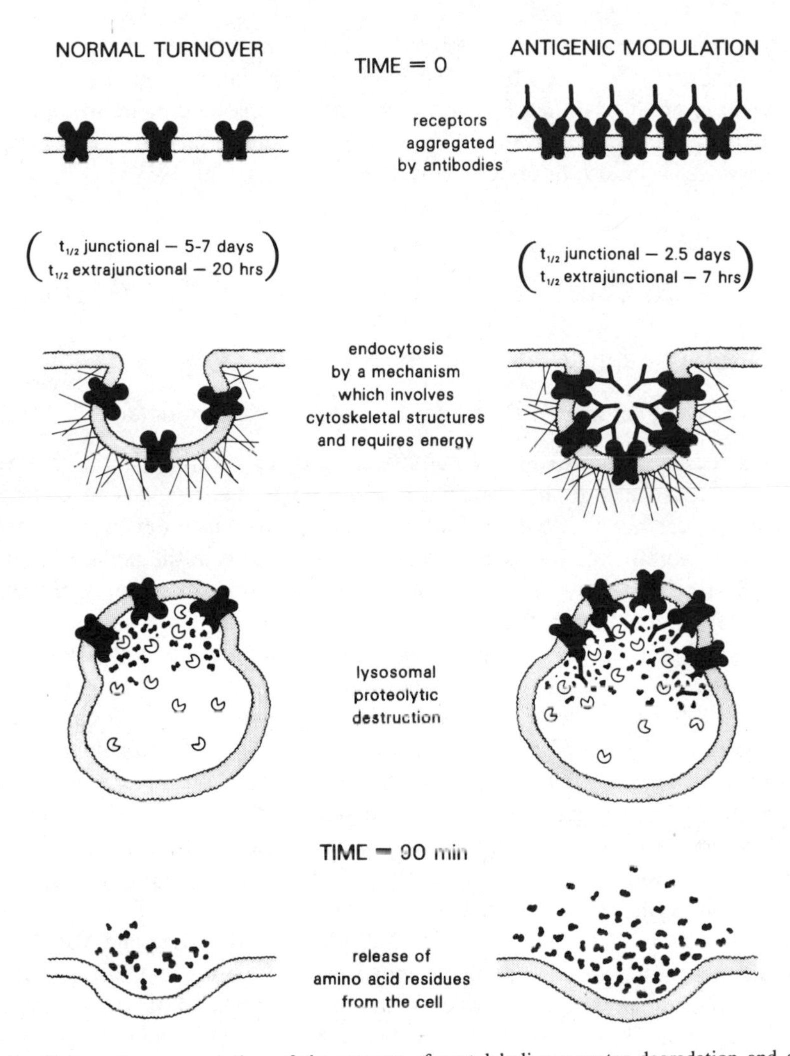


*Figure 6.* Schematic representation of the process of acetylcholine receptor degradation and antigenic modulation in cultured muscle cells. From Lindstrom (1984).

this process, an effect which has been termed "antigenic modulation" and which accounts for a 2- to 3-fold enhancement of the rate of degradation of acetylcholine receptors during the autoimmune disease myasthenia gravis (Heinemann *et al.*, 1977, 1978; Kao and Drachman, 1977; Merlie *et al.*, 1976, 1979a,b; Reiness *et al.*, 1978; Stanley and Drachman, 1978; Lindstrom and Einarson, 1979; Conti-Tronconi *et al.*, 1981a; Drachman *et al.*, 1982). Endocytosis of receptors involves cytoskeletal elements and requires metabolic energy (Berg and Hall, 1974; Devreotes and Fambrough, 1975, 1976; Fambrough, 1979). Although receptors are eventually transported into the lysosomal compartment for their ultimate destruction (Libby *et al.*, 1980; Fumagalli *et al.*, 1982a), inhibition of lysosomal proteases does not affect the rate at which internalization occurs (Libby *et al.*, 1980). Thus, the internalization process and the process of lysosomal proteolysis do not appear to be tightly coupled. The first receptor breakdown products are excreted in the medium about 90 min after initiation of the internalization process (Figure 6; reviewed by Fambrough, 1979).

## III. FUNCTIONAL ASPECTS OF THE ACETYLCHOLINE RECEPTOR

### A. Electrophysiology of the Postsynaptic Membrane

#### 1. Macroscopic Changes in Postsynaptic Membrane Conductance

In this section, we will briefly summarize the early experimental observations that have led to an electrophysiological description of the changes in postsynaptic membrane conductance induced by binding of the neurotransmitter.

During neurotransmission, acetylcholine is released from the nerve terminal primarily as discrete amounts ("quanta") from presynaptic vesicles fusing with the nerve membrane (Nickel and Potter, 1970; Heuser *et al.*, 1979). However, nonquantal release of acetylcholine from nerve terminals has also been described (reviewed by Tauc, 1982). Each quantum contains 6000–10,000 molecules of acetylcholine (Kuffler and Yoshikami, 1975). The release of a single quantum of acetylcholine causes a small subthreshold depolarization of the muscle membrane ~0.5–1.0 mV in amplitude (termed "miniature endplate potential") after a lag period of about 2 msec during which the neurotransmitter diffuses across the 50 nm synaptic cleft (del Castillo and Katz, 1954; Katz and Miledi, 1965). Stimulation of the nerve results in the synchronous release of approximately 300 quanta of acetylcholine. This causes the concentration of acetylcholine in the synaptic cleft to rise in less than 1 msec up to $\sim 3 \times 10^{-4}$ M over a background level of $\sim 10^{-8}$ M (Kuffler and Yoshikami, 1975; Katz and Miledi, 1977). This results through the summation of individual miniature endplate potentials in depolarization of the muscle membrane, which represents the initial trigger for muscle contraction (Fatt and Katz, 1951; del Castillo and Katz, 1954). The response is related to the second power of the dose suggesting that two molecules of acetylcholine are required to open a postsynaptic channel (Adams, 1975; Peper *et al.*, 1975; Dionne *et al.*, 1978; Dreyer *et al.*, 1978). This squared relationship provides a tight coupling between dose and response over a narrow range of acetylcholine concentrations allowing effective control for the rapid initiation and termination of postsynaptic signals. Thus, the postsynaptic membrane operates as an efficient amplification device.

Electrophysiological measurements (Takeuchi and Takeuchi, 1960; Lewis, 1979; Adams *et al.*, 1980) and ion flux experiments (Huang *et al.*, 1978) in solutions of various ionic composition have shown the postsynaptic channel to be cation-selective. The endplate current during *in vivo* neuromuscular transmission is carried mainly by sodium and potassium ions (Takeuchi and Takeuchi, 1960; Jenkinson and Nicholls, 1961; Takeuchi, 1963; Ritchie and Fambrough, 1975; Catterall, 1975; Huang *et al.*, 1978). Detailed studies of the permeability of the endplate conductance showed that the permeability sequence of alkali metal ions through the channel parallels their aqueous mobility sequence (Adams *et al.*, 1980). Hence, such ions experience the channel as a neutral water-filled pore through which they can pass by free diffusion. Divalent cations, however, have a permeability sequence opposite from their mobility sequence and, therefore, must experience some interaction with groups inside the channel (Adams *et al.*, 1980). Large organic cations of the size of agonists, such as glucosamine, were also found to pass through the acetylcholine receptor channel. From the size cut-off of such permeant organic cations, it was estimated that the smallest cross section of the open channel must be at least as large as a 6.5 × 6.5-Å square (Dwyer *et al.*, 1980).

In 1957, Katz and Thesleff observed that the amplitude of the conductance response of the muscle slowly decreases during continuous exposure to acetylcholine (Katz and Thesleff, 1957). This closing of postsynaptic channels during prolonged exposure to the agonist is known as "desensitization." The rate of desensitization depends on the dose of agonist (Katz and Thesleff, 1957; Rang and Ritter, 1970; Magazanik and Vyskocil, 1973; Scubon-Mulieri and Parsons, 1978; Feltz and Trautmann, 1982) and is enhanced by membrane hyperpolarization and divalent cations (Magazanik and Vyskocil, 1970) as well as by temperature (Magazanik and Vyskocil, 1975). Katz and Thesleff (1957) predicted that this refractory, desensitized state of the acetylcholine receptor would be coupled to an increased binding affinity for the neurotransmitter. Subsequent pharmacological evidence demonstrated that an increased affinity for the agonist is indeed characteristic for the desensitized state of the receptor (discussed in detail below). However, in spite of extensive studies for more than two decades by numerous laboratories, the physiological significance of acetylcholine receptor desensitization is still not understood.

Fluctuation analysis of the voltage noise during application of a steady low dose of acetylcholine to frog muscle preparations yielded initial information regarding the conductance and lifetime of single postsynaptic channels. Katz and Miledi (1972) determined that, under their conditions, the elementary acetylcholine-induced current arises from a conductance change of the order of 100 p*S* which lasts for about 1 msec at 20°C and produces a minute depolarization of ~0.3 μV. Hence, during a miniature endplate potential, about 2000 channels are opened. The unit depolarization of 0.3 μV is associated with a net charge transfer of 10–14 C, equivalent to the passage of approximately 50,000 univalent cations per millisecond. Anderson and Stevens (1973) analyzed the conductance fluctuations of the voltage-clamped frog muscle preparation and found a unit conductance of about 32 p*S* in Ringer's solution and a mean channel open time of 6.5 msec at 8°C and 0 mV. Their findings indicated that the closing of open receptor channels is the rate-limiting step in the decay of the postsynaptic current. This process occurred, under their conditions, with a single time constant. They proposed that individual postsynaptic channels open rapidly to a specific conductance which remains constant for an exponentially distributed duration. In addition, the channel lifetime was found to be weakly voltage-dependent such that it is prolonged *e*-fold for each 100 mV of membrane hyperpolarization (Magleby and Stevens, 1972a; Sheridan and Lester, 1975). This voltage dependence could be accounted for by a 50-Debye unit change in the dipole moment of the receptor–transmitter complex during the shift from the open to the closed conformation of the channel (Magleby and Stevens, 1972b).

### 2. Recordings of Single Acetylcholine Receptor Channels

Technical advances in the mid-1970's provided the resolution necessary to record the opening and closing of single extrajunctional acetylcholine receptor channels in denervated frog muscle fibers (Neher and Sakmann, 1976). These studies confirmed the notion that the mature acetylcholine receptor channel can fluctuate between two conductance states, an open state of fixed conductance and a closed state. The mean conductance of the open state was found to be ~28 p*S*, close to the value previously estimated from fluctuation analysis (Neher and Sakmann, 1976). The single-channel conductance appears to be independent of the nature of the cholinergic agonist and of the membrane potential over a wide voltage range (Neher and Sakmann, 1976; Lewis, 1979). Gigohm-seal patch-clamping, which involves the sealing of a patch of muscle membrane over the tip of an electrode followed by excision of the sealed patch from the rest of the membrane, has enabled measurement of the single-channel conductance over a wide range of applied voltages and salt concentrations up to concentrations at which saturation of the channel conductance to sodium ions could be observed (Horn and Patlak, 1980). Hamill and Sakmann (1981) reported in uninnervated embryonic rat muscle two independent classes of acetylcholine-activated conductance states of 25 p*S* and 35 p*S*, respectively. Each of these conductance states could fluctuate between the closed state and a common "substate" of 10 p*S*. This heterogeneity in single-channel conductance may reflect modifications of the acetylcholine receptor molecule during the early stages of synapse formation. In addition, one may speculate that the 10 p*S* conductance state reflects the singly liganded receptor.

In contrast to the single-channel conductance, the mean channel open time appears to depend on the nature of the agonist. Neher and Sakmann (1976) reported values of 45 ± 3 msec for suberyldicholine, 26 ± 5 msec for acetylcholine and 11 ± 2 msec for carbamylcholine using extrajunctional acetylcholine receptors from frog muscle at 8°C and − 120 mV applied voltage (for the structures of these agonists see Figure 4). When carbamylcholine and acetylcholine were applied simultaneously, channels opened by combined activation had a mean channel open time characteristic of carbamylcholine (Trautmann and Feltz, 1980). This observation may be explained by the hypothesis that a channel stays open as long as the agonist remains bound. The mean channel open time would then be proportional to the affinity of the agonist for the binding site. Hence, if two molecules of agonist have to be bound in order to keep the channel open, the mean channel open time will depend on the agonist with lower affinity, in this case carbamylcholine. However, this hypothesis may be an oversimplification, since an analysis in the microsecond time range of current fluctuations through single acetylcholine receptor channels indicates that channels may close briefly and open again during the time for which the receptor remains occupied by agonist (Colquhoun and Sakmann, 1981).

Records of single channels obtained in the presence of substantial concentrations of agonist reveal burst behavior; i.e., during long time intervals, the channel remains in the desensitized state from which it occasionally recovers, fluctuating rapidly between the closed and open state before relaxing again into the desensitized state (Sakmann *et al.*, 1980). From an analysis of the temporal distribution of bursts and intervals between the bursts, Sakmann *et al.* (1980) have resolved the desensitization process into two components: a fast component with an apparent rate constant of about 2 $sec^{-1}$, and a slow component with an apparent rate constant of about 0.2 $sec^{-1}$. The rate constants for recovery from the desensitized state were determined to be ~5 $sec^{-1}$ and 0.03 $sec^{-1}$ for the two components, respectively. This resolution of the desensitization process into two distinct components has subsequently been confirmed by pharmacological studies using rapid spectroscopic techniques (Neubig and Cohen, 1980; Walker *et al.*, 1981b, 1982) and by electrophysiological studies using nerve-released acetylcholine (Magleby and Pallotta, 1981) or iontophoretically applied agonists (Feltz and Trautmann, 1982).

## B. Pharmacology of the Nicotinic Acetylcholine Receptor

### 1. Agonist-Induced State Transitions of the Acetylcholine Receptor

After the initial prediction by Katz and Thesleff in 1957 that desensitization of the acetylcholine receptor would be accompanied by an increase in binding affinity for the agonist, transitions between low- and high-affinity states have been demonstrated by a variety of techniques which monitor ligand binding to the receptor. These techniques include fluorescent spectroscopy (Grünhagen *et al.*, 1977; Barrantes, 1978; Quast *et al.*, 1978, 1979; Heidmann and Changeux, 1979a,b), electron spin resonance (Weiland *et al.*, 1976), inhibition of the initial rate of neurotoxin binding (Weber *et al.*, 1975; Lee *et al.*, 1977; Weiland and Taylor, 1979; Sine and Taylor, 1979), and rapid-mixing techniques monitoring the binding of radioactive ligands (Boyd and Cohen, 1980a,b; Neubig *et al.*, 1982). In each case, it was evident that binding of an agonist to the native membrane-bound acetylcholine receptor results in a slow conformational transition which is characterized by an approximately 300-fold increase of affinity for the agonist. The major conformational transitions underlying acetylcholine receptor function are schematically represented in Figure 7. Binding measurements by equilibrium dialysis reveal a homogeneous population of high-affinity acetylcholine-binding sites (Neubig and Cohen, 1979). Representative values for the apparent dissociation constants of the low- and high-affinity conformation were obtained by Boyd and Cohen (1980a) from studies on the kinetics of binding of [$^3$H]acetylcholine and [$^3$H]carbamylcholine to acetylcholine receptors in membrane vesicles from the electric organ of *Torpedo*. At 4°C, the apparent dissociation constants of the low- and high-affinity conformations for acetylcholine were determined to be 800 nM and 2 nM, respectively, and for carbamylcholine these values were 30 μM and 25 nM, respectively. Other investigators have reported similar values, although these values may vary slightly between different laboratories (Barrantes, 1978; Weiland and Taylor, 1979; Heidmann and Changeux, 1979a).

State transitions in the acetylcholine receptor molecule and agonist-induced cation permeability were correlated in elegant studies by Sine and Taylor (1979) using cultured $BC_3H1$ muscle cells. Agonists were shown to elicit a permeability increase to $^{22}Na^+$ ions that slowly decreased in a way closely paralleled by the conversion of the acetylcholine receptor from an initially low-affinity state to a high-affinity conformation. A similar correlation was demonstrated for the binding of agonists to acetylcholine receptors in membranes from *Torpedo* electric organ (Neubig *et al.*, 1982). Thus, the

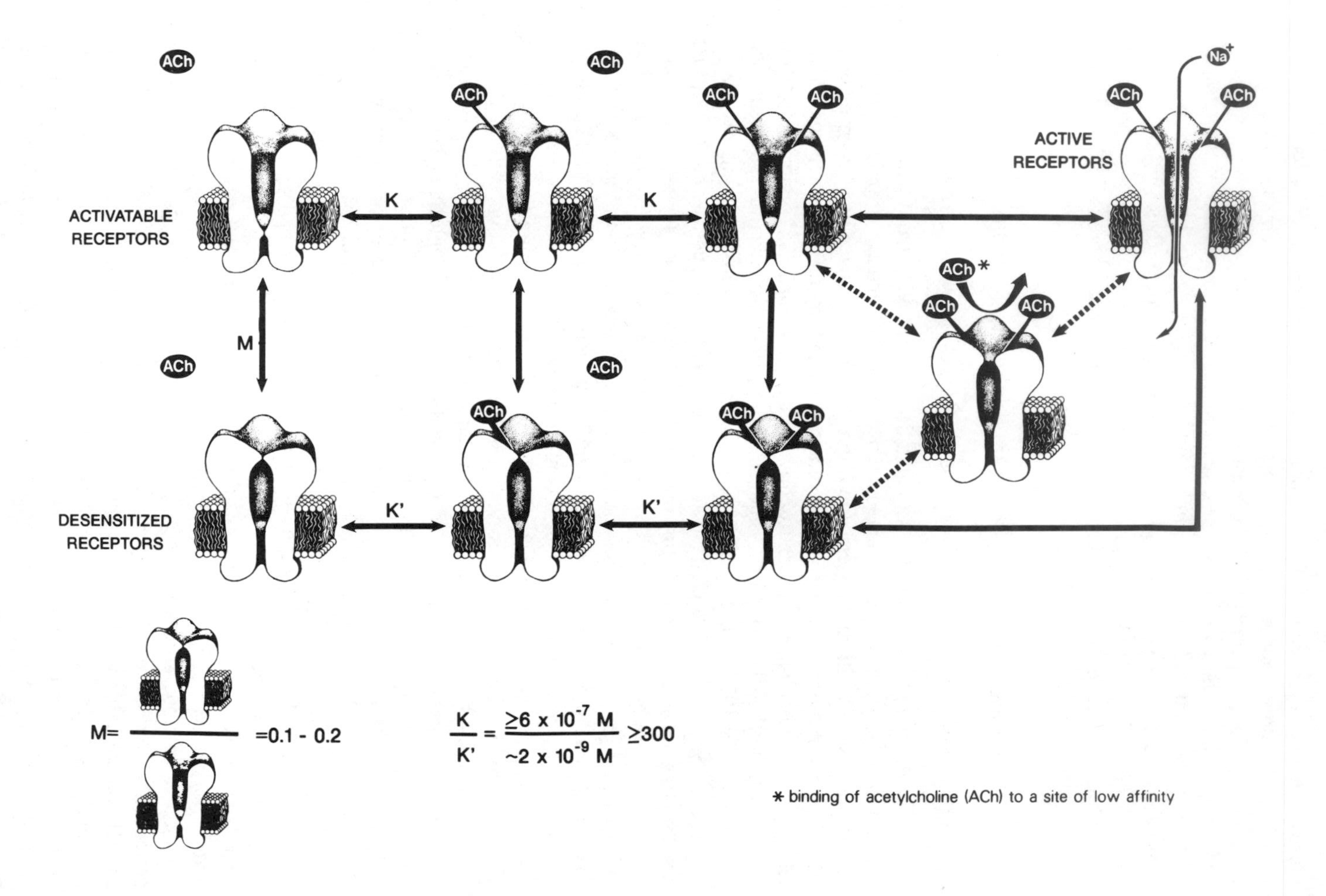

ACh
ACh*
Na+
ACTIVE RECEPTORS
ACTIVATABLE RECEPTORS
DESENSITIZED RECEPTORS
K
K'
M
M= =0.1 - 0.2
$\frac{K}{K'} = \frac{\geq 6 \times 10^{-7}\ M}{\sim 2 \times 10^{-9}\ M} \geq 300$
* binding of acetylcholine (ACh) to a site of low affinity

physiologically observed refractory state of the receptor was shown to correspond with the pharmacologically characterized high-affinity conformation. In $BC_3H1$ muscle cells (but not in *Torpedo*), agonist binding was found to involve a moderate degree of homotropic cooperativity which could be accounted for by assuming that both agonist-binding sites of the acetylcholine receptor go through a concerted conformational transition, which precludes the formation of hybrid species binding agonist with low affinity to one site and with high affinity to the other (Sine and Taylor, 1979).

Agonist-controlled cation translocation shows a second-order dependence on agonist concentration whether assayed electrophysiologically (Adams, 1975; Peper *et al.*, 1975; Sheridan and Lester, 1977; Dionne *et al.*, 1978; Dreyer *et al.*, 1978) or by tracer flux experiments (Sine and Taylor, 1979; Neubig and Cohen, 1980; Cash *et al.*, 1981; Neubig *et al.*, 1982; Walker *et al.*, 1982), indicating that two acetylcholine binding sites are required for activation of the acetylcholine receptor (Figure 7). Neubig *et al.* (1982) estimated that in the absence of agonist less than one channel in $10^7$ is open, and that in the case of the singly liganded receptor, the probability of channel opening is less than 0.03% of that of the doubly liganded receptor. Further evidence that the doubly-liganded receptor dominates the control of channel opening comes from experiments that correlate the fractional occupancy of ligand-binding sites by cobratoxin with inhibition of acetylcholine receptor function. A linear decline of acetylcholine-mediated cation permeability on addition of increasing amounts of cobratoxin would indicate that a singly liganded site controls opening of the acetylcholine receptor channel. However, if the doubly liganded form of the acetylcholine receptor controls the opening of the cation channel, the concentration dependence of the inhibition of acetylcholine receptor function by cobratoxin should be parabolic. Such predicted parabolic decline of the cholinergic response as a function of the cobratoxin concentration has been observed by Sine and Taylor (1980) for the acetylcholine receptor of the $BC_3H1$ muscle cell line and by Anholt *et al.* (1982) for the acetylcholine receptor from *Torpedo* electric organ after reconstitution in lipid vesicles. Attempts to measure the inhibition of receptor function by neurotoxin in native membrane vesicles from *Torpedo* electric organ were complicated due to the dense packing of acetylcholine receptors in these membranes resulting in a limited internal volume per receptor for the flux of cations. Moore *et al.* (1979) observed a decline of cation efflux from electric organ membranes only after approximately 70% of the ligand-binding

*Figure 7.* Schematic representation of the major conformational transitions that underlie acetylcholine receptor function. ACh designates acetylcholine. K and K′ indicate the intrinsic equilibrium dissociation constants for the agonist of the resting and desensitized states of the receptor, respectively. M is the allosteric constant describing the isomerization equilibrium of the interconversion between unliganded, activatable, and desensitized acetylcholine receptors. The scheme indicates that the doubly liganded receptor is able to open a cation-selective channel. Desensitization of the receptor is accompanied by closing of the channel and an increase in binding affinity for the agonist (indicated in the figure by thickening and shortening of the bars connecting ACh with the receptor). Desensitization is a two-step process which may involve a transient intermediate state in which acetylcholine binds to a low-affinity site (indicated in the figure and further explained in the text). The nature of the various closed states of the channel as indicated in the figure is merely illustrative and is, in reality, unknown. For actual values for *Torpedo californica* of the constants K, K′, and M, see Weiland and Taylor (1979) or Boyd and Cohen (1980a).

sites were blocked with toxin. Similar volume limitations on the agonist-controlled cation flux through acetylcholine receptor channels have been observed by other laboratories (Neubig and Cohen, 1980; Lindstrom *et al.*, 1980c). Therefore, responses that are directly proportional to the amount of active receptor can be measured in receptor-rich electric organ membranes only at low concentrations of agonist and after blocking a substantial fraction of the total [$^{125}$I]α-bungarotoxin-binding sites with toxin. As will be discussed below, reconstitution of acetylcholine receptors in lipid vesicles can, to a large extent, circumvent this limitation (Anholt *et al.*, 1982).

Whereas acetylcholine receptor activation requires the binding of two molecules of acetylcholine, evidence suggests that a single molecule of bound agonist is sufficient to cause desensitization of the receptor (Sine and Taylor, 1979; Neubig *et al.*, 1982; Walker *et al.*, 1982; Figure 7). Furthermore, the slow conformational transition of the acetylcholine receptor from the low-affinity into the high-affinity, desensitized state has been resolved into two steps (Neubig and Cohen, 1980; Walker *et al.*, 1981b, 1982; Hess *et al.*, 1982). Analysis of signals from fluorescent ligands (Heidmann and Changeux, 1980) and correlation of permeability measurements with binding parameters (Neubig and Cohen, 1980; Walker *et al.* 1981b, 1982) indicated the involvement of a low-affinity acetylcholine-binding component in the initiation of the desensitization process of the acetylcholine receptor from *Torpedo* (Figure 7). These studies are in agreement with electrophysiological observations using nerve-released acetylcholine (Magleby and Pallotta, 1981) or iontophoretically applied agonists (Feltz and Trautmann, 1982). The latter study indicates that at the frog neuromuscular junction, the time constant of the fast phase of desensitization is 11–12 sec and 4–5 min for the slow phase, regardless of the concentration or the nature of the agonist. In these studies, the recovery from the desensitized state was also found to be a two-step process. Moreover, the onset and offset of the desensitization process were coupled such that a fast recovery was observed when desensitization had lasted only a few seconds, but a slow phase of recovery was prominent when receptors had been kept in the desensitized state for several minutes (Feltz and Trautmann, 1982). Faster time constants for the desensitization process were reported for the acetylcholine receptor from *Torpedo*; e.g., Walker *et al.* (1981b) reported a half-lifetime of ~300 msec for the fast step and 6–7 sec for the slow step. Using a rapid quench-flow technique, Walker *et al.* (1982) measured the agonist-controlled influx of $^{86}Rb^{+}$ through acetylcholine receptor channels in reconstituted vesicles. They observed that the half-maximal rate of the fast phase of the desensitization process is characterized by a carbamylcholine concentration at which both ligand-binding sites for receptor activation are expected to be saturated with agonist. From these observations they suggest that the low-affinity binding component which signals the onset of the desensitization process may represent a third binding site for acetylcholine on the receptor molecule distinct from the two neurotoxin-binding sites that mediate receptor activation. The extensive structural homologies between the subunits of the acetylcholine receptor, which were discussed above, raise the possibility that additional acetylcholine-binding sites may be present on the molecule, which may participate in the desensitization process. Such putative agonist-binding sites may resemble the ligand-binding sites on the α-subunit which control channel opening but may have a lower affinity for acetylcholine. Sites with very low affinity for some cholinergic ligands have been detected on the acetylcholine receptor, but their physiological significance for receptor activation or desensitization is, at best, unclear (Conti-Tronconi *et al.*, 1982d; Dunn and Raftery, 1982).

In the absence of agonist, the majority of acetylcholine receptors exist in the low-affinity activatable conformation. However, a fraction of the nonliganded receptors will preexist in the desensitized conformation (Figure 7). The population distribution of activatable and desensitized receptors is at equilibrium characterized by an allosteric constant, frequently indicated as "M." The value of M has been determined for the acetylcholine receptors from *Torpedo* electric organ and from the $BC_3Hl$ murine smooth muscle cell line. In the case of the receptor from *Torpedo* between 10% (Weiland and Taylor, 1979) and 20% (Barrantes, 1978; Heidmann and Changeux, 1979a; Boyd and Cohen, 1980a) of the receptor population preexists in the desensitized conformation (M = 0.1–0.2). However, in the case of the $BC_3Hl$ muscle cell line, the value of M was found to be $\sim 10^{-4}$ (Sine and Taylor, 1982), indicating that here only one in every 10,000 receptors preexists in the desensitized conformation. This observation accounts for the fact that at high agonist concentrations, desensitization goes to completion in the case of the *Torpedo* receptor (Epstein and Racker, 1978; Lindstrom *et al.*, 1980c; Moore and Raftery 1980; Popot *et al.*, 1981; Walker *et al.*, 1981b; Hess *et al.*, 1982; Anholt *et al.*, 1982), but results in the maximal steady-state desensitization of only 75% of the receptor population in the case of the $BC_3Hl$ receptor (Sine and Taylor, 1979). Boyd and Cohen (1980a) reported the $K_d$ of acetylcholine for the high-affinity conformation of the acetylcholine receptor to be 1.4 nM and its association rate $3 \times 10^7$ $M^{-1}sec^{-1}$ at 4°C. The average dwell time of acetylcholine in the binding site of the desensitized acetylcholine receptor calculated from these data is 24 sec. This dwell time is short relative to the lifetime of the desensitized conformation of the unliganded receptor ($\sim$7 min in the system of Boyd and Cohen, 1980a). Hence, at steady state, low concentrations of agonist can cause receptor desensitization with great efficiency. Since the recovery time from the desensitized state is very slow (Boyd and Cohen, 1980a) and since the value of M is relatively high for *Torpedo* receptor (Weiland and Taylor, 1979; Heidmann and Changeux, 1979a; Boyd and Cohen, 1980a), all of the acetylcholine receptors can be effectively locked in the refractory state after an initial brief (millisecond) transient ion-flux response. However, in the case of the $BC_3H1$ receptor, steady-state flux measurements can still be made at high agonist concentrations (Sine and Taylor, 1979) since 25% of the total receptor population remains active under these conditions due to the intrinsically low value of the allosteric constant M (Sine and Taylor, 1982). The relatively slow recovery time from the desensitized state of acetylcholine receptors in muscle (Rang and Ritter, 1970; Scubon-Mulieri and Parsons, 1977; Feltz and Trautmann, 1982) and the observation that here desensitization can go to completion (Scubon-Mulieri and Parsons, 1977; Magleby and Pallotta, 1981; Feltz and Trautmann, 1982) suggest that the kinetics of the state transitions which occur in receptors from striated muscle may resemble those of the *Torpedo* electric organ receptor more closely than those of the acetylcholine receptor from the murine $BC_3Hl$ muscle cell line.

## 2. Competitive Inhibitors of the Cholinergic Response

Classical antagonists, such as *d*-tubocurarine (curare; Figure 4) and gallamine, block the cholinergic response by competing with the neurotransmitter for the ligand-binding sites on the acetylcholine receptor. Application of curare to the endplate causes a decrease in the amplitude of the postsynaptic potential by reducing the frequency of channel opening events without altering the conductance characteristics of the open channel. This effect can be overcome by an excess of acetylcholine illustrating the competitive nature of the inhibition (Jenkinson, 1960).

In contrast to agonists, conformational transitions in the acetylcholine receptor molecule have not been readily detectable with antagonists (Weber *et al.*, 1975; Weiland *et al.*, 1976; Weiland and Taylor, 1978; Sine and Taylor, 1981). The binding of competitive antagonists to the acetylcholine receptor from *Torpedo* is characterized by two distinct components: *d*-tubocurarine has a high-affinity binding site with a $K_d$ of ~33 nM and low-affinity binding site with $K_d$ of ~8 μM (Neubig and Cohen, 1979). Binding of curare to both populations of binding sites can be abolished by pretreatment with an excess of α-bungarotoxin, or it can be competitively inhibited by carbamylcholine. A similar distinction of some reversible antagonists in affinity for the two ligand binding sites has been observed by Sine and Taylor (1981) for the receptor in $BC_3Hl$ cells. They observed that the higher the apparent binding affinity for the antagonist, the less pronounced is the separation in dissociation constants for the two antagonist-binding sites. In line with this trend, curarimimetic snake neurotoxins which bind with high affinity (a $K_d$ of the order of $10^{-11}$ M) to the acetylcholine receptor do not distinguish between the sites in terms of binding affinity (Witzemann and Raftery, 1978; Nathanson and Hall, 1980; Sine and Taylor, 1981). Since these toxins may interact with the receptor at multiple sites, their binding may, as a result, be less sensitive to conformational alterations at a single site. Sine and Taylor (1981) further compared the concentration dependence of antagonist binding with inhibition of the agonist-elicited cation permeability, and showed that occupation by the antagonist of one of the two ligand-binding sites on the receptor was sufficient to block the response. Antagonists appeared to exert their inhibitory effect primarily through occupation of the site for which they had highest affinity. These observations are consistent with the notion that liganding of both acetylcholine-binding sites is required for receptor activation. Thus far, it has not yet been established which of the two α-subunits carries the high-affinity site and which one carries the low-affinity site for antagonists like curare. It is of interest that the coral diterpenoid lophotoxin appears to inhibit acetylcholine receptor function in $BC_3H_1$ cells via a preferential blockade at the ligand binding site with low affinity for curare (Culver, *et al.*, 1984).

Although *d*-tubocurarine has been considered a competitive antagonist, it can induce desensitization of the receptor (Krodel *et al.*, 1979; Cohen, 1978) and a non-competitive effect of curare on the endplate has also been reported (Katz and Miledi, 1978). Moreover, application of the high-resolution patch-clamp technique to measurements of single acetylcholine receptor channels in cultured rat myotubes has revealed that curare can act as a weak agonist causing the opening of very short-lived receptor channels (Lecar *et al.*, 1982; Morris *et al.*, 1982; Trautmann, 1982). The histogram of the open time distribution of these channels could be fitted by two exponentials, the slower component corresponding to an open time of ~1–2 msec and the faster component ranging between 0.25–0.7 msec (Lecar *et al.*, 1982; Trautmann, 1982). Trautmann (1982) observed that the mean channel open time was dependent on the concentration of curare. The fast component of the channel open time distribution may arise from binding of a single molecule of curare to the high-affinity binding site, whereas, the doubly-liganded receptor may give rise to the slow component. However, this interpretation should be viewed with caution, since curare appears to block its own channels in a noncompetitive fashion, thus complicating the interpretation of the mean channel open time (Trautmann, 1982). The unambiguous conclusion emerging from these studies is that the distinction between agonists and antagonists may be less sharply defined than previously accepted. In fact, snake neurotoxins may well be the only blockers of the acetylcholine-binding site and, hence, the cholinergic response that do not display secondary noncompetitive or agonist-like effects.

### 3. *Noncompetitive Inhibitors of the Cholinergic Response*

In addition to the competitive antagonists, a heterogeneous group of noncompetitive inhibitors of the cholinergic response has been identified. These compounds include, among others, aminated local anesthetics such as procaine, lidocaine, and tetracaine (Steinbach, 1968: Kordas, 1970; Katz and Miledi, 1975; Weiland *et al.*, 1977; Neher and Steinbach, 1978; Koblin and Lester, 1979), as well as proadifen, di- and trimethisoquin (Figure 4; Krodel *et al.*, 1979; Heidmann and Changeux, 1979b, 1981; Waksman *et al.*, 1980), the toxin histrionicotoxin isolated from the Columbian frog *Dendrobates histrionicus* and its perhydro derivative (Figure 4; Eldefrawi *et al.*, 1977, 1980; Elliott *et al.*, 1979; Heidmann and Changeux, 1979b, 1981; Sine and Taylor, 1982), the hallucinogen phencyclidine (Albuquerque *et al.*, 1980a,b; Karpen *et al.*, 1982), and polymixin-type antibiotics (Brown and Taylor 1983).

The inhibitory effect of the aminated local anesthetics on the endplate current manifests itself in the presence of the agonist and is more pronounced at more negative voltages in the case of tertiary amines such as procaine (Neher and Steinbach, 1978; Koblin and Lester, 1979). Single-channel records obtained in the presence of the agonist suberyldicholine and the quaternary lidocaine derivative QX222 (Figure 4) showed that, in the presence of the local anesthetic, single channels were fragmented into bursts of shorter current pulses (Neher and Steinbach, 1978). This flickering of the conductance, together with the voltage dependence and the requirement for the presence of the agonist, has led to the hypothesis that the local anesthetic blocks the agonist-induced conductance by binding to and physically occluding the cation channel (Neher and Steinbach, 1978; Ruff, 1977, 1982). However, a number of compounds such as chlorpromazine (Koblin and Lester, 1979), benzocaine (Ogden *et al.*, 1981), alcohols (Gage *et al.*, 1975), fatty acids (Brisson *et al.*, 1975; Andreasen *et al.*, 1979), various volatile organic solvents (Young *et al.*, 1978), detergents such as Triton X-100 (Brisson *et al.*, 1975), and sodium cholate (Briley and Changeux, 1978) act as voltage-independent noncompetitive blockers (Koblin and Lester, 1979). It has been suggested that these compounds exert their effect by disturbing the interaction between the intramembranous portion of the acetylcholine receptor channel and the surrounding lipid matrix (Koblin and Lester, 1979; Andreasen and McNamee, 1980). Studies of acetylcholine receptor function employing local anesthetics are complicated by the fact that a sharp distinction cannot always be drawn between general effects exerted through the lipid matrix and effects mediated through a specific local anesthetic-binding site (see Koblin and Lester, 1979).

Binding studies have shown that local anesthetics as well as histrionicotoxin appear to enhance rather than to inhibit the binding of agonists and, vice versa, agonists enhance the binding of local anesthetics to the acetylcholine receptor (Kato and Changeux, 1976; Burgermeister *et al.*, 1977; Krodel *et al.*, 1979; Oswald and Changeux, 1981a,b). The net effect of the presence of local anesthetics is to stabilize the acetylcholine receptor in its high-affinity conformation (Weiland *et al.*, 1977; Heidmann and Changeux, 1979b, 1981; Dunn *et al.*, 1981; Sine and Taylor, 1982). Sine and Taylor (1982) have directly measured the effect of histrionicotoxin and the local anesthetics dibucaine and QX314 on the allosteric constant of the acetylcholine receptor in $BC_3Hl$ cells. They demonstrated that the primary effect of the local anesthetic is an ~2000-fold increase in the value of the allosteric constant, M.

Estimates of the stoichiometry of local anesthetic binding are variable. Krodel *et al.*, (1979) reported a stoichiometry of 0.25 local anesthetic-binding sites per agonist-binding site for meproadifen in the presence of carbamylcholine and one local anesthetic-binding site per agonist-binding site in the absence of the agonist. Estimates for the binding of histrionicotoxin vary between 0.7 (Eldefrawi *et al.*, 1978) and two (Eldefrawi *et al.*, 1977) binding sites per agonist-binding site. In addition, 0.25 binding site for perhydrohistrionicotoxin per ligand-binding site has been reported (Elliott and Raftery, 1977). Binding studies are complicated by the hydrophobic nature of these compounds allowing them to partition in the lipid phase of the membrane, which results in a high nonspecific binding (notably in the case of histrionicotoxin; e.g., Eldefrawi *et al.*, 1978), and the observation that at high concentrations, the aminated local anesthetics, which bear resemblance to cholinergic agonists (Figure 4), can bind to the acetylcholine-binding sites as well (Krodel *et al.*, 1979). Sine and Taylor (1982) observed that the local anesthetic-mediated increase in the allosteric constant of the acetylcholine receptor in $BC_3Hl$ cells proceeds in a cooperative fashion consistent with their association with at least two sites per acetylcholine receptor.

Oswald *et al.* (1980) have synthesized a tritiated photoaffinity derivative of the aminated local anesthetic trimethisoquin and have demonstrated that this compound selectively labels the δ-subunit of the membrane-bound acetylcholine receptor from the electric organ of *Torpedo marmorata*. This labeling was enhanced by carbamylcholine and this auxiliary effect of carbamylcholine could be blocked by α-bungarotoxin. Selective labeling of the δ-chain could also be inhibited by unlabeled local anesthetics and histrionicotoxin (Oswald *et al.*, 1980; Saitoh *et al.*, 1980; Oswald and Changeux, 1981a,b). Tryptic fragmentation of the labeled δ-chain indicated that the trimethisoquin derivative was attached to a region of δ-chain located on the synaptic side of the membrane (Wennogle *et al.*, 1981). These studies were extended by the observation by Oswald and Changeux (1981b) that ultraviolet light irradiation could induce the covalent attachment of a variety of noncompetitive blockers to the receptor without the need for prior derivatization. Thus, tritiated trimethisoquin, phencyclidine, and histrionicotoxin were all shown to label predominantly the δ-chain. However, tritiated chlorpromazine displayed less specific labeling and attached to all four subunits. Still, this labeling could be enhanced by carbamylcholine and inhibited by histrionicotoxin (Oswald and Changeux, 1981b). The overall implication arising from these studies is that the δ-chain contains a binding site for local anesthetics. Thus far, it is not known whether this site is located in or near the ion channel or at the periphery of the receptor near the membrane interphase, whether other subunits contribute to the formation of this site, or whether additional local anesthetic-binding sites that are not as readily labeled exist on any of the other receptor subunits. It is of interest that Kaldany and Karlin (1983) observed labeling of all subunits of the acetylcholine receptor by quinacrine mustard with increased labeling of the α and β subunits in the presence of carbamylcholine.

Although the effects of local anesthetics on the cholinergic response have been studied in considerable detail, the physiological significance of local anesthetic binding to the acetylcholine receptor is not clear. Structural homologies between the different subunits of the receptor and structural similarities between acetylcholine and aminated local anesthetics, e.g., QX222 (Figure 4), suggest that local anesthetics may mimic binding of the neurotransmitter to "degenerate" acetylcholine-binding sites distinct from the ligand-binding sites that control receptor activation. The low-affinity acetylcholine-binding site involved in receptor desensitization may be such a degenerate acetylcholine-binding site, and it is tempting to speculate that this site may be identical to the local anesthetic-binding site on the δ-subunit, especially since liganding of this site by either ligand would cause the same effect, namely the allosteric conversion of the acetylcholine receptor to its high-affinity, desensitized conformation.

## IV. STRUCTURE–FUNCTION CORRELATIONS WITHIN THE ACETYLCHOLINE RECEPTOR MOLECULE: RECONSTITUTION STUDIES AS AN EXPERIMENTAL APPROACH

### A. Reconstitution of Acetylcholine Receptors in Model Membranes

#### 1. Incorporation of Acetylcholine Receptors in Lipid Vesicles

Thus far, we have described the biochemical information which has given insights into the structure of the acetylcholine receptor molecule, and electrophysiological and pharmacological results that have yielded important information regarding the function of the acetylcholine receptor. The molecular mechanisms that underlie the conformational transitions, which take place after binding of the agonist, are, however, not yet understood. During the last decade, several laboratories have attempted reconstitution of acetylcholine receptors in model membranes in order to establish systems that would allow investigation of structure–function relationships in the receptor molecule (reviewed by Anholt, 1981; Popot, 1982; McNamee and Ochoa, 1982; Anholt *et al.*, 1983). Reconstitution studies would allow biochemical manipulation of the receptor in detergent solution and subsequent assessment of functional activity after reassembly of the acetylcholine receptor into a model membrane. Thus, one can, in principle, investigate under controlled conditions how structural modifications in the molecule or alterations in lipid environment affect the function of the acetylcholine receptor. Hence, a reconstituted system might give information relevant to the following questions: (1) is the ion channel a part of the purified acetylcholine receptor, (2) which subunits contribute to its structure, and (3) how does binding of acetylcholine to the receptor regulate channel gating.

Early attempts to reconstitute acetylcholine receptors in lipid vesicles were only partially successful (Hazelbauer and Changeux, 1974) and to a large extent not reproducible (Michaelson and Raftery, 1974). Although neurotoxin-binding activity was retained after solubilization of electric organ membranes in detergent, recovery of agonist-induced cation translocation appeared highly problematic (reviewed by Briley and Changeux, 1977). In addition, it became clear that the characteristic agonist-induced conformational transitions between low- and high-affinity states of the receptor were irreversibly lost after disruption of the electric organ membrane by detergent (Sugiyama and Changeux, 1975; Briley and Changeux, 1978).

Significant progress came from experiments by Epstein and Racker (1978), who dissolved electric organ membranes using the anionic dialyzable detergent cholate in the presence of excess crude soybean lipids. Removal of the detergent by dialysis resulted in the incorporation of acetylcholine receptors into lipid vesicles, which exhibited a rapid uptake of $^{22}Na^+$ on addition of carbamylcholine. This response could be blocked by curare, α-bungarotoxin, and local anesthetics. Pre-exposure to agonist in the absence of radioisotope could prevent the subsequent translocation of $^{22}Na^+$, indicating the occurrence of desensitization. These studies led to the realization that the continuous presence of lipids after solubilization of the electric organ membranes is essential to preserve the functional integrity of the cation channel. Several laboratories have subsequently established optimal conditions for the preservation of acetylcholine receptor function in lipid–cholate solution (Huganir *et al.*, 1979; Lindstrom *et al.*, 1980c; Popot *et al.*, 1981; Anholt *et al.*, 1981). Preservation of acetylcholine receptor function after solubilization in the nonionic detergent octylglucoside has also been reported (Paraschos *et al.*, 1982).

Relatively low amounts of lipid were found to be required for the stabilization of the native structure of the acetylcholine receptor in cholate solution in terms of the subsequent recovery of the agonist-induced translocation of cations (Anholt *et al.*, 1981) and the characteristic allosteric state transitions (Heidmann *et al.*, 1980a,b; Criado *et al.*, 1982). Under optimal lipid conditions, the detergent is still present in a 20-fold molar excess over the lipid. Based on this fact, Anholt *et al.* (1981) suggested that the hydrophobic regions of the acetylcholine receptor have a higher affinity for the lipid than for the detergent, allowing them to be surrounded by an annulus of lipid and to be carried as lipoprotein complexes in solution (Figure 8). In this configuration, the planar cyclopentenophenanthrene rings of the cholate may interact with the hydrophobic moieties of these lipoprotein complexes, while the hydroxyl groups attached to the rings interact with the aqueous environment.

Although low concentrations of lipids (2 mg/ml in 2% cholate) are sufficient to protect the native structure of the cation channel in detergent solution after solubilization, it was found that about ten-fold higher lipid concentrations were required to preserve the integrity of the cation channel during the reassembly of lipid and acetylcholine receptors into lipid vesicles (Anholt *et al.*, 1981; Criado *et al.*, 1982; Figure

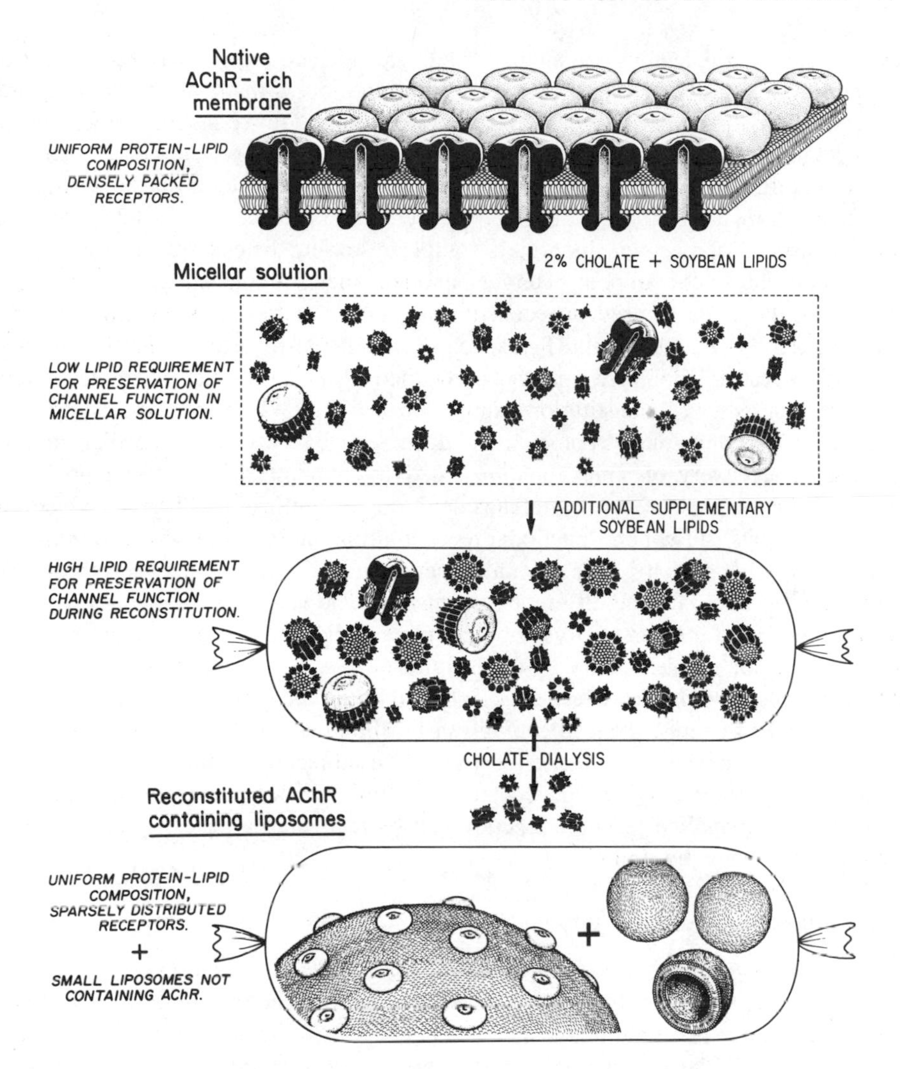


*Figure 8.* Schematic representation of acetylcholine receptor (AChR)–lipid interactions during the solubilization and reconstitution processes. From Anholt *et al.* (1981).

8). When receptors are reconstituted under optimal conditions (lipid:protein > 16:1), they are found to reassemble into lipid vesicles at a constant protein:lipid ratio, such that the packing density of the receptors in the reconstituted membranes is five-fold lower than in the native electric organ membrane (Figure 8; Anholt *et al.* 1981, 1982). Freeze-fracture replicas of reconstituted preparations reveal the presence of two populations of vesicles: large vesicles containing particles which correspond in size to acetylcholine receptors and smaller liposomes free of particles. About 70% of the receptors are oriented with their agonist-binding sites on the external surface of the vesicles. Immune precipitation of reconstituted vesicles with monoclonal antibodies has indicated that receptors in a single vesicle are oriented either all right-side-out or inside-out (Anholt *et al.*, 1981). The reconstitution process that leads to the insertion of receptors at a fixed packing density and orientation in the reconstituted membranes is not yet understood. Anholt *et al.* (1981) discussed the formation of the reconstituted vesicles in terms of steric and electrostatic protein–protein interactions occurring during the vesicle formation process and hypothesized the existence of "pseudocrystalline spherical arrays" of receptor–lipoprotein complexes during the early nucleation events.

In addition to the anionic detergent cholate, the nonionic dialyzable detergent octylglucoside has been used in reconstitution studies of the acetylcholine receptor. Gonzalez-Ros *et al.* (1980) and Paraschos *et al.* (1982) reported solubilization and purification of acetylcholine receptors in octylglucoside in the absence of supplementary lipids and subsequent reconstitution using *Torpedo* lipids. They described the incorporation of acetylcholine receptors in lipid vesicles by octylglucoside dialysis and reported the recovery of some functional activity. Anholt *et al.* (1981), however, observed that the presence of octylglucoside during solubilization of the acetylcholine receptor prevents subsequent functional reconstitution into vesicles of soybean lipids. It is not yet clear which of the several experimental differences between these two approaches accounts for the differences in results. The nonionic detergent Tween 80 has also been used in attempts to solubilize functionally intact acetylcholine receptors in the absence of supplementary lipids (Heidmann *et al.*, 1981; Ruechel *et al.*, 1981; Boheim *et al.*, 1981). However, reassembly of these receptors into sealed vesicles with recovery of agonist-induced cation translocation has not yet been demonstrated. Hence, cholate dialysis remains the most widely used technique for the incorporation of functionally intact acetylcholine receptors into lipid vesicles, since general consensus exists that acetylcholine receptor function can be recovered after exposure to mixed micelles of cholate and lipid.

### 2. *Incorporation of Acetylcholine Receptors in Planar Lipid Bilayers*

In addition to the reconstitution of acetylcholine receptors in lipid vesicles, several laboratories have reported the incorporation of acetylcholine receptors from *Torpedo* electric organ in planar lipid bilayers (Nelson *et al.*, 1980; Schindler and Quast, 1980; Boheim *et al.*, 1981). Planar lipid bilayers can be constructed by spreading two lipid monolayers at the air–water interface of two aqueous compartments separated by a Teflon septum. Subsequent apposition of the two monolayers by sequentially raising the water levels in the two compartments across a small (0.25 mm) aperture in the septum results in the assembly of a planar lipid bilayer. Immersion of an electrode in each compartment allows electrical recordings across this bilayer membrane (Montal and Mueller, 1972; Montal, 1974; reviewed by Montal *et al.*, 1981; Figure 9).

Early attempts to incorporate functionally intact acetylcholine receptors into planar bilayers were unsuccessful, most likely due to the use of organic solvents in the preparation of receptor–lipid complexes (Parisi *et al.*, 1971; Schlieper and De Robertis, 1977). Subsequently, techniques have been developed which circumvent the use of organic solvents. These methods have been more successful and are based either on fusion of reconstituted or native membrane vesicles with preformed lipid bilayers

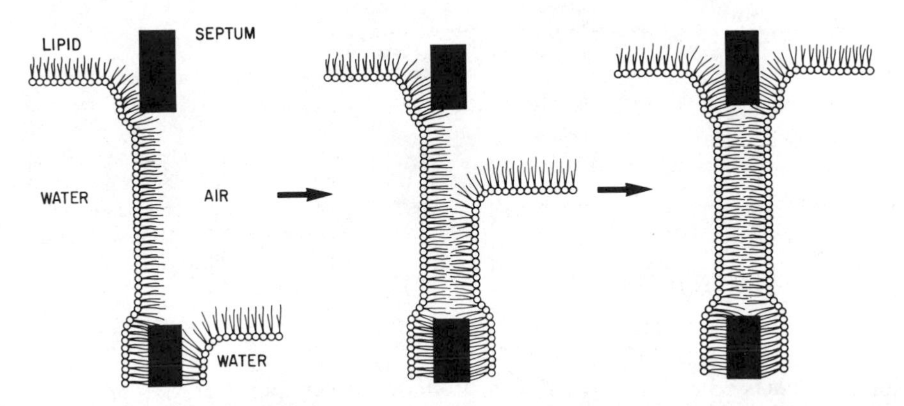


*Figure 9.* Schematic representation of the formation of planar lipid bilayers by apposition of two monolayers across an aperture in a Teflon septum. Membranes are formed by apposition of two monolayers across an aperture in a septum separating two aqueous compartments. The formation of the membrane can be monitored by measuring the amplitude of the capacitative transient, which stabilizes when the hole is completely covered at a capacitance of 0.8 $\mu F/cm^2$ (Montal and Mueller, 1972; Montal, 1974).

(Miller *et al.*, 1976; Boheim *et al.*, 1981) or on the spontaneous formation of monolayers from a suspension of such membrane vesicles at the air–water interface followed by subsequent bilayer assembly (Schindler, 1980; Schindler and Quast, 1980; Nelson *et al.*, 1980).

Planar bilayers derived from native electric organ membranes from *Torpedo marmorata* displayed an increased membrane conductance upon addition of carbamylcholine, which exhibited desensitiztion and could be blocked by α-bungarotoxin and curare. This carbamylcholine-induced membrane permeability was about seven times higher for monovalent cations than for anions. Single acetylcholine receptor channels with a unitary conductance of 20–25 p*S* and a lifetime of 1–3 msec were observed in 0.25 M NaCl containing carbamylcholine (Schindler and Quast, 1980). Similar observations were made by Nelson *et al.* (1980) with planar bilayers generated from purified acetylcholine receptors in reconstituted vesicles. Addition of carbamylcholine gave rise to an increase in both the membrane conductance and the conductance noise. Both the increase in the overall membrane conductance and that the noise could be inhibited by curare. Spontaneous relaxation of the current indicating the occurrence of desensitization was also observed. Fluctuation analysis revealed here a single-channel conductance of 16 p*S* (in 0.1 M NaCl) with a mean channel open time of about 35 msec. This rather long channel lifetime was ascribed to limitations of the time constant of the amplifier. Subsequent application of faster electronics revealed a shorter channel lifetime and enabled a detailed characterization of the single-channel properties (Labarca *et al.*, 1981, 1982, 1984*a,b*; Montal, *et. al.*, 1984). The single-channel conductance was found to be independent of the type or concentration of the agonist. In contrast, the lifetime of the open state of the channel depended on the affinity of the agonist, but not on its concentration, and was longest for suberyldicholine, shortest for carbamylcholine, and intermediate for acetylcholine, in line with

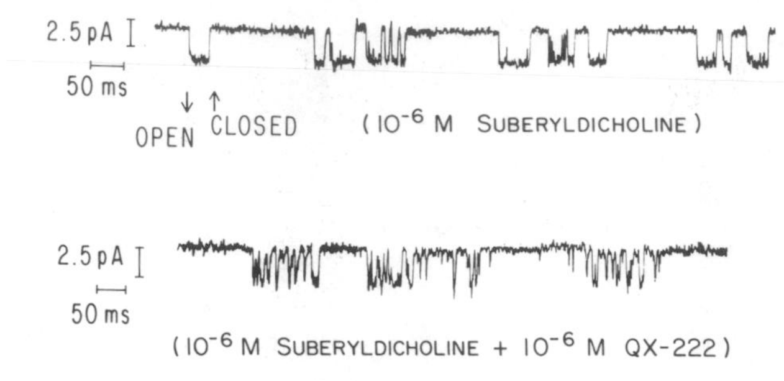


*Figure 10.* Single-channel records obtained from purified acetylcholine receptors in planar lipid bilayers in the presence of suberyldicholine. Notice the flickering of the single-channel conductance observed after addition of the local anesthetic QX222. From Anholt *et al.* (1983), modified from Labarca *et al.* (1982).

single channels recorded at the frog neuromuscular junction (Labarca *et al.*, 1981, 1982, 1984b; Montal, *et al.* 1984; Boheim *et al.*, 1981; Figure 10). In addition, the type of burst kinetics observed by Sakmann *et al.*, (1980) at desensitizing concentrations of agonist with frog muscle preparations, has also been observed in the planar bilayer system. Thus, purified acetylcholine receptors from *Torpedo* electric organ incorporated in planar bilayers appear to behave in a way similar to acetylcholine receptors at the frog neuromuscular junction. Recently, Labarca *et al.* (1984a) have observed that the distribution of channel open times follows a double exponential function. The minor component of this distribution is for each agonist ~5–10 times longer than the predominant short lifetime. The occurrence of the long-lived channels (most likely corresponding to those initially reported by Nelson *et al.*, 1980) was found to be dependent on the prior occurrence of several short-lived events. The nature of these two kinetically distinct states of the receptor is not known.

Although planar lipid bilayers are technically complex systems, they are thus far the only reconstituted system which allows direct observation of the kinetics of opening and closing of single acetylcholine receptor channels at the same high resolution obtained with the patch-clamping technique at muscle endplates (Figure 10). Recently the formation of bilayers over the tip of a patch pipet directly applied to suspensions of reconstituted vesicles containing purified acetylcholine receptors has also been reported and preliminary observations suggest that authentic acetylcholine receptor function can be observed in this system (Suárez-Isla *et al.*, 1983; Tank et al., 1983).

The incorporation of purified acetylcholine receptors into lipid vesicles and the subsequent transformation of these vesicles into planar bilayers enables electrical measurements at high sensitivity and time resolution, thus, providing a system in which the "molecular electrophysiology" of synaptic transmission can be studied at the level of its key element, the isolated acetylcholine receptor.

## B. Functional Studies of Acetylcholine Receptors after Reconstitution in Model Membranes

### 1. Identification of the Functional Unit of the Acetylcholine Receptor

Failure to recover receptor function during early reconstitution attempts raised the question whether the five subunits in the purified receptor complex were sufficient to mediate agonist-induced cation translocation or whether other components of the electric organ membranes might play a functional role. It was also not clear whether the monomeric or the dimeric form of the *Torpedo* receptor represented the minimal molecular entity mediating acetylcholine receptor function.

The realization that the presence of supplementary lipids can prevent denaturation of the ion channel during the purification of receptors by affinity chromatography allowed the isolation and reconstitution of functionally intact acetylcholine receptors. These studies demonstrated that the four different acetylcholine receptor subunits are the only components of the electric organ which are essential for agonist-induced cation translocation. (Huganir *et al.*, 1979; Changeux *et al.*, 1979; Lindstrom *et al.*, 1980c; Popot *et al.*, 1981).

It still remained to be determined whether the monomeric or dimeric form of the acetylcholine receptor represented its functional unit. Miller *et al.* (1978) proposed a functional role for the receptor dimer based on equilibrium efflux measurements using native electric organ membranes. However, Anholt *et al.* (1980) reconstituted purified monomers and dimers separately into vesicles and observed that these forms were indistinguishable in their abilities to mediate agonist-induced channel activation and desensitization. Popot *et al.* (1981) reported the functional reconstitution of acetylcholine receptors after treatment with reducing agents, which results in the virtually complete conversion of dimeric forms into monomers. These studies further supported the notion that the $\alpha_2\beta\gamma\delta$ monomer of the acetylcholine receptor contains all the elements necessary for functional activity. Wu and Raftery (1981) confirmed further the functional equivalence of monomeric and dimeric forms of the acetylcholine receptor using a rapid flux assay which measured the fluorescence quenching of a dye entrapped in reconstituted vesicles as a result of the influx of thallium ions (Wu and Raftery, 1980; Wu *et al.*, 1981). They proposed that the rate of ion flux through the receptor channel should be taken as the only measure for functional integrity of the acetylcholine receptor. However, this is misleading, since the rate of ion flux through the receptor channel will depend on the external conditions, such as the membrane voltage and the ion concentrations on each side of the membrane. In addition, thallium ions have been shown to behave anomalously in terms of their permeation through the receptor channel, having a significantly higher relative permeability compared to sodium ions than aqueous mobility would predict (Adams *et al.*, 1980). Thus, there are grave dangers inherent in applying environment-dependent criteria rather than inherent pharmacological characteristics to the evaluation of functional integrity of the acetylcholine receptor. The records presented by Wu and Raftery (1981) indicate that the initial response is complete within the first 10 msec, during which time a precise determination of their flux rate was precluded due to a poor signal-to-noise ratio. Although these authors reported at saturating concentrations of agonist a flux rate which was two-fold larger for receptor dimers than for monomers, they conclude that within their experimental error, monomers and dimers behave in an identical manner. In addition, Boheim *et al.* (1981) compared the single-channel properties of acetylcholine receptor monomers and dimers in planar lipid bilayers and also concluded that the monomeric form is fully functional. The observation that acetylcholine receptors remain functional after extensive trypsinization which converts dimers into monomers constitutes additional evidence in support of the notion that the monomeric form of the receptor represents its functional unit (Huganir and Racker, 1980).

### 2. Pharmacological Integrity of Reconstituted Acetylcholine Receptors

Reconstituted vesicles containing physiologically active acetylcholine receptors should exhibit an agonist-induced, dose-dependent translocation of cations, which can be blocked by known pharmacological inhibitors, such as snake neurotoxins, *d*-tubocurarine, and local anesthetics, and which can be prevented by preincubation with the neurotransmitter, indicating the occurrence of desensitization. Several laboratories have reported reconstituted systems that satisfy some or all of these requirements (Epstein and Racker, 1978; Huganir *et al.*, 1979; Wu and Raftery, 1979; Lindstrom *et al.*, 1980c; Anholt *et al.*, 1980, 1981, 1982; Gonzalez-Ros *et al.*, 1980; Popot *et al.*, 1981; Walker *et al.*, 1982; Paraschos *et al.*, 1982).

A particular problem inherent in studies with vesicular membrane structures is their finite internal volume. Initial studies on reconstituted receptors have mostly measured agonist-induced cation translocation at equilibrium separating vesicle-entrapped $^{22}Na^+$ or $^{86}Rb^+$ from the external radiosotope by a Millipore filtration assay (Changeux *et al.*, 1979; Wu and Raftery, 1979; Gonzalez-Ros *et al.*, 1980; Popot *et al.*, 1981) or on a Dowex cation exchange resin (Epstein and Racker, 1978; Huganir *et al.*, 1979; Lindstrom *et al.*, 1980c; Anholt *et al.*, 1980, 1981, 1982). In these assays, the cumulative uptake or release of $^{22}Na^+$ or $^{86}Rb^+$ is measured seconds after the addition of agonist, an incubation period which exceeds the time scale of acetylcholine receptor activation (milliseconds), thereby allowing the occurrence of multiple channel openings and the stabilization of acetylcholine receptors in the desensitized conformation. Under these assay conditions, the internal volume of the vesicles rather than acetylcholine receptor desensitization may limit the response and distort actual receptor behavior. Therefore, attempts have been made to prepare vesicles with a large internal volume per receptor. Popot *et al.* (1981) reported the preparation of large vesicles (mean diameter 95 $\pm$ 55 nm) by a technique which employs partial cholate dialysis and subsequent gel filtration. Anholt *et al.* (1982) observed that subjecting a preparation of reconstituted vesicles to a freeze-thaw cycle sealed vesicles which were initially leaky, and caused vesicle fusion. Moreover, inclusion of supplementary cholesterol during the reconstitution procedure greatly enhanced vesicle fusion during the subsequent freeze-thaw cycle to the extent that after freezing and thawing, 50% of the total intravesicular volume was contained in only 3.5% of the vesicles with diameters larger than 140 nm. Limitations of the internal volume of the vesicles on the amplitude of the cumulative $^{22}Na^+$-uptake response were minimal in these preparations and, when tested at a low agonist concentration (2 $\mu$M carbamylcholine) which elicits only about one half to one third of the maximal response, the agonist-induced $^{22}Na^+$ uptake was limited only by desensitization of the acetylcholine receptors and directly proportional to the amount of active receptor (Anholt *et al.*, 1982).

Evidence from ion-flux experiments (Anholt *et al.*, 1982; Walker *et al.*, 1982) and binding studies (Sobel *et al.*, 1980; Criado *et al.*, 1982) indicates that reconstituted receptors, like those in native membranes, assume a conformation with high affinity for the ligand when they desensitize (Figure 11). Rapid kinetic measurements of the agonist-regulated uptake of $^{86}Rb^+$ into reconstituted vesicles using a quench-flow method gave further evidence for the functional integrity of reconstituted receptors in terms of their kinetics of activation and desensitization (Walker *et al.*, 1982). Desensitization of the receptors was here resolved as a biphasic process, each step of which

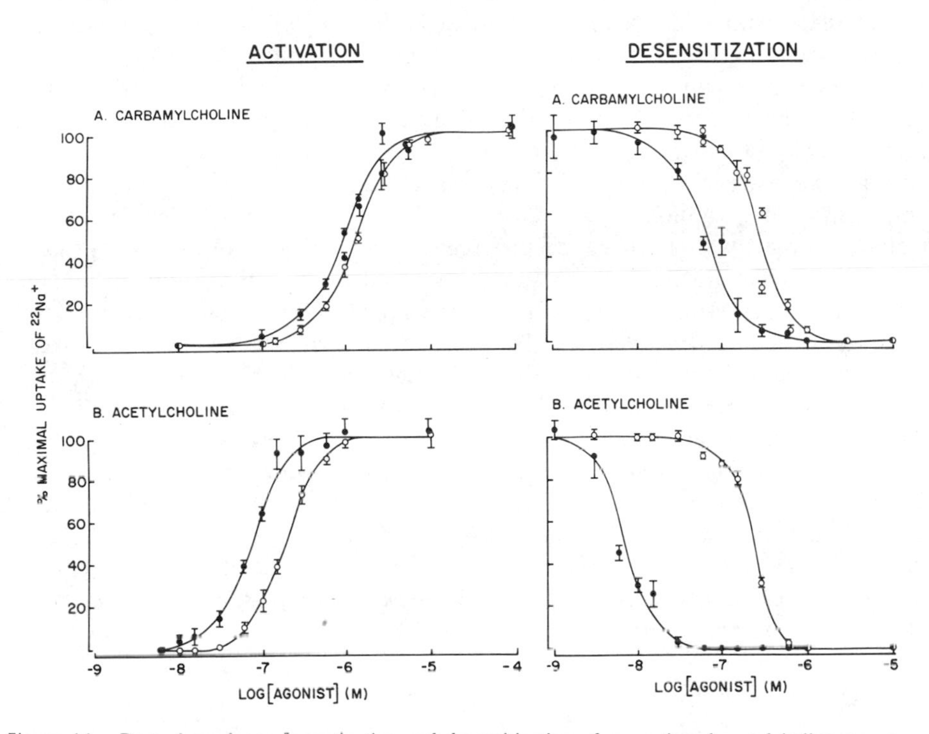


*Figure 11.* Dose dependence for activation and desensitization of reconstituted acetylcholine receptors. Equilibrium measurements were made using a $^{22}Na^+$-uptake assay in which a Dowex cation exchange resin served to separate $^{22}Na^+$ entrapped in the vesicles from external $^{22}Na^+$. Counts at the maximal response were approximately ten-fold higher than the background control. The dose dependence for receptor desensitization was assayed by addition of $10^{-4}$ M carbamylcholine after a preincubation period of at least 30 min with the indicated concentration of agonist in the absence of radioisotope. Assays were performed at 48 nM (●) and 800 nM (○) ligand-binding sites. The dose-response characteristics for activation and desensitization will depend on the receptor concentration when it is high with respect to the dissociation constant for the agonist, since the receptor will, under these conditions, deplete the concentration of free ligand by binding a significant fraction of the total agonist concentration. As a result, the dose-response curve will be displaced depending on the affinity of the receptor for the agonist. The large displacement of the dose-response curves for acetylcholine as compared to those for carbamylcholine reflects the expected higher affinity for acetylcholine. The fact that the dose-response curves for desensitization for each agonist are more sensitive to the receptor concentration than the corresponding dose-response curves for activation indicates that desensitization of the receptor is accompanied by an increase in binding affinity for the receptor. It should be noted that the midpoint of the dose-response curve for receptor activation at low receptor concentrations lies at least an order of magnitude lower when cumulative responses are measured at equilibrium than when initial flux rates are measured. From Anholt *et al.* (1983), adapted from Anholt *et al.* (1982).

displayed a different dependence on the concentration of agonist. The half-maximal rate of the fast phase of desensitization occurred at 1.3 mM carbamylcholine, a concentration at which the two agonist-binding sites that participate in receptor activation are saturated with agonist. Based on this observation, Walker *et al.* (1982) proposed the existence of a third low-affinity binding site for acetylcholine on the receptor molecule, as discussed above. They further observed a hyperbolic concentration dependence for both the slow and fast desensitization processes which suggests that a single site can trigger receptor desensitization. The concentration dependence for acetylcholine receptor activation, on the other hand, was sigmoidal, indicating that two molecules of agonist must bind in order to cause channel opening. This result is in agreement with observations by Anholt *et al.* (1982), who demonstrated a parabolic dependence of the decline of receptor function with the occupancy of binding sites by cobratoxin under conditions in which internal volume limitations are prevented, indicating that the doubly liganded receptor prevails in the control of channel opening.

The collective evidence from the investigations described above indicates that the characteristic pharmacological properties of the acetylcholine receptor are preserved after reconstitution in lipid vesicles. The large internal volume per receptor in these vesicles as compared to native electric organ membranes makes them an attractive system for kinetic studies on acetylcholine receptor function. Preliminary studies indicate that monoclonal antibodies may be particularly useful as probes for structure–function correlations within the receptor molecule (Lindstrom *et al.*, 1981b). Reconstituted vesicles provide a convenient screening system for the identification of such monoclonal antibodies. The combined use of reconstituted vesicles and planar bilayers as model systems for the study of acetylcholine receptor function should provide a powerful approach for the investigation of structure–function relationships within the receptor molecule.

### 3. *Influence of the Lipid Environment on Acetylcholine Receptor Function*

Thus far, only limited information is available with respect to the effect of the lipid environment on acetylcholine receptor function. Gonzalez-Ros *et al.* (1982) reported a detailed analysis of the lipid composition of receptor-rich electric organ membranes and found that these membranes are characterized by a high content of polyunsaturated ethanolamine phosphoglycerides and cholesterol (Table 2). This high level of cholesterol in receptor-rich electric organ membranes agrees with previous reports from other investigators based either on chemical analysis of extracts from electric organ membranes (Popot *et al.*, 1978; Schiebler and Hucho, 1978) or on the cytochemical localization of cholesterol–filipin complexes in postsynaptic membranes from *Torpedo marmorata* by electron microscopy (Perrelet *et al.*, 1982). It is not yet clear to what extent the high cholesterol content of the postsynaptic membranes is important for acetylcholine receptor function. Reconstitution studies using mixtures of different lipid composition have indicated that a mixture of phosphatidylethanolamine and phosphatidylserine (3:1 wt./wt.) supplemented with cholesterol (Dalziel *et al.*, 1980; McNamee *et al.*, 1982) or other neutral lipids, such as α-tocopherol or certain quinones (Kilian *et al.*, 1980), provides an optimal lipid environment in terms of the uptake of $^{22}Na^{+}$ at saturating concentrations of agonist. These studies, however,

*Table 2. Lipid Composition of Acetylcholine Receptor-Rich Electric Organ Membranes*[a]

| Phospholipid class | Molar percentage[b] | Average degree of unsaturation % |
|---|---|---|
| Phosphatidylcholine | 38.4 | 1.1 |
| Phosphatidylethanolamine | 35.8 | 2.9 |
| Phosphatidylserine | 10.9 | 2.4 |
| Other ethanolamine phosphoglycerides | 6.4 | 3.1 |
| Sphingomyelin and lysophosphatidylcholine | 2.3 | |
| Cardiolipin | 3.3 | 1.4 |
| Phosphatidic acid | 2.9 | |
| Phospholipid–cholesterol = 1.10 ± 0.27 (molar ratio) | | |
| Lipid–protein = 0.43 ± 0.12 (weight ratio) | | |

[a] Values are cited from Gonzalez-Ros *et al.* (1982).
[b] For comparison, soybean lipids contain approximately 40% phosphatidylcholine, 29% phosphatidylethanolamine, 14% monophosphoinositide, 4% phosphatidylserine, and small amounts of lysophospholipids (O'Brien *et al.*, 1977).

measured the cumulative uptake of $^{22}Na^+$ at equilibrium. Changes in the volume of the vesicles as a result of alterations in the lipid composition may, in fact, be responsible for apparent increases in functional activity of the reconstituted receptors (Anholt *et al.*, 1982; Ochoa *et al.*, 1982). In addition to altering the size of the vesicles, changes in lipid composition may affect the extent of incorporation of receptors during reconstitution as well as the orientation of the acetylcholine receptors in the reconstituted vesicles (Anholt *et al.*, 1982). Such artifactual effects on the process of vesicle formation and the structure of the resulting vesicles greatly complicate studies of the influence of the lipid composition on receptor function. This has been clearly demonstrated in studies by Ochoa *et al.* (1982) who, in agreement with Anholt *et al.* (1982), demonstrated that part of the stimulatory effects of cholesterol on receptor function could be accounted for by an increase in the size of the reconstituted vesicles. In addition, they observed that vesicles prepared from only dioleoylphosphatidylcholine displayed virtually no recovery of acetylcholine receptor function, which was most likely due to their very small size and/or the exclusion of a large number of the receptors from the reconstituted membranes. Criado *et al.* (1982) reported that, under their conditions, addition of cholesterol greatly enhanced the incorporation of acetylcholine receptors in reconstituted membranes during the reassembly of the receptors with synthetic phospholipids.

To date, there is no evidence that acetylcholine receptors have an absolute requirement for a specific lipid component. Any potential effects of the lipid composition on acetylcholine receptor function will most likely consist of shifts in the allosteric constant, i.e., alterations in the population distribution of low- and high-affinity conformations of unliganded receptors. Although no evidence for alterations in this isomerization equilibrium was evident in the studies of Ochoa *et al.* (1982), precedence for alterations in the value of M comes from studies on the effect of phospholipases on receptor function (Andreasen *et al.*, 1979) as well as studies using other membrane-perturbing agents such as local anesthetics (Weiland *et al.*, 1977; Heidmann and Changeux, 1979b, 1981; Dunn *et al.*, 1981; Sine and Taylor, 1982). A study by

Andreasen and McNamee (1980) which evaluates the effects of free fatty acids partitioned in electric organ membranes on acetylcholine-induced cation permeability is of particular interest in this context. Petroselenic acid, a fatty acid which has a melting point of 30°C, inhibited receptor function above its melting temperature, but was ineffective below its melting point, indicating that lipid fluidity is a prerequisite for the inhibition of receptor function by free fatty acids. In addition, none of the free fatty acids tested affected the ligand-binding properties of the receptor. The authors concluded that inhibition by fatty acids is due to disruption of protein–lipid interactions in the membrane without causing the stabilization of the acetylcholine receptor in its desensitized conformation (Andreasen and McNamee, 1980).

Clearly, more documentation is required before the physiological significance of effects of the lipid environment on the function of the acetylcholine receptor can be unambiguously evaluated.

### 4. *Functional Studies after Dissociation of the Acetylcholine Receptor Complex into Its Component Subunits*

The ultimate aim in reconstitution studies of the acetylcholine receptor would be the dissociation of the complex into its subunits and subsequent reassembly of the isolated components into a functional form. Since the subunits of the acetylcholine receptor can, at present, only be dissociated under denaturing conditions, the technology to realize this goal still seems remote. However, some preliminary studies along this line have been reported.

Huganir *et al.* (1981) reported a preliminary study in which the β-subunit of the acetylcholine receptor was separated from the other subunits using nonionic detergents and urea. The other subunits remained associated as a largely inactive complex, although some α-bungarotoxin-binding activity could be recovered.

Haggerty and Froehner (1981) reported the isolation of α-subunits by preparative gel electrophoresis in sodium dodecyl sulfate SDS and their partial renaturation after removal of the detergent, which manifested itself as low-affinity α-bungarotoxin binding with a $K_d$ approximately $10^4$-fold lower than that observed for native acetylcholine receptor. This α-bungarotoxin-binding activity could be displaced by cholinergic ligands, although with a 50- to 100-fold lower affinity than that found for native receptor. These observations have been confirmed by Gershoni *et al.* (1982) who reported that the isolated α-subunit could be electrophoretically transferred from an acrylamide gel in SDS onto a nylon filter and visualized by autoradiography with [$^{125}$I]α-bungarotoxin.

Tzartos and Changeux (1982) reported the recovery of α-bungarotoxin-binding activity of the isolated α-subunit with a $K_d$ in the order of 0.5 nM and a half-life time of about 3 hr for the receptor–toxin complex. Interestingly, the continuous presence of lipids was essential for the recovery of this toxin-binding activity, which is usually unaffected by the presence and absence of lipids. It is possible that the supplementary lipids in this case substitute for interactions with hydrophobic domains of neighboring subunits which in the native receptor complex may stabilize the native conformation of the α-subunit. However, in these studies, competition for the α-bungarotoxin-binding site by cholinergic ligands was not recovered indicating that re-formation of the acetylcholine-binding site *sensu stricto* had not been achieved.

Although some progress has been made in these preliminary studies, it is evident that the dissociation and complete functional reassociation of the component subunits of the acetylcholine receptor remain a challenging problem. This is not surprising in light of the observation that the assembly of acetylcholine receptor subunits during *in vivo* synthesis appears to be a complex process, in which maturation of the toxin-binding site requires a 15- to 30-min posttranslational period (Merlie and Sebbane, 1981; Merlie *et al.*, 1982).

## V. SUMMARY

The identification of snake neurotoxins as specific high-affinity probes for the acetylcholine-binding site and the recognition of fish electric organs as rich sources of acetylcholine receptors have enabled considerable progress in the biochemical characterization of the acetylcholine receptor primarily from *Torpedo* electric organ.

The acetylcholine receptor is a 250-K transmembrane protein composed of five homologous glycopeptides in the subunit stoichiometry $\alpha_2\beta\gamma\delta$. Acetylcholine receptors from mammalian muscle appear to be structurally similar to electric organ receptors. The acetylcholine receptor occurs in the electric organ of *Torpedo* as a dimeric complex formed through a disulfide bridge between the δ-subunits of adjacent monomers.

The bulk of the receptor protein protrudes from the synaptic side of the membrane and all the subunits are transmembrane polypeptides. Although the exact mechanism of their assembly is still unknown, they are synthesized from separate mRNAs and inserted independently across the membranes of the endoplasmic reticulum. Their membrane insertion appears to be mediated through a signal sequence at the N-terminal end of the nascent chains. Several posttranslational events seem to be required for the assembly and maturation of the functional complex. The mature junctional acetylcholine receptors appear to be immobilized at the postsynaptic membrane by a complex cytoskeletal architecture, involving a prominent 43,000-dalton protein.

The α-subunits of the receptor form all or part of the acetylcholine-binding sites which participate in acetylcholine receptor activation. Liganding of both of these sites is necessary for the transient opening of a cation-selective, transmembrane channel during synaptic transmission. Prolonged exposure to the ligand results in closing of this cation channel, a process known as "desensitization." Desensitization of the acetylcholine receptor is accompanied by a conformational transition in which the affinity of the receptor for agonists is increased several hundred-fold. Evidence indicates that a single molecule of acetylcholine can trigger the conversion of the receptor into this high-affinity, desensitized conformation. Desensitization appears to be a biphasic process and may involve the binding of acetylcholine to a low-affinity acetylcholine-binding site, the physical nature of which is, at present, not clear.

The functional reconstitution of the acetylcholine receptor in lipid vesicles and planar lipid bilayers has been achieved, and reconstitution studies have indicated the $\alpha_2\beta\gamma\delta$ monomeric form of the acetylcholine receptor from *Torpedo* to be its functional entity. It has been demonstrated that reconstituted acetylcholine receptors behave pharmacologically like receptors in native membranes.

Studies on cloned cDNAs encoding acetylcholine receptor subunits, the application of monoclonal antibodies as structural and functional probes, and the use of reconstituted systems to investigate structure–function correlations within the receptor molecule are expected to be key factors in future investigations of the regulation of postsynaptic signal transduction at the molecular level.

ACKNOWLEDGMENTS

We thank Drs. F. J. Barrantes, T. Claudio, J.-P. Changeux, S. Heinemann, P. Labarca, M. G. McNamee, J. Patrick, J.-L. Popot, and S. J. Tzartos for communicating to us information prior to publication. We thank Jan Littrell for secretarial assistance during the preparation of the manuscript. Work from the authors' laboratories was supported by grants from the National Institutes of Health (EY-02084 and RR07011 to M.M. and N.S., 11323 to J.L.), the Office of Naval Research (N00014-79-0798 to M.M. and J.L.), Department of the Army Medical Research (17-82-C221 to M.M.), John Simon Gugenheim Foundation (to M.M.), and the Muscular Dystrophy Association, McKnight Foundation, and California and Los Angeles Chapters of the Myasthenia Gravis Foundation (to J.L.).

## REFERENCES

Adams, D. J., Dwyer, T. M., and Hille, B., 1980, The permeability of endplate channels to monovalent and divalent metal cations, *J. Gen. Physiol.* **75:**493–510.

Adams, P. R., 1975, An analysis of the dose–response curve at voltage-clamped frog endplates, *Pfluegers Arch.* **360:**145–153.

Albuquerque, E. X., Tsai, M.-C., Aronstam, R. S., Witkop, B., Eldefrawi, A. T., and Eldefrawi, M. E., 1980a, Phencyclidine interactions with the ionic channel of the acetylcholine receptor and electrogenic membrane, *Proc. Natl. Acad. Sci. USA* **77:**1224–1228.

Albuquerque, E. X., Tsai, M.-C., Aronstam, R. S., Eldefrawi, A. T., and Eldefrawi, M. E., 1980b, Sites of action of phencyclidine. II. Interaction with the ionic channel of the nicotinic receptor, *Mol. Pharmacol.* **18:**167–178.

Anderson, M. J., and Cohen, M. W., 1977, Nerve-induced and spontaneous redistribution of acetylcholine receptors on cultured muscle cells, *J. Physiol.* **268:**757–773.

Anderson, C. R., and Stevens, C. F., 1973, Voltage-clamp analysis of acetylcholine produced end-plate current fluctuations at frog neuromuscular junction, *J. Physiol.* **235:**655–691.

Anderson, D. J., and Blobel, G., 1981, *In vitro* synthesis, glycosylation and membrane insertion of the four subunits of Torpedo acetylcholine receptor, *Proc. Natl. Acad. Sci. USA* **78:**5598–5602.

Anderson, D. J., Walter, P., and Blobel, G., 1982, Signal recognition protein is required for the integration of acetylcholine receptor δ-subunit, a transmembrane glycoprotein, into the endoplasmic reticulum membrane, *J. Cell Biol.* **93:**501–506.

Anderson, D. J., Blobel, G., Tzartos, S., Gullick, W., and Lindstrom, J., 1983, Transmembrane orientation of an early biosynthetic form of acetylcholine receptor δ subunit determined by proteolytic dissection in conjunction with monoclonal antibodies, *J. Neurosci.* **3:**1773–1784.

Andreasen, T. J., and McNamee, M. G., 1980, Inhibition of ion permeability control properties of acetylcholine receptor from *Torpedo californica* by long-chain fatty acids, *Biochemistry* **19:**4719–4726.

Andreasen, T. J., Doerge, D. R., and McNamee, M. G., 1979, Effects of phospholipase $A_2$ on the binding and ion permeability control properties of the acetylcholine receptor, *Arch. Biochem. Biophys.* **194:**468–480.

Anholt, R., Lindstrom, J., and Montal, M., 1980, Functional equivalence of monomeric and dimeric forms of purified acetylcholine receptor from *Torpedo californica* in reconstituted lipid vesicles, *Eur. J. Biochem.* **109:**481–487.

Anholt, R., Lindstrom, J., and Montal, M., 1981, Stabilization of acetylcholine receptor channels by lipids in cholate solution and during reconstitution in vesicles, *J. Biol. Chem.* **256:**4377–4387.

Anholt, R., 1981, Reconstitution of acetylcholine receptors in model membranes, *Trends Biochem. Sci.* **6:**288–291.

Anholt, R., Fredkin, D. R., Deerinck, T., Ellisman, M., Montal, M., and Lindstrom, J., 1982, Incorporation of acetylcholine receptors into liposomes: Vesicle structure and acetylcholine receptor function, *J. Biol. Chem.* **257:**7122–7134.

Anholt, R., Montal, M., and Lindstrom, J., 1983, Incorporation of acetylcholine receptors in model membranes: An approach aimed at studies of the molecular basis of neurotransmission, in: *Peptide and Protein Reviews,* Vol. 1 (M. Hearn, ed.), Marcel Dekker, New York, pp. 95–137.

Axelrod, D., Ravdin, P., Koppel, D. E., Schlessinger, J., Webb, W. W., Elson, E. L., and Podleski, T. R., 1976, Lateral motion of fluorescently labeled acetylcholine receptors in membranes of developing muscle fibers, *Proc. Natl. Acad. Sci. USA* **73:**4594–4598.

Axelrod, D. P., Ravdin, P. M., and Podleski, T. R., 1978, Control of acetylcholine receptor mobility and distribution in cultured muscle membrane. A fluorescence study, *Biochim. Biophys. Acta* **511:**23–28.

Ballivet, M., Patrick, J., Lee, J., and Heinemann, S., 1982, Molecular cloning of cDNA coding for the γ-subunit of *Torpedo* acetylcholine receptor, *Proc. Natl. Acad. Sci. USA* **79:**4466–4470.

Barnard, E. A., Miledi, R., and Sumikawa, K., 1982, Translation of exogenous messenger RNA coding for nicotinic acetylcholine receptors produces functional receptors in Xenopus oocytes, *Proc. R. Soc. Lond.* **B215:**241–246.

Barrantes, F. J., 1978, Agonist-mediated changes of the acetylcholine receptor in its membrane environment, *J. Mol. Biol.* **124:**1–26.

Barrantes, F. J., 1982, Interactions of the membrane-bound acetylcholine receptor with the nonreceptor peripheral peptide, *in: Neuroreceptors,* Walter de Gruyter and Co., Berlin and New York, pp. 315–328.

Barrantes, F. J., Neugebauer, D.-Ch., and Zingsheim, H. P., 1980, Peptide extraction by alkaline treatment is accompanied by rearrangement of the membrane bound acetylcholine receptor from *Torpedo marmorata, FEBS Lett.* **112:**73–78.

Barrantes, F. J., Mieskes, J., and Wallimann, T., 1983, Creatine kinase activity in the *Torpedo* electrocyte and in the nonreceptor, peripheral ν proteins from acetylcholine receptor-rich membranes, *Proc. Natl. Acad. Sci. USA* **80:**5440–5444.

Bartholdi, M., Barrantes, F. J., and Jovin, T. M., 1981, Rotational molecular dynamics of the membrane-bound acetylcholine receptor revealed by phosphorescence spectroscopy, *Eur. J. Biochem.* **120:**389–397.

Ben-Haim, D., Landau, E. M., and Silman, I., 1973, The role of a reactive disulphide bond in the function of the acetylcholine receptor at the frog neuromuscular junction, *J. Physiol.* **234:**305–325.

Bennett, M. V. L., Wurzel, M., and Grundfest, H., 1961, The electrophysiology of electric organs of marine electric fishes. I. Properties of electroplaques of *Torpedo nobiliana, J. Gen. Physiol.* **44:**757–804.

Beranek, R., and Vyskocil, F., 1967, The action of tubocurarine and atropine on the normal and denervated rat diaphragm, *J. Physiol.* **188:**53–66.

Berg, D. K., and Hall, Z. W., 1974, Fate of α-bungarotoxin bound to acetylcholine receptors of normal and denervated muscle, *Science* **184:**473–475.

Berg, D. K., and Hall, Z. W., 1975a, Loss of α-bungarotoxin from junctional and extrajunctional acetylcholine receptors in rat diaphragm muscle *in vivo* and in organ cultures, *J. Physiol.* **252:**771–789.

Berg, D. K., and Hall, Z. W., 1975b, Increased extrajunctional acetylcholine sensitivity produced by chronic post-synaptic neuromuscular blockage, *J. Physiol.* **244:**659–676.

Berg, D. K., Kelly, R. B., Sargent, P. B., Williamson, P., and Hall, Z. W., 1972, Binding of α-bungarotoxin to acetylcholine receptors in mammalian muscle, *Proc. Natl. Acad. Sci. USA* **69:**147–151.

Bevan, S., and Steinbach, J. H., 1977, The distribution of α-bungarotoxin binding sites on mammalian skeletal muscle developing *in vivo, J. Physiol.* **267:**195–213.

Birks, R., Katz, B., and Miledi, R., 1960, Physiological and structural changes at the amphibian myoneural junction in the course of nerve degeneration, *J. Physiol.* **150:**145–168.

Boheim, G., Hanke, W., Barrantes, F. J., Eibl, H., Sakmann, B., Fels, G., and Maelicke, A., 1981, Agonist-activated ionic channels in acetylcholine receptor reconstituted into planar lipid bilayers, *Proc. Natl. Acad. Sci. USA* **78:**3586–3590.

Boulter, J., and Patrick, J., 1977, Purification of an acetylcholine receptor from a nonfusing muscle cell line, *Biochemistry* **16:**4900–4908.

Boyd, N. D., and Cohen, J. B., 1980a, Kinetics of binding of [$^3$H]-acetylcholine and $^3$[H]-carbamylcholine to *Torpedo* postsynaptic membranes: Slow conformational transitions of the cholinergic receptor, *Biochemistry* **19:**5344–5353.

Boyd, N. D., and Cohen, J. B., 1980b, Kinetics of binding of [$^3$H]-acetylcholine to *Torpedo* postsynaptic membranes: Association and dissociation rate constants by rapid mixing and ultrafiltration, *Biochemistry* **19:**5353–5358.

Briley, M., and Changeux, J.-P., 1977, Isolation and purification of the nicotinic acetylcholine receptor and its functional reconstitution into a membrane environment, *Int. Rev. Neurobiol.* **20:**31–59.

Briley, M. S., and Changeux, J.-P., 1978, Recovery of some functional properties of the detergent-extracted cholinergic receptor protein from *Torpedo marmorata* after reintegration into a membrane environment, *Eur. J. Biochem.* **84:**429–439.

Brisson, A., Devaux, P. F., and Changeux, J. P., 1975, Effet anesthésique local de plusieurs composés liposolubles sur la réponse de l'électroplaque de Gymnote á la carbamylcholine et sur la liaison de l'acétylcholine au récepteur cholinergique de Torpille, *C.R. Acad. Sci. (Paris)* **280D:**2153–2156.

Brockes, J. P., Berg, D. K., and Hall, Z. W., 1975, The biochemical properties and regulation of acetylcholine receptors in normal and denervated muscle, *Cold Spring Harbor Symp. Quant. Biol.* **40:**253–262.

Brown, D. R., and Taylor, P., 1983, The influence of antibiotics on agonist occupation and functional states of the nicotinic acetylcholine receptor, *Mol. Pharmacol.*, **23:**8–16.

Burden, S. J., 1982, Identification of an intracellular postsynaptic antigen at the frog neuromuscular junction, *J. Cell Biol.* **94:**521–530.

Burden, S. J., Sargent, P. B., and McMahan, U. J., 1979, Acetylcholine receptors in regenerating muscle accumulate at original sites in the absence of the nerve, *J. Cell Biol.* **82:**412–425.

Burgermeister, W., Catterall, W., and Witkop, B., 1977, Histrionicotoxin enhances agonist-induced desensitization of acetylcholine receptor, *Proc. Natl. Acad. Sci. USA* **74:**5754–5758.

Cartaud, J., Benedetti, E., Sobel, A., and Changeux, J.-P., 1978, A morphological study of the cholinergic receptor protein from *Torpedo marmorata* in its membrane environment and in its detergent extracted purified form, *J. Cell Sci.* **29:**313–337.

Cartaud, J., Sobel, A., Rousselet, A., Devaux, P., and Changeux, J.-P., 1981, Consequences of alkaline treatment for the ultra-structure of the acetylcholine receptor-rich membranes from *Torpedo marmorata* electric organ, *J. Cell Biol.* **90:**418–426.

Cartaud, J., Oswald, R., Clément, G., and Changeux, J.-P., 1982, Evidence for a skeleton in acetylcholine receptor-rich membranes from *Torpedo marmorata* electric organ, *FEBS Lett.* **145:**250.

Cash, D. J., Aoshima, H., and Hess, G. P., 1981, Acetylcholine-induced cation translocation across cell membranes and inactivation of the acetylcholine receptor: Chemical kinetic measurements in the millisecond time region, *Proc. Natl. Acad. Sci. USA* **78:**3318–3322.

Catterall, W. A., 1975, Sodium transport by the acetylcholine receptor of cultured muscle cells, *J. Biol. Chem.* **250:**1775–1781.

Chang, H. W., and Bock, E., 1977, Molecular forms of the acetylcholine receptor. Effects of calcium ions and a sulfhydryl reagent on the occurrence of oligomers, *Biochemistry* **16:**4513–4520.

Chang, C. C., and Huang, M. C., 1975, Turnover of junctional and extrajunctional acetylcholine receptors of the rat diaphragm, *Nature* **253:**643–644.

Chang, C. C., and Lee, C.-Y., 1963, Isolation of neurotoxins from the venom of *Bungarus multicinctus* and their modes of neuromuscular blocking action, *Arch. Int. Pharmacodyn. Ther.* **144:**241–257.

Changeux, J.-P., 1981, The acetylcholine receptor: An allosteric membrane protein, *Harvey Lect.* **75:**85–254.

Changeux, J.-P., Heidmann, T., Popot, J., and Sobel, A., 1979, Reconstitution of a functional acetylcholine regulator under defined conditions, *FEBS Lett.* **105:**181–187.

Clark, D. G., Macmurchie, D. D., Elliott, E., Wolcott, R. G., Landel, A. M., and Raftery, M. A., 1972, Elapid neurotoxins, purification, characterization, and immunochemical studies of α-bungarotoxin, *Biochemistry* **11:**1663–1668.

Claudio, T., Ballivet, M., Patrick, J., and Heinemann, S., 1983, *Torpedo californica* acetylcholine receptor 60,000 dalton subunit μ nucleotide sequence of cloned cDNA, deduced amino acid sequence, subunit structural predictions, *Proc. Natl. Acad. Sci. USA,* **80:**111–115.

Cohen, J. B., 1978, Ligand binding properties of membrane bound cholinergic receptors of *Torpedo marmorata,* in: *Molecular Specialization and Symmetry in Membrane Function* (A. K. Solomon and M. Karnosky, eds.), Harvard University Press, Cambridge, Massachusetts, pp. 99–128.

Colquhoun, D., and Sakmann, B., 1981, Fluctuations in the microsecond time range of the current through single acetylcholine receptor ion channels, *Nature* **294:**464–466.

Conti-Tronconi, B., Brigonzi, A., Fumagalli, G., Sher, M., Cosi, V., Piccolo, G., and Clementi, F., 1981a, Antibody-induced degradation of acetylcholine receptor in myasthenia gravis: Clinical correlates and pathogenetic significance, *Neurology* **31:**1440–1444.

Conti-Tronconi, B., Tzartos, S., and Lindstrom, J., 1981b, Monoclonal antibodies as probes of acetylcholine receptor structure. II. Binding to native receptor, *Biochemistry* **20:**2181–2191.

Conti-Tronconi, B. M., Hunkapiller, M. W., Lindstrom, J. M., and Raftery, M. A., 1982a, Subunit structure of the acetylcholine receptor from *Electrophorus electricus, Proc. Natl. Acad. Sci. USA* **79:**6489–6493.

Conti-Tronconi, B. M., Hunkapiller, M. W., Lindstrom, J. M., and Raftery, M. A., 1982b, Amino acid sequence homology between α-subunits from *Torpedo* and *Electrophorus* acetylcholine receptor, *Biochem. Biophys. Res. Commun.* **106:**312–318.

Conti-Tronconi, B. M., Dunn, S. M. J., and Raftery, M. A., 1982c, Functional stability of *Torpedo* acetylcholine receptor. Effects of protease treatment, *Biochemistry* **21:**893–899.

Conti-Tronconi, B. M., Dunn, S. M. J., and Raftery, M. A., 1982d, Independent sites of low and high affinity for agonists on *Torpedo californica* acetylcholine receptor, *Biochem. Biophys. Res. Commun.* **107:**123–129.

Conti-Tronconi, B. M., Gotti, C. M., Hunkapiller, M. W., and Raftery, M. A., 1982e, Mammalian muscle acetylcholine receptor: A supramolecular structure formed by four related proteins, *Science* **218:**1227–1229.

Criado, M., and Barrantes, F. J., 1982, Effects of periodate oxidation and glycosidases on structural and functional properties of the acetylcholine receptor and the non-receptor, peripheral ν-polypeptide ($M_r$ 43,000), *Neurochem. Int.,* **4:**289–302.

Criado, M., Eibl, H., and Barrantes, F. J., 1982, Effects of lipids on acetylcholine receptor. Essential need of cholesterol for maintenance of agonist-induced state transitions in lipid vesicles, *Biochemistry* **21:**3622–3629.

Culver, P., Fenical, W., and Taylor, P., 1984, Lophotoxin irreversibly inactivates the nicotinic acetylcholine receptor by preferential association at one of the two primary agonist sties. *J. Biol Chem.* **259:**3763–3770.

Dale, H. H., 1934, Chemical transmission of the effects of nerve impulses, *Br. Med. J.* **1:**835–841.

Dalziel, A. W., Rollins, E. S., and McNamee, M. G., 1980, The effect of cholesterol on agonist-induced flux in reconstituted acetylcholine receptor vesicles, *FEBS Lett.* **122:**193–196.

Damle, V. N., and Karlin, A., 1978, Affinity labeling of one of two α-neurotoxin binding sites in acetylcholine receptor from *Torpedo californica, Biochemistry* **17:**2039–2045.

Damle, V. N., and Karlin, A., 1980, Effects of agonists and antagonists on the reactivity of the binding site disulfide in acetylcholine receptor from *Torpedo californica, Biochemistry* **19:**3924–3932.

Damle, V. N., McLaughlin, M., and Karlin, A., 1978, Bromoacetylcholine as an affinity label of the acetylcholine receptor from *Torpedo californica, Biochem. Biophys. Res. Commun.* **84:**845–851.

Davis, C., Gordon, A., and Diamond, I., 1982, Specificity and localization of the acetylcholine receptor kinase, *Proc. Natl. Acad. Sci. USA* **79:**3666–3670.

Del Castillo, J., and Katz, B., 1954, Quantal components of the endplate potential, *J. Physiol.* **124:**560–573.

Del Castillo, J., and Katz, B., 1954, Quantal components of the endplate potential, *J. Physiol.* **124:**560–573.

Dennis, M. J., Ziskind-Conhaim, L., and Harris, A. J., 1981, Development of neuromuscular junctions in rat embryos, *Dev. Biol.* **81:**266–279.

Devillers-Thiery, A., Changeux, J.-P., Paroutaud, P., and Strosberg, A., 1979, The amino-terminal sequence of the 40,000 molecular weight subunit of the acetylcholine receptor protein from *Torpedo marmorata, FEBS Lett.* **104:**99–105.

Devillers-Thiery, A., Giraudat, J., Bentaboulet, M., and Changeux, J.-P., 1983, Complete mRNA coding sequence of the acetylcholine binding subunit of *Torpedo marmorata* acetylcholine receptor: A model for the transmembrane organization of the polypeptide chain. *Proc. Natl. Acad. Sci. USA* **80:**2067–2071.

Devreotes, P. N., and Fambrough, D. M., 1975, Acetylcholine receptor turnover in membranes of developing muscle fibers, *J. Cell Biol.* **65:**335–358.

Devreotes, P. N., and Fambrough, D. M., 1976, Turnover of acetylcholine receptors in skeletal muscle, *Cold Spring Harbor Symp. Quant. Biol.* **40:**237–251.

Diamond, J., and Miledi, R., 1962, A study of foetal and newborn rat muscle fibres, *J. Physiol.* **162:**393–408.

Dionne, V. E., Steinbach, J. H., and Stevens, C. F., 1978, An analysis of the dose–response relationship of voltage-clamped frog neuromuscular junctions, *J. Physiol.* **281**:421–444.

Drachman, D. B., 1981, The biology of myasthenia gravis, *Annu. Rev. Neurosci.* **4**:195–225.

Drachman, D. B., Adams, R., Josifek, L., and Self, S., 1982, Functional activities of autoantibodies to acetylcholine receptors and the clinical severity of myasthenia gravis, *New Eng. J. Med.* **307**:769–775.

Dreyer, F., Peper, K., and Sterz, R., 1978, Determination of dose–response curves by quantitative iono-phoresis at the frog neuromuscular junction, *J. Physiol.* **281**:395–419.

Dunn, S. J. M., and Raftery, M. A., 1982, Activation and desensitization of Torpedo acetylcholine receptor: Evidence for separate binding sites. *Proc. Natl. Acad. Sci. USA* **79**:6757–6761.

Dunn, M. J., Blanchard, S. G., and Raftery, M. A., 1981, Effects of local anesthetics and histrionicotoxin on the binding of carbamoylcholine to membrane-bound acetylcholine receptor, *Biochemistry* **20**:5617–5624.

Dwyer, T. M., Adams, D., and Hille, B., 1980, The permeability of the endplate-channel to organic cations in frog muscle, *J. Gen. Physiol.* **75**:469–492.

Einarson, B., Gullick, W., Conti-Tronconi, B., and Lindstrom, J., 1982, Subunit composition of fetal calf muscle nicotinic acetylcholine receptor, *Biochemistry* **21**:5295–5302.

Eldefrawi, A. T., Eldefrawi, M. E., Albuquerque, E. X., Oliveira, A. C., Mansour, N., Adler, M., Daly, J. W., Brown, G. G., Burgermeister, W., and Witkop, B., 1977, Perhydrohistrionicotoxin: A potential ligand for the ion conductance modulator of the acetylcholine receptor, *Proc. Natl. Acad. Sci. USA* **74**:2172–2176.

Eldefrawi, M. E., Eldefrawi, A. T., Mansour, N. A., Daly, J. W., Witkop, B., and Albuquerque, E. X., 1978, Acetylcholine receptor and ionic channel of *Torpedo* electroplax: Binding of perhydrohistrion-icotoxin to membrane and solubilized preparations, *Biochemistry* **17**:5474–5484.

Eldefrawi, M. E., Aronstam, R. S., Bakry, N. M., Eldefrawi, A. T., and Albuquerque, E. X., 1980, Activation, inactivation, and desensitization of acetylcholine receptor channel complex detected by binding of perhydrohistrionicotoxin, *Proc. Natl. Acad. Sci. USA* **77**:2309–2313.

Elliott, J., and Raftery, M. A., 1977, Interactions of perhydrohistrionicotoxin with postsynaptic membranes, *Biochem. Biophys. Res. Commun.* **77**:1347–1353.

Elliott, J., Dunn, S. M. J., Blanchard, S. G., and Raftery, M. A., 1979, Specific binding of perhydro-histrionicotoxin to *Torpedo* acetylcholine receptor, *Proc. Natl. Acad. Sci. USA* **76**:2576–2579.

Engel, A. G., Tsujihata, M., Lindstrom, J., and Lennon, V., 1976, The motor end-plate in myasthenia gravis and in experimental autoimmune myasthenia gravis. A quantitative ultrastructural study, *Ann. N.Y. Acad. Sci.* **274**:60–79.

Engel, A. G., Lindstrom, J. M., Lambert, E. H., and Lennon, V. A., 1977a, Ultrastructural localization of the acetylcholine receptor in myasthenia gravis and its experimental autoimmune model, *Neurology* **27**:307–315.

Engel, A., Lambert, E. M., and Howard, G., 1977b, Localization of acetylcholine receptors, antibodies, and complement at endplates of patients with myasthenia gravis, *Mayo Clin. Proc.* **52**:267–280.

Engel, A. G., Sakakibara, H., Sahashi, K., Lindstrom, J. M., Lambert, E. M., and Lennon, V. A., 1979, Passively transferred experimental autoimmune myasthenia gravis, *Neurology* **29**:179–188.

Epstein, M., and Racker, E., 1978, Reconstitution of carbamylcholine-dependent sodium ion flux and desensitization of the acetylcholine receptor from *Torpedo californica, J. Biol. Chem.* **253**:6660–6662.

Fairclough, R. H., Finer-Moore, J., Love, R. A., Kristofferson, D., Desmeules, P. J. and Stroud, R. M., 1984, Subunit organization and structure of an acetylcholine receptor. *Cold Spring Harbor Symposia* **48.** *Molecular Neurobiology,* in press.

Fambrough, D. M., 1979, Control of acetylcholine receptors in skeletal muscle, *Physiol. Rev.* **59**:165–227.

Fambrough, D. M., and Devreotes, P. N., 1978, Newly synthesized acetylcholine receptors are located in the Golgi apparatus, *J. Cell Biol.* **76**:237–244.

Fambrough, D. M., Bayne, E. K., Gardner, J. M., Anderson, M. J., Wakshull, E., and Rotundo, R. L., 1982, Monoclonal antibodies to skeletal muscle cell surface, in: *Neuroimmunology* (J. Brockes, ed.), Plenum, New York, pp. 49–89.

Fatt, P., and Katz, B., 1951, An analysis of the end-plate potential recorded with an intra-cellular electrode, *J. Physiol.* **115**:320–370.

Feltz, A., and Trautmann, A., 1982, Desensitization at the frog neuromuscular junction: A biphasic process, *J. Physiol.* **322**:257–272.

Fertuck, H. C., and Salpeter, M. M., 1974, Localization of $^{125}$I-labeled α-bungarotoxin binding at mouse motor endplates, *Proc. Natl. Acad. Sci. USA* **71:**1376–1378.

Fertuck, H. C., and Salpeter, M. M., 1976, Quantitation of junctional and extrajunctional acetylcholine receptors by electron microscope autoradiography after $^{125}$I-αbungarotoxin binding at mouse neuromuscular junctions, *J. Cell Biol.* **69:**144–158.

Flynn, D. D., Kloog, Y., Potter, L. T., and Axelrod, J., 1982, Enzymatic methylation of the membrane-bound nicotinic acetylcholine receptor, *J. Biol. Chem.* **257:**9513–9517.

Fox, G. O., and Richardson, G. P., 1979, The developmental morphology of *Torpedo marmorata:* Electric organ—Electrogenic phase, *J. Comp. Neurol.* **185:**293–316.

Frank, E., and Fischbach, G. D., 1979, Early events in neuromuscular junction formation *in vitro:* Induction of acetylcholine receptor clusters in the postsynaptic membrane and morphology of newly formed synapses, *J. Cell Biol.* **83:**143–158.

Froehner, S. C., 1981, Identification of exposed and buried determinants of the membrane-bound acetylcholine receptor from *Torpedo californica, Biochemistry* **20:**4905–4915.

Froehner, S. C., Karlin, A., and Hall, Z. W., 1977a, Affinity alkylation labels two subunits of reduced acetylcholine receptor from mammalian muscle, *Proc. Natl. Acad. Sci. USA* **74:**4685–4688.

Froehner, S. C., Reiness, C. G., and Hall, Z. W., 1977b, Subunit structure of the acetylcholine receptor from denervated rat skeletal muscle, *J. Biol. Chem.* **252:**8589–8596.

Froehner, S. C., Gulbrandsen, V., Hyman, C., Jeng, A. Y., Neubig, R. R., and Cohen, J. B., 1981, Immunofluorescence localization at the mammalian neuromuscular junction of the $M_r$ 43,000 protein of *Torpedo* postsynaptic membranes, *Proc. Natl. Acad. Sci. USA* **78:**5230–5234.

Froehner, S. C., Wray, B. E., and Sealock, R., 1983, Ultrastructural localization of 43K protein in Torpedo postsynaptic membranes with monoclonal antibodies, *Soc. for Neurosci. Abstr.,* 168.5, p. 578.

Fumagalli, G., Engel, A., and Lindstrom, J., 1982a, Ultrastructural aspects of acetylcholine receptor turnover at the normal end-plate and in autoimmune myasthenia gravis, *J. Neuropathol. Exp. Neurol.* **41:**567–579.

Fumagalli, G., Engel, A. G., and Lindstrom, J., 1982b, Estimation of acetylcholine receptor degradation rate by external gamma counting *in vivo, Mayo Clin. Proc.,* **57:**758–764.

Gage, P. W., McBurney, R. N., and Schneider, G. T., 1975, Effects of some aliphatic alcohols on the conductance change caused by a quantum of acetylcholine at the toad end-plate, *J. Physiol.* **244:**409–429.

Gardner, J. M., and Fambrough, D. M., 1979, Acetylcholine receptor degradation measured by density labeling: Effects of cholinergic ligands and evidence against recycling, *Cell* **16:**661–674.

Gershoni, J. M., Palade, G. E., Hawrot, E., Klimowicz, D. W., and Lentz, T. L., 1982, Analysis of α bungarotoxin binding to *Torpedo* acetylcholine receptor by electrophoretic transfer techniques, *J. Cell Biol.* **95:**422a.

Giraudat, J., Devillers-Thiery, A., Auffray, C., Rougeon, F., and Changeux, J. P., 1982, Identification of a cDNA clone coding for the acetylcholine binding subunit of *Torpedo marmorata* acetylcholine receptor, *EMBO J.* **1:**713–717.

Gomez, C., Richman, D., Berman, P., Burres, S., Arnason, B., and Fitch, F., 1979, Monoclonal antibodies against purified nicotinic acetylcholine receptor, *Biochem. Biophys. Res. Commun.* **88:**575–582.

Gonzalez-Ros, J. M., Paraschos, A., and Martinez-Carrion, M., 1980, Reconstitution of functional membrane-bound acetylcholine receptor from isolated *Torpedo californica* receptor protein and electroplax lipids, *Proc. Natl. Acad. Sci. USA* **77:**1796–1800.

Gonzalez-Ros, J. M., Llanillo, M., Paraschos, A., and Martinez-Carrion, M., 1982, Lipid environment of acetylcholine receptor from *Torpedo californica, Biochemistry* **21:**3467–3474.

Gordon, A. S., Davis, C. G., and Diamond, I., 1977a, Phosphorylation of membrane proteins at a cholinergic synapse, *Proc. Natl. Acad. Sci. USA* **74:**263–267.

Gordon, A. S., Davis, C. G., Milfay, D., and Diamond, I., 1977b, Phosphorylation of acetylcholine receptor by endogenous membrane protein kinase in receptor-enriched membranes of *Torpedo californica, Nature* **267:**539–540.

Gordon, A. S., Milfay, D., Davis, C. G., and Diamond, I., 1979, Protein phosphatase activity in acetylcholine receptor-enriched membranes, *Biochem Biophys. Res. Commun.* **87:**876–883.

Gordon, A. S., Davis, C. G., Milfay, D., Kaur, J., and Diamond, I., 1980, Membrane-bound protein kinase activity in acetylcholine receptor-enriched membranes, *Biochim. Biophys. Acta* **600:**421–431.

Gotti, C., Conti-Tronconi, B. M., and Raftery, M. A., 1982, Mammalian muscle acetylcholine receptor purification and characterization, *Biochemistry* **21:**3148–3154.
Grunhagen, H. H., Iwatsubo, M., and Changeux, J.-P., 1977, Fast kinetic studies on the interaction of cholinergic agonists with the membrane-bound acetylcholine receptor from *Torpedo marmorata* as revealed by quinacrine fluorescence, *Eur. J. Biochem.* **80:**225–242.
Gullick, W. J., and Lindstrom, J. M., 1982a, The antigenic structure of the acetylcholine receptor from *Torpedo californica, J. Cell. Biochem.,* **19:**223–230.
Gullick, W. J., and Lindstrom, J. M., 1982b, Structural similarities between acetylcholine receptors from fish electric organs and mammalian muscle, *Biochemistry,* **21:**4563–4569.
Gullick, W. J., and Lindstrom, J. M., 1983a, Mapping the binding of monoclonal antibodies to the acetylcholine receptor from *Torpedo californica, Biochemistry* **22:**3312–3320.
Gullick, W. J., and Lindstrom, J. M., 1983b, Comparison of the subunit structure of acetylcholine receptors from muscle and electric organ of *Electrophorus electricus, Biochemistry* **22:**3801–3807.
Gullick, W. J., Tzartos, S., and Lindstrom, J., 1981, Monoclonal antibodies as probes of acetylcholine receptor structure. I. Peptide mapping, *Biochemistry* **20:**2173–2180.
Guy, H. R., 1984, A structural model of the acetylcholine receptor channel based on partition energy and helix packing calculations. *Biophys. J.,* **45:**249–261.
Gysin, R., Wirth, M., and Flanagan, S. D., 1981, Structural heterogeneity and subcellular distribution of nicotinic synapse-associated proteins, *J. Biol. Chem.* **256:**11373–11376.
Haggerty, J. G., and Froehner, S. C., 1981, Restoration of $^{125}$I-α-Bungarotoxin binding activity to the α-subunit of *Torpedo* acetylcholine receptor isolated by gel electrophoresis in sodium dodecyl sulfate, *J. Biol. Chem.* **256:**8294–8297.
Hall, Z. W., Lubit, B. W., and Schwartz, J. H., 1981, Cytoplasmic actin in postsynaptic structures at the neuromuscular junction, *J. Cell Biol.* **90:**789–792.
Hamill, O. P., and Sakmann, B., 1981, Multiple conductance states of single acetylcholine receptor channels in embryonic muscle cells, *Nature* **294:**462–464.
Hamilton, S. L., McLaughlin, M., and Karlin, A., 1979, Formation of disulfide-linked oligomers of acetylcholine receptor in membrane from *Torpedo* electric tissue, *Biochemistry* **18:**155–163.
Hanover, J. A., and Lennarz, W. J., 1981, Transmembrane assembly of membrane and secretory glycoproteins, *Arch. Biochem. Biophys.* **211:**1–19.
Hartzell, H. C., 1982, Physiological consequences of muscarinic receptor activation, *Trends Pharmacol. Sci.* **4:**213–214.
Hartzell, H. C., and Fambrough, D. M., 1972, Acetylcholine receptors. Distribution and extrajunctional density in rat diaphragm after denervation correlated with acetylcholine sensitivity, *J. Gen. Physiol.* **60:**248–262.
Hartzell, H. C., and Fambrough, D. M., 1973, Acetylcholine receptor production and incorporation into plasma membranes of developing muscle fibers, *Dev. Biol.* **30:**153–165.
Hazelbauer, G. L., and Changeux, J.-P., 1974, Reconstitution of a chemically excitable membrane, *Proc. Natl. Acad. Sci. USA* **71:**1479–1483.
Heidmann, T., and Changeux, J.-P., 1979a, Fast kinetic studies on the interaction of a fluorescent agonist with the membrane-bound acetylcholine receptor from *Torpedo marmorata, Eur. J. Biochem.* **94:**255–279.
Heidmann, T., and Changeux, J.-P., 1979b, Fast kinetic studies on the allosteric interactions between acetylcholine receptor and local anesthetic binding sites, *Eur. J. Biochem.* **94:**281–296.
Heidmann, T., and Changeux, J.-P., 1980, Interaction of a fluorescent agonist with the membrane-bound acetylcholine receptor from *Torpedo marmorata* in the millisecond time range: Resolution of an intermediate conformational transition and evidence for positive cooperative effects, *Biochem. Biophys. Res. Commun.* **97:**889–896.
Heidmann, T., and Changeux, J.-P., 1981, Stabilization of the high affinity state of the membrane-bound acetylcholine receptor from *Torpedo marmorata* by noncompetitive blockers. Evidence for dual interaction and pharmacological selectivity, *FEBS Lett.* **131:**239–244.
Heidmann, T., Sobel, A., and Changeux, J.-P., 1980a, Conservation of the kinetic and allosteric properties of the acetylcholine receptor in its Na-cholate soluble 9S form: Effect of lipids, *Biochem. Biophys. Res. Commun.* **93:**127–133.

Heidmann, T., Sobel, A., Popot, J.-L., and Changeux, J.-P., 1980b, Reconstitution of a functional acetylcholine receptor: Conservation of the conformational and allosteric transitions and recovery of the permeability response: Role of lipids, *Eur. J. Biochem.* **110:**35–55.

Heidmann, T., Cuisinier, J. B., and Changeux, J.-P., 1981, Conservation des propriétés allostériques de la protéine réceptrice de l'acétylcholine en solution détergente sans addition de lipides, *C. R. Acad. Sci. (Paris)* **292**(série III):13–15.

Heinemann, S., Bevan, S., Kullberg, R., Lindstrom, J., and Rice, J., 1977, Modulation of the acetylcholine receptor by anti-receptor antibody, *Proc. Natl. Acad. Sci. USA* **74:**3090–3094.

Heinemann, S., Merlie, J., and Lindstrom, J., 1978, Modulation of acetylcholine receptor in rat diaphragm by antireceptor sera, *Nature* **274:**65–68.

Hess, G. P., Pasquale, E. B., Walker, J. W., and McNamee, M. G., 1982, Comparison of acetylcholine receptor-controlled cation flux in membrane vesicles from *Torpedo californica* and *Electrophorus electricus:* Chemical kinetic measurements in the millisecond region, *Proc. Natl. Acad. Sci. USA* **79:**963–967.

Heuser, J. E., and Salpeter, S. R., 1979, Organization of acetylcholine receptors in quick-frozen, deep-etched, and rotary-replicated *Torpedo* postsynaptic membrane, *J. Cell Biol.* **82:**150–173.

Heuser, J. E., Reese, T. S., Dennis, M. J., Jan, Y., Jan, L., and Evans, L., 1979, Synaptic vesicle exocytosis captured by quick freezing and correlated with quantal transmitter release, *J. Cell Biol.* **81:**275–300.

Holtzman, E., Wise, D., Wall, J., and Karlin, A., 1982, Electron microscopy of complexes of isolated acetylcholine receptor, biotinyl-toxin, and avidin, *Proc. Natl. Acad. Sci. USA* **79:**310–314.

Horn, R., and Patlak, J., 1980, Single channel currents from excised patches of muscle membrane, *Proc. Natl. Acad. Sci. USA* **77:**6930–6934.

Huang, L. M., Catterall, W. A., and Ehrenstein, G., 1978, Selectivity of cations and nonelectrolytes for acetylcholine-activated channels in cultured muscle cells, *J. Gen. Physiol.* **71:**397–410.

Hucho, F., Bandini, G., and Suárez-Isla, B. A., 1978, The acetylcholine receptor as part of a protein complex in receptor-enriched membrane fragments from *Torpedo californica* electric tissue, *Eur. J. Biochem.* **83:**335–340.

Huganir, R. L., and Greengard, P., 1983, cAMP-dependent protein kinase phosphorylates the nicotinic acetylcholine receptor, *Proc. Natl. Acad. Sci. USA* **80:**1130–1134.

Huganir, R. L., and Racker, E., 1980, Endogenous and exogenous proteolysis of the acetylcholine receptor from *Torpedo californica, J. Supramol. Struct.* **14:**215–221.

Huganir, R. L., Schell, M. A., and Racker, E., 1979, Reconstitution of the purified acetylcholine receptor from *Torpedo californica, FEBS Lett.* **108:**155–160.

Huganir, R. L., Coronado, R., Silverman, D. H., and Racker, E., 1981, Structure and function of the nicotinic acetylcholine receptor, *Abstracts, VII Internatl. Biophys. Congress and III Pan American Biochem. Congress,* Mexico City, Mexico, p. 258.

James, R. W., Kato, A. C., Rey, M.-J., and Fulpius, B. W., 1980, Monoclonal antibodies directed against the neurotransmitter binding site of nicotinic acetylcholine receptor, *FEBS Lett.* **120:**145–148.

Jenkinson, D. H., 1960, The antagonism between tubocurarine and substances which depolarize the motor end-plate, *J. Physiol.* **152:**309–324.

Jenkinson, D. M., and Nicholls, J. G., 1961, Contracture and permeability changes produced by acetylcholine in depolarized denervated muscle, *J. Physiol.* **159:**111–127.

Kaldany, R. R. J., and Karlin, A., 1983, Reaction of quinacrine mustard with the acetylcholine receptor from *Torpedo californica*. Functional consequences and sites of labeling, *J. Biol. Chem.* **258:**6232–6242.

Kao, I., and Drachman, D. B., 1977, Myasthenic immunoglobulin accelerates acetylcholine receptor degradation, *Science* **196:**527–529.

Karlin, A., 1980, Molecular properties of nicotinic acetylcholine receptors, in: *The Cell Surface and Neuronal Function* (G. Poste, G. Nicolson, and C. Cotman, eds.), Elsevier/North-Holland Biomedical Press, New York, pp. 191–260.

Karlin, A., Weill, C. L., McNamee, M. G., and Valderrama, R., 1975, Facets of the structures of acetylcholine receptors from *Electrophorus* and *Torpedo, Cold Spring Harbor Symp. Quant. Biol.* **40:**203–210.

Karlsson, E., Arnberg, H., and Eaker, D., 1971, Isolation of the principal neurotoxins of two *Naja naja* subspecies, *Eur. J. Biochem.* **21:**1–16.

Karpen, J. W., Aoshima, H., Abood, L. G., and Hess, G. P., 1982, Cocaine and phencyclidine inhibition of the acetylcholine receptor: Analysis of the mechanisms of action based on measurements of ion flux in the millisecond-to-minute time region, *Proc. Natl. Acad. Sci. USA* **79:**2509–2513.

Kato, G., and Changeux, J.-P., 1976, Studies on the effect of histrionicotoxin on the monocellular electroplax from *Electrophorus electricus* and on the binding of $^3$H-acetylcholine to membrane fragments from *Torpedo marmorata, Mol. Pharmacol.* **12:**92–100.

Katz, B., 1966, *Nerve, Muscle and Synapse,* McGraw-Hill, New York.

Katz, B., and Miledi, R., 1965, The measurement of synaptic delay, and the time course of acetylcholine release at the neuromuscular junction, *Proc. Roy. Soc. (London) B* **161:**483–496.

Katz, B., and Miledi, R., 1972, The statistical nature of the acetylcholine potential and its molecular components, *J. Physiol.* **224:**665–699.

Katz, B., and Miledi, R., 1975, The effect of procaine on the action of acetylcholine at the neuromuscular junction, *J. Physiol.* **249:**269–284.

Katz, B., and Miledi, R., 1977, Transmitter leakage from motor nerve endings, *Proc. Roy. Soc. (London) B* **196:**59–72.

Katz, B., and Miledi, R., 1978, A re-examination of curare action at the motor endplate, *Proc. Roy. Soc. (London) B* **203:**119–133.

Katz, B., and Thesleff, S., 1957, A study of the desensitization produced by acetylcholine at the motor end-plate, *J. Physiol.* **138:**63–80.

Kemp, G., Morley, B., Dwyer, D., and Bradley, R. J., 1980, Purification and characterization of nicotinic acetylcholine receptors from muscle, *Memb. Biochem.* **3:**229–257.

Kilian, P. L., Dunlap, C. R., Mueller, P., Schell, M. A., Huganir, R. L., and Racker, E., 1980, Reconstitution of acetylcholine receptor from *Torpedo californica* with highly purified phospholipids: Effects of α-tocopherol, phylloquinone, and other terpenoid quinones, *Biochem. Biophys. Res. Commun.* **93:**409–414.

Kistler, J., and Stroud, R. M., 1981, Crystalline arrays of membrane-bound acetylcholine receptor, *Proc. Natl. Acad. Sci. USA* **78:**3678–3682.

Kistler, J., Stroud, R. M., Klymkowsky, M. W., Lalancette, R. A., and Fairclough, R. H., 1982, Structure and function of an acetylcholine receptor, *Biophys. J.* **37:**371–383.

Kloog, Y., Flynn, D., Hoffman, A. R., and Axelrod, J., 1980, Enzymatic carboxymethylation of the nicotinic acetylcholine receptor, *Biochem. Biophys. Res. Commun.* **97:**1474–1480.

Klymkowsky, M. W., and Stroud, R. M., 1979, Immunospecific identification and three-dimensional structure of membrane-bound acetylcholine receptor from *Torpedo californica, J. Mol. Biol.* **128:**319–334.

Klymkowsky, M. W., Heuser, J. E., and Stroud, R. M., 1980, Protease effects on the structure of acetylcholine receptor membranes from *Torpedo californica, J. Cell Biol.* **85:**823–838.

Koblin, D. D., and Lester, H. A., 1979, Voltage-dependent and voltage-independent blockade of acetylcholine receptors by local anesthetics in *Electrophorus* electroplaques, *Mol. Pharmacol.* **15:**559–580.

Kordas, M., 1970, The effect of procaine on neuromuscular transmission, *J. Physiol.* **209:**689–699.

Krenz, W.-D., Tashiro, T., Waechtler, K., Whittaker, V. P., and Witzemann, V., 1980, Aspects of the chemical embryology of the electromotor system of *Torpedo marmorata* with special reference to synaptogenesis, *Neuroscience* **5:**617–624.

Krodel, E. K., Beckman, R. A., and Cohen, J. B., 1979, Identification of a local anesthetic binding site on postsynaptic membranes isolated from *Torpedo marmorata* electric tissue, *Mol. Pharmacol.* **15:**294–312.

Kuffler, S. W., and Yoshikami, D., 1975, The number of transmitter molecules in a quantum, *J. Physiol.* **251:**465–482.

Labarca, P., Lindstrom, J., and Montal, M., 1981, Channel properties of the purified acetylcholine receptor (AChR) in planar lipid bilayers, *Abstracts, VII International Biophys. Congress and III Pan-American Biochem. Congress,* Mexico City, Mexico, p. 258.

Labarca, P., Lindstrom, J., and Montal, M., 1982, Studies on the properties of the purified acetylcholine receptor reconstituted in planar lipid bilayers, *Biophys. J.* **37:**170a.

Labarca, P., Lindstrom, J., and Montal, M., 1984a, Two kinetically coupled open states of the acetylcholine receptor channel, *J. Neurosci.* **4:**502–507.

Labarca, P., Lindstrom, J. and Montal, M 1984b, Acetylcholine receptor in planar lipid bilayers. Characterization of the channel properties of the purified nicotinic acetylcholine receptor from *Torpedo californica* reconstituted in planar lipid bilayers, *J. Gen. Physiol.* **83.**

Lecar, H., Morris, E. C., and Wong, B. S., 1982, Single channel recording of weak cholinergic agonists, *Biophys. J.* **37:**313a.

Lee, C. Y., 1970, Elapid neurotoxins and their mode of action, *Clin. Toxicol.* **3:**457–472.

Lee, T., Witzemann, V., Schimerlik, M., and Raftery, M. A., 1977, Cholinergic ligand induced affinity changes in *Torpedo californica* acetylcholine receptor, *Arch. Biochem. Biophys.* **183:**57–63.

Lennon, V. A., Thompson, M., and Chen, J., 1980, Properties of nicotinic acetylcholine receptors isolated by affinity chromatography on monoclonal antibodies, *J. Biol. Chem.* **255:**4395–4398.

Lewis, C. A., 1979, Ion-concentration dependence of the reversal potential and the single channel conductance of ion channels at the frog neuromuscular junction, *J. Physiol.* **286:**417–445.

Libby, P., Bursztajn, S., and Goldberg, A. L., 1980, Degradation of the acetylcholine receptor in cultured muscle cells: Selective inhibitors and the fate of undegraded receptors, *Cell* **19:**481–491.

Linden, D., and Fambrough, D., 1979, Biosynthesis and degradation of acetylcholine receptors in rat skeletal muscles. Effects of electrical stimulation, *Neuroscience* **4:**527–538.

Lindstrom, J. M., 1978, Biochemical studies of receptors: Solubilization, purification, characterization and studies with antibodies, in: *Neurotransmitter Receptor Binding* (H. I. Yamamura, S. J. Enna, and M. J. Kuhar, eds.), Raven Press, New York, pp. 91–111.

Lindstrom, J., 1979, Autoimmune response to acetylcholine receptor in myasthenia gravis and its animal model, *Adv. Immunol.* **27:**50.

Lindstrom, J., 1984, Acetylcholine receptors: Structure, function, synthesis, destruction and antigenicity, in: *Myology,* Chap. 27 (A. Engel and B. Banker, eds.), McGraw-Hill, New York, in press.

Lindstrom, J., and Einarson, B., 1979, Antigenic modulation and receptor loss in EAMG, *Muscle Nerve* **2:**173–179.

Lindstrom, J., and Engel, A., 1981, Myasthenia gravis and the nicotinic cholinergic receptor, in: *Receptor Regulation,* Ser. B, Vol. 13, Chapman and Hall, London, pp. 161–214.

Lindstrom, J. M., and Lambert, E. H., 1978, Content of acetylcholine receptor and antibodies bound to receptor in myasthenia gravis, experimental autoimmune myasthenia gravis and in Eaton-Lambert Syndrome, *Neurology* **28:**130–138.

Lindstrom, J., and Patrick, J., 1974 Purification of the acetylcholine receptor by affinity chromatography, in: *Synaptic Transmission and Neuronal Interaction,* 26th Annual Meeting, Soc. Gen. Physiologists, Woods Hole, Massachusetts, September 1972 (M. V. L. Bennett, ed.), Raven Press, New York, pp. 191–216.

Lindstrom, J., Einarson, B., and Merlie, J., 1978, Immunization of rats with polypeptide chains from *Torpedo* acetylcholine receptor causes an autoimmune response to receptors in rat muscle, *Proc. Natl. Acad. Sci. USA* **75:**769–773.

Lindstrom, J., Merlie, J., and Yogeeswaran, G., 1979a, Biochemical properties of acetylcholine receptor subunits from *Torpedo californica, Biochemistry* **18:**4465–4470.

Lindstrom, J., Walter, B., and Einarson, B., 1979b, Immunochemical similarities between subunits of acetylcholine receptors from *Torpedo, Electrophorus* and mammalian muscle, *Biochemistry* **18:**4470–4480.

Lindstrom, J., Gullick, W. J., Conti-Tronconi, B., and Ellisman, M., 1980a, Proteolytic nicking of the acetylcholine receptor, *Biochemistry* **19:**4791–4795.

Lindstrom, J., Cooper, J., and Tzartos, S., 1980b, Acetylcholine receptors from *Torpedo* and *Electrophorus* have similar subunit structures, *Biochemistry* **19:**1454–1458.

Lindstrom, J., Anholt, R., Einarson, B., Engel, A., Osame, M., and Montal, M., 1980c, Purification of acetylcholine receptors, reconstitution into lipid vesicles, and study of agonist-induced cation channel regulation, *J. Biol. Chem.* **255:**8340–8350.

Lindstrom, J., Einarson, B., and Tzartos, S., 1981a, Production and assay of antibodies to acetylcholine receptors, *Meth. Enzymol.* **74:**432–460.

Lindstrom, J., Tzartos, S., and Gullick, W., 1981b, Structure and function of acetylcholine receptors studied using monoclonal antibodies, *Ann. N.Y. Acad. Sci.* **377:**1–19.

Lindstrom, J. M., Cooper, J. F., and Swanson, L. W., 1983, Purification of acetylcholine receptors from the muscle of *Electrophorus electricus, Biochemistry* **22:**3796–3800.

Lo, M. M. S., Garland, P. B., Lamprecht, J., and Barnard, E. A., 1980, Rotational mobility of the membrane-bound acetylcholine receptor of *Torpedo* electric organ measured by phosphorescence depolarization, *FEBS Lett.* **111**:407–412.

Lo, M. M. S., Dolly, O. J., and Barnard, E. A., 1981, Molecular forms of the acetylcholine receptor from vertebrate muscles and *Torpedo* electric organ: Interactions with specific ligands, *Eur. J. Biochem.* **116**:155–163.

Loewi, O., 1921, Über humorale übertragbarkeit der Herznervenwirkung, *Pfluegers Arch. Ges. Physiol.* **189**:239–242.

Magazanik, L. G., and Vyskocil, F., 1970, Dependence of acetylcholine desensitization on the membrane potential of frog muscle fibre and on the ionic changes in the medium, *J. Physiol.* **210**:507–518.

Magazanik, L. G., and Vyskocil, F., 1973, Desensitization at the motor endplate, in: *Drug Receptors* (H. P. Rang, ed.), MacMillan, London, pp. 105–119.

Magazanik, L. G., and Vyskocil, F., 1975, The effect of temperature on desensitization kinetics at the post-synaptic membrane of the frog muscle fibre, *J. Physiol.* **249**:285–300.

Magleby, K. L., and Pallotta, B. S., 1981, A study of desensitization of acetylcholine receptors using nerve-released transmitter in the frog, *J. Physiol.* **316**:225–250.

Magleby, K. L., and Stevens, C. F., 1972a, The effect of voltage on the time course of end-plate currents, *J. Physiol.* **223**:151–171.

Magleby, K. L., and Stevens, C. F., 1972b, A quantitative description of end-plate currents, *J. Physiol.* **223**:173–197.

Martinez-Carrion, M., Sator, V., and Raftery, M. A., 1975, The molecular weight of an acetylcholine receptor isolated from *Torpedo californica, Biophys. Res. Commun.* **65**:129–137.

McNamee, M. G., and Ochoa, E. L. M., 1982, Reconstitution of acetylcholine receptor function in model membranes, *Neuroscience* **7**:2305–2319.

McNamee, M. G., Ellena, J. F., and Dalziel, A. W., 1982, Lipid–protein interactions in membranes containing the acetylcholine receptor, *Biophys. J.* **37**:103–104.

Mebs, D., Narita, K., Iwanaga, S., Samejima, Y., and Lee, C. Y., 1971, Amino acid sequence of α-bungarotoxin from the venom of *Bungarus multicinctus. Biochem. Biophys. Res. Commun.* **44**: 711–716.

Mellinger, J., Belbenoit, P., Ravaille, M., and Szabo, T., 1978, Electric organ development in *Torpedo marmorata* chondrichthyes, *Dev. Biol.* **67**:167–188.

Mendez, B., Valenzuela, P., Martial, J. A., and Baxter, J. D., 1980, Cell free synthesis of acetylcholine receptor polypeptides, *Science* **209**:695–697.

Merlie, J. P., and Sebbane, R., 1981, Acetylcholine receptor subunits transit a precursor pool before acquiring αbungarotoxin binding activity, *J. Biol. Chem.* **256**:3605–3608.

Merlie, J. P., Changeux, J.-P., and Gros, F., 1976, Acetylcholine receptor degradation measured by pulse chase labeling, *Nature* **264**:74–76.

Merlie, J. P., Changeux, J.-P., and Gros, F., 1978, Skeletal muscle acetylcholine receptor purification, characterization and turnover in muscle cell cultures, *J. Biol. Chem.* **253**:2882–2891.

Merlie, J. P., Heineman, S., Einarson, B., and Lindstrom, J. M., 1979a, Degradation of acetylcholine receptor in diaphragms of rats with experimental autoimmune myasthenia gravis, *J. Biol. Chem.* **254**:6328–6332.

Merlie, J. P., Heinemann, S., and Lindstrom, J. M., 1979b, Acetylcholine receptor degradation in adult rat diaphragms in organ culture and the effect of anti-acetylcholine receptor antibodies, *J. Biol. Chem.* **254**:6302–6327.

Merlie, J. P., Hofler, J. G., and Sebbane, R., 1981, Acetylcholine receptor synthesis from membrane polysomes, *J. Biol. Chem.* **256**:6995–6999.

Merlie, J. P., Sebbane, R., Tzartos, S., and Lindstrom, J., 1982, Inhibition of glycosylation with tunicamycin blocks assembly of newly synthesized acetylcholine receptor subunits in muscle cells, *J. Biol. Chem.* **257**:2694–2701.

Meyer, D. I., Krause, E., and Dobberstein, B., 1982, Secretory protein translocation across membranes—the role of the "docking protein " *Nature* **297**:647–650.

Michaelson, D. M., and Raftery, M. A., 1974, Purified acetylcholine receptor: Its reconstitution to a chemically excitable membrane, *Proc. Natl. Acad. Sci. USA* **71**:4768–4772.

Michler, A., and Sakmann, B., 1980, Receptor stability and channel conversion in the subsynaptic membrane of the developing mammalian neuromuscular junction, *Dev. Biol.* **80:**1–17.

Miller, C., Arvan, P., Telford, J. N., and Racker, E., 1976, Calcium-induced fusion of proteoliposomes: Dependence on transmembrane osmotic gradient, *J. Membr. Biol.* **30:**271–282.

Miller, D. L., Moore, H.-P. H., Hartig, P. R., and Raftery, M. A., 1978, Fast cation flux from *Torpedo californica* membrane preparations: Implications for a functional role for acetylcholine receptor dimers, *Biochem. Biophys. Res. Commun.* **85:**632–640.

Mishina, M., Kurosaki, T., Tobimatsu, T., Morimoto, Y., Noda, M., Yamamoto, T., Terao, M., Lindstrom, J., Takahashi, T., Kuno, M. and Numa, S., 1984, Expression of functional acetylcholine receptor from cloned cDNAs, *Nature,* **307:**604–608.

Mochly-Rosen, C., and Fuchs, S., 1981, Monoclonal anti-acetylcholine receptor antibodies directed against the cholinergic binding site, *Biochemistry* **20:**5920–5924.

Momoi, M. Y., and Lennon, V. A., 1982, Purification and biochemical characterization of nicotinic acetylcholine receptor of human muscle, *J. Biol. Chem.* **257:**12757–12764.

Montal, M., 1974, Formation of bimolecular membranes from lipid monolayers, *Meth. Enzymol.* **32:**545–556.

Montal, M., and Mueller, P., 1972, Formation of bimolecular membranes from lipid monolayers and a study of their electrical properties, *Proc. Natl. Acad. Sci. USA* **69:**3561–3566.

Montal, M., Darszon, A., and Schindler, H., 1981, Functional reassembly of membrane proteins in planar lipid bilayers, *Quart. Rev. Biophys.* **14:**1–79.

Montal, M., Labarca, P., Fredkin, D. R., Suárez-Isla, B. A. and Lindstrom, J.,1984, Channel properties of the purified acetylcholine receptor from *Torpedo californica* reconstituted in planar lipid bilayer membranes, *Biophys. J.* **45:**165–174.

Moore, H.-P. H., and Raftery, M. A., 1979, Ligand-induced interconversion of affinity states in membrane-bound acetylcholine receptor from *Torpedo californica.* Effects of sulfhydryl and disulfide reagents, *Biochemistry* **18:**1907–1911.

Moore, H.-P. H., and Raftery, M. A., 1980, Direct spectroscopic studies of cation translocation by *Torpedo* acetylcholine receptor on a time scale of physiological relevance, *Proc. Natl. Acad. Sci. USA* **77:**4509–4513.

Moore, H.-P., Hartig, P. R., and Raftery, M. A., 1979, Correlation of polypeptide composition with functional events in acetylcholine receptor-enriched membranes from *Torpedo californica, Proc. Natl. Acad. Sci. USA* **76:**6265–6269.

Morris, C. E., Jackson, M. B., Lecar, H., Wong, B. S., and Christian, C. N., 1982, Activation of individual acetylcholine channels by curare in embryonic rat muscle, *Biophys. J.* **37:**19a.

Nachmansohn, D., 1953, Metabolism and function of the nerve cell, *Harvey Lect.* **49:**57–99.

Nathanson, N. M., and Hall, Z. W., 1979, Subunit structure and peptide mapping of junctional and extrajunctional acetylcholine receptors from rat muscle, *Biochemistry* **18:**3392–3401.

Nathanson, N. M., and Hall, Z. W., 1980, *In situ* labeling of Torpedo and rat muscle acetylcholine receptor by a photoaffinity derivative of α-bungarotoxin, *J. Biol. Chem.* **255:**1698–1703.

Neher, E., and Sakmann, B., 1976, Single channel currents recorded from membrane of denervated frog muscle fibers, *Nature* **260:**799–802.

Neher, E., and Steinbach, J. H., 1978, Local anesthetics transiently block currents through single acetylcholine receptor channels, *J. Physiol.* **277:**153–176.

Nelson, N., Anholt, R., Lindstrom, J., and Montal, M., 1980, Reconstitution of purified acetylcholine receptors with functional ion channels in planar lipid bilayers, *Proc. Natl. Acad. Sci. USA* **77:**3057–3061.

Neubig, R. R., and Cohen, J. B., 1979, Equilibrium binding of $^3$H-tubocurarine and $^3$H-acetylcholine by Torpedo postsynaptic membranes: Stoichiometry and ligand interactions, *Biochemistry* **18:**5464–5475.

Neubig, R. R., and Cohen, J. B., 1980, Permeability control by cholinergic receptors in *Torpedo* postsynaptic membranes: Agonist dose–response relations measured at second and millisecond times, *Biochemistry* **19:**2770–2779.

Neubig, R. R., Krodel, E. K., Boyd, N. D., and Cohen, J. B., 1979, Acetylcholine and local anesthetic binding to *Torpedo* nicotinic postsynaptic membranes after removal of nonreceptor peptides, *Proc. Natl. Acad. Sci. USA* **76:**690–694.

Neubig, R. R., Boyd, N. D., and Cohen, J. B., 1982, Conformations of *Torpedo* acetylcholine receptor associated with ion transport and desensitization, *Biochemistry* **21:**3460–3467.

Neugebauer, D.-Ch., and Zingsheim, H. P., 1982, Structural changes in alkaline-treated postsynaptic membranes from *Torpedo marmorata* are not due to lipid hydrolysis, *Biochim. Biophys. Acta* **684:**272–276.

Nghiem, H. O., Cartaud, J., Dubreuil, C., Kordeli, C., Buttin, G., and Changeux, J. P., 1983, Production and characterization of a monoclonal antibody directed against the 43,000 dalton $\nu_1$ polypeptide from *Torpedo marmorata* electric organ, *Proc. Natl. Acad. Sci. USA* **80:**6403–6407.

Nickel, E., and Potter, L. T., 1970, Synaptic vesicles in freeze-etched electric tissue of *Torpedo, Brain Res.* **23:**95–100.

Noda, M., Takahashi, H., Tanabe, T., Toyosato, M., Furotani, Y., Hirose, T., Asai, M., Inayama, S., Miyata, T., and Numa, S., 1982, Primary structure of α-subunit precursor of *Torpedo californica* acetylcholine receptor deduced from cDNA sequence, *Nature* **299:**793–797.

Noda, M., Takahashi, H., Tanabe, T., Toyosato, M., Kikyotani, S., Hirose, T., Asai, M., Takashima, H., Inayama, S., Miyata, T., and Numa, S., 1983a, Primary structures of β- and δ-subunit precursors of *Torpedo californica* acetylcholine receptor deduced from cDNA sequences, *Nature* **301:**251–255.

Noda, M., Takahashi, H., Tanabe, T., Toyosato, M., Kikyotani, S., Furutani, Y., Hirose, T., Takashima, H., Inayama, S., Miyata, T., and Numa, S., 1983b, Structural homology of *Torpedo californica* acetylcholine receptor subunits, *Nature* **302:**528–532.

Noda, M., Furutani, Y., Takahashi, H., Toyosato, M., Tanabe, T., Schimizu, S., Kikyotani, S., Kayano, T., Hirose, T., Inayama, S., and Numa, S., 1983c, Cloning and sequence analysis of calf cDNA and human genomic DNA encoding α-subunit precursor of muscle acetylcholine receptor, *Nature* **305:**818–823.

O'Brien, D. F., Costa, L. F., and Ott, R. A., 1977, Photochemical functionality of rhodopsin-phospholipid recombinant membranes, *Biochemistry* **16:**1295–1303.

Ochoa, E. L. M., Dalziel, A. W., and McNamee, M. G., 1983, Reconstitution of acetylcholine receptor function in lipid vesicles of defined compositions, *Biochim. Biophys. Acta,* **727:**151–162.

Ogden, D. C., Siegelbaum, S. A., and Colquhoun, D., 1981, Block of acetylcholine-activated ion channels by an uncharged local anaesthetic, *Nature* **289:**596–598.

Oswald, R. E., and Changeux, J.-P., 1981a, Selective labeling of the δ-subunit of the acetylcholine receptor by a covalent local anesthetic, *Biochemistry* **20:**7166–7174.

Oswald, R., and Changeux, J.-P., 1981b, Ultraviolet light-induced labeling by noncompetitive blockers of the acetylcholine receptor from *Torpedo marmorata, Proc. Natl. Acad. Sci. USA* **78:**3925–3929.

Oswald, R. E., and Changeux, J. P., 1982, Crosslinking of αbungarotoxin to the acetylcholine receptor from *Torpedo marmorata* by ultraviolet light irradiation, *FEBS Lett.* **139:**225–229.

Oswald, R., Sobel, A., Waksman, G., Roques, B., and Changeux, J.-P., 1980, Selective labeling by [$^3$H]-trimethisoquin azide of polypeptide chains present in acetylcholine receptor-rich membranes from *Torpedo marmorata, FEBS Lett.* **111:**29–34.

Paraschos, A., Gonzales-Ros, J., and Martinez-Carrion, M., 1982, Acetylcholine receptor from *Torpedo:* Preferential solubilization and efficient reintegration into lipid vesicles, *Biochim. Biophys. Acta* **691:**249–260.

Parisi, M., Rivas, E., and De Robertis, E., 1971, Conductance changes produced by acetylcholine in lipidic membranes containing a proteolipid from *Electrophorus, Science* **172:**56–57.

Peper, K., Dryer, F., and Mueller, K.-D., 1975, Analysis of cooperativity of drug-receptor interaction by quantitative iontophoresis at frog motor end plates, *Cold Spring Harbor Symp. Quant. Biol.* **40:**187–192.

Perrelet, A., Garcia-Segura, L.-M., Singh, A., and Orci, L., 1982, Distribution of cytochemically detectable cholesterol in the electric organ of *Torpedo marmorata, Proc. Natl. Acad. Sci. USA* **79:**2598–2602.

Popot, J.-L., 1982, Functional studies of purified integral membrane proteins reintegrated into lipid vesicles: The case of the acetylcholine receptor, in: *Méthodologie des Liposomes/Liposome Methodology* (L. D. Leserman and J. Barbet, eds.), INSERM Symposia Series 107, pp. 93–124 and 147–154.

Popot, J.-L., Demel, R. A., Sobel, A., Van Deenen, L. L. M., and Changeux, J.-P., 1978, Interaction of the acetylcholine (nicotinic) receptor protein from *Torpedo marmorata* electric organ with monolayers of pure lipids, *J. Biochem.* **85:**27–42.

Popot, J.-L., Cartaud, J., and Changeux, J.-P., 1981, Reconstitution of a functional acetylcholine receptor: Incorporation into artificial lipid vesicles and pharmacology of the agonist-controlled permeability changes, *Eur. J. Biochem.* **118:**203–214.

Porter, C. W., and Barnard, E. A., 1975, Distribution and density of cholinergic receptors at the motor endplates of a denervated mouse muscle, *Exp. Neurol.* **48:**542–556.

Potter, L. T., and Smith, D. S., 1977, Postsynaptic membranes in the electric tissue of Narcine, *Tissue Cell* **9:**585–644.

Prives, J., and Bar-Sagi, D., 1982, Effect of tunicamycin, an inhibitor of protein glycosylation, on the functional properties of acetylcholine receptors in cultured muscle cells, *J. Cell Biol.* **95:**416a.

Prives, J. M., and Olden, K., 1980, Carbohydrate requirement for expression and stability of acetylcholine receptor on the surface of embryonic muscle cells in culture, *Proc. Natl. Acad. Sci. USA* **77:**5263–5267.

Pumplin, D. W., and Fambrough, D. M., 1982, Turnover of acetylcholine receptors in skeletal muscle, *Annu. Rev. Physiol.* **44:**319–335.

Quast, U., Schimerlik, M., Lee, T., Witzemann, V., Blanchard, S., and Raftery, M. A., 1978, Ligand-induced conformation changes in *Torpedo californica* membrane-bound acetylcholine receptor, *Biochemistry* **17:**2405–2414.

Quast, U., Schimerlik, M. I., and Raftery, M. A., 1979, Ligand-induced changes in membrane-bound acetylcholine receptor observed by ethidium fluorescence. 2. Stopped-flow studies with agonists and antagonists, *Biochemistry* **18:**1891–1901.

Raftery, M. A., Hunkapiller, M. W., Strader, C. D., and Hood, L. E., 1980, Acetylcholine receptor: Complex of homologous subunits, *Science* **208:**1454–1457.

Rang, H. P., and Ritter, J. M., 1970, On the mechanism of desensitization at cholinergic receptors, *Mol. Pharmacol.* **6:**357.

Rang, H. P., and Ritter, J. M., 1971, The effect of disulfide bond reduction on the properties of cholinergic receptors in chick muscle, *Mol. Pharmacol.* **7:**620–631.

Ravdin, P. M., and Berg, D. K., 1979, Inhibition of neuronal acetylcholine sensitivity by α-toxins from *Bungarus multicinctus* venom, *Proc. Natl. Acad. Sci. USA* **76:**2072–2076.

Reiness, C. G. and Hall, Z. W., 1977, Electrical stimulation of denervated muscles reduces incorporation of methionine into the ACh receptor, *Nature* **268:**655–657.

Reiness, C. G., and Weinberg, C. B., 1981, Metabolic stabilization of acetylcholine receptors at newly formed neuromuscular junctions in rat, *Dev. Biol.* **84:**247–254.

Reiness, C. G., Weinberg, C. B., and Hall, Z. W., 1978, Antibody to acetylcholine receptor increases the degradation of junctional and extrajunctional receptors in adult muscles, *Nature* **274:**68–70.

Reynolds, J. A., and Karlin, A., 1978, Molecular weight in detergent solution of acetylcholine receptor from *Torpedo californica*, *Biochemistry* **17:**2035–2038.

Ritchie, A. K., and Fambrough, D. M., 1975, Ionic properties of the acetylcholine receptor in cultured rat myotubes, *J. Gen. Physiol.* **65:**751–767.

Ross, M. J., Klymkowsky, M. W., Agard, D. A., and Stroud, R. M., 1977, Structural studies of a membrane-bound acetylcholine receptor from *Torpedo californica, J. Mol. Biol.* **116:**635–659.

Rotundo, R. L., and Fambrough, D. M., 1982, Synthesis, transport and fate of acetylcholinesterase and acetylcholine receptors in cultured muscle, in: *Membranes in Growth and Development,* Alan R. Liss, New York, pp. 259–286.

Ruechel, R., Watters, D., and Maelicke, A., 1981, Molecular forms and hydrodynamic properties of acetylcholine receptor from electric tissue, *Eur. J. Biochem.* **119:**215–223.

Ruff, R. L., 1977, A quantitative analysis of local anaesthetic alteration of miniature end-plate currents and end-plate current fluctuations, *J. Physiol.* **264:**89–124.

Ruff, R. L., 1982, The kinetics of local anesthetic blockade of end-plate channels, *Biophys. J.* **37:**625–631.

Sahashi, K., Engel, A. G., Lindstrom, J. M., Lambert, E. M., and Lennon, V. A., 1978, Ultrastructural localization of immune complexes (IgG and $C_3$) at the endplate in experimental autoimmune myasthenia gravis, *J. Neuropathol. Exp. Neurol.* **37:**212–223.

Saitoh, T., and Changeux, J.-P., 1980, Phosphorylation *in vitro* of membrane fragments from *Torpedo marmorata* electric organ: Effect on membrane solubilization by detergents, *Eur. J. Biochem.* **105:**51–62.

Saitoh, T., and Changeux, J.-P., 1981, Change in state of phosphorylation of acetylcholine receptor during maturation of the electromotor synapse in *Torpedo marmorata* electric organ, *Proc. Natl. Acad. Sci. USA* **78:**4430–4434.

Saitoh, T., Wennogle, L. P., and Changeux, J. P., 1979, Factors regulating the susceptibility of the acetylcholine receptor protein to heat inactivation, *FEBS Lett.* **108:**489–494.

Saitoh, T., Oswald, R., Wennogle, L. P., and Changeux, J.-P., 1980, Conditions for the selective labeling of the 66,000 dalton chain of the acetylcholine receptor by the covalent non-competitive blocker 5-azido-[$^3$H]trimethisoquin, *FEBS Lett.* **116:**30–36.

Sakmann, B., and Brenner, H. R., 1978, Change in synaptic channel gating during neuromuscular development, *Nature* **276:**401–402.

Sakmann, B., Patlak, J., and Neher, E., 1980, Single acetylcholine-activated channels show burst-kinetics in presence of desensitizing concentrations of agonist, *Nature* **286:**71–73.

Sanes, J. R., Marshall, L. M., and McMahan, U. J., 1978, Reinnervation of muscle fiber basal lamina after removal of myofibers: Differentiation of regenerating axons at original synaptic sites, *J. Cell Biol.* **78:**38–165.

Sargent, P. B., Hedges, B. E., Tsavaler, L., Clemmons, L., Tzartos, S., and Lindstrom, J., 1984, The structure and transmembrane nature of the acetylcholine receptor in amphibian skeletal muscle as revealed by crossreacting monoclonal antibodies, *J. Cell Biol.,* (in press.)

Schiebler, W., and Hucho, F., 1978, Membranes rich in acetylcholine receptor: Characterization and reconstitution to excitable membranes from exogenous lipids, *Eur. J. Biochem.* **85:**55–63.

Schiebler, W., Bandini, G., and Hucho, F., 1980, Quaternary structure and reconstitution of acetylcholine receptor from *Torpedo californica, Neurochem. Int.* **2:**281–290.

Schindler, H., 1980, Formation of planar bilayers from artificial or native membrane vesicles, *FEBS Lett.* **122:**77–79.

Schindler, H., and Quast, U., 1980, Functional acetylcholine receptor from *Torpedo marmorata* in planar membranes, *Proc. Natl. Acad. Sci. USA* **77:**3052–3056.

Schlieper, P., and De Robertis, E., 1977, Lipid bilayers and liposomes in reconstitution experiments with cholinergic proteolipid from *Torpedo* electroplax, *Biochem. Biophys. Res. Commun.* **75:**886–894.

Schmidt, J., and Raftery, M. A., 1973, A simple assay for the study of solubilized acetylcholine receptors, *Anal. Biochem.* **52:**349–354.

Scubon-Mulieri, B., and Parsons, R. L., 1977, Desensitization and recovery at the frog neuromuscular junction, *J. Gen. Physiol.* **69:**431–447.

Scubon-Mulieri, B., and Parsons, R. L., 1978, Desensitization onset and recovery at the potassium-depolarized frog neuromuscular junction are voltage sensitive, *J. Gen. Physiol.* **71:**285–299.

Sealock, R., 1982a, Cytoplasmic surface structure in postsynaptic membranes from electric tissue visualized by tannic-acid-mediated negative contrasting, *J. Cell Biol.* **92:**514–522.

Sealock, R., 1982b, Visualization at the mouse neuromuscular junction of a submembrane structure in common with *Torpedo* postsynaptic membranes, *J. Neurosci.* **2:**918–923.

Sebbane, R., Clokey, G., Merlie, J. P., Tzartos, S., and Lindstrom, J., 1983, Characterization of the mRNA for mouse muscle acetylcholine receptor α-subunit by quantitative translation *in vitro, J. Biol. Chem.* **258:**3294–3303.

Sheridan, R. E., and Lester, H. A., 1975, Relaxation measurements on the acetylcholine receptor, *Proc. Natl. Acad. Sci. USA* **72:**3496–3500.

Sheridan, R. E., and Lester, H. A., 1977, Rates and equilibria at the acetylcholine receptor of *Electrophorus* electroplaques, *J. Gen. Physiol.* **70:**187–219.

Shorr, R. G., Dolly, J. O., and Barnard, E. A., 1978, Composition of acetylcholine receptor protein from skeletal muscle, *Nature* **274:**283–284.

Shorr, R. G., Lyddiatt, A., Lo, M. M. S., Dolly, J. O., and Barnard, E. A., 1981, Acetylcholine receptor from mammalian skeletal muscle: Oligomeric forms and their subunit structures, *Eur. J. Biochem.* **116:**143–153.

Sine, S., and Taylor, P., 1979, Functional consequences of agonist-mediated state transitions in the cholinergic receptor, *J. Biol. Chem.* **254:**3315–3325.

Sine, S. M., and Taylor, P., 1980, The relationship between agonist occupation and the permeability response of the cholinergic receptor revealed by bound cobra α-toxin, *J. Biol. Chem.* **255:**10144–10156.

Sine, S. M., and Taylor, P., 1981, Relationship between reversible antagonist occupancy and the functional capacity of the acetylcholine receptor, *J. Biol. Chem.* **256:**6692–6699.

Sine, S. M., and Taylor, P., 1982, Local anesthetics and histrionicotoxin are allosteric inhibitors of the acetylcholine receptor, *J. Biol. Chem.* **257:**8106–8114.

Sokolovsky, M., and Bartfai, T., 1981, Biochemical studies on muscarinic receptors, *Trends Biochem. Sci.* **6:**303–305.

Smilowitz, H., Hadjian, R. A., Dwyer, J., and Feinstein, M. B., 1981, Regulation of acetylcholine receptor phosphorylation by calcium and calmodulin, *Proc. Natl. Acad. Sci. USA* **78:**4708–4712.

Sobel, A., Weber, M., and Changeux, J.-P., 1977, Large-scale purification of the acetylcholine-receptor protein in its membrane-bound and detergent-extracted forms from *Torpedo marmorata* electric organ, *Eur. J. Biochem.* **80:**215–224.

Sobel, A., Heidmann, T., Hofler, J., and Changeux, J.-P., 1978, Distinct protein components from *Torpedo marmorata* membranes carry the acetylcholine receptor site and the binding site for local anesthetics and histrionicotoxin, *Proc. Natl. Acad. Sci. USA* **75:**510–514.

Sobel, A., Heidmann, T., Cartaud, J., and Changeux, J.-P., 1980, Reconstitution of a functional acetylcholine receptor, *Eur. J. Biochem.* **110:**13–33.

Stanley, E. F., and Drachman, D. B., 1978, Effect of myasthenic immunoglobulin on acetylcholine receptors of intact mammalian neuromuscular junctions, *Science* **200:**1285–1287.

Steinbach, A. B., 1968, Alteration by xylocaine (lidocaine) and its derivatives of the time course of the end plate potential, *J. Gen. Physiol.* **52:**144–161.

Steinbach, J. H., 1981, Developmental changes in acetylcholine receptor aggregates at rat skeletal neuromuscular junctions, *Dev. Biol.* **84:**267–276.

Steinbach, J. H., Merlie, J., Heinemann, S., and Bloch, R., 1979, Degradation of junctional and extrajunctional acetylcholine receptors by developing rat skeletal muscle, *Proc. Natl. Acad. Sci. USA* **76:**3547–3551.

Stephenson, F. A., Harrison, R., and Lunt, G. G., 1981, The isolation and characterization of the nicotinic acetylcholine receptor from human skeletal muscle, *Eur. J. Biochem.* **115:**91–97.

St. John, P. A., Froehner, S. C., Goodenough, D. A., and Cohen, J. B., 1982, Nicotinic postsynaptic membranes from *Torpedo:* Sidedness, permeability to macromolecules, and topography of major polypeptides, *J. Cell Biol.* **92:**333–342.

Strader, C. D., and Raftery, M. A., 1980, Topographic studies of *Torpedo* acetylcholine receptor subunits as a transmembrane complex, *Proc. Natl. Acad. Sci. USA* **77:**5807–5811.

Strader, C. B. D., Revel, J.-P., and Raftery, M. A., 1979, Demonstration of the transmembrane nature of the acetylcholine receptor by labeling with anti-receptor antibodies, *J. Cell Biol.* **83:**499–510.

Strader, C. D., Lazarides, E., and Raftery, M. A., 1980, The characterization of actin associated with postsynaptic membranes from *Torpedo californica, Biochem. Biophys. Res. Commun.* **92:**365–373.

Suárez-Isla, B. A., and Hucho, F., 1977, Acetylcholine receptor: -SH group reactivity as indicator of conformational changes and functional states, *FEBS Lett.* **75:**65–69.

Suárez-Isla, B. A., Wan, K., Lindstrom, J., and Montal, M., 1983, Single-channel recordings from purified acetylcholine receptors reconstituted in bilayers formed at the tip of patch pipets, *Biochemistry* **22,**2319–2323.

Sugiyama, H., and Changeux, J.-P., 1975, Interconversion between different states of affinity for acetylcholine of the cholinergic receptor protein from *Torpedo marmorata, Eur. J. Biochem.* **55:**505–515.

Sumikawa, K., Houghton, M., Emtage, J. S., Richards, B. M., and Barnard, E. A., 1981, Active multi-subunit ACh receptor assembly by translation of heterologous mRNA in *Xenopus* oocytes, *Nature* **292:**862–864.

Sumikawa, K., Barnard, E. A., and Dolly, J. O., 1982a, Similarity of acetylcholine receptors of denervated, innervated and embryonic chicken muscles, *Eur. J. Biochem.* **126:**473–479.

Sumikawa, K., Houghton, M., Smith, J. C., Bell, L., Richards, B. H., and Barnard, E. A., 1982b, The molecular cloning and characterization of cDNA coding for the α-subunit of the acetylcholine receptor, *Nucleic Acid Res.* **10:**5809–5822.

Takeuchi, N., 1963, Some properties of conductance changes at the end-plate membrane during the action of acetylcholine, *J. Physiol.* **167:**128–140.

Takeuchi, A., and Takeuchi, N., 1960, On the permeability of end-plate membrane during the action of transmitter, *J. Physiol.* **154:**52–67.

Tank, D. W., Huganir, R. L., Greengard, P., and Webb, W. W., 1983, Patch-recorded single channel currents of the purified and reconstituted *Torpedo* acetylcholine receptor, *Proc. Natl. Acad. Sci. USA* **80:**5129–5133.

Tarrab-Hazdai, R., Geiger, B., Fuchs, S., and Amsterdam, A., 1978, Localization of acetylcholine receptor in excitable membrane from the electric organ of *Torpedo:* Evidence for exposure of receptor antigenic sites on both sides of the membrane, *Proc. Natl. Acad. Sci. USA* **75:**2497–2501.

Tauc, L., 1982, Nonvesicular release of neurotransmitter, *Physiol. Rev.* **62:**857–891.

Teichberg, V. I., Sobel, A., and Changeux, J.-P., 1977, *In vitro* phosphorylation of the acetylcholine receptor, *Nature* **267:**540–542.

Trautmann, A., 1982, Curare can open and block ionic channels associated with cholinergic receptors, *Nature* **298:**272–275.

Trautmann, A., and Feltz, A., 1980, Open time of channels activated by binding of two distinct agonists, *Nature* **286:**291–293.

Tzartos, S. J., and Changeux, J.-P., 1982, Lipid-dependent recovery of high affinity binding of α-bungarotoxin to the purified α-subunit from *Torpedo* acetylcholine receptors, *Abstracts, Society for Neuroscience, 12th Annual Meeting,* Minneapolis, Minnesota, 91.4, p. 334.

Tzartos, S. J., and Lindstrom, J. M., 1980, Monoclonal antibodies used to probe acetylcholine receptor structure: Localization of the main immunogenic region and detection of similarities between subunits, *Proc. Natl. Acad. Sci. USA* **77:**755–759.

Tzartos, S. J., Rand, D. E., Einarson, B. L., and Lindstrom, J. M., 1981, Mapping of surface structures of Electrophorus acetylcholine receptor using monoclonal antibodies, *J. Biol. Chem.* **256:**8635–8645.

Tzartos, S. J., Seybold, M. E., and Lindstrom, J. M., 1982, Specificities of antibodies to acetylcholine receptors in sera from myasthenia gravis patients measured by monoclonal antibodies, *Proc. Natl. Acad. Sci. USA* **79:**188–192.

Tzartos, S., Langeberg, L., Hochschwender, S., and Lindstrom, J., 1983, Demonstration of a main immunogenic region on acetylcholine receptors from human muscle using monoclonal antibodies to human receptor, *FEBS Lett.* **158:**116–118.

Vandlen, R. L., Wu, W. C.-S., Eisenach, J. C., and Raftery, M. A., 1979, Studies of the composition of purified *Torpedo californica* acetylcholine receptor and of its subunits, *Biochemistry* **18:**1845–1854.

Verdenhalven, Y., Bandini, G., and Hucho, F., 1982, Acetylcholine receptor-rich membranes contain an endogenous protease regulated by peripheral membrane protein, *FEBS Lett.* **147:**168–170.

Vincent, A., 1980, Immunology of acetylcholine receptors in relation to myasthenia gravis, *Physiol. Rev.* **60:**726–824.

Waksman, G., Oswald, R., Changeux, J.-P., and Roques, B. P., 1980, Synthesis and pharmacological activity on *Electrophorus electricus* electroplaque of photoaffinity labelling derivatives of the non-competitive blockers di- and tri-methisoquin, *FEBS Lett.* **111:**23–28.

Walker, J. W., Lukas, R. J., and McNamee, M. G., 1981a, Effects of thio-group modifications on the ion permeability control and ligand binding properties of *Torpedo californica* acetylcholine receptor, *Biochemistry* **20:**2191–2199.

Walker, J. W., McNamee, M. G., Pasquale, E., Cash, D. J., and Hess, G. P., 1981b, Acetylcholine receptor inactivation in *Torpedo californica* electroplax membrane vesicles. Detection of two processes in the millisecond and second time regions, *Biochem. Biophys. Res. Commun.* **100:**86–90.

Walker, J. W., Takeyasu, K., and McNamee, M. G., 1982, Activation and inactivation kinetics of *Torpedo californica* acetylcholine receptor in reconstituted membranes, *Biochemistry* **21:**5384–5389.

Walkinshaw, M. D., Saenger, W., and Maelicke, A., 1980, Three-dimensional structure of the long neurotoxin from cobra venom, *Proc. Natl. Acad. Sci. USA* **77:**2400–2404.

Walter, P., and Blobel, G., 1981a, Translocation of proteins across the endoplasmic reticulum. II. Signal recognition protein (SRP) mediates the selective binding to microsomal membranes of *in vitro* assembled polysomes synthesizing secretory protein, *J. Cell Biol.* **91:**551–556.

Walter, P., and Blobel, G., 1981b, Translocation of proteins across the endoplasmic reticulum. III. Signal recognition protein (SRP) causes signal sequence-dependent and site-specific arrest of chain elongation that is released by microsomal membranes, *J. Cell Biol.* **91:**557–561.

Walter, P. Ibrahimi, I., and Blobel, G., 1981, Translocation of proteins across the endoplasmic reticulum. I. Signal recognition protein (SRP) binds to *in vitro*-assembled polysomes synthesizing secretory protein, *J. Cell Biol.* **91:**545–550.

Weber, M., David-Pfeuty, T., and Changeux, J.-P., 1975, Regulation of binding properties of the nicotinic receptor protein by cholinergic ligands in membrane fragments from *Torpedo marmorata, Proc. Natl. Acad. Sci. USA* **72:**3443–3447.

Weiland, G., and Taylor, P., 1979, Ligand specificity of state transitions in the cholinergic receptor: Behavior of agonists and antagonists, *Mol. Pharmacol.* **15:**197–212.

Weiland, G., Georgia, B., Wee, V. T., Chignell, C. F., and Taylor, P., 1976, Ligand interactions with cholinergic receptor-enriched membranes from *Torpedo:* Influence of agonist exposure on receptor properties, *Mol. Pharmacol.* **12:**1091–1105.

Weiland, G., Georgia, B., Lappi, S., Chignell, C. F., and Taylor, P., 1977, Kinetics of agonist-mediated transitions in state of the cholinergic receptor, *J. Biol. Chem.* **252:**7648–7656.

Weill, C. L., McNamee, M. G., and Karlin, A., 1974, Affinity-labeling of purified acetylcholine receptor from *Torpedo californica, Biochem. Biophys. Res. Commun.* **61:**997–1003.

Weinberg, C. G., and Hall, Z. W., 1979, Antibodies from patients with myasthenia gravis recognize determinants unique to extrajunctional acetylcholine receptors, *Proc. Natl. Acad. Sci. USA* **76:**504–508.

Wennogle, L. P., and Changeux, J.-P., 1980, Transmembrane orientation of proteins present in acetylcholine receptor-rich membranes from *Torpedo marmorata* studied by selective proteolysis, *Eur. J. Biochem.* **106:**381–393.

Wennogle, L. P., Oswald, R., Saitoh, T., and Changeux, J.-P., 1981, Dissection of the 66,000 dalton subunit of the acetylcholine receptor, *Biochemistry* **20:**2492–2497.

Wise, D. S., Karlin, A., and Schoenborn, B. P., 1979, An analysis by low-angle neutron scattering of the structure of the acetylcholine receptor from *Torpedo californica* in detergent solution, *Biophys. J.* **28:**473–496.

Wise, D. S., Schoenborn, B. P., and Karlin, A., 1981a, Structure of acetylcholine receptor dimer determined by neutron scattering and electron microscopy, *J. Biol. Chem.* **256:**4124–4126.

Wise, D. S., Wall, J., and Karlin, A., 1981b, Relative locations of the β- and δ-chains of the acetylcholine receptor determined by electron microscopy of isolated receptor trimer, *J. Biol. Chem.* **256:**12624–12627.

Witzemann, V., and Raftery, M. A., 1978, Affinity directed crosslinking of acetylcholine receptor polypeptide components in post-synaptic membranes, *Biochem. Biophys. Res. Commun.* **85:**1623–631.

Witzemann, V., Muchmore, D., and Raftery, M. A., 1979, Affinity-directed cross-linking of membrane-bound acetylcholine receptor polypeptides with photolabile α-bungarotoxin derivatives, *Biochemistry* **18:**5511–5518.

Wolosin, J. M., Lyddiatt, A., Dolly, J. O., and Barnard, E. A., 1980, Stoichiometry of the ligand-binding sites in the acetylcholine-receptor oligomer from muscle and from electric organ, *Eur. J. Biochem.* **109:**495–505.

Wu, W. C.-S., and Raftery, M. A., 1979, Carbamylcholine-induced rapid cation efflux from reconstituted membrane vesicles containing purified acetylcholine receptor, *Biochem. Biophys. Res. Commun.* **89:**26–35.

Wu, W. C.-S., and Raftery, M. A., 1981, Functional properties of acetylcholine receptor monomeric and dimeric forms in reconstituted membranes, *Biochem. Biophys. Res. Commun.* **99:**436–444.

Wu, W. C.-S., Moore, H.-P. H., and Raftery, M. A., 1981, Quantitation of cation transport by reconstituted membrane vesicles containing purified acetylcholine receptor, *Proc. Natl. Acad. Sci. USA* **78:**775–779.

Young, A. P., Brown, F. F., Halsey, M. J., and Sigman, D. S., 1978, Volatile anesthetic facilitation of *in vitro* desensitization of membrane-bound acetylcholine receptor from *Torpedo californica, Proc. Natl. Acad. Sci. USA* **75:**4563–4567.

Zingsheim, H. P., Neugebauer, D.-Ch., Barrantes, F. J., and Frank, J., 1980, Structural details of membrane-bound actylcholine receptor from *Torpedo marmorata, Proc. Natl. Acad. Sci. USA* **77:**952–956.

Zingsheim, H. P., Barrantes, F. J., Frank, J., Haenicke, W., and Neugebauer, D.-Ch., 1982a, Direct structural localization of two toxin-recognition sites on an ACh receptor protein, *Nature* **299:**81–84.

Zingsheim, H.-P., Neugebauer, D.-Ch., Frank, J., Haenicke, W., and Barrantes, F. J., 1982b, Dimeric arrangement and structure of the membrane-bound acetylcholine receptor studied by electron microscopy, *Eur. Mol. Biol. Org. J.* **1:**541–547.

# 37

# Structural Distinctions among Acetylcholinesterase Forms

Terrone L. Rosenberry

## I. INTRODUCTION

Acetylcholinesterase (EC 3.1.1.7) is associated primarily with cells involved in cholinergic synaptic transmission, but it is also found in a variety of other neuronal and a few nonneuronal cells, like erythrocytes, in which its function is unclear (Nachmansohn, 1959). The catalytic properties of acetylcholinesterase have been studied intensively for more than 40 years, and details about its specificity and catalytic mechanism and features that distinguish it from the similar enzyme cholinesterase (EC 3.1.1.8) can be found in many reviews (Froede and Wilson, 1971; Rosenberry, 1975; Massoulié and Bon, 1982). Although no role in synaptic transmission for cholinesterase has yet been demonstrated, it is remarkable that both these enzymes exist in multiple forms with striking structural parallels (see Massoulié and Bon, 1982). This review will focus on the protein structures of various acetylcholinesterase forms and on the relationship of these structures to cellular localization.

In 1969, Massoulié and Rieger demonstrated that fresh extracts of eel electric organ contained multiple forms of acetylcholinesterase that could be distinguished by their sedimentation coefficients. This report led to similar investigations of many nerve and muscle cells, and as many as six distinct forms could be identified in extracts of mammalian skeletal muscle and bovine superior cervical ganglia (Bon *et al.*, 1979). Hall (1973) also found multiple forms in extracts of rat diaphragm. One of these forms, a 16 S species, was associated primarily with sections of the diaphragm that are rich in neuromuscular junctions. This 16 S form decreased to low levels on denervation. Hall's observations were confirmed and extended in studies in other laboratories as

*Terrone L. Rosenberry* • Department of Pharmacology, Case Western Reserve University, Cleveland, Ohio 44106.

noted below, but they were particularly significant for two reasons. They suggested that differences in the various acetylcholinesterase forms were related to their cellular localizations, and they indicated that the levels of at least one form that was localized at skeletal neuromuscular junctions could be maintained by innervation of the muscle.

Acetylcholinesterase forms have been divided into two classes on the basis of their structural properties. Asymmetric forms (A) show a high degree of asymmetry by hydrodynamic criteria (Massoulié *et al.*, 1971) and contain an elongated "tail" structure in electron micrographs (Rieger *et al.*, 1973; Dudai *et al.*, 1973; see Rosenberry, 1976). The tail structure is composed of subunits that are largely collagen-like, as described in greater detail below. All other acetylcholinesterase forms have been grouped in the class of globular forms (G) largely on the basis that they are devoid of a collagen-like tail. However, this class is heterogeneous and includes both soluble forms and membrane-bound forms that bind nonionic detergents when extracted into solution (Massoulié, 1980). Molecular weight estimates have indicated that the asymmetric and globular forms are oligomeric assemblies constructed primarily of catalytic subunits (Bon *et al.*, 1976; Bon *et al.*, 1979). Aside from the noncatalytic tail subunits in the asymmetric forms (and an unusual 100K component in torpedo asymmetric enzyme noted below), no other noncatalytic subunits have been detected. The molecular mass of the catalytic subunits (70–80K) appears similar in all mammalian and fish electric organ tissues (Rosenberry, 1982; Massoulié and Bon, 1982), although apparently nonidentical catalytic subunits with a larger mass (100–110K) have been identified in both asymmetric and globular forms from avian muscle and brain (Allemand *et al.*, 1981; Rotundo, 1984). These larger subunits appeared very susceptible to trypsin degradation to subunit masses similar to those found in mammals and electric organs (Allemand *et al.*, 1981). Individual asymmetric and globular forms can be denoted by a subscript which designates the number of catalytic subunits in the oligomer (Bon *et al.*, 1979), thus $G_4$ refers to a tetramer in the globular class.

## II. ASYMMETRIC ACETYLCHOLINESTERASE FORMS CONTAIN COLLAGEN-LIKE TAIL STRUCTURES

Acetylcholine has an extremely high turnover rate with acetylcholinesterase, and this catalytic efficiency has permitted the detection of various sedimenting forms even when they are present in minute amounts in crude cellular extracts. Although asymmetric forms are the primary acetylcholinesterase species in fish electric organs and in certain adult chicken muscles (posterior latissimus dorsi; Sketelj *et al.*, 1978; Silman *et al.*, 1979), these forms tend to be minor acetylcholinesterase components in most mammalian tissue extracts including those from skeletal muscle, their most abundant mammalian source. A detailed analysis of the distribution of acetylcholinesterase forms in rat diaphragm is given below. Despite their low abundance, asymmetric forms are of great interest in skeletal muscle because they are considerably enriched in the acetylcholinesterase localized at skeletal neuromuscular junctions; the 16 S form identified in junctional regions by Hall as noted in the introduction is an asymmetric $A_{12}$ form. This localization contributes to the characteristics of neurotransmission at electric organ synapses and skeletal neuromuscular junctions, an extremely rapid process that is mediated by "nicotinic" acetylcholine receptors. Asymmetric acetylcholinesterase is highly concentrated in such synapses and insures that the lifetime of acetylcholine in the synaptic cleft will be less than a few hundred microseconds (see Rosenberry, 1979). Asymmetric forms are present in smaller amounts in autonomic ganglia, in axons of motor nerves (Fernandez *et al.*, 1979), and in smooth and cardiac muscles and their innervating nerves (Skau and Brimijoin, 1980). In contrast, these forms, while prevalent in the central nervous systems of lower vertebrates, are nearly nonexistent in those of higher vertebrates (Rodriguez-Borrajo *et al.*, 1982; Massoulié and Bon, 1982). The relationship of acetylcholinesterase structure to synaptic function in systems largely devoid of asymmetric forms is less clear, particularly in those systems where acetylcholine synaptic transmission is slower and is generally mediated by "muscarinic" receptors.

Asymmetric acetylcholinesterases are membrane bound in all tissues, and high ionic strength solvents are required for their extraction. This observation suggests that electrostatic forces are responsible for their membrane interaction, and evidence that these forms are localized in the extracellular basement membrane is presented below.

## A. *Acetylcholinesterase from Fish Electric Organs Provides a Structural Model for the Asymmetric Forms*

Examination of the protein structure of a particular acetylcholinesterase form requires the isolation of that form in milligram amounts, and it for this reason that the electric organs of several *Torpedo* species and of the eel *Electrophorus electricus* (Nachmansohn and Lederer, 1939) have provided the only asymmetric forms to date that have been purified and characterized in detail. For example, electric organs from the eel contain approximately 50–100 mg of acetylcholinesterase per kg fresh tissue (the size of organs from one typical eel), while rat diaphragm contains about 0.1 mg of this enzyme per kg muscle. The electric organs are phylogenetically derived from muscle (see Nachmansohn, 1959; Fessard, 1958) and contain large amounts of actin and myosin that are not organized in striated structures (Amsterdam *et al.*, 1975). The unusually large quantity of acetylcholinesterase in these organs derives from the fact that an electric organ cell or electroplaque is a giant syncytium which receives thousands of nerve terminals at synapses on its innervated face, and these synapses are remarkably like skeletal neuromuscular junctions in their structure, pharmacology, and acetylcholinesterase content.

Acetylcholinesterase forms from electric organs were isolated and characterized in several laboratories following the introduction of useful affinity chromatography resins (Dudai and Silman, 1974a; Rosenberry, 1975; Rosenberry *et al.*, 1982). Fresh 1 M NaCl extracts of eel electric organ contained primarily 18 S ($A_{12}$) and 14 S ($A_8$) with a small amount of 8 S ($A_4$) acetylcholinesterase (Massoulié and Rieger, 1969). Following purification from such extracts, about 60–70% of the enzyme corresponded to the $A_{12}$ species and most of the remainder, to the $A_8$ (McCann and Rosenberry, 1977). Virtually no other forms could be detected in fresh extracts from this organ. However, when eel electric organ was removed from the animal and stored, only 11

S ($G_4$) acetylcholinesterase could be detected in subsequent extracts (see Rosenberry, 1975). Treatment of the crude or purified $A_{12}$ and $A_8$ forms with trypsin or other proteases also led to their conversion to a $G_4$ form (Massoulié and Rieger, 1969; Massoulié and Bon, 1982), indicating that the conversion in stored tissue probably arose from an endogenous protease.

## 1. Subunit Assembly in Acetylcholinesterase Forms from Eel Electric Organ

The polypeptide compositions of the eel $A_{12}$ and $A_8$ forms and the eel $G_4$ form derived from autolysis of stored tissue, following their purification by affinity chromatography, are compared in Figure 1. The banding patterns in gels 1–3 refer to the $G_4$ enzyme and gels 4–6, to the $A_{12}$ plus $A_8$ preparation. The compositions revealed by these patterns are complex, but interpretation is aided by noting that disulfide bonding between polypeptides is prevalent. Three states of disulfide reduction can be defined for either enzyme preparation. All disulfide bonds were intact in gels 1 and 4

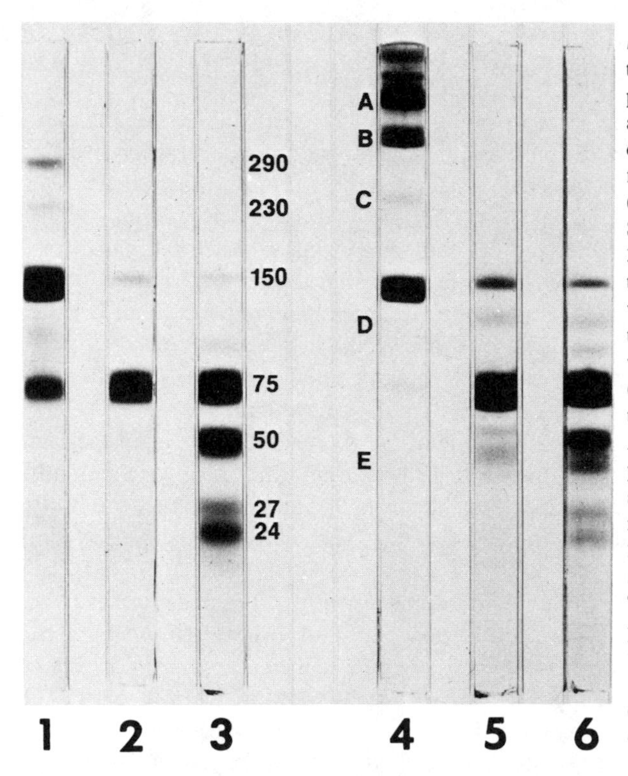


*Figure 1.* Polyacrylamide gel electrophoresis of $G_4$ (gels 1–3) and $A_{12}$ plus $A_8$ (gels 4–6) acetylcholinesterase from eel electric organ in sodium dodecylsulfate demonstrating disulfide bonding between polypeptides (Rosenberry and Richardson, 1977). Samples (25 μg protein) were run on 3.5% cylindrical gels, and polypeptide bands were detected by staining with Coomassie Blue. The bands from the $G_4$ enzyme in gels 1–3 are labeled with their apparent molecular masses (K); additonal bands generated from the $A_{12}$ plus $A_8$ enzyme are labeled A–E. Gels 1 and 4, nonreduced; samples were incubated directly in 1% sodium dodecylsulfate buffer without reducing agent at 50°C for 1 hr. Gels 2 and 5, selective partially reduced; samples were incubated in 100 mM Tris-chloride, 100 mM dithiothreitol, pH 8.0, at 25°C for 30 min, mixed with 500 mM *N*-ethylmaleimide for 15 min, and, following dialysis and concentration, incubated in 1% sodium dodecylsulfate buffer as for gels 1 and 4. Gels 3 and 6, completely reduced; samples were incubated directly in 1% sodium dodecylsulfate buffer containing 100 mM dithiothreitol at 50°C for 1 hr. Further experimental details in Rosenberry and Richardson (1977).

and completely reduced in gels 3 and 6; and only disulfide bonds accessible to dithiothreitol in the *absence* of a denaturant were reduced in gels 2 and 5. The primary species in the $G_4$ enzyme prior to reduction (gel 1) was a 150K dimer of catalytic subunits. In contrast, the $A_{12}$ plus $A_8$ preparation showed two additional major bands A and B that were greater than 300K (gel 4). When the $A_{12}$ and $A_8$ forms in the preparation were individually resolved by sucrose-gradient centrifugation, the A band was found only in $A_{12}$ while the B band was derived only from $A_8$ (McCann and Rosenberry, 1977). Furthermore, if the mixed or resolved $A_{12}$ and $A_8$ preparations were reacted with [$^{32}$P]- or [$^{3}$H]diisopropylfluorophosphate to provide a stoichiometric label for the catalytic subunits prior to electrophoresis, one half of the catalytic subunits were in the 150K dimer band and the remaining one half in the sum of the A and B bands in all cases.

Reduction carried out in the absence of denaturants in gels 2 and 5 resulted in the conversion of all the disulfide-linked catalytic subunit oligomers in bands A, B, and 150K dimer to 75K catalytic subunit monomers. A similar conversion could be obtained at dithiothreitol concentrations as low as 1 mM, after which the reduced sulfhydryl groups could be alkylated with [$^{14}$C]*N*-ethylmaleimide. The amount of radioactive label incorporated into the 75K gel band under these conditions corresponded to 1.1 sulfhydryl groups per catalytic subunit for the $G_4$ enzyme and 1.0 for the $A_{12}$ enzyme (Rosenberry, 1975; P. Barnett and T. L. Rosenberry, unpublished observations). The reduction was quite selective for *intersubunit* disulfide bonds, as is noted below. From these data, it can be concluded that each catalytic subunit in bands A, B, and 15OK dimer is linked to the oligomeric structure through a single disulfide bond.

After complete disulfide reduction, many of the polypeptides obtained from the $G_4$ enzyme (gel 3) and the $A_{12}$ plus $A_8$ enzyme mixture (gel 6) were similar. In addition to the 75K intact catalytic subunit, both gels showed 50K, 27K and 23.5K bands that were previously identified as catalytic subunit fragments with the $G_4$ enzyme (Rosenberry *et al.*, 1974). The smaller amounts of these fragments relative to the 75K band in gel 6 suggested that the proteolytic degradation which generated the fragments was minimized in the rapidly extracted and purified $A_{12}$ plus $A_8$ preparation. A new banding region E was apparent on gel 6 but absent on gel 3, and the polypeptides in this region corresponded to the noncatalytic tail subunits that are discussed below. The formation of 50K, 27K, and 23.5K fragments appears unique to enzyme from eel electric organ and was not apparent, for example, in acetylcholinesterase purified from *Torpedo* electric organ or human erythrocytes (Lee *et al.*, 1982a; Rosenberry and Scoggin, 1984). Thus, this fragmentation appeared to have little functional significance. However, it did provide useful criteria for assessing the selectivity of the reduction in the absence of denaturants. No fragments (linked by *intrasubunit* disulfides) were released by this reduction, and the distribution of [$^{14}$C]*N*-ethylmaleimide label following the reduction among the fragments indicated that very few intrasubunit half-cystine residues were labeled (Rosenberry, 1975; Rosenberry and Richardson, 1977).

The nonreduced oligomeric structures A, B, and 150K dimer were isolated, and complete reduction of the isolated samples demonstrated that the noncatalytic subunit bands E were associated only with A and B prior to reduction (Rosenberry and Rich-

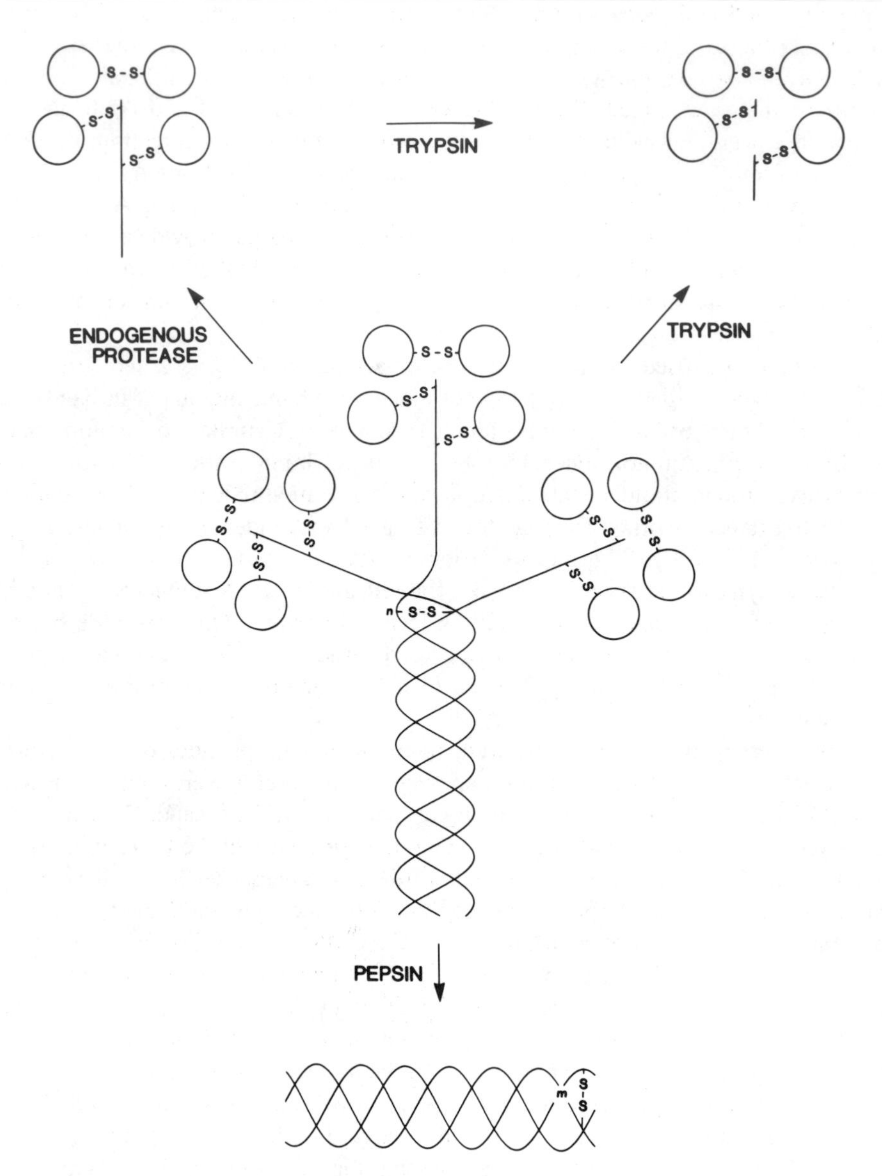


*Figure 2.* Schematic representations of $A_{12}$ eel acetylcholinesterase and the patterns of degradation shown by three different proteases (Rosenberry *et al.*, 1980). The $A_{12}$ form can be degraded to $G_4$ forms either by an endogenous protease in stored electric organ tissue or by trypsin. The $G_4$ forms are not quite equivalent in that trypsin also cleaves between the disulfide linkages to the tail subunit. Pepsin degrades the catalytic subunits and the noncollagenous domain of the tail subunits to give triple helical tail subunit fragments.

ardson, 1977). This information together with that in the preceding paragraphs suggested the schematic structure shown in Figure 2 for the $A_{12}$ eel enzyme. The basic structure of 12 75K catalytic subunits (circles) arranged in three tetrameric groups associated with a tail component was originally proposed by Rieger *et al.* (1973) to account for the apparent molecular weights of the observed $A_{12}$, $A_8$, and $A_4$ eel enzyme forms. The data outlined above indicate the following additional points. Within each tetramer, two catalytic subunits are directly linked by a single disulfide bond while the remaining two are each covalently attached by a single disulfide bond to one noncatalytic tail subunit. Disulfide bonds also crosslink the tail subunits directly to each other. Oligomer A observed prior to disulfide reduction thus corresponds to six catalytic and three tail subunits linked through disulfide bonds. Proteolytic degradation of the $A_{12}$ enzyme by an endogenous protease or trypsin releases $G_4$ forms, as indicated in Figure 2, but leaves residual asymmetric forms; loss of one $G_4$ generates an $A_8$ form, while removal of two $G_4$ tetramers leaves an $A_4$ (Massoulié *et al.*, 1971). The observation that oligomer B is derived from the $A_8$ form indicates that it corresponds to four catalytic, two tail, and one residual tail subunit (McCann and Rosenberry, 1977). Although Figure 2 is schematic, it is noteworthy that a similar structure for stored preparations of the $A_{12}$ enzyme has been observed by electron microscopy (Cartaud *et al.*, 1975).

## 2. *The Tail Subunits from $A_{12}$ and $A_8$ Eel Acetylcholinesterase*

Figure 2 represents each noncatalytic tail subunit of the $A_{12}$ eel enzyme as two-domain polypeptides, a domain that interacts with a corresponding segment in the other two tail subunits to form a collagen-like triple helix, and a noncollagen-like region that includes the disulfide bonds to the catalytic subunits. The eel $G_4$ enzyme generated from the $A_{12}$ and $A_8$ forms by endogenous protease retained most, if not all, of the noncollagenous domain of the tail subunits, as indicated in Figure 2. Selective reduction of this $G_4$ enzyme in nondenaturants followed by alkylation with [$^{14}$C]*N*-ethylmaleimide not only labeled one sulfhydryl group on the catalytic subunits, as noted above, but it also labeled a peptide with a mass of about 8K (Barnett and Rosenberry, 1978). This peptide could be isolated by gel-exclusion chromatography on Sepharose CL-6B in the presence of 6 M guanidine hydrochloride, and its labeling stoichiometry was two sulfhydryls per $G_4$ tetramer (Rosenberry, 1975). In contrast, the $G_4$ generated by trypsin did not contain an 8K fragment, and a nonreduced gel profile like that in gel 1 (Figure 1) showed about equal amounts of 150K dimer and 75K monomer (T. L. Rosenberry, P. Barnett, and C. Mays, unpublished observations; Anglister and Silman, 1978).

The collagen-like domains of the noncatalytic tail subunits were isolated following pepsin digestion, as indicated in Figure 2. Pepsin digestion of tissue or tissue extracts is frequently employed to solubilize collagen-like proteins or to selectively degrade noncollagenous regions in these proteins, and treatment of eel $A_{12}$ plus $A_8$ acetylcholinesterase with 20–100 μg/ml pepsin in 0.5 M acetic acid at 15°C for 6 hr generated several polypeptide fragments (Mays and Rosenberry, 1981). Polyacrylamide gel electrophoresis in sodium dodecylsulfate of the pepsin-digested sample followed by Coomassie Blue staining revealed only two discrete polypeptide fragment bands larger than pepsin. Both the larger band of 72K, an apparent trimer denoted $F_3$, and the smaller band of 48 K, $F_2$, represented polypeptides linked by disulfide bonds, as both bands were converted to a rather broad band $F_1$ of about 24K following exposure to dithiothreitol. Two observations indicated that these polypeptide fragments were not derived from catalytic subunits; the fragments were absent in pepsin digests of the eel $G_4$ enzyme, and the fragments were not labeled by exposure of the $A_{12}$ and $A_8$ preparation to the catalytic site reagent [$^3$H]diisopropylfluorophosphate prior to pepsin digestion. Gel exclusion chromatography of the pepsin digest of $A_{12}$ plus $A_8$ on Sepharose CL-6B in 1 M NaCl demonstrated that the F fragments eluted primarily as aggregates in the void volume where they were well separated from residual small catalytic subunit fragments. The circular dichroism spectra and mean residue ellipticities of the aggregate pool and of acid-soluble calf skin collagen were virtually identical and consistent with literature values for collagen triple helices. Thus, prior to denaturation the F fragments are predominantly triple helical.

The intersubunit disulfide linkages in the collagen-like domain $F_3$ appear close to the end(s) of the pepsin-resistant fragment polypeptide chains. Continued pepsin action for longer periods or at higher temperatures resulted in the appearance of $F_2$ dimers and finally $F_1$ monomers as the predominant species on sodium dodecylsulfate gels prior to reduction, without significant change in the size of the reduced $F_1$ fragments (Mays and Rosenberry, 1981). Pepsin conversion of $F_3$ to $F_2$ was particularly accelerated between 25 and 30°C, suggesting that the structure in the disulfide linkage region undergoes thermal destabilization in this temperature range. The question of whether the three tail subunits that give rise to $F_1$ fragment polypeptides are identical remains open.

The isolation of the intact noncatalytic tail subunits has been achieved following dissociation of the $A_{12}$ and $A_8$ enzyme subunits by reduction in a denaturing solvent (Rosenberry *et al.*, 1980, 1982). Approximately five sulfhydryl groups per tail subunit were labeled with [$^{14}$C]*N*-ethylmaleimide under these conditions. The isolated tail subunit pool was eluted from a Sepharose Cl-6B column in 6 M guanidine hydrochloride at a volume that corresponded to a 38,000-mol. wt. polypeptide relative to protein standards. Gel analyses in sodium dodecylsulfate indicated that this pool contained a predominant band corresponding to 41K (according to noncollagen standards) or 30K (according to collagen standards). Since the tail subunits contain both collagen-like and non-collagen-like domains, a precise molecular weight is difficult to assign from these data.

Data on the amino acid compositions of the noncatalytic tail subunits of eel $A_{12}$ and $A_8$ acetylcholinesterase is summarized in Table 1. The compositions of the pepsin-resistant aggregate (the collagen-like domain) and the 8K residual tail subunit fragment isolated from the eel $G_4$ enzyme (the noncollagenous domain) are compared with the composition of the intact tail subunit pool. The amino acid composition for intact subunits can be estimated as an appropriate weighted average of the compositions of the two isolated domains. This estimate is in good agreement with the observed composition, suggesting that no distinctive regions of the intact subunits are omitted

*Table 1. Comparison of the Amino Acid Mole Percentages of Intact and Fragmented Tail Subunits of $A_{12}$ Plus $A_8$ Acetylcholinesterase from Eel Electric Organ*[a]

| Amino acid | Collagen-like domain | Noncollagenous domain | Intact subunits | |
|---|---|---|---|---|
| | | | Observed | Calculated |
| HYP | 5.0 | 0.0 | 4.2 | 3.8 |
| ASP | 5.0 | 11.4 | 6.7 | 6.5 |
| THR | 2.3 | 4.2 | 2.4 | 2.7 |
| SER | 5.9 | 4.5 | 5.2 | 5.6 |
| GLU | 9.1 | 11.9 | 9.8 | 9.7 |
| PRO | 8.7 | 13.8 | 11.0 | 9.9 |
| GLY | 27.2 | 6.0 | 23.3 | 22.3 |
| ALA | 3.4 | 5.7 | 3.5 | 3.9 |
| VAL | 4.3 | 6.9 | 4.7 | 4.9 |
| MET | 2.8 | 2.6 | 2.0 | 2.8 |
| ILE | 2.2 | 4.3 | 2.4 | 2.7 |
| LEU | 5.2 | 9.9 | 6.5 | 6.3 |
| TYR | 1.7 | 1.8 | 1.5 | 1.7 |
| PHE | 1.2 | 4.0 | 2.0 | 1.9 |
| HYL | 5.3 | 0.0 | 4.6 | 4.1 |
| LYS | 2.3 | 4.5 | 2.6 | 2.8 |
| HIS | 2.2 | 1.1 | 1.8 | 1.9 |
| ARG | 4.1 | 5.0 | 4.8 | 4.3 |
| Half-CYS | | 2.2 | 1.4 | |

[a] Uncorrected 20-hr hydrolyses at 110°C in 6 N-hydrochloric acid containing 80 mM mercaptoethanol under argon. The samples corresponding to each domain are described in the text. The theoretical amino acid composition was calculated from the data in columns 2 and 3 assuming a 32,000-dalton intact subunit, composed of a 24,000-dalton collagen-like domain (100 mean residue weight) and an 8000-dalton noncollagenous domain (110 mean residue weight). See Rosenberry *et al.* (1980) for details.

in the two domains. This suggestion is supported by the satisfactory agreement of the molecular weight estimates of the intact subunits with the sum of the estimated molecular weights of these two domains. About 35% of the proline residues and 70% of the lysine residues in the collagen-like domain were hydroxylated, a modification typical of collagen-like polypeptides. Furthermore, basic amino acid hydrolysates revealed that most of the hydroxylysine residues were linked to 2-*O*-α-D-glucopyranosyl-*O*-β-D-galactopyranose (Rosenberry *et al.*, 1980), a characteristic of basement membrane collagens. The 27% glycine content of the pepsin-resistant fragments is somewhat higher than that of the intact tail subunits, but this percentage is sufficiently below the 33% expected for a complete collagen triple helical structure to indicate the presence of additional short noncollagenous sequences within the collagen-like domain.

The data presented above on the structure of eel acetylcholinesterase forms are in excellent agreement with the observations of Anglister and Silman (1978). These workers as well as others (Bon and Massoulié, 1978; see Rosenberry, 1982) have also examined the degradation of eel $A_{12}$ and $A_8$ forms by bacterial collagenase. Collagenase degradation occurred in two steps, with the initial formation of modified forms $A'_{12}$, $A'_8$, and $A'_4$. The sedimentation coefficient of each of these forms was 2–3 S larger than the corresponding A form but the Stokes radius was smaller, indicating that the structures had become much less asymmetric and had lost 60–100K of molecular mass. Prolonged digestion with collagenase or digestion at higher temperatures resulted in extensive formation of $G_4$ enzyme from the A′ forms. Only the noncatalytic tail subunits appeared cleaved in these steps, and 50–60% of the initial hydroxyproline and hydroxylysine content of the $A_{12}$ plus $A_8$ enzyme was lost in the isolated $A'_{12}$ plus $A'_8$ preparation (Anglister and Silman, 1978). These observations indicated that collagenase had removed a major portion of the tail structure corresponding to the collagenous domains of the tail subunits. Similar conclusions were reached following collagenase digestion of purified torpedo electric organ $A_{12}$ plus $A_8$ acetylcholinesterase (Lee and Taylor, 1982). Although neither intact asymmetric forms nor globular forms from eel or torpedo are dissociated by reduction under nondenaturing conditions (Rosenberry, 1975; Bon and Massoulié, 1976), Lee and Taylor (1982) further demonstrated that the $A'_{12}$ plus $A'_8$ torpedo enzyme was dissociated to a $G_4$ form following reduction under these conditions. Thus, the first stage of collagenase digestion appears to remove much of the collagenous region below the intersubunit disulfide linkages in the tail structure in Figure 2, while continued digestion degrades the structure containing these linkages and releases $G_4$ forms similar to those generated by the endogenous protease in eel electric organ tissue.

### 3. *Subunit Assembly of Asymmetric Forms in Torpedo Electric Organ Appears Slightly More Complex Than in Eel*

Although the hydrodynamic properties of torpedo asymmetric forms are very similar to those of eel, the subunit composition and assembly of the asymmetric forms in *Torpedo californica* were more complex (Lee *et al.*, 1982a; Lee and Taylor, 1982). A third type of subunit, a 100K species that was neither catalytic nor collagen-like but quite sensitive to trypsin, was present in both $A_{12}$ and $A_8$ forms. This subunit appeared to substitute in a random, nonstoichiometric way for only those catalytic subunits that were linked to the tail structure by disulfide bonds. As in the eel asymmetric forms, only about one half of the catalytic subunits could be disulfide-linked to the tail structure; but in *Torpedo,* about one third of these catalytic subunit sites were substituted with the 100K polypeptide. Furthermore, intersubunit disulfide linkages among catalytic and 100K subunits in the *Torpedo* enzyme appeared more complex than the single linkages indicated in Figure 2. The function of the 100K polypeptide is unknown, but Lee and Taylor (1982) suggest that it may be analogous to a procollagen or noncollagenous basement membrane peptide that is largely replaced by catalytic subunits during biosynthetic assembly.

## B. *Asymmetric Acetylcholinesterase Forms in Other Tissues Have Hydrodynamic and Aggregation Properties Similar to the Electric Organ Forms*

Although asymmetric acetylcholinesterases from chicken and mammalian tissues have not been isolated, several close similarities in hydrodynamic and aggregation properties suggest that structural information obtained for asymmetric electric organ forms is directly applicable to chicken and mammalian forms. Correspondences of these properties thus provide extremely useful criteria for the characterization of tiny amounts of native asymmetric forms in partially purified or even crude solutions. The molecular weights deduced from hydrodynamic data for $A_{12}$, $A_8$, and $A_4$ forms, and a collagenase-generated $G_4$ form from bovine superior cervical ganglia were very similar to those for corresponding forms from eel and torpedo electric organs (Bon *et al.*, 1979). Inspection of the molecular weights indicated a tail structure with a molecular weight slightly in excess of 100,000, again consistent with the electric organ forms. The sedimentation coefficients of asymmetric forms and their collagenase-derived products in rat muscle (Younkin *et al.*, 1982) and in chicken muscle (following trypsin modification; Allemand *et al.*, 1981) also paralleled those of the electric organ forms (Massoulié and Bon, 1982).

The solubility and aggregation properties of asymmetric forms from chicken and mammalian tissues also follow those of the electric organ forms. High ionic strength (1 M NaCl) is necessary to extract asymmetric forms from tissues. Extraction was sometimes improved by inclusion of nonionic detergents (Hall, 1973; Dudai and Silman, 1974b), substitution of 2 M $MgCl_2$ for 1 M NaCl (Dudai and Silman, 1974b; Lwebuga-Mukasa *et al.*, 1976; Bulger *et al.*, 1982), or inclusion of EDTA (Rodriguez-Borrajo *et al.*, 1982). Extracted asymmetric forms from eel electric organ have long been known to undergo reversible aggregation at an ionic strength less than about 0.3 (Millar and Grafius, 1965; Rosenberry, 1975), and typical 50–90 S aggregates have been characterized by electron microscopy (Cartaud *et al.*, 1978). Aggregation appeared to involve exclusively the collagenous domains in tail structures of the asymmetric forms, and, thus, it is not surprising that neither the $G_4$ forms nor the A′ forms generated by collagenase showed any tendency to aggregate (Bon and Massoulié, 1978). Acylation of the asymmetric forms with maleic anhydride (Dudai and Silman, 1973) or acetic anhydride (Cartaud *et al.*, 1978) also totally blocked aggregation. Chicken and mammalian asymmetric forms precipitated rather than aggregated from most extracts when the ionic strength was reduced (Bon *et al.*, 1979). However, this precipitation appeared to arise from other components in the crude extracts; purified eel asymmetric forms precipitated in identical fashion when added to these extracts. Collagenase and acetic anhydride blocked precipitation of asymmetric forms from bovine superior ganglia and rat and chicken skeletal muscles under the same conditions in which they abolished aggregation of purified electric organ asymmetric forms (Bon *et al.*, 1979; Massoulié and Bon, 1982).

## C. *Asymmetric Acetylcholinesterases Appear to Be Localized in the Extracellular Basement Membrane Matrix in Skeletal Muscle*

Considerable evidence indicates that the *in situ* membrane interaction of asymmetric forms arises from their collagen-like tail structures. Collagenase and maleic anhydride, two agents that blocked the tail-mediated aggregation of extracted asymmetric forms, solubilized acetylcholinesterase from eel electric organ tissue (Dudai and Silman, 1974b). Collagenase as well as other proteases also were efficient in digesting the extracellular basement membrane at rat and frog skeletal neuromuscular junctions, and during this process released most of the junctional acetylcholinesterase

(Hall and Kelly, 1971; Betz and Sakmann, 1973; Sketelj and Brzin, 1979). Dudai and Silman (1974b) noted these reports and suggested that the collagen-like subunits are responsible for an *in situ* localization of the asymmetric forms in the extracellular basement membrane matrix at these synapses. This suggestion was supported by Lwebuga-Mukasa *et al.* (1976), who reported a torpedo membrane fraction that was enriched in both acetylcholinesterase and hydroxyproline. It was also supported by striking histochemical evidence involving frog neuromuscular junctions obtained by McMahan *et al.* (1978), who demonstrated that a significant fraction of the junctional acetylcholinesterase remains associated with residual basement membrane following destruction of the nerve and muscle cells.

Basement membranes do not contain phospholipid but are rich in both collagen-like and noncollagenous proteins (Kefalides *et al.*, 1979). Some of the noncollagenous proteins, like laminin (Kleinman *et al.*, 1981), may bind directly to basement membrane collagens. In other cases, disulfide bonds may link these two types of protein. For example, a citrate-soluble fraction of lens capsule basement membrane was composed of two components, collagenous filament structures and noncollagenous globular structures, that apparently could be dissociated by disulfide bond reduction (Olsen *et al.*, 1973). Basement membranes, in addition, contain sulfated glycosaminoglycan (Kanwar *et al.*, 1981), and Bon *et al.* (1978) have suggested that glycosaminoglycans participate in the aggregation and basement membrane attachment of asymmetric acetylcholinesterases. These workers found that partially purified $A_{12}$ enzyme from eel electric organ gradually lost its ability to aggregate and that various polyanions, including *in situ* levels of endogenous chondroitin sulfate from the electric organ, could restore aggregation. Vigny *et. al.* (1983) report no disaggregation of aggregated $A_{12}$ and $A_8$ AChE by chondroitinase ABC but a partial disaggregation by heparinase that was partly overcome by the addition of heparin sulfate proteoglycan. The eel $A_{12}$ enzyme purified to homogeneity by affinity chromatography in our laboratory has not been observed to lose its aggregating ability, and further work is required to demonstrate that these purified preparations retain sulfated glycosaminoglycans linked either covalently or noncovalently to acetylcholinesterase tail subunits.

## III. GLOBULAR ACETYLCHOLINESTERASE OCCURS AS SOLUBLE AND AMPHIPATHIC FORMS

In contrast to the asymmetric forms with their distinct collagen-like tail structures, the class of globular acetylcholinesterases is defined less specifically than all other acetylcholinesterase forms that lack a collagen-like tail. Globular forms are more widely distributed than asymmetric forms and can be further subclassified into two groups: forms that interact with nonionic detergents following extraction and presumably correspond to integral membrane proteins *in situ,* and forms that do not bind detergent and presumably are soluble proteins (see Massoulié, 1980). Distinction between these groups was drawn in striking fashion by Lazar and Vigny (1980), who examined the hydrodynamic properties of forms extracted from a neuroblastoma multisympathetic ganglion hybrid cell line T28. They found cellular $G_4$ and $G_1$ forms that interacted with Triton X-100 and secreted $G_4$ and $G_1$ forms that had identical sedimentation coefficients in the presence or absence of Triton X-100 and, thus, were presumed to bind no detergent. Soluble globular acetylcholinesterase forms appear to be secreted by nerve and muscle cells in general, but they also can be generated *in vitro* by the digestion of either asymmetric or detergent-binding globular forms with proteases (see Massoulié and Bon, 1982). The soluble $G_4$ form produced from asymmetric forms by endogenous proteases in stored eel electric organ tissue was the first acetylcholinesterase to be isolated and characterized in detail (see above and Rosenberry, 1975). Neither this $G_4$ form nor the parent asymmetric forms bound radiolabeled detergent (Millar *et al.*, 1978) or showed shifts in sedimentation coefficient on introduction of nonionic detergents (Bon *et al.*, 1978). Soluble $G_2$ and $G_1$ forms from eel also have been generated by degradation involving sonication or extensive autolysis in partially purified extracts (Bon and Massoulié, 1976), and trypsin has been shown to convert asymmetric forms from bovine superior cervical ganglia into soluble $G_4$, $G_2$, and $G_1$ forms (Vigny *et al.*, 1979). The generation of soluble forms from detergent-binding forms by proteases is described below.

The detergent-binding globular acetylcholinesterase forms have been referred to as a "hydrophobic" protein class (Massoulié and Bon, 1982) and as an "amphipathic" class (Rosenberry and Scoggin, 1984). In this review, we will use the term amphipathic, which has been applied to proteins with distinct hydrophilic and hydrophobic domains that can be separated by protease cleavage (Macnair and Kenny, 1979; Dailey and Strittmatter, 1981; Frielle *et al.*, 1982). The hydrophobic domain anchors the protein in a phospholipid membrane while the hydrophilic domain generally is oriented exclusively on one side of the membrane (Engelman and Steitz, 1981). Integral membrane proteins that have been termed amphipathic exhibit several common characteristics that can serve as criteria for this class of proteins: (1) the intact protein requires detergent for extraction from membranes, binds to detergent micelles only, aggregates but does not generally precipitate when detergent is removed, and can be reconstituted into liposomes; (2) the larger hydrophilic domain is oriented on one side of the membrane only, is solubilized by protease cleavage generally with full retention of associated enzymatic activities, and following protease cleavage shows no aggregation or detergent- or liposome-binding properties; and (3) the hydrophobic domain is quite small (generally $< 10{,}000$ daltons) and is located at the N- or C-terminus of the primary sequence. In the following section, we examine several characteristics of purified human erythrocyte acetylcholinesterase, the first detergent-binding acetylcholinesterase form to which these criteria have been applied in detail.

### A. *Human Erythrocyte Acetylcholinesterase is an Amphipathic Form*

Human erythrocyte membranes are a relatively rich source of mammalian acetylcholinesterase because this enzyme comprises about 0.01% of the protein in erythrocyte membranes. Enzyme activity is retained quite well in outdated erythrocytes from blood banks, and these outdated stocks are a convenient starting material for enzyme purification. Although its function in erythrocytes is unclear, the erythrocyte enzyme appears to be an excellent model for other amphipathic acetylcholinesterase forms. Many features of this enzyme are also shown by globular forms in other tissues, and the erythrocyte enzyme has been observed to have close structural relationships to several other mammalian acetylcholinesterases by immunochemical criteria (Fambrough *et al.*, 1982; Rosenberry and Sheldon, unpublished observations).

Previous studies by Brodbeck and his colleagues have established that detergents are required for the extraction of human erythrocyte acetylcholinesterase from erythrocyte membranes and that the extracted enzyme corresponds to a $G_2$ form (Ott *et al.*, 1975, 1982). These workers purified the enzyme essentially to homogeneity by affinity chromatography and showed that the purified enzyme aggregated when depleted of detergents. The aggregates consisted of discrete oligomers of 3–7 protomeric $G_2$ forms which interconverted quite slowly on storage (Ott *et al.*, 1975; Ott and Brodbeck, 1978). Polyacrylamide gel electrophoresis in sodium dodecylsulfate under the three states of protein disulfide reduction shown in Figure 1 revealed that the purified enzyme was a 150–160K dimer prior to disulfide reduction, and that reduction under either nondenaturing or denaturing conditions converted the enzyme to 70–80K monomers (Bellhorn *et al.*, 1970; Ott *et al.*, 1975; Grossmann and Leifländer, 1975; Niday *et al.*, 1977; Rosenberry and Scoggin, 1984). The subunit assembly pattern, thus, is quite similar to that previously observed for the $G_4$ acetylcholinesterase from eel electric organ in gels 1–3 of Figure 1 except that in most preparations no evidence of subunit fragmentation following complete reduction was seen (Rosenberry and Scoggin, 1984). Alkylation of the enzyme reduced in the absence of denaturants with [$^{14}$C]*N*-ethylmaleimide resulted in 1.7 labeled groups per catalytic subunit, somewhat larger than the incorporation values noted above for the eel enzyme (Rosenberry and Scoggin, 1984). However, it appeared somewhat more difficult to direct the reduction exclusively to the intersubunit disulfide bond in the erythrocyte enzyme, and the incorporation data were considered consistent with only a single sulfhydryl group on each catalytic subunit involved in intersubunit disulfide bonding (as depicted in Figure 2 for the eel enzyme). Of greater importance, no additional peptide like the 8K tail subunit fragment noted above was labeled under these conditions. This observation indicates that the catalytic subunits in the $G_2$ erythrocyte enzyme are linked by a single disulfide directly to each other rather than through two disulfides and an intervening peptide.

Reduction of the erythrocyte enzyme in the absence of denaturants resulted not only in the generation of monomers on sodium dodecylsulfate gels but also in the dissociation of the native $G_2$ dimer to active $G_1$ monomers when a nonionic detergent was present (Rosenberry and Scoggin, 1984). Dissociation under these conditions is observed with all $G_2$ acetylcholinesterase forms investigated to date but does not occur with native tetrameric or higher oligomeric forms (Rosenberry, 1975; Bon and Massoulié, 1976; Vigny *et al.*, 1979; Lee *et al.*, 1982b). The hydrodynamic properties of the erythrocyte $G_2$ and $G_1$ forms in the presence of Triton X-100 are shown in Table 2, along with those of the papain-digested $G_2$ enzyme discussed below. Massoulié and his colleagues have stressed that shifts in sedimentation coefficients and Stokes radii ($R_s$) in the presence and absence of detergents permit identification of detergent-binding acetylcholinesterase forms (Lazar and Vigny, 1980; Massoulié and Bon, 1982), and both the $G_2$ and $G_1$ erythrocyte enzymes showed large shifts in these parameters when Triton X-100 was added to purified preparations previously depleted of detergent

*Table 2. Hydrodynamic and Detergent-Binding Properties of Human Erythrocyte Acetylcholinesterase*[a]

| Form | $S_{20,w}$ | $R_s$ (nm) | Detergent-binding ($g$ per $g$ protein) | Protein molecular weight |
|---|---|---|---|---|
| $G_2$ | 6.5 ± 0.2 | 7.4 ± 0.2 | 0.63 ± 0.03 | 160,000 ± 8000 |
| $G_1$ | 4.3 ± 0.1 | 6.2 ± 0.1 | 0.75 ± 0.07 | 85,000 ± 6000 |
| Papain-digested $G_2$ | 6.8 ± 0.2 | 5.8 ± 0.1 | | 157,000 ± 5000 |

[a] Purified enzyme was obtained as a $G_2$ form, and samples were converted to $G_1$ by reduction under nondenaturing conditions. Papain digestion was conducted as outlined in Figure 3. Triton X-100 (1.0%) was included in measurements of sedimentation coefficients ($S_{20,w}$) and Stokes radii ($R_s$) with the nondigested enzymes. $S_{20,w}$ values were corrected for detergent binding. Triton X-100 was deleted in the $S_{20,w}$ and $R_s$ estimates for the papain-disaggregated enzyme, but these estimates did not differ significantly from those obtained in the presence of 1% Triton X-100 ($S_{20,w}$ = 6.8, $R_s$ = 5.9 ± 0.1). Tabulated means were averages of 3–5 determinations and are listed with standard errors. Detergent binding was measured in 0.1% [$^3$H]Triton X-100 by an affinity chromatography procedure, and values were estimated by a least squares regression analysis for several fractions. Molecular weights were estimated from the Svedberg equation assuming a partial specific volume for protein of 0.715 (see Bon *et al.*, 1976) and do not include bound detergent. See Rosenberry and Scoggin (1984) for further experimental details.

(Rosenberry and Scoggin, 1984). Both form similar aggregates in the absence of detergents that are characterized by average $R_s$ values of 9.8 nm, in contrast to the values of 7.4 nm for $G_2$ and 6.2 nm for $G_1$ in the presence of Triton X-100 in Table 2. Disaggregation by Triton X-100 occurred over a narrow range of detergent concentrations near the critical micelle concentration of 0.016% (Wiedmer *et al.*, 1979; Rosenberry and Scoggin, 1984), and this observation suggested that the erythrocyte enzyme was interacting with Triton X-100 only when it formed micelles. Binding studies with [$^3$H]Triton X-100 confirmed that significant amounts of Triton X-100 interacted with the enzyme only above the detergent critical micelle concentration (Rosenberry and Scoggin, 1984). The amounts of [$^3$H]Triton X-100 that bound to both the $G_2$ and the $G_1$ enzyme are listed in Table 2 and are consistent with previously reported binding values for proteins that interact only with Triton X-100 micelles (see Rosenberry and Scoggin, 1984). Furthermore, the $G_1$ enzyme bound about the same amount of detergent per mass of protein as the $G_2$ enzyme, indicating that each *subunit* in both the $G_2$ and $G_1$ forms interacts with a single Triton X-100 micelle.

Membrane-bound acetylcholinesterase activity in human erythrocytes is localized exclusively on the extracellular surface (see Steck, 1974). The one criterion typical of amphipathic membrane proteins that erythrocyte acetylcholinesterase meets only partially is its solubilization from the membrane by proteases. A survey of proteases to determine those most effective in cleaving the hydrophilic and hydrophobic domains of the erythrocyte enzyme revealed that papain and pronase completely converted the purified aggregate to a nonaggregating 6.8 S form (Dutta-Choudhury and Rosenberry, 1984). These two proteases were also most efficient in solubilizing the enzyme from intact red cells in the absence of detergent. However, generally less than 10% of the enzyme was solubilized under these conditions, perhaps because the cleavage site is relatively inaccessible in the enzyme bound to the erythrocyte membrane. Cleavage of the two domains at high concentrations of the purified aggregate occurred most readily with papain, and the papain-digested enzyme was investigated in detail (Dutta-

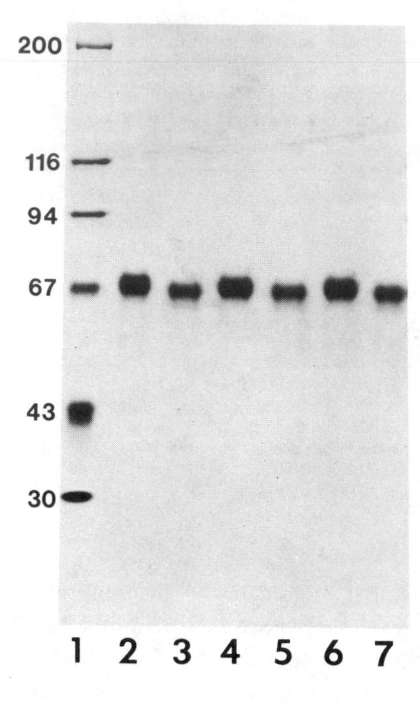


*Figure 3.* Polyacrylamide gel electrophoresis in sodium dodecylsulfate of human erythrocyte acetylcholinesterase before and after papain digestion (Dutta-Choudhury and Rosenberry, 1984). Papain (2.0 mg) linked to Sepharose CL-4B (1.0 ml) was activated with cysteine, washed, and mixed with acetylcholinesterase (2.9 mg) in 10.5 ml (5 mM sodium phosphate, 1 mM edrophonium, pH 7) at 5°C for 45 min. The supernatant was concentrated and chromatographed on Sepharose CL-4B, and fractions containing digested acetylcholinesterase were concentrated. Samples (0.2 μg protein) were reduced with 40 mM dithiothreitol in sodium dodecylsulfate and run on 5–13% gradient slab gels, and polypeptide bands were detected by silver staining. Lane 1, polypeptide standards (myosin, β-galactosidase, phosphorylase b, bovine serum albumin, ovalbumin, and carbonic anhydrase with molecular masses in kilodaltons indicated in order from top). Lanes 2, 4, and 6, triplicate runs of nondigested acetylcholinesterase. Lanes 3, 5, and 7, triplicate runs of the digested enzyme. Apparent molecular masses were 72K for the nondigested and 70K for the digested enzymes. Further experimental details in Dutta-Choudhury and Rosenberry (1984).

Choudhury and Rosenberry, 1984). Papain digestion was accompanied by relatively small losses in enzyme activity (10–50%), particularly when edrophonium chloride, a competitive inhibitor with high affinity for the active site, was included. The sedimentation coefficient and Stokes radius of the digested enzyme (Table 2) were not altered in the presence of 1% Triton X-100 within experimental error, confirming that it corresponded only to the hydrophilic domain. Furthermore, papain digestion blocked the reconstitution of $G_2$ enzyme into liposomes (B. H. Kim and T. L. Rosenberry, unpublished observations). Molecular weight estimates in Table 2 indicated that the papain-digested enzyme was less than 5% smaller than the detergent-binding $G_2$ form and, thus, the hydrophobic domain removed by papain appears quite small. The location of this domain in the primary sequence of the catalytic subunit was investigated by gel electrophoresis in sodium dodecylsulfate. When precautions were taken to prevent further protease degradation in the sodium dodecylsulfate, papain-digested samples fully reduced in this denaturant migrated slightly faster than nondigested $G_2$ enzyme reduced and run in parallel (Figure 3). The apparent molecular mass difference in the catalytic subunits in the two samples was about 2K. A quantitative correlation of the faster migrating band with nonaggregating enzyme was made by fractionating papain-digested enzyme on Sepharose CL-4B in the absence of detergent, labeling the digested pool ($R_s$ = 5.8 nm) with [$^3$H]diisopropylfluorophosphate, and confirming that the labeled enzyme corresponded both to a nonaggregated 6.8 S component on a sucrose gradient in the absence of detergent and to the 70K band on the gel. The apparent mass difference between the nondigested and digested catalytic subunits sets an upper limit for the mass of the hydrophobic domain removed by papain. Since the remaining hydrophilic domain corresponds to 70K of contiguous primary sequence, the hydrophobic domain must be at or very near either the N- or C-terminus of the catalytic subunit.

An assignment of the hydrophobic domain to the C-terminus of the catalytic subunits was made following studies in which the N-terminal amino acid of the erythrocyte enzyme was labeled by reduction methylation prior to papain digestion (Haas and Rosenberry, 1984). The labeled N-terminal amino acid was identified as glutamate, and papain digestion did not remove this labeled amino acid from the catalytic subunit even though it converted the reductively methylated enzyme to a nonaggregating form that appeared identical to unlabeled papain-digested enzyme.

### B. Comparison of Human Erythrocyte Acetylcholinesterase to Globular Acetylcholinesterases in Other Tissues That Bind Detergent

$G_2$ forms are the predominant globular membrane-bound enzyme forms in human erythrocytes and torpedo electric organ tissue, and the $G_2$ enzyme from torpedo is similar to the erythrocyte enzyme in many characteristics that indicate an amphipathic structure (Viratelle and Bernhard, 1980; Bon and Massoulié, 1980; Lee *et al.*, 1982b; see Massoulié and Bon, 1982). $G_2$ forms are also prevalent in human skeletal muscle (Carson *et al.*, 1979; M. C. Sheldon and T. L. Rosenberry, unpublished observations) but are quite minor compared to $G_1$ and $G_4$ forms in other mammalian tissues, particularly in the central nervous system. Part of the cellular $G_1$ and $G_4$ acetylcholinesterase in T28 cultures was extracted by homogenization in the absence of nonionic detergent, but the remainder required detergent for extraction (Lazar and Vigny, 1980). The crude enzyme extracted by detergent aggregated when sedimented in the absence of detergent. Surprisingly, the cellular enzyme extracted in the absence of detergent nonetheless interacted somewhat with detergent as indicated by a decrease in apparent sedimentation coefficient of 0.5–1.2 S when sedimented in the presence of detergent. Bovine brain $G_1$ and $G_4$ forms appeared primarily membrane bound and interacted with nonionic detergents following extraction (Grassi *et al.*, 1982), but after partial purification neither bovine brain $G_1$ and $G_4$ nor rat brain $G_4$ enzyme (M. B. Hodges and T. L. Rosenberry, unpublished observations) aggregated in detergent-free sucrose gradients. The bovine brain $G_1$ and $G_4$ forms, however, appeared to be amphipathic as pronase digestion abolished their detergent interaction (Grassi *et al.*, 1982). The predominant $G_2$ enzyme form extracted from chicken muscle appeared to bind detergent but also did not aggregate in the absence of detergent (Allemand *et al.*, 1981). Thus, neither a detergent requirement for extraction nor aggregation in the absence of detergent appears to be a feature common to all globular amphipathic acetylcholinesterases, and structural distinctions among this enzyme class remain to be pursued.

## IV. RELATIONSHIPS AMONG ACETYLCHOLINESTERASE FORMS

The distribution of acetylcholinesterase activity among asymmetric as well as amphipathic and soluble globular forms has been investigated in a variety of tissues by several laboratories. This chapter will not provide a comprehensive review of the reported distributions or of their changes during development, but these topics have been covered in detail in a recent review by Massoulié and Bon (1982). To illustrate the cellular distribution of these forms, however, a summary of a recent study of adult rat diaphragm (Younkin *et al.*, 1982) will be presented. This tissue is typical of mammalian skeletal muscle in possessing a substantial number of asymmetric and globular forms, and it has the added advantage that the endplate-rich region which contains virtually all the neuromuscular junctions in the muscle can be separated from the nonendplate region by dissection.

## A. Acetylcholinesterase Forms in Rat Diaphragm

### 1. Localization and Cellular Orientation

Endplate and nonendplate regions of the diaphragm were separated and each region was extracted in sequential fashion. Extraction was first conducted with low ionic strength detergent buffer (1.0% Triton X-100 in 10 mM sodium phosphate, pH 7, with protease inhibitors) to solubilize both amphipathic and soluble globular enzyme forms and then with high ionic strength detergent buffer (previous buffer plus 1.0 M NaCl) to solubilize asymmetric forms. This sequential extraction significantly improves the quantitation of individual forms on sucrose gradients, particularly in the 8.5–12.5 S range. The distribution of globular forms was similar in endplate and nonendplate regions and consisted of roughly equal amounts of 4 S ($G_1$) and 10 S ($G_4$) and a small amount of 6.5 S ($G_2$) enzyme. The 16 S ($A_{12}$) form was the predominant asymmetric form in both regions, but a substantial amount of 12.5 S ($A_8$) enzyme also was present, particularly in the nonendplate region. In addition to the extracted globular and asymmetric forms, a significant fraction of the enzyme activity was not solubilized during the sequential extraction procedure. This fraction was termed "nonextractable" and appeared to be derived from the asymmetric pool. It was solubilized by collagenase, and the solubilized enzyme gave a sedimentation pattern identical to the collagenase-digested asymmetric pool. To further characterize the localization of these forms in diaphragm, echothiophate, a cationic phosphorylating agent that reversibly inactivates enzyme activity, was applied to the muscle prior to the separation of endplate and nonendplate regions. This agent penetrates muscle cell membranes poorly at 0°C and, thus, selectively inactivates enzyme forms on the external cell surface.

A summary of the distribution of the rat diaphragm forms is given in Table 3. According to these data, about 21% of the total acetylcholinesterase in the diaphragm was specifically associated with endplates. Only 8% of the total globular forms were localized at endplates, but asymmetric forms were distributed about equally between endplates and the remainder of the muscle. Thus, although in this muscle asymmetric forms were not localized exclusively at endplates, they were highly concentrated in this very small portion of the muscle cell surface. About one half of the total globular forms were intracellular, but this distribution was not uniform. $G_1$ was overwhelmingly intracellular, as noted below, while the majority of $G_4$ was externally oriented in the sarcolemma. Nonextractable enzyme comprised only 9% of the total, but this fraction was of particular interest both because it was the one most concentrated at endplates and because almost all the nonextractable endplate enzyme was externally oriented. This fraction together with extractable asymmetric forms in fact comprised most (84%)

*Table 3. The Distribution of the Molecular Forms of Acetylcholinesterase in Rat Diaphragm*[a]

| | Percent whole muscle AChE activity | | |
|---|---|---|---|
| | Nonendplate | Endplate-specific | Whole muscle |
| Globular forms | | | |
| Intracellular | 33.9 ± 3.3 | 2.6 ± 0.7 | 36.5 ± 3.0 |
| External | 28.6 ± 1.9 | 2.6 ± 2.0 | 31.2 ± 2.2 |
| Asymmetric forms | | | |
| Intracellular | 4.9 ± 0.5 | 2.2 ± 0.3 | 7.1 ± 0.4 |
| External | 8.1 ± 0.9 | 8.2 ± 1.7 | 16.3 ± 1.3 |
| Nonextractable | | | |
| Intracellular | 0.8 ± 0.2 | 0.4 ± 0.1 | 1.2 ± 0.2 |
| External | 2.3 ± 0.3 | 5.3 ± 0.8 | 7.6 ± 0.8 |
| Total | | | |
| Intracellular | 39.6 ± 3.6 | 5.2 ± 0.9 | 44.8 ± 2.9 |
| External | 39.0 ± 2.3 | 16.1 ± 1.0 | 55.2 ± 2.9 |

[a] The data shown are averages from seven experiments. Nonendplate enzyme activity per mg muscle was assumed constant throughout the dissected endplate and nonendplate regions of the diaphragm, and endplate-specific activity was determined by subtracting the nonendplate activity per mg from the total activity per mg in the endplate region. Sequential extractions were carried out after selectively inactivating external enzyme by exposing segments of diaphragm at 0°C to 1.25 μM echothiophate for 1 hr. The enzyme present in each fraction after this procedure was interpreted to be intracellular. Total activity was measured after reactivating each fraction with pralidoxime. External enzyme was calculated by subtracting intracellular from total enzyme. Values shown are the mean ± standard error. See Younkin *et al.* (1982) for details.

of the externally oriented endplate-specific pool, the pool presumably of paramount functional importance because it hydrolyzes acetylcholine released from nerve terminals. Nonextractable endplate enzyme may contribute significantly to the acetylcholinesterase exclusively localized in the endplate basement membrane that was identified by McMahan *et al.* (1978).

The use of echothiophate to determine intracellular and external enzyme in Table 3 was justified by two observations: (1) there were clearly two phases to the inactivation process, a rapid phase that was interpreted as inactivation of external enzyme and a slow phase which was considered to be inactivation of intracellular enzyme, and (2) a differential inactivation of individual enzyme forms was observed that ranged from 93% for the nonextractable endplate enzyme noted above to 27% for the $G_1$ form in the nonendplate region. Caution must be used in interpreting these percentages quantitatively, because the time required for a polar compound like echothiophate to reach fibers deep within the diaphragm results in some inactivation of intracellular enzyme. However, one can conclude from Table 3 that a significant fraction of the asymmmetric forms were intracellular, particularly in the nonendplate region, and, thus, that these forms were assembled inside the cell. This conclusion was even more compelling when the external acetylcholinesterase in cultured embryonic rat myotubes was evaluated by several procedures and nearly 60% of the total asymmetric enzyme was found to be intracellular (Brockman *et al.*, 1982). The percentage of intracellular asymmetric enzyme appears to vary considerably among cell types, however, for in cultured myotubes from the mouse muscle cell line C2, virtually all the asymmetric enzyme was external, much of it in focal patches on the cell surface (Inestrosa *et al.*, 1982). The distribution of enzyme forms in Table 3 should be taken as a qualitative indication of acetylcholinesterase form distributions in skeletal muscle, for the quantitative details vary considerably among species and muscle types (Massoulié and Bon, 1982).

### 2. Effects of Denervation

When many mammalian skeletal muscles are denervated, both endplate-specific and nonendplate acetylcholinesterase decrease dramatically within several days (Guth *et al.*, 1964, 1967; Drachman, 1972; Hall, 1973; Davey and Younkin, 1978; Collins and Younkin, 1982; for an exception in rabbit muscle see Massoulié and Bon, 1982). Reinnervation of muscles at their original junctional sites resulted in reappearance of the $A_{12}$ form, while reinnervation at muscle sites previously free of junctions resulted in reappearance of this form both at the new sites and at the (nonreinnervated) endplates of the original junctions (Guth and Zalewski, 1963; Vigny *et al.*, 1976; Weinberg and Hall, 1979; Lømo and Slater, 1980). The appearance of asymmetric forms in mixed cultures of muscle and nerve has been used as a specific biochemical marker of functional nerve–muscle contacts (Koenig and Vigny, 1978; Rubin *et al.*, 1979), but this approach is limited by observations that some primary muscle cultures and established muscle cell lines produce asymmetric forms in the absence of nerve cells (see above and Massoulié and Bon, 1982). Several laboratories are currently focusing on the mechanisms by which nerve regulates the levels of acetylcholinesterase forms in general and asymmetric forms in particular. While a review of the pertinent literature is far beyond the scope of this article, it appears that regulation is mediated both by the activity (electrical and/or mechanical) generated in muscle by nerve (Lømo and Slater, 1980; Rubin *et al.*, 1979; Rieger *et al.*, 1980; Brockman *et al.*, 1983) and by trophic factors delivered to muscle by nerve (Davey *et al.*, 1979; Fernandez *et al.*, 1980).

The most rapid change in enzyme levels after denervation of rat diaphragm was a large decrease in the predominantly intracellular $G_1$ form in the nonendplate region (McLaughlin and Bosmann, 1976; Carter and Brimijoin, 1981; Collins and Younkin, 1982). The initial decrease in this form suggested a rapid change in the intracellular metabolism of the enzyme rather than an increased internalization and degradation of the predominantly $G_4$ sarcolemmal enzyme. Furthermore, experiments in which acetylcholinesterase was labeled *in vivo* with echothiophate indicated that the turnover of nonendplate enzyme was unaffected by denervation (Younkin, 1981). These results led to the hypothesis that the electromechanical activity generated in muscle by nerve influences the nonendplate enzyme by increasing its rate of synthesis, and strong support for this hypothesis has come from a recent study which showed that both globular and asymmetric enzyme forms were synthesized more rapidly in actively fibrillating rat myotubes than in inactive (tetrodotoxin-treated) cells (Brockman *et al.*, 1984). Endplate-specific asymmetric forms in diaphragm decreased to about one fourth of their innervated levels following denervation, and this loss was accompanied by the transient appearance of an $A_4$ form (Collins and Younkin, 1982). *In vivo* studies that utilized echothiophate as a label detected an accelerated loss of endplate-specific enzyme 18–42 hr after denervation (Younkin, 1981). These observations indicated that accelerated degradation played an important role in the rapid loss of endplate-specific asymmetric forms 1–2 days after denervation, but it was unclear whether the prolonged decrease in these forms after denervation was due to a persistent acceleration of their turnover. Since the rapid loss occurred at about the same time as nerve terminal degeneration (Miledi and Slater, 1970), it is possible that increased degradation of endplate-specific enzyme was a transient event caused by increased proteolysis associated with nerve terminal degeneration. The fact that electromechanical activity increased the synthesis of the asymmetric forms in cultured myotubes (Brockman *et al.*, 1984) suggests that decreased synthesis may contribute to the persistent decrease of endplate-specific asymmetric forms in denervated muscle.

## B. Biosynthesis of Acetylcholinesterase Forms

A fundamental question that challenges current biosynthesis studies is the following: does the polymorphism of acetylcholinesterase forms observed in a single tissue reflect transcriptional or posttranslational processing of a single catalytic subunit gene product, or do these forms arise independently from related but distinct genes? Very few direct genetics experiments to test this question have been conducted. An apparent single structural gene locus *Ace* for acetylcholinesterase in *Drosophila melanogaster* has been identified (Hall and Kankel, 1976; Greenspan *et al.*, 1980). In a preliminary note, Dudai (1977) reported the presence of both asymmetric and globular enzyme forms in *Drosophila* (see Zingde *et al.*, 1983), and he identified mutant flies heterozygous for a small deficiency in the *Ace* locus. These mutants showed a normal distribution of forms but contained only one half the activity of normal flies, an observation consistent with the possibility that all forms derive from this single gene locus.

Several observations on acetylcholinesterase metabolism are consistent with the possibility that all globular and asymmetric forms in a cell arise from posttranslational processing of a single catalytic subunit precursor. The catalytic subunit sizes and turnover numbers of enzyme forms from a given tissue are similar (Vigny *et al.*, 1978; Viratelle and Bernhard, 1980; Lee *et al.*, 1982a,b), and protease-mediated cleavages of amphipathic forms could generate soluble globular forms that are either secreted or assembled into asymmetric forms. In cultured mammalian cells in which acetylcholinesterase was totally inactivated by an organophosphate, newly synthesized enzyme activity reappeared first in the $G_1$ fraction, then in $G_4$, and finally in $A_{12}$ (Rieger *et al.*, 1976; Koenig and Vigny, 1978). The $G_1$ form in rat diaphragm was noted above to be more intracellular than the oligomeric globular or asymmetric forms in this muscle, and similar studies with inactivating or protecting agents that do not penetrate the cell membranes of the neuronal hybrid T28 cells (Lazar and Vigny, 1980; Taylor *et al.*, 1981) or PC12 cells (Inestrosa *et al.*, 1981) indicated that greater than 90% of the $G_1$ forms were intracellular. This percentage was far higher than those for oligomeric globular or asymmetric forms in these cells. Parallel results were obtained with avian cells except that the intracellular form was $G_2$ rather than $G_1$ (Rotundo and Fambrough, 1979, 1980; Taylor *et al.*, 1981). During both development *in vivo* and transition from log to stationary phase in tissue culture, the percentage of total acetylcholinesterase found as $G_1$ decreases several-fold (Massoulié and Bon, 1982). These observations suggest that intracellular soluble $G_1$ (or $G_2$ in avians) may serve as a precursor to membrane-bound $G_4$, secreted $G_4$, and/or asymmetric forms (see Massoulié and Bon, 1982; Younkin *et al.*, 1982).

The intracellular $G_1$ or $G_2$ pool shows a rapid rate of turnover in cultured cells. Experiments involving selective organophosphate inactivation and the use of protein synthesis inhibitors in cultured chick embryo muscle cells indicated that acetylcholinesterase synthesized in the intracellular pool was transported over a 2- to 3-hr period to two alternative destinations. A small fraction appeared in the plasma membrane, where it slowly turned over with a half-life of about 50 hr, while the majority was secreted into the extracellular medium (Rotundo and Fambrough, 1980). Similar results were obtained in a heavy isotope-labeling study of T28 cells (Lazar *et al.*, 1984), where a 3.5-hr half-life for intracellular $G_1$ and a 40-hr half-life for cellular $G_4$ were observed. Secreted enzyme activity that appeared in the medium could not account for the majority of the rapidly lost $G_1$ activity, suggesting that a considerable portion of the intracellular $G_1$ pool may have been degraded without further processing.

An alternative to posttranslational processing of a common catalytic subunit precursor is the transcription of similar but distinct acetylcholinesterase mRNAs from a common genome. Several classes of immunoglobulins occur in membrane-bound and secreted forms. The membrane-bound forms are distinguished from the secreted forms by the presence of a short, largely hydrophobic C-terminal extension of the heavy chain subunits. Alternative mRNA transcripts corresponding to the two forms are made from a single gene by appropriate splicing patterns at the 3′ end of the RNA during transcription (Rogers *et al.*, 1981). The differences between the two mRNAs suggest that developmental control of the site at which poly(A) is added to gene transcripts determines the relative levels of the two immunoglobulin forms. Distinct alternative acetylcholinesterase mRNA transcripts would likely give rise to different primary amino acid sequences in the catalytic subunits of the various enzyme forms. The only tissue currently providing sufficient purified amounts of both globular and asymmetric forms to test for such differences is torpedo electric organ. Catalytic subunit peptide maps of the $A_{12}$ plus $A_8$ forms appeared to differ slightly from those of the $G_2$ amphipathic form, but the differences were equivocal (Lee *et al.*, 1982b). More compelling evidence that these torpedo asymmetric and globular catalytic subunits do not have a precursor–product relationship was obtained from the fractionation of tryptic digests by high-pressure liquid chromatograpy (Doctor *et al.*, 1983). At least four distinct peptides were obtained from the catalytic subunits of each class of forms which could not be attributed to carbohydrate adducts. Although this evidence tends to rule out the hypothesis that all acetylcholinesterase forms derive from a single catalytic subunit polypeptide precursor, it does not invalidate parallel assembly processes for asymmetric and globular amphipathic forms from distinct $G_1$ pools, an important point noted by Massoulié and Bon (1982).

## REFERENCES

Allemand, P., Bon, S., Massoulié, J., and Vigny, M., 1981, The quaternary structure of chicken acetylcholinesterase and butyrylcholinesterase; effect of collagenase and trypsin, *J. Neurochem.* **36**:860–867.

Amsterdam, A., Lamed, R., and Silman, L., 1975, Actomyosin from electric organ tissue of electric eel, *Isr. J. Med. Sci.* **11:**1183.

Anglister, L., and Silman, I., 1978, Molecular structure of elongated forms of electric eel acetylcholinesterase, *J. Mol. Biol.* **125:**293–311.

Barnett, P., and Rosenberry, T. L., 1978, A residual subunit fragment in the conversion of 18S to 11S acetylcholinesterase, *Fed. Proc.* **36:**485.

Bellhorn, M. B., Blumenfeld, O. O., and Gallop, P. M., 1970, Acetylcholinesterase of human erythrocyte membrane, *Biochem. Biophys. Res. Commun.* **39:**267–273.

Betz, W., and Sakmann, B., 1973, Effects of proteolytic enzymes on function and structure of frog neuromuscular junctions, *J. Physiol. (London)* **230:**673–688.

Bon, S., and Massoulié, J., 1976, Molecular forms of *Electrophorus* acetylcholinesterase. The catalytic subunits: Fragmentation, intra- and intersubunit disulfide bonds, *FEBS Lett.* **71:**273–278.

Bon, S., and Massouilié, J., 1978, Collagenase sensitivity and aggregtion properties of *Electrophorus* acetylcholinesterase, *Eur. J. Biochem.* **89:**89–94.

Bon, S., and Massoulié, J., 1980, Collagen-tailed and hydrophobic components of acetylcholinesterase in *Torpedo marmorata* electric organ, *Proc. Natl. Acad. Sci. USA* **77:**4464–4468.

Bon, S., Huet, M., Lemonnier, M., Rieger, F., and Massoulié, J., 1976, Molecular forms of *Electrophorus* acetylcholinesterase. Molecular weight and composition, *Eur. J. Biochem.* **68:**523–530.

Bon, S., Cartaud, J., and Massoulié, J., 1978, The dependence of acetylcholinesterase aggregation at low ionic strength upon a polyanionic component, *Eur. J. Biochem.* **85:**1–14.

Bon, S., Vigny, M., and Massoulieé, J., 1979, Asymmetric and globular forms of acetylcholinesterase in mammals and birds, *Proc. Natl. Acad. Sci. USA* **76:**2546–2550.

Brockman, S. K., Przybylski, R. J., and Younkin, S. G., 1982, Cellular localization of the molecular forms of acetylcholinesterase in cultured embryonic rat myotubes, *J. Neurosci.* **2:**1775–1785.

Brockman, S. K., Younkin, L. H., and Younkin, S. G., 1984, The effect of spontaneous electromechanical activity on the metabolism of acetylcholinesterase in cultured embryonic rat myotubes, *J. Neurosci* **4:**131–140.

Bulger, J. E., Randall, W. R., Nieberg, P. S., Patterson, S. T., McNamee, M. G., and Wilson, B. W., 1982, Regulation of acetylcholinesterase forms in quail and chicken muscle cultures, *Dev. Neurosci.* **5:**474–483.

Carson, S., Bon, S., Vigny, M., Massoulié, J., and Fardeau, M., 1979, Distribution of acetylcholinesterase molecular forms in neural and nonneural sections of human muscle, *FEBS Lett.* **97:**348–352.

Cartaud, J., Rieger, F., Bon, S., and Massoulié, J., 1975, Fine structure of electric eel acetylcholinesterase, *Brain Res.* **88:**127–130.

Cartaud, J., Bon, S., and Massoulié, J., 1978, *Electrophorus* acetylcholinesterase. Biochemical and electron microscope characterization of low ionic strength aggregates, *J. Cell Biol.* **77:**315–322.

Carter, J. L., and Brimijoin, S., 1981, Effects of acute and chronic denervation on the release of acetylcholinesterase and its molecular forms in rat diphragms, *J. Neurochem.* **36:**1018–1025.

Collins, P. L., and Younkin, S. G., 1982, Effect of denervation on the molecular forms of acetylcholinesterase in rat diaphragm, *J. Biol. Chem.* **257:**13638–13644.

Dailey, H. A., and Strittmatter, P., 1981, Orientation of the carboxyl and $NH_2$ termini of the membrane-binding segment of cytochrome $b_5$ on the same side of phospholipid bilayers, *J. Biol. Chem.* **256:**3951–3955.

Davey, B., and Younkin, S. G., 1978, Effect of nerve stump length on cholinesterase in denervated rat diaphragm, *Exp. Neurol.* **59:**168–175.

Davey, B., Younkin, L. H., and Younkin, S. G., 1979, Neural control of skeletal muscle cholinesterase: A study using organ-cultured rat muscle, *J. Physiol. (London)* **289:**501–515.

Doctor, B. P., Camp, S., Genetry, M. K., Taylor, S. S., and Taylor, P., 1983, Antigenic and structural differences in the catalytic subunits of the molecular forms of acetylcholinesterase, *Proc. Natl. Acad. Sci. USA* **80:**5767–5771.

Drachman, D. B., 1972, Neurotrophic regulation of muscle cholinesterase: Effects of botulinum toxin and denervation, *J. Physiol. (London)* **226:**619–627.

Dudai, Y., 1977, Molecular forms of acetylcholinesterase in normal and mutant *Drosophila, Isr. J. Med. Sci.* **13:**944.

Dudai, Y., and Silman, I., 1973, The effect of $Ca^{2+}$ on interaction of acetylcholinesterase with subcellular fractions of electric organ tissue from the electric eel, *FEBS Lett.* **30:**49–52.

Dudai, Y., and Silman, I., 1974a, Acetylcholinesterase, *Meth. Enzymol.* **34:**571–580.

Dudai, Y., and Silman, I., 1974b, The effects of solubilization procedures on the release and molecular state of actylcholinesterase from electric organ tissue, *J. Nuerochem.* **23:**1177–1187.

Dudai, Y., Herzberg, M., and Silman, I., 1973, Molecular structures of acetylcholinesterase from electric organ tissue of the electric eel, *Proc. Natl. Acad. Sci. USA* **70:**2473–2476.

Dutta-Choudhury, T. A., and Rosenberry, T. L., 1984, Human erythrocyte acetylcholinesterase is an amphipathic protein whose short membrane-binding domain is removed by papain digestion, *J. Biol. Chem.* **259:**5653–5660.

Engelman, D. M., and Steitz, T. A., 1981, The spontaneous insertion of proteins into and across membranes: The helical hairpin hypothesis, *Cell* **23:**411–422.

Fambrough, D. M., Engel, A. G., and Rosenberry, T. L., 1982, Acetylcholinesterase of human erythrocytes and neuromuscular junctions: Homologies revealed by monoclonal antibodies, *Proc. Natl. Acad. Sci. USA* **79:**1078–1082.

Fernandez, H. L., Duell, M. J., and Festoff, B. W., 1979, Cellular distribution of 16S acetylcholinesterase, *J. Neurochem.* **32:**581–585.

Fernandez, H. L., Patterson, M. R., and Duell, M. J., 1980, Neurotrophic control of 16S acetylcholinesterase from mammalian skeletal muscle in organ culture, *J. Neurobiol.* **11:**557–570.

Fessard, A., 1958, Les Organes Electrique, in: *Traite de Zoology,* Vol. XIII (P. P. Grasse, ed.), Masson, Paris, pp. 1143–1238.

Frielle, T., Brunner, J., and Curthoys, N. P., 1982, Isolation of the hydrophobic membrane binding domain of rat renal γ-glutamyl transpeptidase selectively labeled with 3-trifluoromethyl-3-(*m*-[$^{125}$I]iodophenyl)diazirine, *J. Biol. Chem.* **257:**14979–14982.

Froede, H. C., and Wilson, I. B., 1971, Acetylcholinesterase, in: *The Enzymes,* Vol. V, 3rd Ed. (P. D. Boyer, ed.), Academic Press, New York, pp. 87–114.

Grassi, J., Vigny, M., and Massoulié, J., 1982, Molecular forms of acetylcholinesterase in bovine caudate nucleus and superior cervical ganglion: Solubility properties and hydrophobic character, *J. Neurochem.* **38:**457–469.

Greenspan, R. J., Finn, J. A., Jr., and Hall, J. C., 1980, Acetylcholinesterase mutants in Drosophila and their effects on the structure and function of the central nervous system, *J. Comp. Neurol.* **189:**741–774.

Grossmann, H., and Liefländer, M., 1975, Affinitätschromatographische reinigung der acetylcholinesterase aus menschlichen erythrozyten, *Hoppe-Seyler's Z. Physiol. Chem.* **356:**663–669.

Guth, L., and Zalewski, A. A., 1963, Disposition of cholinesterase following implantations of nerve into innervated and denervated muscle, *Exp. Neurol.* **7:**316–326.

Guth, L., Albers, R. W., and Brown, W. C., 1964, Quantitative changes in cholinesterase activity of denervated muscle fibers and sole plates, *Exp. Neurol.* **10:**236–250.

Guth, L., Brown, W. C., and Watson, P. K., 1967, Studies on the role of nerve impulses and acetylcholine release in the regulation of the cholinesterase activity of muscle, *Exp. Neurol.* **18:**443–452.

Haas, R., and Rosenberry, T. L., 1984, Identification of the N-terminal amino acid in human erythrocyte acetylcholinesterase by radiolabelled reductive methylation and amino acid analysis, submitted for publication.

Hall, J. C., and Kankel, D. R., 1976, Genetics of acetylcholinesterase in *Drosophila melanogaster, Genetics* **83:**517–535.

Hall, Z. W., 1973, Multiple forms of acetylcholinesterase and their distribution in endplate and non-endplate regions of rat diaphragm muscle, *J. Neurobiol.* **4:**343–361.

Hall, Z. W., and Kelly, R., 1971, Enzymatic detachment of endplate acetylcholinesterase from muscle, *Nature New Biol.* **232:**62–63.

Inestrosa, N. C., Reiness, C. G., Reichardt, L. F., and Hall, Z. W., 1981, Cellular localization of the molecular forms of acetylcholinesterase in rat pheochromocytoma PC12 cells treated with nerve growth factor, *J. Neurosci.* **1:**1260–1267.

Inestrosa, N. C., Silberstein, L., and Hall, Z. W., 1982, Association of the synaptic form of acetylcholinesterase with extracellular matrix in cultured mouse muscle cells, *Cell* **29:**71–79.

Kanwar, Y. S., Hascall, V. C., and Farquhar, M. G., 1981, Partial characterization of newly synthesized proteoglycans isolated from the glomerular basement membrane, *J. Cell Biol.* **90:**527–532.

Kefalides, N. A., Alper, R., and Clark, C. C., 1979, Biochemistry and metabolism of basement membranes, *Int. Rev. Cytol.* **61:**167–228.

Kleinman, H. K., Klebe, R. J., and Martin, G. R., 1981, Role of collagenous matrices in the adhesion and growth of cells, *J. Cell Biol.* **88:**473–485.

Koenig, J., and Vigny, M., 1978, Neural induction of the 16S acetylcholinesterase in muscle cell cultures, *Nature (London)* **271:**75–77.

Lazar, M., and Vigny, M., 1980, Modulation of the distribution of acetylcholinesterase molecular forms in a murine neuroblastoma multisympathetic ganglion cell hybrid cell line, *J. Neurochem.* **35:**1067–1079.

Lazar, M., Salmeron, E., Vigny, M., and Massoulié, J., 1984, Heavy isotope-labeling study of the metabolism of monomeric and tetrameric acetylcholinesterase forms in the murine neuronal-like T28 hybrid cell line, *J. Biol. Chem.* **259:**3703–3713.

Lee, S. L., and Taylor, P., 1982, Structural characterization of the asymmetric (17 + 13) S species of acetylcholinesterase from *Torpedo*. II. Component peptides obtained by selective proteolysis and disulfide bond reduction, *J. Biol. Chem.* **257:**12292–12301.

Lee, S. L., Heinemann, S., and Taylor, P., 1982a, Structural characterization of the asymmetric (17 + 13) S forms of acetylcholinesterase from *Torpedo*. I. Analysis of subunit composition, *J. Biol. Chem.* **257:**12283–12291.

Lee, S. L., Camp, S. J., and Taylor, P., 1982b, Characterization of a hydrophobic, dimeric form of acetylcholinesterase from *Torpedo, J. Biol. Chem.* **257:**12302–12309.

Lømo, T., and Slater, C. R., 1980, Control of junctional acetylcholinesterase by neural and muscular influences in the rat, *J. Physiol. (London)* **303:**191–202.

Lwebuga-Mukasa, J. S., Lappi, S., and Taylor, P., 1976, Molecular forms of acetylcholinesterase from *Torpedo californica:* Their relationship to synaptic membranes, *Biochemistry* **15:**1425–1434.

Macnair, R. D. C., and Kenny, A. J., 1979, Proteins of the kidney microvillar membrane. The amphipathic form of dipeptidyl peptidase IV, *Biochem. J.* **179:**379–395.

Massoulié, J., 1980, The polymorphism of cholinesterases and its physiological significance, *Trends Biochem. Sci.* **5:**160–164.

Massoulié, J., and Bon, S., 1982, The molecular forms of cholinesterase and acetyhlcholinesterase in vertebrates, *Annu. Rev. Neurosci.* **5:**57–106.

Massoulié, J., and Rieger, F., 1969, L'acétylcholinestérase des organes électriques de poissons (torpille et gymnote); complexes membranaires, *Eur. J. Biochem.* **11:**441–455.

Massoulié, J., Rieger, F., and Bon, S., 1971, Espéces acetylcholinesterasiques globulaires et allongées des organes électriques de poisson, *Eur. J. Biochem.* **21:**542–551.

Mays, C., and Rosenberry, T. L., 1981, Characterization of pepsin-resistant collagen-like tail subunit fragments of 18S and 14S acetylcholinesterase from *Electrophorus electricus, Biochemistry* **20:**2810–2817.

McCann, W. F. X., and Rosenberry, T. L., 1977, Identification of discrete disulfide-linked oligomers which distinguish 18S from 14S acetylcholinesterase, *Arch. Biochem. Biophys.* **183:**347–352.

McLaughlin, J., and Bosmann, H. B., 1976, Molecular species of acetylcholinesterase in denervated rat skeletal muscle, *Exp. Neurol.* **52:**263–271.

McMahan, U. J., Sanes, J. R., and Marshall, L. M., 1978, Cholinesterase is associated with the basal lamina at the neuromuscular junction, *Nature (London)* **271:**172–174.

Miledi, R., and Slater, C. R., 1970, On the degeneration of rat neuromuscular junctions after nerve section, *J. Physiol. (London)* **201:**507–526.

Millar, D. B., and Grafius, M. A., 1965, Reversible aggregation of acetylcholinesterase, *Biochim. Biophys. Acta* **110:**540–547.

Millar, D. B., Christopher, J. P., and Burrough, D. O., 1978, Evidence that eel acetylcholinesterase is not an integral membrane protein, *Biophys. Chem.* **9:**9–14.

Nachmansohn, D., 1959, *Chemical and Molecular Basis of Nerve Activity,* Academic Press, New York (revised edition with E. Neumann, 1975).

Nachmansohn, D., and Lederer, E., 1939, Sur la biochemie de la cholinesterase. I. Preparation de l'enzyme: Groupements-SH, *Bull. Soc. Chim. Biol. (Paris)* **21:**797–808.

Niday, E., Wang, C. S., and Alaupovic, P., 1977, Studies on the characterization of human erythrocyte acetylcholinesterase and its interaction with antibodies, *Biochim. Biophys. Acta* **469:**180–193.

Olsen, B. R., Alper, R., and Kefalides, N. A., 1973, Structural characterization of a soluble fraction from lens-capsule basement membrane, *Eur. J. Biochem.* **38:**220–228.

Ott, P., and Brodbeck, U., 1978, Multiple molecular forms of acetylcholinesterase from human erythrocyte membranes, *Eur. J. Biochem.* **88:**119–125.

Ott, P., Jenny, B., and Brodbeck, U., 1975, Multiple molecular forms of purified human erythrocyte acetylcholinesterase, *Eur. J. Biochem.* **57:**469–480.

Ott, P., Lustig, A., Brodbeck, U., and Rosenbusch, J. P., 1982, Acetylcholinesterase from human erythrocyte membranes: Dimers as functional units, *FEBS Lett.* **138:**187–189.

Rieger, F., Bon, S., and Massoulié, J., 1973, Observation par microscopie électronique des formes allongées et globulaires de l'acétylcholinestérase de gymnote *(Electrophorus electricus), Eur. J. Biochem.* **34:**539–547.

Rieger, F., Favre-Bauman, A., Benda, P., and Vigny, M., 1976, Molecular forms of acetylcholinesterase: Their *de novo* synthesis in mouse neuroblastoma cells, *J. Neurochem.* **27:**1059–1063.

Rieger, F., Koenig, J.,and Vigny, M., 1980, Spontaneous contractile activity and the presence of the 16 S form of acetylcholinesterase in rat muscle cells in culture: Reversible suppressive action of tetrodotoxin, *Dev. Biol.* **76:**358–365.

Rodriguez-Borrajo, C., Barat, A., and Ramirez, G., 1982, Solubilization of collagen-tailed molecular forms of acetylcholinesterase from several brain areas in different vertebrate species, *Neurochem. Int.* **4:**563–568.

Rogers, J., Choi, E., Souza, L., Carter, C., Word, C., Keuhl, M., Eisenberg, D., and Wall, R., 1981, Gene segments encoding transmembrane carboxyl termini of immunoglobulin gamma chains, *Cell* **26:**19–27.

Rosenberry, T. L., 1975, Acetylcholinesterase, *Adv. Enzymol.* **43:**103–218.

Rosenberry, T. L., 1976, Acetylcholinesterase, in: *The Enzymes of Biological Membranes,* Vol. 4 (A. Martonosi, ed.), Plenum Press, New York, pp. 331–363.

Rosenberry, T. L., 1979, Quantitative simulation of endplate currents at neuromuscular junctions based on the reaction of acetylcholine with acetylcholine receptor and acetylcholinesterase, *Biophys. J.* **26:**263–290.

Rosenberry, T. L., 1982, Acetylcholinesterase: The relationship of protein structure to cellular localization, in: *Membranes and Transport,* Vol. 2 (A. Martonosi, ed.), Plenum Press, New York, pp. 339–348.

Rosenberry, T. L., and Richardson, J. M., 1977, Structure of 18S and 14S acetylcholinesterase. Identification of collagen-like subunits that are linked by disulfide bonds to catalytic subunits, *Biochemistry* **16:**3550–3558.

Rosenberry, T. L., and Scoggin, D. M., 1984, Structure of human erythrocyte acetylcholinesterase. Characterization of intersubunit disulfide bonding and detergent interaction, *J. Biol. Chem.* **259:**5643–5660.

Rosenberry, T. L., Chen, Y. T., and Bock, E., 1974, Structure of 11 S acetylcholinesterase: Subunit composition, *Biochemistry* **13:**3068–3079.

Rosenberry, T. L., Barnett, P., and Mays, C., 1980, The collagen-like subunits of acetylcholinesterase from the eel *Electrophorus electricus, Neurochem. Int.* **2:**135–147.

Rosenberry, T. L., Barnett, P., and Mays, C., 1982, Acetylcholinesterase, *Meth. Enzymol.* **82:**325–339.

Rotundo, R. L., 1984, Purification and properties of the hydrophobic, membrane-bound form of acetylcholinesterase from chicken brain, *J. Biol. Chem.* (in press).

Rotundo, R. L., and Fambrough, D. M., 1979, Molecular forms of chicken embryo acetylcholinesterase *in vitro* and *in vivo, J. Biol. Chem.* **254:**4790–4799.

Rotundo, R. L., and Fambrough, D. M., 1980, Synthesis, transport and fate of acetylcholinesterase in cultured chick embryo muscle cells, *Cell* **22:**583–594.

Ruben, L. L., Schuetze, S. M., and Fischbach, G. D., 1979, Accumulation of acetylcholinesterase at newly formed nerve-muscle synapses, *Dev. Biol.* **69:**46–58.

Silman, L., Lyles, J. M., and Barnard, E. A., 1979, Intrinsic forms of acetylcholinesterase in skeletal muscle, *FEBS Lett.* **94:**166–170.

Skau, K. A., and Brimijoin, S., 1980, Multiple molecular forms of acetylcholinesterase in rat vagus nerve, smooth muscle, and heart, *J. Neurochem.* **35:**1151–1154.

Sketelj, J., and Brzin, M., 1979, Attachment of acetylcholinesterase to structures of the motor endplate, *Histochemistry* **61:**239–248.

Sketelj, J., McNamee, M. G., and Wilson, B. W., 1978, Effect of denervation on the molecular forms of acetylcholinesterase in normal and dystrophic chicken muscles, *Exp. Neurol.* **60:**624–629.

Steck, T. L., 1974, The organization of proteins in the human red blood cell membrane, *J. Cell Biol.* **62:**1–19.

Taylor, P., Rieger, F., Shelanski, M. L., and Green, L. A., 1981, Cellular localization of the multiple molecular forms of acetylcholinesterase in cultured neuronal cells, *J. Biol. Chem.* **256:**3827–3830.

Vigny, M., Koenig, J., and Rieger, F., 1976, The motor end-plate specific form of acetylcholinesterase: Appearance during embryogenesis and re-innervation of rat muscle, *J. Neurochem.* **27:**1347–1353.

Vigny, M., Bon, S., Massoulié, J., and Leterrier, F., 1978, Active-site catalytic efficiency of acetylcholinesterase molecular forms in *Electrophorus, Torpedo,* rat and chicken, *Eur. J. Biochem.* **85:**317–323.

Vigny, M., Bon, S., Massoulié, J., and Gisiger, V., 1979, The subunit structure of mammalian acetylcholinesterase: Catalytic subunits, dissociating effect of proteolysis and disulphide reduction on the polymeric forms, *J. Neurochem.* **33:**559–565.

Vigny, M., Martin, G. R., and Grotendorst, G. R., 1983, Interactions of asymmetric forms of acetylcholinesterase with basement membrane components, *J. Biol. Chem.* **258:**8795–8798.

Viratelle, O. M., and Bernhard, S. A., 1980, Major component of acetylcholinesterase in *Torpedo* electroplax is not basal lamina associated, *Biochemistry* **19:**4999–5007.

Weinberg, C. B., and Hall, Z. W., 1979, Junctional form of acetylcholinesterase restored at nerve-free endplates, *Dev. Biol.* **68:**631–635.

Wiedmer, T., DiFrancesco, C., and Brodbeck, U., 1979, Effects of amphiphiles on structure and activity of human erythrocyte membrane acetylcholinesterase, *Eur. J. Biochem.* **102:**59–64.

Younkin, S. G., 1981, Turnover of acetylcholinesterase in innervated and denervated rat diaphragm, *11th Neuroscience Meeting, Soc. Neurosci.,* Vol. 7, Abst. 249.12, p. 766.

Younkin, S. G., Rosenstein, C. C., Collins, P. L., and Rosenberry, T. L., 1982, Cellular localization of the molecular forms of acetylcholinesterase in rat diaphragm, *J. Biol. Chem.* **257:**13630–13637.

Zingde, S., Rodrigues, V., Joshi, S. M., and Krishnan, K. S., 1983, Molecular properties of *Drosophila* acetylcholinesterase, *J. Neurochem.* **41:**1243–1252.

# 38

# The Gastric H,K-ATPase

L. D. Faller, A. Smolka, and G. Sachs

## I. INTRODUCTION

The secretion of hydrochloric acid by the stomach is a particularly remarkable example of biological active transport. A luminal pH of 0.8 is generated. Since the average cytosolic pH is 7.7, the proton gradient across the gastric mucosa is more than a million-fold. Ion gradients found across other mammalian tissues are smaller by several orders of magnitude, raising the question of whether the gastric pump works by the same mechanism as the more extensively studied ATP-driven Na,K- and Ca-pumps.

To understand how an ion is translocated against an unfavorable thermodynamic gradient, it is necessary to know the structure and arrangement of the pump peptides, the number and location of the ion-binding sites, their physical relationship to the substrate-binding sites, and how the energy released in ATP hydrolysis is coupled to transport. In an earlier review (Sachs *et al.*, 1982), studies of the gastric proton pump through 1981 were reviewed. The salient points are summarized here without citing experimental evidence.

The site of acid secretion is the parietal cell. ATP is both a necessary and sufficient energy source for active proton transport. In the process, protons are exchanged electroneutrally for $K^+$ ions. The enzyme that couples ATP hydrolysis to proton transport is localized in the microvilli of the canaliculus of the secreting parietal cell, and appears to be a multimer of nonidentical $1.0 \times 10^5$-dalton polypeptides. The molecular weight of $3.2 \times 10^5$ determined by irradiation inactivation with electrons suggested a trimeric structure. Mechanistically, the enzyme belongs to the class of ATPases that form a covalent phosphoenzyme intermediate. Both arginyl and histidyl residues have been implicated in the catalytic mechanism, and $Mg^{2+}$ is required for phosphorylation. The dependence of the hydrolysis rate on $K^+$ is complex. Qualitatively, luminal $K^+$

---

*L. D. Faller, A. Smolka, and G. Sachs* • Center for Ulcer Research and Education, Wadsworth Veterans Administration Center, Los Angeles, California 90073; and School of Medicine, University of California at Los Angeles, Los Angeles, California 90024.

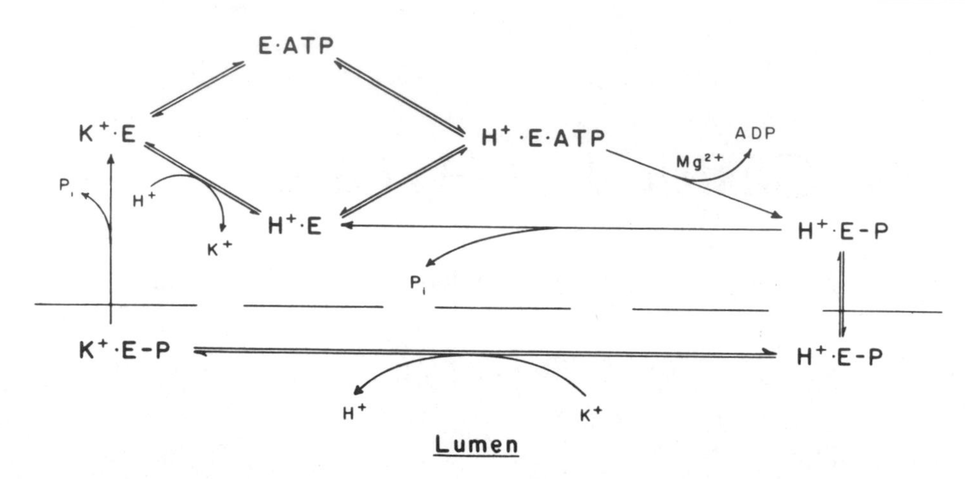


*Figure 1.* A schematic summarizing knowledge of the gastric ATPase reaction mechanism through 1981. The model emphasizes the $Mg^{2+}$ requirement for phosphorylation of the enzyme and the sidedness of $K^+$ in accelerating dephosphorylation and inhibiting phosphorylation. $K^+$ is exchanged across the membrane (dashed line) for protons.

activates the enzyme by increasing the dephosphorylation rate, and cytosolic $K^+$ inhibits the gastric ATPase by decreasing the phosphorylation rate. The effects of $K^+$ are antagonized by protons. The gastric enzyme also catalyzes the hydrolysis of phosphate esters, such as *p*-nitrophenylphosphate (*p*NPP), but protons are not transported. A mechanism synthesizing the experimental evidence through 1981 is shown in Figure 1.

Two additional catalytic activities of the gastric enzyme, ATP–ADP exchange and oxygen-18 exchange, have been described since an earlier review (Sachs *et al.*, 1982) appeared. Three new approaches to the study of the H,K-ATPase have been particularly informative. Monoclonal antibodies to the ATPase have provided more precise information on the site of acid secretion and the structure of the enzyme. The isolation of vesicles containing the enzyme from stimulated animals has demonstrated differences between resting and secreting tissues. Studies utilizing site-selective reagents have given insight into the number of active sites and the mechanism of catalysis. In general, the new results are compatible with the conclusions drawn from earlier studies, but they suggest that the scheme in Figure 1 provides only a partial explanation of gastric acid secretion.

## II. LOCALIZATION OF THE H,K-ATPase WITHIN THE PARIETAL CELL

The parietal cell undergoes dramatic morphological changes when stimulated to secrete acid. The tubulovesicular appearance of the intracellular membrane is largely replaced by microvillous canaliculi. These extended spaces have been shown to be

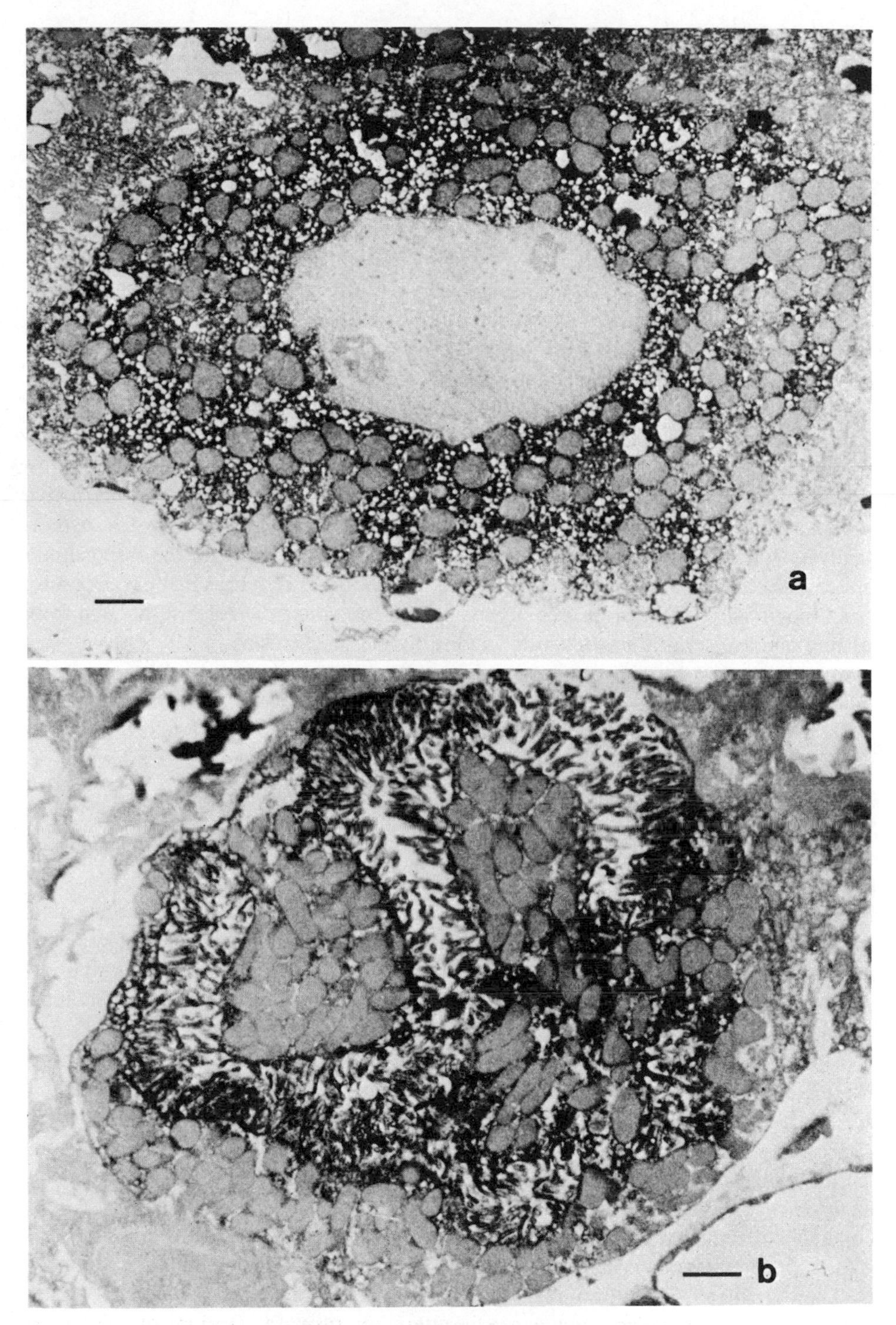


*Figure 2.* Electron micrographs of rabbit parietal cells stained indirectly by the immunoperoxidase technique with a monoclonal antibody against the H,K-ATPase. (a) The tubulovesicles of a parietal cell from a cimetidine-inhibited rabbit are labeled by the antibody. (b) In the case of a stimulated rabbit parietal cell, the same antibody labels open structures called secretory canaliculi.

the sites of acid secretion by their accumulation of basic dyes and to be the location of the H,K-ATPase by labeling of the microvilli with a polyclonal antibody against the microsomal enzyme. Since monoclonal antibodies are specific for a single antigenic determinant, they are a more sensitive and convincing probe of antigen location and structure. Two monoclonal antibodies against the H,K-ATPase have been shown to react selectively with the parietal cells of gastric mucosae (Smolka *et al.*, 1983). Figure 2a is an immunoelectron micrograph of a rabbit parietal cell labeled with a monoclonal antibody to the H,K-ATPase after inhibition of acid secretion by cimetidine. The tubulovesicles of the resting parietal cell are labeled. Figure 2b depicts a parietal cell from an uninhibited animal. In this case, secretory canaliculi dominate and are labeled by the antibody. Two interpretations of the formation of the secretory membrane are consistent with the observed localization of the ATPase in the tubulovesicles of resting cells and in the microvilli of the canaliculi of secreting cells. In one, the six- to tenfold increase in apical membrane surface area seen upon stimulation is attributed to fusion of the tubulovesicles with apical membrane (Forte *et al.*, 1977). The alternative proposal is that the tubulovesicles are collapsed canaliculi which are swollen osmotically to form the open, microvillous structures occupying much of the intracellular space in secreting parietal cells (Berglindh *et al.*, 1980). Differences that have recently been found between the properties of membranes containing the ATPase isolated from resting and secreting tissues seem to favor a fusion mechanism.

## III. DISCOVERY OF A PATHWAY FOR THE TRANSPORT OF $K^+$ SALTS IN MEMBRANE VESICLES ISOLATED FROM SECRETING TISSUES

Much of the work on acid secretion has employed vesicles in the microsomal fraction isolated from gastric mucosae. These vesicles accumulate protons when ATP is added in the presence of $K^+$. However, their permeability to $K^+$ is limited, and an additional requirement for sustained acid accumulation is either preincubation of the vesicles with $K^+$ or addition of an ionophore (Figure 3a). Recently, a redistribution of $K^+$-ATPase activity from the microsomal fraction into a denser fraction containing larger vesicles was found when a secretagogue was administered before sacrifice (Wolosin and Forte, 1981a). The most striking difference between the vesicles from stimulated tissue and the traditional microsomal vesicles is shown in Figure 3b. Acid accumulation is ionophore independent in the former (Wolosin and Forte, 1981b,c). Rapid and selective permeability of the vesicles to $K^+$ salts was demonstrated. From the ionophore requirements for dissipation of the hydrogen ion gradient and a kinetic analysis of the effects of $K^+$ and $Cl^-$ ions on the rate of $H^+$ accumulation, it was concluded that the pathway for KCl transport in vesicles isolated from stimulated tissue is electroneutral (Wolosin and Forte, 1983). The biochemical and functional differences between vesicles isolated from resting and stimulated tissue support the fusion model of secretory membrane formation, since the alternative proposal, swelling of a collapsed canaliculus, does not involve mixing of membrane structural components. It is hy-

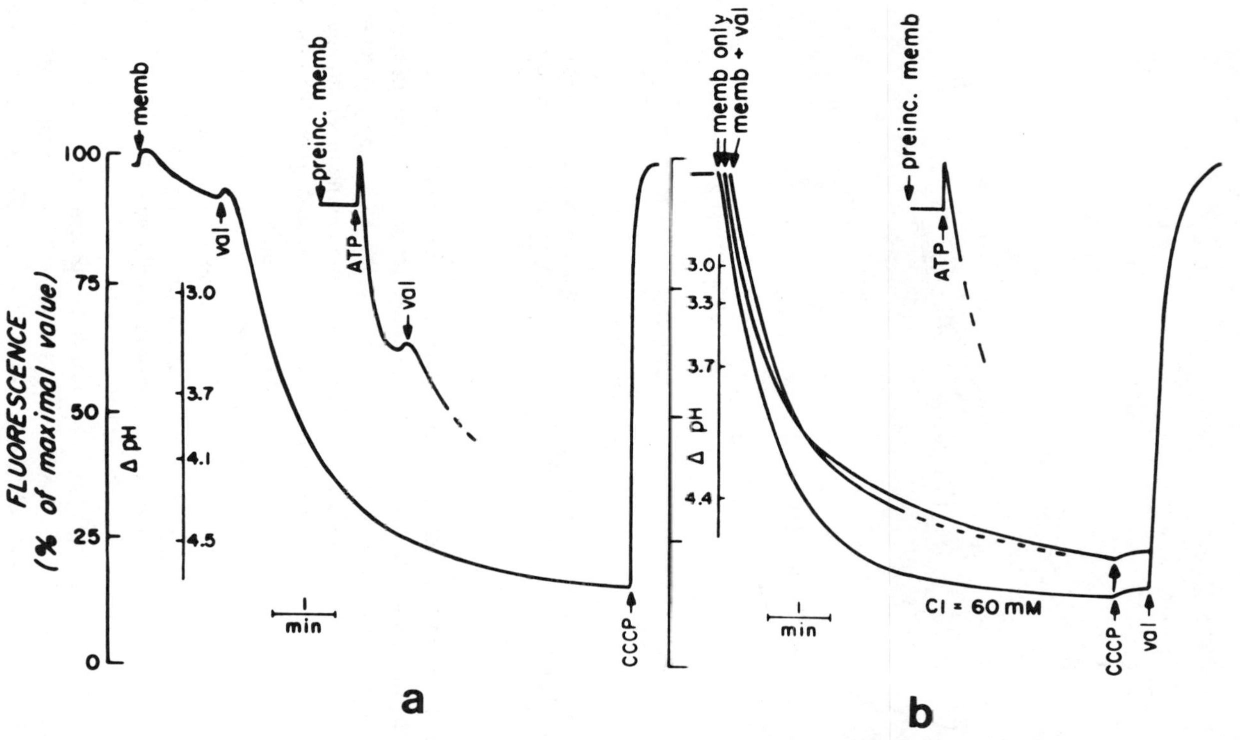


*Figure 3.* ATP-driven acidification of gastric vesicles measured by the accumulation of a weak base, acridine orange. (a) Addition of an ionophore, valinomycin, is required for acid accumulation by vesicles in the microsomal fraction. (b) Acid accumulation by heavier vesicles isolated from stimulated tissue is ionophore independent.

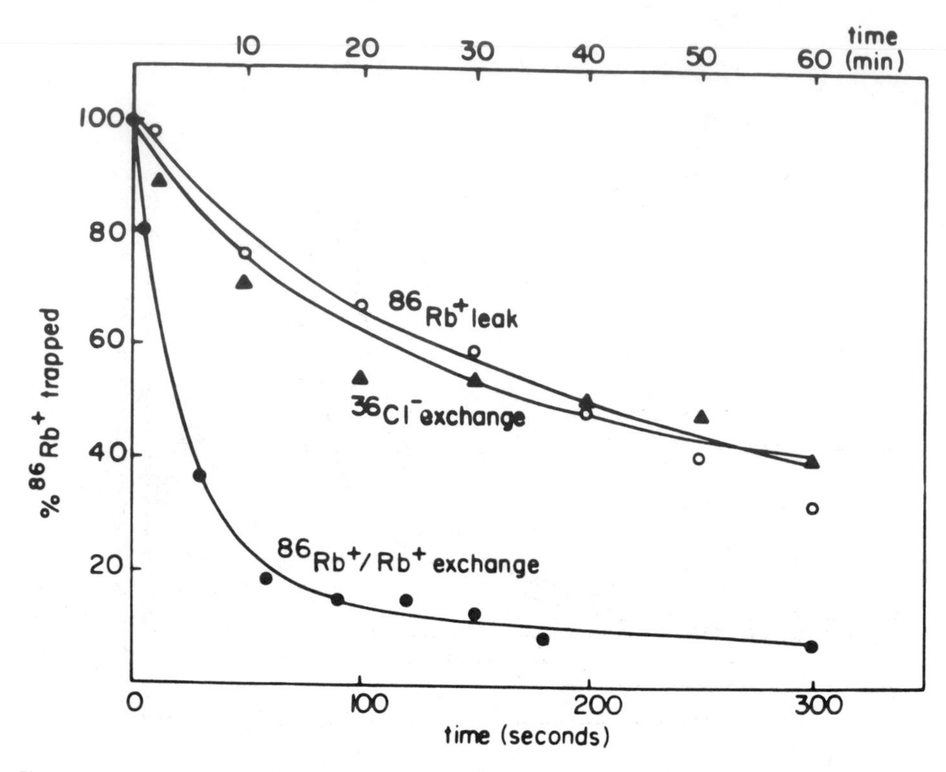


*Figure 4.* The movement of isotopes across hog gastric microsomal membranes. The time course of both RbCl movement and $^{36}Cl^-$:$Cl^-$exchange is slow (upper time scale), but $^{86}Rb^+$:$Rb^+$ exchange is fast (lower time scale).

pothesized that microsomal vesicles, which contain the H,K-ATPase, derive from the tubulovesicles that dominate the resting parietal cell. Stimulation results in fusion of these tubulovesicles with apical membrane containing the KCl transport pathway. Therefore, vesicles derived from the expanded secretory membrane of stimulated cells contain both components of the proton pump and do not require preincubation with KCl, or addition of an ionophore, for sustained acid accumulation. Support for this theory comes from immunopurification of membranes from resting and stimulated tissues using monoclonal antibodies (J. M. Wolosin, personal communication). Membranes from stimulated tissue contained actin, in addition to the 100K band characteristic of the H,K-ATPase. Filamentous actin has been implicated in the morphological changes that accompany secretion (Okamoto *et al.*, 1983).

Studies with isolated gastric glands confirm that stimulation of acid secretion involves activation of $K^+$ and $Cl^-$ pathways either inactive or absent in the apical membrane of the resting parietal cell (Malinowska *et al.*, 1983). One reservation to the proposal that stimulation involves activation of a KCl cotransport pathway is evidence that a $K^+$ pathway is intrinsic to the H,K-ATPase. In Figure 4, the rate of net $^{86}$RbCl flux across microsomal vesicles is compared with the rates of $^{86}$Rb:Rb exchange and $^{36}$Cl:Cl exchange (Schackmann *et al.*, 1977). The rapid cation:cation exchange is inhibited by vanadate ions, which selectively inhibit the H,K-ATPase as discussed in a subsequent section. To account for this observation and more recent experiments with charged dyes that indicate the existence of $K^+$ conductances in membranes isolated from resting tissue, an alternative model which attributes stimulation to insertion or activation of a $Cl^-$ pathway in the secreting membrane has been proposed (Rabon *et al.*, 1983a).

## IV. STRUCTURE OF THE H,K-ATPase

There is general agreement that most of the protein in purified gastric microsomal vesicles migrates with a molecular weight of approximately $1.0 \times 10^5$ in sodium dodecyl sulfate–polyacrylamide gel electrophoresis (SDS–PAGE). Nevertheless, tryptic digests of the membrane protein have been interpreted as showing three nonidentical subunits, and isoelectric focusing (IEF) patterns seem to confirm that the 100K band is heterogeneous. Recently, the solubilized 100K band was analyzed by ultracentrifugation, N-terminal amino acid determination, acylation of ε-$NH_2$ groups with citraconic acid, and binding to concanavalin A (Peters *et al.*, 1982). No evidence of heterogeneity was found, but any differences in the hydrodynamic properties of the subunits, or their reactivity toward the reagents employed, could be too small to detect. Two protein bands with p*I* values close to those previously reported were found by the same laboratory upon isoelectric focusing (Schrijen, 1981). However, both tryptic digestion and IEF are subject to a variety of artifacts. The best evidence for heterogeneity of the 100K band comes from probing the IEF peptide pattern with monoclonal antibodies (Smolka *et al.*, 1983). Figure 5a shows that one monoclonal antibody against the H,K-ATPase reacts specifically with the protein band centered at pH 8.5 on IEF gels. Two-dimensional electrophoresis indicates that this band has a molecular weight of approximately $8.0 \times 10^5$ and corresponds to a shoulder on the 100K band amounting to about 10% of the total protein. The diffuse major protein band centered at pH 6.2 can be resolved into distinct peaks by using a shallower pH gradient. Figure 5b shows that a second monoclonal antibody against the H,K-ATPase reacts almost exclusively with the more acidic peak. Therefore, at the level of antigenic determinants, the H,K-ATPase is composed of different subunits.

Radiation inactivation with electrons gave a functional molecular weight of $3.2 \times 10^5$ for both the ATPase and *p*NPPase activities of the gastric enzyme (Saccomani *et al.*, 1981). A subsequent molecular weight determination using gamma irradiation under similar experimental conditions gave a value of $(3.7 \pm 0.4) \times 10^5$, in reasonable agreement with the earlier value for inactivation of ATPase activity. However, the effective molecular weight found for *p*NPPase activity was one third lower (Schrijen *et al.*, 1983). Inactivation of ATPase activity by ultraviolet irradiation gives nearly twice the target size for inactivation of *p*NPPase activity (Chang *et al.*, 1977; Schrijen, 1981). A monomeric molecular weight approximately one third larger than indicated

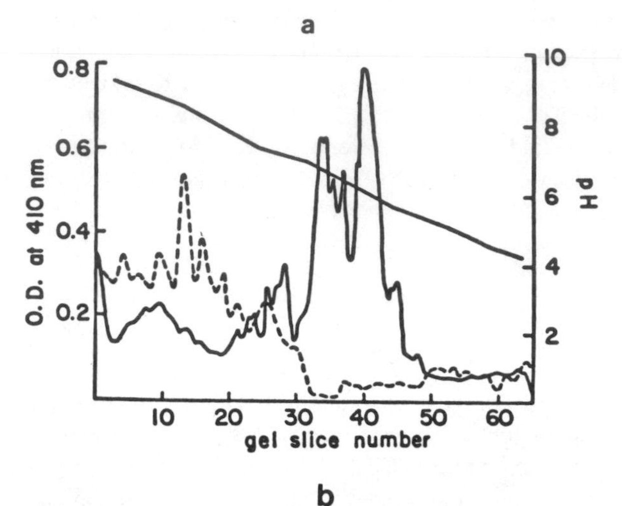


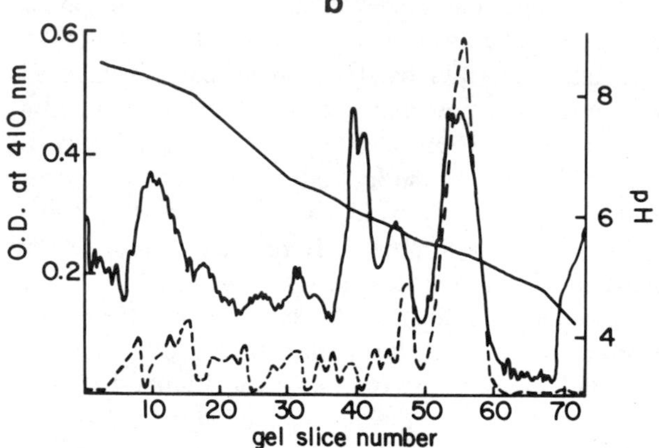


*Figure 5.* Photometric scans of isoelectric focusing patterns of hog gastric mucosal H,K-ATPase, resolved on polyacrylamide gels. The solid lines show the peptide distribution and the broken lines depict the binding of monoclonal antibodies to those peptides eluted from parallel gels and measured at 410 nm by an enzyme-linked immunoabsorbent assay. The diagonal lines are the pH gradients. (a) One antibody reacts specifically with the protein band centered at pH 8.5. (b) Another antibody binds specifically to the p*I* 5.4 subunit.

by SDS–PAGE has been found by sedimentation velocity measurements (Peters *et al.*, 1982). The sedimentation constant increased with time when the detergent used to solubilize the protein was removed by dialysis, and a molecular weight of $5.0 \times 10^5$ was inferred from sedimentation equilibrium measurements for a metastable equilibrium of the aggregating subunits.

Both heterotrimeric (Saccomani *et al.*, 1981) and homotetrameric (Peters *et al.*, 1982) structures have been proposed for the gastric ATPase. Given the reported variability in the molecular weights of the subunit(s) and the multimer, a choice between these and other possible subunit structures for the gastric ATPase cannot be made. Similarly, judgment must be reserved on the suggestion that one or more of the subunits required for ATPase activity are unnecessary for *p*NPPase activity (Chang *et al.*, 1977; Schrijen, 1981), in view of the differing relative molecular weights for ATPase and *p*NPPase activity found using different methods of inactivation. Differences in the reactivity of the chemically modified gastric enzyme with ATP and *p*NPP were explained in an earlier review (Sachs *et al.*, 1982). Different effects of site-selective reagents on the ATPase and *p*NPPase activities of the gastric enzyme have also been found.

## V. CATALYTIC PROPERTIES OF THE H,K-ATPase

The functional dependence of the rate of ATP hydrolysis by the gastric enzyme on ATP concentration is complex and has been interpreted as evidence for two nucleotide sites. In the absence of $K^+$, the Michaelis constants for high-affinity and low-affinity interaction of the substrate with the enzyme are 0.4 μM and 50 μM, respectively. Under these conditions, dephosphorylation of the enzyme is rate limiting, so that measurements of the amount of acid-stable phosphoenzyme formed yield the active site stoichiometry. The maximum amount of acid-stable phosphoenzyme formed is 1.5 nmoles per mg protein determined by the Lowry method. Surprisingly, this level is reached with 5 μM ATP, enough to fill only the high-affinity substrate site. One possible explanation is that the low-affinity nucleotide site is an effector site.

Phosphorylation has recently been shown to occur on the β-carboxyl group of an

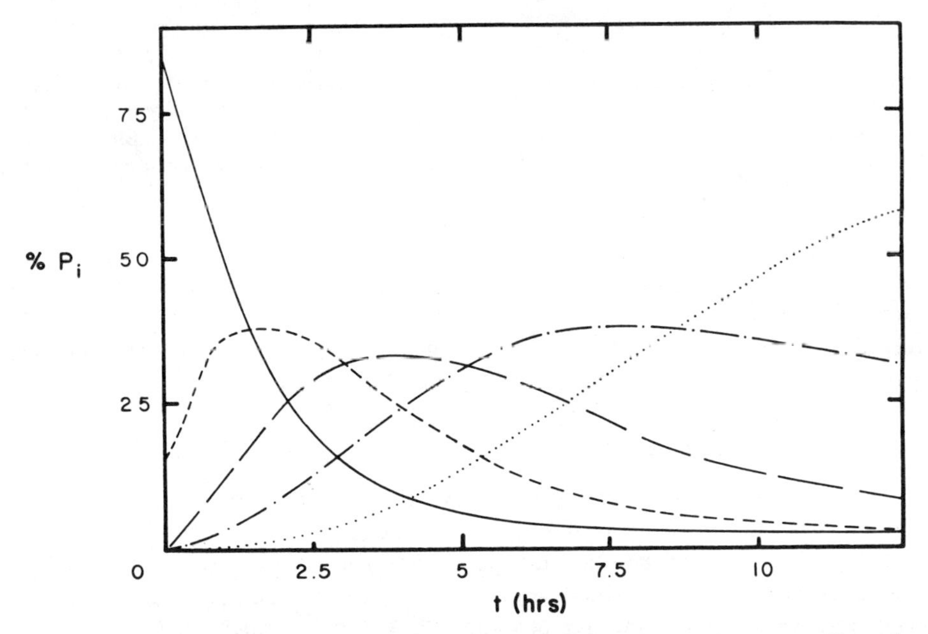


*Figure 6.* Catalysis of oxygen-18 exchange between inorganic phosphate and solvent water. The nuclear magnetic resonance frequency of phosphorus-31 is chemically shifted by oxygen-18. In a 117.5 kGauss magnetic field, the five possible inorganic phosphate species are completely resolved. The percentage of inorganic phosphate containing 4 (—), 3 (---), 2 (– –), 1 (-.-), or 0(· · ·) oxygen-18 atoms is plotted against time. No isotope exchange occurred in the absence of enzyme, or in the presence of a metal chelator, over the time period shown.

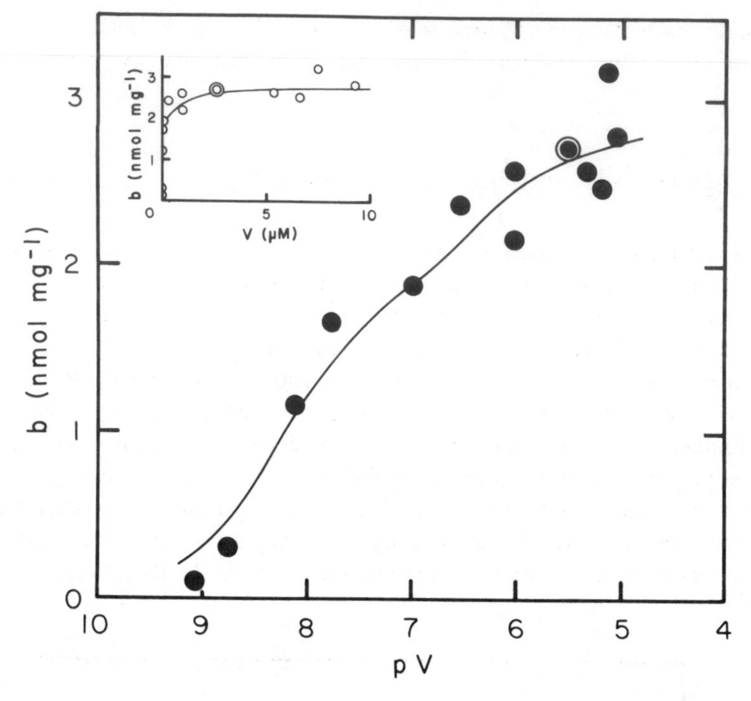


*Figure 7.* Vanadate binding to gastric microsomal vesicles. Vanadium-48 radiolabeled vanadate binding was measured by the rapid sedimentation method. Binding extended over a 10,000-fold range of free vanadate concentrations (semilogarithmic plot, $pV = -\log V$) indicating two classes of binding sites. The inset shows that saturation occurred with a stoichiometry of $2.8 \pm 0.3$ nmol vanadate bound (b) per mg of protein determined by the Lowry method.

aspartic acid residue (Walderhaug *et al.*, 1983), forming a relatively high-energy carboxylic-phosphoric anhydride intermediate. In the presence of $Mg^{2+}$ ions, the gastric enzyme also catalyzes isotopic exchange of carbon-14 between ADP and ATP, which demonstrates the reversibility of the partial reaction

$$\text{ATP} + \text{E} \rightleftharpoons \text{ATP} \cdot \text{E} \overset{Mg^{2+}}{\rightleftharpoons} \text{ADP} \cdot \text{ECOOP} \rightleftharpoons \text{ADP} + \text{ECOOP}$$

and has been interpreted as confirming the existence of a high-energy (ADP-sensitive) phosphoenzyme form (Rabon *et al.*, 1982a). More recently, catalysis of isotopic exchange of oxygen-18 between inorganic phosphate and solvent water has been demonstrated (Faller *et al.*, 1983). Figure 6 shows the time course of appearance and disappearance of each of the five possible $[^{18}O]P_i$ species. Since $Mg^{2+}$ is required, the reaction presumably involves reversal of the last step in the catalytic sequence and implies the existence of a lower energy phosphoenzyme form:

$$\text{HOH} + \text{E-P} \overset{Mg^{2+}}{\rightleftharpoons} \text{E} \cdot \text{P} \rightleftharpoons \text{E} + \text{P}_i$$

Two phosphoenzyme forms have been postulated before to explain biphasic dephosphorylation kinetics in the presence of $K^+$ (Wallmark *et al.*, 1980). However, $K^+$ sensitivity is probably not related to the intermediates in the partial reactions, since ATP–ADP exchange and oxygen-18 exchange are both accelerated by concentrations of $K^+$ that activate the overall hydrolysis reaction.

The inference that there are two nucleotide sites seems to be confirmed by studies of vanadate inhibition of the gastric ATPase (Faller *et al.*, 1983). Vanadate binds to the enzyme with two affinities as shown in Figure 7. Figure 8 shows that vanadate inhibition of ATPase activity is also biphasic. The binding and kinetic data are quantitatively consistent with inhibition by competitive high-affinity binding of ATP and vanadate to one nucleotide site and competitive lower-affinity binding to a second nucleotide site. Comparing Figures 7 and 8, half inhibition of ATPase activity corresponds to approximately 1.5 nmoles/mg of vanadate bound, and 3 nmol $mg^{-1}$ of vanadate must be bound to completely inhibit ATPase activity. One implication of the difference between the stoichiometry of phosphoenzyme formation (1.5 nmoles/mg) and saturable vanadate binding (3 nmoles/mg) is that the low-affinity nucleotide site is not an effector site, but a catalytic site. The more intriguing implication is that hydrolysis at the low-affinity nucleotide site is not catalyzed via a phosphoenzyme intermediate.

In contrast to ATPase activity, the functional dependence of the *p*NPPase activity of the gastric enzyme on substrate concentration is simple and suggests that *p*NPP is

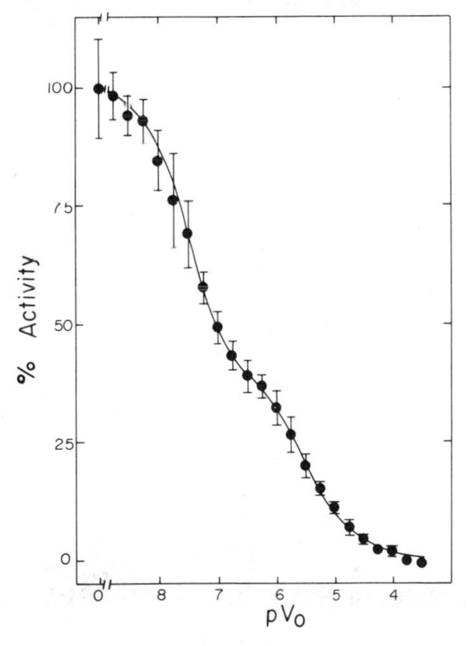


*Figure 8.* Vanadate inhibition of $K^+$-stimulated gastric ATPase activity. The semilogarithmic plot shows that inhibition occurs with two apparent inhibition constants, each accounting for loss of about half the activity.

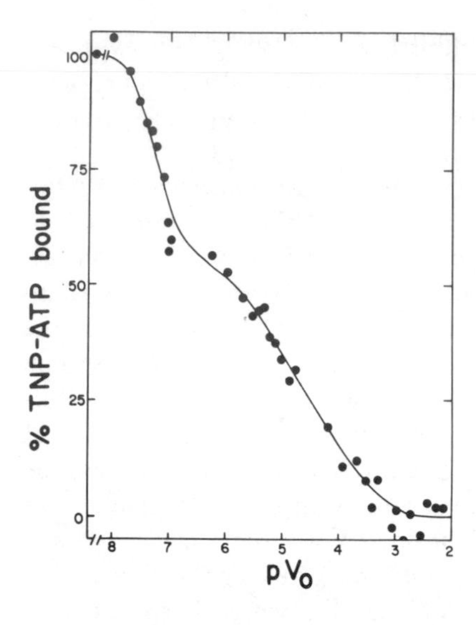


*Figure 9.* Competitive binding of vanadate and TNP-ATP (see text) to the gastric H,K-ATPase. Quenching of the fluorescence of the bound ATP analogue was measured. Two apparent vanadate affinities are evident in this semilogarithmic plot.

hydrolyzed at a single site (Forte *et al.*, 1981). Vanadate inhibition of *p*NPPase activity is compatible with this conclusion, since inhibition is monophasic and can be explained by competitive binding of vanadate and *p*NPP at the high-affinity nucleotide site (Faller *et al.*, 1983).

Binding studies with another reversible inhibitor of the gastric enzyme, the flu-

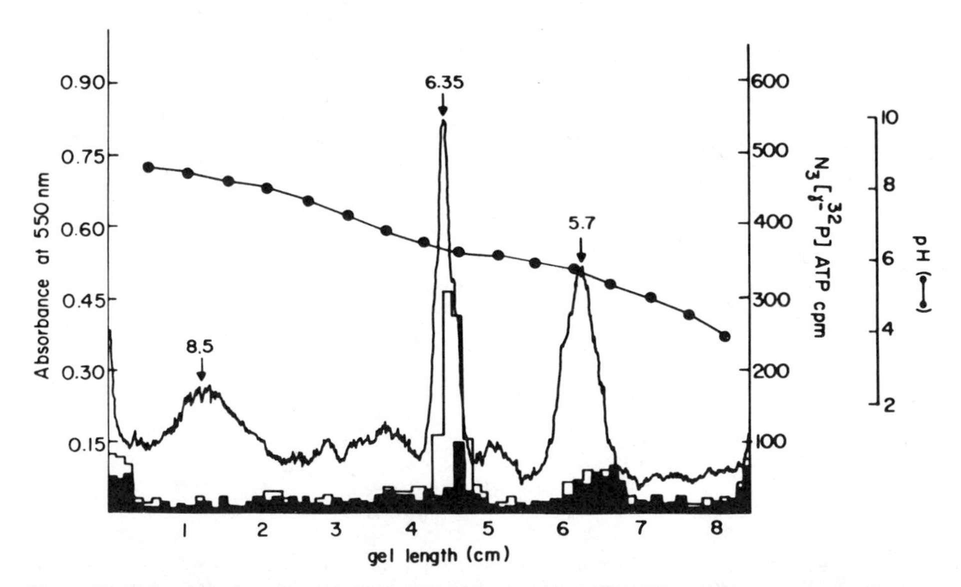


*Figure 10.* Photolabeling of gastric H,K-ATPase by 8-azido[γ-$^{32}$P]ATP. □ In the absence and ■ in the presence of cold ATP. Ampholines pH range 3.5–11 were used to resolve the proteins by isoelectric focusing on polyacrylamide gel. Incorporation of phosphorus-32 occurs largely on the p*I* 6.35 polypeptide.

orescent ATP analogue 2′,3′-*O*-(2,4,6-trinitrocyclohexadienylidine)-ATP (TNP-ATP), also suggest that there are two nucleotide sites (Sartor *et al.*, 1982). Figure 9 shows that TNP-ATP and vanadate ions bind competitively to two sites. Similar displacement curves were obtained with ADP, ATP, and β,γ-methylene ATP. In the absence of divalent cations, the stoichiometry of TNP-ATP binding was found to be 3 nmoles/mg Lowry protein.

Binding studies with a third reversible inhibitor of the gastric enzyme, adenylyl imido diphosphate (AMPPNP), have also been interpreted as evidence for two nucleotide sites (Van de Ven *et al.*, 1981). The maximum stoichiometry of AMPPNP binding observed in the absence of added ligands was 3 nmoles/mg Lowry protein. Titration with $Mg^{2+}$ reduced the number of ATP analogue binding sites by about half. The data were explained by a model in which binding to two independent, identical AMPPNP sites becomes strongly anticooperative when $Mg^{2+}$ is added.

8-Azido ATP ($N_3$ATP) is a photoaffinity analogue of ATP, which irreversibly inhibits the gastric enzyme. Photolysis of $N_3$[α-$^{32}$P]ATP resulted in incorporation of 1.4 nmoles/mg of the radiolabel into gastric vesicles (Saccomani *et al.*, 1983). Using $N_3$[γ-$^{32}$P]ATP, 3.5 ± 0.4 nmole $^{32}$P was incorporated per mg Lowry protein, but half the counts were chased by $K^+$, which accelerates dephosphorylation, and only half the ATPase activity was irreversibly lost when the enzyme was assayed in the presence of $K^+$ (Faller *et al.*, 1982). $K^+$-*p*NPPase activity was not inhibited by photolabeling with $N_3$ATP. A simple explanation, consistent with the vanadate binding and inhibition data, is that one nucleotide site is preferentially phosphorylated and a second site is photolabeled by $N_3$[γ-$^{32}$P]ATP. Figure 10 is an isoelectric focusing pattern of the $N_3$[γ-$^{32}$P]ATP photolabeled gastric enzyme. Phosphoenzyme is lost in this procedure, so the label is due to the covalently bound nucleotide. Comparing Figures 10 and 5b, the more basic peak into which the p*I* 6.2 band can be split by shallower pH gradients is labeled by $N_3$ATP, in contrast to the monoclonal antibody which labeled the more acidic peak.

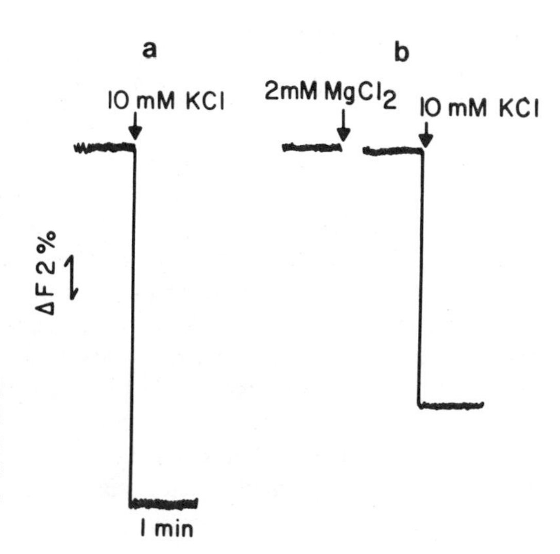


*Figure 11.* Fluorescence quenching induced by $K^+$. FITC-labeled H,K-ATPase was titrated with KCl in the (a) absence and in the (b) presence of $Mg^{2+}$. This result is the most direct evidence for a $K^+$-conformer of the gastric enzyme.

Fluorescein isothiocyanate (FITC) has also been used to probe the number of nucleotide sites and conformational states of the gastric H,K-ATPase (Jackson *et al.*, 1983). FITC, another irreversible inhibitor of the ATPase, labels the ε-$NH_2$ group of lysine residues. The stoichiometry of FITC binding was found to be 1.5 nmoles/mg Lowry protein. $K^+$-ATPase activity was almost completely inhibited by FITC, but *p*NPPase activity was unaffected. ATP protected against inhibition, suggesting that the reagent reacts at or near a nucleotide site. The change in fluorescence caused by titration of the fluorophore-labeled enzyme with KCl (Figure 11) is the most direct evidence to date for a $K^+$-conformation of the gastric enzyme.

## VI. MODEL

Two nucleotide sites, one an effector site, were originally proposed to explain the substrate dependence of the hydrolysis and phosphorylation reactions of the gastric ATPase. The results obtained with site-selective reagents are not inconsistent with a two-site model in which catalysis at the lower-affinity nucleotide site occurs without formation of an acid-stable phosphoenzyme intermediate. The site stoichiometry of 1.5 nmoles/mg of Lowry protein obtained with these reagents implies a molecular weight of $4.7 \times 10^5$, if correction is made for 42% overestimation of protein in gastric microsomal vesicles by the Lowry method (Peters *et al.*, 1982). This value is an upper limit, because only part of the protein in gastric microsomes migrates as a 100K band in SDS–PAGE. Using published estimates of the percentage of total protein in the 100K band gives a range of molecular weights from $3.3–4.2 \times 10^5$. The experimentally determined molecular weights are 3.2 and $3.7 \times 10^5$. More precise data are needed before speculating on the subunit structure of the enzyme. However, labeling of different peaks on IEF gels with different monoclonal antibodies and/or site-selective reagents points to at least two subunits with two types of nucleotide sites on different polypeptide chains. One apparently inconsistent observation is that 1.5 nmoles/mg of FITC completely inhibited ATPase activity, although *p*NPPase activity was unaffected. Since *p*NPP appears to react at the high-affinity site, a possible explanation is that FITC attachment near the lower-affinity site prevents ATP turnover at both sites. Most of the fluorescence in IEF gels of FITC-labeled enzyme was associated with the same peptide peak covalently labeled by 8-azido ATP.

Both monovalent and divalent cations affect the functional and conformational state of the gastric enzyme. Molecular weight values measured either in the presence of $Mg^{2+}$, or on samples purified in the presence of $Mg^{2+}$, were compared in preceding paragraphs. A 20% higher molecular weight has been measured in the absence of $Mg^{2+}$ (Schrijen *et al.*, 1983). The effects of $K^+$ on the fluorescence of the FITC-labeled enzyme and of $Mg^{2+}$ on AMPPNP binding to the enzyme have been cited. TNP-ATP appears to bind homogeneously in the absence of $Mg^{2+}$, but to two distinct classes of sites in its presence. $K^+$ has an analogous effect on vanadate binding (Faller *et al.*, 1983). These effects are poorly understood and could explain some of the seemingly irreconcilable data that have been reported. However, more striking is the consistency of the results obtained using a variety of chemical reagents and structural

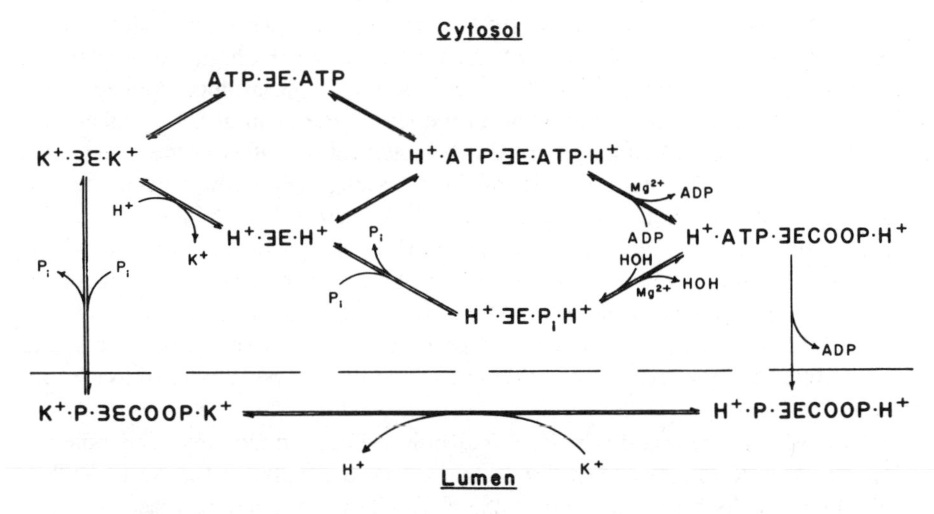


*Figure 12.* Model summarizing the major mechanistic features of the gastric H,K-ATPase. A carboxyl group can be reversibly phosphorylated by ATP or inorganic phosphate in the presence of $Mg^{2+}$. Dephosphorylation is accelerated by a $K^+$-induced change in the enzyme conformation (∋∈). The enzyme is modeled as a dimer because of structural evidence for dissimilar subunits and functional evidence for a second catalytic site that hydrolyzes ATP without forming an acid stable phosphoenzyme intermediate.

probes of the gastric enzyme. Figure 12 is a schematic diagram incorporating those mechanistic features of the H,K-ATPase for which there is reasonable evidence at the present time.

## VII. COMPARISON WITH OTHER TRANSPORT ATPases

The enzymes that couple ATP hydrolysis to metabolite transport across biological membranes may or may not form a covalent phosphoenzyme as an intermediate in the catalytic mechanism. The proton translocating enzymes of thylakoid, mitochondrial, and bacterial membranes are not phosphorylated. Transport enzymes that do react via a phosphoenzyme intermediate are the Na,K-ATPase and the Ca-ATPase found in mammalian tissues. The gastric enzyme is grouped with these latter enzymes, because a phosphoenzyme can be isolated. Mechanistic studies of the H,K-ATPase have been interpreted using schemes proposed for the Na,K-ATPase, with $H^+$ playing the role of $Na^+$.

Recent controversy about the Na,K-ATPase mechanism has focused on the number of nucleotide sites (Norby, 1983). Two sites were originally proposed to explain acceleration of the hydrolysis reaction at high ATP concentrations, and additional evidence for two nucleotide sites has been obtained (Skou and Esmann, 1981; Schuurmans Stekhoven *et al.*, 1981; Koepsell *et al.*, 1982). Alternatively, the complex substrate dependence of the catalytic reaction can be explained by a single site which has different affinities for ATP in different conformations of the enzyme (Smith *et al.*, 1980; Moczydlowski and Fortes, 1981b). The one-site/two-state model was proposed to explain complete inhibition of N,K-ATPase activity by high-affinity vanadate (Cantley *et al.*, 1978) and TNP–ATP binding with a single affinity (Moczydlowski and Fortes, 1981a). Vanadate inhibition of the Ca-ATPase can also be explained by a single site, since no inhibition of the $Ca^{2+}$-dependent activities of the enzyme was found (Pick, 1982). The biphasic inhibition and binding curves found using the same reagents to study the H,K-ATPase (Figures 7–9) cannot be explained by rapidly interconverting states of a single site, because binding of vanadate to the higher-affinity state would shift the equilibrium between them and result in inhibition of all ATPase activity and displacement of all TNP-ATP with a single apparent vanadate affinity. The possibility that the gastric preparation contains two enzymes cannot be excluded but, in addition to copurifying both enzymes, would have to be stimulated to the same extent by $K^+$ ions.

The H,K-ATPase also differs from the Na,K-ATPase structurally. The molecular weight of the H,K-subunits is similar to that of the α-subunit of the Na,K-ATPase, and the high-affinity nucleotide site on the H,K-ATPase mimics the reactivity of the Na,K-ATPase α-subunit. However, the H,K-enzyme does not have a lower molecular weight subunit resembling the β-subunit of the Na,K-ATPase. There is evidence for a 100K subunit which catalyzes ATP hydrolysis without forming a phosphoenzyme intermediate. The gastric enzyme may thus be evolutionarily related to the proton translocating enzymes of plant, bacterial, and mitochondrial origin as well as to the mammalian Na,K- and Ca-ATPases.

## VIII. PROBLEMS AND FUTURE RESEARCH

A coherent and consistent explanation of gastric acid secretion is beginning to emerge, although much of the evidence is circumstantial. For example, no unequivocal demonstration of membrane fusion or of two nucleotide sites on different subunits exists. One approach to the latter question is to show that different polypeptides are phosphorylated and photolabeled. Phosphoenzyme is unstable in the usual basic application to IEF gels, but preliminary experiments with acid applications are promising, showing phosphorylation of peptides with p*I* between 5.5–6.0 (Smolka, 1982). The most convincing approach would be to isolate the individual polypeptides by immunochromatography and study their reactivity with ATP and *p*NPP individually. A promising step in this direction is the development of anti-ATPase subunit monoclonal antibodies (Smolka, 1982) and solubilization of the H,K-ATPase in active form from gastric microsomes (Soumarmon *et al.*, 1983; Rabon *et al.*, 1983b). Another approach is to use a different type of probe. Oxygen-18 exchange can be used to study the mechanisms of enzymes that catalyze hydrolysis of oxygen–phosphorus bonds. Preliminary measurements of oxygen-18 exchange kinetics by NMR show promise of demonstrating the presence or absence of alternative hydrolytic pathways. So far, involvement of the postulated low-affinity nucleotide site in proton transport has not been demonstrated. Vesicles isolated from stimulated and resting tissues generate the same ΔpH of 4–5 units, several units less than the physiological proton gradient. One possibility is that the methods used to quantify proton gradients systematically underestimate the true value (Sack and Spenney, 1982). Alternatively, two hydrolytic pathways functioning sequentially to generate the enormous ion gradient across the gastric mucosa could be uncoupled during the preparation of vesicles. A maximum $H^+$/ATP stoichiometry of 1 can account for the physiological proton gradient, yet measurements of the $K^+$-dependent component of this ratio give values ranging from 1.5–2.0 (Rabon *et al.*, 1982b). Clearly, much remains to be learned about the mechanism of acid secretion and about the enzyme that catalyzes this important transport reaction.

## REFERENCES


Berglindh, T., DiBona, D. R., Ito, S., and Sachs, G., 1980, Probes of parietal cell function, *Am. J. Physiol.* **238:**G165–G176.

Cantley, L. C., Jr., Cantley, L. G., and Josephson, L., 1978, A characterization of vanadate interactions with the (Na,K)-ATPase, *J. Biol. Chem.* **253:**7361–7368.

Chang, H., Saccomani, G., Rabon, E., Schackmann, R., and Sachs, B., 1977, Proton transport by gastric membrane vesicles, *Biochim. Biophys. Acta* **464:** 313–327.

Faller, L., Jackson, R., Malinowska, D., Mukidjam, E., Rabon, E., Saccomani, G., Sachs, G., and Smolka, A., 1982, Mechanistic aspects of gastric ($H^+$ + $K^+$)-ATPase, *Ann. N.Y. Acad. Sci.* **402:**146–163.

Faller, L., Sachs, G., and Elgavish, G., 1983, P-31 NMR studies of O-18 disappearance from labeled inorganic phosphate catalyzed by gastric H,K-ATPase, *Fed. Proc.* **42:**1936.

Faller, L. D., Rabon, E., and Sachs, G.,1983, Vanadate binding to the gastric H,K-ATPase and inhibition of the enzyme's catalytic and transport activities, *Biochemistry* **22:**4676–4685.

Forte, T. M., Machen, T. E., and Forte, J. G., 1977, Ultrastructural changes in oxyntic cells associated with secretory function: A membrane-recycling hypothesis, *Gastroenterology* **73:**941–955.

Forte, J. G., Poulter, J. L., Dykstra, R., Rivas, J., and Lee, H. C., 1981, Specific modification of gastric $K^+$-stimulated ATPase activity by thimerosal, *Biochim. Biophys. Acta* **644:**257–265.

Jackson, R. J., Mendlein, J., and Sachs, G., 1983, Interaction of fluorescein isothiocyanate with the ($H^+$ + $K^+$)-ATPase, *Biochim. Biophys. Acta* **731:**9–15.

Koepsell, H., Hulla, F. W., and Fritzsch, G., 1982, Different classes of nucleotide binding sites in the ($Na^+$ + $K^+$)-ATPase studied by affinity labeling and nucleotide-dependent SH-group modifications, *J. Biol. Chem.* **257:**10733–10741.

Malinowska, D. H., Cuppoletti, J., and Sachs, G., 1983, Cl-requirement of acid secretion in isolated gastric glands, *Am. J. Physiol.* **245:**G573–G581.

Moczydlowski, E. G., and Fortes, P. A. G., 1981a, Characterization of 2′,3′-*O*-(2,4,6-trinitrocyclohexadienylidine) adenosine 5′-Triphosphate as a fluorescent probe of the ATP site of sodium and potassium transport adenosine triphosphatase, *J. Biol. Chem.* **256:**2346–2356.

Moczydlowski, E. G., and Fortes, P. A. G., 1981b, Inhibition of sodium and potassium adenosine triphosphatase by 2′,3′-*O*-(2,4,6-trinitrohexadienylidine) adenine nucleotides, *J. Biol. Chem.* **256:**2357–2366.

Norby, J. G., 1983, Ligand interactions with the substrate site of Na,K-ATPase: Nucleotides, vanadate and phosphorylation, *Curr. Top. Membr. Trans.* **19:**281–314.

Okamoto, C., Wolosin, J. M., Forte, T. M., and Forte, J. G., 1983, Topographical fluorescence microscopy of oxyntic cell microfilaments, *Biophys. J.* **41:**87a.

Peters, W. H. M., Fleuren-Jakobs, A. M. M., Schrijen, J. J., De Pont, J. J. H. H. M., and Bonting, S. L., 1982, Studies on ($K^+$ + $H^+$)-ATPase V. Chemical composition and molecular weight of the catalytic subunit, *Biochim. Biophys. Acta* **690:**251–260.

Pick, U., 1982, The interaction of vanadate ions with the Ca-ATPase from sarcoplasmic reticulum, *J. Biol. Chem.* **257:**6111–6119.

Rabon, E. C., Sachs, G., Mardh, S., and Wallmark, B., 1982a, ATP/ADP exchange activity of gastric ($H^+$ + $K^+$)-ATPase, *Biochim. Biophys. Acta* **688:**515–524.

Rabon, E. C., McFall, T. L., and Sachs, G., 1982b, The gastric [H,K]ATPase: $H^+$/ATP stoichiometry, *J. Biol. Chem.* **257:**6296–6299.

Rabon, E. C., Cuppoletti, J., Malinowska, D., Smolka, A., Helander, H. F., Mendlein, J., and Sachs, G., 1983a, Proton secretion by the gastric parietal cell, *J. Exp. Biol.* **106:**119–133.

Rabon, E. C., Gunther, R. D., and Soumarmon, A., 1983b, Extraction and solubilization of the hog gastric ATPase, *The Physiologist* **26:**A–109.

Saccomani, G., Sachs, G., Cuppoletti, J., and Jung, C. Y., 1981, Target molecular weight of the gastric ($H^+$ + $K^+$)-ATPase functional and structural molecular size, *J. Biol. Chem.* **256:**7727–7729.

Saccomani, G., Cole, L., and Mukidjam, E., 1983, Interaction of the photo-affinity label 8-azido ATP with the gastric ($H^+$ + $K^+$)-ATPase, *Fed. Proc.* **42:**1936.

Sachs, G., Koelz, H. R., Berglindh, T., Rabon, E., and Saccomani, G., 1982, Aspects of gastric proton-transport ATPase, in: *Membranes and Transport,* Vol. 1 (Anthony N. Martonosi, ed.), Plenum Publishing, New York, pp. 633–643.

Sack, I., and Spenney, I. G., 1982, Aminopyrine accumulation by mammalian gastric glands: An analysis of the technique, *Am. J. Physiol.* **243:**G313–G319.

Sartor, G., Mukidjam, E., Faller, L., Saccomani, G., and Sachs, G., 1982, Nucleotide probes of gastric ATPase, *Biophys. J.* **37:**375a.

Schackmann, R., Schwartz, A., Saccomani, G., and Sachs, G., 1977, Cation transport by gastric $H^+$ : $K^+$ ATPase, *J. Membr. Biol.* **32:**361–381.

Schrijen, J. J., 1981, Structure and mechanism of gastric ($K^+$ + $H^+$)-ATPase, Dissertation, University of Nijmegen, Nijmegen, The Netherlands.

Schrijen, J. J., Van Groningen-Luyben, W. A. H. M., Nauta, H., De Pont, J. J. H. H. M., and Bonting, S. L., 1983, Studies on ($K^+$ + $K^+$)-ATPase VI. Determination of the molecular size by radiation inactivation analysis, *Biochim. Biophys. Acta* **731:**329–337.

Schuurmans Stekhoven, F. M. A. H., Swarts, H. G. P., De Pont, J. J. H. H. M., and Bonting, S. L., 1981, Studies on ($Na^+$ + $K^+$)-activated ATPase XLV. Magnesium induces two low-affinity non-phosphorylating nucleotide binding sites per molecule, *Biochim. Biophys. Acta* **649:**533–540.

Skou, J. C., and Esmann, M., 1981, Eosin, a fluorescent probe of ATP-binding to the ($Na^+$ + $K^+$)-ATPase, *Biochim. Biophys. Acta* **647:**232–240.

Smith, R. L., Zinn, K., and Cantley, L. C., 1980, A study of the vanadate-trapped state of the (Na,K)-ATPase: Evidence against interacting nucleotide site models, *J. Biol. Chem.* **255:**9852–9859.

Smolka, A., 1982, A study of the proton translocating adenosine triphosphatase of the gastric mucosa using monoclonal antibodies, Dissertation, The University of Alabama in Birmingham, Birmingham, Alabama.

Smolka, A., Helander, H. F., and Sachs, G., 1983, Monoclonal antibodies against gastric $H^+$ + $K^+$ ATPase, *Am. J. Physiol.* **245:**G589–G596.

Soumarmon, A., Grelac, F., and Lewin, M. J. M., 1983, Solubilization of active ($H^+$ + $K^+$)-ATPase from gastric membrane, *Biochim. Biophys. Acta* **732:**579–585.

Van De Ven, F. J. M., Schrijen, J. J., De Pont, J. J. H. H. M., and Bonting, S. L., 1981, Studies on ($K^+$ + $H^+$)-ATPase III. Binding of adenylyl imidodiphosphate, *Biochim. Biophys. Acta* **640:**487–499.

Walderhaug, M. O., Saccomani, G., Wilson, T. H., Briskin, D., Leonard, R. T., Sachs, G., and Post, R. L., 1983, Aspartyl residue may be phosphorylated in a variety of membrane-bound adenosine triphosphatases, *Fed. Proc.* **42:**1275A.

Wallmark, B., Stewart, H. B., Rabon, E., Saccomani, G., and Sachs, G., 1980, The catalytic cycle of gastric ($H^+$ + $K^+$)-ATPase, *J. Biol. Chem.* **255:**5313–5319.

Wolosin, J. M., and Forte, J. G., 1981a, Changes in the membrane environment of the ($K^+$ + $H^+$)-ATPase following stimulation of the gastric oxyntic cell, *J. Biol. Chem.* **256:**3149–3152.

Wolosin, J. M ., and Forte, J. G., 1981b, Functional differences between $K^+$-ATPase rich membranes isolated from resting or stimulated rabbit fundic mucosa, *FEBS Lett.* **125:**208–212.

Wolosin, J. M., and Forte, J. G., 1981c, Isolation of the secreting oxyntic cell apical membrane-identification of an electroneutral KCl symport, in: *Membrane Biophysics: Structure and Function in Epithelia* (M. A. Dinno and A. B. Callahan, eds.), Alan R. Liss, New York, pp. 189–204.

Wolosin, J. M., and Forte, J. G., 1983, Kinetic properties of the KCl transport at the secreting apical membrane of the oxyntic cell, *J. Membr. Biol.* **71:**195–207.

# 39

# $H^+$-Translocating ATPase and Other Membrane Enzymes Involved in the Accumulation and Storage of Biological Amines in Chromaffin Granules

Sally E. Carty, Robert G. Johnson, and Antonio Scarpa

## I. INTRODUCTION

Within the adrenal medulla, catecholamines accumulate into and are stored within a highly specialized subcellular organelle, the chromaffin granule. Since their isolation almost 30 years ago, an interdisciplinary approach from pharmacologists, transport and cellular physiologists, electron microscopists, and anatomists has helped to delineate the salient aspects of the structure, composition, and function of these granules. One of the reasons for such profound interest is that the adrenal chromaffin granule has a common embryologic origin with the adrenergic neurotransmitter granules of the central and peripheral nervous system, suggesting that the two amine storage organelles may have functional properties in common.

Rapidly accumulating data indicate that the chromaffin granule is not, as had been thought for many years, a passive storage organelle. The identification of the electrochemical gradient for protons across the chromaffin granule membrane and the recognition that a membrane proton-pumping ATPase is responsible for the generation and maintenance of this gradient, have led to a wide range of experiments which have

*Sally E. Carty, Robert G. Johnson, and Antonio Scarpa* • Department of Biochemistry and Biophysics, University of Pennsylvania School of Medicine, Philadelphia, Pennsylvania 19104.

defined its integral role in the mechanism of amine accumulation. Coupled with advances such as the ability to form functional chromaffin ghosts devoid of endogenous catecholamines and gradients, and the isolation and characterization of most of the major proteins of the chromaffin granule membrane and matrix space, the chromaffin granule is now the best-understood subcellular organelle containing a hormone or neurotransmitter.

Several excellent descriptive reviews of various aspects of chromaffin granule composition, biogenesis, and function have appeared in the literature in the last several years (Winkler and Westhead, 1980; Winkler and Carmichael, 1982; Winkler, 1977; Njus and Radda, 1978; Njus *et al.*, 1981; Phillips, 1982; Johnson *et al.*, 1982d). The purpose of the present review is to describe in a general format the unique features of the chromaffin granule as they relate to its role in amine homeostasis, with particular focus on the bioenergetics and thermodynamics of enzymatic proton translocation and catecholamine transport.

## II. ISOLATION OF CHROMAFFIN GRANULES AND PREPARATION OF CHROMAFFIN GHOSTS

Chromaffin granules are routinely isolated by most investigators from adrenal glands of slaughterhouse animals such as the cow or pig. Immediately after the death of the animal, the cortex of the adrenal gland is removed by dissection and the medulla is minced and then homogenized at 4°C in isotonic buffered sucrose. Since the chromaffin granule is denser than the other organelles in the homogenate, isolation of highly purified granules by differential and density gradient centrifugation can be easily achieved.

Differential centrifugation of the homogenate produces a crude chromaffin granule fraction which, although rich in chromaffin granule constituents, is contaminated by large amounts of mitochondria and lysosomes. Further purification of the crude fraction is generally required and is accomplished by a variety of density gradient centrifugation techniques. The purest granules are prepared by ultracentrifugation of the crude granule fraction through hypertonic step gradients of 1.6 M sucrose, as described by Smith and Winkler (1967). Unfortunately, this method is inadequate for some experimentation since some lysis of the granules occurs upon resuspension in isotonic media.

There are three well-characterized isotonic purification procedures: (1) ultracentrifugation on step gradients of Ficoll–$D_2O$–sucrose (Trifaro and Dworkind, 1970), (2) ultracentrifugation on mechanically generated continuous gradients of metrizamide–sucrose (Morris and Schovanka, 1977), and (3) standard speed centrifugation on continuous gradients of Percoll–sucrose (Carty *et al.*, 1980). Of these, the isotonic Percoll method is the technique of choice. It produces the purest granule fraction, the yield is high, and there is no need for ultracentrifugation, mechanical gradient generation, or pelleting of granules. Preparation of a highly purified chromaffin granule suspension can thus be achieved rapidly and relatively inexpensively, and with minimal insult to the integrity of the organelle.

In intact chromaffin granules, any attempt to elucidate the membrane enzymatic properties and transport phenomena is hampered by the presence of the internal matrix rich in ATP, catecholamines, ions, amine-binding sites, and other components such as $Mg^{2+}$ and ascorbate. Matrix content may be eliminated by the use of chromaffin ghosts, generally prepared by one of two methods: (1) hypoosmotic lysis of purified granules followed by revesicularization of the washed membranes in isotonic media with extensive dialysis (Schuldiner *et al.*, 1978; Johnson *et al.*, 1981), or (2) application of the crude granule fraction to a hypoosmotic Sephadex G-50 column, or to a solution of glycerol, followed by density gradient ultracentrifugation of the membranes on a Ficoll or sucrose $D_2O$ step gradient with subsequent recovery of the re-formed ghosts (Phillips, 1974a; Deupree *et al.*, 1982; Njus and Radda, 1979). The first technique produces greater yield and less mitochondrial and lysosomal contamination, and offers the substantial advantage of allowing revesicularization in a variety of ionic and buffered media. Thus, depending on the selection of intravesicular and extravesicular media, ghosts can be formed so as to generate and/or maintain transmembrane proton, electrical, or ionic gradients as required by the choice of the experimental conditions. The development of this technique for preparation of purified chromaffin ghosts devoid of endogenous ions, amines, and nucleotides gives the necessary accuracy and flexibility to the study of energy-linked function and amine transport in the chromaffin granule membrane.

## III. THE COMPOSITION OF THE CHROMAFFIN GRANULE

The chromaffin granule is the best-studied subcellular organelle containing a neurotransmitter or hormone. Extensive analysis has been performed on the membrane lipids and proteins and also the components of the intragranular matrix space. The lipid composition of the membrane has been identified and most of the major proteins of the membrane and matrix space have been purified to homogeneity. The granule possesses lipids, proteins, substrates, ions, and enzymes clearly differentiated from those of other identified organelles such as mitochondria, lysosomes, or endoplasmic reticulum.

Several excellent and comprehensive reviews are available (Pletscher, 1974; Winkler, 1976, 1977, 1982; Winkler and Carmichael, 1982). This section will present an overview of the composition of the chromaffin granule so that the transport function and properties relating to transmembrane proton gradients may be appreciated (see Table 1).

### A. Membrane Proteins

The chromaffin granule maintains the lowest protein–lipid ratio of any isolated subcellular organelle. In fact, the only mammalian membrane with a lower ratio is that of the Schwann cell, which functions primarily as an electrical insulator. Approximately one-fifth of the total protein of the chromaffin granule is localized within the membrane. These proteins, a majority of which possess enzymatic activity, can be separated into at least 20 discrete bands with SDS electrophoresis. Two of the bands, identified as dopamine-β-hydroxylase and cytochrome $b_{561}$, comprise over 40% of the membrane protein. The $F_1$ $H^+$-translocating ATPase and NADH (acceptor) oxidoreductase have also been isolated to homogeneity.

Table 1. *Composition of the Chromaffin Granule*

| | Percent dry weight | nmol/mg protein | Concentration (M) |
|---|---|---|---|
| I. Protein | 42 | 1800 | |
| A. Membrane | 8 | | |
| 1. Enzymes | | | |
| NADH oxidoreductase | | | |
| Dopamine-β-hydroxylase | 2 | | |
| $F_1$ $H^+$-translocating ATPase | | 6.5 | |
| Cytochrome $b_{561}$ | 1.5 | | |
| Phosphatidylinositol kinase | | | |
| 2. Structural | | | |
| Glycoprotein | | | |
| Synaptin | | | |
| B. Soluble | 34 | | |
| 1. Enzymes | | | |
| Dopamine-β-hydroxylase | 1.8 | | |
| 2. Other | | | |
| Chromogranin | 13.6 | | |
| Enkephalin | | | |
| Polypeptides | | | |
| II. Lipid | | | |
| A. Phospholipid | 15 | 480 | |
| B. Cholesterol | 4.5 | 290 | |
| III. Catecholamine | 19 | 2500 | 0.500 |
| A. Dopamine | 0.2 | 20 | 0.004 |
| B. Norepinephrine | 5.2 | 680 | 0.126 |
| C. Epinephrine | 13.7 | 1800 | 0.360 |
| IV. Nucleotides | 16.8 | 800 | 0.1250 |
| A. ATP | 11.8 | 560 | 0.0875 |
| B. GTP | 1.5 | 76 | 0.0125 |
| C. UTP | 0.8 | 43 | 0.0068 |
| D. ADP | 1.4 | 66 | 0.010 |
| E. AMP | 0.8 | 38 | 0.006 |
| V. Ascorbate | 0.5 | 63 | 0.013 |
| VI. Metal ions | | | |
| A. Cations | | | |
| 1. $H^+$ | | | 0.00001 |
| 2. $Ca^{2+}$ | 0.12 | 76 | 0.014 |
| 3. $Mg^{2+}$ | 0.02 | 21 | 0.005 |
| 4. $Na^+$ | | 47.8 | 0.006 |
| 5. $K^+$ | | 23.0 | 0.012 |
| 6. $Cu^{2+}$ | 0.0016 | 0.6 | 0.0001 |
| 7. $Fe^{2+}$ | 0.0023 | 1.0 | 0.00028 |
| B. Anions | | | |
| 1. $Cl^-$ | | | |
| 2. $PO_4^{3-}$ | | | |
| VII. Mucopolysaccharides | 0.6 | | |
| VIII. Gangliosides | 0.8 | 0.015 | |

### 1. *Dopamine-β-hydroxylase*

This copper-containing enzyme is a mixed-function oxidase responsible for the β-hydroxylation of dopamine to form norepinephrine. Approximately 50% of its activity is localized within the membrane (requiring detergent for removal), while the remaining activity is associated with the cytosol (Skotland and Ljones, 1979; Rosenberg and Lovenberg, 1980). Composed of monomer subunits each containing a copper moiety, its native functional form is a tetramer with a molecular weight of 290,000 (Wallace *et al.*, 1973; Craine *et al.*, 1973). Although indirect evidence exists to suggest that the enzyme can be reduced directly by cytosolic ascorbate or other reducing equivalents (Grouselle and Phillips, 1982; Zaremba and Hogue-Angeletti, 1981), β-hydroxylation must proceed on the matrix side of the membrane. The $K_m$ of the enzyme for dopamine has not been determined under physiologic conditions, i.e., pH 5.5. At higher pH values (pH 6.4–7.2), the $K_m$ ranges from 1.67–5.0 mM, depending upon the experimental conditions (Goldstein *et al.*, 1968; Craine *et al.*, 1973).

### 2. *$F_1$ $H^+$-Translocating ATPase*

The physicochemical properties, activity, and function of this ATPase are discussed in detail in Section V. Briefly, the enzyme consists of an $F_1$ subunit (with α, β, and γ chains) and a dicyclohexylcarbodiimide-binding $F_0$ portion (Apps and Glover, 1978; Apps and Schatz, 1979; Buckland *et al.*, 1979; Apps *et al.*, 1980b; Sutton and Apps, 1981). The protein has similar subunit composition, fingerprinting, and antibody-binding characteristics to the mitochondrial $F_1$ ATPase (Apps and Schatz, 1979). However, significant differences exist with respect to electrophoretic mobility, antibody binding, and aurovertin fluorescence, denoting nonidentity of composition and structure (Apps and Schatz, 1979).

### 3. *Cytochrome $b_{561}$*

This enzyme comprises up to 15% of the membrane protein and its heme prosthetic portion imparts a characteristic pink color to the isolated granule. The cytochrome has been purified to homogeneity with a molecular weight of 35,000 (Silsand and Flatmark, 1974; Abbs and Phillips, 1980; Duong and Fleming, 1982; Flatmark and Gronberg, 1981). The high content of hydrophobic amino acids in conjunction with results of labeling experiments suggest that it is a transmembrane protein (Abbs and Phillips, 1980; Duong and Fleming, 1982).

Chromomembrin B, a previously identified major band on SDS gels, is now thought to be equivalent to the cytochrome $b_{561}$. The organization and possible function of the electron-transfer components within the membrane of the chromaffin granule are listed in Section VII.

### 4. *NADH (Acceptor) Oxidoreductase*

The protein utilizing cytosolic NADH as a substrate has been purified to homogeneity. Its molecular weight is 55,000, and it is thought to be a transmembrane protein (Zaremba and Hogue-Angeletti, 1982).

#### 5. *Phosphatidylinositol Kinase*

Phosphorylation of phosphatidylinositol has been reported by several groups (Buckley *et al.*, 1971; Phillips, 1973; Muller and Kirshner, 1975). The site of the phosphorylation is on the outer (cytosolic) surface of the bimolecular leaflet. However, since no diphosphoinositide is detectable in isolated chromaffin granules, the physiological role of the diphosphoinositide is not substantiated. The enzyme has not been isolated.

#### 6. *Structural*

Several proteins without enzymatic activity have been identified: (1) α-actinin, an integral membrane protein which binds actin in isolated systems (Jockusch *et al.*, 1977), (2) synaptin, which has similar properties to that isolated from the brain (Bock and Helle, 1977), and (3) several glycoproteins (in addition to dopamine-β-hydroxylase; Eagles *et al.*, 1975; Zinder *et al.*, 1978; Huber *et al.*, 1979; Cahill and Morris, 1979; Abbs and Phillips, 1980).

### B. *Soluble Proteins*

Eighty percent of the proteins of the chromaffin granule are soluble and are found within the intragranular matrix space. The proteins comprising the largest percentage of this fraction have been identified, but most of the remainder are unidentified small polypeptides. Only one, dopamine-β-hydroxylase, has appreciable enzymatic activity.

#### 1. *Dopamine-β-hydroxylase*

One-twentieth of the soluble protein is contributed by dopamine-β-hydroxylase. The only structural difference between the soluble and membrane bound forms appears to be a minute hydrophobic moiety which is too small to alter the amino acid composition (Skotland and Flatmark, 1979). Immunologically, the two forms are equivalent (Hortnagl *et al.*, 1973; Helle *et al.*, 1978).

#### 2. *Enkephalin*

The matrix space of the chromaffin granule contains several proteins with opiate activity in the body. To date, three enkephalins have been identified: Met-enkephalin, leu-enkephalin, and met-enkephalin [Arg-6 Phe] (Viveros *et al.*, 1979, 1980; Stern et al., 1979, 1981; Livett *et al.*, 1981).

#### 3. *Chromogranin A*

The chromogranins make up 40% of the proteins of chromaffin granule lysate (Winkler, 1976). They exist as a group of proteins with a molecular weight of 74,000–81,000 and migrate to the acidic region on isoelectric electrophoresis. Their structural configuration is that of a random coil. Their physiologic role remains conjectural (Winkler and Carmichael, 1982).

### 4. Small Polypeptides

A series of small polypeptides (in addition to the enkephalins), ranging in molecular weight from 300–5000, have been observed on SDS gels (Roda and Hogue-Angeletti, 1979). Their precise identification as prohormones, hormones, peptide precursors, or simple degradative proteins is an area of current investigation.

## C. Lipids

The phospholipid profile of the chromaffin granule is unique in that approximately 13–18% of the total phospholipid is constituted by lysophosphatidylcholine. The high percentage of this phospholipid does not appear to be *post mortem* or isolation artifact. The remainder of the phospholipids are phosphatidylcholine (~26%) and phosphatidylethanolamine (~30%), with sphingomyelin (~11%) and phosphatidylserine plus phosphatidylinositol (10%) making up the remainder (Winkler and Carmichael, 1982).

The chromaffin granule contains threefold more gangliosides than do microsomal membranes (Geissler *et al.*, 1977). The major gangliosides of chromaffin granules include two hematosides similar in structure to $G_{M_3}$ (Price *et al.*, 1975).

## D. Storage Components

### 1. Catecholamines

Almost 20% of the dry weight of the chromaffin granule consists of the dopamine, norepinephrine, and epinephrine which are contained within the intragranular space (Winkler, 1976). Quantitatively, this would correspond to 2.5 μmoles/mg protein or a matrix concentration of 500 mM if all the endogenous amines were free in solution (Winkler, 1976).

The most probable intracellular sites of catecholamine biosynthesis are as follows. Dopamine is formed from L-Dopa in the cytosol through decarboxylation by the enzyme amino acid decarboxylase. It then accumulates across the chromaffin granule membrane via a specific amine transporter driven by the electrochemical proton gradient (see Section VI). The efficiency of subsequent β-hydroxylation of dopamine to norepinephrine is reflected in the very low content of dopamine within the granule (<1% of total catecholamines). Because epinephrine is the product of *N*-methylation of norepinephrine by the cytosolic enzyme phenylethanolamine-*N*-methyltransferase, it is generally accepted that in some cells norepinephrine is released from the granule, methylated, and then reaccumulated as epinephrine. Morphological and histochemical evidence indicates that two distinct types of cells exist: norepinephrine cells which lack the enzyme PNMT, and epinephrine cells which contain all the biosynthetic enzymes (Coupland and Hopwood, 1966; Winkler, 1969; Terland *et al.*, 1979).

### 2. Nucleotides

Nucleotides are present within the intragranular space in approximately one-fourth the quantity of the catecholamines (Winkler, 1976). The predominant nucleotide is

ATP, although significant amounts of GTP and UTP are also present. A number of studies have addressed the issue of the catecholamine–ATP ratio yielding values ranging from 2–12 (for review see Winkler and Carmichael, 1982). Variations have been documented to occur with granule subpopulations (new vs. old, norepinephrine vs. epinephrine-containing granules), seasonal variation in catecholamine and ATP content, and as a function of isolation procedure. The significance of the precise quantitation of this ratio has never been determined.

### 3. *Ascorbate*

The adrenal medulla has the highest ascorbic acid content of any mammalian tissue. One-third of this is localized to the chromaffin granule (Ingebretsen *et al.*, 1980). It is present in a concentration of approximately 63 nmoles/mg protein or one-fortieth of the catecholamine content, which would correspond to a concentration of approximately 13 mM if all the ascorbate were free in solution. The ascorbate has been measured to be in the reduced form in excess of 98% (Ingebretsen *et al.*, 1980; Terland and Flatmark, 1975; Sen and Sharp, 1980).

### 4. *Ions*

Multiple cations can be measured in the lysate of chromaffin granules. Calcium is found in the greatest amount, ranging from 75–100 nmoles/mg protein (Borowitz *et al.*, 1965; Borowitz, 1967, 1969; Phillips, 1981; Serck-Hannsen and Christiansen, 1973), with magnesium approximately one third to one fourth of that (Borowitz *et al.*, 1965; Phillips and Allison, 1977). The sodium content is approximately twice that of potassium (Johnson, unpublished observations). Most of the iron in the membrane is confined to the cytochrome $b_{561}$, and likewise, the copper to dopamine-β-hydroxylase (Phillips and Allison, 1977). The proton concentration is $10^{-5.5}$ as measured by distribution of radiochemically labeled methylamine, fluorescence amines, and [$^{31}$P]-NMR (see Section IV-2). No accurate matrix values for anions, including chloride and phosphate, have been reported.

### 5. *Mucopolysaccharides*

Eighty percent of the total mucopolysaccharides are solubilized in the matrix space (Geissler *et al.*, 1977). The predominant glycosaminoglycans (for both membrane-associated and soluble forms) are chondroitin-4-sulfate (61%), chondroitin-6-sulfate (35%), and heparin sulfate (4%; Margolis and Margolis, 1973).

## E. *Topography*

Careful analysis of the composition of the lipids and proteins of the chromaffin granule membrane has led to assignment of their spacial orientation within the lipid bilayers (for review see Winkler and Westhead, 1980; Winkler and Carmichael, 1982). This is shown schematically in Figure 1.

As with other biological membranes there is a high degree of asymmetry in the lipid–protein distribution. The lipid moiety is composed of a bimolecular leaflet of

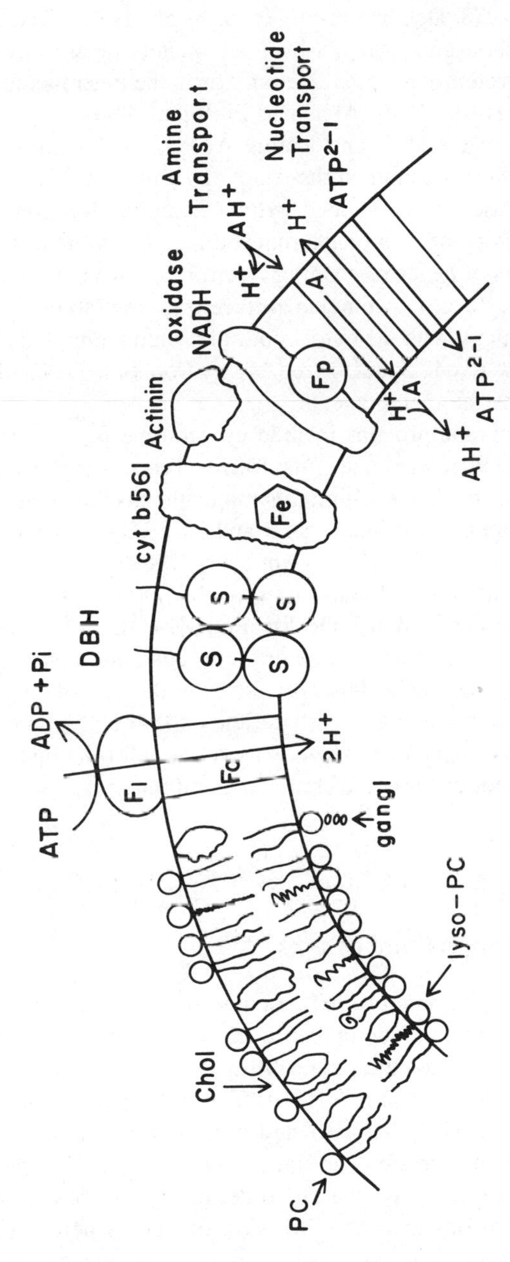


*Figure 1.* Schematic representation of protein and lipid distribution within the membrane of chromaffin granules. More details in the text.

phospholipids. The majority of lysolecithin is localized to the inner leaflet of the bilayer, and most of the phosphatidylethanol is in the outer leaflet (Buckland *et al.*, 1978; Voyta *et al.*, 1978; DeOliveira-Filgueiras *et al.*, 1981). Distribution of the other phospholipids and cholesterol has not been adequately determined. Most of the gangliosides and glycoproteins are accessible only from the inner membrane leaflet (Eagles *et al.*, 1975; Huber *et al.*, 1979; Abbs and Phillips, 1980).

The $F_1$ portion of the $H^+$-translocating ATPase is thought to be external to the outer membrane surface based upon the following evidence: (1) antibody binding, (2) dependence of ATPase on ATP added extravesicularly (Johnson *et al.*, 1982a), (3) loss of enzyme activity after pronase digestion of the granule (Abbs and Phillips, 1980), and (4) results of negative-staining electron microscopy, which revealed knobs projecting from the cytosolic membrane surface presumed to be the ATPase (Schmidt *et al.*, 1982). By analogy with the mitochondrial proton pump, the proton translocating portion of the ATPase ($F_0$) is characterized as a hydrophobic DCCD-binding lipoprotein complex which spans the membrane.

Other transmembrane proteins include cytochrome $b_{561}$, which may actually be anisotropically oriented toward the outside membrane surface (Abbs and Phillips, 1980). Localization of the heme within this molecule has not been established. NADH oxidoreductase is a transmembrane protein and its catalytic portion is exposed to the cytosolic surface (Zaremba and Hogue-Angeletti, 1982).

The precise orientation and association of flavoprotein with other enzymes have yet to be determined. Localization of the dopamine-β-hydroxylase within the membrane is controversial as well. Most investigators now conclude that the enzyme does not completely span the membrane; however, since reduction of its copper moiety can proceed directly via extragranular ascorbate, the exposure of a portion of the molecule to the cytosol cannot be excluded. There is as yet no information about the orientation of the amine or nucleotide carrier within the membrane.

## IV. THE ELECTROCHEMICAL $H^+$ GRADIENT

### A. Membrane Permeability to Ions

Sensitive spectrophotometric and potentiometric techniques coupled with the appropriate use of ionophores or lipophilic membrane-permeable ions have permitted extensive investigation of the permeability of the chromaffin granule membrane to small ions (Johnson and Scarpa, 1976). The diffusion of the cations $Na^+$, $K^+$, $H^+$, $Ca^{2+}$, $Mg^{2+}$, and $Mn^{2+}$ into isolated chromaffin granules was virtually negligible, indicating that these cations are quite impermeable. The passive permeability of $H^+$ across the membrane was over one order of magnitude less than that across the inner mitochondrial membrane. The diffusion of anions across the membrane was higher than that for cations with relative permeabilities in the order $SCN^- > I^- > Br^- > Cl^- > SO_4^- > F^- >$ isethionate, $PO_3^{2-}$ (Johnson and Scarpa, 1976; Casey *et al.*, 1976; Dolais-Kitabgi and Perlman, 1975; Phillips, 1977a).

The observed low proton conductivity of the chromaffin granule membrane prompted attempts to measure the intragranular pH and later the transmembrane potential difference of the granules after isolation and storage. Since the granule matrix is too small to accommodate a standard pH or potential microelectrode, more indirect methods were developed to characterize $H^+$ flux across the membrane.

## B. Measurement of the ΔpH

The difference between the intragranular pH and the pH of the medium in which isolated chromaffin granules are suspended (termed the ΔpH) has been measured by (1) [$^{14}$C]methylamine distribution, (2) quenching of fluorescent amines, and (3) [$^{31}$P]-NMR. Consistently, the intragranular matrix space has been found to be quite acidic by all these techniques.

Methylamine is a weak base ($pK_a$ = 10.6) that can only permeate biologic membranes in the deprotonated, uncharged form. Although, at equilibrium, the concentration of uncharged methylamine will be the same on both sides of a membrane, if one side of the membrane is more acidic, methylamine will become protonated on that side and will be effectively trapped in the acidic medium. When exposed to an organelle with an acidic interior relative to the external medium, the total equilibrium methylamine concentration will thus be much higher inside than outside. At near physiologic pH values, the ratio of internal to external methylamine closely approximates the ratio of internal to external $H^+$, i.e., of the ΔpH. The ΔpH across the membrane of an organelle can thus be accurately and reproducibly quantitated by radioassay of the distribution of tracer amounts of [$^{14}$C]methylamine (for review see Rottenberg, 1980). A small deviation from ideal methylamine distribution according to the ΔpH occurs because of methylamine binding to membrane and matrix components, but this deviation is confined to, at most, the equivalent of 0.2 pH units (Johnson and Scarpa, 1976).

When freshly isolated chromaffin granules are suspended in buffered sucrose, KCl, NaCl, Na–isethionate, or choline–Cl media at pH 7.0 and incubated with 10 μM [$^{14}$C]methylamine, a transmembrane ΔpH of 1.43 units is measured (Johnson and Scarpa, 1976). When the pH of sucrose medium is varied from 7.9 to 5.4, there is a proportional decrease in the ΔpH, which is equal to zero at an external pH of 5.5 (Figure 2). The measurements are consistent with an intragranular pH of 5.5, and additional experiments have shown that this acidic intragranular pH is independent of the external pH and of the ionic composition of the media (Johnson and Scarpa, 1976). The acidic chromaffin granule interior is further evidence for an extremely low membrane conductance to protons and has been confirmed by numerous investigators to be 5.5–5.7 (Salama *et al.*, 1979; Bashford *et al.*, 1976; Ritchie, 1975; Pollard *et al.*, 1979).

The endogenous ΔpH of isolated granules can be perturbed by protonophores and large concentrations of ammonia (and other primary amines). It decays only very slowly with time. Isolated chromaffin granules at 4°C have been found to maintain a ΔpH of nearly 1.4 units 48 hr after isolation (Johnson *et al.*, 1978). Ammonia and other primary amines (such as methylamine, catecholamines, and pharmacologic amines) can permeate the chromaffin granule membrane in the uncharged form as described above. If present in the incubation medium in large amounts (generally >0.5 mM), reprotonation of accumulated amine will occur, removing $H^+$ from the intravesicular

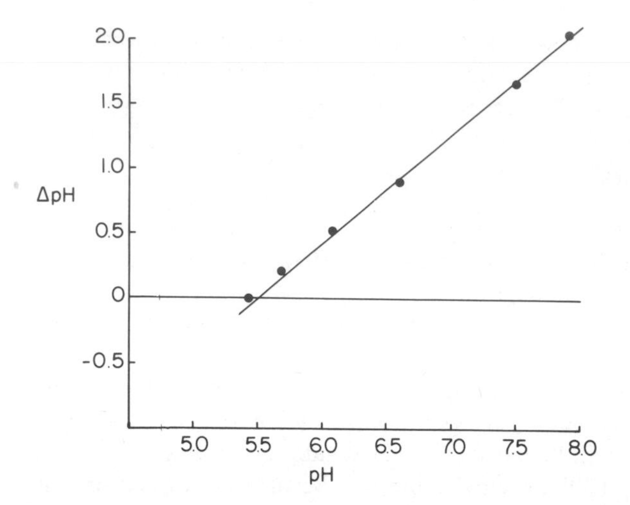


*Figure 2.* Measurement of ΔpH between the intravesicular space of chromaffin granules and suspending media of varying pHs. The medium consisted of 0.25 M sucrose buffered with 20 mM each of tris (hydroxymethyl)aminomethane(Tris) and 2-[*N*-morpholino]ethanesulfonate (MES) at the pHs indicated in the abscissa, and 2 mg protein/ml of chromaffin granules. Tracer amounts of $^3H_2O$ and [$^{14}$C]methylamine or [$^{14}$C]polydextran were also added to measure intravesicular space and ΔpH as described previously (Johnson and Scarpa, 1976).

solution. As the buffering capacity of the matrix space is exceeded, alkalinization of the intragranular space with stepwise decrease in the ΔpH will result. From comparison of the changes in intravesicular pH and labeled methylamine accumulation in the matrix, a very high intragranular buffering capacity (approximately 300 μmol $H^+$/pH unit × g dry weight; see Johnson *et al.*, 1978) can be estimated.

Ionophores such as FCCP, nigericin, and A23187 perturb the ΔpH in predictable ways (Johnson *et al.*, 1978). FCCP catalyzes electrogenic proton movement through a membrane down the electrochemical gradient for $H^+$. When added to a suspension of chromaffin granules, it allows protons to leak out, slowly collapsing the ΔpH. Nigericin, which catalyzes electroneutral $K^+$–$H^+$ exchange, collapses the ΔpH of granules suspended in KCl media but slightly increases the ΔpH of granules in $K^+$-free media. The ionophore A23187, when capable of electroneutral $Ca^{2+}$–$H^+$ exchange in the ratio 1 : 2 (Reed and Lardy, 1972), can collapse the ΔpH if granules are suspended in medium containing $Ca^{2+}$ (Johnson *et al.*, 1978).

Certain primary amines such as 9-aminoacridine and atebrin maintain a high native fluorescence when free in solution and can accumulate into chromaffin granules by the same mechanism as methylamine. Once trapped in the intragranular space, their fluorescence is quenched. By measuring the degree of quenching as well as the internal and external water spaces, the ΔpH can be calculated (Rottenberg, 1980). The measurement of chromaffin granule ΔpH using fluorescent dyes has produced values in good agreement with those obtained using [$^{14}$C]methylamine distribution (Salama *et al.*, 1979; Bashford *et al.*, 1976).

Because of the unusually high ATP content of the isolated chromaffin granule, [$^{31}$P]-NMR can be used as another probe for measuring intragranular pH. Since the $pK_a$ of the γ-P of ATP is within the physiologic range, the resonance of the γ-P is shifted according to the pH of the environment in which the ATP is located. Comparison of the shifts obtained in intact granules with those of model suspensions approximating them has made possible the determination of not only intragranular composition but also intragranular pH, with results comparable to those of the two other techniques

(Ritchie, 1975; Casey *et al.*, 1977a,b; Njus *et al.*, 1978; Pollard *et al.*, 1979). Although [$^{31}$P]-NMR has proved to date to be chiefly a confirmatory method, it is potentially a powerful tool in the study of ΔpH *in situ* or *in vivo*.

The physiologic role of the acidic intragranular pH is still speculative but may serve a variety of biological functions of the granules. These include: (1) the maintenance of stored catecholamines in unoxidized form, (2) the provision of optimum pH environment for DβH activity, (3) the driving force for catecholamine accumulation and storage, and (4) the provision of driving force for transport of other substrates present within the chromaffin granule. The evidence to date for these concepts is presented in succeeding sections.

## C. Measurement of the ΔΨ

The potential difference of charge across the membrane of the chromaffin granule is termed the ΔΨ (positive inside), and has been measured using both radioisotopic and spectrophotometric tracers. Thiocyanate ($SCN^-$) and tetramethyl-phenylphosphonium ($TPMP^+$) are lipophilic ions that can freely permeate membranes in the charged form to distribute according to the transmembrane Nernst potential (Rottenberg, 1980). When incubated with a suspension of chromaffin granules or ghosts, the ratio of internal to external [$^{14}$C]-$SCN^-$ or [$^3$H]-$TPMP^+$ will give an exact measure of the potential, positive inside (ΔΨ, mV) or potential, negative inside (−ΔΨ, mV), respectively. Aspecific binding of [$^{14}$C]-$SCN^-$ accounts for an error in calculation of only <6 mV; pH-dependent binding of [$^3$H]-$TPMP^+$ to chromaffin membranes, however, is appreciable (up to 20 mV), and this factor must be controlled for in quantitative measurements (Johnson and Scarpa, 1979). These radioisotopic methods have been used with great success to reproducibly and sensitively quantitate the chromaffin granule ΔΨ, and have been independently verified in other organelles large enough to permit potential electrodes (Johnson and Scarpa, 1979; Holz, 1979; Pollard *et al.*, 1976; Phillips and Allison, 1978).

When freshly isolated chromaffin granules were suspended in sucrose in the absence of permeant anions and in the presence of 10 μM [$^3$H]-$TPMP^+$, a resting ΔΨ of −10 to −20 mV was measured (Figure 3). Addition of MgATP produced a rapid and marked increase in the ΔΨ which reached a plateau of +80 mV, as measured by distribution of 10 μM [$^{14}$C]-$SCN^-$. The potential could be sustained for over 40 min and had no effect on the ΔpH (not shown). Under these experimental conditions, the observed MgATP-dependent generation of a large positive ΔΨ could only be caused by inward movement of positive charge. Addition of the ionophore FCCP, which allows $H^+$ ions to leak down their electrochemical gradient, evoked a rapid reversal of the ΔΨ to −80 mV; this value was exactly equal to the diffusion potential for $H^+$ contributed by the acidic intragranular space (ΔpH = 1.3; see below). These phenomena are consistent with the generation of a positive ΔΨ by inward translocation of $H^+$ in a process driven by ATP hydrolysis. The MgATP-stimulated generation of a ΔΨ across the chromaffin granule membrane has been reported by a number of other investigators (Johnson and Scarpa, 1979; Holz, 1979; Pollard *et al.*, 1976; Phillips and Allison, 1978; Bashford *et al.*, 1975; Scherman and Henry, 1980).

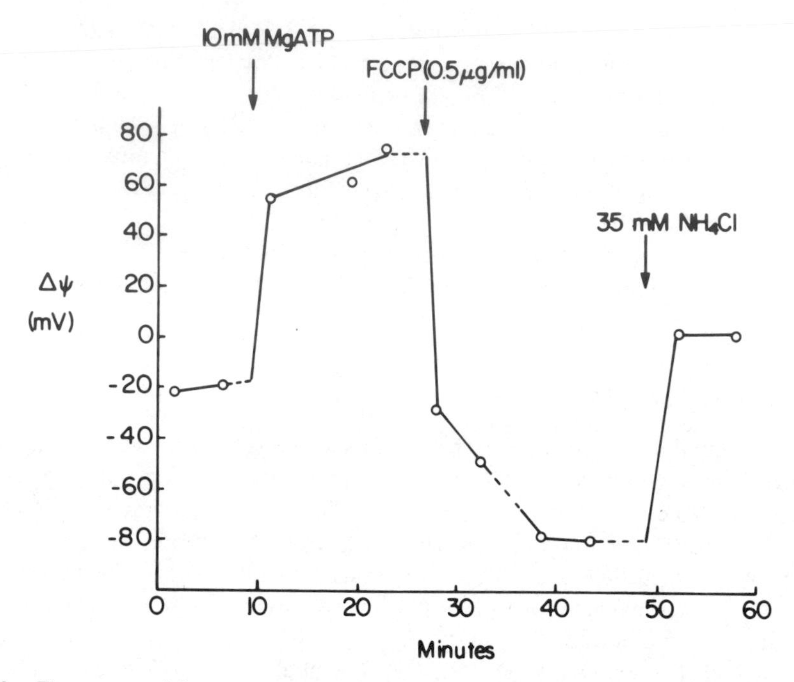


*Figure 3.* Time course of the measurement of $\Delta\psi$ in isolated chromaffin granules. The medium contained 0.27 M sucrose, 30 mM Tris-maleate (pH 6.8), and 10.8 mg chromaffin granule protein/ml. To one chamber thermostated at 24°C, [$^{14}$C]-$SCN^-$ and $^3H_2O$ were added to measure the membrane potential (positive inside), to another, [$^3$H]tetramethylphenylphosphonium ($TPMP^+$) and [$^{14}$C]polydextran to measure the membrane potential (negative inside). ATP, carbonyl cyanide *p*-trifluoromethoxyphenylhydrazone (FCCP) and $NH_4^+Cl$ were added where and at the concentrations indicated. More details in Johnson and Scarpa (1979).

Once generated, the $\Delta\Psi$ can be perturbed by several compounds including FCCP, $SCN^-$ (and other lipophilic anions), and $Cl^-$ (Johnson and Scarpa, 1979). Although the ionophore FCCP is quite potent (1.0 μg/ml), $SCN^-$, $ClO_4^-$, and small permeable anions such as $Cl^-$ must be present in the external medium in large amounts (1–60 mM) to collapse the $\Delta\Psi$. All exert their effect in a dose-dependent fashion. The generation of a $\Delta\Psi$ can be inhibited partially or completely by DCCD and alkyltin; these compounds, however, act indirectly by inhibition of $H^+$-translocating ATPase activity (see Section V).

Several membrane permeable potential-sensitive dyes, namely ANS, Oxonol-V, and di-S-$C_3$-(5), have been used to measure $\Delta\Psi$ in isolated chromaffin granules (Salama *et al.*, 1979; Bashford *et al.*, 1975; Scherman and Henry, 1980). All can be used to monitor changes in the $\Delta\Psi$ by changes in fluorescence and/or transmittance. The cyanine dye di-S-$C_3$-(5) was found to have a linear optical response to the Nernst potential for $H^+$ and $K^+$ over the range $-60$ to $+90$ mV in isolated chromaffin granules. Measurements of $\Delta\Psi$ using this dye agreed well with those obtained using the radiochemical distribution method (Salama *et al.*, 1979).

The physiologic role of the $\Delta\Psi$ in chromaffin granule homeostasis is only partially understood. The $\Delta\Psi$ is known to contribute to the driving force for catecholamine accumulation and storage. It is also thought to promote accumulation of ATP; these concepts are presented in succeeding sections. Other functions of the $\Delta\Psi$ such as in transport of other substrates, in stimulus–secretion coupling, or in interaction with the $e^-$-transport chain remain purely conjectural.

## D. Measurement of the $\Delta\bar{\mu}_{H^+}$

According to the Nernst equation, the $\Delta\Psi$ across a membrane (mV) may be calculated as follows:

$$\Delta\Psi = (2.3\ RT/F)\log\ \{([^{14}C]\text{-}SCN^-)_{in}/([^{14}C]\text{-}SCN^-)_{out}\}$$

At room temperature, the constant (2.3 $RT/F$) is equal to 58.8. The diffusion potential for $H^+$ (in mV) is expressed as:

$$\Delta pH = 58.8\ \log\ \{([^{14}C]\text{-}MA^-)_{in}/([^{14}C]\text{-}MA^-)_{out}\}$$

The $\Delta pH$ and $\Delta\Psi$ can be related by the chemiosmotic theory outlined by Mitchell in other energy-transducing organelles (Mitchell, 1961, 1968):

$$\Delta\mu_{H^+} = \Delta\Psi - 58.8\ \Delta pH$$

where $T = 23°C$, $\Delta pH$ is defined as $pH_{in} - pH_{out}$ as in mitochondria, and $\Delta\bar{\mu}_{H^+}$ is the electrochemical proton gradient (Mitchell, 1968). The $\Delta\bar{\mu}_{H^+}$ across the chromaffin granule membrane has been measured by both radiochemical and spectrophotometric techniques, most reliably by the distribution of [$^{14}$C]methylamine and [$^{14}$C]-$SCN^-$ in parallel and identically treated samples (Johnson and Scarpa, 1979; Johnson *et al.*, 1979, 1981). In this laboratory, the $\Delta\bar{\mu}_{H^+}$ consistently measures 180–210 mV in intact freshly isolated granules suspended at pH 7.4 with MgATP. Considering the large magnitude of this gradient, which seems likely to exist under physiologic conditions, it is not surprising that the $\Delta\bar{\mu}_{H^+}$ plays a pivotal role in amine transport, as will be discussed in Section VI.

# V. THE $H^+$-TRANSLOCATING ATPase

The discovery of an endogenous proton gradient ($\Delta pH$, inside acidic) and of an ATP-stimulated electrical gradient ($\Delta\Psi$, inside positive) suggested that a $H^+$-translocating ATP hydrolase might exist in the chromaffin granule membrane. Evidence for the existence of a proton-pumping ATPase, now generally accepted, was based on fulfillment of four minimal criteria: (1) Measurement of ATP-stimulated $H^+$ translocation in ghost preparations devoid of endogenous gradients, (2) elucidation of a fixed stoichiometry between ATP hydrolysis and $H^+$ translocation, (3) demonstration of ATP synthesis from ADP and $P_i$ in the presence of artificially magnified $H^+$ gradients, and (4) inhibition of proton translocation by known inhibitors of other $H^+$-

pumping ATPases. Additional evidence included: (5) isolation and physicochemical characterization of the ATPase, and (6) demonstration that antibodies to mitochondrial $F_1$ ATPase also bind to the chromaffin granule proton pump.

## A. Generation of a $\Delta\bar{\mu}_{H^+}$

The characteristics of the $H^+$-translocating ATPase are best studied using purified chromaffin ghosts, which are devoid of the endogenous ion, amine, and ATP gradients of intact granules and which can be formed in a variety of ionic media (see Section II). To evaluate proton translocation, purified ghosts were formed and suspended at neutral pH in two different types of ionic media, and changes in the ΔpH and ΔΨ were measured using the radioisotopic methods described in Section IV. In the first medium (Na-isethionate), the addition of MgATP should result in vectorial $H^+$ translocation inward. Since no permeable anion (or cation) was present in this medium, however, the initial generation of a ΔΨ (inside positive) would build up a large diffusion potential, positive inside, which would rapidly limit further $H^+$ influx. The overall result would be generation of a large ΔΨ and a negligible ΔpH. Conversely, in the second type of medium (KCl), the rapid influx of the permeable anion $Cl^-$ in response to an initial inwardly positive ΔΨ would collapse any membrane potential and allow large-scale $H^+$ influx to ensue. The overall result in KCl medium would be the generation of a large ΔpH and a minimal ΔΨ.

These considerations were borne out by experimental results (Figure 4; Johnson *et al.*, 1979). When MgATP was added to ghosts in Na-isethionate medium, a large ΔΨ was rapidly generated and was maintained for over 30 min. The ΔΨ was entirely abolished by the addition of permeable anions or by the electrogenic proton ionophore FCCP. When MgATP was added to ghosts in KCl medium, on the other hand, a large ΔpH was rapidly generated and persisted for over 30 min. The addition of 1–80 mM $NH_4^+$ completely collapsed the ΔpH in a dose-dependent fashion (ammonia is thought to permeate the membrane as $NH_3$ and realkalinize the ghost interior by buffering $H^+$ as $NH_4^+$). Other experiments with ion-selective ionophores such as nigericin and valinomycin were consistent with this explanation and have excluded the possibility that $H^+$ flux may be secondary to primary ATP-dependent translocation of another ion (Johnson and Scarpa, 1976). These data provided strong support for the hypothesis that ATP hydrolysis is linked to proton translocation in the chromaffin granule membrane.

The chemiosmotic hypothesis of Mitchell (1968) states that the electrochemical proton gradient across a biologic membrane is composed of interconvertible electrical and concentration components according to the relationship

$$\Delta\bar{\mu}_{H^+} = \Delta\Psi - Z\Delta pH$$

where $Z = 2.3\ RT/F$. The experiments cited above (Johnson *et al.*, 1979) and others (Johnson *et al.*, 1978; Phillips, 1977a; Schuldiner *et al.*, 1978; Ingebretsen and Flatmark, 1979; Bashford *et al.*, 1976; Holz, 1978; Knoth *et al.*, 1980; Njus and Radda, 1979; Kanner *et al.*, 1980; Scherman and Henry, 1979; Johnson and Scarpa, 1979; Apps *et al.*, 1980a) have conclusively demonstrated that, in chromaffin ghosts, addition

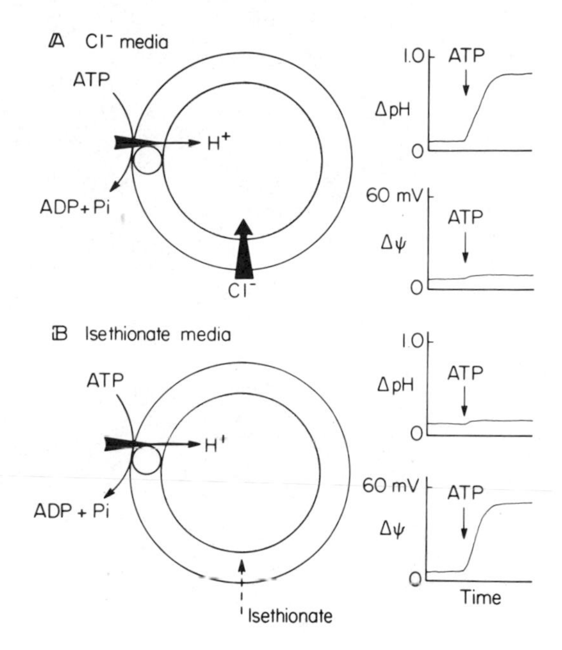


*Figure 4.* Schematic illustration of ATP-dependent generation of (A) ΔpH in media containing $Cl^-$, and (B) Δψ in media containing isethionate.

of MgATP can produce a ΔpH alone, a ΔΨ alone, or both, depending upon the ionic composition of the medium. They provide convincing evidence for the existence of an electrogenic proton pump. Extrapolation of the *in vitro* results to cytosolic conditions of pH, $[Cl^-]$, and [ATP] suggests that *in situ,* the chromaffin granule ATPase may maintain a transmembrane $\Delta\bar{\mu}_{H^+}$ of over 200 mV.

## B. Stoichiometry of $H^+$: ATP

Coupling of ATP hydrolysis and anisotropic $H^+$ movement across the chromaffin granule membrane will occur in a fixed ratio if the ATPase activity observed is truly that of a proton pump. Experiments to determine the precise stoichiometry between ATP hydrolysis and $H^+$ translocation were technically difficult to design and interpret due to several factors: (1) hydrolysis of ATP itself results in net pH change of the medium at most values of pH and $[Mg^{2+}]$, (2) movement of $H^+$ through other coupled transport systems often occurs, and (3) adenylate kinase activity is ubiquitous in most subcellular preparations. However, through the meticulous choice of initial pH and $Mg^{2+}$ concentrations, the utilization of chromaffin ghosts devoid of endogenous $H^+$ gradients, the use of continuous potentiometric pH monitoring, and of reserpine (to inhibit possible movement of minute amounts of catecholamine), most of these technical difficulties were overcome.

As is illustrated in Figure 5, under the conditions specified, any pH change in the medium reflected actual translocation of $H^+$ across the chromaffin ghost membrane (Johnson *et al.*, 1982a). Addition of ATP to the reaction mixture (Figure 5A) resulted in a large time-dependent alkalinization of the external medium consistent with $H^+$

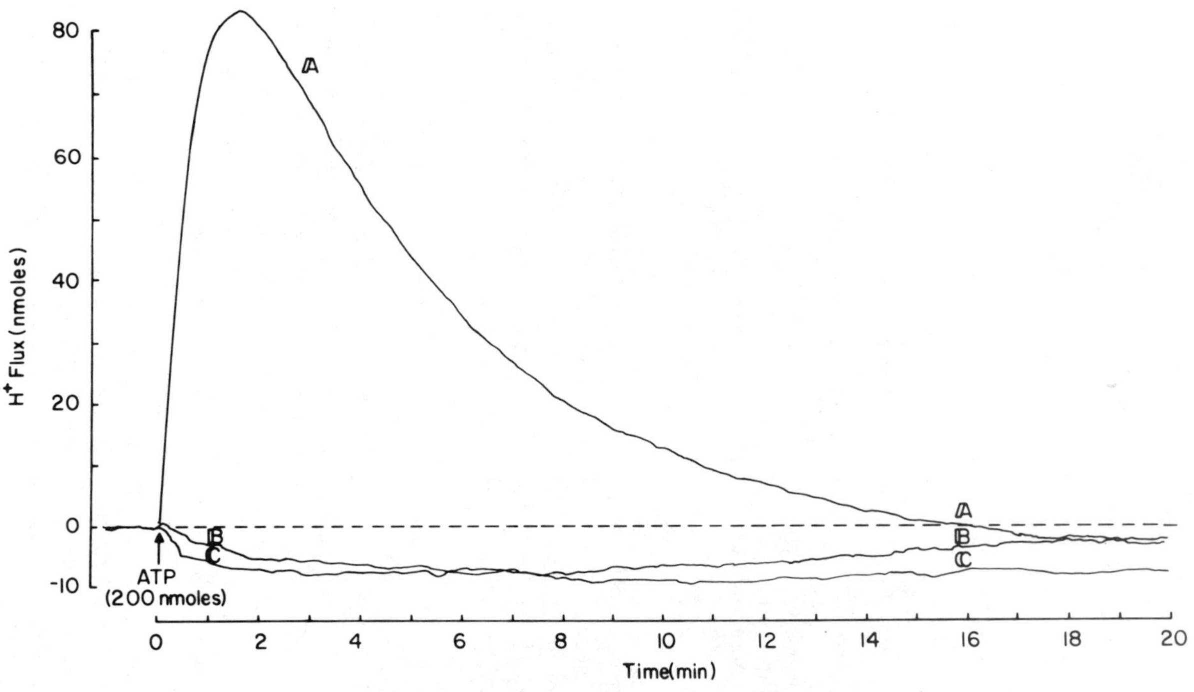


*Figure 5.* ATP-driven $H^+$ translocation into chromaffin ghosts. ATP-driven $H^+$ flux across the membrane of chromaffin granule ghosts was measured by following the pH changes of the medium. Ghosts (1.35 mg protein) re-formed in 185 mM KCl and 1 mM MES (pH 6.10) were incubated in a medium consisting of 185 mM KCl, 500 μM MES, 1.7 mM $MgSo_4$, and 10 μg valinomycin. The pH was adjusted to 6.10 with HCl, and the mixture (3.0 ml final volume) was allowed to equilibrate 5–10 min until a stable baseline was achieved. MgATP (200 nmoles) was added where indicated by the arrow. The recordings shown are (A) no other additions, (B) plus 7.5 μg FCCP, (C) plus 0.83 μmole trimethyltin (TMT). Temperature was 22°C. From Johnson *et al.* (1982a).

translocation into the intravesicular space. In the presence of FCCP (a proton ionophore; Figure 5B) or trimethyltin (an ATPase inhibitor; Figure 5C), no pH changes were recorded. Precise measurement of ATP hydrolysis under identical conditions permitted calculation of the $H^+$: ATP stoichiometry, which ranged from 1.52–1.75. These compared favorably with other values in the literature calculated less directly (Flatmark and Ingebretsen, 1977; Njus *et al.*, 1978). Based on this data, the ideal $H^+$: ATP stoichiometry is most likely 2.0, although nonintegral proton translocation, for example, that secondary to an enzyme involved in proton translocation which may have a $pK_a$ near the pH of the reaction medium, cannot be excluded.

## C. Reversal of the ATPase

Thermodynamic principles dictate that for an $H^+$-translocating ATPase, generation of ATP should proceed in the presence of ADP and $P_i$ by reversal of the pump through an imposed proton gradient. One laboratory has synthesized up to 4 nmoles ATP over 10 sec using the $\Delta\bar{\mu}_{H^+}$ across the chromaffin ghost membrane (Roisin *et al.*, 1980) as the driving force for synthesis. The synthesis was inhibited by proton ionophores, ATP inhibitors, and modulation of the magnitude of the $\Delta\bar{\mu}_{H^+}$.

## D. Physicochemical Properties of the ATPase

Recently, data are accumulating which provide a more complete characterization and understanding of the properties of the chromaffin granule ATPase. A compilation of these measurements is presented in Table 2.

### 1. $K_m$

The $K_m$ of the ATPase is 69 μM (Johnson *et al.*, 1982a), which is far below the cytosolic ATP concentration. This fact suggests that *in situ* ATPase activity is not affected by changes in cytosolic ATP concentration over the physiologic range.

### 2. $V_{max}$

The large variability that exists in measurements of the $V_{max}$ of the ATPase in isolated granules and membranes can probably be ascribed to different methods of preparation, varying experimental temperatures, and the unsuitability of some protein assays utilized. The $V_{max}$ was consistently measured in this laboratory to be 110–130 nmoles ATP/mg protein per min at 10°C (Johnson *et al.*, 1982a). The observed threefold difference (Table 2) between the basal ATPase rate and that in the presence of FCCP indicates that the ATPase is sensitive to changes in the $\Delta\Psi$, since the membrane potential falls from +60 mV to −90mV as the ATPase rate rises (Johnson *et al.*, 1982a). The many measurements of $V_{max}$ in membranes (for review see Johnson *et al.*, 1982a) have ranged from 150–250 nmoles ATP/min per mg protein, and the threefold difference between these values and the granular $V_{max}$ probably reflects the fact that only 30% of granule protein is contributed by membrane protein and the remaining 70% by soluble proteins of the matrix space.

Table 2. *Properties of the Chromaffin Granule $H^+$ ATPase*

| Property | Measurement | Reference |
|---|---|---|
| Specificity | ATP = ITP > GTP > UTP > CTP | Kirshner, 1962 |
| $K_m$ | 69 μM | Johnson *et al.*, 1982a |
| $V_{max}$ (granules) | 20, + FCCP 60 | Bashford *et al.*, 1976 |
| | 91, + FCCP 167 | Pazoles *et al.*, 1980 |
| | 10, + FCCP 45 | Johnson *et al.*, 1982a |
| pH optimum | 6.4 | Muller and Kirsher, 1975 |
| | 6.5 | Bashford *et al.*, 1976 |
| | 7.4 | Taugner, 1971 |
| | 7.3 | Johnson *et al.*, 1982a |
| Stoichiometry ($H^+$: ATP) | 2 | Njus *et al.*, 1978 |
| | 1.53 | Flatmark and Ingebretson, 1977 |
| | 1.65 | Johnson *et al.*, 1982a |
| Molecular weight | 400,000 | |
| | α-subunit, 51,000 | Apps and Glover, 1978 |
| | β-subunit, 50,000 | Apps and Glover, 1978 |
| | γ-subunit, 28,000 | Apps and Glover, 1978 |
| Inhibitors | Alkyl tin, $K_{i(50)}$ = 10 nM/mg prot | Johnson *et al.*, 1982a |
| | DCCD, $K_{i(50)}$ = 80 | Bashford *et al.*, 1976 |
| | DCCD, $K_{i(50)}$ = 300 | Johnson *et al.*, 1982a |

### 3. *pH Optimum*

The $H^+$ ATPase exhibited a broad pH dependence with a maximum activity at pH 7.3. Indirect evidence is accumulating that the $F_1$ portion of the ATPase is spacially oriented on the cytosolic side of the membrane. Since the cytosolic pH is probably 7.4 (as opposed to pH 5.5 in the intragranular space), the ATPase would be functioning optimally *in vivo*.

### 4. *Inhibitors*

Several agents which selectively inhibit $H^+$ ATPases in mitochondria, chloroplasts, chromatophores, and bacteria have been tested for their effect upon the ATPase activity of fragmented chromaffin granule membranes. Each of these ATPases has been shown to include an $F_1$ portion responsible for ATP hydrolysis and an $F_0$ portion which is conceptualized as the lipoprotein-lined channel through which proton conductance proceeds. In many of these systems, the ATPase can donate electrons to the membrane electron transport chain and vice versa. The application of several of these agents to the chromaffin granule system produced the following results:

1. Rotenone and antimycin A (electron transport chain inhibitors), oligomycin and aurovertin (ATPase inhibitors in mitochondria and chloroplasts), and reserpine (biogenic amine transport inhibitor) failed to significantly influence the ATPase activity of chromaffin granule membranes (Johnson *et al.*, 1982a; Apps and Schatz, 1979).

2. DCCD and alkyltin (trimethyltin), which are known to specifically inhibit the ATPases of mitchondria, chloroplasts, and *E. coli*, were excellent inhibitors of the

chromaffin granule ATPase. No analogues of DCCD were more effective as inhibitors than DCCD on a mole-to-mole basis (Johnson *et al.*, 1982a).

3. While ETOH had no effect on ATPase activity, the solvent DMSO (dimethyl sulfoxide) nonspecifically stimulated activity by 154% in isolated membranes (Johnson *et al.*, 1982a).

### E. Isolation and Reconstitution of the ATPase

The ATPase has been isolated from the chromaffin granule membrane with a 60-fold increase in activity and 60% yield (Apps and Schatz, 1979). The ATPase closely resembles the mitochondrial $F_1$ ATPase of bovine heart or adrenal medulla with respect to subunit composition, molecular weight on gel filtration, immunologic crossreaction against $F_1$ ATPase of bovine heart mitochondria, and proteolytic fingerprints of the three largest subunits. The chromaffin granule ATPase, however, reacted differently from the mitochondrial $F_1$ ATPase by quantitative complement fixation.

There have been two attempts at reconstitution of the chromaffin granule ATPase from bile acid extracts. In the first, the reconstituted preparation in native lipid or in soybean phospholipid resulted in ATPase activity which was inhibited by DCCD and mildly stimulated by the uncoupler S-13 (Buckland *et al.*, 1979). In the second, ATPase activity of the reconstituted fraction was sensitive to triphenyltin and DCCD (Giraudat *et al.*, 1980). No quantitative measurements of electrochemical proton gradient generation were reported in either study.

## VI. AMINE ACCUMULATION

The accumulation of biogenic amines into chromaffin granules is a pivotal factor in the physiological control of amine homeostasis. First, it allows the >10,000-fold concentration of amines prior to physiologic release. Second, it protects stored amines from oxidation and/or degradative cytosolic enzymes. Third, it permits specific control, at the membrane level, of the species and amount of biogenic amine to be stored. Fourth, it provides for biosynthesis of norepinephrine from cytosolic dopamine via membrane-bound DβH.

In recognition of its biologic importance, the investigation of amine accumulation into the chromaffin granule has proceeded at a great rate for over two decades and a very large body of empirical data has been gathered. It is only recently, however, that the discovery of the existence of an endogenous pH gradient and $H^+$-translocating ATPase in intact, isolated granules has suggested new directions of study. These, in turn, have rapidly led to new hypotheses better able to account for experimental observations. Because the vast majority of existing data predate these recent discoveries and lack any documentation of factors now known to control amine accumulation (such as pH, ΔpH, ΔΨ, the presence of ATP, medium composition, etc.), earlier data can only be interpreted with caution. The development of purified membrane and re-formed ghost preparations has occurred in concert with the need for careful modulation

Table 3. Specificity of Amine Accumulation[a]

| Investigator | Pletscher *et al.*, 1973 | DaPrada *et al.*, 1975 | Slotkin and Kirshner, 1971 | Phillips, 1974 |
|---|---|---|---|---|
| Preparation | Ghosts | Ghosts | Granules | Ghosts |
| Measurements | nmol/mg protein/ 15 min | nmol/mg protein/ 30 min | nmol/100μg catechols/ 20 min | nmole/mg protein/ 10 min |
| Concentration | 45 μM | 45 μM | 100 μM | 100 μM |
| Dopamine | 7.9 | 77.5 | | |
| Norepinephrine | 2.8 | 72.4 | 1.98 | 77 |
| Epinephrine | 3.1 | 62.4 | 2.1 | 60 |
| β-Phenylethylamine | | 12.0 | 0.6 | |
| Dopa | | | 0.006 | |
| Tyramine | | 36.3 | 2.73 | |
| Tryptamine | | 11.9 | | |
| Serotonin | 8.9 | 70 | 8.1 | 70 |

[a] Temperature was 37°C for all measurements.

of electrochemical variables and has indeed proven to be a powerful and elegant tool in the investigation of the properties of amine accumulation into chromaffin granules.

## A. Physicochemical Properties

Amine accumulation into the chromaffin granule is structurally specific. The results of several studies of specificity (Pletscher *et al.*, 1973; DaPrada *et al.*, 1975; Slotkin and Kirshner, 1971; Phillips, 1974b; Carty *et al.*, 1984; Table 3) suggest that hydroxylation of the catechol ring is required for accumulation, e.g., β-phenylethylamine, that β-hydroxylation limits uptake, e.g., dopamine vs. epinephrine and norepinephrine, and that carboxylation renders the uptake unfavorable, e.g., dopa. In addition, 5-hydroxytryptamine (serotonin) has a greater specificity for uptake than do the catecholamines and their derivatives, even though this biogenic amine is not found in the adrenal medulla *in vivo*. L-Epinephrine maintains a threefold stereospecificity with respect to D-epinephrine (Phillips, 1974b). Amines which are not accumulated to a large extent, e.g., β-phenylethylamine and tyramine, can, nevertheless, act as competitive inhibitors of epinephrine accumulation (Slotkin and Kirshner, 1971), suggesting that amine binding to the carrier is not necessarily related to the rate of accumulation. The structures of the principal accumulated biogenic amines are given in Figure 6.

Temperature dependence of [$^{14}$C]epinephrine uptake has been measured with a $Q_{10}$ of 2.8 from 20–30°C (Kirshner, 1962) and a $Q_{30}$ of 6.9 from 0–30°C (Slotkin, 1975a) in isolated granules. No measurements of temperature dependence in ghosts or in relation to chemiosmotic gradient formation have been reported.

The uptake of catecholamines is saturable, although there is considerable controversy as to specific values for $K_m$ and $V_{max}$ and a wide range of measurements have been reported (35–300 μM) (Slotkin, 1975a; Jonasson *et al.*, 1964; Lundborg, 1966; Ramu *et al.*, 1983). Recognizing the inaccuracies inherent in use of intact granules laden with endogenous catecholamines, a good many investigators have utilized ghost preparations for kinetic measurements, with even more widely varying results (12–2500 μM; Phillips, 1974b; DaPrada *et al.*, 1975; Ingebretsen and Flatmark, 1979; Kanner *et al.*, 1979; Knoth *et al.*, 1981; Scherman and Henry, 1981; Deupree *et al.*, 1982). Most of these data were obtained without documentation of the external pH, internal pH, and/or chemiosmotic driving force for amine accumulation (see Section VI-C). Utilizing a new and extremely sensitive on-line amperometric technique (Hayflick *et al.*, 1981), our laboratory has recently measured the $K_m$ and $V_{max}$ of accumulation for several biogenic amines under precisely defined conditions and these data are presented in Table 4 (Carty *et al.*, 1984).

The rapid velocity of amine accumulation and its specificity and affinity for various biogenic amines, together with the existence of specific inhibitors (Section VI-B), are an indication that amine accumulation is catalyzed by the operation of a specific transporter molecule in the chromaffin granule membrane. Although no successful attempt at isolation of the carrier has been reported, reconstitution of carrier activity has been achieved after solubilization of the entire chromaffin membrane (Maron *et al.*, 1979). Moreover, a tentative identification of the transporter polypeptide in chromaffin membranes has been made using photoaffinity labeling with 4-azido-3-nitrophenyl-azo-(5-hydroxytryptamine) (Gabizon *et al.*, 1982). The information concerning the molecular mechanism of the carrier and its turnover number, charge (if any), and asymmetry is extremely sparse.

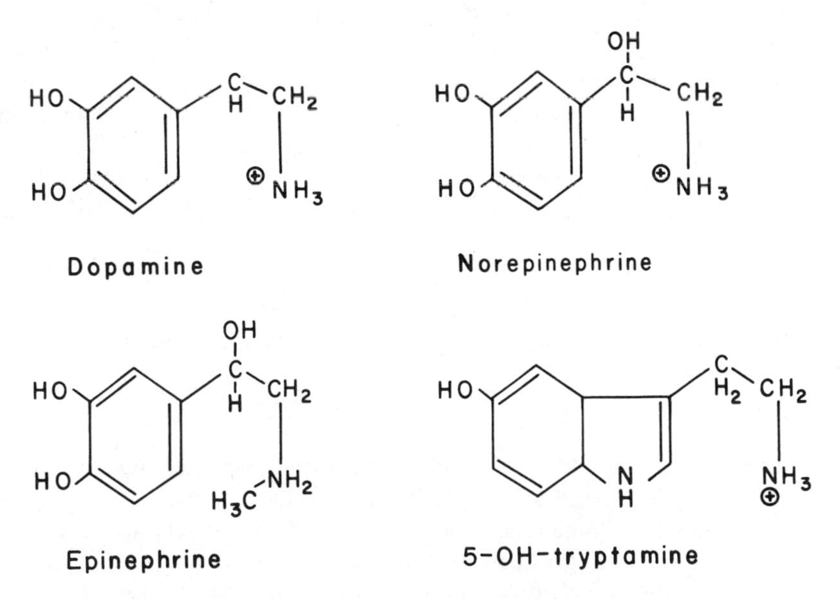


*Figure 6.* Structure of various biological amines.

Table 4. $K_m$ and $V_{max}$ of Biogenic Amines[a]

| Amine | $K_m$ (μM) | $V_{max}$ (nmol/min × mg protein) | Number of experiments |
|---|---|---|---|
| Dopamine | 14.6 ± 3.6 | 12.2 ± 3.0 | 9 |
| ℓ-Norepinephrine | 30.0 ± 7.5 | 15.1 ± 4.6 | 7 |
| ℓ-Epinephrine | 32.3 ± 10.8 | 15.0 ± 5.4 | 6 |
| 5-Hydroxytryptamine | 3.5 ± 1.1 | 7.0 ± 2.2 | 8 |

[a] Purified ghosts were formed and incubated in buffered isotonic KCl. Four mM mgATP was present. The external pH was 7.0 and internal pH was 6.0 ± 0.1. The membrane potential was 0 to −2 mV. Temperature was 37°C. Values are accurate to ±1 S.D.

## B. Inhibitors

Reserpine is the prototypical competitive inhibitor of specific carrier-mediated amine accumulation (Slotkin and Kirshner, 1971; Jonasson *et al.*, 1964; Johnson and Scarpa, 1979; Stitzel, 1977). Structurally, it consists of a substituted indoleamine with a bulky, extremely lipophilic side chain which confers irreversibility of inhibition to the molecule. Reserpine is thought to bind specifically to the biogenic amine transporter protein of the granule membrane. It reportedly has a $K_{i(50)}$ of $10^{-7}$ M for epinephrine (Kirshner, 1962), although since exogenous reserpine solubilizes almost completely in the apolar membrane phase, concentrations have no physiologic relevance and reserpine content is best expressed in μg/mg protein. In this laboratory, 1 μg reserpine/mg protein consistently produces >95% inhibition of catecholamine uptake into isolated chromaffin granules. Reserpine does not inhibit nonspecific, noncarrier-mediated movement of amines across the membranes of granules or ghosts (Section VI-C), and is without effect on $H^+$-pumping ATPase activity.

Reversible competitive inhibitors of amine accumulation include harmine ($K_i$ = 2.8 μM) and harmaline ($K_i$ = 3.9 μM), which have structures intermediate between 5-hydroxytryptamine and reserpine. Harmine is particularly noteworthy in that it is probably not accumulated and is also a potent inhibitor of monoamine oxidase (Slotkin, 1975b). Recently, a light-sensitive specific inhibitor of amine uptake, 4-azido-3-nitrophenyl-azo-(5-hydroxytryptamine), has been described with a reversible $K_{i(50)}$ of 2 μM in the dark that converts to an irreversible $K_{i(50)}$ of 4 μM upon illumination (Gabizon *et al.*, 1982).

Compounds such as FCCP, nigericin, trimethyltin, DCCD, $SCN^-$, and $NH_4^+$, which secondarily inhibit amine accumulation by altering the transmembrane proton and electrical gradients, have been discussed in Sections V and VI.

## C. Coupling to the $\Delta\bar{\mu}_{H^+}$

The discovery that the internal pH of isolated chromaffin granules is acidic (Section IV) and that ATP hydrolysis causes inward proton translocation (Section V) suggested to many researchers that amine accumulation was linked not to ATP hydrolysis itself but to the existence of the electrochemical proton gradient across the granule membrane. This concept was indirectly supported by several observations: (1) the pH optima of

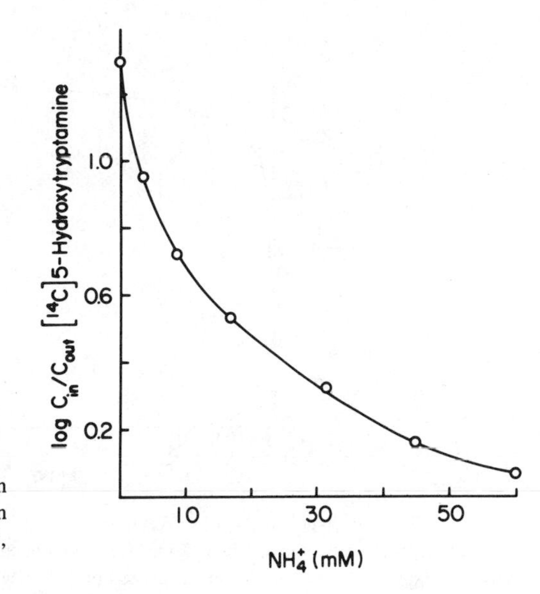


*Figure 7.* Dose-dependent effect of $NH_4^+$ upon [$^{14}$C]5-hydroxytryptamine (serotonin) uptake in chromaffin granules. From Johnson *et al.* (1981), wherein more details can be found.

the ATPase and of amine accumulation were disparate (Taugner, 1972), (2) the measured ratio of catecholamine accumulated to ATP hydrolyzed ranged from 1–350 (Phillips, 1974b; Taugner, 1971), and (3) ATP addition to granules in sucrose medium resulted in amine accumulation, while in chloride medium it caused complete granule lysis (Trifaro and Poisner, 1967; Lishajko, 1971; Izumi *et al.*, 1975). Direct evidence was supplied both qualitatively by experiments with isolated granules and quantitatively by experiments using re-formed ghosts.

### 1. Accumulation in Response to a ΔpH

A fruitful series of experiments in this laboratory has shown that the rate and extent of biogenic amine accumulation into chromaffin granules and ghosts are directly proportional to the magnitude of the transmembrane pH gradient (ΔpH; Johnson and Scarpa, 1979; Johnson *et al.*, 1979, 1981). When isolated granules suspended in sucrose medium at neutral pH were incubated with small amounts of radiolabeled biogenic amine, rapid kinetic amine accumulation was observed. If the magnitude of the resting ΔpH was reduced or collapsed by varying concentrations of $NH_4^+$, the rate and extent of subsequent amine uptake were reduced in a directly proportional fashion, as illustrated in Figure 7. The development of a highly purified chromaffin ghost preparation, devoid of endogenous amines and matrix proteins and capable of generating a large ΔpH upon incubation with MgATP, made possible quantitative steady-state measurements of ΔpH-dependent uptake. Under these conditions, amine accumulation was observed to undergo a rapid initial phase followed by a plateau of maximal accumulation (Figure 8). Collapse of the ΔpH after attainment of maximal accumulation (arrow in Figure 8) resulted in nearly complete efflux of accumulated amines, indicating that true steady-state conditions were present. Quantitative simul-

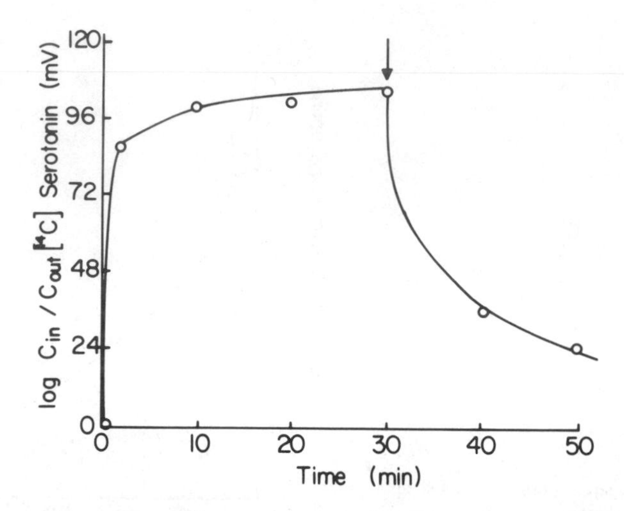


*Figure 8.* Time-resolved uptake of [$^{14}$C]serotonin in the presence of electrochemical proton gradient. Chromaffin ghosts were formed in 185 mM Na-isethionate, 20 mM ascorbate, and 10 mM Tris-maleate buffer at pH 7. The reaction medium consisted of 185 mM Na-isethionate and 30 mM Tris-maleate, pH 7.0, and contained either [$^{14}$C]methylamine and $^{3}H_2O$ or [$^{14}$C]-$SCN^-$, as well as 8 mM MgATP and chromaffin ghosts (1 mg protein/ml). The incubation volumes were 9.0 ml, 30 mM for the measurements of $\Delta pH$ and $\Delta\psi$ (not shown), or [$^{14}$C]serotonin and $^{3}H_2O$. $(NH_4)_2SO_4$ and FCCP (8.3 μg/ml) were added to the reaction medium after 30 min had elapsed (arrow). At the times indicated, 1.2 ml samples were removed and centrifuged 7 min in an Eppendorf desk microcentrifuge. Supernatants and pellets were assayed for relative radioactivity. Temperature was 24°C. From Johnson *et al.*, (1981).

taneous measurements of the $\Delta pH$ and the steady-state $\Delta\bar{\mu}_A$ (transmembrane amine gradient) then revealed a constant ratio of 0.38 ($\Delta pH : \Delta\bar{\mu}_A$; Figure 9, curve A). The $H^+$ : amine ratio was unaffected by changes in the pH of the external medium over the range 6–8. These experiments indicate that transport most probably occurs via a membrane carrier protein driven by the magnitude of the $\Delta pH$, which possesses a p$K_a$ outside the physiologic range. The $\Delta pH$-dependent uptake of biogenic amine has been observed by numerous investigators in both isolated granules and re-formed ghosts (Hayflick *et al.*, 1981; Taugner, 1972; Schuldiner *et al.*, 1978; Phillips and Allison, 1978; Johnson *et al.*, 1979; Ingebretsen and Flatmark, 1979).

Accumulation of biogenic amines into chromaffin granules and ghosts in response to a $\Delta pH$ actually occurs by two distinct mechanisms. The first, already described, is a rapid, species-specific, saturable, temperature-dependent process which is competitively inhibited by reserpine. Direct evidence that this process is carrier-mediated now includes the recent identification of the transporter polypeptide (mol. wt. 45,000) by photoaffinity labeling (Gabizon *et al.*, 1982). The second mechanism of pH-dependent amine accumulation differs from the first in that it is nonspecific, unsaturable, nontemperature-dependent, reserpine-insensitive, and quite slow. In this process, the neutral form of a biogenic or pharmacologic amine, which exists in p$K_a$-dependent equilibrium with charged forms, is able to permeate the chromaffin membrane bilayer much as do $NH_4^+$ or methylamine (Section IV). As protonation occurs in the acidic

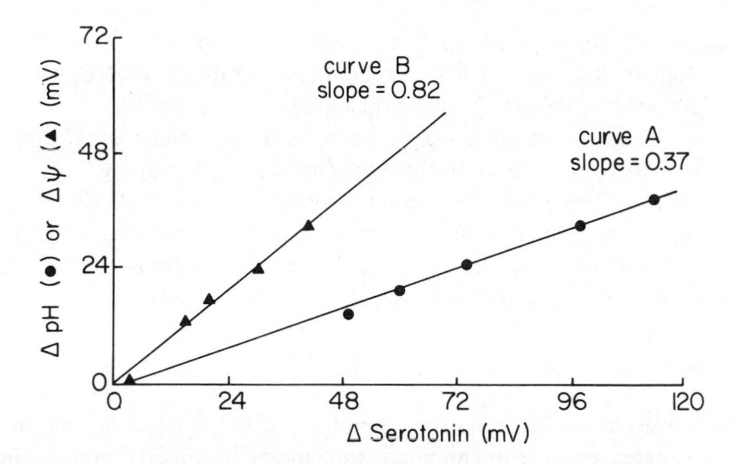


*Figure 9.* Relationship between ΔpH or Δψ and steady-state accumulation of [$^{14}$C]serotonin. (A) ΔpH vs. [$^{14}$C]serotonin accumulation. Chromaffin ghosts, formed in 185 mM Na-isethionate, 20 mM ascorbate, and 4 mM Tris-maleate at pH 7.0, were suspended in an incubation medium of 185 mM Na-isethionate, 30 mM Tris-maleate at pH 6.95, 8 mM MgATP, $^3H_2O$, 60 mM NaSCN, and varying concentrations of $(NH_4)_2SO_4$ (0–30 mM). Half the samples contained [$^{14}$C]methylamine, and [$^{14}$C]serotonin was added to the other half when 10 min incubation had elapsed. Reaction volumes were 1.2 ml. The reaction was ended at 20 min by centrifugation, and samples were processed. The Δψ under these conditions was found to be slightly negative (−5 mV). Temperature was 24°C. (B) Δψ vs. [$^{14}$C]serotonin accumulation. Chromaffin ghosts were formed and suspended (1.4 mg of protein/ml) in 30 mM Tris-maleate at pH 6.94, containing 8 mM MgATP, $^3H_2O$, 30 mM $(NH_4)_2SO_4$, and various concentrations of NaSCN (0–60 mM). Experimental conditions were otherwise identical with those of (A) except that [$^{14}$C]-$SCN^-$ was substituted for methylamine. The ΔpH under these conditions was found to be 0.05 units. From Johnson *et al.*, (1981).

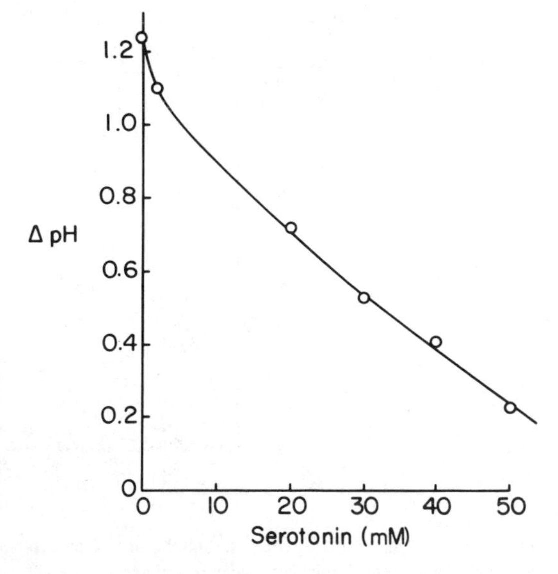


*Figure 10.* ΔpH as a function of extravesicular amine concentration in chromaffin granules. Freshly isolated chromaffin granules were suspended at a concentration of 1.7 mg protein/ml in 1.2 ml medium containing 0.32 M sucrose, 20 mM Tris-maleate (pH 6.8) containing [$^{14}$C]methylamine, $^3H_2O$, and 5-hydroxytryptamine at the concentrations indicated in the figure. ΔpH was measured as described in Figure 2. Temperature was 23°C.

vesicle, interior amines become trapped inside and eventually distribute in a transmembrane gradient that mimics the pH gradient. Similarly to $NH_4^+$, when large amounts (<500 μM) of biogenic or pharmacologic amine are added to a suspension of granules or ghosts possessing a ΔpH, slow dose-dependent alkalinization of the vesicle interior occurs once the buffering capacity of the internal medium is exceeded. This can be measured as a reduction in the magnitude of the ΔpH (Figure 10). Nonspecific, noncarrier-mediated uptake of amines has been observed not only in chromaffin granules and ghosts possessing a pH, but also in liposomes wherein a pH gradient has been artificially imposed across the membrane (Deamer *et al.*, 1972).

## 2. *Accumulation in Response to a* $\Delta\psi$

In addition to dependence on the magnitude of the ΔpH, biogenic amine accumulation in isolated chromaffin granules and ghosts is directly proportional to the magnitude of the transmembrane electrical gradient ($\Delta\psi$; Johnson and Scarpa, 1979). Isolated granules suspended at neutral pH in sucrose medium were found to maintain a resting membrane near zero, but generated a $\Delta\psi$ (inside positive) of 80 mV or more upon addition of MgATP. The establishment of a $\Delta\psi$ was found to stimulate kinetic amine accumulation quite markedly over that observed in the presence of a ΔpH alone (Figure 11). Using a highly purified ghost preparation formed in Na-isethionate medium

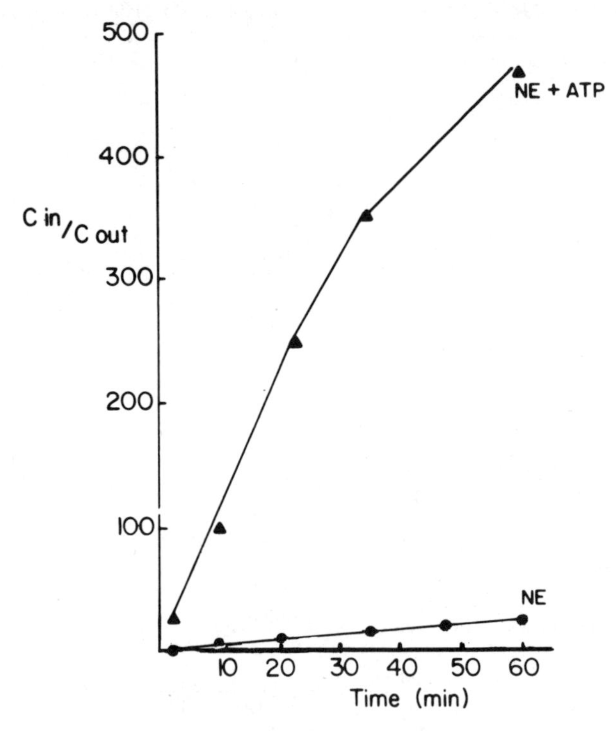


*Figure 11.* Time dependence of the uptake of [$^{14}$C]norepinephrine (NE) expressed as the ratio of internal to external concentration in the presence and absence of 10 mM MgATP. From Johnson and Scarpa (1979).

and capable of generating a large $\Delta\psi$ with a negligible $\Delta pH$ upon ATP addition, amine accumulation in the face of a $\Delta\psi$ alone could be quantitatively studied (Johnson *et al.*, 1981). Under these circumstances, biogenic amine accumulation varied as the magnitude of the $\Delta\psi$ with a constant ratio of 0.76 ($\Delta\psi : \Delta\bar{\mu}_A$) (Figure 9, curve B). The ratio was unchanged by variation in the pH of the external medium over the range 6–8. Uptake was completely inhibited by reserpine. These experiments demonstrate that carrier-mediated transport of biogenic amines into chromaffin granules is driven with a fixed, pH-insensitive stoichiometry by the transmembrane $\Delta\psi$ as well as by the transmembrane $\Delta pH$. The $\Delta\psi$-stimulated accumulation of biogenic amines into chromaffin granules and ghosts has been reported by a number of other workers in the field (Njus and Radda, 1979; Holz, 1978; Scherman and Henry, 1979; Kanner *et al.*, 1980).

### 3. Accumulation in Response to a $\Delta\bar{\mu}_{H^+}$

The discovery that amine accumulation into isolated chromaffin ghosts could be driven by either the $\Delta pH$ or the $\Delta\psi$ prompted several investigations into elucidation of the precise contributions of the $\Delta\bar{\mu}_{H^+}$ to amine uptake (Scherman and Henry, 1981; Johnson *et al.*, 1979; Knoth *et al.*, 1980, 1981; Apps *et al.*, 1980a). The relationship between the $\Delta\bar{\mu}_{H^+}$ and $\Delta\bar{\mu}_A$ under conditions in which the gradients are in equilibrium with each other, is predicted by the chemiosmotic hypothesis, which states that the energy for substrate transport systems is derived from the $\Delta\bar{\mu}_{H^+}$ generated by the $H^+$-translocating ATPase. The coupling between the transported amine and $H^+$ approaches equilibrium, i.e., net transport vanishes, whenever

$$\Delta\bar{\mu}_A - n\Delta\bar{\mu}_{H^+} = 0$$

where $n$ is defined as the stoichiometry of $H^+$ efflux (since the diffusion gradient for protons in the chromaffin granule is outward) to amine influx. It is the value $n$, the ratio of $H^+$–amine antiport, which determines the relative contribution of the $\Delta pH$ and $\Delta\psi$ to the driving force.

In chromaffin ghosts formed to permit the independent variation of the magnitudes of the $\Delta pH$ and $\Delta\psi$ between 0–0.85 pH unit and 0–55 mV, respectively, a broad range of values for the $\Delta\bar{\mu}_{H^+}$ could be generated. Moreover, different combinations of $\Delta pH$ and $\Delta\psi$ values were able to produce the same magnitude of the $\Delta\bar{\mu}_{H^+}$. Utilizing this experimental design, rigorous measurements have shown that the fastest and quantitatively largest amine accumulation into purified chromaffin ghosts is observed whenever both the $\Delta pH$ and $\Delta\psi$ are present (Johnson *et al.*, 1981). Only when the driving force was expressed as $\Delta\psi - 2Z\Delta pH$, however, was a first order relationship observed between the magnitude of the driving force and amine accumulation [ratio of $\Delta\psi - 2Z\Delta pH$ (mV) $= \Delta\bar{\mu}_A$ (mV) $= 0.96 \pm 0.08$]. These results are consistent with data obtained from the study of amine uptake in the presence of a $\Delta pH$ or $\Delta\psi$ alone, which showed that, for a given magnitude of the driving force ($\Delta pH$ or $\Delta\psi$; see Figure 9), the accumulation of biogenic amines in the presence of the $\Delta pH$ alone is twice that observed in the presence of a $\Delta\psi$ of the same magnitude.

The physiological implications of this relationship can be explored (Figure 12). If it is assumed that the internal pH of the chromaffin granule is 5.5 and the cytosolic

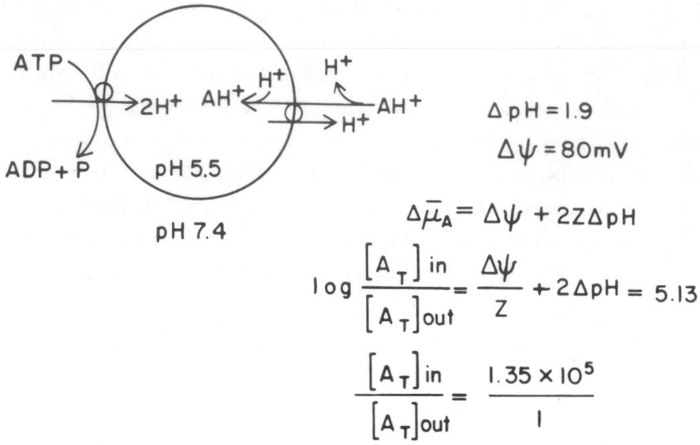

ATP
ADP + P
2H+ AH+ H+
H+
AH+
H+
pH 5.5
pH 7.4

$\Delta pH = 1.9$

$\Delta\psi = 80 mV$

$$\Delta\bar{\mu}_A = \Delta\psi + 2Z\Delta pH$$

$$\log \frac{[A_T]in}{[A_T]out} = \frac{\Delta\psi}{Z} + 2\Delta pH = 5.13$$

$$\frac{[A_T]in}{[A_T]out} = \frac{1.35 \times 10^5}{1}$$

*Figure 12.* Schematic representation of catecholamine uptake into chromaffin granules according to the proton-motive force. More details in the text.

pH 7.4, then a ΔpH of 1.9 pH units may exist *in vivo*. Isolated chromaffin granules can generate a membrane potential of up to 80 mV. Substituting values into the equation $\Delta\bar{\mu}_A = \Delta\psi + 2Z\Delta pH$, a transmembrane catecholamine gradient of 135,000 to 1 can be predicted. Generation of this enormous (by biologic standards) catecholamine gradient is based solely upon the existence of a $\Delta\bar{\mu}_{H^+}$ and an amine transporter molecule able to cycle in response to the $\Delta\bar{\mu}_{H^+}$. If the cytosolic concentration of catecholamines is 2 μM as has been suggested by indirect measurements (Perlman and Sheard, 1982), then the intragranular concentration would correspond to 270 mM, a value which approximates the concentration previously calculated from measurements of catecholamine content (see Table 1).

## D. Net Accumulation of Biogenic Amines

The existence of an active transport process is best indicated by the demonstration of net substrate accumulation. Despite the extensive and thorough characterization of biogenic amine accumulation into chromaffin granules and ghosts which is present in the literature, nearly all the data were gathered under experimental conditions which did not permit discrimination between amine uptake and amine exchange. Every major study utilizing the distribution of radiolabeled amines to measure amine transport has thus been limited by: (1) exchange of endogenous nonlabeled amines for labeled amines; (2) exchange of intravesicular labeled amines with extravesicular labeled amines; (3) possible oxidation or metabolism of exogenous amines, especially at site of radiolabel; and (4) possible redistribution of gradients during preparative and/or final centrifugations.

Two studies have clearly demonstrated net biogenic amine accumulation in a preparation of purified chromaffin ghosts, using both high-pressure liquid chromatography for the measurement of extravesicular catecholamines after filtration of the granules (Johnson *et al.*, 1982c) and on-line kinetic, amperometric detection of catecholamine flux by a glassy carbon electrode (Hayflick *et al.*, 1981; Johnson *et al.*,

1982c). In both studies net accumulation of catecholamines was found to occur by a carrier-mediated, pH-dependent mechanism whose kinetics are similar to those measurable by radiolabeled distribution methods.

The amperometric technique permits very sensitive and accurate measurements of transport kinetics and has already been used to determine $V_{max}$ and $K_m$ for a variety of biogenic and pharmacologic amines (Carty *et al.*, 1984). Moreover, it promises to be a powerful tool in elucidating the pH dependence and electrical charge of the membrane amine transporter (Johnson *et al.*, 1984).

## VII. THE ELECTRON-TRANSPORT CHAIN

The chromaffin granule contains a number of soluble and membrane-associated components which can participate in oxidation–reduction reactions. The soluble redox mediators include ascorbate (at an effective concentration of 13 mM), catecholamines ($<$ 500 mM), and dopamine-β-hydroxylase. The membrane redox carriers are NADH (acceptor) oxidoreductase, flavoprotein, cytochrome $b_{561}$, and dopamine-β-hydroxylase. In addition to the intrinsic electron carriers, the local environment of the chromaffin granule is also rich in reducing equivalents since the cytosolic concentration of ascorbate approaches 2 mM (Ingebretsen *et al.*, 1980; Terland and Flatmark, 1975). Because the range of redox potentials of the intragranular and membrane components spans the thermodynamic spectrum from $-320$ to $+600$ mV, it is evident that the chromaffin granule membrane is well equipped to participate in a number of possible electron-transfer reactions. Despite intensive investigation, however, the reaction sequence of electron transfer is not well understood, transmembrane electron fluxes are only in the initial stages of documentation, terminal electron acceptors are not yet identified, and the physiological role of electron transfer and coupling reactions remains in the earliest phases of conceptualization.

### A. Organization

The thermodynamic properties of chromaffin granule redox mediators, taken together with the best available evidence to date permit the formulation of a plausible scheme for electron flow, illustrated in Figure 13. Membrane electron flow can best be described in a series of two transfer reactions.

In the first sequence, electrons are transferred from cytosolic NADH to the membrane enzyme NADH oxidase (Bashford *et al.*, 1976). Flavoprotein is probably a cofactor in this reaction, based upon analogy with other pyridine nucleotide oxidase interactions. Despite evidence that NADH oxidase is a transmembrane protein (Zaremba and Hogue-Angeletti, 1982), and that a favorable thermodynamic couple exists for the subsequent transfer of $e^-$ from the NADH oxidase–flavoprotein to dehydroascorbate, no NADH-stimulated transmembrane electron flux or NADH-stimulated dehydroascorbate reduction has been demonstrated (Terland and Flatmark, 1980).

In the second sequence, electrons are transferred from matrix or cytosolic ascorbate to membrane cytochrome $b_{561}$ (Terland and Flatmark, 1973). There are three components with possible cytochrome $b_{561}$ oxidase activity: (1) semidehydroascorbate,

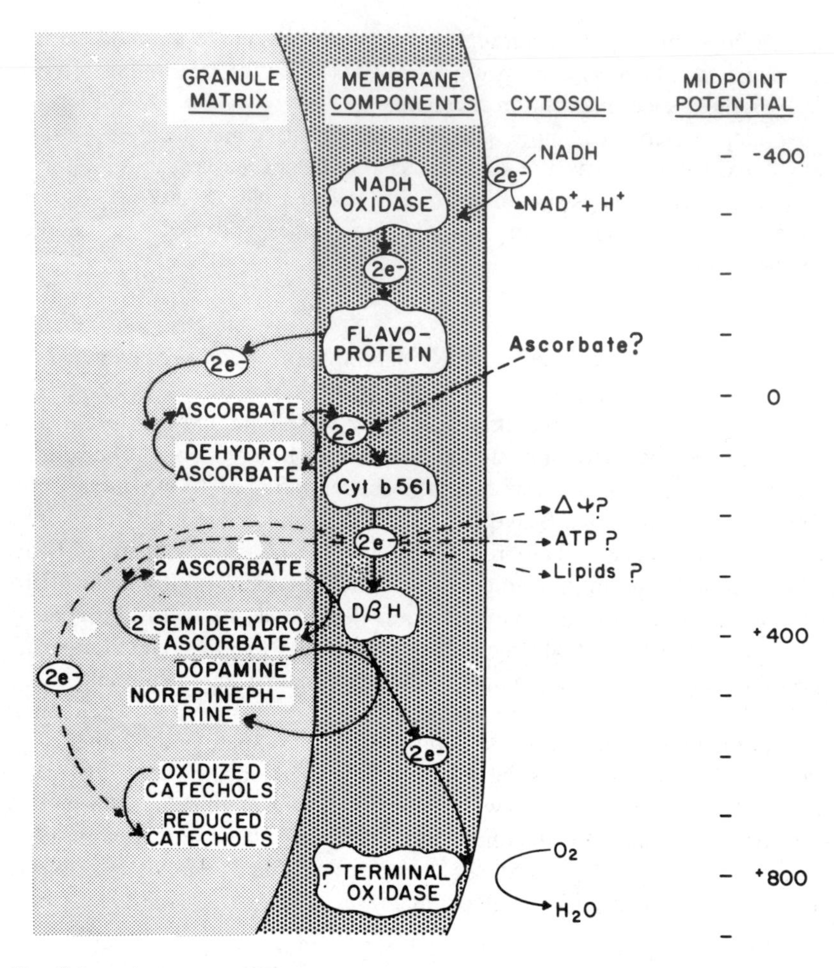


*Figure 13.* Schematic representation of the electron-transport chain of chromaffin granules. More details in the text.

(2) matrix-oxidized catecholamines, and (3) matrix or membrane-bound dopamine-β-hydroxylase. Semidehydroascorbate, a high midpoint potential free radical compound, is formed during the one electron reduction of the copper molecules within dopamine-β-hydroxylase. As a strong oxidizing agent, semidehydroascorbate in interaction with the cytochrome would maintain matrix ascorbate in the fully reduced form. Due to the transmembrane nature of the cytochrome $b_{561}$, reduction of the cytochrome by cytosolic ascorbate has been postulated (Njus *et al.*, 1983). Transfer of electrons to intragranular semidehydroascorbate could constitute an effective shuttle mechanism for transfer of reducing equivalents across the chromaffin granule membrane.

*In vitro,* purified dopamine-β-hydroxylase and oxidized catecholamines are efficiently reduced by ascorbate. Within the chromaffin granule, however, electron flow from cytochrome $b_{561}$ (which in the membrane maintains an 8 : 1 ratio with that of dopamine-β-hydroxylase) may be preferential.

Mixed function oxidase activity exists in the form of dopamine-β-hydroxylase. However, no terminal oxidase activity has been demonstrated in purified chromaffin granule membrane preparations.

## B. Physicochemical Properties

Investigation of the properties of the redox components of intact chromaffin granules has been hampered by (1) the enormous quantities of ascorbate and catecholamines within the intragranular space of the chromaffin granule which serve as redox buffers, (2) contamination of isolated granule preparations by mitochondria which provide a source of cytochrome *c* and oxidase activity, and (3) the lack of experimental inhibitors of NADH oxidase or cytochrome $b_{561}$ oxidoreductase activity. Most of the reliable information to date as to the properties of the electron-transport proteins has been gained from studies utilizing either highly purified chromaffin membranes (generally prepared after the method of Terland and Flatmark, 1980), or the isolated proteins themselves.

### 1. NADH (Acceptor) Oxidoreductase

The enzyme, a transmembrane protein, has been isolated to homogeneity and has a molecular weight of 55,000 (Zaremba and Hogue-Angeletti, 1982). The basal rate of NADH oxidation within purified chromaffin granule membranes is 35 nmoles/min per mg protein, and increases tenfold in the presence of an exogenous electron acceptor such as ferricyanide (Terland and Flatmark, 1973). Neither NADPH nor succinate is a substrate. The enzyme activity is inhibited by PHMB (*p*-hydroxymerubenzoate) but not by rotenone (Terland and Flatmark, 1973). No NADH -dehydroascorbate, NADH-catecholamine-quinone, or NADH-glyceraldehyde-oxidase activities have been detected (Terland and Flatmark, 1980).

### 2. Flavoprotein

A broad spectral band at 450 nm is visualized in an oxidized-minus-reduced spectrum of chromaffin granule membranes (Figure 14); this absorption spectrum is consistent with a flavoprotein. Using a partially purified membrane preparation, Pollard *et al.* (1973) measured an enzymatic activity of 2.1 nmoles/mg of protein; removal of mitochondrial contaminants, however, significantly reduced the presumed flavoprotein content. At this time, therefore, the existence of a flavoprotein within the chromaffin granule membrane is not well substantiated.

### 3. Cytochrome $b_{561}$

The only cytochrome within the chromaffin granule membrane has oxidized-minus-reduced absorption maxima at 561 (A), 532 (β), and 430 nm (Soret, Figure 14; Pollard *et al.*, 1973; Bank, 1965; Spiro and Ball, 1961; Duong and Fleming, 1982; Flatmark and Gronberg, 1981; Flatmark and Terland, 1971). It is an intrinsic membrane protein present at 2.29 nmoles/mg of protein within the chromaffin granule membrane and has been purified to homogeneity with reported molecular weights of 20,500–35,000

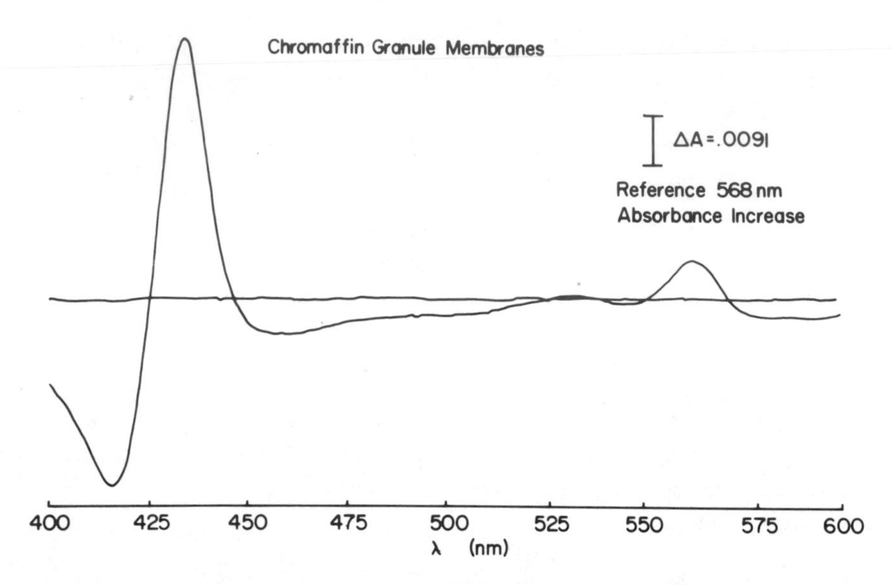


*Figure 14.* The oxidized-minus-reduced spectrum of chromaffin granule membrane. The oxidized-minus-reduced spectrum was obtained on a Johnson Foundation double beam spectrophotometer using a low temperature (−80°C) attachment. The reference wavelength was held constant at 468 nm and the measure wavelength was scanned from 600 to 400 nm, and the differential absorbance readout at 60 Hz was stored in a microprocessor and plotted. Highly purified chromaffin granule fragmented membranes were suspended at a concentration of 0.18 mg/ml in 50 mM potassium phosphate buffer at pH 6.6. From Johnson and Scarpa (1981).

(Abbs and Phillips, 1980; Duong and Fleming, 1982; Flatmark and Gronberg, 1981; Silsand and Flatmark, 1974). The following properties have been determined: (1) the molar extinction coefficient (ε) of the reduced-minus-oxidated spectrum at 561–575 nm is 26.8 $nM^{-1}$ $cm^{-1}$ (Silsand and Flatmark, 1974), (2) the prosthetic heme group is a protoheme IX and, in fact, contains all of the heme within the membrane (Terland *et al.*, 1974), (3) CO binds to the enzyme weakly, suggesting that under the appropriate conditions (probably denaturing) the enzyme is auto-oxidizable (Terland *et al.*, 1974), and (4) the midpoint potential is 140 mV at pH 7.0 (Flatmark and Terland, 1971).

Although its spectra and molecular weight are quite similar to those of the mitochondrial cytochrome *b*, the midpoint potential and amino acid composition render the cytochrome $b_{561}$ of the chromaffin granule quite distinct from the mitochondrial enzyme (Duong and Fleming, 1982; Silsand and Flatmark, 1974).

Ascorbate can reduce the cytochrome within isolated chromaffin granule membranes; the $K_m$ for ascorbate in this reaction was found to be 0.34 mM (Flatmark and Terland, 1971). NADH is not capable of reducing cytochrome $b_{561}$ (Terland and Flatmark, 1980). There are no known inhibitors of the cytochrome and no evidence exists to implicate the cytochrome in vectorial $H^+$ fluxes. Vectorial electron movement in a nonphysiologic direction, i.e., from the matrix to the cytosol, has been demonstrated using ghosts loaded with ascorbate and with cytochrome *c* as an external electron acceptor. It was presumed that the electron flow proceeded through the cytochrome $b_{561}$ (Njus *et al.*, 1983).

### 4. Terminal Oxidase

Although mixed-function oxidase activity exists in the form of dopamine-β-hydroxylase, there is no spectrophotometrically resolvable terminal oxidase. The existence of a terminal oxidase (other than dopamine-β-hydroxylase) was postulated from early experiments using only partially purified membrane preparations (Flatmark and Terland, 1971). Subsequent investigation, however, has revealed that in highly purified membrane preparations devoid of mitochondrial or lysosomal contamination, oxygen consumption in the absence of dopamine-β-hydroxylase activity is minimal (Terland and Flatmark, 1980). Therefore, at present, there is no evidence for the existence of a terminal oxidase, and consequently no documentation that the electron transfer components constitute a "respiratory chain" *per se*.

## C. Physiologic Role

The physiologic function of electron transport within the chromaffin granule is unknown. Indeed, in the final analysis, not one but several electron transport sequences may exist. Based upon the known properties of the chromaffin granule, several reasonable mechanisms can be advanced which existing evidence cannot prove or refute:

1. Electron flow from cytochrome $b_{561}$ to semidehydroascorbate, generated through reduction of dopamine-β-hydroxylase by ascorbate, may serve to regenerate the ascorbate in the granule matrix. If re-reduction of the cytochrome then occurred at the other (cytosolic) membrane face, the transmembrane, anisotropic movement of electrons through the cytochrome would constitute an effective shuttle to keep intragranular ascorbate in the reduced form. It is hypothesized that the semidehydroascorbate formed in the cytosol would be reduced by mitochondrial semidehydroascorbate reductase (Njus *et al.*, 1983).

2. Catecholamines maintain biologic activity only in the reduced form. Interaction of oxidized catecholamines (midpoint potential +645 mV) with electron carriers may result in direct transfer of reducing equivalents and regeneration of biologically active catecholamines.

3. Dopamine-β-hydroxylase may accept electrons directed from the electron-transport chain rather than from ascorbate. This may have important implications for the membrane-bound-to-free dopamine-β-hydroxylase.

4. Several of the membrane redox components may yet be identified as $H^+$ vectorial carriers, thereby contributing to the generation and maintenance of the $\Delta\bar{\mu}_{H^+}$.

5. The energy derived from electron movement may be directly coupled to ATP synthesis or transport.

6. In other biologic systems, the cytochrome *b* has been implicated in lipid metabolism, functioning to maintain unsaturated lipid in the reduced form.

7. Electron-transport activity virtually identical to that of the chromaffin granule membrane has been identified recently in the membrane of the platelet serotonin granule (Johnson and Scarpa, 1981). Given the disparity in embryologic origin, content, and physiologic activity of these storage granules, the possession of a common electron-

transport chain by both suggests that its function is coupled to functions or properties common to both organelles.

A systematic effort will be required to elicit the fundamental properties of electron transfer through study of isolated redox proteins, reconstituted systems, and purified membrane preparations. In addition, an active search for other electron-transporting components which may be in equilibrium with the recognized electron carriers (such as iron sulfur proteins) may lead to a more elegant documentation of the sequence of electron flow in the chromaffin granule membrane.

## VIII. OTHER TRANSPORT SYSTEMS

### A. ATP Accumulation

The unusually high intragranular ATP concentration (approaching 200 mM if all ATP were free in solution) constitutes an *in situ* transmembrane ATP gradient of 100 to 1. This has prompted one laboratory in particular to investigate the properties of nucleotide transport in the intact chromaffin granule using radiolabeled compounds (Aberer *et al.*, 1978; Weber and Winkler, 1981; Weber *et al.*, 1983; Kostron *et al.*, 1977b). While most of the conclusions advanced probably possess physiologic significance, it should be noted that all of the experiments to date have been performed using intact chromaffin granules; therefore, exchange of labeled for the large pool of endogenous amines cannot be excluded. In addition, quantitation of the kinetic parameters ($K_m$ and $V_{max}$) was probably altered by the leakage of endogenous nucleotides into the incubation medium, competition by endogenous substrates, etc. No net uptake of nucleotides has yet been demonstrated.

The known properties of nucleotide uptake into isolated chromaffin granules are listed in Table 5. The nucleotide carrier maintains a rather low affinity for ATP (and less for other nucleotides), suggesting that at cytosolic concentrations of ATP transport of the nucleotide may be submaximal. The $V_{max}$ for ATP uptake is approximately one eighth of that for catecholamines under similar conditions (Weber and Winkler, 1981). Nucleotide accumulation is independent of the magnitude of the $\Delta$pH, but is sensitive to the magnitude of the transmembrane potential (Aberer *et al.*, 1978). This relationship to the $\Delta\mu_{H^+}$ would be predicted for a highly negatively charged compound moving in the face of a positive potential. The specificity of the nucleotide carrier is quite broad (Weber and Winkler, 1981), which may help to explain the existence of GTP and UTP within the intragranular space (9% and 5% of matrix nucleotides, respectively). The broad specificity allows transport of several compounds structurally quite dissimilar from the nucleotides but each possessing two negative charges: $SO_4^{2-}$, $HPO_4^{2-}$, and $PEP^{2-}$ (phosphoenolpyruvate); in fact, the $K_m$ and $V_{max}$ for $SO_4^{2-}$ accumulation are similar to those for ATP (Table 5). ATP uptake is inhibited by atractyloside but at several orders of magnitude greater than that needed for inhibition of the mitochondrial nucleotide shuttle. Inhibition by phenylglyoxal is thought to indicate involvement of an arginine residue in a functionally active position on the putative transport molecule (Weber *et al.*, 1983).

While distinct similarities between the $F_1$ ATPase of the chromaffin granule and

*Table 5. Properties of Nucleotide Transport into Isolated Chromaffin Granules*

| | | References |
|---|---|---|
| $K_m$ | 0.9–1.4 mM | Weber and Winkler, 1981; Aberer *et al.*, 1978 |
| $V_{max}$ | 476 nmol/min/mg protein at 37°C | Weber and Winkler, 1981 |
| pH Dependence | None | Aberer *et al.*, 1978 |
| Driving force | $\Delta\Psi$ | Aberer *et al.*, 1978 |
| Specificity | ATP – GTP – UTP $\gg$ ADP $\gg$ AMP | Weber and Winkler, 1981 |
| Analogues | $SO_4^{2-}$ ($K_m$ = 0.4 mM, $V_{max}$ = 380 nmol/mg protein/min)<br>$PO_4^{3-}$<br>Phosphoenolpyruvate (PEP) | Weber *et al.*, 1983 |
| Inhibitors $K_{i(50\%)}$ | $SO_4^{2-}$ (5 mM)<br>$PO_4^{3-}$ (10 mM)<br>PEP (1 mM)<br>Atractyloside (0.1 mM)<br>Carboxyatractyloside (0.4 mM)<br>Phenylglyoxal (5 mM) | Weber *et al.*, 1983 |

that of the mitochondrion have been documented, it appears that nucleotide carriers for these subcellular organelles have different values for $K_m$, $V_{max}$, specificity, inhibitors, and coupling. Only inhibition by phenylglyoxal is common to both (Cross, 1981).

The function of intragranular ATP stores is still conjectural. It is generally accepted that the interaction of ATP with catecholamines and metal ions within the intragranular space acts to effectively lower the osmolality of the matrix space. In contradistinction to the ATP within synaptic vesicles, however, there is no evidence that released ATP is an effector at a target receptor, or that it is important in reaccumulation of amines across the cell membrane.

## B. Ascorbate Accumulation

The adrenal medulla contains the highest tissue level of ascorbate within the body (4.1 μmoles/g wet weight; Hornig, 1975; Ingebretsen *et al.*, 1980). The ascorbate is localized predominantly within the chromaffin granules, wherein the intragranular concentration would approach 13 mM if all matrix ascorbate existed free in solution. This compares with a cytosolic concentration of approximately 2 mM (Terland and Flatmark, 1975). Over 99% of the ascorbate within the intragranular space is in the fully reduced form (Ingebretsen *et al.*, 1980). One percent or less of fully oxidized ascorbate (dehydroascorbate) is present, with free radical ascorbate (semidehydroascorbate) existing as a fraction of this amount (Ingebretsen *et al.*, 1980).

The impetus for the study of ascorbate transport arose from the supposition that if ascorbate is the physiologic electron donor for dopamine-β-hydroxylase, then for each mole of dopamine hydroxylated one mole of ascorbate must be oxidized. Since the intragranular matrix ratio of catecholamines to ascorbate is 25 to 1, it follows then that a mechanism must exist by which intragranular dehydroascorbate (or semidehydroascorbate) is either replaced by or regenerated to ascorbate. Electron flux through the cytochrome $b_{561}$ and/or direct reduction of membrane-bound dopamine-β-hydroxylase by cytosolic ascorbate are two mechanisms which probably occur *in vivo* and could contribute to ascorbate reduction. However, they are probably not capable of maintaining the ascorbate <99% reduced.

The experiments to date indicate that, in fact, ascorbate is not transported across the chromaffin granule membrane (Tirrell and Westhead, 1978, 1979). Moreover, no ascorbate–dehydroascorbate exchange, i.e., ascorbate influx coupled to dehydroascorbate efflux, has been measured. Dehydroascorbate, on the other hand, does accumulate at a slow rate of 0.07 nmole/mg protein per min. The uptake is independent of MgATP, displays nonsaturability, and is stimulated only threefold from 0–30°C. Its properties are consistent with a noncarrier-mediated diffusion process similar to that observed in the red blood cell and mitochondrion (Hughes and Maton, 1968), wherein dehydroascorbate is thought to permeate the apolar region of the membrane while ascorbate is excluded due to its much lower solubility in the apolar phase. Therefore, the present data exclude dehydroascorbate or ascorbate transport as significantly contributing to maintenance of reduced ascorbate within the chromaffin granule matrix.

## C. Calcium Accumulation

The calcium content of isolated chromaffin granules ranges between 58–125 nmoles/mg protein, corresponding to an intragranular concentration of 18–31 mM if all the calcium were free in solution (Borowitz *et al.*, 1965; Borowitz, 1967, 1969; Phillips, 1981; Serck-Hannsen and Christiansen, 1973). However, most of the intragranular calcium is probably bound with high affinity to a matrix storage complex for the following reasons: (1) in artificial solutions of ATP and catecholamines, the addition of calcium profoundly increases the stability of the high molecular weight complexes formed (Ingebretsen *et al.*, 1980), (2) incubation of isolated granules with EDTA fails to decrease the intragranular calcium content (Borowitz *et al.*, 1965; Serck-Hannsen and Christiansen, 1973), and (3) addition of a calcium ionophore (which can only equilibrate the calcium gradient across the membrane) results in significant calcium accumulation into isolated granules (Johnson and Scarpa, 1976).

Several studies have shown that $^{45}Ca^{2+}$ can be accumulated into chromaffin granules (Table 6), but no *net* uptake of calcium into isolated granules or ghosts has been demonstrated. Only one out of the four studies to date has attempted to measure the transport of calcium directly across the chromaffin granule membrane (Phillips and Allison, 1977; Krieger-Brauer and Gratzl, 1982). The others resorted to density-gradient distribution of calcium after incubation with a crude granule preparation (Hausler *et al.*, 1981; Kostron *et al.*, 1977a; Niedermaier and Burger, 1981). Table 6 lists the known properties of so-called calcium uptake.

The physiologic function of the intragranular calcium is speculative. Certainly *in vitro* model system experiments suggest that calcium may stabilize the catecholamine–ATP complexes, or it may simply accumulate bound to ATP. It has been

Table 6. *Properties of Calcium Uptake into Isolated Chromaffin Granules or Ghosts*

| Investigator | $V_{max}$ (nmoles/min/mg) | Temperature | $K_m$ (μM) | ATP dependence | Saturable | $Na^+$ dependence |
|---|---|---|---|---|---|---|
| Kostron *et al.*, 1977a,b | 1.01 | 37 | 400 | No | ? | No |
| Phillips, 1981 | 28.0 | 37 | 38 | No | Yes | Yes |
| Hausler *et al.*, 1981 | 0.57 | 20 | ? | Yes | ? | No |
| Niedermaier and Burger, 1981 | 2.71 | 20 | ? | No | ? | No |
| Krieger-Brauer and Gratzl, 1982 | 1.20 | 37 | ? | No | No | Yes |

proposed that the chromaffin granule accumulates the cytosolic calcium which increases in the chromaffin cell during stimulus-secretion coupling, thereby acting in tandem with the plasma membrane to regulate the intracellular calcium concentration. Alternatively, the chromaffin granule may function as a long-term homeostatic regulator of cellular calcium rather than influence each cycle of calcium depolarization–repolarization.

## IX. CONCLUSION: BIOGENIC AMINE TRANSPORT INTO OTHER ORGANELLES

Evidence is rapidly accumulating to suggest that the maintenance of a transmembrane $\Delta\bar{\mu}_{H^+}$ may be a generally operative mechanism for the physiologic accumulation of neurotransmitters or hormones into their subcellular storage organelles. Techniques and methodologies perfected for investigation of the chromaffin granule are now being applied to the analysis of the composition, ATPase activity, amine uptake, and transmembrane ion gradients of a variety of less easily isolated amine-containing subcellular organelles. In isolated platelet dense granules and human pheochromocytoma granules, accumulation of serotonin and of catecholamines, respectively, has been shown to proceed via a $\Delta\bar{\mu}_{H^+}$-dependent carrier-mediated mechanism virtually identical to that of biogenic amine accumulation into bovine adrenal chromaffin granules (Carty *et al.*, 1981; Johnson *et al.*, 1982b; Wilkins and Salganicoff, 1981; Rudnick *et al.*, 1980). $H^+$-translocating ATPase activity has recently been measured in anterior pituitary dense granules containing prolactin and growth hormone (Carty *et al.*, 1982) and in insulin-secretory granules isolated from a pancreatic islet cell tumor (Hutton and Peshavaria, 1982). Existence of a $\Delta\bar{\mu}_{H^+}$ has been reported in isolated anterior pituitary dense granules (Carty *et al.*, 1982), posterior pituitary granules (Russell and Holz, 1981), and pancreatic insulin granules (Hutton, 1982). Moreover, both the histamine-containing granules of isolated mast cells (Johnson *et al.*, 1980) and synaptic vesicles isolated from *Torpedo* electric organ (Michaelson and Angel, 1980) have been shown to possess markedly acidic interiors. The common presence of an electrochemical proton gradient in these various organelles of widely differing composition, function, and embryologic origin is striking.

Although the role of the $\Delta\bar{\mu}_{H^+}$ in biogenic amine transport into chromaffin granules and platelet dense granules is becoming well defined, its function in anterior pituitary, neurohypophyseal, acetylcholine-containing, and pancreatic β-cell granules is still enigmatic since these organelles contain few or no biogenic amines *in vivo*. The existence of a $\Delta\bar{\mu}_{H^+}$ may, in fact, assume a different purpose than to act as the driving force for amine accumulation. For example, amine uptake into anterior pituitary granules does not proceed through a reserpine-sensitive transporter (Carty *et al.*, 1982); rather, amines probably distribute in response to the ΔpH by a nonspecific mechanism, with the granule amine content reflecting only the amount and type of cytosolic amines. It can be hypothesized that the $\Delta\bar{\mu}_{H^+}$ may provide the driving force for uptake and/or homeostasis of both acetylcholine (Parsons and Koenigsberger, 1980; Toll and Howard, 1980) and peptide hormones in their storage organelles.

No longer thought of as a passive amine storage organelle, the dynamic contributions of the chromaffin granule to amine homeostasis have become increasingly well characterized during the last 5 years. Use of chromaffin granules as the basis for future investigations in other storage organelles should permit the detailed clarification of such enigmas as (1) the role of lysolipids in membrane integrity and function, (2) the implications of electron transport in transmembrane redox function, (3) the characterization of ion and nucleotide transport and homeostasis, (4) the identification, synthesis, transport, and maintenance of small peptides within the matrix storage space, (5) the interaction of membranes and transporters with a variety of pharmacologic agents, (6) the relationship of transmembrane ion gradients to secretory or postsecretory events, and (7) the elucidation of the molecular mechanism of neurotransmitter or hormone carrier function in terms of specificity, stoichiometry, turnover, and coupling.

ACKNOWLEDGMENTS

The experimental work outlined in this review was supported by Grant HL-18708 from NIH. Many thanks are due to several collaborators who have participated over the years in the various phases of this project: M. Beers, N. Carlson, S. Hayflick, A. Pallant, D. Pfister, and G. Salama. The skillful help of Dan Brannen in the preparation of the manuscript is also greatly appreciated.

## REFERENCES

Abbs, M. T., and Phillips, J. H., 1980, Organization of the proteins of the chromaffin granule membrane, *Biochim. Biophys. Acta* **595**:200–221.

Aberer, W. H., Kostron, H., Huber, E., and Winkler, H., 1978, A characterization of the nucleotide uptake by chromaffin granules of bovine adrenal medulla, *Biochem. J.* **172**:353–360.

Apps, D. K., and Glover, L. A., 1978, Isolation and characterization of magnesium adenosinetriphosphatase from the chromaffin granule membrane, *FEBS Lett.* **85**:254–257.

Apps, D. K., and Schatz, G., 1979, An adenosine triphosphatase isolated from chromaffin granule membranes is closely similar to $F_1$-adenosine triphosphatase of mitochondria, *Eur. J. Biochem.* **100**:411–419.

Apps, D. K., Pryde, J. G., and Phillips, J. H., 1980a, Both the transmembrane pH gradient and the membrane potential are important in the accumulation of amines by resealed chromaffin-granule "ghosts," *FEBS Lett.* **111**:386–390.

Apps, D. K., Pryde, J. G., Sutton, R., and Phillips, J. H., 1980b, Inhibition of adenosine triphosphatase, 5-hydroxytryptamine transport, and proton translocation activities of resealed chromaffin granule ghosts, *Biochem. J.* **190**:273–282.

Bank, S. P., 1965, The adenosine-triphosphatase activity of adrenal chromaffin granules, *Biochem. J.* **95**:490.

Bashford, C. L., Radda, G. K., and Ritchie, G. A., 1975, Energy-linked activities of the chromaffin granule membrane, *FEBS Lett.* **50**:21–24.

Bashford, C. L., Cassey, R. P., Radda, G. K., and Ritchie, G. A., 1976, Energy coupling in adrenal chromaffin granules, *Neuroscience* **1**:399–412.

Bock, E., and Helle, K. B., 1977, Localization of synoptin on synaptic vesicle membranes, synaptosomal plasma membranes, and chromaffin granule membranes, *FEBS Lett.* **82**:175–178.

Borowitz, J. L., 1967, Calcium binding by subcellular fractions of bovine adrenal medulla, *J. Cell. Comp. Physiol.* **69**:305–310.

Borowitz, J. L., 1969, Effect of acetylcholine on the subcellular distribution of $^{45}Ca^{++}$ on bovine adrenal medulla, *Biochem. Pharmacol.* **18**:715–723.

Borowitz, J. L., Fuiva, K., and Weiner, N., 1965, Distribution of metals and catecholamines in bovine adrenal medulla sub-cellular fractions, *Nature* **205:**42–43.

Buckland, R. M., Radda, G. K., and Shennon, K. D., 1978, Accessibility of phospholipids in the chromaffin granule membrane, *Biochim. Biophys. Acta* **513:**321–337.

Buckland, R. M., Radda, G. K., and Wakefield, E. M., 1979, Reconstitution of the MgATPase of the chromaffin granule membrane, *FEBS Lett.* **103:**323–327.

Buckley, J. T., Lefebre, Y. A., and Hawthorne, J. A., 1971, Identification of an actively phosphorylated component of adrenal medulla chromaffin granules, *Biochim. Biophys. Acta* **239:**517–519.

Cahill, A. L., and Morris, S. J., 1979, Soluble and membrane lectin-binding glycoproteins of the chromaffin granule, *J. Neurochem.* **32:**855–867.

Carty, S. E., Johnson, R. G., and Scarpa, A., 1980, The isolation of intact chromaffin granules using isotonic Percoll density gradients, *Anal. Biochem.* **106:**438–445.

Carty, S. E., Johnson, R. G., and Scarpa, A., 1981, Serotonin transport in isolated platelet granules: Coupling to the electrochemical proton gradient, *J. Biol. Chem.* **256:**11244–11250.

Carty, S. E., Johnson, R. G., and Scarpa, A., 1982, Electrochemical proton gradient in dense granules isolated from anterior pituitary, *J. Biol. Chem.* **257:**7269–7273.

Carty, S. E., Johnson, R. G., Vaugh T., Pallant, A., and Scarpa, A., 1984, Kinetic parameters of amine accumulation into chromaffin ghosts, *Eur. J. Biochem.* (submitted).

Casey, R. P., Njus, D., Radda, G. K., and Sehr, P. A., 1976, Adenosine triphosphate-evoked catecholamine release in chromaffin granules, *Biochem. J.* **158:**583–588.

Casey, R. P., Njus, D., Radda, G. K., and Sehr, P. A., 1977a, Active proton uptake by chromaffin granules: Observation by amine distribution and phosphorus-31 nuclear magnetic resonance techniques, *Biochemistry* **16:**972–977.

Casey, R. P., Njus, D., Radda, G. K., and Sehr, P. A., 1977b, Adenosine triphosphate-evoked catecholamine release in chromaffin granules. Osmotic lysis as a consequence of proton translocation, *Biochem. J.* **158:**583–588.

Coupland, R. E., and Hopwood, D., 1966, The mechanism for differential staining reaction for adrenaline- and noradrenaline-storing granules in tissue fixed in gluteraldehyde, *J. Anat.* **100:**227–243.

Craine, J. E., Daniels, G. H., and Kaufman, S., 1973, Dopamine-β-hydroxylase, the subunit structure and anion activation of the bovine adrenal enzyme, *J. Biol. Chem.* **248:**7838–7844.

Cross, R. L., 1981, The mechanism and regulation of ATP synthesis by $F_1$ -ATPases, *Annu. Rev. Biochem.* **50:**681–714.

DaPrada, M., Obrist, R., and Pletscher, A., 1975, Discrimination of monoamine uptake by membranes of adrenal chromaffin granules, *Br. J. Pharmacol.* **53:**257–265.

Deamer, D. N., Prince, R., and Crofts, A. R., 1972, The response of fluorescent amines to pH gradients across liposome membranes, *Biochim. Biophys. Acta* **274:**323–335.

DeOliveira-Filgueiras, O. M., van den Bosch, H., Johnson, R. G., Carty, S. E., and Scarpa, A., 1981, Phospholipid composition of some amine containing storage granules, *FEBS Lett.* **129:**309–313.

Deupree, J. D., Weaver, J. A., and Downs, D. A., 1982, Catecholamine content of chromaffin granule "ghosts" isolated from bovine adrenal glands, *Biochim. Biophys. Acta* **714:**471–478.

Dolais-Kitabgi, J., and Perlman, R. L., 1975, The stimulation of catecholamine release from chromaffin granules by valinomycin, *Mol. Pharmacol.* **11:**745–750.

Duong, L. T., and Fleming, P. J., 1982, Isolation and properties of cytochrome $b_{561}$ from bovine adrenal chromaffin granules, *J. Biol. Chem.* **257:**8561–8564.

Eagles, P. A. M., Johnson, L. N., and van Horn, C., 1975, The distribution of concanavaline A receptor sites on the membrane of chromaffin granules, *J. Cell Soc.* **19:**33–34.

Flatmark, T., and Gronberg, M., 1981, Cytochrome $b_{561}$ of the bovine adrenal chromaffin granules, *Biochem. Biophys. Res. Commun.* **99:**292–301.

Flatmark, T., and Ingebretsen, O. C., 1977, ATP dependent proton translocation in resealed chromaffin granule ghosts, *FEBS Lett.* **78:**53–56.

Flatmark, T., and Terland, O., 1971, Cytochrome $b_{561}$ of the bovine adrenal chromaffin granule: A high potential *b*-type cytochrome, *Biochim. Biophys. Acta* **253:**487–491.

Gabizon, R., Yetinson, T., and Schuldiner, S., 1982, Photoinactivation and identification of the biogenic amine transporter in chromaffin granules from bovine adrenal medulla, *J. Biol. Chem.* **257:**15145–15150.

Geissler, D., Martinek, A., Margolis, R. U., Margolis, R. K., Shrivanek, J. A., Ledeen, R., Konig, P., and Winkler, H., 1977, Composition and biogenesis of complex carbohydrates of ox adrenal chromaffin granules, *Neuroscience* **2:**685–693.

Giraudat, J., Roisin, M., and Henry, J-P., 1980, Solubilization and reconstitution of the adenosine 5′-triphosphate dependent proton translocase of bovine chromaffin granule membrane, *Biochemistry* **19:**4499–4505.

Goldstein, M., Joh, T. H., and Gravey, T. Q., 1968, Kinetic studies of the enzyme dopamine-β hydroxylation reaction, *Biochemistry* **7:**2724–2730.

Grouselle, M., and Phillips, J. H., 1982, Reduction of membrane bound dopamine-β-hydroxylase from the cytoplasmic surface of the chromaffin granule membrane, *Biochem. J.* **202:**759–770.

Hausler, R., Burger, A., and Niedermaier, W., 1981, Evidence for an inherent, ATP-stimulated uptake of calcium into chromaffin granules, *Naunyn-Schmiedeberg's Arch. Pharmacol.* **315:**255–267.

Hayflick, S., Johnson, R. G., Carty, S. E., and Scarpa, A., 1981, Kinetic and quantitative measurements of catecholamine transport in chromaffin ghosts using a catecholamine electrode, *Anal. Biochem.* **126:**58–66.

Helle, K. B., Serck-Hanssen, G., and Boch, E., 1978, Complexes of chromogranin A and dopamine β-hydroxylase among the chromogranins of the bovine adrenal medulla, *Biochim. Biophys. Acta* **533:**396–407.

Holz, R. W., 1978, Evidence that catecholamine transport into chromaffin vesicles is coupled to vesicle membrane potential, *Proc. Natl. Acad. Sci. USA* **75:**5190–5194.

Holz, R. W., 1979, Measurement of membrane potential of chromaffin granules by the accumulation of triphenyl methylphosphonium cations, *J. Biol. Chem.* **254:**6703–6709.

Hornig, D., 1975, Distribution of ascorbic acid, metabolites, and analogs in man and animals, *Ann. N.Y. Acad. Sci.* **258:**103–118.

Hortnagl, H., Winkler, H., and Lochs, H., 1973, Membrane proteins of chromaffin granules. Dopamine-beta-hydroxylase, a major constituent, *Biochem. J.* **129:**187–195.

Huber, E., Konig, P., Schuler, G., Aberer, W., Plattner, H., and Winkler, H., 1979, Characterization and topography of the glycoproteins of adrenal chromaffin granules, *J. Neurochem.* **32:**35–417.

Hughes, R. E., and Maton, S. C., 1968, The passage of vitamin C across the erythrocyte membrane, *Br. J. Haematol.* **14:**247–253.

Hutton, J. C., 1982, The internal pH and membrane potential of the insulin-secretory granule, *Biochem. J.* **204:**171–178.

Hutton, J. C., and Peshavaria, M., 1982, Proton-translocating $Mg^{2+}$-dependent ATPase activity in insulin-secretory granules, *Biochem. J.* **204:**161–170.

Ingebretsen, O. C., and Flatmark, T., 1979, Active and passive transport of dopamine in chromaffin granule ghosts isolated from bovine adrenal medulla, *J. Biol. Chem.* **254:**3833–3839.

Ingebretsen, O. C., Terland, O., and Flatmark, T., 1980, Subcellular distribution of ascorbate in bovine adrenal medulla. Evidence for accumulation in chromaffin granules against a concentration gradient, *Biochim. Biophys. Acta* **628:**182–189.

Izumi, F., Oka, M., Morita, K., and Azuma, H., 1975, Catecholamine releasing factor in bovine adrenal medulla, *FEBS Lett.* **56:**73–76.

Jockusch, B. M., Burger, M. M., DaPrada, M., Richards, J. G., Chaponnier, C., and Gabbiani, G., 1977, Alpha-actin attached to membranes of secretory vesicles, *Nature* **270:**628–629.

Johnson, R. G., and Scarpa, A., 1976, Ion permeability of isolated chromaffin granules, *J. Gen. Physiol.* **68:**601–631.

Johnson, R. G., and Scarpa, A., 1979, Protonmotive force and catecholamine transport in isolated chromaffin granules, *J. Biol. Chem.* **254:**3750–3760.

Johnson, R. G., and Scarpa, A., 1981, The electron transport chain of serotonin dense granules of platelets, *J. Biol. Chem.* **256:**11966–11969.

Johnson, R. G., Carlson, N., and Scarpa, A., 1978, ΔpH and catecholamine distribution in isolated chromaffin granules, *J. Biol. Chem.* **253:**15120–15121.

Johnson, R. G., Pfister, D., Carty, S. E., and Scarpa, A., 1979, Biological amine transport in chromaffin ghosts, *J. Biol. Chem.* **254:**10963–10972.

Johnson, R. G., Carty, S. E., Fingerhood, B., and Scarpa, A., 1980, The internal pH of mast cell granules, *FEBS Lett.* **120:**75–79.

Johnson, R. G., Carty, S. E., and Scarpa, A., 1981, Proton : substrate stoichiometries during active transport of biogenic amines in chromaffin ghosts, *J. Biol. Chem.* **256:**5773–5780.

Johnson, R. G., Beers, M. F., and Scarpa, A., 1982a, $H^+$ ATPase of chromaffin granules: Kinetics, regulation, and stoichiometry, *J. Biol. Chem.* **257:**10701–10707.

Johnson, R. G., Carty, S. E., and Scarpa, A., 1982b, Catecholamine transport and energy-linked function of chromaffin granules isolated from a human pheochromocytoma, *Biochim. Biophys. Acta* **716:**366–376.

Johnson, R. G., Hayflick, S., Carty, S. E., and Scarpa, A., 1982c, Net uptake of catecholamines into isolated chromaffin granules demonstrated by a novel polarographic technique, *FEBS Lett.* **141:**63–67.

Johnson, R. G., Carty, S. E., and Scarpa, A., 1982d, The electrochemical proton gradient and catecholamine accumulation into isolated chromaffin granules and ghosts, in: *Membranes and Transport: A Critical Review* (A. Martonosi, ed.), Plenum, New York, pp. 237–244.

Johnson, R. G., Carty, S. E., and Scarpa, A., 1984, Kinetic parameters of uptake of various biological amines in isolated chromaffin granules, manuscript in preparation.

Jonasson, J., Rosengren, E., and Waldeck, B., 1964, Effects of some pharmacologically active amines on the uptake of arylalkylamines by adrenal medullary granules, *Acta Physiol. Scand.* **60:**136–140.

Kanner, B. I., Fishkes, H., Maron, R., Sharon, I., and Schuldiner, S., 1979, Reserpine as a competitive and reversible inhibitor of the catecholamine transporter of bovine chromaffin granules, *FEBS Lett.* **100:**175–178.

Kanner, B. I., Sharon, I., Maron, R., and Schuldiner, S., 1980, Electrogenic transport of biogenic amines in chromaffin granule membrane vesicles, *FEBS Lett.* **111:**83–86.

Kirshner, N., 1962, Uptake of catecholamines by a particulate fraction of the adrenal medulla, *J. Biol. Chem.* **237:**2311–2317.

Knoth, J., Handloser, K., and Njus, D., 1980, Electrogenic epinephrine transport in chromaffin granule ghosts, *Biochemistry* **19:**2938–2942.

Knoth, J., Isaacs, J. M., and Njus, D., 1981, Amine transport in chromaffin granule ghosts, *J. Biol. Chem.* **256:**6541–6543.

Kostron, H., Winkler, H., Geissler, D., and Konig, P., 1977a, Uptake of calcium by chromaffin granules *in vitro, J. Neurochem.* **23:**487–493.

Kostron, H., Winkler, H., Peer, L. J., and Konig, P., 1977b, Uptake of adenosine triphosphate by isolated adrenal chromaffin granules: A carrier-mediated transport, *Neuroscience* **2:**159–166.

Krieger-Brauer, H., and Gratzl, M., 1982, Uptake of $Ca^{++}$ by isolated secretory vesicles from adrenal medulla, *Biochim. Biophys. Acta* **691:**61–70.

Lishajko, F., 1971, Studies on catecholamine release and uptake in adreno-medullary storage granules, *Acta Physiol. Scand.* **362**(Suppl.)**:**3–39.

Livett, B. G., Dean, D. M., Whelan, L. G., Udenfriend, S., and Rossier, J., 1981, Co-release of enkephalin and catecholamine from cultured adrenal chromaffin cells, *Nature* **289:**317–319.

Lundborg, P., 1966, Uptake of metaraminol by the adrenal medullary granules, *Acta Physiol. Scand.* **67:**423–429.

Margolis, R. U., and Margolis, R. K., 1973, Isolation of chondroitin sulfate and glycopeptides from chromaffin granules of adrenal medulla, *Biochem. Pharmacol.* **22:**2195–2197.

Maron, R., Fishkes, H., Kanner, B. I., and Schuldiner, S., 1979, Solubilization and reconstitution of the catecholamine transporter from bovine chromaffin granules, *Biochemistry* **18:**4781–4785.

Michaelson, D. M., and Angel, I., 1980, Determination of ΔpH in cholinergic synaptic vesicles: Its effect on storage and release of acetylcholine, *Life Sci.* **27:**39–44.

Mitchell, P., 1961, Coupling of phosphorylation to electron and hydrogen transfer by a chemi-osmotic type of mechanism, *Nature* **191:**144–148.

Mitchell, P., 1968, *Chemiosmotic Coupling and Energy Transduction,* Glynn Research, Bodmin, Cornwall.

Morris, S. J., and Schovanka, I., 1977, Some physical properties of adrenal medulla chromaffin granules isolated by a new continuous iso-osmotic density gradient method, *Biochim. Biophys. Acta* **464:**53–64.

Muller, T. W., and Kirshner, N., 1975, ATPase and phosphatidylinositol kinase activities of adrenal chromaffin vesicles, *J. Neurochem.* **24:**1155–1161.

Nichols, J. W., and Deamer, D. W., 1976, Catecholamine uptake and concentration by liposomes maintaining pH gradients, *Biochim. Biophys. Acta* **455:**269–271.

Niedermaier, W., and Burger, A., 1981, Two different ATP-dependent mechanisms for calcium uptake into chromaffin granules and mitochondria, *Naunyn-Schmiedeberg's Arch. Pharmacol.* **316**(1)**:**69–80.

Njus, D., and Radda, G. K., 1978, Bioenergetic processes in chromaffin granules: A new perspective on some old problems, *Biochim. Biophys. Acta* 219–244.

Njus, D., and Radda, G. K., 1979, A potassium ion diffusion potential causes adrenaline uptake into chromaffin granule "ghosts," *Biochem. J.* **180:**579–585.

Njus, D., Sehr, P. A., Radda, G. K., Ritchie, G. M., and Seeling, R. J., 1978, Phosphorous-31 nuclear magnetic resonance studies of active proton translocation in chromaffin granules, *Biochemistry* **17:**4337–4343.

Njus, D., Knoth, J., and Zallakian, M., 1981, Proton-linked transport in chromaffin granules, *Curr. Top. Bioenerg.* **13:**107–145.

Njus, D., Knoth, J., Cook, C., and Kelly, P. M., 1983, Electron transfer across the chromaffin granule membrane, *J. Biol. Chem.* **258:**27–30.

Parsons, S. M., and Koenigsberger, R., 1980, Specific stimulated uptake of acetylcholine by *Torpedo* electric organ synaptic vesicles, *Proc. Natl. Acad. Sci. USA* **77:**6234–6238.

Pazoles, C. J., Creutz, C. E., Ramu, A. and Pollard, H. B., 1980, Permeant anion activation of Mg, ATPase activity in chromaffin granules, *J. Biol. Chem.* **255:**7863–7869.

Perlman, R. L., and Sheard, B. E., 1982, Estimation of the cytoplasmic catecholamine concentration in pheochromocytoma cells, *Biochim. Biophys. Acta* **719:**334–340.

Phillips, J. H., 1973, Phosphatidyl kinase, a component of the chromaffin granule membrane, *Biochem. J.* **136:**579–587.

Phillips, J. H., 1974a, Transport of catecholamines by resealed chromaffin-granule "ghosts," *Biochem. J.* **144:**311–318.

Phillips, J. H., 1974b, Steady-state kinetics of catecholamine transport by chromaffin granule "ghosts," *Biochem. J.* **144:**319–325.

Phillips, J. H., 1977a, 5-Hydroxytryptamine transport by the bovine chromaffin granule membrane, *Biochem. J.* **170:**673–679.

Phillips, J. H., 1977b, Passive ion permeability of the chromaffin granule membrane, *Biochem. J.* **186:**289–297.

Phillips, J. H., 1981, Transport of $Ca^{++}$ and $Na^{+}$ across the chromaffin granule membrane, *Biochem. J.* **200:**99–107.

Phillips, J. H., 1982, Dynamic aspects of chromaffin granule structure, *Neuroscience* **7:**1595–1609.

Phillips, J. H., and Allison, Y. P., 1977, The distribution of calcium, magnesium, copper, and iron in the bovine adrenal medulla, *Neuroscience* **2.**147–152.

Phillips, J. H., and Allison, Y. P., 1978, Proton translocation by the bovine chromaffin-granule membrane, *Biochem. J.* **170:**661–672.

Pletscher, A., DaPrada, M., Steffen, H., Lutold, B., and Berneis, K. H., 1973, Mechanism of catecholamine accumulation in adrenal chromaffin granules, *Brain Res.* **62:**317–326.

Pletscher, A., DaPrada, M., Berneis, K. H., Steffen, H., Lutold, B., and Weder, H. G., 1974, Molecular organization of amine storage organelles of blood platelets and adrenal medulla, *Adv. Cytol. Pharmacol.* **2:**257–264.

Pollard, H. B., Miller, A., and Cox, G. C., 1973, Synaptic vesicles: Structure of chromaffin granule membranes, *J. Supramol. Struct.* **1:**295–306.

Pollard, H. B., Zinder, O., Hoffman, P. G., and Nikodejevic, O., 1976, Regulation of the transmembrane potential of isolated chromaffin granules by ATP, ATP analogs, and external pH, *J. Biol. Chem.* **251:**4544–4550.

Pollard, H. B., Shindo, H., Creutz, C. E., Pazoles, C. T., and Cohen, J. S., 1979, Internal pH and state of ATP in adrenergic chromaffin granules determined by $^{31}P$ nuclear magnetic resonance spectroscopy, *J. Biol. Chem.* **254:**1170–1177.

Price, H., Kinder, S., and Ledeen, R., 1975, Structure of gangliosides from bovine adrenal medulla, *Biochemistry* **14:**1512–1518.

Ramu, A., Levine, M., and Pollard, H., 1983, *Proc. Natl. Acad. Sci. USA* **80:**2107–2111.

Reed, P. W., and Lardy, H. A., 1972, A23187: A divalent cation ionophore, *J. Biol. Chem.* **247:**6970–6977.

Ritchie, G. A., 1975, Ph.D. thesis, University of Oxford, Oxford, England.

Roda, L. G., and Hogue-Angeletti, R. A., 1979, Peptides in the adrenal medulla chromaffin granule, *FEBS Lett.* **107:**393–397.

Roisin, M. P., Scherman, D., and Henry, J.-P., 1980, Synthesis of ATP by an artificially imposed electrochemical proton gradient in chromaffin granule ghosts, *FEBS Lett.* **115:**143–146.

Rosenberg, R. C., and Lovenberg, W., 1980, Dopamine-β-hydroxylase, *Essays Neurochem. Neuropharmacol.* **4:**163–209.
Rottenberg, H., 1980, The measurement of membrane potential and ΔpH in cells, organelles, and vesicles, *Meth. Enzymol.* **5:**547–569.
Rudnick, G., Fishkes, H., Nelson, P. J., and Schuldine, S., 1980, Evidence for two distinct serotonin transport systems in platelets, *J. Biol. Chem.* **255:**3638–3641.
Russell, J. T., and Holz, R. W., 1981, Measurement of ΔpH and membrane potential in isolated neurosecretory vesicles from bovine neurohypophyses, *J. Biol. Chem.* **256:**5950–5953.
Salama, G., Johnson, R. G., and Scarpa, A., 1979, Spectrophotometric measurements of transmembrane potential and pH gradients in chromaffin granules, *J. Gen. Physiol.* **75:**109–140.
Scherman, D., and Henry, J.-P., 1979, Effet du potential trans-membranaire sur le transport de la noradrenaline par les granules chromaffines, *C. R. Acad. Sci. (Paris)* **289:**911–914.
Scherman, D., and Henry, J.-P., 1980, Oxonol-V as a probe of chromaffin granule membrane potentials, *Biochim. Biophys. Acta* **599:**150–166.
Scherman, D., and Henry, J.-P., 1981, pH-dependence of the ATP-driven uptake of noradrenaline by bovine chromaffin-granule ghosts, *Eur. J. Biochem.* **116:**535–539.
Schmidt, W., Winkler, H., and Plattner, H., 1982, Adrenal chromaffin granules: Evidence for an ultrastructural equivalent of the proton pumping ATPase, *Eur. J. Cell Biol.* **27:**96–104.
Schuldiner, S., Fishkes, H., and Kanner, B. L., 1978, Role of a transmembrane pH gradient in epinephrine transport by chromaffin granule membrane vesicles, *Proc. Natl. Acad. Sci. USA* **75:**3713–3716.
Sen, R., and Sharp, R. R., 1980, The soluble components of chromaffin granules, a carbon-13 NMR survey, *Biochim. Biophys. Acta* **630:**447–458.
Serck-Hannsen, G., and Christiansen, E. N., 1973, Uptake of calcium in chromaffin granules of bovine adrenal medulla stimulated *in vitro, Biochim. Biophys. Acta* **307:**404–414.
Silsand, T., and Flatmark, T., 1974, Purification of cytochrome $b_{561}$, an integral heme protein of the adrenal chromaffin granule membrane, *Biochim. Biophys. Acta* **395:**257–266.
Skotland, T., and Flatmark, T., 1979, On the amphiphilic and hydrophobic forms of dopamine-β-monooxygenase in bovine adrenal medulla, *J. Neurochem.* **31:**1861–1863.
Skotland, T., and Ljones, T., 1979, Dopamine-β-hydroxylase: Structure, mechanism, and properties of the enzyme bound copper, *Inorg. Perspect. Biol. Med.* **2:**151–180.
Slotkin, T. A., 1975a, Maturation of the adrenal medulla. III. Practical and theoretical considerations of the age-dependent alterations in kinetics of incorporation of catecholamines and non-catecholamines, *Biochem. Pharmacol.* **24:**89–97.
Slotkin, T. A., 1975b, Structure-activity relationships for the reserpine-like actions of derivatives of β-carboline *in vitro, Life Sci.* **15:**439–454.
Slotkin, T. A., and Kirshner, N., 1971, Uptake, storage, and distribution of amines in bovine adrenal medullary vesicles, *Mol. Pharmacol.* **7:**581–592.
Smith, A. D., and Winkler, H., 1967, A simple method for the isolation of adrenal chromaffin granules on a large scale, *Biochem. J.* **103:**480–482.
Spiro, M. J., and Ball, E. G., 1961, Studies on the respiratory enzymes of the adrenal gland, *J. Biol. Chem.* **236:**225–229.
Stern, A. S., Lewis, R. V., Kimura, S., Rossier, J., Gerber, L. D., Brink, L., Stern, S., and Udenfriend, S., 1979, Isolation of the opioid heptopeptide Met-enkephalin ($Arg^6Phe^7$) from the bovine adrenal medullary granules and striatum, *Proc. Natl. Acad. Sci. USA* **76:**6680–6683.
Stern, A. S., Jones, B. N., Shively, J. E., Stern, S., and Udenfriend, S., 1981, Two adrenal opioid polypeptides: Proposed intermediates in the processing of proenkephalin, *Proc. Natl. Acad. Sci. USA* **78:**1962–1966.
Stitzel, R., 1977, The biological fate of reserpine, *Pharmacol. Rev.* **28:**179–205.
Sutton, R., and Apps, D. K., 1981, Isolation of a DCCD binding protein from bovine chromaffin granule membranes, *FEBS Lett.* **130:**103–106.
Taugner, G., 1971, The membrane of catecholamine storage vesicles of adrenal medulla. Catecholamine fluxes and ATPase activity, *Naunyn-Schmiedeberg's Arch. Pharmacol.* **270:**392–406.
Taugner, G., 1972, The membrane of catecholamine storage vesicles of adrenal medulla. Uptake and release of noradrenaline in relation to the pH and the concentration of steric configuration of the amine present in medium, *Naunyn-Schmiedeberg's Arch. Pharmacol.* **274:**299–314.

Terland, O., and Flatmark, T., 1973, NADH (NADPH) acceptor oxidoreductase activities of the bovine adrenal chromaffin granules, *Biochim. Biophys. Acta* **305:**206–218.

Terland, O., and Flatmark, T., 1975, Ascorbate as a natural constituent of chromaffin granules from the bovine adrenal medulla, *FEBS Lett.* **59:**52–56.

Terland, O., and Flatmark, T., 1980, Oxidoreductase activities of chromaffin granule ghosts isolated from the bovine adrenal medulla, *Biochim. Biophys. Acta* **597:**318–330.

Terland, O., Silsand, T., and Flatmark, T., 1974, Cytochrome $b_{561}$ as the single heme protein of the bovine adrenal chromaffin granule membrane, *Biochim. Biophys. Acta* **359:**253–256.

Terland, O., Flatmark, T., and Kryvi, H., 1979, Isolation and characterization of noradrenaline storage granules of bovine adrenal medulla, *Biochim. Biophys. Acta* **553:**460–468.

Tirrell, J. G., and Westhead, E. W., 1978, Ascorbate uptake and metabolism by adrenal medullary granules, in: *Catecholamines: Basic and Clinical Frontiers* (E. Usdine, I. H. Kopin, and J. Barchas, eds.), Pergamon Press, New York, pp. 181–186.

Tirrell, J. G., and Westhead, E., 1979, The uptake of ascorbic acid and dehydroascorbic acid by chromaffin granules of the adrenal medulla, *Neuroscience* **4:**181–186.

Toll, L., and Howard, B. D., 1980, Evidence that an ATPase and a protonmotive force function in the transport of acetylcholine into storage vesicles, *J. Biol. Chem.* **255:**1787–1789.

Trifaro, J. M., and Dworkind, J., 1970, A new and simple method for the isolation of adrenal chromaffin granules by means of an isotonic density gradient, *Anal. Biochem.* **34:**403–412.

Trifaro, J. M., and Poisner, A. M., 1967, The role of ATP and ATPase in the release of catecholamines from the adrenal medulla. II. ATP evoked fall in optical density of isolated chromaffin granules, *Mol. Pharmacol.* **3:**572–580.

Viveros, O. H., Diliberto, E. J., Hazum, E., and Chang, K. J., 1979, Opiate like materials in the adrenal medulla: Evidence for storage and secretion with catecholamines, *Mol. Pharmacol.* **16:**1101–1108.

Viveros, O. H., Diliberto, E. J., Hazum, E., and Chany, K. J., 1980, Enkephalin as possible adreno-medullary hormones: Storage, secretion, and regulation of synthesis, in: *Advances in Biochemical Psychopharmacology,* Vol. 22 (E. Costa and M. Trabucchi, eds.), Raven Press, New York, pp. 191–204.

Voyta, J. C., Slakey, L. L., and Westhead, E. W., 1978, Accessibility of lysolecithin in catecholamine secretory vesicles to acyl CoA: lysolecithin acyl transferase. *Biochem. Biophys. Res. Commun.* **80:**413–417.

Wallace, E. G., Krantz, M. J., and Lovenberg, W., 1973, Dopamine-β-hydroxylase: A tetrameric glycoprotein, *Proc. Natl. Acad. Sci. USA* **70:**2253–2255.

Weber, A., and Winkler, H., 1981, Specificity and mechanisms of nucleotide uptake by adrenal chromaffin granules, *Neuroscience* **6:**2269–2276.

Weber, A., Westhead, E. W., and Winkler, H., 1983, Specificity and properties of the nucleotide carrier in chromaffin granules from bovine adrenal medulla, *Biochem. J.* **210:**789–794.

Wilkins, J. A., and Salgonicoff, L., 1981, Participation of a transmembrane proton gradient in 5-hydroxytryptamine transport by platelet dense granules and dense-granule ghosts, *Biochem. J.* **198(1):**113 123.

Winkler, H., 1969, Isolierung and Charakterisierung von chromaffinen Noradrenalin-Granula aus Schweine-Nebennierenmark, *Naunyn-Schmiedebergs Arch. Pharmakol.* **263:**340–357.

Winkler, H., 1976, The composition of adrenal chromaffin granules: An assessment of controversial results, *Neuroscience* **1:**65–80.

Winkler, H., 1977, The biogenesis of adrenal chromaffin granules, *Neuroscience* **2:**657–683.

Winkler, H., 1982, The proteins of catecholamine-storing organelles, *Scand. J. Immunol.* **15**(Suppl. 9):75–96.

Winkler, H., and Carmichael, S. W., 1982, *The Secretory Granule* (E. Poisner and J. Trifaro, eds.), Elsevier Biomedical Press, Amsterdam.

Winkler, H., and Westhead, E., 1980, The molecular organization of adrenal chromaffin granules, *Neuroscience* **5:**1803–1823.

Zaremba, S., and Hogue-Angelleti, R. A., 1981, Transmembrane nature of chromaffin granule dopamine-β-monooxygenase, *J. Biol. Chem.* **256:**12310–12315.

Zaremba, S., and Hogue-Angelleti, R. A., 1982, NADH: (acceptor) oxidoreductase from bovine adrenal medulla chromaffin granules, *Arch. Biochem. Biophys.* **219:**297–305.

Zinder, O., Hoffman, P. G., Bonner, W. M., and Pollard, H. B., 1978, Comparison of chemical properties of purified plasma membranes and secretory vesicle membranes from the bovine adrenal medulla, *Cell Tissue Res.* **188:**153–170.

# 40

# Hexose Transport and Its Regulation in Mammalian Cells

*Jeffrey E. Pessin and Michael P. Czech*

## I. INTRODUCTION

The uptake of solutes across the cell-surface membrane can occur by either an active or passive transport mechanism. Active transport mechanisms are characterized by the uptake of solutes against their concentration gradient at the expense of metabolic energy. Active transport of hexoses occurs in two major tissues in mammals, kidney and intestine, and will not be dealt with in this review. Passive transport can be subdivided into the two categories of simple diffusion or facilitative diffusion. In both cases, the movement of solutes across the cell membrane is driven solely by the concentration gradient between the intracellular and extracellular environment without any metabolic energy being required. Net uptake ceases when the concentration of solutes between the inside and outside of the cell has reached equilibrium. Facilitative diffusion differs from simple diffusion in that the former process is mediated by membrane-bound proteins which exhibit a high degree of specificity and whose activity is competitively inhibited with appropriate analogues. The difference in transport rate between simple and facilitative diffusion is dramatically exemplified by the permeability of D-glucose across synthetic lipid bilayers with a permeability coefficient of $10^{-9}$–$10^{-10}$ cm/sec (Lidgard and Jones, 1975; Jung, 1971a), whereas for the intact erythrocyte, the permeability coefficient is approximately $10^{-4}$ cm/sec (Jung, 1971b).

The molecular mechanism governing the facilitative diffusion of solutes across the cell-surface membrane has not been clearly defined to date. Natural peptide ionophores, however, have been invaluable tools in providing useful information under

*Jeffrey E. Pessin and Michael P. Czech* • Department of Biochemistry, University of Massachusetts Medical Center, Worcester, Massachusetts 01605.

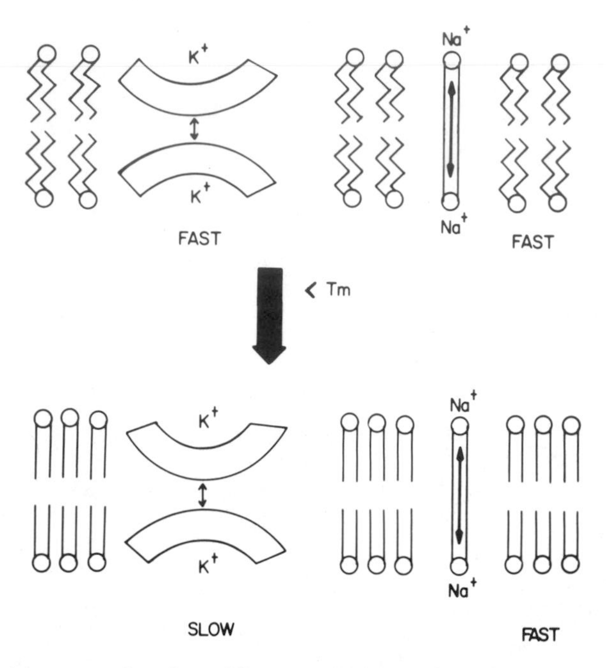


*Figure 1.* Schematic representation of two different mechanisms by which ionophores catalyze the transfer of solutes across biological membranes. Cage-carrier ionophores such as valinomycin diffuse or rotate across the phospholipid bilayer and thereby facilitate the diffusion of bound ions (upper left). The activity of this process is greatly diminished when the bilayer is in a liquid–crystalline phase (lower left). In contrast, the fixed channel catalyzed mode of solute transport exemplified by gramicidins is relatively insensitive to the physical state of the bilayer (upper and lower right). Reprinted with permission from Czech (1980).

*in vitro* conditions. Two extreme models of facilitative solute transport are illustrated by the cage carrier (valinomycin) and fixed-channel (gramicidin) concepts as schematically drawn in Figure 1. Valinomycin specifically binds to the solute being transported, e.g., $K^+$ or $Rb^+$, and diffuses or rotates within the membrane bilayer to effect transport in the direction of the electrochemical gradient (Finkelstein and Cass, 1968; Shemyakin *et al.*, 1969; Eisenman *et al.*, 1968). This type of carrier has been found to be extremely sensitive to the molecular motion of the phospholipid bilayer. Transport catalyzed by this cage-carrier mechanism is greatly inhibited upon converting the phospholipid bilayer from a less ordered liquid–crystalline phase to a more ordered gel phase (Krasne *et al.*, 1971). In contrast, fixed-channel type transporters such as gramicidin facilitate the movement of cations, e.g., $Na^+$, by creating a rigid pore through the bilayer (Hladky and Haydon, 1970; Urry, 1971; Tosteson *et al.*, 1968). This mode of transport is virtually unaffected by the physical state of the phospholipid bilayer (Krasne *et al.*, 1971). Thus, the difference in temperature sensitivity of the cage-carrier or channel-forming modes of transport can provide insight into the molecular mechanism of membrane transporters.

## A. Facilitative D-Glucose Transport

Facilitative D-glucose transport systems are found almost universally throughout all animal tissues (Plagemann and Richey, 1974) and are especially abundant in rabbit reticulocytes (Albert, 1982) and human erythrocytes (Barnett *et al.*, 1973; Taverna and Langdon, 1973; Lin *et al.*, 1974; Lin and Spudich, 1974a; Lienhard *et al.*, 1977; Jung and Rampal, 1977; Basketter and Widdas, 1977). All the facilitative D-glucose transport systems are characterized by a high degree of stereospecificity for hexose and pentose monosaccharides in the pyranose ring form (Barnett *et al.*, 1973; Le Fevre, 1961). The C1 chair conformation, in which the hydroxyl groups are in an equitorial position, appears to be the ring structure most effectively transported (Lacko *et al.*, 1972). For example, D-glucose has a $K_m$ value for uptake into human erythrocytes of approximately 2 mM (Le Fevre, 1961; Lacko *et al.*, 1972), while L-glucose has essentially no appreciable transport into human erythrocytes with a $K_m$ greater than 3 M (Le Fevre, 1961). Several common monosaccharides used for transport measurements are 3-*O*-methylglucose and 2-deoxyglucose, both having $K_m$ in the range of 1–5 mM (Le Fevre, 1961). The use of 3-*O*-methylglucose to measure transport in intact

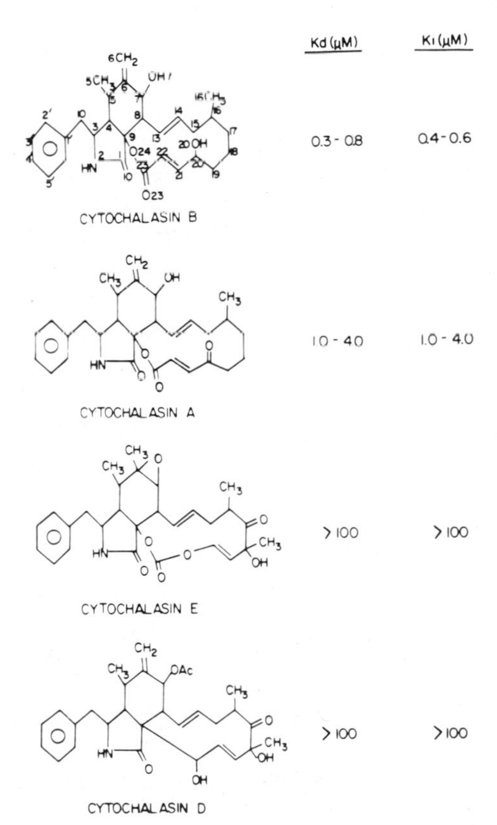


*Figure 2.* The chemical structure and biochemical activities of several cytochalasins. The $K_d$ values represent the high-affinity binding to the D-glucose transporter in human erythrocytes and the $K_i$ values represent the inhibition of D-glucose transport activity. These data were obtained from Griffin *et al.* (1982).

cells has become widespread because it cannot be metabolized (Plagemann and Richey, 1974) and, thus, is a true measure of transport activity. In contrast, 2-deoxyglucose is phosphorylated by hexokinase (Renner *et al.*, 1972), and only under conditions where hexokinase activity is low can this analogue really be used effectively. Another unique feature of all facilitative D-glucose transport systems in mammalian cells is their sensitivity to cytochalasin B. This fungal metabolite (Binder and Tamm, 1973) is a potent reversible inhibitor of D-glucose transport activity (Bloch, 1973; Taverna and Langdon, 1973; Mizel and Wilson, 1974; Taylor and Gagneja, 1973; Jung and Rampal, 1977) and has a $K_i$ of D-glucose transport inhibition equal to its $K_d$ of binding, approximately 0.1 μM (Lin and Spudich, 1974a; Jung and Rampal, 1977; Sogin and Hinkle, 1978, 1980a; Baldwin *et al.*, 1979). Figure 2 illustrates the structure of cytochalasin B as well as several cytochalasin B analogues. Studies on the competition of cytochalasin B binding and D-glucose transport inhibition have revealed that the C-20–C-23 region plays a major role in its binding and inhibition properties (Rampal *et al.*, 1980). Diffraction studies also indicate that cytochalasin B binding to the D-glucose transporter occurs through hydrogen bonding at the N-2, O-7, and O-23 positions (Griffin *et al.*, 1982). The hydrophobic region from C-13–C-19 appears to be essential in binding, and is thought to act as an anchor in a hydrophobic domain of the D-glucose transporter (Griffin *et al.*, 1982).

## B. Assay Methodology

One of the original methods for assaying the rate of D-glucose transport took advantage of the fact that water permeability through membranes is approximately three orders of magnitude greater than D-glucose permeability in human erythrocytes (Rendi, 1964). This approach was originally reported by Sen and Widdas (1962) and later by Wilbrandt (1978), who preloaded human erythrocyte ghosts with high concentrations of D-glucose. These vesicles were then rapidly diluted into a low osmotic buffer, which causes a rapid influx of water and swelling of the membranes. Subsequently, a slower phase of facilitative D-glucose exit occurs, with the concomitant movement of water and decrease in membrane size. The changes in membrane vesicle size can be measured spectrophotometrically as a function of light scattering. This method has a distinct advantage in that very rapid and accurate kinetic analysis can be performed by use of stop-flow type instrumentation (Carruthers and Melchoir, 1983).

A more commonly used assay methodology has been to measure the flux of radiolabeled D-glucose or D-glucose analogues in cells or membrane vesicles. This technique requires a method to separate the free ligand from the intravesicular ligand. This is accomplished by either rapidly filtering the cells or vesicles in suspension or washing surface-attached cultured monolayer cells at reduced temperatures. A novel method for separating adipocytes from free ligand was developed by Gliemann *et al.* (1972). They took advantage of the fact that adipocytes have a density less than one, and will float on top of a silicone oil layer.

As opposed to measuring D-glucose transport activity in intact cells or membranes, a method for the reconstitution of stereospecific D-glucose uptake into phospholipid vesicles has been developed. For the human erythrocytes, membrane proteins were solubilized in Triton X-100 or β-D-octylglucopyranoside and reconstituted into synthetic phospholipid vesicles by dialysis (Kasahara and Hinkle, 1976, 1977). These reconstituted vesicles exhibited initial rates of D-glucose uptake approximately 30 times faster than for L-glucose (Kasahara and Hinkle, 1976). A similar approach has been devised for the reconstitution of the adipocyte D-glucose transporter (Shanahan and Czech, 1977; Carter-Su *et al.*, 1980), but differed in that cholate was used instead of Triton X-100 and the formation of vesicles was accomplished on a Sephadex G50 column. A freeze–thaw cycle and brief sonication step was also employed to enlarge the volume of the reconstituted vesicles which facilitates the assay of transport activity by rapid filtration. Reconstitution of the D-glucose transporter into planar bilayer phospholipids has also been used to study the permeability properties of D-glucose (Nickson and Jones, 1977; Jones and Nickson, 1978).

All the above methods take advantage of the known transporter stereospecificities and the inhibitory properties of agents such as cytochalasin B on the D-glucose transporter. In order to demonstrate that the assay of choice is measuring facilitative D-glucose transport activity, a means to determine the amount of simple diffusion is required. This is generally accomplished by measuring the uptake or efflux of substrate in the presence and absence of cytochalasin B or by comparing the D-glucose transport activity with that of L-glucose.

A third assay technique has been developed to specifically determine the number of D-glucose transporters in membranes. In this assay, the saturable high-affinity binding of [$^3$H]cytochalasin B to membranes is determined as a function of increasing concentrations of cytochalasin B (Taverna and Langdon, 1973; Plagemann and Richey, 1974; Lienhard *et al.*, 1977; Lin and Snyder, 1977; Jung *et al.*, 1980). The binding of [$^3$H]cytochalasin B to nontransporter membrane components is corrected for by determining the binding in the absence and presence of D-glucose, which specifically blocks the binding of [$^3$H]cytochalasin B to the transporter. This method may be used as a molecular marker for the identification of the D-glucose transporter in membrane fractions when a close correlation between cytochalasin B binding and D-glucose transporter inhibition can be demonstrated. The relative high-affinity binding of cytochalasin B to the D-glucose transporter which is inhibited by D-glucose has been shown to accurately reflect the number of D-glucose transporters in a given preparation (Taverna and Langdon, 1973; Lin *et al.*, 1974; Lin and Spudich, 1974a; Lienhard *et al.*, 1977; Jung and Rampal, 1977; Basketter and Widdas, 1977; Sogin and Hinkle, 1978; Baldwin *et al.*, 1979; Jung *et al.*, 1980).

## II. HUMAN ERYTHROCYTE D-GLUCOSE TRANSPORT SYSTEM

Over the past several years, extensive efforts have been made to identify and purify the stereospecific D-glucose transporter from human erythrocytes. These endeavors have focused on the use of specific reagents that are expected to covalently bind solely to the transport protein (Taverna and Langdon, 1973; Mullins and Langdon, 1980a,b; Carter-Su *et al.*, 1982; Shanahan, 1982), to nonspecific reagents that covalently bind to many membrane components (Jung and Carlson, 1975; Batt *et al.*, 1976; Lienhard *et al.*, 1977; Shanahan and Jacquez, 1978), and to purification of specific membrane components which exhibit stereospecific D-glucose transport activity when reconstituted into phospholipid vesicles (Kasahara and Hinkle, 1977; Kahlenberg and Zala, 1977; Goldin and Rhoden, 1979; Phutrakul and Jones, 1979; Lundahl *et al.*, 1981). Despite this rather intensive effort, the identity of the native D-glucose transporter from human erythrocytes has not been unequivocally determined. Some of these methods will be described below and the reader is referred to several reviews for a more detailed description (Plagemann and Richey, 1974; Hokin, 1981; Jones and Nickson, 1981; Baldwin and Lienhard, 1981).

## A. Affinity Labeling

Differential labeling of human erythrocyte ghosts with fluorodinitrobenzene (FDNB) or impermeant glutathione-maleimide has been one approach to covalently label the D-glucose transporter (Eady and Widdas, 1973; Batt *et al.*, 1975, 1976; Jung and Carlson, 1975; Lienhard *et al.*, 1977; Shanahan and Jacquez, 1978). This method is based upon the covalent labeling of membrane protein with [$^{14}$C]-FDNB in the presence of reagents which potentiate the labeling reaction, D-glucose or 2-deoxyglucose (Eady and Widdas 1973; Jung and Carlson, 1975), compared to the labeling of [$^{3}$H]-FDNB in the presence of reagents which inhibit the labeling reaction, ethylidene glucose, cytochalasin B, or D-maltose (Eady and Widdas 1973; Shanahan and Jacquez, 1978). Protein components in the $M_r$ 180,000–200,000 (Jung and Carlson, 1975; Shanahan and Jacquez, 1978), $M_r$ 90,000 (Shanahan and Jacquez, 1978), and $M_r$ 45,000–70,000 (Batt *et al.*, 1975, 1976; Lienhard *et al.*, 1977) have been identified as possible candidates for the D-glucose transporter in this manner. However, when cytochalasin B is used as a protecting reagent, there is a general consensus that the protein in the $M_r$ 45,000–70,000 range is specifically labeled.

Direct affinity labeling of the human erythrocyte D-glucose transporter has been attempted with maltosyl isothiocyanate (MITC), which is an impermeant reversible inhibitor of D-glucose uptake (Mullins and Langdon, 1980a,b). Using [$^{14}$C]-MITC, Mullins and Langdon demonstrated that the inhibition of D-glucose transport activity was linearly related to the labeling of a protein of $M_r$ 88,000. This covalent labeling was protected by maltose and cytochalasin B but not by L-glucose or sucrose. Complete inhibition of D-glucose transport activity required $3.9 \times 10^5$ molecules of MITC bound/cell, consistent with previous studies indicating $3 \times 10^5$ D-glucose transporters/cell (Jones and Nickson, 1981). These data not only indicated that the human erythrocyte D-glucose transporter is a component in the Band 3 region (nomenclature according to Fairbanks *et al.*, 1971) but that prolonged incubation even at 4°C resulted in proteolysis to lower-molecular weight species in the Band 4.5 region, $M_r$ 45,000–70,000.

Recently, a method to photoaffinity label the D-glucose transporter with [$^{3}$H]cytochalasin B has been developed (Carter-Su *et al.*, 1982; Shanahan, 1982). This method takes advantage of an intrinsic property of the system, that upon irradiation of membranes or cells with high-intensity ultraviolet light in the presence of

[$^3$H]cytochalasin B labeling of the D-glucose transporter ensues. The [$^3$H]cytochalasin B photoaffinity labeling was shown to follow all the characteristics expected of the stereospecific D-glucose transporter with a high degree of specificity (Carter-Su *et al.*, 1982). When this method was applied to freshly isolated intact human erythrocytes, only the protein of $M_r$ 45,000–70,000 was covalently labeled in a D-glucose sensitive fashion. Further, no evidence was found for the involvement of any higher-molecular weight components, indicating that at the very minimum the protein of $M_r$ 45,000–70,000 is the fundamental D-glucose sensitive cytochalasin B binding component.

## B. *Purification*

The most convincing evidence that the human erythrocyte D-glucose transporter is a heterogeneous glycoprotein of $M_r$ 45,000–70,000 comes from the purification of this protein to homogeneity (Kahlenberg and Zala, 1977; Kasahara and Hinkle, 1977; Sogin and Hinkle, 1978; Baldwin *et al.*, 1979). Peripheral membrane proteins were extracted from isolated ghosts by an alkaline-EDTA treatment. The resulting particulate fraction was then solubilized in nonionic detergents, Triton X-100 (Sogin and Hinkle, 1978; Baldwin *et al.*, 1979), or β-D-octylglucopyranoside (Kasahara and Hinkle, 1976; Baldwin *et al.*, 1982). This extract was then chromatographed on a DEAE-cellulose column and reconstituted into phospholipid vesicles. The purified and reconstituted D-glucose transporter was shown to stereospecifically transport D-glucose and bind cytochalasin B with a stoichiometry of approximately 0.8 mole cytochalasin B/mole of transporter (Baldwin and Baldwin, 1981; Baldwin *et al.*, 1982). In sodium dodecyl sulfate polyacrylamide gels, the D-glucose transporter stained positive with periodic acid-Schiff reagent (Steck, 1974), indicating this protein to be a glycoprotein. Further, treatment with endo-B-galactosidase reduced the apparent heterogeneity and shifted the average molecular weight from $M_r$ 54,000 to $M_r$ 46,000 (Baldwin *et al.*, 1982; Gorga *et al.*, 1979). Polyclonal antibodies generated against the purified protein of $M_r$ 45,000–70,000 were shown to immunoprecipitate the D-glucose transport activity and did not crossreact with any higher-molecular weight protein components (Baldwin and Lienhard, 1980; Sogin and Hinkle, 1980b). These antibodies also immunoprecipitated a similar molecular weight protein in L-929 and Hela cells. Thus, these results, coupled with the affinity labeling data (Carter-Su *et al.*, 1982; Shanahan, 1982), provide convincing evidence that the native functional unit of the D-glucose transport system in human erythrocytes is a heterogeneous glycoprotein in the Band 4.5 region, $M_r$ 45,000–70,000.

There is uncertainty as to exactly what the native stoichiometry of the D-glucose transporter in the human erythrocyte membrane may be. Although it is clear that the functional transporter unit in reconstituted vesicles is a monomeric glycoprotein of $M_r$ 45,000–70,000, Sogin and Hinkle (1978), using freeze-fracture electron microscopy, have indicated that the D-glucose transporter may exist as a dimer. Jung *et al.* (1980) employed radiation-inactivation studies in which the target analysis indicated that the D-glucose transporter could be tetrameric in nature. Direct chemical analysis of this protein has been very difficult to obtain, most probably due to the extremely hydrophobic nature of the protein. Nevertheless, it has been shown to contain approximately

17% carbohydrate by weight (5% neutral sugars, 7% glucosamine, and 5% sialic acid). Amino acid composition analysis also does not indicate any unusual properties, with 41% nonpolar residues (Sogin and Hinkle, 1978).

## III. REGULATION OF THE D-GLUCOSE TRANSPORTER IN CULTURED CELLS

In contrast to the human erythrocyte which expresses D-glucose transport activity at a constant level, other cells can regulate their facilitative D-glucose transport activity in response to a variety of physiological stimuli. Hamster and chicken embryo fibroblasts maintained in D-glucose-deficient culture medium for 18–24 hr (starved cells) have greatly elevated levels (10 to 20-fold) of D-glucose transport activity compared with that of normally maintained (fed) cultures (Martineau *et al.*, 1972; Kalckar and Ullrey, 1973; Kletzien and Perdue, 1975b; Ullrey *et al.*, 1975; Christopher *et al.*, 1976a,b; Christopher, 1977). This increase in D-glucose transport activity appears to be directly regulated by D-glucose since D-fructose substitution into the culture medium is not able to depress the increase in D-glucose transport activity (Christopher *et al.*, 1976b; Christopher, 1977). Cultured cells transformed by several viruses (sarcoma and polyoma viruses) have also been shown to exhibit elevated levels of D-glucose transport activity (Kawai and Hanafusa, 1971; Weber, 1973; Venuta and Rubin, 1973; Kalckar *et al.*, 1973; Kalckar and Ullrey, 1973; Hatanaka, 1974, 1976; Kletzien and Perdue, 1974; Dolberg *et al.*, 1975). The increased rate of D-glucose transport uptake into these cells appears to require *de novo* protein synthesis, in that protein synthesis inhibitors can block the starvation-induced increase in D-glucose transport activity (Martineau *et al.*, 1972; Kletzien and Perdue, 1975a,b; Christopher *et al.*, 1976a–c; Christopher, 1977). Tunicamycin has also been demonstrated to inhibit the transformation-induced elevation of D-glucose transport activity, indicating that the transport protein(s) in these cells are asparagine-linked glycoproteins (Olden *et al.*, 1974). Kinetic analysis of monosaccharide uptake in starved or transformed cultured cells has demonstrated that the elevated rate of D-glucose transport activity occurs by an increase in the $V_{max}$ for uptake without any change in the $K_m$ of binding (Weber, 1973; Kletzien and Perdue, 1974; Dolberg *et al.*, 1975; Franchi *et al.*, 1978; Innui *et al.*, 1980). However, Christopher *et al.* (1976a) have shown that both starved and fed chicken embryo fibroblasts have at least two transport systems, a low affinity ($K_m$ 6–8 mM) system and a higher affinity ($K_m$ 1–3 mM) system for D-glucose uptake. In general, there are two possible mechanisms by which one can envision an increase in D-glucose transport activity. There could be an increase in the number of D-glucose transporters which can occur by (1) an increase in protein synthesis and/or a decrease in protein degradation, and (2) a movement of preformed intracellular transporters to the cell-surface membrane. Alternatively, there could be an increase in the specific activity of the preexisting D-glucose transporters, either through direct covalent modification or via a modification of an allosteric regulator. Scatchard analysis of [$^3$H]cytochalasin B binding to both starved (Pessin *et al.*, 1982) and transformed (Salter and Weber, 1979) chicken embryo fibroblast plasma membranes has demonstrated a 6- to 10-fold increase in the number of D-glucose-specific cytochalasin B binding sites. Further, antibodies raised against the purified human erythrocyte D-glucose transporter were shown to specifically precipitate a protein of $M_r$ 41,000 and to a lesser degree a protein of $M_r$ 82,000 (Salter *et al.*, 1982). The amount of both the $M_r$ 41,000 and 82,000 proteins increased in direct proportion to the increase in D-glucose transport activity.

Photoaffinity labeling of starved and fed chicken embryo fibroblast plasma membranes with [$^3$H]cytochalasin B has specifically identified two proteins of $M_r$ 46,000 and 52,000 (Figure 3). The covalent labeling of both proteins was highly sensitive to the presence of 500 mM D-glucose compared to the labeling in the presence of 500 mM D-sorbitol (a nontransportable sugar alcohol used as an osmotic control). Differ-

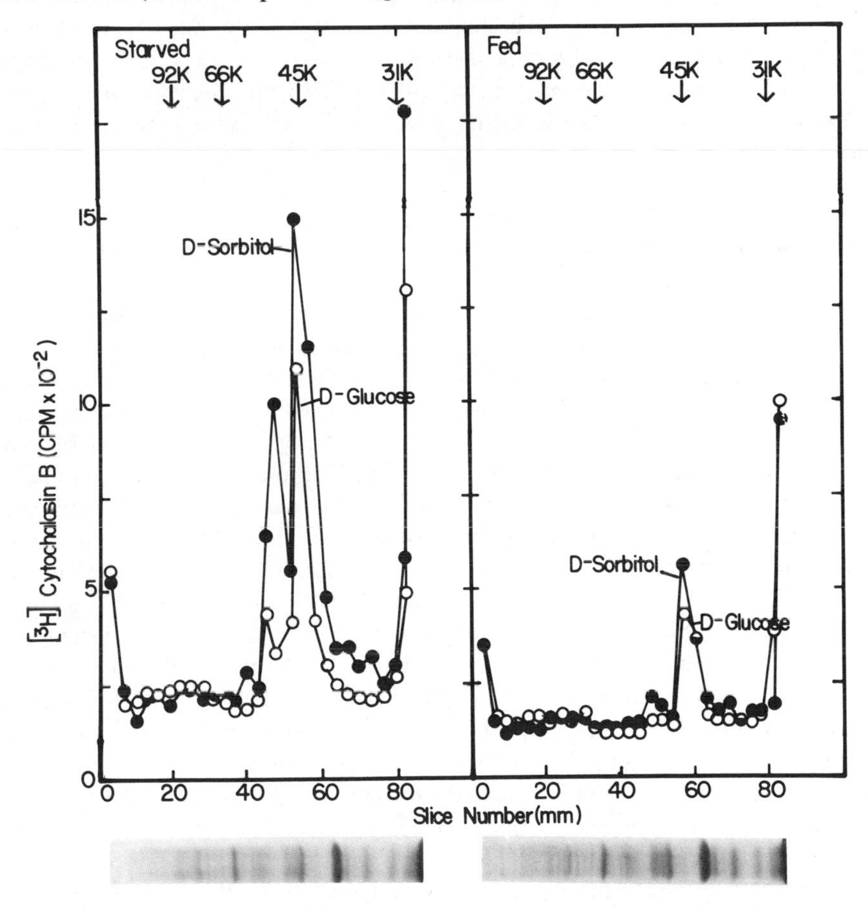


*Figure 3.* Plasma membranes were isolated from control or D-glucose-starved chicken embryo fibroblasts. The membranes were incubated with 0.5 μM [$^3$H]cytochalasin B in the presence of 500 mM D-sorbitol (●) or 500 mM D-glucose (○). D-Sorbitol is a nontransportable sugar alcohol and was used as an osmotic control. After equilibrium binding was reached, the samples were irradiated with a high-intensity ultraviolet light which results in covalent labeling of the D-glucose transporter (Pessin *et al.*, 1982). The proteins were then separated by SDS-gel electrophoresis and the amount of covalently bound [$^3$H]cytochalasin B was determined by scintillation counting. The Coomassie brilliant blue gel profile is shown at bottom. Reprinted with permission from Pessin *et al.* (1982).

ences in the D-glucose sensitivities of the [$^3$H]cytochalasin B photoaffinity labeling between the proteins of $M_r$ 52,000 and 46,000 is consistent with high and low affinity D-glucose transport systems found in the intact cells (Christopher *et al.*, 1976a). The D-glucose-specific [$^3$H]cytochalasin B affinity labeling in the starved cell plasma membrane was increased approximately ten-fold, consistent with Scatchard analysis of binding (Pessin *et al.*, 1982). A significant amount of [$^3$H]cytochalasin B was found at the gel dye front but was not D-glucose sensitive and represents the covalent labeling of a protein of $M_r$ 18,000 (unpublished results).

Although different molecular-weight species have been identified by different techniques in both starved and transformed chicken embryo fibroblasts, the data appear to be consistent with an increase in the number of D-glucose transport proteins after cell starvation or viral transformation. Clearly, further studies are needed to distinguish the relationships of these various proteins and the molecular mechanism(s) by which an increase in the steady-state levels of these proteins on the cell-surface membrane is maintained.

## IV. INSULIN REGULATION OF D-GLUCOSE TRANSPORT ACTIVITY

Unlike the long-term regulation of glucose transport observed in cultured cells, insulin action on D-glucose transport activity in both muscle and fat cells is extremely rapid in onset (maximal activation occurs in 10–15 min at 37°C) and is readily observed in the presence of protein synthesis inhibitors. The rat adipocyte is extraordinarily sensitive to insulin action and has probably been most employed to study the regulation of D-glucose transport activity. These cells are readily prepared from pieces of adipose tissue by digestion with crude bacterial collagenase–protease mixtures (Rodbell, 1964). However, one problem with this system has been the difficulty in measuring initial rates of D-glucose transport due to the small cytoplasmic volume relative to the large cell surface area (Angel and Farkas, 1970). In spite of this difficulty, the isolated rat adipocyte has been shown to contain a facilitative D-glucose transport system with a $K_m$ of approximately 1–5 mM (Czech *et al.*, 1974; Czech, 1976a; Vinten *et al.*, 1976; Vinten, 1978; Whitesell and Gliemann, 1979). Adipocytes were also shown to have a D-glucose transport system kinetically analogous to that of erythrocytes and cultured cells in respect to substrate stereospecificity (Plagemann and Richey, 1974; Czech, 1976a; Holman and Reis, 1982) and [$^3$H]cytochalasin B-binding properties (Czech *et al.*, 1973; Vinten, 1978; Wardzala *et al.*, 1978). Further, insulin was shown to stimulate the $V_{max}$ of D-glucose transport 5- to 10-fold, without any alteration in the $K_m$ of binding (Czech, 1976a; Vinten *et al.*, 1976; Vinten, 1978; Whitesell and Gliemann, 1979) similar to that found for starved and transformed fibroblasts in culture (Weber, 1973; Kletzien and Perdue, 1974; Dolberg *et al.*, 1975; Franchi *et al.*, 1978; Innui *et al.*, 1980).

### A. Insulin Binding and D-Glucose Transport Activation

The initial step in insulin action is the necessary binding of the hormone to specific receptors on the cell surface (Roth, 1973). Rat adipocytes possess more receptors than are required to generate a maximal biological response, indicating the existence of spare receptors. Thus, less than 5% of the total cell-surface receptors must be occupied in order to observe a maximal effect (Kono and Barham, 1971b; Meuli and Froesch, 1977; Davidson and Frank, 1980; Fehlmann and Freychet, 1981). The stimulation of D-glucose transport activity has been shown not to occur with the same time course as insulin binding to its receptor. It has been well established that when, at 37°C, adipocytes are incubated with insulin concentrations which result in a fractional receptor occupancy greater than the spare receptor threshold, there is an initial lag of 20–40 sec before any stimulation of D-glucose transport activity can be observed (Haring *et al.*, 1978, 1981; Whitesell and Gliemann, 1979; Ciaraldi and Olefsky, 1979). This lag period, before maximal D-glucose transport activation occurs, has been shown to be time, temperature, and energy dependent (Whitesell and Gliemann, 1979; Ciaraldi and Olefsky, 1979; Haring *et al.*, 1981). Deactivation of D-glucose transport activity is also a much slower process than is the dissociation of insulin from the adipocyte insulin receptor (Ciaraldi and Olefsky, 1980). However, this appears to be dependent on the concentration of insulin used to reach steady-state binding (Ciaraldi and Olefsky, 1982a). The half-time of insulin dissociation at 37°C varies slightly, with a $t_{1/2}$ ranging from 8–12 min over the insulin concentration range of 0.3–10 ng/ml used to obtain equilibrium binding, whereas the half-time of D-glucose transport deactivation is 12 min at 0.3 ng/ml insulin, but increases to 25 min at 0.5 ng/ml and 60 min at 10 ng/ml. The lack of correlation of insulin receptor occupancy and D-glucose transport activation is thought to be consistent with the hypothesis that insulin generates a chemical signal or mediator and it is the level of this signal which is dependent on receptor occupancy (Ciaraldi and Olefsky, 1982a). Thus, the level of mediator generated would be proportional to receptor occupancy until the spare receptor level is reached. At this point, a decrease in receptor occupancy would only indirectly affect the deactivation process by no longer generating the mediator. The rate of deactivation would then be dependent on the rate of decay of the chemical mediator.

### B. Insulinomimetic Agents

One key missing piece to the problem of insulin action is the identity of events occurring immediately in response to insulin receptor occupancy. It is not known whether one or multiple signals are generated, although a multiple signal process would be perhaps more consistent with the diverse effects of insulin which can occur with different dose–response relationships (Fain *et al.*, 1966). It is not within the scope of this chapter to review the extensive literature related to the molecular basis of insulin action and possible second messengers and the reader is referred to several reviews (Czech, 1977, 1980, 1981). Several of these findings, however, may be quite relevant to the regulation of hexose transport.

The insulin-sensitive adipocyte D-glucose transport system is extremely responsive to a wide variety of membrane perturbants. Treatment of adipocytes with such diverse agents as oxidants (Czech *et al.*, 1974a–c; Czech, 1976b; Mukherjee and Lynn, 1977; May and de Haen, 1979a,b), polyamines (Lockwood *et al.*, 1971; Lockwood and East, 1974), vanadate (Dubyak and Kleinzeller, 1980), hypertonicity (Clausen *et al.*, 1974), sulfhydryl reagents (Carter and Martin, 1969; Minemura and Crofford, 1969), lectins (Czech and Lynn, 1973; Cuatrecasas, 1973b; Czech *et al.*, 1974c; Katzen and Soderman, 1975), antireceptor antibodies (Flier *et al.*, 1975, 1976a,b; Kahn *et al.*, 1976, 1977, 1978b; Jarrett *et al.*, 1976), proteases (Kono and Barham, 1971a), phospholipases (Rodbell, 1966; Blecher, 1968), and neuraminidase (Rosenthal and Fain, 1971; Cuatrecasas and Illiano, 1971) mimics the action of insulin to stimulate D-glucose transport activity. Due to the broad specificity that characterizes this group of reagents, their value as supportive evidence for any specific hypothesis on insulin action is rather limited. Nevertheless, certain characteristics of D-glucose transport activation have emerged from these types of studies.

The insulin-like activity exhibited by oxidants has led to the hypothesis that membrane redox reactions may be involved in D-glucose transport activation (Czech *et al.*, 1974b; May and de Haen, 1979b). This hypothesis has been supported by the finding that insulin and other hormones which have insulin-like activity in adipocytes stimulate $H_2O_2$ production (Mukherjee *et al.*, 1978; de Haen *et al.*, 1980; Muchmore *et al.*, 1981). Further, exogeneously added $H_2O_2$ mimics many of the effects of insulin including activation of D-glucose transport activity (Czech *et al.*, 1974d; Ciaraldi and Olefsky, 1982b) whereas reductants inhibit the D-glucose transport activation (Czech *et al.*, 1974d; Czech, 1976c). This thiol redox hypothesis has remained viable, although no direct data have been established to prove or refute the concept. Recently, Muchmore *et al.* (1982) have reported that under a variety of conditions, an inverse relationship exists between lipolysis and $H_2O_2$ production. However, they demonstrated that the end product of lipolysis, glycerol, had no effect on $H_2O_2$ production but free fatty acids were inhibitory. Further, in the absence of any significant effects of insulin on glycerol production there was a direct inverse relationship with the decrease in free fatty acids and rise in $H_2O_2$ production. These results indicate that $H_2O_2$ production is related to the level of free fatty acids and as a consequence renders its role as an insulin receptor second messenger unlikely.

It has also been suggested that insulin action results in a rapid increase in intracellular free calcium pools (Bihler, 1972, 1974; Clausen *et al.*, 1974; Fraser and Russell, 1975; Kissebah *et al.*, 1975; Schudt *et al.*, 1976; Clausen, 1977a,b). Several reports have shown that calcium is required for several cellular responses to insulin (Holoszy and Narahara, 1966, 1967; Gould and Chaudry, 1970; Bihler, 1972; Jacobs and Krahl, 1973; Bonne *et al.*, 1977, 1978). However, the incomplete inhibition of insulin action in the absence of calcium (Rihan *et al.*, 1967; Letarte and Reynold, 1969) has decreased the impact of these findings. Although it has been reported that the calcium ionophore A23187 mimicked the insulin-induced activation of D-glucose transport activity (Reeves, 1975; Clausen and Martin, 1977), this has not been consistently observed (Grinstein and Erlij, 1976; Bonne *et al.*, 1977). Further, the ionophore A23187 in rat adipocytes was shown to cause an inhibition of glycogen synthetase activity (Lawrence and Larner, 1978), whereas insulin is known to stimulate its activity. Recently, the involvement of magnesium rather than calcium has been implicated in the activation of the D-glucose transporter in soleus muscle (Hall *et al.*, 1982). Due to the confusing and often contradictory literature on the involvement of divalent cations in insulin action, it is very difficult to accommodate any unified hypothesis.

In recent years, evidence has accumulated that insulin may activate or release an intrinsic plasma membrane factor which can stimulate pyruvate dehydrogenase activity (Seals and Jarett, 1980; Seals and Czech, 1980, 1981; Kiechle *et al.*, 1981; Saltiel *et al.*, 1981). Release of this putative second messenger has been shown to occur from isolated plasma membranes treated with insulin and could be mimicked by treating the membranes with low concentrations of trypsin (Seals and Jarett, 1980; Seals and Czech, 1980), concanavalin A, or insulin-receptor antibodies (Seals and Jarett, 1980). This factor has been shown to be heat-stable and trypsin-sensitive with a molecular weight of 1000–4000 on Sephadex G-25 (Kiechle *et al.*, 1981; Seals and Czech, 1981; Saltiel *et al.*, 1981). Further studies are needed to purify and examine its effects on other physiological actions of insulin.

## C. Insulin-Receptor Aggregation

An interesting class of agents used to mimic insulin action are those that have multivalent binding sites for the adipocyte plasma membrane. The plant lectins, concanavalin A, wheat germ agglutinin, and *Lens culinaris* agglutinin have all been shown to stimulate D-glucose uptake in rat adipocytes (Czech and Lynn, 1973; Cuatrecasas and Tell, 1973). These lectins were shown to bind the cell-surface insulin receptor (Cuatrecasas, 1973a,b), which suggests that their action may be mediated via the insulin receptor itself. However, it was subsequently shown that adipocyte rendered unresponsive to insulin action by trypsinization had full responsiveness to these lectins (Czech and Lynn, 1973). It thus appears that a nonreceptor mechanism mediates lectin action perhaps by initiating patching of lectin-bound sites on the cell surface as a first step.

Circulating antibodies from patients with a rare form of insulin resistance and acanthosis nigricans have been shown to inhibit [$^{125}$I]insulin binding to adipocytes and mimic insulin action (Flier *et al.*, 1975, 1976a,b; Jarrett *et al.*, 1976; Kahn *et al.*, 1976, 1977, 1978b). Polyclonal antibodies generated against the purified insulin receptor have also been shown to activate D-glucose transport activity in adipocytes (Jacobs *et al.*, 1978). However, antibodies raised in rabbits against adipocyte plasma membranes mimicked insulin action but did not inhibit [$^{125}$I]insulin binding (Pillion and Czech, 1978; Pillion *et al.*, 1979). For these antibodies, the association with the insulin-binding site on the insulin receptor is not necessary for initiation of the biologic response (Pillion and Czech, 1978). These antibodies are similar to the lectins in that D-glucose transport can be readily activated in cells which exhibit little or no insulin binding to the insulin receptor (Pillion *et al.*, 1979).

An important implication of these studies is that lateral aggregation of membrane proteins may be an obligatory feature of their insulinomimetic action. This concept was further addressed by preparing monovalent Fab antibody fragments which were found to be devoid of biological activity in adipocytes (Kahn *et al.*, 1978b; Pillion *et al.*, 1979). Upon addition of anti-Fab antibodies, which restores multivalency of the Fab fragments by crosslinking them, activation of D-glucose transport activity was observed. In addition, Kahn *et al.* (1978a) have reported potentiation of the action of submaximal doses of insulin by addition of antiinsulin antibody, indicating aggregation of receptors or other components may be involved in insulin action itself. It should be emphasized, however, that although membrane crosslinking or aggregation phenomena appear to be involved in D-glucose transport activation, there is no direct evidence that such events play a role in the action by insulin itself.

Studies by Jarret and Smith (1974, 1975, 1977) have shown that the insulin receptor exists as clusters on the adipocyte cell-surface membrane when monitored by their ability to be bound to an insulin–ferritin complex. These clusters do not appear to be aggregated in response to insulin binding and action since immobilization of the cell surface by crosslinking did not change the cluster distribution observed by the ferritin–insulin (Jarett and Smith, 1977). Interestingly, cytochalasin B was shown to initiate the disaggregation of receptor cluster concomitant with its ability to inhibit D-glucose transport activity (Jarett and Smith, 1979). Cytochalasin D, a cytochalasin B analogue which does not inhibit D-glucose transport but does alter cell morphology and dynamics, failed to modulate the distribution of the ferritin–insulin receptor cluster. It was proposed, therefore, that insulin receptor clusters may constitute D-glucose transporter pores in the cell-surface membrane (Jarett and Smith, 1979), and that insulin binds to a macromolecular complex containing the D-glucose transport system. Cytochalasin B could thus inhibit transport activity by dissociating the pore-forming complex. However, no further support for this idea has been forthcoming.

### D. *Structural Relationships between the D-Glucose Transporter and Insulin Receptor*

The development of a reliable system for measuring the stereospecific D-glucose transport activity in reconstituted phospholipid vesicles (see Section IV-A) has allowed chromatography of the detergent-solubilized adipocyte membrane proteins. These types of studies have indicated that the adipocyte D-glucose transporter in cholate migrates with an apparent Stokes radius of 60–80 Å on Sepharose 6B and is therefore large enough to span the plasma membrane (Carter-Su *et al.*, 1980). Sequential use of immobilized lectin chromatography followed by gel filtration on Sepharose 6B yielded a fraction which appeared to be enriched by 25 to 50-fold in its ability to catalyze D-glucose uptake in reconstituted vesicles. However, these data also demonstrated that none of the major membrane proteins in the rat adipocytes comprises the D-glucose transporter and that the transporter is present in the membrane in very small quantities (<0.08%). It is interesting that the D-glucose transport system itself does not bind to immobilized concanavalin A since this lectin markedly activates D-glucose transport in the intact adipocytes (Czech and Lynn, 1973; Cuatrecasas, 1973a,b; Katzen and Soderman, 1975).

Over the past several years, methods have been developed to specifically and covalently label the insulin receptor with [$^{125}$I]insulin (Oppenheimer and Czech, 1983). This type of crosslinking methodology has demonstrated that the insulin receptor is a heterotetromeric disulfide-linked complex, composed of two protein subunits of $M_r$ 125,000 and two of $M_r$ 90,000 (Massague *et al.*, 1980). The ability to affinity label the insulin receptor and to assay the D-glucose transport activity of solubilized proteins using the reconstitution method has allowed a direct comparison of the chromatographic properties of these two membrane components. When [$^{125}$I]insulin was crosslinked to the insulin receptor in intact membranes before detergent solubilization, nearly all the labeled receptor eluted from a hydroxylapatite column with 0.1 M phosphate while no D-glucose transport activity was found (Carter-Su *et al.*, 1981). Elution of the D-glucose transporter with higher salt was accompanied by very small amounts of labeled receptor. Similar results were also found with DEAE-cellulose chromatography. Further, it has been observed that Triton X-100-solubilized insulin receptor binds very tightly to immobilized lectins (Cuatrecasas, 1973b; Katzen and Soderman, 1975) while the D-glucose transporter does not (Carter-Su *et al.*, 1980).

### E. Identification of the Insulin-Sensitive D-Glucose Transporter

Recently, a rabbit antibody generated against the human erythrocyte D-glucose transporter has been used to label the D-glucose transporter from adipocyte cell plasma membranes and low-density microsomal membranes after transfer from sodium dodecyl sulfate polyacrylamide gels to nitrocellulose paper (Wheeler *et al.*, 1982). These results have indicated that the insulin-sensitive D-glucose transporter in rat adipocytes is of $M_r$ 45,000. However, this cross-reacting antibody does not react with the native adipocyte D-glucose transporter and only recognized the sodium dodecyl sulfate-denatured protein with poor efficiency. There also appears to be a discrepancy between the quantitation of the D-glucose transporter as measured with this antibody compared to the cytochalasin B-binding analysis. Insulin induced a four-fold increase in the plasma membrane cytochalasin B sites, whereas the antibody indicated only a 1.3-fold increase. [$^{3}$H]cytochalasin B photoaffinity labeling of rat adipocyte low-density microsomes has also indicated one D-glucose-sensitive labeled protein at approximately $M_r$ 46,000 (Shanahan *et al.*, 1982). We have further demonstrated that this protein of $M_r$ 46,000 displays all the characteristics one would expect of the insulin-sensitive D-glucose transporter (unpublished results).

It thus appears that the activities, as well as the proteins of the insulin receptor and of the D-glucose transporter, are distinct and separate entities. The insulin-induced D-glucose transport activation must therefore involve some type of membrane transduction mechanism by which the insulin-generated signal can interact with the effector system. In all these studies, however, it has not been ruled out that the D-glucose transport system may be associated with the insulin receptor by noncovalent means which are disrupted by detergent solubilization.

## V. MECHANISM OF INSULIN ACTIVATION OF THE D-GLUCOSE TRANSPORTER

A central problem in the area of D-glucose transport regulation by insulin is whether the actual number of transport systems in the cell-surface membrane changes during activation. Two general models of increased D-glucose transport activity are consistent with the observed enhancement of $V_{max}$ with no change in $K_m$. Insulin action could either convert the preexisting D-glucose transporters from a less active to a more active (efficient) state or increase the number of D-glucose transporters in the cell-surface membrane. In this latter hypothesis, the characteristics of the D-glucose transport system, e.g., temperature dependence, modification by protein reagents, etc., should be identical in basal- or insulin-activated cells. The former hypothesis predicts that the structure or the membrane microenvironments of the D-glucose transporter are altered by insulin, and therefore basal and insulin-stimulated D-glucose transport activity may exhibit different properties (Czech, 1980).

To address this question, [$^3$H]cytochalasin B binding has been used as previously described in human erythrocytes (Taverna and Langdon, 1973; Plagemann and Richey, 1974; Lienhard *et al.*, 1977; Lin and Snyder, 1977; Jung *et al.*, 1980) to assess the absolute number of D-glucose transporters in isolated plasma membranes from rat adipocytes (Wardzala *et al.*, 1978; Karnieli *et al.*, 1981). These studies demonstrated an increase in the D-glucose-sensitive fraction of [$^3$H]cytochalasin B-binding sites in plasma membranes prepared from insulin-treated compared to control cells. These studies were further extended and the increased amount of [$^3$H]cytochalasin B binding in the plasma membrane was quantitatively equal to the loss of [$^3$H]cytochalasin B binding in a low-density microsome fraction (Karnieli *et al.*, 1981). Similar results have also been found in isolated rat diaphragm muscle cells (Wardzala and Jeanrenaud, 1981).

Identical conclusions have also been drawn based upon reconstitution of detergent-solubilized D-glucose transport activity after separation of subcellular membranes on sucrose gradients (Suzuki and Kono, 1980; Kono *et al.*, 1981; Gorga and Lienhard, 1982). These results have suggested that the D-glucose transport system in muscle and fat cells is recycled between the plasma membrane and membranes in an unidentified intracellular site that migrate in a low-density fraction. According to this hypothesis, insulin treatment induces the translocation or redistribution of the D-glucose transporters from the intracellular site to the plasma membrane. These studies also demonstrated that this putative recruitment process itself is energy-dependent and that uncouplers of oxidative phosphorylation (DNP, KCN, $NaN_3$, etc.) inhibit this process (Kono *et al.*, 1981).

These results explain the long-standing observation that energy uncouplers inhibit the insulin-induced increase in D-glucose transport activity at concentrations which do not effect insulin binding in rat adipocytes (Kono *et al.*, 1977a,b; Chandramouli *et al.*, 1977; Ciaraldi and Olefsky, 1980; Siegal and Olefsky, 1980; Vega *et al.*, 1980; Haring *et al.*, 1981). In soleus muscle, however, uncouplers of oxidative phosphorylation have been shown to not only inhibit the insulin-induced increase in D-glucose transport activity but also to inhibit insulin binding to the insulin receptor in direct proportion to the level of intracellular ATP (Yu and Gould, 1977).

A similar insulin-induced redistribution of the insulin-like growth factor II (IGF-II) receptor has also been observed in rat adipocytes (Oppenheimer *et al.*, 1983). Plasma membranes prepared from insulin-treated cells exhibited a 1.5-fold increase in the number of IGF-II receptors compared to controls whereas the low-density microsome fractions from insulin-treated cells had a 40% decrease, without any change in receptor affinity. These results are qualitatively similar to the insulin-induced membrane redistribution of the D-glucose transporter. However, Scatchard analysis of IGF-

II binding to intact adipocytes demonstrated that insulin treatment causes a ten-fold increase in the apparent affinity of the IGF-II receptor, without any change in the total number of cell-surface receptors (Oppenheimer *et al.*, 1983; King *et al.*, 1982). Thus, the binding of IGF-II to isolated membranes does not accurately reflect the effect of insulin on the intact adipocyte. Based upon this finding, it is clear that the number of D-glucose transporters must be determined in the intact adipocyte rather than in isolated membrane fractions.

At least two instances of divergent properties of control vs. insulin-stimulated D-glucose transport activity have been reported. Several groups have found a different temperature sensitivity of insulin-activated 3-*O*-methylglucose transport activity compared with control transport activity in intact cells (Kono *et al.*, 1977a; Czech, 1976a; Amatruda and Finch, 1979; Whitesell and Gliemann, 1979), although others have failed to make this observation (Vinten, 1978; Olefsky, 1978). Secondly, in plasma membranes prepared from basal and insulin-stimulated cells, the $K_d$ of [$^3$H]cytochalasin B binding to the D-glucose transporter was found to be approximately 98 nM. However, the $K_d$ of [$^3$H]cytochalasin B binding to the intracellular D-glucose transporter was 140 nM in basal cells but after insulin treatment decreased to that value observed in the plasma membrane (Simpson *et al.*, 1981).

The differences in temperature sensitivities of basal and insulin-stimulated D-glucose transport activities have recently been addressed by Ezaki and Kono (1982). Rat adipocyte basal D-glucose transport activity has been shown to be elevated at 15–25°C compared to the D-glucose transport activity at 37°C (Vega and Kono, 1979). It was subsequently found that incubation of rat adipocytes at reduced temperatures would stimulate the basal D-glucose transport activity in a time-dependent manner which could be blocked by the addition of energy uncouplers (Ezaki and Kono, 1982). They concluded that apparent discrepancies in the temperature coefficient of D-glucose transport activity is affected by the insulinomimetic effect of low temperatures and not a direct result of insulin treatment on the transporter activity itself.

## VI. SUMMARY AND CONCLUSIONS

A large and extensive effort has been made to understand the physical properties and the metabolic regulation of the facilitative D-glucose transporter in a number of cell types. One of the major problems facing researchers in this area is the extremely low abundance of this system in cells that contain physiologically regulated D-glucose transporters. This problem is further complicated by the fact that no true substrate is generated from this activity, but only a vectorial displacement occurs, making assay conditions very time-consuming and laborious.

Despite these formidable difficulties, substantial progress has been made in our understanding of the D-glucose transport process. We now know that in cultured cells the increased rate of D-glucose uptake occurs by an increase in the number of D-glucose transport systems in the cell-surface membrane. In cultured cells, this is probably primarily due to an increase in D-glucose transporter synthesis and/or decrease in degradation. Obviously, we still have a long way to go to understand the mechanism by which this occurs. In the future, it will be necessary to study the detailed molecular life cycle of the D-glucose transporter to understand what role transformation and cell starvation play.

In adipocyte and muscle, the available evidence implicates a rapid recruitment of D-glucose transporters from a low-density membrane fraction, presumably intracellular, to the plasma membrane in response to insulin, which is energy-dependent but protein synthesis independent. However, direct demonstration that the low-density membranes containing hexose transporters are indeed intracellular in locus is lacking. Studies are clearly needed to determine the true cellular locations of these low-density microsome D-glucose transporters and the means by which the transporters redistribute into the different membrane fractions. The approaches we have outlined here will hopefully, in the next several years, provide insight to these very interesting and fundamental problems of membrane biochemistry.

ACKNOWLEDGMENTS

We wish to thank Judith A. Kula for her excellent secretarial assistance.

## REFERENCES

Albert, S. G., 1982, The effect of insulin on glucose transport in rabbit erythrocytes and reticulocytes, *Life Sci.* **31:**265–271.

Amatruda, J. M., and Finch, E. D., 1979, Modulation of hexose uptake and insulin action by cell membrane fluidity, *J. Biol. Chem.* **254:**2619–2625.

Angel, A., and Farkas, J., 1970, Structural and chemical compartments in adipose cells, in: *Hormone and Metabolic Research, Supplement 2. Adipose Tissue, Regulation and Metabolic Functions* (B. Jeanrenaud and D. Hepp, eds.), Academic Press, New York, pp. 152–161.

Baldwin, S. A., and Baldwin, J. M., 1981, The stoichiometry of cytochalasin B binding to the human erythrocyte glucose transporter, *Fed. Proc.* **40:**1983.

Baldwin, S. A., and Lienhard, G. E., 1980, Immunological identification of the human erythrocyte monosaccharide transporter, *Biochem. Biophys. Res. Commun.* **94:**1401–1408.

Baldwin, S. A., and Lienhard, G. E., 1981, Glucose transport across plasma membranes: Facilitated diffusion systems, *TIBS* **6:**208–211.

Baldwin, S. A., Baldwin, J. M., Gorga, F. R., and Lienhard, G. E., 1979, Purification of the cytochalasin B binding component of the human erythrocyte monosaccharide transport system, *Biochim. Biophys. Acta* **552:**183–188.

Baldwin, S. A., Baldwin, J. M., and Lienhard, G. E., 1982, Monosaccharide transport of the human erythrocyte: Characterization of an improved preparation, *Biochemistry* **21:**3836–3849.

Barnett, J. E. G., Holmon, G. D., and Munday, K. A., 1973, Structural requirement for binding to the sugar-transport system of the human erythrocyte, *Biochem. J.* **131:**211–221.

Basketter, D. A., and Widdas, W. F., 1977, Competitive inhibition of hexose transfer in human erythrocytes by cytochalasin B, *J. Physiol.* **265:**39P.

Batt, E. R., Abbott, R. E., and Schachter, D., 1975, Two types of sulfhydryl groups involved in erythrocyte hexose transport, *Fed. Proc.* **34:**250.

Batt, E. R., Abbott, R. E., and Schachter, D., 1976, Impermeant maleimides, *J. Biol. Chem.* **251:**7184–7190.

Bihler, I., 1971, Ionic effects in the regulation of sugar transport in muscle, in: *The Role of Membranes in Metabolic Regulation* (M. A. Mehlman and R. W. Hanson, eds.), Academic Press, New York, pp. 411–422.

Bihler, I., 1974, Mechanisms regulating the membrane transport of sugars in the myocardium, in: *Recent Advances in Studies on Cardiac Structure and Metabolism* (N. S. Dhalla, ed.), University Park Press, Baltimore, pp. 209–216.

Binder, M., and Tamm, C., 1973, The cytochalasins: A new class of biologically active microbial metabolites, *Angew. Chem.* **12:**370–380.

Blecher, M., 1968, Action of insulin on a glucose transport mechanism in the plasma membrane of the isolated adipose cell. Participation of membrane phospholipids and cyclic adenosine monophosphate in the transport processes, *Gumma Symp. Endocrinol.* **5:**145–161.

Bloch, R., 1973, Inhibition of glucose transport in the human erythrocyte by cytochalasin B, *Biochemistry* **12:**4779–4801.

Bonne, D., Belhadj, O., and Cohen, P., 1977, Modulation by calcium of the insulin action and of the insulin-like effect of oxytocin on isolated rat adipocytes, *Eur. J. Biochem.* **75:**101–105.

Bonne, D., Belhadj, O., and Cohen, P., 1978, Calcium as modulator of the hormonal-receptors-biological-response coupling system. *Eur. J. Biochem.* **86:**261–266.

Carruthers, A., and Melchior, D. L., 1983, Asymmetric or symmetric cytosolic modulation of human erythrocyte hexose transfer, *Biochim. Biophys. Acta* **727:**421–434.

Carter, J. R., and Martin, D. B., 1969, The effect of sulfhydryl blockade on insulin action and glucose transport in isolated adipose tissue cells, *Biochim. Biophys. Acta* **177:**521–526.

Carter-Su, C., Pillion, D. J., and Czech, M. P., 1980, Reconstituted D-glucose transport from the adipocyte plasma membrane: Chromatographic resolution of transport activity from membrane glycoproteins using immobilized concanavalin A, *Biochemistry* **19:**2374–2385.

Carter-Su, C., Pilch, P. F., and Czech, M. P., 1981, Chromatographic resolution of insulin receptor from insulin-sensitive D-glucose transporter of adipocyte plasma membranes, *Biochemistry* **20:**216–221.

Carter-Su, C., Pessin, J. E., Mora, R., Gitomer, W., and Czech, M. P., 1982, Photoaffinity labeling of the human erythrocyte D-glucose transporter, *J. Biol. Chem.* **257:**5419–5425.

Chandramouli, V., Milligan, M., and Carter, J. R., Jr., 1977, Insulin stimulation of glucose transport in adipose cells. An energy-dependent process, *Biochemistry* **16:**1151–1158.

Christopher, C. W., 1977, Hexose transport regulation in cultured hamster cells, *J. Supramol. Struct.* **6:**485–494.

Christopher, C. W., Kohlbacher, M. S., and Amos, H., 1976a, Derepression and carrier turnover: Evidence for two regulators in animal cells, *Biochem. J.* **158:**439–450.

Christopher, C. W., Colby, W. W., and Ullrey, D., 1976b, Transport of sugars in chick-embryo fibroblasts, *J. Cell Physiol.* **89:**683–692.

Christopher, C. W., Ullrey, D., Colby, W., and Kalckar, H. M., 1976c, Paradoxical effects of cycloheximide and cytochalasin B on hamster cell hexose uptake, *Proc. Natl. Acad. Sci. USA* **73:**2429–2433.

Ciaraldi, T. P., and Olefsky, J. M., 1979, Coupling of insulin receptors to glucose transport: A temperature-dependent time lag in activation of transport, *Arch. Biochem. Biophys.* **193:**221–231.

Ciaraldi, T. P., and Olefsky, J. M., 1980, Relationship between deactivation of insulin-stimulated glucose transport and insulin dissociation in isolated rat adipocytes, *J. Biol. Chem.* **255:**327–330.

Ciaraldi, T. P., and Olefsky, J. M., 1982a, Kinetic relationship between insulin receptor binding and effects on glucose transport in isolated rat adipocytes, *Biochemistry* **21:**3475–3480.

Ciaraldi, T. P., and Olefsky, J. M., 1982b, Comparison of the effects of insulin and $H_2O_2$ on adipocyte glucose transport, *J. Cell Physiol.* **110:**323–328.

Clausen, T., 1977a, Calcium, glucose transport and insulin action, in: *Biochemistry of Membrane Transport,* FEBS Symposium No. 42 (G. Semenza and E. Carafoli, eds.), Springer-Verlag, New York.

Clausen, T., 1977b, The role of calcium in the action of insulin, in: *Membrane Proteins,* FEBS 11th Meeting Copenhagen, Vol. 45, Symposium A4.

Clausen, T., and Martin, B. R., 1977, The effect of insulin on the washout of [$^{45}$Ca]calcium from adipocytes and soleus muscle of the rat, *Biochem. J.* **164:**251–255.

Clausen, T., Elbrink, J., and Martin, B. R., 1974, Insulin controlling calcium distribution in muscle and fat cells, *Acta Endocrinol.* **77**(Suppl. 191)**:**137–143.

Cuatrecasas, P., 1973a, Interaction of wheat germ agglutinin and concanavalin A with isolated fat cells, *Biochemistry* **12:**1312–1323.

Cuatrecasas, P., 1973b, Interaction of concanavalin A and wheat germ agglutinin with the insulin receptor of fat cells and liver, *J. Biol. Chem.* **248:**3528–3534.

Cuatrecasas, P., and Illiano, G., 1971, Membrane sialic acid and the mechanism of insulin action in adipose tissue cells. Effect of digestion with neuraminidase, *J. Biol. Chem.* **246:**4938–4946.

Cuatrecasas, P., and Tell, G. P. E., 1973, Insulin-like activity of concanavalin A and wheat germ agglutinin-direct interactions with insulin receptors, *Proc. Natl. Acad. Sci. USA* **70:**485–489.

Czech, M. P., 1976a, Regulation of the D-glucose transport system in isolated fat cells, *Mol. Cell. Biochem.* **11:**51–63.

Czech, M. P., 1976b, Differential effect of sulfhydryl reagents on activation and deactivation of the fat cell hexose transport system, *J. Biol. Chem.* **251:**1164–1170.

Czech, M. P., 1976c, Current status of the thiol redox model for the regulation of hexose transport by insulin, *J. Cell Physiol.* **89:**661–668.

Czech, M. P., 1977, Molecular basis of insulin action, *Annu. Rev. Biochem.* **46:**359–384.

Czech, M. P., 1980, Insulin action and the regulation of hexose transport, *Diabetes* **29:**399–409.

Czech, M. P., 1981, *Insulin Action: Second Messengers, Handbook of Diabetes Mellitus; Islet Cell Function/Insulin Action* (M. Brownlee, ed.), Garland Press, New York, pp. 117–149.

Czech, M. P., and Lynn, W. S., 1973, Stimulation of glucose metabolism by lectins in isolated white fat cells, *Biochim. Biophys. Acta* **217:**386–397.

Czech, M. P., Lynn, D. G., and Lynn, W. S., 1973, Cytochalasin B-sensitive 2-Deoxyglucose transport in adipose cell ghosts, *J. Biol. Chem.* **248:**3636–3641.

Czech, M. P., Lawrence, J. C., Jr., and Lynn, W. S., 1974a, Hexose transport in isolated brown fat cells. A model system for investigating insulin action on membrane transport, *J. Biol. Chem.* **249:**5421–5427.

Czech, M. P., Lawrence, J. C., and Lynn, W. S., 1974b, Evidence for electron transfer reactions involved in the $Cu^{2+}$-dependent thiol activation of fat cell glucose utilization, *J. Biol. Chem.* **249:**1001–1006.

Czech, M. P., Lawrence, J. C., and Lynn, W. S., 1974c, Activation of hexose transport by concanavalin A in isolated brown fat cells. Effects of cell surface modification with neuraminidase and trypsin on lectin and insulin action, *J. Biol. Chem.* **249:**7499–7505.

Czech, M. P., Lawrence, J. C., and Lynn, W. S., 1974d, Evidence for the involvement of sulfhydryl oxidation in the regulation of cell hexose transport by insulin, *Proc. Natl. Acad. Sci. USA* **71:**4173–4177.

Davidson, M. B., and Frank, H. J. L., 1980, Decreased spare hepatic receptors for insulin: Possible importance for insulin action, *Diabetes* **29**(Suppl. 2):39A.

Dolberg, D. S., Bassham, J. A., and Bissell, M. J., 1975, Selective inhibition of the facilitated mode of sugar uptake by cytochalasin B in cultured chick fibroblasts, *Exp. Cell Res.* **96:**129–137.

Dubyak, G. R., and Kleinzeller, A., 1980, The insulinomimetic effects of vanadate in isolated rat adipocytes, *J. Biol. Chem.* **255:**5306–5312.

Eady, R. P., and Widdas, W. F., 1973, The use of sugars and fluorodinitrobenzene (FDNB) to differentially label red cell membrane components involved in hexose transfers, *Quart. J. Exp. Physiol.* **58:**59–66.

Eisenman, G., Ciani, S. M., and Szabo, G., 1968, Some theoretically expected and experimentally observed properties of lipid bilayer membranes containing neutral molecular carries of ions, *Fed. Proc.* **27:**1289–1304.

Ezaki, O., and Kono, T., 1982, Effects of temperature on basal and insulin-stimulated glucose transport activities in fat cells, *J. Biol. Chem.* **257:**14306–14310.

Fain, J. N., Kovacev, V. P., and Scow, R. O., 1966, Antilipolytic effect of insulin in isolated fat cells of the rat, *Endocrinology* **78:**773–778.

Fairbanks, G., Steck, T. L., and Wallach, D. F. H., 1971, Electrophoretic analysis of the major polypeptides of the human erythrocyte membrane, *Biochemistry* **10:**2606–2616.

Fehlmann, M., and Freychet, P., 1981, Insulin and glucagon stimulation of ($Na^+$-$K^+$)-ATPase transport activity in isolated rat hepatocytes, *J. Biol. Chem.* **256:**7449–7453.

Finkelstein, A., and Cass, A., 1968, Permeability and electrical properties of thin lipid membranes, *J. Gen. Physiol.* **52:**145–173.

Flier, J. S., Kahn, C. R., Roth, J., and Bar, R. S., 1975, Antibodies that impair insulin receptor binding in an unusual diabetic syndrome with severe insulin resistance, *Science* **190:**63–65.

Flier, J. S., Kahn, C. R., Jarrett, D. B., and Roth, J., 1976a, The immunology of the insulin receptor, *Immunol. Commun.* **5:**361–373.

Flier, J. S., Kahn, C. R., Jarrett, D. B., and Roth, J., 1976b, Characterization of antibodies to the insulin receptor; a cause of insulin-resistant diabetes in man, *J. Clin. Invest.* **58:**1442–1449.

Franchi, A., Silvestre, P., and Pouyssegur, 1978, "Carrier activation" and glucose transport in Chinese hamster fibroblasts, *Biochem. Biophys. Res. Commun.* **85:**1526–1534.

Fraser, T., and Russell, M. D., 1975, Is insulin's second messenger calcium? *Proc. Roy. Soc. Med.* **68:**785–791.

Gliemann, J., Osterlind, K., Vinten, J., and Gammeltoff, S., 1972, A procedure for measurement of distribution space in isolated fat cells, *Biochim. Biophys. Acta* **286:**1–9.

Goldin, S. M., and Rhoden, V., 1979, Reconstitution and "transport specificity fractionation" of the human erythrocyte glucose transport system, *J. Biol. Chem.* **253:**2575–2583.

Gorga, J. C., and Lienhard, G. E., 1982, Insulin stimulation of glucose transport in adipocytes, *Fed. Proc.* **41:**627.

Gorga, F. R., Baldwin, S. A., and Lienhard, G. E., 1979, The monosaccharide transporter from human erythrocytes is heterogeneously glycosylated, *Biochem. Biophys. Res. Commun.* **91:**995–961.

Gould, M. K., and Chaudry, I. H., 1970, The action of insulin on glucose uptake by isolated rat soleus muscle, I. Effects of cations, *Biochim. Biophys. Acta* **215:**247–249.

Griffin, J. F., Rampal, A. L., and Jung, C. Y., 1982, Inhibition of glucose transport in human erythrocytes by cytochalasins: A model based on diffraction studies, *Proc. Natl. Acad. Sci. USA* **79:**3759–3763.

Grinstein, S., and Erlij, D., 1976, Action of insulin and cell calcium: Effect of ionophore A23187, *J. Membr. Biol.* **29:**313–328.

de Haen, C., Muchmore, D. B., and Little, S. A., 1980, Stimulation of intracellular $H_2O_2$ production in rat epididymal adipocytes by insulin, insulin fragments, and other hormones and growth factors with insulin-like activities, *Insulin Chemistry, Structure and Function of Insulin and Related Hormones* (D. Brandenburg and A. Wollmer, eds.), Walter de Gruyter and Co., New York, pp. 461–468.

Hall, S., Keo, L., Yu, K. T., and Gould, M. K., 1982, Effect of ionophore A23187 on basal and insulin-stimulated sugar transport by rat soleus muscle, *Diabetes* **31:**846–850.

Haring, H. U., Kemmler, W., Renner, R., and Hepp, H. D., 1978, Initial lagphase in the action of insulin on glucose transport and cAMP levels in fat cells, *FEBS Lett.* **95:**177–180.

Haring, H. U., Biermann, E., and Kemmler, W., 1981, Coupling of insulin binding and insulin action on glucose transport in fat cells, *Am. J. Physiol.* **240:**E556–E565.

Hatanka, M., 1974, Transport of sugars in tumor cell membranes, *Biochim. Biophys. Acta* **355:**77–104.

Hatanaka, M., 1976, Saturable and nonsaturable process of sugar uptake: Effect of oncogenic transformation in transport and uptake of nutrients, *J. Cell Physiol.* **89:**745–750.

Hladky, S. B., and Haydon, D. A., 1970, Discreteness of conductance change in bimolecular lipid membranes in the presence of certain antibiotics, *Nature* **225:**451–453.

Hokin, L. E., 1981, Reconstitution of "carriers" in artificial membranes, *J. Membr. Biol.* **60:**77–93.

Holloszy, J. O., and Narahara, H. H., 1966, Enhanced permeability to sugar associated with muscle contraction, *J. Gen. Physiol.* **50:**551–562.

Holloszy, J. O., and Narahara, H. H., 1967, Studies in tissue permeability: X. Changes in permeability to 3-*O*-methylglucose associated with contraction of isolated frog muscle, *J. Biol. Chem.* **240:**3493–3500.

Holman, G. D., and Reis, W. D., 1982, Side-specific analogues for the rat adipocyte sugar transport system, *Biochim. Biophys. Acta* **685:**78–86.

Innui, K.-I., Tillotson, L. G., and Isselbacher, K. J., 1980, Hexose and amino acid transport by chicken embryo fibroblasts infected with temperature-sensitive mutant of Rous sarcoma virus, *Biochim. Biophys. Acta* **598:**616–627.

Jacobs, B. O., and Krahl, M. E., 1973, The effects of divalent cations and insulin on protein synthesis in adipose cells, *Biochim. Biophys. Acta* **319:**410–415.

Jacobs, S., Chang, K., and Cuatrecasas, P., 1978, Antibodies to purified insulin receptors have insulin-like activity, *Science* **200:**1283–1285.

Jarett, L., and Smith, R. M., 1974, Electron microscopic demonstration of insulin receptors on adipocyte plasma membranes utilizing a ferritin-insulin conjugate, *J. Biol. Chem.* **249:**7024–7031.

Jarett, L., and Smith, R. M., 1975, Ultrastructural localization of insulin receptors on adipocytes, *Proc. Natl. Acad. Sci. USA* **72:**3526–3530.

Jarett, L., and Smith, R. M., 1977, The natural occurrence of insulin receptors in groups on adipocyte plasma membranes as demonstrated with monomeric ferritin-insulin, *J. Supramol. Struct.* **6:**45–59.

Jarett, L., and Smith, R. M., 1979, Effect of cytochalasin B and D on groups of insulin receptors and on insulin action in rat adipocytes, *Clin. Invest.* **6:**571–579.

Jarrett, D. B., Roth, J., Kahn, C. R., and Flier, J. S., 1976, Direct method for detection and characterization of cell surface receptors for insulin by means of $^{125}$I-labeled autoantibodies against the insulin receptors, *Proc. Natl. Acad. Sci. USA* **73:**4115–4119.

Jones, M. N., and Nickson, J. K., 1978, Electrical properties and glucose permeability of bilayer lipid membranes on incorporation of erythrocyte membrane extracts, *Biochim. Biophys. Acta* **509:**260–271.

Jones, M. N., and Nickson, J. K., 1981, Monosaccharide transport proteins of the human erythrocyte membrane, *Biochim. Biophys. Acta* **650:**1–20.

Jung, C. Y, 1971a, Evidence of high stability of the glucose transport carrier function in human red cell ghosts extensively washed in various media, *Arch. Biochem. Biophys.* **146:**215–226.

Jung, C. Y., 1971b, Permeability of bimolecular membranes made from lipid extracts of human red cell ghosts to sugars, *J. Membr. Biol.* **5:**200–214.

Jung, C. Y., and Carlson, L. M., 1975, Glucose transport carrier in human erythrocyte membranes, *J. Biol. Chem.* **250:**3217–3220.

Jung, C. Y., and Rampal, A. L., 1977, Cytochalasin B binding sites and glucose transport carrier in human erythrocyte ghosts, *J. Biol. Chem.* **252:**5456–5463.

Jung, C. Y., Hsu, T. L., Hah, J. S., Cha, C., and Haas, M. N., 1980, Glucose transport carrier of human erythrocytes, *J. Biol. Chem.* **253:**361–364.

Kahlenberg, A., and Zala, C. A., 1977, Reconstitution of D-glucose transport in vesicles composed of lipids and intrinsic protein (zone 4.5) of the human erythrocyte membrane, *J. Supramol. Struct.* **7:**287–300.

Kahn, C. R., Flier, J. S., Bar, R. S., Archer, J. A., Gorden, P., Martin, M. M., and Roth, J., 1976, The syndromes of insulin resistance and acanthosis nigricans, Insulin-receptor disorders in man, *N. Engl. J. Med.* **294:**739–745.

Kahn, C. R., Baird, K., Flier, J. S., and Jarrett, D. B., 1977, Effects of autoantibodies to the insulin receptor on isolated adipocytes. Studies of insulin binding and insulin action, *J. Clin. Invest.* **60:**1094–1106.

Kahn, C. R., Baird, K., Baird, R., Jarrett, D. B., and Flier, J. S., 1978a, Direct demonstration that receptor cross-linking or aggregation is important in insulin action, *Proc. Natl. Acad. Sci. USA* **75:**4209–4213.

Kahn, C. R., Baird, K., Jarrett, D. B., and Flier, J. S., 1978b, Monovalent anti-receptor antibodies regain insulinomimetic actions when crosslinked by a second antibody, *Diabetes* **27**(Suppl. 2)**:**449.

Kalckar, H. M., and Ullrey, D., 1973, Two distinct types of enhancement of glucose uptake into hamster cells: Tumor-virus transformation and hexose starvation, *Proc. Natl. Acad. Sci. USA* **70:**2502–2504.

Kalckar, H. M., Ullrey, D., Kijomoto, S., and Hakomori, S., 1973, Carbohydrate catabolism and the enhancement of uptake of galactose in hamster cells transformed by polyoma virus, *Proc. Natl. Acad. Sci. USA* **70:**839–843.

Karnieli, E., Zarnowski, M. J., Hissin, P. J., Simpson, I. A., Salans, L. B., and Cushman, S. W., 1981, Insulin-stimulated translocation of glucose transport systems in the isolated rat adipose cell, *J. Biol. Chem.* **256:**4772–4777.

Kasahara, M., and Hinkle, P. C., 1976, Reconstitution of D-glucose transport catalyzed by a protein fraction from human erythrocytes in sonicated liposomes, *Proc. Natl. Acad. Sci. USA* **73:**396–400.

Kasahara, M., and Hinkle, P. C., 1977, Reconstitution and purification of the D-glucose transporter from human erythrocytes, *J. Biol. Chem.* **252:**7384–7390.

Katzen, H. M., and Soderman, D. D., 1975, Interaction of carbohydrate binding sites on concanavalin A-agarose with receptors on adipocytes studied by buoyant density method, *Biochemistry* **14:**2293–2298.

Kawai, S., and Hanafusa, H., 1971, The effects of reciprocal changes in temperature on the transformed state of cells infected with a Rous sarcoma virus mutant, *Virology* **46:**470–479.

Kiechle, F. L., Jarett, L., Kotagal, N., and Popp, D. A., 1981, Partial purification from rat adipocyte plasma membranes of a chemical mediator which stimulates the action of insulin on pyruvate dehydrogenase, *J. Biol. Chem.* **256:**2945–2951.

King, G. L., Rechler, M. M., and Kahn, C. R., 1982, Interactions between the receptors for insulin and the insulin-like growth factors on adipocytes, *J. Biol. Chem.* **257:**10001–10006.

Kissebah, A. H., Hope-Gill, H., Vydelingum, N., Tulloch, B., Clarke, P., and Fraser, T. R., 1975, Mode of insulin action, *Lancet* **1:**144–147.

Kletzien, R. F., and Perdue, J. F., 1974, Sugar transport in chick embryo fibroblasts, *J. Biol. Chem.* **249:**3375–3382.

Kletzien, R. F., and Perdue, J. F., 1975a, Regulation of sugar transport in chick embryo fibroblasts infected with a temperature-sensitive mutant of RSV, *Cell* **6:**513–520.

Kletzien, R. F., and Perdue, J. F., 1975b, Induction of sugar transport in chick embryo fibroblasts by hexose starvation, *J. Biol. Chem.* **250:**593–600.

Kono, T., and Barham, F. W., 1971a, Insulin-like effects of trypsin on fat cells. Localization of the metabolic steps and the cellular site affected by the enzymes, *J. Biol. Chem.* **246:**6204–6209.

Kono, T., and Barham, F. W., 1971b, The relationship between the insulin-binding capacity of fat cells and the cellular response to insulin, *J. Biol. Chem.* **246:**6210–6216.

Kono, T., Robinson, F. W., Sarver, J. A., Vega, F. V., and Pointer, R. A., 1977a, Action of insulin in fat cells. Effects of low temperature, uncouplers of oxidative phosphorylation, and respiratory inhibitors, *J. Biol. Chem.* **252:**2226–2233.

Kono, T., Vega, F. V., Raines, K. B., and Shumway, S. J., 1977b, Deactivation of the once stimulated sugar transport reaction in fat cells, *Fed. Proc.* **36:**341.

Kono, T., Suzuki, K., Dansey, L. E., Robinson, F. W., and Blevins, T. L., 1981, Energy-dependent and protein synthesis-independent recycling of the insulin-sensitive glucose transport mechanism in fat cells, *J. Biol. Chem.* **256:**6400–6407.

Krasne, S., Eisenman, G., and Szabo, G., 1971, Freezing and melting of lipid bilayers and the mode of action of nonactin, valinomycin, and gramicidin, *Science* **174:**412–415.

Lacko, L., Wittke, B., and Kromphardt, H., 1972, Zur kinetik der glucose-aufnahme in erythrocyten effekt der trans-konzentration, *Eur. J. Biochem.* **25:**447–454.

Lawrence, J. C., Jr., and Larner, J., 1978, Effects of insulin, methoxamine, and calcium and glycogen synthase in rat adipocytes, *Mol. Pharamacol.* **14:**1079–1091.

Le Fevre, P. G., 1961, Sugar transport in the red blood cells: Structure-activity relationships in substrates and antagonists, *Pharmacol. Rev.* **13:**39–45.

Letarte, J., and Reynold, A. E., 1969, Ionic effects on glucose transport and metabolism by isolated mouse fat cells incubated with or without insulin. I. Lack of effect of medium $Ca^{2+}$, $Mg^{2+}$ or $PO_4{}^{3-}$. *Biochim. Biophys. Acta* **183:**350–356.

Lidgard, G. P., and Jones, M. N., 1975, D-Glucose permeability of black lipid membranes modified by human erythrocyte membrane fractions, *J. Membr. Biol.* **21:**1–10.

Lienhard, G. E., Gorga, F. R., Orasky, J. E., and Zoccol, M. A., 1977, Monosaccharide transport system of the human erythrocyte: Identification of the cytochalasin B binding component, *Biochemistry* **16:**4921–4926.

Lin, S., and Snyder, C. E., Jr., 1977, High affinity cytochalasin B binding to red cell membrane proteins which are unrelated to sugar transport, *J. Biol. Chem.* **252:**5464–5471.

Lin, S., and Spudich, J. A., 1974a, Binding of cytochalasin B to a red cell membrane protein, *Biochem. Biophys. Res. Commun.* **61:**1471–1476.

Lin, S., and Spudich, J. A., 1974b, Biochemical studies on the mode of action of cytochalain B, *J. Biol. Chem.* **249:**5578–5783.

Lin, S., Santi, D. V., and Spudich, J. A., 1974, Biochemical studies on the mode of action of cytochalasin B, *J. Biol. Chem.* **249:**2268–2274.

Lockwood, D. H., and East, L. E., 1974, Studies of the insulin-like actions of polyamines on lipid and glucose metabolism in adipose tissue cells, *J. Biol. Chem.* **249:**7717–7722.

Lockwood, D. H., Lipsky, J. J., Meronk, F., Jr., and East, L. E., 1971, Actions of polyamines on lipid and glucose metabolism of fat cells, *Biochem. Biophys. Res. Commun.* **44:**600–617.

Lundahl, P., Acevedo, F., Froman, G., and Phutrakul, S., 1981, The stereospecific D-glucose transport activity of cholate extracts from human erythrocyte membranes, *Biochim. Biophys. Acta* **644:**101–107.

Martineau, R., Kohlbacher, M. S., Shaw, S. N., and Amos, H., 1972, Enhancement of hexose entry into chick fibroblasts by starvation: Differential effect on galactose and glucose, *Proc. Natl. Acad. Sci. USA* **69:**3407–3411.

Massague, J., Pilch, P. F., and Czech, M. P., 1980, Electrophoretic resolution of three major insulin receptor structures with unique subunit stoichiometries, *Proc. Natl. Acad. Sci. USA* **77:**7137–7141.

May, J. M., and de Haen, C., 1979a, Insulin-stimulated intracellular hydrogen peroxide production in rat epididymal fat cells, *J. Biol. Chem.* **254:**2214–2220.

May, J. M., and de Haen, C., 1979b, The insulin-like effect of hydrogen peroxide on pathways of lipid synthesis in rat adipocytes, *J. Biol. Chem.* **254:**9017–9021.

Meuli, C., and Froesch, E. R., 1977, Insulin and nonsupressible insulin-like activity (NSILA-S) stimulate the same glucose transport system via two separate receptors in rat heart, *Biochem. Biophys. Res. Commun.* **75:**689–695.

Minemura, T., and Crofford, O. B., 1969, Insulin-receptor interaction in isolated fat cells. I. The insulin-like properties of *p*-chloromercuribenzene sulfonic acid, *J. Biol. Chem.* **244:**5181–5188.

Mizel, S. B., and Wilson, L., 1974, Inhibition of the transport of several hexoses in mammalian cells by cytochalasin B, *J. Biol. Chem.* **247:**4102–4105.

Muchmore, D. B., Little, S. A., and de Haen, C., 1981, A dual mechanism of action of ocytocin in rat epididymal fat cells, *J. Biol. Chem.* **256:**365–372.

Muchmore, D. B., Little, S. A., and de Haen, C., 1982, Counterregulatory control of intracellular hydrogen peroxide production by insulin and lipolytic hormones in isolated rat epididymal fat cells: A role of free fatty acids, *Biochemistry* **21:**3886–3892.

Mukherjee, S. P., and Lynn, W. S., 1977, Reduced nicotinamide adenine dinucleotide phosphate oxidase in adipocyte plasma membrane and its activation by insulin. Possible role in the hormone's effects on adenylate cyclase and the hexose monophosphate shunt, *Arch. Biochem. Biophys.* **184:**69–76.

Mukherjee, S. P., Lane, R. H., and Lynn, W. S., 1978, Endogenous hydrogen peroxide and peroxidative metabolism in adipocytes in response to insulin and sulfhydryl reagents, *Biochem. Pharmacol.* **27:**2589–2594.

Mullins, R. E., and Langdon, R. G., 1980a, Maltosyl isothiocyanate: An affinity label for the glucose transporter of the human erythrocyte membrane. 1. Inhibition of glucose transport, *Biochemistry* **19:**1199–1205.

Mullins, R. E., and Langdon, R. G., 1980b, Maltosyl isothiocyanate: An affinity label for the glucose transporter of the human erythrocyte membrane. 2. Identification of the transporter, *Biochemistry* **19:**1205–1211.

Nickson, J. K., and Jones, M. N., 1977, Reconstitution of the monosaccharide-transport system of the human erythrocyte membrane, *Biochem. Trans.* **5:**147–149.

Olden, K., Pratt, R. M., Jaworski, C., and Yamada, K. M., 1974, Evidence for role of glycoprotein carbohydrates in membrane transport: Specific inhibition by tunicamycin, *Proc. Natl. Acad. Sci. USA* **76:**791–795.

Olefsky, J. M., 1978, Mechanisms of the ability of insulin to activate the glucose-transport system in rat adipocytes, *Biochem. J.* **172:**137–145.

Oppenheimer, C. L., and Czech, M. P., 1983, Affinity labeling of receptors, in: *Growth and Maturation Factors,* John Wiley and Sons, New York.

Oppenheimer, C. L., Pessin, J. E., Massague, J., Gitomer, W., and Czech, M. P., 1983, Insulin action rapidly modulates the affinity of the insulin-like growth factor II receptor, *J. Biol. Chem.* **258:**4824–4830.

Pessin, J. E., Tillotson, L. G., Yamada, K., Gitomer, W., Carter-Su, C., Mora, R., Isselbacher, K. J., and Czech, M. P., 1982, Identification of the stereospecific hexose transporter from starved and fed chicken embryo fibroblasts, *Proc. Natl. Acad. Sci. USA* **79:**2286–2290.

Phutrakul, S., and Jones, M. N., 1979, The permeability of bilayer lipid membranes on the incorporation of erythrocyte membrane extracts and the identification of the monosaccharide transport proteins, *Biochim. Biophys. Acta* **550:**188–200.

Pillion, D. J., and Czech, M. P., 1978, Antibodies against intrinsic adipocyte plasma membrane proteins activate D-glucose transport independent of interaction with insulin binding sites, *J. Biol. Chem.* **253:**3761–3764.

Pillion, D. J., Grantham, J. R., and Czech, M. P., 1979, Biological properties of antibodies against rat adipocyte intrinsic membrane proteins. Dependence on multivalency for insulin-like activity, *J. Biol. Chem.* **254:**3211–3220.

Plagemann, P. G. W., and Richey, D. P., 1974, Transport of nucleosides, nucleic acid bases, choline and glucose by animal cells in culture, *Biochim. Biophys. Acta* **344:**263–305.

Rampal, A. L., Pinkofsky, H. B., and Jung, C. Y., 1980, Structure of cytochalasin B binding sites in human erythrocyte membranes, *Biochemistry* **19:**679–683.

Reeves, J. P., 1975, Calcium-dependent stimulation of 3-*O*-methylglucose uptake in rat thymocytes by the divalent cation ionophore A 23187, *J. Biol. Chem.* **250:**9428–9430.

Rendi, R., 1964, Water extrusions in isolated subcellular fractions, *Biochim. Biophys. Acta* **84:**694–706.

Renner, E. D., Plagemann, P. G. W., and Bernlohr, R. W., 1972, Permeation of glucose by simple and facilitated diffusion by Novikoff rat hepatoma cell in suspension culture and relationship of glucose metabolism, *J. Biol. Chem.* **247:**5765–5776.

Rihan, Z., Jarrett, R. J., and Keen, H., 1967, EDTA and insulin: A study of the effect of salts of EDTA upon insulin action *in vivo* and *in vitro, Diabetologia* **3:**449–452.

Rodbell, M., 1964, Metabolism of isolated fat cells. 1. Effects of hormones on glucose metabolism and lipolysis, *J. Biol. Chem.* **239:**375–380.

Rodbell, M., 1966, Metabolism of isolated fat cells. 1. The similar effects of phospholipase C, and of insulin on glucose and amino acid metabolism, *J. Biol. Chem.* **241:**130–139.

Rosenthal, J. W., and Fain, J. N., 1971, Insulin-like effect of clostridial phospholipase C, neuraminidase, and other bacterial factors on brown fat cells, *J. Biol. Chem.* **246:**5888–5895.

Roth, J., 1973, Peptide hormone binding to receptors: A review of direct studies *in vitro, Metabolism* **22:**1059–1073.

Salter, D. W., and Weber, M. J., 1979, Glucose-specific cytochalasin B binding is increased in chicken embryo fibroblasts transformed by Rous sarcoma virus, *J. Biol. Chem.* **254:**3554–3561.

Salter, D. W., Baldwin, S. A., Lienhard, G. E., and Weber, M. J., 1982, Proteins antigenically related to the human erythrocyte glucose transporter in normal and Rous sarcoma virus-transformed chicken embryo fibroblasts, *Proc. Natl. Acad. Sci. USA* **79:**1540–1544.

Saltiel, A., Jacobs, S., Siegel, M., and Cuatrecasas, P., 1981, Insulin stimulates the release from liver plasma membranes of a chemical modulator of pyruvate dehydrogenase, *Biochem. Biophys. Res. Commun.* **102:**1041–1047.

Schudt, C., Gaertner, U., and Pette, D., 1976, Insulin action on glucose transport and calcium fluxes in developing muscle cells *in vitro, Eur. J. Biochem.* **68:**103–111.

Seals, J. R., and Czech, M. P., 1980, Evidence that insulin activates in intrinsic plasma membrane protease in generating a secondary chemical mediator, *J. Biol. Chem.* **255:**6529–6531.

Seals, J. R., and Czech, M. P., 1981, Characterization of a pyruvate dehydrogenase activator released by adipocyte plasma membranes in response to insulin, *J. Biol. Chem.* **256:**2894–2899.

Seals, J. R., and Jarett, L., 1980, Activation of pyruvate dehydrogenase by direct addition of insulin to an isolated plasma membrane/mitochondria mixture: Evidence for generation of insulin's second messenger in a subcellular system, *Proc. Natl. Acad. Sci. USA* **77:**77–81.

Sen, A. K., and Widdas, W. F., 1962, Determination of the temperature and pH dependence of glucose transfer across the human erythrocyte membrane measured by glucose exit, *J. Physiol.* **160:**392–403.

Shanahan, M. F., 1982, Cytochalasin B, *J. Biol. Chem.* **257:**7290–7293.

Shanahan, M. F., and Czech, M. P., 1977, Purification and reconstitution of the adipocyte plasma membrane D-glucose transport system, *J. Biol. Chem.* **252:**8341–8343.

Shanahan, M. F., and Jacquez, J. A., 1978, Differential labeling of components in human erythrocyte membranes associated with the transport of glucose, *Membr. Biochem.* **1:**239–267.

Shanahan, M. F., Olson, S. A., Weber, M. J., Lienhard, G. E., and Gorga, J. C., 1982, Photolabeling of glucose-sensitive cytochalasin B binding proteins in erythrocyte, fibroblasts and adipocyte membranes, *Biochem. Biophys. Res. Commun.* **107:**38–43.

Shemyakin, M. M., Ovchinnikov, Y. A., Ivanov, V. I., Antonov, V. K., Vinogradova, E. I., Shkrob, A. M., Malenkov, G. G., Evstratov, A. V., Laine, I. A., Melnik, E. I., Ryabova, I. D., 1969, Cyclodepsipeptides as chemical tools for studying ionic transport through membranes, *J. Membr. Biol.* **1:**402–430.

Siegal, J., and Olefsky, J. M., 1980, Role of intracellular energy in insulin's ability to activate 3-*O*-methylglucose transport by rat adipocytes, *Biochemistry* **19:**2183–2190.

Simpson, I. A., Wheeler, T. J., Sogin, D. C., Hinkle, P. C., and Cushman, S. W., 1981, Characterization of intracellular glucose transport systems and their insulin-induced translocation to the plasma membrane in the rat adipose cell using [$^3$H]cytochalasin B and a rabbit antibody against the human erythrocyte glucose transporter, *J. Cell Biol.* **91:**413a.

Sogin, D. C., and Hinkle, P. C., 1978, Characterization of the glucose transporter from human erythrocytes, *J. Supramol. Struct.* **8:**447–453.

Sogin, D. C., and Hinkle, P. C., 1980a, Binding of cytochalasin B to human erythrocyte glucose transporter, *Biochemistry* **19:**5417–5420.

Sogin, D. C., and Hinkle, P. C., 1980b, Immunological identification of the human erythrocyte glucose transporter, *Proc. Natl. Acad. Sci. USA* **77:**5725–5729.

Steck, T. L., 1974, The organization of proteins in the human red blood cell membrane, *J. Cell Biol.* **62:**1–19.

Suzuki, K., and Kono, T., 1980, Evidence that insulin causes translocation of glucose transport activity to the plasma membrane from an intracellular storage site, *Proc. Natl. Acad. Sci. USA* **77:**2542–2545.

Taverna, R. D., and Langdon, R. G., 1973, Reversible association of cytochalasin B with the human erythrocyte membrane, *Biochim. Biophys. Acta* **323:**207–219.

Taylor, N. F., and Gagneja, G. L., 1973, A model for the mode of action of cytochalasin B inhibition of D-glucose transport in the human erythrocyte, *Can. J. Biochem.* **53:**1078–1084.

Tosteson, D. C., Andreoli, T. E., Tieffenberg, M., and Cook, P., 1968, The effects of macrocyclic compounds on cation transport in sheep red cells and thin and thick lipid membranes, *J. Gen. Physiol.* **51:**373.

Ullrey, D., Gammon, M. T., and Kalckar, H. M., 1975, Uptake patterns and transport enhancements in cultures of hamster cells deprived of carbohydrates, *Arch. Biochem. Biophys.* **167:**410–416.

Urry, D. W., 1971, The gramicidin A transmembrane channel: A proposed (L,D) helix, *Proc. Natl. Acad. Sci. USA* **68:**672–676.

Vega, F. V., and Kono, T., 1979, Sugar transport in fat cells: Effects of mechanical agitation, cell-bound insulin, and temperature, *Arch. Biochem. Biophys.* **192:**120–127.

Vega, F. V., Key, R. J., Jordan, J. E., and Kono, T., 1980, Reversal of insulin effects in fat cells may require energy for deactivation of glucose transport but not for deactivation of phosphodiesterase, *Arch. Biochem. Biophys.* **203:**167–173.

Venuta, S., and Rubin, H., 1973, Sugar transport in normal and Rous sarcoma virus-transformed chick-embryo fibroblasts, *Proc. Natl. Acad. Sci. USA* **70:**653–657.

Vinten, J., 1978, Cytochalasin B inhibition and temperature dependence of 3-*O*-methylglucose transport in fat cells, *Biochim. Biophys. Acta* **511:**259–273.

Vinten, J., Gliemann, J., and Sterlind, K., 1976, Exchange of 3-*O*-methylglucose in isolated fat cells, *J. Biol. Chem.* **254:**794–800.

Wardzala, L. J., and Jeanrenaud, B., 1981, Potential mechanism of insulin action on glucose transport in the isolated rat diaphragm, *J. Biol. Chem.* **256:**7090–7093.

Wardzala, L. J., Cushman, S. W., and Salans, L. B., 1978, Mechanism of insulin action on glucose transport in the isolated rat adipose cell, *J. Biol. Chem.* **253:**8002–8005.

Weber, M. J., 1973, Hexose transport in normal and in Rous sarcoma virus-transformed cells, *J. Biol. Chem.* **248:**2978–2983.

Wheeler, T. J., Simpson, I. A., Sogin, D. C., Hinkle, P. C., and Cushman, S. W., 1982, Detection of the rat adipose cell glucose transporter with antibody against the human red cell glucose transporter, *Biochem. Biophys. Res. Commun.* **105:**89–95.

Whitesell, R. R., and Gliemann, J., 1979, Kinetic parameters of transport of 3-*O*-methylglucose and glucose in adipocytes, *J. Biol. Chem.* **254:**5276–5283.

Wilbrandt, W., 1978, *Cell Membrane Receptors for Drug and Hormones: A Multidisciplinary Approach* (R. W. Straub and L. Bolis, eds.), Raven Press, New York, pp. 243–249.

Yu, K. T., and Gould, M. K., 1977, Insulin-stimulated sugar transport and [$^{125}$I]insulin binding by rat soleus muscle: Permissive effect of ATP, *Biochem. Biophys. Res. Commun.* **77:**203–210.

41

# The Bacterial Phosphoenolpyruvate:Sugar Phosphotransferase System

Norman D. Meadow, Maria A. Kukuruzinska, and Saul Roseman

## I. INTRODUCTION

The bacterial phosphoenolpyruvate:glycose phosphotransferase system, or PTS, plays a key role in several important physiological processes. Those PTS functions thus far identified include transport of PTS sugar substrates across the cytoplasmic membrane coupled with their phosphorylation, chemotaxis toward PTS sugar substrates, and regulation of the synthesis of enzymes and permeases required for the catabolism of certain non-PTS sugars. The latter function is achieved primarily by regulating both adenylate cyclase and the respective non-PTS sugar permeases.

The PTS is a complex system of interacting cytoplasmic and integral membrane proteins. This complexity may be required for the PTS to perform its many functions, the most extensively studied of which is the translocation/phosphorylation of PTS sugars. Figure 1 schematically illustrates the two glucose uptake systems of the enteric bacteria *Escherichia coli* and *Salmonella typhimurium*. The overall reaction is

$$\text{Phosphoenolpyruvate}_{in} + \text{sugar}_{out} \rightleftharpoons \text{sugar phosphate}_{in} + \text{pyruvate}_{in} \qquad (1)$$

As illustrated in Figure 1, Reaction (1) usually requires four proteins. Two of the proteins are nonspecific for the sugar substrates, and are called general proteins (Enzyme I and HPr), while two are sugar-specific. The latter comprise the Enzyme II complexes (III/II-B or II-A/II-B). Variants of this general format are indicated below.

---

*Norman D. Meadow, Maria A. Kukuruzinska, and Saul Roseman* • Department of Biology and the McCollum-Pratt Institute, The Johns Hopkins University, Baltimore, Maryland 21218. The authors were supported by Program Project Grant CA21901 from the National Cancer Institute of the NIH. Contribution 1247 from the McCollum–Pratt Institute.

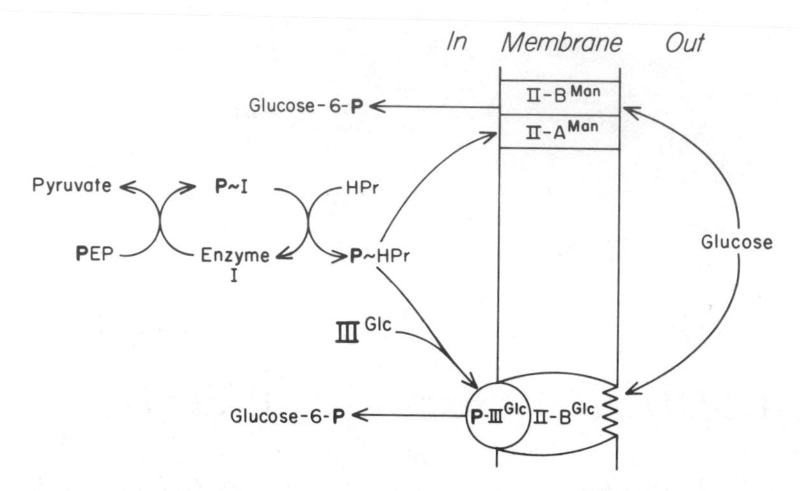


*Figure 1.* Schematic representation of the phosphotransferase system (Weigel *et al.*, 1982c). The phosphoryl group is sequentially transferred from phosphoenolpyruvate to Enzyme I to HPr and then to one of the sugar-specific proteins, II-$A^{Man}$ (an integral membrane protein) or $III^{Glc}$ (a soluble and/or peripheral membrane protein). The integral membrane proteins, II-$B^{Man}$ and II-$B^{Glc}$, are the sugar receptors and catalyze the transfer of the phosphoryl group from II-$A^{Man}$ and $III^{Glc}$ to the sugar (glucose) concomitant with the translocation of the sugar across the membrane. The II-$A^{Man}$/II-$B^{Man}$ complex is designated $II^{Man}$, while $II^{Glc}$ denotes the $III^{Glc}$/II-$B^{Glc}$ complex. Methyl α-D-glucopyranoside is phosphorylated by $II^{Glc}$, while mannose, 2-deoxyglucose, and several other sugars are phosphorylated by $II^{Man}$. The positions of the genes for these and other PTS proteins are given below. Positions on the *E. coli* map are taken from Bachman (1983) and are followed by an "Ec"; those on the *S. typhimurium* map are from Sanderson and Roth (1983) and are followed by an "St." NM represents not mapped; NS, not studied. Alternate symbols for the genes are shown in parentheses and are also followed by letters designating the species. A discussion of nomenclature of PTS genes is found in Rephaeli and Saier (1980). *ptsH* (*ctr, Hpr,* Ec; *carB,* St) for HPr, 52 min Ec, 49 min St. *ptsI* (*ctr,* Ec; *carA,* St) for Enzyme I, 52 min Ec, 49 min St. *crr* (*tgs, gsr,* Ec) for $III^{Glc}$, 52 min Ec, 48 min St. *iex* (*crr,* Ec) unknown product, 52 min Ec, NS, St. *ptsG* (*car, CR, gpt, gptA, tgl, umg,* Ec; *glu, gpt, cat,* St) for Enzyme II-$B^{Glc}$, 24 min Ec, 25 min St. *ptsM* (*gptB, mpt, pel, ptsX,* Ec; *manA,* St) for the $II^{Man}$ complex, 40 min Ec, NM, St. *mtlA* for Enzyme II-$B^{Mtl}$, 81 min Ec, 78 min St. *bglC* (*bglB,* Ec) for Enzyme II-$B^{Bgl}$ (the β-glucoside permease), 83 min Ec, NS, St.

The size and complexity of the recent literature make it difficult to present a detailed survey of the entire field. In this brief review, we therefore focus on three topics: the biochemistry of the PTS proteins, regulation of the PTS, and regulation of non-PTS permeases by the PTS. We shall refer primarily to studies with *E. coli* and *S. typhimurium*. The last review from this laboratory appeared in 1976 (Postma and Roseman). Other reviews have been published in recent years (Dills *et al.*, 1980; Cordaro, 1976; Postma, 1982; Saier, 1977; Saier and Moczydlowski, 1978; Silhavy *et al.* 1978).

## II. ENZYME I

Enzyme I, the first protein in the PTS reaction sequence, catalyzes the transfer of the phosphoryl group from phosphoenolpyruvate to HPr, the second protein of the sequence:

$$\text{phosphoenolpyruvate} + \text{HPr} \xrightleftharpoons{\text{Enzyme I, } Mg^{2+}} \text{P} \sim \text{HPr} + \text{pyruvate} \qquad (2)$$

Enzyme I is the key enzyme in the overall reaction pathway because it initiates the whole chain of phosphoryl transfer reactions.

Enzyme I has been purified to homogeneity from *E. coli* (Robillard *et al.*, 1979; Waygood and Steeves, 1980) and *S. typhimurium* (Waygood *et al.*, 1977; Weigel *et al.*, 1982c). One procedure for isolating Enzyme I from *E. coli* is based on methods that this laboratory developed to isolate the protein from *S. typhimurium* (Waygood *et al.*, 1977; Weigel *et al.*, 1982c), while the other procedure utilizes hydrophobic interaction chromatography (Robillard *et al.*, 1979). However, the latter method yields a preparation significantly contaminated with ribonucleic acid (Misset *et al.*, 1980). Subsequently, the hydrophobic column method was modified, although no conclusive evidence for purity was given (Brouwer *et al.*, 1982; Misset and Robillard, 1982). The purification of Enzyme I from *S. typhimurium* was recently reported (Weigel *et al.*, 1982c). The method takes advantage of two properties of the enzyme. First, it undergoes a temperature-dependent change in molecular weight, and second, it is phosphorylated by incubation with phosphoenolpyruvate (PEP) and $MgCl_2$ which changes its charge and chromatographic properties. This technique provides the protein in good yield (20%) and the preparation stands up to rigorous tests of purity (Chrambach and Rodbard, 1971).

Estimates of the molecular weight of pure, dissociated Enzyme I from *E. coli* range from 67,000–84,000 (Robillard *et al.*, 1979; Misset *et al.*, 1980; Waygood and Steeves, 1980), and are based on standard procedures for gel filtration chromatography and polyacrylamide gel electrophoresis in the presence of sodium dodecyl sulfate (SDS–PAGE). In contrast, the monomer molecular weight of Enzyme I from *S. typhimurium* is around 58,000. This value was obtained from extensive sedimentation equilibrium studies and gel filtration measurements (Kukuruzinska *et al.*, 1982) under denaturing conditions (6 M guanidinium chloride). Moreover, SDS–PAGE (Weigel *et al.*, 1982c) indicates a molecular weight value of 60,000–62,000. Since Enzyme I from *S. typhimurium* appeared to have a molecular weight significantly lower than the values reported for the *E. coli* enzyme, crossed immunoelectrophoresis was used to see if the isolated *S. typhimurium* Enzyme I was a proteolytic product formed during purification. The analysis suggests that the isolated *S. typhimurium* protein represents native, intact Enzyme I (Kukuruzinska *et al.*, 1982). It should be noted that Mattoo and Waygood (1983) report that Enzyme I from *S. typhimurium* has an immunological reactivity which is quantitatively different from Enzyme I from *E. coli* to antiserum prepared against the *E. coli* enzyme. Since both proteins have the same specific activity, this implies that the Enzymes I from the two species differ structurally (Mattoo and Waygood, 1983).

Recent experiments from this laboratory provide evidence that Enzyme I subunits from *S. typhimurium* are identical. Exhaustive dansylation of the enzyme reveals only an $NH_2$-terminal methionine, and amino terminal sequence analysis reveals a single amino acid at each of the first 17 residues. In addition, the protein migrates as a single band during both SDS–PAGE and isoelectric focusing (Weigel *et al.*, 1982c).

Studies from a number of laboratories (Waygood *et al.*, 1977, 1979; Robillard *et al.*, 1979; Misset *et al.*, 1980; Saier *et al.*, 1980; Waygood and Steeves, 1980; Hoving *et al.*, 1981, 1982; Misset and Robillard, 1982) indicate that Enzyme I undergoes reversible, temperature- and concentration-dependent association. Gel filtration chromatography under nondenaturing conditions shows that, at 4–6°C, the enzyme migrates as a protein of 70,000 (apparent) molecular weight; however, at room temperature, its apparent molecular weight increases to about 130,000 (Waygood *et al.*, 1977; Waygood and Steeves, 1980; Weigel *et al.*, 1982c). Association of monomers of Enzyme I from *S. typhimurium* has been examined in detail by sedimentation equilibrium experiments (Kukuruzinska *et al.*, 1982). At 8°C, in the absence of substrates and cofactor, and with an initial protein concentration of 0.5 mg/ml, Enzyme I self-associates, although weakly, favoring primarily the monomer. Nonlinear least squares analysis of the weight-average molecular weight distributions could not distinguish between monomer–dimer and isodesmic reaction mechanisms. However, rigorous examination of number-average and Z-average molecular weight distributions, as well as studies on Enzyme I association using optical scanning chromatography and elution gel chromatography (Valdes and Ackers, 1979), support a monomer–dimer reaction mechanism (unpublished results in collaboration with Dr. G. Ackers). The latter mechanism agrees with published kinetic data (Waygood *et al.*, 1979; Misset *et al.*, 1980; Saier *et al.*, 1980; Hoving *et al.*, 1981). Preliminary scanning gel chromatography studies indicate that, at 21°C and a concentration of 2 mg/ml, the association of Enzyme I monomers is strong and is not significantly affected by $Mg^{2+}$. This is in contrast to results reported for the *E. coli* enzyme by one group of investigators (Misset *et al.*, 1980).

Enzyme I monomer is a globular molecule with a frictional ratio very close to 1 (Kukuruzinska *et al.*, 1982). The sedimentation coefficient of Enzyme I in buffer and at 4°C is 3.36 S ($S_{20,w}$ = 5.59 S), which approximates the value reported for *E. coli* Enzyme I (Waygood and Steeves, 1980). The partial specific volume, determined from the density of the protein, is 0.72 ml/g (Kukuruzinska *et al.*, 1982). Upon isoelectric focusing, Enzyme I forms a sharp stationary band with a pI of 4.5. Amino acid compositions for both *E. coli* and *S. typhimurium* Enzyme I appear similar. The tryptophan content of Enzyme I is low since the extinction coefficients are 4.4 and 4.0 for 10 mg/ml solutions of the *E. coli* and *S. typhimurium* proteins, respectively (Waygood and Steeves, *et al.*, 1980; Weigel *et al.*, 1982c). Originally, we reported that *S. typhimurium* Enzyme I contained four residues of tryptophan per monomer (Weigel *et al.*, 1982c) More recently (unpublished results in collaboration with Dr. R. Levine), second derivative ultraviolet spectroscopy coupled with statistically weighted multicomponent analysis (Levine and Federici, 1982) indicates that there are only two tryptophan residues per monomer of Enzyme I from *S. typhimurium*.

The optimum pH for the activity of Enzyme I is at or near the physiological pH (Waygood *et al.*, 1979; Weigel *et al.*, 1982c); it is very sensitive to inhibition by sulfhydryl reagents; and when pure, Enzyme I is fairly labile (Postma and Roseman, 1976; Weigel *et al.*, 1982c). The stability of the protein increases in the presence of $Mg^{2+}$ and phosphoenolpyruvate (Misset *et al.*, 1980; Saier *et al.*, 1980; Waygood and Steeves, 1980; Hoving *et al.*, 1982).

Enzyme I exhibits ping-pong kinetics in phosphorylating HPr, and it catalyzes isotope exchange between phosphoenolpyruvate and pyruvate (Saier *et al.*, 1980; Hoving *et al.*, 1981; Weigel *et al.*, 1982c). For the *S. typhimurium* enzyme, the $K_m$ value for the interaction with HPr is 4.5 μM, while values of 0.2 and 0.4 mM for the interaction with phosphoenolpyruvate have been reported (Saier *et al.*, 1980; Weigel *et al.*, 1982c). The *E. coli* enzyme has $K_m$ values for HPr and phosphoenolpyruvate of 9 μM and 0.18 mM, respectively (Waygood and Steeves, 1980). Enzyme I requires $Mg^{2+}$ ion for activity with a $K_m$ of 0.53 mM for *S. typhimurium* Enzyme I (Weigel *et al.*, 1982c), but divalent ions such as $Mn^{2+}$ and $Co^{2+}$ ($K_m$ of 0.05 mM in each case) can substitute for $Mg^{2+}$. This has been confirmed by isotope exchange studies (Saier *et al.*, 1980). However, at high $Co^{2+}$ concentration, Enzyme I activity can be inhibited ($K_i$ = 0.7 mM; Weigel *et al.*, 1982c).

Preliminary studies from this and other laboratories (Simoni *et al.*, 1973b; Stein *et al.*, 1974; Postma and Roseman, 1976) with the *S. typhimurium* and *Staphylococcus aureus* enzymes show that phosphorylation of HPr proceeds via a phospho-Enzyme I intermediate. Preliminary studies with *S. aureus* (Stein *et al.*, 1974) show that a preparation of phosphorylated Enzyme I could serve as a phosphoryl donor to HPr, and that it could phosphorylate the sugar analogue isopropyl 1-thio-β-D-galactopyranoside in the presence of $Mg^{2+}$, HPr, and the relevant sugar-specific proteins. More recent experiments (Weigel *et al.*, 1982a) with homogeneous *S. typhimurium* enzyme show that, with excess phosphoenolpyruvate, close to 1 mole of phosphoryl group is incorporated per mole of Enzyme I monomer. In contrast, others report that one phosphoryl group is incorporated per dimer of *E. coli* Enzyme I (Misset and Robillard, 1982) according to pyruvate burst and isotope-exchange studies. It has been suggested that the mechanism of phosphorylation of Enzyme I is accompanied by a stereospecific proton transfer (Hoving *et al.*, 1983).

Kinetic studies from this and other laboratories indicate that the monomer is not catalytically active (Waygood *et al.*, 1977, 1979; Misset *et al.*, 1980; Saier *et al.*, 1980; Hoving *et al.*, 1981, 1982; Weigel *et al.*, 1982a). From experiments on both [$^{14}$C]pyruvic acid exchange (between phosphoenolpyruvate and pyruvate) and the rate of phosphorylation of the enzyme, it appears that a divalent metal ion must be bound to the enzyme to render it active (Hoving *et al.*, 1982). Phosphorylation of Enzyme I requires $Mg^{2+}$; yet, transfer of the phosphoryl group from phospho-Enzyme I to HPr proceeds in the presence of 20 mM EDTA (Weigel *et al.*, 1982a). The rate of phosphorylation of Enzyme I from *S. typhimurium* is biphasic (Weigel *et al.*, 1982a), which suggests that the enzyme is phosphorylated only in an associated state; a similar conclusion has been reached from kinetic studies with the *E. coli* enzyme (Saier *et al.*, 1980; Hoving *et al.*, 1981, 1982).

The apparent equilibrium constant for phosphorylation of Enzyme I from *S. typhimurium* at very low concentrations of substrate (phosphoenolpyruvate and pyruvate) is about 1.5, and the reaction appears independent of pH in the range of 6.5–8.0 (Weigel *et al.*, 1982a). This value is different from the one reported for the *E. coli* enzyme, measured using isotope exchange between PEP and pyruvate, which was found to be about 40 (Hoving *et al.*, 1982). However, the accuracy of the latter result has been questioned (Weigel *et al.*, 1982a). When the equilibrium constant is measured at high concentrations of substrate, only about 50% of the enzyme is phosphorylated

(Weigel *et al.*, 1982a), and this degree of phosphorylation changes only slightly with large changes in the ratio of phosphoenolpyruvate to pyruvate. From these results, the authors (Weigel *et al.*, 1982a) conclude that the reaction (phosphoenolpyruvate + Enzyme I $\rightleftharpoons$ phospho-Enzyme I + pyruvate) is likely to occur as written at low substrate concentrations, but may be much more complex at high substrate concentrations.

Recently, phospho-Enzyme I from *S. typhimurium* was isolated in homogeneous form (Weigel *et al.*, 1982a). From the pH stability of the phospho-enzyme and its sensitivity to hydrolysis in the presence of pyridine and hydroxylamine, it is concluded that the phosphoryl group is linked to position N-3 in the imidazole ring of a histidine residue. In addition, 3-phospho-histidine was isolated after alkaline hydrolysis of $^{32}$P-labeled phospho-Enzyme I. Phospho-Enzyme I is the true intermediate in Reaction (2) since it can transfer its phosphoryl group to pyruvate (to form phosphoenolpyruvate) and to HPr (to form phospho-HPr). Furthermore, in the presence of HPr and appropriate sugar-specific proteins, phospho-Enzyme I donates its phosphoryl group to methyl α-glucoside (to form sugar-phosphate). Phospho-Enzyme I has a standard free energy of hydrolysis of about −14.5 kcal/mole, which approximates that of phosphoenolpyruvate, thus placing the phosphate transfer potential of phospho-Enzyme I among the highest of known biological phosphoryl derivatives (Weigel *et al.*, 1982a).

As noted below, the structural gene for Enzyme I (*ptsI*) has been cloned onto a plasmid.

## III. HPr

HPr is the second protein that participates in the phosphoryl transfer reactions of the PTS. It interacts with phospho-Enzyme I and accepts 1 mole of phosphoryl group per mole of protein (Postma and Roseman, 1976). HPr has been purified to homogeneity from *E. coli* (Anderson *et al.*, 1971), *S. typhimurium* (Beneski *et al.*, 1982b), *S. aureus* (Simoni *et al.*, 1973b), *Bacillus subtilis* (Marquet *et al.*, 1976), and *Mycoplasma capricolum* (Ullah and Cirillo, 1976). HPr from all of these organisms consists of a single polypeptide chain of low molecular weight, ranging from 8000 to slightly more than 9000. The *E. coli* and *S. typhimurium* proteins appear identical in their amino acid composition (Weigel *et al.*, 1982b) and can substitute for one another in the heterologous sugar phosphorylation assay system, whereas HPr from *S. aureus* substitutes poorly for HPr from *S. typhimurium* (Postma and Roseman, 1976). The complete amino acid sequences for HPr from *S. aureus* and *S. typhimurium* are now available (Bayreuther *et al.*, 1977; Weigel *et al.*, 1982b), and except for 13 residues, 12 of which appear in the first 31 residues and cluster around the active site (histidine), the amino acid sequences of the two proteins are quite different. The *predicted* secondary structures of these proteins, however, show major regions of homology (Weigel *et al.*, 1982b).

The entire *pts* region of the *E. coli* chromosome has been cloned into specialized λ-transducing phages (Britton *et al.*, 1983; D. Saffen and A. Wong, unpublished results). These were subcloned into pBR322 plasmids (D. Saffen, unpublished results) and the structural genes which specify Enzyme I *(ptsI)*, HPr *(ptsH)*, and $III^{Glc}$ (*crr;* see Figure 1) have been isolated. The *ptsH* gene is completely sequenced and the predicted amino acid sequence for *E. coli* HPr has been determined (T. Doering and D. Saffen, unpublished results). This sequence disagrees at several residues with the sequence actually determined for *S. typhimurium* HPr (Weigel *et al.*, 1982b). The latter result has been reexamined (D. Powers and S. Roseman, unpublished results), and the corrected *S. typhimurium* sequence agrees completely with the predicted sequence for *E. coli*. Therefore, the HPr proteins from the two different species are identical.

HPr from *S. typhimurium* has a molecular weight of 9017, a value obtained from the complete amino acid sequence (Weigel *et al.*, 1982b). The amino acid composition is unusual in that it lacks tryptophan, tyrosine, and cysteine residues. One report claims that a fraction of a mole of tyrosine is present per mole of HPr (Roossien *et al.*, 1979) while another (Dooijewaard *et al.*, 1979) claims that HPr is closely associated with a polysaccharide. Both claims have been disputed (Weigel *et al.*, 1982b). HPr from *S. typhimurium* contains two histidine residues, one of which is phosphorylated in the phosphotransfer reaction, and two methionine residues. The $NH_2$-terminal methionine has been successfully derivatized with fluorescent, spin-labeled, and radioactive compounds (Grill *et al.*, 1982; Hildenbrand *et al.*, 1982). HPr from *S. aureus* has a molecular weight of 8600 (Simoni *et al.*, 1973b), and contains two tyrosine residues and only one histidine residue, which also is phosphorylated in the PTS reaction.

Initially, HPr was considered to be resistant to heating, but later it was shown that heating the pure protein to 100°C resulted in deamidation and loss of PTS activity (Anderson *et al.*, 1971). It is not known if this deamidation takes place at specific sites. Heat-treated extracts have two species of HPr (HPr-1 and HPr-2) which contain one and two fewer amide nitrogens, respectively, and have reduced (20–80%) specific activity. Nevertheless, pure HPr is quite stable at room temperature and can be stored frozen in water.

The phosphoryl group in phospho-HPr is linked to the N-1 position of the imidazole ring of a histidine residue (Anderson *et al.*, 1971) which, in *S. typhimurium,* is residue 15 of the protein. In *S. aureus,* the same linkage is formed, a conclusion supported by PMR measurements (Simoni *et al.*, 1973b; Schrecker *et al.*, 1975). The phosphoryl bond was identified by three criteria: pH stability, sensitivity to hydrolysis in the presence of pyridine and hydroxylamine, and identification of the phosphorylated amino acid after alkaline hydrolysis of $^{32}P$-labeled protein. In *S. aureus,* the equilibrium constant for the phosphorylation of HPr from phosphoenolpyruvate by Enzyme I is about 10 (Simoni *et al.*, 1973b), which is very close to the value obtained with the *S. typhimurium* protein (Weigel *et al.*, 1982a).

Phosphorylated HPr is quite unstable; the phosphoryl group has a half-life of 13 min at pH 8.0 and 37°C (Weigel *et al.*, 1982a). At higher pH values, its stability increases as expected, and at pH 9.2, the half-life is about 1 hr. As in the case of phospho-Enzyme I, the phosphate transfer potential of phospho-HPr is among the highest of known biological phosphate derivatives, since the standard free energy of hydrolysis of the phosphoprotein is $-13$ kcal/mole, slightly below that of phospho-Enzyme I and phosphoenolpyruvate (Weigel *et al.*, 1982a).

The enteric bacteria can compensate for loss of the protein HPr. This was first noted with *ptsH* mutants which show an unexpectedly high rate of reversion (Cordaro and Roseman, 1972; Cordaro, 1976). The revertants are, in fact, pseudorevertants. That is, another protein is generated (Waygood *et al.*, 1975) called "pseudo-HPr" which substitutes for HPr and permits growth of the cells on PTS sugars. Pseudo-HPr has not yet been isolated in homogeneous form.

## IV. $III^{Glc}$

### A. Isolation and Characterization

Earlier in this review, we noted that a given sugar-specific Enzyme II complex catalyzes the transfer of the phosphoryl group of phospho-HPr to its sugar substrate. In addition to lipid and divalent cation, Enzyme II complexes generally contain a *pair* of sugar-specific proteins (Postma and Roseman, 1976). An exception to this rule has recently been reported; in *E. coli*, the mannitol-specific Enzyme II consists of only a single integral membrane protein (Jacobson *et al.*, 1979; Lee *et al.*, 1981), unlike its counterpart from *S. aureus* (Simoni *et al.*, 1968) which is composed of two proteins.

The sugar receptor in an Enzyme II complex is an integral membrane protein called Enzyme II-B, while the protein which accepts the phosphoryl group from phospho-HPr may be either an integral membrane protein, called II-A, or a soluble protein, called III. An Enzyme II complex from *E. coli*, now designated $II^{Man}$ (Stock *et al.*, 1982), was the first to be dissociated into its corresponding $II\text{-}A^{Man}$ and Enzyme $II\text{-}B^{Man}$ components (Kundig and Roseman, 1971). The III-type proteins were then discovered in extracts of cells of *S. aureus* (Simoni *et al.*, 1973b), and the III protein specific for lactose ($III^{Lac}$) was isolated in homogeneous form (Hays *et al.*, 1973). It directly accepts a phosphoryl group from homogeneous phospho-HPr, and the equilibrium constant for the reaction was determined.

In *E. coli* and *S. typhimurium*, there are two Enzyme II complexes capable of taking up and phosphorylating glucose. One of these is the $II^{Man}$ complex (described briefly above), which has an affinity and capacity for glucose similar to those for mannose (Stock *et al.*, 1982; Curtis and Epstein, 1975). Also, $II^{Man}$ serves to take up other gluco- or manno-analogues such as 2-deoxy-D-glucose and *N*-acetylglucosamine, but not methyl α-D-glucopyranoside. The other Enzyme II complex in enteric bacteria is much more glucose specific. It is called the $II^{Glc}$ complex, and is composed of a soluble component, $III^{Glc}$, and a membrane component, Enzyme $II\text{-}B^{Glc}$, specific for glucose and methyl α-glucoside (Stock *et al.*, 1982).

$III^{Glc}$, because of its regulatory role, is one of the most important PTS proteins. A significant part of this review will be concerned with it and its physiological functions. The protein was discovered in this laboratory about 10 years ago (Kundig, 1974), but was purified to homogeneity only recently (from *S. typhimurium;* Meadow and Roseman, 1982). The gene from *E. coli* which codes for $III^{Glc}$ has been cloned (Meadow *et al.*, 1982b) and sequenced (D. Saffen, unpublished results). The information from the DNA sequence confirms and extends that obtained from the protein itself.

$III^{Glc}$ consists of a single polypeptide chain; its molecular weight as measured by SDS–PAGE is about 20,000 (Meadow and Roseman, 1982) and that from the DNA sequence is 18,162. Like HPr from the enteric bacteria, $III^{Glc}$ contains no cysteine, tryptophan, or tyrosine. Electrophoresis in polyacrylamide gels under nondenaturing conditions separates $III^{Glc}$ into two distinct species which were isolated (Meadow and Roseman, 1982) and named for their relative mobilities, $III^{Glc}_{slow}$ and $III^{Glc}_{fast}$. Amino-terminal sequence analysis showed that $III^{Glc}_{fast}$ is a processed form lacking the $NH_2$-terminal heptapeptide of $III^{Glc}_{slow}$. Further, the DNA sequence shows that $III^{Glc}_{slow}$ is the primary transcript; it lacks only the $NH_2$-terminal methionine found in the DNA sequence.

A standard procedure for determining the molecular weights of proteins is based on their relative electrophoretic mobilities under denaturing conditions, usually in sodium dodecyl sulfate containing polyacrylamide gels (SDS–PAGE). $III^{Glc}_{fast}$ does not behave as expected. While it is a processed form of $III^{Glc}_{slow}$ and, therefore, of somewhat lower molecular weight, $III^{Glc}_{fast}$ migrates more slowly than $III^{Glc}_{slow}$ in SDS–PAGE (Meadow and Roseman, 1982).

$III^{Glc}_{slow}$ is the major $III^{Glc}$ species in *E. coli* and *S. typhimurium* with $III^{Glc}_{fast}$ present only in trace quantities. In addition, $III^{Glc}_{slow}$ is the form active in sugar phosphorylation, at least of methyl α-glucoside. Homogeneous phospho-$III^{Glc}_{slow}$ was isolated and the following were studied (Meadow and Roseman, 1982):

$$\text{phospho-HPr} + III^{Glc}_{slow} \rightleftharpoons \text{phospho-}III^{Glc}_{slow} + \text{HPr} \tag{3}$$

$$\text{phospho-}III^{Glc}_{slow} + \begin{matrix}\text{methyl}\\ \alpha\text{ glucoside}\end{matrix} \xrightarrow{\text{Enzyme II-B}^{Glc}} \begin{matrix}\text{methyl}\\ \alpha\text{-glucoside}\\ \text{6-phosphate}\end{matrix} + III^{Glc}_{slow} \tag{4}$$

Some of the results obtained with phospho-$III^{Glc}_{slow}$ are as follows: (1) using a phospho-HPr-generating system (Enzyme I, HPr, and phosphoenolpyruvate), a maximum of 1 mole of phosphoryl group can be introduced per mole of $III^{Glc}_{slow}$ (or $III^{Glc}_{fast}$); (2) the phosphoryl group appears to be linked at the N-3 position of a histidine moiety in $III^{Glc}_{slow}$, similar to the linkage of the phosphoryl group of $III^{Lac}$ from *S. aureus* (Hays *et al.*, 1973) and Enzyme I from *S. typhimurium* (Weigel *et al.*, 1982a), but not like that of phospho-HPr where the linkage is at the N-1 position (Anderson *et al.*, 1971); (3) Reaction (3) is very rapid and demonstrably reversible with a $K'_{eq}$ of about 0.1; and (4) Reaction (4) was directly demonstrated (without Enzyme I, PEP, and HPr), which shows that phospho-$III^{Glc}_{slow}$ is a true intermediate in the sugar phosphorylation reaction.

The kinetics of Reaction (3) can be studied using a phospho-HPr-generating system (Enzyme I, HPr, and phosphoenolpyruvate) by measuring the rate of formation of pyruvate using lactate dehydrogenase (Waygood *et al.*, 1979). $III^{Glc}_{slow}$ is phosphorylated too rapidly to obtain accurate initial velocities; $III^{Glc}_{fast}$ also is an effective phosphoryl group acceptor, and may be as effective as $III^{Glc}_{slow}$. An accurate comparison of the phosphorylation of the two species of $III^{Glc}$ requires determination of the respective equilibrium constants and initial velocities for Reaction (3).

$III^{Glc}_{fast}$ is thus a good phosphoryl group acceptor, but in contrast, it is a poor phosphoryl donor (at least in Reaction (4), so poor that its $K_m$ and $V_{max}$ values could not be adequately determined. It was estimated to be about 2–3% as active as phospho-$III^{Glc}_{slow}$. These results suggest that the $NH_2$-terminal heptapeptide of phospho-$III^{Glc}_{slow}$ is required for its binding to the membrane protein Enzyme II-$B^{Glc}$. This idea can be tested with a chemically modified form of $III^{Glc}_{slow}$, derivatized with fluorescein isothiocyanate at its N-terminal amino group (Jablonski *et al.*, 1983).

The protease responsible for processing $III^{Glc}_{slow}$ to $III^{Glc}_{fast}$ is localized in the membrane fraction of crude extract of *S. typhimurium* (Meadow and Roseman, 1982, 1983), and is resistant to extraction by Triton X-100. It is active in the presence of EDTA, but is stimulated fivefold by calcium (5 mM) or magnesium (10 mM) ions. The pH optimum of the enzyme is in the range 6.6–7.0, and its activity is inhibited (50%) by 0.5 M KCl. Phospho-$III^{Glc}_{slow}$ is cleaved at the same rate as $III^{Glc}_{slow}$.

In addition to its localization in the membrane, the site where PTS-mediated regulation occurs, the proteolytic processing of $III^{Glc}_{slow}$ has the following important characteristics. (1) As mentioned above, the processing specifically alters the phosphotransfer properties of $III^{Glc}$. The processed protein, $III^{Glc}_{fast}$, remains a good acceptor of a phosphoryl group, but its ability to participate in the phosphorylation of sugar is severely reduced. (2) $III^{Glc}_{fast}$ is found in bacterial cells at low levels. (3) The protease is a highly specific, endopeptidase-like enzyme. It cleaves between the seventh and eighth amino acids of $III^{Glc}_{slow}$. Using synthetic heptapeptide (courtesy of Drs. Akira Komoriya and C. Anfinsen) as a marker for the electrophoretic separation of the products of the incubation of $III^{Glc}_{slow}$ with *S. typhimurium* membranes, Meadow and Roseman (1983) showed that the expected heptapeptide was liberated, rather than smaller peptides or amino acids. (4) Both products of the processing are stable to further incubation, $III^{Glc}_{fast}$ for at least 12 hr at 37°C in crude extract, the heptapeptide for at least 18 hr at 37°C in the presence of membranes. These properties suggest that the processing of $III^{Glc}_{slow}$ may play a role in PTS-mediated repression.

Other forms of $III^{Glc}$ may exist. Another group (Scholte *et al.*, 1981) has reported the purification of $III^{Glc}$ from *S. typhimurium*. The preparation was considered to be about 90% pure, and albumin was added to prevent loss of $III^{Glc}$ during purification. The molecular weight was also found to be about 20,000 by SDS–PAGE and two forms of $III^{Glc}$ were observed on electrophoresis under nondenaturing conditions. The authors suggest (Scholte *et al.*, 1981) that the two electrophoretically distinguishable forms result from aggregation of the monomeric form of $III^{Glc}$, although the two forms were not separated and assayed independently. It seems possible that the two forms are either closely related or identical to the two species (fast and slow) discussed above, but this interpretation has not been proven.

The $III^{Glc}$ preparation isolated by Scholte *et al.* (1981) is only slightly active in the PTS assay for methyl α-glucoside phosphorylation. When the $III^{Glc}$ is added at "saturating" concentrations, the rate of methyl α-glucoside phosphorylation increases only about twofold over that of the control (assay mixture lacking $III^{Glc}$). By contrast, the $III^{Glc}_{slow}$ isolated by Meadow and Roseman (1982) stimulates methyl α-glucoside phosphorylation about 100-fold (Figure 4). One interpretation of this apparent difference in the activity of the two preparations is that the material isolated by Scholte *et al.* (1981) consists primarily of $III^{Glc}_{fast}$. Alternatively, the problem may arise from differences between the methods of assay, and to explain this interpretation, a discussion of the assay and kinetics of PTS proteins is required.

## B. Kinetic Studies with $III^{Glc}$

The kinetics of phosphate transfer from phosphoenolpyruvate to sugar by the PTS are exceedingly complex *in vitro* and even more so *in vivo* (Schachter, 1973). For this reason, there is disagreement over some of the nomenclature, such as "Enzyme III" vs. "Factor III" or "III," and more importantly, over the conditions for their assay.

Are all of the PTS proteins enzymes or are some of them substrates? This is not a trivial question because the answer affects the design of assay conditions and the interpretation of results. In a formal sense, HPr can be considered an enzyme because it catalyzes the transfer of a phosphoryl group (from phospho-Enzyme I to III or II-A-type proteins) and in this process remains unchanged. It is more difficult to make this argument in the case of a protein such as $III^{Glc}$ because the overall reaction it catalyzes,

$$\text{phospho-HPr} + \text{sugar} \xrightarrow{\text{III}^{\text{Glc}},\ \text{Enzyme II-B}^{\text{Glc}}} \text{sugar phosphate} + \text{HPr} \qquad (5)$$

requires yet another protein, Enzyme $II\text{-}B^{Glc}$, which does not appear to be phosphorylated. In our view, however, neither HPr nor the III or II-A proteins are enzymes. Instead, they are phosphocarrier proteins and act as substrates of Enzymes I and II-B. This conclusion is based partly on analogy with the proteins of the electron transport chain, and partly on the *in vivo* concentrations and kinetics of the PTS proteins.

1. In the sense that they act as "carriers," albeit phosphocarriers, some of the PTS proteins resemble the proteins of the electron transport chain. Are cytochrome *c*; the iron sulfur proteins (and in fact ubiquinone) enzymes? An accepted definition is as follows: "The initial and final steps of electron transfer in these chains are catalyzed by discrete enzymes that can be separated from the other components and studied by conventional methods of enzymology. . . . The subsequent steps of electron transfer in the mitochondrial respiratory chain are catalyzed by electron-transfer carriers, both nonprotein (ubiquinone) and protein. Although each protein carrier, being a catalytically active protein, satisfies the most all-embracing definition of an enzyme, many do not readily fit in with the scheme of enzyme nomenclature, since they catalyze hydrogen or electron transfer from another enzyme to a third enzyme. Moreover, since much is known about the electron-carrying center of these enzymes, it is more appropriate to classify them on the basis of chemical structure of the prosthetic groups. . . ."* According to this analogy, I and $II\text{-}B^{Glc}$ are enzymes, whereas the other PTS proteins are not.

2. The properties of the PTS proteins *in vivo* are defined by their relative concentrations. HPr might be considered an enzyme, for instance, if it were less concentrated than its putative substrates, Enzyme I and $III^{Glc}$. The *in vivo* concentrations of

* From *Enzyme Nomenclature 1978,* 1979, Academic Press, New York, pp. 593–594.

proteins can be estimated from published levels in crude extract assuming that 1 mg of cellular water contains 0.25 mg of dry weight and that dry weight is 50% protein (Roberts *et al.*, 1963). Thus, the concentrations of the soluble PTS proteins in wild-type *S. typhimurium* grown on glucose are (μM): Enzyme I monomer, 6 (Weigel *et al.*, 1982c), 10 (Mattoo and Waygood, 1983); HPr, 106 (Beneski *et al.*, 1982b), 50 (Mattoo and Waygood, 1983); $III^{Glc}$, 15 (Meadow and Roseman, 1982), 50 (Nelson *et al.*, 1983). Mattoo and Waygood (1983) have reported that the molar ratio of HPr to Enzyme I monomer ranges from 7–25 in both *S. typhimurium* and *E. coli*. Based on a recent report (Erni *et al.*, 1982), we estimate that the "concentration" of Enzyme $II\text{-}B^{Glc}$ is about 4 μM, although this is a spurious calculation since Enzyme $II\text{-}B^{Glc}$ is an integral membrane protein and cannot be reported in the usual units of concentration.

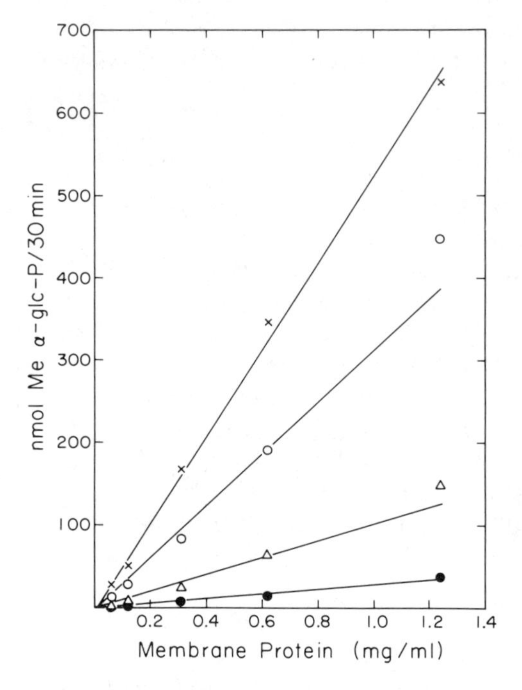


*Figure 2.* The dependence of the rate of sugar phosphorylation by the PTS on the quantity of Enzyme $II\text{-}B^{Glc}$ at various concentrations of $III^{Glc}_{slow}$. The assays were conducted as described in Waygood *et al.* (1979). Incubations of 0.1 ml volume contained Tris-HCl buffer, pH 8.0, 50 mM; potassium fluoride, 12 mM; magnesium chloride, 5 mM; phosphoenolpyruvate, potassium salt, 10 mM; methyl [U-$^{14}$C]-α-glucoside, sp. act. 219 dpm/nmol, 10 mM; Enzyme I, homogeneous, 6 units; HPr, homogeneous, 2 μM. A unit is defined as the quantity of enzyme which will catalyze the formation of 1 μmole of sugar phosphate in 30 min at 37°C. The source of Enzyme $II\text{-}B^{Glc}$ was washed membranes from *S. typhimurium* SB1687 (*ptsM*$^-$) added in quantities of from 0.006–0.124 μgm; they are plotted on the abscissa, however, as the concentration of membrane protein in mg/ml. $III^{Glc}$ was from step 2 of procedure B (Meadow and Roseman, 1982) and the preparation contained 91% $III^{Glc}_{slow}$ and 9% $III^{Glc}_{fast}$. It was used at final concentrations (of $III^{Glc}_{slow}$) of: x—x, 29 μM; ○—○ 7.2 μM; △—△, 1.8 μM; ●—●, 0.45 μM. The incubation was for 20 min at 37°C and the rates were calculated on a 30-min basis. The reaction rate was constant for at least 45 min. Me α-glc-P is methyl α-glucoside 6-phosphate.

Thus, the *in vivo* concentrations of the PTS proteins also support the view that HPr and $III^{Glc}$ are not enzymes, but substrates for Enzymes I and $II\text{-}B^{Glc}$, respectively.

3. Finally, we consider the conditions for assaying the PTS proteins *in vitro*. According to the views presented above, in a complete assay mixture containing optimal concentrations of phosphoenolpyruvate, sugar, $Mg^{2+}$, etc., the rate of sugar phosphorylation should be directly proportional to the quantity of either Enzyme I or $II\text{-}B^{Glc}$, whichever is rate limiting. A linear relationship is, in fact, obtained when the activity of Enzyme $II\text{-}B^{Glc}$ is measured in washed membranes, and the relationship is independent of the concentration of $III^{Glc}$ (Figure 2).

The rate of sugar phosphorylation vs. concentration of a given phosphocarrier protein should resemble an hyperbola if the protein behaves in typical Michaelis–Menten fashion, i.e., as a putative substrate. Indeed, this is the case for HPr (Anderson *et al.*, 1971; Waygood *et al.*, 1979), $III^{Lac}$ (Simoni *et al.*, 1973a), and $III^{Glc}$ (Meadow and Roseman, 1982; Meadow *et al.*, 1982a). This point is further emphasized in Figure 3; apparent hyperbolas are obtained when activity is plotted as a function of the concentration of $III^{Glc}$ (actually phospho-$III^{Glc}$) at any level of Enzyme $II\text{-}B^{Glc}$ (over a 20-fold concentration range).

The data in Figures 2 and 3 clearly demonstrate that: (1) the sugar phosphorylation reaction *cannot* be "saturated" with Enzyme $II\text{-}B^{Glc}$ (Figure 2) and, thus, (2) the "activity" of a fixed quantity of $III^{Glc}$ is always proportional to the quantity of Enzyme $II\text{-}B^{Glc}$ (Figure 3). From these results, it is clear that the experiments in the literature

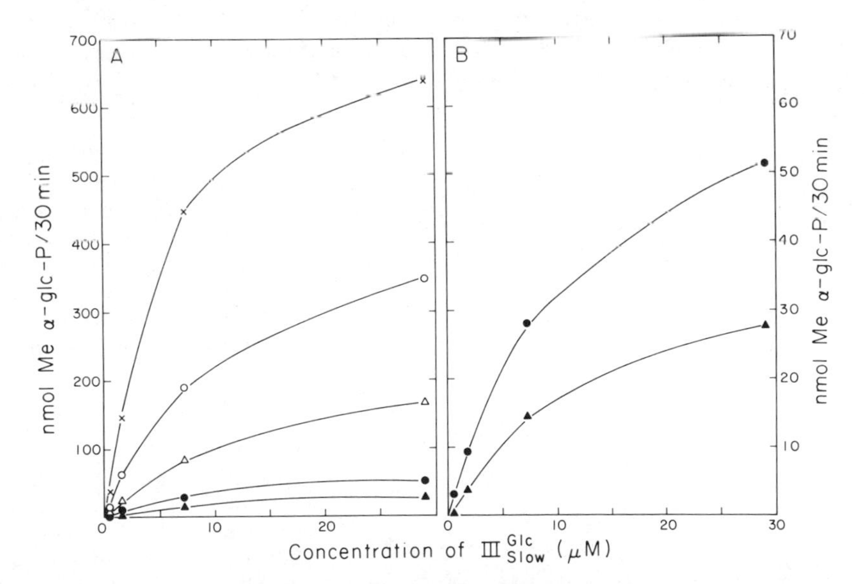


*Figure 3.* The dependence of the rate of sugar phosphorylation by the PTS on the concentration of $III^{Glc}_{slow}$ using various quantities of Enzyme $II\text{-}B^{Glc}$. (A) The data from Figure 2 replotted as a function of the concentration of $III^{Glc}_{slow}$ (from 0.45–29.0 μM). The quantities of SB1687 membranes are: x—x, 0.124 μgm; ○—○, 0.062 μgm; △—△, 0.031 μgm; ●—●, 0.012 μgm; ▲—▲, 0.006 μgm. (B) The lowest two curves from (A) replotted on a 10-fold expanded ordinate in order to better visualize the shape of the curves. Me α-glc-P is methyl α-glucoside 6-phosphate.

which give $III^{Glc}$ as *activities* rather than in units of mass cannot be reproduced, or even interpreted, unless Enzyme $II\text{-}B^{Glc}$ is used at rate-limiting levels and the precise quantity of Enzyme $II\text{-}B^{Glc}$ activity is specified. The same problems arise in the many papers which report HPr as an activity.

It is important to emphasize what we have stated previously (Waygood *et al.*, 1979), that is, since HPr and proteins such as $III^{Glc}$ are substrates, their *relative* concentrations can be measured by *in vitro* assays if their concentrations are set at or below their respective $K_m$ values. Under these conditions, even though the catalytic rate-limiting protein is either Enzyme I or $II\text{-}B^{Glc}$, the rate of sugar phosphorylation is also approximately proportional to the concentration of the phosphocarrier protein. *However, it is absolutely essential that the appropriate enzyme be used at rate-limiting levels* (Waygood *et al.*, 1979).

This point is illustrated in Figure 4. Identical concentrations of $III^{Glc}_{slow}$ were added to incubation mixtures containing several ratios of Enzyme $I{:}II\text{-}B^{Glc}$ with the quantity of $II\text{-}B^{Glc}$ held constant. When the ratio was high, with $II\text{-}B^{Glc}$ the rate-limiting enzyme, as much as 220 nmoles of methyl α-glucoside were phosphorylated in 30 min. When the ratio was low, with Enzyme I rate limiting, only 20 nmoles of methyl α-glucoside were phosphorylated. The rate of methyl α-glucoside phosphorylation thus showed a 10-fold difference, depending on whether Enzyme I or $II\text{-}B^{Glc}$ was rate limiting. Furthermore, when the same ratios of activity of Enzyme I:Enzyme $II\text{-}B^{Glc}$ were tested, but with their absolute quantities reduced 10-fold, and methyl α-glucoside at 1 mM, the same results were obtained (data not shown).

But there is an even more important effect. Whereas, there is a 30- to 50-fold difference in activity between $III^{Glc}_{slow}$ and $III^{Glc}_{fast}$, and between $III^{Glc}$ from a wild-type strain and a *crr*⁻ mutant (Meadow and Roseman, 1982; Meadow *et al.*, 1982a) when

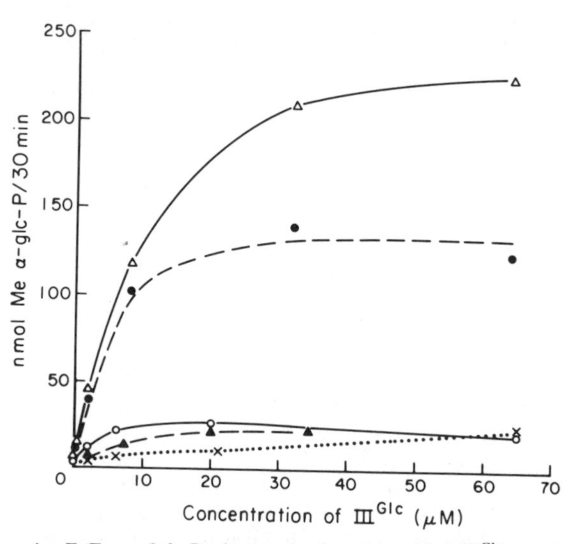


*Figure 4.* Effect of the ratio of Enzyme I to Enzyme $II\text{-}B^{Glc}$ on the rate of methyl α-glucoside phosphorylation at various concentrations of wild-type and mutant $III^{Glc}$. The assays were conducted as described in Figure 2 using homogeneous soluble PTS proteins and 37 μg (0.22 units, a unit is defined in Figure 2) of membranes from *S. typhimurium* SB1687 in *all* tubes. △—△, $III^{Glc}_{slow}$ from *S. typhimurium* LT-2 (wild type); Enzyme I, 2.2 units; ratio, $E_I/E_{II}$ = 10. ●--●, $III^{Glc}_{slow}$, as above; Enzyme I, 0.22 units; ratio, $E_I/E_{II}$ = 1. ○—○, $III^{Glc}_{slow}$, as above; Enzyme I, 0.044 units; ratio, $E_I/E_{II}$ = 0.2. ▲--▲, $III^{Glc}_{fast}$ from *S. typhimurium* LT-2; Enzyme I, 0.044 units; ratio, $E_I/E_{II}$ = 0.2. x · · · x, $III^{Glc}_{slow}$ from *S. typhimurium* SB1796 (*crr*⁻); Enzyme I, 0.044 units; ratio, $E_I/E_{II}$ = 0.2. Background values (no added $III^{Glc}$) are *not* subtracted. Note that there is little or no difference in the activities of the three species of $III^{Glc}$ (wild-type $III^{Glc}_{slow}$ and $III^{Glc}_{fast}$ and *crr*⁻ mutant $III^{Glc}_{slow}$) when Enzyme I is rate limiting.

the assays are done with Enzyme II-B$^{Glc}$ rate limiting; this large difference disappears almost entirely when Enzyme I is rate limiting (Figure 4).

Thus, when Enzyme I is rate limiting, the activities of wild-type $III^{Glc}_{slow}$, wild-type $III^{Glc}_{fast}$, and $III^{Glc}_{slow}$ from a *crr*$^-$ mutant cannot be distinguished. We believe that unfavorable conditions for assaying $III^{Glc}_{slow}$ have led to reports of relatively inactive wild-type $III^{Glc}_{slow}$ (Scholte *et al.*, 1981) with about the same activity as protein isolated from *crr*$^-$ cells (Scholte *et al.*, 1982).

## V. SUGAR RECEPTOR (II-B) PROTEINS

### A. Purification

The integral membrane proteins of the PTS, the II-B proteins, are enzymes catalyzing the following reaction:

$$\begin{matrix}\text{phospho-III} \\ \text{or phospho-II-A}\end{matrix} + \text{sugar} \xrightarrow{\text{II-B}} \begin{matrix}\text{III} \\ \text{or II-A}\end{matrix} + \text{sugar phosphate} \qquad (6)$$

Enzyme II-B$^{Man}$ was the first of these enzymes to be purified to single band on SDS–PAGE and studied. The II-B$^{Man}$, when separated from the II-A protein of the II$^{Man}$ complex and from lipid, requires phosphatidylglycerol or cardiolipin, and divalent cation for reconstitution (Kundig and Roseman, 1971). Most interestingly, the purified II-B$^{Man}$, divalent cation, lipid, and II-A had to be mixed in that specific sequence for maximum reconstitution of activity. Also, $Ca^{2+}$, which is not active with the intact PTS, was an effective cation in the reconstitution of II-B$^{Man}$. The criteria used for judging the purity of the Enzyme II-B$^{Man}$ preparations are no longer recognized as sufficiently rigorous (Chrambach and Rodbard, 1971). One of the other findings of this paper requires reexamination, that is, that there are three different II-A proteins, active with II-B$^{Man}$, each of which is specific for a different sugar. At the time of these studies, the II$^{Glc}$ complex (III$^{Glc}$/Enzyme II-B$^{Glc}$) system had not been properly defined, and it was not known that the II$^{Glc}$ and II$^{Man}$ systems had overlapping sugar specificities at high substrate concentrations (Stock *et al.*, 1982).

Erni *et al.* (1982) purified Enzyme II-B$^{Glc}$ from membranes of *S. typhimurium*, using a neutral detergent (polydisperse octyl oligooxyethylene) as the solubilizing agent. The purified enzyme catalyzes the phosphorylation of methyl α-glucoside with a specific activity 100 times higher than that of crude membranes. The enzyme requires the soluble PTS proteins (including III$^{Glc}$), and phosphatidylglycerol stimulates its activity 20-fold. The molecular weight of the purified monomer by SDS-PAGE is 40,000, but it appears to form dimers or trimers in nondenaturing detergent during gel filtration chromatography or analytical ultracentrifugation.

Enzyme II-B$^{Mtl}$ from *E. coli* has also been purified (Jacobson *et al.*, 1979, 1983). Purified enzyme is active in both phosphoenolpyruvate-dependent mannitol phosphorylation and mannitol 1-phosphate:mannitol transphosphorylation. It functions directly with phospho-HPr without a III-type intermediate (see above). The purification factor is approximately 230-fold and activity is stimulated (approximately 2.5-fold) by phospholipid and the detergent Lubrol PX. The molecular weight of II-$B^{Mtl}$, estimated by SDS–PAGE, is approximately 60,000. Crosslinking experiments failed to reveal any but very large multimers, but these experiments were considered inconclusive. The gene for the enzyme was cloned (Lee and Saier, 1983) and the DNA sequence predicted a primary translation product of 637 amino acids (molecular weight = 67,893). The criteria used by Jacobson *et al.* (1979) for judging the purity of their preparations do not meet the recommendations of Chrambach and Rodbard (1971).

## B. General Properties

Curtis and Epstein (1975), in genetic and transport experiments, recognized two different PTS systems for glucose uptake in *E. coli* cells. These observations were confirmed and extended in a number of studies: Rephaeli and Saier, 1978; Waygood *et al.*, 1979; Postma and Stock, 1980; Postma, 1981; and Scholte and Postma, 1981). The two systems ($II^{Glc}$ and $II^{Man}$) were characterized systematically by Stock *et al.* (1982), who measured the transport *(in vivo)* and phosphorylation *(in vitro)* of a number of sugars by wild-type and various *ptsG* and *ptsM* mutant strains of both *E. coli* and *S. typhimurium*. Mutants defective in the *ptsG* gene have low II-$B^{Glc}$ activity, while mutants defective in *ptsM* have low $II^{Man}$ activity. All *ptsM* mutants are thought to be of a single class and to map at one locus in both the *E. coli* and *S. typhimurium* genomes. However, recent unpublished studies by C. Shea *et al.* show that there are two classes of *ptsM* mutants, one with mutations in II-$B^{Man}$, and the other in II-$A^{Man}$ in addition, the gene which encodes for II-$B^{Man}$ has been cloned into a pBR322 plasmid.

Wild-type cells generally contain both the $II^{Glc}$ and $II^{Man}$ systems. When high concentrations of sugars are used, the substrate specificities of $II^{Glc}$ and $II^{Man}$ overlap (Stock *et al.*, 1982). Thus, assays for the two systems require carefully defined conditions. Such assay conditions have been reported (Waygood *et al.*, 1979; Stock *et al.*, 1982). Stock *et al.* (1982) also give conditions for assaying each system *in vivo* by measuring the rate of transport of appropriate analogues. At low concentrations, the effective substrates for $II^{Man}$ are glucose, mannose, and 2-deoxyglucose; whereas $II^{Glc}$ is effective only with glucose, and both methyl α- and β-glucosides. Thioglucosides are also specific substrates for II-$B^{Glc}$ (Scholte and Postma, 1981).

In the study by Stock *et al.* (1982), the kinetics of sugar uptake by whole cells were compared to the rate of sugar phosphorylation by membranes prepared from the same batch of cells. In general, the *in vivo* and *in vitro* results agreed, except that the $K_m$ for methyl α-glucoside transport (170 μM) was much higher than the $K_m$ for its phosphorylation (6 μM). The $V_{max}$ of transport and phosphorylation of methyl α-glucoside agreed, but there was an apparent discrepancy between corresponding $V_{max}$ values for the transport and phosphorylation of 2-deoxyglucose which were 400 and 72 μmoles per gram cells, dry weight, per min, respectively. As one explanation, the authors suggested that the II-$A^{Man}$ and II-$B^{Man}$ proteins may have been partially separated by fragmentation of the membranes during rupture of the cells. However, to repeat, PTS kinetics are extremely complex, and there is no reason *a priori* for kinetic results obtained *in vitro* and *in vivo* to agree (Schachter, 1973).

In an earlier kinetic study, Rephaeli and Saier (1978) determined $K_m$ values for glucose, mannose, and methyl α-glucoside, and these compare well to those reported by Stock *et al.* (1982). Later, Rephaeli and Saier (1980) measured the inhibition by various sugars of [$^{14}$C]mannose uptake by cells of a *ptsG* mutant of *S. typhimurium*. They reported relative substrate affinities for the Enzyme II$^{Man}$ in decreasing order: glucose, mannose, 2-deoxyglucose, *N*-acetylglucosamine, glucosamine, *N*-acetyl mannosamine, mannosamine, and fructose.

Are the II-B proteins themselves phosphorylated when they catalyze Reaction 6? The evidence is conflicting. Attempts to show the formation of labeled phospho-II-B$^{Man}$ have failed (Kundig and Roseman, unpublished results). Of course, this could have been due to gross instability of the putative phosphoprotein.

On the basis of kinetic analyses, Rose and Fox (1971) concluded that, in *E. coli* Enzyme II for β-glucosides functions *via* a "ping-pong" mechanism and presumably is phosphorylated. In contrast, in *S. aureus,* Enzyme II-B$^{Lac}$ functions *via* a sequential mechanism and, presumably is not phosphorylated (Hays *et al.*, 1973; Schachter, 1973). Recently, Begley *et al.* (1982) investigated the stereochemical course of the PTS sugar phosphorylation reaction *in vitro*. Using an incubation mixture containing chiral phosphoenolpyruvate, dialyzed crude extract from a strain of *S. typhimurium* that overproduces Enzyme I, HPr, and III$^{Glc}$, and unwashed membranes from a strain of *E. coli* mutated in Enzyme II$^{Man}$ (ML308, *manA* 58), they showed that the phosphoryl group is transferred an odd number of times during formation of methyl α-glucoside 6-phosphate. Since there are three phosphoryl transfers between phosphoenolpyruvate and phospho-III$^{Glc}$ (Weigel *et al.*, 1982a; Meadow and Roseman, 1982), Begley *et al.* (1982) conclude that there must be two additional transfer reactions between phospho-III$^{Glc}$ and sugar phosphate, and therefore, that the Enzyme II-B$^{Glc}$ is itself phosphorylated. These results must be interpreted with caution, however. The reaction conditions used by Begley *et al.* (1982) require a 5-hr incubation at 37°C in what is essentially crude extract in order to get sufficient product to analyze. If any other reactions of the phosphoryl group occur during this time and involve an odd number of phosphoryl transfers, their conclusions are compromised. Since their results showed clearly that 100% of the phosphoryl group in the product (methyl α-glucoside 6-phosphate) was transferred an odd number of times, any postulated side reaction must occur quantitatively.

In the past, the question of the tightness of the coupling between translocation and phosphorylation by the PTS has given rise to some controversy. Do Enzyme II complexes catalyze translocation in the absence of phosphorylation, that is, carry out facilitated diffusion (Postma and Roseman, 1976)? Obligatory coupling was reported by Simoni and Roseman (1973) and Saier *et al.* (1973) whereas translocation in the absence of phosphorylation was observed by Gachelin (1970), Kornberg and Riordan (1976), Postma (1976), and Solomon *et al.* (1973). Interpretation of the results is complicated by residual Enzyme I activity in the PTS mutants used for the studies, and also by uptake of PTS sugars at low rates by non-PTS permeases with overlapping specificities. For instance, several substrates of Enzyme II$^{Man}$ (glucose, mannose, and 2-deoxyglucose) are also substrates of the galactose permeases (Postma, 1976).

Postma and Stock (1980) approached this problem by using strains of *S. typhimurium* which carried deletions for the soluble PTS proteins and mutations in various non-PTS permeases, such as galactose permease. They concluded that the PTS does not catalyze a significant uptake of sugars in the absence of phosphorylation. (Uptake through $II^{Man}$ and $II^{Glc}$ were less than 0.03% and 0.2% of that of wild-type strains, respectively.) Postma (1981) has isolated mutant strains of *S. typhimurium* in which transport and phosphorylation by the $II^{Glc}$ complex are uncoupled. Genetic and biochemical analysis of these strains suggests that the mutation is in, or at least very close to, the *ptsG* gene. Phenotypically, this new mutation allows growth on glucose by strains carrying deletions of both the *pts* operon and the galactose permease. At high concentrations of glucose (1%), the growth rate of the mutant cells is comparable to that of wild-type cells. The apparent affinity for the uptake of glucose is approximately 1000 times lower for mutant cells than for wild-type cells. Membranes from the mutant cells cannot phosphorylate methyl α-glucoside. The molecular lesion caused by the mutation is unknown.

## C. Organization of the Enzymes II-B in Membranes

In studies with membrane vesicles of *S. typhimurium,* Beneski *et al.* (1982a) showed that the $II\text{-}B^{Glc}$ protein and the $II^{Man}$ complex were oriented quite differently in the membrane. $II\text{-}B^{Glc}$ recognizes only intravesicular phospho-$III^{Glc}$, and uses this substrate to take up methyl α-glucoside from the outside. In sharp contrast, the $II^{Man}$ complex phosphorylates 2-deoxyglucose when phospho-HPr is either intra- or extravesicular. From these results, the authors suggest that $II\text{-}B^{Glc}$ is oriented asymmetrically in the membrane, while the $II^{Man}$ complex is symmetric.

As noted above, the $II\text{-}B^{Mtl}$ gene has been cloned and sequenced (Lee and Saier, 1983). Computer analysis of this primary product predicts that the amino terminal half of $II\text{-}B^{Mtl}$ is intramembranal while the carboxyl terminal half has the properties of a soluble protein.

Evidence for multiple functional domains on Enzyme $II\text{-}B^{Mtl}$ was obtained by analysis of a large number of mutations in the protein (Leonard and Saier, 1981). The mutants were assayed for several Enzyme II-mediated functions: fermentation, chemotaxis, transport, phosphoenolpyruvate-dependent mannitol phosphorylation, and mannitol:mannitol 1-phosphate-dependent transphosphorylation. The mutants were grouped into four classes, each of which exhibited rather specific loss of some of the functions but not others.

## D. Exchange Transphosphorylation

In a previous review (Postma and Roseman, 1976), "exchange transphosphorylation" was defined as a vectorial process where:

$$\text{sugar}_{1\,(out)} + \text{sugar}_2\text{ phosphate}_{(in)} \rightleftharpoons \text{sugar}_{2\,(out)} + \text{sugar}_1\text{ phosphate}_{(in)} \quad (7)$$

That this process might occur was first suggested by results reported by Egan and Morse (1966), prior to characterization of the PTS in *S. aureus*. They showed that *S. aureus* cells accumulated a variety of sugars, both mono- and disaccharides, as unknown derivatives (later shown to be phosphate esters). When the cells were preloaded with one sugar, which accumulated as sugar phosphate, and a second sugar was added, the first sugar was expelled from the cell while the second was taken up. Subsequently, Stock showed (see Postma and Roseman, 1976) that, in *S. typhimurium*, methyl α-glucoside outside of the cell does not exchange with free methyl α-glucoside inside, but only with the phosphate ester as shown in Reaction (7). As noted in Postma and Roseman (1976), the PTS could be the mediator of such an exchange, and it might proceed with only Enzyme II-B or require other components of the PTS.

To our knowledge, Rose and Fox (1971) conducted the first direct test of exchange transphosphorylation. Using membranes containing the Enzyme $II^{Bgl}$ complex (for β-glucosides), they reported that the membranes did not catalyze transphosphorylation between β-glucosides in the presence or absence of HPr.

Other results have been obtained by several laboratories. Saier and Newman (1976) found that the Enzymes $II^{Mtl}$ in membrane preparations from *S. typhimurium* and *Spirochaeta aurantia* catalyze a transfer of the phosphoryl group from mannitol 1-phosphate to mannitol. This report was soon followed by a series of papers on the mannitol system in *S. aurantia* (Saier *et al.*, 1977a), the $II^{Glc}$ system in *E. coli* (Saier *et al.*, 1977b), the $II^{Man}$ system in *S. typhimurium* (Rephaeli and Saier, 1978, 1980), the $II^{Glc}$, $II^{Man}$, and $II^{Mtl}$ systems in *E. coli* and *S. typhimurium* and the $II^{Lac}$ systems in *S. aureus* (Saier *et al.*, 1977c), the fructose-specific system in *Bacillus subtilis* (Perret and Gay, 1979), and the glucose-specific system in *Streptococcus faecalis* (Hüdig and Hengstenberg, 1980).

The basic observations of these reports are as follows: (1) Saier and co-workers, using appropriate mutant strains, concluded that only the Enzymes II-B, as defined in this review, are necessary to catalyze transphosphorylation. For instance, in a strain of *S. typhimurium* deleted for Enzyme I, HPr, and $III^{Glc}$, the membranes were more active than membranes from wild-type cells in transferring the phosphoryl group from glucose 6-phosphate to methyl α-glucoside. A *ptsG* mutant, on the other hand, did not carry out this reaction. (2) Under optimum conditions, the transphosphorylation reaction of a particular Enzyme II-B is much slower than its corresponding phosphoenolpyruvate-dependent phosphorylation reaction. The rate of transphosphorylation was about 1–2% of the rate of phosphoenolpyruvate-dependent phosphorylation when tested with membranes from *S. typhimurium* using the following combinations: methyl α-glucoside/glucose 6-phosphate, mannose/glucose 6-phosphate, mannitol/mannitol 1-phosphate, sorbitol/sorbitol 6-phosphate (Saier *et al.*, 1977c). Membranes from *S. aureus* gave a similarly low rate when assayed for transphosphorylation between galactose 6-phosphate and thiomethyl β-galactoside. Relative rates may be even less than 1–2% since at least some of the phosphoenolpyruvate-dependent phosphorylations were conducted at less than optimal sugar concentrations. (3) Transphosphorylation reactions require concentrations of sugar phosphate greater than 10 mM, and concentrations of free sugar below 0.1 mM. Higher concentrations of the free sugar substrate inhibit the reaction. (4) When purified, $II\text{-}B^{Mtl}$ is reconstituted into phospholipid vesicles, it also catalyzes exchange transphosphorylation (Leonard and Saier, 1983).

Kinetic studies on transphosphorylation yield different interpretations. Rephaeli and Saier (1978, 1980) found a Bi–Bi sequential mechanism, and thus state that a phosphorylated enzyme intermediate is not involved. Transphosphorylation in membranes from *B. subtilis* (Perret and Gay, 1979) and *S. faecalis* (Hüdig and Hengstenberg, 1980) were found to have a ping-pong mechanism, which involves a phosphorylated enzyme intermediate. These conflicting results are discussed by Rephaeli and Saier (1980).

Saier and Schmidt (1981), using *E. coli* membrane vesicles, report a transphosphorylation reaction that is nonvectorial. Robillard and Lageveen (1982) claim, however, that this reaction does not occur and that the observations of Saier and Schmidt (1981) were due to leaky vesicles. The vectorial exchange transphosphorylation reaction, if it occurs in whole cells and vesicles, might regulate the PTS since the exchange does not result in net uptake of sugar phosphate. This point is further considered in the section on regulation.

## VI. REGULATION OF THE PTS

### A. Introduction

Even the most cursory examination indicates that the PTS must be highly regulated. The primary reaction it catalyzes, Reaction (1), is the coupled uptake and phosphorylation of PTS sugars. The apparent equilibrium constant for this reaction is about $10^8$ (Weigel *et al.*, 1982a). At thermodynamic equilibrium, therefore, if the sugar concentration outside of the cell is 1 μM, the concentration of sugar phosphate within the cell would be 100 M! In fact, with metabolizable sugars, there is initially a rapid uptake of sugar followed by a rapid decline in rate. An example of such kinetics is found in *Streptococcus lactis* cells (Mason *et al.*, 1981). When glucose is added to starved cells, uptake is very rapid for 1.0–1.5 min, then practically stops for several minutes and then resumes at a slow but steady rate. The report by Mason *et al.* (1981), along with unpublished material (kindly provided by Dr. Waggoner) is considered by Weigel *et al.* (1982a).

When nonmetabolizable analogues of the PTS sugars are added to whole cells, again one sees a burst of uptake followed by a rapid decline in the rate until a steady state is attained. These kinetics are illustrated by the uptake of thiomethyl β-galactoside through the lactose PTS of *S. aureus* (Simoni and Roseman, 1973). Here, the steady-state level of intracellular sugar phosphate is seven orders of magnitude below the value estimated from the equilibrium constant, and barely above the external concentration of free sugar.

It is possible that the cell has an initial store of PEP which is rapidly depleted on addition of sugar. This is not the mechanism for PTS regulation, however, as has been shown by Mason *et al.* (1981) who measured phosphoenolpyruvate and its precursors in intact *S. lactis* cells (see also Weigel *et al.*, 1982a). In addition, when *S. typhimurium* membrane vesicles are supplied with phosphoenolpyruvate, biphasic uptake kinetics are observed. Uptake of PEP is much more rapid than sugar uptake by these vesicles, and the intravesicular pool of phosphoenolpyruvate is maintained at a high level throughout the experiment (Liu and Roseman, 1983a,b).

The suggested mechanisms for regulation of the PTS are largely speculative, but the problem is sufficiently important to be considered in detail in the following sections.

## B. Regulation via Enzyme I

Since Enzyme I is the first protein in the PTS pathway, it is an excellent candidate as a regulator, and two possible mechanisms are considered by Weigel *et al.* (1982a).

First, association–dissociation of Enzyme I, discussed earlier, could regulate the PTS. If, as we originally proposed (Waygood *et al.*, 1977), Enzyme I monomer is inactive, several regulatory schemes are plausible. The balance of catalytically active and inactive species of Enzyme I, during uptake and at the steady state, would, in turn, be regulated by metabolites or other proteins.

Second, another mechanism for regulation *via* Enzyme I concerns the absolute concentrations and ratios of intercellular phosphoenolpyruvate and pyruvate. From the apparent $K'_{eq}$ for the phosphorylation of Enzyme I by phosphoenolpyruvate, the ratio of phospho-I to free Enzyme I would necessarily be close to the ratio of phosphoenolpyruvate to pyruvate. This potential mechanism is considered in detail by Weigel *et al.* (1982a).

## C. Regulation via Acetate Kinase

Fox and Roseman (1983) suggest that acetate kinase may play an important role in regulating the PTS. This enzyme may link the PTS to the Krebs cycle, and thus to the concentration of key metabolites such as acetyl-SCoA, ATP and ADP, GTP and GDP, etc. Historically, acetate kinase is an important enzyme because its discovery led to the discovery of Coenzyme Å (Lipmann, 1944). The enzyme catalyzes the following reaction:

$$\text{ATP (or GTP)} + \text{acetate} \rightleftharpoons \text{acetyl phosphate} + \text{ADP (or GDP)} \qquad (8)$$

The equilibrium constant for this reaction is 5–6 $\times$ $10^{-3}$ M. In other words, the reaction proceeds primarily to the left. Nevertheless, in bacterial cells, acetate kinase is thought to be the first enzyme in the pathway of acetate metabolism. The literature on the enzyme is extensive (reviewed in Fox, 1983). The points most relevant to the present discussion are: (1) when acetate kinase is incubated with ATP (or GTP) or acetyl-phosphate, a phosphoenzyme is formed, and the phosphoryl group is linked to a carboxyl group in the protein, i.e., the phosphoprotein is an acyl phosphate, (2) the phosphoprotein can donate its phosphoryl group to acetate or to ADP, (3) whether or not the phosphoenzyme is an obligatory intermediate in the above reaction is highly controversial, and (4) previous preparations of acetate kinase from *E. coli* were not greater than about 20% pure.

The enzyme has now been obtained in homogeneous form (Fox and Roseman, 1983). This preparation accepts a phosphoryl group from ATP, and the phosphoprotein donates the phosphoryl group to both acetate and ADP, as described with the cruder preparations. The phosphoprotein also has the properties of an acyl phosphate.

Acetate kinase interacts with Enzyme I of the PTS, a conclusion based on the following results: (1) ATP or GTP and acetate kinase substitute for phosphoenolpyruvate in the phosphorylation of glucose by a complete PTS assay mixture. The rate of the reaction is about 20% of the rate observed when phosphoenolpyruvate was used.

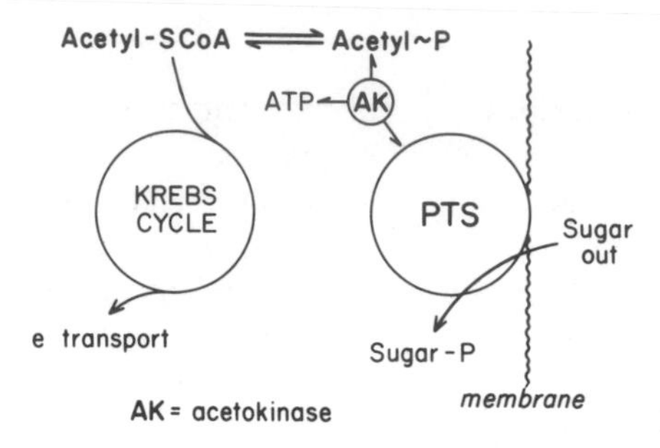


*Figure 5.* A model for the interaction among the PTS proteins, acetokinase, and the Krebs cycle.

(2) Homogeneously pure proteins catalyze partial reactions, for instance, ATP phosphorylates $III^{Glc}$ in the presence of pure acetate kinase, Enzyme I, and HPr. In addition, these phosphotransfers are reversible; phospho-$III^{Glc}$ phosphorylates acetate kinase in the presence of Enzyme I and HPr. (3) Finally, phospho-Enzyme I directly transfers its phosphoryl group to acetate kinase.

The significance of these observations is summarized in Figure 5. Since the level of phosphoacetate kinase is determined by the relative levels of ATP and ADP, GTP and GDP, and by the concentration of acetate, acetyl-SCoA, etc., acetate kinase may link sugar transport by the PTS with a key metabolic cycle, the Krebs cycle.

## D. Regulation of the Activity of the Enzymes II

Some Enzymes II of the PTS are sensitive to sulfhydryl reagents, but only under specific conditions. Haguenauer-Tsapis and Kepes (1973, 1977a) showed that methyl α-glucoside transport in *E. coli* is inhibited by *N*-ethylmaleimide; inactivation is greatly enhanced by substrate or by inhibitors of the synthesis of phosphoenolpyruvate. When the PTS from *N*-ethylmaleimide-treated cells was fractionated into its soluble and membrane-bound components, the Enzyme II-$B^{Glc}$ was found to be inactivated. The rate of inactivation was directly proportional to the rate of inhibition of sugar transport by the PTS (Haguenauer-Tsapis and Kepes, 1977a).

Even though Enzyme I is known to have sensitive sulfhydryl groups (Weigel *et al.*, 1982c), it is inactivated only slightly under the conditions used by Haguenauer-Tsapis and Kepes (1973). The function of Enzyme I, however, is important; its loss, in a temperature-sensitive mutant (Bourd *et al.*, 1971), leads to enhanced inactivation by *N*-ethylmaleimide (Haguenauer-Tsapis and Kepes, 1977b). Thus, the inactivation by Enzyme II-$B^{Glc}$ is prevented, not by the synthesis of phosphoenolpyruvate *per se,* but by transfer of the phosphoryl group from phosphoenolpyruvate to proteins of the PTS. Haguenauer-Tsapis and Kepes (1977b) postulate that unmasking of essential thiol groups in Enzyme II-$B^{Glc}$ is regulated by passage of the enzyme between "energized and deenergized" states; these might be phosphorylated and dephosphorylated states of the enzyme. (One might also suggest that the same result could be obtained by phosphorylation and dephosphorylation of $III^{Glc}$ bound to Enzyme II-$B^{Glc}$.)

Extending these studies, Haguenauer-Tsapis and Kepes (1980) obtained information about the symmetry of insertion of Enzyme II in the membrane. They found that *N*-ethylmaleimide inactivated both Enzyme II-B$^{Glc}$ and Enzyme II$^{Bgl}$ in a substrate-dependent manner. The sulfhydryl reagents *p*-chloromercuribenzoic acid and *p*-mercuriphenylsulfonic acid, however, inhibited only the activity of Enzyme II$^{Bgl}$. Enzyme II-B$^{Glc}$ was resistant to the two mercurials, unless the permeability barrier of the cells was destroyed by toluene treatment. Thus, the essential thiols of Enzyme II$^{Bgl}$ appear to be oriented outward, while those of Enzyme II-B$^{Glc}$ are inward. Both proteins obviously have an asymmetric orientation in the membrane and they do not appear to undergo a flip-flop when they function. Other evidence for an asymmetric orientation of Enzyme II-B$^{Glc}$ is mentioned above.

Regulation of the activity of the PTS by the state of cellular metabolism has been known for some time (Hagihira *et al.*, 1963; Hoffee *et al.*, 1964). Del Campo *et al.* (1975) found that the uptake of methyl α-glucoside by intact cells of *E. coli* was inhibited by such substrates as succinate which stimulate cellular respiration. They conclude that inhibition is mediated by an "energy rich state of the membrane." This effect was confirmed by Reider *et al.* (1979) who demonstrated it in both *E. coli* membrane vesicles and cells. In aerobically grown cells, carbonylcyanide *m*-chlorophenylhydrazone, dinitrophenol, or cyanide (which uncouple oxidative phosphorylation and dissipate the energized state of the membrane) all stimulate uptake of methyl α-glucoside. In membrane vesicles, respiration of lactate or ascorbate inhibits uptake of methyl α-glucoside, and cyanide reverses the inhibition. Reider *et al.* (1979) also conclude that membrane energization affects the activity of the PTS.

Robillard and Konings (1981) studied the effects of oxidative metabolism on both solubilized membrane fragments and inverted vesicles prepared from *E. coli*. In inverted vesicles, the Enzymes II are accessible to soluble PTS protein and substrates added to the medium. Robillard and Konings (1981) found that the rate of methyl α-glucoside phosphorylation is greatly inhibited by oxidizable substrates (ascorbate-phenazine methosulfate, NADH, succinate), but only when the methyl α-glucoside is at a low concentration (approximately 5 μM). Further, they found inhibition of PTS activity not only by oxidizable substrates, but by a variety of oxidants with redox potentials of $-208$ mV or higher. The oxidant they tested most extensively was ferricyanide, which increases the $K_m$ for methyl α-glucoside phosphorylation by at least 100-fold, with no change in $V_{max}$. The fact that the inhibition is a $K_m$ effect explains why it is observed only at low concentrations of methyl α-glucoside. Since the ferricyanide effect protects Enzyme II-B$^{Glc}$ against inhibition by *N*-ethylmaleimide, Robillard and Konings (1981) conclude that the oxidation effect is, in fact, an –SH to S–S conversion. From other results, they also conclude that ferricyanide may protect the same –SH group that methyl α-glucoside sensitizes to *N*-ethylmaleimide (Haguenauer-Tsapis and Kepes, 1973, 1977a).

Stock *et al.* (1982) measured the effects of uncouplers and oxidizable substrates on both the initial rates and steady-state levels of methyl α-glucoside in cells of *S. typhimurium*. They found that the initial rate of uptake is stimulated by the uncouplers carbonylcyanide *m*-chlorophenylhydrazone, dinitrophenol, and cyanide. The effects of oxidizable substrates, however, are variable. Cells were tested only with the substrate that had been the carbon source during growth. Although both glycerol and lactate decrease the rate of uptake by 40% and 25%, respectively, the effects are smaller than those reported by others. Furthermore, melibiose causes a 1.4-fold increase in the initial rate of uptake. Scholte and Postma (1981) also found that a metabolizable sugar such as galactose had no effect on the rate of methyl α-glucoside uptake by cells of *S. typhimurium*. Stock *et al.* (1982) explained these results as effects on the phosphoenolpyruvate levels of the cells. But, because it was under experimental control, variation in the concentration of phosphoenolpyruvate does not explain the results of Robillard and Konings (1981) and perhaps Reider *et al.* (1979).

## E. Regulation of Methyl α-Glucoside Transport in Membrane Vesicles

One potential mechanism for regulating the PTS was alluded to earlier, i.e., exchange transphosphorylation, Reaction (7). $Sugar_1$ may or may not be the same as $sugar_2$. If they are different, then a different sugar phosphate accumulates inside the cell, but with no change in the total intracellular concentration of sugar phosphate. If $sugar_1$ and $sugar_2$ are the same, then there is no change at all, qualitatively or quantitatively, in the internal sugar phosphate. Nevertheless, exchange transphosphorylation may control the PTS by competing with the phosphoenolpyruvate-driven reaction for the relevant Enzyme II-B, which is the rate-limiting protein in the sugar transport process (Beneski *et al.*, 1982a). In the limiting case, the transphosphorylation reaction would stop sugar uptake entirely. This is an attractive idea, since it predicts that the product of the transport process, sugar phosphate, gradually inhibits and finally stops accumulation of more sugar phosphate.

There are several difficulties with this simple idea. As already mentioned, transphosphorylation, at least *in vitro,* functions at a fraction of the rate of the phosphoenolpyruvate-driven reaction. In addition, transphosphorylation requires very high concentrations of sugar phosphate and very low concentrations of free sugar. As noted earlier, however, the PTS is regulated within the first minutes of uptake, long before the extracellular sugar concentration is reduced, and before internal sugar phosphate rises to the high concentrations required for significant transphosphorylation.

The hypothesis that exchange transphosphorylation is responsible for shutting down net sugar uptake has been tested directly (Liu and Roseman, 1983b). Membrane vesicles isolated from *S. typhimurium* take up methyl α-glucoside exclusively as the sugar phosphate, and this reaction requires exogenous phosphoenolpyruvate. A careful kinetic study of the uptake process shows that an initial rapid uptake rate, designated $v_1$, persists for about 6 sec, followed by a slower rate, $v_2$. Toluene-treated vesicles, which are permeable to small molecules, show only the $v_1$ rate, which persists for the time course of the experiment (1.2 min). In the intact vesicles, the occurrence of the slower rate, $v_2$, is not explained by limiting quantities or rates of transport of phosphoenolpyruvate, or by inhibition by pyruvate and its catabolites. In toluene-treated vesicles, sugar phosphate did inhibit the rate of sugar phosphorylation, but only at very high concentrations, much higher than the 1–2 mM methyl α-glucoside 6-phosphate accumulated in intact vesicles at the transition from $v_1$ to $v_2$.

To test the hypothesis that regulation of the PTS (the change from $v_1$ to $v_2$) is effected by exchange transphosphorylation, the vesicles were preloaded with $^{3}$H-labeled methyl α-glucoside 6-phosphate, and were then exposed to $^{14}$C-labeled methyl α-glucoside. The uptake of $^{14}C$ was followed, along with the concentration of $^{3}H$ in the vesicles. Exchange transphosphorylation requires loss of $^{3}H$ coupled to uptake of $^{14}C$ on an equimolar basis. In contrast, the phosphoenolpyruvate-driven reaction would result in no change in the $^{3}H$ content of the vesicles, but only in the usual continuous uptake of $^{14}C$-sugar over the time course of the incubation, 1.2 min. In fact, only the latter result was obtained, which indicates that under these conditions at least, exchange transphosphorylation does not significantly compete with net uptake. Similar experiments need to be done under other conditions, such as the steady state, before we can conclude that exchange transphosphorylation is not a significant regulatory mechanism in membrane vesicles.

In sum, we can conclude that the PTS is stringently regulated, but that the mechanism(s) for this regulation are not yet clear. We have tried to summarize current theories about this important process, and perhaps some of these will prove to be involved. We would argue that the uptake of sugar by bacterial cells is too important a phenomenon to be controlled by only one or two regulatory mechanisms, and that under different conditions of cell growth, different modes of regulation will be found to control the rate of sugar uptake *via* the PTS.

## VII. PTS REGULATION OF NON-PTS SYSTEMS: PTS-MEDIATED REPRESSION

### A. The *crr* and *iex* Genes

Observations on the "glucose effect," or diauxic growth, date back to the last century (for review, see Magasanik, 1970). As an aside, it is interesting to note that Monod's studies on diauxic growth led directly to the Jacob and Monod postulate of mRNA.

The basic observation is that when cells are placed in a medium containing two carbon sources, they will preferentially grow on one until it is depleted from the medium. At this point, the cells enter a brief stationary phase and begin to induce the synthesis of the catabolic enzymes and permeases required for utilization of the second compound. The cells then grow on the second compound.

Gershanovitch *et al.* (1967), who first proposed that the PTS plays a role in this regulation, showed that synthesis of the inducible enzyme systems that catabolize certain non-PTS compounds is repressed in Enzyme I and HPr mutants. Several laboratories confirmed this phenomenon (Fox and Wilson, 1968; Wang and Morse, 1968; Tyler and Magasanik, 1970). In a series of papers on this PTS-mediated repression (Saier *et al.*, 1976; Saier and Roseman, 1976a,b), the following conclusions were drawn: (1) PTS sugars are used in preference to non-PTS sugars. (2) The catabolism of lactose, maltose, melibiose, and glycerol is sensitive to PTS-mediated repression. (3) A major mechanism of PTS-mediated repression is inhibition of uptake of the particular non-PTS solute, an effect called inducer exclusion (Magasanik, 1970). Repression is enhanced in Enzyme I (*ptsI*) or HPr (*ptsH*) mutants, and is most severe in strains with the lowest activities of Enzyme I or HPr. (4) A functional Enzyme II complex for the repressing PTS sugar is required for PTS-mediated repression to occur, but there is no requirement for uptake or phosphorylation of appreciable quantities of the repressing PTS sugar. It has not been determined whether or not PTS-mediated repression requires uptake and phosphorylation of trace quantities of the PTS sugar. (5) PTS-mediated repression is overcome by a single mutation, called *crr* (carbohydrate repression resistance), which eliminates the PTS effect when it is introduced into Enzyme I or HPr mutants as well as into wild-type cells, i.e., *crr*⁻ mutants are more resistant to the repressive effects of PTS sugars than are wild-type cells. The *crr* locus is closely linked to the *pts* operon but is not part of it (Cordaro *et al.*, 1974; Cordaro, 1976). (6) The only biochemical defect detected in *crr*⁻ strains is a great reduction or absence of the activity of $III^{Glc}$ in crude extracts. Thus, the *crr* gene appears either to code for or to regulate the synthesis of the protein, $III^{Glc}$, and this became a central question in the study of $III^{Glc}$, and is discussed in detail below.

Cyclic AMP is a critical regulator of the induction of catabolic enzyme systems in enteric bacteria (Postma and Roseman, 1976; Postma, 1982; Scholte and Postma, 1980). It is well established that the PTS regulates the activity of adenylate cyclase (Peterkofsky and Gazdar, 1975, 1978; Harwood *et al.*, 1976; Feucht and Saier, 1980). Furthermore, such PTS sugars as glucose, when added to the growth medium, cause release of cAMP from the cells, resulting in lower intracellular cAMP levels (Makman and Sutherland, 1965).

Thus, the PTS represses the synthesis of certain non-PTS catabolic systems in two ways: it inhibits both the uptake of inducing molecules and the activity of adenylate cyclase (Saier *et al.*, 1976).

In addition to the *crr* gene, Kornberg and co-workers (Kornberg and Watts, 1979; Kornberg *et al.*, 1980; Parra *et al.*, 1983) have described another gene designated *iex* (*i*nducer *ex*clusion) which is involved in PTS-mediated regulation of the utilization of non-PTS sugars. This gene, mapped using specialized λ-transducing phages, is located close to the promoter of the *ptsH* gene (Britton *et al.*, 1983). In contrast to *crr* mutants (*tgs* or *gsr* in Kornberg's nomenclature), *iex* mutants are not deficient in the activities of either $III^{Glc}$ or adenylate cyclase. However, the repressive effects of all PTS sugars on the utilization of non-PTS sugars are abolished in *iex*⁻ strains. The product of the *iex* locus remains to be determined.

## B. *crr* Is the Structural Gene for $III^{Glc}$

Both biochemical and genetic methods were used to show that the *crr* gene codes for $III^{Glc}$:

1. $III^{Glc}$ was isolated from an *S. typhimurium crr*⁻ mutant (SB1796). As with wild-type proteins, $III^{Glc}$ from the mutant occurs in two electrophoretically distinguishable forms, $III^{Glc}_{slow}$ and $III^{Glc}_{fast}$. The studies described below pertain to the $III^{Glc}_{slow}$ protein. It is altered in its chemical and physical properties relative to those of wild-type $III^{Glc}_{slow}$ (Meadow *et al.*, 1982a,b). It tends to aggregate under conditions where wild-type $III^{Gls}_{slow}$ does not. Most importantly, its activity in the complete PTS sugar

phosphorylation assay was much lower (5–10%) than that of the wild-type protein. The reduction in activity was an effect on the $V_{max}$ of Enzyme II-B$^{Glc}$, not on the $K_m$. The fact that $III^{Glc}_{slow}$ from the *crr*⁻ mutant has any activity at all in the sugar phosphorylation assay implies that it can accept a phosphoryl group. This is the case; $III^{Glc}_{slow}$ from the SB1796 mutant accepts 1 mole of phosphate per mole of protein, as does $III^{Glc}_{slow}$ from wild-type cells. In short, $III^{Glc}_{slow}$ from the *crr*⁻ mutant is a good phosphoryl acceptor, but a poor donor.

Of the four *S. typhimurium crr*⁻ strains assayed by immunological methods (SB1796, SB1798, SB1799, and SB2026), three had relatively normal levels of III$^{Glc}$, while the fourth (SB2026) had no detectable cross-reactive material. Strain SB2026 may possess a nonsense mutation close to the start of the gene. The conclusion from these studies (Meadow *et al.*, 1982a,b) is that *crr*⁻ strains of *S. typhimurium* possess an altered $III^{Glc}_{slow}$ molecule in approximately normal quantities.

2. Genetic evidence is also consistent with the interpretation that the *crr* gene is the structural gene for III$^{Glc}$. The *crr* gene from *E. coli* was cloned into a plasmid as part of a 1.3 kilobase fragment. The plasmid transformed *crr*⁻ strains to *crr*⁺ and, in addition, directed the synthesis of III$^{Glc}$, which was detected by immunological methods (Meadow *et al.*, 1982b). Sequencing of the cloned gene provides further evidence that the protein encoded by the plasmid is III$^{Glc}$ (D. Saffen, unpublished results). The first 30 amino acid residues predicted from the DNA sequence of the cloned *E. coli* gene are identical to the residues of the amino acid sequence of *S. typhimurium* $III^{Glc}_{slow}$.

3. Finally, use of the technique of $\gamma\delta$ mutagenesis of the plasmid (Guyer, 1978) demonstrates simultaneous loss of the *crr*⁺ phenotype and modification of the III$^{Glc}$ protein (D. Saffen, unpublished results). The mutagenized plasmids do not transform the *crr*⁻ phenotype and either produce no protein which crossreacts with anti-III$^{Glc}$ serum or yield peptides of molecular weight lower than III$^{Glc}$ which react with the antiserum. Thus, abolition of gene function coincides with modification of the III$^{Glc}$ protein, the classical method for demonstrating a gene–protein product relationship.

Another recent report (Scholte *et al.*, 1982) relating the *crr* gene product to III$^{Glc}$ differs from those of Meadow *et al.* (1982a,b) in two significant respects. First, using immunological methods similar to those used by Meadow *et al.* (1982a), Scholte *et al.* (1982) found that most of the *crr*⁻ mutants they tested had either little or no III$^{Glc}$ protein while two strains contained from 20–30% of wild-type levels. Second, in the complete PTS sugar phosphorylation assay, the III$^{Glc}$ isolated from these latter two *crr*⁻ strains had the same activity as wild-type III$^{Glc}$. These two findings, considered together, are more consistent with the interpretation that the *crr* gene regulates the levels of III$^{Glc}$ rather than serves as the structural gene. Scholte *et al.* (1982), however, also observed that III$^{Glc}$ from *crr*⁻ strains tends to aggregate when compared to III$^{Glc}$ from wild-type strains. On the basis of this latter difference alone, Scholte *et al.* (1982) concluded that the *crr* gene is probably the structural gene for III$^{Glc}$.

Whether mutant III$^{Glc}$ is approximately as active (Scholte *et al.*, 1982) or is far less active (Meadow *et al.*, 1982a,b) than wild-type III$^{Glc}$ is an important question. We suggest that this apparent discrepancy is an artifact of the assay conditions (Figure 4), where the normally high activity of wild-type $III^{Glc}_{slow}$ is not expressed.

## C. Mechanism of Regulation of Non-PTS Transport Systems

Early quantitative measurements of repression of the lactose operon by glucose, i.e., the glucose effect, showed that the degree of repression depended on when glucose was added to the cells, which were growing under inducing conditions (Magasanik, 1970). As stated by Magasanik (1970), "The fact that preinduction can partly overcome the effect of glucose suggested . . . that the acquisition of the β-galactoside permease makes it possible for the cells to become induced in the presence of glucose."

Relief from repression by prior induction has recently been extended from glucose to PTS sugars in general and from the lactose operon to other non-PTS operons, such as glycerol, maltose, and melibiose (Nelson *et al.*, 1982; Saier *et al.*, 1982; Mitchell *et al.*, 1982). Saier *et al.* (1982) call the phenomenon "desensitization" of the non-PTS permeases and adenylate cyclase, and show that including cyclic AMP in the growth medium leads to such "desensitization." This effect of cyclic AMP was extensively studied by Pastan and Perlman (1970), who came to essentially the same conclusions. While Saier *et al.* (1982) did not directly measure PTS and non-PTS proteins, but only the activity of transport systems in whole cells, their results are consistent with those described below.

Nelson *et al.* (1982) used various techniques to alter the levels of the glycerol and maltose permeases in *S. typhimurium*. Under their conditions, the transport of 2-deoxyglucose via the $II^{Man}$ system, and $III^{Glc}$ (measured immunologically), varied only slightly. As the non-PTS permeases increased in the cell, the capacity of PTS sugars to inhibit non-PTS permeases decreased. Using *crr* mutants, in which regulation is essentially lost, and assuming that *crr* codes for $III^{Glc}$ (which had not yet been established), Nelson *et al.* (1982) conclude that the PTS protein that regulates non-PTS permeases is $III^{Glc}$.

In the third study along these lines, Mitchell *et al.* (1982) investigated the lactose and melibiose systems in *S. typhimurium* into which the lactose operon had been inserted on an episome. Three strains were used, one containing wild-type PTS proteins, one lacking Enzyme I, and one lacking Enzyme I, HPr, and $III^{Glc}$ (a deletion). The PTS and non-PTS proteins were quantitated under various conditions of growth and induction, and the following conclusions were drawn: (1) methyl α-glucoside inhibits influx, efflux, and exchange of galactosides via the permease. Inhibition does not require either the generation of metabolic energy or energy coupling to the permeases (see also Winkler and Wilson, 1967; Boniface and Koch, 1967). In other words, the PTS inhibits the non-PTS permeases directly. It should be emphasized that inhibition does not involve competition for the permeases between PTS and non-PTS sugars (see also Winkler and Wilson, 1967; Boniface and Koch, 1967); functional $II\text{-}B^{Glc}$ is required. (2) The degree of inhibition depends directly on the *ratio* of PTS proteins to the permeases. This is consistent with the idea that one or more PTS proteins interact stoichiometrically with the permease. (3) Finally, the manner in which inducer exclusion is expressed depends on the routes available for the non-PTS sugar to exit from the cell.

The view held currently by most workers is that regulation of the non-PTS permeases (and very likely adenylate cyclase) is effected through the PTS protein

$III^{Glc}$. This idea stems from the work discussed above, namely, that PTS-mediated regulation is controlled by the *crr* gene product, and that the gene product is $III^{Glc}$. What is the mechanism for this regulation? We proposed the mechanism shown in Figure 6, in which there is a stoichiometric interaction between $III^{Glc}$ and the permeases, leading to at least partial inhibition of the permeases. Phospho-$III^{Glc}$, in contrast, is either inactive, or is a positive effector of the permeases (Roseman, 1972, 1977; Postma and Roseman, 1976). Thus, critical factors in regulation could be the ratio of phospho-$III^{Glc}$ to $III^{Glc}$, or their absolute levels, if they compete for the permeases. The idea has been tested in a number of whole-cell experiments (Saier and Feucht, 1980; Scholte and Postma, 1981; Postma and Roseman, 1976; Saier, 1977; Postma, 1982). In each case, the authors conclude that their evidence is indeed consistent with the hypothesis.

More direct evidence has recently been published. Membrane vesicles were prepared from *E. coli,* and soluble PTS proteins were shocked into them (Dills *et al.*, 1982). Uptake of lactose was partially inhibited in vesicles containing partially purified $III^{Glc}$ but not HPr or Enzyme I; this inhibition was prevented by adding phosphoenolpyruvate to the vesicles. Similar but not identical results were obtained by Misko (1983). Vesicles were prepared from a strain of *S. typhimurium* deleted for *ptsH, ptsI,* and *crr,* but which contained the lactose permease (Mitchell *et al.*, 1982). Homogeneous $III^{Glc}_{slow}$ or $III^{Glc}_{fast}$, with or without homogeneous Enzyme I and HPr, was incorporated into the vesicles during preparation (Beneski *et al.*, 1982a). $III^{Glc}_{slow}$ was found to partially inhibit the lactose permease while $III^{Glc}_{fast}$ had less of an effect. When Enzyme I was incorporated into the vesicles, in the presence or absence of HPr, the results were not easily interpretable. The model shown in Figure 6 may be too simple, and may require modification.

Evidence directly in accord with the model was obtained from two other sets of experiments. Osumi and Saier (1982), using a strain of *E. coli* carrying a plasmid that overproduces the lactose permease, investigated the binding of $III^{Glc}$ to membranes.

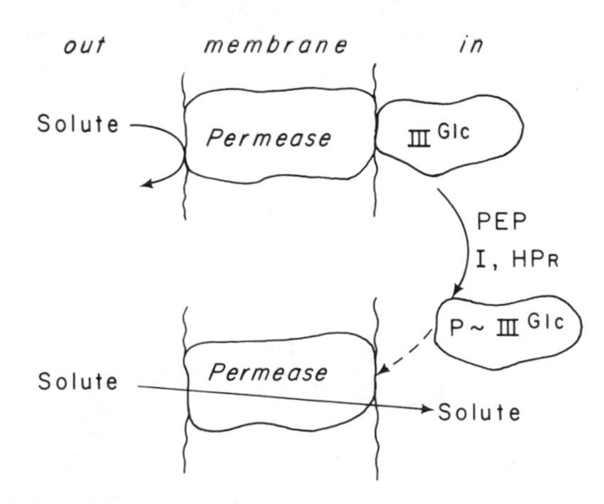


*Figure 6.* A model for the mechanism of inducer exclusion. The solute is a non-PTS compound.

The authors reported the following results: (1) in most experiments, specific binding to the membranes is about fourfold greater than nonspecific. (2) Specific binding is observed in the presence of substrates of the lactose permease, but not in their absence or in the presence of maltose. Other sugars were not tested. Binding is greater in the presence of more effective substrates of the lactose permease. (3) Membranes from the parental *E. coli* strain, which do not contain a functional *lacY* gene, show only nonspecific binding. When treated with *N*-ethylmaleimide, a reagent known to inactivate the lactose permease, membranes from the plasmid-carrying strain show only nonspecific binding of $III^{Glc}$. When membranes were protected with substrate during treatment with *N*-ethylmaleimide, they only partially lost the specific binding of $III^{Glc}$. (4) Finally, the membranes did not bind $III^{Glc}$ when the solution contained Enzyme I, HPr, and phosphoenolpyruvate. Osumi and Saier (1982) conclude that their results agree with the model shown in Figure 6.

Nelson *et al.* (1983) confirmed and extended these results. Using *E. coli,* they measured $III^{Glc}$ bound to membranes by rocket immunoelectrophoresis. They obtained essentially the same results as Osumi and Saier (1982). In addition, Nelson *et al.* (1983) found that 1.0–1.3 moles of $III^{Glc}$ bind per mole of lactose carrier, and the dissociation constant varies between 16 and 5 μM. Furthermore, when Enzyme I, HPr, and phosphoenolpyruvate are added to the system, binding does not exceed background levels. Not only is a galactoside required for $III^{Glc}$ binding to the membranes, but $III^{Glc}$ affects the binding of the galactosides, lowering the $K_d$ for one of these compounds about fourfold.

Nelson *et al.* (1983) conducted two additional kinds of experiments. One measured the effect of $III^{Glc}$ on the function of the lactose carrier in natural membrane vesicles. The methods used were counterflow (thiodigalactoside for lactose) or exchange (lactose for lactose). They report relatively small inhibitions, but in this kind of experiment, exchange and counterflow are so rapid that initial kinetics are difficult to obtain. In their second experiment, using liposomes containing homogeneous lactose carrier protein, they measured both binding of $III^{Glc}$ to the lactose carrier and the effect of $III^{Glc}$ on the function of the carrier. These experiments are, undoubtedly, the most important of any of those described because the effects of $III^{Glc}$ on natural membrane vesicles may have been indirect and not the result of a direct interaction between $III^{Glc}$ and the lactose carrier. Unfortunately, there was only a slight effect of $III^{Glc}$ on the purified lactose carrier reconstituted into liposomes.

In summary, then, all published experiments agree with the model shown in Figure 6. However, unequivocal evidence that proves or disproves this model remains to be provided. It seems likely that the model is too simple and may have to be modified.

## VIII. PROSPECTUS

This review has covered only a fraction of the recent literature on the PTS. That much has been accomplished by the combined efforts of various laboratories is obvious. That much remains to be done is equally obvious. Although Kundig *et al.* (1964), in the very first paper on the subject, recognized that the PTS was a multiprotein phosphocarrier system and Simoni *et al.* (1967) showed that it was involved in the transport of its sugar substrates, more questions have been raised than answered. How do the soluble PTS proteins interact with one another and with the II-B proteins in the membrane? Do they form a multiprotein complex with the II-B proteins, or do one or more of the soluble proteins shuttle on and off, carrying the phosphoryl group to the membrane? What precisely happens to the II-B proteins during sugar translocation? When, precisely, are the sugars phosphorylated? How is the PTS regulated? How does it regulate other transport systems and adenylate cyclase? How is the synthesis of the PTS proteins regulated?

These and many more questions remain to be answered. The growing interest in and work on the PTS should answer many of these questions in the next decade. Of course, it is equally likely that each answer will lead to other questions; that, at least, is the hope of these reviewers.

ACKNOWLEDGMENTS

We wish to thank Mr. Terrance Monmaney for his assistance in editing, Mrs. Dorothy Regula for preparation of the manuscript, and Mr. Jules Shin for the original data which appear in Section IV-B.

## REFERENCES

Anderson, B., Weigel, N., Kundig, W., and Roseman, S., 1971, Sugar transport. III. Purification and properties of a phosphocarrier protein (HPr) of the phosphoenolpyruvate-dependent phosphotransferase system of *Escherichia coli*, *J. Biol. Chem.* **246:**7023–7033.

Bachmann, B. J., 1983, Linkage map of *Escherichia coli* K-12, Edition 7, *Microbiol. Rev.* **47:**180–230.

Bayreuther, K., Raufuss, H., Schrecker, C., and Hengstenberg, H., 1977, The phosphoenolpyruvate-dependent phosphotransferase system of *Staphylococcus aureus*. I. Amino acid sequence of phosphocarrier protein HPr, *Eur. J. Biochem.* **75:**275–286.

Begley, G. S., Hansen, D. E., Jacobson, G. R., and Knowles, J. R., 1982, Stereochemical course of the reactions catalyzed by the bacterial phosphoenolpyruvate:glucose phosphotransferase system, *Biochemistry* **21:**5552–5556.

Beneski, D. A., Misko, T. P., and Roseman, S., 1982a, Sugar transport by the bacterial phosphotransferase system. Preparation and characterization of membrane vesicles from mutant and wild type *Salmonella typhimurium*, *J. Biol. Chem.* **257:**14565–14575.

Beneski, D. A., Nakazawa, A., Weigel, N., Hartman, P. E., and Roseman, S., 1982b, Sugar transport by the bacterial phosphotransferase system. Isolation and characterization of a phosphocarrier protein HPr from wild type and mutants of *Salmonella typhimurium*, *J. Biol. Chem.* **257:**14492–14498.

Boniface, J., and Koch, A. L., 1967, The interaction between permeases as a tool to find their relationship on the membrane, *Biochim. Biophys. Acta* **135:**756–770.

Bourd, G. I., Bol'shakova, T. N., Saprykina, T. P., Klyucheva, V. V., and Gershanovitch, V. N., 1971, Reduction in biosynthesis rate for RNA and protein in a thermosensitive *E. coli* K12 mutant defective in the Roseman phosphotransferase system, *Mol. Biol.* **5:**384–389.

Britton, P., Boronat, A., Hartley, D. A., Jones-Mortimer, M. C., Kornberg, H. L., and Parra, F., 1983, Phosphotransferase-mediated regulation of carbohydrate utilization in *Escherichia coli* K12. Location of the *gsr* (*tgs*) and *iex* (*crr*) genes by specialized transduction, *J. Gen. Microbiol.* **129:**349–358.

Brouwer, M., Elferink, M. G. L., and Robillard, G. T., 1982, Phosphoenolpyruvate-dependent fructose phosphotransferase system of *Rhodopseudomonas sphaeroides:* Purification and physicochemical and immunochemical characterization of a membrane-associated Enzyme I, *Biochemistry* **21:**82–88.

Chrambach, A., and Rodbard, D., 1971, Polyacrylamide gel electrophoresis, *Science* **172:**440–451.

Cordaro, C., 1976, Genetics of the bacterial phosphoenolpyruvate:glycose phosphotransferase system, *Annu. Rev. Genet.* **10:**341–359.

Cordaro, J. C., and Roseman, S., 1972, Deletion mapping of the genes coding for HPr and Enzyme I of the phosphoenolpyruvate:sugar phosphotransferase system in *Salmonella typhimurium, J. Bacteriol.* **112:**17–29.

Cordaro, J. C., Anderson, R. P., Grogan, E. W., Jr., Wenzel, D. J., Engler, M., and Roseman, S., 1974, Promoter-like mutation affecting HPr and Enzyme I of the phosphoenolpyruvate:sugar phosphotransferase system in *Salmonella typhimurium, J. Bacteriol.* **120:**245–252.

Curtis, S. J., and Epstein, W., 1975, Phosphorylation of D-glucose in *Escherichia coli* mutants defective in glucosephosphotransferase, mannosephosphotransferase, and glucokinase, *J. Bacteriol.* **122:**1189–1199.

del Campo, F. F., Hernandez-Asensio, M., and Ramirez, J. M., 1975, Transport of α-methyl glucoside in mutants of *Escherichia coli* K12 deficient in $Ca^{2+}$, $Mg^{2+}$-activated adenosine triphosphatase, *Biochem. Biophys. Res. Commun.* **63:**1099–1105.

Dills, S. S., Apperson, A., Schmidt, M. R., and Saier, M. H., Jr., 1980, Carbohydrate transport in bacteria, *Microbiol. Rev.* **44:**385–418.

Dills, S. S., Schmidt, M. R., and Saier, M. H., Jr., 1982, Regulation of lactose transport by the phosphoenolpyruvate-sugar phosphotransferase system in membrane vesicles of *Escherichia coli, J. Cell. Biochem.* **18:**239–244.

Dooijewaard, G., Roossien, F. F., and Robillard, G. T., 1979, *Escherichia coli* phosphoenolpyruvate dependent phosphotransferase system. Copurification of HPr and α1-6 glucan, *Biochemistry* **18:**2990–2996.

Egan, J. B., and Morse, M. L., 1966, Carbohydrate transport in *Staphylococcus aureus* III. Studies of the transport process, *Biochim. Biophys. Acta* **112:**63–73.

Erni, B., Trachsel, H., Postma, P. W., and Rosenbusch, J. P., 1982, Bacterial phosphotransferase system. Solubilization and purification of the glucose-specific enzyme II from membranes of *Salmonella typhimurium, J. Biol. Chem.* **257:**13726–13730.

Feucht, B. U., and Saier, M. H., Jr., 1980, Fine control of adenylate cyclase by the phosphoenolpyruvate:sugar phosphotransferase systems in *Escherichia coli* and *Salmonella typhimurium, J. Bacteriol.* **141:**603–610.

Fox, D. K., 1983, The purification and characterization of acetate kinase from *Salmonella typhimurium,* Ph.D. dissertation, The Johns Hopkins University, University Microfilms International, Ann Arbor, Michigan.

Fox, D. K., and Roseman, S., 1983, Interaction between the PEP:glycose phosphotransferase system (PTS) and acetate kinase of *Salmonella typhimurium, Fed. Proc.* **42:**1942.

Fox, C. F., and Wilson, G., 1968, The role of a phosphoenolpyruvate-dependent kinase system in β-glucoside catabolism in *Escherichia coli, Proc. Natl. Acad. Sci. USA* **59:**988–995.

Gachelin, G., 1970, Studies on the α-methylglucoside permease of *Escherichia coli.* A two step mechanism for the accumulation of α-methylglucoside 6-phosphate, *Eur. J. Biochem.* **16:**342–357.

Gershanovitch, V. N., Bourd, G. I., Jorovitzkaya, N. V., Skavronskaya, A. G., Klyucheva, V. V., and Shabolenko, V. P., 1967, β-Galactosidase induction in cells of *Escherichia coli* not utilizing glucose, *Biochim. Biophys. Acta* **134:**188–190.

Grill, H., Weigel, N., Gaffney, B. J., and Roseman, S., 1982, Sugar transport by the bacterial phosphotransferase system. Radioactive and electron paramagnetic resonance labeling of the *Salmonella typhimurium* phosphocarrier protein (HPr) at the $NH_2$-terminal methionine, *J. Biol. Chem.* **257:**14510–14517.

Guyer, M. S., 1978, The γδ sequence of F is an insertion sequence, *J. Mol. Biol.* **126:**347–365.

Hagihira, H., Wilson, T. H., and Lin, E. C. C., 1963, Studies on the glucose-transport system in *Escherichia coli* with α-methylglucoside as substrate, *Biochim. Biophys. Acta* **78:**505–515.

Haguenauer-Tsapis, R., and Kepes, A., 1973, Changes in accessibility of the membrane bound transport enzyme glucose phosphotransferase of *E. coli* to protein group reagents in presence of substrate or absence of substrate or absence of energy source, *Biochem. Biophys. Res. Commun.* **54:**1335–1341.

Haguenauer-Tsapis, R., and Kepes, A., 1977a, Unmasking of an essential thiol during function of the membrane bound enzyme II of the phosphoenolpyruvate glucose phosphotransferase system of *Escherichia coli, Biochim. Biophys. Acta* **465:**118–130.

Haguenauer-Tsapis, R., and Kepes, A., 1977b, The role of enzyme I in the unmasking of an essential thiol of the membrane-bound enzyme II of the phosphoenolpyruvate-glucose phosphotransferase system of *Escherichia coli, Biochim. Biophys. Acta* **469:**211–215.

Haguenauer-Tsapis, R., and Kepes, A., 1980, Different sidedness of functionally homologous essential thiols in two membrane-bound phosphotransferase enzymes of *Escherichia coli* detected by permeant and nonpermeant thiol reagents, *J. Biol. Chem.* **255:**5075–5081.

Harwood, J. P., Gazdar, C., Prasad, C., and Peterkofsky, A., 1976, Involvement of the glucose Enzymes II of the sugar phosphotransferase system in the regulation of adenylate cyclase by glucose in *Escherichia coli, J. Biol. Chem.* **251:**2462–2468.

Hays, J. B., Simoni, R. D., and Roseman, S., 1973, Sugar transport. V. A trimeric lactose-specific phosphocarrier protein of the *Staphylococcus aureus* phosphotransferase system, *J. Biol. Chem.* **248:**941–956.

Hildenbrand, K., Brand, L., and Roseman, S., 1982, Sugar transport by the bacterial phosphotransferase system. Nanosecond fluorescence studies of the phosphocarrier protein (HPr) labeled at the $NH_2$-terminal methionine, *J. Biol. Chem.* **257:**14518–14525.

Hoffee, P., Englesberg, E., and Lamy, F., 1964, The glucose permease system in bacteria, *Biochim. Biophys. Acta* **79:**337–350.

Hoving, H., Lolkema, J. S., and Robillard, G. T., 1981, *Escherichia coli* phosphoenolpyruvate-dependent phosphotransferase system: Equilibrium kinetics and mechanism of enzyme I phosphorylation, *Biochemistry* **20:**87–93.

Hoving, H., Koning, J. H., and Robillard, G. T., 1982, *Escherichia coli* phosphoenolpyruvate-dependent phosphotransferase system: Role of divalent metals in the dimerization and phosphorylation of enzyme I, *Biochemistry* **21:**3128–3135.

Hoving, H., Nowak, T., and Robillard, G. T., 1983, *Escherichia coli* phosphoenolpyruvate-dependent phosphotransferase system: Stereospecificity of proton transfer in the phosphorylation of enzyme I from (Z)-phosphoenolbutyrate, *Biochemistry* **22:**2832–2838.

Hüdig, H., and Hengstenberg, W., 1980, The bacterial phosphoenolpyruvate dependent phosphotransferase system (PTS). Solubilisation and kinetic parameters of the glucose-specific membrane bound enzyme II component of *Streptococcus faecalis, FEBS Lett.* **114:**103–106.

Jablonski, E. G., Brand, L., and Roseman, S., 1983, Sugar transport by the bacterial phosphotransferase system. Preparation of a fluorescein derivative of the glucose-specific phosphocarrier protein $III^{Glc}$ and its binding to the phosphocarrier protein HPr, *J. Biol. Chem.* **258:**9690–9699.

Jacobson, G. R., Lee, C. A., and Saier, M. H., Jr., 1979, Purification of the mannitol-specific enzyme II of the *Escherichia coli* phosphoenolpyruvate:sugar phosphotransferase system, *J. Biol. Chem.* **254:** 249–252.

Jacobson, G. R., Lee, C. A., Leonard, J. E., and Saier, M. H., Jr., 1983, Mannitol-specific enzyme II of the bacterial phosphotransferase system. I. Properties of the purified permease, *J. Biol. Chem.* **258:**10748–10756.

Kornberg, H. L., and Riordan, C., 1976, Uptake of galactose into *Escherichia coli* by facilitated diffusion, *J. Gen. Microbiol.* **94:**75–89.

Kornberg, H. L., and Watts, P. D., 1979, *tgs* and *crr* genes involved in catabolite inhibition and inducer exclusion in *Escherichia coli, FEBS Lett.* **104:**313–316.

Kornberg, H. L., Watts, P. D., and Brown, K., 1980, Mechanism of "inducer exclusion" by glucose, *FEBS Lett.* **117** (Suppl.)**:**K28–K36.

Kukuruzinska, M. A., Harrington, W. F., and Roseman, S., 1982, Sugar transport by the bacterial phosphotransferase system. Studies on the molecular weight and association of enzyme I, *J. Biol. Chem.* **257:**14470–14476.

Kundig, W., 1974, Molecular interactions in the bacterial phosphoenolpyruvate-phosphotransferase system (PTS), *J. Supramol. Struct.* **2:**695–714.

Kundig, W., and Roseman, S., 1971, Sugar transport. II. Characterization of constitutive membrane-bound Enzymes II of the *Escherichia coli* phosphotransferase system, *J. Biol. Chem.* **246:**1407–1418.

Kundig, W., Ghosh, S., and Roseman, S., 1964, Phosphate bound to histidine as an intermediate in a novel phospho-transferase system, *Proc. Natl. Acad. Sci. USA* **52:**1067–1074.

Lee, C. A., and Saier, M. H., Jr., 1983, Mannitol-specific enzyme II of the bacterial phosphotransferase system. III. The nucleotide sequence of the permease gene, *J. Biol. Chem.* **258:**10761–10767.

Lee, C. A., Jacobson, G. R., and Saier, M. H., Jr., 1981, Plasmid-directed synthesis of enzymes required for D-mannitol transport and utilization in *Escherichia coli, Proc. Natl. Acad. Sci. USA* **78:**7336–7340.

Leonard, J. E., and Saier, M. H., Jr., 1981, Genetic dissection of catalytic activities of the *Salmonella typhimurium* mannitol enzyme II, *J. Bacteriol.* **145:**1106–1109.

Leonard, J. E., and Saier, M. H., Jr., 1983, Mannitol-specific Enzyme II of the bacterial phosphotransferase system. II. Reconstitution of vectorial transphosphorylation in phospholipid vesicles, *J. Biol. Chem.* **258:**10757–10760.

Levine, R. L., and Federici, M. M., 1982, Quantitation of aromatic residues in proteins: Model compounds for second-derivative spectroscopy, *Biochemistry* **21:**2600–2606.

Lipmann, F., 1944, Enzymatic synthesis of acetyl phosphate, *J. Biol. Chem.* **155:**55–70.

Liu, K. D. F., and Roseman, S., 1983a, Kinetic properties and regulation of methyl α-glucoside uptake by *Salmonella typhimurium* membrane vesicles, *Fed. Proc.* **42:**1941.

Liu, K. D. F., and Roseman, S., 1983b, Kinetic characterization and regulation of phosphoenolpyruvate-dependent methyl α-D-glucopyranoside transport by *Salmonella typhimurium* membrane vesicles, *Proc. Natl. Acad. Sci. USA,* **80:**7142–7145.

Magasanik, B., 1970, Glucose effects: Inducer exclusion and repression, in: *The Lactose Operon* (J. R. Beckwith and D. Zipser, eds.), Cold Spring Harbor Press, Cold Spring Harbor, New York, pp. 189–219.

Makman, R. S., and Sutherland, E. W., 1965, Adenosine 3′,5′-phosphate in *Escherichia coli, J. Biol. Chem.* **240:**1309–1314.

Marquet, M., Creignou, M., and Dedoner, R., 1976, The phosphoenolpyruvate:methyl α-D-glucoside phosphotransferase system in *Bacillus subtilis* Marburg 168: Purification and identification of the phosphocarrier protein (HPr), *Biochimie* **58:**435–441.

Mason, P. W., Carbone, D. P., Cushman, R. A., and Waggoner, A. S., 1981, The importance of inorganic phosphate in regulation of energy metabolism of *Streptococcus lactis, J. Biol. Chem.* **256:**1861–1866.

Mattoo, R. L., and Waygood, E. B., 1983, Determination of the levels of HPr and enzyme I of the phosphoenolpyruvate-sugar phosphotransferase system in *Escherichia coli* and *Salmonella typhimurium, Can. J. Biochem. Cell Biol.* **61:**29–37.

Meadow, N. D., and Roseman, S., 1982, Sugar transport by the bacterial phosphotransferase system. Isolation and characterization of a glucose-specific phosphocarrier protein ($III^{Glc}$) from *Salmonella typhimurium, J. Biol. Chem.* **257:**14526–14537.

Meadow, N. D., and Roseman, S., 1983, A protease in *S. typhimurium* membranes which processes $III^{Glc}$, a protein of the phosphotransferase system, *Fed. Proc.* **42:**1813.

Meadow, N. D., Rosenberg, J. M., Pinkert, H. M., and Roseman, S., 1982a, Sugar transport by the bacterial phosphotransferase system. Evidence that *crr* is the structural gene for the *Salmonella typhimurium* glucose-specific phosphocarrier protein $III^{Glc}$, *J. Biol. Chem.* **257:**14538–14542.

Meadow, N. D., Saffen, D. W., Dottin, R. P., and Roseman, S., 1982b, Molecular cloning of the *crr* gene and evidence that it is the structural gene for $III^{Glc}$, a phosphocarrier protein of the bacterial phosphotransferase system, *Proc. Natl. Acad. Sci. USA* **79:**2528–2532.

Misko, T. P., 1983, Studies on the transport and regulatory functions of the phosphoenolpyruvate:glycose phosphotransferase system in *Salmonella typhimurium,* Ph.D. dissertation, The Johns Hopkins University, University Microfilms International, Ann Arbor, Michigan.

Misset, O., and Robillard, G. T., 1982, *Escherichia coli* phosphoenolpyruvate-dependent phosphotransferase system: Mechanism of phosphoryl-group transfer from phosphoenolpyruvate to HPr, *Biochemistry* **21:**3136–3142.

Misset, O., Brouwer, M., and Robillard, G. T., 1980, *Escherichia coli* phosphoenolpyruvate-dependent phosphotransferase system. Evidence that the dimer is the active form of enzyme I, *Biochemistry* **19:**883–890.

Mitchell, W. J., Misko, T. P., and Roseman, S., 1982, Sugar transport by the bacterial phosphotransferase system. Regulation of other transport systems (lactose and melibiose), *J. Biol. Chem.* **257:**14553–14564.

Nelson, S. O., Scholte, B. J., and Postma, P. W., 1982, Phosphoenolpyruvate : sugar phosphotransferase system-mediated regulation of carbohydrate metabolism in *Salmonella typhimurium, J. Bacteriol.* **150:**604–615.

Nelson, S. O., Wright, J. K., and Postma, P. W., 1983, The mechanism of inducer exclusion. Direct interaction between purified $III^{Glc}$ of the phosphoenolpyruvate:sugar phosphotransferase system and the lactose carrier of *Escherichia coli, Eur. Mol. Biol. Org. J.* **2:**715–720.

Osumi, T., and Saier, M. H., Jr., 1982, Regulation of lactose permease activity by the phosphoenolpyruvate : sugar phosphotransferase system: Evidence for direct binding of the glucose-specific enzyme III to the lactose permease, *Proc. Natl. Acad. Sci. USA* **79:**1457–1461.

Parra, F., Jones-Mortimer, M. C., and Kornberg, H. L., 1983, Phosphotransferase mediated regulation of carbohydrate utilization in *Escherichia coli* K12. The nature of the *iex* (*crr*) and *gsr* (*tgs*) mutations, *J. Gen. Microbiol.* **129:**337–348.

Pastan, I., and Perlman, R., 1970, Cyclic adenosine monophosphate in bacteria, *Science* **169:**339–344.

Perret, J., and Gay, P., 1979, Kinetic study of a phosphoryl exchange reaction between fructose and fructose 1-phosphate catalyzed by the membrane-bound enzyme II of the phosphoenolpyruvate-fructose 1-phosphotransferase system of *Bacillus subtilis, Eur. J. Biochem.* **102:**237–246.

Peterkofsky, A., and Gazdar, C., 1975, Interaction of Enzyme I of the phosphoenolpyruvate:sugar phosphotransferase system with adenylate cyclase of *Escherichia coli, Proc. Natl. Acad. Sci. USA* **72:**2920–2924.

Peterkofsky, A., and Gazdar, C., 1978, The *Escherichia coli* adenylate cyclase complex: Activation by phosphoenolpyruvate, *J. Supramol. Struct.* **9:**219–230.

Postma, P. W., 1976, Involvement of phosphotransferase system in galactose transport in *Salmonella typhimurium, FEBS Lett.* **61:**49–53.

Postma, P. W., 1981, Defective enzyme II-$B^{Glc}$ of the phosphoenolpyruvate:sugar phosphotransferase system leading to uncoupling of transport and phosphorylation in *Salmonella typhimurium, J. Bacteriol.* **147:**382–389.

Postma, P. W., 1982, Regulation of sugar transport in *Salmonella typhimurium, Ann. Microbiol. (Inst. Pasteur)* **133A:**261–267.

Postma, P. W., and Roseman, S., 1976, The bacterial phosphoenolpyruvate:sugar phosphotransferase system, *Biochim. Biophys. Acta* **457:**213–257.

Postma, P. W., and Stock, J. B., 1980, Enzymes II of the phosphotransferase system do not catalyze sugar transport in the absence of phosphorylation, *J. Bacteriol.* **141:**476–484.

Reider, E., Wagner, E. F., and Schweiger, M., 1979, Control of phosphoenolpyruvate-dependent phosphotransferase-mediated sugar transport in *Escherichia coli* by energization of the cell membrane, *Proc. Natl. Acad. Sci. USA* **76.**5529–5533.

Rephaeli, A. W., and Saier, M. H., Jr., 1978, Kinetic analyses of the sugar phosphate:sugar transphosphorylation reaction catalyzed by the glucose enzyme II complex of the bacterial phosphotransferase system, *J. Biol. Chem.* **253:**7595–7597.

Rephaeli, A. W., and Saier, M. H., Jr., 1980, Substrate specificity and kinetic characterization of sugar uptake and phosphorylation, catalyzed by the mannose enzyme II of the phosphotransferase system in *Salmonella typhimurium, J. Biol. Chem.* **255:**8585–8591.

Roberts, R. B., Cowie, D. B., Abelson, P. H., Bolton, E. T., and Britten, R. J., 1963, Studies of biosynthesis in *Escherichia coli,* Carnegie Institution of Washington Publication 607, Washington, D.C., pp. 5, 15.

Robillard, G. T., and Konings, W. N., 1981, Physical mechanism for regulation of phosphoenolpyruvate-dependent glucose transport activity in *Escherichia coli, Biochemistry* **20:**5025–5032.

Robillard, G. T., and Lageveen, R. G., 1982, Non-vectorial phosphorylation by the bacterial PEP-dependent phosphotransferase system is an artifact of spheroplast and membrane vesicle preparation procedures, *FEBS Lett.* **147:**143–148.

Robillard, G. T., Dooijewaard, G., and Lolkema, J., 1979, *Escherichia coli* phosphoenolpyruvate dependent phosphotransferase system. Complete purification of Enzyme I by hydrophobic interaction chromatography, *Biochemistry* **18:**2984–2989.

Roossien, F. F., Dooijewaard, G., and Robillard, G. T., 1979, The *Escherichia coli* phosphoenolpyruvate-dependent phosphotransferase system: Observation of heterogeneity in the amino acid composition of HPr, *Biochemistry* **18:**5793–5797.

Rose, S. P., and Fox, C. F., 1971, The β-glucoside system of *Escherichia coli*. II. Kinetic evidence for a phosphoryl-enzyme II intermediate, *Biochem. Biophys. Res. Commun.* **45:**376–380.

Roseman, S., 1972, Carbohydrate transport in bacterial cells, in: *Metabolic Transport,* Vol. VI (L. E. Hokin, ed.), Academic Press, New York, pp. 41–89.

Roseman, S., 1977, The transport of sugars across bacterial membranes, in: *Biochemistry of Membrane Transport, FEBS-Symposium No. 42* (G. Semenza and E. Carafoli, eds.), Springer-Verlag, New York, pp. 582–597.

Saier, M. H., Jr., 1977, Bacterial phosphoenolpyruvate:sugar phosphotransferase systems: Structural, functional, and evolutionary interrelationships, *Bacteriol. Rev.* **41:**856–871.

Saier, M. H., Jr., and Feucht, B. U., 1980, Regulation of carbohydrate transport activities in *Salmonella typhimurium:* Use of the phosphoglycerate transport system to energize solute uptake, *J. Bacteriol.* **141:**611–617.

Saier, M. H., Jr., and Moczydlowski, E. G., 1978, The regulation of carbohydrate transport in *Escherichia coli* and *Salmonella typhimurium,* in: *Bacterial Transport* (B. P. Rosen, ed.), Marcel Dekker, New York, pp. 103–125.

Saier, M. H., Jr., and Newman, M. J., 1976, Direct transfer of the phosphoryl moiety of mannitol 1-phosphate to [$^{14}$C]mannitol catalyzed by the enzyme II complexes of the phosphoenolpyruvate:mannitol phosphotransferase systems in *Spirochaeta aurantia* and *Salmonella typhimurium, J. Biol. Chem.* **251:**3834–3837.

Saier, M. H., Jr., and Roseman, S., 1976a, Sugar transport. The *crr* mutation: Its effect on repression of enzyme synthesis, *J. Biol. Chem.* **251:**6598–6605.

Saier, M. H., Jr., and Roseman, S., 1976b, Sugar transport. Inducer exclusion and regulation of the melibiose, maltose, glycerol, and lactose transport systems by the phosphoenolpyruvate:sugar phosphotransferase system, *J. Biol. Chem.* **251:**6606–6615.

Saier, M. H., Jr., and Schmidt, M. R., 1981, Vectorial and nonvectorial transphosphorylation catalyzed by enzymes II of the bacterial phosphotransferase system, *J. Bacteriol.* **145:**391–397.

Saier, M. H., Jr., Bromberg, F. G., and Roseman, S., 1973, Characterization of constitutive galactose permease mutants in *Salmonella typhimurium, J. Bacteriol.* **113:**512–514.

Saier, M. H., Jr., Simoni, R. D., and Roseman, S., 1976, Sugar transport. Properties of mutant bacteria defective in proteins of the phosphoenolpyruvate:sugar phosphotransferase system, *J. Biol. Chem.* **251:**6584–6597.

Saier, M. H., Jr., Cox, D. F., and Moczydlowski, E. G., 1977a, Sugar phosphate:sugar transphosphorylation coupled to exchange group translocation catalyzed by the enzyme II complexes of the phosphoenolpyruvate:sugar phosphotransferase system in membrane vesicles of *Escherichia coli, J. Biol. Chem.* **252:**8908–8916.

Saier, M. H., Jr., Feucht, B. U., and Mora, W. K., 1977b, Sugar phosphate:sugar transphosphorylation and exchange group translocation catalyzed by the enzyme II complexes of the bacterial phosphoenolpyruvate:sugar phosphotransferase system, *J. Biol. Chem.* **252:**8899–8907.

Saier, M. H., Jr., Newman, M. J., and Rephaeli, A. W., 1977c, Properties of a phosphoenolpyruvate:mannitol phosphotransferase system in *Spirochaeta aurantia, J. Biol. Chem.* **252:**8890–8898.

Saier, M. H., Jr., Schmidt, M. R., and Lin, P., 1980, Phosphoryl exchange reaction catalyzed by enzyme I of the bacterial phosphoenolpyruvate:sugar phosphotransferase system. Kinetic characterization, *J. Biol. Chem.* **255:**8579–8584.

Saier, M. H., Jr., Keeler, D. K., and Feucht, B. U., 1982, Physiological desensitization of carbohydrate permeases and adenylate cyclase to regulation by the phosphoenolpyruvate:sugar phosphotransferase system in *Escherichia coli* and *Salmonella typhimurium*. Involvement of adenosine cyclic 3′,5′-phosphate and inducer, *J. Biol. Chem.* **257:**2509–2517.

Sanderson, K. E., and Roth, J. R., 1983, Linkage map of *Salmonella typhimurium,* Edition VI, *Microbiol. Rev.* **47:**410–453.

Schachter, H., 1973, On the interpretation of Michaelis constants for transport, *J. Biol. Chem.* **248:**974–976.

Scholte, B. J., and Postma, P. W., 1980, Mutation in the *crp* gene of *Salmonella typhimurium* which interferes with inducer exclusion, *J. Bacteriol.* **141:**751–757.

Scholte, B. J., and Postma, P. W., 1981, Competition between two pathways for sugar uptake by the phosphoenolpyruvate-dependent sugar phosphotransferase system in *Salmonella typhimurium, Eur. J. Biochem.* **114:**51–58.

Scholte, B. J., Schuitema, A. R., and Postma, P. W., 1981, Isolation of III$^{Glc}$ of the phosphoenolpyruvate-dependent glucose phosphotransferase system of *Salmonella typhimurium, J. Bacteriol.* **148:**257–264.

Scholte, B. J., Schuitema, A. R. J., and Postma, P. W., 1982, Characterization of Factor III$^{Glc}$ in catabolite repression-resistant (*crr*) mutants of *Salmonella typhimurium, J. Bacteriol.* **149:**576–586.

Schrecker, O., Stein, R., Hengstenberg, W., Gassner, M., and Stehlik, D., 1975, The *Staphylococcal* PEP dependent phosphotransferase system, proton magnetic resonance (PMR) studies on the phosphoryl carrier protein HPr: Evidence for a phosphohistidine residue in the intact phospho-HPr molecule, *FEBS Lett.* **51:**309–312.

Silhavy, T. J., Ferenci, T., and Boos, W., 1978, Sugar transport systems in *Escherichia coli,* in: *Bacterial Transport* (B. P. Rosen, ed.), Marcel Dekker, New York, pp. 127–169.

Simoni, R. D., and Roseman, S., 1973, Sugar transport. VII. Lactose transport in *staphylococcus aureus, J. Biol. Chem.* **248:**966–976.

Simoni, R. D., Levinthal, M., Kundig, F. D., Kundig, W., Anderson, B., Hartman, P. E., and Roseman, S., 1967, Genetic evidence for the role of a bacterial phosphotransferase system in sugar transport, *Proc. Natl. Acad. Sci. USA* **58:**1963–1970.

Simoni, R. D., Smith, M., and Roseman, S., 1968, Resolution of a Staphylococcal phosphotransferase system into four protein components and its relation to sugar transport, *Biochem. Biophys. Res. Commun.* **31:**804–811.

Simoni, R. D., Hays, J. B., Nakazawa, T., and Roseman, S., 1973a, Sugar transport. VI. Phosphoryl transfer in the lactose phosphotransferase system of *Staphylococcus aureus, J. Biol. Chem.* **248:**957–965.

Simoni, R. D., Nakazawa, T., Hays, J. B., and Roseman, S., 1973b, Sugar transport. IV. Isolation and characterization of the lactose phosphotransferase system in *Staphylococcus aureus, J. Biol. Chem.* **248:**932–940.

Solomon, E., Miyai, K., and Lin, E. C. C., 1973, Membrane translocation of mannitol in *Escherichia coli* without phosphorylation, *J. Bacteriol.* **114:**723–728.

Stein, R., Schrecker, O., Lauppe, H. F., and Hengstenberg, H., 1974, The *Staphylococcal* PEP dependent phosphotransferase system: Demonstration of a phosphorylated intermediate of the enzyme I component, *FEBS Lett.* **42:**98–100.

Stock, J. B., Waygood, E. B., Meadow, N. D., Postma, P. W., and Roseman, S., 1982, Sugar transport by the bacterial phosphotransferase system. The glucose receptors of the *Salmonella typhimurium* phosphotransferase system, *J. Biol. Chem.* **257:**14543–14552.

Tyler, B., and Magasanik, B., 1970, Physiological basis of transient repression of catabolic enzymes in *Escherichia coli, J. Bacteriol.* **102:**411–422.

Ullah, A., and Cirillo, V., 1976, *Mycoplasma* phosphoenolpyruvate-dependent sugar phosphotransferase system: Purification and characterization of the phosphocarrier protein, *J. Bacteriol.* **127:**1298–1306.

Valdes, R., Jr., and Ackers, G. K., 1979, Study of protein subunit association equilibria by elution gel chromatography, *Meth. Enzymol.* **61:**125–142.

Wang, R. J., and Morse, M. L., 1968, Carbohydrate accumulation and metabolism in *Escherichia coli* I. Description of pleiotropic mutants, *J. Mol. Biol.* **32:**59–66.

Waygood, E. B., and Steeves, T., 1980, Enzyme I of the phosphoenolpyruvate:sugar phosphotransferase system of *Escherichia coli*. Purification to homogeneity and some properties, *Can. J. Biochem.* **58:**40–48.

Waygood, E. B., Cordaro, J. C., and Roseman, S., 1975, Pseudo-HPr, a substitute for HPr in the PEP:sugar phosphotransferase system, *Proc. Can. Fed. Biol. Soc.* **18:**115.

Waygood, E. B., Weigel, N., Nakazawa, A., Kukuruzinska, M., and Roseman, S., 1977, Purification and properties of Enzyme I of the PEP:glycose phosphotransferase system (PTS), *Proc. Can. Fed. Biol. Soc.* **20:**54.

Waygood, E. B., Meadow, N. D., and Roseman, S., 1979, Modified assay procedures for the phosphotransferase system in enteric bacteria, *Anal. Biochem.* **95:**293–304.

Weigel, N., Kukuruzinska, M. A., Nakazawa, A., Waygood, E. B., and Roseman, S., 1982a, Sugar transport by the bacterial phosphotransferase system. Phosphoryl transfer reactions catalyzed by enzyme I of *Salmonella typhimurium, J. Biol. Chem.* **257:**14477–14491.

Weigel, N., Powers, D. A., and Roseman, S., 1982b, Sugar transport by the bacterial phosphotransferase system. Primary structure and active site of a general phosphocarrier protein (HPr) from *Salmonella typhimurium, J. Biol. Chem.* **257:**14499–14509.

Weigel, N., Waygood, E. B., Kukuruzinska, M. A., Nakazawa, A., and Roseman, S., 1982c, Sugar transport by the bacterial phosphotransferase system. Isolation and characterization of enzyme I from *Salmonella typhimurium, J. Biol. Chem.* **257:**14461–14469.

Winkler, H. H., and Wilson, T. H., 1967, Inhibition of β-galactoside transport by substrates of the glucose transport system in *Escherichia coli, Biochim. Biophys. Acta* **135:**1030–1051.

# 42

# The Maltose–Maltodextrin-Transport System of *Escherichia coli* K-12

*Howard A. Shuman and Nancy A. Treptow*

## I. INTRODUCTION

Gram-negative bacteria inhabit a wide variety of environments and have evolved different strategies for capturing useful substances from the external milieu. Some active transport systems for sugars and amino acids are composed of a single polypeptide species. These systems usually function as secondary transporters that are energized by the electrochemical proton gradient. Examples of these are the permeases for β-galactosides and glycerol-3-phosphate. In contrast to these "simple" systems, many growth substrates are transported by multicomponent systems that include a water-soluble substrate-binding protein in the periplasmic space, as well as proteins in the cytoplasmic membrane. These systems function as primary active transport systems that pump substrates into the cell at the expense of chemical energy. They are energized by an as yet undefined compound that is derived from the high-energy phosphoester pool of the cell (Berger, 1973; Berger and Heppel, 1974).

The mechanism by which these systems translocate substrates across the membrane and couple this process to chemical energy remains almost totally obscure. The dominant reason for studying these systems as models for active transport is the ability to use the genetic approach to analyze the structures that make up the systems, define the interactions among these proteins, and to ask specific questions about the mechanism of active transport.

*Howard A. Shuman and Nancy A. Treptow* • Department of Microbiology, College of Physicians and Surgeons, Columbia University, New York, New York 10032.

The maltose–maltodextrin transport system of *Escherichia coli* is an example of such a multicomponent system. It catalyzes the accumulation of maltose [glucose α(1–4) glucopyranoside] and oligomers of glucose in α(1–4) linkage up to seven units in length. Initial genetic and physiological experiments have provided a description of the overall characteristics of this system. More recent biochemical and genetic studies have been able to provide a more detailed description of the individual components and their interactions. This has made it possible to think of testable models for the organization and operation of this and related transport systems.

This chapter is a summary of recent results that are relevant to the organization of this transport system in the membrane and its operation as a pump.

## II. MALTOSE AND MALTODEXTRIN CATABOLISM IN *Escherichia coli* K-12

In wild-type *Escherichia coli,* there are eight genes that are specifically required for the efficient utilization of maltose and maltodextrins. These eight genes are located in two loci on the standard *E. coli* map (Schwartz, 1966). One group of genes is located at the *malA* locus. These genes are called *malT, malP,* and *malQ*. The *malT* gene product functions as a positive regulatory protein that stimulates expression of the other *mal* genes in the presence of maltose (Debarbouille *et al.*, 1978). The *malP* and *malQ* genes code for the soluble, cytoplasmic enzymes, maltodextrin phosphorylase and amylomaltase. The remaining five genes are located at the *malB* locus. These

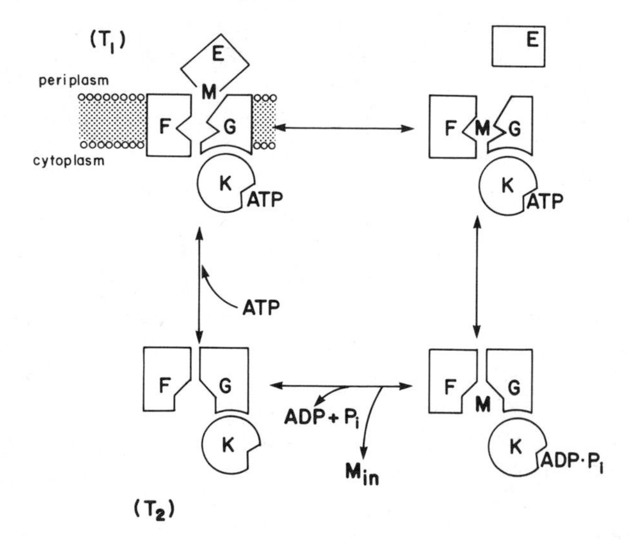


*Figure 1.* A model for the active transport of maltose and maltodextrins across the cytoplasmic membrane of *Escherichia coli*. In this model, the MalF, MalG, and MalK proteins form a complex in the inner membrane. Substrate molecules bound to MBP interact with this complex from the external periplasmic face of the membrane. M indicates a molecule of substrate. See text for a detailed description of this model.

genes are called *malE, malF, malG, malK,* and *lamB*. Each of these five genes codes for a protein which is a component of the active transport system (Silhavy *et al.*, 1979; Raibaud *et al.*, 1979). These five genes are arranged in two operons that are transcribed in opposite directions. The *malE* gene codes for a water-soluble maltose-binding protein (MBP) that is found in the periplasmic space (Kellermann and Szmelcman, 1974). The *malF, malG,* and *malK* genes code for proteins that are found in the inner, cytoplasmic membrane, and the *lamB* gene codes for a protein that is found in the outer membrane of the cell and also serves as the receptor protein for the bacteriophage lambda (Figure 1; Randall-Hazelbauer and Schwartz, 1973).

## III. GENERAL PROPERTIES OF MALTOSE AND MALTODEXTRIN TRANSPORT

Bacterial mutants which lack amylomaltase (*malQ*⁻) can accumulate large concentrations of maltose. This compound is transported by a single system that is made up of the five products of the *malB* locus. The absence of secondary transport systems is in contrast to the multiplicity of transport systems for other sugars and amino acids. This has simplified genetic and physiological experiments.

The normal substrates for this system are glucose polymers up to seven residues in length. Longer dextrins cannot serve as growth substrates but can competitively inhibit maltose transport. The ability of the long dextrins to interact with the cell envelope is mediated by the *lamB* gene product (Ferenci, 1980a). These dextrins most likely occur in nature as breakdown products of starch and amylose degradation. Other nonmetabolizable compounds related to maltose have been shown to be substrates for the transport system. Radioactive methyl-α-maltoside and 5-thiomaltose were prepared and shown to be actively accumulated by this system (Ferenci, 1980b). Addition of these compounds to cultures that were induced for the maltose system resulted in severe growth inhibition even in the presence of other metabolizable carbon sources. The selective toxicity of these compounds might be used to select mutants that are defective in this transport system.

The apparent affinity of the maltose-transport system is in the micromolar range. Half-maximal transport activity is observed at external concentrations between 0.2 and 2.0 μM for all substrates. The concentration gradient that can be formed in a cell which is unable to metabolize substrate is approximately $10^4$ : 1 (Szmelcman *et al.*, 1976). The exit of accumulated radioactive maltose is much slower than the initial rate of uptake and is not mediated by the products of the *malB* region (Boos *et al.*, 1981). Exit of unmodified maltose is not detectable in unenergized cells. In cells that are given glycolytic substrates such as glucose or fructose, some of the internal maltose is acetylated. The acetylmaltose rapidly leaves the cell by an uncharacterized mechanism. It is clear from studies with *malB* mutant strains that the acetylation reaction and exit of acetylmaltose are not mediated by the products of the *malB* locus, and are most likely unrelated to the normal pathway of substrate entry. A plausible role for the acetylation reaction may be some sort of detoxification (Boos *et al.*, 1981). At the present time, this idea is difficult to test since no mutants have been isolated which lack the ability to acetylate maltose. In view of these results, it seems that this system functions asymmetrically in the inward direction only. A similar observation has been made for the β-methylgalactoside transport system (Parnes and Boos, 1973).

As mentioned above, there is very little information concerning the energy source for active transport. Treatment of cells with substances that decrease the pool of high-energy phosphoesters such as arsenate completely abolishes substrate entry. Other treatments which abolish the proton gradient without depleting the pool of high-energy phosphate (such as proton ionophores in a strain that lacks $BF_1$ ATPase) do not result in a significant loss of transport activity. A second series of experiments carried out by Berger also indicate the involvement of high-energy phosphate rather than the proton gradient in energization of periplasmic component-containing transport systems (Berger, 1973; Berger and Heppel, 1974).

## IV. PROPERTIES OF THE INDIVIDUAL COMPONENTS

### A. The LamB Protein

The LamB protein was first identified as the receptor for bacteriophage lambda. It had been known that the ability of bacteria to adsorb lambda phage was dramatically increased if cells were grown in the presence of maltose. Furthermore, it was shown that the *malT* gene product was required for lambda sensitivity and phage adsorption (Schwartz, 1967). Finally, the structural gene for the receptor protein itself was shown to map within the *malB* locus. Randall-Hazelbauer and Schwartz (1973) were able to solubilize a lambda-inactivating activity from whole cells and purify the activity to apparent homogeneity. They showed that the protein was able to cause DNA ejection from whole phage and that this protein was the product of the *lamB* gene. The phage-receptor properties of this protein have been reviewed in detail elsewhere (Schwartz, 1980). The role of the LamB protein in maltose and maltodextrin transport was not recognized immediately. This was the case because mutants that lacked this protein can still grow on maltose as a carbon source. The $Mal^+$ phenotype of *lamB* mutants was always evaluated at maltose concentrations commonly used in standard media (0.5–1.0 mM). It was not until the properties of these mutants were examined at much lower maltose concentrations that the role of the LamB protein became apparent (Szmelcman and Hofnung, 1975). Below $10^{-4}$ M maltose, the absence of the LamB protein has a pronounced effect on the rate of maltose entry. For substrates longer than maltotriose, the requirement for the LamB protein is absolute. Thus, *lamB* mutants are able to grow on high (mM) concentrations of maltose, but are unable to grow on any concentration of polymers longer than maltotriose ($Dex^-$ phenotype; Szmelcman *et al.*, 1976).

LamB protein is found in the outer membrane of the cell envelope. When bacteria are grown on maltose, there are approximately $10^5$ copies of the LamB polypeptide per cell. These are probably arranged as a trimer in the membrane (Nakae and Ishii, 1982; Neuhaus, 1982). That the LamB protein spans the width of the outer membrane is indicated by the fact that the protein is accessible from the outside of the cell to both phages and antibodies; in addition, it is accessible to the periplasmic face of the membrane since it has been shown to interact noncovalently with the peptidoglycan layer (Gabay and Yasunaka, 1980), and also with periplasmic maltose-binding protein (see below). Studies with monoclonal antibodies have demonstrated that at least two antigenic sites within the 70-amino-acid-residue carboxy-terminal region are located at the cell surface (Gabay *et al.*, 1983).

Purified LamB protein has been shown to form aqueous channels when it is reconstituted in phospholipid vesicles or bilayers (Boehler-Kohler *et al.*, 1979; Nakae and Ishii, 1980). The transport properties of this pore indicate that the channel has some specificity for maltose and dextrins (Luckey and Nikaido, 1980). The rates of permeation of maltose and other disaccharides were measured. It was found that the rate of maltose permeation was 40 times faster than that of sucrose and eight times faster than that of cellobiose. Also, the rate of maltose permeation was 12 times faster than that of maltoheptaose. These results suggest that the LamB protein forms a pore that has a certain degree of stereospecificity. In agreement with these results, it has been shown that in the absence of other *malB*-encoded proteins, the LamB protein can act as a binding protein for dextrins that are too long to be transported into the cell, as well as for branched starch molecules (Ferenci *et al.*, 1980). The binding of the LamB protein to various sugars was measured and found to be specific for maltooligosaccharides and the apparent affinity increased with increasing chain length. For example, the apparent affinity for maltose was 14 mM and for maltodecaose was found to be 0.075 mM. It should be pointed out that the LamB protein can also serve to transport other sugars *in vivo* such as glucose and lactose (Heuzenroeder and Reeves, 1980). Despite these transport and binding properties of the LamB protein itself, there is evidence that the role of the LamB protein in maltose and maltodextrin transport across the outer membrane is to allow external substrate molecules direct access to the maltose-binding protein. This may be the result of the physical interaction of the maltose-binding protein and the LamB protein. The evidence for this interaction will be described below.

The nucleotide sequence of the *lamB* gene has been determined (Clement and Hofnung, 1981). The 5′ extremity encodes a 25-amino-acid "signal sequence" which is required for the initiation of LamB export to the outer membrane. This sequence is removed during the process of localization. Mutations which prevent initiation of export but have no effect on gene expression map within the region that codes for a hydrophobic core within the signal sequence (Emr and Silhavy, 1980).

Many other mutations which affect different LamB functions have been characterized, but the only mutations that have been sequenced are those that result in decreased ability of the LamB protein to adsorb the lambdoid phage K10. These mutations occur at codons 154 and 155 (Roa and Clement, 1980). Apparently, these residues are exposed at the external surface of the outer membrane.

Structural features of the LamB protein that have become apparent from the nucelotide sequence are α-helical regions composed of negatively and positively charged residues. The α-helical regions that contain these pairs may correspond to parts of the polypeptide which span the hydrophobic core of the outer membrane. These are reasonable candidates for membrane-spanning regions since there are no obvious stretches of hydrophobic residues that could span this membrane.

## B. The Maltose-Binding Protein

The maltose-binding protein was identified as a water-soluble protein, released by the classical osmotic-shock procedure of Neu and Heppel (1965), that could bind radioactive maltose (Kellerman and Szmelcman, 1974). This protein was purified from shock fluid either by conventional protein-purification techniques or by affinity chromatography on crosslinked amylose (Ferenci and Klotz, 1978). The affinity of the MBP for all maltooligosaccharides is in the micromolar range and does not seem to depend on the length of the particular dextrin. Ligand-binding results in a decrease in the intrinsic fluorescence of MBP (Szmelcman *et al.*, 1976). This protein seems to have only a single binding site for ligand and exists as a monomer. In fully induced cells, there are approximately $3 \times 10^4$ monomers per cell. It is possible to calculate that under these conditions the concentration of MBP in the periplasmic space is close to millimolar (Dietzel *et al.*, 1978).

If the maltose-binding protein is absent from cells due to deletion of the *malE* gene, it is not possible to detect translocation of substrate into the cell even at external maltose concentrations of up to 25 mM (Shuman, 1982). This argues that this protein plays an essential role in the transport of substrates across the cytoplasmic membrane. It has been proposed that the components of the transport system in the cytoplasmic membrane recognize the MBP–maltose complex rather than free maltose (see below).

The maltose-binding protein also functions as the chemoreceptor for maltose during chemotaxis. The component of the chemotaxis system that interacts with the MBP is the methyl-accepting chemotaxis protein, Tar (Koiwai and Hayashi, 1979). The maltose-binding protein, therefore, functions in substrate transport across both the outer membrane and the cytoplasmic membrane and also interacts with the chemotaxis system. It, therefore, has the ability to interact specifically with the LamB protein, the Tar protein, and the MalF–MalG–MalK complex (see below).

## C. The MalF and MalK Proteins

The extensive characterization of the maltose-binding protein and the LamB protein has been possible because assays exist for each of these proteins and both are present in large amounts when cells are grown on maltose. The study of the remaining components is more difficult because there is no functional assay for any of them. Many unsuccessful attempts to detect these proteins have been made. Comparison of the protein patterns obtained when cell fractions from wild-type and appropriate mutant strains were subjected to two-dimensional electrophoresis did not permit identification of the MalF or MalG proteins (Silhavy, Boos, and Schwartz, unpublished) but did allow detection of MalK (Bavoil and Nikaido, 1981).

In order to identify and characterize the products of the *malF* and *malK* genes, *malF–lacZ* and *malK–lacZ* gene fusions were constructed. These gene fusions are constructed so that the amino-terminal region of β-galactosidase is replaced by an amino-terminal fragment of the desired gene product (in this case, either MalF or MalK). These fusions produce hybrid proteins that retain β-galactosidase activity. Originally, two classes of *malF–lacZ* gene fusions were found. The first class contained only a small portion of the *malF* gene. The hybrid β-galactosidase produced by this class is found exclusively in the cytoplasm of the cell. The second class of *malF–lacZ* fusions contained almost all of the *malF* gene. The hybrid β-galactosidase produced by this class is located in the cytoplasmic membrane of the cell. This suggested that if the MalF protein itself is located in the cytoplasmic membrane, then the portion of the *malF* gene present in the fusion is sufficient to localize this protein to the cytoplasmic membrane (Silhavy *et al.*, 1976).

The *malK–lacZ* gene fusion was constructed in a manner that resulted in a complex fusion which contains the entire *malK* gene, a small portion of the *lamB* gene, and the *lacZ* gene. The hybrid protein produced by this fusion is also found in the cytoplasmic membrane. The role of the small portion of *lamB* sequence in determining the location of the hybrid protein is not known (Emr and Silhavy, 1980).

The hybrid protein product from the gene-fusion strains could be identified after electrophoresis of membrane extracts on polyacrylamide gels containing SDS. In addition, it was shown that these proteins could be precipitated by anti-β-galactosidase antibodies.

The molecular weight of the MalF hybrid protein is approximately 150,000 and that of MalK hybrid is approximately 170,000 (Silhavy *et al.*, 1976; Emr and Silhavy, 1980). This indicates that both of the hybrids contain substantial portions of the MalF and MalK proteins, respectively.

The MalF and MalK proteins were identified using antibodies that recognized these proteins. The antibodies were prepared by immunizing rabbits with purified MalF and MalK hybrid protein. Some of the antibodies produced by the rabbits would recognize *lacZ*-coded β-galactosidase sequences but others would recognize sequences coded for by the *malF* or *malK* gene. The MalF- or MalK-specific antibodies could then be used to detect the MalF and MalK proteins in wild-type cells.

The hybrid proteins have been purified by a simple procedure based on the fact that each is the largest major polypeptide species in the membranes of the respective fusion strain.

Rabbits were then immunized with each hybrid protein. The immune response was followed by monitoring the ability of serum samples to precipitate the appropriate hybrid protein. After it was determined that antibodies were being produced, serum samples were used to detect the MalF and MalK proteins in immune precipitation experiments (Shuman *et al.*, 1980; Shuman and Silhavy, 1981).

This was accomplished by labeling cells with radioactive amino acids, preparing whole cell extracts and adding the appropriate antibody to the extracts. After incubating this mixture, the immune complexes were collected and subjected to electrophoresis on SDS–polyacrylamide gels. Radioactive antigens which had combined with the antibody were then detected in the gels by fluorography. The MalF and MalK proteins were detected as maltose-inducible proteins that were precipitated by the appropriate antibody preparation. In addition, the MalF protein was absent in *malF* mutants and the MalK protein was absent in *malK* mutant strains. The various *malB* mutant strains were grown in maltose to induce expression of any *malB* region genes which were not inactivated. Maltose entry was made possible by providing the *malB* mutant strains with a secondary maltose transport system (Shuman and Beckwith, 1979). Finally, it was shown that the MalF and MalK proteins crossreacted immunologically with their respective hybrid proteins. The molecular weights of these two proteins have been estimated by their electrophoretic mobility on SDS–polyacrylamide gels. Both the MalF and MalK proteins have an apparent molecular weight of approximately 41,000–43,000 (Shuman *et al.*, 1980; Shuman and Silhavy, 1981). The nucleotide sequence of the *malF* gene indicates that the estimate for the MalF protein is in error. The nucleotide sequence of the *malF* gene codes for a protein of 56,964 mol. wt. (S. Froshauer, personal communication). The sequence of the *malK* gene predicts a molecular weight of 40,700 for the MalK protein, which is in good agreement with the observed molecular weight of this protein (Gilson *et al.*, 1982a; Bavoil *et al.*, 1980).

The location of the MalF and MalK proteins was determined by fractionating radioactively labeled wild-type cells and testing each fraction for the presence of the proteins by immune precipitation. The MalF protein was shown to be a cytoplasmic membrane protein because it sedimented with the total membrane fraction but could be solubilized by Triton X-100 in the absence of EDTA (Shuman *et al.*, 1980). The MalK protein was shown to be located in the envelope fraction in wild-type cells. When the location of this protein was examined in different *malB* mutant strains, it was found that in *malG* mutants, the MalK protein was not located in the membrane fraction but could be recovered from the cytoplasm (Shuman and Silhavy, 1981). In addition, it has been shown that this protein can be solubilized from wild-type membranes by Triton X-100 (Bavoil *et al.*, 1980), and can be removed from membranes by sonication (Shuman, unpublished). These results are interpreted to indicate that the MalK protein is peripherally associated with the inside surface of the cytoplasmic membrane via an interaction with the MalG protein. This location of the MalK protein implies that it is unlikely to participate in substrate translocation or binding-protein substrate recognition. A more reasonable role for the MalK protein would be coupling metabolic energy to substrate transport.

Only very crude estimates of the number of MalF and MalK protein molecules per cell have been made. Based on the amount of radioactivity present in the MalF and MalK bands after immunoprecipitation from radioactively labeled extracts, there appear to be approximately 500–1000 copies of each of these proteins per cell This is at least 10–50 times less than the amount of maltose-binding protein and LamB protein per cell (Shuman *et al.*, 1980; Shuman and Silhavy, 1981). Alternate estimates for the amount of MalK protein based on dye intensity yielded a value of 10,000 MalK molecules per cell. These two estimates were determined on different strains of *E. coli;* the discrepancy in the number of MalK polypeptides may be due to differences in the two strains (Bavoil *et al.*, 1980).

The sequence of the *malF* gene indicates that there are large stretches of hydrophobic amino acid residues. Although there is not a "signal sequence" in the classical sense for secreted proteins, there is an amino-terminal region which contains positively charged residues. One of the hydrophobic stretches follows this positively charged region. It is tempting to speculate that these regions are involved in determining the membrane location of this protein. Another hydrophobic stretch (32 residues) is located in the middle part of the sequence (S. Froshauer, personal communication).

The *malK* sequence does not code for any long stretches of hydrophobic amino acids. There is no evidence for a "signal sequence" at any location within the gene

(Gilson *et al.*, 1982a). This is not inconsistent with the proposed location of this protein on the inner surface of the cytoplasmic membrane. Extensive homology was discovered between the *malK* gene and the *hisP* gene. The *hisP* gene codes for a component of the histidine-transport system of *Salmonella typhimurium* (Gilson *et al.*, 1982b). This homology most likely reflects the similarity in the function and/or structure of these two proteins. Two regions of both the *malK* and *hisP* genes code for sequences that have been found in a number of unrelated ATP-requiring proteins and enzymes (Higgins, personal communication). These sequences have been shown to occur in regions which are involved in nucleotide binding in the cases of myosin, phosphofructokinase, and adenylate kinase (Walker *et al.*, 1982). This observation suggests strongly that the MalK and HisP proteins contain an adenine nucleotide-binding fold. It is tempting to speculate that this binding site may be involved in the coupling of active transport to the hydrolysis of a high-energy phosphoester bond in ATP.

### D. The MalG Protein

The MalG protein has been identified with the aid of a plasmid which expresses the *malF* and *malG* genes from an artificial promoter (Ptac). Strains that contain this plasmid (pHS2) produce approximately 100-fold more MalF protein than fully induced wild-type cells. In addition, it has been possible to show that cells harboring pHS2 produce a cytoplasmic membrane protein of molecular weight 24,000 which is absent in an isogenic strain that contains a *malG* amber mutation on pHS2. Further experiments to confirm that this polypeptide is the product of the *malG* are underway. The role of the MalG protein in determining the location of the MalK protein implies that MalG should also be located in the cytoplasmic membrane (Shuman and Silhavy, 1981). Furthermore, a *malG* mutation has been reported which exhibits a Mal$^+$ Dex$^-$ phenotype. This mutation may be defective in the translocation of dextrins across the cytoplasmic membrane but may still be able to transport maltose (Wandersmann *et al.*, 1979). At the present time, there is no way to estimate the number of MalG polypeptides in wild-type bacteria.

## V. INTERACTIONS BETWEEN THE MALTOSE-BINDING PROTEIN AND THE MEMBRANE COMPONENTS

The major characteristic that distinguishes the multicomponent system from the $H^+$-symport monocomponent active transport systems is the utilization of a periplasmic substrate-binding protein. There is evidence that the periplasmic components play an essential role in some systems (Shuman, 1982), or an auxillary one in others (Robbins and Rotman, 1975). In order to determine if the maltose-binding protein is essential for substrate translocation across the cytoplasmic membrane, we constructed a strain that contains an internal nonpolar deletion of the MBP structural gene, *malE*. This deletion was transferred to a strain that expresses the *mal* regulon constitutively. Under these conditions, all of the *mal* regulon proteins are present in near fully induced levels, with the exception of MBP. If any maltose were translocated across the cytoplasmic membrane, it would be rapidly metabolized by the intracellular enzymes and the cells would grow. This strain fails to grow even when the external maltose concentration is as high as 25 mM. This indicates that, under these conditions, no detectable substrate translocation occurs in the absence of MBP. We interpret these results to indicate that MBP must directly interact with one of the components in the cytoplasmic membrane during substrate translocation (Shuman, 1982).

Revertants of the *malE* deletion strain that regained the ability to grow on maltose were obtained after mutagenesis with ultraviolet light. One of these revertants contains a mutation in the *malG* gene in addition to the *malE* deletion. This strain can accumulate maltose independently of MBP but cannot transport dextrins. Active transport is still sensitive to energy poisons and stereospecific for maltose. In contrast to wild-type cells, transport occurs equally well in whole cells and osmotically sensitive spheroplasts. These results indicate that in the revertant strain there is a recognition site for maltose formed by the cytoplasmic membrane components. It is likely that this site exists in wild-type bacteria as well, but is only available to substrates bound to MBP.

When MBP is replaced genetically by introducing the *malE*$^+$ gene into the revertant strain, transport of maltose is completely *inhibited* and the resulting strain is Mal$^-$. This result suggests that the MBP aberrantly interacts with the cytoplasmic membrane components so that transport cannot occur. Since the *malG* mutation in the revertant strain may only indirectly influence the interaction of wild-type MBP with the membrane components, it is not yet possible to conclude that MBP is interacting with the MalG protein. The strict requirement of MBP for maltose transport in wild-type cells and its complete inhibition of transport in the *malG* MBP-independent mutant do, however, indicate that MBP interacts with some part of the MalF–G–K complex during active transport. A precise description of this interaction will require further genetic and biochemical characterization of these proteins. The maltose-binding protein also participates in transport across the outer membrane. Although the LamB protein itself can act as a facilitator for the diffusion of maltose and maltodextrins *in vitro* and has a measurable affinity for these compounds, these properties do not adequately explain the role of the LamB protein in maltose and maltodextrin transport (Ferenci and Boos, 1980).

As a result of genetic experiments, Wandersmann *et al*. (1979) proposed that, in wild-type bacteria, there is an interaction between the maltose-binding protein and the LamB protein. The result of this interaction is that external substrates have direct access to periplasmic maltose-binding protein without an apparent permeability barrier. A strong argument for the physiological relevance of this interaction is based on the properties of a class of *malE* mutations that produce a defective maltose-binding protein. These mutant strains produce MBP that is not defective in ligand-binding *in vitro*, but the strains exhibit a Mal$^+$ Dex$^-$ phenotype *in vivo*. That is, growth on high concentrations of maltose is unaffected, but the mutant strains are unable to grow on dextrins. In fact, maltotetraose is unable to inhibit maltose transport in these cells, even though the mutant MBP has a higher affinity for maltotetraose than for maltose when measured *in vitro*. These results imply that in wild-type cells, penetration of dextrins into the periplasm is not solely dependent on LamB and that, in this class of *malE* mutants, dextrins cannot gain access to the periplasm because the mutant MBP does not interact properly with the LamB protein (Wandersmann *et al*., 1979).

Recently, it has been possible to demonstrate an MBP–LamB interaction *in vitro*. A column of MBP immobilized on Sepharose was prepared. When detergent extracts of outer membrane were applied to this column, the LamB protein was selectively retained. The adsorbed LamB could be eluted with high salt (Bavoil and Nikaido, 1981). These two sets of independent data suggest strongly that, in wild-type cells, there is a physical interaction between the maltose-binding protein and the LamB protein that is necessary for the transport of dextrins into the periplasmic space. Further evidence could be obtained to strengthen this hypothesis if it were possible to obtain $Dex^+$ revertants of the *malE* mutants that mapped in the *lamB* gene. In addition, one would predict that a Sepharose MBP column prepared with MBP from the $Dex^-$ mutant strain should not retain LamB protein.

Recently, it has been possible to reconstitute active transport of maltose with purified MBP and $Ca^{2+}$-permeabilized cells that contain a deletion of the *malE* gene. It was shown that the reconstituted transport activity was kinetically similar to transport activity in wild-type cells. This indicates that the reconstituted MBP is interacting faithfully with both the LamB protein (this is required at maltose concentration less than $10^{-4}$ M) and the plasma-membrane components. It should be possible to use this reconstitution system to study the details of the interactions of MBP with membrane components of this system (Brass *et al.*, 1981, 1983).

## VI. A MODEL FOR THE OPERATION OF THE MALTOSE–MALTODEXTRIN TRANSPORT SYSTEM

It is possible to formulate a working model for the operation of the maltose–maltodextrin transport system. This model takes into account the following points:

1. The LamB protein provides external substrate molecules direct access to periplasmic MBP. This could be the result of the binding site on LamB for α(1–4)-linked glucosides, the capacity of the LamB protein to form an aqueous channel with specificity for α(1–4)-linked glucosides, the interaction between the LamB protein and MBP, or some combination of all three phenomena.
2. The maltose-binding protein is essential for detectable substrate translocation across the cytoplasmic membrane. This implies an interaction between the maltose-binding protein and the membrane components.
3. The membrane components form a substrate recognition site or a transmembrane channel that is available only to ligand that is bound to MBP.
4. Exit of substrate does not occur via the same components that mediate entry.

### A. Transport across the Outer Membrane

Transport across the outer membrane is not an active process that requires energy; the concentration of *free* substrate is never higher within the periplasm than in the medium. The total amount of substrate, both free and bound to MBP, however, may be as high as the molar concentration of MBP or 1 mM. Substrate molecules can gain access to MBP using two independent routes: the LamB route and the porin route.

The LamB route is exclusively used for dextrins and is kinetically predominant at maltose concentrations below $10^{-4}$ M. The porin route can operate for maltose and maltotriose at high concentrations. It has been proposed that the LamB protein interacts with MBP so that as substrate molecules pass through the LamB protein, they gain access to the binding site on MBP. Transport across the outer membrane *via* LamB protein can be considered to be a two-step process. In the absence of substrate, MBP and LamB are bound together; when a molecule of substrate is added, it traverses whatever distance may be necessary to gain access to the MBP-binding site and binds to MBP. The conformational change of MBP that is observed upon substrate-binding results in disruption of the LamB–MBP interaction and the MBP–ligand complex is discharged into the periplasm. When substrate molecules traverse the outer membrane via the OmpC and OmpF porin proteins, they encounter free MBP in the periplasmic space.

## B. Transport across the Cytoplasmic Membrane

Transport across the cytoplasmic membrane is an active process that requires energy and MBP. The properties of this transport system can be accommodated by models in which the MalF, MalG, and MalK proteins form a complex in the membrane that exists in two conformations, $T_1$ and $T_2$. According to the pore model, $T_1$ corresponds to the open conformation and $T_2$ to the closed conformation. In an alternative, recognition-site model, the two states differ with respect to the orientation of a substrate-recognition site; $T_1$ represents the conformation in which the site is exposed to the periplasm and $T_2$ represents the state that is exposed to the cytoplasm. The requirement for MBP can be explained by both types of models if the $T_2$ state predominates in the absence of MBP–substrate complex and only $T_1$ can interact with MBP–substrate complex.

The energy associated with the binding and/or hydrolysis of a compound such as ATP which contains a high-energy phosphoester could be used to interconvert the $T_1$ and $T_2$ forms in a cyclic manner. This would result in unidirectional transfer of substrate from MBP to the cytoplasm (see Figure 1).

The existence of a substrate-recognition site formed by the MalF and MalG proteins is suggested by the properties of the mutant strain in which stereospecific active transport takes place in the absence of MBP. In terms of the model, MBP-independent transport in the mutant would be the result of an increase in the $T_1$ conformation in the absence of MBP. This corresponds either to an open pore or to a substrate-recognition site exposed to the periplasm. Since the MBP-independent transport that is observed is competitively inhibited exclusively by $\alpha$(1–4) glucosides, the existence of a nonspecific pore seems less likely than a stereospecific recognition site. Since the mutation which results in exposure of this site is in *malG*, it seems likely that the MalG protein makes up part of this site.

The fact that MBP *inhibits* transport of maltose in the *malG*, MBP-independent mutant can be explained most easily by a substrate-recognition model. In this model, the affinity of the wild-type recognition site for free maltose is so low that transport does not occur. In the MBP-independent mutant, however, the affinity for free maltose is raised sufficiently so that transport can occur in the absence of MBP. At the same time, the affinity for MBP–maltose is also raised, but is so high that transport can no longer take place. This model predicts that the interaction of MBP with membranes from the *malG* mutant should have a higher affinity than the interaction with wild-type membranes. Experiments designed to test this prediction are currently in progress.

The proposed location of the MalK protein on the inner surface of the cytoplasmic membrane makes it an attractive candidate for the site of energy coupling. Binding of a high-energy compound (such as ATP) and its subsequent hydrolysis by the MalK protein could drive the conformational changes in MalF and MalG which correspond to the interconversion of the $T_1$ and $T_2$ states.

It should be possible to test various aspects of these models by studying the properties of this transport system in subcellular vesicles or in a reconstituted system with purified components.

ACKNOWLEDGMENTS

The work in the authors' laboratory is supported by Grant AI-19276 from the National Institutes of Health. We thank Christopher Higgins for pointing out the presence of the adenine nucleotide fold in the MalK and HisP protein sequences. It is a pleasure to acknowledge the skills and cheerful assistance of Ms. Irma Portalatin with the preparation of this manuscript.

## REFERENCES

Bavoil, P., and Nikaido, H., 1981, Physical interaction between the phage λ receptor protein and the carrier-immobilized maltose binding protein of *E. coli*, *J. Biol. Chem.* **256:**11385–11388.

Bavoil, P., Hofnung, M., and Nikaido, H., 1980, Identification of a cytoplasmic membrane associated component of the maltose transport system of *Escherichia coli*, *J. Biol. Chem.* **255:**8366–8369.

Berger, E. A., 1973, Different mechanisms of energy coupling for the active transport of proline and glutamine in *Escherichia coli*, *Proc. Natl. Acad. Sci. USA* **70:**1514–1518.

Berger, E. A., and Heppel, L. A., 1974, Different mechanisms for energy coupling for the shock-sensitive and shock-resistant amino acid permeases of *Escherichia coli*, *J. Biol. Chem.* **249:**7747–7755.

Boehler-Kohler, B. A., Boos, W., Dieterle, R., and Benz, R., 1979, Receptor for bacteriophage lambda of *Escherichia coli* forms larger pores in black lipid membranes than matrix protein (porin), *J. Bacteriol.* **138:**33–39.

Boos, W., Ferenci, T., and Shuman, H. A., 1981, Formation and excretion of acetylmaltose after accumulation in *E. coli*, *J. Bacteriol.* **146:**725–732.

Brass, J. M., Boos, W., and Hengge, R., 1981, Reconstitution of maltose transport in *malB* mutants of *E. coli* through calcium-induced disruptions of the outer membrane, *J. Bacteriol.* **146:**10–17.

Brass, J. M., Ehmann, U., and Bukau, B., 1984, Reconstitution of maltose transport in *Escherichia coli:* Conditions affecting import of maltose binding protein into the periplasm of calcium treated cells, *J. Bacteriol.*, in press.

Clement, J. M., and Hofnung, M., 1981, Gene sequence of the λ receptor, an outer membrane protein of *E. coli* K-12, *Cell* **27:**507–514.

Debarbouille, M., Shuman, H. A., Silhavy, T. J., and Schwartz, M., 1978, Dominant constitutive mutations in *malT*, the positive regulator gene of the maltose regulon in *Escherichia coli*, *J. Mol. Biol.* **124:**359–371.

Dietzel, I., Kolb, V., and Boos, W., 1978, Pole cap formation in *Escherichia coli* following induction of the maltose binding protein, *Arch. Mikrobiol.* **118:**207–218.

Emr, S. D., and Silhavy, T. J., 1980, Mutations affecting localization of an *Escherichia coli* outer membrane protein, the bacteriophage lambda receptor, *J. Mol. Biol.* **141:**63–90.

Ferenci, T., 1980a, Methyl-α-maltoside and 5-thiomaltose: Analogs of maltose transported by the *Escherichia coli* maltose transport system, *J. Bacteriol.* **144:**7–11.

Ferenci, T., 1980b, The recognition of maltodextrins by *Escherichia coli, Eur. J. Biochem.* **108:**631–636.

Ferenci, T., and Boos, W., 1980, The role of the *Escherichia coli* λ receptor in the transport of maltose and maltodextrins, *J. Supramol. Struct.* **13:**101–116.

Ferenci, T., and Klotz, U., 1978, Affinity chromatographic isolation of the periplasmic maltose binding protein of *Escherichia coli, FEBS Lett.* **94:**213–217.

Ferenci, T., Schwentorat, M., Ullrich, S., and Vilmart, J., 1980, Lambda receptor in the outer membrane of *Escherichi coli* as a binding protein for maltodextrins and starch polysaccharides, *J. Bacteriol.* **142:**521–526.

Gabay, J., and Yasunaka, K., 1980, Interaction of the LamB protein with the peptidoglycan layer in *E. coli* K-12, *Eur. J. Biochem.* **104:**13–18.

Gabay, J., Benson, S., and Schwartz, M., 1983, Genetic mapping of antigenic determinants on a membrane protein, *J. Biol. Chem.* **258:**2410–2414.

Gilson, E., Nikaido, H., and Hofnung, M., 1982a, Sequence of the *malK* gene of *E. coli* K-12, *Nucleic Acids Res.* **10:**7449–7458.

Gilson, E., Higgins, C. F., Hofnung, M., Ferro-Luzzi Arnes, and Nikaido, H., 1982b, Extensive homology between membrane associated components of histidine and maltose transport systems of *Salmonella typhimurium* and *Escherichia coli, J. Biol. Chem.* **257:**9915–9918.

Heuzenroeder, M. W., and Reeves, P., 1980, Periplasmic maltose binding protein confers specificity on the outer membrane maltose pore of *Escherichia coli, J. Bacteriol.* **141:**431–435.

Kellermann, O., and Szmelcman, S., 1974, Active transport of maltose in *Escherichia coli* K-12: Involvement of a "periplasmic maltose binding protein," *Eur. J. Biochem.* **47:**139–149.

Koiwai, O., and Hayashi, H., 1979, Studies on bacterial chemotaxis: VI. Interaction of maltose receptor with membrane bound chemosensing component, *J. Biochem. (Tokyo)* **86:**27–34.

Luckey, M., and Nikaido, H., 1980, Specificity of diffusion channels produced by λ-phage receptor protein of *Escherichia coli, Proc. Natl. Acad. Sci. USA* **77:**167–171.

Nakae, T., and Ishii, J. M., 1980, Permeability properties of *Escherichia coli* outer membrane containing pore forming proteins: Comparison between lambda receptor protein and porin for saccharide permeation, *J. Bacteriol.* **142:**735–740.

Nakae, T., and Ishii, J. M., 1982, Molecular weights and subunit structure of LamB proteins, *Ann. Microbiol. (Inst. Pasteur)* **133A:**21–25.

Neu, H. C., and Heppel, L., 1965, The release of enzymes from *E. coli* by osmotic shock and during the formation of spheroplasts, *J. Biol. Chem.* **240:**3685–3692.

Neuhaus, J. M., 1982, The receptor protein of phage λ: Purification, characterization and preliminary electrical studies in planar lipid bilayers, *Ann. Microbiol. (Inst. Pasteur)* **133A:**27–32.

Parnes, J. R., and Boos, W., 1973, Undirectional transport activity mediated by the galactose binding protein of *Escherichia coli, J. Biol. Chem.* **248:**4436–4445.

Raibaud, O., Roa, M., Braun-Breton, C., and Schwartz, M., 1979, Structure of the *malB* region in *Escherichia coli* K-12 I. Genetic map of the *malK-lamB* operon, *Mol. Gen. Genet.* **174:**241–248.

Randall-Hazelbauer, L. L., and Schwartz, M., 1973, Isolation of the bacteriophage lambda receptor from *Escherichia coli* K-12, *J. Bacteriol.* **116:**1436–1446.

Roa, M., and Clement, J. M., 1980, Location of a phage binding site on an outer membrane protein, *FEBS Lett.* **121:**127–129.

Robbins, A., and Rotman, B., 1975, Evidence for binding protein-independent substrate translocation by the β-methyl galactoside transport system of *Escherichia coli* K-12, *Proc. Natl. Acad. Sci. USA* **72:**423–427.

Schwartz, M., 1966, Location of the maltose A and B loci on the genetic map of *Escherichia coli, J. Bacteriol.* **92:**1083–1089.

Schwartz, M., 1967, Sur l'existence chez *Escherichia coli* K-12 d'une regulation commune a la biosynthése des recepteurs du bacteriophage lambda et au metabolisme du maltose, *Ann. de l'Institut Pasteur* **113:**685–704.

Schwartz, M., 1980, Interaction of phages with their receptor proteins, in: *Virus Receptors,* Vol. 7 (L. L. Randal and L. Philipson, eds.), Chapman and Hall, London, pp. 61–94.

Schwartz, M., Kellerman, O., Szmelcman, S., and Hazelbauer, G., 1976, Further studies on the binding of maltose to the maltose binding protein of *Escherichia coli, Eur. J. Biochem.* **71:**167–170.

Shuman, H. A., 1982, Active transport of maltose in *Escherichia coli* K-12: Role of the periplasmic maltose binding protein and evidence for a substrate recognition site in the cytoplasmic membrane, *J. Biol. Chem.* **257:**5455–5461.

Shuman, H. A., and Beckwith, J. R., 1979, Mutants of *Escherichia coli* K-12 that allow transport of maltose via the β-galactoside transport system, *J. Bacteriol.* **137:**365–373.

Shuman, H. A., and Silhavy, T. J., 1981, Identification of the *malK* gene product: A peripheral membrane component of the *E. coli* maltose transport system, *J. Biol. Chem.* **256:**560–562.

Shuman, H. A., Silhavy, T. J., and Beckwith, J., 1980, Labeling proteins with β-galactosidase by gene fusion: Identification of a cytoplasmic membrane component of the *Escherichia coli* maltose transport system, *J. Biol. Chem.* **255:**168–174.

Silhavy, T. J., Casadaban, M., Shuman, H. A., and Beckwith, J. R., 1976, Conversion of β-galactosidase to a membrane-bound state by gene fusion, *Proc. Natl. Acad. Sci. USA* **73:**3423–3427.

Silhavy, T. J., Brickman, E., Bassford, P., Casadaban, M. J., Shuman, H. A., Schwartz, V., Guarente, L., Schwartz, M., and Beckwith, J. R., 1979, Structure of the *malB* region in *Escherichia coli* K-12 II. Genetic map of the *malE,F,G* operon, *Mol. Gen. Gen.* **174:**249–259.

Szmelcman, S., and Hofnung, M., 1975, Maltose transport in *Escherichia coli* K-12: Involvement of the bacteriophage lambda receptor, *J. Bacteriol.* **124:**112–118.

Szmelcman, S., Schwartz, M., Silhavy, M., and Boos, W., 1976, Maltose transport in *Escherichia coli* K-12: A comparison of transport kinetics in wild-type and λ resistant mutants with the dissociation constants of the maltose binding protein as measured by fluorescence quenching, *Eur. J. Biochem.* **65:**13–19.

Walker, J. E., Saraste, M., Runswick, M. J., and Gay, N. J., 1982, Distantly related sequences in the α and β subunits of ATP synthase, myosin, kinases and other ATP-requiring enzymes and a common nucleotide binding fold, *Eur. Mol. Biol. Org. J.* **1:**945–951.

Wandersmann, C., Schwartz, M., and Ferenci, T., 1979, Mutants of *Escherichia coli* impaired in the transport of maltodextrins, *J. Bacteriol.* **140:**1–13.

43

# Bacterial Amino Acid Transport Systems

Robert Landick, Dale L. Oxender, and Giovanna Ferro-Luzzi Ames

## I. INTRODUCTION

The roles of the cell membrane as both a permeability barrier to foreign substances and as a selective filter which admits nutrient molecules have been recognized as fundamentally important to animal physiology since the first decade of this century. It was not until the 1950's, however, that microbiologists recognized that bacteria possess membrane transport systems, distinct from their metabolic enzymes, that mediate the uptake of amino acids and other nutrients. The work of Cohen and Monod (1957) established modern research on bacterial membrane transport. While much of this early work focused on transport of sugars, Gale provided the first evidence for accumulation of amino acids by bacteria during this era (Gale, 1947, 1954).

Today, amino acid transport systems for almost all of the amino acids have been recognized and studied. This review will tabulate information on these systems, but it is not intended to be a comprehensive treatment of all amino acid transport systems. For additional information on other systems, the reader is referred to recent reviews on the subject (Oxender, 1972, 1974, 1975; Slayman, 1973; Boos, 1974; Halpern, 1974; Simoni and Postma, 1975; Wilson, 1978; Iaccarino *et al.*, 1980; Anraku, 1980; Furlong and Schellenberg, 1980).

The existence of active amino acid transport systems has been deduced or inferred from a variety of biochemical and genetic approaches. In some instances, the existence

*Robert Landick* • Department of Biological Sciences, Stanford University, Stanford, California 94305. *Dale L. Oxender* • Department of Biological Chemistry, The University of Michigan, Ann Arbor, Michigan 48109. *Giovanna Ferro-Luzzi Ames* • Department of Biochemistry, University of California at Berkeley, Berkeley, California 94720.

and role of specific amino acid transport proteins is well documented while, in other cases, the existence of a given amino acid transport system has been inferred from kinetic analysis of amino acid uptake measurements. The two best characterized amino acid transport systems are the high-affinity periplasmic binding protein-(BP) dependent transport system for leucine, isoleucine, and valine (LIV-I) system in *E. coli* (Piperno and Oxender, 1966, 1968; Nakane *et al.*, 1968; Penrose *et al.*, 1970; Rahmanian and Oxender, 1972; Rahmanian *et al.*, 1973; Harrison *et al.*, 1975; Quay *et al.*, 1975a,b, 1977, 1978; Anderson *et al.*, 1976; Oxender and Quay, 1976a–c; Quay and Oxender, 1976, 1977, 1979, 1980a,b; Anderson and Oxender, 1977, 1978; Oxender *et al.*, 1977, 1980a,b; Landick *et al.*, 1980, 1983; Daniels *et al.*, 1980, 1981; Landick and Oxender, 1981; Wood, 1975; Guardiola *et al.*, 1974a,b; Templeton and Savageau, 1974a,b; Yamato and Anraku, 1977, 1980); and the high-affinity histidine transport system in *S. typhimurium* (Ames, 1964, 1972; Ames and Roth, 1968; Ames and Lever, 1970, 1972; Rosen and Vasington, 1971; Lever, 1972; Kustu and Ames, 1973; Shaltiel *et al.*, 1973; Ames and Spudich, 1976; Ames *et al.*, 1977; Robertson *et al.*, 1977; Ames and Nikaido, 1978; Kustu *et al.*, 1979; Noel *et al.*, 1979; Manuck and Ho, 1979; Ho *et al.*, 1980; Ardeshir and Ames, 1981; Ardeshir *et al.*, 1981; Higgins and Ames, 1981, 1982; Higgins *et al.*, 1982a,b; Gilson *et al.*, 1982b). These systems both fall in the class of high-affinity, periplasmic, BP-dependent systems and possess remarkable similarities. This review will focus on these two examples of amino acid transport systems. While this approach will necessarily limit treatment of the various mechanisms and components that are found in different amino acid transport systems, we hope it will give a comprehensive and focused picture of the chosen model systems.

We begin by considering definitions of amino acid transport and the different types of systems which have been recognized and then proceed to a tabulation of the multiple amino acid transport systems found in bacteria and a detailed examination of the protein components and genetic organization of the histidine and LIV-I systems. The sequence of events which leads from the synthesis of transport system components to their final assembly in the inner membrane and periplasm is discussed. This is followed by a review of the energization and reconstitution of amino acid transport systems. We conclude with discussions of how amino acid transport systems are regulated and how amino acid transport systems may have evolved.

## II. CLASSES OF TRANSPORT SYSTEMS

The terms *transport system* and *permease* will be considered synonymous and taken to mean the entire assembly of protein components which are required for a given transport process. The individual components will be referred to as *transport proteins* whether they are membrane-bound or soluble periplasmic proteins.

Transport of solutes across membranes is usually divided into three classes: (1) *passive diffusion,* meaning diffusion through the lipid–protein bilayer without specific interaction, (2) *facilitated diffusion,* meaning passage through the membrane by specific binding to and release from a carrier protein without requiring energy, and (3) *active transport,* meaning passage of a solute through the membrane by binding to and release from a carrier protein in a process coupled to metabolic energy. While passive diffusion and facilitated diffusion require that the solute always move to dissipate a concentration gradient, active transport is usually taken to mean that the solute is accumulated against a concentration gradient. This review will be limited to amino acid transport systems that fulfill the criteria of active transport.

At least four distinct mechanisms of bacterial active transport have been demonstrated: (1) sodium-independent, membrane-bound transport systems which require only membrane components, (2) sodium-dependent, membrane-bound transport systems which require only membrane components and sodium ions, (3) periplasmic BP-dependent transport systems which require both periplasmic and membrane-bound components, and (4) cation-stimulated ATPase systems which are multicomponent assemblies and are thought to be responsible for creating a transmembrane proton gradient (Mitchell, 1970). Of these mechanisms, only numbers (1)–(3) have been shown to function for amino acid transport and will be discussed here. A fifth, somewhat distinct mechanism of transport is *group translocation* which is best exemplified by the phosphoenolpyruvate-utilizing phosphotransferase systems that transport various carbohydrates (Postma and Roseman, 1976; Dills *et al.*, 1980). There are no known examples of group translocation involving amino acids in bacteria.

## A. Multiplicity of Amino Acid Transport Systems

Table 1 lists the various amino acid transport systems that have been described in the literature. In compiling this list, we have given consideration to the criteria used in identifying systems and the arguments presented above. Rather than compile a comprehensive list, we have attempted to present those systems which are based on the strongest evidence. Where conflicts exist in reports about a given system or class of systems, we have used the nomenclature or values from the best documented evidence. References to other interpretations are included. The table contains the information available on *E. coli* and *S. typhimurium;* these two organisms are extremely similar from the biological and genetic points of view and, in many cases, the results obtained with one species also apply to the other. Limited information is also given on *P. aeruginosa* where amino acid transport systems similar to those found in *E. coli* and *S. typhimurium* have been characterized.

It is obvious from this table that each amino acid may be concentrated from the extracellular medium by several different permeases or transport systems with overlapping specificities. Leucine enters by at least three systems: LIV-I, LS, and low-affinity, membrane-bound transport system for leucine, isoleucine, and valine (LIV-II; Piperno and Oxender, 1966, 1968; Nakane *et al.*, 1968; Penrose *et al.*, 1970; Rahmanian and Oxender, 1972; Rahmanian *et al.*, 1973; Harrison *et al.*, 1975; Quay *et al.*, 1975a,b, 1977, 1978; Anderson *et al.*, 1976; Oxender and Quay, 1976a–c; Quay and Oxender, 1976, 1977, 1979, 1980a,b; Anderson and Oxender, 1977, 1978; Oxender *et al.*, 1977, 1980a,b; Landick *et al.*, 1980, 1983; Daniels *et al.*, 1980, 1981; Landick and Oxender, 1981; Wood, 1975; Guardiola *et al.*, 1974a,b; Templeton and Savageau, 1974a,b; Yamato and Anraku, 1977, 1980); histidine, by five systems: aromatic, high-affinity histidine, LAO, and two kinetically distinct residual systems

Table 1. Bacterial Amino Acid Transport Systems

| System name | Comments | Substrates | K(μM) | Organism characterized in | Genes | References[a] |
| --- | --- | --- | --- | --- | --- | --- |
| Arginine-specific | Uses the Arg-BP, repressed by L-arginine or L-ornithine | L-Arginine, L-ornithine, D-arginine | 0.12 | *E. coli* | *abpS*, Arg-BP, gene, min 50 | 1–7 |
| | | | | | *abpR*, Regulatory gene, min 50, mutation allows growth on D-arginine | |
| General aromatic | General aromatic amino acid transport system | L-Phenylalanine, L-tyrosine, L-tryptophan, L-histidine | | *S. typhimurium* | | 8 |
| | | | 0.1–0.5 | *S. typhimurium* | *aroP*, Min 3 | 9,10 |
| | | | 0.1–0.5 | *E. coli* | *aroP*, Min 2 | 11,12 |
| | | | 0.1–0.5 | *P. aeruginosa* | | 13 |
| Asparagine-I | Repressed by asparagine | L-Asparagine, L-aspartic acid | 3.5 | *E. coli* | | 14 |
| Asparagine-II | Necessary for utilization of asparagine as a nitrogen source | L-Asparagine | 80 | *E. coli* | | 14 |
| Aspartate-specific | A membrane-bound transport system | L-Aspartic acid | 3.7 | *E. coli* | | 15,16 |
| Cystine | A binding protein-dependent system that uses Cys-BP | L-Cystine, diamino-pimelic acid | 0.3 | *E. coli* | | 17–20 |
| | | | 14 | *S. typhimurium* | | 21 |
| DAG | A membrane-bound system for small amino acids; mutation renders cells resistant to D-serine and D-cycloserine D-cycloserine | Glycine, L-alanine, D-alanine, L-serine, D-serine D-serine | 3–5<br>2<br>2 | *E. coli* | *dag*, Min 83 | 22–27,11,93 |

| | | | | | | |
|---|---|---|---|---|---|---|
| Low-affinity glutamate, aspartate | A membrane-bound system which is inhibited by β-hydroxyaspartate | L-Glutamic acid, L-aspartic acid, β-hydroxy-aspartic acid | 5<br>5 | *E. coli* | | 16,28 |
| High-affinity glutamate, aspartate | A binding protein-dependent system which uses glutamate, aspartate BP | L-Glutamic acid, L-aspartic acid | 0.5<br>0.5 | *E. coli* | | 16,29–35 |
| Glutamate-specific transport system | A membrane-bound, sodium-dependent transport system | L-Glutamate L-methyl-glutamate | 1.5 | *E. coli* | | 16,36–40 |
| Glutamine | Inhibited by glutamylhydrazide and glutamylhydrazone, repressed by glutamine, derepressed by nitrogen starvation | L-Glutamine | 0.05 | *E. coli*<br>*S. typhimurium* | | 40–45<br>46,47 |
| Histidine | A binding protein-dependent system which uses the His-BP | L-Histidine, D-histidine | 0.01 | *S. typhimurium* | *hisJ*, Min 48.5, *hisP*, min 48.5, *hisQ*, min 48.5, *hisM*, min 48.5, *duaH*, min 48.5; mutation allows cells to grow on D-histidine as a nitrogen source | 48–53 |
| | | | | *E. coli* | | 54 |
| LAO | A binding protein-dependent system which uses the LAO-BP and the *hisPQM* components | L-Lysine, L-arginine, L-ornithine, L-histidine | | *S. typhimurium* | *argT*, Min 48.5, *hisP*, min 48.5, *hisQ*, min 48.5, *hisM*, min 48.5 | 46,53–57 |

*(continued)*

Table 1. (Continued)

| System name | Comments | Substrates | K(μM) | Organism characterized in | Genes | References[a] |
|---|---|---|---|---|---|---|
| LIV-I | A binding protein-dependent system which uses the *livGH* components, repressed by L-leucine | L-Alanine, L-leucine, L-isoleucine, L-valine, L-threonine | 30<br>0.2–0.6 | *E. coli* | *livJ*, Min 74.5, *livH*, min 74.5, *livG*, min 74.5, *livR*, Min 20, a regulatory gene, mutation derepresses LIV-I transport | 58–63 |
| | | | | *S. typhimurium* | | 64 |
| | | | | *P. aeruginosa* | | 65,66 |
| Leucine-specific (LS) | A binding protein-dependent system that uses the LS-BP and *livGH* components, repressed by L-leucine | L-Leucine, D-leucine | 0.2–0.6 | *E. coli* | *livK*, Min 74.5, *livG*, min 74.5, *livH*, min 74.5, *lstR*, min 20, a regulatory gene; mutation allows cells to use D-leucine as a source of L-leucine | 58,59,63,67 |
| LIV-II | A membrane-bound system that is only weakly repressed by leucine | L-Leucine,<br>L-isoleucine,<br>L-valine | 10–20<br>5<br>7 | *E. coli* | *livP*, Min 75 | 58,59,68–71 |
| | | | | *P. aeruginosa* | | 66 |
| Methionine-I | | L-Methionine, D-methionine | 0.075 | *E. coli* | *metD*, Min 5 | 72–74 |
| | | | | *S. typhimurium* | *metD*, Min 6 | 47,75 |
| Methionine-II | | L-Methionine, D-methionine | | *E. coli* | | 72–74 |
| Phenylalanine-specific | | L-Phenylalanine | 1 | *S. typhimurium* | *pheP*, Min 15–35 | 48,49 |
| | | | | *E. coli* | | 12,79 |

| | | | | | | |
|---|---|---|---|---|---|---|
| Proline PP-I | A membrane-bound, high-affinity system, induced by proline, catabolite repressed, required for growth on proline as sole carbon or nitrogen source | L-Proline, L-hydroxy-proline | 2 | *S. typhimurium* | *putP*, Min 22 | 77–80 |
| | | | | | *putA*, Min 22, gene for a PP-I system repressor | |
| | | | | *E. coli* | | 81–87 |
| | | | | *P. aeruginosa* | | 13 |
| Proline PP-II | Low-affinity, membrane-bound system, induced by amino acid starvation | L-Proline | 300 | *S. typhimurium* | *proP*, Min 92 | 77,88,89 |
| | | | | *E. coli* | | 83 |
| Proline PP-III | Functions only at high osmolarity (0.3 M NaCl) | L-Proline | | *S. typhimurium* | *proU*, Min 59 | 90 |
| Tryptophan-specific I | Inducible by tryptophan | L-Tryptophan | 0.1 | *E. coli* | *trpP*, Min 69 | 11,12,95 |
| | | | | *S. typhimurium* | | 48,49 |
| Tryptophan-specific II | Constitutively expressed | L-Tryptophan | 1 | *E. coli* | | 12 |
| | | | | *S. typhimurium* | | 48,49 |
| Tryptophan-specific III ($T_3A$) | Catabolite repressed, induced by tryptophan | L-Tryptophan | 10 | *E. coli* | *tnaB*, Min 83 | 91,92,94,95 |
| Tyrosine-specific | | L-Tyrosine | 1 | *S. typhimurium* | | 48,49 |
| | | | | *E. coli* | *tyrP*, Min 42 | 12,76 |

[a] (1) Rosen, 1971, (2) Rosen, 1973b, (3) Rosen, 1973a, (4) Cells, 1981, (5) Cells, 1977, (6) Cells, 1982, (7) Cells *et al.*, 1973, (8) Kreischman *et al.*, 1973, (9) Ames, 1964, (10) Ames and Roth, 1968, (11) Piperno and Oxender, 1968, (12) Brown, 1971, (13) Kay and Gronlund, 1969, (14) Willis and Woolfolk, 1975, (15) Kay, 1971, (16) Schellenberg and Furlong, 1977, (17) Leive and Davis, 1965a, (18) Leive and Davis, 1965b, (19) Berger and Heppel, 1972, (20) Oshimaa *et al.*, 1974, (21) Baptist and Kredich, 1977, (22) Lee *et al.*, 1975, (23) Lombardi and Kaback, 1975, (24) Cosloy, 1973, (25) Kaback and Kostellow, 1968, (26) Wargel *et al.*, 1970, (27) Wargel *et al.*, 1971, (28) Barash and Halpern, 1975, (29) Willis and Furlong, 1975, (30) Furlong and Schellenberg, 1980, (31) Schellenberg, 1978, (32) Miner and Frank, 1974, (33) Barash and Halpern, 1975, (34) Schellenberg and Furlong, 1977, (35) Aksamit *et al.*, 1975, (36) Frank and Hopkins, 1969, (37) Kahane *et al.*, 1975, (38) Tsuchiya *et al.*, 1977, (39) Britten and McClure, 1962, (40) MacDonald *et al.*, 1977, (41) Weiner and Heppel, 1971, (42) Weiner *et al.*, 1971, (43) Willis *et al.*, 1975, (44) Masters and Hong, 1981b, (45) Plate, 1979, (46) Kustu *et al.*, 1979, (47) Ayling, 1981, (48) Ames, 1964, (49) Ames and Roth, 1968, (50) Ames and Nikaido, 1978, (51) Ames *et al.*, 1977, (52) Masters and Hong, 1981b, (53) Higgins *et al.*, 1982a, (54) Ardeshir and Ames, 1981, (55) Ardeshir *et al.*, 1981, (56) Higgins and Ames, 1981, (57) Higgins and Ames, 1982, (58) Rahmanian *et al.*, 1973, (59) Anderson and Oxender, 1977, (60) Oxender *et al.*, 1977, (61) Anderson and Oxender, 1978, (62) Oxender *et al.*, 1980a, (63) Landick *et al.*, 1980, (64) Kiritani, 1974, (65) Hoshino and Nishio, 1982, (66) Hoshino and Kageyama, 1982, (67) Oxender *et al.*, 1980b, (68) Wood, 1975, (69) Guardiola *et al.*, 1974b, (70) Yamato and Anraku, 1980, (71) Iaccarino *et al.*, 1978, (72) Kadner, 1974, (73) Kadner and Watson, 1974, (74) Kadner, 1977, (75) Ayling *et al.*, 1979, (76) Whipp *et al.*, 1980, (77) Wood, 1981, (78) Ratzkin *et al.*, 1978, (79) Wood *et al.*, 1979, (80) Menzel, 1980, (81) Kaback and Stadtman, 1966, (82) Kaback and Deuel, 1969, (83) Morikawa *et al.*, 1974, (84) Amanuma *et al.*, 1977. (85) Rowland and Tristam, 1974, (86) Motojima *et al.*, 1978, (87) Condamine, 1971, (88) Anderson *et al.*, 1980, (89) Menzel and Roth, 1980, (90) Csonka, 1982, (91) Boezi and Demoss 1961, (92) Burrous and Demoss, 1963, (93) Robbins and Oxender, 1973, (94) Edwards and Yudkin, 1982, (95) V. Stewart, personal communication.

(Ames, 1964; Ames and Roth, 1968; Ames and Lever, 1970; Kustu *et al.*, 1979; Higgins *et al.*, 1982a); and proline, by at least three systems: PP-I, PP-II, and PP-III (Kaback and Stadtman, 1966; Kaback and Deuel, 1969; Morikawa *et al.*, 1974; Amanuma *et al.*, 1977; Wood, 1981; Ratzkin *et al.*, 1978; Wood *et al.*, 1979; Menzel, 1980; Rowland and Tristam, 1974; Motojima *et al.*, 1978; Condamine, 1971; Kusaka *et al.*, 1976; Anderson *et al.*, 1980; Menzel and Roth, 1980; Csonka, 1982). Glutamate may enter bacteria by as many as five routes and aspartate and arginine by at least three (Furlong and Schellenberg, 1980; Schellenberg and Furlong, 1977; Ames, 1972; Rosen, 1973a; Celis, 1977, 1981, 1982; Higgins and Ames, 1981).

Determining the existence of multiple amino acid transport systems for a given substrate should involve several approaches (Ames, 1972). The existence of system multiplicity has often been predicated almost entirely on kinetic analysis alone. On occasion, multiple transport systems have been postulated solely on the basis of biphasic transport kinetics. While the demonstration of biphasic Lineweaver–Burk plots is suggestive of multiple transport systems, it is by no means conclusive. It is important to keep in mind that the Michaelis–Menten assumptions are most likely not fulfilled by complex transport processes even if the individual components may obey the equation. It is, therefore, not appropriate to base arguments for multiple routes of entry on this type of analysis alone. A variety of other explanations for two or more apparent $K_m$s of amino acid uptake are possible. Carrier proteins may be subject to allosteric effects which alter their affinity for substrates as a function of substrate concentration. Other cellular processes may respond to changes in the concentration of a given amino acid and alter the properties of carrier proteins indirectly by changing, for instance, the transmembrane electrochemical potential; and, on occasion, impurities in the radioactively labeled amino acids used in uptake measurements can lead to spurious conclusions about transport kinetics. Given these potential pitfalls, multiphasic kinetics should not be used alone as the only argument for multiple transport systems.

Analogue inhibition analysis can provide alternative criteria for multiple transport systems (Ames, 1964, 1974) and has been used extensively in studies of animal cell transport (Oxender and Christensen, 1963; Christensen, 1975, 1982). With this method, multiple systems are revealed when some component of uptake remains after inhibition with a large concentration of a structural analogue. For example, only 80–90% of L-leucine uptake at 0.5 μM substrate can be inhibited by 200 μM L-threonine. The 10–20% remaining is uptake by the LS transport system while the threonine inhibitable fraction is due to uptake by the LIV-I system. Since it avoids relying on the validity of kinetics assumptions, the analogue inhibition strategy is a valuable technique for identifying multiple transport systems.

The most definitive method for defining bacterial amino acid transport systems is mutational analysis. If mutants in a given transport system can be isolated by selection or identified by screening, other systems with overlapping specificities can be readily identified. Furthermore, once mutants are obtained, identification of the protein components affected by the mutation and molecular cloning of the gene(s) for the defective component(s) are possible. This type of genetic approach has been exploited in our analysis of the leucine and histidine transport systems and has led to rapid advances in our understanding of how these systems are assembled and how they function.

Although a precise definition of all the amino acid transport systems in *E. coli* or *S. typhimurium* is not yet possible, a general picture has emerged. Bacterial amino acid transport systems usually transport a class of amino acids, such as basic amino acids or branched-chain amino acids, but this specificity is not as broad as is found for amino acid transport systems in eukaryotic microorganisms such as yeast or *Neurospora* or in higher eukaryotes. *Neurospora* and yeast, for instance, each possess a transport system which transports all amino acids, a so-called general amino acid permease (Pall, 1969; Grenson and Hon, 1972; Seaston *et al.*, 1973, 1976; Rao *et al.*, 1975; DeBusk and DeBusk, 1980; Ogilvie-Villa *et al.*, 1981).

In addition to systems transporting multiple amino acids, most amino acids are also transported by systems which are specific for a single amino acid. No such systems have been found, so far, in eukaryotes. However, unlike eukaryotic cells, bacteria are able to synthesize all 20 amino acids and therefore adapt to a wide variety of environments which may be deficient in one or more amino acids. Any one amino acid may utilize either membrane-bound systems or periplasmic BP-dependent systems, or both.

## B. Membrane-Bound Systems

The best-characterized example of a sodium-independent membrane-bound system is the proline transport system (PP-I in Table 1) in *E. coli* and *S. typhimurium* (Kaback and Stadtman, 1966; Kaback and Deuel, 1969; Morikawa *et al.*, 1974; Amanuma *et al.*, 1977; Wood, 1981). The best-characterized example of a sodium-dependent, membrane-bound system is the glutamate transport system of *E. coli* (Britten and McClure, 1962; Frank and Hopkins, 1969; Schellenberg and Furlong, 1977; Kahane *et al.*, 1975; MacDonald *et al.*, 1977; Tsuchiya *et al.*, 1977). Sodium-dependent transport of glutamate has also been observed in the halophilic genus *Halobacter* (Stevenson, 1966) and in the marine genus *Pseudomonad* (Wong *et al.*, 1980). The sodium-dependent, membrane-bound glutamate transport system is one of three distinct *E. coli* systems which transport glutamate. It is distinguished from the other systems by its resistance to osmotic shock and inhibition by the amino acid analog α-methylglutamate. Sodium-dependent amino acid transport is relatively rare in bacteria, unlike mammalian cells where transport of many solutes is linked with the movement of sodium ions. Early in the study of amino acid transport, some systems were erroneously thought to be sodium-dependent because inclusion of sodium ion in uptake assay and wash buffers partially blocked dumping of amino acid pools when cells were treated with cold wash buffer (Piperno and Oxender, 1968; Britten and McClure, 1962). Once the phenomenon of cold-shock-induced amino acid pool loss was widely appreciated and experimental protocols were modified to include room temperature washes, most of these effects disappeared.

## C. Periplasmic BP-Dependent Systems

In addition to uptake by membrane-bound systems, some amino acids, as well as some ions and sugars, are transported by more complicated systems which include both periplasmic and membrane-bound components. The periplasmic components are solute BPs which bind specific substrates or sets of substrates with very high affinity. When the periplasmic BPs are removed from the cell by the osmotic shock procedure of Neu and Heppel (1965) or by lysozyme–ethylenediamine tetraacetic acid (EDTA) treatment, transport of amino acids by these high-affinity systems is no longer observed. Hence, these systems have been termed periplasmic BP-dependent transport systems or shock-sensitive systems. Bacterial periplasmic BPs that bind arginine [lysine, arginine, ornithine-binding protein (LAO-BP) and Arg-BP] (Neu and Heppel, 1965; Wilson and Holden, 1969a,b; Rosen, 1971, 1973a,b; Celis *et al.*, 1973; Celis, 1977, 1981, 1982; Kustu *et al.*, 1979; Wong *et al.*, 1980; Ardeshir *et al.*, 1981; Higgins *et al.*, 1982a), aspartic acid [glutamate, aspartate-BP] (Miner and Frank, 1974; Willis and Furlong, 1975; Aksamit *et al.*, 1975; Schellenberg, 1978), cystine [cystine-BP] (Berger and Heppel, 1972; Oshimaa *et al.*, 1974), glutamic acid [glutamate, aspartate-BP] (Miner and Frank, 1974; Willis and Furlong, 1975; Schellenberg, 1978), glutamine [glutamine-BP] (Weiner and Heppel, 1971; Weiner *et al.*, 1971; Kreischman *et al.*, 1973; Schellenberg and Furlong, 1977; Schellenberg, 1978; Kustu *et al.*, 1979; Celis, 1982), histidine [histidine-BP] (Rosen and Vasington, 1971; Ames and Lever, 1970, 1972; Lever, 1972; Shaltiel *et al.*, 1973; Ames and Spudich, 1976; Robertson *et al.*, 1977; Ames and Nikaido, 1978; Manuck and Ho, 1979; Ho *et al.*, 1980), isoleucine [leucine, isoleucine, valine-binding protein (LIV-BP), also referred to in other publications as leucine, isoleucine, valine, threonine-binding protein (LIVT-BP) and leucine, isoleucine, valine, alanine, threonine-binding protein (LIVAT-BP) and isoleucine-preferring, leucine, valine-binding protein (ILV-BP), also referred to in other publications as isoleucine-preferring, leucine, valine, threonine-binding protein (ILVT-BP] (Piperno and Oxender, 1966; Anraku and Heppel, 1967; Nakane *et al.*, 1968; Anraku, 1968a,b; Penrose *et al.*, 1970; Furlong *et al.*, 1973; Amanuma and Anraku, 1974; Amanuma *et al.*, 1976; Willis *et al.*, 1974; Antonov *et al.*, 1974, 1976; Oxender and Quay, 1976c; Quay *et al.*, 1977; Anderson and Oxender, 1977; Ovchinnikov *et al.*, 1977; Oxender *et al.*, 1977; Hoshino and Kageyama, 1980; Daniels *et al.*, 1980; Landick and Oxender, 1981; Hoshino and Nishio, 1982; Landick *et al.*, 1983), leucine [LIV-BP leucine-specific binding protein (LS-BP), also referred to in other publications as leucine-binding protein (LBP), and ILV-BP] (Piperno and Oxender, 1966; Anraku and Heppel, 1967; Nakane *et al.*, 1968; Anraku, 1968a,b; Penrose *et al.*, 1970; Furlong and Weiner, 1970; Weiner and Heppel, 1971; Amanuma and Anraku, 1974; Amanuma *et al.*, 1976; Willis *et al.*, 1974; Antonov *et al.*, 1974, 1976; Oxender and Quay, 1976c; Quay *et al.*, 1977; Anderson and Oxender, 1977; Ovchinnikov *et al.*, 1977; Oxender *et al.*, 1977, 1980b; Hoshino and Kageyama, 1980; Daniels *et al.*, 1980, 1981; Landick and Oxender, 1981; Hoshino and Nishio, 1982; Landick *et al.*, 1983), lysine [LAO-BP] (Wilson and Holden, 1969a,b; Kustu *et al.*, 1979), phenylalanine [phenylalanine-BP] (Klein *et al.*, 1970; Kuzaya *et al.*, 1971), and valine [LIV-BP and ILV-BP] (Piperno and Oxender, 1966; Anraku and Heppel, 1967; Nakane *et al.*, 1968; Anraku, 1968a,b; Penrose *et al.*, 1970; Furlong *et al.*, 1973; Amanuma and Anraku, 1974; Amanuma *et al.*, 1976; Willis *et al.*, 1974; Antonov *et al.*, 1974, 1976; Oxender and Quay, 1976c; Quay *et al.*, 1977; Anderson and Oxender, 1977; Ovchinnikov *et al.*, 1977; Oxender *et al.*, 1977; Hoshino and Kageyama, 1980; Daniels *et al.*, 1980;

Landick and Oxender, 1981; Hoshino and Nishio, 1982; Landick *et al.*, 1983) have been identified and characterized. In addition to BPs, periplasmic BP-dependent transport systems require the presence of proteins in the inner membrane to interact with the BPs and accomplish the actual transport event.

## III. NATURE OF PROTEIN COMPONENTS OF BP-DEPENDENT TRANSPORT SYSTEMS

### A. Periplasmic Components

The properties of periplasmic BPs are remarkably similar. Periplasmic amino acid BPs bind amino acids with high affinity. Equilibrium dialysis measurements usually give $K_d$ values between $10^{-6}$ and $10^{-7}$. Stopped-flow binding studies on the arabinose-BP have shown that the pseudo-first-order rate constant for arabinose binding decreases as the fraction of arabinose-BP containing bound arabinose increases (Miller *et al.*, 1980). This variable affinity, which has also been observed with the LIV-BP (Amanuma *et al.*, 1976), may explain why removal of substrate from BPs usually requires dialysis or gel filtration in the presence of 2–3 M guanidinium HCl or urea. The physiological significance and biochemical explanation of this variable binding are not yet understood. Most amino acid BPs are quite stable and can withstand treatment at 100°C for several minutes or with detergents such as sodium dodecyl sulfate (SDS) without losing binding activity. This stability may reflect a structural organization which resists denaturation and/or easily refolds to active form. BPs must resist degradation in the periplasmic space where proteases readily degrade many proteins and where pH cannot be readily controlled by the cell. Significant conformational differences, as monitored by proton-NMR or quenching of tryptophan fluorescence, between BPs freed of substrate and their bound forms have been demonstrated (glutamine-BP, Weiner and Heppel, 1971; Krieschman *et al.*, 1973; histidine-BP, Robertson *et al.*, 1977; Manuck and Ho, 1979; Ho *et al.*, 1980; S. Zukin, personal communication; LIV-BP, R. Landick, unpublished results). Early attempts to demonstrate this conformational shift with the LIV-BP (Penrose *et al.*, 1970) were unsuccessful because it was not appreciated that simple equilibrium dialysis against substrate-free buffer will not remove substrate bound by some bacterial periplasmic BPs.

Complete amino acid sequences are now available for the histidine-binding protein (his-BP), LAO-BP, LS-BP, and LIV-BP (Ovchinnikov *et al.*, 1977; Higgins and Ames, 1981; Hogg, 1981; R. Landick and D. L. Oxender, manuscript in preparation) as well as for the non-amino acid BPs: ara-BP, gal-BP, and sulfate-BP (Hogg and Hermondson, 1977; Isihara and Hogg, 1980; Mahoney *et al.*, 1981). Comparison of these sequences has not revealed significant conservations except between BPs coded by closely linked genetic loci, e.g., LAO-BP and his-BP. The conservation between the LAO-BP and his-BP is about 70% while the conservation between the LS-BP and LIV-BP is approximately 80%. In the case of the LAO-BP and his-BP pair the pattern of conservation suggests clearly which portions of the BPs are involved in substrate binding and which ones in transport (Higgins and Ames, 1981).

X-ray crystal structure determination has revealed that typical BPs which bind carbohydrate (arabinose-BP), ions (sulfate-BP), and amino acids (LIV-BP) all crystallize in the $P2_12_12_1$ space group with four molecules per unit cell (Quiocho *et al.*, 1979). Like the arabinose-BP (Quiocho *et al.*, 1977), the LIV-BP folds into a bilobate structure in which the cleft between the two domains contains the substrate-binding site (Saper and Quiocho, 1983). It is likely that the other amino acid-BPs share this bilobate structure. For the arabinose-BP and LIV-BP, the N-terminal region of the protein folds to form one domain while the C-terminal region folds to form the other. One possible model for the two domains is that one is important for substrate binding and the second, for interaction with the membrane component(s). This idea is consistent with the finding of two genetically distinct sites, one for substrate binding and one for membrane protein interaction, on the histidine-BP (Kustu and Ames, 1974; Higgins and Ames, 1981). Thus, we envisage that the BP is triggered to release its substrate from its binding site upon interaction of the other site with the membrane component(s). This type of model will be discussed further in a later section of this review.

### B. Membrane Components

Isolation and characterization of the membrane components of these systems have not progressed as far as has the study of the BPs, but a tentative picture of their roles is now emerging. The best-characterized membrane components are those of the histidine transport systems of which there are three, P protein, Q protein, and M protein (Higgins *et al.*, 1982a). Three potential proteins have also been identified for the LIV-I transport system (*livG, livH, livM;* Anderson and Oxender, 1977; R. Landick, T. Su, P. Nazos, and D. L. Oxender, manuscripts in preparation). A carbohydrate transport system, the maltose transport system, also appears to have three membrane-bound components: *malF, malG,* and *malK* (Shuman *et al.*, 1980; Bavoil *et al.*, 1980; Shuman and Silhavy, 1981; Shuman, 1982). While it has not yet been demonstrated that all of these proteins are, in fact, membrane-bound or membrane-associated components, it has been suggested that two components may be required to form a transport channel or binding site in the membrane, and that the third component may function either to act as a link between the periplasmic component, bound substrate, and the membrane components (Kustu and Ames, 1974; Ames and Spudich, 1976; Higgins *et al.*, 1982a), or in coupling transport to a source of energy (Shuman, 1982). The complete nucleotide sequences of all three of these genes for the histidine genes have been determined (Higgins *et al.*, 1982a).

## IV. GENETIC AND PHYSICAL MAPS OF THE LIV-I AND HISTIDINETRANSPORT GENES

Both leucine and histidine are transported into *Salmonella* and *E. coli* by more than one route, but in each case, the high-affinity, periplasmic BP-dependent system has been characterized most extensively. Each system is composed of both soluble, periplasmic proteins and membrane-bound proteins which are coded for by genes that are situated together in a cluster on the chromosome. These two gene clusters, *liv* at min 74.5 in *E. coli* and *his* at min 47 in *S. typhimurium*, are remarkably similar. Figure 1 shows a schematic representation of the *his* and *liv* gene clusters.

The four genes responsible for the high-affinity histidine permease, together with a regulatory locus, *dhuA*, constitute an operon located at min 47 on the *S. typhimurium* chromosome (Higgins *et al.*, 1982a; see Figure 1). The *hisJ* gene encodes the his-BP (J) which has been purified and characterized (Ames and Lever, 1972; Lever, 1972). The *hisQ* and *hisM* genes code for very hydrophobic, basic proteins (Q and M) which are probably integral membrane proteins (Higgins *et al.*, 1982a). The *hisP* gene codes for an inner membrane-bound protein (P) which is basic and very hydrophilic (Ames and Nikaido, 1978; Higgins *et al.*, 1982a). The P protein is presumed, on the basis of genetic evidence, to interact directly with the periplasmic J protein during histidine transport (Kustu and Ames, 1974; Ames and Nikaido, 1978). The Q, M, and P proteins are also thought to interact with each other, on the basis of complementation studies and, thus, possibly form a multicomponent membranous complex. Models suggesting mechanisms of action for this system are presented in a later section.

Four genes are also responsible for the high-affinity leucine permease and constitute an operon at minute 74.5 on the *E. coli* chromosome which is surprisingly similar to the *S. typhimurium* histidine transport operon. The *livK* gene, which codes for the LS-BP, is the first gene in the operon and is followed by three genes which code for proteins synthesized in relatively lower amounts. The first of these genes, *livH*, encodes a basic hydrophobic protein (H). The last two genes in the operon, *livM* and *livG*, code for the M and G protein. The H, M, and G proteins are thought to be membrane-associated proteins or integral membrane proteins. It is quite likely that the H, M, and G leucine transport proteins function in a membrane complex mechanistically identical to the Q, M, P histidine transport protein complex.

Of particular interest are the findings (1) that the Q, M, and P histidine transport proteins are also essential for an arginine transport system which requires the LAO-

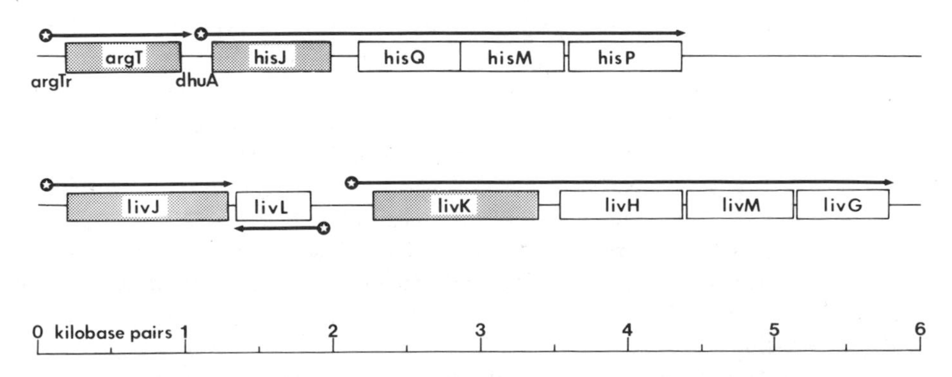


*Figure 1.* Organization of the histidine and LIV-I transport genes. Stars indicate the positions at which transcription of the genes originates. The arrows indicate the length of each transcript. Transcription of the *livL* gene occurs from the opposite strand in and in opposite direction from the other transport genes.

BP (Kustu and Ames, 1973; Kustu *et al.*, 1979; Higgins and Ames, 1981) and (2) that the H, M, and G leucine transport proteins are also required for a branched-chain amino acid transport system which requires the LIV-BP. The gene coding for the LAO protein (*argT*) and its regulatory locus are adjacent to the histidine transport operon and form a separate monocistronic operon (Higgins and Ames, 1982). Likewise, the gene coding for the LIV-BP (*livJ*) and its regulatory region form a monocistronic operon which is near the leucine-specific transport operon (R. Landick and D. L. Oxender, manuscript in preparation). The LAO-BP is believed to interact with the P protein during arginine transport in a manner analogous to that of the J protein (his-BP) during histidine transport. The his-BP and LAO-BP are 70% homologous while the LIV-BP and LS-BP are 80% homologous. Each pair of genes is believed to have arisen by duplication of an ancestral gene coding for an amino acid BP.

The only striking difference between these two transport gene clusters is the *livL* gene which is located between *livJ* and *livK* in the *liv* cluster. *livL* has been characterized on the basis of DNA-sequencing studies and by gene fusion to *lacZ* to create a hybrid protein. These experiments have revealed that *livL* is transcribed in the opposite orientation to the other *liv* genes and that it can be translated into a basic, hydrophilic protein with a mass of 18.7 K. There is no known function for the *livL* protein, but it has been identified as an *in vitro* transcription–translation product of several plasmids containing the *livL* gene. It is possible that the *livL* protein may play a role in the regulation of *liv* gene expression.

## V. ASSEMBLY OF TRANSPORT COMPONENTS

After synthesis, the components of the LIV-I and histidine transport systems must be assembled into their proper locations in either the periplasm or cytoplasmic membrane. Like all other known periplasmic proteins, LAO-BP, his-BP, LIV-BP, and LS-BP are synthesized as precursor molecules with N-terminal signal peptides of 22 (*his*) or 23 (*liv*) amino acids (Oxender *et al.*, 1980b; Daniels *et al.*, 1980; Higgins and Ames, 1981; Higgins *et al.*, 1982a). These sequences are shown in Table 2. A comparison between these two pairs of evolutionarily related signal sequences reveals that they have diverged from each other faster than the mature portion of the BPs. Thus, his-BP and LAO-BP signal sequences are 41% conserved vs. 70% conservation in the mature sequences, and the LIV-BP and LS-BP signal sequences are 47% conserved vs. 80% conservation in the mature sequences. This difference indicates less selective pressure on the signal sequence composition which is in agreement with the widely held belief that the hydrophobic nature of the signal rather than a precise sequence of amino acids is required for protein secretion. This view is further supported by a comparison of the his-BP and LAO-BP signal sequences with the LIV-BP and LS-BP signal sequences where 23% is the maximum conservation of sequence which can be produced by various alignments.

Secretion of the LS-BP has been studied in considerable detail (Oxender *et al.*, 1980b; Daniels *et al.*, 1980, 1981; Landick *et al.*, 1983). The intact precursor of the

Table 2. N-terminal Sequences of LIV-I and Histidine-Transport Proteins

| N-terminal sequence | | | | | | | | | | | | | | | | | | | | | | | | | | Transport protein |
|---|---|---|---|---|---|---|---|---|---|---|---|---|---|---|---|---|---|---|---|---|---|---|---|---|---|---|
| | | | | | | | | | | | | | | | | | | | | | | Processing site[a] ↓ | | | | |
| | Met | Lys | Lys | Leu | Ala | Leu | Ser | Leu | Ser | Leu | Val | Leu | Ala | Phe | Ser | Ser | Ala | Thr | Ala | Ala | Phe | Ala | Ala | Ile | Pro | His-BP |
| | Met | Lys | Lys | Thr | Val | Leu | Ala | Leu | Ser | Leu | Leu | Ile | Gly | Leu | Gly | Ala | Thr | Ala | Ala | Ser | Tyr | Ala | Ala | Leu | Pro | LAO-BP |
| Met | Asn | Ile | Lys | Gly | Lys | Ala | Leu | Leu | Ala | Gly | Leu | Ile | Ala | Leu | Ala | Phe | Ser | Asn | Met | Ala | Leu | Ala | Glu | Asp | Ile | LIV-BP |
| Met | Lys | Arg | Asn | Ala | Lys | Thr | Ile | Ile | Ala | Gly | Met | Ile | Ala | Leu | Ala | Ile | Ser | His | Thr | Ala | Met | Ala | Asp | Asp | Ile | LS-BP |
| Met | Leu | Tyr | Gly | Phe | Ser | Gly | Val | Ile | Leu | Gln | Gly | Ala | Ile | Val | Thr | Leu | Glu | Leu | Ala | Leu | Ser | Ser | Val | Val | Leu | *hisQ* Protein |
| Met | Ile | Glu | Ile | Ile | Gln | Glu | Tyr | Trp | Lys | Ser | Leu | Leu | Trp | Thr | Asp | Gly | Tyr | Arg | Phe | Thr | Gly | Val | Ala | Ile | Thr | *hisM* Protein |
| Met | Met | Ser | Glu | Asn | Lys | Leu | His | Val | Ile | Asp | Leu | His | Lys | Arg | Tyr | Gly | Gly | His | Glu | Val | Leu | Lys | Gly | Val | Ser | *hisP* Protein |
| Met | Ser | Glu | Gln | Phe | Leu | Tyr | Phe | Leu | Gln | Gln | Met | Phe | Asn | Gly | Val | Thr | Leu | Gly | Ser | Thr | Tyr | Ala | Leu | Ile | Ala | *livH* Protein |
| Met | Trp | Ser | His | Ser | Pro | Cys | Asp | Ser | Gly | Ala | Ala | Gly | Asp | Ala | Asp | Arg | Tyr | Ser | Gly | Ser | Pro | Gly | Gly | Gly | Glu | *livM* Protein |
| Met | Lys | Leu | Thr | Ile | Ile | Arg | Leu | Glu | Lys | Phe | Ser | Asp | Gln | Asp | Arg | Ile | Asp | Leu | Ala | Lys | Asp | Leu | Gly | Arg | Glu | *livL* Protein |

[a] The processing site refers only to the first four sequences, the periplasmic His-BP, LAO-BP, LIV-BP, and LS-BP. The remaining N-terminal sequences are shown to illustrate their lack of a consensus signal sequence for secretion.

LS-BP has been synthesized *in vitro* and is processed to the mature form (Daniels *et al.*, 1980) by the leader peptidase enzyme characterized by Zwizinski and Wickner (1980). *In vivo,* the LS-BP can also be detected as a precursor during pulse-chase labeling experiments with $^{35}$Smethionine (Daniels, 1981; Landick *et al.*, 1984). This result suggests that the LS-BP precursor is synthesized as a complete polypeptide chain prior to being processed by the leader peptidase and is at odds with the cotranslational secretion model as postulated by Blobel and Dobberstein (1975a,b) and extended to bacterial periplasmic proteins by Davis and co-workers (Ron *et al.*, 1966; Smith *et al.*, 1977, 1978). Randall and co-workers have recently shown that maltose-BP may be processed either posttranslationally or cotranslationally, but never before the polypeptide chain is 80% complete (Josefsson and Randall, 1981). Secretion of some proteins, such as the maltose-BP, appears to be primarily cotranslational while secretion of others, presumably including the LS-BP, is primarily posttranslational. Secretion of periplasmic BPs has been found to be dependent on the inner membrane electrochemical potential (Daniels *et al.*, 1981; Enequist *et al.*, 1981; Landick *et al.*, 1983). Processing of the LS-BP can be blocked by membrane potential dissipating agents such as carbonylcyanide-*m*-chlorophenylhydrazone (CCCP) or valinomycin (Daniels *et al.*, 1981). When secretion is blocked by treating spheroplasts with valinomycin, the LS-BP precursor builds up in the cytoplasmic membrane. The trapped precursor can be processed and released into the periplasm in a completely posttranslational fashion by resuspending valinomycin-treated spheroplasts in the absence of KCl (Daniels *et al.*, 1981). These conditions create a transient membrane potential which is apparently sufficient to stimulate processing and secretion of the LS-BP. Based on these data, we envisage a secretory process that takes advantage of the transmembrane electrochemical field to align the signal sequence across the membrane and allow the signal peptidase, which is located on the outside of the cytoplasmic membrane, to cleave off the signal sequence at the correct location. The inserted signal peptide loop appears to be a structural requirement for initiation of secretion. This conformation may be achieved by an incomplete nascent chain on the ribosome or by a freely soluble protein which would account for the observation of both posttranslational and cotranslational processing of periplasmic BPs. This secretory model is shown in Figure 2.

In addition to the secretory process, periplasmic BPs must fold into the proper conformation once they are released into the periplasm. This conformation is conserved among periplasmic nutrient BPs which have been examined (Quiocho *et al.*, 1979). This conformation has two domains, one of which includes primarily the N-terminal portion of the protein and one of which includes primarily the C-terminal portion. The two domains are connected by a hinge region. In the case of the LS-BP, this conformation is resistant to trypsin, even though the protein contains 27 lysines and nine arginines (Oxender *et al.*, 1980b). This trypsin-resistant conformation is specific to the mature form of the protein; the LS-BP precursor is rapidly degraded by trypsin. We believe that the trypsin resistance of the LS-BP reflects a conformation which may be generally resistant to protease degradation because the periplasmic space contains several proteolytic activities (Sveedhara-Swamy and Goldberg, 1982), and no degradation of the LS-BP occurs in the periplasm. Small deletions in the C-terminal region of the LS-BP render it susceptible to periplasmic protease degradation so that none of

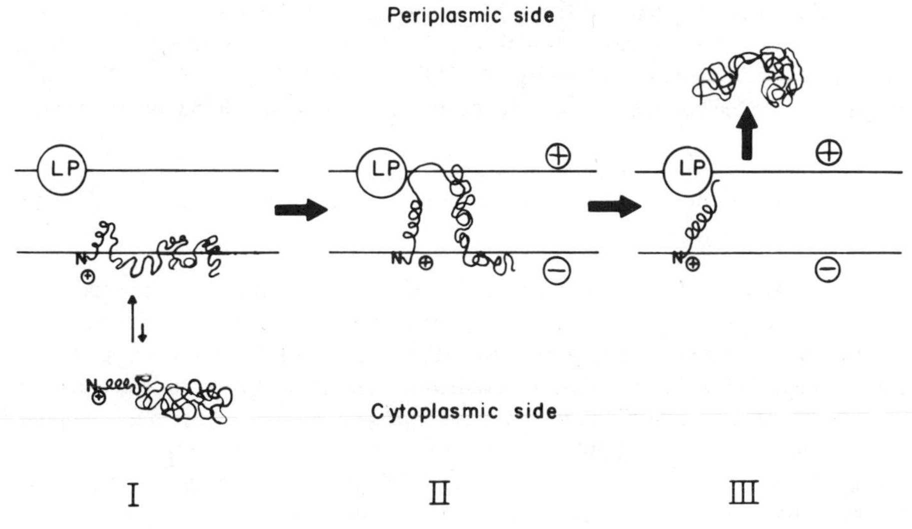


*Figure 2.* Model for the secretion of the LS-BP.

the mutated protein can be detected in the periplasm, even though it is processed and secreted normally (Landick *et al.*, 1984).

The three membrane proteins of the histidine transport system, two potential membrane proteins of the LIV-I system, and the *livL* protein do not have typical signal sequences at their N-termini (see Table 2). The signal sequence is apparently important for transport of the proteins into the periplasm or into the outer membrane, but a cleavable signal is not required for proteins to be inserted into the cytoplasmic membrane. This is also the case for other membrane proteins (Gay and Walker, 1981a,b; Young *et al.*, 1981; Gilson *et al.*, 1982a; Higgins *et al.*, 1982a). The lactose permease gene codes for what might possibly be a signal peptide, but no processing occurs (Ehring *et al.*, 1980; Buchel *et al.*, 1980). Whether the *liv* and *his* membrane transport proteins are inserted cotranslationally, individually inserted posttranslationally, or assembled in the cytoplasm and then inserted as a complex has not been determined. Because these proteins are highly basic, hydrophobic, and produced in exceedingly small quantities, experiments to answer these questions will be difficult. We anticipate that overproduction of the membrane components by recombinant DNA methods will facilitate these studies.

## VI. ENERGIZATION AND RECONSTITUTION OF AMINO ACID TRANSPORT

Amino acid transport, as we have defined it here, is accumulation of amino acids against a concentration gradient and, therefore, necessarily requires a source of energy. A precise description of the mechanism of energy coupling to active transport is not yet possible; instead, we will discuss the current status of the energetics of amino acid transport. Comprehensive reviews on this subject have been recently published (Simoni and Postma, 1975; Booth and Hamilton, 1980; Hunt and Hong, 1981b). We will also briefly describe the current status of the reconstitution of amino acid transport systems.

### A. Membrane-Bound, Osmotic Shock-Resistant Systems: Energization and Reconstitution

It is now generally accepted that membrane-bound, osmotic shock-resistant amino acid transport systems, such as the proline PP-I system, utilize primarily electrochemical proton-motive force to energize transport (Berger, 1973; Berger and Heppel, 1974; Booth and Hamilton, 1980). The mechanism of energy coupling for these transport systems has been studied in membrane vesicle preparations by addition of compounds which either stimulate or inhibit generation of a proton-motive force (Kaback and Milner, 1970; Kaback, 1972, 1974; Futai, 1974; Ramos and Kaback, 1977a,b). In general, D-lactate is an efficient energy donor for stimulating amino acid transport in vesicle preparations, perhaps because of its rapid translocation into membrane vesicles (Nichols and Hamilton, 1976). D-Lactate is converted to pyruvate by the membrane-bound lactate dehydrogenase which supplies electrons to the respiratory chain in the form of NADH. To a lesser extent, succinate, L-lactate, D,L-α-hydroxybutyrate, and NADH also stimulate osmotic shock-resistant amino acid transport in vesicle preparations. An artificial electron donor system composed of phenazine methosulfate and ascorbate is capable of stimulating transport even more effectively than D-lactate (Konings *et al.*, 1971). In each case, this stimulation may be explained by generation of a proton-motive force which, in turn, drives membrane-bound, osmotic shock-resistant amino acid transport systems. In membrane vesicles, D-lactate or other energy sources can stimulate transport of alanine, aspartate, cysteine, glutamate, glycine, histidine, isoleucine, leucine, lysine, phenylalanine, proline, tryptophan, tyrosine, serine, and valine (Kaback, 1972).

The precise mechanism that couples the proton-motive force to these transport systems is not yet established. The proton-motive force is generally thought to be composed of two components: (1) the actual proton gradient or $\Delta pH$, and (2) the membrane potential or $\Delta\Psi$. Either or both of these components may be coupled to amino acid transport. Ramos and Kaback (1977b) have suggested that the mechanism may be different depending on the nature of the substrate transported. Negatively charged amino acids such as glutamate may be taken up by electrogenic proton symport (cotransport of protons and glutamate molecules), whereas neutral or positively charged amino acids may be taken up by a membrane potential-driven mechanism. Amino acid transport also may be indirectly coupled to the proton gradient by sodium ion cotransport as in the case of sodium-dependent glutamate transport (Britten and McClure, 1962; Frank and Hopkins, 1969; Kahane *et al.*, 1975; MacDonald *et al.*, 1977; Schellenberg and Furlong, 1977; Tsuchiya *et al.*, 1977).

Reconstitution of membrane-bound, osmotic shock-resistant amino acid transport systems from purified protein components may allow a better definition of the mechanism of energy coupling. The *E. coli* lactose carrier has been reconstituted in this manner (Foster *et al.*, 1982). In this case, it has been suggested that either the membrane potential ($\Delta\Psi$) or the proton gradient ($\Delta$pH) may energize lactose transport. $\Delta\Psi$ may be coupled to lactose transport through a net negative charge at the carrier protein's active site (Schuldiner *et al.*, 1975; Kaback, 1976). Thus, the membrane potential would increase the rate of outward translocation of the free carrier and decrease its lifetime at the inner surface of the membrane. $\Delta$pH may be coupled to lactose transport by cotransport of lactose and protons (lactose–proton symport). Studies on the reconstituted lactose carrier have provided convincing support for these ideas and have even been extended to suggest a role for histidine residues in energy coupling to the proton gradient (Garcia *et al.*, 1982; Patel *et al.*, 1982). Because purified lactose carrier protein was used, these experiments demonstrate that no other proteins are required for coupling of lactose transport to either $\Delta\Psi$ or $\Delta$pH. Further work will be necessary to extend these proposals (if warranted) to membrane-bound, osmotic shock-resistant amino acid transport systems.

## B. Energization of BP-Dependent Systems

The mechanism of energy coupling in BP-dependent, osmotic shock-sensitive amino acid transport systems is even less well understood. Both phosphate bond energy and the membrane potential appear to be required for this class of transport system.

The requirement for phosphate bond energy is most clearly evident from two observations: (1) an *E. coli* mutant that is unable to synthesize ATP via oxidative phosphorylation can energize BP-dependent transport of glutamine by metabolism of glucose but not lactate, while the parent strain can utilize either substrate, and (2) arsenate, which rapidly depletes cellular ATP levels (Klein and Boyer, 1972), blocks BP-dependent transport of glutamine when either glucose or lactate is the carbon source (Berger, 1973). Subsequent investigations have suggested a phosphate bond energy requirement for the BP-dependent transport of arginine, histidine, ornithine (Berger and Heppel, 1974; Gutowski and Rosenberg, 1976), methionine (Kadner and Winkler, 1975), leucine, isoleucine, and valine (Oxender, 1972; Wood, 1975). Initially, it was thought that ATP might directly phosphorylate the BPs of osmotic shock-sensitive systems, but no correlation between ATP levels and the extent of osmotic shock-sensitive transport has been found (Plate *et al.*, 1974; Lieberman and Hong, 1976) and, in disagreement with the earlier view, it is now generally thought that some other high-energy phosphate bond metabolite must act as the primary energy coupling factor (Lieberman and Hong, 1976). Based on the observation that an *E. coli* mutant unable to synthesize acetyl phosphate (*pta*) was unable to use pyruvate to stimulate BP-dependent glutamine transport in the presence of cyanide and fluoride, Hong and co-workers proposed that acetyl phosphate may be the intermediate metabolite (Hong *et al.*, 1979). Attempts to obtain more direct evidence for the role of acetyl phosphate as an intermediate have generally been unsuccessful. No change in acetyl phosphate levels was observed when *E. coli* cells were treated with arsenate (Hong and Hunt, 1980). At this point, the available evidence indicates that it is unlikely that acetyl phosphate is an intermediate in the energization of osmotic shock-sensitive transport systems.

Several lines of evidence suggest that phosphate bond energy, alone, is not sufficient for energization of amino acid transport systems involving BPs. Proton ionophores or anaerobiosis which prevent the generation of a proton-motive force can inhibit BP-dependent transport by as much as 60% (Berger, 1973; Berger and Heppel, 1974; Cowell, 1974). In a proton ATPase-negative mutant, treatment of cells with colicin K results in a significant decline of BP-dependent glutamine uptake even though ATP levels rise (Plate *et al.*, 1974). Temperature-sensitive *E. coli* mutants, which lack a membrane potential but have normal proton gradients and ATP levels, exhibit reduced BP-dependent glutamine uptake (Lieberman and Hong, 1976; Plate, 1976; Lieberman *et al.*, 1977). Thus, it appears that the membrane potential, in addition to high-energy phosphate bond energy, has a function in the activation of BP-dependent amino acid transport. This function might be a true contribution to the energization of transport or it might be simply to orient membrane protein components of the transport systems into their proper conformation.

## C. *Reconstitution of BP-Dependent Amino Acid Transport*

Ultimately, reconstitution from purified protein components may play an important role in defining the mechanism and energization of BP-dependent amino acid transport systems. To date, however, reconstitutions of BP-dependent systems, even in spheroplasts or membrane vesicles, have been plagued with problems. The large number of controls necessary to rule out artifactual results has made these experiments difficult. In spite of the difficulties, several investigators have reported the successful reconstitution of BP-dependent amino acid transport. Wilson and Holden (1969b) first reported reconstitution of BP-dependent amino acid transport by addition of partially purified arginine-BP to osmotically shocked cells. Anraku also reported stimulation of leucine transport by addition of partially purified LIV-BP and other protein factors to osmotically shocked cells (Anraku *et al.*, 1973). Successful reconstitutions of BP-dependent phosphate (Gutowski and Rosenberg, 1976) and ribose (Robb and Furlong, 1980) transport in spheroplasts have also been reported. More recently, Hong and co-workers have reported successful reconstitution of BP-dependent glutamine transport by addition of glutamine-BP to both spheroplasts (Masters and Hong, 1981a) and membrane vesicles (Hunt and Hong, 1981a). In the latter work, incorporation of NAD into the membrane vesicles and addition of sodium pyruvate to the assay were required for maximal stimulation. Even under this best situation, however, the reconstitution achieved was more than an order of magnitude lower than that achieved with an equal amount of spheroplasts (based on protein content). Correction for the differences in protein content between spheroplasts and membrane vesicles would result in a much greater difference. Even these low levels of reconstituted uptake, however, have been used for recent studies to define the energy requirements of BP-dependent amino acid transport (Hunt and Hong, 1983a,b). These recent experiments have suggested that two classes for compounds (succinate and those which can be metabolized to pyruvate) are capable of stimulating BP-dependent transport of glutamine in membrane vesicles (Hunt and Hong, 1983a). Evidence implicating a role for glutamine-BP tryptophan and histidine residues in the interaction on glutamine-BP with membrane-bound components has also been found with reconstitution experiments (Hunt and Hong, 1983b). Most recently, studies on the reconstitution of BP-dependent transport of glutamine in membrane vesicles have led to the conclusion that redox energy and a membrane potential may both be required to energize BP-dependent active transport (J.-s. Hong, personal communication).

## VII. POSSIBLE MODELS FOR AMINO ACID TRANSPORT

With the availability of the complete nucleotide sequence of all of the genes of the histidine transport operon, with the nucleotide sequence of the LIV-I genes nearing completion, and with the identification of most of the proteins involved in these two permeases, it is now possible to formulate tentative models for the molecular mechanism of transport by these systems (Higgins *et al.*, 1982a). Two possible alternatives are presented here, both using as examples the histidine transport system and taking into account all the knowledge available for that system. The general architecture of the complex of proteins is the same in two models (Figure 3), but in one case, the membrane-bound components bear no specificity for transport, while in the other case, one or more specific substrate-binding sites are present in the membrane-bound components. In the "pore" model (shown in Figure 3A), histidine binds to J (his-BP) causing a conformational change, thus allowing interaction of the J–histidine complex with P. Additional conformational changes in the hypothetical P,Q,M complex allow a pore to be formed within or between the proteins of the complex (or both). Histidine is able to diffuse through such a pore to the cell interior. This model proposes that the specificity of the transport system is dictated by a single specific component, the his-BP, J. The specificity is transmitted to the system as a whole by a specific interaction

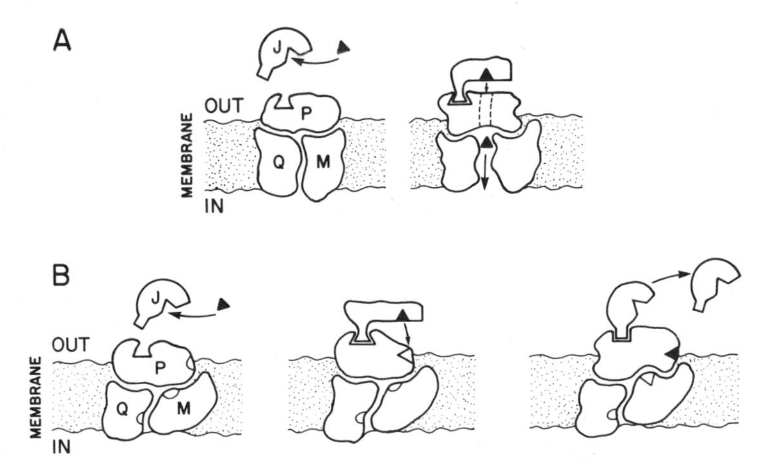


*Figure 3.* Models for the mechanism of amino acid transport systems. (A) The "pore" model, (B) the "binding-site" model.

between J and P. Presumably, any substrate able to bind to J (and therefore causing the prescribed conformational change) will be transported through an unspecific pore in the membrane. Alternatively, the "binding-site" model (shown in Figure 3B) involves an identical initial step which would be followed, upon interaction of the J–histidine complex with the membrane proteins, by a sequential "activation" of one or more "histidine-binding sites" which would transport histidine in a cascade-like manner through the membrane. This second model postulates a more active role for the membrane-bound proteins, involving a histidine-binding site on each of the membrane components (although, of course, not all of the membrane components need to have such a site). Both models postulate that the interaction of the J–histidine complex with the membrane-bound proteins should be such that histidine is very close to the pore or to the next histidine-binding site. This arrangement is assumed to be necessary to avoid diffusion of the released histidine into the periplasm as soon as it is released from the periplasmic component. In other words, it is possible that the most probable function of the BP is that of "trapping" the substrate in a bound form and delivering it directly to the next site, without allowing it to return to a freely diffusible state. It is also possible to envisage an intermediate situation where a pore is formed which has such a shape and/or disposition of amino acid residues that only histidine or a substrate with a closely related shape can pass through it.

It is too early to distinguish between the two models of Figure 3. However, we can predict that if the binding site model is correct, transport mutants should exist which have a defective binding site in a membrane protein thus resulting in an *altered* spectrum of specificity for transport. Several mutants in membrane components of the histidine permease have been characterized, which indeed present an altered spectrum of specificity, thus tentatively supporting the "binding-site" model (Higgins *et al.*, 1982a). The existence of a substrate-binding site on a membrane component has been postulated also for another, non-amino acid transport system (Shuman, 1982). It should be noted that, although not represented in Figure 3, an energy-coupling mechanism, such as described in the previous section, must be involved in the process of concentrative uptake.

## VIII. REGULATION OF AMINO ACID TRANSPORT

Amino acid transport systems are regulated to meet the nutrient requirements of bacteria. Low-affinity transport systems which typically have high $V_{max}$s may function primarily for concentration of amino acids from the growth medium, while high-affinity systems may function primarily as scavengers to prevent the loss of amino acids which are synthesized intracellularly and leak out through the inner membrane (Ames, 1972; Landick and Oxender, 1981). Since wild type bacteria can synthesize all 20 amino acids, they may have a great need for transport systems to recapture amino acids that they have expended energy to synthesize. For example, it has been calculated (Brenner and Ames, 1971) that, for the synthesis of one molecule of histidine, the potential energy corresponding to 41 ATP molecules is sacrificed. Thus, it should be more desirable from the point of view of cellular economy that the cell be able to transport one molecule of histidine from the medium or recapture a previously synthesized one rather than synthesize a new histidine molecule *de novo* from glucose. This is achieved through the evolution of high-affinity transport systems which efficiently concentrate any small amounts of environmentally available substrate. Measurement of transport in growing cells, under physiological conditions, reveals an interesting measurable parameter, the *limit concentration,* which is the minimal external concentration required by the permease in order to supply the cell with enough amino acid to carry on protein synthesis without utilizing any of the biosynthetic amino acid (Ames and Roth, 1968; Ames, 1972); the lower this value, the more effective is the uptake mechanism with respect to growth. For example, the limit concentration for histidine is $1.5 \times 10^{-7}$ M (Ames, 1964). While the histidine and LIV-I transport systems certainly fulfill the role of high-affinity amino acid transport systems, all of their substrate amino acids can also be transported by other, higher $K_m$ and $V_{max}$ systems (see Table 1). The LIV-I system, in particular, may play a central role in fine tuning the intracellular concentration of leucine.

## A. Regulation of the Histidine Transport System

The histidine transport system has been shown to be regulated by nitrogen availability: cells respond to nitrogen limitation by increased expression of the histidine transport genes (among others; Kustu *et al.*, 1979; Higgins and Ames, 1982). This system is not affected by the level of histidine in the medium. In addition it is regulated independently from the histidine biosynthetic operon as its expression is unaffected by the presence of regulatory mutations affecting histidine biosynthesis. Thus, it seems that this system must depend on regulatory mechanisms which do not involve a typical transcription attenuation mechanism like those described for biosynthetic operons (Yanofsky, 1981). In agreement with this it was found that the nucleotide sequence of neither regulatory region, *dhuA* or *argT*, displays any of the features typical of the regulatory regions of amino acid biosynthetic operons (Higgins and Ames, 1982). Among the features of interest found in these two regions are several dyad symmetries. Of particular significance may be a dyad symmetry found in each region, the two bearing considerable homology to each other. In both cases, the dyad symmetry overlaps the promoter site (−10 and −35 regions) and retains a stretch of high homology at the "foot" of the hypothetical stem-loop structure which each of them can form. It is not known what, if any, is the function of this structure.

One might expect to find regulatory features which must be involved in the response of these operons to nitrogen availability. A pseudo-mirror symmetry was in fact found in each of these regions, and by model building and computer analysis (cylindrical projection kindly performed by S.-H. Kim) it was possible to determine that the distribution of functional groups in such pseudo-symmetries does indeed correspond to a true twofold symmetry in the minor groove of the DNA double helix and, thus, would be able to respond to a regulatory protein that displays dyad symmetry. Because extensive homology was found between these mirror symmetries and one occurring in the regulatory region of *glnA*, the gene coding for glutamine synthetase and known to be under nitrogen regulation, it was postulated that these structures are involved in nitrogen regulation (Higgins and Ames, 1982). If this were true, it may be that these sequences are recognition sites for interaction with protein(s) produced by genes involved in responding to nitrogen availability (*ntrA*, *ntrB*, and *ntrC*; McFarland *et al.*, 1981).

## B. Regulation of the LIV-I Transport System

LIV-I transport is directly regulated by the concentration of L-leucine in the growth medium (Quay and Oxender, 1976, 1980a; Landick *et al.*, 1980). When cells are grown in the absence of leucine, LIV-I transport is derepressed. Addition of leucine reverses this derepression, but no other amino acid can accomplish this reversal. Thus, the LIV-I transport system, though it functions for uptake of L-leucine, L-isoleucine, L-valine, L-threonine, L-alanine, and D-leucine, is regulated by the single amino acid, L-leucine. This control, however, is mediated by mechanisms which involve several other cellular components. Table 3 shows the various mutations which are known to affect LIV-I gene expression.

Repression by exogenous leucine is a different mode of regulation than that found for other amino acid transport systems, such as the proline PP-II system (see Table 1). Here, amino acid transport activity is induced when the given amino acid is present in the growth medium. Most carbohydrate transport systems are also regulated in this fashion (Dills *et al.*, 1980).

Transcription attenuation of biosynthetic operon expression has been documented for the *trp, his, ilv, leu,* and *phe* biosynthetic operons and the mechanism for this attenuation is reasonably well understood (Yanofsky, 1981). Over the past 7 years, a variety of results have been obtained which suggest that, like the biosynthetic operons, expression of the LIV-I transport gene cluster is controlled by transcription attenuation. Most notable is the derepression of LIV-I transport in strains with mutations in *rho* or *leuS* (Quay *et al.*, 1975a, 1977; Quay and Oxender, 1976, 1977; 1980b; Landick *et al.*, 1980). Recently, this speculation has been strengthened by examination of the DNA sequence preceding the *livJ* gene and studies on the *in vitro* transcription of this gene. We have found that transcription of *livJ* originates 104 bases prior to the beginning of the structural gene, and that the leader region of this transcript potentially codes for a small peptide with tandem leucine codons. On the basis of these and other results we have proposed a model for transcription attenuation of *livJ* (Landick, 1984). In this model, transcription termination by the product of the *rho* gene is coupled to translation of the small leader peptide which is, in turn, controlled by the availability of charged leucyl-tRNA. This model is quite different from the transcription attenuators found in the biosynthetic operons where transcription termination is coupled to translation without the involvement of the *rho* gene product. Further work will be required to verify the model for *livJ* transcription attenuation, but it now appears that this LIV-I transport gene is controlled by a unique transcription attenuator. Similar transcription attenuation of the *livK* gene has not been documented.

A strong case for the involvement of regulatory proteins can be made for LIV-I transport gene expression. Mutations with repressor-like phenotypes have been found which derepress LIV-I and map at a location far removed from the LIV-I transport

*Table 3. LIV-I Transport System Regulatory Factors*

| Gene | Gene product | Phenotypic effect of the mutation on transport | References |
|---|---|---|---|
| *rho* | Transcription termination factor Rho | Elevated level of leucine transport | Quay and Oxender (1977), Quay *et al*. (1978), Landick *et al*. (1980) |
| *leuS* | Temperature sensitive leucyl-tRNA synthetase | Elevated leucine transport at the nonpermissive temperature | Quay *et al*. (1975a), Quay and Oxender (1976), Quay and Oxender (1980b) |
| *hisT* | Enzyme which converts uridine to pseudouridine in tRNA | Unable to derepress leucine transport. Higher concentration of charged leucyl-tRNA | Oxender and Quay (1976a), Quay *et al*. (1978) |
| *relA* | A ribosome-associated protein which synthesizes ppGpp from GTP in response to amino acid starvation | Unable to derepress leucine transport | Quay and Oxender (1979) |
| *livR* | Unknown | Elevated leucine transport. Primary effect on LIV-BP | Anderson *et al*. (1976), Landick *et al*. (1980) |
| *lstR* | Unknown | Elevated leucine transport. Primary effect on the LS-BP | Anderson *et al*. (1976) |

genes (Anderson *et al.*, 1976; Landick *et al.*, 1980). These loci, *livR* and *lstR*, map at min 22 on the *E. coli* chromosome and are described in Table 3. Palindromic sequences are located surrounding the origins of transcription for all four transport promoters. Palindromic sequences have been found at the binding sites for other known repressor proteins (Gilbert *et al.*, 1974; Maniatis *et al.*, 1975; Gunsalus and Yanofsky, 1980). In the case of the *livJ* gene transcript, the palindromic symmetry is located directly overlapping the known origin of transcription as determined by sequencing of an *in vitro* synthesized *livJ* transcript.

### C. Control of Membrane Protein Synthesis

Present evidence suggests that both the specific amino acid BPs, his-BP and LS-BP, and their respective membrane components are encoded on a single polycistronic mRNA. The membrane components, however, are synthesized in quantities ten-fold lower than the BPs (Ames and Nikaido, 1978). Secondary structure in the mRNA may be responsible for controlling the level of messenger for the membrane component genes to accomplish this lowered synthesis. Some of this reduction may also be due to a weaker ribosome-binding site for the membrane components, but a strong case can be made for regulation of mRNA levels as a mechanism for lowering the expression of these genes. This is best illustrated in the *his* transport regulon where a large stem-loop structure is present in the mRNA between the *hisJ* and *hisQ* genes (Higgins *et al.*, 1982b). Extremely similar structures are also present in the mRNA sequences between the *hisG* and *hisD* biosynthetic genes and between the *lamB* and *molA* genes in the *malK–lamB* operon and others. These structures, which exhibit very high homology may be involved in regulating the differential expression of the individual genes in these operons (Higgins *et al.*, 1982b). The stem-loop structure between *hisJ* and *hisQ* is shown in Figure 4. This structure may function as a recognition site for mRNA processing. Recent data indicate that it is not involved in transcription termination (Stern *et al.*, 1984). A small stem-loop structure is found between *livK* and *livH* in the LIV-I regulon (shown in Figure 4). This structure contains a GC-rich stem and is followed by three Ts, however, it bears no homology to the structure found in the histidine operon. Such a structure may be a typical transcription termination signal. It is unlikely, however, that this termination signal separates the leucine transport genes into two operons both because no promoter sequence is present preceding *livH* and because mutations in *livK* are polar on expression of the downstream *livH* gene. It is possible that regulation in the *his* transport operon utilizes mRNA processing, while the regulation between *livK* and *livH* involves transcriptional attenuation.

## IX. EVOLUTIONARY RELATIONSHIPS AMONG PERIPLASMIC SYSTEMS

A comparison of the characteristics of the known periplasmic systems (including non-amino acid ones) suggests that the underlying mechanism of transport is possibly

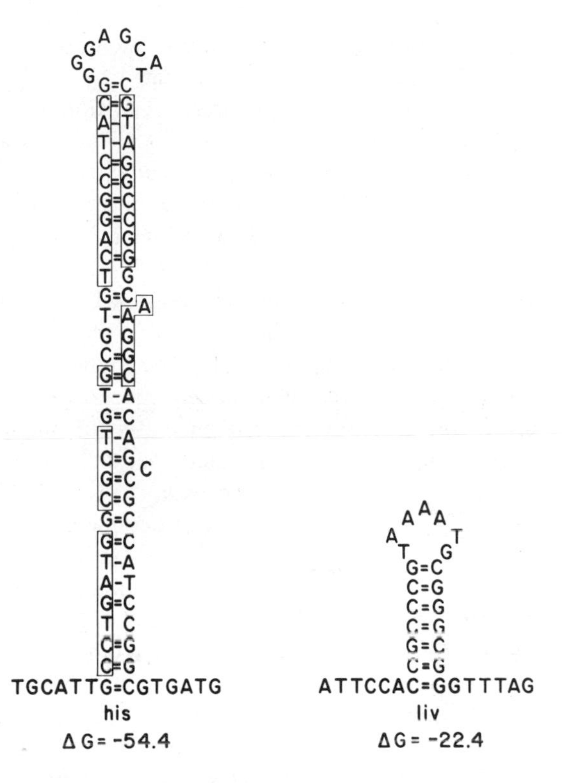


*Figure 4.* Secondary structures in the mRNA between the *hisJ* and *hisQ* genes and the *livK* and *livH* genes. In this figure the DNA sequence is shown. Thymine residues (T) will appear as uracil residues (U) in mRNA synthesized from the DNA sequence shown here. The free energies of stabilization for the stem loops were computed using the rules of Tinoco *et al.* (1973). Free energies are given in kcal/mole. The regions of sequence conservation with the stem-loop structures found in the *his* biosynthetic and *malK–lamB* operons are enclosed in boxes.

the same for all of them (Ames and Higgins, 1983). The physical composition of all of the permeases which have been extensively studied, i.e., histidine, LIV-I, maltose, and galactose, is similar, involving one or more periplasmic components and two or more membrane components. Where sufficient genetic information is available, it is clear that the genes coding for the components are always closely linked on the chromosome, possibly forming an operon in all cases (two divergent operons are known for the maltose transport system). In at least two cases (histidine and galactose), interaction may occur between the periplasmic BP and a membrane-bound component. In addition, it is interesting that the tandem duplication and divergent evolution of the gene coding for the periplasmic component have occurred in more than one system. This has been shown clearly for the *hisJ* and *argT* genes in the histidine system. A remarkably analogous situation exists for LIV-I where genes *livJ* and *livK* also probably originate by duplication of an ancestral gene. The incomplete characterization of most other systems does not allow a generalization to be formulated as far as duplications of genes for periplasmic components is concerned. However, it is possible that the genes for the galactose- and arabinose-BPs also originated as the result of a duplication. These proteins are antigenically related, they share a small but definite homology, and their genes are located quite close to each other on the chromosome.

Considering the complexity of the organization of the periplasmic systems, it is possible that they have all originated by duplication and divergence from a single ancestral system, perhaps already containing a duplication of the periplasmic component, rather than having arisen independently and then acquired some similarity by convergent evolution. Each system arising by duplication would have evolved different specificity while the same basic architecture would have been retained. If the above were true, it might be possible to uncover sequence homologies between proteins of unrelated systems. In agreement with this hypothesis is the fact that the sequence of one of the membrane-bound components of the histidine transport system, the P protein, has a clearly significant homology with the sequence of a membrane-bound component of the maltose-transport system, the MalK protein (Gilson *et al.*, 1982b). However, comparison of all the available sequences of periplasmic BPs has shown only marginally significant homologies between proteins from several completely unrelated systems (Ferro-Luzzi Ames, Farrah, and Doolittle, unpublished results). A more sophisticated statistical analysis will be necessary before the significance of such homologies can be determined. In favor of a structural and functional relationship between these proteins are also the elegant studies of Quiocho and his collaborators on the X-ray structures of several periplasmic transport proteins, which showed in all cases strongly similar two-domain structures, even though the substrates transported by the proteins analyzed are quite different from each other (leucine-isoleucine-valine, arabinose, and sulfate; Quiocho *et al.*, 1977, 1979; Miller *et al.*, 1980).

A problem arising if we are to conclude that all periplasmic systems arose from an ancestral system which already contained a duplication of the periplasmic components is that homology between the two periplasmic components of one system is greater than the homology between those same components and their hypothetical duplicated versions as they appear in a system of completely different specificity. For example, the LAO-BP resembles the his-BP more than either the LAO-BP or the his-BP resembles the LS-BP or LIV-BP. A reasonable explanation is that the LAO-BP and his-BP are constrained in their evolution by their need to interact with the same membrane component (P), while the LS-BP and LIV-BP presumably need to interact with their own specialized membrane component. In support of this hypothesis is the finding that within the his-BP and LAO-BP are segments which are much more highly conserved (>90% homology) than the overall sequence (70% homology) and which are believed to be involved in forming that domain of the molecule which is responsible for the interaction with the membrane protein P (Figure 3). A similar pattern of sequence homology and divergence is seen for the LIV-BP and LS-BP. These two proteins contain two long stretches (>30 amino acids) of absolute sequence conservation in the N-terminal half of the protein while the C-terminal half of the protein is only about 60% conserved vs. 80% overall conservation. Interestingly, the location of conserved and diverged sequences for the LS-BP and LIV-BP differs significantly from those observed for the his-BP and LAO-BP. Therefore, each would have evolved as a "package" independently from the other periplasmic systems. Regardless of the exact sequence of duplication, it is quite likely that present-day amino acid transport systems have evolved from an ancestral amino acid transport system with much broader specificity but which utilized a similar transport mechanism and a similar arrangement of protein components.

## REFERENCES

Aksamit, R. R., Howlett, B. J., and Koshland, D. E., 1975, Soluble and membrane-bound aspartate-binding activities in *Salmonella typhimurium*, *J. Bacteriol.* **123:**1000–1005.

Amanuma, H., and Anraku, Y., 1974, Transport of sugars and amino acids in bacteria. XII. Substrate specificities of the branched chain amino acid-binding protein of *Escherichia coli*, *J. Biol. Chem. (Tokyo)* **76:**1165–1173.

Amanuma, H., Itoh, J., and Anraku, Y., 1976, Transport of amino acids and sugars. XVII. On the existence and nature of substrate amino acids bound to purified branched chain amino acid-binding proteins of *Escherichia coli*, *J. Biol. Chem. (Tokyo)* **79:**1167–1182.

Amanuma, H., Motojima, K., Yamaguchi, A., and Anraku, Y., 1977, Solubilization of a functionally active proline carrier from membrane of *Escherichia coli* with an organic solvent, *Biochem. Biophys. Res. Commun.* **74:**366–373.

Ames, G. F.-L., 1964, Uptake of amino acids in *Salmonella typhimurium*, *Arch. Biochem. Biophys.* **104:**1–18.

Ames, G. F.-L., 1972, Components of histidine transport, in: *Membrane Research*, First ICN-UCLA Symp. Mol. Biol. (C. F. Fox, ed.), Academic Press, New York, pp. 409–426.

Ames, G. F.-L., 1974, Two methods for the assay of amino acid transport in bacteria, *Meth. Enzymol.* **32:**843–856.

Ames, G. F.-L., and Higgins, C. F., 1983, The organization, mechanism of action, and evolution of periplasmic transport systems, *Trends Biochem. Sci.* **8:**97–100.

Ames, G. F.-L., and Lever, J. E., 1970, Components of histidine transport: Histidine-binding proteins and *hisP* protein, *Proc. Natl. Acad. Sci. USA* **66:**1096–1103.

Ames, G. F.-L., and Lever, J. E., 1972, The histidine binding protein J is a component of histidine transport, *J. Biol. Chem.* **247:**4309–4316.

Ames, G. F.-L., and Nikaido, K., 1978, Identification of a membrane protein as a histidine transport component in *Salmonella typhimurium*, *Proc. Natl. Acad. Sci. USA* **75:**5447–5451.

Ames, G. F.-L., and Roth, J. R., 1968, Histidine and aromatic permeases of *Salmonella typhimurium*, *J. Bacteriol.* **96:**1742–1749.

Ames, G. F.-L., and Spudich, E. N., 1976, Protein–protein interaction in transport: Periplasmic histidine-binding protein J interacts with P protein, *Proc. Natl. Acad. Sci. USA* **73:**1887–1891.

Ames, G. F.-L., Noel, K. D., Taber, H., Spudich, E. N., Nikaido, K., Afong, J., and Ardeshir, F., 1977, Fine-structure map of the histidine transport genes in *Salmonella typhimurium*, *J. Bacteriol.* **129:**1289–1297.

Anderson, J. J., and Oxender, D. L., 1977, *Escherichia coli* mutants lacking binding proteins and other components of the branched-chain amino acid transport systems, *J. Bacteriol.* **130:**384–392.

Anderson, J. J., and Oxender, D. L., 1978, Genetic separation of high- and low-affinity transport systems for branched chain amino acids in *Escherichia coli*, *J. Bacteriol.* **136:**168–174.

Anderson, J. J., Quay, S. C., and Oxender, D. L., 1976, Mapping of two loci affecting the regulation of branched chain amino acid transport in *Escherichia coli*, *J. Bacteriol.* **126:**80–90.

Anderson, R. R., Menzel, R., and Roth, J., 1980, Biochemistry and regulation of a second L-proline transport system in *Salmonella typhimurium*, *J. Bacteriol.* **141:**1071–1076.

Anraku, Y., 1968a, Transport of sugars and amino acids. I. Purification and specificity of the galactose- and leucine-binding proteins, *J. Biol. Chem.* **243:**3116–3122.

Anraku, Y., 1968b, Transport of sugars and amino acids in bacteria. II. Properties of galactose- and leucine-binding proteins, *J. Biol. Chem.* **243:**3128–3135.

Anraku, Y., 1980, Transport and utilization of amino acids in bacteria, in: *Microorganisms and Nitrogen Sources* (J. W. Payne, ed.), John Wiley and Sons, New York, pp. 9–34.

Anraku, Y., and Heppel, L. A., 1967, On the nature of the changes induced in *Escherichia coli* by osmotic shock, *Fed. Proc. Fed. Am. Soc. Exp. Bio.* **26:**393.

Anraku, Y., Kobayashi, H., Amanuma, H., and Yamaguchi, A., 1973, Transport of sugars and amino acids in bacteria. VII. Characterization of the reaction of restoration of active transport mediated by binding protein, *J. Biochem. (Tokyo)* **74:**1249–1261.

Antonov, V. K., Arsen'eva, E. L., Gavrilova, N. A., Ginodman, L. M., and Krylova, Y. I., 1974, Novyisposob vydeleniia i nekotorye svoistva leitsinsviazyraivschchego belka iz kishechnoi palochki, *Biokhimiya* **38:**1088–1091.

Antonov, V. K., Alexandrov, S. L., and Vorotyntseva, T. I., 1976, Reversible association as a possible regulatory mechanism for controlling the activity of the nonspecific leucine-binding protein from *Escherichia coli, Adv. Enzyme Reg.* **14:**269.

Ardeshir, F., and Ames, G. F.-L., 1981, Cloning of the histidine transport genes from *Salmonella typhimurium* and characterization of an analogous system in *Escherichia coli, J. Supramol. Struct.* **13:**117–130.

Ardeshir, F., Higgins, C. F., and Ames, G. F.-L., 1981, Physical map of the *Salmonella typhimurium* histidine transport operon: Correlation with the genetic map, *J. Bacteriol.* **147:**401–409.

Ayling, P. D., 1981, Methionine sulfoxide is transported by high-affinity methionine and glutamine transport systems in *Salmonella typhimurium, J. Bacteriol.* **148:**514–520.

Ayling, P. D., Mojica, A. T., and Klopotowski, T., 1979, Methionine transport in *Salmonella typhimurium:* Evidence for at least one low-affinity transport system, *J. Gen. Microbiol.* **114:**227–246.

Baptist, E. W., and Kredich, N. M., 1977, Regulation of L-cystine transport in *Salmonella typhimurium, J. Bacteriol.* **131:**111–118.

Barash, H., and Halpern, Y. S., 1975, Purification and properties of glutamate binding protein from the periplasmic space of *Escherichia coli* K-12, *Biochim. Biophys. Acta* **386:**168–180.

Bavoil, P., Hofnung, M., and Nikaido, H., 1980, Identification of a cytoplasmic membrane-associated component of the maltose transport system of *Escherichia coli, J. Biol. Chem.* **255:**8366–8369.

Berger, E. A., 1973, Different mechanisms of energy coupling for the active transport of proline and glutamine in *Escherichia coli, Proc. Natl. Acad. Sci. USA* **70:**1514–1518.

Berger, E. A., and Heppel, L. A., 1972, A binding protein involved in the active transport of cystine and diaminopimelic acid in *Escherichia coli, J. Biol. Chem.* **247:**7684–7694.

Berger, E. A., and Heppel, L. A., 1974, Different mechanisms of energy coupling for the shock-sensitive and shock-resistant amino acid permeases of *Escherichia coli, J. Biol. Chem.* **249:**7747–7755.

Blobel, G., and Dobberstein, B., 1975a, Transfer of proteins across membranes. I. Presence of proteolytically processed and unprocessed nascent immunoglobulin light chains on membrane-bound ribosomes of murine myeloma, *J. Cell. Biol.* **67:**835–851.

Blobel, G., and Dobberstein, B., 1975b, Transfer of proteins across membranes. II. Reconstitution of functional rough microsomes from heterologous components, *J. Cell. Biol.* **67:**852–862.

Boezi, J. A., and Demoss, R. D., 1961, Properties of a tryptophan transport system in *Escherichia coli, Biochim. Biophys. Acta* **49:**471–484.

Boos, W., 1974, Bacterial transport, *Annu. Rev. Biochem.* **43:**123–146.

Booth, I. R., and Hamilton, W. A., 1980, Energetics of bacterial amino acid transport, in: *Microorganisms and Nitrogen Sources* (J. W. Payne, ed.), J. Wiley and Sons, New York, pp. 171–207.

Brenner, M., and Ames, B. N., 1971, The histidine operon and its regulation, in: *Metabolic Regulation* (H. J. Vogel, ed.), Academic Press, New York, pp. 349–387.

Britten, R. J., and McClure, F. T., 1962, The amino acid pool of *Escherichia coli, Bacteriol. Rev.* **26:**292–299.

Brown, K. D., 1971, Maintenance and exchange of the aromatic amino acid pool in *Escherichia coli, J. Bacteriol.* **106:**70–81.

Buchel, D. E., Gronenborn, B., and Muller-Hill, B., 1980, Sequence of the lactose permease gene, *Nature* **283:**541–545.

Burrous, S. E., and Demoss, R. D., 1963, Studies on the tryptophan permease in *Escherichia coli, Biochim. Biophys. Acta* **73:**623–637.

Celis, R. T. F., 1977, Properties of an *Escherichia coli* K12 mutant defective in the transport of arginine and ornithine, *J. Bacteriol.* **130:**1234–1243.

Celis, R. T. F., 1981, Chain terminating mutants affecting a periplasmic binding protein involved in the active transport of arginine and ornithine in *Escherichia coli, J. Biol. Chem.* **256:**773–779.

Celis, R. T. F., 1982, Mapping of two loci affecting the synthesis and structure of a periplasmic protein involved in arginine and ornithine transport in *Escherichia coli* K-12, *J. Bacteriol.* **151:**1314–1319.

Celis, R. T. F., Rosenfeld, H. J., and Maas, W. K., 1973, A mutant of *Escherichia coli* K12 deficient in the transport of basic amino acids, *J. Bacteriol.* **116:**619–626.

Christensen, H. N., 1975, Recognition sites for material transport and information transfer, *Curr. Top. Membr. Trans.* **6:**227–258.

Christensen, H. N., 1982, The analog inhibition strategy for discriminating similar transport systems: Is it succeeding?, in: *Membranes and Transport,* Vol. 2 (A. Martonosi, ed.), Plenum Press, New York, pp. 145–151.

Cohen, G. N., and Monod, J., 1957, Bacterial permeases, *Bacteriol. Rev.* **21:**169–194.

Condamine, H., 1971, Sur la régulation de la production de proline chez *E. coli* K-12, *Ann. Inst. Pasteur (Paris)* **120:**126–143.

Cosloy, S. D., 1973, D-Serine transport system in *Escherichia coli* K-12, *J. Bacteriol.* **114:**679–684.

Cowell, J. L., 1974, Energization of glycylglycine transport in *Escherichia coli, J. Bacteriol.* **120:**139–146.

Csonka, L. A., 1982, A third L-proline permease in *Salmonella typhimurium* which functions in media of elevated ionic strength, *J. Bacteriol.* **151:**1433–1443.

Daniels, C. J., 1981, Synthesis and processing of the periplasmic leucine transport proteins of *Escherichia coli,* Ph.D. thesis, The University of Michigan, Ann Arbor, Michigan.

Daniels, C. J., Anderson, J. J., Landick, R. C., and Oxender, D. L., 1980, The *in vitro* synthesis and processing of the branched chain amino acid binding proteins, *J. Supramol. Struct.* **14:**305–311.

Daniels, C. J., Bole, D. G., Quay, S. C., and Oxender, D. L., 1981, Role for membrane potential in the secretion of protein into the periplasm in *Escherichia coli, Proc. Natl. Acad. Sci. USA* **78:**5396–5400.

DeBusk, R. M., and DeBusk, D. E., 1980, Physiological and regulatory properties of the general amino acid transport system of *Neurospora crassa, J. Bacteriol.* **143:**188–197.

Dills, S. S., Apperson, A., Schmidt, M. R., and Saier, M. H., 1980, Carbohydrate transport in bacteria, *Microbiol. Rev.* **44:**385–418.

Edwards, R. M., and Yudkin, M. D., 1982, Location of the gene for the low-affinity tryptophan-specific permease of *E. coli, Biochem. J.* **204:**617–619.

Ehring, R., Beyreuther, K., Wright, J. K., and Overath, P., 1980, *In vitro* and *in vivo* products of the *Escherichia coli* lactose permease gene are identical, *Nature* **283:**537–540.

Enequist, H. G., Hirst, T. R., Harayama, S., Hardy, S. J. S., and Randall, L. L., 1981, Energy is required for maturation of exported proteins in *Escherichia coli, Eur. J. Biochem.* **116:**227–233.

Foster, D. L., Garcia, M. L., Newman, M. J., Patel, L., and Kaback, H. R., 1982, Lactose–proton symport by purified *lac* carrier protein, *Biochemistry* **21:**5634–5638.

Frank, L., and Hopkins, I., 1969, Sodium-stimulated transport of glutamate in *Escherichia coli, J. Bacteriol.* **100:**329–336.

Furlong, C. E., and Schellenberg, G. D., 1980, Characterization of membrane proteins involved in transport, in: *Microoganisms and Nitrogen Sources* (J. W. Payne, ed.), John Wiley and Sons, New York, pp. 89–123.

Furlong, C. E., and Weiner, J. H., 1970, Purification of a leucine-specific binding protein from *Escherichia coli, Biochem. Biophys. Res. Commun.* **38:**1076–1083.

Furlong, C. E., Cirakoglu, C., Willis, R. C., and Santy, P. A., 1973, A simple preparative polyacrylamide disc gel electrophoresis apparatus: Purification of three branched chain amino acid binding proteins from *Escherichia coli, Anal. Biochem.* **51:**297–311.

Futai, M., 1974, Orientation of membrane vesicles from *Escherichia coli* prepared by different procedures, *J. Membr. Biol.* **15:**15–28.

Gale, E. F., 1947, The assimilation of amino acids by bacteria. 1. The passage of certain amino acids across the cell wall and their concentration in the internal environment of *Streptococcus faecalis, J. Gen. Microbiol.* **1:**53–76.

Gale, E. F., 1954, The accumulation of amino acids within staphylococcal cells, in: *Active Transport and Secretions,* Symposia of the Society for Experimental Biology, Vol. VIII, Academic Press, New York, pp. 242–253.

Garcia, M. L., Patel, L., Padan, E., and Kaback, H. R., 1982, Mechanism of lactose transport in *Escherichia coli* membrane vesicles: Evidence for the involvement of histidine residues in the response of the *lac* carrier to the proton electrochemical gradient, *Biochemistry* **21:**5800–5805.

Gay, N. J., and Walker, J. E., 1981a, The *atp* operon: Nucleotide sequence of the region encoding the α-subunit of *Escherichia coli* ATP-synthase, *Nucleic Acid Res.* **9:**2187–2194.

Gay, N. J., and Walker, J. E., 1981b, The *atp* operon: Nucleotide sequence of the promoter and the genes for the membrane proteins, and the delta subunit of *Escherichia coli* ATP-synthase, *Nucleic Acid Res.* **9:**3919–3926.

Gilbert, W., Maizels, N., and Maxam, A., 1974, Sequences of controlling regions of the lactose operon, *Cold Spring Harbor Symp. Quant. Biol.* **38:**845–855.

Gilson, E., Higgins, C. F., Hofnung, M., Ames, G. F.-L., and Nikaido, H., 1982a, Extensive homology between membrane-associated components of histidine and maltose transport systems of *Salmonella typhimurium* and *Escherichia coli, J. Biol. Chem.* **257:**9915–9918.

Gilson, E., Nikaido, H., and Hofnung, M., 1982b, Sequence of the *malK* gene in *Escherichia coli* K-12, *Nucleic Acids Res.* **10:**7449–7458.

Grenson, M., and Hon, C., 1972, Ammonia inhibition of the general amino acid permease and its suppression in NADPH-specific glutamate dehydrogenaseless mutants of *Saccharomyces cerevisiae, Biochem. Biophys. Res. Commun.* **48:**749–756.

Guardiola, J., DeFelice, M., Klopotowski, T., and Iaccarino, M., 1974a, Mutation affecting the different transport systems for isoleucine, leucine, and valine in *Escherichia coli* K-12, *J. Bacteriol.* **117:**382–392.

Guardiola, J., DeFelice, M., Klopotowski, T., and Iaccarino, M., 1974b, Multiplicity of isoleucine, leucine, and valine transport systems in *Escherichia coli* K-12, *J. Bacteriol.* **117:**393–405.

Gunsalus, R. P., and Yanofsky, C., 1980, Nucleotide sequence and expression of *Eshcerichia coli trpR:* The structural gene for the *trp* aporepressor, *Proc. Natl. Acad. Sci. USA* **77:**7117–7121.

Gutowski, S. J., and Rosenberg, H., 1976, Energy coupling to active transport in anaerobically grown mutants of *Escherichia coli* K-12, *Biochem. J.* **154:**731–734.

Halpern, Y. S., 1974, Genetics of amino acid transport in bacteria, *Annu. Rev. Genet.* **8:**103–133.

Harrison, L. I., Christensen, H. N., Handlogten, M. E., Oxender, D. L., and Quay, S. C., 1975, Transport of L-4-azaleucine in *Escherichia coli, J. Bacteriol.* **122:**957–965.

Higgins, C. F., and Ames, G. F.-L., 1981, Two periplasmic binding proteins which interact with a common membrane receptor show extensive homology: Complete nucleotide sequences, *Proc. Natl. Acad. Sci. USA* **78:**6038–6042.

Higgins, C. F., and Ames, G. F.-L., 1982, Regulatory regions of two transport operons under nitrogen control: Nucleotide sequences, *Proc. Natl. Acad. Sci. USA* **79:**1083–1087.

Higgins, C. F., Haag, P. D., Nikaido, K., Ardeshir, F., Garcia, G., and Ames, G. F.-L., 1982a, Complete nucleotide sequence and identification of membrane components of the histidine transport operon of Salmonella typhimurium, *Nature* **298:**723–727.

Higgins, C. F., Ames, G. F.-L., Barnes, W. M., Clement, J. M., and Hofnung, M., 1982b, A novel intercistronic regulatory element of prokaryotic operons, *Nature* **298:**760–762.

Ho, C., Giza, Y., Takahashi, S., Ugen, K. E., Cotlam, P. F., and Dowd, S. R., 1980, A proton nuclear magnetic resonance investigation of histidine binding protein J of *Salmonella typhimurium:* A model for transport of L-histidine across the cytoplasmic membrane, *J. Supramol. Struct.* **13:**131–143.

Hogg, R. W., 1981, The amino acid sequence of the histidine binding protein of *Salmonella typhimurium, J. Biol. Chem.* **256:**1935–1939.

Hogg, R. W., and Hermondson, M. A., 1977, Amino acid sequence of the L-arabinose-binding protein from *Escherichia coli* B/r, *J. Biol. Chem.* **252:**5135–5141.

Hong, J.-s., and Hunt, A. G., 1980, The role of acetyl phosphate in active transport, *J. Supramol. Struct.* Suppl. 4, Abst. No. 189, p. 77.

Hong, J.-s., Hunt, A. G., Masters, P. S., and Lieberman, M. A., 1979, Requirement for acetyl phosphate for the binding protein-dependent transport systems in *Escherichia coli, Proc. Natl. Acad. Sci. USA* **76:**1213–1217.

Hoshino, T., and Kageyama, M., 1980, Purification and properties of a binding protein for branched-chain amino acids in *Pseudomonas aeruginosa, J. Bacteriol.* **141:**1055–1063.

Hoshino, T., and Kageyama, M., 1982, Mutational separation of transport systems for branched-chain amino acids in *Pseudomonas aeruginosa, J. Bacteriol.* **151:**620–628.

Hoshino, T., and Nishio, K., 1982, Isolation and characterization of a *Pseudomonas aeruginosa* PAO mutant defective in the strucutral gene for the LIVAT-binding protein, *J. Bacteriol.* **151:**729–736.

Hunt, A. G., and Hong, J.-s., 1981a, The reconstitution of binding protein-dependent active transport of glutamine in isolated membrane vesicles from *Escherichia coli, J. Biol. Chem.* **256:**11988–11991.

Hunt, A. G., and Hong, J.-s., 1981b, The energetics of osmotic shock-sensitive active transport in *Escherichia coli,* in: *Membranes and Transport,* Vol. 2 (A. Martonosi, ed.), Plenum Press, New York, pp. 9–13.

Hunt, A. G., and Hong, J.-s., 1983a, Properties and characterization of binding protein dependent transport of glutamine in isolated membrane vesicles of *Escherichia coli, Biochemistry* **22:**844–850.

Hunt, A. G., and Hong, J.-s., 1983b, Involvement of histidine and tryptophan residues of glutamine binding protein with membrane-bound components of the glutamine transport system of *Escherichia coli, Biochemistry* **22:**851–854.

Iaccarino, M., Guardiola, J., and DeFelice, M., 1978, On the permeability of biological membranes, *J. Membr. Sci.* **3:**287–302.

Iaccarino, M., Guardiola, J., and DeFelice, M., 1980, Genetics of amino acid transport, in: *Microorganisms and Nitrogen Sources* (J. W. Payne, ed.), J. Wiley and Sons, New York, pp. 125–151.

Isihara, H., and Hogg, R. W., 1980, Amino acid sequence of the sulfate-binding protein from *Salmonella typhimurium* LT2, *J. Biol. Chem.* **255:**4614–4618.

Josefsson, L.-G., and Randall, L. L., 1981, Processing *in vivo* of precursor maltose-binding protein in *Escherichia coli* occurs post-translationally as well as cotranslationally, *J. Biol. Chem.* **256:**2504–2507.

Kaback, H. R., 1972, Transport mechanisms in isolated bacterial cytoplasmic membrane vesicles, in: *Membrane Research,* 1st ICN-UCLA Symp. Mol. Biol. (C. F. Fox, ed.), Academic Press, New York, pp. 473–501.

Kaback, H. R., 1974, Transport studies in bacterial membrane vesicles, *Science* **186:**882–892.

Kaback, H. R., 1976, Molecular biology and energetics of membrane transport, *J. Cell. Physiol.* **89:**575–594.

Kaback, H. R., and Deuel, F., 1969, Proline uptake by disrupted membrane preparations from *Escherichia coli, Arch. Biochem. Biophys.* **132:**118–129.

Kaback, H. R., and Kostellow, A. B., 1968, Glycine uptake in *Escherichia coli, J. Biol. Chem.* **243:**1384–1389.

Kaback, H. R., and Milner, L. S., 1970, Relationship of a membrane bound D-(-)-lactic dehydrogenase to amino acid transport in isolated bacterial membrane preparations, *Proc. Natl. Acad. Sci. USA* **66:**1008–1012.

Kaback, H. R., and Stadtman, E. R., 1966, Proline uptake by an isolated cytoplasmic membrane preparation of *Escherichia coli, Proc. Natl. Acad. Sci. USA* **55:**920–927.

Kadner, R. J., 1974, Transport systems for L-methionine in *Escherichia coli, J. Bacteriol.* **117:**232–241.

Kadner, R. J., 1977, Transport and utilization of D-methionine and other methionine sources in *Escherichia coli, J. Bacteriol.* **129:**207–216.

Kadner, R. J., and Watson, W. J., 1974, Methionine transport in *Escherichia coli:* Physiological and genetic evidence for two uptake systems, *J. Bacteriol.* **119:**401–409.

Kadner, R. J., and Winkler, H. H., 1975, Energy coupling for methionine transport in *Escherichia coli, J. Bacteriol.* **123:**985–991.

Kahane, S., Marcus, M., Barash, H., and Halpern, Y. S., 1975, Sodium-dependent glutamate transport in membrane vesicles of *Escherichia coli* K-12, *FEBS Lett.* **560:**235–239.

Kay, W. W., 1971, Two aspartate transport systems in *Escherichia coli, J. Biol. Chem.* **246:**7373–7382.

Kay, W. W., and Gronlund, A. F., 1969, Amino acid transport on *Pseudomonas aeruginosa, J. Bacteriol.* **97:**273–281.

Kiritani, K., 1974, Mutants of *Salmonella typhimurium* defective in transport of branched chain amino acids, *J. Bacteriol.* **120:**1093–1101.

Klein, W. L., and Boyer, P. D., 1972, Energization of active transport by *Escherichia coli, J. Biol. Chem.* **247:**7257–7265.

Klein, W. L., Dahms, A. S., and Boyer, P. D., 1970, The nature of the coupling of oxidative energy to amino acid transport, Abstract No. 540, *Fed. Proc. Fed. Am. Soc. Exp. Biol.* **29:**341.

Konings, W. N., Barnes, E. M., and Kaback, H. R., 1971, Mechanism of active transport in isolated membrane vesicles. III. The coupling of reduced phenazine methosulfate to the concentrative uptake of β-galactosides and amino acids, *J. Biol. Chem.* **246:**5857–5861.

Kreischman, G. P., Robertson, D. E., and Ho, C., 1973, PMR studies of the substrate induced conformational change of the glutamine binding protein from *E. coli, Biochem. Biophys. Res. Commun.* **53:**18–23.

Kusaka, I., Hayakawa, K., Kanai, K., and Fukui, S., 1976, Isolation and characterization of hydrophobic proteins (H proteins) in the membrane fraction of *Bacillus subtilus.* Involvement in membrane biosynthesis and the formation of biochemically active membrane vesicles by combining H proteins with lipids, *Eur. J. Biochem.* **71:**451–458.

Kustu, S. G., and Ames, G. F.-L., 1973, The *hisP* protein, a known histidine transport component in *Salmonella typhimurium,* is also an arginine transport component, *J. Bacteriol.* **166:**107–113.

Kustu, S. G., and Ames, G. F.-L., 1974, The histidine-binding protein J, a histidine transport component, has two different functional sites, *J. Biol. Chem.* **249:**6976–6983.

Kustu, S. G., McFarland, N. C., Hui, S. P., Esmon, B., and Ames, G. F.-L, 1979, Nitrogen control in *Salmonella typhimurium:* Co-regulation of synthesis of glutamine synthetase and amino acid transport systems, *J. Bacteriol.* **138:**218–234.

Kuzaya, H., Bromwell, K., and Guroff, G., 1971, The phenylalanine-binding protein of *Comanonas sp.* (ATCC 11299a), *J. Biol. Chem.* **246:**6371–6380.

Landick, R., 1984, Regulation of LIV-I transport system gene expression, in: *Microbiology 1984* (D. Schlessinger, ed.), American Society for Microbiology, Washington, D.C., in press.

Landick, R. C., and Oxender, D. L., 1981, Bacterial periplasmic binding proteins, in: *Membranes and Transport,* Vol. 2 (A. Martonosi, ed.), Plenum Press, New York, pp. 81–91.

Landick, R., Anderson, J. J., Mayo, M. M., Gunsalus, R. P., Mavromara, P., Daniels, C. J., and Oxender, D. L., 1980, Regulation of high affinity leucine transport in *Escherichia coli, J. Supramol. Struct.* **14:**527–537.

Landick, R. C., Daniels, C. J., and Oxender, D. L., 1983, Assays for the role of membrane potential in the secretion of proteins in bacteria, in: *Methods in Enzymology* (B. Fleischer and S. Fleischer, eds.), **97:**146–153.

Landick, R., Duncan, J. R., Copeland, B., Nazos, P., and Oxender, D. L., 1984, Secretion and degradation of mutant leucine-specific binding protein molecules containing C-terminal deletions, *J. Cellular Biochem.,* in press.

Lee, M., Robbins, J. C., and Oxender, D. L., 1975, Transport properties of merodiploids covering the *dagA* locus in *Escherichia coli* K-12, *J. Bacteriol.* **122:**1001–1005.

Leive, L., and Davis, B. D., 1965a, The transport of diaminopimelate and cystine in *Escherichia coli, J. Biol. Chem.* **240:**4263–4369.

Leive, L., and Davis, B. D., 1965b, Evidence for a gradient of exogenous and endogenous diaminopimelate in *Escherichia coli, J. Biol. Chem.* **240:**4370–4376.

Lever, J. E., 1972, Purification and properties of a component of histidine transport in *Salmonella typhimurium:* The histidine-binding protein J, *J. Biol. Chem.* **247:**4317–4326.

Lieberman, M. A., and Hong, J.-s., 1976, Energization of osmotic shock-sensitive transport systems in *Escherichia coli* requires more than ATP, *Arch. Biochem. Biophys.* **172:**312–315.

Lieberman, M. A., Simon, M., and Hong, J.-s., 1977, Characterization of *Escherichia coli* mutant incapable of maintaining a transmembrane potential, *J. Biol. Chem.* **252:**4056–4067.

Lombardi, F. J., and Kaback, H. R., 1975, Mechanisms of active transport in isolated bacterial membrane vesicles. VIII. The transport of amino acids by membranes prepared from *Escherichia coli, J. Biol. Chem.* **247:**7844–7857.

MacDonald, R. E., Lanyi, J. K., and Greene, R. V., 1977, Sodium-stimulated glutamate uptake in membrane vesicles of *Escherichia coli:* The role of ion gradients, *Proc. Natl. Acad. Sci. USA* **74:**3156–3170.

Mahoney, W. C., Hogg, R. W., and Hermondson, M. A., 1981, The amino acid sequence of the D-galactose-binding protein from *Escherichia coli* B/r, *J. Biol. Chem.* **256:**4350–4356.

Maniatis, T., Ptashne, M., Backman, K., Kleit, D., Flashman, S., Jeffery, A., and Mauer, R., 1975, Recognition sequences of repressor and polymerase in the operators of phage lambda, *Cell* **5:**109–113.

Manuck, B. A., and Ho, C., 1979, High resolution nuclear magnetic resonance studies of histidine-binding proteins J of *Salmonella typhimurium.* An investigation of substrate and membrane interaction sites, *Biochemistry* **18:**566–573.

Masters, P. S., and Hong, J.-s., 1981a, Reconstitution of binding protein dependent active transport of glutamine in spheroplasts of *Escherichia coli, Biochemistry* **20:**4900–4904.

Masters, P. S., and Hong, J.-s., 1981b, Genetics of the glutamine transport system of *Escherichia coli, J. Bacteriol.* **147:**805–819.

McFarland, N., McCarter, L., Artz, S., and Kustu, S., 1981, Nitrogen regulatory locus "*glnR*" of enteric bacteria is composed of cistrons *ntrB* and *ntrC:* Identification of their protein products, *Proc. Natl. Acad. Sci. USA* **78:**2135–2139.

Menzel, R., 1980, The biochemistry and genetics of proline degradation in *S. typhimurium*, Ph.D. thesis, University of California, Berkeley, California.

Menzel, R., and Roth, J., 1980, Identification and mapping of a second proline permease in *Salmonella typhimurium, J. Bacteriol.* **141:**1064–1070.

Miller, D. M., Olson, J. S., and Quiocho, F. A., 1980, The mechanism of sugar binding to the periplasmic receptor for galactose chemotaxis and transport in *Escherichia coli, J. Biol. Chem.* **255:**2465–2471.

Miner, K. M., and Frank, L., 1974, Sodium-stimulated glutamate transport in osmotically shocked cells and membrane vesicles of *Escherichia coli, J. Bacteriol.* **117:**1093–1098.

Mitchell, P., 1970, Membranes of cells and organelles: Morphology, transport, and metabolism, *Symp. Soc. Gen. Microbiol.* **29:**121–166.

Morikawa, A., Suzuki, H., and Anraku, Y., 1974, Transport of sugars and amino acids in bacteria. VIII. Properties and regulation of the active transport reaction of proline in *Escherichia coli, J. Biol. Chem. (Tokyo)* **75:**229–241.

Motojima, K., Yamamoto, I., and Anraku, Y., 1978, Proline transport carrier defective mutants of *Escherichia coli* K12: Properties and mapping, *J. Bacteriol.* **136:**5–9.

Nakane, P. K., Nichoalds, G. E., and Oxender, D. L., 1968, Cellular localization of leucine-binding protein from *Escherichia coli, Science* **161:**182–183.

Neu, H. C., and Heppel, L. A., 1965, The release of enzymes from *Escherichia coli* by osmotic shock and during the formation of spheroplasts, *J. Biol. Chem.* **240:**3685–3692.

Nichols, W. W., and Hamilton, W. A., 1976, The transport of D-lactate by membrane vesicles from *Poracoccus denitrificans, FEBS Lett.* **65:**107–110.

Noel, D., Nikaido, K., and Ames, G. F.-L., 1979, A single amino acid substitution in a histidine-transport protein drastically alters its mobility in sodium dodecyl sulfate–polyacrylamide gel electrophoresis, *Biochemistry* **18:**4159–4165.

Ogilvie-Villa, S., DeBusk, R. M., and DeBusk, A. G., 1981, Characterization of 2-aminoisobutyric acid transport in *Neurospora crassa:* A general amino acid permease-specific substrate, *J. Bacteriol.* **147:**944–948.

Oshimaa, R. B., Willis, R. C., Furlong, C. E., and Schnieder, J. A., 1974, Binding assays for amino acids. The utilization of a cystine-binding protein from *Escherichia coli* for the determination of acid-soluble cystine in small physiological samples, *J. Biol. Chem.* **249:**6033–6039.

Ovchinnikov, Y. A., Aidanova, N. A., Grinkevich, V. A., Arzamazova, N. M., and Movoz, I. N., 1977, The primary structure of a Leu, Ile, and Val (LIV)-binding protein from *Escherichia coli, FEBS Lett.* **78:**313–316.

Oxender, D. L., 1972, Membrane transport, *Annu. Rev. Biochem.* **41:**777–814.

Oxender, D. L., 1974, Membrane transport proteins, in: *Biomembranes*, Vol. 5 (L. A. Manson, ed.), Plenum Press, New York, pp. 25–79.

Oxender, D. L., 1975, Genetic approaches to the study of transport, in: *Biological Transport,* Chapter VI (H. N. Christensen, ed.), Benjamin Press, New York, pp. 214–231.

Oxender, D. L., and Christensen, H. N., 1963, Evidence for two types of mediation of neutral amino acid transport in Ehrlich cells, *J. Biol. Chem.* **238:**3686–3699.

Oxender, D. L., and Quay, S. C., 1976a, Regulation of leucine transport and binding proteins in *Escherichia coli, J. Cell. Physiol.* **89:**517–521.

Oxender, D. L., and Quay, S. C., 1976b, Isolation and characterization of membrane binding proteins, in: *Methods in Membrane Biology,* Vol. 6, Chapter IV (E. D. Korn, ed.), Plenum Press, New York, pp. 183–242.

Oxender, D. L., and Quay, S. C., 1976c, Binding proteins and membrane transport, *Ann. N. Y. Acad. Sci.* **264:**358–374.

Oxender, D. L., Anderson, J. J., Mayo, M. M., and Quay, S. C., 1977, Leucine binding protein and regulation of transport in *Escherichia coli, J. Supramol. Struct.* **6:**419–431.

Oxender, D. L., Anderson, J. J., Daniels, C. J., Landick, R., Gunsalus, R. P., Zurawski, G., Selker, E., and Yanofsky, C., 1980a, Structural and functional analysis of cloned DNA containing genes responsible for branched-chain amino acid transport in *Escherichia coli, Proc. Natl. Acad. Sci. USA* **77:**1412–1416.

Oxender, D. L., Anderson, J. J., Daniels, C. J., Landick, R., Gunsalus, R. P., Zurawski, G., and Yanofsky, C., 1980b, Amino-terminal sequence and processing of the precursor of the luecine-specific binding protein, and evidence for conformational differences between the precursor and mature form, *Proc. Natl. Acad. Sci. USA* **77:**2005–2009.

Pall, M. L., 1969, Amino acid transport in *Neurospora crassa*. I. Properties of two amino acid transport systems, *Biochem. Biophys. Acta* **173:**113–129.

Patel, L., Garcia, M. L., and Kaback, H. R., 1982, Direct measurement of lactose/proton symport in *Escherichia coli* membrane vesicles: Further evidence for the involvement of histidine residue(s), *Biochemistry* **21:**5805–5810.

Penrose, W. R., Zand, R., and Oxender, D. L., 1970, Reversible conformational changes in a leucine binding protein from *Escherichia coli, J. Biol. Chem.* **245:**1432–1437.

Piperno, J. R., and Oxender, D. L., 1966, Amino acid-binding protein released from *Escherichia coli* by osmotic shock, *J. Biol. Chem.* **241:**5732–5743.

Piperno, J. R., and Oxender, D. L., 1968, Amino acid transport systems in *Escherichia coli* K12, *J. Biol. Chem.* **243:**5914–5920.

Plate, C. A., 1976, Mutant of *Escherichia coli* defective in response to colicin K and active transport, *J. Bacteriol.* **125:**467–474.

Plate, C. A., 1979, Requirement for membrane potential in active transport of glutamine by *Escherichia coli, J. Bacteriol.* **137:**221–225.

Plate, C. A., Suit, J. L., Jetten, A. M., and Luria, S. E., 1974, Effect of colicin K on a mutant of *Escherichia coli* deficient in $Ca^{+2}$, $Mg^{+2}$-activated adenosine triphosphatase, *J. Biol. Chem.* **249:**6138–6143.

Postma, P. W., and Roseman, S., 1976, The bacterial phosphoenolpyruvate:sugar phosphotransferase system, *Biochem. Biophys. Acta* **457:**213–257.

Quay, S. C., and Oxender, D. L., 1976, Regulation of branched-chain amino acid binding proteins in *Escherichia coli, J. Bacteriol.* **127:**1225–1238.

Quay, S. C., and Oxender, D. L., 1977, Regulation of amino acid transport in *Escherichia coli* by transcription termination factor rho, *J. Bacteriol.* **139:**1024–1029.

Quay, S. C., and Oxender, D. L., 1979, The *relA* locus specifies a positive effector for branched-chain amino acid transport regulation, *J. Bacteriol.* **137:**1059–1062.

Quay, S. C., and Oxender, D. L., 1980a, Regulation of membrane transport, in: *Biological Regulation and Development,* Vol. 2 (R. Goldberger, ed.), Plenum Press, New York, pp. 413–436.

Quay, S. C., and Oxender, D. L., 1980b, Role of $tRNA^{leu}$ in branched-chain amino acid transport, in: *tRNA: Biological Aspects* (D. Söll, J. Abelson, and P. R. Schimmel, eds.), Cold Spring Harbor Press, Cold Spring Harbor, New York, pp. 481–491.

Quay, S. C., Kline, E. L., and Oxender, D. L., 1975a, Role of leucyl-tRNA synthetase in regulation of branched-chain amino acid transport, *Proc. Natl. Acad. Sci. USA* **72:**3921–3924.

Quay, S. C., Oxender, D. L., Tsuyumu, S., and Umbarger, H. E., 1975b, Separate regulation of transport and biosynthesis of leucine, isoleucine, and valine in bacteria, *J. Bacteriol.* **122:**994–1000.

Quay, S. C., Dick, T. E., and Oxender, D. L., 1977, Role of transport systems in amino acid metabolism: Leucine toxicity and the branched-chain amino acid transport systems, *J. Bacteriol.* **129:**1257–1265.

Quay, S. C., Lawther, R. P., Hatfield, G. W., and Oxender, D. L., 1978, Branched-chain amino acid transport regulation in mutants blocked in tRNA maturation and transcription termination, *J. Bacteriol.* **134:**683–686.

Quiocho, F. A., Gilliland, G. L., and Phillips, G. N., 1977, The 2.8-Å resolution structure of the L-arabinose-binding protein from *Escherichia coli, J. Biol. Chem.* **252:**5142–5149.

Quiocho, F. A., Meador, W. E., and Pflugrath, J. W., 1979, Preliminary crystallographic data on receptors for transport and chemotaxis of *Escherichia coli:* D-galactose and maltose binding proteins, *J. Mol. Biol.* **133:**181–184.

Rahmanian, M., and Oxender, D. L., 1972, Derepressed leucine transport activity in *Escherichia coli, J. Supramol. Struct.* **1:**55–59.

Rahmanian, M., Claus, D. R., and Oxender, D. L., 1973, Multiplicity of leucine transport systems in *Escherichia coli* K12, *J. Bacteriol.* **116:**1258–1266.

Ramos, S., and Kaback, H. R., 1977a, The electrochemical proton gradient in *Escherichia coli* membrane vesicles, *Biochemistry* **16:**848–853.

Ramos, S., and Kaback, H. R., 1977b, The relationship between the electrochemical proton gradient and active transport in *Escherichia coli* membrane vesicles, *Biochemistry* **16:**855–859.

Rao, E. Y. T., Rao, T. K., and DeBusk, A. G., 1975, Isolation and characterization of a mutant of *Neurospora crassa* deficient in general amino acid permease activity, *Biochem. Biophys. Acta* **413:**45–51.

Ratzkin, B., Grabnar, M., and Roth, J., 1978, Regulation of a major proline permease gene of *Salmonella typhimurium, J. Bacteriol.* **133:**737–739.

Rhoads, D. B., and Epstein, W., 1977, Energy coupling to net $K^+$ transport in *Escherichia coli* K12, *J. Biol. Chem.* **252:**1394–1401.

Robb, F. T., and Furlong, C. E., 1980, Reconstitution of binding protein dependent ribose transport in spheroplasts derived from a binding protein negative *Escherichia coli* K12 mutant and from *Salmonella typhimurium, J. Supramol. Struct.* **13:**183–190.

Robbins, J. C., and Oxender, D. L., 1973, Transport systems for alanine, serine, and glycine, *J. Bacteriol.* **116:**12–18.

Robertson, D. E., Kroon, P. A., and Ho, C., 1977, Nuclear magnetic resonance and fluorescence studies of substrate induced conformational changes of histidine-binding protein J of *Salmonella typhimurium, Biochemistry* **16:**1443–1451.

Ron, E. Z., Kohler, R. E., and Davis, B. D., 1966, Polysomes extracted from *Escherichia coli* by freeze-thaw-lysozyme lysis, *Science* **153:**1119–1120.

Rosen, B. P., 1971, Basic amino acid transport in *Escherichia coli, J. Biol. Chem.* **246:**3653–3662.

Rosen, B. P., 1973a, Basic amino acid transport in *Escherichia coli:* Properties of canavanine-resistant mutants, *J. Bacteriol.* **116:**627–635.

Rosen, B. P., 1973b, Basic amino acid transport in *Escherichia coli*. II. Purification and properties of an arginine specific binding protein, *J. Biol. Chem.* **248:**1211–1218.

Rosen, B. P., and Vasington, F. D., 1971, Purification and characterization of a histidine-binding protein from *Salmonella typhimurium* LT-2 and its relationship to the histidine permease system, *J. Biol. Chem.* **246:**5351–5356.

Rowland, I., and Tristam, H., 1974, Specificity of the *Escherichia coli* proline transport system, *J. Bacteriol.* **123:**871–877.

Saper, M. A. and Quiocho, F. A., 1983, Leucine, isoleucine, valine-binding protein from *Escherichia coli:* Structure at 3.0 Å resolution and location of the binding site, *J. Biol. Chem.* **258:**11057–11062.

Schellenberg, G. D., 1978, The multiplicity of glutamate and aspartate transport systems in *Escherichia coli,* Ph.D. dissertation, University of California, Riverside, California.

Schellenberg, G. D., and Furlong, C. E., 1977, Resolution of the multiplicity of the glutamate and aspartate transport systems of *Escherichia coli, J. Biol. Chem.* **352:**9055–9064.

Schuldiner, S., Kung, H., Weil, R., and Kaback, H. R., 1975, Differentiation between binding and transport of dansylgalactosides in *Escherichia coli, J. Biol. Chem.* **250:**3679–3682.

Seaston, A., Inkson, C., and Eddy, A. A., 1973, The absorption of protons with specific amino acids and carbohydrates by yeast, *Biochem. J.* **154:**1031–1043.

Seaston, A., Carr, G., and Eddy, A. A., 1976, The concentration of glycine by preparations of the yeast *Saccharomyces carlsbergensis* depleted of adenosine triphosphate: Effects of proton gradients and uncoupling agents, *Biochem. J.* **169:**210–218.

Shaltiel, S., Ames, G. F.-L., and Noel, K. D., 1973, Hydrophobic chromatography in the purification of the histidine-binding protein J from *Salmonella typhimurium, Arch. Biochem. Biophys.* **159:**174–179.

Shuman, H. A., 1982, Active transport of maltose in *Escherichia coli* K12: Role of the periplasmic maltose-binding protein and evidence for a substrate recognition site in the cytoplasmic membrane, *J. Biol. Chem.* **257:**5455–5461.

Shuman, H. A., and Silhavy, T. J., 1981, Identification of the *malK* gene product: A peripheral membrane component of the *Escherichia coli* maltose transport system, *J. Biol. Chem.* **256:**560–562.

Shuman, H. A., Silhavy, T. J., and Beckwith, J. R., 1980, Labeling of proteins with β-galactosidase by gene fusion: Identification of a cytoplasmic membrane component of the *Escherichia coli* maltose transport system, *J. Biol. Chem.* **255:**168–174.

Simoni, R. D., and Postma, P. W., 1975, The energetics of amino acid transport, *Annu. Rev. Biochem.* **44:**523–544.

Slayman, C. W., 1973, The genetic control of membrane transport, *Curr. Top. Membr. Trans.* **4:**1–175.

Smith, W. P., Tai, P.-C., Thompson, R., and Davis, B. D. 1977, Extracellular labeling of nascent polypeptides traversing the membrane of *Escherichia coli, Proc. Natl. Acad. Sci. USA* **74:**2830–2834.

Smith, W. P., Tai, P.-C., and Davis, B. D., 1978, Nascent peptide as sole attachment of polysomes to membranes in bacteria, *Proc Natl. Acad. Sci. USA* **75:**814–817.

Stern, M. F., Ames, G. F.-L., Smith, N. H., Robinson, C. E., and Higgins, C. F., 1984, Repetitive extragenic palindromic (REP) sequences: a major component of the bacterial genome, *Cell* (in press).

Stevenson, J., 1966, The specific requirement for sodium chloride for the active uptake of L-glutamate by *Halobacterium salinarium, Biochem. J.* **99:**257–260.

Sveedhara-Swamy, K. H., and Goldberg, A. L., 1982, Subcellular distribution of various proteases in *Escherichia coli, J. Bacteriol.* **149:**1027–1033.

Templeton, B. A., and Savageau, M. A., 1974a, Transport of biosynthetic intermediates: Regulation of homoserine and threonine uptake in *Escherichia coli, J. Bacteriol.* **120:**114–120.

Templeton, B. A., and Savageau, M. A., 1974b, Transport of biosynthetic intermediates: Homoserine and threonine uptake in *Escherichia coli, J. Bacteriol.* **117:**1002–1009.

Tinoco, I., Borer, P. N., Dengler, B., Levine, M. D., Uhlenbeck, O. C., Crothers, D. M., and Gralla, J., 1973, Improved estimation of secondary structure in ribonucleic acids, *Nature* **246:**40–41.

Tsuchiya, T., Hasan, S. M., and Raven, J., 1977, Glutamate transport driven by an electrochemical gradient of sodium ions in *Escherichia coli, J. Bacteriol.* **131:**848–853.

Wargel, R. J. C., Shadur, C. A., and Neuhaus, F. C., 1970, Mechanism of D-cycloserine action: Transport systems for D-alanine, D-cycloserine, alanine, and glycine, *J. Biol. Chem.* **103:**778–788.

Wargel, R. J. C., Shadur, C. A., and Neuhaus, F. C., 1971, Mechanism of D-cycloserine action: Transport mutants for D-alanine, D-cycloserine, alanine, and glycine, *J. Bacteriol.* **105:**1028–1035.

Weiner, J. H., and Heppel, L. A., 1971, A binding protein for glutamine and its relation to active transport in *Escherichia coli, J. Biol. Chem.* **246:**6933–6941.

Weiner, J. H., Furlong, C. E., and Heppel, L. A., 1971, A binding protein for L-glutamine and its relation to active transport in *Escherichia coli, Arch. Biochem. Biophys.* **142:**715–717.

Whipp, M. J., Halsall, D. M., and Pittard, A. J., 1980, Isolation and characterization of an *Escherichia coli* K12 mutant defective in tyrosine- and phenylalanine-specific transport systems, *J. Bacteriol.* **143:**1–7.

Willis, R. C., and Furlong, C. E., 1975, Purification and properties of a periplasmic glutamate-aspartate binding protein from *Escherichia coli* K-12 strain W3092, *J. Biol. Chem.* **250:**2574–2580.

Willis, R. C., and Woolfolk, C. A., 1975, L-Asparagine uptake in *Escherichia coli, J. Bacteriol.* **123:**937–945.

Willis, R. C., Morris, R. G., Cirakoglu, C., Schellenberg, G. D., Gerber, N. H., and Furlong, C. E., 1974, Preparations of the periplasmic binding proteins from *Salmonella typhimurium* and *Escherichia coli, Arch. Biochem. Biophys.* **161:**64–75.

Willis, R. C., Iwata, K. K., and Furlong, C. E., 1975, Regulation of glutamine transport in *Escherichia coli, J. Bacteriol.* **122:**1032–1037.

Wilson, D. B., 1978, Cellular transport mechanisms, *Annu. Rev. Biochem.* **47:**933–965.

Wilson, O. H., and Holden, J. T., 1969a, Arginine transport and metabolism in osmotically shocked cells of *Escherichia coli* W, *J. Biol. Chem.* **244:**2737–2742.

Wilson, O. H., and Holden, J. T., 1969b, Stimulation of arginine transport in osmotically shocked *Escherichia coli* W cells by purified arginine-binding protein fractions, *J. Biol. Chem.* **244:**2743–2749.

Wong, P. T. S., Thompson, J., and McCleod, R. A., 1980, Nutrition and metabolism of marine bacteria. XVII. Ion-dependent retention of $\alpha$-aminoisobutyric acid and its relation to $Na^+$-dependent transport in a marine pseudomonad, *J. Biol. Chem.* **244:**1016–1025.

Wood, J. M., 1975, Leucine transport in *Escherichia coli:* The resolution of multiple transport systems and their coupling to metabolic energy, *J. Biol. Chem.* **250:**4477–4485.

Wood, J. M., 1981, Genetics of L-proline utilization in *Escherichia coli, J. Bacteriol.* **146:**895–901.

Wood, J. M., Zwadorny, D., Lohmeier, E., and Weiner, J. H., 1979, Characterization of an inducible porter for L-proline catabolism by *Escherichia coli, Can. J. Biochem.* **57:**1328–1330.

Yamato, I., and Anraku, Y., 1977, Transport of sugars and amino acids in bacteria. XVIII. Properties of an isoleucine transport carrier in the cytoplasmic membrane vesicles of *Escherichia coli, J. Biol. Chem (Tokyo)* **81:**1517–1523.

Yamato, I., and Anraku, Y., 1980, Genetic and biochemical studies of transport systems for branched-chain amino acids in *Escherichia coli:* Isolation and properties of mutants defective in leucine-repressible activities, *J. Bacteriol.* **144:**36–44.

Yanofsky, C., 1981, Attenuation in the control of expression of bacterial operons, *Nature* **289:**751–758.

Young, I. G., Rogers, B. L., Campbell, H. D., Jawornowski, A., and Shaw, D. C., 1981, Nucleotide sequence coding for the respiratory NADH dehydrogenase of *Escherichia coli:* UUG initiation codon, *Eur. J. Biochem.* **116:**165–170.

Zwizinski, C., and Wickner, W., 1980, Purification and characterization of a leader (signal) peptidase from *Escherichia coli, J. Biol. Chem.* **255:**7973–7977.

# 44

# The Iron-Transport Systems of *Escherichia coli*

Volkmar Braun

## I. TYPES OF TRANSPORT SYSTEMS IN *E. coli*

Cells of *E. coli* like those of the other gram-negative bacteria are surrounded by two membranes, the outer membrane and the cytoplasmic membrane. Between these two membranes is situated the murein or peptidoglycan layer which confers rigidity to the cells. It is part of the periplasmic space from which water-soluble proteins can be released without release of the cytoplasmic proteins by a treatment called osmotic shock. Solutes which are taken up by the cells have therefore to be translocated through two membranes, the periplasmic space and the murein layer. The latter two are not considered to form a permeability barrier.

### A. Uptake through the Outer Membrane

The outer membrane protects cells of *E. coli* against detrimental substances, for example, the bile salts and hydrolytic enzymes in the gut. For small hydrophilic substrates, it does not form a permeability barrier. Proteins that are present in high amounts, in the order of $5 \times 10^4$ copies per $\mu m^2$, form essentially water-filled channels through which solutes diffuse as long as their molecular weight is less than about 600 (Nikaido and Nakae, 1979; Nikaido, 1982). The rate of diffusion is mainly determined by the size of the solutes. These proteins, called porins, show some preference for cations (Benz *et al.*, 1979). However, there are also more specific uptake processes through the outer membrane. Under phosphate-limiting growth conditions, a protein is synthesized that forms a channel which favors diffusion of inorganic phosphate and

*Volkmar Braun* • Mikrobiologie II, Universität Tübingen, D-7400 Tübingen, West Germany.

organic phosphate compounds (Argast and Boos, 1980; Overbeeke and Lugtenberg, 1982). In addition, growth on maltose induces an outer membrane protein, the *lamB* gene product, that facilitates the uptake of maltose and of larger maltodextrins (Szmelcman and Hofnung, 1975; Szmelcman *et al.*, 1976; Braun and Krieger-Brauer, 1977). The LamB protein exerts configurational specificity since it binds maltodextrins (Luckey and Nikaido, 1980; Ferenci, 1980; Ferenci *et al.*, 1980). The maltose-binding protein in the periplasm physically interacts with the LamB protein of the outer membrane (Bavoil and Nikaido, 1981) and, thus, contributes to the specificity of the LamB pore for maltodextrins (Wandersman *et al.*, 1979; Ferenci and Boos, 1980; Heuzenroeder and Reeves, 1980).

Another system probably comparable to the translocation of maltodextrins across the outer membrane is that for the uptake of nucleosides and deoxynucleosides. Their rate of uptake is accelerated by the presence of the Tsx protein (Hantke, 1976). Exceptions are cytidine and deoxycytidine whose uptake rates are not altered by the Tsx protein. This observation was supported by the finding that synthesis of the Tsx protein was controlled by the same genes that also regulated the various nucleoside and deoxynucleoside-transport systems (Krieger-Brauer and Braun, 1980).

A third type of uptake through the outer membrane strictly depends on highly specific receptor proteins. They will be described in Section II-B.

Extensive reviews on the structure, function, and biosynthesis of the outer membrane have appeared recently (DiRienzo *et al.*, 1978; Osborn and Wu, 1980; Lugtenberg and van Alphen, 1983).

## B. Uptake through the Cytoplasmic Membrane

Diffusion is certainly not an important uptake process in the case of microorganisms which usually live in an environment where some or all of the required substrates are present in very low concentration. However, glycerol is taken up into *E. coli* by facilitated diffusion for which a protein has been identified in the cytoplasmic membrane. Nutrients are usually taken up by active transport. The active transport systems reside in the cytoplasmic membrane. Energy is provided by the various electron-transport chains, by ATP or PEP hydrolysis. The membrane potential derived from the electron-transport chains or from ATP hydrolysis can be used directly for the transport of solutes. Examples are the uptake of lactose and proline into *E. coli*.

Some sugars, like maltose and galactose, have first to be bound to so-called binding proteins in the periplasmic space from where they are donated to the transport systems in the cytoplasmic membrane. The energy-providing process for the transport that depends on binding proteins is still not clear.

Some sugars, for example, glucose, are phosphorylated during the transport, and PEP is the energy-rich compound which provides the phosphate group. For more detailed information on these types of transport, the reader is referred to a book in which recent developments have been discussed by experts (Martonosi, 1982).

As far as is known, transport of iron follows none of the mechanisms outlined above for the uptake of substrates through the outer and cytoplasmic membrane of *E. coli*. The peculiarities of iron transport will now be discussed.

## *II. PECULIARITIES OF THE IRON-TRANSPORT SYSTEMS*

The oxidation reduction potential of $Fe^{3+}$, $Fe^{2+}$ spans an extremely wide scale (from +300 mV to −500 mV). It depends on the mode of the iron binding and the surrounding provided by the proteins. It is, therefore, the most versatile and the most abundant catalytically active metal ion.

In *E. coli,* iron plays an essential role in heme and nonheme iron proteins of the electron-transport chains, and as part of a number of enzymes in the intermediary metabolism, for example, in glutamate synthase and in ribotide reductase. *E. coli* also contains a small amount of ferritin (Bauminger *et al.*, 1980; Yariv *et al.*, 1981) whose physiological function is unknown, and which is composed of two chromophores, a polynuclear iron compound and a *b*-type cytochrome. Of the 800 μg of iron per gram cells, only 5 μg is stored as ferritin.

### *A. Requirement of Siderophores*

At pH7, iron(III) forms polymeric hydroxy–aquo complexes. The free ion concentration in equilibrium with the polymer is in the range of $10^{-12}$ μM (Table 1). Since *E. coli* cells require about 0.1 μM iron for growth, the iron concentration available at aerobic growth conditions would be 11 orders of magnitude too low (1000 iron ions for $10^8$ cells in 1 ml)! Therefore, *E. coli,* as most other microorganisms, synthesizes and excretes strong iron chelators which, after scavenging the free iron ions, are transported into the cells. These iron-chelating agents are called siderophores (see the review of Neilands, 1981) to account for their function as iron-transport vehicles. However, this term does not imply that they function like ionophores. For example, valinomycin complexes potassium ions and is thereby converted from a hydrophilic to a more hydrophobic molecule. Potassium is then shuttled through a

*Table 1. Free-Iron(III) Concentrations*

| Ligand | $Fe^{3+}(H_2O)_6$ (μM) |
|---|---|
| Enterochelin | $10^{-29}$[a] |
| Aerobactin | $10^{-17}$[a] |
| Deferriferrichrome | $10^{-19}$[a] |
| Deferrichrysin | $10^{-20}$ |
| Citrate[b] | |
| Human transferrin | $10^{-18}$ |
| Human lactoferrin | $<10^{-18}$ |
| $(Fe(OOH))_n$ | $10^{-12}$ |
| Concentration for growth | $10^{-1}$ |

[a] The iron concentrations were calculated from the formation constants for a solution which is 1 μM in iron(III) and 10 μM in ligand, at pH 7.4. From Raymond *et al.* (1982).

[b] Formation constants determined for the iron–citrate complex are $10^{-25}$ for the unprotonated and $10^{-12}$ for the protonated form. From Sillen and Martell (1964).

lipid bilayer membrane by a pure diffusion process. Gramicidin, on the other hand, forms a pore through a lipid bilayer. In contrast to these passive diffusion processes, the siderophores are taken up by active transport catalyzed by a number of proteins.

Four siderophores are known to be used for the uptake of iron(III) by *E. coli*. They are listed in Table 1 together with the free iron concentrations which are present in equilibrium with the iron complexes. The iron concentrations have been estimated from the formation constants of the iron complexes. It is evident from these figures that these complexing agents extract iron very effectively from the growth medium.

The function of siderophores as iron carriers is supported by two findings. First, they are only synthesized at growth-limiting iron concentrations in the medium. Under these conditions, the concentration of enterochelin (synonymous with enterobactin) can reach 100 mg/l (Young, 1976), and that of aerobactin, 50–200 mg/l (Braun, 1981; Braun *et al.*, 1982). Second, for each iron complex, there exists a specific transport system which was proven by the electrophoretic identification of transport proteins and by the isolation of mutants which became deficient in the transport of only one iron siderophore complex.

*Enterochelin (enterobactin)* is the most widely distributed siderophore among Enterobacteriaceae. It has been found in *E. coli, S. typhimurium, Enterobacter (Aerobacter) aerogenes* (Rosenberg and Young, 1974), *Shigella sonnei, Klebsiella pneumoniae* (Perry and San Clemente, 1979), and in *Enterobacter cloacae* (van Tiel-Menkveld *et al.*, 1982). Its biosynthesis starts from chorismic acid, the common precursor of phenylalanine, tyrosine, and tryptophan. In a number of steps, chorismic acid is converted to 2,3-dihydroxybenzoic acid, to which L-serine is added. Three molecules of the resulting amide then form a cyclic triester (Figure 1). Since the appearance of the review by Rosenberg and Young (1974), some progress was made in relating gene products with enzymatic activities without deciphering the biochemistry of the whole reaction sequence from dihydroxybenzoic acid to enterochelin (Greenwood and Luke, 1976; Woodrow *et al.*, 1979). It is thought that an enzyme complex catalyzes the reactions. If the complex were membrane-bound, enterochelin could immediately be excreted and thus be separated from the cytoplasmic degradative enzyme, entero-

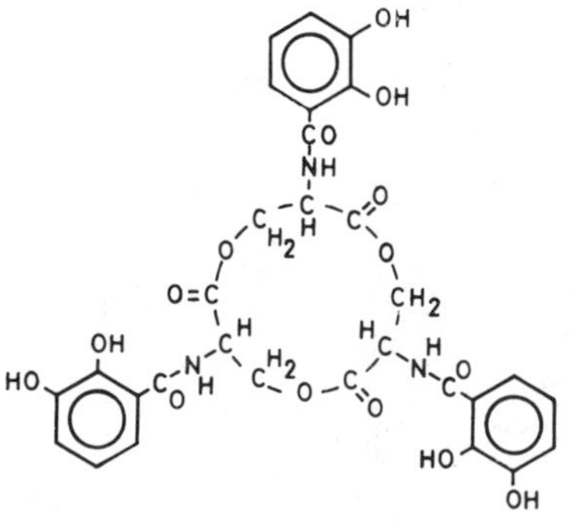

*Figure 1.* Structure of enterochelin (enterobactin).

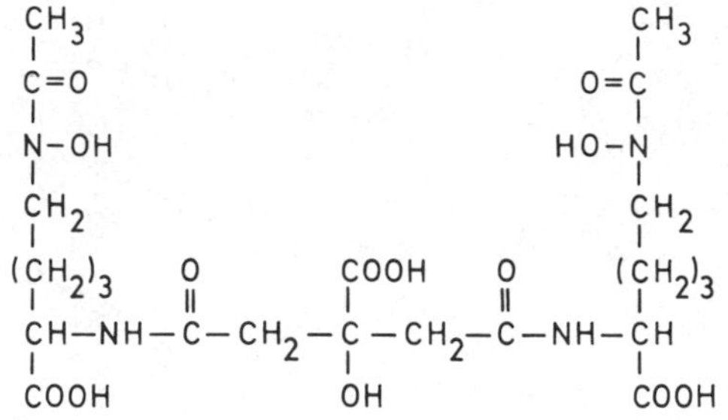

```
CH3                                             CH3
 |                                               |
C=O                                            O=C
 |                                               |
N-OH                                          HO-N
 |                                               |
CH2                                             CH2
 |                                               |
(CH2)3      O          COOH       O           (CH2)3
 |          ||          |         ||             |
CH—NH—C—CH2—C—CH2—C—NH—CH
 |                      |                        |
COOH                    OH                      COOH
```

*Figure 2.* Structure of aerobactin.

chelin esterase (see Section II-C). This mechanism would also keep the cytoplasm free of enterochelin that could inhibit iron-dependent reactions by withdrawing iron from enzymes. Enterochelin has a very high affinity for $Fe^{3+}$ (Table 1). Enterochelin belongs to the iron ligands of the catecholate type.

The hydroxamate iron ligands seem to occur more frequently in nature than the catecholates. *Aerobactin* was originally isolated from *Aerobacter* (now named *Eneterobacter) aerogenes* (Gibson and Magrath, 1969). It consists of two *N*-hydroxy-*N*-acetyl-L-lysine residues linked by citrate (Figure 2). Regarding its biosynthesis, hydroxylation of L-lysine to ε-*N*-hydroxylysine was achieved in a cell-free system (Murray *et al.*, 1977; Parniak *et al.*, 1979). *N*-Hydroxylation was optimal when the supernatant fraction was combined with the membrane fraction suggesting a membrane-bound synthesis, and immediate excretion of the iron ligand as discussed above for enterochelin. Iron is bound by the hydroxamate groups and probably by the central carboxylate and hydroxyl group of the citrate residue (Harris *et al.*, 1979a,b). Aerobactin has recently been found in strains of *E. coli* (Braun, 1981; Warner *et al.*, 1981) bearing certain ColV plasmids (Williams, 1979; Stuart *et al.*, 1980), in two *Salmonella* strains and in one *Arizona* strain (Warner *et al.*, 1981), in *Shigella flexneri* (Payne, 1980), and in *Enterobacter cloacae* (van Tiel-Menkveld *et al.*, 1982).

*Ferrichrome* is another iron–hydroxamate complex taken up by *E. coli* and *S. typhimurium* via a highly specific transport system, although the ligand is only synthesized by certain fungi and not by bacteria (Neilands, 1981). It consists of a cyclic hexapeptide composed of three consecutive residues of δ-*N*-acetyl-L-δ-*N*-hydroxyornithine and three glycine residues (Figure 3). One or two glycine residues can be replaced by serine residues. These compounds have been named ferricrocin and ferrichrysin (Zähner *et al.*, 1963; Figure 3). Structures based on X-ray analyses have been elucidated for ferrichrome (van der Helm *et al.*, 1980) and for ferrichrysin (Norrestam *et al.*, 1975). The antibiotic albomycin is taken up by the same transport system as ferrichrome and its derivatives. Albomycin contains only four of the six amino acids of ferrichrysin. The glycine residue is lacking, and the second serine residue is replaced by an amino acid with a serine grouping in which one β-hydrogen

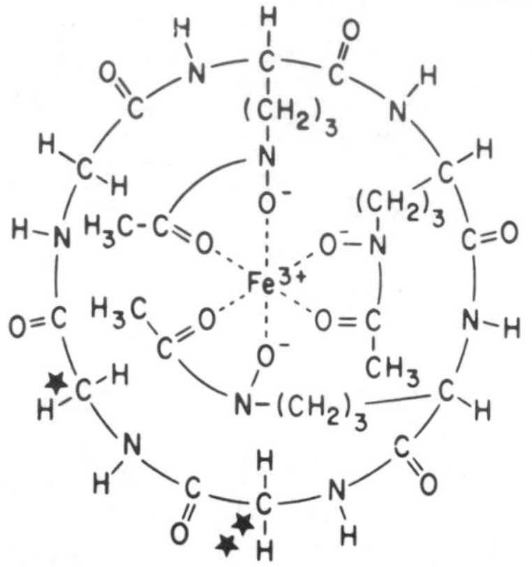

Ferrichrome.

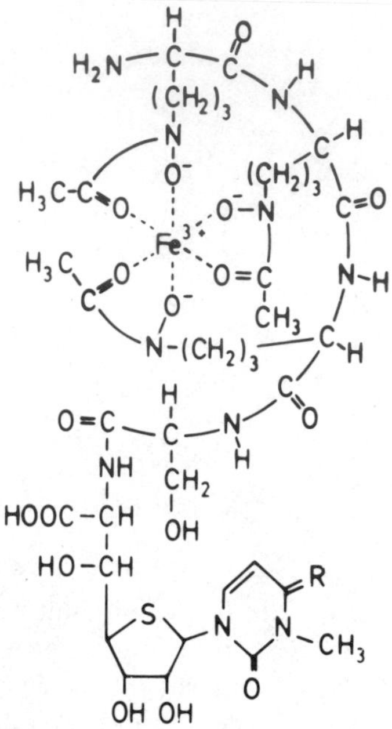

Albomycin $\delta_1$ R = O

Albomycin $\delta_2$ R = N–C(=O)–NH$_2$

Albomycin ε R = NH

*Figure 3.* Structure of ferrichrome and of albomycin. The stars indicate the glycine residue (*) replaced in ferricrocin by a serine residue or by two serine residues (* and **) in ferrichrysin.

atom is replaced by a thioribosylpyrimidine group (Benz *et al.*, 1982). The transport system apparently tolerates the lack of a cyclic hexapeptide and the addition of a large side chain. For further discussion of albomycin transport, see Section II-D.

The fourth characterized transport system in *E. coli* uses *citrate* as iron ligand (Frost and Rosenberg, 1973; Woodrow *et al.*, 1978, 1979; Hussein *et al.*, 1981; Wagegg and Braun, 1981). For induction of the transport system and for measuring transport, the ratio of citrate to iron used is usually 1000 : 1 which favors the formation of the iron dicitrate complex (Spiro *et al.*, 1967; Hussein *et al.*, 1981). The conformation of citrate was determined by X-ray analysis (Glusker, 1980) but the geometry of the $Fe^{3+}$ complex is not known.

## B. Receptor Proteins at the Cell Surface

The iron-siderophore-transport systems share with the vitamin $B_{12}$-transport system the requirement for receptor proteins in the outer membrane. This property distinguishes them from all the other transport systems studied so far in bacteria. These

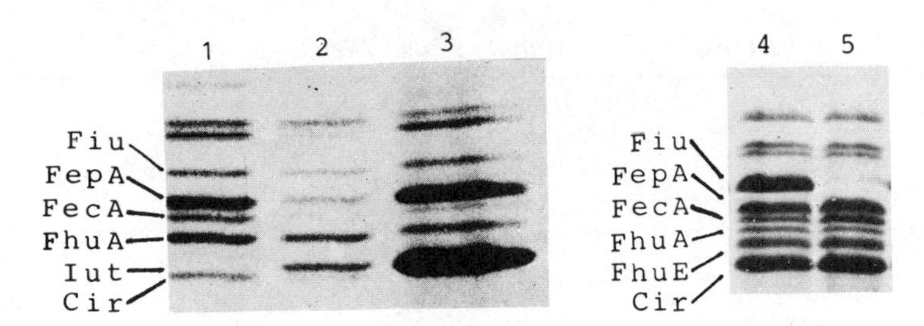


*Figure 4.* *E. coli* outer membrane proteins separated by electrophoresis on polyacrylamide gels in the presence of 0.1% sodium dodecyl sulfate. Only the section containing the proteins regulated by iron (74K to 83K) is shown. For comparison see Figures 5 and 10. The nomenclature is explained in Table 2, except for Fiu which denotes the 83K protein (see also Table 3). The iron supply was limited in the cultures 1, 3, 4, 5, and sufficient in culture 2. The following strains were used: lanes 1–3, wild-type *E. coli* from the stool of a typhoid patient which contains the ColV-K229 plasmid that determines the Iut receptor protein for the uptake of ferric-aerobactin and for cloacin (Braun *et al.*, 1982). Note the very strong expression of the Iut protein under iron limitation shown in lane 3. The outer membrane proteins applied to lane 1 are derived of a mutant which is devoid of the ColV plasmid. Lanes 4 and 5 contain outer membranes of *E. coli* K12. Synthesis of the Fiu protein was abolished in one strain (lane 5) by insertion of Mu (Ap,*lac*). See also Table 3.

proteins are recognized by their strongly enhanced production at iron-limiting growth conditions. They are all of similar size, ranging from 74,000–83,000 (74–83K). They were considered to be related to certain iron-siderophore-transport systems, when it was found that the proteins also serve as binding sites for certain bacteriophages and colicins. Phage- or colicin-resistant mutants are usually devoid of the proteins, and concomitantly are entirely transport inactive for one type of iron-siderophore. These mutants can easily be isolated. One drop of a phage suspension, or of a colicin solution on a nutrient agar plate seeded with the strain to be studied results in the appearance of spontaneous resistant colonies in the zone of growth inhibition or of cell lysis. The genetically homogeneous population of one resistant colony can then be tested for the loss of an outer membrane protein and the lack of transport of an iron–siderophore complex. The electrophoretic system with the best resolution in the 74–83K region is that developed by Lugtenberg *et al.*, (1975). In the years 1975 and 1976, all of the outer membrane proteins related to iron-siderophore transport were identified except the one for the recently discovered ferric aerobactin-transport system. The proteins as they appear after separation by polyacrylamide gel electrophoresis in the presence of 0.1% sodium dodecyl sulfate are shown in Figure 4, and their participation in iron transport is listed in Table 2. The early literature has been reviewed repeatedly and only more recent developments will be discussed here (Braun and Hantke, 1977, 1981, 1982; Kadner and Bassford, 1978; Braun, 1978; Konisky, 1979; Neilands, 1982).

The iron–siderophore complexes prevent killing of cells by the phages and colicins indicating that they bind the same receptor proteins. Mutants become deficient in the uptake of the same agents that have been grouped together by the binding studies. However, the involvement of an outer membrane protein has to be proved by showing that a single protein has been lost by the mutation. Additional genetic studies have to be performed to corroborate the relation of a gene with a certain protein, and of the protein with a receptor function. Here, it is particularly important to rule out deletions

Table 2. *Iron-Transport Proteins in the Outer Membrane of Escherichia coli*

| Designation | Size | Siderophore | Additional receptor activity | References[a] |
|---|---|---|---|---|
| FhuA (TonA) | 78K | Ferrichrome<br>Ferrichrysin<br>Ferricrocin | Phage T1, T5, Ø80, Colicin M | Hantke and Braun (1975b) |
| FhuE | 76K | Coprogen | None | Hantke, unpublished |
| Iut[b] | 74.5K | Aerobactin | Cloacin | Bindereif *et al.* (1982), Greval *et al.* (1982), van Tiel-Menkveld *et al.* (1982) |
| FepA | 81K | Enterochelin | Colicin B<br>Colicin D | Braun *et al.* (1976), Hancock *et al.* (1976), Pugsley and Reeves (1977), Ichihara and Mizushima (1978) |
| FecA | 80K | Citrate | None | Hancock *et al.* (1977), Pugsley and Reeves (1977), Wagegg and Braun (1981) |

[a] Only those references are listed where the proteins were identified by gel electrophoresis. Several reports described competition between iron-siderophores and phage or colicin adsorption to cells, for example, Luckey *et al.* (1975).
[b] Designation follows the proposal of Williams and Warner (1980) for transport-deficient mutants on the ColV-K30 plasmid. All the proteins have been named after their structural genes to provide a uniform nomenclature.

covering several genes, or mutations that exert polar regulatory effects on the expression of a number of consecutive genes. It has generally been found that almost all spontaneous mutants resistant to a phage or a colicin have lost the outer membrane protein. This observation may have several reasons: (1) the frequency of spontaneous mutations leading to deletions is rather high, (2) the export of the proteins through the cytoplasmic membrane and the periplasm into the outer membrane may restrict the number of possible point mutations, (3) point mutations would have to be located in a small region of the protein exposed at the cell surface in order to be effective, and (4) other point mutations leading to the destabilization of the conformation of a protein in solution may be ineffective for proteins embedded in a membrane that provides tight associations with other membrane components. However, the picture is far from being simple since the structural requirements for being exported, deposited in the outer membrane, and active as receptor seem to be complex. Unprocessed proteins still bearing the signal peptide (see Figure 5) or truncated proteins lacking one third of the carboxy-terminal amino acid sequence have been found in the outer membrane of mutants (Bremer *et al.*, 1982). In addition, fused proteins consisting of parts of a membrane protein and a cytoplasmic protein are deposited according to the normal location of the complete membrane protein (see review of Michaelis and Beckwith, 1982).

Point mutants devoid of only one of several receptor functions have been obtained. For example, *fhuA* mutants were isolated that expressed, in normal amounts, a protein with the electrophoretic mobility of the FhuA protein. The strains were resistant to phage T1, to a host-range T1 variant (T1h), and to colicin M, but they were sensitive to phage T5. However, the number of plaques formed by T5 was only 0.5–4.0%

compared with the number obtained on the parent strain. Interestingly, the mutants were entirely devoid of ferrichrome transport and resistant to the structurally similar antibiotic, albomycin. Another mutant was resistant to phage T1, completely sensitive to T1h, and partially sensitive to albomycin and colicin M (Hantke and Braun, 1978). These data indicate that localized small mutations affect all agents that bind to the receptor proteins. Since the extent by which they are affected differs, one can conclude that they all bind to a narrow region of the receptor protein, but the stereochemistry determined by the amino acid side chains is not identical for all of the agents. A similar conclusion can be drawn from the studies on an altered receptor protein for enterochelin and the colicins B and D (McIntosh *et al.*, 1979). The mutated protein has a greater electrophoretic mobility, functions in iron-enterochelin uptake, but renders cells resistant to the colicins. The receptor protein extracted from the membrane is also more thermolabile with regard to the receptor function for iron-enterochelin than for colicin B (Hollifield and Neilands, 1978).

The iron-transport systems are particularly interesting with regard to the translocation of the compounds across the outer membrane. Two questions have to be answered. What is the function of the receptor proteins, and how are the iron-siderophores taken up through the outer membrane? The porins allow the passage of hydrophilic compounds up to a mol. wt. 600–800 (see Section I-A). The iron complexes of ferrichrome (mol. wt. 740) and of enterochelin (mol. wt. 720) are close to the exclusion limit of the porins. However, ferric-aerobactin (mol. wt. 620) and the ferric–dicitrate complex (mol. wt. 443) should be able to pass through the channels formed by the porins. Yet, receptor-deficient mutants are transport-inactive. Even when concentrations of iron-siderophores as high as 0.1 mM are supplied, growth of receptorless mutants is only weakly supported. One could argue that binding to the receptor protein is a necessary prerequisite for the uptake into the cytoplasmic membrane, just as certain amino acids and sugars must first be bound to the binding proteins of the periplasm from where they are donated to the transport proteins in the cytoplasmic membrane. To test this notion, methods were developed to bypass the receptor proteins. Cells were treated with pronase under conditions where they remained viable (Wookey *et al.*, 1981). They became sensitive to the antibiotic actinomycin which is otherwise excluded from entry into the cells by the permeability barrier imposed by the outer membrane. Despite the destruction of the FhuA receptor protein, the uptake rate of iron as ferrichrome complex was even faster than into untreated cells. In addition, transport-deficient mutants devoid of the FhuA receptor protein took up ferrichrome at a rate comparable to that of the parent strain. The apparent $K_m$ value for transport with the receptor was 0.06 μM, and after pronase treatment, 0.7 μM. The $K_m$ into the receptorless, pronase-treated mutants was 0.45 μM. This shows that affinity of the transport system for ferrichrome is largely determined by the outer membrane receptor since the $K_m$ of the subsequent step is higher by an order of magnitude. The $V_{max}$ values were less influenced by the receptor protein [$V_{max}$ (*p*moles/mg cell dry weight) = 110 for the untreated parent, 50 for the pronase-treated parent, and 30 for the pronase-treated mutant]. Additional mutants, which were mapped close to, but outside, the *fhuA* gene, remained transport-deficient after pronase treatment. The products of these genes were therefore assigned to the cytoplasmic membrane (see also Section II-C).

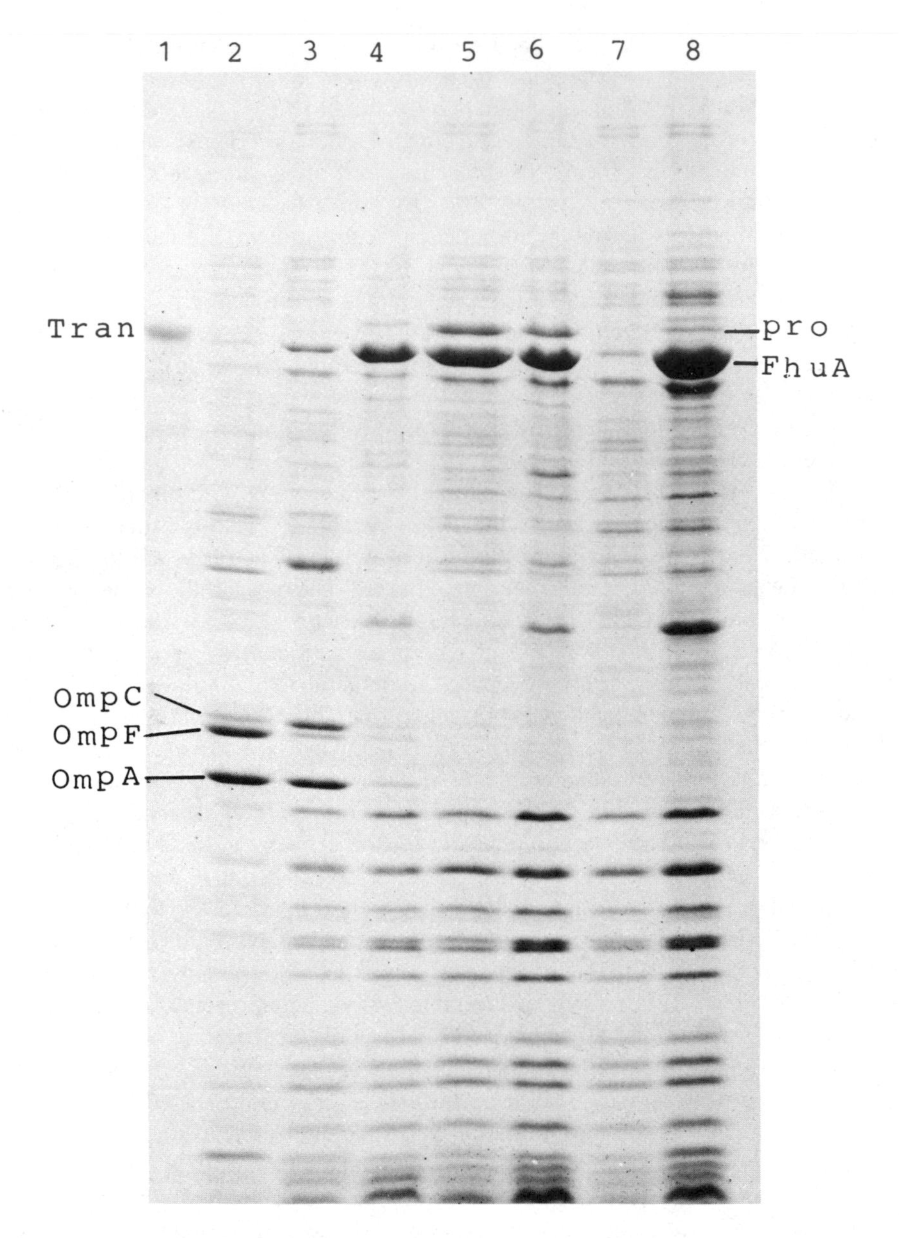


*Figure 5.* *E. coli* K-12 strain superproducing the FhuA (TonA) protein. Strain JE5615 pHK232 was constructed by H. Kraut, National Institute of Genetics, Misima, Japan, by cloning the gene region around *fhuA* from the plasmid pLC19-19 of the Clarke and Carbon collection into the temperature-sensitive multicopy plasmid pSY343 of Seiichi Jasuda, same institute as above. Lane 1, transferrin standard, mol. wt. 80K; lane 2, *E. coli* K-12 grown at 37°C to an absorbance at 578 nm (A) of 0.6, corresponding to about $5 \times 10^8$ cells/ml; lane 3, JE5615 pHK232 grown at 27°C to A = 0.5; lane 4, JE5615 pHK232 shifted at A = 0.05 from 27–37°C and cultured up to A = 0.65; lane 5, *E. coli* HO830 pHK232 grown at 27°C to A = 0.55; lane 6, *E. coli* HO830 pHK232 shifted at A = 0.2 to 37°C and grown to A = 0.35; lane 7, *E. coli* HO831 pHK232 grown at 27°C to A = 0.6; lane 8, *E. coli* HO831 pHK232 shifted at A = 0.15 from 27–37°C

Another way to bypass the outer membrane receptor protein is to convert cells into spheroplasts. By treatment of cells with lysozyme in a buffer containing 30% sucrose to counterbalance the osmotic pressure of the cytoplasm, round cells are formed which have lost part of the outer membrane. Mutants devoid of the FhuA function transported ferrichrome (Wookey and Rosenberg, 1978; Weaver and Konisky, 1980), mutants lacking the FepA function took up ferric-enterochelin (Wookey and Rosenberg, 1978), while *fepB* mutants remained transport-inactive. These consistent results obtained with different methods demonstrate that the receptor proteins are only required for the translocation of iron-siderophores across the outer membrane, and that the uptake through the cytoplasmic membrane takes place without participation of the outer membrane receptors. The receptors apparently function as primary binding sites. In natural environments, they extract the scarce iron-siderophore molecules from the environment and thus concentrate them at the cell surface. This assumption is supported by the low $K_m$ values of the transport systems which are determined by the receptor proteins. Ideas about their release from the strong binding to the receptor proteins and about their translocation across the outer membrane will be discussed in Section II-D. The function of the receptor proteins in natural environments seems, thus, to be clear. It is not understood why the hydrophilic iron-siderophores do not enter cells via the channels formed by the porins when they are supplemented experimentally at rather high concentrations (0.1 mM). This concentration is sufficient to support growth of strains auxotrophic for an amino acid. For this comparison, one has to take into consideration that the amount of iron per cell (4–10 nmoles/mg dry weight; McIntosh and Earhart, 1977; Hartmann and Braun, 1981) is less than 1% of the amount of an amino acid in the proteins (roughly 250 nmoles/mg cell dry weight). Mutants devoid of the porin proteins Ia and Ib were not impaired in ferrichrome transport (Coulton and Braun, 1979). However, a third protein, Ic (PhoE), was synthesized which could have replaced the porin function of Ia (OmpF) and Ib (OmpC).

To gain further insights into the transport steps after the binding to the receptor, it will be necessary to insert isolated native receptor proteins into black lipid membranes, or into liposomes, in order to test whether they render lipid layers specifically permeable to the respective iron-siderophores. Up to now, such studies were hampered by the difficulties in isolating these proteins in pure form. They are minor proteins usually present in small amounts; they are also intrinsic membrane proteins to which lipopolysaccharide and other proteins are strongly bound. First attempts for their isolation have been made (Braun *et al.*, 1973; Braun *et al.*, 1976; Hollifield and Neilands, 1978). Genetic methods will have to be used to improve the yield of these

---

and grown to A = 0.45. OmpC, OmpF, and OmpA are the major outer membrane proteins of *E. coli* K-12 strains. The strains containing the multicopy plasmic pHK232 synthesize already as much (lane 3) or more of the FhuA protein (lanes 5 and 7) at 27°C compared with the Omp proteins, which upon deregulation of plasmid replication at 37°C are made in great abundance (lanes 4, 6, and 8). The term *pro* denotes the proform of the FhuA protein. Cell envelopes containing outer membrane and cytoplasmic membrane were dissolved in Tris-HCl buffer containing 0.2% sodium dodecylsulfate (SDS) and mercaptoethanol at 98°C. The proteins were separated by SDS electrophoresis in polyacrylamide gels and stained with Coomassie blue. Data from H. Hoffmann, this institute.

proteins. An example of a successful attempt is given in Figure 5 where the FhuA protein became the dominant protein in a strain with the *fhuA* gene cloned on a multicopy plasmid.

### C. Functions of Genes Assigned to the Cytoplasmic Membrane; Release of Iron from the Siderophores; Modification of Siderophores

Only the proteins of the outer membrane have been identified and related to certain iron-siderophore-transport systems. No report has been published that described the assignment of additional gene products to the cytoplasmic membrane. The reason is that the expression of these proteins is too low to be detected among the many proteins of the cytoplasmic membrane. The method of choice is to clone the structural genes on multicopy plasmids and study their expression in chromosome-deficient minicells or in ultraviolet-treated maxicells. This approach revealed the arrangement of four gene loci (Figure 6). Mutants in any one of the *fhuB,C,D* genes are transport-deficient for ferric-aerobactin (Braun *et al.*, 1982) and for ferrichrome. Two additonal proteins apart from the FhuA protein were identified in this way. Only the *fhuC* gene product has been localized in the membrane fraction. Some of the mutants formerly designated *fhuB* or *fhuC* (Kadner *et al.*, 1980) may, in fact, be *fhuD* mutants. This gene arrangement was established by the combination of a number of methods: mutagenesis by transposons, cleavage by restriction enzymes, complementation analysis, functional studies, and the determination of the gene products.

The participation of these genes in ferrichrome and ferric-aerobactin transport was derived from functional studies. Since *fhuB,C,D* mutants, whose outer membranes were made permeable by treatment with pronase or by conversion into spheroplasts,

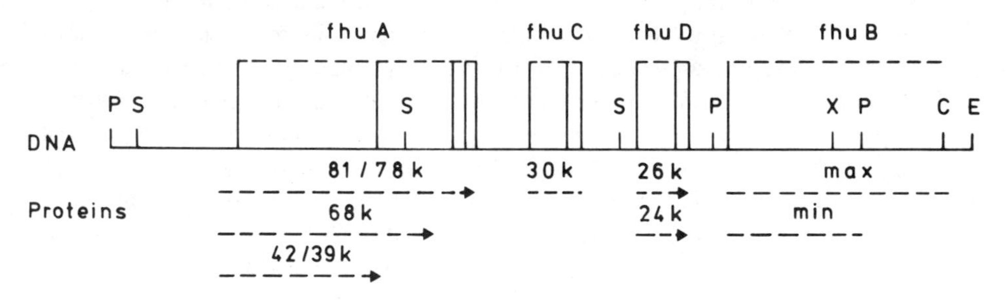


*Figure 6.* Arrangement of the *fhu* gene locus involved in ferric hydroxamate uptake. E, C, P, X, S denote cleavage sites of the restriction enzymes *EcoRI*, *ClaI*, *PstI*, *SalI*, and *XhoI*. The long vertical bars indicate insertion sites of the transposon Tn5 which resulted in inactivation of the genes. Cloned fragments in pBR322 expressed in minicell polypeptides of 81K and 78K in the case of the *fhuA* gene where the 81K protein is the precursor of the 78K protein (see also Figure 5), a 30K protein determined by the *fhuC* gene, and two 26/24K proteins specified by *fhuD*, presumably also the precursor–product pair. No protein derived of the *fhuB* locus could be identified. The broken line indicates the maximum and minimum sizes of the presumed FhuB protein derived from deletion mapping. Truncated proteins, 68K, 42/39K, 24K, formed by Tn5 mutants are also shown from which the direction of transcription, indicated by the arrows, was deduced. Data from L. Fecker, this institute.

were still transport-inactive, it was concluded that these proteins are involved in transport processes after the translocation of the iron-siderophores through the outer membrane. It is unknown whether these proteins are all active in the transport step across the cytoplasmic membrane, or whether some of them are required for the release of ferric-iron from the siderophores by reduction to the ferrous form, or for the modification of the ligand after delivery of the iron ion to the cells.

It was shown for ferrichrome that the ligand is inactivated at the same rate as the iron is transported into the cell (Hartmann and Braun, 1980). It was suggested from the incorporation of radioactive acetate that one of the three hydroxamate groups of the siderophore was acetylated. Thus, binding of iron was inhibited. The modified siderophore was found in the medium. Modification took place in an isolated membrane fraction (Schneider *et al.*, 1981). A reducing cofactor like NADH or NADPH had to be added to achieve modification. These observations were consistent with the conclusion that the iron is released in the membrane fraction by reduction. Mutants in the *tonB* (see Section II-D) and *fhu* gene regions exhibited the same *in vitro* activity as wild-type strains. Apparently, the functions for reduction and modification of ferrichrome, if there are specific ones, are not encoded in these gene regions. The same reaction sequence was obtained with a membrane fraction of *S. typhimurium*. Separation of the iron and the ligand in the cytoplasmic membrane is also indicated by the transport activity of negatively-charged ferrichrome derivatives. The transport rates of iron with ferricrocinyl succinate and with ferrichrysinyl disuccinate were the same as with the uncharged ferrichrome (Coulton *et al.*, 1979; Schneider *et al.*, 1981). In the frame of the known transport mechanisms, uptake of a neutral molecule and of charged molecules by the same transport system through the cytoplasmic membrane is difficult to imagine. The same argument applies to the uptake of iron via aerobactin controlled by the *fhuB,C,D* genes (Braun *et al.*, 1982), since this iron complex also bears two negative charges.

The antibiotic activity of albomycin requires the thioribonucleoside moiety to be taken up into the cytoplasm. Indeed, it was found that the $^{35}$S-labeled part of the molecule stays in the cells, whereas the [$^{3}$H]acetyl-labeled part of the molecule is released into the medium (Hartmann and Braun, 1979). The sulfur label was concentrated 500-fold within the cell. It was not incorporated into nucleic acids or proteins. Potential cleavage sites for the release of the antibiotically active part from the iron-siderophore carrier are the peptide bonds (Figure 3). Mutants were obtained which still took up albomycin but were less sensitive or resistant to the antibiotic. One was mapped close to the *pyrD* gene on the *E. coli* chromosome (L. Zimmerman, unpublished result). They presumably lack peptidase activity since one class of mutants was more resistant when grown on tryptone–yeast extract medium which probably suppresses peptidase activity. In the resistant mutants, albomycin serves as iron carrier. In contrast to ferrichrome, it bears a negatively charged carboxy group which supports the conclusions made above that charged and uncharged siderophores are substrates of the transport system determined by the *fhu* locus.

For the uptake of ferric-dicitrate, a gene region called *fecB* was identified in addition to the *fecA* gene that controls the outer membrane receptor protein. It is unknown how many genes are located at the *fecB* locus. It is assumed that the *fecB* entD fes(B) entF fep entC entA entG entB entE

*Figure 7.* Sequence of genes involved in enterochelin biosynthesis (*entA,B,C,D,E,F,G*), ferric-enterochelin transport (*fep*), and release of iron from enterochelin [*fes*(B)]. The genes are organized in at least six transcriptional units at 13 min of the 100 min linkage map of *E. coli* K-12 (Laird and Young, 1980).

gene product resides in the cytoplasmic membrane. Woodrow *et al.* (1978) have mapped mutations in ferric-dicitrate transport at 7 min of the 100 min linkage map of *E. coli*. They found that some alleles were cotransducible with the *argF* gene. Later, it was shown that the allele which was cotransducible with *argF* determined an outer membrane receptor protein (Wagegg and Braun, 1981; Hussein *et al.*, 1981). Since ten times more iron was taken up than citrate was found associated with the cells, it was concluded that citrate is largely excluded from the transport (Hussein *et al.*, 1981). As in the case of the ferric-hydroxamates, it is likely that the iron complex is already dissociated at the cytoplasmic membrane. There is no transport system for citrate in *E. coli* K-12 (Hall, 1982).

The genes of ferric-enterochelin transport and of enterochelin biosynthesis map at 13 min on the *E. coli* chromosome (Rosenberg and Young, 1974; Hancock *et al.*, 1977; Pugsley, 1977; McIntosh *et al.*, 1979; Laird *et al.*, 1980; Laird and Young, 1980). The arrangement of the genes is given in Figure 7. Although complementation studies revealed only one gene *(fep)* for transport (Laird *et al.*, 1980), physiological evidence suggests more than one gene. The outer membrane receptor protein is determined by at least one gene and this has been designated *fepA*. Additional transport-negative mutants were isolated suggesting at least one further gene (tentatively designated *fepB*) that specifies a function for the uptake of ferric-enterochelin through the cytoplasmic membrane. The most direct evidence comes from transport studies into spheroplasts, discussed in Section II-B. Spheroplasts of *fepB* mutants are unable to take up ferric-enterochelin (Wookey and Rosenberg, 1978).

An additional function claimed to be required for ferric-enterochelin uptake is the enterochelin esterase which hydrolyzes the serine ester bonds and yields the monomer, dimer, and trimer of 2,3-dihydroxybenzoyl serine (Rosenberg and Young, 1974). It consists of the two subunits A and B. Mutants in the *fesB* gene accumulated ferric-enterochelin, but grew poorly on media with ferric-enterochelin as sole iron source. Iron is apparently not released from the intracellular ferric–enterochelin complex since the precipitated cell pellet had the pink color of ferric-enterochelin. It was concluded that the ligand had to be hydrolyzed in order to release the iron. This notion gained support by the very low reduction potential of ferric-enterochelin ($-750$ mV vs. the normal hydrogen electrode at pH 7) which seemed to be too low to be physiologically reducible (Harris *et al.*, 1979a,b). However, this concept has been challenged on several grounds. It is unclear whether ferric-enterochelin (Langmann *et al.*, 1972) or enterochelin (Greenwood and Luke, 1978) is the substrate of the esterase. Furthermore, synthetic derivatives of enterochelin lacking the triester ring supported growth of mutants of *S. typhimurium* and of *E. coli* which were unable to synthesize enterochelin (Hollifield and Neilands, 1978; Venuti *et al.*, 1979; Heidinger *et al.*, 1983).

A rather extensive study about the growth-promoting activity of a group of enterochelin analogues listed in Figure 8 on a set of *E. coli* strains with mutations in various genes for iron transport revealed the following: the cyclic derivatives MECAM, $Me_3$MECAM, and TRIMCAM, and the linear compound LICAMS supported growth under iron-limiting conditions (Heidinger *et al.*, 1983). The *fepB* and *fesB* gene products were required but the *fepA* gene product was dispensable. Regarding the function determined by the *fepB* gene, which was tentatively assigned to the cytoplasmic membrane, it seems to be certain that uptake of the analogues into or across this membrane follows the route of ferric-enterochelin. The same conclusion can be drawn for the *fesB* function. However, no hydrolysis of ester bonds in the analogues is possible and the assumption that, instead, the amide bonds may be hydrolyzed is very unlikely in the case of the methylated amide nitrogen of $Me_3$MECAM. The same result with regard to the *fepB* and *fesB* functions was obtained with another enterochelin analogue, *cis*-1,5,9-*tris*(2,3-dihydroxybenzamido)cyclododecane (Hollifield and Neilands, 1978). Since the *fesB* function is required but cannot act as an esterase, it must have another function, for example, release of iron by reduction of $Fe^{3+}$ to $Fe^{2+}$. It is also not excluded that the studies with the analogues revealed an additional mutation in the various *fes* mutants used which was not recognized in the experiments employing ferric-enterochelin.

Release of iron from enterochelin, MECAM, and $Me_3$MECAM was also achieved with the use of a ferric-siderophore reductase system obtained from *Bacillus subtilis*

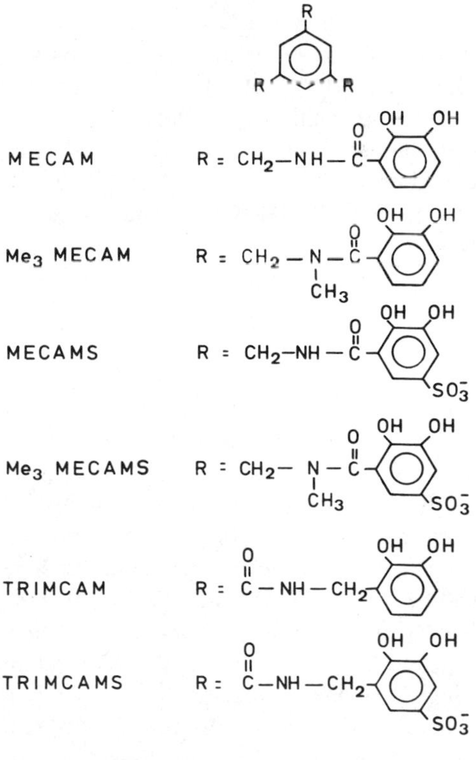

*Figure 8.* Synthetic enterochelin analogues. MECAM 1,3,5-*N,N′,N″-tris*-(2,3-dihydroxybenzoyl)-triaminomethylbenzene, $Me_3$MECAM (*N*-methyl derivative of MECAM), MECAMS and $Me_3$MECAMS (2,3-dihydroxy-5-sulfonyl derivatives), TRIMCAM 1,3,5-*tris*(2,3-dihydroxybenzoylcarbomido)-benzene, TRIMCAMS (2,3-dihydroxy-5-sulfonyl derivative of TRIMCAM), LICAMS 1,5,10-*N,N′,N″-tris* (5-sulfo-2,3-dihydroxybenzoyl)-triazedecane, LICAMC (4-carboxy derivative), Diisp-3,4 LICAM 1,10-*N,N″*-di(isopropyl)-*N,N′,N″-tris* (2,3-dihydroxylbenzoyl)-1,5,10-triazadecane.

WB2802 (Lodge *et al.*, 1980). The nonhydrolytic reductive mobilization of iron from these complexes demonstrates that the thermodynamically difficult reduction mentioned above is possible without concomitant destruction of the ligand. Even glutathione is able to abstract iron from ferric-enterochelin yet the pH value has to be below 6 (Hamed *et al.*, 1982). At this pH, protonation of one or more of the catechol oxygen atoms changes the ligand field of the coordinated iron. In 50% aqueous methanol, a solution of low dielectric, iron(II) has been observed in enterochelin without glutathione present (Hider *et al.*, 1979). An ambient low pH and low dielectric could be provided on the surface of the cytoplasmic membrane or at an enzyme, possibly at the protein of the *fes* gene. Under such conditions, reduction of iron(III) to iron(II) followed by release of iron(II) would not be a thermodynamic problem any longer. It is likely that such a mechanism is realized in ferric-enterochelin transport.

In contrast to the cyclododecane derivative, the active analogues listed in Figure 8 did not require the outer membrane receptor protein determined by the gene *fepA* (Heidinger *et al.*, 1983). These analogues find their way through the outer membrane by various routes. Some specificity was encountered since growth of double and triple mutants in receptor proteins related to iron-transport systems was found to be reduced. In addition, the sulfonated derivatives of the cyclic analogues were inactive showing restraint of the compounds accepted by the ferric-enterochelin transport system. Their inactivity may result from the additional negative charge. This argument is, however, invalid for the pair LICAMS–LICAMC, where the compound sulfonated at the 5-position was active, whereas the compound carboxylated at the 4-position was inactive. Therefore, activity may depend on the charge and the stereochemistry of the compounds. The conformation of the compounds around the iron seems to be the most important factor. They are all centered around the iron in an octahedral array. Their specific recognition apparently occurs at the side of the molecule composed of the iron and three catechol rings. Direct evidence for this conclusion came from competition experiments. Ferric-MECAM inhibited binding of colicin B to the common receptor protein, but MECAMS sulfonated at the 5-position of the 2,3-dihydroxybenzoyl ring was inactive.

## D. The TonB and ExbB Functions

The *tonB* gene was originally defined on the basis that *tonB* mutants became resistant to phage T1 *(Tone)*. Wang and Newton (1971) found that *tonB* mutants of *E. coli* have a ten-fold higher apparent Michaelis–Menten constant for iron uptake than *tonB*$^+$ cells. They probably measured iron uptake via enterochelin synthesized by their strains and via citrate present in the medium. Later studies showed, for all defined iron uptake systems via aerobactin, ferrichrome, enterochelin, and citrate, absolute dependence on a functional *tonB* gene. Furthermore, transport of vitamin $B_{12}$ and the sensitivity of cells to some colicins, designated B, D, G, H, I, M, Q, $S_1$, and V, depend on an active *tonB* gene. The common denominator of all of the *tonB*-dependent agents is that they strictly require receptor proteins at the cell surface to which they bind and from where they are taken up into the cell. But not all receptor-dependent uptake systems rely on the *tonB* function. There are many phages and colicins which infect cells via receptor proteins independent of the *tonB* gene product for example, phage T5, BF23, T2, T6,2, 434, the pili-dependent small DNA and RNA phages, and the group A colicins A, E1, E2, E3, K, and L). Even the uptake of agents which bind to the same receptor protein may either be *tonB* dependent or *tonB*-independent. Therefore, different mechanisms for the uptake of substances are used in the steps following binding to the receptor proteins. The iron complexes, vitamin $B_{12}$, and the colicins stay bound at the receptor proteins in energy-deprived cells (see reviews of Braun and Hantke, 1981; Konisky, 1979, 1982), but this property is not unique for the *tonB*-dependent uptake systems via receptor proteins. Also, the group A colicins are bound to but not taken up into deenergized cells. However, the *tonB*-dependent substances remain adsorbed to the receptors in *tonB* mutants regardless of whether they are energized or not (Hantke and Braun, 1978; Braun *et al.*, 1980; Reynolds *et al.*, 1980). These data indicate that release of the substances of the receptor proteins requires energy, and that for a certain group of substances, the *tonB* gene product somehow mediates the energy of the cell produced in the cytoplasm and the cytoplasmic membrane to the outer membrane receptor proteins.

This view was for the first time derived from the study of the adsorption of phage T1 and Ø80 to sensitive cells (Hancock and Braun, 1976). These phages adsorb reversibly to unenergized cells and to *tonB* mutants as if they would probe the suitability of cells for phage multiplication. Only metabolically active cells are infected. Irreversible adsorption with concomitant release of the DNA and its active uptake occurs only by energized $tonB^+$ cells. An energized cytoplasmic membrane is required since uncouplers prevent infection. Energy can be derived from ATP hydrolysis or from electron-transport chains. Similar conclusions were drawn in a study about vitamin $B_{12}$ uptake (Reynolds *et al.*, 1980). In *btuC* mutants devoid of the transport step through the cytoplasmic membrane, cyanocobalamin was accumulated in the periplasmic space. Uptake into the periplasmic space was dependent on a functional *tonB* gene product and on the proton-motive force. It was concluded that the dissociation rate of the vitamin $B_{12}$ receptor complex is increased by the *tonB* gene product and the proton-motive force.

If the TonB function is required for the translocation of certain substances across the outer membrane, as suggested by the above data, experimental conditions which allow the bypassing of the outer membrane receptor proteins should also abolish the need for the TonB function. Indeed, when FhuA-receptorless strains of *E. coli* were treated by osmotic shock, they became sensitive to colicin M and their killing occurred independent of the TonB function (Braun *et al.*, 1980). Spheroplasts prepared from FhuA-receptorless strains of *E. coli* transported iron as ferrichrome complex independent of the TonB function (Weaver and Konisky, 1980). Both treatments render the outer membrane permeable to certain compounds which normally require the receptor proteins for the transfer across the intact membrane. For the uptake of ferric-enterochelin into spheroplasts, the TonB function was still required (Wookey and Rosenberg, 1978) but these results could not be reproduced by others (Weaver and Konisky, 1980). On the contrary, when the requirement for the outer membrane protein FepA was avoided by another experimental approach, transport of ferric-enterochelin became *tonB*-independent. Mutants that required the percursor 2,3-dihydroxybenzoate for enterochelin synthesis could grow in low-iron media supplemented with 2,3-dihydroxybenzoate (Frost and Rosenberg, 1975). Growth was independent of the TonB function as shown with *tonB* mutants, provided that the concentration of 2,3-dihydroxybenzoate was low (2–5 μM). Increasing concentrations of 2,3-dihydroxybenzoate in the medium reduced the growth rate. Growth promotion by 2,3-dihydroxybenzoate required enterochelin synthesis but exogenously supplied enterochelin did not act as a growth factor. The authors postulated that iron may be carried across the outer membrane as 2,3-dihydroxybenzoate complex. Enterochelin excreted from the cytoplasm into the periplasmic space takes over the iron and transports it across the cytoplasmic membrane. With higher concentrations of 2,3-dihydroxybenzoate, higher amounts of enterochelin are synthesized which are excreted into the medium. The extremely high affinity of enterochelin for iron leads to the formation of the ferric–enterochelin complex. The complex cannot be taken up from the medium into the cells because its transfer across the outer membrane requires the TonB function. Later, it was shown by iron-transport experiments that not only the requirement for the TonB function but also for the outer membrane receptor protein was bypassed with the provision of low concentrations of 2,3-dihydroxybenzoate in the medium (Hancock *et al.*, 1977). In addition, transport of iron as enterochelin complex across the cytoplasmic membrane was demonstrated since *fepB* and *fesB* mutants were transport-negative (Hancock *et al.*, 1977). The latter study strongly supported the original proposal of Frost and Rosenberg (1975) which was put forward at a time when the participation of the FepA receptor protein in ferric-enterochelin transport was not known.

Another way to bypass the TonB function was the isolation of host-range mutants of phage T1 (Gratia, 1964; Franklin *et al.*, 1965; Hantke and Braun, 1978). The phage mutants productively infect *tonB* mutants with high efficiency. The uncoupling of the adsorption from the TonB function and the energy of the cell is demonstrated by the fact that these phage mutants bind irreversibly to the receptor of isolated outer membranes (Hantke and Braun, 1978). These data also show that it is the adsorption process of the wild-type phage which requires the TonB function, and that DNA uptake through the outer and the cytoplasmic membrane is TonB-independent.

The biochemical mechanism(s) by which the TonB function exerts its effect on the various processes that rely on outer membrane receptor proteins remains to be elucidated. The reader is referred to a paper in which viewpoints different from the one discussed in this chapter were proposed (Wookey, 1982).

The function controlled by the *exbB* gene is even more elusive than the one of the *tonB* gene. Mutants in the *exbB* gene show a similar phenotype as mutants in the *tonB* gene. Originally, *exb* mutants were found by their property to hyperexcrete an inhibitor of the colicins B, I, and often also V (Guterman and Luria, 1969). It was later shown that the inhibitor is enterochelin (Guterman, 1971). Subsequent studies revealed that *exbB* strains require higher concentrations of iron in the medium for growth. Hyperexcretion of enterochelin 250-fold, compared with *exbB*$^+$ strains, is the consequence of iron starvation. *exbB* mutants, unlike *tonB* mutants are sensitive to colicin B, provided no enterochelin is synthesized (*aroB* mutants; Hantke and Zimmermann, 1981). However Pugsley and Reeves (1976) found insensitivity to colicins B and D in the absence of enterochelin synthesis, and to the other group B colicins in the presence of enterochelin synthesis. The latter observation may come from the suppression of receptor synthesis when the iron supply provided by enterochelin is sufficient (see also Section II-E).

Ferric-enterochelin formed in the growth medium protects cells against colicin B. *exbB* mutants, like *tonB* mutants, do not transport vitamin $B_{12}$ (Hantke and Zimmermann, 1981). Both types of mutants are deficient in ferrichrome transport and are insensitive to colicin M and albomycin. The relation between *tonB* and *exbB* is not clear. Mutations in *exbB* map at a different position (58 min) than *tonB* mutations (27 min). The TonB function participates in more processes than the ExbB function. Furthermore, mutations of the *tonB* gene completely abolish *tonB*-related processes, whereas the activity of some of the *exbB*-related functions are only reduced. To gain more insights in the mechanisms of the activity of the *tonB* and *exbB* genes, identification of the gene products, their location within the cell, and *in vitro* systems will be required. The *tonB* gene has been cloned, and the gene product has been ascribed to polypeptides of 36,000 (Postle and Reznikoff, 1979) and 40,000 molecular weight (Plastow and Holland, 1979). The latter authors provided evidence that the protein resides in the cytoplasmic membrane. Such a location has tacitly been assumed by most workers in the field and would fit into the proposed function as a coupler between the energy state of the cytoplasmic membrane and outer membrane receptors. Two physical assumptions can be made for such a mode of action. Either a soluble substance is produced breaching the gap between inner and outer membranes, or the communication between the two membranes takes place at junctions (fusion points) of the two membranes. If this is the way the *tonB* gene product functions, it can be foreseen that a simpler cell-free membrane system that is still active will be difficult to obtain. From the few and sometimes even conflicting data on the phenotype of *exbB* mutations, it is even more difficult to deduce its function and mode of action.

## *E. Regulation of the Iron-Transport Systems*

Regulation of transcription by a metal ion is an interesting phenomenon and iron is a good candidate to study the underlying mechanism(s). Iron depletion of *E. coli* leads to the synthesis of the enzymes for enterochelin formation and enterochelin degradation, to aerobactin synthesis, and to the derepression of the transport systems. Bryce and Brot (1971) have estimated that synthesis of 2,3-dihydroxybenzoylserine, the precursor of enterochelin, sets in at an iron concentration of $22 \times 10^{-19}$ moles per cell which is 2.2 mM assuming 1 $\mu m^3$ cellular volume. This figure was later confirmed by McIntosh and Earhart (1977) who deduced from their data a regulatory iron level of 2.55 mM. It is important to note that this figure comprises the total amount of iron per cell. The regulatory effective concentration of iron is certainly much lower since most of the iron is built into proteins and, above all, into redox enzymes which presumably have no direct regulator effect on transcription.

In addition to the determination of biosynthetic enzyme activities, the concentration of siderophores released by the cells into the growth medium, and the rates of iron transport at various iron concentrations in the growth medium, the amount of outer membrane proteins can be measured by scanning the stained proteins and by counting the radioactively labeled proteins after their separation on polyacrylamide gels. Besides the proteins listed in Table 2, enhanced synthesis of outer membrane proteins with molecular weights of 83,000 (83K), 76,000 (76K), and 74,000 (74K) occurs at iron-limiting growth conditions (Figure 4). No iron transport systems were related to these proteins. However, there is evidence that mutants lacking the 76,000 protein cannot take up ferric coprogen (K. Hantke, unpublished results). An immediate increase in the rate of synthesis of the FepA protein (81K) was observed after addition of 50 μM dipyridyl to a cell culture pulse-labeled with [$^3$H]leucine (Boyd and Holland, 1979). Basically the same technique was used by Klebba *et al.* (1982), who found a coordinate control of the 83K, 81K, and 74K proteins.

Casadaban and Cohen (1979) developed a method that allows the study of transcriptional regulation of any gene. They inserted into phage Mu the structural genes of the *lac* operon and a gene conferring ampicillin resistance. Phage Mu can be integrated into most genes, rendering cells ampicillin-resistant. The expression of the *lac* genes comes under the control of the gene in question. When Mudl (Ap,*lac*) is inserted into genes whose expression is controlled by the cellular iron concentration, synthesis of β-galactosidase is regulated by iron (Hantke, 1981; Worsham and Konisky, 1981). Determination of β-galactosidase activity allows a quantitative determination of the transcription of iron-regulated genes. The results obtained in this laboratory are compiled in Table 3. They include the genes for the receptor proteins of ferric-enterochelin (*fepA*) and of ferrichrome (*fhu*), the outer membrane receptor protein of colicin I (*cir*), the genes of the 83K and 76K proteins, the *fecB* gene for ferric-citrate transport, and the *exbB* and *exbC* genes. Furthermore, Mu (Ap,*lac*) insertion into the ColV plasmic abolished aerobactin synthesis and reduced ferric-aerobactin transport (Braun and Burkhardt, 1982). With regard to the expression of the genes of the outer membrane proteins, the data agree with the semiquantitative determinations of the proteins separated from each other by polyacrylamide gel electrophoresis. Concerning the expression of the *cir* gene, Worsham and Konisky (1981) obtained similar figures. It can be seen that transcription of these genes responds to iron deprivation (figures with sufficient iron supply vs. figures under iron limitation in the presence of dipyridyl), but the extent of regulation varies considerably. At present, no data based on this method are available on the iron-regulated *ent, fes,* and *fepB* genes. They would be of considerable interest because the genes seem to be organized into at least six transcriptional units all localized at 13 min of the linkage map of the *E. coli* chromosome (Laird and Young, 1980; see Figure 7). In contrast to the *fep, fes* genes, the *fhu* genes comprising the *fhuA, fhuB, fhuC,* and *fhuD* genes (Figure 6) seem to be contiguous and regulated to the same extent (compare the expression of β-galactosidase in *fhuA* and in *fhuB*). The *fhuB* gene, being located downstream of *fhuA,* is somewhat less expressed as in other operons. A regulatory element has been determined in front of the *fhuA* gene (V. Braun, unpublished result).

The availability of these strains with *lacZ* operon fusions made it possible to isolate mutants that expressed β-galactosidase independent of the iron supply. By mutagenizing a *fhuA* : : Mu(Ap,*lac*) strain, a mutant was obtained which not only overproduced β-galactosidase in the *fhuA* gene but in all the other iron-related genes that contained Mu(Ap,*lac*) (Hantke, 1981). The mutation was designated *fur* for ferric

Table 3. Iron-Regulated Gene Expression, Mu(Ap,lac)

| Gene[a] | Specific activity of β-galactosidase | | | |
|---|---|---|---|---|
| | *fur*$^+$ | | *fur* | |
| | $+Fe^{3+}$ | $-Fe^{3+}$ | $+Fe^{3+}$ | $-Fe^{3+}$ |
| *fhuA* | 70 | 220 | 280 | 280 |
| *fhuB* | 40 | 120 | | |
| *fhuE* | 2 | 49 | 23 | 34 |
| *ColV* | 6 | 70 | 220 | 360 |
| *fecB*[b] | 29 | 139 | 167 | |
| *fepA* | 6 | 160 | 18 | 220 |
| *cir* | 20 | 240 | 300 | 400 |
| *fiu*[c] | 11 | 201 | 79 | 82 |
| *tonB* | 26 | 67 | | |
| *tonB*[d] | 8 | 15 | | |
| *exbB* | 75 | 170 | | |
| *exbB*[d] | 15 | 200 | | |
| *exbC* | 14 | 147 | | |
| *exbC*[d] | 6 | 26 | | |

[a] Genes into which Mu(Ap,*lac*) was inserted (operon fusions). If not indicated otherwise, cells were grown aerobically in tryptone–yeast extract medium with no addition, or with 0.2 mM 2,2′-dipyridyl ($-Fe^{3+}$). The experiments with *ColV* insertion mutants employed nutrient broth to which either 1 mM citrate or 0.2 mM 2,2′-dipyridyl ($-Fe^{3+}$) was added. The *tonB* insertion mutants were grown in tryptone–yeast extract medium to which 50 μM $FeCl_2$ or 0.2 mM 2,2′-dipyridyl ($-Fe^{3+}$) were added.
[b] The *fecB* insertion mutants were grown in nutrient broth or in nutrient broth supplemented with 1 mM citrate and 50 μM 2,2′-dipyridyl ($-Fe^{3+}$).
[c] *fiu* designates the gene that controls synthesis of the 83K protein.
[d] Cells were grown under anaerobic conditions. β-Galactosidase activity was determined after 2–3 hr logarithmic growth and expressed in units (amount of enzyme that will hydrolyze 1 nmol of *o*-nitrophenyl-β-D-galacto-pyranoside per min per mg cell protein at 28°C). Data were taken from Hantke (1981), Hantke and Zimmermann (1981), Braun and Burkhardt (1982), and Hantke and Zimmermann (unpublished results) (*fiu, fecB, fhuE, exbC, tonB*). *fur*$^+$ denotes wild-type gene; *fur*, the mutated gene.

uptake regulation. The *fur* mutation affected the various genes under iron regulation to a different extent (Table 3). The overall picture that emerged from these studies is that a chromosomal gene locus regulates all these iron-related genes including the genes for the ferric-aerobactin receptor and for aerobactin synthesis encoded on the ColV plasmid (Braun and Burkhardt, 1982). Regulation by the *fur* locus is superimposed on the individual regulatory devices of each of the iron-transport systems. The iron-transport systems are probably negatively regulated by the product(s) of the *fur* locus since, in merodiploids containing a mutated and a nonmutated *fur* locus, the fur phenotype, overproduction of the iron functions, is not observed (Hantke, 1982). It is proposed that the *fur* locus determines a repressor protein that is active with iron bound to it. The iron–repressor complex binds to the control regions of the iron-related structural genes and shuts off their expression. When the regulatory active portion of the intracellular iron is reduced, fewer active iron-repressor complexes are formed and the iron-related genes are turned on. In the *fur* mutants, no active repressor protein is formed and the part of regulation exerted by the *fur* gene product does not take place. In most of the iron-transport systems, lack of the *fur* activity leads to a nearly constitutive expression of the iron-transport functions. The *fepA* gene is the least *fur*-regulated locus among the iron-transport genes tested (Figure 9).

The fur phenotype was also described in a mutant of *S. typhimurium*. Uptake of ferric-enterochelin and ferrichrome, intracellular degradation of ferric-enterochelin, synthesis of enterochelin, and of three outer membrane proteins were high, irrespective of the concentration of iron supplied in the growth medium (Ernst *et al.*, 1978). The *fur* mutation in *S. typhimurium* has not been mapped so that it is unclear whether it is the same as in *E. coli*. The latter was mapped in the *lac* gene region (Hantke, 1982).

Since regulation of the ferric-citrate transport system shows some peculiarities, it will be discussed separately. The activity of the transport system is increased when cells are grown in the presence of 1 mM citrate (Frost and Rosenberg, 1973). Concomitant with the increase of the citrate-mediated iron-transport rate, the outer membrane protein FecA is formed (Wagegg and Braun, 1981). However, there is no citrate-transport system in *E. coli* K-12, and citrate supplied in the growth medium is no carbon source. These data suggested that the ferric–citrate complex does not enter the cytoplasm in order to induce the transport system. This assumption was corroborated by the following observations (Hussein *et al.*, 1981). Fluorocitrate and phosphocitrate induced ferric-citrate transport, although they supported iron uptake only very poorly. Synthesis of the FecA outer membrane protein was induced in *fecB* or *tonB* mutants which were devoid of ferric-citrate transport. The intracellular citrate and (overall) iron concentrations were 10–100 times higher than the external concentrations required for induction of the transport system. Once induced, the system still required the inducing threshold concentration of 0.1 mM citrate to remain induced. Apparently, the ferric-citrate-transport system does not transport the inducer so that lower concentrations in the medium suffice to maintain synthesis of the transport proteins. Such an autocatalytic mechanism is usually observed in transport systems. For example, 1 mM

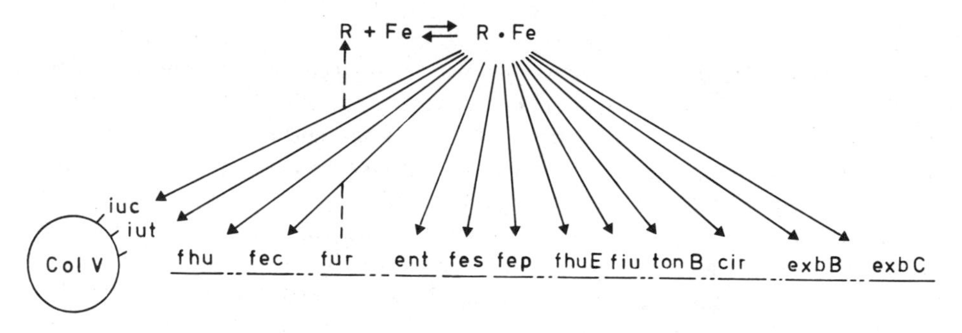


*Figure 9.* Gene loci of *E. coli* controlled by the *fur* gene. The model postulates that *fur* determines a protein R which is converted to a repressor by binding of iron. The repressor–iron complex R · Fe binds to the regulatory regions of the loci *fhu* (*fhuA,B,C,D*) at 3 min, *fec* (*fecA*, *fecB*) at 7 min, *fhuE* and *fiu* (76K and 83K protein, respectively) at 17 min, *tonB* at 27 min, *cir* (colicin I receptor) at 45 min, and to *exbB* and *exbC* at 65 and 58 min, respectively, and shuts off their expression. R also regulates the genes for aerobactin synthesis, ferric-aerobactin transport, and release of iron from aerobactin (*iuc* and *iut* loci) encoded by the ColV plasmid. The genes *fhuE* and *fiu* were mapped by K. Hantke (unpublished results). The *fur* locus seems to be close to *fec* but its precise location has still to be determined.

methylthiogalactoside is required to initially induce the lactose-transport system. However, 5 μM methylthiogalactoside is sufficient to maintain the induced state. Citrate alone does not induce the transport system since after withdrawal of iron by a strong chelator like deferri-ferrichrome, citrate does not induce the ferric-citrate transport system of a strain unable to take up ferrichrome. Thus, the conclusion that ferric-citrate transport is induced by exogenous ferric-citrate is based on the facts that there are inducers which are not transported, that transport-deficient strains *(fecB, tonB)* can be induced, that the assumed intracellular ferric-citrate is inactive, and that there is no autocatalysis.

More recent data (Hantke and Zimmerman, this institute, unpublished) support this hypothesis. The *lacZ* gene inserted as part of the phage Mu [Mud1 (Ap,*lac*)] into *fecB* is turned on when cells are grown in the presence of ferric-citrate. As shown in Table 3, induction requires citrate and a limited amount of iron. In tryptone–yeast extract medium, induction is higher than in nutrient broth because the cells grow better in the former medium. The citrate concentration in the tryptone–yeast extract medium is sufficient to induce the ferric-citrate transport system. The fact that β-galactosidase of Mu (Ap, *lac*) in *fecB* can also be induced normally in *tonB* and *exbB* mutants clearly supports the conclusion that the transport system across the cytoplasmic membrane is not involved in its induction.

In contrast to the *fecB*::Mu(Ap,*lac*) strains, *fecA*::Mu(Ap,*lac*) strains show no induction of β-galactosidase. It is concluded that a functional FecA outer membrane protein is required for induction. Either the FecA protein occupied with ferric-citrate transmits a signal into the cells which leads to induction or, more likely, ferric-citrate is transported into the periplasm and to the outer surface of the cytoplasmic membrane from where it interacts with a component that spans the cytoplasmic membrane and that acts in the occupied form directly as inducer, or releases an inducing substance into the cytoplasm. Exogenous induction has been claimed before for the induction of the transport systems for glucose 6-phosphate in *E. coli* (Dietz, 1976) and for C4-dicarboxylic acid anions of *Azotobacter vinelandii* (Reuser and Postma, 1973). The mechanism of induction has not been revealed for any of the systems. It is tempting to speculate that in the case of ferric-citrate a periplasmic-binding protein is involved. Binding proteins can exert two functions. They can donate their substrate, for example, maltose, either to the transport system or to the chemotactic system (Hazelbauer and Parkinson, 1977). In the latter case, a complex reaction sequence is initiated that finally leads to the swimming of *E. coli* toward higher concentrations of the substrate. In the case of ferric-citrate transport, a processing system across the cytoplasmic membrane different from the transport system would lead to the induction of the chromosomal genes.

There is evidence that modification of tRNAs is altered at iron-restricted growth conditions. The tRNAs of the aromatic amino acids and two minor serine-tRNAs lack the 2-methylthio group on $\Delta^2$-isopentenyl-adenosine located adjacent to the 3′ end of the anticodon (Buck and Griffith, 1982). Expression of the *trp* operon was enhanced by a factor of 2–3 in an iron-deplete strain of *E. coli* where the dominant controlling element *trpR* was not functioning. The undermodified phe-tRNA showed a reduced translocation of poly-U-directed poly-Phe synthesis. This observation was taken as evidence for a less efficient attenuation of transcription of aromatic amino acid operons in the presence of undermodified tRNAs. These results indicate a modulation of the expression of aromatic amino acid operons by the iron status of the cells. A coordinate regulation of the biosynthesis of enterochelin and of the aromatic amino acids would make sense because they share the common precursor chorismic acid. However, as long as no data on the regulatory effect of tRNA modification on enterochelin synthesis are available, it is difficult to estimate the value of this suggestion.

## III. IRON SUPPLY AND VIRULENCE OF *E. coli*

Pathogenicity of a microorganism depends on many factors such as production of endo- and exotoxins, adherence to and invasion into tissues, resistance to complement and to phagocytosis, and resistance against hydrolytic degradation, to mention only the common ones. It is also well documented that diseases leading to iron excess in the serum, like various kinds of anemia, hepatitis, hemochromatosis, porphyria, malaria, or bartonellosis, increase the incidence of microbial infections in man (Weinberg, 1978).

Microorganisms have to multiply in order to become pathogenic for eukaryotes. Growth is severely restricted by the low concentration of available iron ions. In the body fluids of vertebrates, iron is either bound to transferrin or to lactoferrin. Different figures are given in the literature about the concentration of free iron ions in equilibrium with transferrin that is usually only 20–35% saturated. Bullen *et al.* (1976) estimated a concentration of $10^{-12}$ μM at 20% saturation, and Weinberg (1974), $6 \times 10^{-9}$ μM at 25% saturation (see also Table 1). Apart from these differences, it becomes obvious that growth of microorganisms in body fluids is severely restricted by iron limitation. The affinity of lactoferrin for ferric ions is 260 times higher than that of transferrin (Aisen and Leibman, 1972), so that the free iron ion concentration in the presence of a surplus of these proteins is more than eight orders of magnitude below the concentration required for microbial growth. The concentration of transferrin in human serum is about 30 μM (2.4 mg/ml), and that of lactoferrin is about 0.01 μM (0.8 μg/ml). However, in human milk, large quantities of lactoferrin (2–6 mg/ml) and small amounts of transferrin (10–15 μg/ml) are present. Lactoferrin has a protective effect against the proliferation of microorganisms in milk and in other secretory fluids, for example, on the mucosa of the respiratory tract and the gut (Arnold *et al.*, 1977; Masson *et al.*, 1966). In addition to transferrin and lactoferrin, iron is stored in hemoglobin, ferritin, and hemosiderin, besides the many enzymes which contain iron. Among these iron proteins, only hemoglobin may add substantially to the proliferation of *E. coli* under conditions when it is released from the erythrocytes. Such conditions are all kinds of anemia or bacterial infections that lead to lysis of erythrocytes. The virulence-enhancing effect of hemolysin secreted by hemolytic *E. coli* strains (Goebel and Hedgpeth, 1982) may be due to the provision of increased amounts of hemoglobin iron (Lingwood and Ingram, 1982; Welch *et al.*, 1981).

Hemoglobin has a synergistic effect on the lethality of intraperitoneal or subcutaneous inocula of *E. coli*. Recently, a new function for haptoglobin has been suggested

(Eaton *et al.*, 1982). Haptoglobin binds to hemoglobin. The complex has a dissociation constant of $2 \times 10^{-7}$ M. It is rapidly cleared by the reticuloendothelial system. Upon infections, haptoglobin in plasma strongly increases. Hemoglobin (20 mg) administered together with $2.5 \times 10^7$ cells of an *E. coli* strain into the peritoneum of rats led to more than 80% mortality under conditions in which the rats survived the infection without added hemoglobin. Concomitant supply of haptoglobin (30 mg) abolished entirely the effect of hemoglobin. Since addition of 2.3 mg of ferric ammonium citrate to the *E. coli* inoculum increased the mortality to the same extent as hemoglobin, the effect of hemoglobin was ascribed to its iron content. This notion could be supported by growth studies in iron-poor synthetic medium where ferric-ammonium citrate or hemoglobin but not the hemoglobin–haptoglobin complex strongly accelerated the growth rate of *E. coli*. The hemoglobin in contrast to the hemoglobin–haptoglobin complex was apparently proteolytically degraded, and iron in the released heme supported growth. The authors concluded that prevention of the use of hemoglobin iron may be an important function of haptoglobin, besides its role in the clearance of hemoglobin.

There are numerous reports which beyond doubt establish the important role of iron in the outcome of a bacterial infection. It seems that iron is the most important nutritional factor. The term "nutritional immunity" has been coined to account for this fact (Kochan, 1973). The injection of enough iron into the peritoneum of young guinea pigs to saturate the circulating transferrin (5 mg Fe/kg) either in the form of ferric-ammonium citrate, lysed red cells, or hematin, greatly enhanced the virulence of *E. coli* 0111 (Bullen *et al.*, 1968). The $LD_{100}$ (dose that kills 100% of the animals) was $10^8$ bacteria without added iron and $10^3$ bacteria with iron. Furthermore, intravenous injection of $4 \times 10^8$ *E. coli* 0141 did not cause pyelonephritic lesions in rats, yet administration of four doses of iron sorbitol citrate (25 mg/kg) after injection of the bacteria increased the number of bacteria in the kidney 100- to 1000-fold and led to the formation of lesions (Fletcher and Goldstein, 1970).

Conversion of a virulent *S. typhimurium* strain into an avirulent one was achieved by mutation in the pathway of enterochelin synthesis in the step following chorismic acid (Yancey *et al.*, 1979). The mutant was unable to grow in complement-inactivated serum. Addition of enterochelin relieved growth inhibition in serum and in the peritoneum of the mouse. A single dose of 300 μg of enterochelin injected together with the bacteria resulted in the same $LD_{50}$ of the mutant as of the virulent parent strain. It is generally assumed that siderophores released by pathogenic strains can acquire enough iron from transferrin to support growth. This has been shown in synthetic media and in complement-inactivated human serum where iron was present in the form of partially saturated transferrin (Rogers, 1973; Kochan *et al.*, 1977; Tidmarsh and Rosenberg, 1981). Addition of exogenous enterochelin stimulated growth of *S. paratyphi* B and *S. typhimurium* mutants unable to synthesize their own enterochelin. However, Mellencamp *et al.* (1981) failed to observe acquisition of iron from transferrin or enterochelin saturated with radioactive iron. They grew serum-resistant (virulent) and serum-sensitive (avirulent) strains of *E. coli* and *S. typhimurium* in bovine serum. Although the virulent strains multiplied, and the avirulent strains grew when [$^{55}$Fe]enterochelin was added, both types of strains did not take up iron. The authors concluded that the bacteria used intracellular iron stores for growth. Bacteria depleted of iron by repeated growth on iron-poor synthetic medium ceased growth after a few generations.

The recently discovered iron-transport system directed by aerobactin was also correlated with the ability of such strains to kill mice (Williams and Warner, 1980). Young adult mice were inoculated intraperitoneally with $5 \times 10^6$ cells of *E. coli* K-12 *entA* derivatives which were unable to synthesize enterochelin. Transfer of the ColV plasmid bearing the aerobactin-transport system converted the nonpathogenic *E. coli* to a pathogenic one which killed the mice within 34 hr. Mutants in the aerobactin-transport system did not cause death. It is, therefore, likely that the virulence conferred by the ColV plasmid was due to the iron-transport system. Additional virulence factors encoded by some ColV plasmids may have added to the pathogenicity of the K-12 strains, such as serum resistance (Binns *et al.*, 1979) and adhesion to tissues (Clancy and Savage, 1981), but the virulence-enhancing effect of the ferric-aerobactin-transport system seems to be established.

Two mechanisms have been suggested for the bacteriostatic effect of serum. Enterochelin-specific immunoglobulins of the IgA isotype were isolated from normal human serum (Moore *et al.*, 1980), and it was shown that they specifically inhibit uptake of $^{55}Fe^{3+}$ via enterochelin into *E. coli* (Moore and Earhart, 1981). The antibodies did not inhibit $^{55}Fe^{3+}$ uptake directed by ferrichrome or by citrate. The question arises how enterochelin-specific antibodies are formed and how they come into the serum. Molecules as small as enterochelin are usually not immunogenic. It could be exposed to the immune system coupled to a large molecule, for example, a protein involved in the excretion or the uptake of enterochelin. Furthermore, antibodies directed against the colitose (3,6-dideoxy-L-glucose) residue of the *E. coli* 0111 lipopolysaccharide proved to reduce growth in normal horse serum (Fitzgerald and Rogers, 1980). Colitose abolished growth inhibition in horse serum and in the peritoneum of mice. It also reduced the frequency and the rate of survival of mice infected peritoneally with $10^8$ *E. coli* 0111. Based on previous work which showed that bacteriostasis resulted from a decrease in enterochelin production and not from interference with ferric-enterochelin uptake (Rogers, 1973), the authors concluded that the lipopolysaccharide-specific antibodies of this particular strain prevent release of enterochelin. Funakoshi *et al.* (1982) concluded from their data that serum IgA prevents siderophore synthesis by *E. coli* by the action of specific anti-*E. coli* antibodies. Additional experiments with purified antibodies will have to be performed to show directly in a defined synthetic medium that excretion of enterochelin is actually reduced or totally abolished.

Host iron-binding proteins do, in fact, induce the iron-transport systems of *E. coli*. Enterochelin was isolated from the peritoneal washings of guinea pigs lethally infected with *E. coli* 0111 K58 H2 (Griffith and Humphreys, 1980). *E. coli* recovered from the peritoneum contained the altered transfer ribonucleic acids discussed in Section II-E, and outer membrane proteins were very strongly enhanced (see Figure 10). Their concentration reaches those of the major outer membrane proteins (see Section I-A). The degree of derepression of the iron-related proteins in the virulent strain is at least as high as the derepression obtained with the genetically deregulated avirulent *E. coli* K-12 *fur* mutants under *in vitro* conditions (compare Figures 4 and 10).

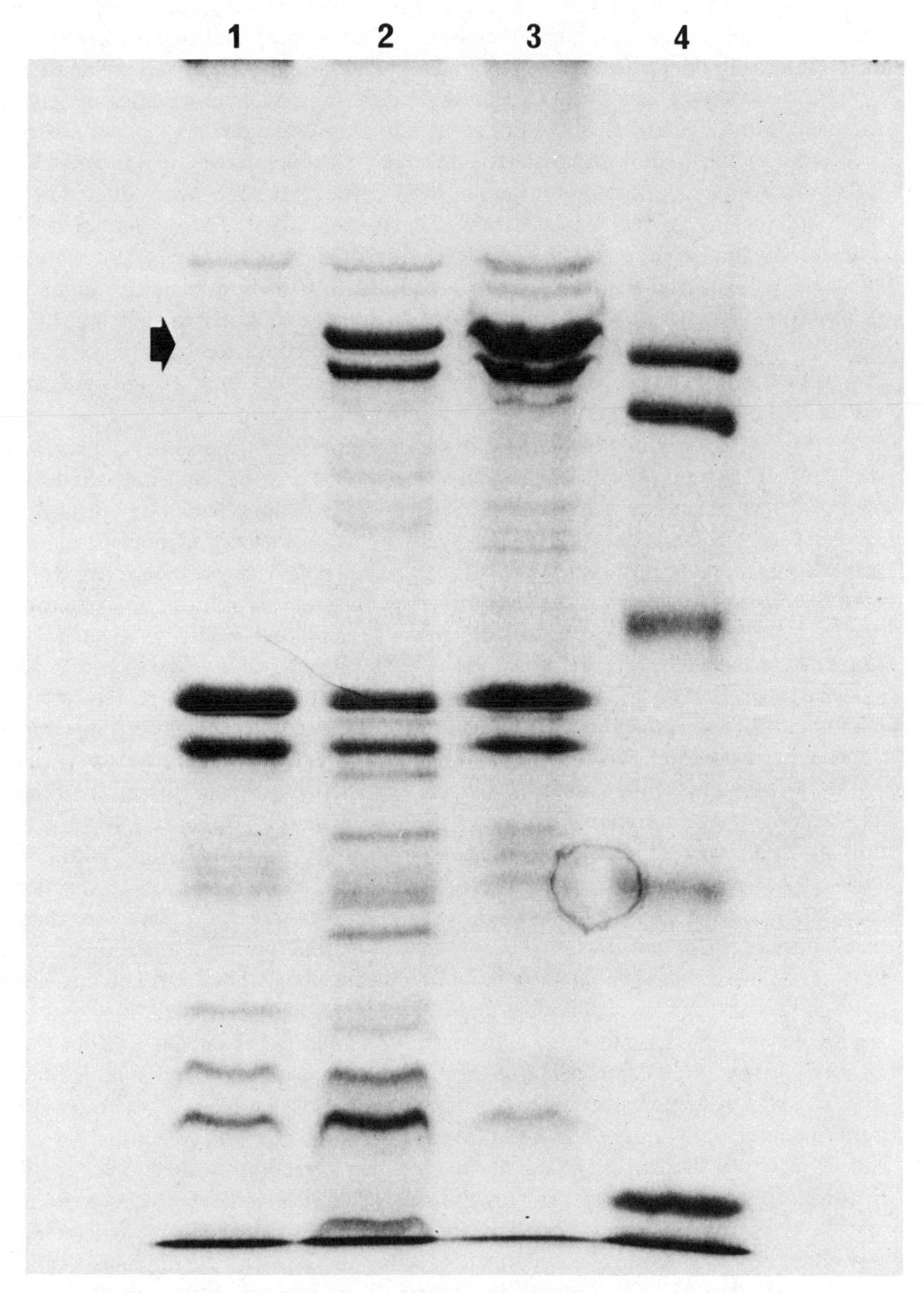


*Figure 10.* Polyacrylamide gel electrophoresis of the outer membrane proteins of *E. coli* 0111 (1) grown in trypticase soy broth, (2) grown in trypticase soy broth containing 0.5 mg/ml ovotransferrin, (3) recovered from the peritoneal washings of lethally infected guinea-pigs, (4) molecular weight standards from top to bottom 78K, 66K, 45K, 26K, 17K. The arrow indicates the proteins formed at iron limitation (compare with Figures 4 and 5). From Griffiths *et al.* (1982), with permission.

Milk is the medium of choice to study growth inhibition of lactoferrin in secretory fluids. A rather complex picture emerged from numerous studies. Lactoferrin alone exerts little inhibitory effect; however, in combination with secretory immunoglobulin A, it exerted a strong bacteriostatic effect on *E. coli* strains of enteropathogenic serotype (Stephens *et al.*, 1980). Rabbit antisera raised against enteropathogenic strains of *E. coli* had a strong bacteriostatic activity together with lactoferrin. It was specific for the serotype of the strain used for immunization. The antibodies were directed against the *O*-side chains of lipopolysaccharide (Rogers and Synge, 1978). Inhibition of *E. coli* 0111 by human milk and by a mixture of purified lactoferrin and *E. coli* antibody could be abolished by the addition of iron (Bullen, 1976; lactoferrin of human milk is 11–43% saturated with iron). These observations are similar to those made with secretory and serum immunoglobulins (IgA), where a concerted bacteriostatic action of transferrin and lactoferrin on a number of strains including *E. coli* was noted (Funakoshi *et al.*, 1982).

However, Dolby and Stephens (private communication) observed a discrepancy between experiments using isolated human milk IgA, lactoferrin, and whole milk. Milk, irrespective of the geographical location of the donor (British or West African mothers), was bacteriostatic for all serotypes of *E. coli* tested, pathogens and commensals alike, and also for other Enterobacteriaceae. The bacteriostatic activity of IgA was O-antigen-specific and was mainly if not exclusively directed against enteropathogenic strains, leaving commensals of the same serotype largely unaffected. They suggested that there might be an additional iron-related factor in whole milk which still has to be discovered. A single method for isolating lactoferrin, secretory immunoglobulin A, and lysozyme from human milk was described recently so that studies under defined conditions will be possible (Boesman-Finkelstein and Finkelstein, 1982).

Suppression of enteropathogenic *E. coli,* which is a major source of infantile gastroenteritis, by the combined action of lactoferrin and IgA (IgG and IgM) in human milk is certainly of importance. As discussed by Bullen (1976), feeding with human milk confers resistance to infection by pathogenic *E. coli*. This broad protection is especially effective until the infant has developed its own repertoire of immune defense mechanisms. The infant obtains the iron-unsaturated lactoferrin and the specific immunoglobulins with the mother's milk and they pass undenatured into the gut. Under these conditions, addition of iron to baby food is certainly not advisable, and there are no reasons for such measures with healthy children either (Betke, 1981).

The balanced situation between the infecting microorganisms having elaborate iron-transport systems via excreted siderophores and the human host with its strong iron-binding proteins can be shifted in favor of the host. The iron content of the serum and of secretory fluids can be actively reduced as a reaction to infections. The deprivation of microbial invaders of growth-essential iron is achieved by a decrease of intestinal absorption, increase of transferrin and lactoferrin synthesis, and increased deposition of iron in storage compartments such as the liver and the reticuloendothelial system (Weinberg, 1978). Reduction of plasma iron seems to pose no problem since it turns over ten times daily and the amount of plasma iron is only 0.4% of the amount of storage iron. Two examples will illustrate this point: First, $10^5$ *E. coli* injected into the peritoneum of NMRJ mice led within 10 hr to the reduction of total serum iron from 353 ± 65 to 38 ± 62 μg/100 ml (Ganzoni and Puschmann, 1977). Accumulation of lactoferrin at sites of infection in humans is well established. Second, the concentration of lactoferrin in the synovial fluid of arthritic patients was increased from 14 nM in noninflammatory cases to 330 nM in inflamed cases (Bennett *et al.*, 1973).

## IV. OUTLOOK

The most interesting aspects of the iron-transport systems into *E. coli* are the following: Do the data collected, so far, really indicate an energy-dependent transport through the outer membrane? Are the receptor proteins also involved in the actual translocation step across the outer membrane or are they merely primary binding sites? As long as all genes and their products have not been identified, localized to one of the two membranes or to the periplasm, and their functions determined, the present model is still subject to substantial alterations. If the proposed model turns out to be basically correct, transport of iron-siderophores (and of vitamin $B_{12}$) sheds an entirely new light on the function of the outer membrane. For these substrates, the outer membrane would exert similar transport properties as the cytoplasmic membrane has for most of the other substrates. It would form a strict permeability barrier which can only be overcome by specific transport systems. However, one has to be aware of the fact that iron once taken up into the cell is "metabolized" further. Therefore, there is no equilibrium of free iron outside and inside the cell whose formation is catalyzed by the transport system. Outside the cells, iron is incorporated into siderophores and inside the cells, into proteins. In addition, the charge of the ion, $Fe^{3+}$ outside and $Fe^{2+}$ inside, can be altered.

If the transport across the outer membrane requires energy, the source of energy is of great interest. Are energy-rich bonds hydrolyzed or do the transport proteins (the receptor proteins) switch between different conformations. Since neutral and negatively charged molecules are transported by the same system, it is unlikely that a membrane potential is directly involved. Since point mutations affect uptake of the small iron-siderophores and of the large colicin proteins, the primary binding for both substrates should be close to the cell surface. How are the two very different kinds of substrates released again from receptor proteins and taken up through the outer membrane? With regard to the specificity of uptake across the outer membrane and the cytoplasmic membrane, it is of interest that for each of the iron-hydroxamate complexes, ferrichrome, aerobactin, and coprogen, a different receptor protein in the outer membrane is used, but the additional functions determined by the *fhuB,C,D* genes are common for all three iron-hydroxamates. The latter functions presumably reside in the cytoplasmic membrane. Furthermore, the rather different iron-catecholate substrates listed in Figure 8 are all accepted by the *fepB* and *fesB* gene products which again indicates that the transport steps after the outer membrane tolerate rather different structures. If the iron-siderophores are not taken up into the cytoplasm but rather are dissociated in the cytoplasmic membrane, the enzyme systems determined by the *fepB, fesB* genes on the one hand and by the *fhuB,C,D* genes on the other hand would exert a rather broad specificity for iron-catecholates and for iron-hydroxamates, respectively. Their specificity would be directed toward the immediate surrounding of iron, the catecholate, and the hydroxamate bondings. These are the questions future research will have to be focused on. In addition, the mode of action of the functions determined by the *tonB*, *exbB*, and *exbC* genes will have to be unraveled.

The third area of interest to be studied is the mode of transcriptional regulation exerted by iron. Research might be guided by the rather advanced knowledge about iron-dependent regulation of the synthesis of diphtheria toxin (Pappenheimer, 1977). Furthermore, the induction of the citrate-dependent iron-transport system poses problems separate from the hydroxamate- and catecholate-dependent iron-transport systems. Since the iron-dicitrate-transport system is induced under experimental conditions where iron and citrate are not transported in measurable quantities, one could expect a new mechanism of transcriptional regulation which is active across the cytoplasmic membrane.

Another field of research will have to deal with the mechanisms by which higher organisms reduce the iron supply of infecting microorganisms. The formation of antibodies which inhibit iron assimilation of *E. coli* is at present actively studied. But it will be of equal importance to understand what kind of mechanisms trigger the reduction of the iron concentrations in body fluids upon infections.

## REFERENCES


Aisen, A., and Leibman, A., 1972, Lactoferrin and transferrin: A comparative study, *Biochim. Biophys. Acta* **257**:314–323.

Argast, M., and Boos, W., 1980, Co-regulation in *Escherichia coli* of a novel transport system for *sn*-glycerol-3-phosphate and outer membrane protein Ic(e,E) with alkaline phosphatase and phosphate binding protein, *J. Bacteritol.* **143**:142–150.

Arnold, R. R., Cole, M. F., and McGhee, J. R., 1977, A bactericidal effect for human lactoferrin, *Science* **197**:163–265.

Bauminger, E. R., Cohen, S. G., Dickson, D. P. E., Levy, A., Ofer, S., and Jariv, J. Y., 1980, Mössbauer spectroscopy of *Escherichia coli* and its iron storage protein, *Biochim. Biophys. Acta* **623**:237–242.

Bavoil, P., and Nikaido, H., 1981, Physical interaction between the phage λ receptor protein and the carrier-immobilized maltose-binding protein of *Escherichia coli, J. Biol. Chem.* **256**:11385–11388.

Bennett, J. L., Jr., Eddie-Quartey, A. C., and Holt, P. J. L., 1973, Lactoferrin—an iron binding protein in synovial fluid, *Arthr. Rheum.* **16**:186–190.

Benz, R., Janko, K., and Läuger, P., 1979, Ionic selectivity of pores formed by the matrix protein (porin) of *Escherichia coli, Biochim. Biophys. Acta* **551**:238–247.

Benz, G., Schröder, T., Kurz, J., Wünsch, C., Karl, W., Steffens, G., Pfitzner, J., and Schmidt, D., 1982, Konstitution der Desferriform der Albomycine, *Angew. Chem. Suppl.* 1322–1335.

Betke, K., 1981, Eisenprophylaxe beim normalen Säugling? *Pädiat. Pädol.* **16**:115–119.

Bindereif, A., Braun, V., and Hantke, K., 1982, The cloacin receptor of ColV-bearing *Escherichia coli* is part of the $Fe^{3+}$-aerobactin transport system, *J. Bacteriol.* **150**:1472–1475.

Binns, M. M., Davies, D. L., and Hardy, K. G., 1979, Cloned fragments of the ColV, I-K94 specifying virulence and serum resistance, *Nature* **279**:778–781.

Boesman-Finkelstein, M., and Finkelstein, R. A., 1982, Sequential purification of lactoferrin, lysozyme and secretory immunoglobulin A from human milk, *FEBS Lett.* **144**:1–5.

Boyd, A., and Holland, I. B., 1979, Regulation of the synthesis of surface protein in the cell cycle of *E. coli* B/r, *Cell* **18**:287–296.

Braun, V., 1978, Structure–function relationships of the gram-negative bacterial cell envelope, in: *Relations between Structure and Function in the Prokaryotic Cell* (R. Y. Stanier, H. J. Rogers, and B. J. Ward, eds.), Cambridge University Press, pp. 111–138.

Braun, V., 1981, *Escherichia coli* cells containing the plasmid ColV produce the iron ionophore aerobactin, *FEMS Microbiol. Lett.* **11**:225–228.

Braun, V., and Burkhardt, R., 1982, Regulation of the ColV plasmid-determined iron(III)-aerobactin transport system in *Escherichia coli, J. Bacteriol.* **152**:223–231.

Braun, V., and Hantke, K., 1977, Bacterial receptors for phages and colicins as constituents of specific transport systems, in: *Microbial Interactions: Receptors and Recognition,* Series B, Vol. 3 (J. L. Reissig, ed.), Chapman and Hall, London, pp. 101–130.

Braun, V., and Hantke, K., 1981, Bacterial cell surface receptors, in: *Organization of Prokaryotic Cell Membranes,* Vol. II (B. K. Ghosh, ed.), CRC Press, Boca Raton, pp. 1–74.

Braun, V., and Hantke, K., 1982, Receptor-dependent transport systems in *Escherichia coli* for iron complexes and vitamin B12, in: *Membranes and Transport,* Vol. 2 (A. N. Martonosi, ed.), Plenum Press, New York and London, pp. 107–113.

Braun, V., and Krieger-Brauer, H. J., 1977, Interrelationship of the phage λ receptor protein and maltose transport in mutants of *Escherichia coli* K12, *Biochim. Biophys. Acta* **469**:89–98.

Braun, V., Schaller, K., and Wolff, H., 1973, A common receptor protein for phage T5 and colicin M in the outer membrane of *Escherichia coli, Biochim. Biophys. Acta* **328**:87–97.

Braun, V., Hancock, R. E. W., Hantke, K., and Hartmann, A., 1976, Functional organization of the outer membrane of *Escherichia coli:* Phage and colicin receptors as components of iron uptake systems, *J. Supramol. Struct.* **5**:37–58.

Braun, V., Frenz, S., Hantke, K., and Schaller, K., 1980, Penetration of colicin M into cells of *Escherichia coli, J. Bacteriol.* **142**:162–168.

Braun, V., Burkhardt, R., Schneider, R., and Zimmermann, L., 1982, Chromosomal genes for ColV plasmid-determined $Fe^{3+}$-aerobactin transport in *Escherichia coli, J. Bacteriol.* **152**:553–559.

Bremer, E., Cole, S. T., Hindennach, I., Henning, U., Beck, E., Kurz, C., and Schaller, K., 1982, Export of a protein into the outer membrane of *Escherichia coli* K12. Stable incorporation of the OmpA protein requires less than 193 amino-terminal amino acid residues, *Eur. J. Biochem.* **122**:220–231.

Bryce, G. F., and Brot, N., 1971, Iron transport in *Escherichia coli* and its relation to the repression of 2,3-dihydroxy-*N*-benzoyl-L-serine synthetase, *Arch. Biochem. Biophys.* **142**:399–406.

Buck, M., and Griffith, E., 1982, Iron mediated methylthiolation of tRNA as a regulator of operon expression in *Escherichia coli, Nucleic Acids Res.* **10**:2609–2623.

Bullen, J. J., 1976, Iron-binding proteins and other factors in milk responsible for resistance to *Escherichia coli,* in: *Acute Diarrhoea in Childhood,* Ciba Foundation Symposium 42, Elsevier/Excerpta Medica/North-Holland, Amsterdam, pp. 149–169.

Bullen, J. J., Leigh, L. C., and Rogers, H. J., 1968, The effect of iron compounds on the virulence of *Escherichia coli* for guinea pigs, *Immunology* **15**:581–588.

Casadaban, M. J., and Cohen, S. N., 1979, Lactose genes fused to exogenous promoters in one step using a Mu-*lac* bacteriophage: *In vivo* probe for transcriptional control sequences, *Proc. Natl. Acad. Sci. USA* **76**:4530–4533.

Clancy, J., and Savage, D. C., 1981, Another colicin V phenotype: *In vitro* adhesion of *Escherichia coli* to mouse intestinal epithelium, *J. Infect. Immunol.* **32**:343–352.

Coulton, J. W., and Braun, V., 1979, Protein II* influences ferrichrome-iron transport in *Escherichia coli* K12, *J. Gen. Microbiol.* **110**:211–220.

Coulton, J. W., Naegeli, H.-U., and Braun, V., 1979, Iron supply of *Escherichia coli* with polymer-bound ferricrocin, *Eur. J. Biochem.* **99**:39–47.

Dietz, G. W., Jr., 1976, The hexose phosphate transport system of *Escherichia coli,* in: *Advances in Enzymology,* Vol. 44 (A. Meister, ed.), John Wiley, New York, pp. 237–259.

DiRienzo, J. M., Nakamura, K., and Inouye, M., 1978, The outer membrane proteins of gram-negative bacteria: Biosynthesis, assembly and functions, *Annu. Rev. Biochem.* **47**:481–532.

Eaton, J. W., Brandt, P., Mahoney, J. R., and Lee, J. T., Jr., 1982, Haptoglobin: A natural bacteriostat, *Science* **215**:691–693.

Ernst, J. F., Bennett, R. L., and Rothfield, L. I., 1978, Constitutive expression of the iron-enterochelin and ferrichrome uptake systems in a mutant strain of *Salmonella typhimurium, J. Bacteriol.* **135**:928–934.

Ferenci, T., 1980, The recognition of maltodextrins by *Escherichia coli, Eur. J. Biochem.* **108:**631–636.

Ferenci, T., and Boos, W., 1980, The role of the *Escherichia coli* λ receptor in the transport of maltose and maltodextrins, *J. Supramol. Struct.* **13:**101–116.

Ferenci, T., Schwentorat, M., Ullrich, S., and Vilmart, J., 1980, Lambda receptor in the outer membrane of *Escherichia coli* as a binding protein for maltodextrins and starch polysaccharides, *J. Bacteriol.* **142:**521–526.

Fitzgerald, S. P., and Rogers, H. J., 1980, Bacteriostatic effect of serum: Role of antibody to lipopolysaccharide, *Infect. Immunol.* **27:**302–308.

Fletcher, J., and Goldstein, E., 1970, The effect of parenteral iron preparations on experimental pyelonephritis, *Arch. J. Exp. Pathol.* **51:**280–285.

Franklin, N. C., Dove, W. F., and Yanofsky, C., 1965, The linear insertion of a prophage into the chromosome of *E. coli* shown by deletion mapping, *Biochem. Biophys. Res. Commun.* **18:**910–923.

Frost, G. E., and Rosenberg, H., 1973, The inducible citrate-dependent iron transport system in *Escherichia coli* K12, *Biochim. Biophys. Acta* **330:**90–101.

Frost, G. E., and Rosenberg, H., 1975, Relationship between the *tonB* locus and iron transport in *Escherichia coli, J. Bacteriol.* **124:**704–712.

Funakoshi, S., Doi, T., Nakajima, T., Suyama, T., and Tokuda, M., 1982, Antimicrobial effect of human serum IgA, *Microbiol. Immunol.* **26:**227–239.

Ganzoni, A. M., and Puschmann, M., 1977, Iron status and host defense, in: *Proteins of Iron Metabolism* (E. B. Brown, P. Aisen, J. Fielding, and R. C. Crichton, eds.), Grune & Stratton, New York, pp. 427–432.

Gibson, F., and Magrath, D. J., 1969, The isolation and characterization of a hydroxamic acid (aerobactin) formed by *Aerobacter aerogenes* 62-1, *Biochim. Biophys. Acta* **192:**175–184.

Glusker, J. P., 1980, Citrate conformation and chelation: Enzymatic implications, *Acc. Chem. Res.* **13:**345–352.

Goebel, W., and Hedgpeth, J., 1982, Cloning and functional characterization of the plasmid-encoded hemolysin determinant of *Escherichia coli, J. Bacteriol.* **151:**1290–1298.

Gratia, J.-P., 1964, Résistance á la colicine B chez *Eschérichia coli.* Relations de spécificité entre colicine B, I, et phage T1. Etude génétique, *Ann. Inst. Pasteur (Paris)* **107:**132–151.

Greenwood, K. T., and Luke, R. K. J., 1976, Studies on the enzymatic synthesis of enterochelin in *Escherichia coli* K-12. Four polypeptides involved in the conversion of 2,3-dihydroxy-benzoate to enterochelin, *Biochim. Biophys. Acta* **454:**285–287.

Greenwood, K. T., and Luke, R. K. J., 1978, Enzymatic hydrolysis of enterochelin and its iron complex in *Escherichia coli* K-12. Properties of enterochelin esterase, *Biochim. Biophys. Acta* **525:**209–218.

Greval, K. K., Warner, P. J., and Williams, P. H., 1982, An inducible outer membrane protein involved in aerobactin-mediated iron transport by ColV strains of *Escherichia coli, FEBS Lett.* **140:**27–30.

Griffith, E., and Humphreys, J., 1980, Isolation of enterochelin from the peritoneal washings of guinea pigs lethally infected with *Escherichia coli, Infect. Immunol.* **28:**286–289.

Griffith, E., Stevenson, P., and Joyce, P., 1983, Pathogenic *Escherichia coli* express new outer membrane proteins when growing *in vivo, FEMS Microbiol. Lett.* **16:**95–99.

Guterman, S. K., 1971, Inhibition of colicin B by enterochelin, *Biochem. Biophys. Res. Commun.* **44:**1149–1155.

Guterman, S., and Luria, S., 1969, *Escherichia coli:* Strains that excrete an inhibitor of colicin B, *Science* **164:**1414.

Hall, B., 1982, Chromosomal mutation for citrate utilization by *Escherichia coli* K-12, *J. Bacteriol.* **151:**269–273.

Hamed, M. Y., Hider, R. C., and Silver, J., 1982, The competition between enterobactin and glutathione for iron, *Inorg. Chim. Acta* **66:**13–18.

Hancock, R. E. W., and Braun, V., 1976, Nature of the energy requirement for the irreversible adsorption of bacteriophages T1 and Ø80 to *Escherichia coli* K-12, *J. Bacteriol.* **125:**409–415.

Hancock, R. E. W., Hantke, K., and Braun, V., 1976, Iron transport in *Escherichia coli* K-12: Involvement of the colicin B receptor and of a citrate-inducible protein, *J. Bacteriol.* **127:**1370–1375.

Hancock, R. E. W., Hantke, K., and Braun, V., 1977, Iron transport in *Escherichia coli* K-12. 2,3-dihydroxybenzoate-promoted iron uptake, *Arch. Microbiol.* **114:**231–239.

Hantke, K., 1976, Phage T6-colicin K receptor and nucleoside transport in *Escherichia coli, FEBS Lett.* **70:**109–112.

Hantke, K., 1981, Regulation of ferric iron transport in *Escherichia coli* K-12: Isolation of a constitutive mutant, *Mol. Gen. Gen.* **182**:288–292.

Hantke, K., 1982, Negative control of iron uptake systems in *Escherichia coli, FEMS Microbiol. Lett.* **15**:83–86.

Hantke, K., and Braun, V., 1975a, A function common to iron enterochelin transport and actions of colicins B,I,V in *Escherichia coli, FEBS Lett.* **59**:277–281.

Hantke, K., and Braun, V., 1975b, Membrane receptor dependent iron transport in *Escherichia coli, FEBS Lett.* **49**:301–305.

Hantke, K., and Braun, V., 1978, Functional interaction of the *tonA/tonB* receptor system in *Escherichia coli, J. Bacteriol.* **135**:190–197.

Hantke, K., and Zimmermann, L., 1981, The importance of the *exbB* gene for vitamin B12 and ferric iron transport, *FEMS Microbiol. Lett.* **12**:31–35.

Harris, W. R., Carrano, C. J., and Raymond, K. N., 1979a, Coordination chemistry of microbial iron transport compounds. Isolation, characterization, and formation constants of ferric aerobactin, *J. Am. Chem. Soc.* **101**:2722–2727.

Harris, W. R., Carrano, C. J., Cooper, S. R., Sofen, S. R., Avdeef, A. E., McArdle, J. V., and Raymond, K. N., 1979b, Coordination chemistry of microbial iron transport compounds. 19. Stability constants and electrochemical behavior of ferric enterobactin and model complexes, *J. Am. Chem. Soc.* **101**:6097–6104.

Hartmann, A., and Braun, V., 1979, Uptake and conversion of the antibiotic albomycin by *Escherichia coli* K-12, *Eur. J. Biochem.* **99**:517–524.

Hartmann, A., and Braun, V., 1980, Iron transport in *Escherichia coli:* Uptake and modification of ferrichrome, *J. Bacteriol.* **143**:246–255.

Hartmann, A., and Braun, V., 1981, Iron uptake and iron limited growth of *Escherichia coli* K-12, *Arch. Microbiol.* **130**:353–356.

Hazelbauer, G. L., and Parkinson, J. S., 1977, Bacterial chemotaxis, in: *Microbial Interactions: Receptors and Recognition,* Series B, Vol. 3 (J. L. Reissig, ed.), Chapman and Hall, London, pp. 61–98.

Heidinger, S., Braun, V., Pecoraro, V. L., and Raymond, K. N., 1983, Iron supply to *Escherichia coli* by synthetic analogs of enterochelin, *J. Bacteriol.* **153**:109–115.

Heuzenroeder, M. W., and Reeves, P., 1980, Periplasmic maltose-binding protein confers specificity on the outer membrane maltose pore of *Escherichia coli, J. Bacteriol.* **141**:431–435.

Hider, R. C., Silver, J., Neilands, J. B., Morrison, I. E. G., and Rees, L. V. E., 1979, Identification of iron(II) enterobactin and its possible role in *Escherichia coli* iron transport, *FEBS Lett.* **102**:325–328.

Hollifield, W. C., and Neilands, J. B., 1978, Ferric enterobactin transport system in *Escherichia coli* K-12. Extraction, assay, and specificity of the outer membrane receptor, *Biochemistry* **17**:1922–1928.

Hussein, S., Hantke, K., and Braun, V., 1981, Citrate-dependent iron transport system of *Escherichia coli* K-12, *Eur. J. Biochem.* **117**:431–437.

Ichihara, S., and Mizushima, S., 1978, Identification of an outer membrane protein responsible for the binding of the Fe-enterochelin complex to *Escherichia coli* cells, *J. Biochem.* **83**:137–140.

Kadner, R. J., and Bassford, P. J., Jr., 1978, The role of the outer membrane in active transport, in: *Bacterial Transport* (B. P. Rosen, ed.), Marcel Dekker, New York, pp. 413–462.

Kadner, R. J., Heller, K., Coulton, J. W., and Braun, V., 1980, Genetic control of hydroxamate-mediated iron uptake in *Escherichia coli, J. Bacteriol.* **143**:256–264.

Klebba, P. E., McIntosh, M. A., and Neilands, J. B., 1982, Kinetics of biosynthesis of iron-regulated membrane proteins in *Escherichia coli, J. Bacteriol.* **149**:880–888.

Kochan, I., 1973, The role of iron in bacterial infections with special consideration of host-tubercle bacillus interactions, *Curr. Top. Microbiol. Immunol.* **60**:1–30.

Kochan, I., Krach, J. T., and Wiles, T. I., 1977, Virulence-associated acquisition of iron in mammalian serum by *Escherichia coli, J. Infect. Dis.* **135**:623–632.

Konisky, J., 1979, Specific transport systems and receptors for colicins and phages, in: *Bacterial Outer Membranes. Biogenesis and Functions* (M. Inouye, ed.), John Wiley & Sons, New York, pp. 319–359.

Konisky, J., 1982, Colicins and other bacteriocins with established modes of action, *Annu. Rev. Microbiol.* **36**:125–144.

Krieger-Brauer, H. J., and Braun, V., 1980, Functions related to the receptor protein specified by the *tsx* gene of *Escherichia coli, Arch. Microbiol.* **114**:233–242.

Laird, A. J., and Young, I. G., 1980, Tn*5* mutagenesis of the enterochelin gene cluster of *Escherichia coli, Gene* **11**:359–366.

Laird, A. J., Ribbons, D. W., Woodrow, G. C., and Young, I. G., 1980, Bacteriophage Mu-mediated gene transposition and *in vitro* cloning of the enterochelin gene cluster of *Escherichia coli, Gene* **11**:347–357.

Langmann, L., Young, I. G., Frost, G. E., Rosenberg, H., and Gibson, F., 1972, Enterochelin system of iron transport in *Escherichia coli:* Mutations affecting ferric enterochelin esterase, *J. Bacteriol.* **112**:1142–1149.

Lingwood, M. A., and Ingram, P. L., 1982, The role of alpha haemolysin in the virulence of *Escherichia coli* for mice, *J. Med. Microbiol.* **15**:23–30.

Lodge, J. S., Gaines, C. G., Arceneaux, J. E. L., and Byers, B. R., 1980, Non-hydrolytic release of iron from ferrienterobactin analogs by extracts of *Bacillus subtilis, Biochem. Biophys. Res. Commun.* **17**:1291–1295.

Luckey, M., and Nikaido, H., 1980, Specificity of diffusion channels produced by λ phage receptor protein of *Escherichia coli, Proc. Natl. Acad. Sci. USA* **77**:167–171.

Luckey, M., Wayne, R., and Neilands, J. B., 1975, *In vitro* competition between ferrichrome and phage for the outer membrane T5 receptor complex of *Escherichia coli, Biochem. Biophys. Res. Commun.* **64**:687–693.

Lugtenberg, B., and van Alphen, L., 1983, Molecular architecture and functioning of the outer membrane of *Escherichia coli* and other gram-negative bacteria, *Biochim. Biophys. Acta.* **737**:51–115.

Lugtenberg, B., Meijers, J., Peters, R., van der Hoek, P., and van Alphen, L., 1975, Electrophoretic resolution of the major outer membrane protein of *Escherichia coli* K12 into four bands, *FEBS Lett.* **58**:254–258.

Martonosi, A. N. (ed.), 1982, *Membranes and Transport,* Plenum Press, New York, London.

Masson, P. L., Heremans, J. F., and Dive, C., 1966, An iron-binding protein common to many external excretions, *Clin. Chim. Acta* **14**:735–739.

McIntosh, M. A., and Earhart, C. F., 1977, Coordinate regulation by iron of the synthesis of phenolate compounds and three outer membrane proteins in *Escherichia coli, J. Bacteriol.* **131**:331–339.

McIntosh, M. A., Chenault, S. S., and Earhart, C. F., 1979, Genetic and physiological studies on the relationship between colicin B resistance and ferrienterochelin uptake in *Escherichia coli* K-12. *J. Bacteriol.* **137**:653–657.

Mellencamp, M. W., McCabe, M. A., and Kochan, I., 1981, The growth-promoting effect of bacterial iron for serum-exposed bacteria, *Immunology* **43**:483–491.

Michaelis, S., and Beckwith, J., 1982, Mechanism of incorporation of cell envelope proteins in *Escherichia coli, Annu. Rev. Microbiol.* **36**:435–465.

Moore, D. G., and Earhart, C. F., 1981, Specific inhibition of *Escherichia coli* ferrienterochelin uptake by a normal human serum immunoglobulin, *Infect. Immunol.* **31**:631–635.

Moore, D. G., Yancey, J., Lankford, C. E., and Earhard, C. F., 1980, Bacteriostatic enterochelin-specific immunoglobulin from human serum, *Infect. Immunol.* **27**:418–423.

Murray, G. J., Clark, G. E. D., Parniak, M. A., and Viswanatha, T., 1977, Effect of metabolites on ε-*N*-hydroxylysine formation in cell-free extracts of *Aerobacter aerogenes* 62-1, *Can. J. Biochem.* **55**:625–629.

Neilands, J. B., 1981, Microbial iron compounds, *Annu. Rev. Biochem.* **50**:715–731.

Neilands, J. B., 1982, Microbial envelope proteins related to iron, *Annu. Rev. Microbiol.* **36**:285–309.

Nikaido, H., 1982, Proteins forming large channels in biological membranes, in: *Membranes and Transport,* Vol. 2 (A. N. Martonosi, ed.), Plenum Press, New York and London, pp. 265–270.

Nikaido, H., and Nakae, T., 1979, The outer membrane of gram-negative bacteria, in: *Advances in Microbial Physiology,* Vol. 20 (A. H. Rose and J. G. Morris, eds.), Academic Press, London, pp. 163–250.

Norrestam, R., Stensland, B., and Bränden, C.-J., 1975, On the conformation of cyclic iron-containing hexapeptides: The crystal structure and molecular structure of ferrichrysin, *J. Mol. Biol.* **99**:501–506.

Osborn, M. J., and Wu, H., 1980, Proteins of the outer membrane of gram-negative bacteria, *Annu. Rev. Microbiol.* **34**:369–422.

Overbeeke, N., and Lugtenberg, B., 1982, Recognition site for phosphorus-containing compounds and other negatively charged solutes on the PhoE protein pore of the outer membrane of *Escherichia coli* K-12, *Eur. J. Biochem.* **126**:113–118.

Pappenheimer, A. M., 1977, Diphtheria toxin, *Annu. Rev. Biochem.* **46:**69–94.

Parniak, M. A., Jackson, G. E. D., Murray, G. J., and Viswanatha, T., 1979, Studies on the formation of $N^6$-hydroxylysine in cell-free extracts of *Aerobacter aerogenes* 62-1, *Biochim. Biophys. Acta* **569:**99–108.

Payne, S. M., 1980, Synthesis and utilization of siderophore by *Shigella flexneri, J. Bacteriol.* **143:**1420–1424.

Perry, R. D., and San Clemente, C. L., 1979, Siderophore synthesis in *Klebsiella pneumoniae* and *Shigella sonnei* during iron deficiency, *J. Bacteriol.* **140:**1129–1132.

Plastow, G. S., and Holland, I. B., 1979, Identification of an *Escherichia coli* inner membrane polypeptide specified by a λ-*tonB* transducing bacteriophage, *Biochem. Biophys. Res. Commun.* **90:**1007–1014.

Postle, K., and Reznikoff, W., 1979, Identification of the *Escherichia coli tonB* gene product in minicells containing *tonB* hybrid plasmids, *J. Mol. Biol.* **131:**619–636.

Pugsley, A. P., 1977, Map location of the *cbr* gene coding for production of the outer membrane receptor for ferrienterochelin and colicins B and D in *Escherichia coli* K-12, *FEMS Microbiol. Letts.* **2:**275–277.

Pugsley, A. P., and Reeves, P., 1976, Characterization of colicin B-resistant mutants of *Escherichia coli* K-12: Colicin resistance and the role of enterochelin, *J. Bacteriol.* **127:**218–228.

Pugsley, A. P., and Reeves, P., 1977, The role of colicin receptors in the uptake of ferrienterochelin by *Escherichia coli* K-12, *Biochem. Biophys. Res. Commun.* **74:**903–911.

Raymond, K. N., Harris, W. R., Carrano, C. J., and Weitl, F. L., 1982, The synthesis, thermodynamic behaviour, and biological properties of metal-iron-specific sequestering agents for iron and the actinides, in: *The Biological Chemistry of Iron* (H. B. Dunford, D. Dolphin, K. N. Raymond, and L. Sieker, eds.), D. Reidel, Dortrecht, Boston, London, pp. 85–105.

Reuser, A. J. J., and Postma, P. W., 1973, The induction of translocators for di- and tricarboxylic-acid anions in *Azotobacter vinelandii, Eur. J. Biochem.* **33:**584–592.

Reynolds, P. R., Mottur, G. P., and Bradbeer, C., 1980, Transport of vitamin B12 in *Escherichia coli.* Some observations of the roles of the gene products of *btuC* and *tonB, J. Biol. Chem.* **255:**4313–4319.

Rogers, H. J., 1973, Iron binding catechols and virulence in *Escherichia coli, Infect. Immunol.* **7:**445–456.

Rogers, H. J., and Synge, C., 1978, Bacteriostatic effect of human milk on *Escherichia coli:* The role of IgA, *Immunology* **34:**19–28.

Rosenberg, H., and Young, I. G., 1974, Iron transport in the enteric bacteria, in: *Microbial Iron Metabolism* (J. B. Neilands, ed.), Academic Press, New York, London, pp. 67–82.

Schneider, R., Hartmann, A., and Braun, V., 1981, Transport of the iron ionophore ferrichrome in *Escherichia coli* and *Salmonella typhimurium* LT2, *FEMS Microbiol. Lett.* **11:**115–119.

Sillen, L. G., and Martell, A. E., 1964, *Stability Constants of Metal–Ion Complexes,* The Chemical Society, London.

Spiro, T. G., Pape, L., and Saltman, P., 1967, The hydrolytic polymerization of ferric citrate. I. The chemistry of the polymer, *J. Am. Chem. Soc.* **89:**5555–5559.

Stephens, S., Dolby, J. M., Montreuil, J., and Spik, G., 1980, Differences in inhibition of the growth of commensal and enteropathogenic strains of *Escherichia coli* by lactotransferrin and secretory immunoglobulin A isolated from human milk, *Immunology* **41:**597–603.

Stuart, S. J., Greenwood, K. T., and Luke, R. K. J., 1980, Hydroxamate-mediated transport of iron controlled by ColV plasmids, *J. Bacteriol.* **143:**35–42.

Szmelcman, S., and Hofnung, M., 1975, Maltose transport in *Escherichia coli* K-12. Involvement of the bacteriophage lambda receptor, *J. Bacteriol.* **124:**112–118.

Szmelcman, S., Schwartz, M., Silhavy, T. J., and Boos, W., 1976, Maltose transport in *Escherichia coli* K-12. A comparison of transport kinetics in wild-type and λ-resistant mutants with the dissociation constants of the maltose-binding protein as measured by fluorescence quenching, *Eur. J. Biochem.* **65:**13–19.

Tidmarsh, G. F., and Rosenberg, L. T., 1981, Acquisition of iron from transferrin by *Salmonella paratyphi* B, *Curr. Microbiol.* **6:**217–220.

van der Helm, D., Baker, J. R., Eng-Wilmot, D. L., Hossain, M. B., and Loghry, R. A., 1980, Crystal structure of ferrichrome and a comparison with the structure of ferrichrome A, *J. Am. Chem. Soc.* **104:**4224–4231.

van Tiel-Menkveld, G. J., Mentjox-Vervuurt, J. M., Oudega, B., and de Graaf, F., 1982, Siderophore production by *Enterobacter cloacae* and a common receptor protein for the uptake of aerobactin and cloacin DF13, *J. Bacteriol.* **150:**490–497.

Venuti, M. C., Rastetter, W. H., and Neilands, J. B., 1979, 1,3,5-Tris(*N,N'*-2,3-dhydroxybenzoyl)aminomethyl benzene, a synthetic iron chelator related to enterobactin, *J. Med. Chem.* **22:**123–124.

Wagegg, W., and Braun, V., 1981, Ferric citrate transport in *Escherichia coli* requires outer membrane receptor protein FecA, *J. Bacteriol.* **145:**156–163.

Wandersman, C., Schwartz, M., and Ferenci, T., 1979, *Escherichia coli* mutants impaired in maltodextrin transport, *J. Bacteriol.* **140:**1–13.

Wang, C. C., and Newton, A., 1971, An additional step in the transport of iron defined by the *tonB* locus of *Escherichia coli, J. Biol. Chem.* **246:**2147–2151.

Warner, P. J., Williams, P. H., Bindereif, A., and Neilands, J. B., 1981, ColV plasmid-specified aerobactin synthesis by invasive strains of *Escherichia coli, Infect. Immunol.* **33:**540–545.

Weaver, C. H., and Konisky, J., 1980, *tonB*-independent ferrichrome-mediated iron transport in *Escherichia coli* spheroplasts, *J. Bacteriol.* **143:**1513–1518.

Weinberg, E. D., 1974, Iron and susceptibility to infectious disease, *Science* **184:**952–956.

Weinberg, E. D., 1978, Iron and infection, *Microbiol. Rev.* **42:**45–66.

Welch, R. A., Dellinger, E. P., Minshew, A., and Falkow, S., 1981, Haemolysin contributes to virulence of extraintestinal *E. coli* infections, *Nature* **294:**665–667.

Williams, P., 1979, Novel iron uptake system specified by ColV plasmids: An important component in the virulence of invasive strains of *Escherichia coli, Infect. Immunol.* **26:**925–932.

Williams, P., and Warner, P. J., 1980, ColV plasmid-mediated, colicin V-independent iron uptake system of invasive strains of *Escherichia coli, Infect. Immunol.* **29:**411–416.

Woodrow, G. C., Langman, L., Young, I. G., and Gibson, F., 1978, Mutations affecting the citrate-dependent iron uptake system in *Escherichia coli, J. Bacteriol.* **133:**1524–1526.

Woodrow, C. W., Young, I. G., and Gibson, F., 1979, Biosynthesis of enterochelin in *Escherichia coli* K-12. Separation of the polypeptides coded by the *entD, E, F* and *G* genes, *Biochim. Biophys. Acta* **582:**145–153.

Wookey, P., 1982, The *tonB* gene product in *Escherichia coli*. Energy-coupling or molecular processing of permeases, *FEBS Lett.* **139:**145–154.

Wookey, P., and Rosenberg, H., 1978, Involvement of inner and outer membrane components in the transport of iron and in colicin B action in *Escherichia coli, J. Bacteriol.* **133:**661–666.

Wookey, P., Hussein, S., and Braun, V., 1981, Functions in outer and inner membranes of *Escherichia coli* for ferrichrome transport, *J. Bacteriol.* **146:**1158–1161.

Worsham, P. L., and Konisky, J., 1981, Use of *cir-lac* operon fusions to study transcriptional regulation of the colicin Ia receptor in *Escherichia coli* K-12, *J. Bacteriol.* **145:**647–650.

Yancey, R. J., Breeding, S. A. L., and Lankford, C., 1979, Enterochelin (enterobactin): Virulence factor for *Salmonella typhimurium, Infect. Immunol.* **24:**174–180.

Yariv, J., Kalb, J. A., Sperling, R., Bauminger, E. R., Cohen, S. G., and Ofer, S., 1981, The composition and the structure of bacterioferritin of *Escherichia coli, Biochem. J.* **197:**171–175.

Young, I. G., 1976, Preparation of enterochelin from *Escherichia coli, Prep. Biochem.* **6:**123–131.

Zähner, H., Keller-Schierlein, W., Hütter, R., Hess-Leisinger, K., and Deér, A., 1963, Sideramine von Mikroorganismen. Sideramine aus Aspergillaceen, *Arch. Microbiol.* **45:**119–135.

# 45

# Potassium Pathways in *Escherichia coli*

*Adam Kepes*[†], *Jean Meury, and Aline Robin*

## I. INTRODUCTION

Prokaryotes and *Escherichia coli* in particular do not possess (Na, K) ATPase in their plasma membrane like higher eukaryotes. Nevertheless, $K^+$ is the predominant cytoplasmic cation and it is essential for growth. The reason for the absolute requirement for potassium in prokaryotes as in eukaryotes is not entirely understood. One role of potassium is, however, probably specific to bacteria, namely the necessity to maintain an osmotic pressure in the cytoplasm in excess of the osmotic pressure in the medium, so that a hydrostatic pressure, the turgor pressure, keeps the plasma membrane in close contact with the rigid murein layer.

The importance of potassium is underlined by the number of mechanisms which concur in its regulation. This multiplicity of mechanisms and the complexity of the regulation made the study of potassium pathways in prokaryotes relatively unattractive, so much so that the molecular basis of the potassium fluxes in higher organisms has been largely elucidated earlier. Few groups devoted sustained attention to the problems of potassium uptake in prokaryotes. Most of what is presently known of the $K^+$ pathways is due to the work of the group of W. Epstein. They established the map of genes involved in potassium metabolism, characterized the functions of important gene products, and notably, identified and isolated the membrane-bound $K^+$ ATPase (for review see Silver, 1978; Helmer *et al.*, 1982).

In the following, an attempt is made to describe potassium fluxes in bacteria with the aim of illustrating some problems of general importance open to further investigation rather than to give an exhausting account of accomplishments.

*Adam Kepes, Jean Meury, and Aline Robin* • Laboratoire des Biomembranes, Institut Jacques Monod 75251 Paris Cédex 05, France.

## II. ACQUISITION OF POTASSIUM-FREE CELLS: A MECHANICAL DISORGANIZATION OF THE HYDROPHOBIC CONTINUUM

Since potassium is essential for growth, bacteria harvested from a culture possess a normal pool of potassium (0.1–0.3 M) in their cytoplasm. Resuspended in a sodium-based potassium-free medium, after a small initial loss of 15–25%, they lose intracellular potassium at a rate of about 10% per hr. Addition of valinomycin was utilized extensively to deplete potassium fast and thereby to create an electrical potential in membrane vesicles; however, in whole bacteria, due to the barrier posed by the outer membrane against access to the inner membrane, the effect of valinomycin is somewhat delayed and higher concentrations of the ionophore are required. Cells depleted by this method remain permeable to $K^+$ and they are not an optimal starting material to explore the $K^+$ pumps.

One physiological means to deplete $K^+$ would involve the utilization of the $K^+/H^+$ exchanger described by Rosen (Tsuchiya and Rosen, 1976; Brey *et al.*, 1980; Plack and Rosen, 1980; Sorensen and Rosen, 1980). This "antiporter" is believed to be essential for the regulation of intracellular pH and at alkaline pH, it is expected to pump $H^+$ in and $K^+$ out. In reality, potassium efflux was described in Tris-buffered alkaline media at pH values where a $Tris^+/K^+$ exchange could become operative, instead of a $H^+/K^+$ exchange. In order to check this hypothesis, $K^+$ depletion was measured as a function of pH and of the concentration of Tris, keeping the osmotic pressure of the medium constant. Over a wide range of pH, the displacement of potassium correlated well with the concentration of $Tris^+$ ions in the medium. Replacement of Tris buffer by triethanolamine buffer or by imidazole buffer changed the pH of half-maximal $K^+$ displacement in accordance with the respective p*K* values of the buffering species, indicating again a correlation with the concentration of the cationic form of the buffer rather than with the pH (Meury, unpublished).

Another way to deplete cell potassium is to inhibit the energy-generation process necessary for its accumulation. 2,4-Dinitrophenol was utilized to this effect at a concentration of 10 mM (Rhoads *et al.*, 1976). This is 10–50 times the concentration necessary to deplete cytoplasmic solutes accumulated by proton-motive transport systems. At these concentrations, the proton-conducting uncoupler might play the role of a $K^+$ ionophore (Kessler *et al.*, 1976).

The method utilized in the authors' laboratory is the sudden decrease of the osmotic pressure of the cell environment, an osmotic down-shock (Tsapis and Kepes, 1977; Kepes *et al.*, 1977). The method can be utilized after centrifugation by resuspending the pellet in a solution of low osmotic strength or after filtration, by employing very dilute aqueous solutions, or even distilled water for washing. When the osmolarity of both the suspending fluid and the washing solution was varied, half-depleting shocks were obtained with washing fluids whose osmolarities were directly related to the osmolarity of the suspension fluid to which bacteria were previously adapted. The drop of osmotic pressure necessary to attain this result was roughly 0.4 osM. Bacteria adapt to an osmolarity different from that of the culture medium in a few minutes and as shown below, the adaptation is due essentially to the adjustment of the intracellular $K^+$ pool.

The leak of $K^+$ caused by the osmotic shock is, however, not part of a physiological adaptation process. It is due to a momentary breakdown of the osmotic barrier of the plasma membrane, which regains its normal impermeability a few seconds (and maybe in fractions of 1 sec) later. The leak is not specific; sugar pools, amino acid pools, and nucleotides also leak out during the shock, and probably nonpenetrant solutes can be introduced into the cytoplasm if present in the washing fluid (Haguenauer-Tsapis and Kepes, 1980). The leak is restricted to small molecules; no TCA-precipitable material is released. These observations rule out both a physiological regulatory response of the cells to the osmotic shock which should be based on solute recognition and a gross damage of the membrane which should require more time to repair and could release also macromolecules. We therefore proposed a more subtle physical modification of the membrane, caused by the sudden increase of the turgor pressure, due to water uptake as the underlying mechanism. Such a turgor pressure would press the membrane against the rigid murein layer and this anisotropic transmembrane pressure might dislocate the bilayer structure; after a short time of relaxation, the pressure becomes again isotropic and the organization of the bilayer reestablishes the hydrophobic continuum. A support for this relaxation of the bilayer comes from the observation that osmotic shocks are more effective at low temperature than at 37°C although the correlation with the transition temperature of the membrane lipids was not established.

The hypothesis involving a modification of the bilayer structure such that it becomes permeable to hydrophilic solutes under a transient anisotropic pressure is largely speculative; it is based on the known changes of the turgor pressure and on the known effects on the release of solutes. Such anisotropic pressure is difficult to produce experimentally, except in the presence of a rigid envelope. Another similar situation may be the electrostriction effect on planar bilayer membranes, which occurs at transmembrane electrical potentials of the order of 300 mV and causes an instantaneous breakdown of the electric resistance of the membrane (Requena *et al.*, 1975). Speculation along these lines could be extended to the observation of the formation of aqueous channels in artificial bilayers at very low pH. The mechanical effect here envisioned, namely a repulsion of the electric charges of the same sign on the polar head groups on the two sides of the membrane, is geometrically opposite to the supposed effects due to an increase in turgor or in coulombian attraction. Perhaps the three phenomena can be grouped into a class of mechanical effects at an intermolecular or quasimolecular scale having important physicochemical consequences.

## III. POTASSIUM UPTAKE AND ITS MECHANOCHEMICAL SWITCH

The mapping of genes governing the $K^+$ requirement of *Escherichia coli* as established by Epstein and Davies (1970) includes genes coding for two primary potassium pumps. The Kdp pump of micromolar affinity for $K^+$ is repressed in the usual high-potassium media and is synthesized only under $K^+$ starvation. It consists of a $K^+$-dependent ATPase (Epstein *et al.*, 1978; Wieczorek and Altendorf, 1979), three subunits of which have been identified on acrylamide gel electrophoresis (Laimins *et al.*, 1978). The kdp D gene is the regulatory gene of the operon, and derepression is dependent on the low turgor pressure as shown by inserting lac Z gene under its control (Laimins *et al.*, 1981). The second potassium pump, coded by the *trkA* gene, is synthesized constitutively and it has a millimolar affinity for $K^+$. Its direct energy donor is not known; both a transmembrane electrochemical $H^+$ ion gradient and ATP were found to be necessary for its function (Rhoads and Epstein, 1977).

The potassium pool built up by the primary pump in the cytoplasm is linearly related to the osmotic pressure of the medium, while it varies only by a factor of about 2 over a $10^5$-fold change in external potassium (Kepes *et al.*, 1977). As a consequence, the chemical gradient of $K^+$ ions in steady-state conditions can also vary from $2.10^5$ to about 4 corresponding to equilibrium potentials ranging from $-310$ to $-36$ mV. Since the steady-state membrane potential in *Escherichia coli* is usually in the range of $-80$ to $-120$mV, one can conclude that $K^+$ does not distribute between medium and cytoplasm according to Nernst equilibrium. In other words, $K^+$ uptake is not due to an electrophoretic process. Similarly, the electrochemical gradient of $H^+$ ions in *Escherichia coli* is maintained in relatively narrow limits under physiological conditions of the order of $-180$ to $-240$ mV. Therefore, a mechanism of $K^+/H^+$ antiport would create a $K^+$ gradient opposite to that observed, whereas, an $H^+/K^+$ symport appears absurd, since this should represent a primary proton pump. The $K^+$ pool size fails to give an indication as to the energy source of the TrKA pump. If the highest observed $K^+$ gradients are compatible with the $K^+$ ATPase mechanism of the Kdp system with $K^+$/ATP stoichiometry of one, the identification of the energy sources may not help to explain the regulation of the $K^+$ pool; the only significant regulatory mechanism to which it appears to respond so far is the osmotic pressure.

Figure 1 shows the behavior of $K^+$-depleted cells when presented with 1 mM $^{42}K$, with or without a carbon source. The rate of uptake is similar in the two cases, and the $K^+$ pool established differs by a factor of less than two, being higher in the presence of the carbon source, glycerol. In both situations, an increase of the osmotic pressure of the medium by addition of NaCl is immediately followed by a readjustment of the $K^+$ pool to a higher value. The duration and the amplitude of this response are also nearly insensitive to the presence of the carbon source. The same type of response are obtained by the addition of sucrose, causing the same increment in osmolarity; therefore, $Na^+$ or $Cl^-$ ions play no specific role. Instead, the effect of NaCl emphasizes the $K^+$ specificity of the response.

If the presence of the carbon source has only a moderate quantitative effect on the phenomena described so far, it is not so when the chase of $^{42}K$ by excess of $^{40}K$ is considered. In the glycerol-fed suspension, cellular $K^+$ appears as exchangeable, while in the carbon-starved suspension, $K^+$ is not exchangeable.

Since, in the case of many other transported solutes, the steady-state pool is the result of a balance between uptake and efflux (whether efflux is due to a leak or to the reversal of inward transport), it was often assumed that the steady-state $K^+$ pool is also due to such a balance. The carbon-starved suspension of *Escherichia coli* gives the most convincing refutation of this belief, since at the steady state no influx and no efflux are observed. The response to NaCl indicates that the arrest of influx is not due to the insufficiency of available energy, but probably attributable to a regulatory

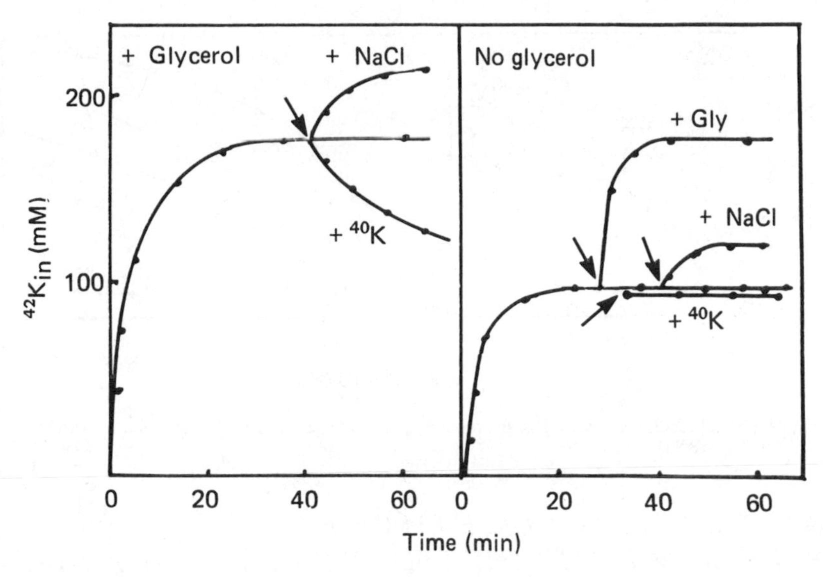


*Figure 1.* Uptake, osmotic adjustment, and chase of $^{42}$K in the presence (left) or absence (right) of a carbon source.

arrest. Thus, the signal that starts the $K^+$ pump is not the actual value of either the environmental or the intracellular osmotic pressure but a change of the difference between the two. The addition of the osmoactive solute causes an immediate outflux of water across the membrane, the cytoplasm tends to shrink, and the turgor pressure decreases; this switches on the $K^+$ pump. Together with the $K^+$ uptake, water is entering the cell, the cytoplasm tends to swell, and the turgor pressure increases. Once it reaches its new value, equal to that which prevailed before the disturbance, the $K^+$ pump is switched off. This on–off operation was made clear by the experiment on the carbon-starved suspension, whereas in the carbon fed cells, the $K^+/K^+$ exchange masked the *off* situation. The decrease in turgor pressure beyond a certain threshold can initiate plasmolysis; at this stage, the shrinkage of the cell is such that the plasma membrane loses contact with the murein and outer membrane. In this respect, it is worth noting that most empirical procedures which aim to plasmolyze *Escherichia coli* are carried out in potassium-free media (Neu and Heppel, 1964), thus depriving the cell of its defense response of pumping potassium in. It is likely that the response of the $K^+$ pump does not require plasmolysis but occurs already with subtle changes in local molecular interaction. The process should be classified as a mechanomolecular action, only a small step away from a mechanochemical transduction. It requires a mechanoreceptor at the molecular level, which can be either distinct or identical with the molecule that carries out the $K^+$-pumping function. Direct kinetic evidence of a similar regulation of the kdp ATPase is not available, but the steady-state regulation makes it likely that it also obeys the same kind of signal. Even the so-called TrKF system seems to be capable of fulfilling a limited osmoregulatory function. Since derepression of the kdp system is also under osmotic control (Laimins *et al.*, 1981), a whole family of mechanoreceptors is conceivable.

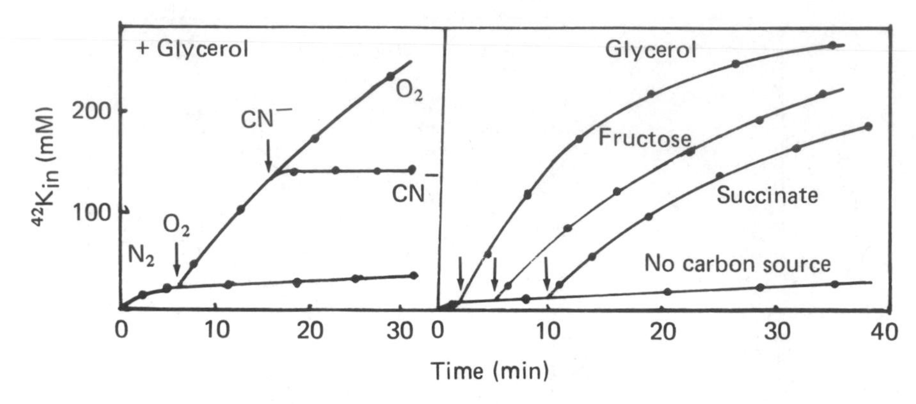


*Figure 2.* $K^+$ exchange in potassium-replete cells under a variety of metabolic conditions.

## IV. POTASSIUM–POTASSIUM EXCHANGE: A METABOLISM-DEPENDENT, ENERGY-INDEPENDENT PROCESS

This exchange can be observed in the context of chase of $^{42}K$ with excess nonradioactive $K^+$ as shown in Figure 1, or by the addition of trace amounts of $^{42}K$ to cells containing a normal amount of nonradioactive potassium without perturbation of any of the prevailing parameters of possible physiological significance. The exchange under these conditions is 2–5 times slower than the rate of initial uptake, suggesting that the two phenomena might not follow the same route. The finding that the rate of exchange follows a Michaelian behavior as a function of extracellular $K^+$ concentration, with an apparent $K_m$ close to that of uptake via the TrKA system, suggested instead that TrKA was responsible for the inward flux during exchange and a leak was only necessary to release the regulatory lock of TrKA.

But Figure 2 shows that exchange only occurs when a carbon source is present and actively metabolized. Indeed, exchange is blocked by anoxia produced either by bubbling nitrogen through cells grown aerobically on glycerol or by the addition of cyanide; bubbling of $O_2$ has permitted the exchange. Under aeration, glucose as well as succinate, a purely oxidative substrate, were able to support the exchange phenomenon; glucuronate uptake in a nonmetabolizing mutant AJ A9 has no effect, although its uptake system is fully active under conditions of carbon starvation.

It was shown in Figure 1 that a carbon source helps the cell to build up a higher $K^+$ pool (maybe by generating anionic metabolites?), and a higher pool should lead to a leak by exceeding a hypothetical threshold of efflux. The experiment represented in Figure 3 makes this supposition unlikely. Potassium uptake was monitored by $^{42}K$ in a carbon-starved suspension and $K^+/K^+$ exchange by repeated 6-min chase experiments with a 50-fold excess of nonradioactive KCl. Chase was very ineffective. Once the steady state was reached, a small amount of glucose was added that could permit only 5–10 min active metabolism. The $K^+$ pool started to increase immediately and, at the same time, chase became efficient even before maximal $K^+$ pool was reached; significantly, chase became inefficient again when glucose was exhausted, well before the $K^+$ pool dropped back to the carbon-starved level. The precision of this experiment is modest because the chase periods are long enough to overlap $K^+$ uptake, plateau, and leak periods, but it is difficult to escape the conclusion, that neither the active uptake nor the high $K^+$ pool, but only the active glucose metabolism, correlates closely with the exchange process. Indeed, if the threshold of leak was exceeded by the increased $K^+$ pool, this should tend to lock rather than unlock the $K^+$ pump and, thereby, should stop the exchange. The only *ad hoc* hypothesis to save the role of the TrKA pump in the exchange process would be to assume that the metabolism of a carbon source sets the threshold by locking the pump at a substantially higher level than during starvation and higher than the threshold of leak, i.e., reverse the order of the two thresholds prevailing under carbon starvation. Figure 4 rules out even this hypothesis. The effects of two energy inhibitors were tested on uptake, exchange, and steady-state pool of $K^+$ in a medium containing glycerol. Both cyanide and the uncoupler CCCP lowered the steady-state $K^+$ pool by 30–50% but did not dramatically deplete the $K^+$ pool. Cyanide decreased the rate of uptake proportionally to the decrease of the pool, while CCCP completely inhibited uptake. Conversely, CCCP decreased the rate of exchange proportionally to the decrease of the $K^+$ pool, while cyanide completely inhibited exchange. Let us remember that CCCP, while abolishing proton-motive energy, permits or even stimulates (in state 4) the $O_2$ uptake. The important

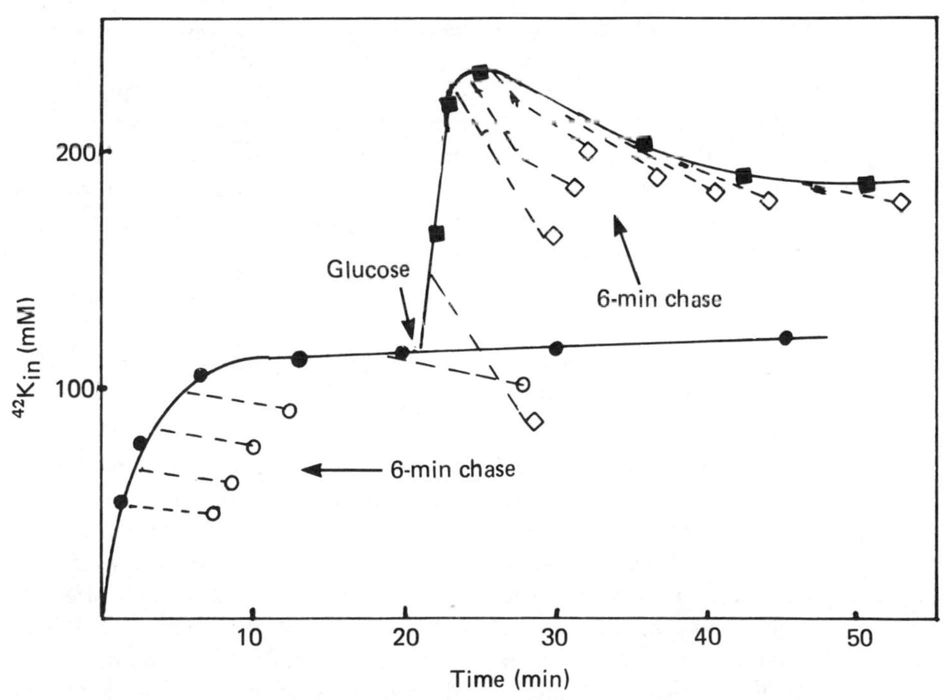


*Figure 3.* Transient activation of $K^+$ exchange (measured by chase) during rapid consumption of added glucose.

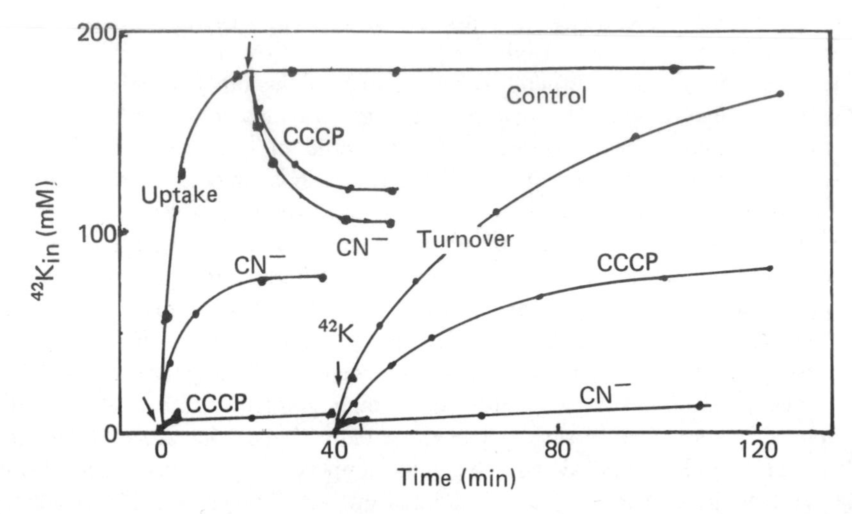


*Figure 4.* Effect of two metabolic inhibitors on uptake, steady-state pool, and exchange of potassium in aerobic suspension with carbon source present.

conclusion of this experiment is that the TrKA pump is not responsible for the inward flux during exchange, because exchange survives the complete inhibition of uptake via TrKA. Its strict dependence on oxidative metabolism is not related to an energy requirement. Since $K^+/K^+$ exchange does not accomplish thermodynamic work, it is plausible that it can operate in the presence of a proton-conducting uncoupler.

The metabolism dependence of the exchange rather suggests the necessity of a cofactor which is a ubiquitous but short-lived metabolic intermediate.

The distinctness of the metabolism-dependent exchange (MDE) from the primary potassium pump is strongly supported by the experiments reported here, but its molecular basis remains to be identified.

## V. THE EFFECT OF THIOL REAGENTS: A REVERSIBLE OPENING OF POTASSIUM-SPECIFIC CHANNELS

The addition of *N*-ethylmaleimide (NEM), *p*-chloromercuribenzoate (*p*CMB), or $HgCl_2$ provoked a massive and complete efflux of cell potassium (Meury *et al.*, 1980). This was reminiscent of the efflux of lactose analogues accumulated by lactose permease, where the key effect was the inactivation of the carrier. The slower effect of the slowly penetrating *p*CMB on the $K^+$ pool, compared to its immediate effect on lactose permease, could be due to a different orientation of the essential –SH. But, as shown in Figure 4, inhibitors of the $K^+$ pump such as CCCP do not cause a similar massive and complete leak of $K^+$. Even more astonishing was the finding that by decreasing the concentrations of NEM not only the leak of $K^+$ became slower but, after some time, $K^+$ was reaccumulated (Meury, unpublished). With optimal concentrations of NEM, 0.5 mM, the addition of 2-mercaptoethanol (2-ME) also resulted in a rapid recovery of the $K^+$ pool, and this was not inhibited by chloramphenicol. Loss of $K^+$ by addition of NEM and recovery after adding 2-ME could be repeated several times. Alkylation is irreversible, so one was led to believe that the essential thiol compound was resynthesized but that it was not a protein. It had to be, therefore, a cofactor of the protein which we believed earlier to be the $K^+$ pump.

It was difficult to reconcile this hypothesis of the inactivation of the $K^+$ pump and its recovery with two observations. First, it was shown that NEM caused a massive leak, even when the $K^+$ pump was stopped by the regulatory switch and the exchange mechanism was inactivated by carbon starvation or inhibited by cyanide; the addition of NEM established both an inward and an outward flux of $K^+$. The second observation was made with cells in which the kdp system was derepressed so that they accumulated $K^+$ from a medium containing $10^{-5}$ M KCl. Addition of NEM provoked a massive loss of $K^+$ similar to that observed with cells utilizing the low-affinity constitutive TrKA system. These experiments strongly suggested that NEM opened a highly permeable potassium channel through which $K^+$ can freely flow both ways, instead of closing a $K^+$ pump which normally assured a purely inward flow. The reversal of the $K^+$ loss could then be attributed to the closing of these channels so that the activity of the $K^+$ pump(s) can accomplish the $K^+$ accumulation.

*N*-ethylmaleimide has multiple effects on bacterial cells including the inactivation of the lactose carrier (Kepes, 1960; Fox and Kennedy, 1965), the glucuronic acid permease (Jimeno-Abendano and Kepes, 1973), and glucose phosphotransferase enzyme II (Haguenauer-Tsapis and Kepes, 1973). Nevertheless, it does not destroy the permeability barrier of the plasma membrane. The solutes accumulated by lactose permease leak out at a rate dependent on the hydrophobicity of the aglycon; i.e., lactose leaks slowly, and TMG faster, in conformity with their respective rate of turnover during the activity of the uptake system. Glucuronic acid leaks out at negligible rates upon addition of NEM; its turnover under the conditions of active uptake goes entirely through the shuttle of its carrier. Some other uptake systems are not sensitive at all to thiol reagents; for example, substrates accumulated by gluconic acid permease (Robin and Kepes, 1975), or by β-methylgalactose permease, remain in the cell when thiol reagents are added. Since among the negative findings quoted above only neutral and anionic sugars were listed, the possibility remains that cations other than $K^+$ and its substitutes $Rb^+$ and $Tl^+$ can also be accommodated by the channel opened by the reaction with NEM (Vorisek and Kepes, 1972). Bakker and Mangerich (1982) showed that protons and triphenyl phosphonium ions exchange against $K^+$ during the opening of the channel although the cell membrane is not permeabilized dramatically.

## VI. THE $K^+$ CHANNEL IS SPECIFICALLY CONTROLLED BY GLUTATHIONE: IT IS A GSH-CONTROLLED CHANNEL (GCC)

In search of the thiol compound which leaves the $K^+$ channel open when removed by NEM and which closes it again during recovery, it was necessary to first eliminate glutathione (GSH), the overwhelmingly abundant thiol compound encountered in *Esch-*

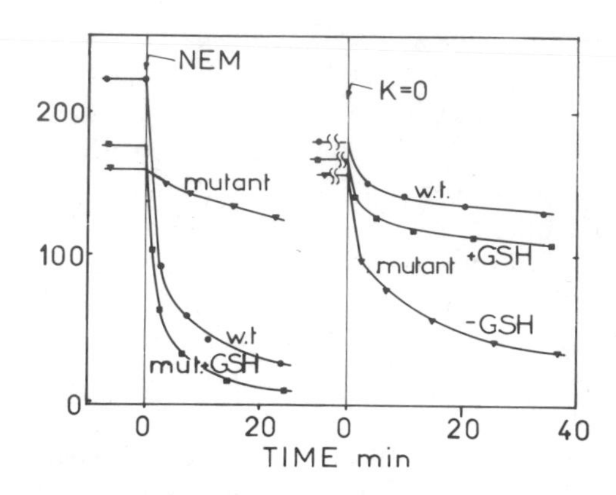


*Figure 5.* Potassium loss by addition of a thiol reagent or by resuspension in a potassium-free medium of wild type, and a GSH-deficient mutant of *Escherichia coli* supplemented or not with GSH.

*erichia coli*. The concentration of glutathione is normally 1–3 mM in the cytoplasm. Mutants of *Escherichia coli* deficient in glutathione synthesis, such as AB7, AB821, and AB830 isolated by Apontoweil and Berends (1975a,b) appeared as the most suitable strains to start the search for the essential thiol. These mutants had growth characteristics undistinguishable from that of the wild type; in particular, they did not require supplements in the growth medium such as glutathione or one of its precursors. Nevertheless, inoculated into media containing 1 mM KCl, they failed to grow, contrasting with the normal growth of the parent strain (Meury and Kepes, 1982). After growth on media containing 20 mM KCl or more, resuspension in a potassium-free medium caused a massive loss of their $K^+$ pool similar to that observed with the $K^+$ leaky strains TrKB, TrKC, or B525 and B526, and very different from the behavior of wild-type strains (Figure 5). When supplemented with glutathione in the growth medium, they regained the capacity to grow on 1 mM KCl and displayed the $K^+$ tight phenotype in a potassium-free medium.

It seemed, therefore, that GSH itself was responsible for the $K^+$ tightness of the bacterial membrane and that, in its absence, the $K^+$ leak occurred probably through the same channels that could be opened by addition of NEM in wild-type bacteria. When the GSH-less mutants were grown on 20 mM KCl, the addition of NEM did not provoke the rapid and massive loss of $K^+$ observed on the wild type, but when grown with glutathione in the medium, the effect of NEM became similar to that observed in the wild type. The concentration of glutathione in the growth medium required to obtain the two effects, i.e., growth on low $K^+$ and sensitivity to NEM, is less than $10^{-5}$ M and appreciable changes can be observed with as little as $10^{-7}$ M GSH. This observation rules out the possibility that the active species could be an impurity contained in GSH. The possibility that the active species is a metabolic product of GSH is made relatively unlikely by the fact that besides disulfide bond formation, no significant metabolic reaction is known to occur with GSH, and that the repair of the $K^+$ leaky phenotype starts immediately upon addition of GSH. This can be shown either as the increase of the $K^+$ pool upon addition of $5.10^{-5}$ M GSH to the mutants suspended in a medium containing 0.1 mM KCl or as the appearance of NEM sensitivity. With this concentration of GSH in the medium, it takes some 30 min to reach full effect, and during this time, a significant accumulation of GSH by the bacteria occurs, achieving concentration ratios of several hundred-fold (Robin, unpublished).

The GSH-deficient mutant, which is impaired in glutathione synthetase and which accumulates γ-glutamyl-cysteine, behaves similarly to the mutant impaired in γ-glutamyl-cysteine synthetase. Thus, the dipeptide cannot relace the tripeptide GSH. In contrast, ophthalmic acid, γ-glutamyl-α-amino isobutyrylglycine, an analogue of glutathione without an –SH group, is capable of replacing glutathione as a supplement in the growth medium or as an additive that restores $K^+$ tightness in nongrowing suspensions of the mutant; of course, ophthalmic acid does not restore NEM sensitivity.

A number of commercially available tripeptides have been tried as substitutes for glutathione. So far, only α-γ glutamyl cysteine $(Gly)_2$, a pentapeptide with a disulfide bond, gave positive results. After incubation with the pentapeptide, the $K^+$ tight phenotype and the NEM sensitivity were restored, probably by an effect of glutathione reductase on the disulfide bond. Therefore, the present tentative conclusion is that only GSH and close analogues of it can be recognized by the $K^+$ leak channel. The SH group is not essential for activity, nor is the free α-carboxyl group of the glutamic acid residue necessary.

## VII. SOME UNSOLVED PROBLEMS

The reason for the existence of a GSH-controlled $K^+$ channel is far from clear, and the physiological gating by GSH has no supporting evidence. One possibility was the involvement of GCC in the metabolism-dependent $K^+$ exchange, a more physiological process than a $K^+$ leak. In a preliminary inquiry, we explored whether the glutathione-supplemented and the glutathione-deprived GSH mutants exhibit the metabolism-dependent $K^+$ exchange to the same extent. The results, represented in Figure 6, gave an affirmative answer to this question. The level of the $K^+$ pool is lower in the unsupplemented cells, but its sensitivity to NEM is very much lower than that of the GSH-supplemented cell. The unsupplemented cells also exhibit a spontaneous nonmetabolism-dependent $K^+$ turnover, which appears upon chase with excess nonradioactive $K^+$. Nevertheless, the $K^+$ exchange flux is accelerated upon addition of glycerol to the same extent as in the GSH-supplemented cells; thus, metabolism-dependent exchange does not depend on the GCC–GSH complex.

In view of this negative conclusion, it cannot be ruled out that GCC is a mere artifact, namely, a glutathione-binding membrane protein which changes its configuration when glutathione is removed. This new configuration can fortuitously yield passage to potassium ions. Of course, the same reasoning applies to the TrKB and TrKC gene products, which could be genuinely unrelated to potassium pathways of *Escherichia coli* but become $K^+$ channels due to genetic damage. In this case, these gene products as well as the GCC could still serve as model systems insofar as they permit new physicochemical, structural, or genetic approaches, less easily available with the physiologically significant channels. In this respect, the glutathione-controlled channel has the advantage of the reversibly controlled opening and closing by exper-

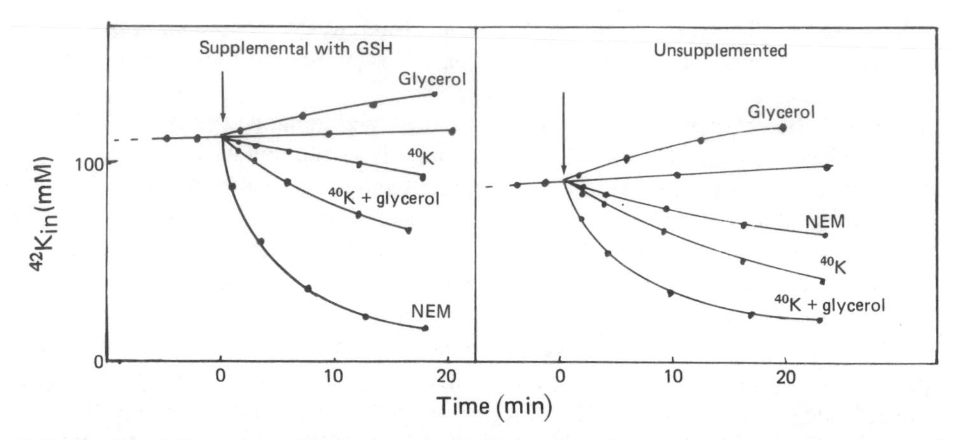


*Figure 6.* Metabolism-dependent $K^+$ exchange (MDE) in supplemented and unsupplemented suspensions of a GSH-deficient mutant. MDE is the difference between the flux provoked by $^{40}$K or by $^{40}$K and glycerol.

imental exposure to glutathione or thiol reagents. Inversely, it is possible that GCC is a physiological cation channel, which can be gated by effectors other than GSH yet to be discovered. It is also possible that the physiological role of GCC is to conduct something other than $K^+$ ions, for example, $Ca^{2+}$ ions.

These questions are not the only ones connected with the $K^+$ pathways of *Escherichia coli*. We showed above that the metabolism-dependent $K^+/K^+$ exchange is also in need of a plausible physiological role, and the elucidation of its molecular mechanism is still in the future. The TrKA system works with an unknown energy source; it is controlled by an ill-understood mechanochemical transduction and maybe for this reason has, so far, not been amenable to function in isolated membrane vesicles.

The $K^+/H^+$ exchanger, which was explored only in inverted membrane vesicles, was entrusted with the role of regulating intracellular pH in alkaline media; by so doing, this electroneutral exchange should transport $H^+$ ions and $K^+$ ions, each against their respective concentration gradients, and needs, therefore, an energy-coupling mechanism. Moreover, if the TrKA system is clearly responsible for adjusting the $K^+$ pool upward in response to an increase in the osmotic pressure of the medium, the downward adjustment following a decrease in osmotic pressure has still to await the identification of a pathway.

## REFERENCES

Apontoweil, P., and Berends, W., 1975a, Glutathione biosynthesis in *Escherichia coli* K12. Properties of the enzymes and regulation, *Biochim. Biophys. Acta* **399:**1–9.

Apontoweil, P., and Berends, W., 1975b, Isolation and initial characterization of glutathione-deficient mutants of *Escherichia coli* K12, *Biochim. Biophys. Acta* **399:**10–22.

Bakker, E. P., and Mangerich, W. E., 1982, *N*-Ethylmaleimide induces $K^+$-$H^+$ antiport activity in *Escherichia coli* K-12, *FEBS Lett.* **140:**177–180.

Brey, R. N., Rosen, B. P. and Sorensen, E. N., 1980, Cation/proton antiport systems in *Escherichia coli*, *J. Biol. Chem.* **255:**39–44.

Epstein, W., and Davies, M., 1970, Potassium-dependent mutants of *Escherichia coli* K-12, *J. Bacteriol.* **101:**836–843.

Epstein, W., Witelaw, V., and Hesse, J., 1978, A $K^+$ transport ATPase in *Escherichia coli, J. Biol. Chem.* **253:**6666–6668.

Fox, C. F., and Kennedy, E. P., 1965, Specific labeling and partial purification of the M protein, a component of the β-galactoside transport system of *Escherichia coli, Proc. Natl. Acad. Sci. USA* **54:**891–899.

Haguenauer-Tsapis, R., and Kepes, A., 1973, Changes in accessibility of the membrane-bound transport enzyme glucose-phosphotransferase of *E. coli* to protein group reagents in presence of substrate or absence of energy source, *Biochem. Biophys. Res. Commun.* **54:**1335–1341.

Haguenauer-Tsapis, R., and Kepes, A., 1980, Different sidedness of functionally homologous essential thiols in two membrane-bound phosphotransferase enzymes of *Escherichia coli* detected by permeant and nonpermeant thiol reagents, *J. Biol. Chem.* **255:**5075–5081.

Helmer, G. L., Laimins, L. A., and Epstein, W., 1982, Mechanisms of potassium transport in bacteria, in: *Membranes and Transport,* Vol. 2 (A. Martonosi, ed.), Plenum Press, New York and London, pp. 123–128.

Jimeno-Abendano, J., and Kepes, A., 1973, Sensitization of D-glucuronic acid transport system of *E. coli* to protein group reagents in presence of substrate or absence of energy source, *Biochem. Biophys. Res. Commun.* **54:**1342–1346.

Kepes, A., 1960, Etudes cinétiques sur la galactoside-permease d'Escherichia coli, *Biochim. Biophys. Acta* **40:**70–84.

Kepes, A., Meury, J., Robin, A., and Jimeno, J., 1977, Some ion transport systems in *E. coli* (transport of potassium and anionic sugars), in: *Biochemistry of Membrane Transport,* FEBS Symposium 42 (G. Semenza and E. Carafoli, eds.), Springer-Verlag, Berlin, Heidelberg, pp. 633–647.

Kessler, R. J., Tyson, C. A., and Green, D., 1976, Mechanism of uncoupling in mitochondria: Uncouplers as ionophores for cycling cations and protons, *Proc. Natl. Acad. Sci. USA* **73:**3141–3145.

Laimins, L. A., Rhoads, D. B., Altendorf, K., and Epstein, W., 1978, Identification of the structural proteins of an ATP-driven potassium transport system in *Escherichia coli, Proc. Natl. Acad. Sci. USA* **75:**3216–3219.

Laimins, L. A., Rhoads, D. B., and Epstein, W., 1981, Osmotic control of Kdp operon expression in *Escherichia coli, Proc. Natl. Acad. Sci. USA* **78:**464–468.

Meury, J., and Kepes, A., 1982, Glutathione and the gated potassium channels of *Escherichia coli, Eur. Mol. Biol. Org. J.* **1:**339–343.

Meury, J., Lebail, S., and Kepes, A., 1980, Opening of potassium channels in *Escherichia coli* membranes by thiol reagents and recovery of potassium tightness, *Eur. J. Biochem.* **113:**33–38.

Neu, H. C., and Heppel, L. A., 1965, The release of enzymes from *Escherichia coli* by osmotic shock and during the formation of spheroplasts, *J. Biol. Chem.* **240:**3685–3692.

Plack, R. H., Jr., and Rosen, B. P., 1980, Cation/proton antiport systems in *Escherichia coli, J. Biol. Chem.* **255:**3824–3825.

Requena, J., Haydon, D. A., and Hladky, S. B., 1975, Lenses and the compression of black lipid membranes by an electric field, *Biophys. J.* **15:**77–81.

Rhoads, D. B., and Epstein, W., 1977, Energy coupling to net $K^+$ transport in *Escherichia coli* K-12*, *J. Biol. Chem.* **252:**1394–1401.

Rhoads, D. B., Water, F. B., and Epstein, W., 1976, Cation transport in *Escherichia coli.* VIII. Potassium transport mutants, *J. Gen. Physiol.* **67:**325–341.

Robin, A., and Kepes, A., 1975, Inducible gluconate pemease in a gluconate kinase deficient mutant of *Escherichia coli, Biochim. Biophys. Acta* **406:**50–54.

Silver, S., 1978, Transport of cations and anions, in: *Bacterial Transport* (B. Rosen, ed.), Marcel Dekker, New York, pp. 221–324.

Sorensen, E. N., and Rosen, B. P., 1980, Effects of sodium and lithium ions on the potassium ion transport systems of *Escherichia coli, Biochemistry* **19:**1458–1462.

Tsapis, A., and Kepes, A., 1977, Transient breakdown of the permeability barrier of the membrane of *Escherichia coli* upon hypoosmotic shock, *Biochim. Biophys. Acta* **469:**1–12.

Tsuchiya, T., and Rosen, B. P., 1976, Calcium transport driven by a proton gradient in inverted membrane vesicles of *Escherichia coli, J. Biol. Chem.* **251:**962–967.

Vorisek, J., and Kepes, A., 1972, Galactose transport in *Escherichia coli* and the galactose-binding protein, *Eur. J. Biochem.* **28:**364–372.

Wieczorek, L., and Altendorf, K., 1979, Potassium transport in *Escherichia coli*. Evidence for a $K^+$-transport adenosine-5′-triphosphatase, *FEBS Lett.* **98:**233–235.

# Index


A23187 ionophore, effect on $Ca^{2+}$-induced $K^{+}$ transport, 195, 206, 209
Acetate kinase, involvement in regulation of PTS, 543
Acetic anhydride, effects on acetylcholinesterase, 413
Acetylcholine, 335–401
Acetylcholine receptor, 335, 401
  affinity labeling, 348
  amino acid sequence, 340
  antigenic structure, 345
  assembly, 355
  degradation, 358
  electrophysiology in postsynaptic membrane, 362
  functional unit, 376
  incorporation into lipid bilayers, 374
  inhibitors of cholinergic response, 368–369
  in muscle, 351
  lipids, effects of, 380
  pharmacology of agonists, 363
  reconstitution into lipid vesicles, 371
  structural aspects, 337
  subunit topography, 343
Acetylcholinesterase, 403–429
  affinity chromatography, 405, 416
  biosynthesis, 422
  diaphragm (rat), 420
  mRNA processing (*in vivo*), 424
  neuromuscular junctions, 403, 404, 420, 421
  papain, 417, 418
  pronase, 417, 419
  protein–detergent interactions, 417
  red cell forms (human), 415
  reductive methylation, 419
  soluble and amphipathic forms, 414
  subunit composition, 406
  trypsin, 404, 406, 412, 415
Acetylphosphate, 543
Acidic phospholipids, effect on plasma membrane $Ca^{2+}$-ATPase, 241
Active transport, primary, 27
Adamantine diazirine, labeling of $Na^{+}$,$K^{+}$-ATPase by, 73
Adenosine triphosphatase, $Ca^{2+}$-activated
  plasma membrane, 235–248
  sarcoplasmic reticulum, 115–191
Adenosine triphosphatase, $Na^{+}$,$K^{+}$-activated, 35–114, *see also:* $Na^{+}$,$K^{+}$ transporting adenosine triphosphatase
Adrenergic control of the $Na^{+}$,$Ca^{2+}$ cycle in mitochondria, 276
*Aerobacter aerogenes*, 621
Aerobactin, 619, 623, 624, 628, 631, 635
Affinities of ligands, intrinsic and apparent, 5, 14, 23, 30
Affinity labeling, of acetylcholine receptor, 348
  erythrocyte glucose transporter, of, 502, 503, 511
Agonists of acetylcholine receptors, 363
Albomycin, 622, 629
Alkaline phosphatase, possible role in Pi translocation, 303
Alloxan diabetes, 499
Amine accumulation, chromaffin granule, 469, 473, 476, 477, 478
Amino acid binding proteins in bacteria, 579–583, 587
Amino acid transport, bacteria, 577, 615
  assembly of transport system, 590
  classes of transport systems, 578–586
  energization, 593
  evolution, 602
  Liv-I and histidine transport genes, 588
  membrane-bound systems, 585, 588
  models of mechanisms, 597

Amino acid transport (*cont'd*)
periplasmic binding protein (BP)-dependent systems, 585, 587
reconstitution of BP-dependent systems, 596
reconstitution of membrane-bound systems, 594
regulation, 598
Amphipathic proteins, 415, 419
Amphiuma red cells, 197, 200, 203, 209
Antibodies against enterochelins, 642, 644
Antigenic structure of acetylcholine receptor, 345
Antiport, kinetics, 11
Aplysia neurons, 213
Ara-BP, 587, 588
Arsenite, as inhibitor of $Na^+,K^+$-ATPase, 52
Ascorbate, chromaffin granule, 456, 480, 485
Aspartyl-phosphate intermediate in plasma membrane $Ca^{2+}$-ATPase, 236–239
Aspartyl-phosphate intermediate in sarcoplasmic reticulum $Ca^{2+}$-ATPase, 157–191
ATP accumulation, chromaffin granule, 484
ATP-ADP exchange, $Na^+$,-$K^+$-activated ATPase, 48
ATP analogs, as inhibitors of $Na^+,K^+$-ATPase, 60, 74, 75, 82–84
ATPase, $Ca^{2+}$-activated, 117–191
ATPase, $Na^+,K^+$-activated, 35–114, *see:* $Na^+,K^+$ transporting adenosine triphosphatase
ATP-dependent $Ca^{2+}$ transport
in mitochondria, 249–286
in plasma membrane, 235–248, 321–334
in sarcoplasmic reticulum, 115–191
ATP, effect on $Ca^{2+}$-induced $K^+$ transport, 205
*Azotobacter vinelandii,* 639

Bacterial amino acid transport, 577–615
Bacteriophage, 622, 633, 639
Basement membrane (basal lamina), 405, 413, 414, 421
Biosynthesis of acetylcholinesterase, 423, 424
Bromoacetylcholine, 349
α-Bungarotoxin, 349
Butanedione, inhibitor of $Na^+,K^+$-ATPase, 74

$Ca^{2+}$ accumulation
chromaffin granule, 486
sarcoplasmic reticulum, 157–191
$Ca^{2+}$ antagonists, effect on mitochondrial $Ca^{2+}$ transport, 266
$Ca^{2+}$-ATPase, plasma membrane, 235, 236, 237, 238, 239, 240, 241, 242, 243, 245
$Ca^{2+}$-ATPase, sarcoplasmic reticulum, 115–191
$Ca^{2+}$-binding proteins in sarcoplasmic reticulum, 117–118
$Ca^{2+}$ channel, 235
$Ca^{2+}$ cycles, mitochondria, 272–279
$Ca^{2+}$-dependent ATP hydrolysis,
in plasma membrane vesicles, 235–248
in sarcoplasmic reticulum, 157–191
Caffeine, effect on Ca release, 135
$Ca^{2+}$-induced $K^+$ transport, 193–233
Calcitonin, effects on intestinal phosphate (Pi) transport, 309
Calcium, relation to intestinal phosphate transport, 287, 289, 292, 293, 295, 297, 299, 303, 309, 313
Calcium transport in mitochondria, 249–286
adrenaline, effects of, 276
$Ca^{2+}$ antagonists, effect of, 266
$Ca^{2+}$ capacity of mitochondria, 267
$Ca^{2+}$ cycles, 270–279
$Ca^{2+}$ uniporter, 251
lanthanides, effect of, 265
$Na^+,Ca^{2+}$ carrier, 255–262
$Na^{2+}$-independent release of calcium, 262–265
nucleotides and $Mg^{2+}$, effects of, 269
ruthenium red, effect of, 266
Calcium transport in nerve membranes, 326–329
ATP, effects on, 328
$Ca^{2+}$ efflux, 327
$Ca^{2+}$ entry with depolarization, 327
$Ca^{2+}$ influx, 326
internal $Ca^{2+}$ concentration, 327
Calcium transport in sarcoplasmic reticulum, 157–191
Calmodulin
effect on $Ca^{2+}$-induced $K^+$ transport, 205, 219
effect on plasma membrane $Ca^{2+}$-ATPase, 238–242, 245
Calsequestrin, 116, 117, 138, 142
$Ca^{2+}$-pump, 157–191, 194, 219, 235–248
Cardiac glycosides, effect on $Na^+,K^+$-ATPase, 37, 68, 85–88
Cardiac Purkinje fibers, $Ca^{2+}$-induced $K^+$ transport, 216
Cardiac sarcoplasmic reticulum, 137–140
$Ca^{2+}$ release from sarcoplasmic reticulum, 134–137
Carrier, simple, 7
Catecholamines, chromaffin granule, 455
Cation-stimulated ATPases, involvement in amino acid transport, 579–583
$Ca^{2+}$ transport ATPase of plasma membrane, 235–248
acidic phospholipids, 241
calmodulin, 238, 240–241
mechanism, 236–238
purification, 238–240

$Ca^{2+}$ transport ATPase of plasma membrane (*cont'd*)
  reconstitution, 238, 240–242
  trypsin digestion, 240
$Ca^{2+}$ transport ATPase of sarcoplasmic reticulum, 117–191
  activation by $Ca^{2+}$, 160
  amino acid sequence, 123–125
  biosynthesis, 140–142
  $E_1$–P → $E_2$P interconversion, 169
  kinetics of $Ca^{2+}$ transport, 157–191
  lipid annulus, 130
  lipid requirement, 128–132
  oligomers, 125–128
  phosphorylated intermediate, 165
  purification, 122
  reconstitution, 132, 133
  reversal of $Ca^{2+}$ transport, 174–182
  substrate specificity, 164
  tryptic fragments, 123
$Ca^{2+}$ transport in mitochondria, 249–286
$Ca^{2+}$ transport in nerve membranes, 326–329
cDNAs of acetylcholine receptor, 341
Cell-free synthesis of sarcoplasmic reticulum proteins, 142
Chemiosmotic coupling, primary active transport, 27
Chicken embryo fibroblasts, glucose transport, 504–506
Chromaffin cells, 215
Chromaffin ghosts, preparation, 450, 465
$\Delta\mu_{H^+}$, chromaffin granule, 463, 464, 467, 472, 477, 478, 483, 488
Chromaffin granules, 449–495
  ATPase, 449, 458
  isolation and purification, 450
  membrane enzymes, 449, 480
Chromogranin A, 454
Chronic alcoholism, hypophosphatemia, 314
Citrate, 619, 622, 624, 629, 630, 638–639
$Cl^-$ fluxes in nerve membranes, 326
$Cl^-$ permeability, 201, 202
Colicins, 622, 632, 633
Collagen, collagen-like proteins, 404, 408, 409, 411
Collagenase, 411, 412, 413
Competitive inhibitors of acetylcholine receptor, 368
Conductance of postsynaptic membrane, 360
Coprogen, 624
Cotransport systems, kinetics, 18
  apparent affinities, 23
  ordered reaction, 24, 26
Countertransport kinetics, 11
  apparent affinities, 14
Coupling of flows on a simple carrier, 11
Cr(III) arylazido-β-alanyl, labelling of $Na^+,K^+$-ATPase by, 73
Crosslinking studies on $Na^+,K^+$-ATPase, 78
crr gene, the structured gene for $III^{Glc}$ of bacterial PTS, 548
Cultured cells, $Ca^{2+}$-induced $K^+$ transport, 211
Cyclic AMP, effect on intestinal phosphate (Pi) transport, 308
Cytochalasin B, 500, 501, 504–506, 512, 513
Cytochrome $b_{561}$, chromaffin granule, 453, 479, 480, 482

Degradation of acetylcholine receptors, 358
  antigenic modulation, 359
  endocytosis, role in, 359
  lysosomes, role in, 359
  normal turnover, 358
Denervation, effects on acetylcholinesterase, 422
24R, 25-Dihydroxyvitamin $D_3$, 305
1,25-Dihydroxyvitamin $D_3$, 304
Dopamine, 471
Dopamine-β-hydroxylase, 453, 454, 480, 483
Drugs, effects on $Ca^{2+}$-induced $K^+$ transport, 207

*E. coli*, amino acid transport, 579, 583
  iron transport, 617–645
  maltose transport, 561–575
  outer membrane, 564, 565, 571
  periplasmic space, 561, 566
  potassium transport, 653–665
  PTS system, 525
Electric organs of electric fish, acetylcholinesterase, 403, 405, 406
Electrochemical $H^+$ gradient, 458
Electrode measurements of ion transport in nerve membranes, 323
Electron transport chain, chromaffin granule, 479
*Electrophorus electricus*
  acetylcholine receptor, 337
  $Na^+,K^+$-ATPase, 49
Endocrine β cells, 212
Energetics of active transport, 1–33
Energization of amino acid transport, 591, 595
Enkephalin, 454
Enterochelin, 619, 620, 624, 630, 635, 642
  structural analogs, 631, 632
Enzyme I, of bacterial PTS, 524
  regulation, 543
Eosin, labelling and inhibition of $Na^+,K^+$-ATPase by, 63–65
Epinephrine, 471
Erythrocyte acetylcholinesterase, 415
Erythrocyte, D-glucose transport, 501

Erythrocytes, human
  $Ca^{2+}$-induced $K^+$ fluxes, 194–208
*Escherichia coli* K-12, maltose transport, 561, 562
Ethylmercurythiosalicylate (thimerosal), effect on $Na^+,K^+$-ATPase, 91–93
Evolution of amino acid transport in bacteria, 602–604
ExbB function, 623–635

Ferrichrome, 619, 622, 624, 628, 269
Ferritin, 619
Fibroblasts, 211
Fluorescein isothiocyanate, inhibition and labelling of $Na^+,K^+$-ATPase by, 62, 63, 73
Formycin nucleotides and $Na^+,K^+$-ATPase, 60, 62
Fur gene product, 637, 638

β-Galactosidase, hybrid proteins, 566, 567, 568
Gal-BP, 587, 588
Gárdos-effect, 193–233
Gastric acid secretion, stimulation, 432–437, 446–447
Gastric ATPase
  active sites, 439–447
  catalytic mechanism, 431–432, 439–445, 446–447
  isoelectric focusing, 437, 438, 442, 444, 446
  localization, 432–434
  monoclonal antibodies, 432, 433, 434, 436, 437, 444, 446
  partial reactions, 432, 439–441, 446–447
  structure, 431, 437–438
Gastric vesicles, 434–437, 446–447
Gene fusions, 566
Gene structure, of *E. coli* $K^+$ transport, 655–656
  bacterial amino acid transport, 577–615
  bacterial iron transport, 617–652
  PTS, 523–599
Glucocorticoids, effects on intestinal phosphate (Pi) transport, 309, 310, 312
Glucose-6-phosphate, 524
Glucose transport, erythrocytes, 501–504
  affinity labeling, 502
  cultured cells, in, 504
  glucocorticoids, effect of, 499
  mammalian cells, in, 498–508, 510–514
  purification of carrier, 503
Glutamate–aspartate binding protein, 586
Glutamate transport in bacteria, 585
Glutamine binding protein, 586
Glutathione, effect on $K^+$ channel in *E. coli*, 661
Glycoprotein, $Na^+,K^+$-ATPase, 119
Glycoproteins, sarcoplasmic reticulum, 118–119, 120
Glycosaminoglycans, 414

Haptoglobin, 641
Harmaline, 472
$H^+$-ATPase, chromaffin granules, 449, 453, 463, 467
$H^+$/ATP stoichiometry, chromaffin granule, 466
Heart mitochondria, $Ca^{2+}$ cycles, 275
Heart sarcolemma, 235, 242, 244
Hexose transport in animal cells, 497–522
  cultured cells, 504
  erythrocytes, 501
  glucose transporter, 511
  insulin receptor, 509–510
  insulin regulation, 506
$HgCl_2$, effects on K channels in *E. coli*, 660
High affinity $Ca^{2+}$ binding protein, 117, 118, 140, 141
Histamine granules, 488
Histidine binding proteins (his-BP), 586, 587
Histidine transport
  assembly into functional transport system, 590
  components P, Q, M, 588
  gene products, 590
  genes, 589
  regulation, 599
Histrionicotoxin, 349
Hormonal regulation of Pi transport, 303
HPr of bacterial PTS, 528
$H^+$ transport in nerve membranes, 331
Human erythrocytes, glucose transport, 498–504
Hyp mice, defective intestinal Pi transport, 314
Hypoparathyroidism, 314
Hypophosphatemia, 314
25-Hydroxyvitamin $D_3$, 304

iex gene, PTS mediated repression, 547
Inside-out vesicles, (IOV), 198, 206, 207, 220
Insulin, effects on intestinal phosphate (Pi) transport, 310
Insulin, mechanism of action, 506–514
Insulin receptor, 506–511
  glucose transporter, relationship to, 510
Insulin regulation of glucose transport, 506
  insulin binding, 506
  insulinomimetic agents, 507
  insulin receptor, 509
  mechanism, 511
Intestinal phosphate transport, 287–320; *see:* Phosphate transport, intestinal

Intramembrane particles
  $Ca^{2+}$-ATPase in sarcoplasmic reticulum, 120, 121
Intrinsic and apparent affinities, 5
Invertebrate neurons, $Ca^{2+}$-induced $K^+$ transport, 212
Iodonaphthylazide, labelling of $Na^+,K^+$-ATPase by, 73
Ion permeability, chromaffin granule, 458
Ion transport, nerve membrane, 321–334
  $Ca^{2+}$ transport, 326
  $Cl^-$ transport, 326
  $H^+$ transport, 337
  integration of ion fluxes, 331
  $K^+$ fluxes, 326
  methods for study, 322
  $Mg^{2+}$ transport, 329
  Na fluxes, 323–325
Ion transporting ATPases, 35
Iron
  ColV plasmid, 621, 623, 637, 638
  regulation, 635–640
  transport, 619–645
  t-RNA modification, 639
  virulence of *E. coli*, 640–645
Iron transport in *E. coli*, 617–652
  gene functions, 628
  receptor proteins, 622
  regulation, 635
  siderophores, 619
  synthetic enterochelins, 631
  TonB and ExbB functions, 632
  virulence, 640
Iso Bi-Bi enzymes, 18
Isoleucine preferring leucine, valine, threonine binding protein (ILVT-BP), 586

$K^+$ channels, $Ca^{2+}$ regulated, 193–233
$K^+$-dependent ATPase, in *E. coli* $K^+$ transport, 655
Kdp D gene, coding for $K^+$ pump in *E. coli*, 656
Kdp pump in $K^+$ transport in *E. coli*, 655
$K^+$ fluxes in nerve membranes, 326
Kinetics of a simple carrier-mediated transport, 7
$K^+,K^+$ exchange, 45
Krebs cycle, regulation of bacterial PTS, 544
$K^+$ transport, $Ca^{2+}$-induced, 193–233
  calcium dependence, 195
  calmodulin, effect of, 205
  channel properties, 221
  divalent and trivalent cations, effects of, 200
  drugs, effects of, 206
  in cardiac Purkinje fibers, 216

$K^+$ transport, $Ca^{2+}$-induced (*cont'd*)
  in cultured cells, 211
  in invertebrate excitable tissues, 213
  in leucocytes, 209
  in liver cells, 212
  in platelets, 209
  in red cells, 194–208
  in secretory tissues, 211
  in skeletal muscle, 217
  in smooth muscle, 216
  in vertebrate neurons, 214
  membrane potential, effects of, 200
  molecular mechanism of, 218

Lacrimal glands, 211
Lactoferrin, 619, 640, 644
LamB, gene and its products, 563
  monoclonal antibodies against, 565
  mutations in, 565
  nucleotide sequence, 565
  purification and characterization, 564–565
Lanthanum, 200, 216, 265
LAO-BP, 586
Leucine binding protein (LBP), 586
Leucine-isoleucine-valine binding protein (LIV-BP, LIVT-BP or LIVAT-BP), 586
Leucocytes, $Ca^{2+}$-induced $K^+$ transport, 209
Lipid annulus, 130
Lipids, role in $Na^+,K^+$ transport, 84
Liposomes, acetylcholinesterase, 415, 418
Liver cells, $Ca^{2+}$-induced $K^+$ transport, 212
LIV-I gene, 588–589
LIV-I gene products
  assembly into functional transport system, 590
LIV-I transport system
  components *livG, livH, livM*, 588
  regulation
    *hisT* gene, 600
    *leuS* gene, 600
    *livR* gene, 600
    *lstR* gene, 600
    *relA* gene, 600
    *rho* gene, 600
Low angle neutron scattering studies on $Na^+,K^+$-ATPase, 77
Lymphocytes, 209
Lysine–arginine–ornithine binding protein (LAO-BP and Arg-BP), 586
Lysine binding protein (LAO-BP), 586, 587

Macrophage, 210
malE, 563
  deletion of, 566, 569–570

malE or MBP
  interaction with inner membrane components, 569–571
  interaction with lamB, 570–572
  interaction with Tar protein, 566
  purification and characterization, 566
  role in chemotaxis, 566
malF, 566
  antibodies directed against, 566–568
  location in the cell, 566–568
  malF-lacZ gene fusions, 566–567
  malF-lacZ hybrid protein, 566–568
  nucleotide sequence of, 568
malG, 563
  location in the cell, 569
malK, 563
  antibodies directed against, 567–568
  homology with hisP, 569
  location in the cell, 568
  malK-lacZ gene fusions, 566–568
  nucleotide sequence of, 568
malP, 562
malQ, 562
malT, 562
Maltodextrin, active transport, 561, 563, 564
Maltose transport in *E. coli,* 561–575
  energy source for, 561, 564, 569, 572, 573
  exit reaction, 563
  kinetics of, 563
  lamB protein, 564
  malF and malK proteins, 566
  malG protein, 569
  maltose binding protein, 566
  membrane-binding protein interactions, 569
  model of transport, 571–573
  reconstitution with MBP and $Ca^{2+}$ permeabilized cells, 571
Measurements of $\Delta\psi$ and $\Delta pH$, 459, 461
Membrane enzymes, chromaffin granule, 449, 451, 457, 480
Membrane of *E. coli*
  cytoplasmic, 618, 628–632
  outer, 564, 565, 571, 617, 622–628, 636
  receptor proteins, 622, 623–628, 636, 643
Membrane potential, effect on $Ca^{2+}$-induced $K^+$ transport, 200, 201, 203, 213, 221
Meproadifen, 349
Methyl α glucoside, 531, 546
Methyl α glucoside-6-phosphate, 531
Methyl α glucoside transport, 546
$Mg^{2+}$ transport in nerve membranes, 329
Mitochondria, $Ca^{2+}$ transport, 249–286
  $Na^+$-$Ca^{2+}$ carrier, 255–262
  $Na^+$-independent $Ca^{2+}$ transport, 262–269
Motoneurons, 214
Muscle
  acetylcholine receptor, 351
  $Ca^{2+}$-activated $K^+$ fluxes, 214
  junctional and extrajunctional receptor, 353
  $Na^+$ transport, 324
Muscle cell culture, 141–142

N-acetylimidazole, inhibition of $Na^+$,$K^+$-ATPase by, 74
$Na^+$-$Ca^{2+}$ exchange, mitochondria, 255–262
$Na^+$-$Ca^{2+}$ exchange, plasma membrane, 235, 242, 243, 244, 245
$Na^+$-dependent membrane-bound amino acid transport system in bacteria, 579–583, 585
NADH oxidoreductase, chromaffin granule, 453, 480, 483
Na fluxes in nerve membranes, 323
Na gradient driven Pi transport, 299
Na-independent $Ca^{2+}$ transport in mitochondria, 262–269
$Na^+$-independent membrane-bound transport system for amino acids in bacteria, 579–583
$Na^+$,$K^+$ exchange, 43
$Na^+$,$K^+$ transporting adenosine triphosphatase, 35–114
  antibodies to, 38
  ATP–ADP exchange, 48
  ATP binding sites, high and low affinity, 82
  ATP synthesis by, 43
  binding of $Na^+$ and $K^+$ ions, 65–71
  binding of nucleotides, 60, 61, 80–84
  cardiac glycosides, 85–88
  cation fluxes, 35, 40–47
  crosslinking of subunits, 72, 78
  effect of divalent cations, 52
  electrogenic effects, 35, 43
  electron microscopy, 72, 73
  EPR studies, 75
  fluorescence studies, 60–66, 75
  functions, 35, 36
  "half-of-the-sites," behavior of, 77, 79–80
  identity with sodium pump, 37, 38
  incorporation into vesicles, 38, 42, 69
  inhibitors of, 85–93 (see also under names of individual inhibitors)
  $K^+$,$K^+$ exchange, 45
  kinetic studies of phosphorylation and dephosphorylation, 55–58

$Na^+$, $K^+$ transporting adenosine triphosphatase (*cont'd*)
  lipids, 84
  mechanism of hydrolysis of phosphoenzyme, 54, 55
  molecular weight, 75–78
  $Na^+$,$K^+$ exchange, 43
  $Na^+$,$Na^+$ exchange, 43, 47
  neutron scattering, 77
  NMR studies of, 49, 75
  $^{18}O$-exchanges catalyzed by, 49, 53–55
  occlusion of $K^+$, 54, 67–70
  occlusion of $Na^+$, 70, 71
  oligomers, 79
  oligomycin, 90–91
  phosphatase activity, 50, 51
  phosphoenzyme intermediates, 52–55, 58–67
  phosphorylation by ATP, 52, 53
  phosphorylation by orthophosphate, 53, 54
  proteolysis, 52, 61, 69, 70
  purification, 38–40
  radiation inactivation, 76, 77
  reversal of, 43
  solubilization of, 39
  spectroscopic studies, 75
  structure of $Na^+$,$K^+$-ATPase, 71–85
  subunits, 39, 71–74
  thimerosal, 91
  uncoupled $Na^+$ efflux, 45, 46
  vanadate, 88–90
$Na^+$,$Na^+$ exchange, 43
*Narcine brasiliensis,* acetylcholine receptor, 337
*Narcine japonica,* acetylcholine receptor, 337
Na-transport systems in nerve membranes, 324
NBD chloride (7-chloro-4-nitrobenzo-2-oxa-1,3,diazole): inhibitor of $Na^+$,$K^+$-ATPase, 74
Nerve membranes, ion transport, 321–334
*N*-ethylmaleimide, 407, 660
  inhibitor of $Na^+$,$K^+$-ATPase, 51, 52, 60, 70, 71
Neuroblastoma cells, 215
Neuromuscular junctions, 403, 404, 420, 421
Neurons, $Ca^{2+}$-induced K transport, 212
Neutrophils, 209
Nicotine acetylcholine receptor, 335–401
Noncompetitive inhibitors of acetylcholine receptor, 369
Norepinephrine, 471
Nucleotide accumulation, chromaffin granule, 484
Nucleotides, chromaffin granule, 455

Occlusion of $K^+$ in $Na^+$,$K^+$-ATPase, 54, 67
Occlusion of $Na^+$ in $Na^+$,$K^+$-ATPase, 70
$^{18}O$ exchange between Pi and $H_2O$ in $Na^+$,$K^+$-ATPase, 49
Oligomycin
  effect on $Ca^{2+}$-induced K transport, 207
  inhibitor of $Na^+$,$K^+$-ATPase, 90–91
Optical measurement of ion transport in nerve membranes, 323
Osmotic stability, role of $Na^+$,$K^+$-ATPase, 36–38
Ouabain, effect on $Na^+$,$K^+$-ATPase, 86–88
Ox 222, 349

Parathyroid hormone, effect on intestinal Pi transport, 287, 308, 309, 314
Parietal cell
  monoclonal antibodies, 432, 433, 434, 436, 437, 444
  morphology, 431–434
Patch-clamp, 214, 217
P-chloromercurybenzoate (PCMB), effects on $K^+$ channel in *E. coli,* 660
Periplasmic binding protein dependent amino acid transport in bacteria, 579–583, 585
Phage lambda receptor, 563, 564
Phenylalanine-insensitive $Ca^{2+}$-activated alkaline phosphatase, 303
Phenylalanine-sensitive Mg-activated alkaline phosphatase, 303
Phosphate transport, intestinal, 287–320
  absorption, 287–297, 304, 309–314
  active transport of Pi, 288, 290–299, 312
  alkaline phosphatase, role of, 299, 303
  alterations in human disease, 313
  basolateral membrane, flux across, 290–297, 299
  brush-border membrane, flux across, 290–298, 305, 308, 309, 313
  brush-border vesicles, uptake by, 299–303, 306, 307, 312, 314
  calcitonin, effects of, 309, 312
  calcium, relation to, 287, 289, 292, 293, 294, 297, 299, 303, 309, 313
  carrier affinity, 293, 297–303, 305–308, 311
  cyclic AMP, effect of, 302, 303, 308, 309
  diffusion, 289, 290, 293, 296, 311, 312
  1,24R,25-dihydroxyvitamin $D_3$, influence of, 305
  1,25-dihydroxyvitamin $D_3$, stimulation by, 289, 298, 303–305, 307, 312, 313
  24R,25-dihydroxyvitamin $D_3$, influence of, 305
  glucocorticoids, effects of, 309, 310, 312
  hormonal regulation, 303
  25-hydroxyvitamin $D_3$, stimulation by, 298, 299, 304–306

Phosphate transport, intestinal (*cont'd*)
hyperphosphatemia, in, 309, 314
hypophosphatemia, in, 313, 314
insulin, effects of, 310–313
kinetics, 298–300, 306
mucosal surface, transfer across, 290–298, 302, 303
$Na^+$ gradient driven Pi transport, 299–303, 306–308, 311
paracellular pathways, 290, 295, 296
parathyroid hormone effects, 287, 308, 309, 314
passive transport, 288, 290–294, 312
pH, influence of, 291, 292, 300, 301, 312
Pi carrier, characterization of, 302
potential difference, influence of, 290–294, 296, 312
secretion, 296, 297, 309
serosal surface, transfer across, 290–297
sites of phosphate absorption, 288–297
transcellular pathways, 290, 295, 296
vitamin D, regulation by, 289, 293, 295, 297–300, 302, 304–308, 310, 313, 314
vitamin D, stimulation by, 298, 302, 304–308, 313
Phosphatidylinositol kinase, chromaffin granule, 454
Phosphoenolpyruvate:sugar phosphotransferase system (PTS), 523–559
enzyme I, 524
$III^{Glc}$, 530
HPr, 528
regulation, 542
sugar receptor proteins (II-B), 537
Phospholamban, 138–140
Phospholipids, phosphatidylserine, role in regulation of plasma membrane $Ca^{2+}$-ATPase, 238, 239, 240, 241
Phosphorylated intermediate
$Ca^{2+}$-activated ATPase of sarcoplasmic reticulum, 165, 169–172, 177–179
$Na^+,K^+$-ATPase, 52–54, 70
plasma membrane $Ca^{2+}$-ATPase, 236–239
Phosphorylation
cardiac sarcoplasmic reticulum, 138, 139
skeletal sarcoplasmic reticulum, 134
Phosphotransferase system, involvement in bacterial sugar transport, 523–559
Pi carrier
alkaline phosphatase, role of, 303
characterization of, 302
inhibitors, effects of, 302
phosphate binding to, 302
phosphorylation of, 302
Pituitary granules, 488
Plasma membrane, $Ca^{2+}$ transporting ATPase, 235–248; *see:* $Ca^{2+}$ transport ATPase of plasma membrane
Platelets
$Ca^{2+}$-induced $K^+$ transport, 209
dense granules, 488
Polycistronic mRNA for amino acid transport systems in bacteria, 602
Porins, 617, 625, 627
Postsynaptic membrane, electrophysiology, 360
Potassium fluxes in nerve membranes, 326
Potassium transport in *E. coli,* 653–665
gene mapping, 655
glutathione, effect on $K^+$ channel, 661
$K^+/H^+$ antiport, 656
$K^+/K^+$ exchange, 658
mechanochemical switch, 655–657
thiol reagents, 660
Primary active transport, chemiosmotic coupling, 27
apparent affinities, 30
kinetics, 28
Proline transport in bacteria, 585
Propranolol, effect on $Ca^{2+}$-induced $K^+$ transport, 206
Protein $III^{Glc}$
of bacterial PTS, 530
crr gene, 548
Protein subunits, acetylcholinesterase, 406, 416
Proteolipid, in sarcoplasmic reticulum, 120, 132, 133
Proteolysis of $Na^+,K^+$-ATPase, 61
PTS-mediated repression, 547

Quercetin, inhibitor of $Na^+,K^+$-ATPase, 52

Radiation inactivation of $Na^+,K^+$-ATPase, 76
Rat adipocytes, glucose transport, 506–514
Rat diaphragm, acetylcholinesterase, 420
denervation, effects of, 422
localization, 420
Receptor, acetylcholine, 335–401
Reconstitution of acetylcholine receptor, 371–383
dissociation into subunits, 382
identification of functional unit, 376
incorporation into lipid vesicles, 371
incorporation into planar lipid bilayers, 374
pharmacological integrity of reconstituted receptors, 378
Reconstitution of
binding protein-dependent amino acid transport systems, 596

Reconstitution of (*cont'd*)
$Ca^{2+}$-ATPase of plasma membrane, 238, 240, 241, 242
calcium transport, 132, 133
membrane-bound amino acid transport systems, 594
Red cells
$Ca^{2+}$ homeostasis, 194
$Ca^{2+}$-induced $K^+$ transport in animals, 208
Regulation of bacterial phosphoenolpyruvate:sugar phosphotransferase system (PTS), 542
acetate kinase, 543
crr and iex genes, 547
enzyme I, 543
enzyme II, 544
Regulation of hexose transport in animal cells, 497–522
Reserpine, 472
Ruthenium red, effect on mitochondrial $Ca^{2+}$ transport, 266

Salivary glands, 211
Salmonella, 621, 629, 641
Sarcoplasmic reticulum of cardiac muscle, 137–140
Sarcoplasmic reticulum (SR) of skeletal muscle, 115–191
amino acid sequence of $Ca^{2+}$-ATPase, 123–125
asymmetry, 120–122
ATPase–ATPase interactions, 125
biosynthesis of $Ca^{2+}$-ATPase, 140–143
$Ca^{2+}$ release, 134
detergent solubilization, 122, 123, 126
glycoproteins, 118, 119, 133, 138
kinetics of Ca transport, 157–191
lipid annulus, 130
lipid–ATPase interactions, 128
oligomers of $Ca^{2+}$-ATPase, 125–128
phosphorylation of SR proteins, 134
protein composition, 117–120
purification of $Ca^{2+}$-ATPase, 122
reconstitution of $Ca^{2+}$-ATPase, 132
structure, 115, 120–122
Secretory canaliculus, 432, 433, 434, 436
Secretory tissues, $Ca^{2+}$-induced $K^+$ transport, 211
SH group reagents, effects on $Ca^{2+}$-induced $K^+$ transport, 207
*Shigella flexneri*, 621
Showdomycin, inhibitor of $Na^+,K^+$-ATPase, 74
Siderophores, 619, 620, 624
Single channel recordings of acetylcholine receptor, 362
SITS, DIDS, effects on $Ca^{2+}$-induced K transport, 207
Skeletal muscle acetylcholinesterase, 404, 420, 421
Smooth muscle, $Ca^{2+}$-induced $K^+$ transport, 216
Sodium pump (see also $Na^+,K^+$ transporting adenosine triphosphatase), 35–38, 40–48
Squid axon, ion transport, 321–334
Storage, biological amines, 449, 455
*S. typhimurium*
amino acid transport, 579–583
PTS system, 525
Suberyldicholine, 349
Sugar receptor proteins (II-B) of bacterial PTS, 537–547
exchange transphosphorylation, 540
organization in membranes, 540
properties, 538
role in regulation, 544
Sulfate-BP, 587, 588
Surface membrane
of muscle cell, 115–117
Sympathetic neurons, 214
Synthesis and assembly of acetylcholine receptors, 355

Thermodynamics of ligand-protein interactions, 3–5
Thimerosal (ethylmercurythiosalicylate), inhibitor of $Na^+,K^+$-ATPase, 66, 91–93
Thiol reagents, effects on $K^+$ channels in *E. coli*, 661
TonB function, 632–635
*Torpedo californica*, acetylcholine receptor, 337
*Torpedo marmorata*, acetylcholine receptor, 337
*Torpedo nobiliana*, acetylcholine receptor, 337
Transferrin, 619, 640
Transverse tubules (T-systems), 115–117
Triad, 115
3-Trifluoromethyl-3-phenyldiazirine, labeling of $Na^+,K^+$-ATPase, 73
1, 24R, 25 Trihydroxyvitamin $D_3$, 305
Trimethisoquin, 349
trkA gene, coding for $K^+$ pump in *E. coli*, 656
Tryptic fragments of $Ca^{2+}$ transport ATPases, sarcoplasmic reticulum, 123–125
T-system (transverse tubules), 115–117, 135, 137
d-Tubocurarine, 349
Tubulovesicles, gastric, 432, 433, 434, 436

Uni-uni isoenzyme mechanism, 5
Uterus muscle, 216

Valine binding protein (LIV-BP), 586, 587
Vanadate, inhibitor of $Na^+,K^+$-ATPase, 88–90
Vertebrate neurons, $Ca^{2+}$-induced $K^+$ transport, 214

Vitamin D, effect on intestinal Pi transport, 289, 304, 310
electrochemical $Na^+$ gradient, effect on, 307
genomic action, 305
mechanisms, 305
synthesis of carrier protein, 306

Vitamin D-resistant rickets, 314

Whence side, definition, 17
Whither side, definition, 17

X-ray crystal structure of binding proteins, 588